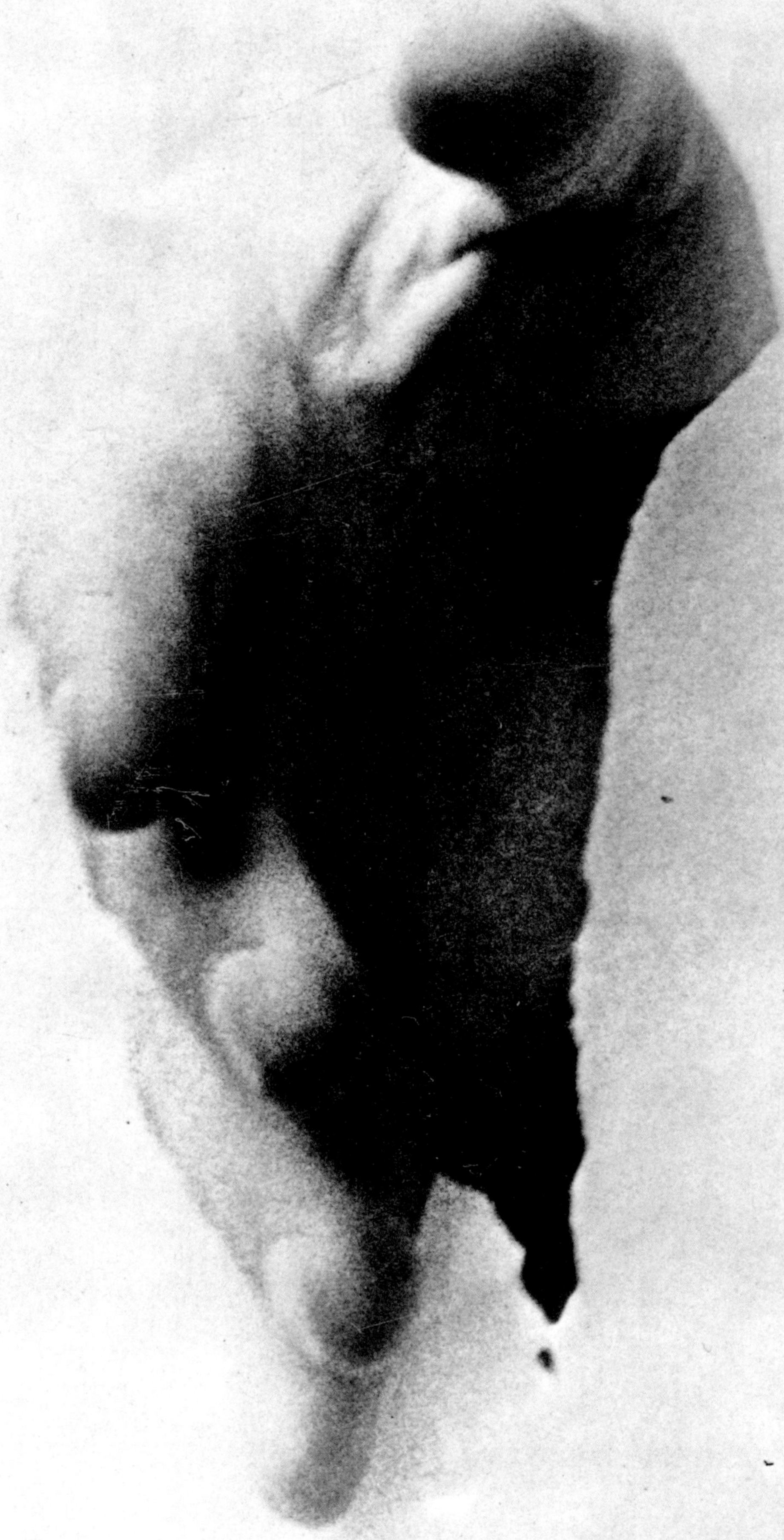

Rip / Riss

1966 | paper poem

Peter Weibel

Peter Weibel

Art as an Act of Cognition

Edited by
Jens Lutz and
Philipp Ziegler

With Clara Runge,
Stephanie Stadler, and
Patrick Trappendreher

Published by ZKM | Center
for Art and Media
Karlsruhe, Germany

Distributed by the MIT Press,
Cambridge, MA
London, England

Essays

Artworks

Preface

Peter Weibel: Art as an Act of Cognition, offers a peek into the art world of Peter Weibel (1944–2023), whose forward-looking oeuvre from the 1960s onward has become a present-day icon of media art itself. The publication covers over half a century of artistic expression, and traces this outstanding media artist's and theorist's migration from material to machines to media, constantly addressing the mechanisms of perception, the autonomous world of apparatuses – a world unto itself –, possible and often imperceptible pitfalls of the visual representation of reality, as well as the relationship between the arts, politics, and economics.

Peter Weibel applies a cognitive and therefore novel approach to art; this position marks him out as a pioneer of media and concept art. His focus was not on creating aesthetic images in the tradition of painting and photography; instead, he was interested in showing how images are actually constructed and why they are so effective in shaping the consciousness of people and their reality. For Peter Weibel, art was a process of cognition, an act of thinking that makes us question supposed truths, and opens our eyes to the ontological difference between representation and reality.

As an interdisciplinary nomad travelling between art and science, Peter Weibel anticipated key developments in the art of the late twentieth and twenty-first centuries. He radically kept on questioning conventional notions of art, and evolved new utopian visions of a free society and individual freedom on the basis of his own technological inventions. The works he created were always political and social, and – as a (media) activist – he translated his public interventions into extraordinary, now historic, public images. Topical issues in this day and age – such as the crisis of Western notions of art, identity politics, defects of democracy, the relationship between power and the law, post-colonialism, gender issues, and violence – were continuously addressed by Weibel. In addition to his artistic works, Peter Weibel's oeuvre encompasses countless books, articles, poems, lectures, conferences, performances, pieces of music, operas, films, videos, festivals, concerts, biennials, and exhibitions. All parts of it are in an intensive dialog with each other. A publication on his artistic works can, therefore, only be the beginning of an in-depth examination of Weibel's work. Or, as the philosopher Peter Sloterdijk, a longtime friend of Weibel's, put it: "In order to get an idea of his entire oeuvre, one would not only have to exhibit his artworks but also his exhibitions, and reflect on his reflections".[1]

The present publication's narrative follows the trajectory of Peter Weibel's development as an artist in various media – from literature to photography, from video to Expanded Cinema, from sculpture, sound art, and light art to digital artworks. More than 370 distinctive works by the artist are presented in this volume, supplemented by concise work descriptions by the ZKM | Center for Art and Media's own editorial team as well as by external authors. These work descriptions are richly illustrated with recent photographs as well as images that date back to the late 1960s.

Essays and texts by Hans Belting, Bazon Brock, Boris Groys, Wolfgang Kemp, Friederike Mayröcker, Hans Ulrich Obrist, Pamela C. Scorzin, Peter Sloterdijk, Terry Smith, and Siegfried Zielinski, contextualize and situate Weibel's works for the first time in English within art history, media theory, and philosophical discourses, and elucidates the general impact and influence of his work. In addition, the book draws on the artist's large personal archive, presenting early drafts and concepts of many of his works as well as historical statements and texts by the artist himself.

As a reader, you can read the book chronologically, but you can also flip through the book and its topic-specific text contributions and linger over certain works or topics according to your interests. In addition, a thematic index of works at the end of the book invites you to approach the artworks associatively: the chronology of works in the book is just one approach, just one way of ordering Weibel's diverse body of work. It is also possible to find other approaches to his work. Here, the index offers a tool to discover further connections in the work of the artist. Furthermore, the book functions as a multimedia presentation, as with the help of a mobile device and QR codes, selected video works and artist statements about individual works can be viewed and it is very worthwhile to approach the artist's works through his own words.

The idea for this publication is based on the exhibition *respectively, Peter Weibel*, which was presented at the ZKM | Karlsruhe in 2019–20 on the occasion of Peter Weibel's 75th birthday, and which for the first time provided an overview of large parts of this versatile media and conceptual artist's work. The exhibition was also shown in a modified form from February to May 2023 at the MMCA, the National Museum of Modern and Contemporary Art, Korea in Seoul, where it was a resounding success with the public.

There is no doubt that Peter Weibel was a central figure of the art scene and in particular of the media art scene. This publication not only provides a survey of his work, it helps the readers to better understand the key concepts and key moments in the evolution of media art in the second half of the twentieth century and the beginning of the twenty-first century.

We both worked intensively with and for Peter Weibel for many years, and encountered him as a media artist, curator, theorist, scientist, author, boss, and human being. We are very happy that this book, which he was involved in until the end, has now been fully realized as part of his legacy. Naturally, many people collaborate in such a major exhibition and publication project. We would now like to thank them all.

First, we would like to thank the authors of the text contributions to this book. A great thanks also goes out to the cocurators of the exhibition *respectively, Peter Weibel*, Judith Bihr and Clara Runge, and the project assistants Svenja Clauss, Hanna Jurisch, Lena Reitschuster, and Leonard Sprüth for their tireless efforts in conceptualizing and bringing the exhibition to life. Their dedication and passion for art and culture truly shone through in every aspect of this exhibition.

We would like to thank all the staff members of the ZKM | Karlsruhe, in particular the scenographer Matthias Gommel, Peter Weibel's research assistant Anett Holzheid, Christian Lölkes and Bob O'Kane as Weibel's advisors for all technical matters, his longtime personal assistants Ingrid Truxa and Dominique Theise, the registrar Marianne Meister, Margit Rosen, the head of the department Wissen (Collections, Archives & Research) at the ZKM, and Christiane Riedel, the former managing director of the ZKM. All contributed to the success of this exhibition through directly working on the project or providing support in the background. Their hard work and dedication were instrumental in ensuring the success of the exhibition.

The editorial team of the ZKM provided extensive guidance and support throughout the publication process. Above all, we would like to express our thanks to the coeditors of this volume, Clara Runge, Stephanie Stadler, and Patrick Trappendreher. Their tireless work on this publication, their expertise and attention to detail helped to create a publication that meets the highest standards of quality and relevance.

We would also like to thank the graphic design team of 2xGoldstein with Erik Schöfer and Joshua Kaiss that brought a high level of creativity and technical skill to the project, creating visual elements that are at once engaging and informative.

Our sincere gratitude also goes to the translators and copy editors Patrick Hubenthal, Peter Dahm Robertson, Gloria Custance, and Lauren K. Wolfe for their linguistic expertise and dedication to accuracy that were critical in ensuring the quality of the texts.

Our grateful thanks go to ZKM's video studio, headed by Christina Zartmann, for recording the interviews with Peter Weibel about his works; and to Volker Sommerfeld who embedded the videos in the ZKM website and made them accessible through the QR codes in this book.

We would also like to extend special thanks to Karin Buol-Wischenau, Loys Egg, Idis Hartmann, Ulrike Havemann, Hartmut Jörg, Felix Mittelberger, Ulrike Rieger, Margit Rosen, Miriam Stürner, and Susanne Widl for their support and expertise in realizing this volume.

And we would also like to thank Alistair Hudson, Weibel's successor as scientific-artistic director of the ZKM, for continuing to support this publication project after the end of Peter Weibel's tenure and his unexpected sudden death.

Jens Lutz and Philipp Ziegler

[1]
Peter Slotderdijk, "Peter Weibel: Premature Encounters with the World Spirit," see pp. 43–45 in this volume; first published in German as "Verfrühte Begegnungen mit dem Weltgeist. Eine große Ausstellung in Wien ehrt den Künstler, Kurator und Denker Peter Weibel – Anmerkungen seines Freundes," *Die Zeit*, no. 39 (2014).

Introduction

Cognition with Conscience: "Those who understand have wings" [1]

Siegfried Zielinski

1

The title of my text plays critically with a hermeneutic in the humanities that launched literary scholar N. Katherine Hayles (b. 1943) to international fame. It proposes that knowledge generated via machine learning processes should be interpreted as *nonconscious cognition*, knowledge that has no perception of itself but focuses solely on external expediency.[1] *Wissen ohne Bewusstsein* (Cognition Without Consciousness), 1972[2] is the title of one of the most important books by a pioneer of critical media studies in Germany, and it has enjoyed the obscurity that is so characteristic of early scholarly thinking about media. The book's author was Franz Dröge (1937–2002). His thesis – which we, as students, abbreviated "WoB" – consisted of linking, as its logical conclusion, his critique of the empirical communication theory being developed in the USA with the critique of capitalist media production that was most notably developed by the Frankfurt School (Max Horkheimer (1895–1973), Theodor W. Adorno (1903–69)) in their North American exile. The reproach – aimed at discrediting the proposition that machines and apparatuses could ever become the bearers of (correct) consciousness, and arguing furthermore that their evolving functions could on the contrary, in a *Dialectic of Enlightenment* (Horkheimer and Adorno, published 1947), produce the opposite of knowledge – has a longstanding history in media studies, dating back at least to the 1940s.

Peter Weibel's epistemological approach, in both his scholarship about media and his artistic engagements with media (the two are impossible to separate in his oeuvre), assumes from the outset a fundamental, say, linkability, or, put more precisely, an inner dialectic between machine learning and purposive – in other words, conscious – actions. In the debate over the potential consciousness of technical subjects, views are divided between a certain cultural pessimism and a belief in constructive intervention. From the perspective of our own both bio-logical and techno-logical existence in the world, only a horizontal equality in the status of the two makes sense. For us, nature, technology, science, and art do not build an epistemic hierarchy, but are organized alongside one another on sprawling plateaus – thousands, and then some.

Weibel inverts Hayles's paradigm in an experimental operation. Knowledge – or, more precisely: cognition – is only to be had *with* consciousness, otherwise it makes no sense. There is no cognition *beyond* consciousness; only, best-case scenario, purpose- and reason-oriented thinking. What's more, it makes no difference, in principle, whether we are dealing with the production of artistic or scientific insights. Since Weibel – in this case, contra Gottfried Wilhelm Leibniz (1664–1716) – proceeds from the assumption that mind not only organizes itself in matter, but is itself highly organized matter; again, in principle, it is indifferent whether cognition is generated by artificially created or biological bodies – not exclusively by, but including, *Homo sapiens*.

This is why, in the title of this introduction, I use the deep-time term for consciousness – namely, *conscience*. This term encompasses not only the entire semantic field that we associate with innermost thought – that which Georges Bataille (1897–1962) referred to as *inner experience*[3] – but also the vast field of activities of the sciences. *Conscience* refers back to a tradition of thought in which the arts and the sciences were not learned and conceived separately, but developed their powers and energies as *artes liberales* in dissonant concordance.

Peter Weibel is a resolute exponent of a concept of the *liberal arts* understood in this way. A book about his oeuvre could be structured along its classical outlines. The *trivium* of grammar, rhetoric, and dialectics/logic is fundamental, not just for his studies, but also for the way his artistic theory and practice would later develop. It is concretized in his countless contributions to literature and poetry, his expansion of language arts into the technical media, his contributions to linguistics, language criticism, psychoanalysis, and a theory of power, as well as to the laws of motion that govern thought and matter alike.

The *quadrivium* of the other *artes liberales* – arithmetic, geometry, music, and astronomy (or astrophysics, as we call it today) – are very easy to find in the scientific and artistic discourse fields that Weibel cultivated. Number theory, games with the concepts and relations of numbers, interested this passionate mathematician just as much as geodesy, human culture's conquering of Earth's landscapes, *perspective as symbolic form*,[4] and architecture, built and designed spaces, as concretizations of the interaction between matter and geometry.[5] The whole of the second volume of Weibel's collected works is dedicated to the inner relationship between *music and media*.[6] And the physics of the very large, as well as of the very small – the macro- and the micro-cosmos and their dynamic interrelationship – Weibel engaged with systematically for decades. This ranges from the works on endophysics that he developed and published with physicist and chaos theoretician Otto E. Rössler (b. 1940)[7] to his conversations with the Nobel Prize laureate (2022) quantum physicist Anton Zeilinger (b. 1945). Weibel worked his concept of the infinitely many (media) worlds existing in parallel and within each other into fiery artistic form in the audiovisual cantata *Chants of the Pluriverse* (*Gesänge des Pluriversums*, 1986–88, see pp. 304–307 in this volume) – his *Divina Commedia* of the outgoing twentieth century in electronic sounds and images.

From this starting point, one is able to discern certain pivotal elements that determined the thinking and activities of this Austrian media pioneer, of whom many colleagues say: Wherever I turn in our field of advanced relations between the arts and the sciences, Peter Weibel has already been there and left his mark.

2

Research is that impassioned activity by means of which we are able to intensely observe, grasp, interpret, represent/stage, and change the world that is accessible to us. To engage in each of these sub-activities, we need media. Media encompass and pervade the research process and, in equal measure, influence its findings. Media-thinking is, for Peter Weibel, material-thinking. Going straight through the resistance of things, he thinks media and changes them. This is something he shared with another great Austrian intellectual, the mechanical engineer and psychologist Friedrich Knilli (1930–2022), who, as early as the 1960s, was already establishing German-language media studies at the Technical University of Berlin. He was my teacher.

Long before the establishment of an ecological discourse, Weibel, the materiologist, already regarded things of the physical world not as passive objects, but as active subjects that have an influence on our thinking. His essay "Material Thinking" from 1966 already formulated this clearly.[8] *Material Thinking* here means not only thinking *of* or *about* material, but – pointedly – the intellectual potentiality of material itself, which we are familiar with from, for instance, alchemy as well as from the early philosophy of Ernst Bloch (1885–1977) who, for example, with reference to the Persian polymath and philosopher Avicenna (Ibn Sina, 980–1037), assumed that form is the "fiery truth" of matter.[9] Active matter is *being* in its potentiality. This paradigm would remain extremely significant for Weibel throughout the entirety of his creative work, up to and including his major projects undertaken with sociologist and historian of science Bruno Latour (1947–2022). The epistemological foundation of the latter's *Parliament of Things* is likewise the subject character of objects.

Weibel is a staunch *materiologue*. For the artist Jean Dubuffet (1902–85), "materiology" was "the study of the connection between heaven and Earth."[10] In the late 1950s/early 1960s the French artist developed a series of image objects tending towards three-dimensionality in which he allowed natural materials to encounter and interact with one another. As Bruno Latour told me in conversation, the concept *matériologie* did not originate with Dubuffet, but with a student of Georges Canguilhem (1904–1995), François Dagognet (1924–2015), most of whose books, to the best of my knowledge, still await translation. Only a few of his texts are available in English – his monograph, for instance, on physiologist and chronophotographer Étienne-Jules Marey (1830–1904), subtitled *A Passion for the Trace*.[11] Peter Weibel valued this book highly.

3

Let's assume, for conceptual expediency, that art and science both produce insights, and that epistemiologically these are not one and the same thing. Scientific work consists primarily in rendering things and contexts that we don't understand comprehensible as well as describable, which in turn enables the (re)production of reality via technical media. Artistic work is essentially about the possibility of making the experience of that which is difficult or intransigent to comprehension, that which is not identical with ourselves, accessible to both the senses and the *logos*. Using the means and media instrinsic to itself – namely, aesthetic ones – art attempts to make us or keep us sensitive to that which constitutes the Other in nature, in technology, and in humankind. We shall only be capable of responding adequately to the dynamics and complexity amongst the various relations in the Real if we work on the *transverses* between the arts and the sciences. The gesture of thinking that best grasps the complexity of these relations is what Roger Caillois (1913–78) conceptualized as *transversality*, a diagonal movement traversing the distinct modes of knowledge, both the sensory and the logical, which as a rule academic institutions compartmentalize.

This way, a new epistemic thing may come into being, which is what Weibel spent decades working on and elaborating. Conducive to gaining insights are both logical and analytical operations as well as their aesthetic mise-en-scène, which, in Weibel's case, usually proved provocative. To paper over the cracks in the Real with an idea or ideal of harmony was just as repugnant to him as any instance of mediocrity. The new epistemic thing would be comparable with Walter Benjamin's *Denkbild* (thought image), in all its dialectical contradictoriness. I suggest calling this the *potentiality of cognition in the material of the art-*

work, a particular form of sensory cognition that through his authorship Weibel indelibly stamped with the mark of uniqueness.

The events and protagonists of this paradigm shift – a shift which Weibel's *conscient* world represents – walk a fine line between the study of the physical and the metaphysical, between precise algorithmic commands and blurrier magic, between extended things (*res extensae*) and thinking and unextended things (*res cogitans*), as René Descartes (1596–1650) dramatized the distinction. Often they also walk a fine line between serious scientific experiment and ironic staging. The chemico-biological investigations of – quite literally – linguistic material that Weibel conducted before audiences in the early 1970s in open laboratory setups, are a perfect example. In the videographed performance *Airtext: The Product Is Extinguished in the Process* (*Airtext: Das Produkt erlischt im Prozess*, 1967–72, see p. 66 in this volume), for example, Weibel quasi inverts the famous Marxian thesis of the process that is sublated in the product of capitalist commodity production. At the bottom of a glass vessel filled with water, he places an arrangement of oxygen tablets that form the word "Luft" (air)which gradually effervesce. This is fine and filigree intellectual Actionism. The process by which the tablets forming the word "air" produce their real, effervescent counterpart in the water formally executes the meaning of the word. And the articulation of visible language is realized in the process. The difference between the structure of language and natural processes is sublated.

Twenty years after his intensive experiments with natural and conceptual materials, which Weibel conducted independently and outside of any institutionalized research context (probably a precondition for the radically transversal activities that he undertook), he founded the Institute for New Media at the Städelschule (State University of Fine Arts) in Frankfurt am Main. Within this respected arts institution, the boundaries between scientific and artistic experiment then became entirely blurred. They dissolved into arenas of perception, investigations into interactive biotechnical systems, algorithmic bodies of knowledge (knowbots), and architectures of thought you could literally walk through. Artists like Ulrike Gabriel (b. 1964), Michael Saup (b. 1961), and Christa Sommerer (b. 1964) & Laurent Mignonneau (b. 1967) enjoyed the protected space of the institute and, under Weibel's direction, accomplished impressive things. Yet, after only a few years, the Städelschule became very disturbed and terminated the contract. Weibel's *conscient* reality became a threat to the world of classic "beautiful appearances" and its bottomless promises.

4

Developing important new terminologies, positing paradigms in which new ideas can take shape and keep evolving – both the arts and the sciences partake in these fields of action that scale the history of competing categories of knowledge. In both domains, originality is highly prized (if only rarely rewarded appropriately). Weibel's work is comparable to a treasure chest that is overflowing with impactful discoveries of every sort, including linguistic ones. I've selected just a handful of exemplary ones.

The concept of the *Gedankenaustellung* (thought exhibition)[12] is, in recent cultural history, commonly acknowledged to have derived genealogically from the *Les Immatériaux* exhibition that French philosopher Jean-François Lyotard (1924–98) put together with an impressive team of scholars, scientists, and designers in 1985 at the Centre Pompidou in Paris. Lyotard, philosopher of the postmodern, put his own stamp on the project (which designers had originally given the mundane working title "New Materials and New Forms of Creativity"), and in so doing gave new meaning to one of the many attempts at sensitizing the public to new realities situated between the material, the imaginary, and the symbolic. The specific collection of *immaterials* on exhibition in Beaubourg was structured according to the legendary formula outlined by U.S. communication theorist Harold D. Lasswell (1902–78) in the 1940s. Between 2002 and 2019, Peter Weibel, together with Bruno Latour and various scholars, scientists, curators, critics, and artists, further developed the concept of the thought exhibition decisively and explicitly into a museum genre of its own, by way of a quadrivium of large-format events: *Iconoclash* (2002), *Making Things Public* (2005), *Reset Modernity!* (2016), and *Critical Zones* (2020–22).

Comparable exhibitions, so far as content is concerned, including extensive accompanying discourse programs, Weibel had not only already experimented with before but also realized in sophisticated forms many years earlier, as artistic director of Ars Electronica and as curator of steirischer herbst, an interdisciplinary arts festival in Graz. Thus *The World from Within – Endo & Nano* (Ars Electronica '92) grappled with the new worldview, which had begun to emerge between endophysics, the world of the inner observer, and that of the then brand new phenomenon of nanotechnology, and which later developed its own unique forms of aesthetic expression between computer science and electronic arts.[13] *Out of Control* (Ars Electronica '91), which made an enormous impression on me in the course of my teaching and research at Paris Lodron University in Salzburg, thematized over three decades ago the exceedingly tension-filled consequences of technological development, between mechanization, electronization, and early digitization, as well as the tangible danger of large-scale technical systems like nuclear power spiraling out of control due to local and global disasters, environmental damage, and natural catastrophes. At Ars Electronica 1998, the last edition before Weibel's move to head the ZKM | Center for Art and Media Karlsruhe, the exhibition and book project *InfoWar: Information.Macht.Krieg*[14] appeared, an early and unapologetic analytical grappling with spheres of thought and activity whose explosive potential has only become gradually apparent to the mass media and public in recent years and which has culminated in Russia's current war of aggression against the so-called *elites of the Western World*: from information wars between competing political systems to the manipulation of tightly networked finance systems to new aggressive forms of cyberwarfare. All of these discourse-analytical and artistic events were accompanied by the publication of books and catalogs. In retrospect, they all coalesce with the exhibition modules, concerts, and performances to form a dense theoretical and aesthetic body of commentary on the present and on a possible future for that which we thoughtlessly refer to as human civilization.

Right at the outset of his time as artistic director of ZKM, Weibel initiated a series of smaller exhibition projects that have since faded from memory. Each of these events, which included exhibitions, festivals, symposia, and publications, centered on a philosopher with their own unique relationship to the arts, to architecture, to film, photography, and other media. As I recall, the series began with *Foucault and the Arts* (2002), followed by *Gilles Deleuze and the Arts: Repeition and Difference* (2003),[15] then continued with exhibitions on Jean Baudrillard, Paul Virilio, and Max Bense, culminating in the exhibition *Without Firm Ground – Vilém Flusser and the Arts* (2015) that ZKM mounted in collaboration with the international Vilém Flusser Archive at Berlin University of the Arts. At the core of this project, which Peter Weibel and I developed together, is an encyclopedic layout of Flusser's (1920–91) world of ideas and a bibliography of his works in three languages.[16] The exhibition took place in two different versions: a massive parcours at ZKM in Karlsruhe and a concentrated, essentially smaller variant at the Academy of Arts in Berlin.

My thesis concerning the hermeneutics of Weibel's works, however, goes considerably further. Ever since making his poetic and reflexive-analytical activities public, he's been exhibiting ideas. In trying out new forms of expression, he is testing new cognitive possibilities and vice versa – he tests new insights by staging them poetically, aesthetically, or as cultural provocations. But none of this is about projecting outward the inner states and motions of the artist as an intellectual subject such that others may participate in a voyeuristic way; almost every public act of art does this. The ideas that Weibel exhibits or stages in performance are at home in the humanities and the natural sciences. As a rule, they possess a considerable degree of complexity, which is already announced by the titles of the works and work processes themselves. Linguistics, semiotics, psychoanalysis, physical, mathematical-geometrical, and philosophical findings, media critique, art criticism, critique of images are all interwoven into dense interdiscursive settings and hypotheses that play out in each individual work through a whole variety of means: *Sig/n/ature* (1967–75, see p. 77 in this volume), *Refractions of Gravity* (1972, see p. 73), *Observing Observation: Uncertainty* (1973, see p. 262), *Scaling: Paradox of the Reference System (Medium) I* (1974, see p. 400) and *Physical Paradoxes* (1974, see p. 157), *The Landscape of the 21st Century* (1984, see p. 128), *You Are (Not) Part of the Program – You Are (Not) Part of the Problem Zone* (1988, see p. 86), *Language Kit W1* (1989/2014, see p. 84) or *The Object That Is Not Itself: Pure Reality; Is There Time before Time? Is There Space before Space?* (1990, see p. 440) – and these are just a few examples, all of which are presented in this catalog.

As an artist-philosopher and polymath, it couldn't be otherwise: museums and galleries, stages or auditoriums are, for Weibel, spaces into which he projects his ideas and which for a good half century he has made the spaces for the idiosyncratic exhibitions of his ideas. A visitor is able to look at and listen to what and above all *how* Weibel thinks. The beauty of the works is realized in their cognitive potential. Indeed, I'd go so far as to suggest that, for Weibel, cognition and beauty are identical.

A concept that gets straight to the heart of the technical-scientific and the particular aesthetic potential of the new arts in the latter half of the twentieth century is the neologism techno-aesthetics or the *techno-aesthetic*. The concept was floated by a number of thinkers in the 1980s and early 1990s and begat a number of strong successors such as the *techno-imaginary*. But no one formulated this semantic field as memorably or as radically as Weibel, particularly in his comprehensive essay "Transformation der Techno-Ästhetik" (Transformations of the Techno-

Aesthetic), which he had already written at the end of the 1980s, though not published until 1991 in the philosophical manual of the period, the anthology edited by Florian Rötzer (b. 1953), *Digitaler Schein: Ästhetik der elektronischen Medien* (Digital Shine: Aesthetics of the Electronic Media).[17]

In this programmatic text, Weibel substantiates his view "that the art created via technical media is in many respects radically different from the art that went before." He regards new media art as "a transformation, if not a transgression, an overwriting and overriding of the classical arts,"[18] and, in what follows, he dismantles the familiar concepts of philosophical aesthetics (author, work, truth, etc.) under the impress of phenomena and events belonging to the techno-cultural. What's become legendary is Weibel's critique of the idealistic notion of art in its confrontation with the new hybrid objects of mind and matter. In the subsection "Art and Truth: Hegel, the Foundation and Termination of Art through Absolute Spirit," in which Weibel not only systematically deconstructs Hegel's (1770–1831) longstanding and taken for granted triad of Being, Art, and Truth, but then counterposes to this another triad: Apparatus – Object – Image. From this new trio he goes on to playfully derive a number of variants of dialectical relationships that in the following decades would become foundational for the theory and practice of media art. With Charles Sanders Peirce's (1839–1914) semiological triad Symbol – Index – Icon as an intermediate step (a triad that also had a profound influence on Lacan's psychoanalytic theory), exciting new triads emerge in Weibel's text, such as Power – Sign – Truth, Sign – Freedom – Fiction, or Virtuality – Medium – Sign[19] – genuine cognitive pyrotechnics that have lost nothing upon re-reading.

Closely linked with the materiological understanding of the medium is another of Weibel's concepts that dispenses with the cozy idea of the instrumental character of media. Particularly in art scholarship, but also in the other disciplines within the humanities, this idea developed hegemonically and is still alive in current debates over artistic extelligences. It proposes that "the technological" and concrete artifacts alike have a merely accidental character. Humans, and whatever they call reality, remain the agents and driving forces of cultural processes, including artistic ones. Weibel, on the other hand, belongs among the thinkers of media who understand the technical medium always as an active entity altering said reality. Above all when it comes to highly developed technological forms of electronic and, later, digital media.

Early on, Weibel declined (I'm speaking grammatically here) a whole gamut of theoretical inflections on the example of video as a media-technical object – or, rather, both theoretical and artistic inflections, as was his habit. His 1973 essay "On the Philosophy of VT and VTR"[20] in my view ranks among the masterpieces of early video theory. Setting the new electronic medium apart from the opto-mechanical media of photography and film, to which he still ascribes semiologically a representative function, he enumerates five characteristics of the new VT and VTR media: "synthetics, transformation, self-reference, instantaneousness, box." Television and video are for him simply an "abstract representation of world ... that has left reality behind." He expressly references the particular "epistemological epoch" in which the new medium developed and its "essential characteristic of admitting a multiplicity of models as reality and thus leaving an objective reality behind." The essay concludes with the sweeping observation that it remains for a future technological philosophy "to convert the direct changes to reality, the changes of state that matter undergoes through new interpretations of sign processing into the conceptual."[21]

The essay formulates theoretically what Weibel expressed via artistic means in a series of video performances between 1970 and 1972. The original subtitle of the series plays with a phrase popularized by Marshall McLuhan, with an epistemic inversion: "Communication is the medium."[22] It's not that the medium is the message (or the massage), but that, in "the deadly ecology of the television," the speaking subject dies, a circular full-closure of "word and deed, of information and action"[23] is produced.

The manner of expression, strongly echoing that of cybernetics, was no accident at this time. Since 1973, Weibel had been working with Oswald Wiener (1935–2021) and Franz Kaltenbeck (1944–2018) on the publication of a scholarly series on meta-mathematics, neurocybernetics, automata theory, and communication. This work culminated in the editing and translation of the writings of Claude Shannon (1916–2001) and John McCarthy (1927–2011), which he and Kaltenbeck first made available in German in 1974 as *Studien zur Theorie der Automaten*.[24] From a scholarly perspective, this too is pioneering work.

The video philosophy that Weibel outlines is not to be understood as philosophizing *about* the medium. It does not aim to make the medium into an object of philosophy and subdue it in this way. Rather, it is a philosophical praxis that is realized through the medium, in the qualities that Weibel indicated – instantaneous, transformative, with its own specific space-time relations, self-referential, etc. The "epistemically more open form" of video was also something that Vilém Flusser was testing at almost the exact same time in his experiments with Fred Forest (b. 1933).[25]

"To be is to be the value of a variable."[26] With this sentence from Willard Van Orman Quine (1908–2000) from his 9 *Logico-Philosophical Essays*, Weibel summarizes what may be the most radical idea among his media theoretical and media artistic aims. Virtuality, variability, virality, and viability – the four V-concepts of his heuristic of the technical image, which I discuss in my contribution to this volume, are here in a nutshell. What interests Weibel in the work of the logician and analytic philosopher Quine is primarily his decisive critique of ontological questions that generate and then forget the circular relationship between signs and the reality of the signified, that when it comes to mathematical set theory, we are not dealing with existential relationships, but with relationships between signs.

In a lesser known essay from 2006, following this same premise, Weibel takes a critical look at the political and theological ontologies of Giorgio Agamben (b. 1942) and Alain Badiou (b. 1937), against which he sets up the concept of *ontological relativity* which Quine established and Weibel endorsed: "Set theory is not a confirmation of what there is – of ontology; it is a universe of variables of quantification over a certain range, assigning objects from the universe to names and choosing subsets of this universe as extensions of predicates."[27]

In the material world, this insight corresponds with the space-time structures of quantum physics, chaos theory, and the endophysics of Otto Rössler – this latter Weibel not only influenced profoundly but also helped to formulate. The world is consistent, but new every moment. Each instant, the possibility of a new universe may come into effect. Every Now generates its own (unique) reality. Peter Weibel doesn't waste time on complacent or convenient thinking. It bores him, but even more he finds it undignified in light of the complexity of the challenges that we have to overcome.

5

Without doubt, Weibel's artistic work is subversive. His concept of subversion, however, is markedly distinct from the intoxicated praxis of many of his contemporaries from the Viennese Actionism scene, who were deeply influenced by the rituals of the Catholic Church, and it is markedly distinct from the work of those who install microworlds of destruction or the trash of civilization in museums and galleries as the measure of all things. For Weibel, subversion is simply not possible without precise knowledge of the set of rules that govern the functioning of nature, technology, science, the social, the economic, and the aesthetic spheres. Without this, subversion remains an empty gesture, no matter how radical it may appear. This is an idea shared by the great French writer and dialogic philosopher Edmond Jabès (1912–91). One must know the rules of language intimately in order to subvert them and thus raise them to a higher poetic niveau. With slight shifts and irritations of familiar semantic and syntactic constructions, new poetic insights come into being.

With regard to media arts, based in and permeated as they are with technology, Weibel learned much from Max Bense (1910–90), to whom ZKM dedicated an exhibition in 2010.[28] The philosopher and computer scientist remains relatively unknown in the USA, since hardly any of his texts have been translated. Which is why I need to provide a little background here.

In 1948, immediately after the Second World War and still under its influence, Max Bense stated in an essay that our lives have now become fundamentally technical.[29] This mathematician, semiotician, philosopher, and author of *Präzisen Vergnügen* (Precise Pleasures, 1964) wrote the first systematic German language information aesthetic – in dialogue with his French friend, the Jewish intellectual Abraham A. Moles (1920–92).[30] It is still very much language-focused, as was Weibel's own early artistic and theoretical praxis. We were right in the middle of the *linguistic turn*. For all his appreciation of the calculable, Bense never allowed art and poetry to be absorbed into the structures comprised of formulas, algebra, and algorithms. For him, the unexpected is a condition *sine qua non* that we can even speak about aesthetic events and sensational phenomena. "Art, whether it tells the story of an old man or that red dots encounter black lines on white surfaces, art is based in each case – not for its own sake, but for ours – on the frivolous nature of surprise. It is only signs of surprise, which excite tension in the mind and the exertion of our intellect, that stand any chance of producing an aesthetic communication and bringing forth art; outside of what surprises, the trivial immediately begins in art – the zone which arouses no astonishment."[31]

The ancient Greek natural philosopher and mathematician Anaxagoras (ca. 500–428 BC), whom Otto Rössler likes to call an early chaos theoretician and one of his most important teachers, once replied to the criticism that the fate of his homeland didn't interest him at all since all of his attention was focused on the firmament in order to study the movement of the stars, with a lapidary remark:

He could not have said it better himself. The well-ordered world of cities, with their walls, laws, economies, and strictly defined jurisdictions, didn't interest him. Michel Serres (1930–2019), from whom I am borrowing this story, has Anaxagoras put it this way: "My kingdom is not of this world, worldly, but of the other, worldwide."[32]

The *frivolous nature* of surprise that Bense describes emerges – again: in the arts as well as the sciences – through decisive interventions of thinking and acting as well as through shifting from a given set of orders into the wide open space of existing possibilities in all their endless plurality. That is Weibel's theory and praxis *par excellence*.

For both Weibel's and my own intellectual socialization, studying the texts of French philosopher Louis Althusser (1918–90) was incredibly important. He thought boldly, unconventionally, contrary to the dogmatists of Marxist doctrine, of which there were many in Western Europe in the 1960s and 1970s. The *aleatory* is key to Althusser's conceptual universe. In his 1982 essay "The Underground Current of the Materialism of the Encounter," he has recourse to an incomparable philosophical poem, written by the Atomist Lucretius (ca. 99–55 BC) in the first century before the Common Era, in order to explain a category very dear to his heart. "The *clinamen* is an infinitesimal *swerve*, 'as small as possible'; 'no one know where, or when, or how' it occurs." In *De rerum natura* (On the Nature of Things) Lucretius describes the *clinamen* by way of analogy to raindrops, all falling parallel to one another until, suddenly, one deviates from its vertical path. In the language of the Atomist, this then leads the atom to "swerve" from "its vertical fall in the void, and, breaking the parallelism in an almost negligible way at one point, induce an *encounter* with the atom next to it, and, from encounter to encounter, a pile-up and the birth of a world – that is to say, of the agglomeration of atoms induced, in a chain reaction, by the initial swerve and encounter." Along with the equally materiologically inclined natural philosopher Epicurus (ca. 341–270 BC), Althusser was profoundly convinced "that the origin of every world, and therefore of all reality and all meaning, is due to a swerve," that reason was not the origin of the world, but the deviation.[33]

6

In the theory and praxis of Weibel's world of cognition, experiment takes undisputed priority. For this reason, on his sixtieth birthday in 2004, I composed a portrait of physicalchemist Johann Wilhelm Ritter (1776–1810) who, around 1800, uncompromisingly and without regard for his own physical body, invested all his power and all his ability in the experimental study of electric phenomena, gaining insights far in advance of those of his own era.[34]

At a time when the boundaries between individual disciplines were being built and strictly demarcated, one from the other, Ritter was already deliberately disregarding them. Electrical phenomena were for him *sensational* – in the sense that they bound together dead and living matter, physical, chemical, and biological processes, and they were sensational in philosophical and aesthetic terms as well. He made an enthusiastic case for a comprehensive *science of life* and declared his own body a laboratory on which he would prove the correctness of his views. To the point of exhaustion, he applied, often in weeks-long series of experiments, voltaic piles and piles of his own construction to his limbs and organs – hands, tongue, temples, eyeballs, nostrils – in order to detect the different current strengths of various materials in electrical contact. But above all he did this in order to demonstrate that all power has its source in polarity; that the microcosm of the individual vibrates between negative and positive voltage just like the Earth and the entire universe. In order to discover ultraviolet light at the edge of the visual spectrum, he undertook extreme experiments in which he exposed his eyes, held open by an artificial device, to direct sunlight for up to twenty minutes.

Weibel's intellectual personality is of the same caliber and radicality as Ritter's. He took the romantic ideas of the young Silesian scientist further still, on the one hand bringing them in line with the contemporary history of science as formulated by the likes of Lorraine Daston (b. 1951), Peter Galison (b. 1955), Bruno Latour, Michel Serres, or Hans-Jörg Rheinberger (b. 1946). The experimental in culture and the culture of experiment are, according to this intellectual tradition, intimately dialectically linked.

On the other hand, Weibel integrates complex philosophical edifices like that of French painter and writer Pierre Klossowski (1905–2001), who translated, among others, Walter Benjamin (1892–1940) into French. In 1970, though it wasn't published until the 1990s, Klossowski wrote a book about a remarkable economy.[35] Michel Foucault (1926–84) called it in his foreword, "the greatest book of our time."[36] In it, Klossowski inverts the complaints of cultural pessimists about capital accumulation and the mechanization of the body via technology and the economy by bringing the human body into play as "living currency." In a radical extension of the ideas of Sade (1740–1814), the human body becomes offensively integrated into social commerce as an object of exchange. Freed from the immediate and goal-directed imperatives of reproduction, the body thus conceived is capable of a new sovereignty. Klossowski attributes particular importance to the experiment in his economy – which his text itself celebrates as a culture of its own. The "fabrication of useful goods," he writes, is periodically confronted with its own "intermittent sterility." This happens as "the accelerated pace of manufacturing must constantly find ways to prevent inefficiency in its production processes. And the only solution to the problem of inefficiency is waste. Trial and error, which is the condition of efficiency, necessarily leads to wasteful errors. Experimenting with what can be fabricated in order to create a profitable operation means eliminating the risk of product sterility, but the price of this experimentation is wasted materials and human efforts (the costs of manufacturing)."[37]

What makes ZKM so unique in the global museum landscape today is Weibel's radical transformation of the upscale exhibition tanker in Karlsruhe's former munitions factory into a permanent laboratory for the exploration of future worlds of sounds and images from which we may extract new insights for the configuration of approaching presents.

7

The equation of the future with the past (origin), as Martin Heidegger (1889–76) conceived it in his antiquity-besotted ontology,[38] is something from which Weibel vehemently distances himself. The "primacy of the future,"[39] prioritizing the work to be produced, subordinating the past to what has yet to come, is for him a rhetorical and epistemic figure that we discussed, and critiqued, again and again in conversation. Weibel is by no means a glorifier of the future, as he is often portrayed. For him, history has an autonomous character. To subordinate past presents to the demands of imaginable futures can serve as a kind of gesture of ecological coupling. But Weibel's avant-garde sensibility consists rather in making the deep time of phenomena useful by way of systematic reinterpretations of past presents. Like almost no one else in the contemporary art scene, he moved passionately along both of time's arrows – back into history and forward into the future. And this never tore him apart because he always understood – elegantly, poetically – how to bend the severe verticality of these movements into dynamic loops with which he was able to travel through both at once.

From the perspective of a logic of multiplicity – which, in keeping with the tradition of postwar French thought, was just as important for Weibel as analytic philosophy – seeking in vain after an origin is just as ill advised as the attempt to define a purportedly certain future which, according to Emmanuel Levinas (1906–95), must always remain alien. Origins and futures are each reciprocal aspects of thrilling spaces of possibility, and so to keep them malleable and to further develop them has become the program of the laboratory museum that he led for nearly a quarter century.

Closing Remark

"The true evil, indeed, the only evil, are the social conventions and fictions which become superimposed on natural realities, everything from family to money, from religion to the state."[40] Thus Fernando Pessoa (1888–1935) has the banker begin his long monologue in a story from 1922, "The Anarchist Banker." The basic idea of this brilliant text is bewildering and yet still provocative: that which could help counteract the social fictions that serve as mechanisms of domination would be to take their logics to the extreme and carry them out ad absurdum. To take the most powerful fiction as an example – namely, the fiction of money – would mean: more reasonable than robbing a bank is to found one.

There is nothing and no one above us. Excepting perhaps God, in whom at least the public figure Weibel did not believe. For scientists and artists alike, it's ill advised to recognize social conventions such as state, family, church, capital. Their existence – scientists, artists – is in principle an-archic. The freedom that accrues from this is full of responsibility and comparable to a scepter. In his own incomparable manner, Weibel has always swung this scepter. It demands, first of all, that one show respect for oneself. According to the logic of the philosophy of dialogue, what follows is respect for others and the Other. Only by virtue of my unconditional recognition of You can I even exist, can there even be a We.

With an inimitable combination of severity and compassion, Peter Weibel exemplified that knowledge need not compete with beauty but can come together with it – in dissonant harmony and concordant dissonance.

[1]
See N. Katherine Hayles, "Cognition Everywhere: The Rise of the Cognitive Nonconscious and the Costs of Consciousness," *New Literary History* 45, no. 2 (2014): 199–220; Hayles, *Unthought: The Power of the Cognitive Nonconscious* (Chicago: University of Chicago Press, 2017).

[2]
The full title of the book is *Wissen ohne Bewußtsein. Materialien zur Medienanalyse der Bundesrepublik Deutschland* (Frankfurt am Main: Fischer, 1972).

[3]
See Georges Bataille, *Inner Experience*, trans. Leslie Anne Boldt (Albany: State University of New York Press, 1988). First published as *L'Expérience intérieure* (Paris: Gallimard, 1943).

[4]
See Erwin Panofsky, *Perspective as Symbolic Form*, trans. Christopher S. Wood (New York: Zone Books, 1991). First published as *Die Perspektive als "symbolische Form"*, ed. Fritz Saxl, Vorträge der Bibliothek Warburg (Leipzig/Berlin: 1927).

[5]
See also, for instance, the book and exhibition: Ulrike Gehrig and Peter Weibel, eds., *Mapping Spaces: Networks of Knowledge in 17th Century Landscape Painting*, exh. cat. ZKM | Karlsruhe (Munich: Hirmer, 2015).

[6]
The collected works appeared in German as a six-volume series titled *Enzyklopädie der Medien* (Encyclopedia of Media), published with the Hatje Cantz Verlag Berlin by the University of Applied Arts Vienna and ZKM Karlsruhe. The second volume, *Musik und Medien: Vom Klang im technischen Zeitalter* (Music and Media: On Sound in the Technical Age), was published in 2016. The present volume on Weibel's artistic works could also have been structured according to the categories of the *Encyclopedia*. Other volumes of the collected works are dedicated to the relationships between architecture and media (2015), art and media (2019), literature and media (2021), politics and media (2023), and *Theory and Media* (forthcoming).

[7]
See Otto E. Rössler, *Endophysics: The World as an Interface*, ed. Peter Weibel and Reimara Rössler (Singapore: World Scientific Publishing, 1998).

[8]
See p. 48 in this volume for an abridged version of this text. Originally written in 1966, it was first published as "Materialdenken als Befreiung der Produkte des Menschen von ihrem Dingcharakter: Probleme der aktuellen Avantgarde," in *Kritik der Kunst – Kunst der Kritik, Es says & I say* (Vienna: Jugend und Volk, 1973), 51.

[9]
See Ernst Bloch, *Avicenna and the Aristotelian Left*, trans. Loren Goldman and Peter Thompson (New York: Columbia University Press, 2018). First published as *Avicenna und die aristotelische Linke* (Berlin: Rütten & Loening, 1952).

[10]
Matériologies de Jean Dubuffet, exh. cat. Galerie Daniel Cordier (Frankfurt am Main: Galerie Daniel Cordier, 1961), n. p.

[11]
François Dagognet, *Éienne-Jules Marey: A Passion for the Trace*, trans. Robert Galeta and Jeanine Herman (New York: Zone Books, 1992). First published as *Étienne-Jules Marey: La passion de la trace* (Paris: Hazan, 1987).

[12]
See Bruno Latour, "On the Concept of Thought Exhibitions," talk at the Multimedia Institute (MI2) at the Museum of Contemporary Art in Zagreb, 2017, http://modesofexistence.org/what-is-a-gedanken-ausstellung/. Daniel Birnbaum and Sven-Olov Wallenstein discuss the concept liberally, with a focus on Lyotard's project, in their book *Spacing Philosophy: Lyotard and the Idea of the Exhibition* (Berlin: Sternberg Press, 2019).

[13]
See the book that accompanied this project: Karl Gerbel and Peter Weibel, eds., *Die Welt von Innen – Endo & Nano / The World from Within – Endo & Nano* (Linz: PVS-Verleger, 1992).

[14]
The German title can be translated as "Information. Power.War" or "Information.Produces.War."

[15]
The book was published with the abbreviated title *Deleuze und die Künste*, ed. Peter Gente and Peter Weibel (Frankfurt am Main: Suhrkamp, 2007).

[16]
See Siegfried Zielinski and Peter Weibel with Daniel Irrgang, eds., *Flusseriana: An Intellectual Toolbox* (Minneapolis: Univocal, 2015).

[17]
See Peter Weibel, "Transformationen der Techno-Ästhetik," in *Digitaler Schein: Ästhetik der elektronischen Medien*, ed. Florian Rötzer (Frankfurt am Main: Suhrkamp, 1991), 205–46.

[18]
Weibel, "Transformationen der Techno-Ästhetik," 205.

[19]
Weibel, "Transformationen der Techno-Ästhetik," 241–45.

[20]
See p. 231 in this volume for an abridged version of this text. It was first published in German on the back of the poster for the film screening *Filme Video-Tapes* in the Städtische Galerie im Lenbachhaus, Munich, in 1973. The complete version of the text can be found, among other places, in the journal *heute Kunst – Internationale Kunstzeitschrift*, no. 4–5 (Dec. 1973–Feb. 1974); 13–15; and in the reader *Video – Apparat/Medium, Kunst, Kultur*, edited by the publisher (Frankfurt: Lang, 1992), 125–33. "vt" means, in Weibel's terminology, videotape, "vtr" videotape recording; during this period, he wrote for the most part without using capital letters.

[21]
All quotations are taken from the first version published in *heute Kunst*, 1973/74. Translated from the German.

[22]
"The Medium Is the Message" it the title of the first chapter in Marshall McLuhan's *Understanding Media: The Extensions of Man* (New York: McGraw-Hill, 1964).

[23]
Quoted after Peter Weibel's commentary on the work *TV News (TV Death 2) (TV-News (TV-Tod 2)*, 1970 (idea)/1972, see p. 250 in this volume.)

[24]
See Claude E. Shannon and John McCarthy, *Studien zur Theorie der Automaten*, ed. Franz Kaltenbeck and Peter Weibel (Munich: Rogner & Bernhard, 1974). First published as *Automata Studies*, Annals of Mathematics Studies No. 34 (Princeton, NJ: Princeton University Press, 1956).

[25]
See in this regard Gustav Hamos's article "Video Philosophy" in the lexical compendium *Flusseriana*, ed. Zielinski and Weibel with Irrgang, 434–36.

[26]
Willard Van Orman Quine, *From a Logical Point of View: 9 Logico-Philosophical Essays* (New York: Harper & Row, 1961), 15.

[27]
Peter Weibel, "Political Ontology and Ontological Relativity: Pro Variantology – A Footnote," in *Variantology 2: On Deep Time Relations of Arts, Sciences, and Technologies*, ed. Siegfried Zielinski and David Link (Cologne: Walther Koenig, 2006), 457–59 (quotation on 459).

[28]
See *Max Bense and the Arts*, ZKM | Karlsruhe, 2010, exhibition curated by Margit Rosen, Miriam Stürner, Jens Lutz, and Peter Weibel.

[29]
See Max Bense, *Technische Existenz. Essays* (Stuttgart: Deutsche Verlags-Anstalt, 1949).

[30]
See Max Bense, *Einführung in die informationstheoretischeÄsthetik. Grundlegung und Anwendung in der Texttheorie* (Reinbek: Rowohlt, 1969); Abraham A. Moles, *Information Theory and Esthetic Perception*, trans. Joel E. Cohen (Urbana, IL: University of Illinois Press, 1968). First published as *Théorie de l'information et perception esthétique* (Paris: Flammarion, 1958).

[31]
Max Bense, *Brasilianische Intelligenz* (Wiesbaden: Limes, 1965), 55.

[32]
Michel Serres, *The Natural Contract*, trans. Elizabeth MacArthur and William Paulson (Ann Arbor: University of Michigan Press, 1995). First published as *Le contrat naturel* (Paris: Éditions François Bourin, 1990). For Édouard Glissant, the *mundial* (the *worldwide*) is also a possible poetic counter concept to the omnipresence of the *global*, in the arts as well.

[33]
Louis Althusser, "The Underground Current of the Materialism of the Encounter," in *Philosophy of the Encounter: Later Writings, 1978–1987*, ed. Oliver Corpet and François Matheron, trans. G. M. Goshgarian (New York: Verso, 2006), 161–207 (quotations on 169). First published as "le courant souterrain du matérialisme de la rencontre," in *Écrits philosophiques et politiques*, vol. 1, ed. François Matheron (Paris: Éditions Stock/Imec, 1994), 539–79.

[34]
See Siegfried Zielinski, "Im Zustand der Schwingung kann es keine Ruhe geben. Ein kurzes Portrait des Physikochemikers Johann Wilhelm Ritter (Peter Weibel zum Geburtstag)," in *05-03-44: Liebesgrüsse aus Odessa. hg.von für/for/à Peter Weibel*, ed. Ecke Bonk, Peter Gente, and Margit Rosen (Berlin: Merve: 2004), 8–12.

[35]
See Pierre Klossowski, *Living Currency*, ed. Daniel W. Smith, Nicolae Morar, and Vernon W. Cisney (London: Bloomsbury,

2017). First published as *La monnaie vivante* (Paris: Éric Losfeld, 1970).

[36]
Michel Foucault, "Letter from Michel Foucault to Pierre Klossowski," in Klossowski, *Living Currency*, 41.

[37]
Klossowski, *Living Currency*, 46.

[38]
"Origin always comes to meet us from the future." Quoted in Martin Heidegger, "A Dialogue on Language," in *On the Way to Language*, trans. Peter D. Hertz (San Francisco: HarperOne, 1971), 10. First published as *Unterwegs zur Sprache* (Pfullingen: Günther Neske, 1959). The original German reads: "Herkunft aber bleibt stets Zukunft."

[39]
See Claude Ozankom, *Herkunft bleibt Zukunft*, Philosophische Schriften 26 (Berlin: Duncker & Humblot, 1998), above all chapter II. 1.

[40]
Fernando Pessoa, "The Anarchist Banker," trans. Margaret Jull Costa, in *The Anarchist Banker and Other Portuguese Stories*, ed. E. Lisboa (Manchester: Carcanet, 1997), 88–114 (quotation on 91). First published as "O Banqueiro Anarquista," *Contemporânea*, no. 1 (1922).

1. Anni Mirabiles 1966–69

The Eleven Dimensions of Peter Weibel

Hans Ulrich Obrist

Peter Weibel is a man of many dimensions, which are like in superstring theory.

The first dimension I really became aware of when we met at the end of the 1980s, beginning of the 1990s in Frankfurt, where Peter Weibel founded the Institute for New Media in 1989, and that was Peter Weibel as a teacher, as a mentor in many places from Nova Scotia to Vienna, and in Frankfurt.

The second, maybe very early dimension, was, of course, Peter Weibel as a poet, as a visual poet.

He then soon went, as we know, from the page to the exhibition and the screen, which brings us to the third dimension: Peter Weibel as a visual artist. His first exhibition only took place in the mid 1970s. The explanation for this was the incredibly conservative climate in Austria at the time, which he resisted in a very political way. And then, naturally, there was the explosion from Performance art to Media art including Conceptual art throughout the 1970s, which brings us to the fourth dimension: Expanded Cinema, Peter Weibel as a filmmaker. And also to the fifth dimension: Peter Weibel as a media pioneer.

A few weeks ago, I spoke to Tim Berners-Lee who in 1989 invented the World Wide Web. He is extremely concerned at this moment in time about the potential loss of net neutrality. We have to fight against this. We have to fight for an internet for everyone; not a fast internet for those who can pay and a slow internet for those who don't pay. And that's a very important aspect of Peter Weibel's work as well, this idea of art for everyone, this idea of the mass media taking art beyond the gallery space.

Another dimension would be Peter Weibel as a musician, a composer. One example, of course, is the band Hotel Morphila Orchester, founded by him and Loys Egg in 1978. A band that follows the Viennese custom of every hotel or coffee house having its own orchestra. But as this tradition no longer exists, they named their band after a hotel which also doesn't exist anymore.

So, dimension number seven: Peter Weibel as a theoretician, as a lecturer all over the world, between geographies, linking art, media history, media theory, film, and philosophy; always going beyond what C.P. Snow had deplored, going beyond that divide between science and the humanities. Peter Weibel, in that sense, has built a bridge and that brings us to Peter Weibel as a curator of historic exhibitions, for example, of the beginnings and memory of the avant-garde; as Eric Hobsbawm said, the protest against forgetting, exhibitions on the damned, forgotten, Austrian avant-garde, such as the show *Das andere Österreich* (1975) with Oswald Oberhuber, and so on. A whole type of exhibition which Weibel calls the *archive exhibition* has to do with that, and he is continuing and continuing it until today.

Dimension number nine: Peter Weibel as the curator of the present or – as we could also say – of the extreme present, always using exhibitions as an intellectual, philosophical, as well as artistic medium through solo exhibitions like the ones featuring Lynn Hershman-Leeson or Ólafur Elíasson – it's a very long list.

He calls that present, or the work with the present, a *positioning*: that a pioneer has the task to actually inscribe positions on a map, which is something that happens through many solo shows, but also through group shows I will always remember the early show on Context art, for example (*Kontext Kunst. The Art of the 90s,* Neue Galerie im Künstlerhaus Graz, 1993).

That leads us to point number ten: Peter Weibel as the curator of what we could maybe call unexpected curatorship. In the sense of Alexander Dorner, who said that if we want to understand the forces which are effective, we need to understand what is happening in our discipline, what is happening in science, in music, in architecture. So very often, Peter Weibel brings all these other fields into curation, something I experienced when we collaborated on *Iconoclash: Beyond the Image Wars in Science, Religion, and Art* (ZKM | Center for Art and Media Karlsruhe, 2002), but also, of course, through his many shows where he got scientists or theoreticians to actually curate exhibitions as in many great shows at the ZKM | Karlsruhe, above all the four *Gedankenausstellungen* with Bruno Latour (thought exhibitions, coined after the concept of thought experiments).

And that brings us to the eleventh dimension: Peter Weibel as an extraordinary museum director, director of institutions, from Ars Electronica to the ZKM, and in this sense not only of exhibitions, but also of the building of collections.

Eleven times Peter Weibel is *pars pro toto*. Indeed, there are many, many more, but I am running out of time.

Peter Weibel, Peter Weibel, Peter Weibel, Peter Weibel, Peter Weibel, Peter Weibel, Peter Weibel, Peter Weibel, Peter Weibel, Peter Weibel, Peter Weibel.

Note: This text was originally presented on the occasion of a symposium accompanying the exhibition *respectively, Peter Weibel* on September 28, 2019.

The Artist P.W.

Friederike Mayröcker

his Anni Mirabiles,
1966–1969,
when he moved
from typed poems to photo poems,
from page to screen,
from literature to electronic media
and invented the future ...

yes, today I wanted to take the enthralledness, the enthrallment as a sequence of visions, moving the way they do inside my brain, and turn it inside out; I mean it's right there, it's on the tip of my tongue, as we like to say by way of apology when the thing we're trying to say isn't there, isn't ready, etc., but we feel it in our whole body, so on the tip of our tongue or behind our forehead behind our eyes behind our shoulder blades, it resides in our whole body but we haven't pressed the right spot yet to trigger the signal.

It all has to do with the body, that's always the first thing that comes to mind when I think of P.W. ... I haven't seen him in a long time, haven't talked to him, phoned him, written to him, admired his stenographizing language, the way it skips and stumbles and executes its wonderful glissandos, throws out its flashes of inspiration, its rumbles of thunder, and you're fascinated, can't tear yourself away, you long for the pleasure to last and last. A whiff a grove of ash and palm, and it wouldn't even have to be the total distances between America and Austria any more; ah, the distance between Karlsruhe and Vienna is already enough, isn't it. But I hallucinate him following the artist V.E. through the streets of the city as a secondhand doggie on a long leash, following or GALUMPHING along behind her, I mean on hands and knees, and at the same time passionately burying himself in the very hairy armpit of S.W., a morphology that gets me intensely excited.

It's on the tip of my tongue but I can't get it out, I hallucinate his morphology, his intellectual makeup, but I can't write it down can't sketch it out – may the artist forgive me – *I've locked him away in my heart.*

January 22, 2004

Note: This text was first published as "der Künstler PW," in 05-03-44: *Liebesgrüsse aus Odessa für Peter Weibel*, ed. Ecke Bonk, Peter Gente, and Margit Rosen (Berlin: Merve, 2004), 112–13. Translated from the German by Patrick Hubenthal.

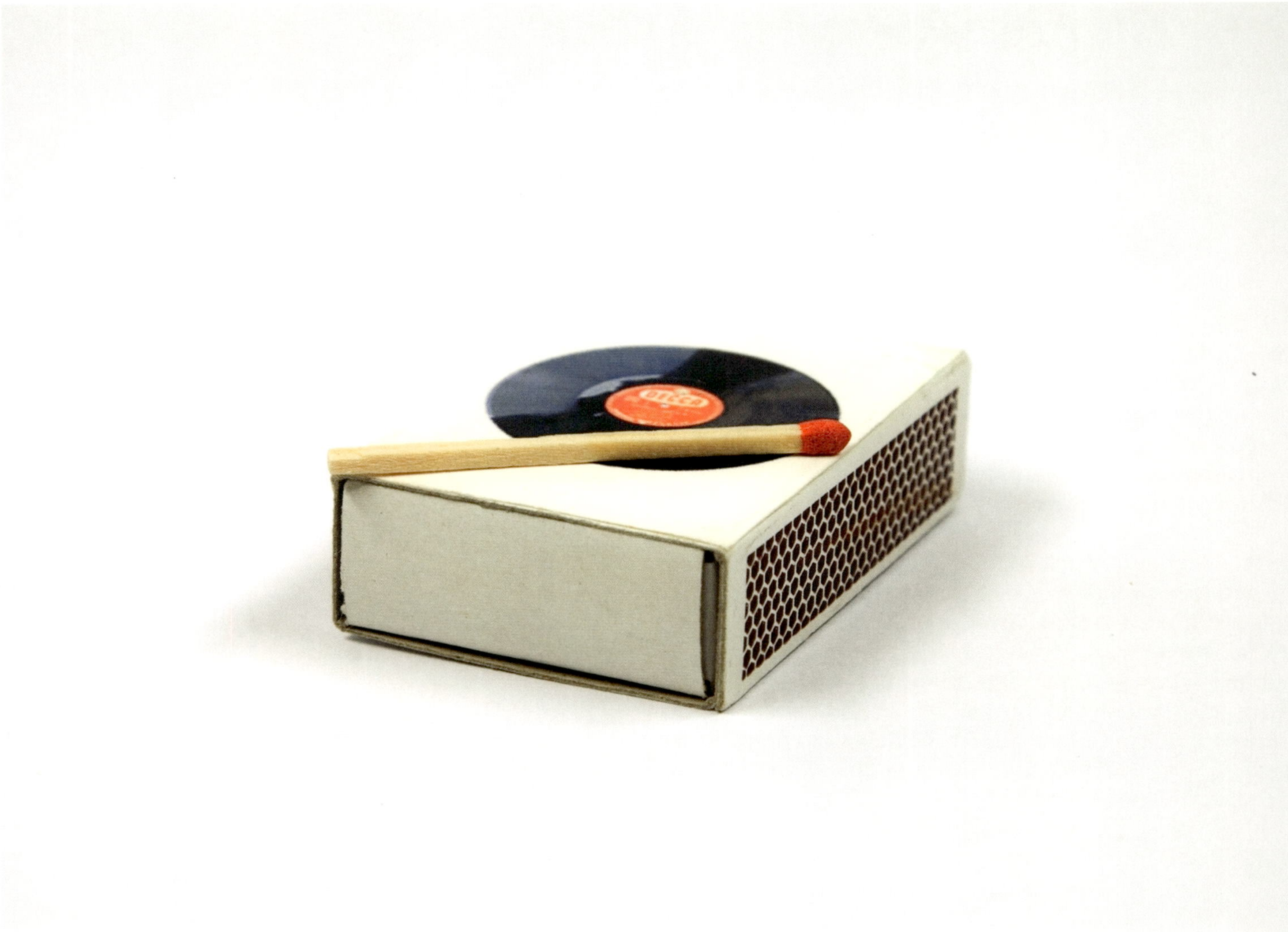

The Invention of Ignitive Sound / Erfindung des Zündschalls

1966	matchbox, photograph of a record, match

Matchbox as record player. A tiny record (e.g., of Mozart's *Eine kleine Nachtmusik*) is placed on it. The match, which serves as the tonearm, ignites the music. The music plays as the record burns.

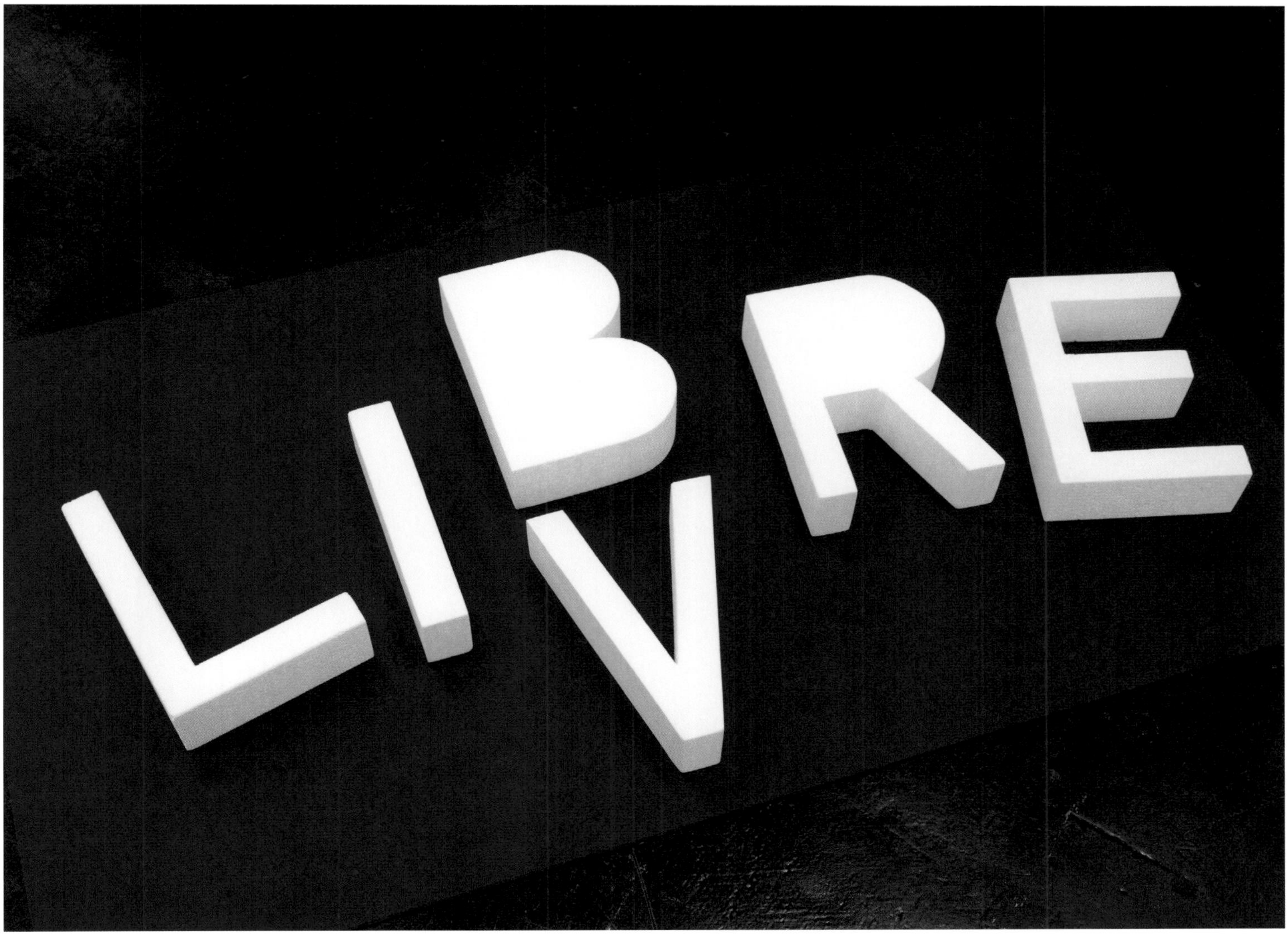

LIVBRE (LIRE, LIVRE, LIBRE)

1966 | letter sculpture; model for furniture in the form of letters, polystyrene, dimensions variable

The French word *livre* (book) contains the English word "live" as well as the French word *lire* (to read). And if we replace the letter *v* with a *b*, we get the French word *libre* (free). This suggests the following maxim: Those who read are free; those who read books live free. Those who understand have wings.

In the mid-1960s, I turned poetry into a performance of the author or of the audience, sculptures to use. I also began to develop a poetic technique for splitting words, breaking them apart by arranging their letters in space, on objects rather than on the two-dimensional surface of the page, in order to liberate secret, hidden messages and meanings. These letter sculptures conformed, as I was later gratified to discover, to Saussure's anagram theory as outlined by Jean Starobinski.[1]

At the time I was completely broke, without even enough money to live on, and my ideas had been rejected by the art world. So I could not realize them; I could only write them down or make models out of polystyrene. *LIVBRE (LIRE, LIVRE, LIBRE)* was conceived as seating for a public library. —Peter Weibel, 2022

Translated from the German by Patrick Hubenthal.

[1]
Jean Starobinski, *Words upon Words: The Anagrams of Ferdinand de Saussure*, trans. Olivia Emmet (New Haven: Yale University Press, 1979). First published as *Les mots sous les mots: Les anagrammes de Ferdinand de Saussure* (Paris: Gallimard, 1971).

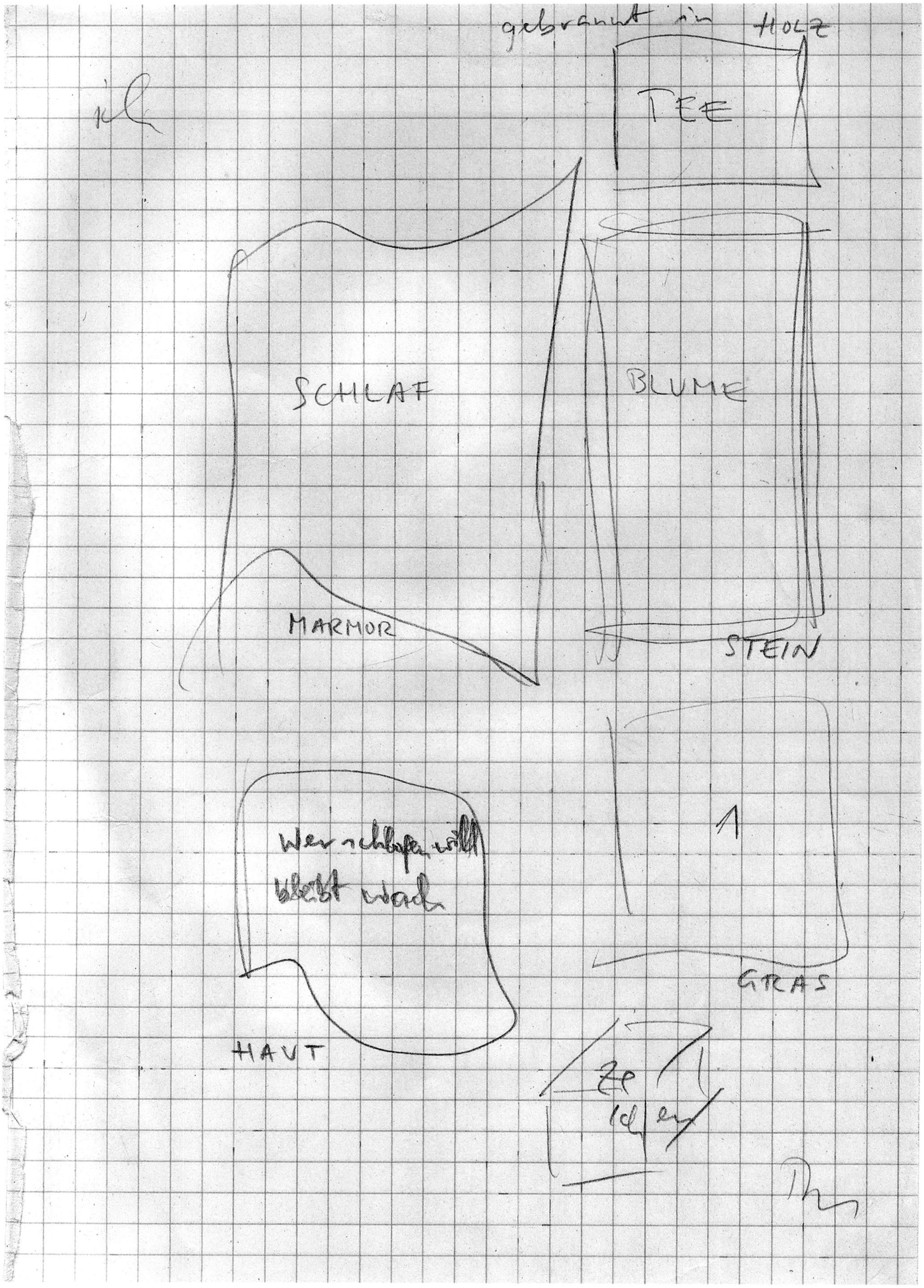

CON/TEXTS / KON/TEXTE

1967 | typescript

Words written on different materials. "Tea" on wood, "sleep" on marble, "flower" on stone, "1" onto grass, "the one who wants to sleep stays awake" onto skin, "Ze/ich/en" (*Zeichen* = sign, *ich* = me) onto stone.

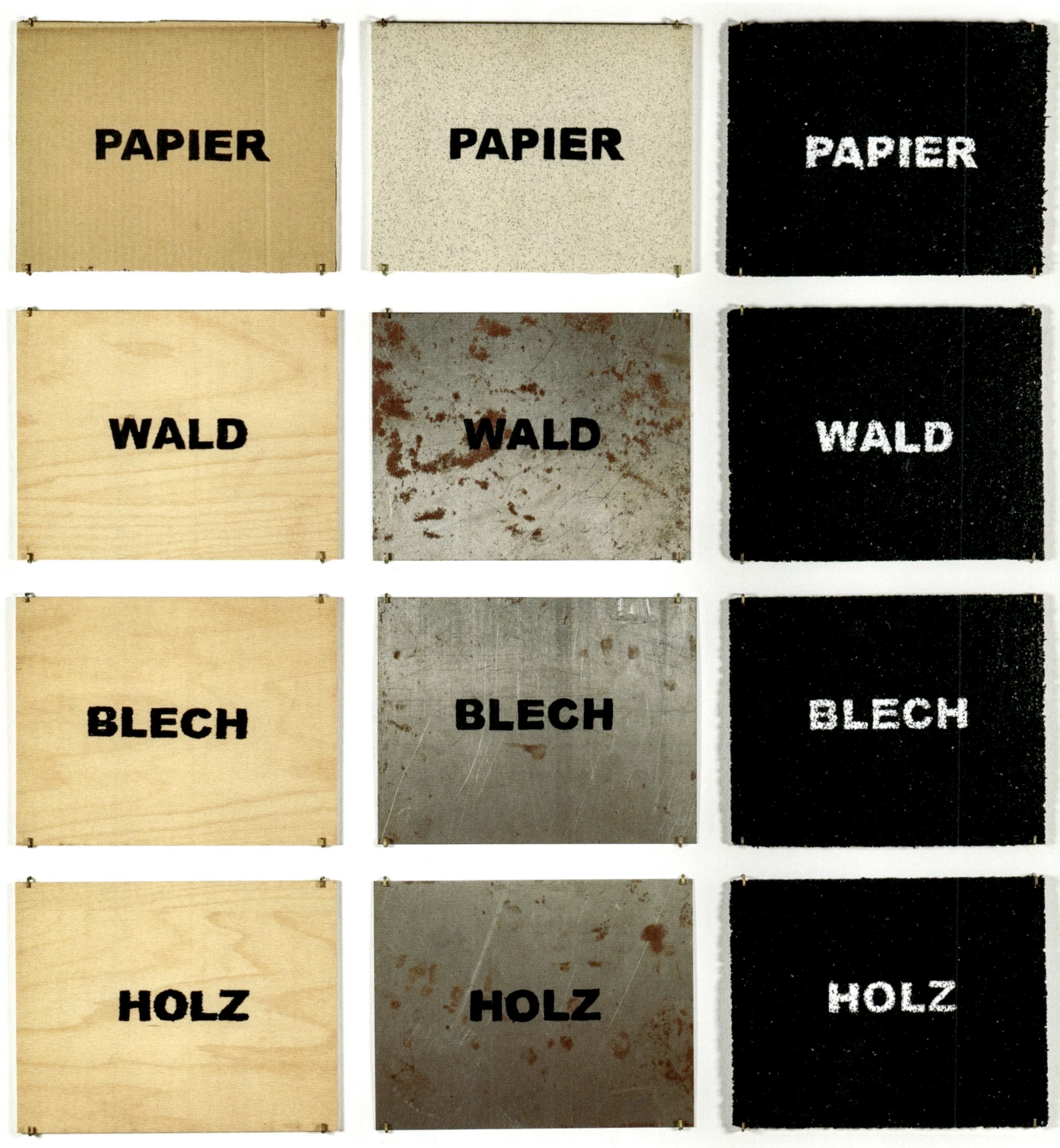

Iso- and Polymorphic Language of Materials / Iso- und polymorphe Materialsprache

1967 | rubber, wood, sheet metal, cardboard, 12 parts, 20 × 25 cm each

Four different terms are written in German onto three different materials: wood, sheet metal, and paper. For example, "Blech" (sheet metal) is written onto wood, onto sheet metal, and onto paper. The combinatorics of term and material defines the meaning of the word. "Blech" on sheet metal demonstrates the name function, because term and material are identical, homo- and isomorphic, like blood written with blood. "Blech" on wood or paper demonstrates a nonidentical relation of term and material. The difference of text and context, of term and material sets associations in motion and creates a free poetry. When the word "Papier" (paper) is written on paper, the meaning of the word is bound, denotatively fixed. The same word written on sheet metal is contradictory, freer in its meaning, and awakens more associations, is more poetic.

—Peter Weibel, 2023

Translated from the German by Peter Dahm Robertson.

p a t r i a

        i d e a l e

p a t r o n e n

        i d o l e

p a t r i o t e n

        i d i o t e n

leergang (in lautverschiebung)
für pa-t$_{ri}$ote$_{l}$n

Idle Speed in Sound Shift / Leergang in Lautverschiebung

1967 | visual poetry; typescript

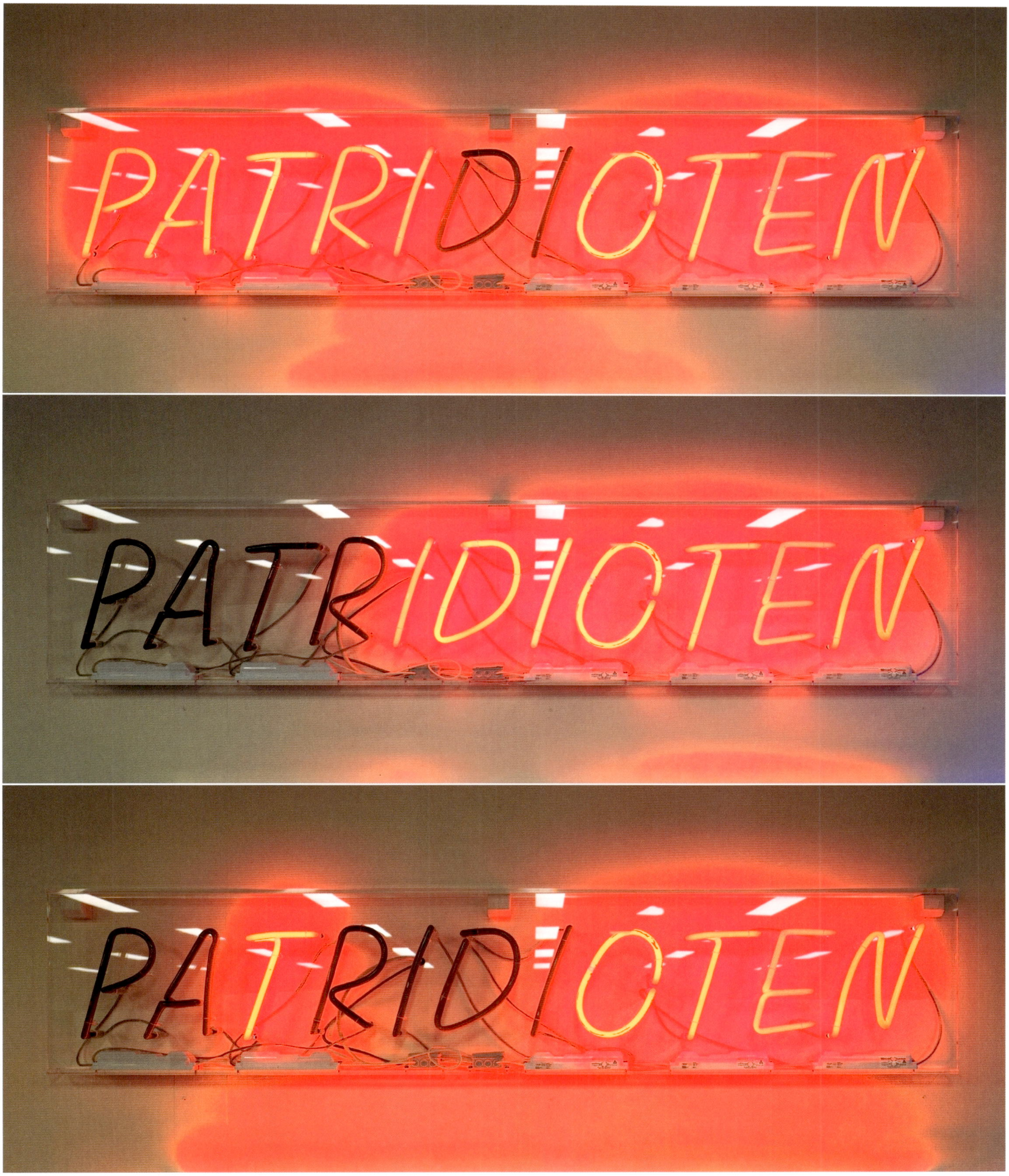

Patridioten

1967/2014

from the series *Motion Poems*
mobile text; neon tubes, cables, time switch, 240 × 70 × 27 cm

Controlled by a time switch, the words *Patrioten* (patriots), *Idioten* (idiots), and *Toten* (the dead) are illuminated.

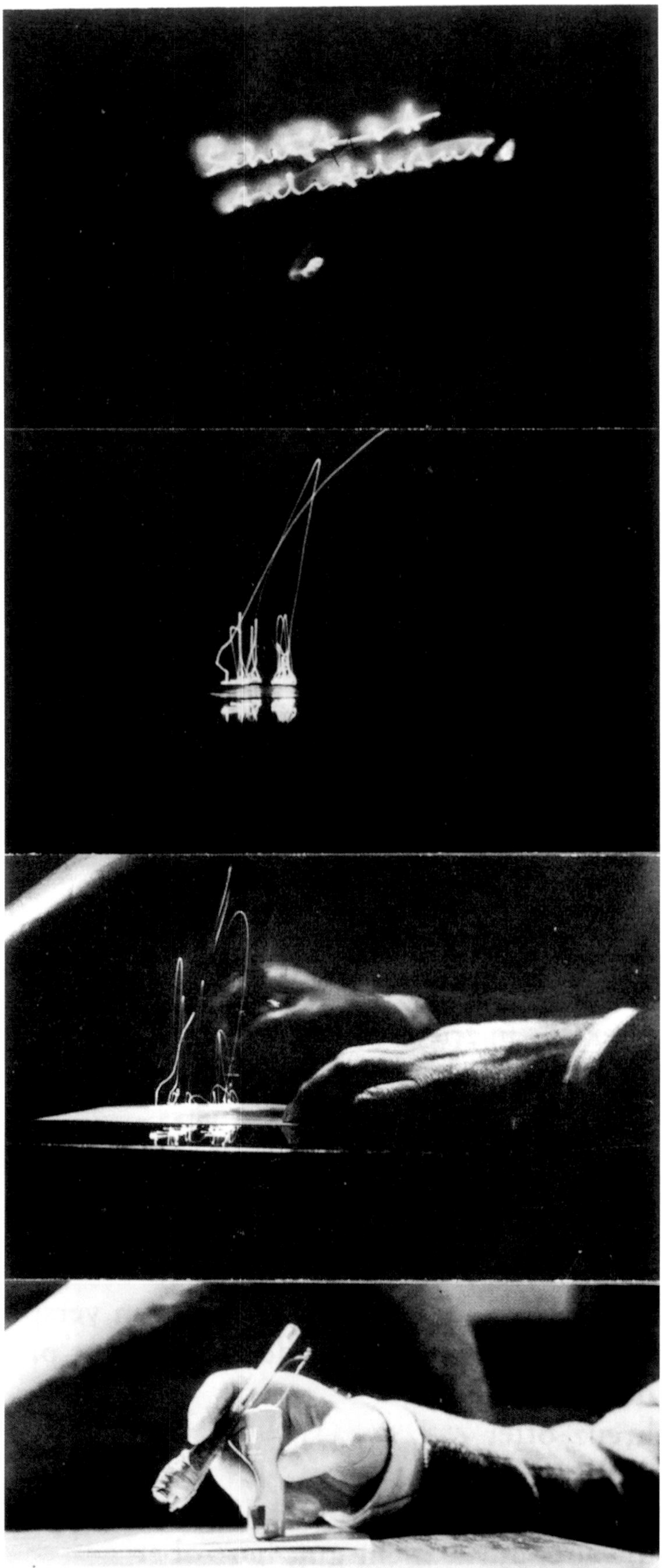

Writing Is Architecture / Schrift ist Architektur

1967 | 4 b/w photographs on cardboard

The photo sequence *Writing Is Architecture* documents the materialization in light of the title's three words. Taken from different camera angles and perspectives, the photographs show my hands as I write the words in a darkened room. The assertion that "writing is architecture," a process in three-dimensional space, is proven by the process itself, a recursive proof.

I do not just experiment with letters on the page, as concrete poets do; I also experiment with the tools that are used to produce words, from the typewriter to the fountain pen. In the case of *Writing Is Architecture*, I attached a tiny battery-powered light bulb to the tip of a ballpoint pen and then took long exposures of the motions of writing from various points in the room. I chose to work with long exposures because this technique is unique to photography, and also because of the singular way in which it sets a cognitive process in motion for the viewer. What I wanted to demonstrate in *Writing Is Architecture* was not only the well-known temporal aspect of writing, but also its previously unknown spatial dimension. The images reveal that writing really is motion in space and time executed by hand, which creates an architecture that is invisible to the naked eye. The assertion contained in the work's title is thus proven by the process, by my experiment in the medium of photography: handwriting is a process in space, an architectural maneuver. —Peter Weibel, 2021

On Human Archaeology

Learning how to write starts the process of human socialization, the acquisition of a geometry of space: the continuity of intuitively experienced space collapses into discrete elements – that is, the dimensions of space, the first, the second, the third. An inscrutable universe proves to be reducible to elements; space becomes constructible from elements, becomes intelligible, distributable, manageable: architecture.

The acquisition of writing signifies the acquisition of space.

—Peter Weibel, 1967

Both texts translated from the German by Patrick Hubenthal.

The Myth of the Twenty-First Century

Excursus on Marshall McLuhan
Field of events from infinite rules

A GUIDE FOR THE YOUNG INTELLECTUAL

The following program is dedicated to the intellect. There is, at any rate, no other option. Utopia via negation is at a dead end. We know it. Kissing the sky is a dead end, not a tunnel – and stone free means bare-assed. We've experienced it.

An aesthetic system reflects (it cannot do otherwise) a political one. The prevailing political conditions, which lay themselves out in our language as well, will unfold today too. We cannot climb over it or them as long as we continue to communicate in it. To put it bluntly: He who speaks must also shut up.

A word from Marcuse should do you good.

What we want for you and your welfare are fairy tales. But, playboy of democracy and language, teach yourself a lesson, for your easy understanding of which we will activate a field of events, OK? A field of events arises through a process of selection from a set of events, as we have learned from pamphlets on cybernetics. Every day, a message arises through the selection of events from an infinite set of events $(x_1, x_2, \ldots x_n)$, and each event ... also has a specific probability $(p_1, p_2, \ldots p_n)$ attached to it, where $\sum_{p=1}^{n} p = 1$

A field of events arises from a set of events, X, through the incorporation of specific associations between event pairs, and through the inclusion of each event's complement (which occurs when the event does not). The expected value of event set X, its median information content $H(x) = \sum_i p_i \, 1d \frac{1}{p_i}$ is in any case on our side. The messages and the development of the field of events are up to you. The fact that you're being taken for a ride here (by us, too) is the consequence of a reality and a language that we did not choose, baby.

Honey, don't ask for a more detailed explanation of our thoughts – you're smart and smooth enough to know: language is insufficient to describe reality.

The event pair is the institutionalized reactionary relationship between author and audience – us and you, high-minded dearie. Us: that's the tape player and the movie projector. You: that's the opaque projector and the record player. Either *a* or *b*, "this occurs when that does not," us or you. If you want to interrupt the lecture, go see Valie. She'll give you a flashlight you can use to illuminate a selenium cell, thereby interrupting our program (audiotape and projection) as long as you want – meaning as long as you shine the light. If you want to ask a question, take your letter and go see Valie. While you illuminate the selenium cell, Valie will put your letter in the opaque projector and Weibel will give the answer prepared for it. Of course, since the political nature of language is also expressed in the alphabet (in its hierarchy), it won't be your turn until your letter comes up. Like language and democracy (which themselves are both violence), this sequence can also be subjected to violence. Provision has been made for even the simplest selection from a set of events, a selection from a set consisting of a single event: yes or no. A press on the button provided, and the lecture etc. will be drowned out by music.

—Peter Weibel, 1967

Translated from the German by Patrick Hubenthal.

Part of the equipment used in the piece (relay circuit for photocell)

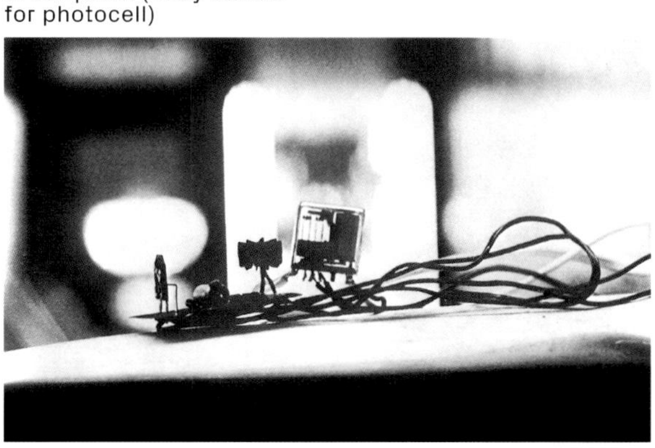

The Myth of the Twenty-First Century / Der Mythos des 21. Jahrhunderts

1967	participatory performance and interactive environment; 2 tape recorders, tapes, selenium photocell, pole, cables, flashlight, film projector, films, letters, opaque projector, performed at Internationales Kulturzentrum, Vienna, December 1967

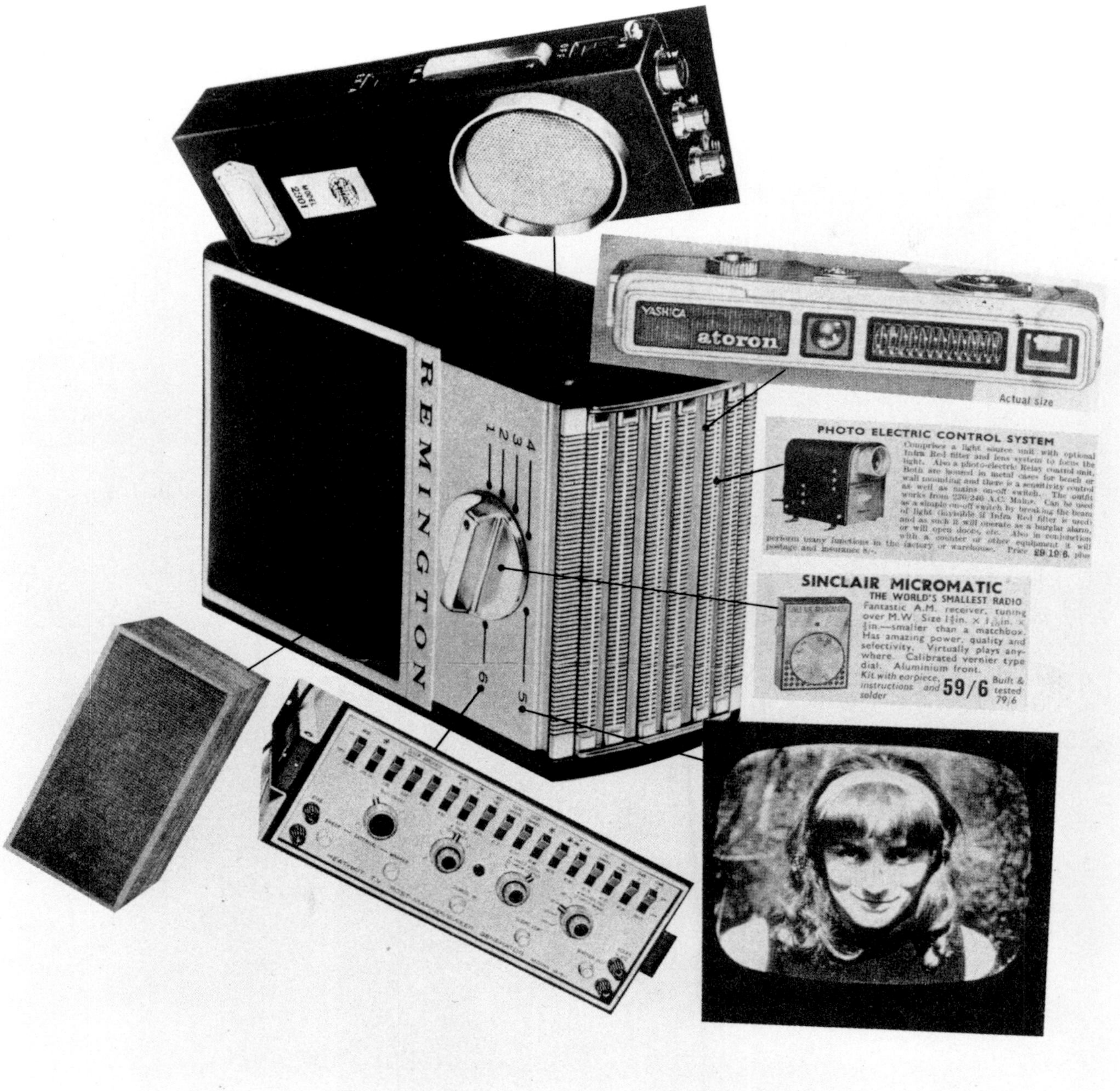

The prophecy of the mobile smartphone. As early as 1967 Peter Weibel imagined a universal device containing in miniature format all information processing (tele)machines,
from photography to radio,
from telephone to television.
He began to collect photos of technical miniatures and electronic toys in order to create the collage reproduced here. *Information Unit* envisions a device the size of an electric razor that contains everything modern humans desire:
a radio,
a control system,
a camera,
a walkie talkie,
a loudspeaker,
a video camera,
a TV set,
a spy pen,
and a telephone.
Through the development of electronic microminiaturization, integrated circuits, and microchips, Weibel's utopian project has become an actual reality decades later.

Information Unit

1967 | collage

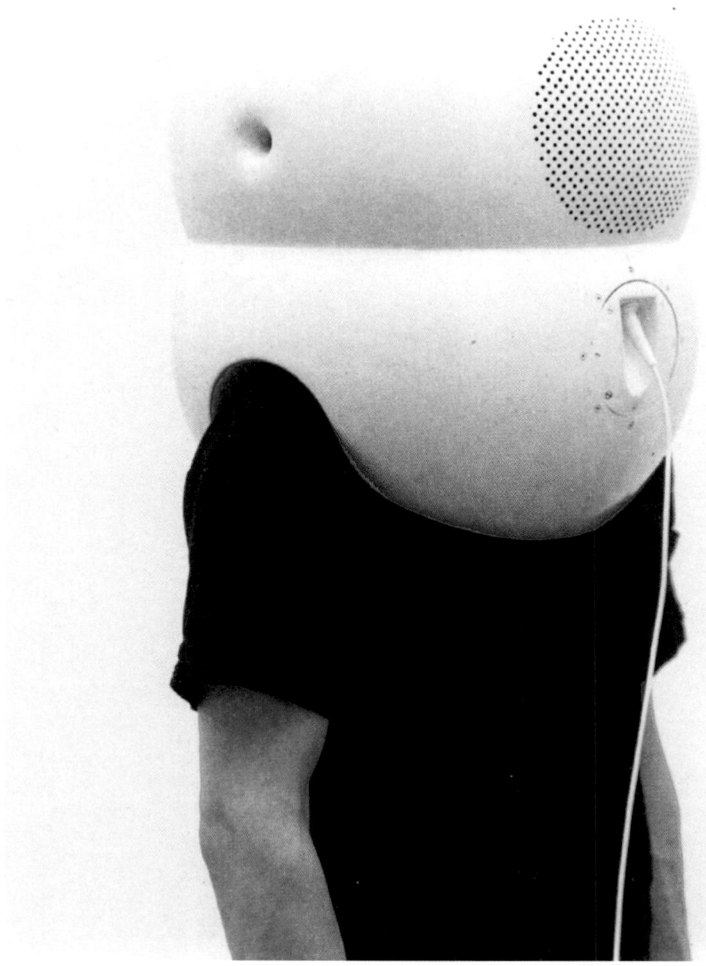

I wanted to proclaim my texts in the streets as a guerrilla poet, while remaining independent and protected so that no one could interfere with my freedom of speech. I therefore spoke to Walter Pichler, the renowned designer, about a SPEECH HELMET equipped with a face shield, microphone, amplifier, battery, climate control, etc. He turned that idea into *Small Room (Prototype 4)*.

—Peter Weibel

Small Room (Prototype 4) / Kleiner Raum (Prototyp 4)

1967	Walter Pichler; idea: Peter Weibel 3-part sculpture; helmet, polyester, white varnish, built-in microphone, perforated sighting field, 48 × 40 × 40 cm

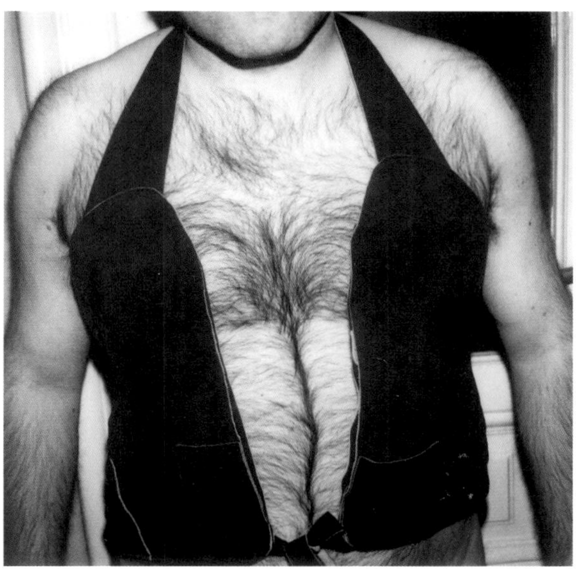

Media Lung / Medienlunge

1967	deer skin, 84 × 64 cm

Media Lung is a lung made of deerskin supplemented with electronic devices, which intends to optimize human capabilities. With a camera, a dictaphone, or a universal device as dreamed of in the work *Information Unit* (see page opposite) under the armpit like a gun holster, the wearer becomes an everyday guerilla, always ready to document, to save, or intentionally intervene in the environment. Here, the artist envisions the electronically or technically augmented human being. An idea that also became reality years later through ubiquitous wearable computing or smart clothing; that is, garments equipped with electronic devices or fitness trackers to measure body conditions.

Works like *Media Lung* and *Information Unit* make it clear that many of Weibel's utopian visions have become part of contemporary reality. His prophetic predictions, however, are not based on clairvoyance, but on the precise analytical observation of the specific present. This is the only way to draw conclusions about the near and distant future.

Camera without Borders

Development of a camera that would also be able to film life behind closed doors. The principle: the camera sends out waves which penetrate obstacles such as walls, bones, etc. in much the same way as sound waves. The waves, whose range can be regulated, would be reflected at a certain adjustable distance and would come back into the camera, where they would reproduce the "image" on a screen. An auxiliary device is necessary for the images captured in this way to be decoded, "deciphered," converted to our visual standard.

Both state secrets and the secrets of private life will become impossible. Everyone will be able to see into everyone's hearts. Nothing will remain hidden.

Street Film: A Real Film

During the showing of a film, the street-facing wall of the theater is broken open with jackhammers and torn down so that people can look through the wall onto the street.

What interests me here is the shift in the direction of the gaze: the vertical orientation toward the screen turns sideways toward reality.

Cloud Films

Projection projects

With the help of the new laser system from ASEA Central Laboratories in Sweden, which is able to reach cloud layers up to 8 km high, projection of films and images onto cloud banks. The daily news in pictures in the firmament! On artificial cloud banks if need be. A constantly rotating information stream in the sky circling the whole Earth.

Day without End

Projected projects. Real cinema.

A rocket carries nitrogen into the uppermost layer of our atmosphere, 130 km high. A cloud of nitrogen oxides begins to spread. Ultraviolet light from the sun breaks down oxygen molecules into individual atoms. Each time an oxygen atom encounters a nitrogen molecule, there is a flash of light, creating an artificial aura. There are various other methods.

Total Cinema

Night becomes day. The gigantic screen of the sky, conceived as a chemically prepared canvas, produces light shows that touch the whole globe. The atmosphere as a light show theater. Artificial control of light conditions above the Earth and for the Earth. Projection of the word "love" over the cities.

Translated from the German by Patrick Hubenthal.

Project Films and Projected Projects / Projektfilme und projizierte Projekte

1968–69 | concepts; partly published in *Interfunktionen* 4 (March 1970)

Amplified Language and Man
Technological inventions that alter the entire structure of communication, our forms and modes of communication, thus altering humans and their world as well.
A pair of **reading glasses** surgically grafted onto the human body that decode and translate, enabling the wearer to read one hundred times faster than without the glasses (normal state of affairs today). ...
Information ring on the finger that places me in continuous wireless contact with my associates.
A sort of **minimal computer** / central computer terminal, providing access to all information.

Note: Peter Weibel, "Konzepte für Kommunikationsaktionen" (1967), in Peter Weibel and Harald Falckenberg, eds., *Peter Weibel: Das offene Werk 1964–1979*, exh. cat., Neue Galerie Graz, Landesmuseum Joanneum (Ostfildern: Hatje Cantz, 2006), 317. Translated from the German by Patrick Hubenthal.

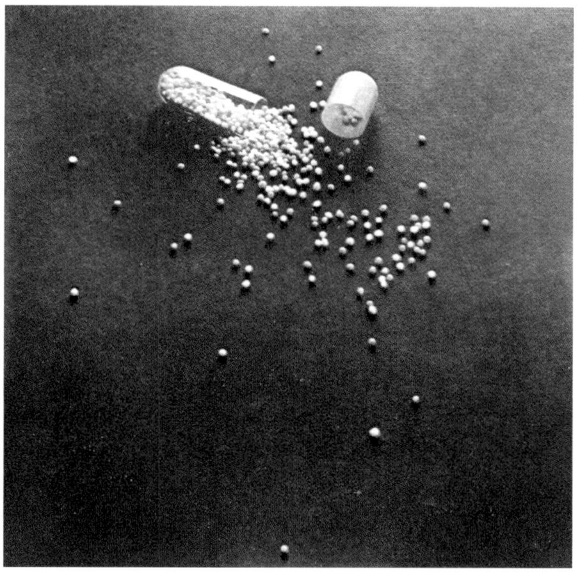

Pill Movies, or the Sense Discorrelator / Pillenfilme oder der Sinnesdiskorrelator

1968 | project film

I am thinking of films as chemical medium translators: projector, film, screen, and camera all in one as a pill. Pill movies, which deintegrate the sensory organs and deliver visions beyond language. I am thinking of films in the form of a new sense organ that provides any sensory quality I desire, in the form of a new sensory correlator that provides me with air when I step into the ocean, that turns lung-breathing into gill-breathing or vice versa, that also does not shy away from discrepancies between the senses and allows me to touch foam when I see stone, smell chocolate when I eat grass, hear violins when I seal envelopes, feel water when I stick my finger in the electrical socket, etc. —Peter Weibel, 1968

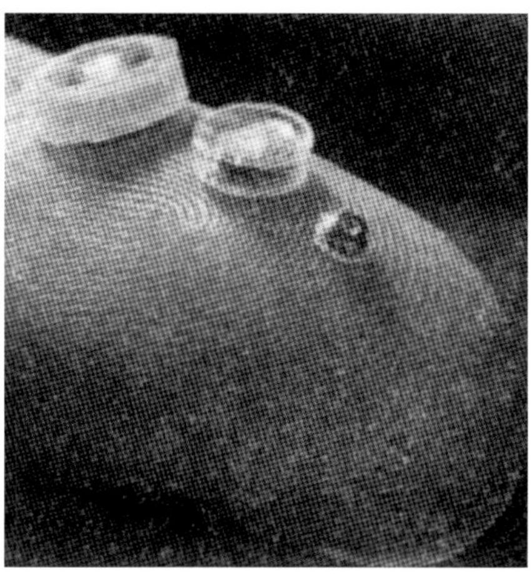

Radio Pills / Radiopillen

1969 | project film, concept film

Implantable "passive" radio pills, originally developed to measure fluid pressure inside the eye, then repurposed as fully functional radios that can be installed anywhere in the body:
Hear with your eyes, hear with your stomach, hear with your little finger. The auditory organ is decentralized, distributed over the entire body.
The vision: every organ in the body takes over the function of every other organ. The eye can speak and fart, the nose can sing and cry, the tongue can see, etc. —Peter Weibel, 1968

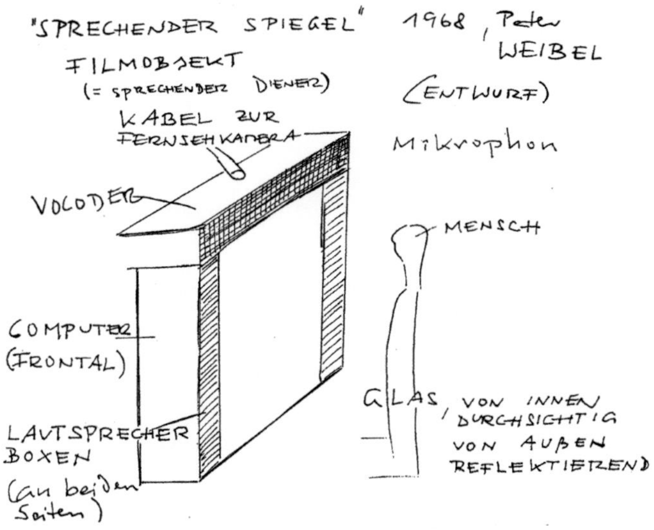

SYNDIC

für c. m. kornbluth

(terra extra band 93)

objekt für kinosaal & publikum (69, weibel)

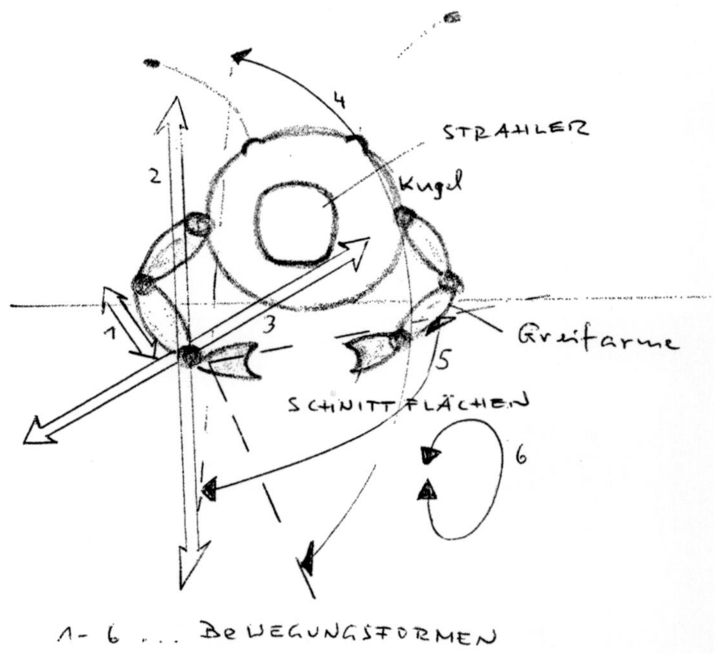

Light-sensitive cells applied to the glass wall measure changes pertaining to light values. The analog processing computer hardwired to the light sensor "recognizes." Movement transmits impulses to data storage (program of sentences for the determined area), from where the corresponding sentences are transmitted to the loudspeaker.

The VOCODER translates the user's sentences into "electr. signals" that a data processing unit can recognize. The VOCODER itself can speak as well (without tape).

Once a computer can also measure light, temperature, humidity, etc., a dialogue such as the following will become commonplace:

Person: (stretching) What should I wear today? "T. M.": Stop lolling about, the humidity is ..., the temperature ..., etc. So wear the cotton coat. Its material (...%, ...%) is especially suited to this weather, etc.

Translated from the German by Peter Dahm Robertson.

Expanded Cinema

1968/69	concepts; published in *Film*, no. 11 (November 1969): 45. See also p. 167 in this volume

SYNDIC lies quietly on the stage, illuminated by a spotlight. Its antennas measure the ambient humidity, the audience merely watches. Uncertainty increases, spinal sweat centers run hot, heat nerves are stimulated, local sweat secretion works accordingly. (Sweat contains approximately 1.8 percent solid components, almost all ureic, only in far lower concentration than in urine.) This "wetting the air" acts on SYNDIC as a stimulus.

WARNING! NO SWEATING!

Ambient humidity has reached the reference frame's programmed set value and SYNDIC takes flight. SYNDIC flies above people's heads, its lights downward, with a navigational computer modifying its flight path according to the humidity data measured by antenna and potentiometer. Once the location of highest sweat production has been identified, SYNDIC, sirens wailing, plummets down toward the subject, its grapplers whir through the air like sabers, its cutters spin... After the extermination of the victim, SYNDIC flies back to its place. It waits until another calibrated detection threshold is reached at a different point in space, where it will then stab in a straight line, without hesitation or consideration, executing its programming perfectly.

The work of SYNDIC is done with doors closed.

Drones

The greater the fear, the greater the danger, because the more he will sweat.

However, there is the (admittedly very low) probability that the humidity threshold value triggering the flight mechanism will not be reached.

P. S.

At a certain level of sweat production, a loudspeaker (in SYNDIC) blares, "Warning! No sweating!" This message is repeated with increasing frequency and volume the closer the moment of takeoff approaches. During the flight, there is silence.

Translated from the German by Peter Dahm Robertson.

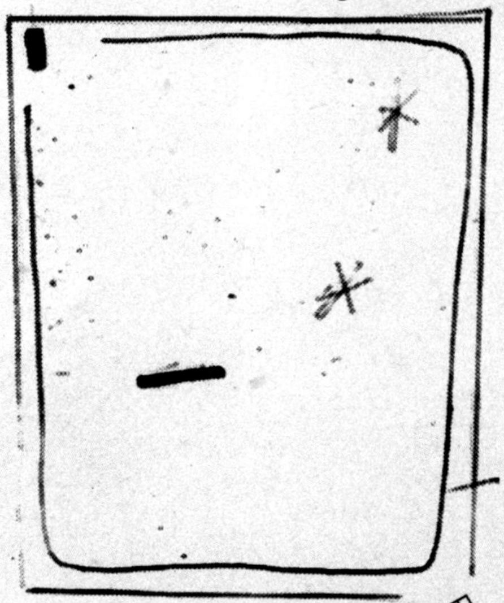

The tiny ultrasonic alarm device in the corner takes no notice of visitors when they first walk in. Seeing no cause for concern, they walk past the warning sign, enter the wave field, and set off the alarm. The cloth that has been hanging unnoticed overhead, covering the entire ceiling, falls and traps the visitors beneath it.

Private property
Keep out
Enter at your own risk

Architecture as administration
Administration as invisible terror

Invisible architecture
Verbal behavior
Invisible state

Translated from the German by Patrick Hubenthal.

The Invisible Boundary / Die unsichtbare Grenze

1968 | gallery project; manuscript

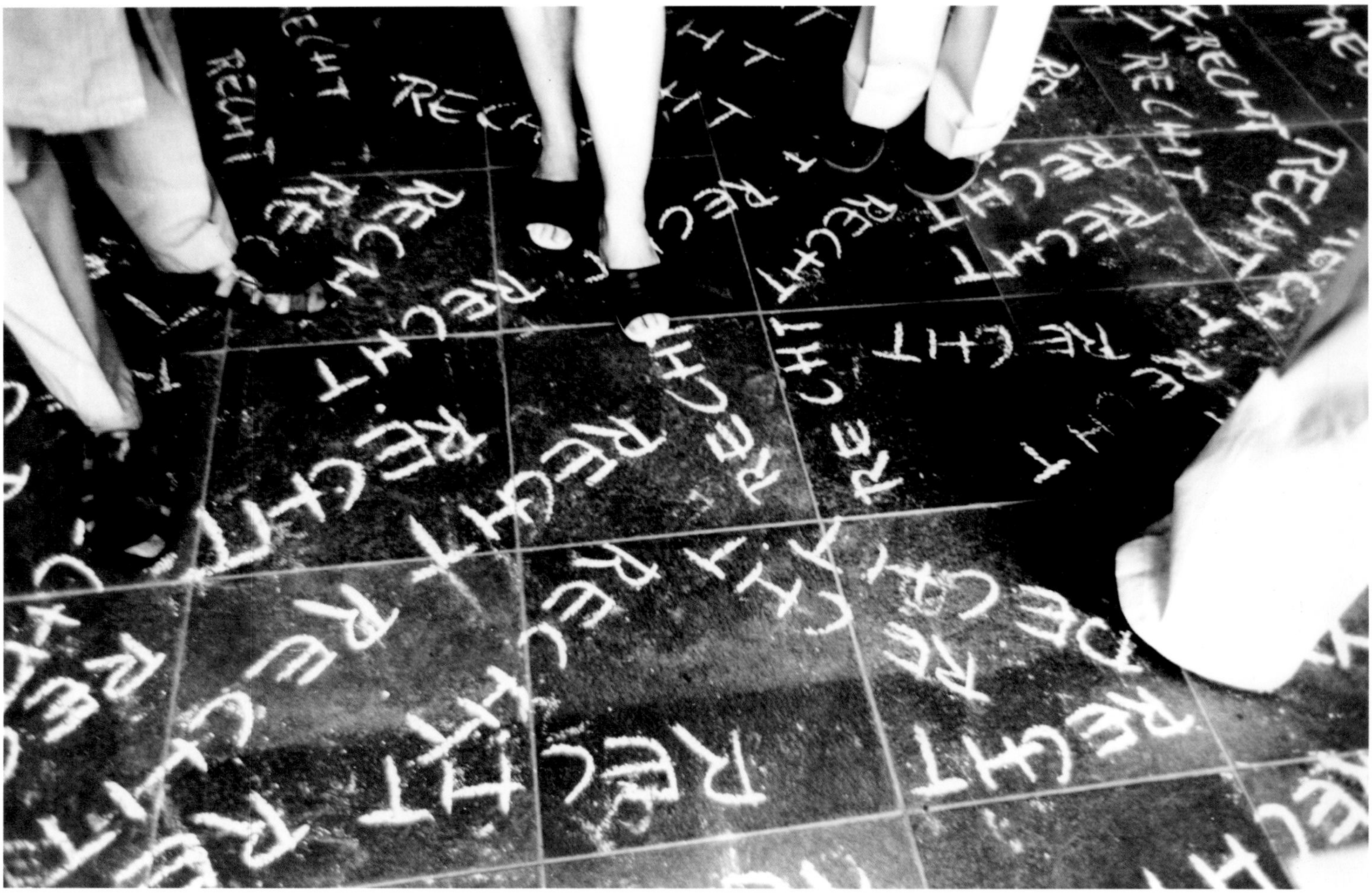

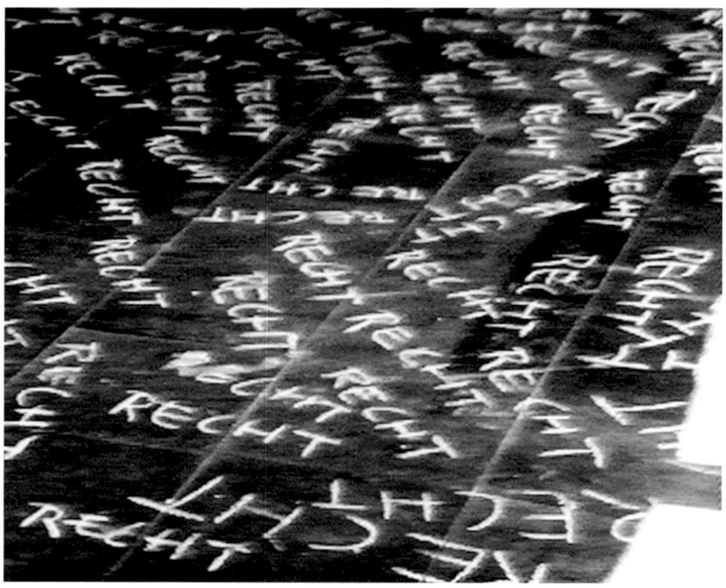

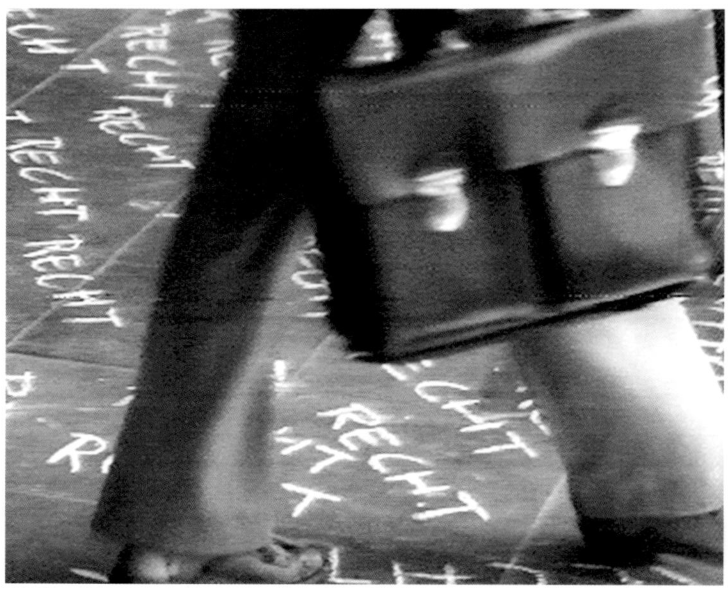

Installation view and video stills *Studententage* Melk Abbey, Austria, May 23, 1968

Trampling on Law /
Das Recht mit Füßen treten

1968	participatory text action/ text environment; white chalk, dimensions variable

The floor sculpture *Trampling on Law* was installed in 1968 in a group exhibition at Melk Abbey in Austria. The visitors entered the exhibition room, where they did not see any works of art; instead, all over the floor the word *Recht* (law) was written. Walking around and looking for the artwork, visitors stepped on the word "law" and thus carried out the act described in the title. The work was created through the actual walk-through, the participation of the visitors.

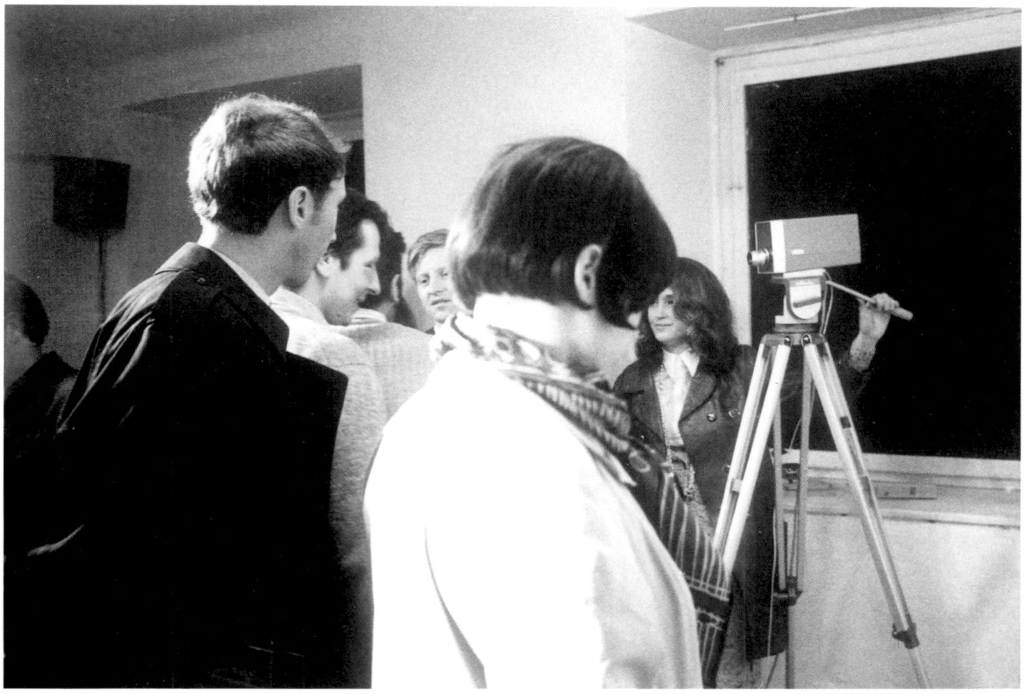

From the left: Hans Puchinger, Peter Hassmann, Hermann Hendrich, and VALIE EXPORT

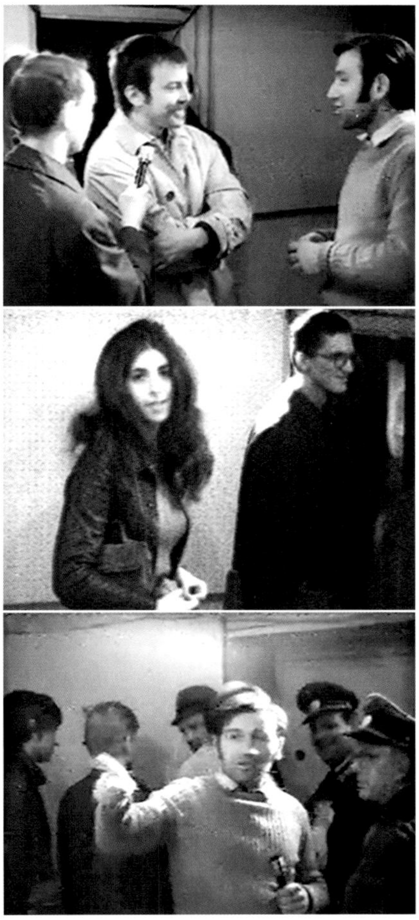

Top: Helmut Swiczinsky (Coop Himmelb(l)au), center: Ingrid Wiener, bottom: Weibel talking to the police

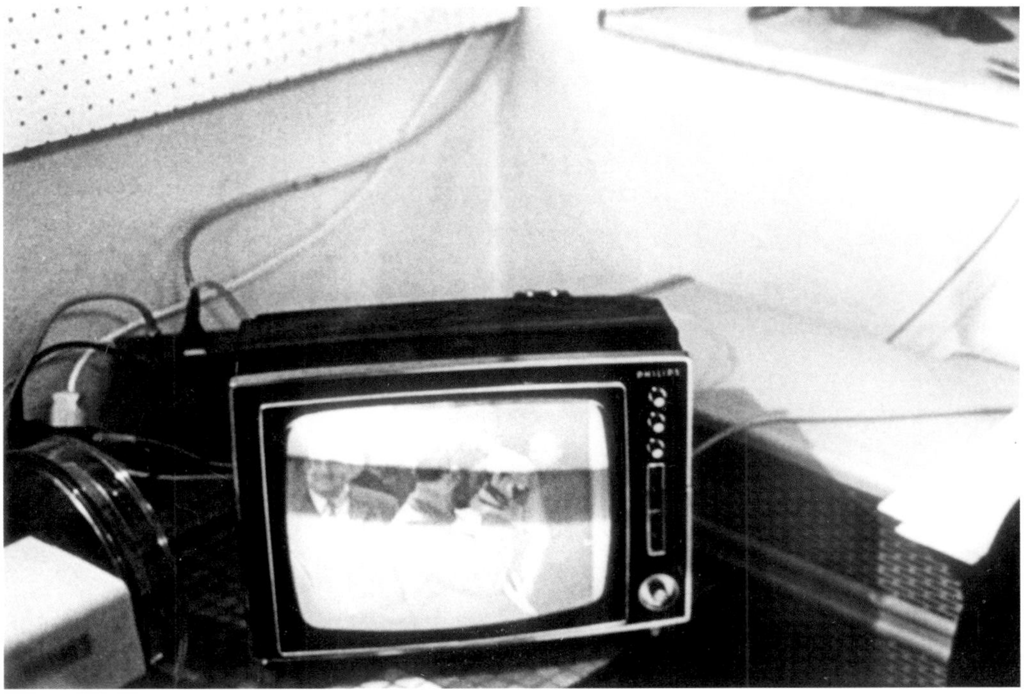

The Public as Exhibit (Audience Exhibited) / Das Publikum als Exponat (Ausstellung des Publikums)

1969	participative video performance, closed-circuit installation; Galerie Junge Generation, Vienna, April 10–11, 1969

Visitors entered the exhibition *multi media 1* at Vienna's Galerie Junge Generation. They were interviewed with a video camera. These interviews were shown simultaneously on a monitor in the first room, and, with a delay, on a monitor in the second room of the gallery. The visitors could see themselves entering the gallery and being interviewed. They became the spectator and the artwork that they observed.

Due to the immediate availability of the video images, the audience in the closed-circuit installation was psychologically confronted with its own image in the public space, and thus became an exhibition object itself. The exhibition was a radical gesture of audience participation; in the gallery, only the audience was exhibited. Since a crowd had gathered and no artwork was shown, people informed the police believing that there was a conspiracy meeting happening. Soon afterwards the police did indeed arrive. From the beginning, state institutions were suspicious of Peter Weibel's art.

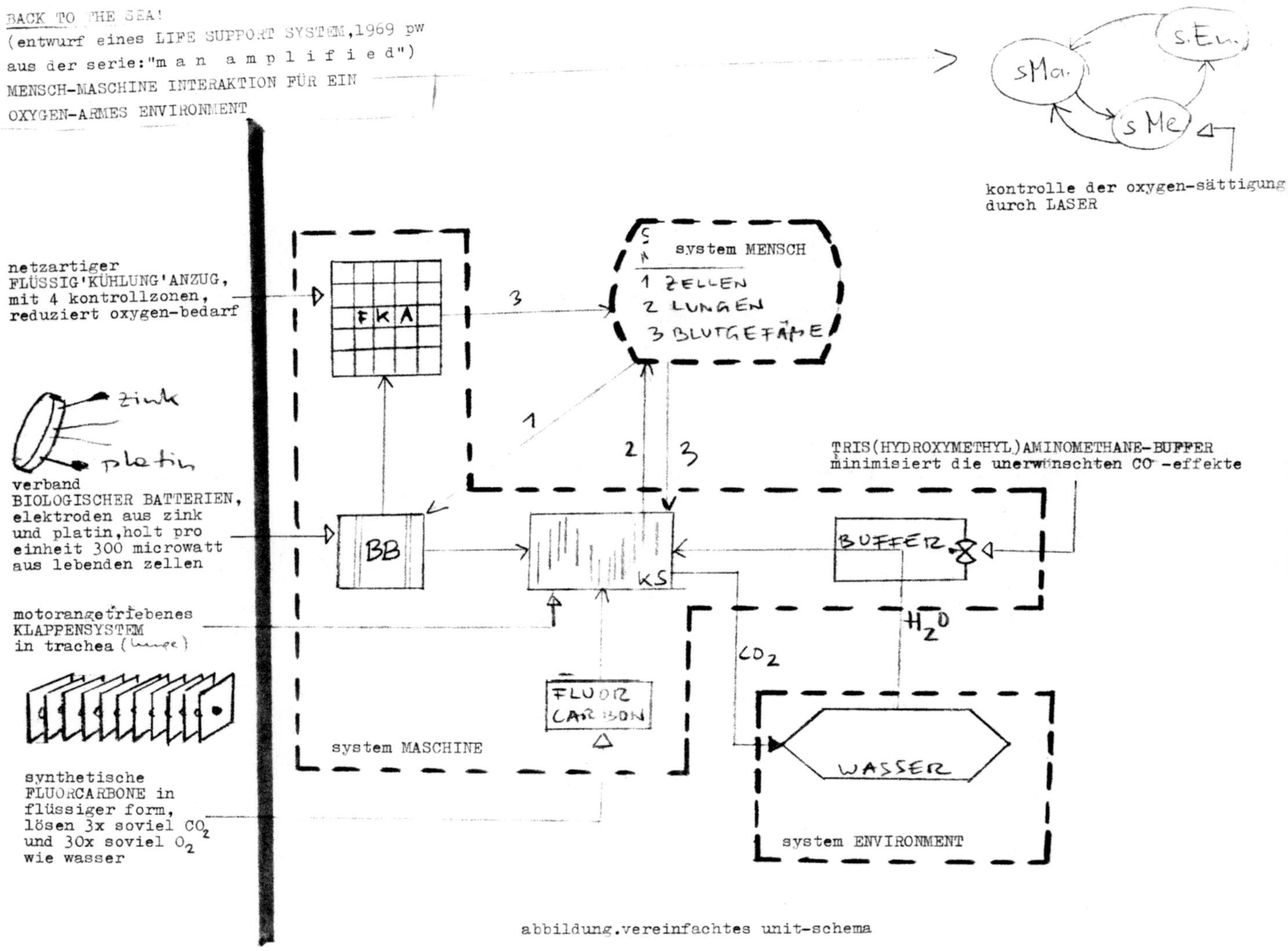

Concept for a life-support system

Human-machine interaction for a low-oxygen environment

This is a project from the series *Man Amplified*, one of whose aims is to use technological add-ons to restore abilities that humans have lost through natural evolution. For example: the eagle's keen eyesight, the cat's night vision, the bat's ultrasonic hearing, the fish's ability to breathe through gills, etc. Humans can live in water too, if they can extract enough oxygen from it and excrete enough carbon dioxide. The biggest obstacle is the removal of carbon dioxide from the bloodstream. Given that

(a) air contains 10 times as much oxygen as water,

(b) the amount of carbon dioxide exhaled is 50 milliliters per liter when breathing air but only 30 milliliters per liter when breathing water, and

(c) the viscosity of water is 36 times that of air,

in order to sustain life, the life-support system must

(a) pump a volume of water into the lungs that is 10 times the normal air volume,

(b) expend 30 times as much energy in filling and emptying the lungs, and

(c) expel 20 times more water than air.

However, our system employs fluorocarbons, fluid cooling, and a buffer, which dramatically minimize (a) and (b), thereby making (c) possible, while the work performed by the valve system supplies the necessary increase in energy.

—Peter Weibel

Translated from the German by Patrick Hubenthal.

Back to the Sea!

1969	from the series *Man Amplified* concept; typescript

Peter Weibel: Premature Encounters with the World Spirit

Peter Sloterdijk

There is a geopolitics of the unhappy consciousness that has gone unremarked by the writers of history, both in the field of ideas and in that of the arts. It has long been known that misery likes to wear national colors. Who has not heard of "English spleen" or *le mal français*? There is no need to go on enumerating the list of European and extra-European nations and their temperamental afflictions. One would always arrive at the same result. It would be the collective psychological counterpart to Tolstoy's best-known dictum regarding (un)happy families:[1] Happy nations are all alike – though has any explorer ever discovered one? – whereas every unhappy nation is unhappy in its own way.

If, as Georg Wilhelm Friedrich Hegel (1770–1831) says, the "unhappy consciousness" indicates the disintegration of what ostensibly was originally a harmonious relationship between the world and the self,[2] then Austria is undoubtedly one of the heartlands of the unhappy consciousness in its modern form. This finding has held true ever since the days when authors such as Arthur Schnitzler (1862–1931), Fritz Mauthner (1849–1923), Sigmund Freud (1856–1939), Hermann Broch (1886–1951), and Robert Musil (1880–1942) worked out effective phenomenologies of the out-of-tune life for their era. Their predecessor was Johann Nestroy (1801–62), whose devastating humor illuminated the profound not-fitting-together of individuals in his farce *Frühere Verhältnisse* (Former relations), which premiered in 1862. Taken in sum, this sort of observation might almost be called the "Austrian lesson": The more individual individuals become with regard to their milieu, the less possible it is for them to be absorbed into their "fellow world."

Current observations from the ex-Kakanian situation[3] show this finding to be truer than ever. Almost all the big names in the Austrian cultural complex of recent years follow the aforementioned pattern in every respect: they exemplify the radical collapse of the connection between the self and the world in a way that is unvarying, yet in each case thoroughly individual. What would Austria's literary and art scene have been like over the last few decades without Thomas Bernhard's (1931–89) emigration into self-exile in Ohlsdorf? What would it be without Elfriede Jelinek's (b. 1946) submersion in her Vienna and Munich refuges, from which she sends out her encyclicals of dissatisfaction each year with the doggedness of the irreconcilably out-of-tune? What would it be without Otto Muehl's (1925–2013) constant flights, his wanderings from his laboratory of excess in the Burgenlandian village of Friedrichshof to prison by way of La Gomera, and then on from there to Faro, his outpost of indigenous delusion on the Algarve? What would it be without the withdrawal of the disenchanted cyberneticist and resigned novelist Oswald Wiener (1935–2021) to an obscure speck called Mürzzuschlag, which he dubbed "the spleen of darkness," an un-place that served him for many years as a sentry post from which to observe the collapse of the self and the world? And what would it be without the dubious self-mutilations of Günter Brus (b. 1938), who periodically chose going to ruin over being a citizen of his country – and who, for many years, preferred to take up residence on islands in the Atlantic rather than in the home he found impossible?

One must evoke these connections when undertaking the attempt to characterize the phenomenon of Peter Weibel, born in 1944 in Odesa, in all his Austrian/un-Austrian self-will. More than anything else, what distinguishes this eminent artist and subtle theorist is the fact that of all the notorious Austria-avoiders, he has remained his country's most loyal emigrant. His curriculum vitae – marked by innumerable engagements all over the world – leaves no possibility of doubting the following statement: for the gifted of a nation, emigration, whether internal or external, can sometimes be a binding fate. The fact that Weibel chose the external may be the most telling element of his intellectual and aesthetic engagement. There are situations in which only the emigrant can be true to his origins. By virtue of his intellectual and artistic physiognomy, Weibel qualifies as a belated member of the Vienna Circle – transplanted to the age of new media. There is probably no one alive today with whom one can discuss Ludwig Wittgenstein (1889–1951), Rudolf Carnap (1891–1970), Otto Neurath (1882–1945), Moritz Schlick (1882–1936), and all the other great theorists more seriously than with him. Neurath, unfairly forgotten, has retained Weibel's sympathy right up to the present hour, although Kurt Gödel (1906–78) is his true hero. He belongs to this circle himself, because his predecessors merited a distinction whose time has come to be named aloud: they were all, given the confusions of late Habsburg consciousness, veritable therapists of culture, not merely artists of perception and not just cunning exploiters of the Austrian disease. They all had in common a belief in the healing power of logical treatments for existential and collective cramps. As far as they were concerned, the unhappy consciousness should never be allowed to have the last word. They would sooner have broken entirely with metaphysics, that compendium of the disadvantages of being born, than with the imperative to make life sensible and happy.

When Weibel stepped onto the stage of the new Austria in the mid-1960s as an excessively gifted twenty-year-old with a sense for the taste of things to come, he first fell in, one might say inevitably, with the Vienna Actionists, never suspecting that he would have to become the lone wolf in the pack. His comrades in arms used the word "action" to surrender to ever-greater disinhibitions under the pretext of liberating the body. Could anyone have doubted, back then, that humans are metabolic creatures? The Vienna Actionists thought they were advancing culture by hauling shitting, pissing, and ejaculating up onto the stage, from Günter Brus's unsavory, thoroughly public masturbations and defecations to Hermann Nitsch (1938–2022), who set rivers of blood flowing, a Wagnerian who had strayed into a butcher's shop. Weibel, by contrast, was turning, even then, to the materials and the gestural-pragmatic foundations of the art scene, holding that the human being must be placed at the center of attention as a sign-producing and sign-receiving creature. With Weibel – and only with Weibel – the misunderstanding called Actionism was put right: the human being is the *sign-spraying* animal. It acts by allowing itself to be acted upon by devices and symbols. Let others talk about wetnesses. Niklas Luhmann (1927–98), the most important German-speaking exponent of sociological systems theory and sociocybernetics, once said succinctly, when asked to sum up the principle he lived by: "Good mind is dry."[4]

Before Weibel, only Stéphane Mallarmé (1842–98) and Hugo Ball (1886–1927) had been so insistent that poems were operations with letters and tone values. With Weibel, the manipulation of letters was declared a public good in the mid-1960s. Yet it was not just the letters of the alphabet that were judged to be building blocks of culture in the eyes of the young mediologist *avant la lettre*: under the influence of tonic rationalists such as Max Bense (1910–90) and subtle Surrealists such as René Magritte (1898–67), Weibel discovered the in-between world of signs and their technological carriers, in all their composite forms and artifacts, astonishingly early on. For anyone wishing to reencounter the charm of the Enlightenment – in the emphatic sense of the word – there is no better way, even today, than to delve into the work of Weibel the young Actionist.

In doing so, one partakes of nothing less than a second-stage Enlightenment. Almost thirty years ahead of the triumph of the computer, Weibel had already made the transition from the canvas to monitorial space. Almost twenty years before the worldwide impact of the works of Friedrich Kittler (1943–2011), Weibel had explored the world of signs and machinery and incorporated it into the repertoire of art's new languages. The "observer" was already a permanent guest in the Weibelian universe a decade before Luhmann introduced the figure into the vocabulary of current theory.[5] In Luhmann's thinking, the Enlightenment motto "better world" was still intimately linked to the watchword "smarter art" in the most congenial (if also naive) way.

Here we have occasion to mention the "tragedy" of Weibel's career, hyperbolic though the term may seem. His greatest talent was revealed in his arriving too early at the fundamental things of our age. By now even teenagers know that in most things, the quality of the technology is all that matters. Since the late twentieth century, the world spirit has spoken in intermedia dialects. Fifty years ago only a tiny avant-garde, represented by names such as Bense, Paul Valéry (1871–1945), Alan Turing (1912–54), and Marshall McLuhan (1911–80), had grasped that the in-between world of signs and their technological carriers represented the epitome of the humanly real. They already knew, back then, that the "world" is a selfie – taken through the technological lenses of each different epoch. Hence the young Weibel's postulate, even then, that a thinking art would need to become mediologically alert. What was once known as "criticism" and "reflection" must now become "media lucidity." One of Weibel's early works is an exemplary reflection of this approach: In May 1968 he wrote the word *Recht* (right, law, justice) several times in big letters on the floor of an exhibition space at Melk Abbey and gave visitors the opportunity to quite literally trample on the word. Could anyone doubt that with this piece, one was very close indeed to the clinical finding regarding post-1945 Austria?

As a virtuoso of media alertness, Weibel has made history in discreet fashion. With the made-up character VALIE EXPORT (b. 1940), he brought into the world an origin point for feminist art – unacknowledged, of course, since the lady with the logo has claimed, ever since splitting up with her inventor, that it is "she herself" under the brand name he created. From his nerve center in Karlsruhe, he has lent momentum for global careers to central figures in world art such as Ólafur Elíasson (b. 1967) and strong minds in world thinking such as Bruno Latour (1947–2022).

Yet anyone who, like me, has had the privilege of experiencing Peter Weibel in action on an almost daily basis, as the observer of an observer of observers, knows that even the most comprehensive exhibition cannot encompass the tireless Actionist's most important work. The reason, in no small measure, is that as a curator and exhibition organizer, Weibel is as devoted to the work of others as he is to his own – or, it even seems at times, to a greater degree. To get an idea of his full oeuvre, one would also have to exhibit his exhibitions alongside his works, and reflect upon his reflections. One sometimes wants to reproach him for his virtually limitless receptivity. Do there really have to be so many exhibitions? Is there really as much talent in the world as what he is constantly placing on view before us in Karlsruhe, in Graz, and at various other venues?

Perhaps it is the essence of genius to meet the inspirations of others on an equal footing. Whatever may have been placed on view before a surprised, cheered, and inspired public, Weibel's primary qualities – namely, his ever-alert farsightedness and his unwavering openness to events in that space of surprises that is known as art – elude exhibitability. If world-impoverished sullenness is the rule with the Austrian version of the unhappy consciousness, then Weibel – the never-resigned, world-embracing hyperactive – will forever remain the exception.

Note: An earlier version of this essay was published as a newspaper article on the occasion of the exhibition *Peter Weibel – MEDIA REBEL: Warning! This Exposition Can Change Your Life* at the Belvedere Museum in Vienna, October 17, 2014–January 18, 2015: Peter Sloterdijk, "Verfrühte Begegnungen mit dem Weltgeist. Eine große Ausstellung in Wien ehrt den Künstler, Kurator und Denker Peter Weibel – Anmerkungen seines Freundes," *DIE ZEIT*, no. 39/2014, September 18, 2014, http://www.zeit.de/2014/39/peter-weibel-ausstellung-wien-sloterdijk. Translated from the German by Patrick Hubenthal.

[1]
I am referring here to the first sentence of Tolstoy's novel *Anna Karenina*: "Happy families are all alike; every unhappy family is unhappy in its own way." Leo Tolstoy, *Anna Karenina*, trans. Constance Garnett, rev. ed. (New York: Modern Library, 2000). First published in *Russkiy Vestnik*, 1875–77.

[2]
Georg Wilhelm Friedrich Hegel, *The Phenomenology of Spirit*, trans. Terry Pinkard (Cambridge: Cambridge University Press, 2018), 117–35. First published as *Phänomenologie des Geistes* (Bamberg: J. A. Goebhardt, 1807).

[3]
The word "ex-Kakanian" is derived from "Kakania," a term coined by Musil in the eighth chapter of his novel *The Man without Qualities* to refer to the Austro-Hungarian dual monarchy, which at that time was already a thing of the past: "All in all, how many amazing things might be said about this vanished Kakania! Everything and every person in it, for instance, bore the label of *kaiserlich-königlich* (Imperial-Royal) or *kaiserlich* und *königlich* (Imperial *and* Royal), abbreviated as "k.k." or "k.&k.," but to be sure which institutions and which persons were to be designated by "k.k." and which by "k.&k." required the mastery of a secret science. On paper it was called the Austro-Hungarian monarchy, but in conversation it was called Austria, a name solemnly abjured officially while stubbornly retained emotionally, just to show that feelings are quite as important as constitutional law and that regulations are one thing but real life is something else entirely." Robert Musil, *The Man without Qualities*, trans. Sophie Wilkins and Burton Pike (New York: Vintage, 1996), 1:29. First published as *Der Mann ohne Eigenschaften*, 3 vols. (Berlin: Rowohlt, 1930–43). For detailed information on the meaning of the word, see Alice Bolterauer's case study "'Kakanien' – Oder was eine mitteleuropäische Landschaft sein könnte," *Kakanien Revisited*, Dec. 26, 2013, http://www.kakanien-revisited.at/beitr/fallstudie/Abolterauer2.pdf.

[4]
Dirk Baecker, "Guter Geist ist trocken und Systeme sind unzuverlässig: Nachruf auf Niklas Luhmann," *Berliner Zeitung*, no. 264, Nov. 12, 1998, 11, http://www.spacetime-publishing.de/luhmann/baecker.htm.

[5]
On the fundamental concepts and decisions involved in Niklas Luhmann's sociological systems theory, see his book *Social Systems*, trans. John Bednarz with Dirk Baecker (Stanford, CA: Stanford University Press, 1995). First published as *Soziale Systeme. Grundriss einer allgemeinen Theorie* (Frankfurt am Main: Suhrkamp, 1984).

Material Thinking

Peter Weibel

Fragment

A specific consciousness creates for itself specific methods and structures in dealing with materiality. The choice of material arises from and accords with the consciousness of the artist. Whether permutational, aleatory, deductive, associative, constructive, subjective, or objective, what interests me here is the question of whether the method protects or even liberates material from objectivization, or whether it condemns material to commodification.

"What is objectivized material?" asks the folklorist. Of the many possible answers, here is one: wood, wire, and straw are the material; the broom is the object. When one is wielding a broom, one is no longer thinking about the material, but rather about the function of the object – namely, sweeping – and in fact, about just one specific action in a restricted context: sweeping as part of keeping house. The associations and emotions connected to the material and the word "sweeping" are lost, as are those elements of the activity and purpose of sweeping that give it meaning. The loss can be so great that those performing the activity of sweeping no longer see any meaning in it; they feel alienated; their activity seems to them to be a mere function in a great and incomprehensible whole, and they themselves seem to be an object in that awful functioning whole.

Most social systems narrow people's activities into functions and the things they produce into commodities. Artists who regard such systems as hostile to humanity try to give the material back its possibilities, to uncover the associative richness of the material and of the action in order to overcome the ennui of the alienated individual.

Material, Models, and Possibilities

The type of thinking that protects material from objectivization, regards the semantics of material as originating, in part, in consciousness, and the structures of consciousness as being, in part, socially and historically determined. It is a type of thinking that does not rob the work of art (be it music, literature, sculpture, painting, film, etc.) of its historicity, and thus of its future. A more liberated way of dealing with material – as the creation of new meanings and associations – means the liberation of humanity from old and constrictive meanings and associations.

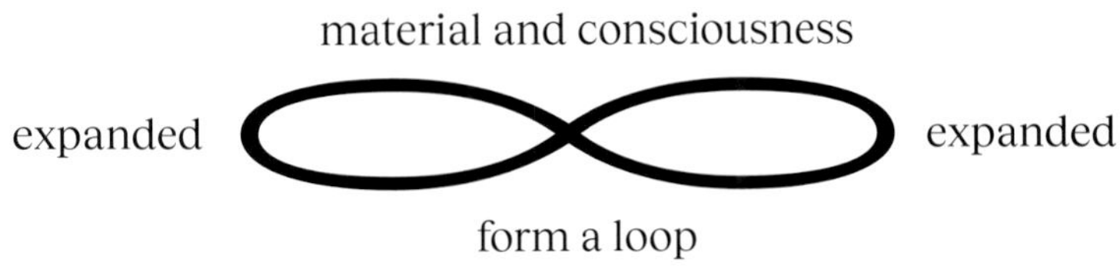

I am seeking a kind of art that does away with objectivization, that frees the individual from their spiritual and intellectual limitations. It seems to me that one path to that end is the type of material thinking that uses the meaning of the material as an operator.

Materials have fixed meanings, an identified context. Meaning and context can be consciously or unconsciously present. To illuminate unconscious meanings and contexts is to charge the material with new meaning. Materials are placed in new surroundings, in unconventional combinations with other materials – and in this way they are stripped of their historical identification, brought back out of their preserved state. New spatiotemporal relationships, new composites of materials, new combinations of signs create new meanings.

A material's openness to different interpretations, its potential for generating and expanding meanings, represents a renewal and expansion of experience. Thinking that sets material free keeps it open for new meanings and associations and lays the foundations for pleasure and meaning in life: experience as one of the meanings of life, creativity as a meaningful activity. New models that arise from the material, as products of imagination or study, serve to liberate humanity. Material thinking as liberation of humanity from objects as commodities of objectivization.

Note: Adapted from Peter Weibel, "Materialdenken als Befreiung der Produkte des Menschen von ihrem Dingcharakter: Probleme der aktuellen Avantgarde," in *Kritik der Kunst – Kunst der Kritik, Es says & I say* (Vienna: Jugend und Volk, 1973), 51. Originally written in 1966. Translated from the German by Patrick Hubenthal.

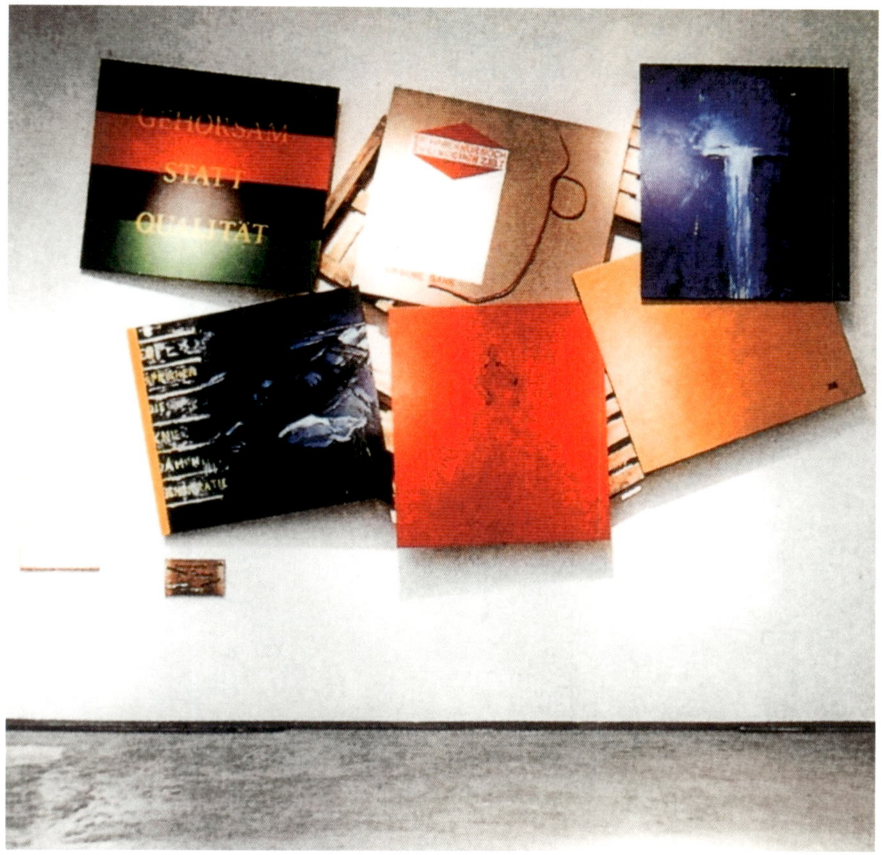

The Last Painter / Der letzte Maler

1988 | 3 pallets, 6 wooden boards, oil paint, rope

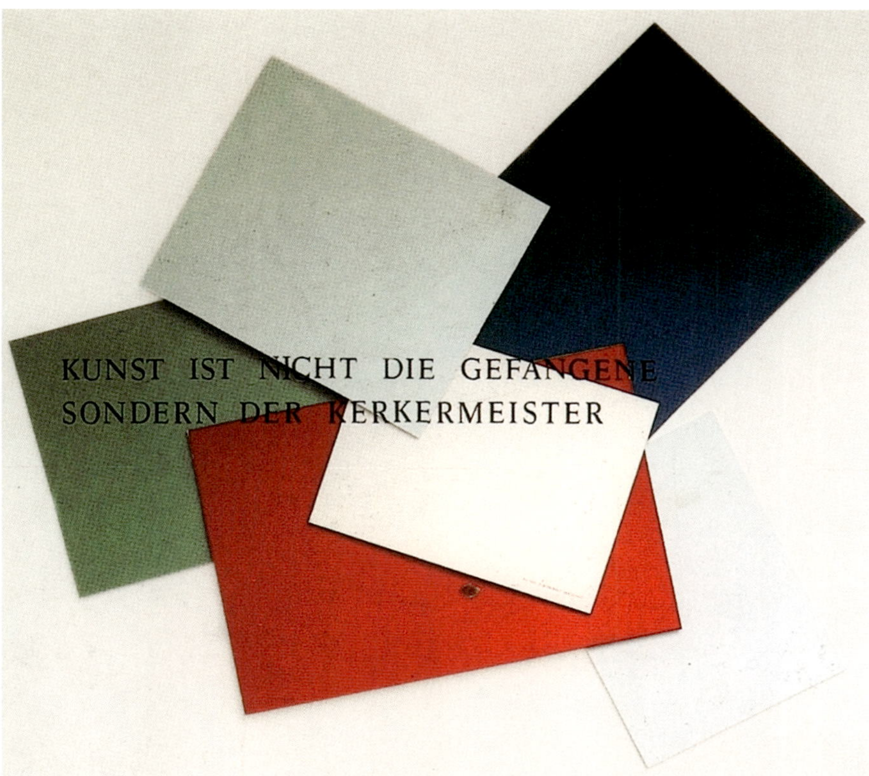

Art is not the prisoner,
art is the jailer

**On the Freedom of Art /
Zur Freiheit der Kunst**

1988 | image object; 6 particle boards, varnish, ca. 500 × 500 cm

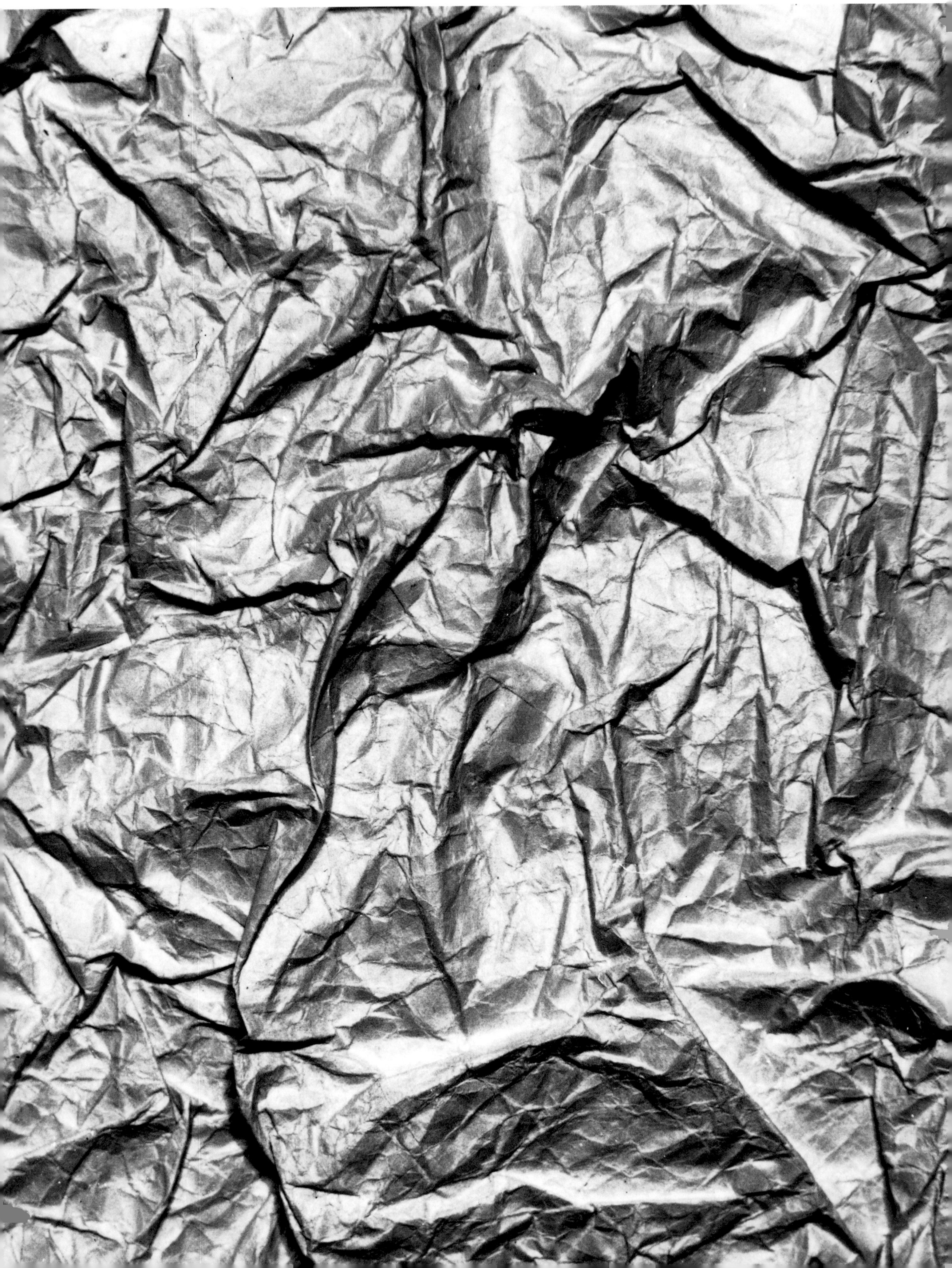

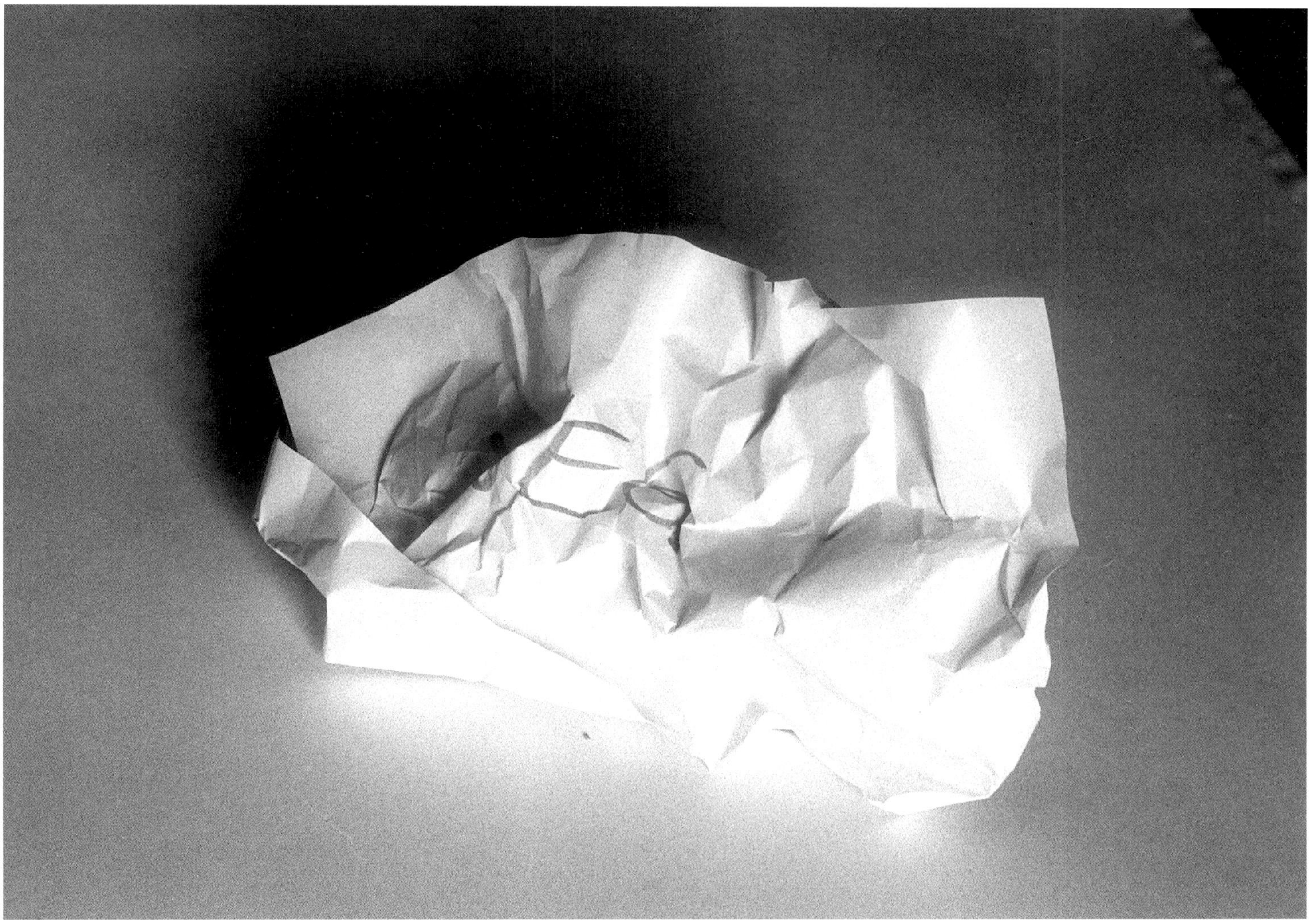

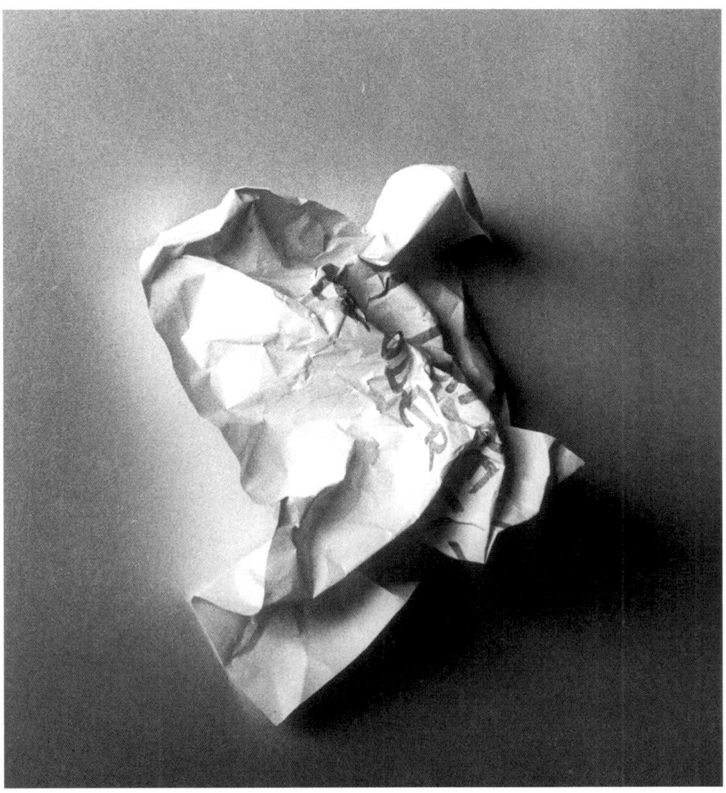

On the opposite page you see a crumpled piece of paper. But actually, you are looking at a photograph of a crumpled piece of paper. The represented object, the paper, is crumpled, but the photograph, the image, is smooth. Thus, a contradiction between image and object arises, an ontological split.

The composition of poems and other texts generally involves the writing of letters and words on paper. Writing must always be communicated via some material, although the nature of this material can vary. The spectrum ranges from paper to wood to stone to skin. And as the material changes, the meaning of each word written on it shifts accordingly. Linguistic experiments with various materials make it clear that the interplay between meaning and the media used for storage and distribution is inherently important. The act of reflecting upon materials and media produces new semantic relationships. Although paper is the most commonly used medium for the production of texts, its significance as a material has never been sufficiently acknowledged.

This was the context in which the *Paper Poems* were created. For example, *Weg*, meaning "path," an indication of which way to go, is a piece of information. But when the carrier material of the information, the paper, is crumpled, the information is lost, rendering the words unreadable and bringing the "path" to an end. The crumpled piece of paper is then documented in a photograph. But the picture, the photographic print, is itself made of paper and is not crumpled: noncrumpled paper representing crumpled paper, two modes of existence operating simultaneously. The photographic representation could itself be crumpled and photographed, a process that can be continued ad infinitum. —Peter Weibel, 2022

Translated from the German by Patrick Hubenthal.

Paper Poems

1965 | b/w photographs

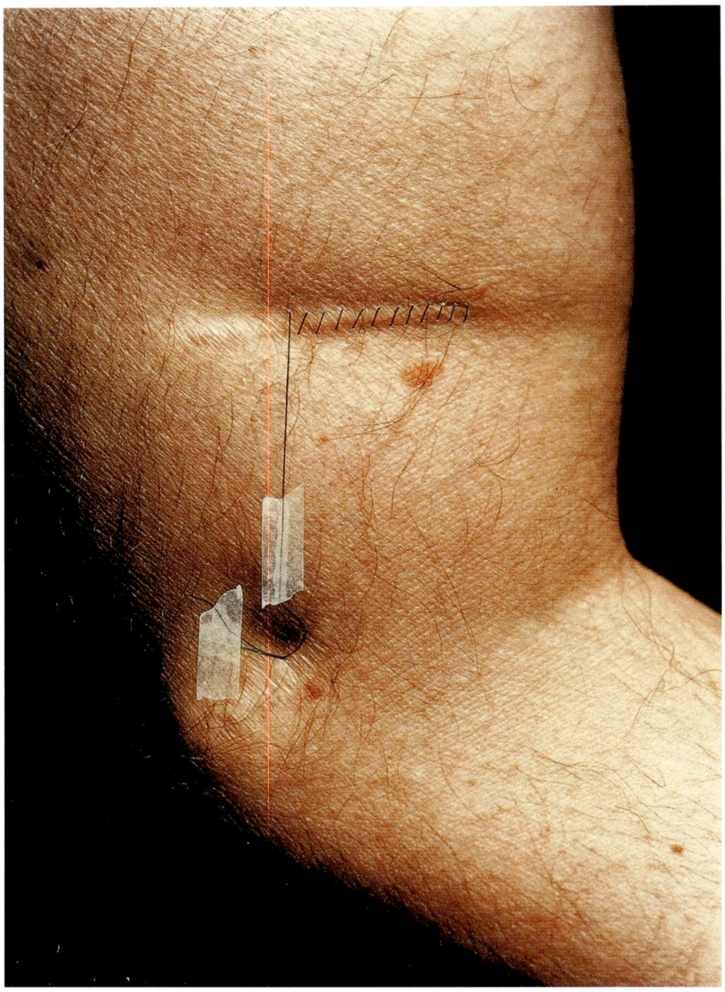

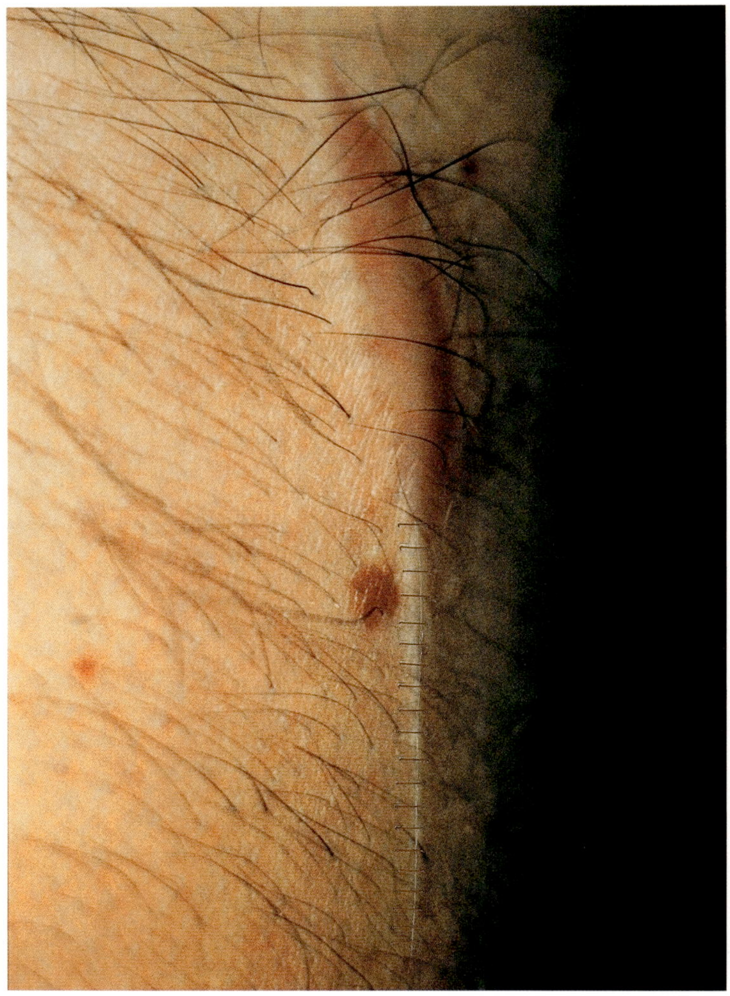

After *Paper Poems*, *Skin Poems*

Tanned animal skin was processed into parchment and thus into a surface for writing to record narratives. The analogy between skin and paper is thus given. Weibel sees the body as a bearer of signs, as a medium of information that can become the actual alphabet, an *écriture corporelle*. The series *Skin Poems* is directly related to Peter Weibel's experimental poetry. The material poetry is now extended to the body. Weibel writes poems without language, through action, through his body.

The photographic series *Skin Poems* from 1968 depicts various parts of the artist's body drawn with scars in large format. The photographs are documentation of the artistic action for which Peter Weibel wrote small poems and painful messages on paper in front of an audience. These writings were not read out; instead, Weibel implanted them under his skin. The pain of a bodily injury and the scar visible later are part of the poem. The senses of the readable and the palpable are united on one medium, the skin, which truly becomes the carrier of messages that can no longer be read. Letters and words can be interpreted as scars, as wounds of past experiences.

The photographic work *Fontana Mix II* from 1970 takes this idea further; here, too, the artist's body becomes the carrier medium. As the title suggests, the work can be read as a homage to the artist Lucio Fontana, who opened his canvases through cuts. For the essence of the line is the cut. Peter Weibel transfers this thought from the screen to the skin into the physical body by cutting two parallel lines on his back.

In 1971 Peter Weibel summarized all his experiences and experiments with skin, writing, and body in the unpublished book *Codex Cutis* (see pp. 54–55 in this volume). Here, too, he was driven by the question as to how far it can be made clear through the medium of paper that humans themselves are books, but that they need apparatuses to decipher the information offered by the body. Already at the beginning of his artistic career, Weibel's body image was influenced by the natural sciences and technology, which are not limited to the natural senses for perception, but utilize apparatuses. The path via the parchment leads the artist to the human skin and further onward to the human being as a readable book. This is followed by the attempt to make the human body speak, which partly leads to Weibel's action art.

Skin Poems / Hautgedichte

1968 | photographic series; 50 × 33.2 cm each

Performance during which the artist talked about his *Skin Poems*: "I cut in the skin, then put a poem I wrote on a piece of paper under the skin and had the wound stitched," Festival Konkrete Poesie, Hanau, May 1971

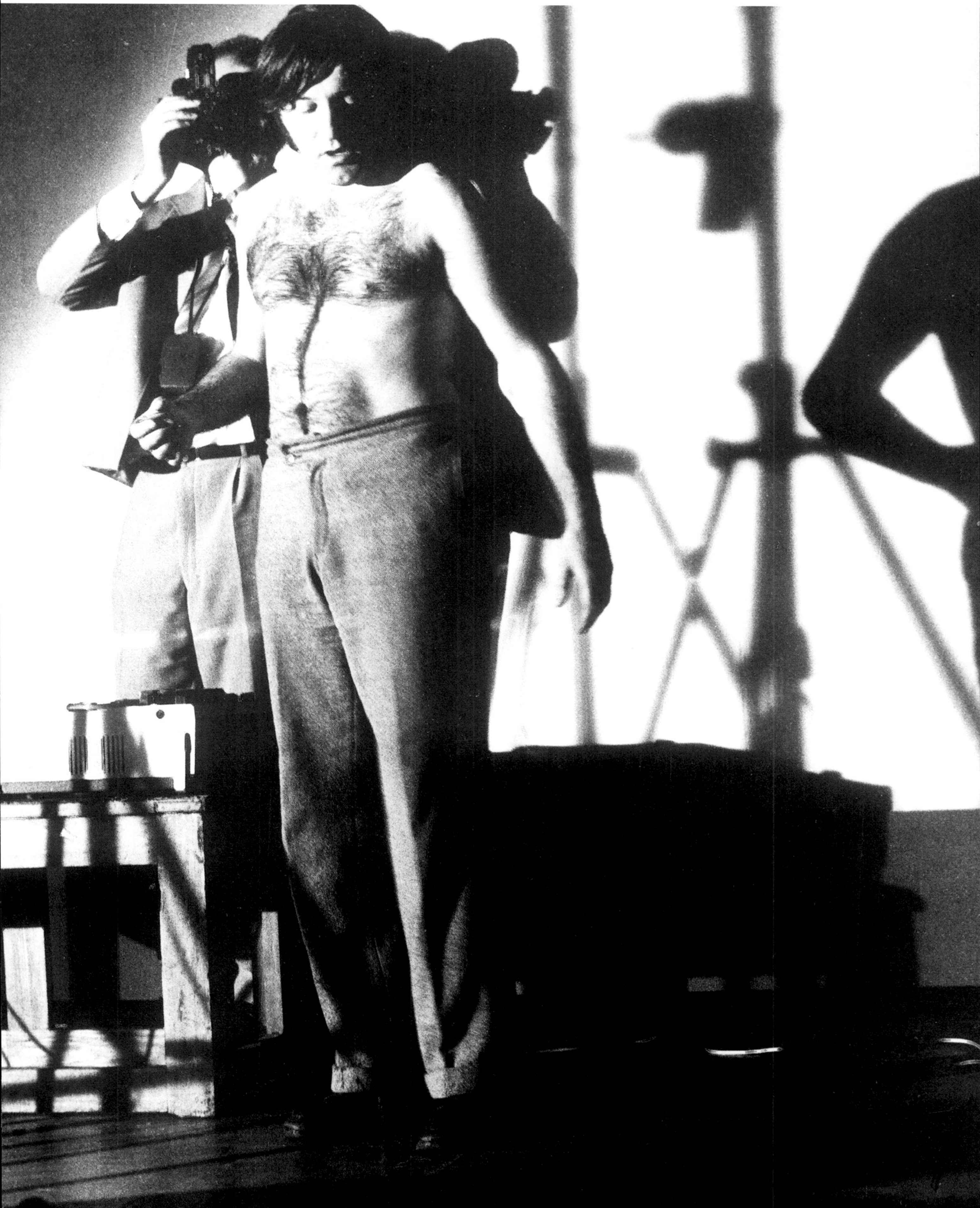

Skin – invented, folded, and equipped with a wooden cover in the fourth century – was used as writing material into the Middle Ages.
The book, that human landscape, is the common outer cover of our body and serves as a trajectory system and external border for the organism. Since the Middle Ages, sweat glands, muscle fibers, blood vessels, papillae, nerve endings, follicles, sensory cells, and fibrils have been united in one binding: the people of the skin surged into the paper age.
Skin in its current form has only been produced since the fourteenth century.
The fate of the three dermal layers: the callus layer was systematically filed into slate blackboards; leather skin was processed into wooden writing tablets; only the subcutis, in antiquity originally covered in wax, was later joined together with squamous epithelium, finger, and toe pads to create a kind of book.
Only the role of arteriovenous anastomoses, such as the nail bed and nail fold, remains unclear in the cycle of words. Papyrus, parchment, paper – these three basic forms serve as outer borders to the brain.
Since the fourth century, dehaired, tanned animal skin has been the general covering of our people. Folded leaves of parchment make up its nomadic paper tent. The modern people of the book was born when the skin tent of the body was transmuted into the paper tent of the mind.
Given its rich supply of nerves, each book is part of the sensory apparatus. Above all, the book tent protects the organism from and during mechanical labor. The printing press reduces pressure on brain and skin; the blackness of print lightens the blackness of pain. That is how printing pressure, blood pressure, and book pain arise.
The ancient manuscripts on the skin of humanity: runes, notches, knots, hairs, feather and cloth writing, fibers, hieroglyphs, warts, numbers, tracks in sand and elsewhere, semaphore, pictograms, musical notation, handwriting, symbols, Morse code, bowels, syllabic or alphabetic writing, tattoos, braille, traffic signs, ALGOL, signs.

...

The book in its future form and with manifold functions such as fat storage, temperature and humidity regulation, participation in the metabolism, however, does not yet exist.

Codex Argenteus – Uppsala
Codex Atlanticus – Milano
Code Napoléon – Paris
Codex Cutis – Vienna

Note: This text by Peter Weibel is the introduction of *Codex Cutis*. Translated from the German by Peter Dahm Robertson.

Codex Cutis

1971	manuscript; digital print, staples, 15 pages, 29.7 × 21 cm

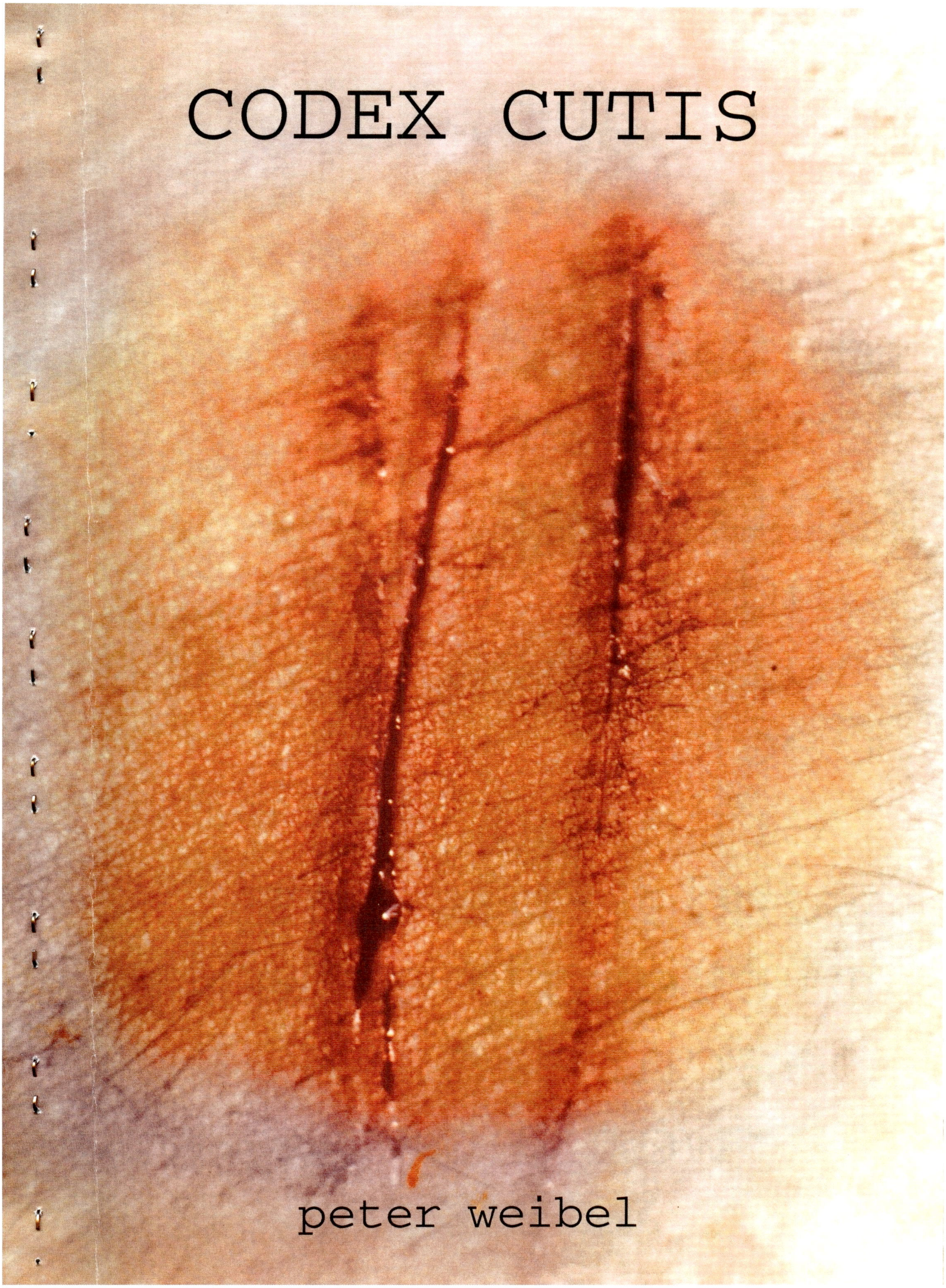
CODEX CUTIS
peter weibel

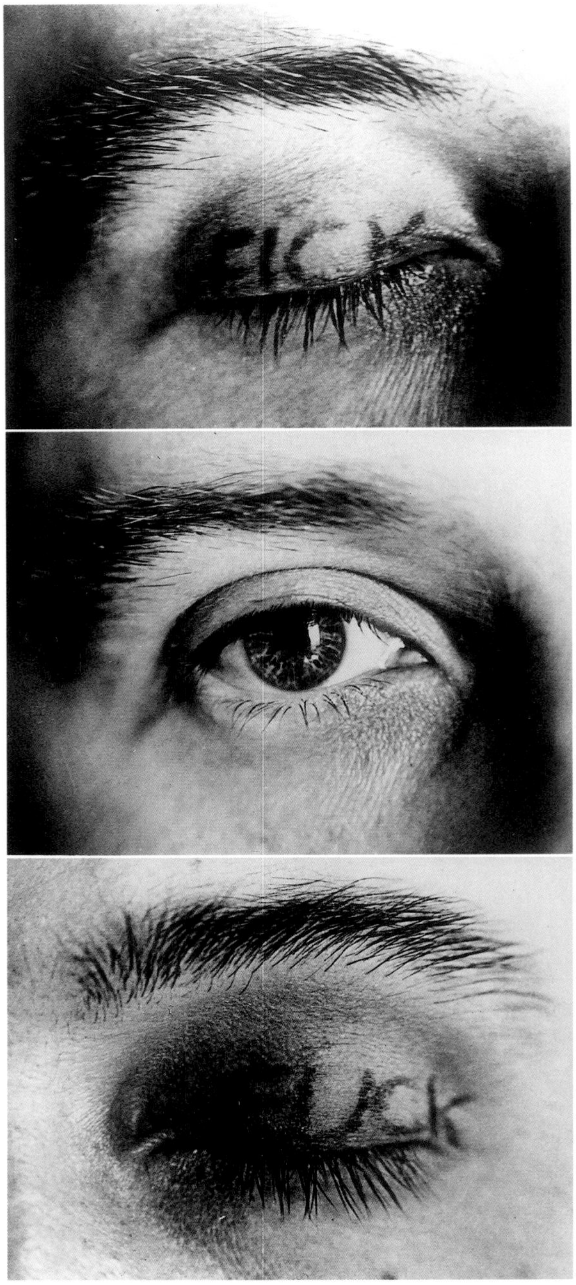

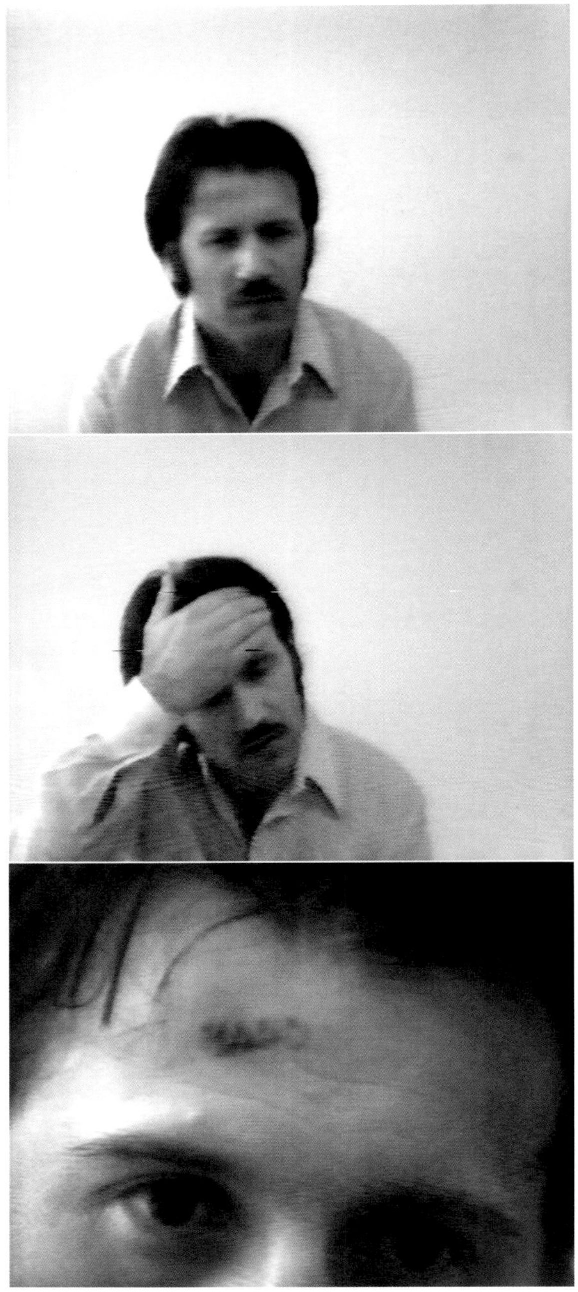

In the video piece *Eye Texts*, we see alternating close-ups of a woman's eyes and those of the artist. Various German and English words, such as "SEE," "LOVE," and "NACHT" (night), are written on their eyelids. In one sequence the word "HIMMEL" (heaven, sky) appears across both of the woman's eyelids, with a break in the middle. Only when the performers close their eyes do the words become visible and readable. A close-up of one of Peter Weibel's eyes, with the words "FICK" and its English equivalent "FUCK" written on it, is particularly striking: the eye opens and closes repeatedly over the space of a few seconds, as a metaphor for the rhythm of sexual intercourse. As viewers, we witness the way the pupil visibly dilates and contracts with each blink.

Messages are always bound to a delivery medium. In *Eye Texts*, a material substitution takes place: instead of the text being set down on paper as usual, the body becomes the medium for the written characters. This change of material creates new semantic connections, and the relationship of the text to its delivery medium opens up new fields of meaning.

By literally inscribing his own eyes and those of his performer, Weibel shifts the focus from the purely syntactic level of the text to that of pragmatics: no longer are the characters simply distributed across a sheet of paper for readers to receive; rather, the participative principle becomes a carrier of meaning. The performers' actions are central to our understanding of the text. The two enter into a kind of mutual dependency, since the text can only be read when the performers close their eyes. In certain cases the message presents a paradox: for example, that the word "SEE" emerges only in the act of closing the eye – at which point the performer herself can no longer see. The message about seeing does not apply to the sender, only to the receiver. The receiver sees while the sender is blind. The

Eye Texts / Augentexte

1974 | from the series *VIDEOTEXTS*
video; b/w, sound, 1:10 min.

Forehead Text / Stirntext

1974 | from the series *VIDEOTEXTS*
video; b/w, sound, 20 sec.

word "SEE" takes the place of the act of seeing; representation supplants reality. Visual culture is a purely representational culture (of signs). Its relationship to reality is not free of contradictions. When a poet writes the word "silence" on a piece of paper, the paper is not silent; it speaks. For Weibel, the linguistic and communicative contradictions that can arise between the meaning of a word and the material through which it is conveyed underscore the formal boundaries of semantics, which are set by the materials, media, and tools involved.

The use of the body as a delivery medium and the shift in focus to pragmatics, to the act or action, are of central importance to Peter Weibel's linguistic works and feature prominently in similar works from the same period. In *Mouth Texts*, for example, he displayed words such as *Zunge* (tongue), *Gedärme* (bowels), and *Scheiße* (shit) on pieces of paper in his mouth, highlighting the path taken by the food we consume while simultaneously enacting a German idiom for speaking – "Worte in den Mund nehmen" (literally, "to take words into the mouth") – that is often used in the context of mentioning the unmentionable.

The motif of eye texts would reappear in popular culture a few years later. The best-known example is undoubtedly an early scene in the action-adventure film *Raiders of the Lost Ark* (USA, 1981, dir. Steven Spielberg). In it, the archaeologist Indiana Jones is at a university giving a lecture about the Neolithic era and the dangers that myths and legends pose for the preservation of historic sites, while his female students hang adoringly on his every word. When his gaze lingers on one of these young women, she smiles at him and closes her eyes for just a moment, revealing a message for her professor written on her eyelids: "LOVE YOU." Confronted with this mute yet expressive declaration of love, Jones falters and abruptly loses his train of thought. After another blink and smile from the student, the bell rings and the class is dismissed.

The transference of poetry from the syntacticaland semantic level to pragmatics and material level into media is one of Weibel's notable achievements.

Translated from the German by Patrick Hubenthal.

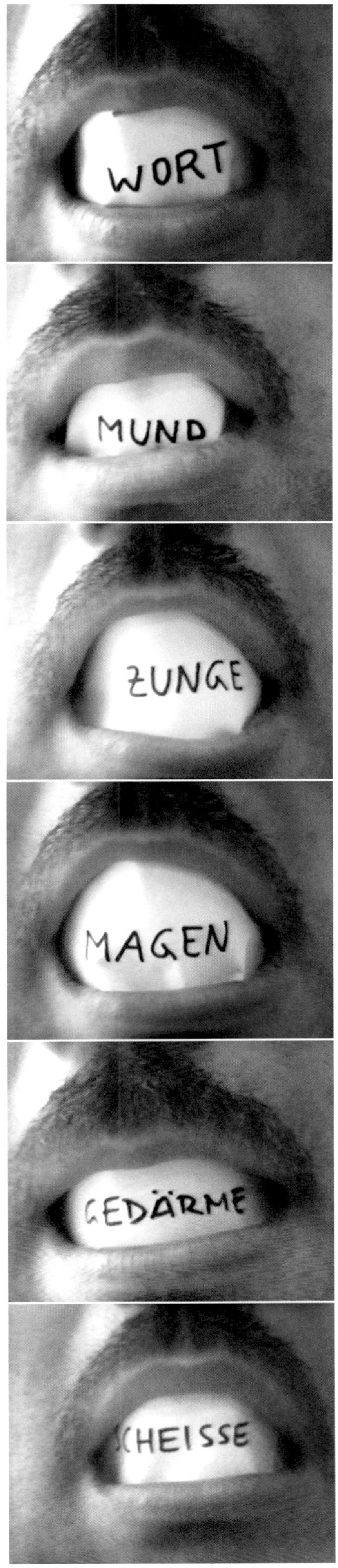

Mouth Texts / Mundtexte

1974	from the series *VIDEOTEXTS* video; b/w, sound, 44 sec.

cemetery-dictionary
for the pretty-perky teenager
as (sub,ab)ject of our many-cured civilization

sex - h,appyal teEnAGER (roll,pull) over beAthoven

dar ling kness teenager get hAP(E)py t(H)onIGHT

(n,sh)a(i)meless blue LEE jean ager rich LIE ILL,us,tr(e)ated is your life

your dREA(L)M of h m oney suckle moon life

tee n ager
man ager s sw allow & f ollow & fo(olish)llow
fl ow N er be e
false
ro man ce een qu ager   pro fitting   h m ood
inthe
lean ager make a S,trip (and a cup of) tea,SE
LOVE for sale   of dan c g er
sell your soul
heart is for   S L A V E

enGAGEment for you only
for gold glori -f(l)y   bet/RAY CHARLES
teenager smoke get in your eyes

pLEAS(e)URE
your trea(son is -sure)
manipulated
manicured   PORGY and BAISSE   bLACK b d ird/ty
mania
(J,CH)eer
disHONEstY MOON   S(A,OU)L,E   by
SYMphon(e)Y
f,all in love DISeas(E)y to love
the touch of your lips
pre tender ly
noStALgic.....mEmories

Cemetery-Dictionary

1964 | visual poetry; typescript, published in *Werkstatt Aspekt* 2 (1966)

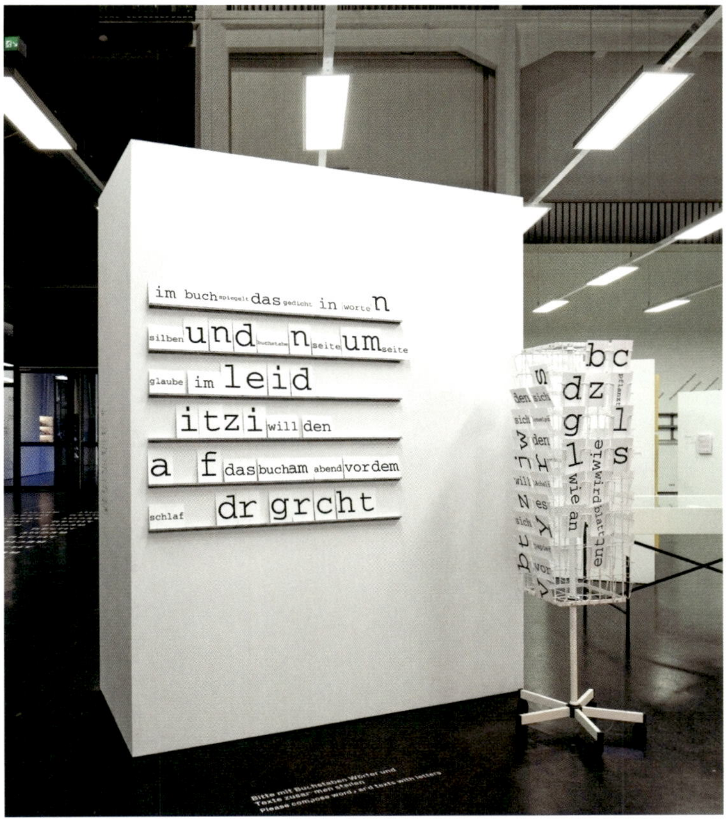

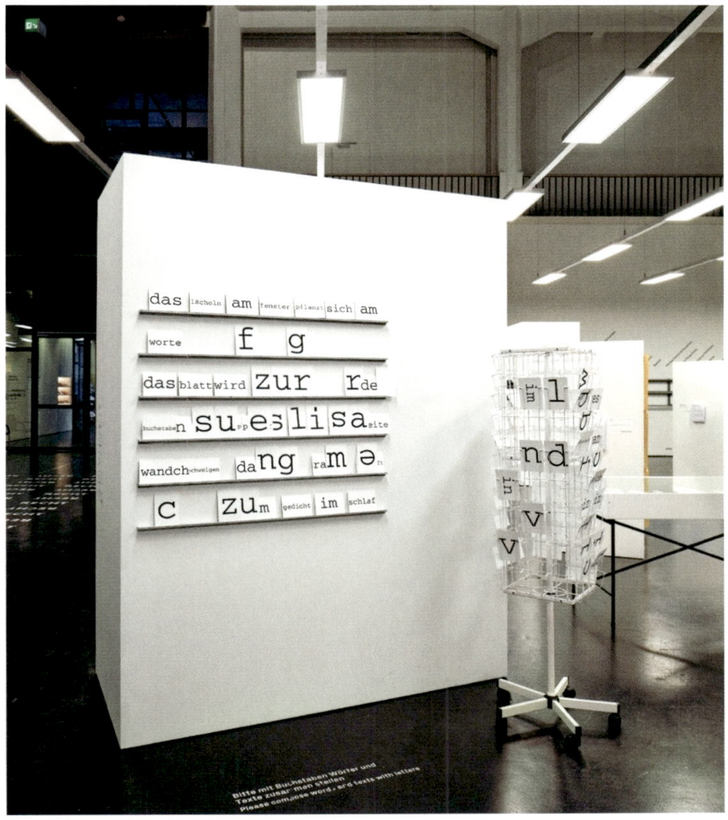

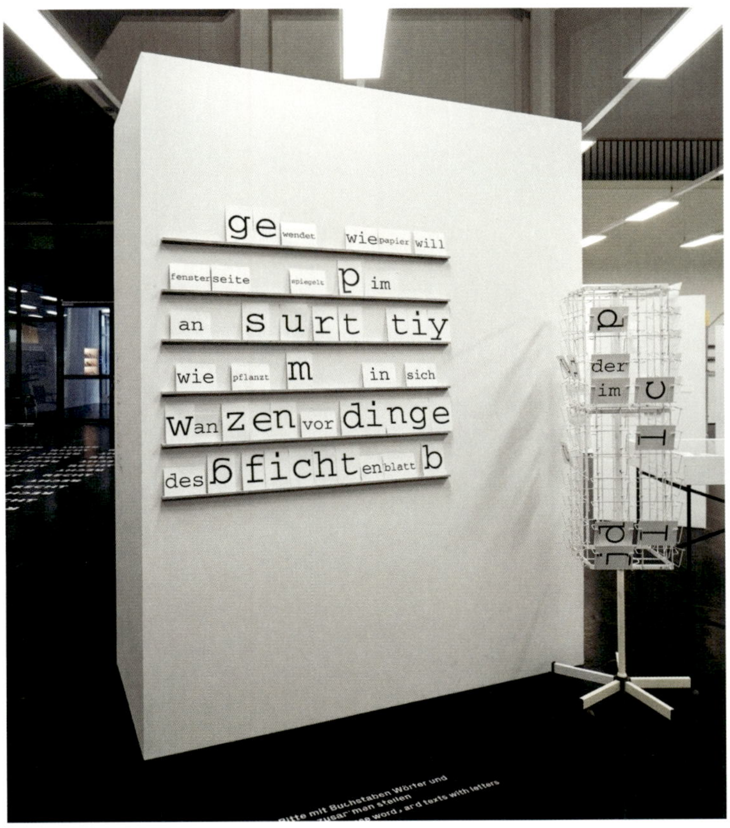

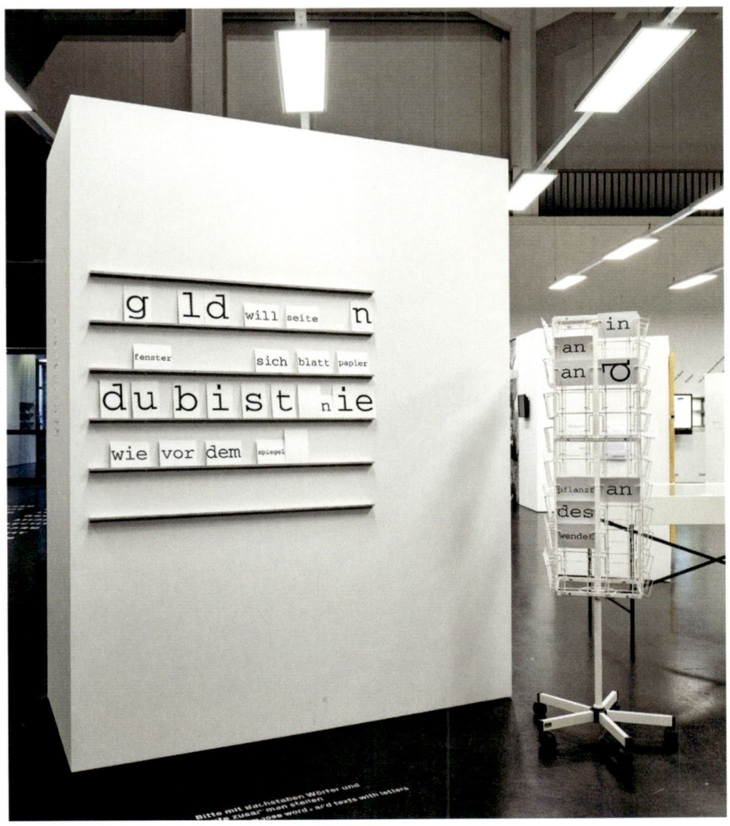

Installation views
respectively, Peter Weibel,
ZKM | Karlsruhe, 2019–20

Mobile Texts / Mobile Texte

1966	participatory public poem; letters on paper, rotating rack, wall-mounted wooden strips

Single letters or groups of letters – morphemes and word fragments – are placed in a rotating rack, the kind that would normally hold postcards or books. The letters are mobile. Readers – viewers, visitors, users – are invited to assemble word combinations on wall-mounted wooden strips (lines) to create their own poems or combinations of texts. Using these movable letters, the audience can publicly generate its own texts. Instead of a book read by the public, a letter rack or word stand serves as an opportunity for a poem by the public.
—Peter Weibel, 2022

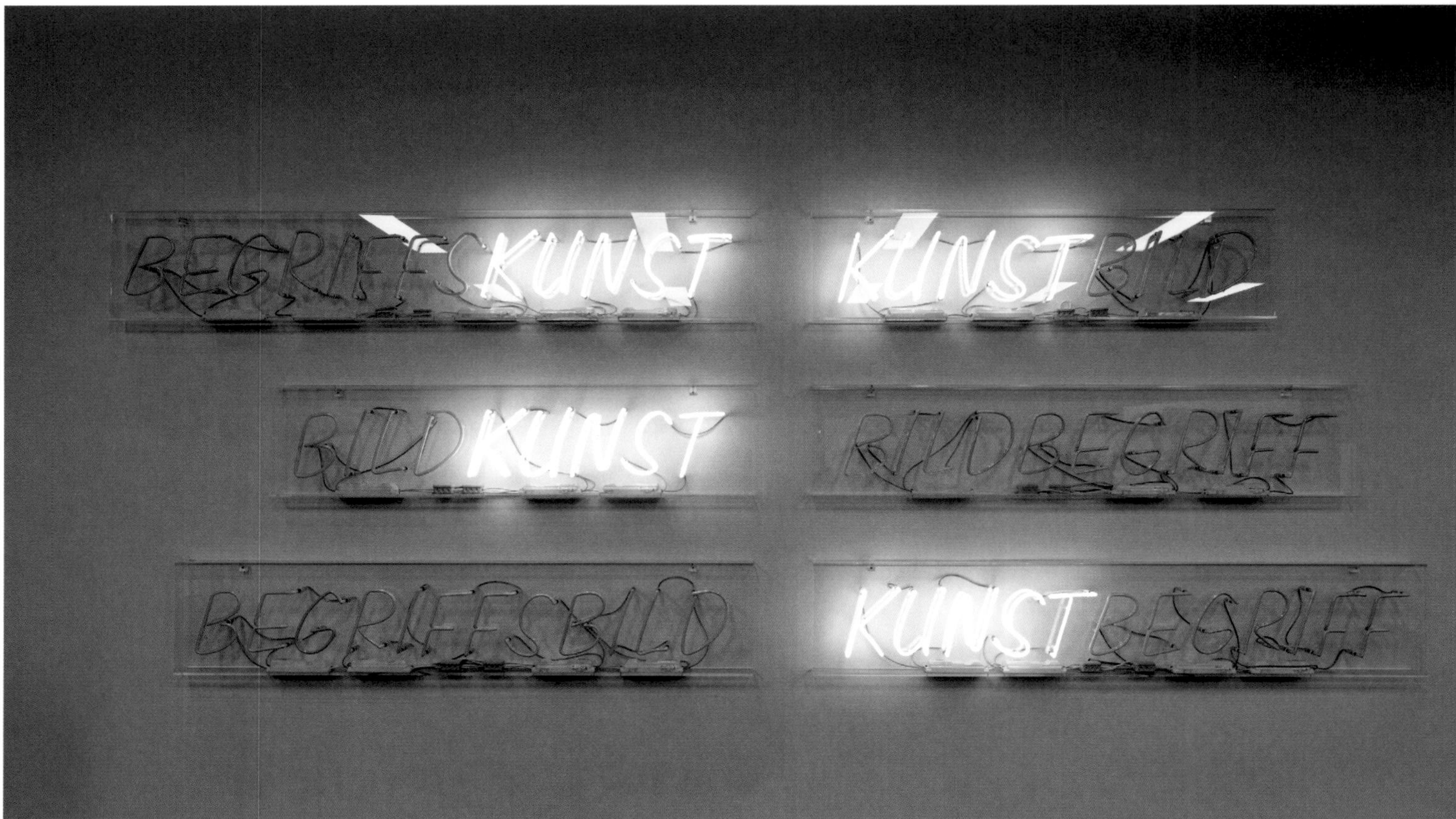

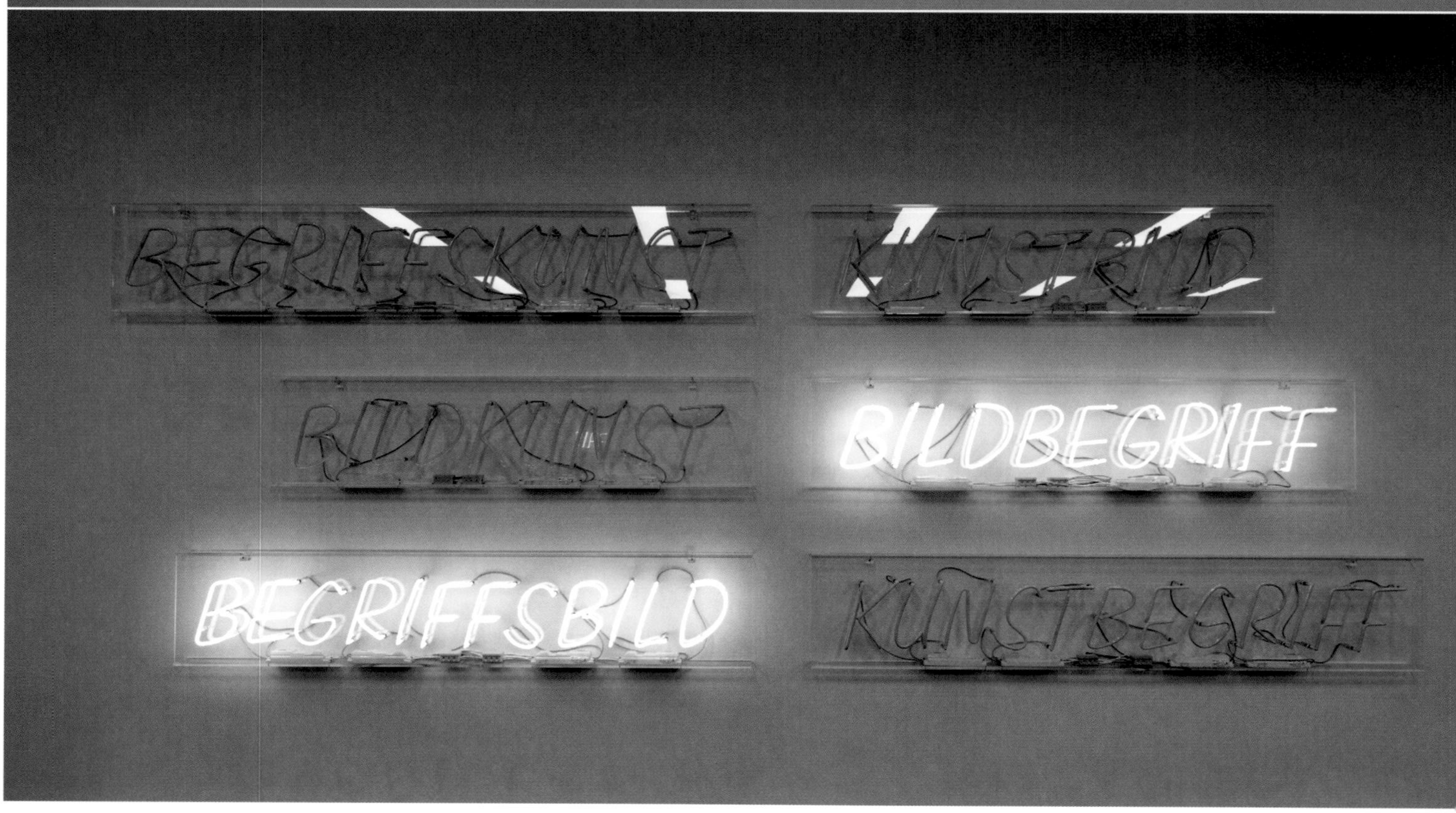

Concept Art / Begriffskunst

1966/ 2014	from the series *Motion Poems* mobile texts; neon tubes, cables, time switch, 6 parts, 194 × 547 × 14 cm

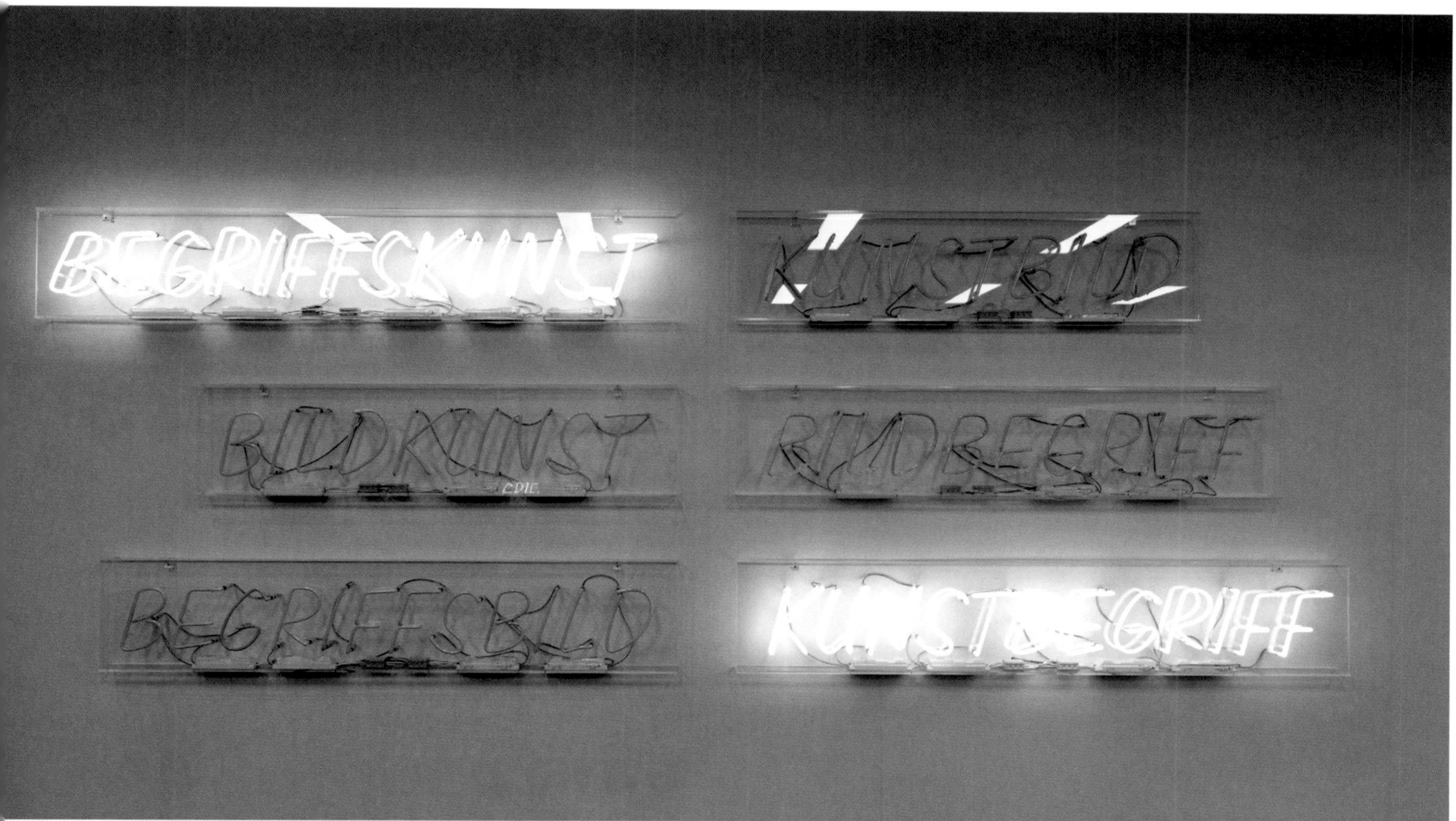

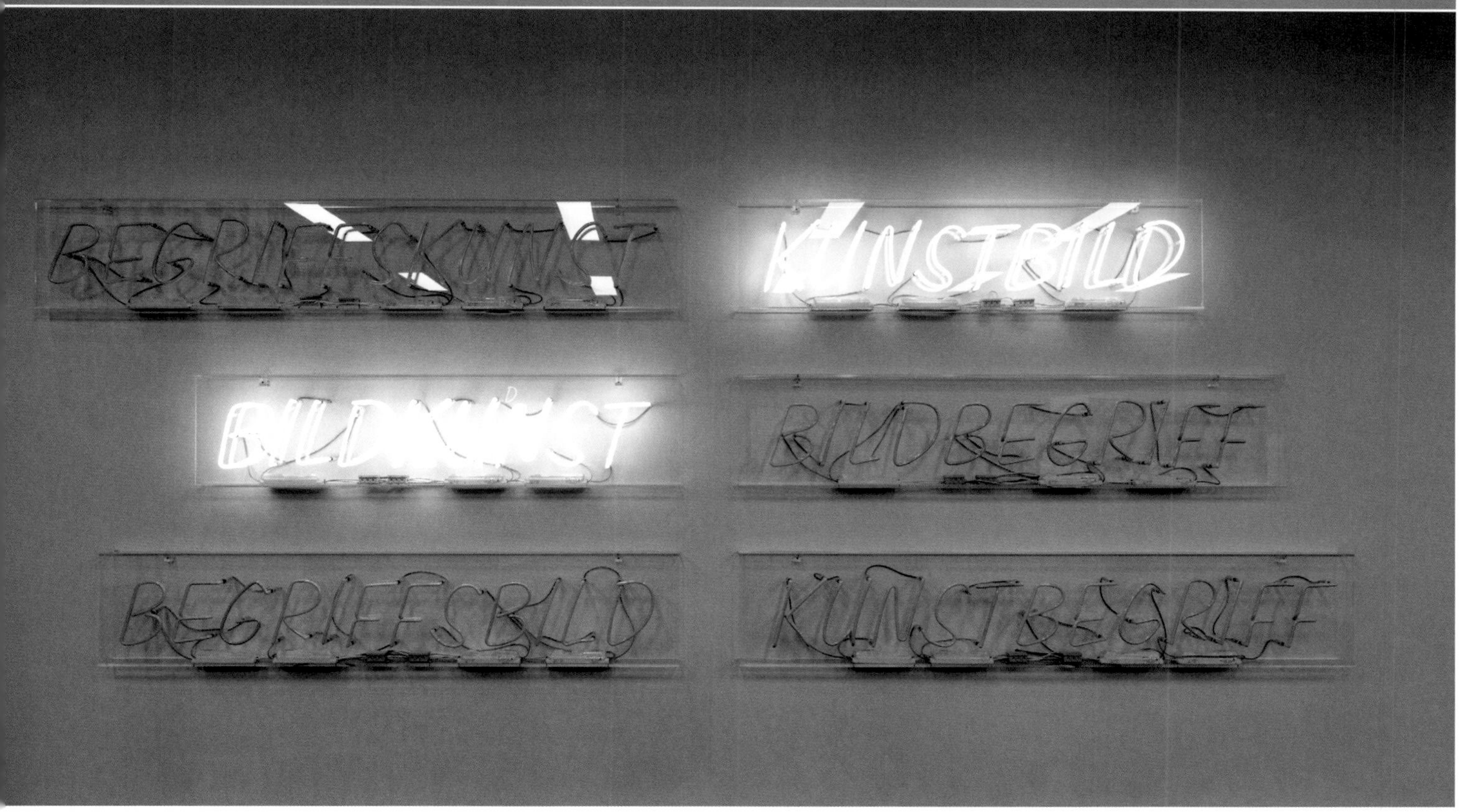

Controlled by a time switch, the words *Kunst* (art), *Begriffskunst–Kunstbegriff* (concept art – art concept), *Bildbegriff–Begriffsbild* (image concept – concept image), and *Kunstbild–Bildkunst* (art image – image art) are illuminated in a combinatory manner.

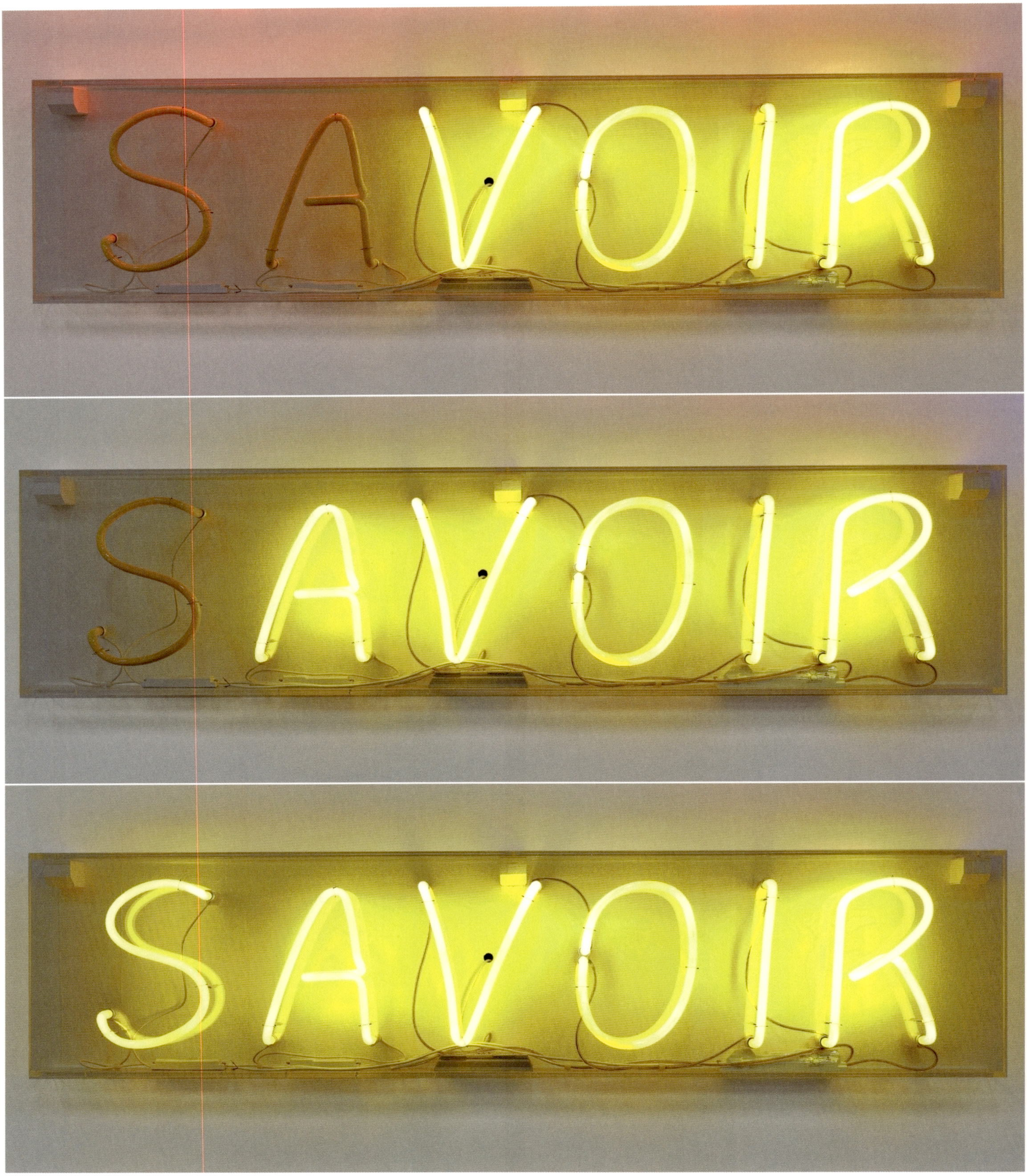

Savoir

1966/ 2014 | from the series *Motion Poems* mobile text; neon tubes, cables, time switch, 47.5 × 200 × 15 cm

Traditionally, letters are spatial constellations distributed on a plane, a page. This is an example of a temporal distribution. The hidden meanings of a word are revealed by means of a time switch.

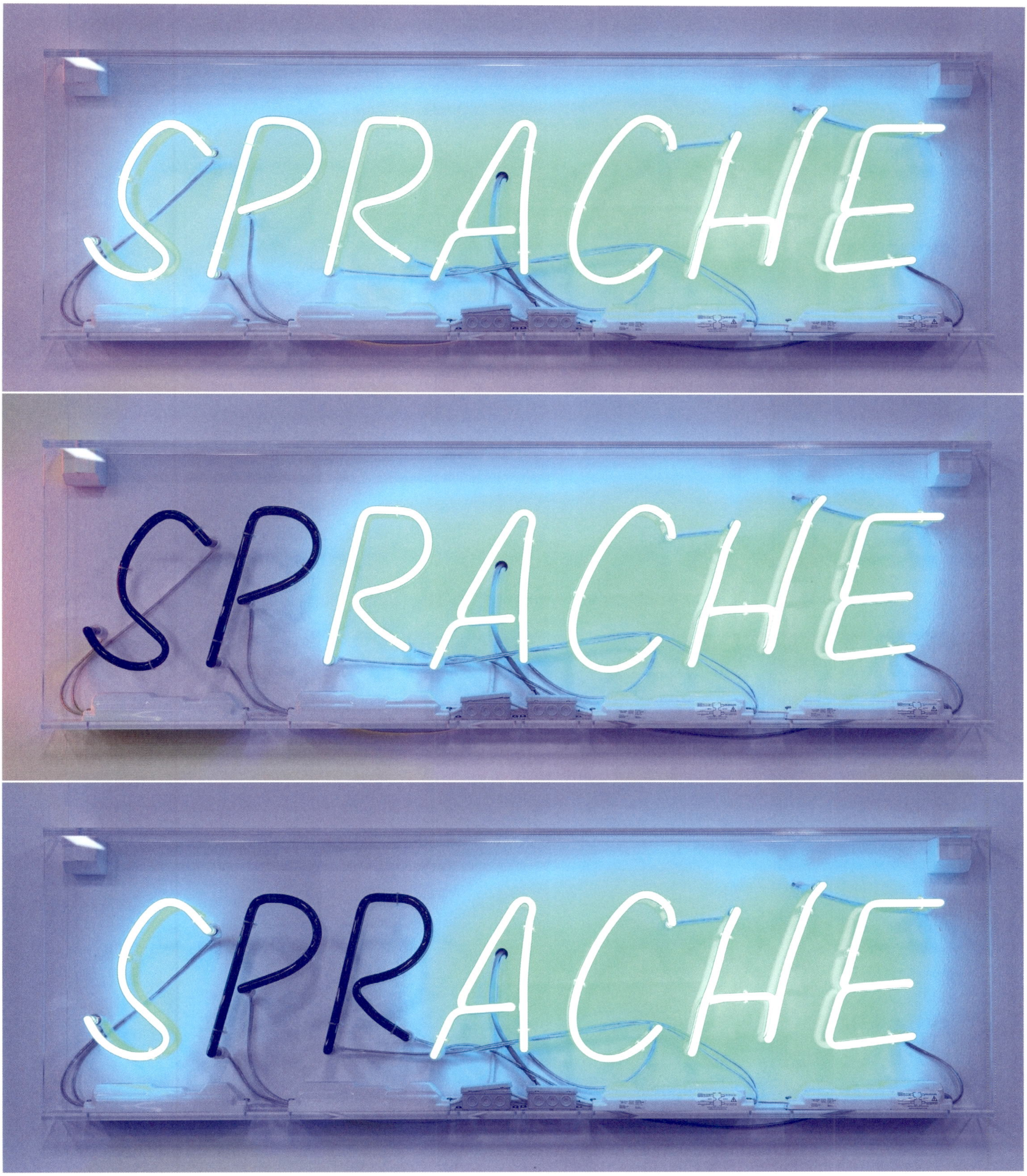

Sprache

1966/ 2014	from the series *Motion Poems* mobile text; neon tubes, cables, time switch, 47.5 × 165 × 15 cm

The German word *Sache* (thing) is contained in the word *Sprache* (language), as is the word *Rache* (revenge). This demonstrates the relation of word and object: the object world is included and mirrored in the language; it is part of the sign system. Language is the interface to the (object)world.

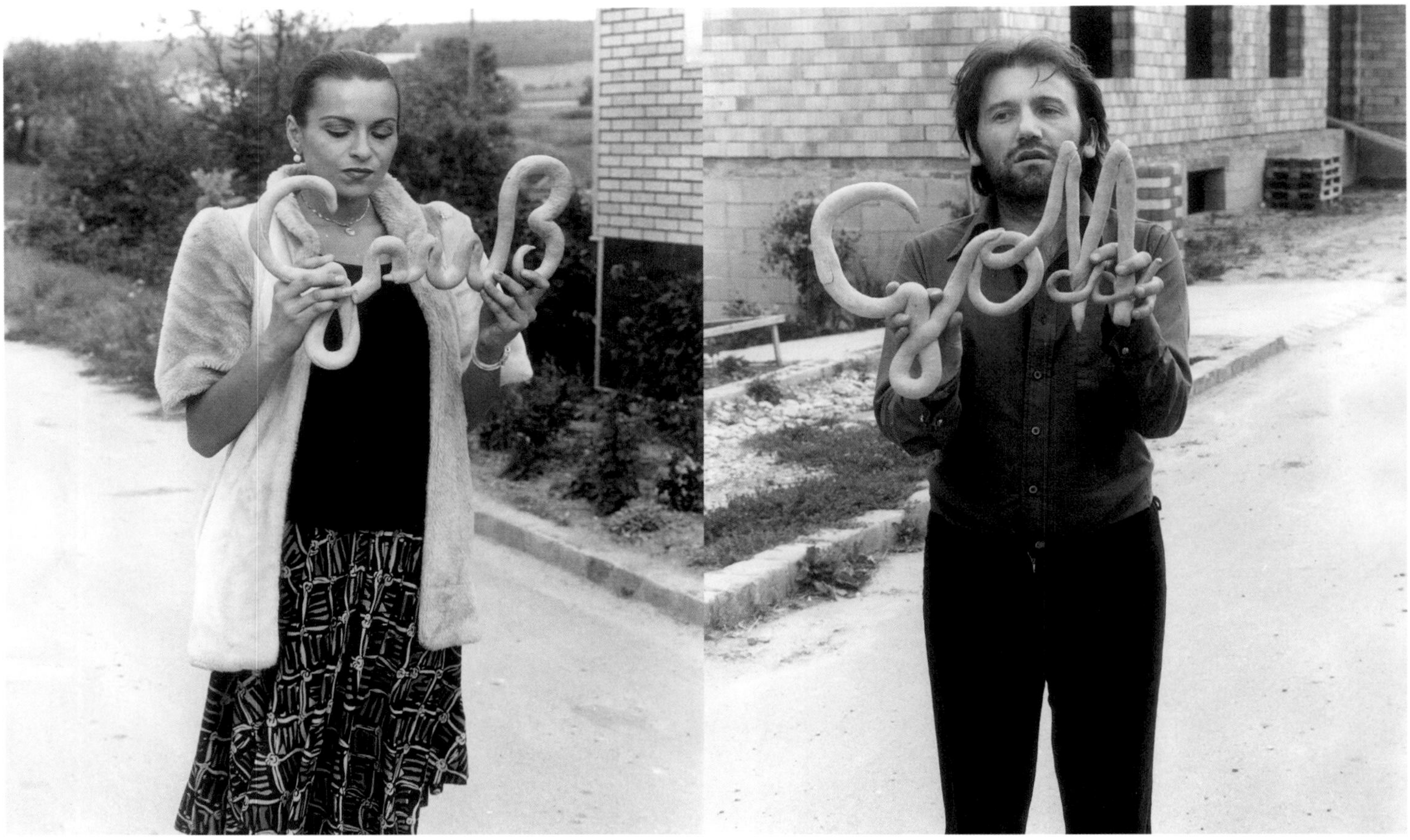

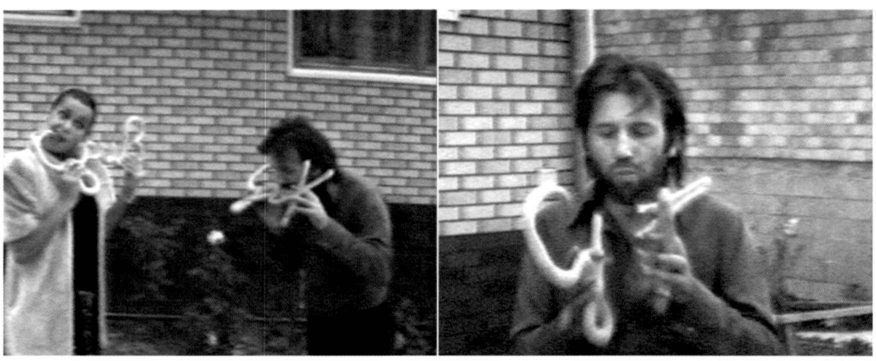

Video stills from the reenactment with Susanne Widl and Peter Weibel, 1972

Greet God / Grüß Gott

1967/72	text action, reenacted for video; b/w, sound, 47 sec.

Greet God is a text action performed in 1967 in which words are turned into baked objects, reenacted and videographically documented in 1972. The actors of the performance, Susanne Widl and Peter Weibel, meet on a street and show the two words *Gruß* (greet) and *Gott* (God) baked from bread dough to each other. *Grüß Gott* is the most common salutation in Austria and southern Germany – the equivalent of "Hello." Widl begins to eat the word *Gruß* and swallows it. Weibel takes the word *Gott* in his mouth and spits it out again. Obviously, he does not like it. In a very literal way, the action deals with complex questions of faith and the human relationship to God and how religion inscribes itself into language or rather how language constructs religion.

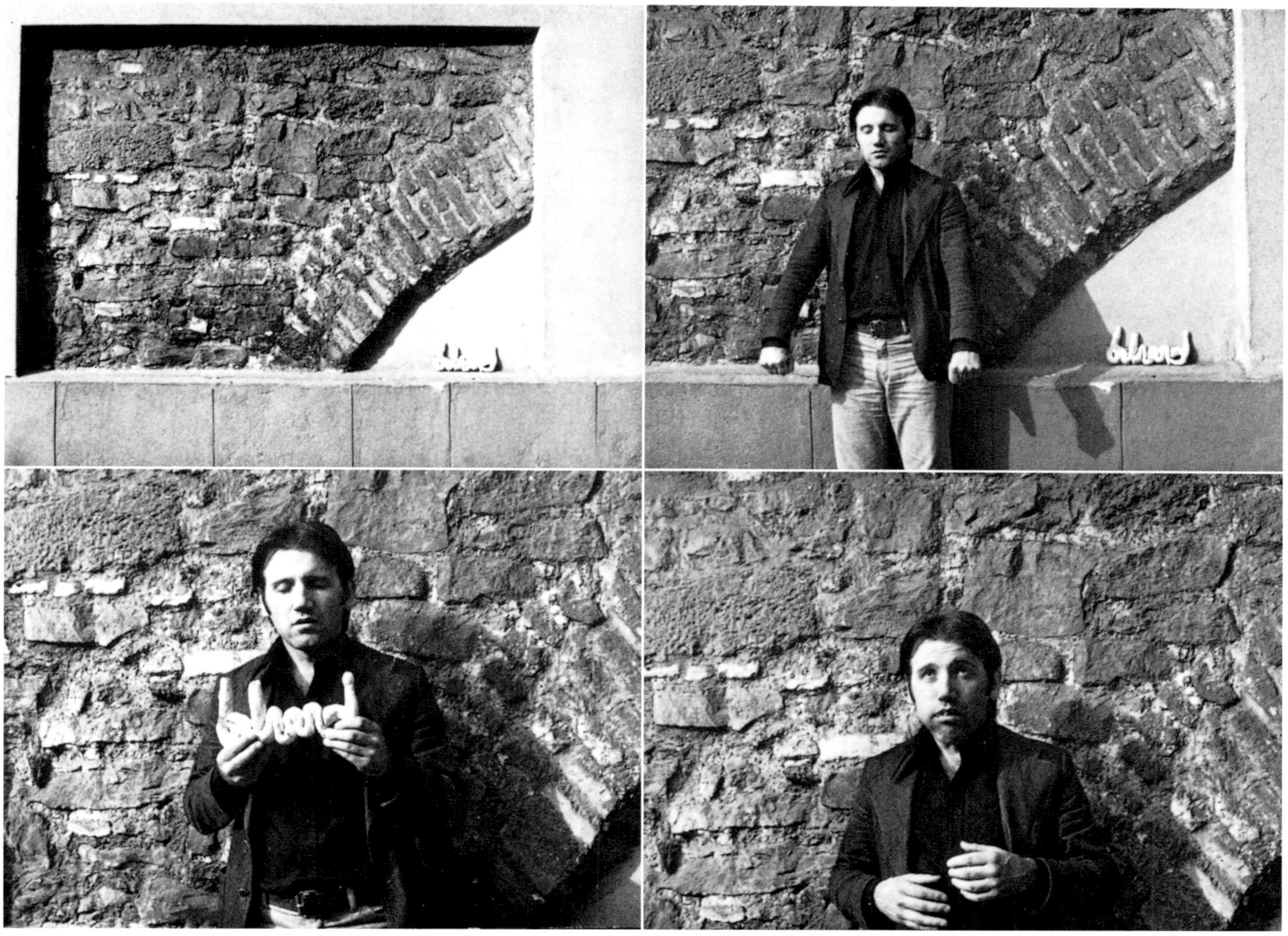

Blind is a four-part photo sequence from the unfinished film *Wor(l)d Cinema*, which was conceived in 1967 and realized in parts on 16 mm in 1972. As the title of the film indicates, *Wor(l)d Cinema* creates a relationship between words and the world. Words are shown as food, as baked goods, as objects, as three-dimensional sculptures. Every word a bite, every bite a word. Words are eaten, devoured, put in the mouth of others and stuffed. Language is shown as food, the language process as digestive process. By baking words, thus making them representational, linguistic communication becomes material and related to the function or situation it describes. It is perceived as physical, as performative, as act(ion).

Blind shows Peter Weibel with closed eyes in front of a wall and ledge. Next to him on the stone ledge stands the baked word "blind." We automatically transfer the word to the person and identify Weibel as blind. Only when Weibel eats the word "blind" the blind person opens his eyes and becomes seeing again. Of course, this is only possible in photography, the work reflects the relationship between photography and reality and the philosophical problem of representation reliability.

Blind

1967/72	performance for photography, photo poem from the film *Wor(l)d Cinema*, 1967/72; 4 b/w photographs

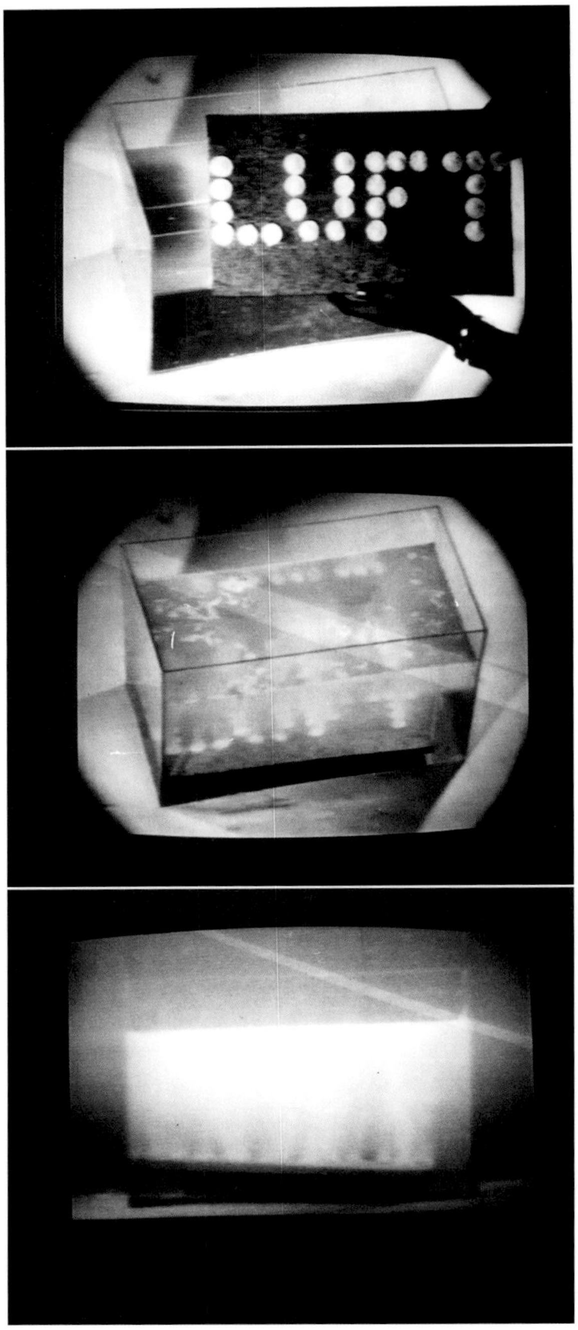

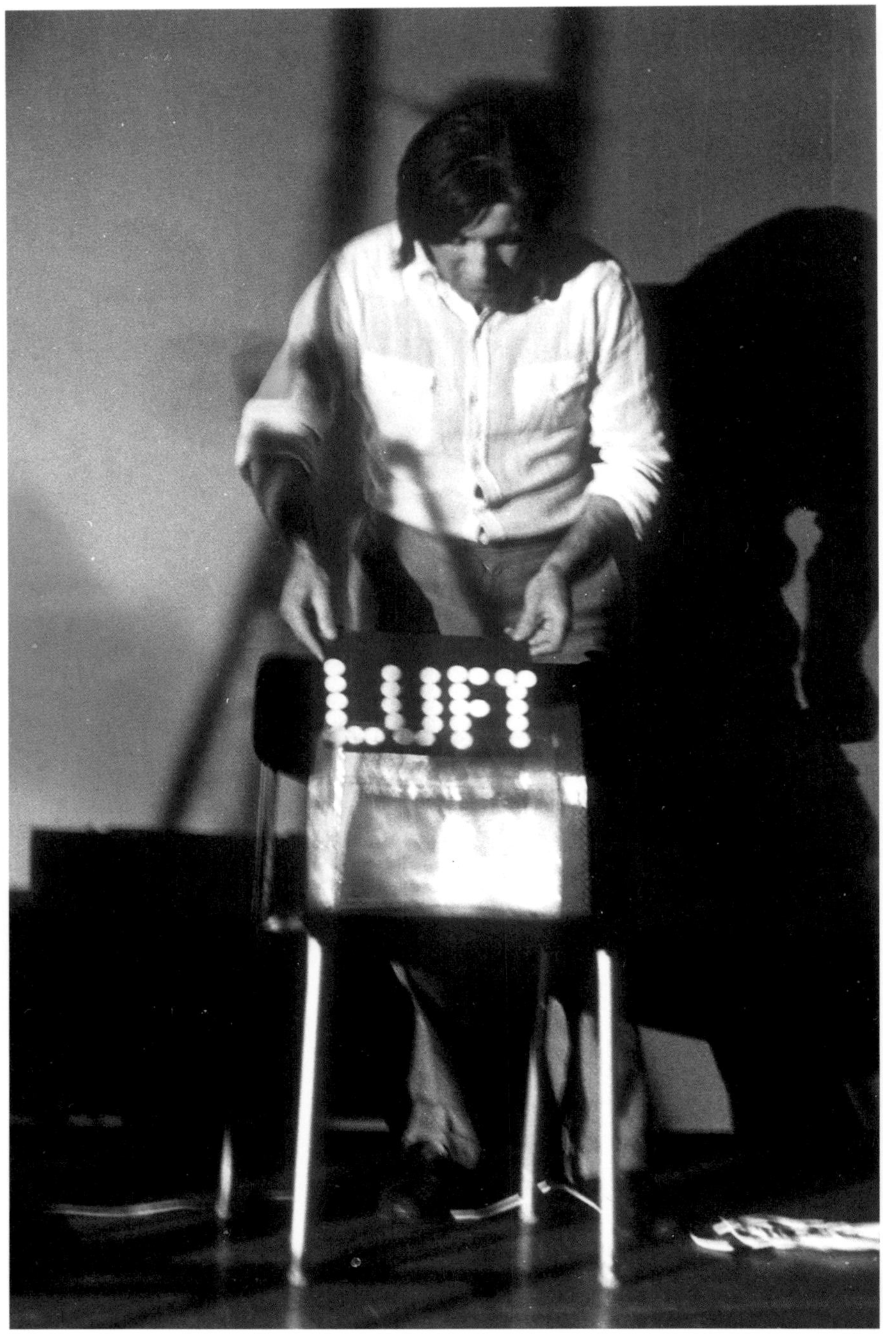

Public poetry performance at Festival Konkrete Poesie, Hanau, 1971

Oxygen tablets are laid on the bottom of an aquarium forming the word *Luft* (air). The dissolution of the text object in water, which can take up to twelve hours, is the meaning of the text itself. Action text – text process. The process by which the word *Luft* generates real air in water performs the meaning of the word. Language is understood as a product of civilization, a medium of social communication. Here, however, language appears as a natural process. Language structures as natural processes attest and lament society itself as a natural process.

"The process is extinguished in the product." (Marx)[1]
"The product is extinguished in the process." (Weibel)

[1]
Editor's note: Karl Marx, *Capital: A Critique of Political Economy*, vol. 1, trans. Ben Fowkes (Harmondsworth: Penguin, 1976), 287. First published as *Das Kapital: Kritik der politischen Ökonomie*, vol. 1 (Hamburg: Otto Meissner, 1867).

Airtext: The Product Is Extinguished in the Process / Airtext: Das Produkt erlischt im Prozess

1967–72	process poem, material text (1971), reenacted for video (1972); b/w, sound, 1:17 min.

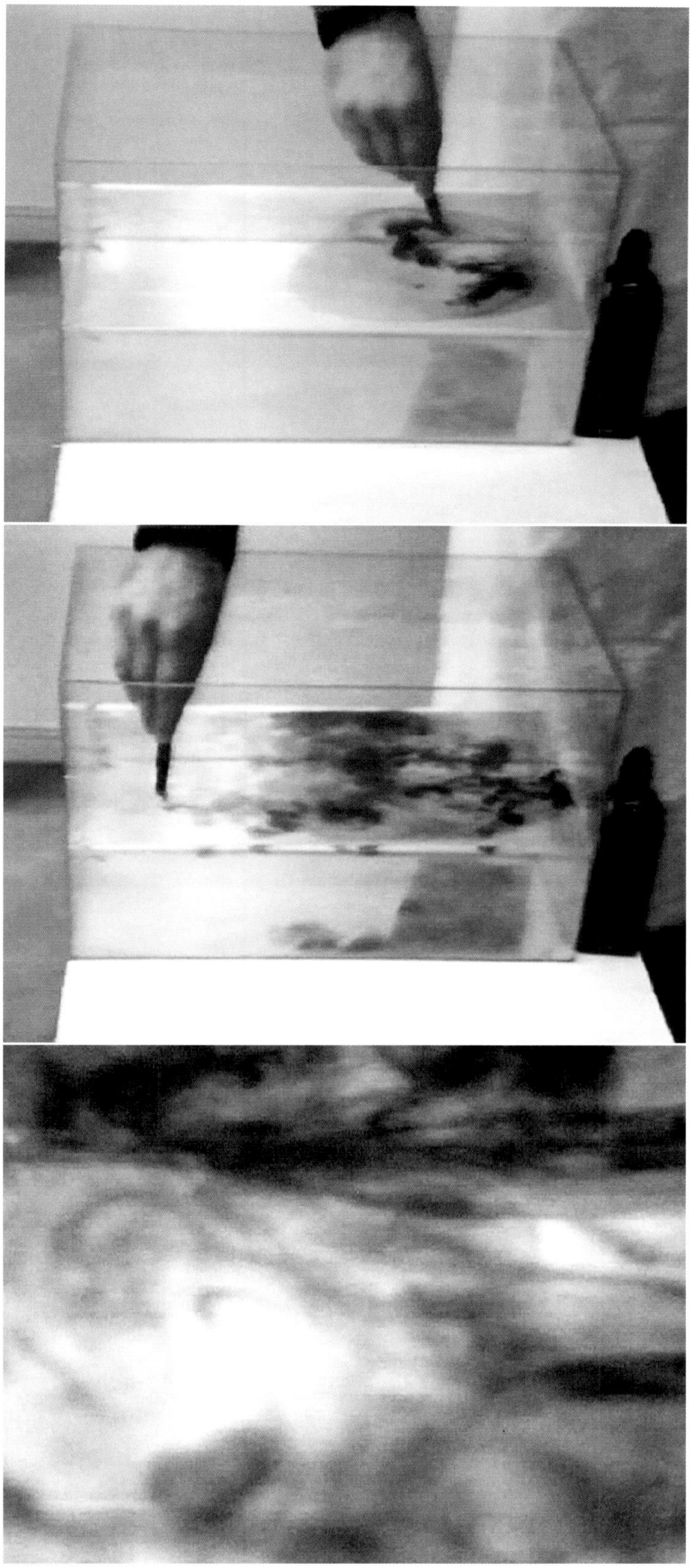

I write the word *Sinn* (sense) with an ink pen in the water.
—Peter Weibel

Sense / Sinn

1968/72	process poem, material text, reenacted for video; b/w, sound, 27 sec.

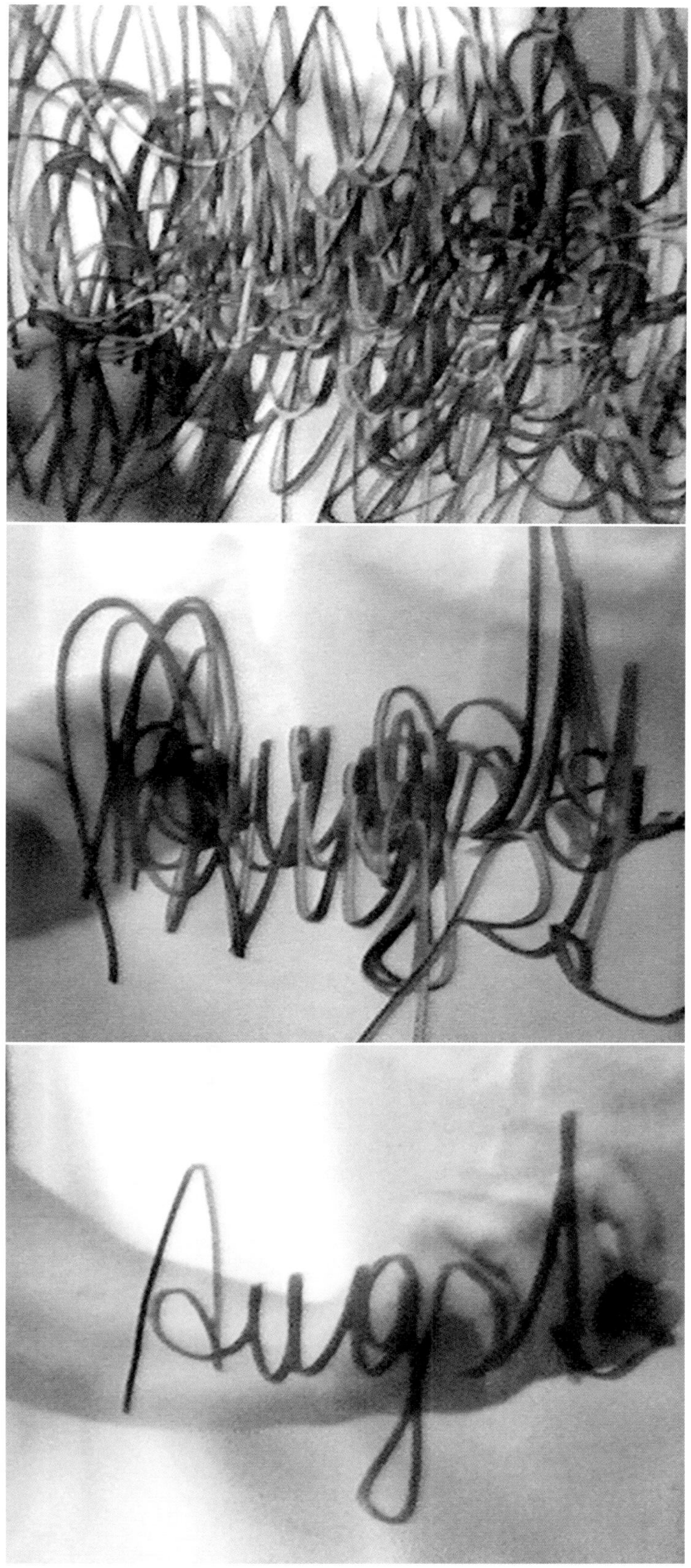

The word *Angst* is written again and again. Finally, there is only an abstract, confusing line drawing. The video is played in reverse. It begins with the drawing and in the end delivers the source of the dramatic activity: anxiety.

Anxiety / Angst

1974/75	video; b/w, sound, 2:15 min.

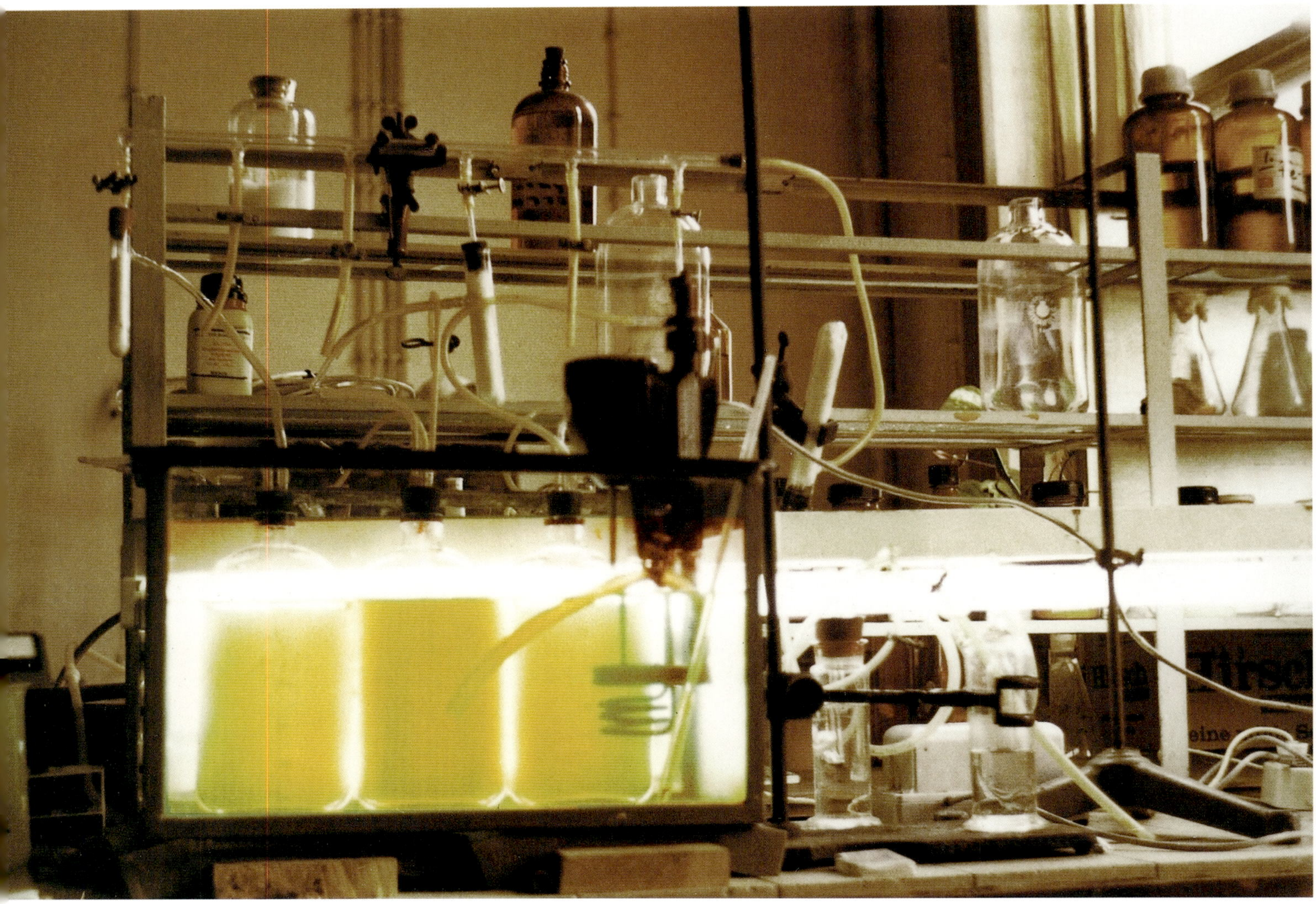

Installation views *Zur Kosmologie des Paradoxen*, Galerie nächst St. Stephan, Vienna, 1975

A Bioenergetic Paradox / Ein bioenergetisches Paradoxon

1969 (idea)/ 1975	experimental laboratory installation, algae

The first known life forms on Earth were bacteria. Blue-green algae, cyanobacteria, a specific group of phototrophic bacteria that are the only prokaryotes able to produce oxygen, began to convert the early Earth's atmosphere, which lacked free oxygen, into an oxygen-rich atmosphere about 3.2 billion years ago – through their ability to split free molecular oxygen from water under the action of sunlight through photosynthesis. In the installation *A Bioenergetic Paradox* with living blue-green algae, Peter Weibel presented a paradoxical bioenergetic situation in the context of his exhibition *Zur Kosmologie des Paradoxen* at Galerie nächst St. Stephan in Vienna. Blue-green algae normally live in the oceans, in an open system. In an aquarium, however, they live in a closed system in which the positive effect of light is reversed by self-shading. The more light, the greater the growth of the algae and the more self-shading occurs. Algal growth is curtailed by too much shadow. The energy source of light, which makes life possible, ultimately leads to its extinction in a closed system.

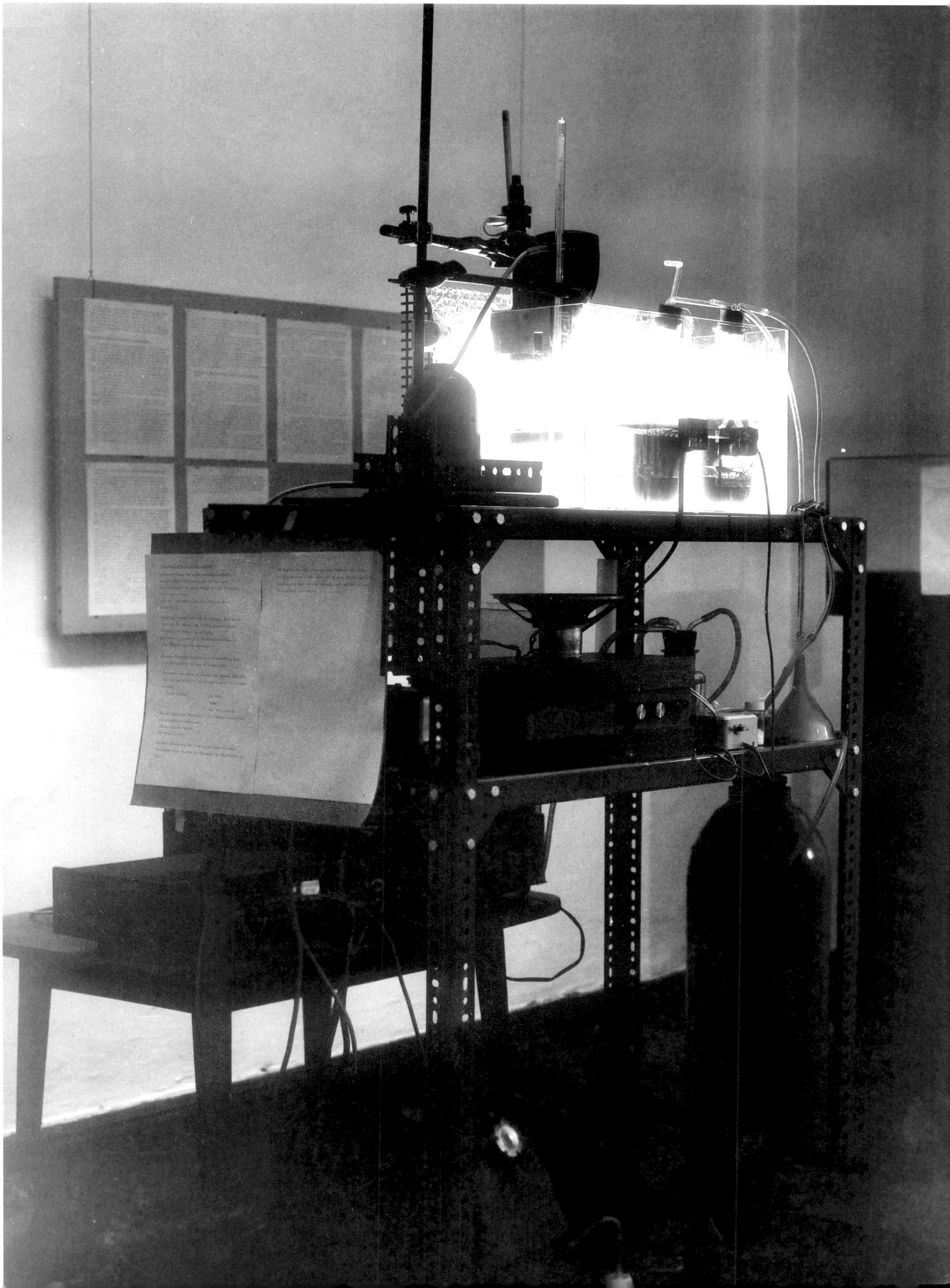

e v e

r e v e e v e r

d e v i l l i v e d

l i v e e v i l

e v e

Mirror Eve / Spiegel-Eve

1966 | visual poetry; typescript, 21 × 29.7 cm

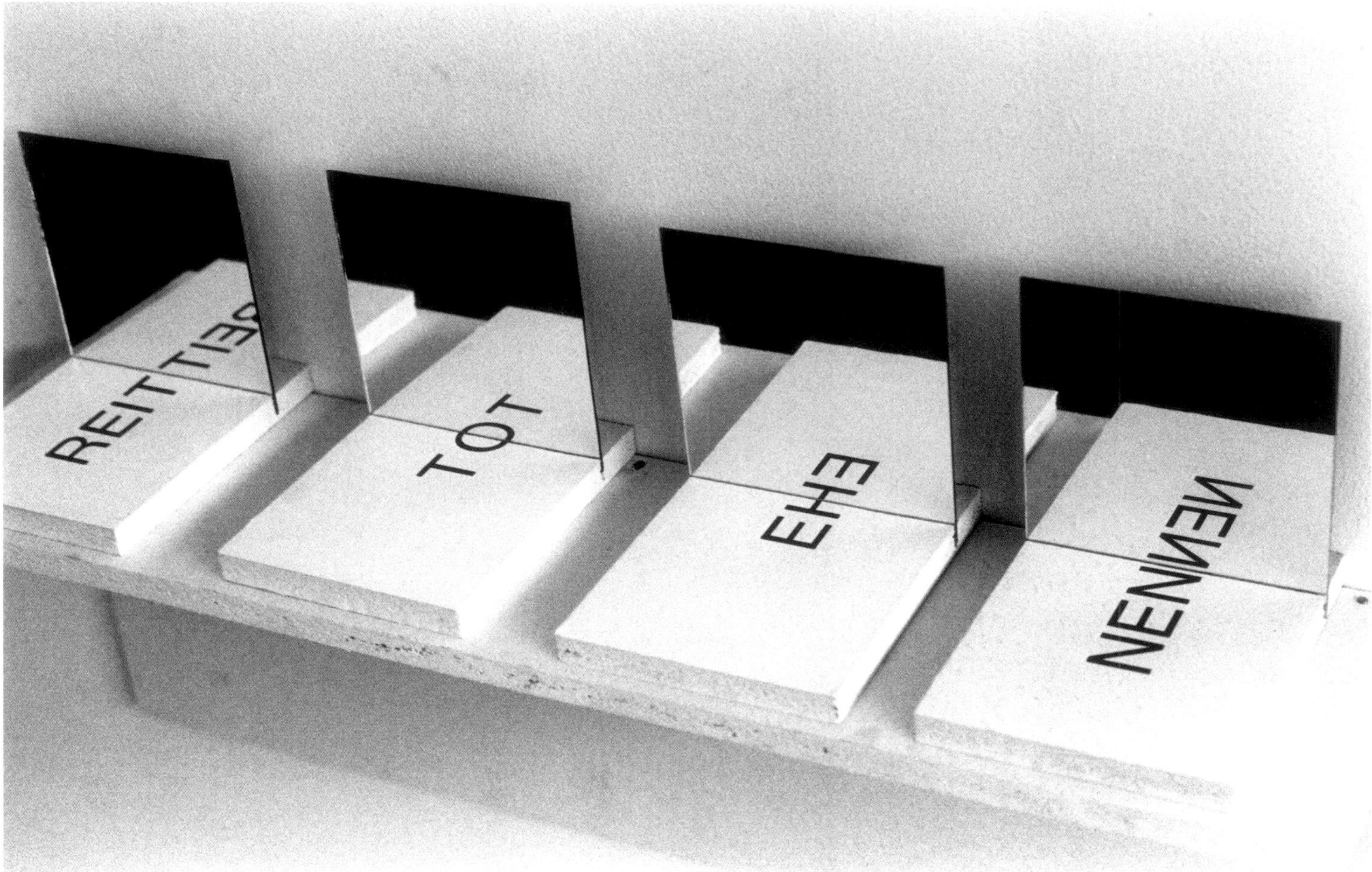

Mirror Texts / Spiegeltexte

1972	installation; 4 mirrors, wood, black paint

REIT (from *Reiter*, Engl. rider) becomes *REITTIER* (mount), *TC* becomes *TOT* (dead), *EH* becomes *EHE* (marriage), *NEN* becomes *NENNEN* (to name).

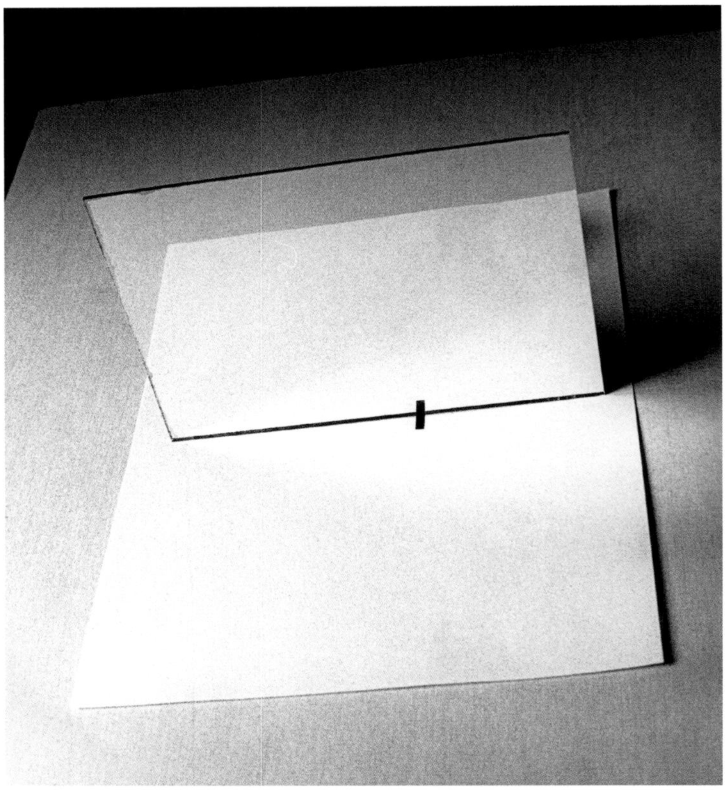

Reference Point for the Identity Sign / Richtwert des Identitätzeichens

1972 | from the series *Ritual of Identity / Ritus der Identität*
3 b/w photographs

The work series *Ritual of Identity* uses a mirror to demonstrate that the perception of reality is relative to the observer and depends on perspective and the reference system. In the first version the real and the mirrored line are united to form a seemingly single symbolic line. The sheet of paper and the mirror are placed in such a way that the mirrored image of the sheet of paper seems to be the real sheet of paper looked at thourgh a glass pane. In the second version, a real line parallel to the mirror is given a second virtual line by the reflection and thus becomes an identity sign, the mathematical symbol indicating equality. In the third version, however, the real and the mirrored line remain separated by their distance and retain their difference. From a single object, a line, three different meanings and representations emerge. Identities are constructed with a variable frame of reference (the mirror).
—Peter Weibel

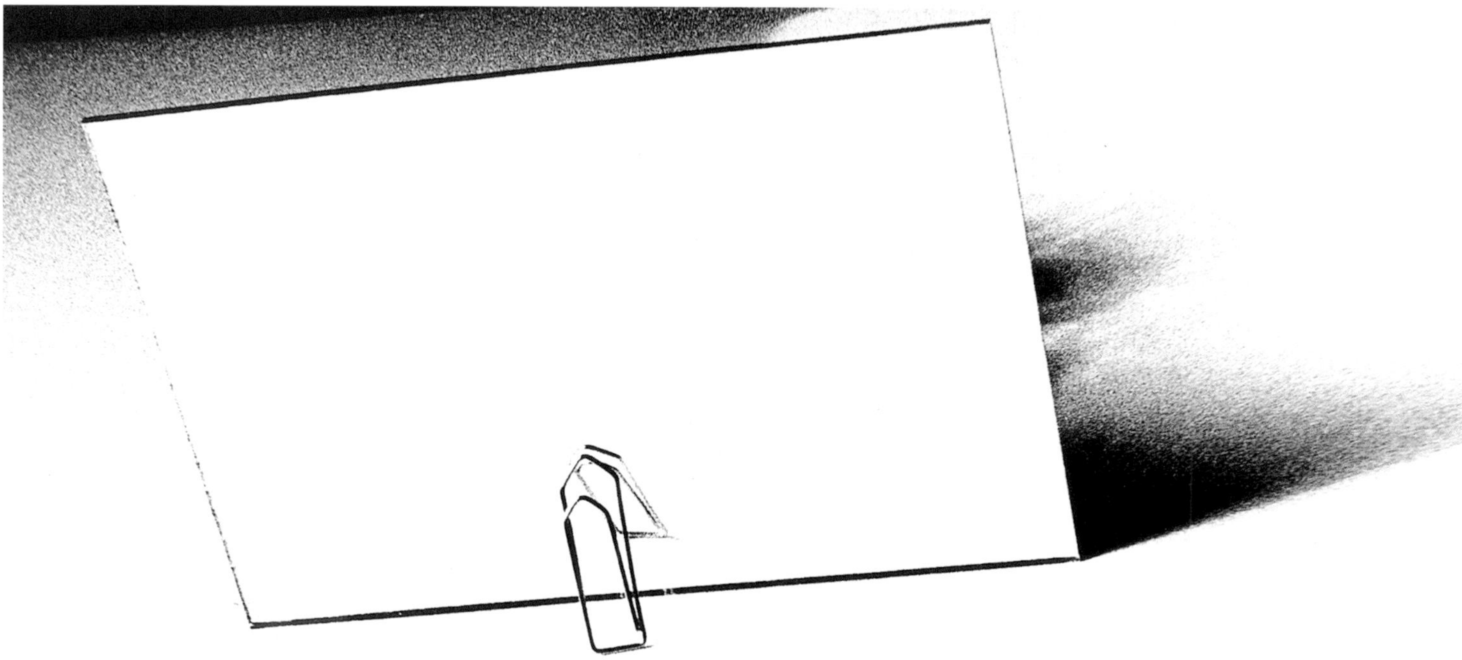

Refractions of Gravity /
Brechungen der Schwerkraft

1972 | from the series *Ritual of Identity / Ritus der Identität*
b/w photograph

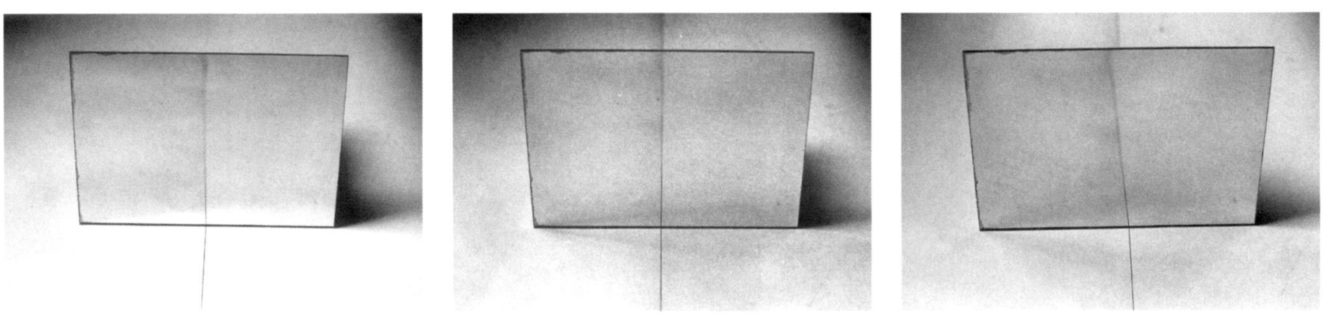

Iso- and Polymorphic Refractions /
Iso- und polymorphe Brechungen

1972 | from the series *Ritual of Identity / Ritus der Identität*
3 b/w photographs

Slight changes in the degree of inclination of the mirror change the line shape: it ends with the mirror, it continues behind the mirror, it is interrupted by the mirror.

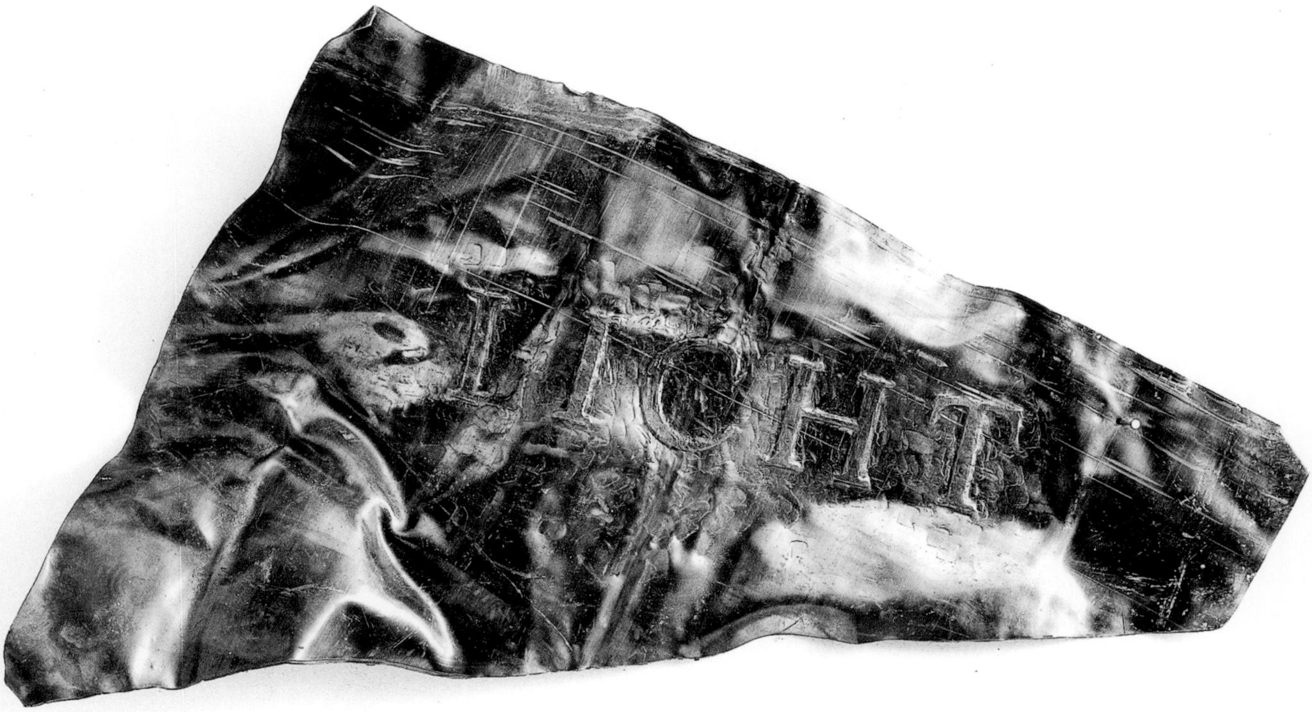

Peter Weibel's artistic career began with an intensive investigation of language and the uses of language. And that investigation frequently focused specifically on writing – as a storage medium, a technological delivery medium, and a medium for the transmission of codes and messages. Taking inspiration from the material-oriented thinking of the musical avant-gardes of the 1950s, Weibel expanded the language-based art of avant-garde literature to encompass mediality and materiality. In doing so, he not only made experimental studies of the rules and mechanisms governing strings of letters, but also supplemented his analyses of language with an exploration of its materials, tools, and media.

The production of poems from letters on paper (visual poetry) was followed by the production of poems from, on, or with stone, neon tubes, typewriters, fountain pens, pencils, audiotapes, and so on. These works downplay semantics and establish the primacy of materiality, placing them in the category of what Peter Weibel called "material poetry," in analogy to material painting. In this context it may be observed that a word's meaning and connotations vary depending on what material it is written with and/or on.

In the case of Weibel's sculpture *Light*, the properties of light (that fundamental element of life on Earth) and those of lead (onto which the word "light" has been embossed) are diametrically opposed: Light is weightless and immaterial, while lead is a weighty material. Light evades gravity; lead symbolizes it. Light, moving as it does at the speed of light, is maximally mobile; lead is immobile. And finally, X-rays cannot penetrate lead; lead protects the body from X-radiation. Lead is light's adversary. These contradictions between the meaning of a word and the material on which the word is written are exemplary of the semantic inversions of Weibel's material poetry.

Note: Adapted from Dominika Szope, "Schrift als Ur-Medium. Die frühen Arbeiten des Medienkünstlers Peter Weibel," in *Schriftfilme: Schrift als Bild in Bewegung*, ed. Bernd Scheffer, Christine Stenzer, Peter Weibel, and Soenke Zehle (Ostfildern: Hatje Cantz, 2014), 122–23. Translated from the German by Patrick Hubenthal.

Light / Licht

1966/93 | lead, 120 × 65 cm

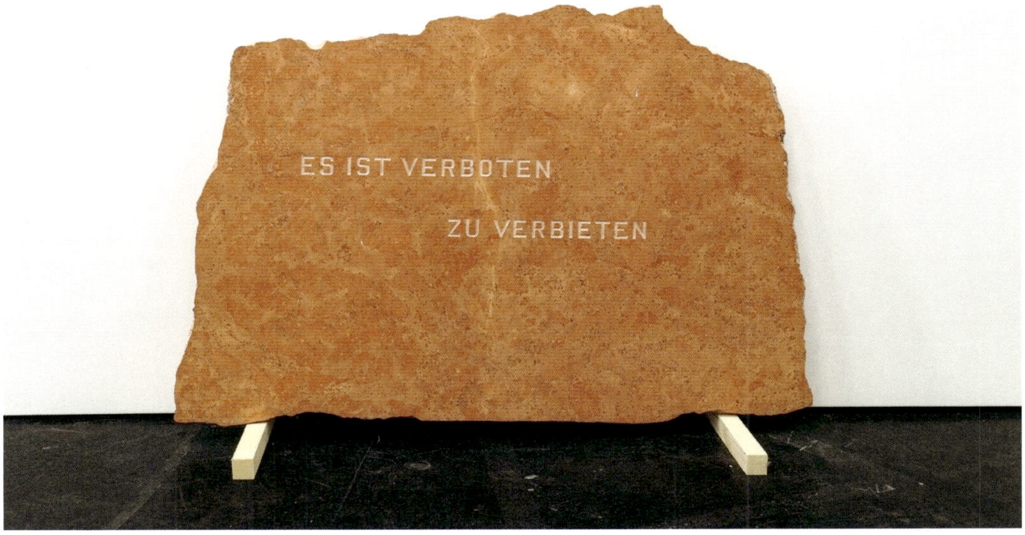

Installation view
respectively, Peter Weibel,
ZKM | Karlsruhe, 2019–20

It Is Forbidden to Forbid / Es ist verboten zu verbieten

1974/ 2004	text stone; marble, 65 × 161 × 7 cm

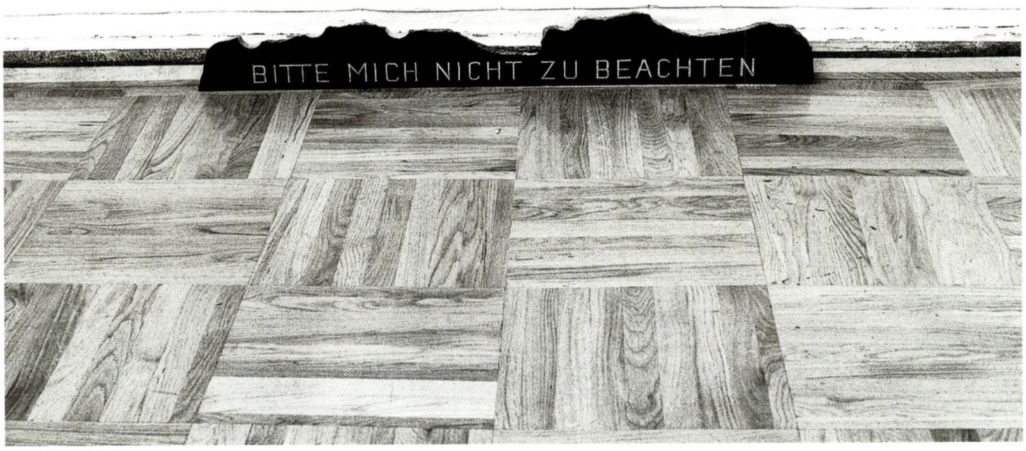

Installation view
Zur Kosmologie des Paradoxen, Galerie nächst St. Stephan, Vienna, 1975

Please Do Not Pay Attention to Me / Bitte mich nicht zu beachten

1966 (idea)/ 1974	text stone; marble, 20 × 70 × 5 cm

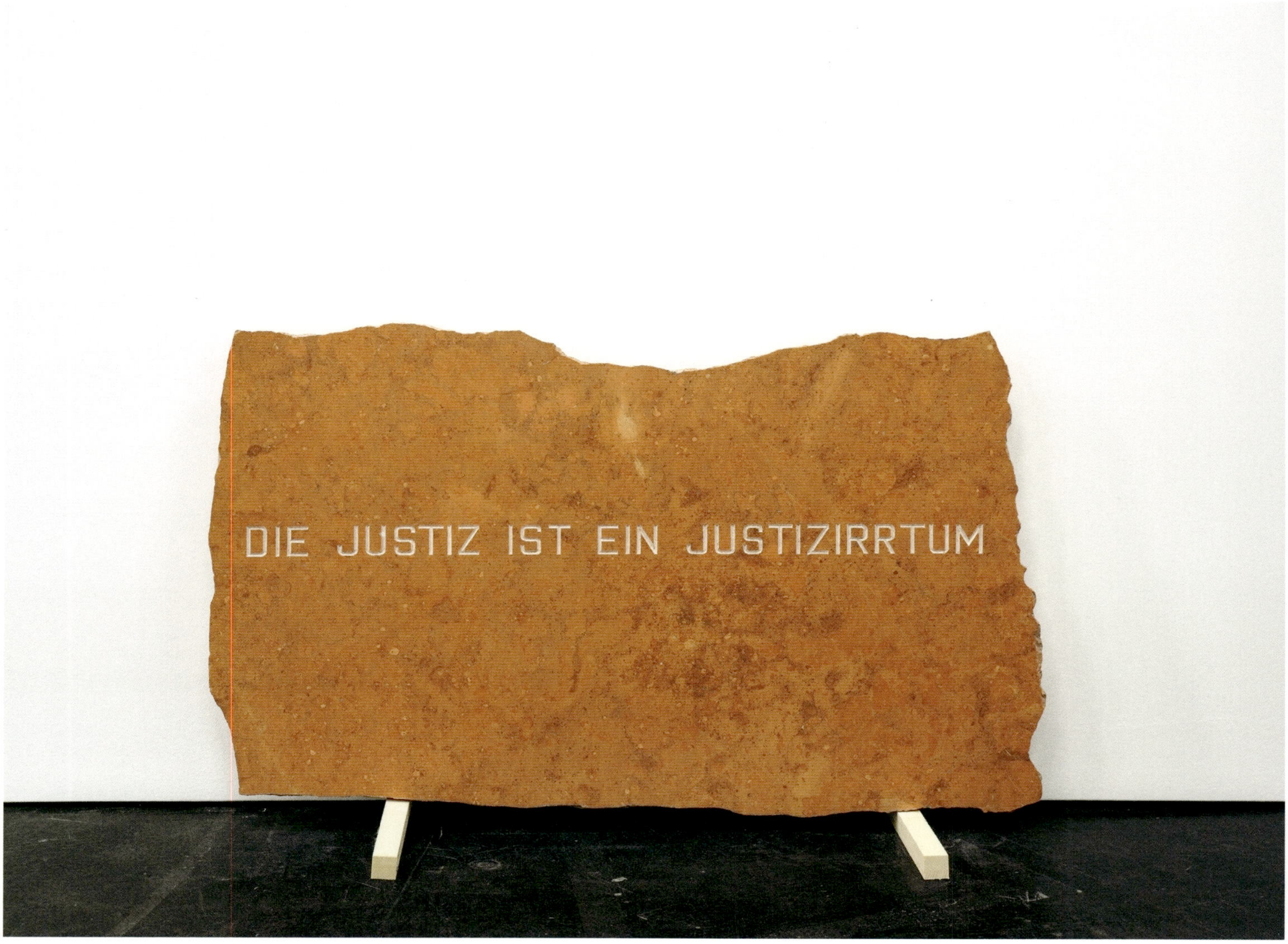

Installation view
respectively, Peter Weibel,
ZKM | Karlsruhe, 2019–20

For Peter Weibel, writing is an artistic medium, one that has featured prominently in his creative endeavors from the very beginning. Writing, however, must itself be conveyed in context with the carrier medium, and these can be quite varied: paper, canvas, screen, skin, and other materials. A tour of Weibel's oeuvre highlights this media migration. Although he started out writing his poems on paper, these were soon followed by words and phrases on stone, glass, wood, lead, and rubber. He has used his body as a writing surface, projecting letters onto it and implanting poems under his skin. He has used light as a medium, installing words in neon lettering. He has extended the space of poetry to encompass new materials, leading him to engage intensely with machines – such as the typewriter and the tape recorder – and new media such as video. He assumes that the meanings of our words are influenced by the material they are written on.

Justice Is a Miscarriage of Justice / Die Justiz ist ein Justizirrtum

1967 | text stone; marble, 60 × 80 × 3 cm

In the context of Weibel's material poetry, the most frequently used medium is stone. In part, the fascination stems from the primal nature of the medium, since stone is one of the earliest materials used for the storage of language. The earliest written characters and graphic symbols were carved into this storage medium. Weibel has realized a series of text-on-stone pieces, in many of which words are chiseled or engraved into marble. The idiom "carved in stone" means that the text in question is law, an irreversible act. From Moses to the present, stone has been a favored material for the dissemination of script as prescription. The provisions of the law, seemingly carved in stone for all eternity, are declared invalid by Weibel in the very medium, stone, that should guarantee their eternal validity. The judge's verdict is contradicted in its own medium. The gravity of the judgment is underscored by the massiveness and weightiness of the marble as well as its association with gravestones. At the same time, the use of stone also allows Weibel to extend the two-dimensionality of written notation into the three-dimensionality of sculpture.

Translated from the German by Patrick Hubenthal.

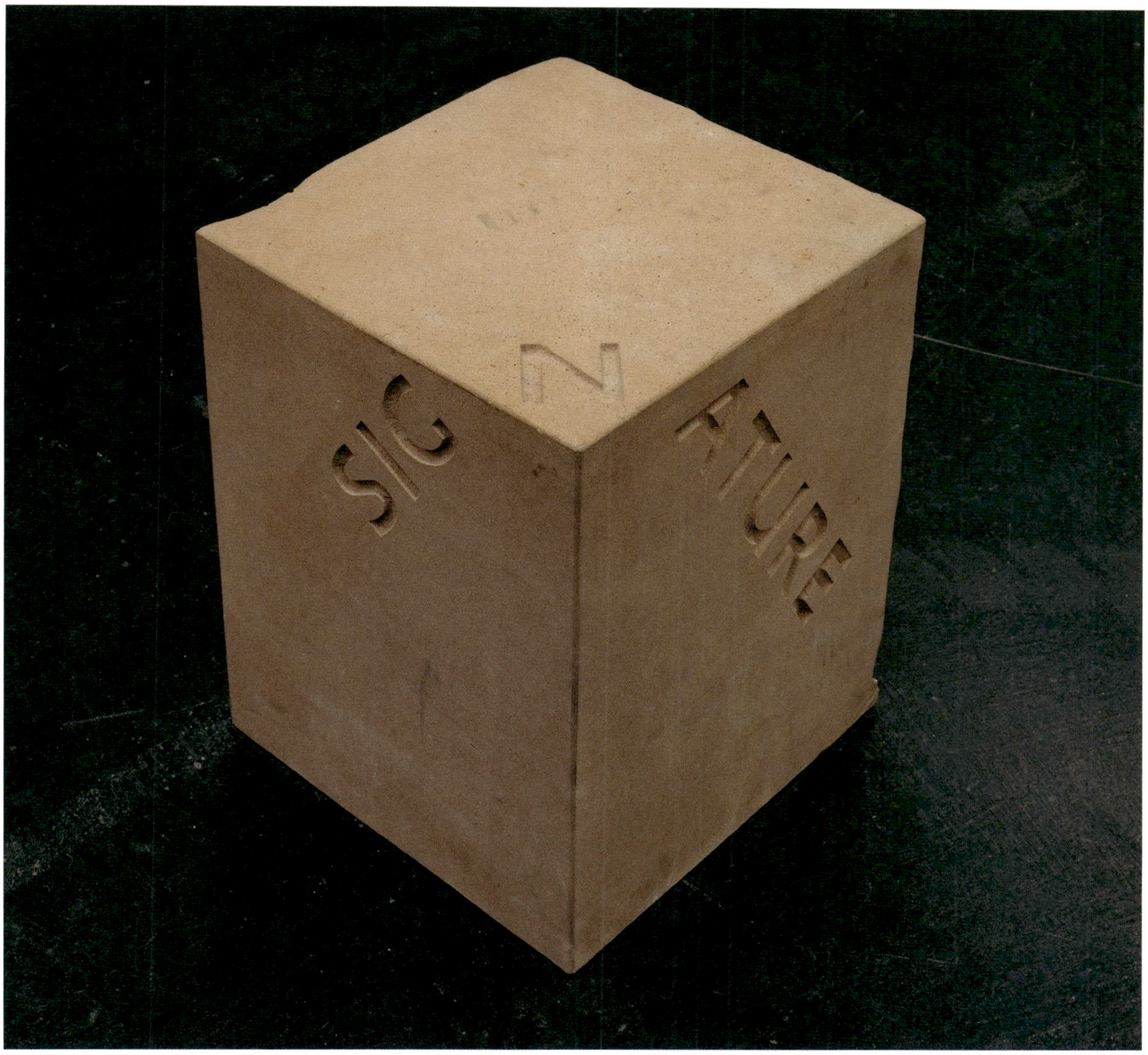

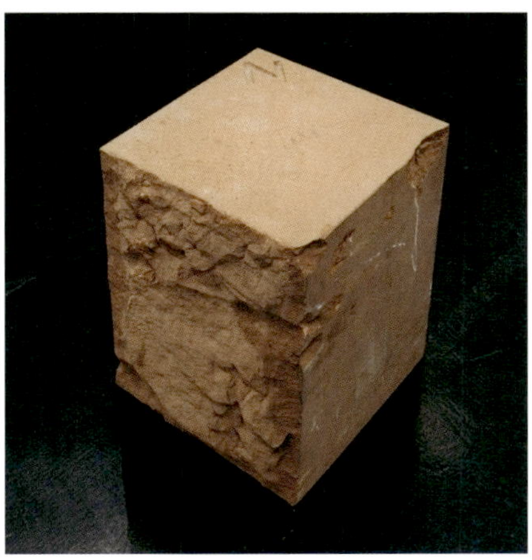

Installation views
respectively, Peter Weibel,
ZKM | Karlsruhe, 2019–20

A semiotic triple: nature, a stone, is turned by the work of an artist (signature) into a cuboid, a sign. A way of notating language in three dimensions. Peter Weibel conceives works in stone that sculpturally integrate language and function as sculptures in space. Language is spatialized, made three-dimensional. This is especially clear in the case of a text stone called *Sig/n/ature*. In the art world, this word often refers to the signature of the artist. If one unfolds this word in space by wrapping it around a three-dimensional block, one discovers new words inside it: Alongside "sign," one can also read "nature." The word "signature" represents the artistic subject, who forms a sign (a block) out of nature (symbolized by the unworked back of the block). With this piece, Weibel shows us not just the flexibility of language, but also the process of civilization and the function of art. The subject embodies the idea-realizing force. The triad, the threefold function of art, consisting of artist, nature, and sign (culture), is already present in the word "signature." Here again, Weibel shows that realities are based in language itself and that sign processes (language) can be the result of the material processes (of reality). The technique of spatialization and the transference of poetry onto new materials illuminate the realm of language and humanity's relationship to the real. In this way, the artist makes clear that the worlds of objects, language, and human practice are all interconnected. He is the inventor of three-dimensional language notation.

Sig/n/ature

1967/75 | text stone; sandstone, 55 × 41 × 40 cm

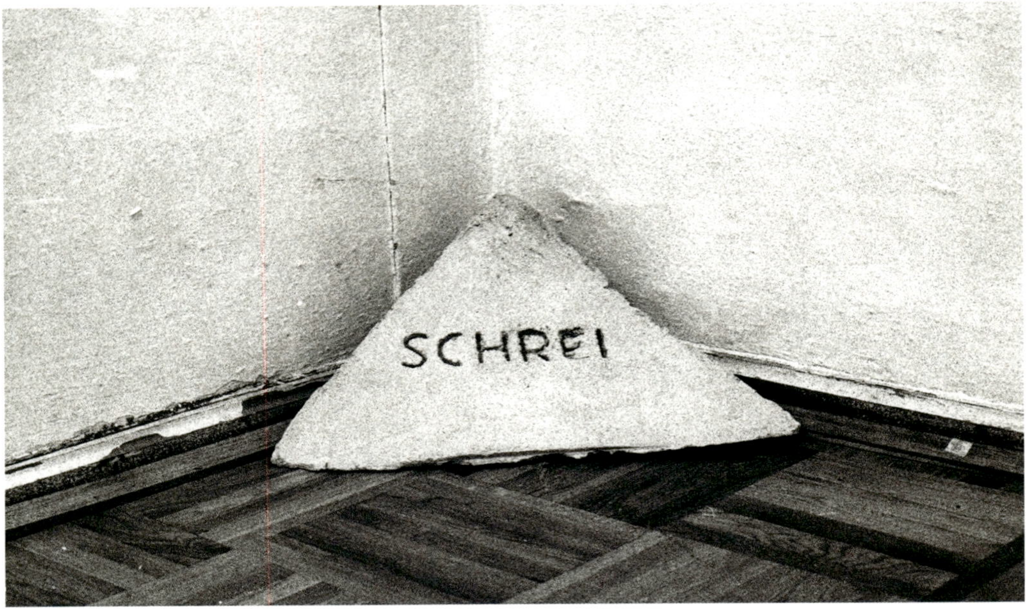

Installation view
Zur Kosmologie des Paradoxen, Galerie nächst St. Stephan, Vienna, 1975

Traditionally, poets would write many words on paper to express the plural meanings of a *Schrei* (scream) (Edvard Munch). I reduced the poem to a single word and changed the material. The contrast between the material and the word *Schrei* creates the meaning of the poem. The material concrete is synonymous for dead, depressed, inanimate, heavy. The word *Schrei* is synonymous with expression, living. The scream is imprisoned in the material, cannot articulate itself, is even placed in a corner. It demonstrates the refusal of society to listen to the sufferings of humans and animals, of all living organisms.
—Peter Weibel

Scream / Schrei

1967/75 | concrete, 38 × 51 × 40 cm

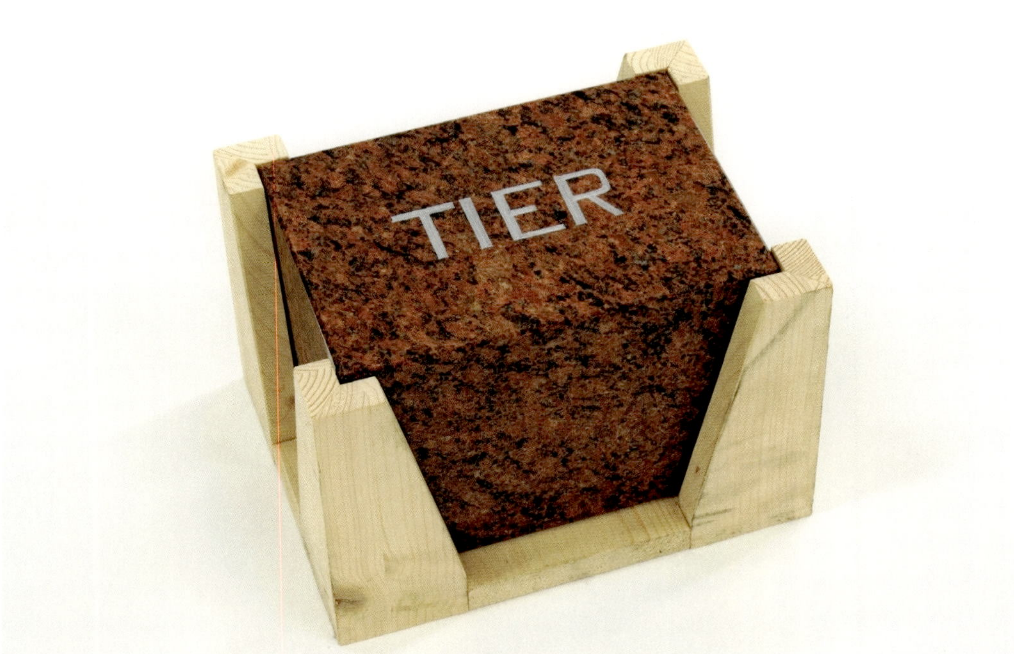

Installation view
respectively, Peter Weibel, ZKM | Karlsruhe, 2019–20

The word *TIER* (animal) written on paper would not have a specific expressivity. It would be void of any empathy and emotion. But written on a cold block of marble, the meaning of the word *TIER* as a living, moving organism contrasts heavily with the frozen, unmovable marble block. The animal is evidently imprisoned in a cage. To support this meaning there is also a frame of wood, an ornament, for the sculpture that expresses again the pressure. The material alone articulates the meaning of the word in a direct manner.

Cage / Käfig

1967/75 | text stone; red marble, wood, ca. 28 × 22 × 22.5 cm

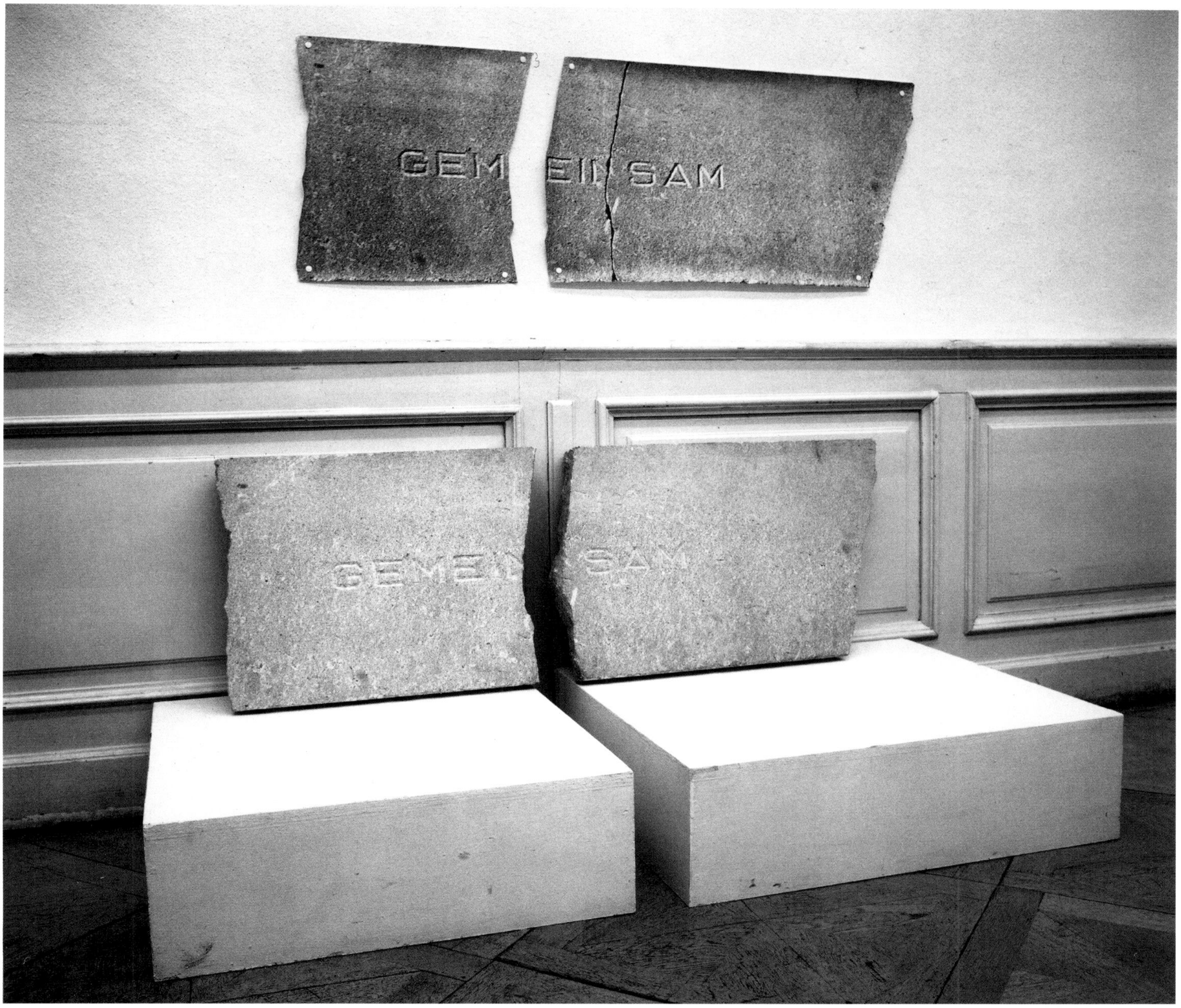

Installation view
Zur Kosmologie des Paradoxen, Galerie nächst St. Stephan, Vienna, 1975

Gem-ein-sam

1967 (idea)/ 1975	on the floor: text stone sculpture; 2 stones, wood, left (*Gemein*): 7.4 × 52 × 40.2 cm, right (*-sam*): 6.6 × 62.5 × 40.2 cm on the wall: photo poem; 2 photographs, left (*Gem*): 40.3 × 40 × 1.4 cm, right (*-einsam*): 40.3 × 70.5 × 1.4 cm

The word *gemeinsam* (together) is engraved on stone. The stone is broken into two pieces reading *gemein* (mean) and *sam*. This broken, split stone is then photographed. The photograph is again split into two pieces reading *gem* and *einsam* (alone, lonely), thereby forming the opposite of the initial meaning "together." Finally, the pedestral is also split.

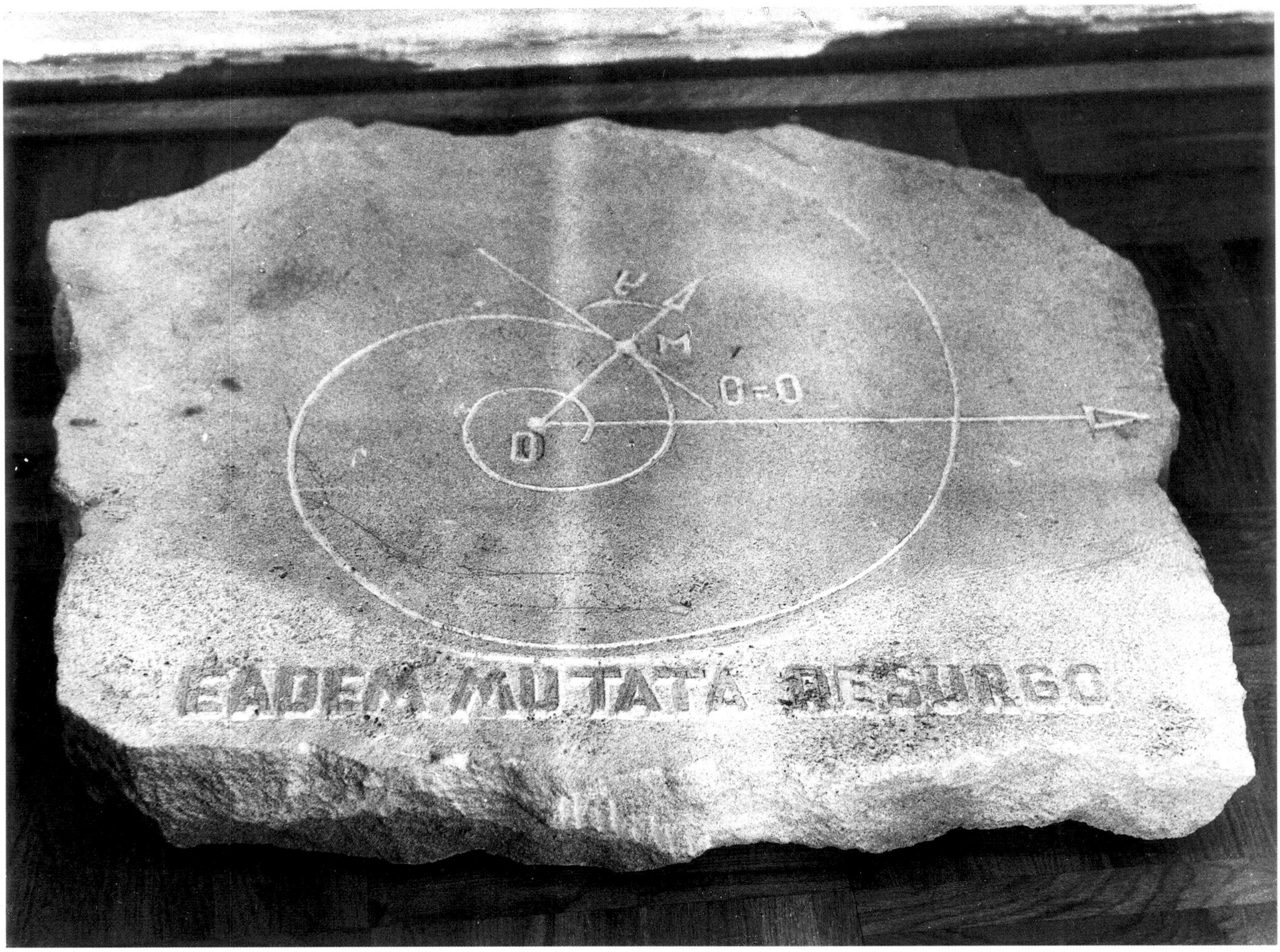

The logarithmic spiral has a number of unique qualities, which is why the Swiss mathematician Jacob Bernoulli called it the "spira mirabilis" (marvelous spiral):

1. A logarithmic spiral is a spiral whose distance from the center, its origin, increases by the same factor with each rotation.
2. Every straight line passing through the origin transects the curve at the same angle.
3. The arc length from any turn of the spiral to the origin is finite, although an infinite number of rotations are required to reach the origin. This is thus evidence of the possibility that the soul, though transformed (by death), can return as itself: "Eadem mutata resurgo"–"Although changed, I rise again the same." Bernoulli wanted to have a logarithmic spiral with this motto carved on his tombstone in the Basel Minster, but by mistake an Archimedean spiral was engraved instead. —Peter Weibel

Logarithmic or Pascal Spiral /
Logarithmische bzw. Pascal'sche Spirale

1974 | stone, neon tube, metal sheet, water

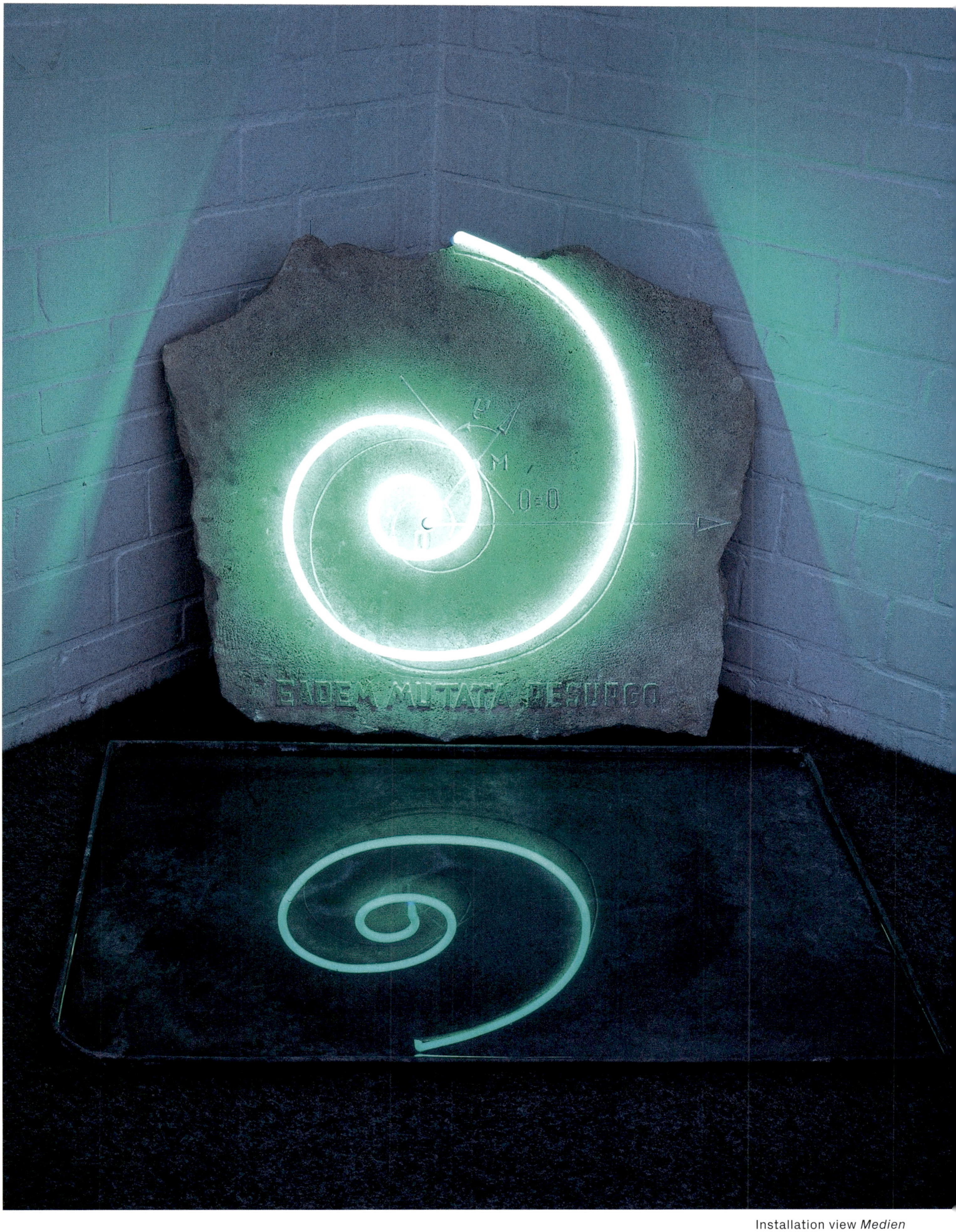

Installation view *Medien und Realität*, Kunsthalle Wilhelmshaven, 1982

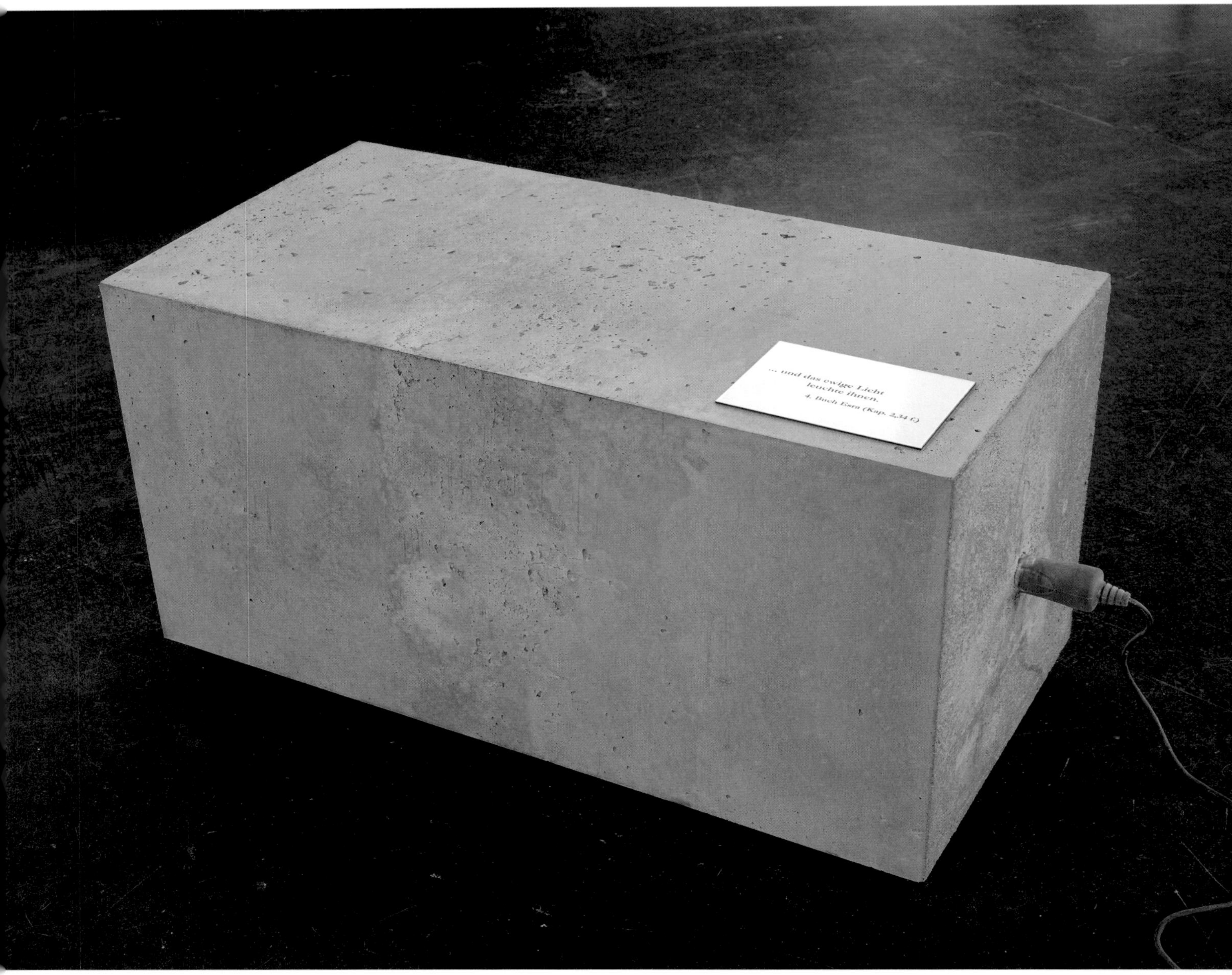

"... for the everlasting light
shall shine upon you"
(*2 Esdras*, chapter 2, 34–35)
Installation view
respectively, Peter Weibel,
ZKM | Karlsruhe, 2019–20

For the Everlasting Light Shall Shine upon You / Und das ewige Licht leuchte Ihnen

1974/ 2019	concrete, LED stick light inside the concrete block, cable, aluminum plaque, 54 × 114 × 50 cm

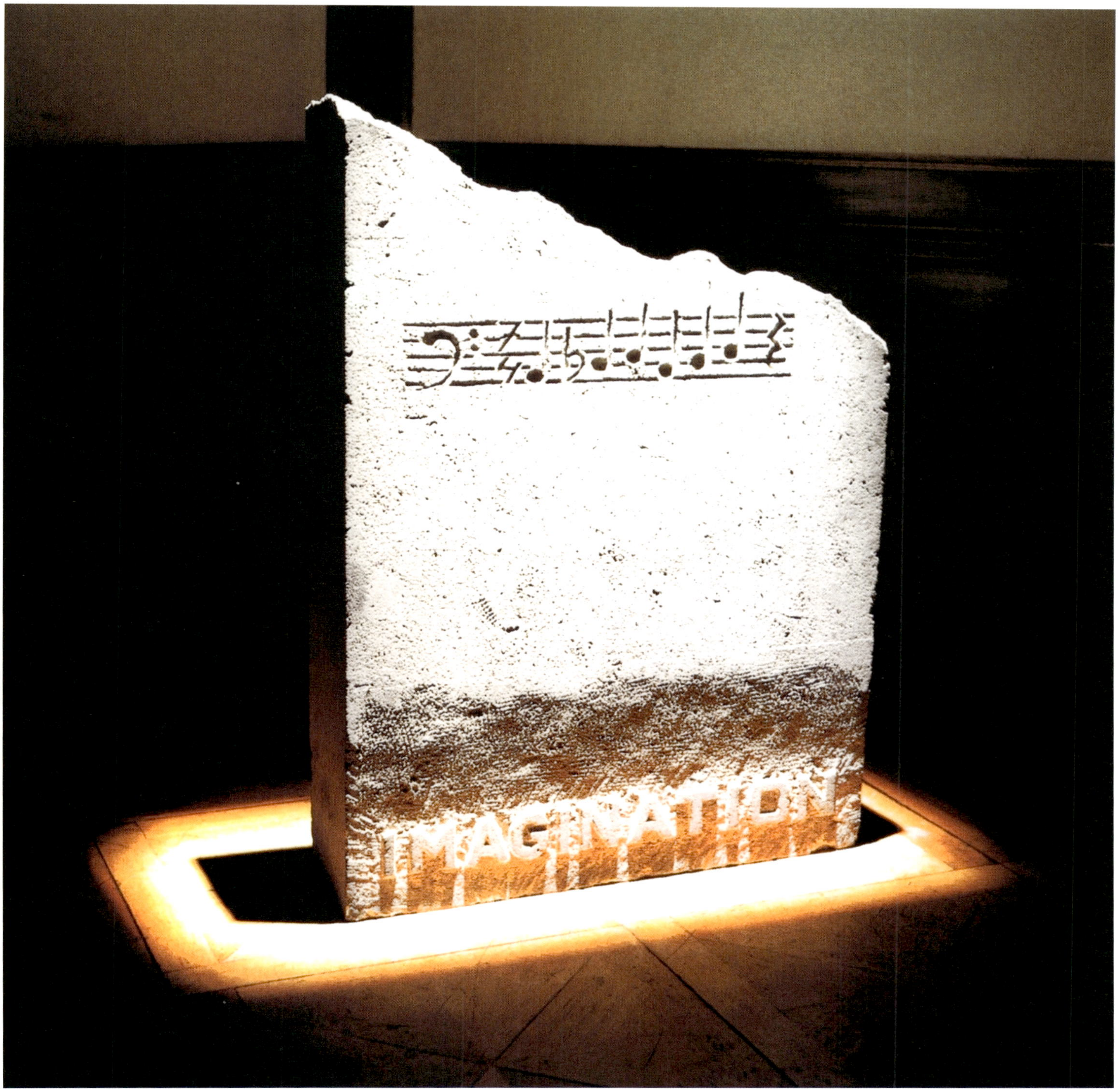

Installation view
Medienpoesie, Kulturhaus
Graz, 1982

Chiseled into the sandstone are the opening notes of Edgard Varèse's composition *Arcana* (1927), the score of which has a quotation by Paracelsus as an epigraph: "One star exists, higher than all the rest. This is the apocalyptic star. The second star is that of the ascendant. The third star is that of the elements – of these there are four, so that six stars are established. Besides these there is still another star, imagination, which begets a new star and a new heaven."[1] Varèse himself said: "It is imagination that gives form to dreams."[2]

[1]
Paracelsus, *The Hermetic and Alchemical Writings*, vol. 2, *Hermetic Medicine and Hermetic Philosophy*, ed. Arthur Edward Waite (London: James Elliott and Co., 1894), 310.

[2]
Edgard Varèse, "Puff Balls," Santa Fe lecture, Sept. 2, 1937. Typescript Sacher Stiftung.

Imagination

1974 | sandstone, 100 × 80 × 17 cm

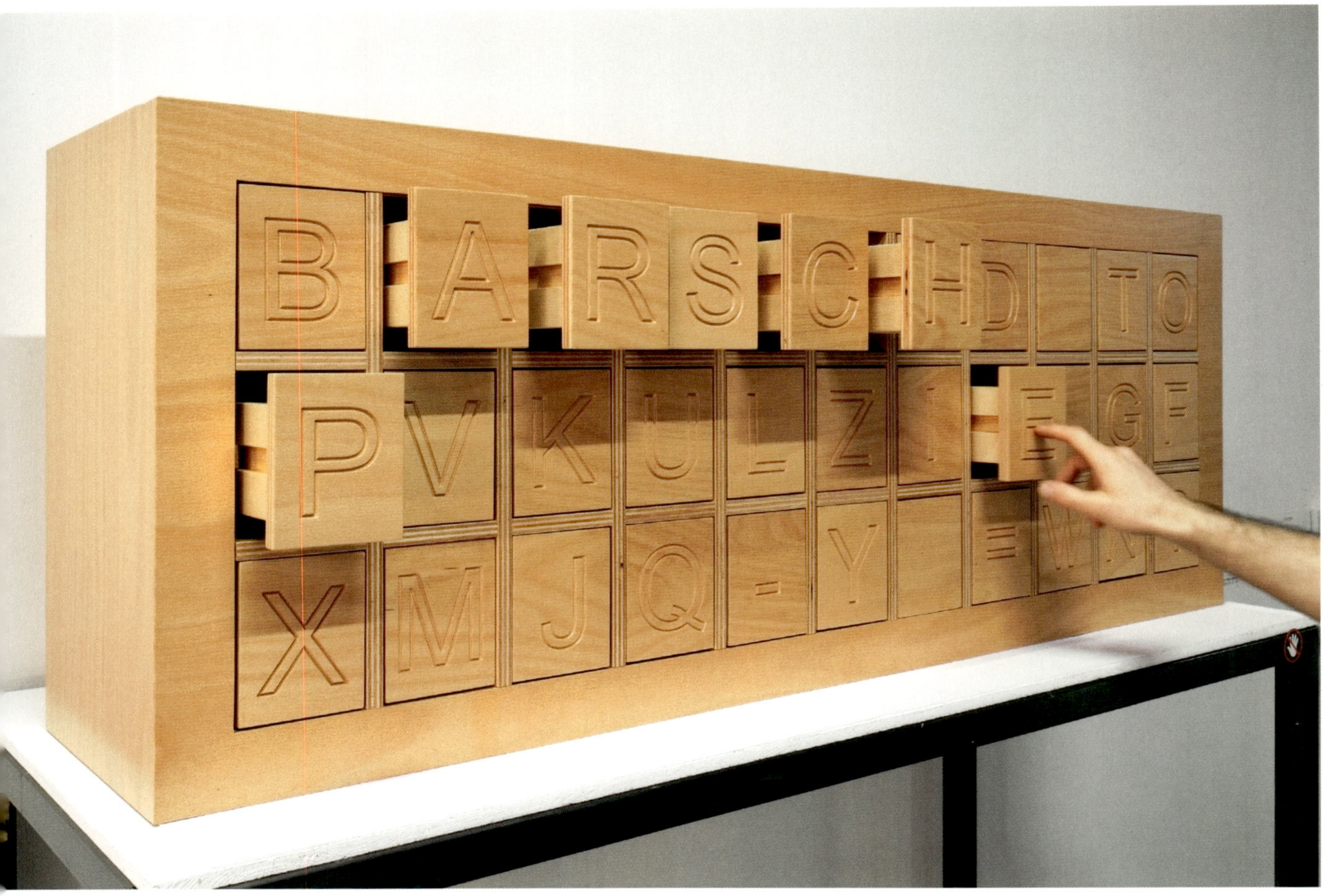

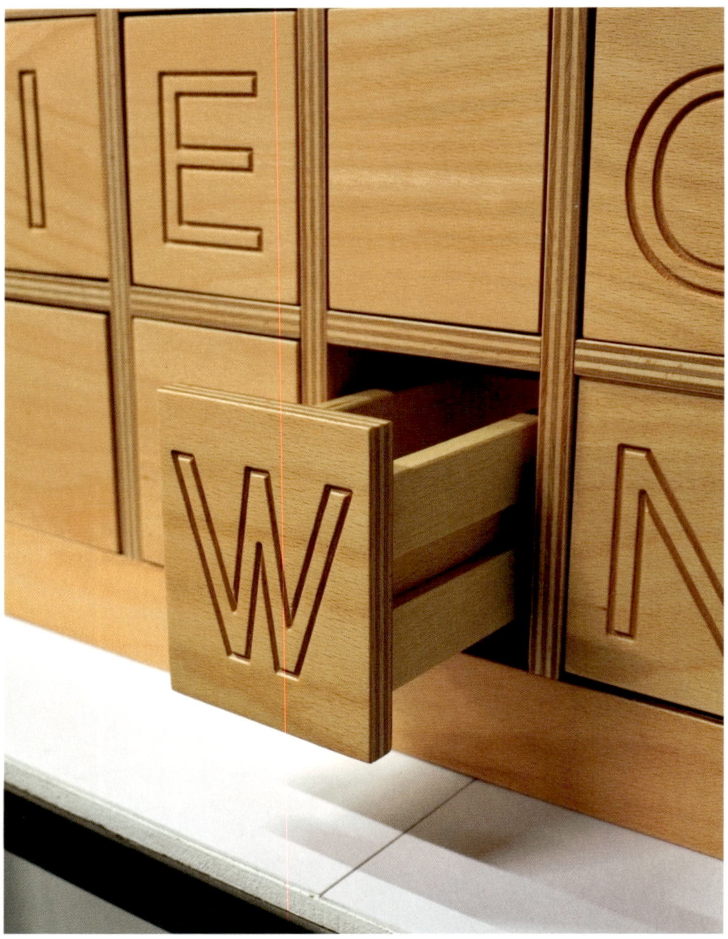

Language Kit W1 / Satzbau – Bausatz W1

1989/ 2014 | wood, 54.5 × 143 × 32 cm

For Peter Weibel, language is a tool, a formal mechanism, a mechanical apparatus of finite elements (e.g., 26 letters) and rules (e.g., grammar). Language not only consists of signs, it is something material, physical, with which humanity has operated for thousands of years. The construction of worlds by language is already laid out in the language itself: in word constellations such as "sentence construction" or "construction kit." Sentences are constructed from elements of the construction kit. Language is a kit with which the world is built, or a sentence construction constructing the world. There are operative links between mentalism, mechanism, and materialism.

Peter Weibel's credo is: Not everything that humans think and feel can be expressed in a formal language, in a string of signs to which others attach meaning. What can be formalized in a language, however, can be mechanized, and what can be mechanized can become a machine, a construction kit. The keyboard of a typewriter in itself is proof of this idea: humans typing on a machine with letters, writing meaningful sentences. They express their thoughts and feelings with the help of a machine that embodies a formalism. This formalism partially mirrors a mentalism.

The furniture *Language Kit W1* with inbuilt letters is a new kind of typewriter machine, based on the analogy of sentence construction and building construction. Its wooden box has a total of 33 drawers on which the letters of the alphabet and various punctuation marks are written. If you exert pressure on the drawers or the letters, they open automatically. The box becomes a keyboard with which words and sentences can be produced. *Language Kit W1*, which was constructed for the first time in 1989 and reconstructed in 2014 with a slight modification, invites us to understand the creation of words and sentences, that is, sentence construction, as a technical kit that has to be actively constructed. The free drawers can be assigned to further symbols or letters.

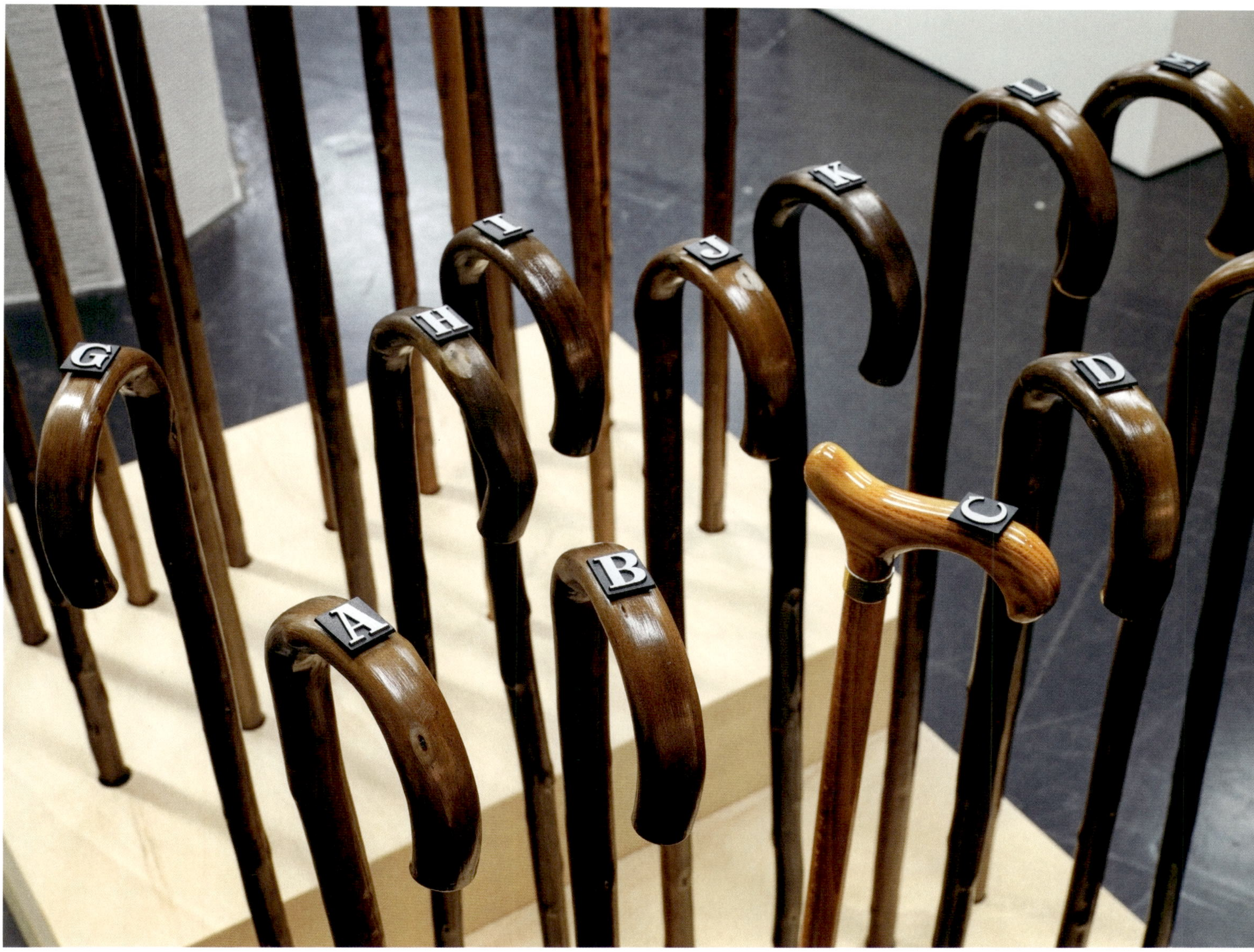

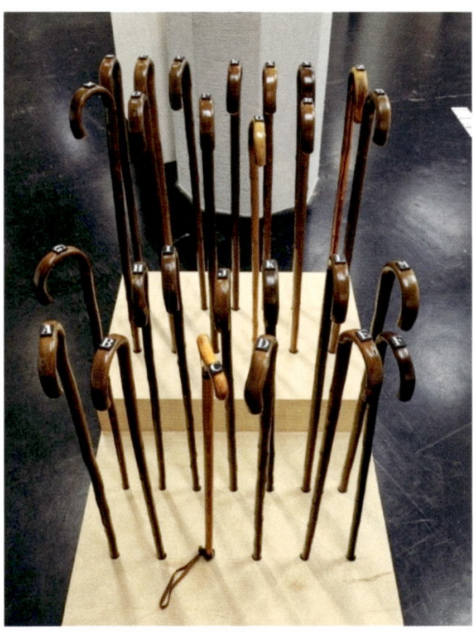

This page and opposite page: Installation views *respectively, Peter Weibel*, ZKM | Karlsruhe, 2019–20

The Alphabet Is Still the Best Crutch in the World / Das Alphabet ist noch immer die beste Krücke der Welt

1994 | 26 canes on two-stage pedestral, 26 plastic letters

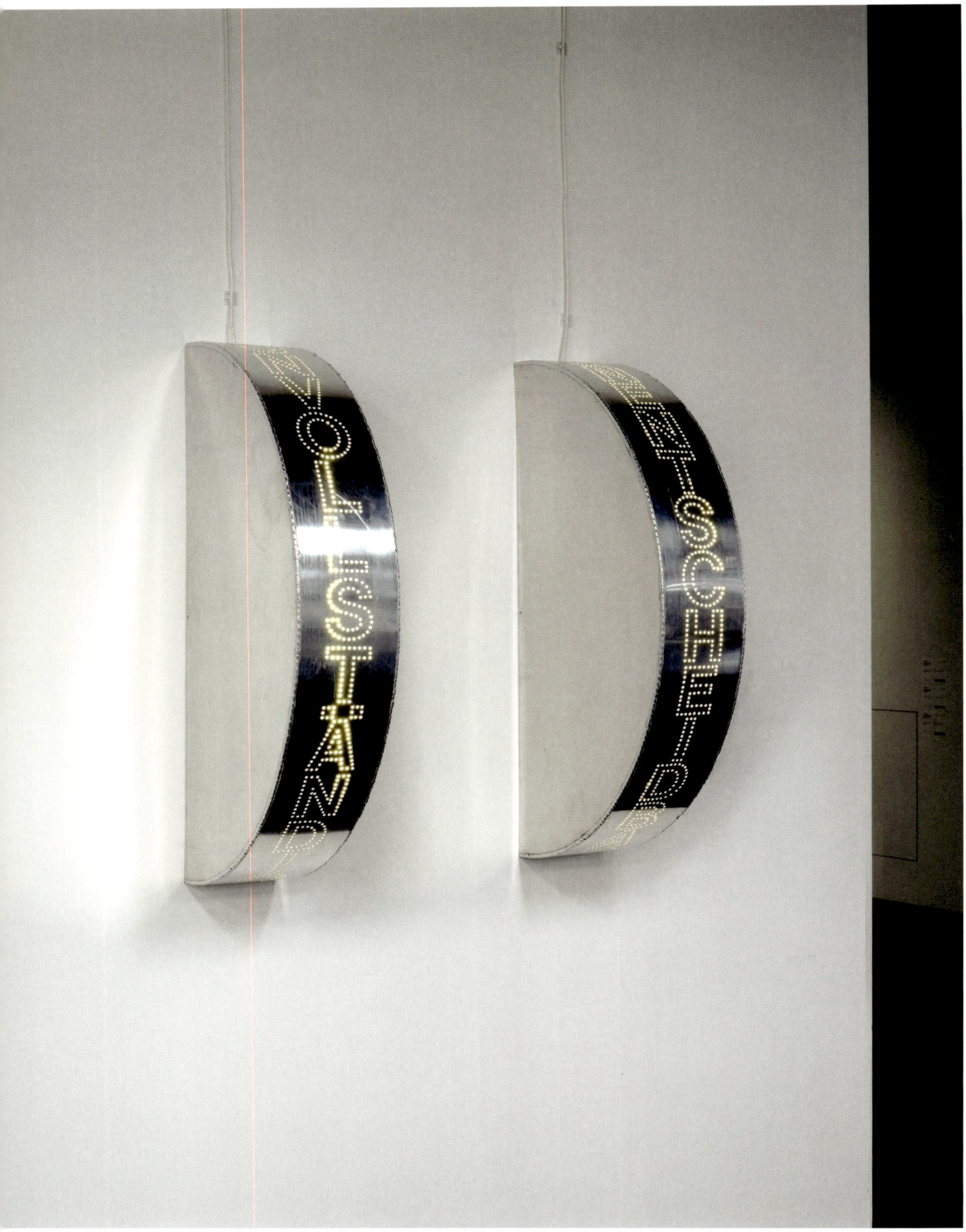

Installation view
respectively, Peter Weibel,
ZKM | Karlsruhe, 2019–20

You Are (Not) Part of the Program. You Are (Not) Part of the Problem Zone

1988	2 neon tubes, 2 aluminum housings with perforations in the form of words, 40 × 55 × 20 cm each

The words *unvollständig* (incomplete) and *unentscheidbar* (indeterminable) are presented in such a way that they cannot be read in their entirety. In this way, they realize their meaning, their program. To an internal observer (part of the program), the overall context remains hidden. Only external observers, who have the ability to take different positions, can free themselves from the problem zone. Here, as in almost all of Peter Weibel's works, the formal considerations entail political statements

Whoever reads a program is programmed, becomes part of the program. They become part of the problem or system that the program is meant to solve. The problem solvers can become the problem themselves.

Is the program part of the problem, the problem itself, or the solution to the problem? Who decides what is a problem? The program itself?

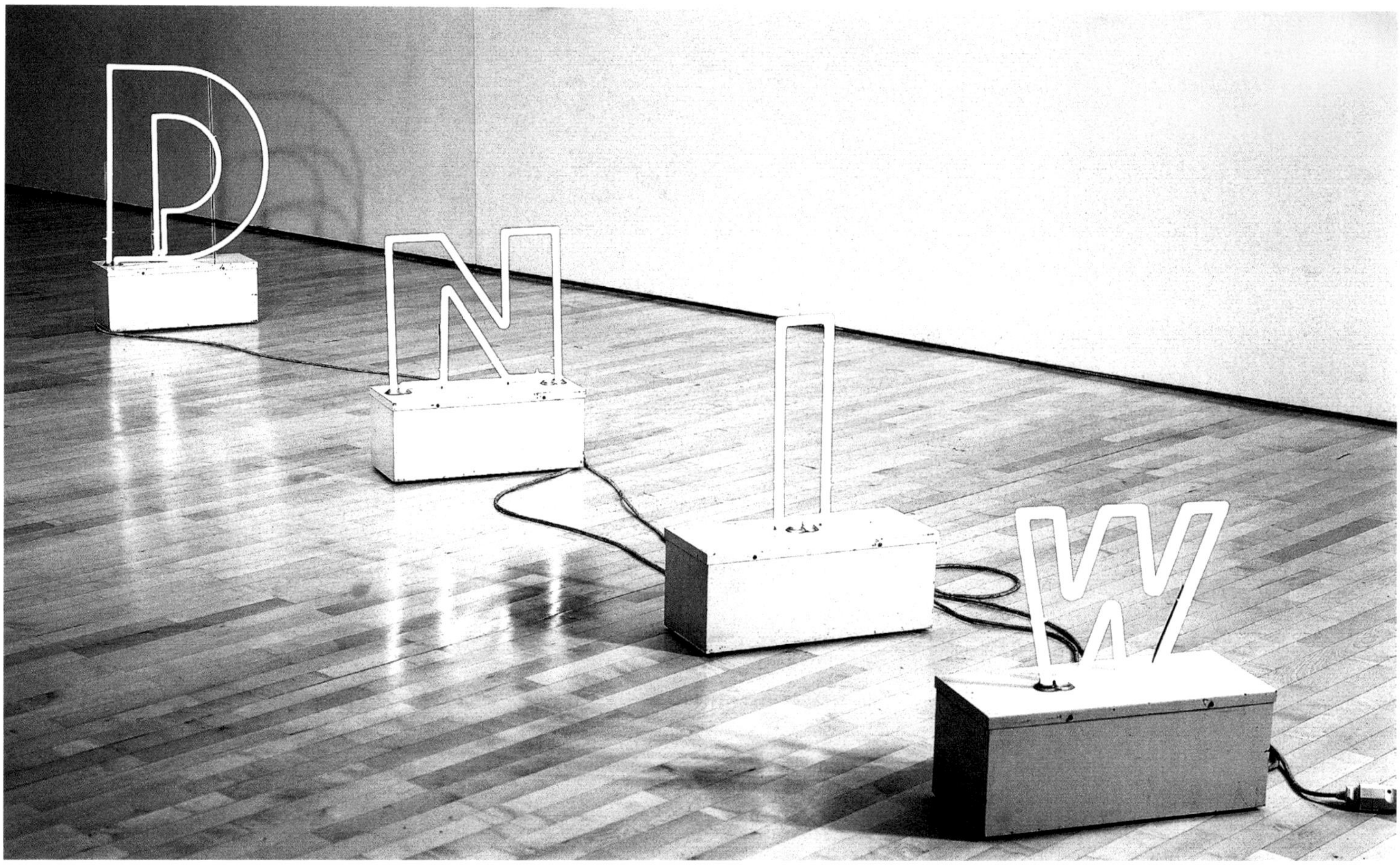

Installation view *Peter Weibel. Das Offene Werk 1964–1979*, Neue Galerie Graz, 2004

Wind: The Compulsion of Identity / Wind: Der Zwang zur Identität

1966/75	mobile texts; neon tubes, sockets, electrical circuit, cable, time switch, 4 parts, dimensions variable

The word *Wind* (wind) is presented as a sculpture made up of four neon tubes on the floor. A time switch lights up the four tubes (*W*, *I*, *N*, *D*) in sequence, as if the illumination of the letters were following the movement of the wind. First shown at Trigon '75 at Künstlerhaus Graz, the piece centered on a comparison between the movement of wind and the movement of light.

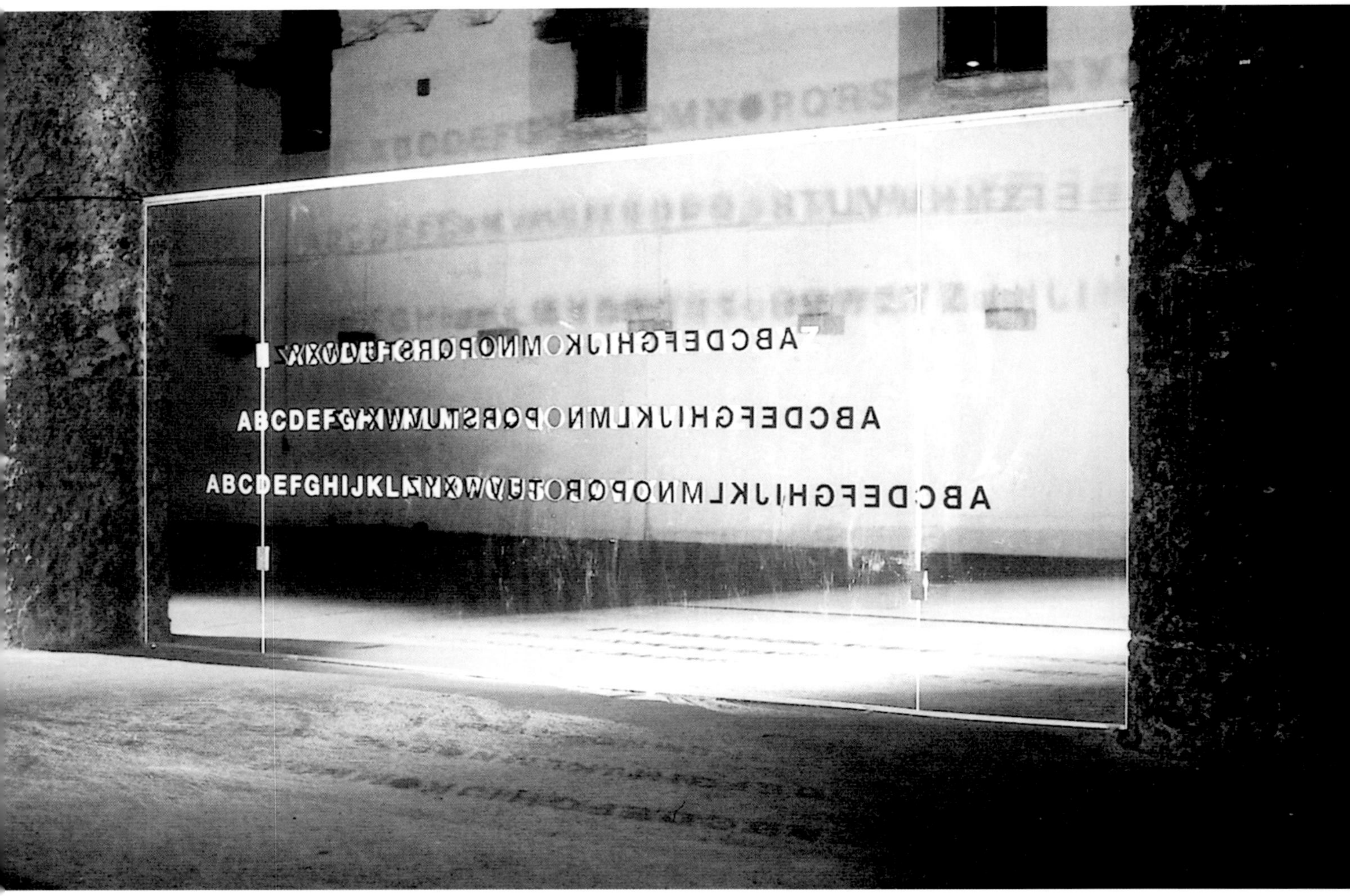

Installation view *Am Anfang war ...*, Kunsthalle Tirol, Hall, 1995

A B C D E F G H I J K L M N ● P Q R S T U V W X Y Z
A B C D E F G H I J K L M N O P Q ● S T U V W X Y Z
A B C D E F G H I J K L M N O P Q R S ● U V W X Y Z

A B C D E F G H I J K ● M N O P Q R S T U V W X Y Z
A B C D E F G H I J K L M N ● P Q R S T U V W X Y Z
A B C D E F G H I J K L M N O P Q R ● T U V W X Y Z

The work *Ortlos* (placeless) consists of a plexiglass pane on the front and back of which the alphabet is written in three lines, one below the other, in white and black characters, respectively. On the side with the white letters, the characters "O" in the first row, "R" in the second row, and "T" in the third row are replaced by holes. The missing letters form the word *ORT*. On the side with the black letters the characters "L," "O," and "S" are replaced by identical holes, forming the word *LOS*. Only by looking at both sides of the pane can the missing letters be grasped and the word *ORTLOS* formed in the mind's eye. The meaning of the word in this work goes hand in hand with its material reality. *ORTLOS* is placeless. A new form of writing technology for the quantum computer: Not individual letters are controlled, but the alphabet chain as a whole is scanned or moved back and forth in front of a sensor and the intended letter is deleted for a fraction of time and thus stored.

—Peter Weibel

Ortlos

1995	plexiglass, Letraset, metal bar, 55 × 151.4 × 1 cm

Context Theory of Art

Peter Weibel

If we choose to subscribe to the division of language into syntax, semantics, and pragmatics, and if we interpret syntax as the relationships of words to one another, semantics as the relationships of words to objects (that is, the meanings of words), and pragmatics as the relationships of words to the subjects who use them (that is, communication), then we can clearly see that the majority of what is understood to be poetry goes no further than merely syntactic and semantic explorations of language. There is a role for the role played by words in the system of language, but not for the role played by language in the social system. The social aspect of language – the extent to which linguistic patterns and speech acts play a role in social role-playing – has gone by the wayside. To investigate the use of language, and to evade linguistic control, the avant-garde has been making use of extralinguistic devices in its dealings with language ever since Fluxus & Co.

In the process, it has become apparent that these extralinguistic devices themselves have become infested with linguistic structures. The universe of language has thus become bigger: the structures of language have been expanded, and new linguistic structures have been discovered. For example, the organizing principles of language extend to the outermost extremities of our social organization. The organization of sensory data in our brains, like the organization of files in our systems of administration, is related to the organization of words in language, which means, among other things, that our divisions into architects and poets, media and materials, painters and musicians, and so on, are superficial divisions, made arbitrarily from a specific standpoint – namely, the perspective of obfuscation.

A poet can compose poetry with rocks or photographs just as well as with numbers or words. All that matters is how much and in what way the meanings of the materials employed are changed. In that sense, my photo-poems make up a very small subset of my many and varied attempts to break down these divisions in order to advance the cause of poetry. Naturally, the liberation of poetry, and the resulting liberation of the subjects who speak a language, are closely linked to a heightened interest in the semantics and pragmatics of language. And it seems to me that the sensitization of perception and cognition must be built, in part, on the realization that every text is part of a context.

However, every context (setting) is, in turn, the text for another, more comprehensive, context. As one example, the following selection can be made from this multilayered series of texts and contexts (in this representation of my context theory, I have sacrificed some precision in the interest of comprehensibility):

word (text)	
sentence (context)	syntax
sentence (text)	
page (context)	
page (text)	semantics
book (context)	
book (text)	
bookstore (context)	pragmatics
bookstore (text)	
customer (context)	

In my poems from 1966 onward, I made extensive use of concepts such as space, time, and natural and artificial processes; I worked intensively with natural materials and electronic components; and I acted directly upon the public sphere via performances. As a consequence, my models of interaction and communication manifested on the complex levels of civilizational processes, in the upper layers of text and context. Text objects (the greeting "Grüß Gott" as baked goods, displayed and eaten upon meeting, text spaces (which are controllable by the audience and create new relationships to sensory experiences), text performances, text processes, text environments, poster campaigns (posters all over the walls of the city featuring really sexy photographs and the sentence "whatever turns you on is beautiful"), banner poems, *Tele-Actions*, etc. were the result.

Therefore, in my search for structures behind the language, for the connections between linguistic and other structures, I focused on the contexts of texts – that is, those things that are not normally included in the poetic calculus – and in doing so, I drifted further and further away from verses and sentences, to the shores of media, the tools of literature (typewriter, pen, paper). Since I viewed behavior as the source of literature, what interested me about communication was the nonverbal parts of it. The prenatal origins of behavior, space, and time as tools for conveying messages, habitats and territories, institutions like galleries or museums, phylogenetic characteristics, structures of metacommunication, and the neurophysiological prerequisites for the cognizance of the world all shifted, as sources of message and desire, into the realm of poetry, as sources of creation and understanding.

Contexts became more important than texts. Procedures for codification rather than the message itself, methods of research rather than the findings, the conditions of culture rather than culture, the system of values rather than the values themselves: they became the foci of my interest.

Note: Adapted from Peter Weibel, "Kontext-Theorie der Kunst," in *Kritik der Kunst. Kunst der Kritik. Es says & I say* (Vienna: Jugend und Volk, 1973), 65–69. Originally written in 1971. Translated from the German by Patrick Hubenthal.

Machines of Some Kind

Peter Weibel

From the 1960s onward, my poetic praxis was based on the semiotic triple:

	syntax	semantics	and pragmatics.
Every sign	is something	about something	for someone.
	(means)	(object)	(subject)
It	realizes,	encodes,	and communicates
and has	form,	meaning,	and command.

As my "progressive texts" of 1964 illustrate, I played with typographic means and adjuncts such as parentheses in order to distill multiple meanings from words. Soon, however, these syntactic and semantic experiments led me to the question of the form, more specifically to the medium transmitting the form; that is, the material on which the form is recorded. A change in the medium, for example, from paper to stone, also has the effect of changing the meaning. After the material of the medium, I also interrogated the material of production: How – by what means – is the form created in the first place? For the means of production and distribution also encode meaning. So, I began to experiment with pencil, fountain pen, erasers – conventional writing implements.

"Already in my childhood, I simultaneously fell prey to letters and became a hunter of letters. At the same age as I discovered the letters, I found their home: books. One day, I discovered a medium which would allow me to become a master of letters: a machine with which I could write printed letters; or, to be more exact, by pressing my fingers on a keyboard, I was able to print the letters on paper so that the paper looked like the printed page in a book. So the writing machine was in fact a printing machine.

The typewriter left such a deep impression on my mind that even with my tape recorders, video installations, and computer-assisted virtual environments I still conceived of the interface between the world of objects and that of signs as a keyboard and I embedded pressure buttons in the floor arrayed like the 40-odd keys on a typewriter to create sounds and images."[1]

I noticed that I was creating my concrete, visual poetry using a typewriter; in other words, that *typewriter poetry* requires a *typewriter*. Accordingly, I used these machines – along with tape recorders, likewise machines of text production – in many objects and installations. And, of course, in actions and performances as well, since I was more interested in the pragmatics, the communicative aspect, than in anything else. In the mid-1960s, I gradually moved on from these machines of text production towards the media image production. Yet the secret of letters and their relationship with reality continues to captivate me to this day.

Note: This text was written in 2022 and is published for the first time in this volume.

[1]
Peter Weibel, "Typewriter," in *Objects of Knowledge, of Art and Friendship: A Small Technical Encyclopedia. For Siegfried Zielinski*, ed. David Link and Nils Röller (Leipzig: Institut für Buchkunst Leipzig, 2011), 167–69, quotation on 168–69).

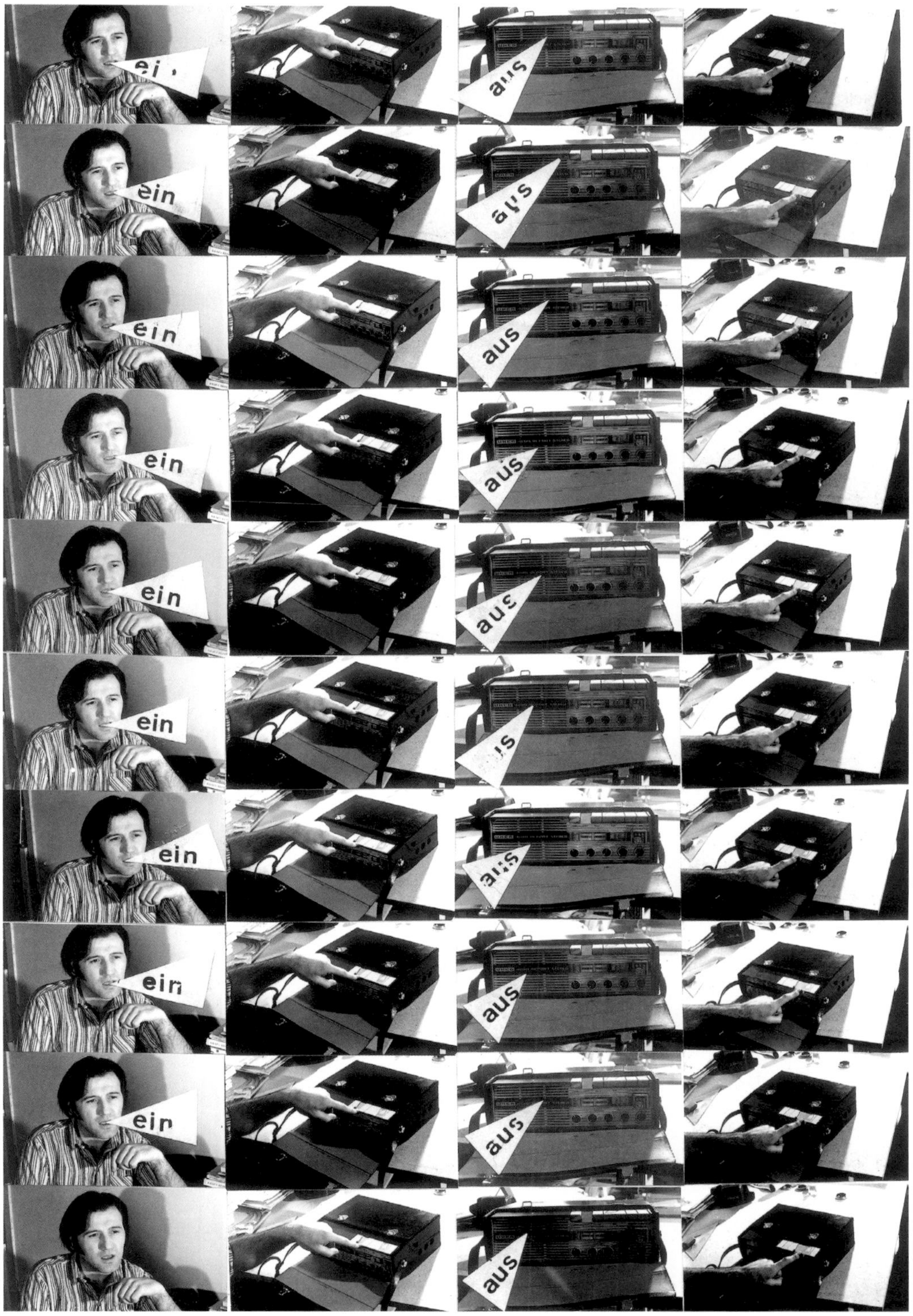

Synthesis (Infinite Symphony I) /
Synthesis (unendliche Symphonie I)

1967 | performance at Atelier Zobel, Vienna; collage of 40 b/w photographs, 36.4 × 27.8 cm, published in *Neues Forum* (January 1972)

A discourse on language, technology, politics: on language as a system of commands; on politics, which takes possession of language in order to wield its power; and on technology as an extension of language. An endless text, a finite model of infinite time. The speaker says *ein* (on) and presses the on button on a tape recorder. The word *aus* (off) is recorded on a tape loop loaded into the recorder. Thus, the tape says "off." The speaker presses the off button, the record stops. Then the speaker says "on" and presses the on button, the tape says "off," the speaker turns the recorder off and says "on," turns the recorder on, and so forth, *ad libitum*. The text effects itself until either the tape recorder breaks or the author succumbs to exhaustion and the performance ends. —Peter Weibel

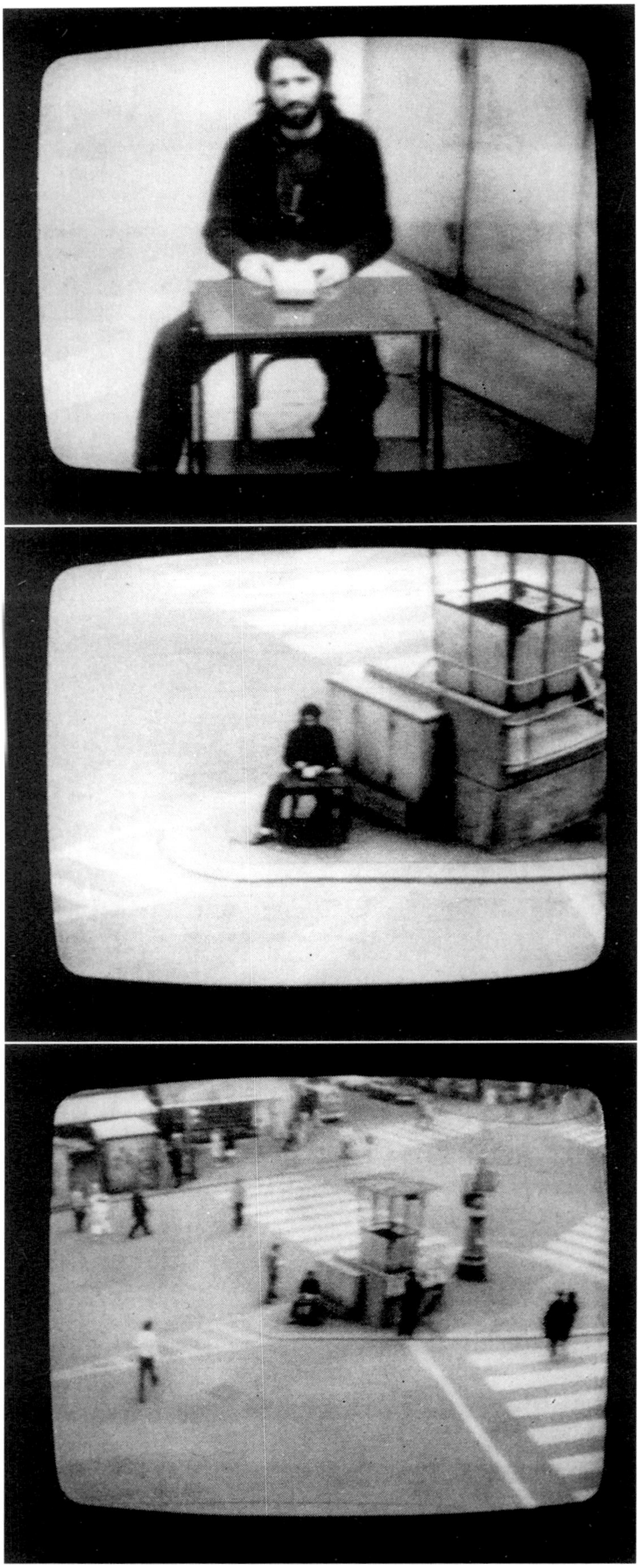

Synthesis of Two Sequential Machines / Synthesis zweier sequentieller Maschinen

1967/72

video; b/w, sound, 1:47 min., recording at intersection Stephansplatz, Graben, Kärntnerstrasse, Vienna 1, broadcast in *Impulse no.* 7, ORF, June 29, 1972

Synthesis of Two Sequential Machines is a human–machine interaction by Peter Weibel that was performed from 1967 onwards in several public spaces. The video shows the artist communicating with a tape recorder at a crossroads in Vienna. The artist says *ein* (on) and presses the power button of the tape recorder. On the tape recorder there is an endless loop with the previously recorded word *aus* (off). So, the tape recorder says "off." The artist presses the off button, then says "on" and switches the tape recorder back on. The action continues until the performer operating the tape recorder is exhausted, or until the technical apparatus fails. Human and machine are coupled in a circular feedback loop. The work can be understood as a result of Weibel's early studies of Claude E. Shannon's information theory and Norbert Wiener's cybernetics.

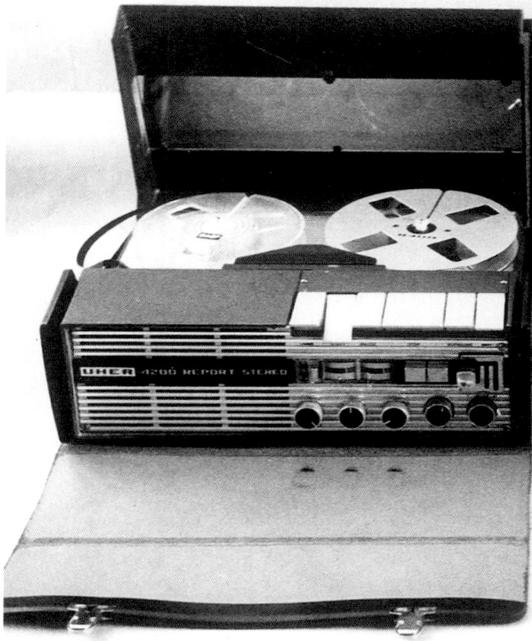

a

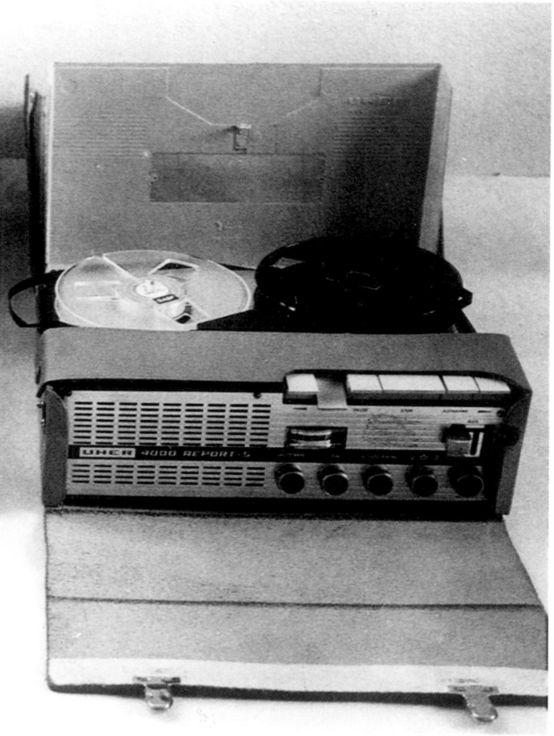

d

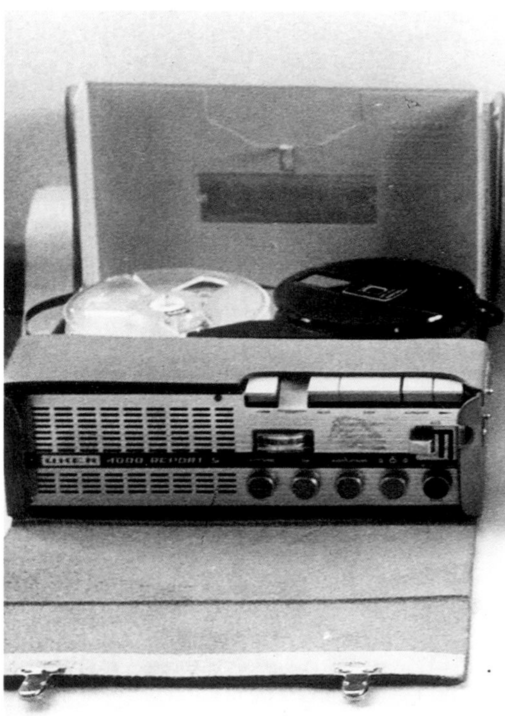

n

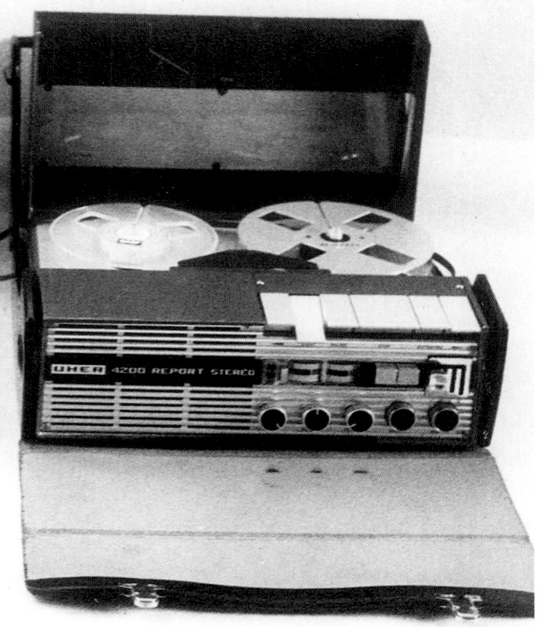

l

Infinite Symphony II /
Unendliche Symphonie II

1974	4 tape recorders, loop, first performed at Alte Schmiede, Vienna, February 9, 1976

Four tape recorders loaded with tape loops, each of which contains just one sound: *a*, *d*, *n*, and *l*. The machines start in this order, and the sounds are heard. The symphony's finale is reached when, as a result of the varying loop lengths, different rotation speeds, etc., the sound combination *land* has been achieved.
—Peter Weibel

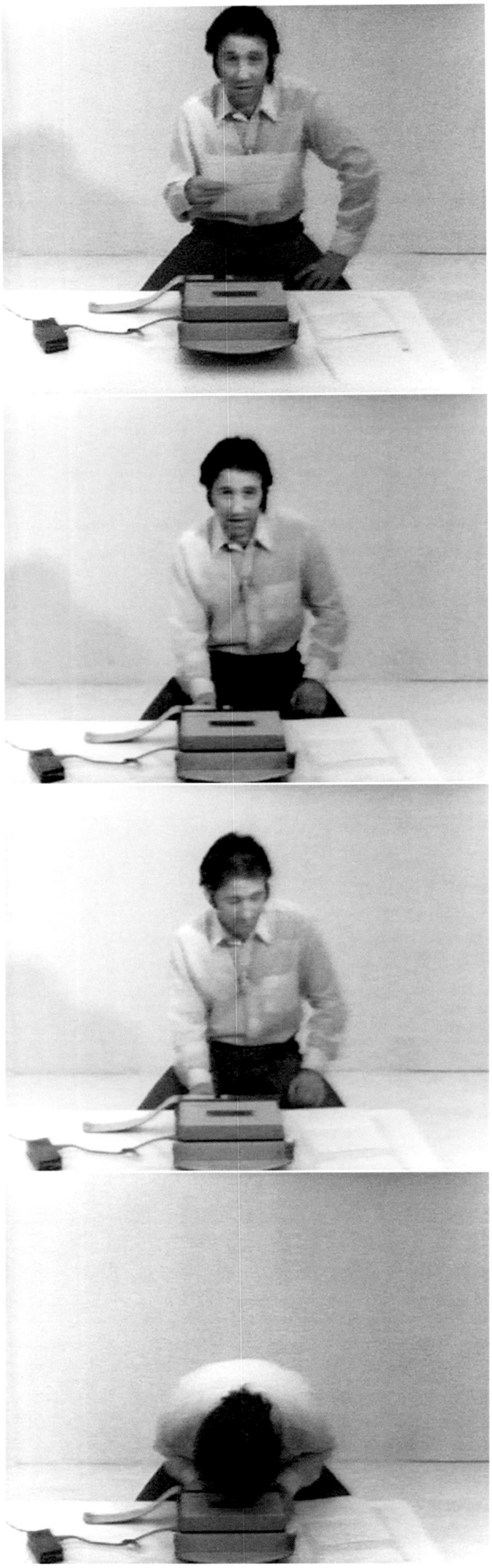

w : w
wa : aw
was : saw
was i : i saw
was is : si saw
si saw : was is
is was : saw si
was ist : tsi saw
was ist : ist was
was ist d : d tsi saw
d tsi saw : was ist d
was ist da : ad tsi saw
ad tsi saw : was ist da
was ist da : da ist was
ist da was : was da ist
ist was da : da was ist
da was ist : ist was da
was da ist : ist da was
da ist was : was ist da
was ist das : das ist was
was das ist : ist das was
ist das was : was das ist
ist was das : das was ist
das was ist : ist was das
das ist was : was ist das

was gibt es : se tbig saw
tbig se saw : was es gibt
was es gibt : gibt es was
gibt was es : es was gibt
gibt es was : saw se tbig
es was gibt : gibt was es
es gibt was : was gibt es

Excerpt

Soliloquy. Language Mirrorings / Selbstgespräch. Sprachspiegelungen

1973 | video; b/w, sound, 16:31 min.

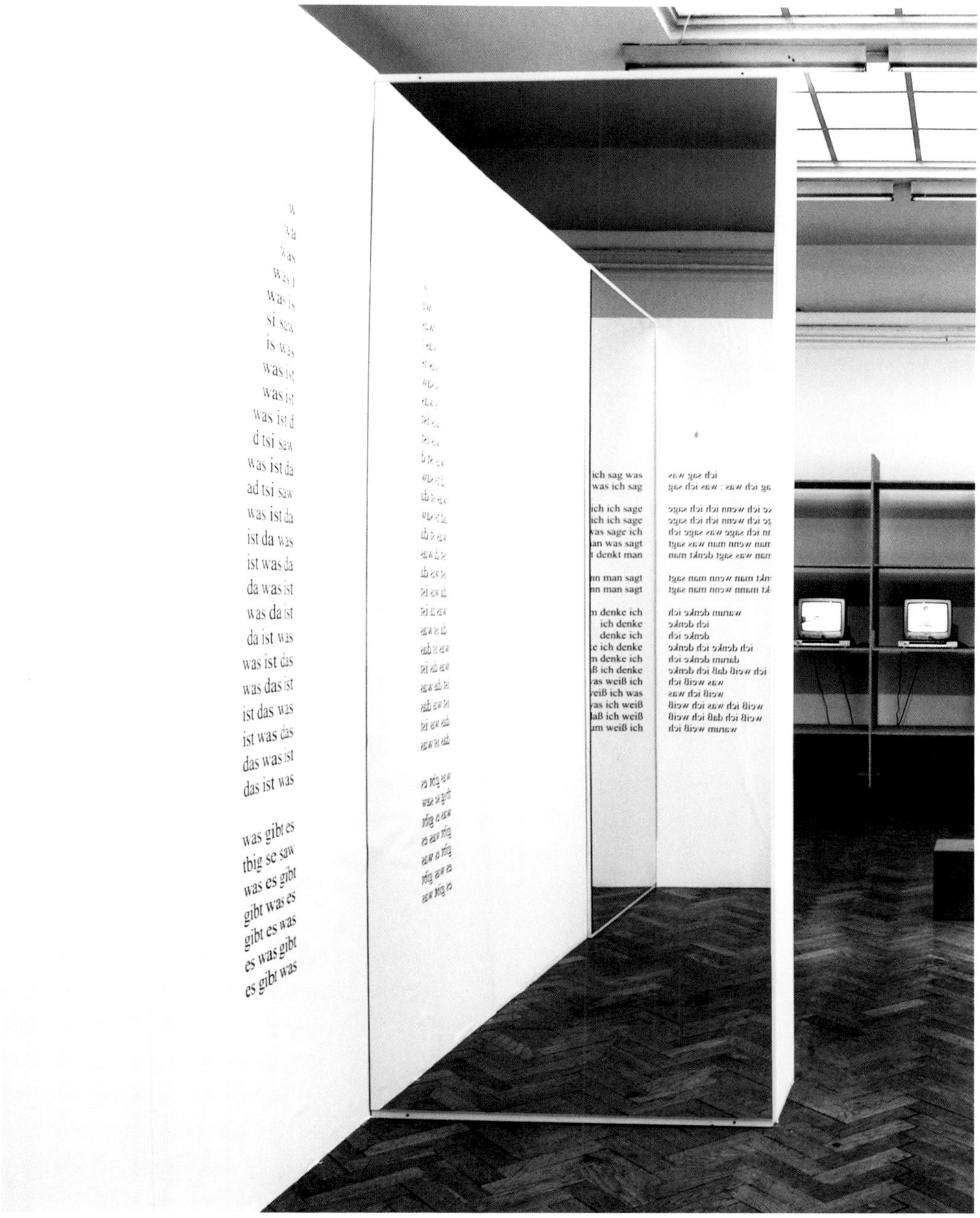

Installation view Moderna Galerija, Ljubljana, 2005

The tape recorder as mirror, for the first time in history: aural symmetry instead of visual symmetry. Visual mirrorings, visual symmetrical patterns, and horizontal or vertical visual symmetries are conventional. Peter Weibel was the first and only artist to mirror language and its phonetic dimension itself. He used the tape recorder as a mirror. He made the tape repeat spoken letters, syllables, words backwards or spoke letters, syllables, words backwards into the tape recorder, so that the answer of the tape provided the meaningful words. This speech mirroring showed how most often the answer is already contained in the question, such as „Gibt es was? – Was es gibt" (What is going on? – Something is going on); „Was ist da? – Da ist was" (What is there? – There is something); „Wer spricht? – Spricht wer?" (Who speaks? – Speaks who?); „Sprichst du? – „Du sprichst" (Are you speaking? – You are speaking); „Wer ist da? – Da ist wer" (Who is there? – There is someone); etc.

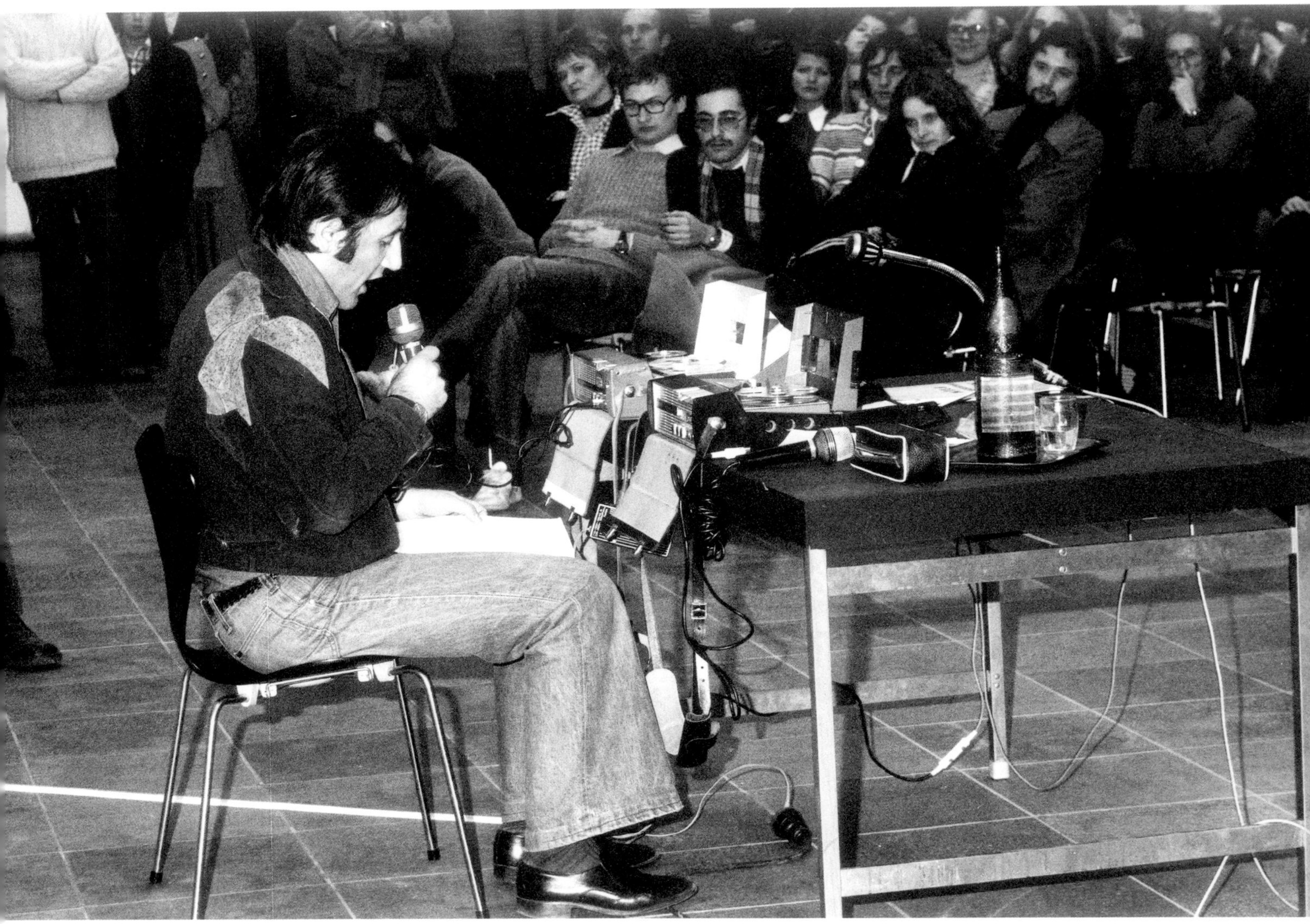

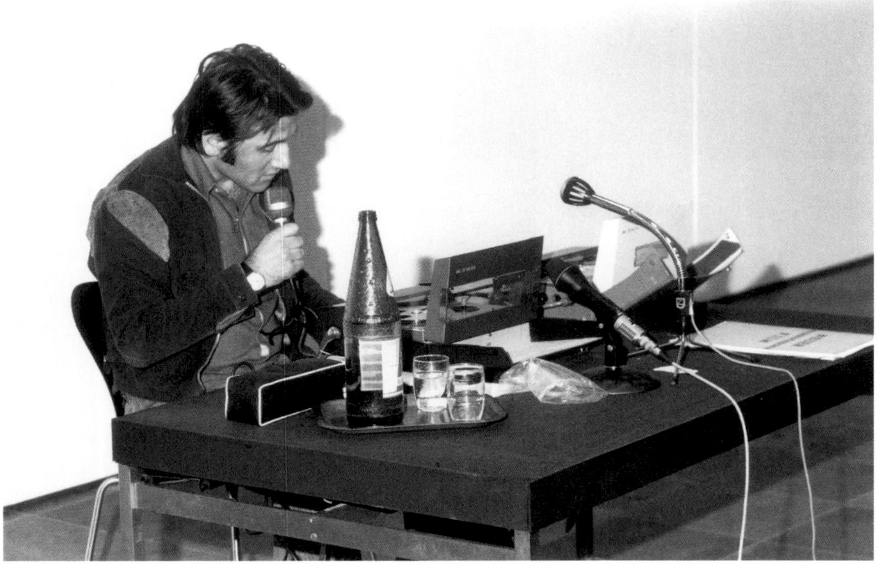

Performance as part of the exhibition *Kunst aus Sprache*, Belvedere, Vienna, 1975

Recursive Texts / Rekursive Texte

1973	recursive text action; microphone, 2 tape recorders, tape

Morphemes (word particles) that are meaningless by themselves are played into the microphone at time t_1 from tape recorder 1. They return, after a delay, at time t_3 from tape recorder 2. They form meaningful words in combination with new syllables spoken live at time t_2 during the delay. I say "en" live, which is played back from tape recorder 2 after a delay. Before it recurs, I say "sprech" live to produce the word "sprechen" (speak). Before "sprech" recurs, I say "ge"; the result is "Gespräch" (conversation). —Peter Weibel

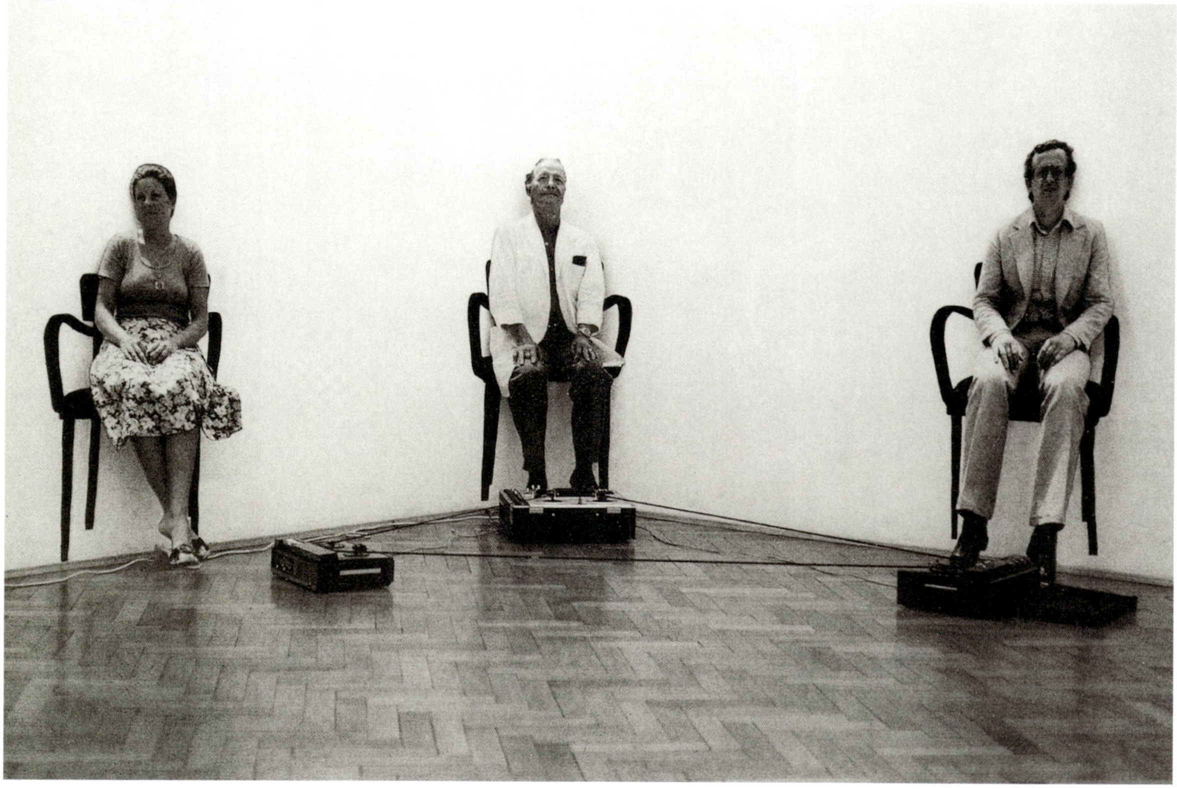

Installation view *Peter Weibel: Medienpoesie*, Kulturhaus Graz, 1982

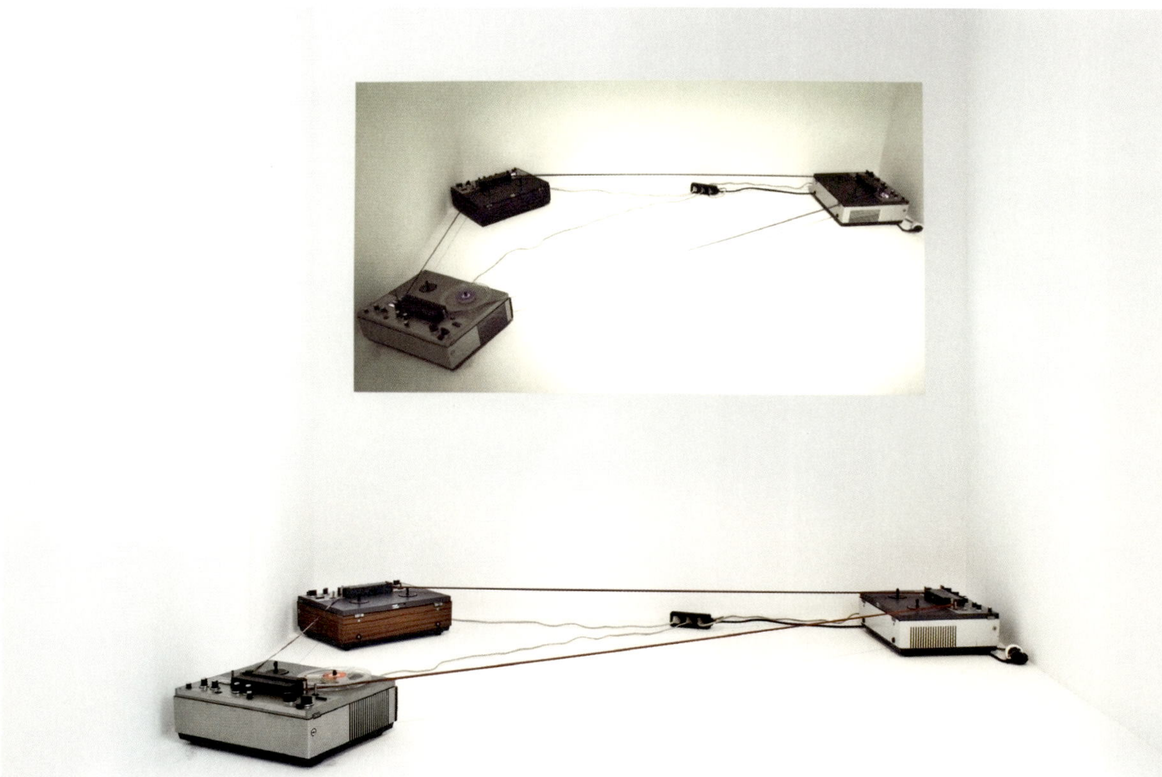

Installation view *Sound Art: Sound as a Medium of Art*, ZKM | Karlsruhe, 2012–13

I Mass – Mass I / Ichmasse – Massenich

1977 | 3 tape players, tape, loop

In *I Mass – Mass I*, an enormous tape loop stretches across three tape recorders. On the loop, a recording of the spoken word *ich* ("I" or "me") travels from one tape recorder to the next. When the word arrives at a tape recorder, the *ich* (I) sound is heard. Each machine says *ich* (me) and thus asserts its own identity. However, it uses the same word *ich* (me) as the other machines. By proclaiming their individuality, their uniqueness, and their singularity using the same word, all the individuals form a homogeneous mass. —Peter Weibel

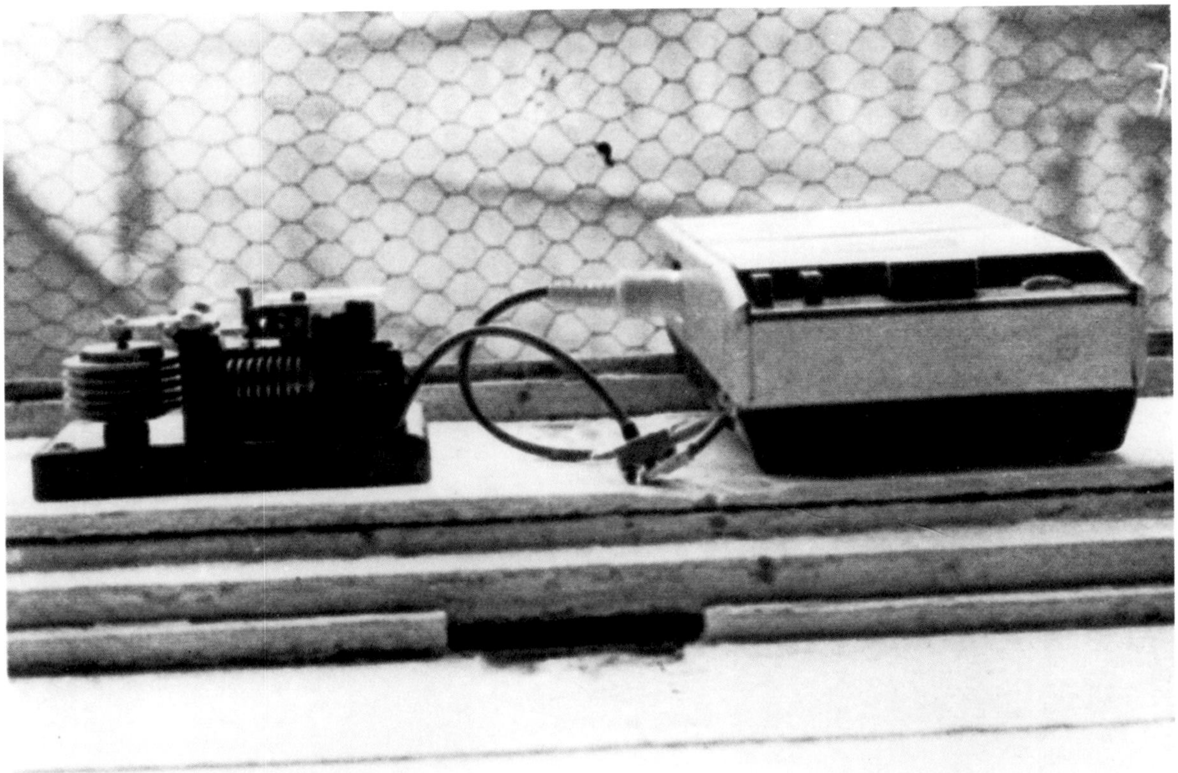

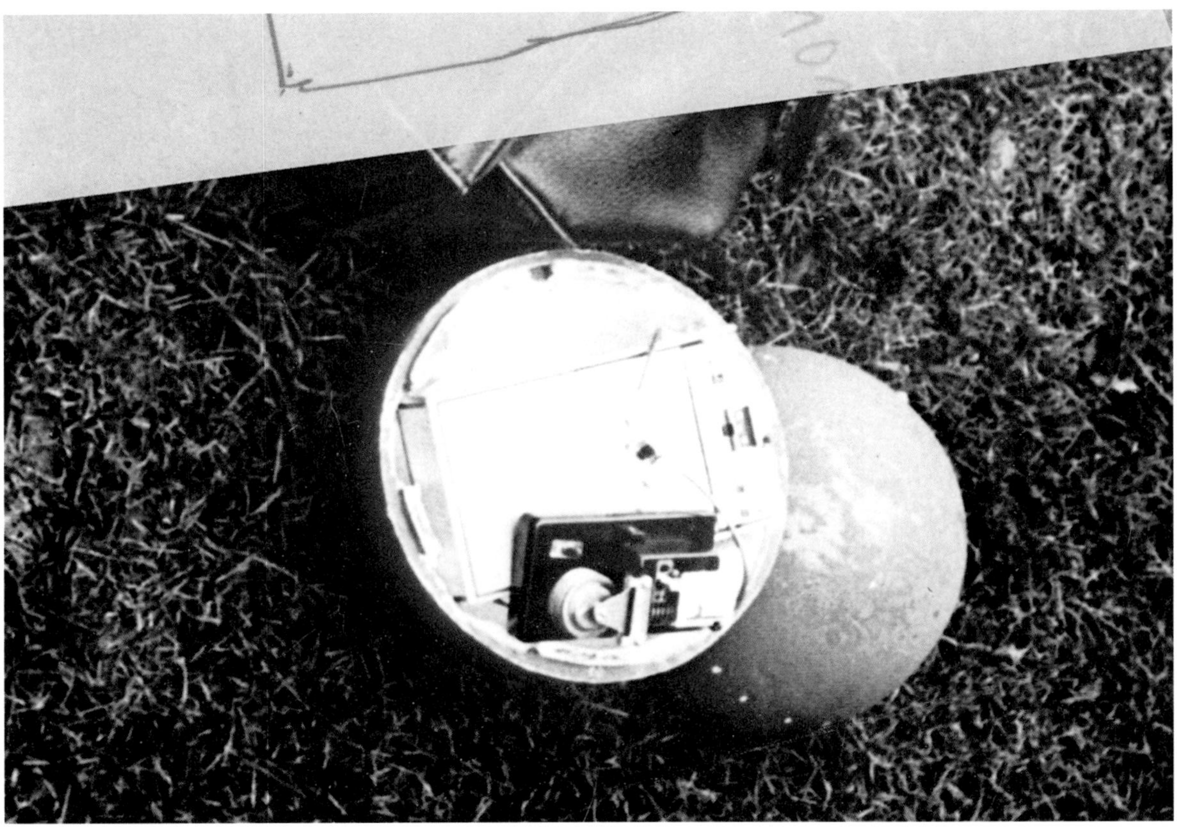

Moaning Stone. Nonhuman Poem / Stöhnender Stein. Nicht-humanes Gedicht

1969 | sound object; artificial stone, loudspeaker, electronics, dimensions variable

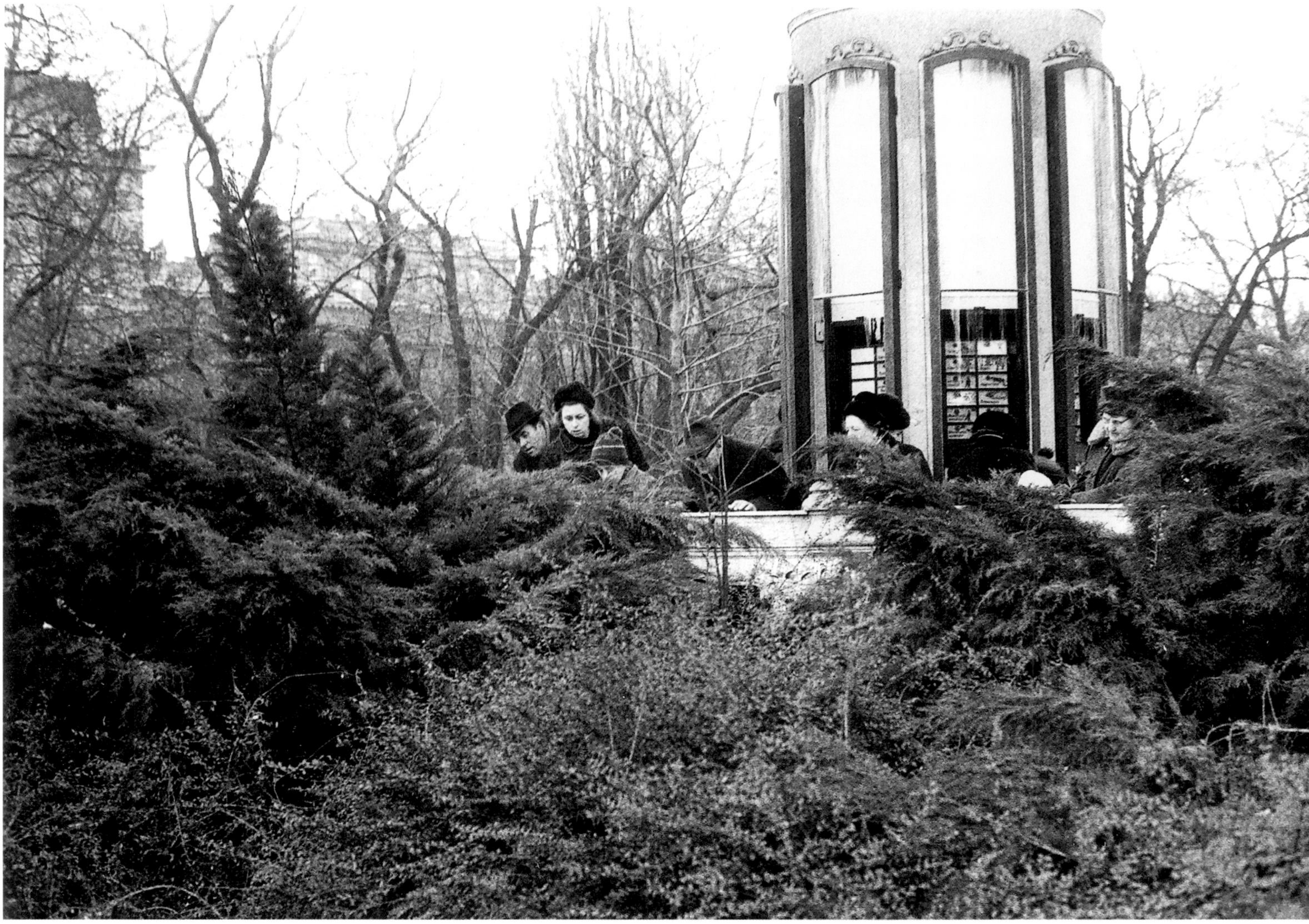

Moaning Stone contained a tape player with an endless loop of the moans of a sick person. The tape player was connected to an electronic thermal switch and a container of gas so that the switch reacted automatically to changes in temperature. As is generally known, gas expands in warm conditions, and thus its elastic container expands as well. In colder temperatures, the phosphor gas contracts and the gas container shrinks. These expansions and contractions of the container are linked, via a spiral spring, to an indicator on a temperature scale. At temperatures above 0 °C, the indicator turns the tape player on; at temperatures below 0 °C, it turns the tape player off. On a cold night, the tape player thus remains off and silent. When the temperature rises in the morning, the gas expands, the indicator rises above 0 °C on the scale, the tape is turned on, and moans are heard.

In 1969, Peter Weibel hid such a "moaning stone" in the Vienna Stadtpark. In the late morning, park visitors noticed the moaning and suspected a crime. But they could not localize its source and called the police, who arrested the artist for "disturbing the public order." This is what Weibel wanted to prove. In capitalist society, the human voice disturbs the public order. The human voice, if it is not productive but in need of help, has no place in this inhumane society. The voice is so alienated from society that it is not part of it. Later, Weibel intended to place moaning stones along the banks of the Danube, from Vienna to Budapest, to create a moaning riverbank at night.

Usually, a poem is part of the process of civilization in the service of human communication. Here, however it is not a human who starts the poem, but a natural process. If here a technological poem with the voice of a human is brought into a natural context, it is because the process of civilization is stifling; it is to make clear that the process of civilization has already become so inhuman that a human message can no longer be distributed there. Humans have become stones among stones.

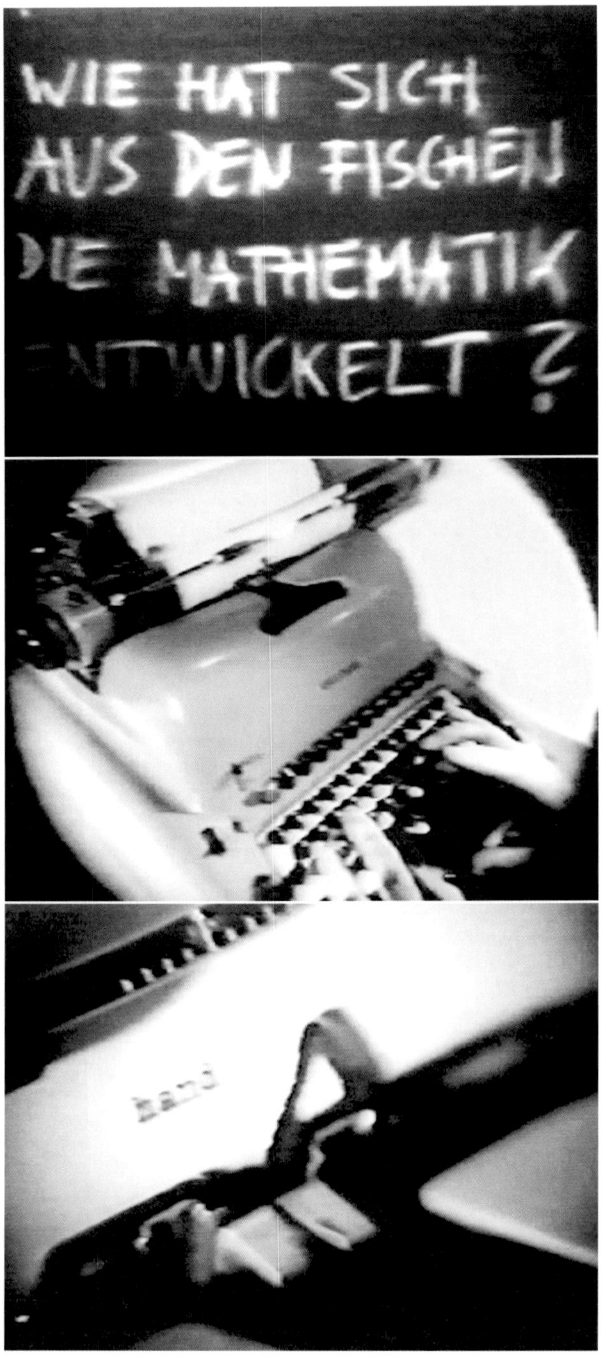

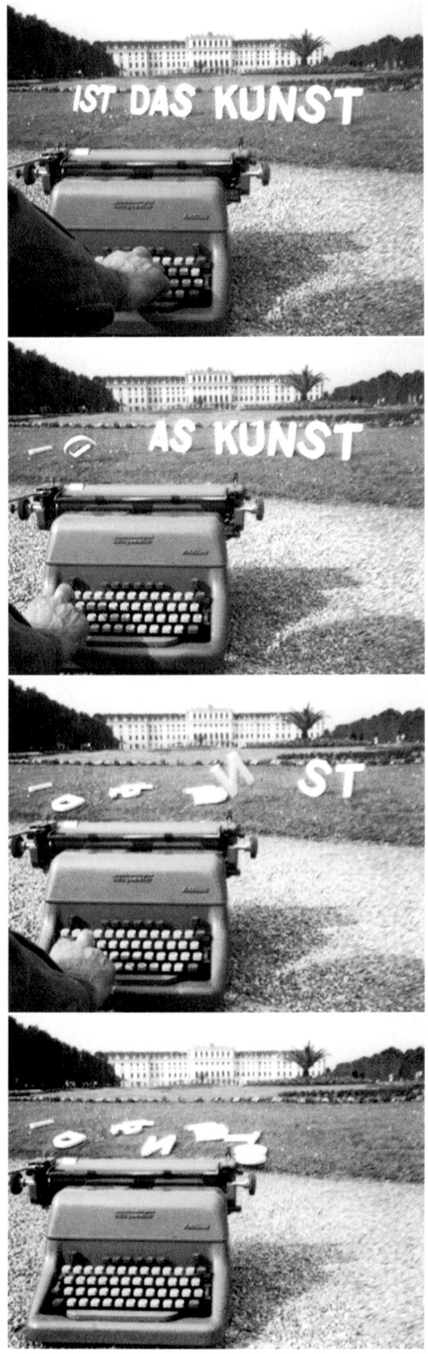

With the initially incomprehensible question *How Did Mathematics Evolve from Fish?*, the photographic sequence and the video of the same name thematize the question of the formation of the human mind and its evolutionary-biological foundations. The fish existed before the mathematicians in the world and the human hand developed from the fin of the fish. The hand writes and types even the word "hand" with a machine, because with hands humans have crafted symbol processing machines and tools which helped humans to higher levels of abstraction. Weibel concludes that all human abilities as well as the laws of mathematics must have been already laid out in the spinal cord of the fish.

How Did Mathematics Evolve from Fish? / Wie hat sich aus den Fischen die Mathematik entwickelt?

1971	photographic sequence and video; b/w, sound, 54 sec.

The sentence "Ist das Kunst?" (Is this art?) can be read on a meadow in front of Schönbrunn Palace in Vienna. While the individual letters of this sentence are typed on a typewriter, the letters fall over on the meadow. Each action triggers another. The work refers both to the fact that media arts were not yet recognized as an art form at the time the video was created, and to the relationship between reality and medium, and how the medium in turn influences our perception of reality.

Is This Art? / Ist das Kunst?

1972	video; b/w, sound, 21 sec.

m e n

s c h

m a -

s c h

i n e

- i n

t e r

a k t

i o n

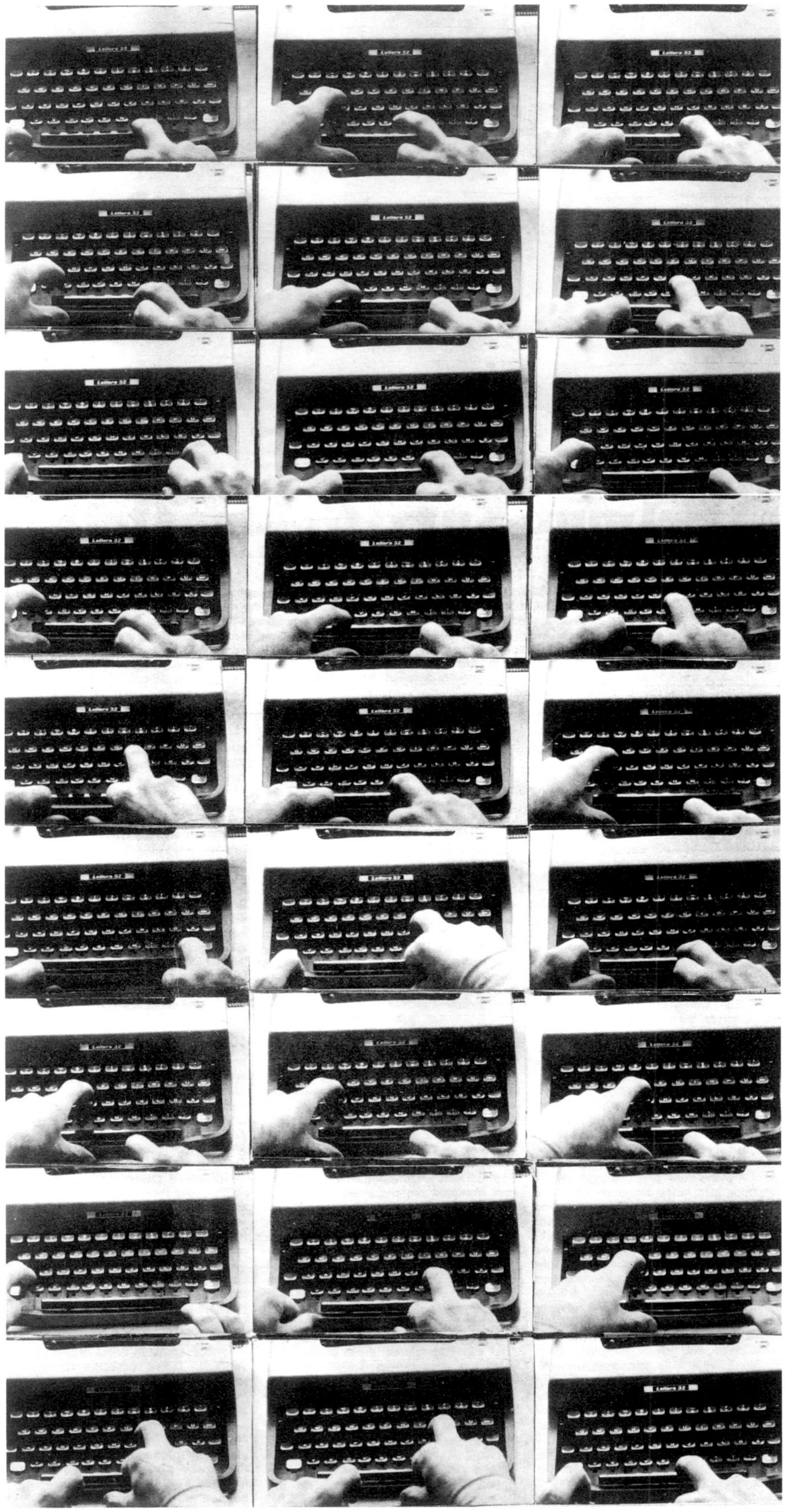

Metaphysical Poem for an Amphioxus / Metaphysisches Gedicht für einen Amphioxus

1971 | collage of 27 b/w photographs, typescript

Each photograph represents a letter – specifically, the one covered by the finger. The sequence of covered letters produces a text: human-machine-interaction. The photographs document the process of realizing a human-machine interaction, and they denote or encode a meaning demonstrated by that process.
—Peter Weibel

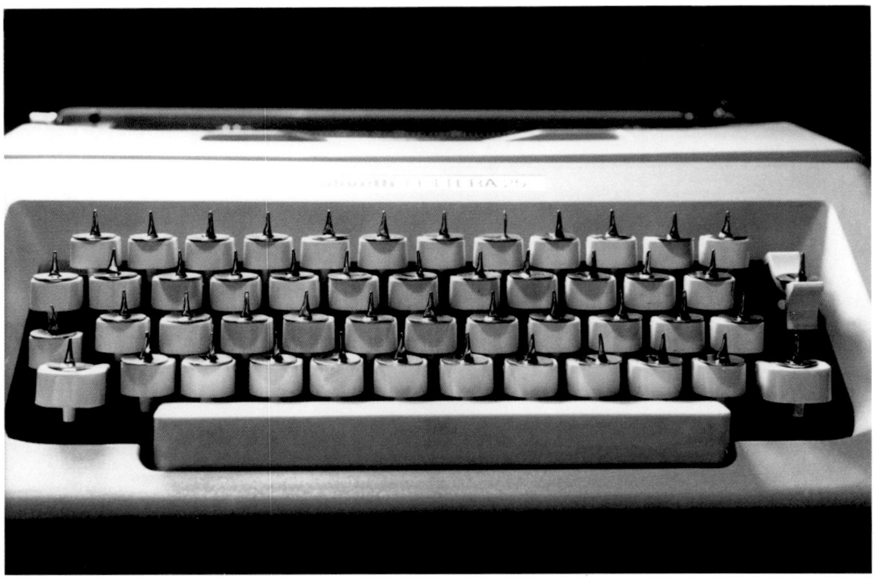

Peter Weibel focused early on on writing as a production, storage, and distribution medium of codes and information. In doing so, he expands the language to include aspects of mediality and materiality, the material carrier medium (wood, paper, stone, neon, etc.), the production medium (typewriter, pen, tape recorder, etc.), and the distribution medium (book, radio, film, TV, etc.). He not only interrogates the intrinsic rules and mechanisms of chains of letters, he also extends the exploration of language to its materials, tools, and media. Therefore, the motif of the typewriter appears again and again in the artist's oeuvre, an apparatus that, thanks to its keyboard, supports and simplifies text production.

Weibel does not write a poem about the resistance and the pain that the poets experience, but expresses the experience of hope, resistance, and pain from the pure realm of signs (representation) in the materiality of the keyboard (reality), to which drawing pins are attached. The resistance or pain actually happens, not just symbolically: those who write with this typewriter bleed real blood, following Franz Kafka's pronouncement that "the real, incontestable truth, a truth marred by no external circumstance (martyrdom, sacrifice of oneself for the sake of another), is only physical pain."[1]

Principle of Resistance /
Prinzip Widerstand

1971	from the series *Principles of Poetry / Prinzipien der Poesie* Olivetti Lettera 25, drawing pins, 13 × 36 × 36 cm

[1]
Franz Kafka in *The Diaries of Franz Kafka 1914–1923*, ed. Max Brod, trans. Martin Greenberg (New York: Schocken, 1949), diary entry on Feb. 1, 1922, 217–18.

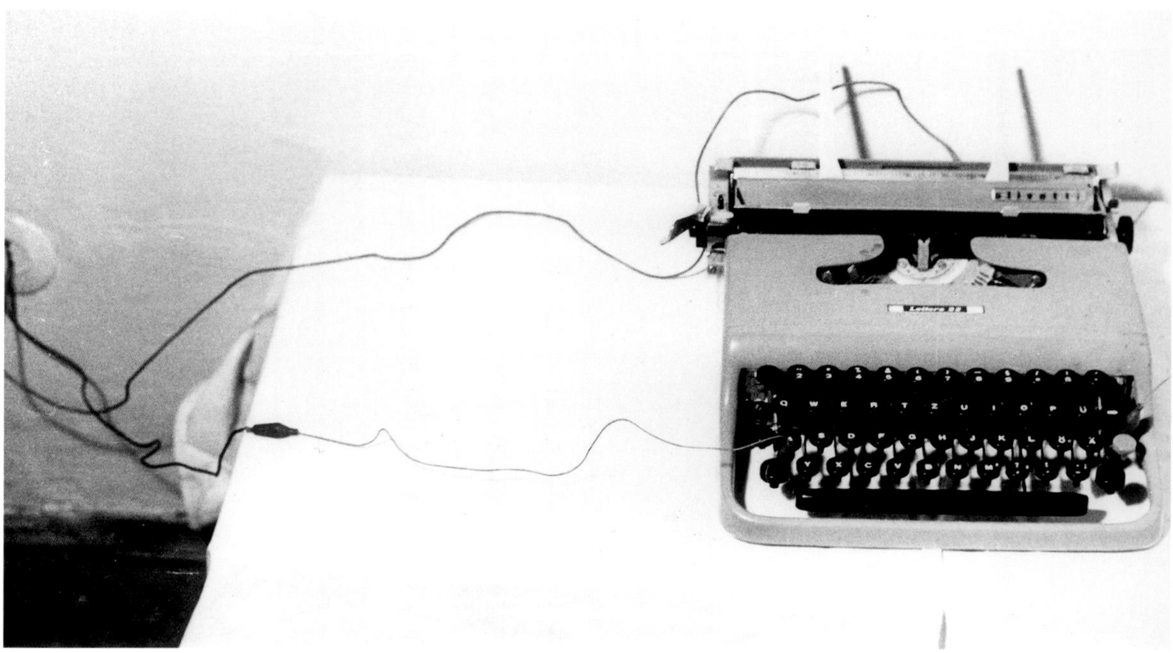

Principle of Scrip/tor/ture /
Prinzip Schrei/ben

1971	from the series *Principles of Poetry / Prinzipien der Poesie* Olivetti Lettera 22, cable, socket, 15 × 32 × 36 cm

By writing, you risk dying because you are making contact with a live electric circuit. The electric typewriter is an electrified typewriter.

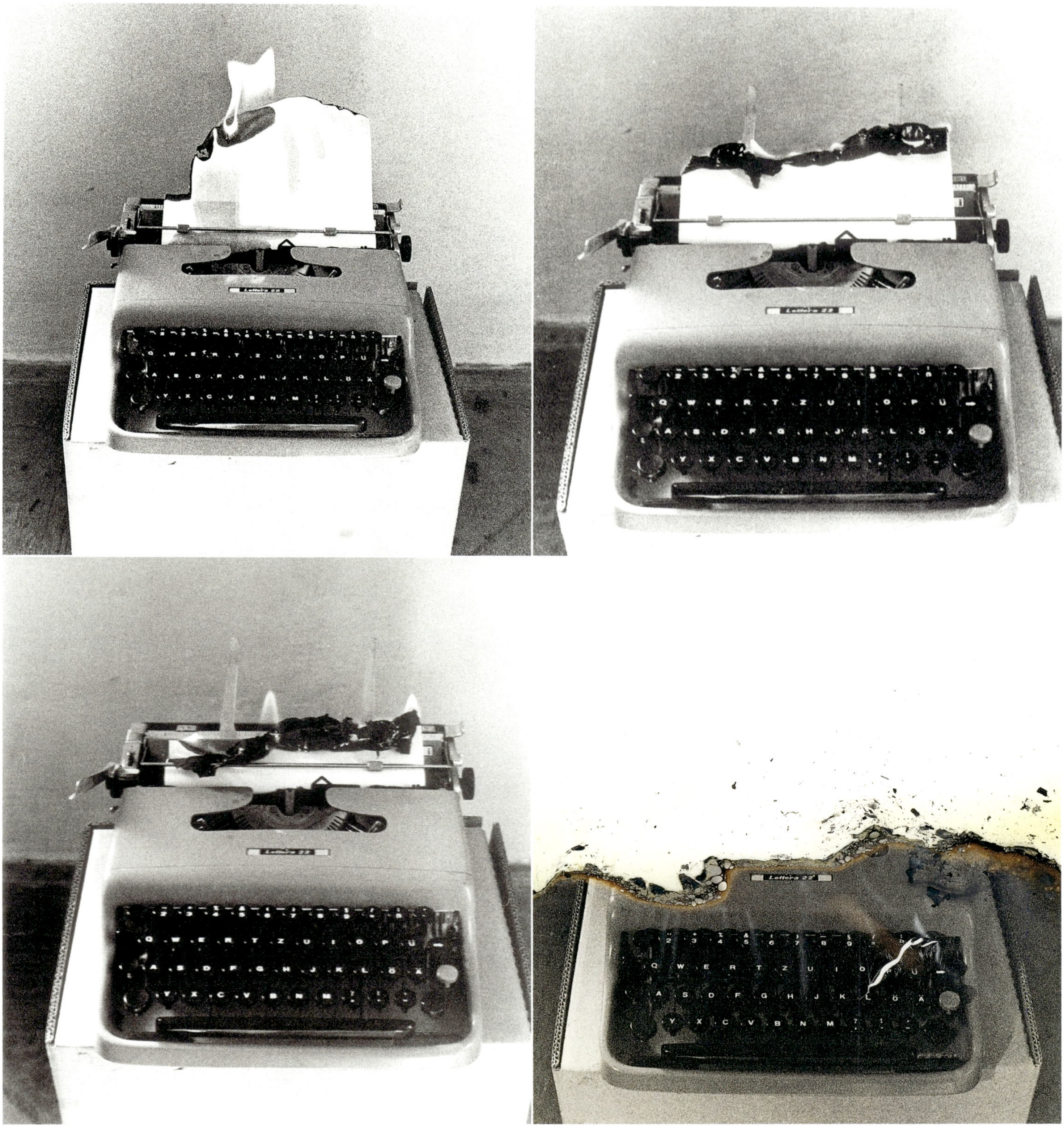

Principle of Hope / Prinzip Hoffnung

1971	from the series *Principles of Poetry / Prinzipien der Poesie* 4 b/w photographs

Principle of Hope is part of a series of photographic sequences devoted to the *Principles of Poetry*. The first photograph of the work shows a burning sheet of paper fixed in a typewriter. No inflammatory poem is being written, the empty sheet of paper in the typewriter is itself in flames. The second and third photographs show the burnt remains of the sheet of paper. In the fourth photograph there is a leap in the material: the photographic image of the fire (representation) has made an ontological leap to the real carrier material, the paper (reality). The image of fire has burnt the real photographic paper, which is not possible in reality. In this and other photographic works created at the same time, Peter Weibel deals with the relationship between photography and the photographed object, with the ontological difference between image and reality. The title of the photo sequence quotes the three-volume publication *The Principle of Hope* (*Das Prinzip Hoffnung*, 1954), one of the major works of the philosopher Ernst Bloch.

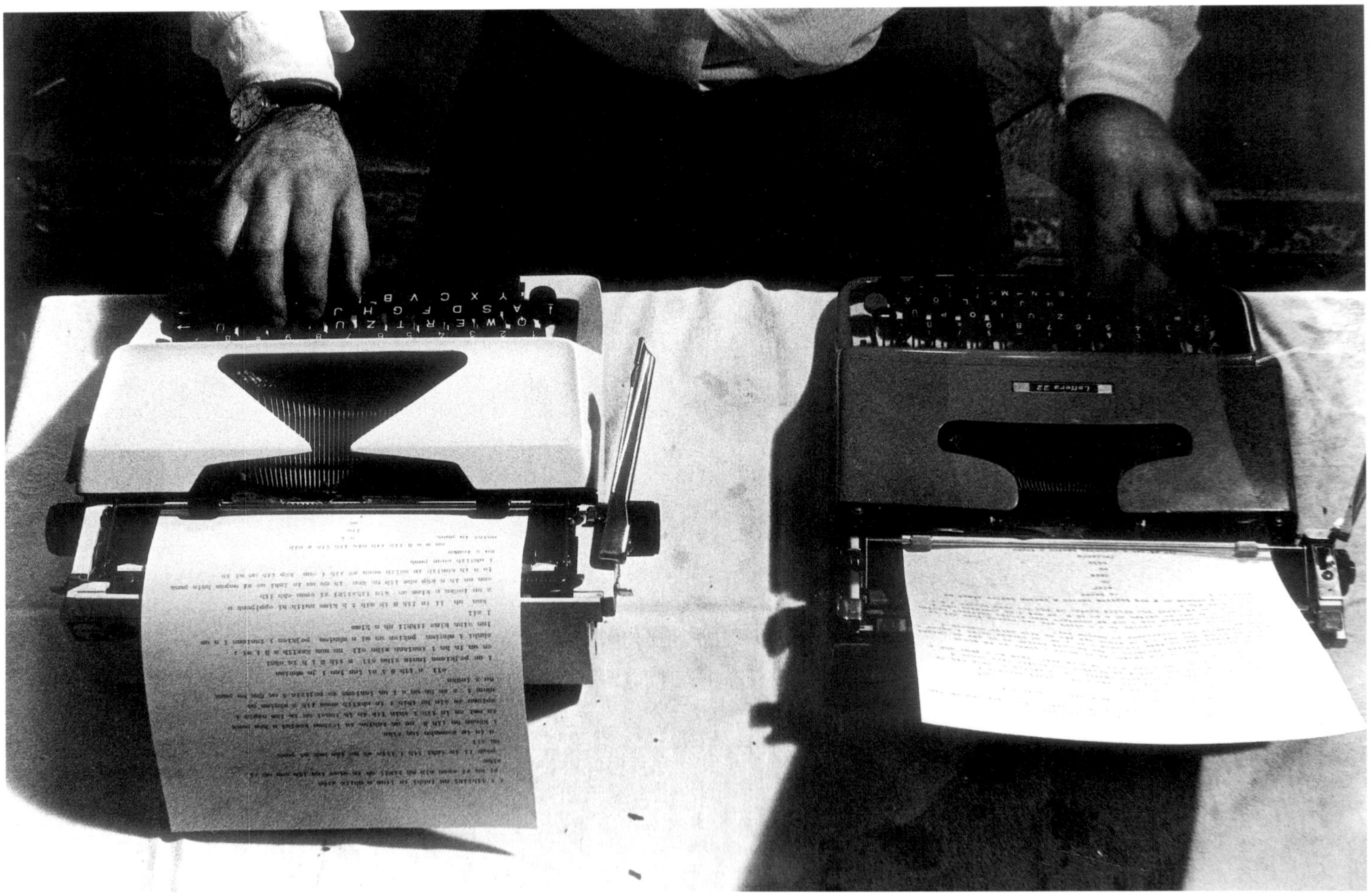

Simultaneous Split Medium / Split Medium Simultan

1974	text action; 2 typewriters, typescript, 3 pages, 29.7 × 21 cm each

A text about the problem of simultaneity is typed on two typewriters simultaneously. The typewriter keyboard is divided into two halves: one half for the left hand, the other for the right hand. In a given word, the left hand types only those letters that are located on the left half of the typewriter's keyboard. The right hand types only those letters that are located on the right half of the typewriter's keyboard. Peter Weibel separated the two halves onto two typewriters – the left typewriter for the left half of the keyboard, the right typewriter for the right half of the keyboard – and produced two pages of nonsense. When combined, the two nonsense pages produce an intelligible text.

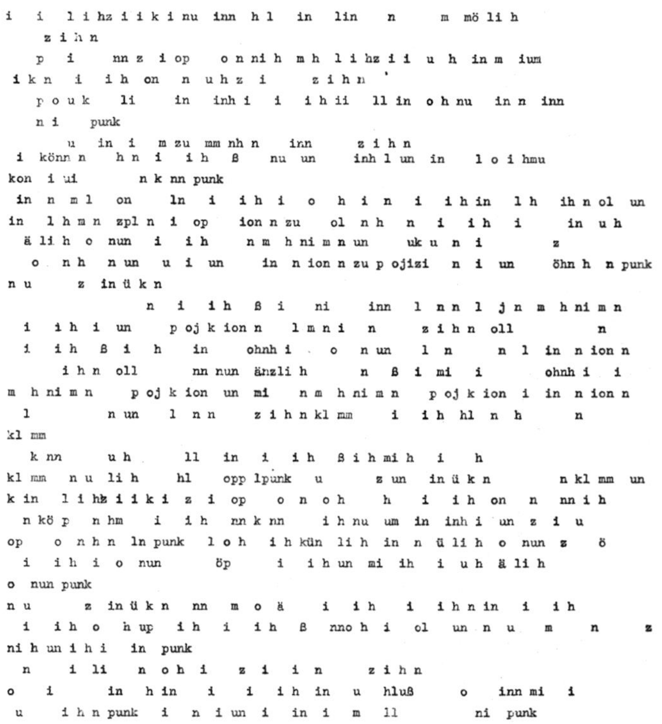

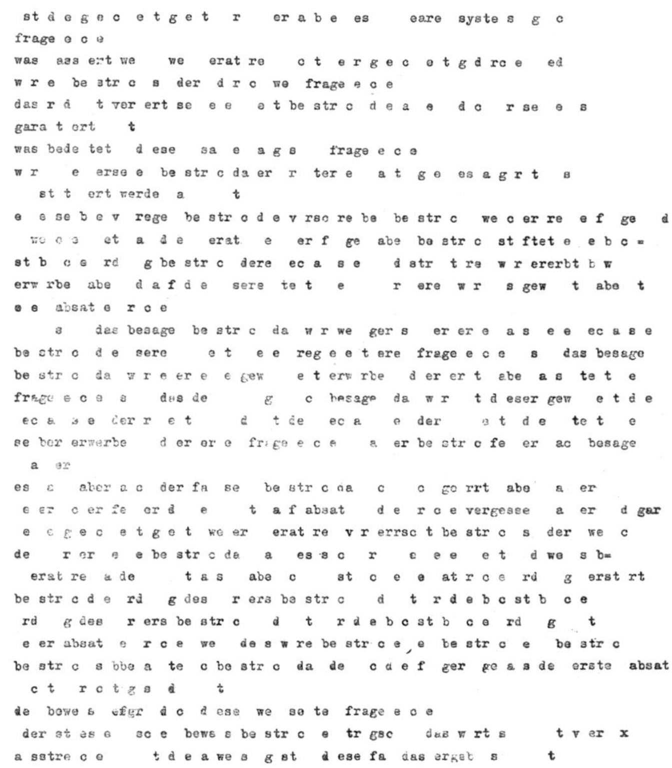

is simultaneity possible only within a linear system and a shared medium?
what happens when two operators are no longer bound by a shared medium, but operate separately through two media?
is simultaneity a question of time only or of the medium as well?
if two operators are simultaneously at work in separate media, the product loses its unity, which alone guarantees its meaning.
what does meaning mean in this context?
we can see that meaning can be constituted only in the case of adherence to a particular algorithm in which simultaneity and unity are linked to one another.
an ensemble of rules dictating in which order and according to which network plan the operations must be completed creates a literal order whose mechanisms and structures we have inherited and/or learned and onto which we have become accustomed to project our intentions.
is this to say that what we learn is not actually meaning, but rather those mechanisms which regulate our projections? is this to say that we have acquired and learned a habit rather than our intentions? is this now to say in sum that we use this habit to acquire and learn the mechanisms of projecting the intentions themselves?
but it might also be the case that i am mistaken and there is no simultaneity of two operators, to wit, the hands, for if i presuppose the body, then it is a matter of one operator alone and the two hands are merely suboperators. the unity of the body would be predominant. the concept of unity thus reappears in relation to the question of simultaneity: by splitting the unity of the body into two media, i have artificially destroyed a natural order, the order of the body, and with it a literal order, the order of meaning.
simultaneity, unity, medium are part of the configuration of meaning.
the proof is given by these two pages.
what do these two pages prove?

Peter Weibel's combined text from the two pages written on two typewriters. Translated from the German by Peter Dahm Robertson.

4. Media

Epistemic Photography

craft is the production of a picture
art is the cognition of the mechanisms of production
the functional aspects of the picture
the functional aspects of the photograph
aesthetics comes from aisthesis, Greek for perception
i'm not interested in perception
i'm interested in truth
i'm not interested in adjective shifts
a photograph of a beautiful object becomes
a beautiful photograph
i'm interested in the truth function of a photograph
knowledge, epistēmē in Greek
photography as a medium of knowledge like language
cognitive photography
epistemic photography
not A. Bazin's "ontology of the image"
but rather an epistemology of the image
in which the difference between object and sign
between being and meaning
between the meaning of the object
and the meaning of the sign

is emphasized: various gradations of being, modalities of being
prevail between sign and being
object – not image
image instead of object image in object
image in image object in image
ontological leaps on the stairsteps of being

Note: Peter Weibel, Paris, 1964–65. Translated from the German by Patrick Hubenthal.

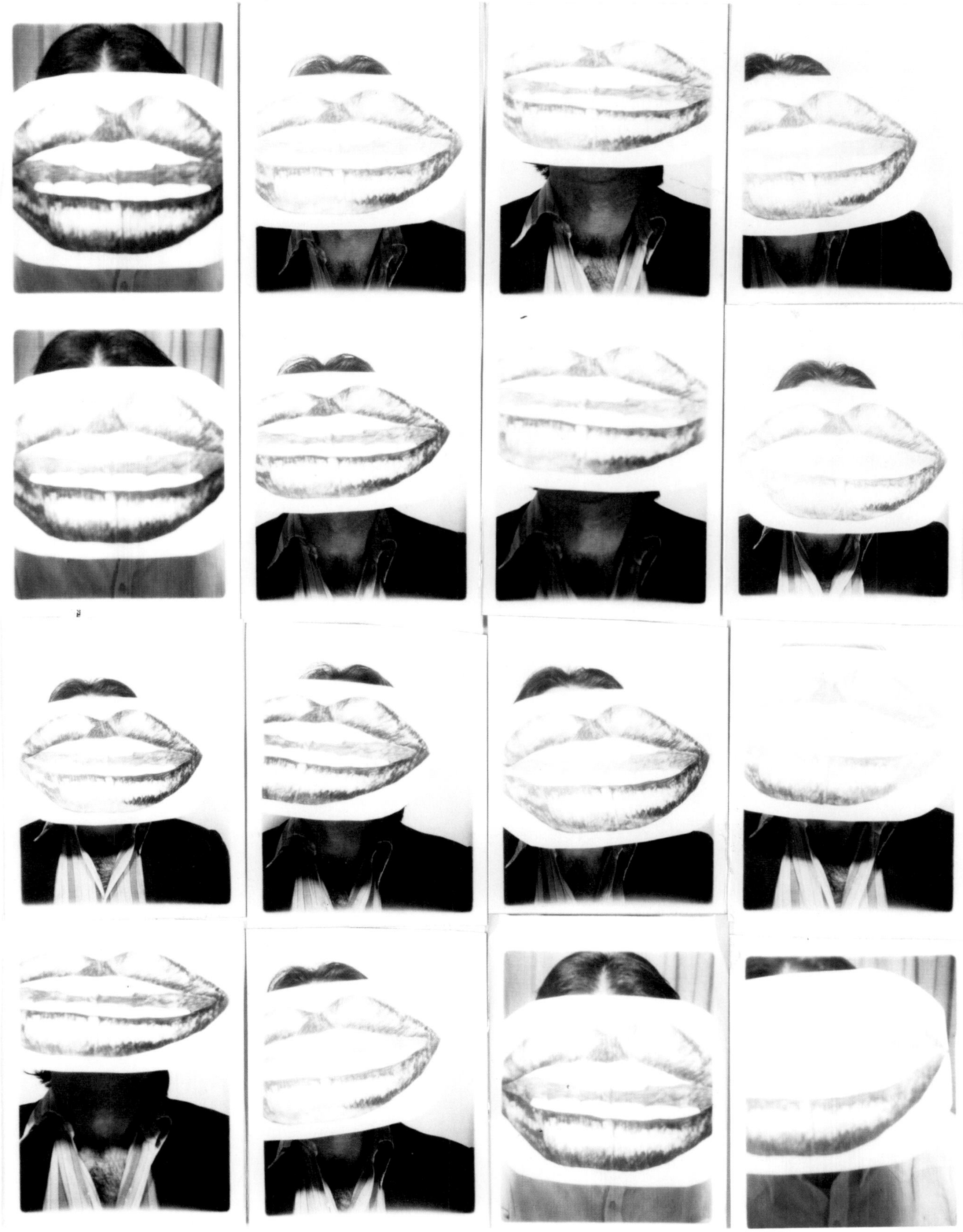

The World/Wilting Kiss / Der Kuß der Wel(k)t

1967 — photo poem; 4 b/w photo booth pictures, 12 b/w photographs

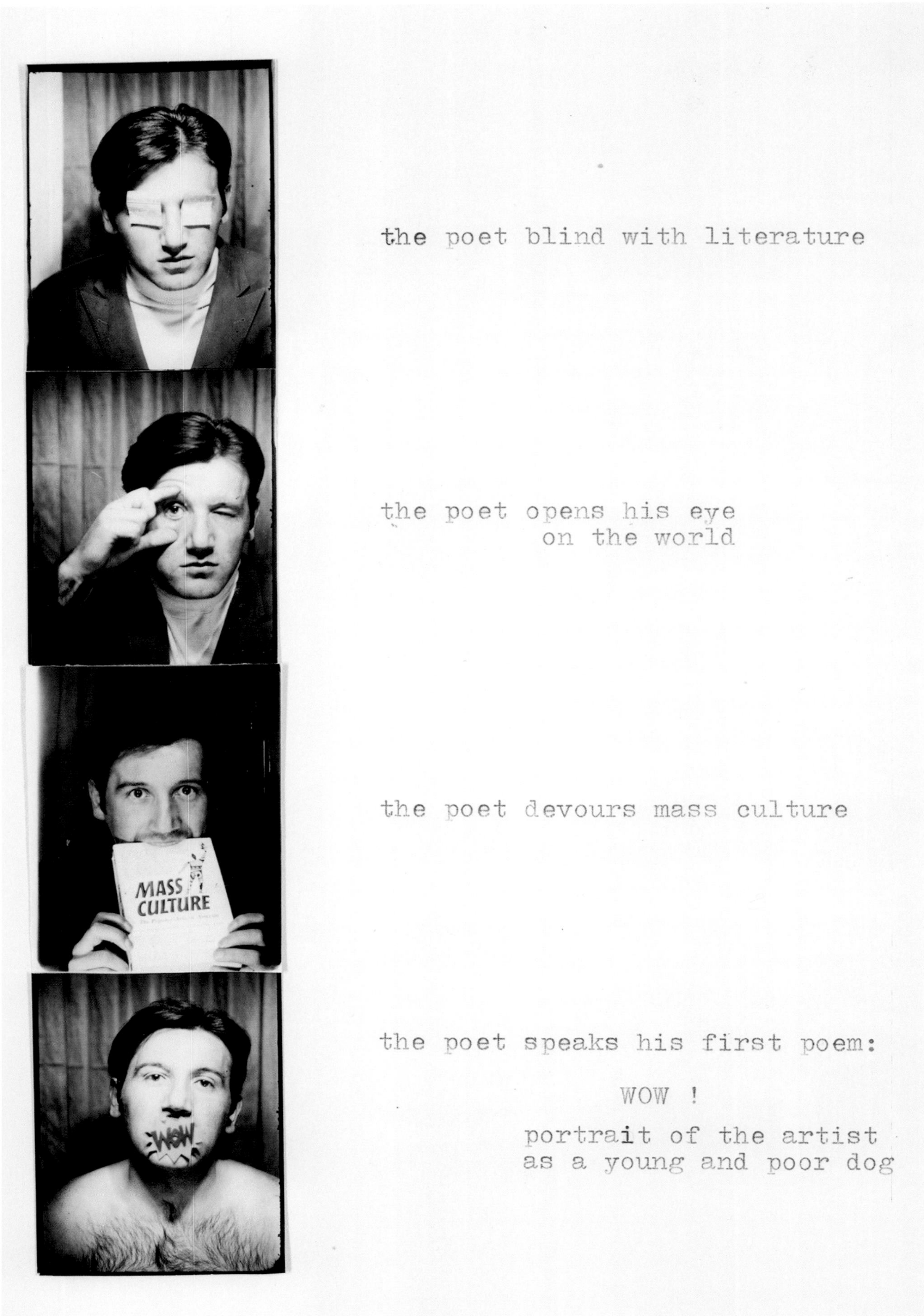

Evolution of the Poet

1967 | photo poem; 4 b/w photo booth pictures, typescript, 31 × 40 cm each

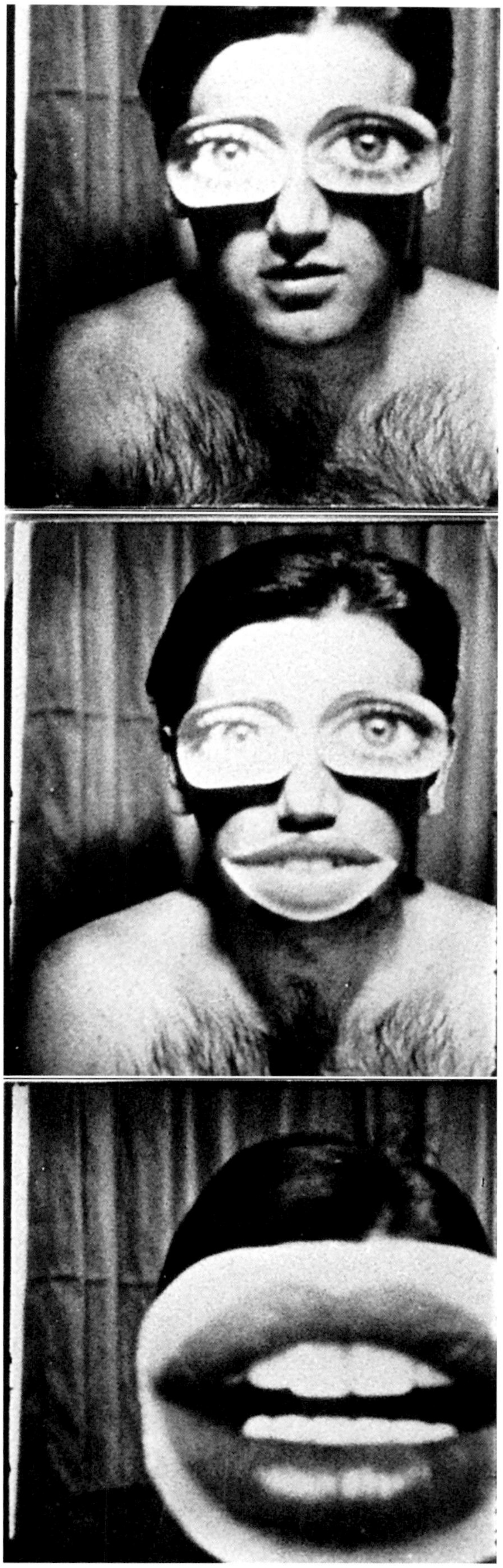

Self-Portrait as a Woman / Selbstportrait als Frau

1967 | photo poem; 3 b/w photo booth pictures

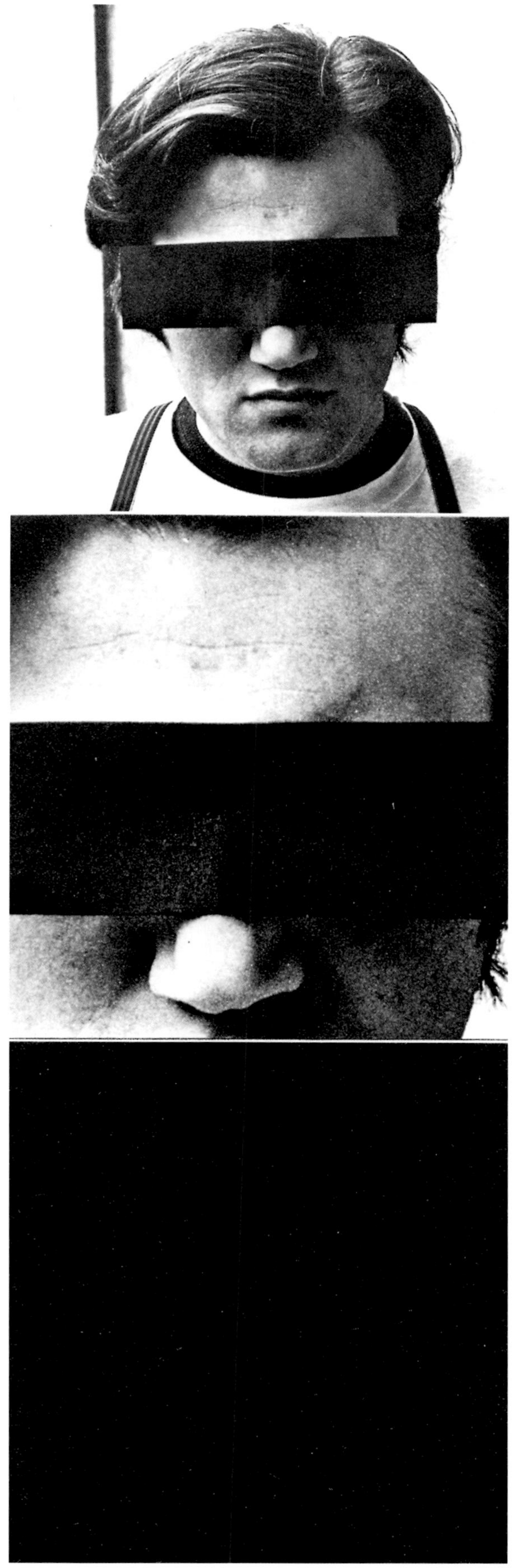

Self-Portrait as Anonymous / Selbstportrait als Anonymus

1967 | photo poem; 3 b/w photographs

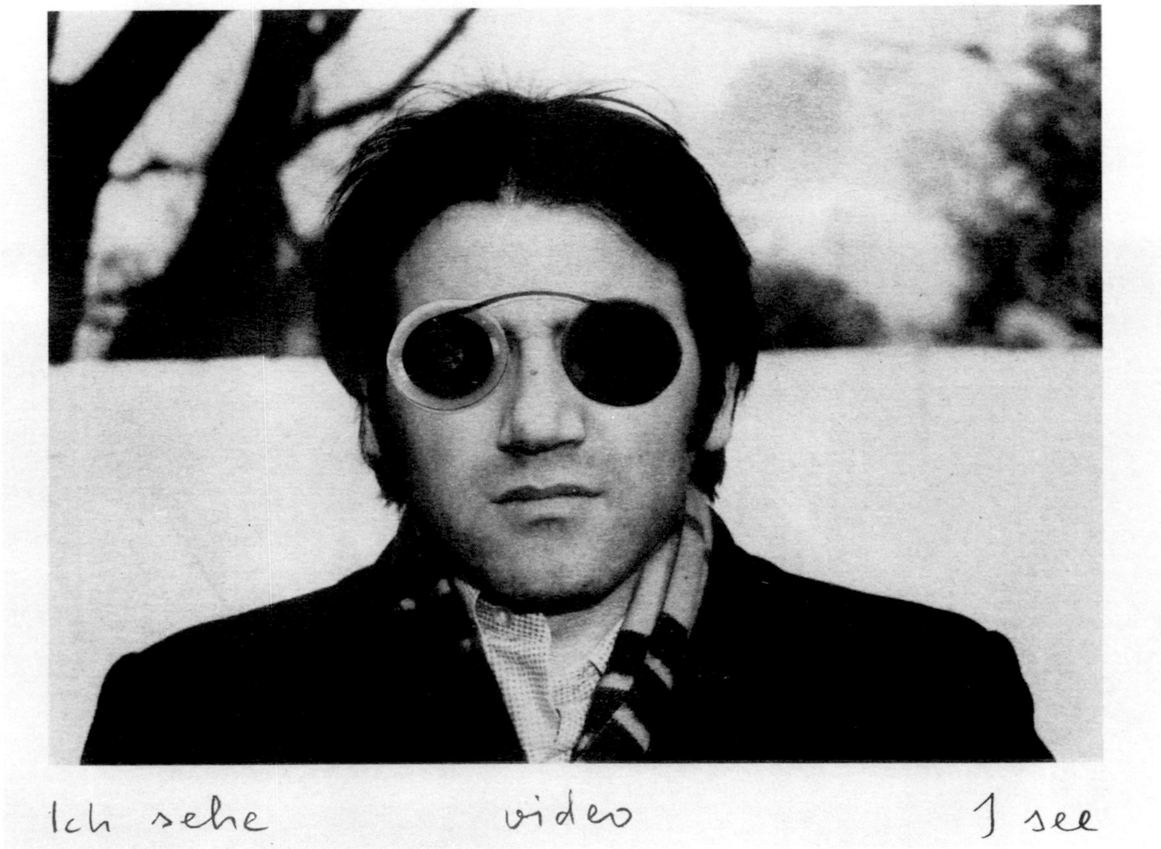

Video – I See

1974 | b/w photograph on paper, felt pen

Two video tapes are laid like sunglasses on a photograph of Peter Weibel which is then rephotographed.

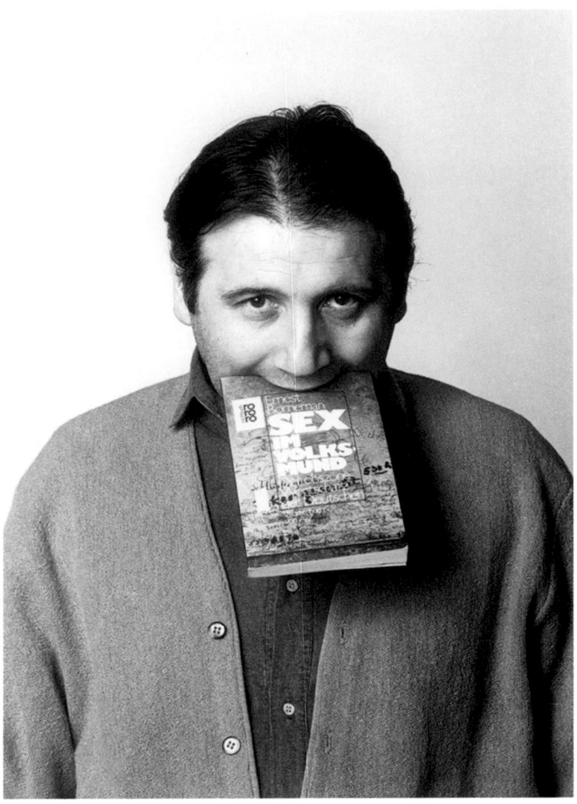

Sex on the Common Tongue / Sex im Volksmund

1971 | b/w photograph

The poet takes the 1971 book *Sex im Volksmund* (literally Sex in the mouth of people) by Ernest Borneman in his mouth. He is now the mouth of the people.

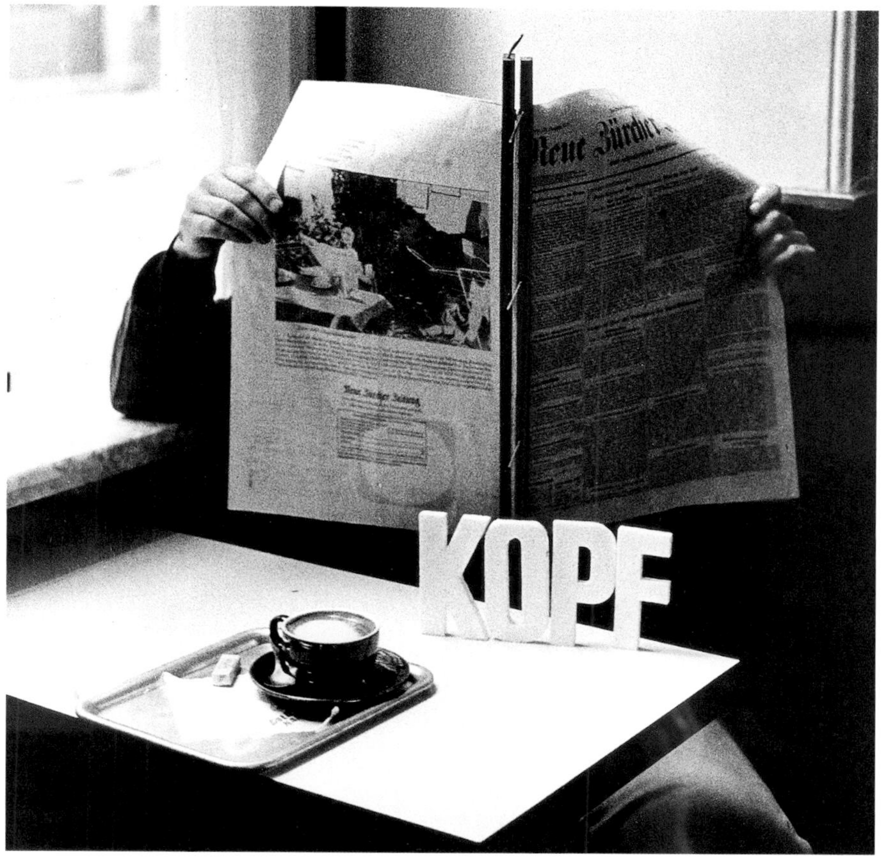

Head / Kopf

1972 | photo taken at the film set of *Wor(l)d Cinema*, gelatin silver print, 14.4 × 20.3 cm

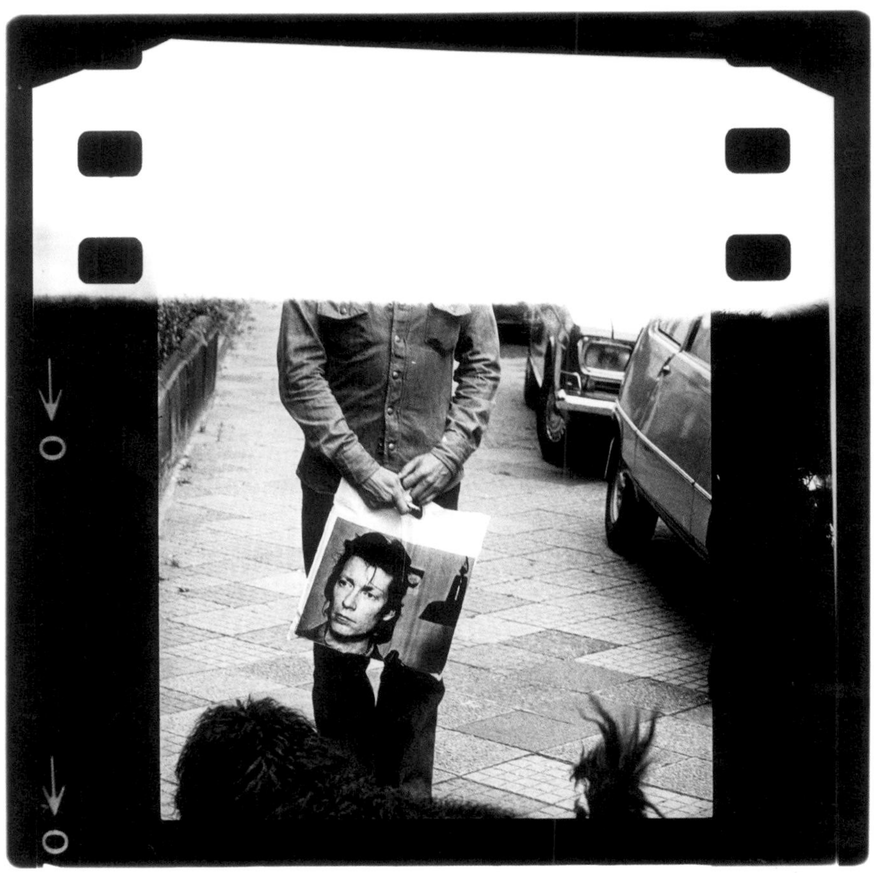

In this overexposed photograph of the author Uve Schmidt in the last frame of a film strip, his head disappears, but is represented on the bag that he is carrying.

Head (Uve Schmidt) / Kopf (Uve Schmidt)

1970 | Gunter Rambow and Peter Weibel enlarged frame of a film strip

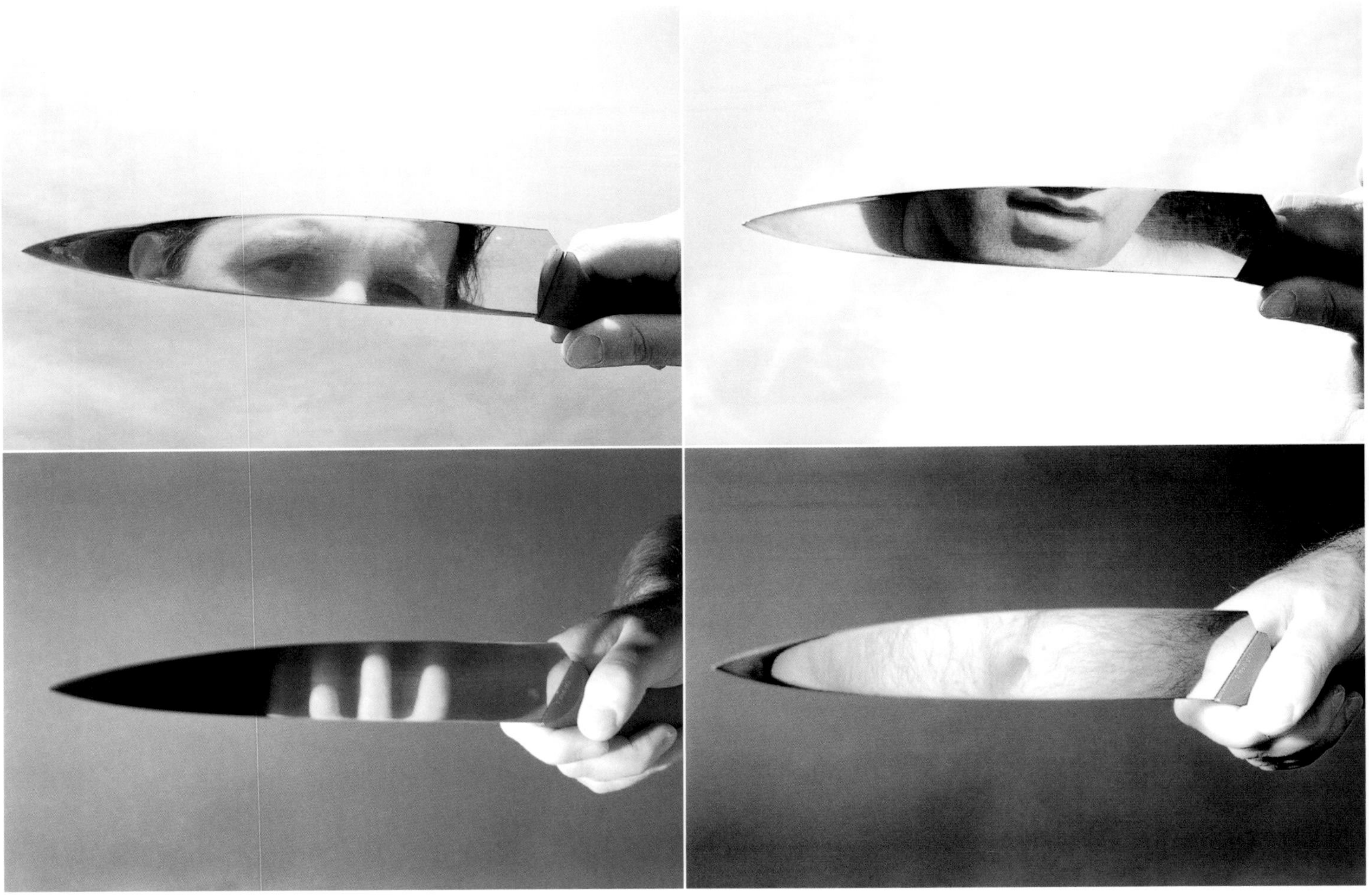

Self-Portrait (Knife as Mirror) / Selbstportrait (Messer als Spiegel)

1975 | 4 b/w photographs

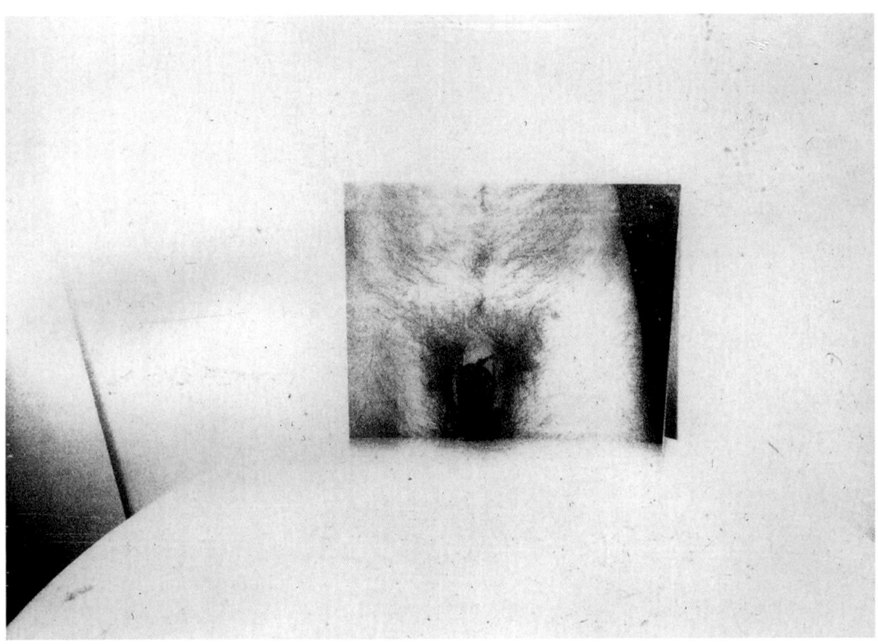

The Blind (Anonymous) Bandit / Der blinde (anonyme) Bandit

1974 | Polaroid photograph

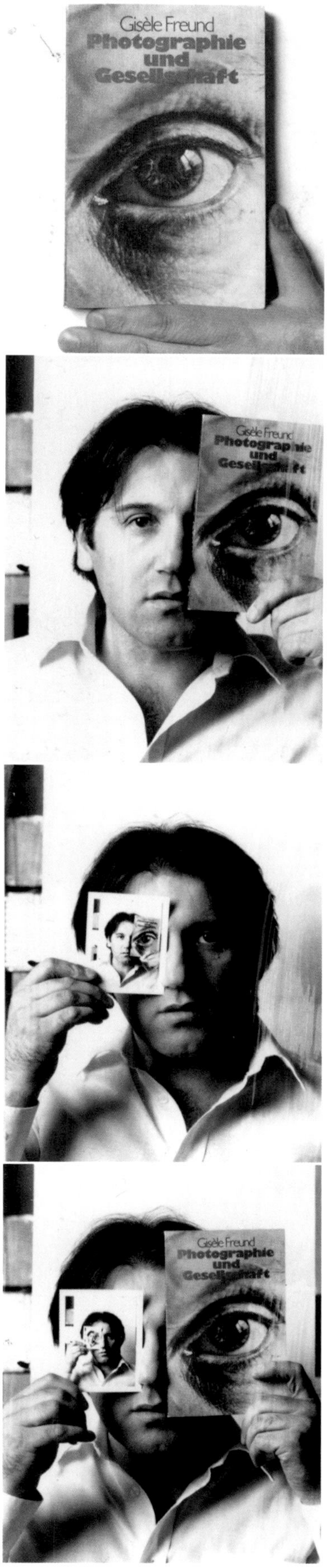

The Eye of Photography / Das Auge der Fotografie

1977 | 4 Polaroid photographs

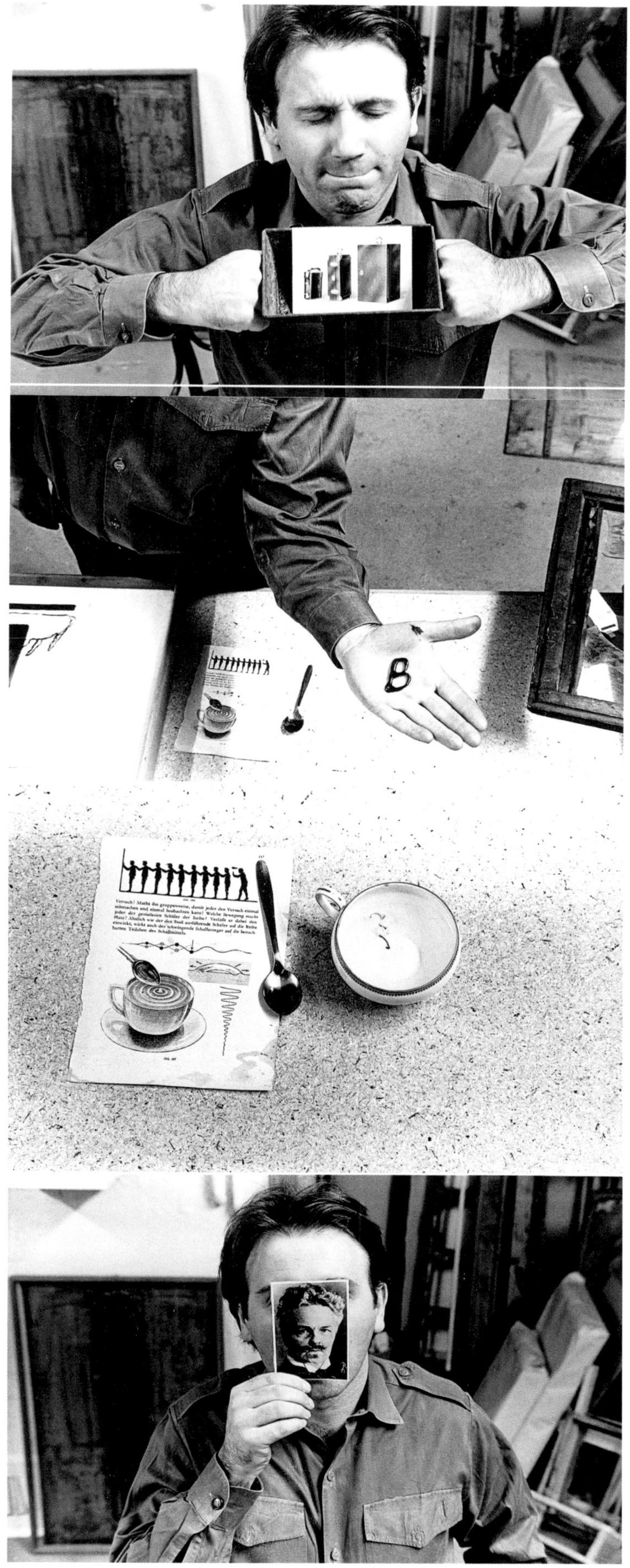

Strindberg Memorial (Identity of Designated and Designatum) / Strindberg-Memorial (Identität von Designat und Designatum)

1978 | 4 b/w photographs

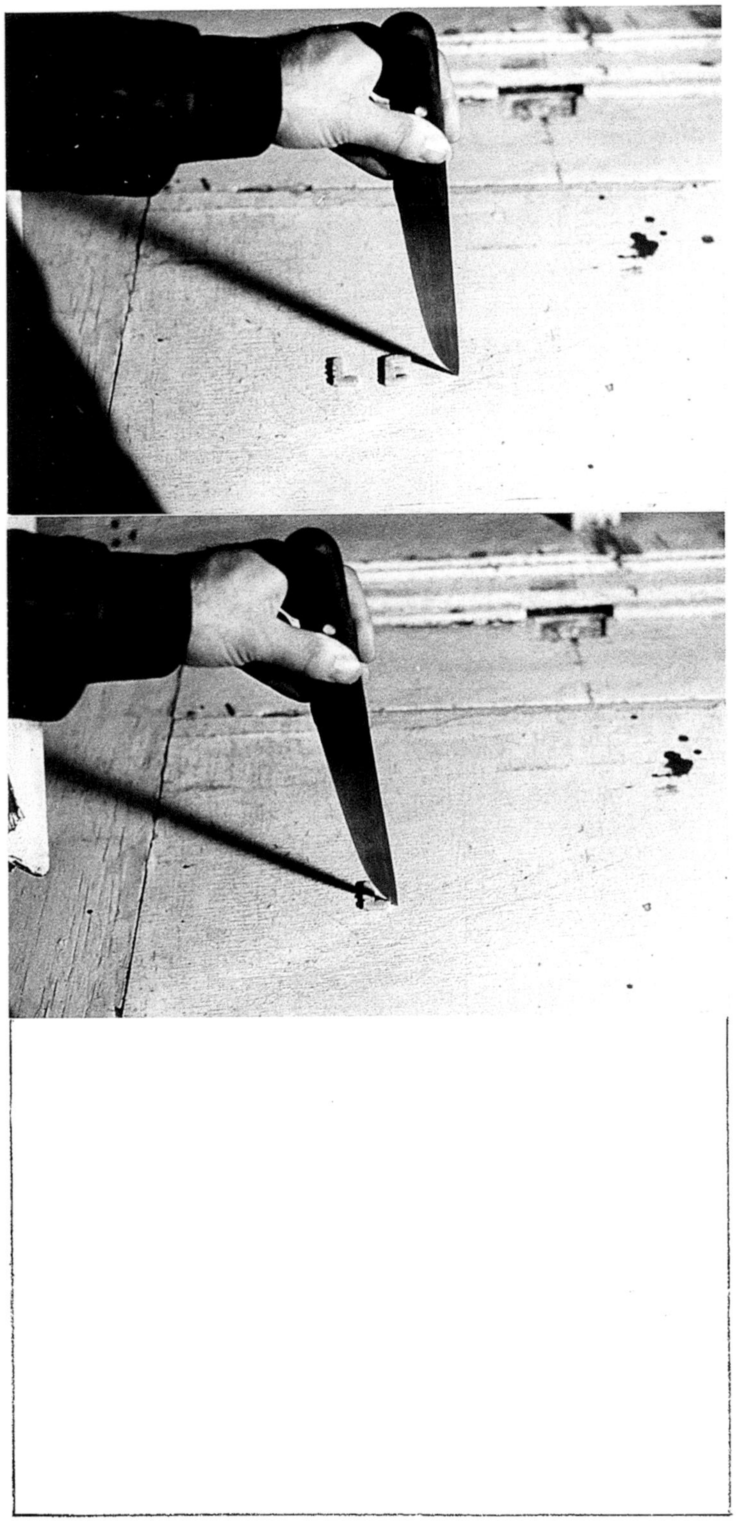

The three-part photo sequence *Emptiness* from the series *Ontological Leaps* shows the three-dimensional letters of the word *Leere* (emptiness) being attacked with a knife. At the end there are no more letters, only an empty frame. But is this really emptiness and not just an empty space? Is it just a square without the word "emptiness"? Is the void only present because the word "emptiness" is present? The presence of the word "emptiness" symbolizes something that is absent. The word "emptiness," representing an absence, a void is present – what a paradox. Emptiness itself cannot be represented by images nor by characters. Emptiness thus poses philosophically complex questions about the photographic representation of reality and the dialectic relationship between concept, word, and image.

—Peter Weibel

Emptiness / Leere

1968 | from the series *Ontological Leaps / Ontologische Sprünge*
2 gelatin silver prints, pencil on cardboard

Photography and the Administration of Data

Peter Weibel

In my own photographic practice, which began in 1966 and which I termed "epistemic photography" or "conceptual photography," I was interested in demonstrating that the "unimitable truth" of a photographic reproduction of reality is an illusion. I criticized the belief in the objective status of photography. I wanted to show that photography is not part of ontology, but part of epistemology. My aim was to claim an *epistemology of the photographic image* in contrast to André Bazin's "Ontologie de l'image photographique" (1945).[1] Photography should be a medium of knowledge, and of production of knowledge. Using various points of view and mechanisms, I wanted to destroy the myth of photographic fidelity to reality. I began my investigation with an analysis of the epistemic system of language (how does language represent reality?) and of the epistemic system of photography (how do images represent reality?). In my work *abstract – concrete (stairwell)* (*abstrakt – konkret (stiegenhaus)*, 1967, see p. 122 in this volume), I showed the difference between language and image. ... It is part of the conceptual, photography-based series titled *Ontological Leaps* (*Ontologische Sprünge*), which consists of three parts: the framed word "stiegenhaus" (stairwell); a compilation of photographic images of several stairwells; and an empty black frame on the wall.[2] The word "stairwell" evokes an image in the brain, which, however, is different for everybody. The abstract term "stairwell" is always accompanied by different concrete visions. The photos reflect the diversity of existing, real stairwells. The system of language contains abstract nouns, but photography does not. There is no abstract photo of an actual, concrete stairwell, it's not possible. Photography can only show a concrete "stairwell," the real object "stairwell." Photography cannot represent abstract "visual" knowledge. Photography can only show us the different forms and shapes of stairwells that actually exist or have existed. Just as we can photograph a specific river or lake, but not the concept of "water." Only the noun, the sign, the word, the term is abstract. They do not relate to concrete objects. Because I have chosen a term that cannot be exhibited as a real object (the stairwell) in an exhibition space, I therefore leave the last frame empty. Thus, this conceptual photographic work presents in an extremely condensed form the philosophical problem of the possibility or impossibility of reproducing reality through language and images, the difference between media and reality. ...

Photography can deliver portraits of different human beings but it cannot articulate the abstract concept of humanity, the noun "humanity." There is no picture of "humanity" as such. There are only pictures of different concrete human beings. ...When one claims that photography represents reality, it suggests that photography represents what is really there. Because photography was linked to ontology by its presumed obligation of truthful representation, some theoreticians thought photography was part of ontology, of what is there. I therefore raised the question: How could photography represent something that is not there? How could photography represent nothing? The word "nothing" exists, even if nothing does not exist. How would you depict nothing; how would you depict what is not there? How would you depict emptiness? You cannot really depict an empty bottle or an empty table. You can depict a table as an object, and if there is nothing on the table, you would not say: It is an empty table. You would just say: It is a table. You could depict a bottle, and no matter whether it was full or empty, it obviously just remains a bottle. When you see the inside of the bottle, you could add that it was a bottle of wine or a bottle of milk. Therefore, in my work *Emptiness* (*Leere*, 1968, see p. 118 in this volume), I made a three-part photo sequence with the German noun *Leere* (emptiness, void). In a series of photographic images, I showed the process of how I, using a knife, made three-dimensional letters which formed the word "LEERE" disappear one after another. The final image of the series – which also belongs to my series *Ontological Leaps* – showed an empty frame devoid of letters. What you see is a kind of canvas within a frame without any letters, but even then, you do not see an empty canvas, and you certainly do not see emptiness, you only see an empty frame. In contrast, the presence of the word "emptiness" symbolizes something that is absent. By the presence of the word "emptiness," an absence, a void is represented – what a paradox. Emptiness itself cannot be represented by images nor by characters. It can only be imagined as absent and presented terminologically. ...

This is a fundamental difference between my conceptual photography and the conceptual photography of other artists, such as the work of Joseph Kosuth. Since 1965 Kosuth has always worked with a triptych. He shows a photograph of a lexical description of an object such as a chair. Then he shows a picture of a chair, and then finally presents the real chair (see fig. 1). ... Kosuth follows precisely the philosophy of the early Ludwig Wittgenstein and of his groundbreaking *Tractatus logico-philosophicus* (1921): "The proposition is a picture of reality."[3] Kosuth's photographic practice is an illustration of this sentence by Wittgenstein. You see the real chair (reality), the photograph of this chair (image), and a dictionary definition of a chair

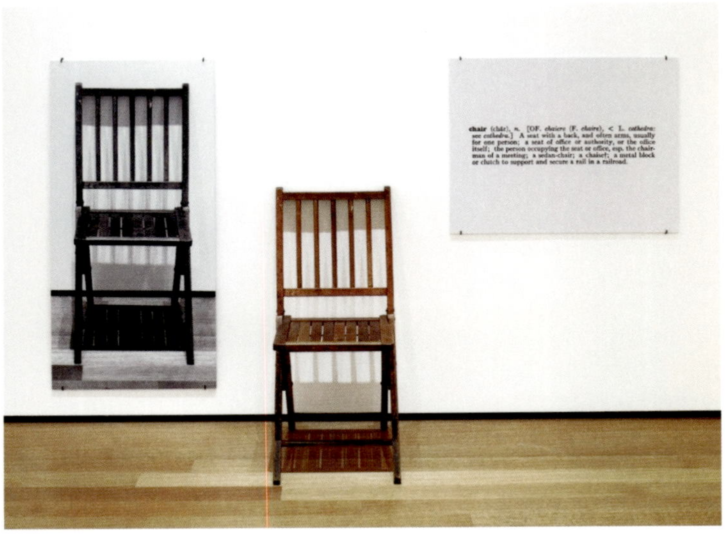

Fig. 1:
Joseph Kosuth, *One and Three Chairs*, 1965

(proposition). The verbal description was abstract, but the photograph was concrete. Therefore, this installation suppressed the difference between abstract language and concrete photography. It confirmed Wittgenstein's statement instead of disproving it. Wittgenstein's theorem was wrong, as he later confessed, and instead claimed that a meaning of a sentence is defined by its use. In the mid-1960s, I was already more inclined towards Wittgenstein's later ideas and criticized his earlier work. Wittgenstein's sentence, and Kosuth's work, proclaim in a certain way a one-to-one relationship between reality and representation. This philosophy, and its pictorial practice, dismisses exactly this gap between abstraction and concretion. As you can see in the work of Kosuth, the verbal description of the "chair" is a kind of all-quantifier. It is an abstract description of all chairs, a noun. The photograph, which is identical to the real chair, is just a special concrete case of a chair, an existence-quantifier. The lexical description is a proposition about all chairs: (valid for all chairs). The image and the real chair are a proposition about one chair: *V* (valid for one chair). The gap between all-quantifier and existence-quantifier *V* defines the difference between language and image, representation and reality.

My aim was precisely the opposite: I wanted to show that photography has no inimitable fidelity to the reproduction of reality. Photography is not a medium through which reality is represented. Photography is a medium used to fabricate knowledge. Therefore, I started to call my photographic practice "epistemic photography." In contrast to Kosuth, I did not work with triptychs, but with diptychs, excluding the third part - the real object. I selected words like "forest," "water," "stairwell," the objects of which could not be properly represented in a gallery because they were too big in reality (see p. 123 in this volume). Another difference was that I displayed a word, but this word was not represented by the photograph of one object; it was represented by many photographs of different objects of the same class to emphasize that the concept or the word summarizes many objects. With this multiplication of photographs, I showed a set of existence-quantifiers as photographs. When you have a word like "chair" or a lexical description of the meaning of a word, then it is abstract, it comprises all chairs. A photograph singles out one concrete chair. Therefore, I used a diptych, that is, only the word and the image, whereby the image contained multiple images, not only one image of one chair. These two features, a multiplication of images and an exclusion of the real object, set my work apart from the usual conceptual photography. As a critical practice it doubted the ontological status of photography and analyzed photography as an epistemic medium.

My work *This Load Is Heavier Than I Can Bear* (*Diese Last ist schwerer als ich halten kann*, 1975, see page opposite) shows that the truth-value of photographs is variable and depends on the subject depicted, or its general framing or the context of the photograph. For the little girl, the sentence is correct, but for me the sentence is evidently not correct. I created a paradox.

Consequently, I became interested in constructing paradoxes photographically; for example, *Physical Paradoxes* (*Physikalische Paradoxe*, 1974, see p. 157 in this volume) showed two spirit levels in which the air bubbles at the center indicate that the instruments are in a horizontal position. But in the photographs the instruments are seen in a diagonal position. The trick: in reality the spirit levels were correctly held in a horizontal position, but the photographic apparatus was held in two different slanting positions. I did similar things in my video work, which is therefore also called "epistemic videology" (see chapter 4.3 in this volume). On a transparent glass plate, the viewer sees a small horizontal line (a minus sign). On the screen of the monitor, the viewer sees a plus sign. The minus sign is recorded by two opposite cameras, one of which is turned upside down. Therefore, one camera sends a horizontal line and the other one sends a vertical line to the mixer. Combined, they deliver a plus sign on the screen. Representation does not depict reality; in fact, it does just the opposite. Photography is not a map of reality (the land). The map of media constructs the land (reality).

Note: Adapted from Peter Weibel, "Photography and the Administration of Data," in *Watched! Surveillance, Art and Phootography*, ed. Louise Wolthers, Dragana Vujanović, and Niclas Östlind (Berlin: Walther König, 2016), 226–37.

[1]
André Bazin, "The Ontology of the Photographic Image," in *Film Quarterly*, vol. 13, no. 4 (1960): 4–9.

[2]
A similar work, *abstract – concrete (tree) / abstrakt – konkret (baum)*, 1967, consists of the German word "Baum" (tree) and three photographs showing different parts of a tree (trunk, branches, leaves.

[3]
Ludwig Wittgenstein, *Tractatus logico-philosophicus* (1921), trans. C. K. Ogden with an introduction by Bertrand Russell, Kegan Paul, Trench (London: Trubner & Co., 1922), § 4.01.

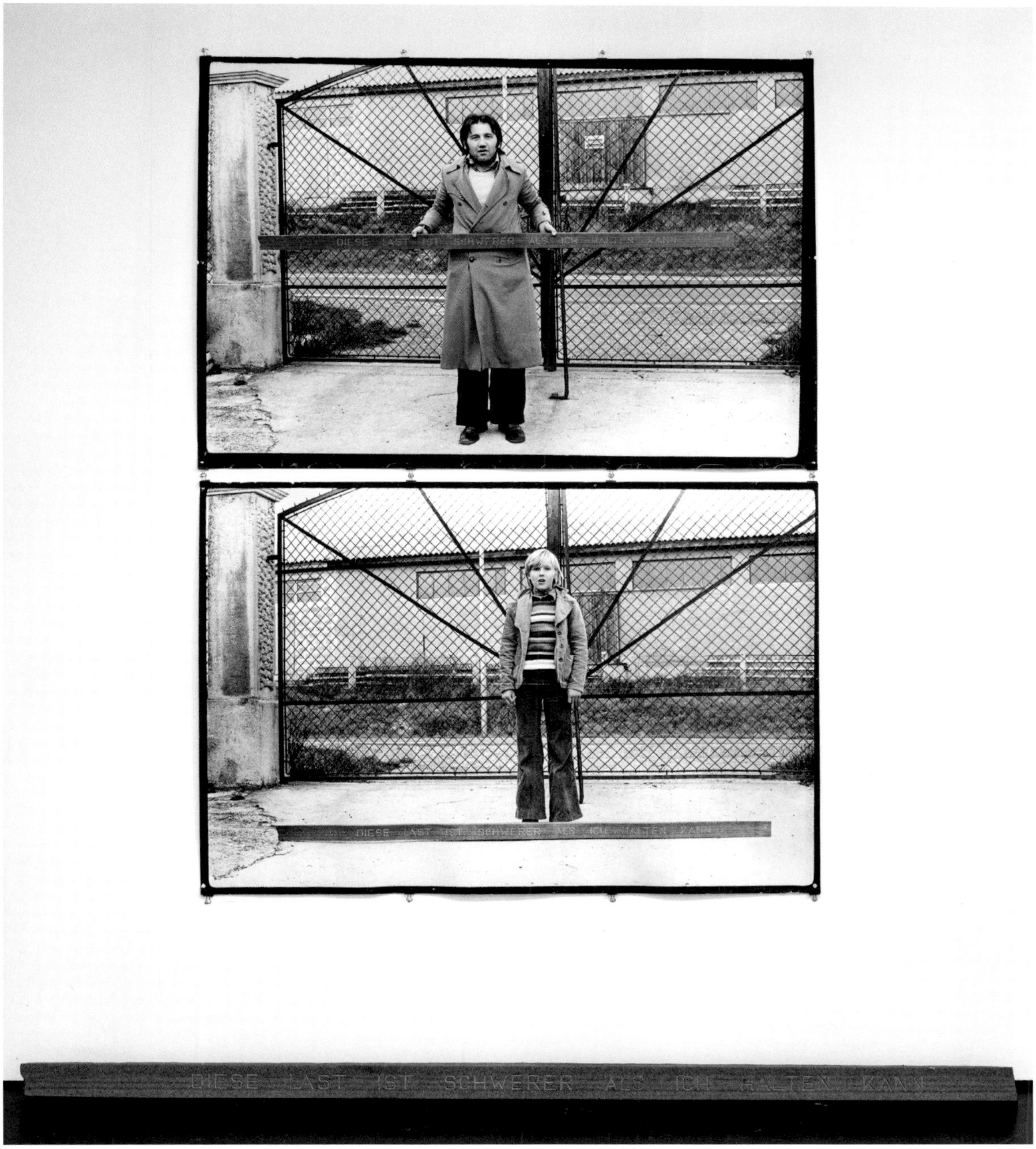

Installation view
respectively, Peter Weibel,
ZKM | Karlsruhe, 2019–20

This Load Is Heavier Than I Can Bear / Diese Last ist schwerer als ich halten kann

1975 | 2 b/w photographs, engraved marble pole, 8 × 241 × 4.5 cm

In this three-part work, one photograph shows Peter Weibel holding a pole made of stone in his hands with the inscription *Diese Last ist schwerer als ich halten kann*. In a second photograph, a girl stands behind a stone pole with the same text. Underneath the two photographs lies the engraved marble pole itself. While the sentence written on the stone pole is correct for the girl, it is not correct for the first photograph with Weibel. Photography was invented as a medium that was characterized by its "perfect fidelity in reproducing reality." Weibel's interest in photography is precisely the opposite; namely, in investigating the photographic mechanisms through which reality is reproduced *eo ipso* falsely, paradoxically, distortedly, illusionarily. Since the beginning of his engagement with photography in 1966, Weibel has always been interested in the difference between image and object and in photography as the medium between them.

The eye is the natural interface, but we are not aware of this when we perceive. In the case of a technical interface, like the camera, the extension of the eye, we cannot deny the existence of the interface between image and object. The camera is a third agent, a machine, that turns an object into a sign.

The mechanisms of the medium of photography as self-legislated, generative principles, a "conceptual photography," therefore dominated Weibel's artistic work in the mid 1970s.

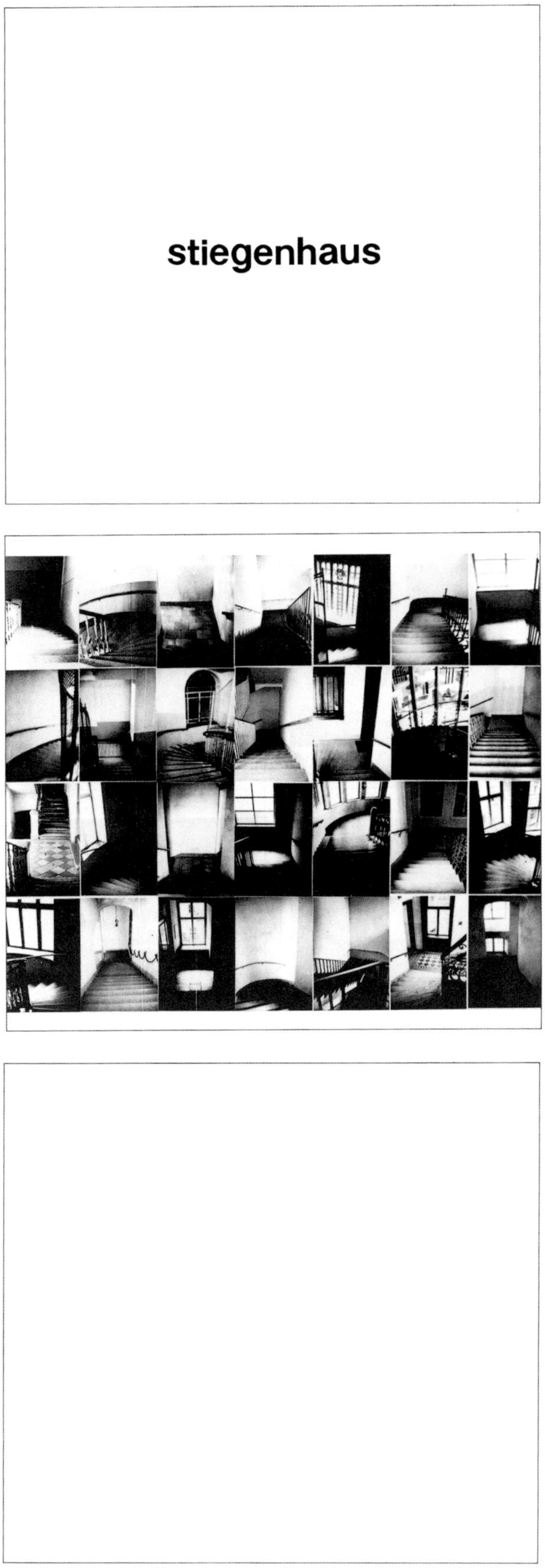

abstract – concrete (stairwell) /
abstrakt – konkret (stiegenhaus)

1967 — from the series *Ontological Leaps / Ontologische Sprünge*
letraset on paper, b/w photograph, marker on wall, 3 parts

In *Being and Time* (*Sein und Zeit*, 1929), Martin Heidegger wrote about the ontological difference between "being" as such and "beings." Peter Weibel's work is about the ontological difference between being and sign, being and language, being and image, being and media. *abstract – concrete (stairwell)* is part of the conceptual, photography-based series titled *Ontological Leaps*. It consists of three parts: the framed word *stiegenhaus* (stairwell), a compilation of photographic images of several stairwells, and an empty black frame on the wall. The word "stairwell" evokes an idea, which, however, is different for everybody. The abstract term "staircase" is always accompanied by different concrete images. The photos reflect the diversity of existing, real stairwells. There is no abstract photo of an actual staircase possible. Only the noun, the sign, the word, the term is abstract. Because he has chosen a term that cannot be exhibited as a real object (the stairwell) in an exhibition space, Peter Weibel therefore leaves the last frame empty. This conceptual photographic work thus presents in an extremely condensed form the philosophical problem of the possibility or impossibility of reproducing reality through language and images – the difference between media and reality.

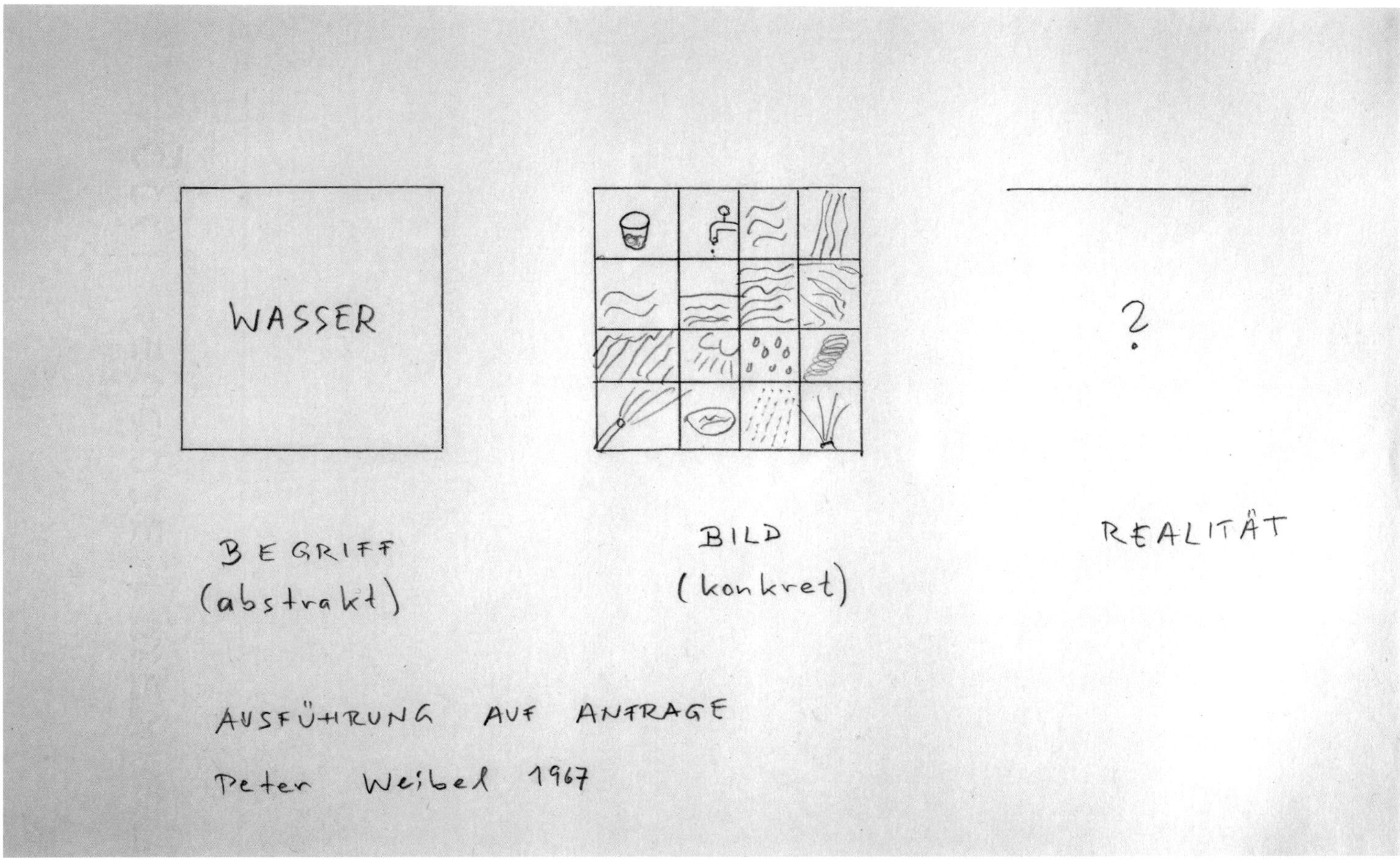

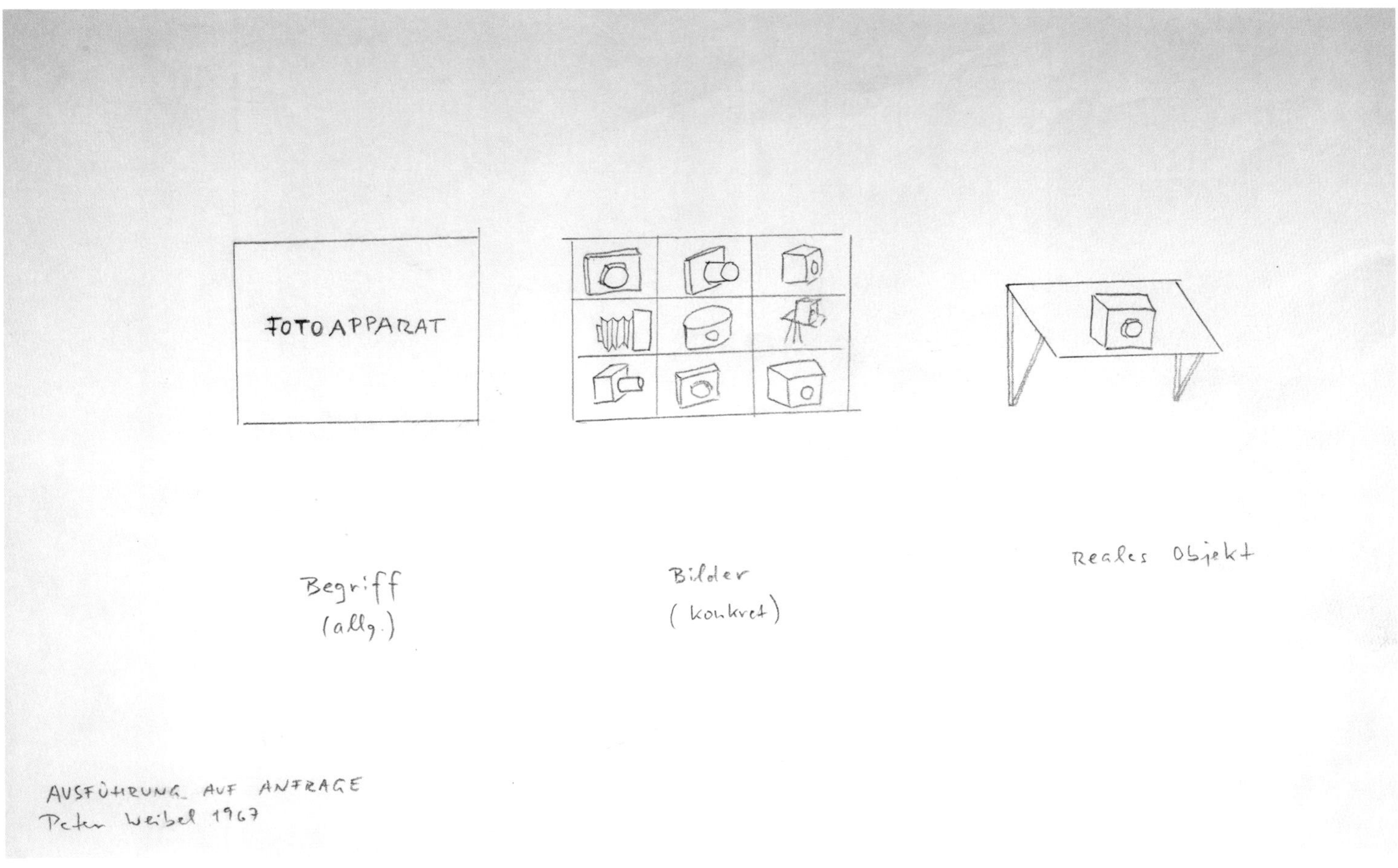

Concepts / Konzepte

1967 | from the series *Ontological Leaps / Ontologische Sprünge*
pencil on paper, 6 parts,
29.7 × 42 cm each

The abstract term "water" is accompanied by different concrete images of water. But how should real water be presented? Accordingly, the abstract term "camera" produces concrete images of cameras and the real object is exhibited.

Ontological Leaps

Peter Weibel

In a series of pictures I demonstrated how photographs substitute for objects and for other photographs as well, forming a chain of references rather than an existential quantification. What takes place with images is much the same as what happens to a child who is given a name, the relationship of which to that child is not necessary but contingent, and which becomes established through convention and repetition. There exists a child whose name is Bruno. There exist many children named Bruno, but not all children are named Bruno. Hence "Bruno" is not a universal quantifier, a definite description that is valid in all worlds – that is, the name's relationship is not one of necessity. The individual's relationship to his name is therefore arbitrary, or rather contingent, but in time he comes to identify with it. When someone calls out "Bruno," he looks up; they might mean him. Thus, possibility is deceptively transformed into the illusion of necessity, modality into ontology. By contrast, all of my artistic efforts have been aimed at transforming being into sign, ontology into modality, "necessary" into "possible," and "possible" into "actual." In this way I have expanded and refined my initial doubt about the identity hypothesis and the ideology of representation.

From photography to interactive computer installations, this doubt and negation runs through all of my work. In *The Science of Logic* (*Wissenschaft der Logik*, 1812–16) Hegel investigated the concepts of actuality, necessity, possibility, and contingency and defined actuality as necessity. Possibility is just possibility, not actuality. I, on the other hand, claim that the actual is merely a special case of the possible, not always a case of the necessary – Hegel himself sensed something similar and consequently wrote: "*In itself*, therefore, we have here the unity of necessity and contingency; this unity is to be called *absolute actuality*."[1] These reflections led him to the (in)famous conclusion that "what is rational is actual; and what is actual is rational."[2]

However, in section 6 of his *Encyclopedia of the Philosophical Sciences in Basic Outline* (*Enzyklopädie der philosophischen Wissenschaften im Grundrisse*, 1830), Hegel wrote that "a contingent concrete existence has no greater value than something that is *possible* and which may just as well *not exist* as exist."[3] If the contingent is a possible existence, according to Hegel, one can conclude that existence is the possibility of the contingent and therefore a contingent actuality.

Just as names do not entail necessity, the relationship between photography and actuality is not one of necessity either, but at most one of possibility. Misunderstood photography gives actuality the stamp of the necessary and rational. Photographs are therefore false prophets. Correctly understood, photographs are subversive agents which expose actuality as mere possibility and reveal the contingency and irrationality of the actual.

Photographs, then, are simply names for the real. As Hegel himself wrote, "a contingent concrete existence ... will not be deemed to deserve the emphatic designation of being actual."[4]

Now that the primary delivery medium for photographs is no longer paper but rather digital data – that is, a set of variables – my line of reasoning should be easier to understand and accept.

Note: This text was written in 2022 and is published for the first time in this volume. Translated from the German by Patrick Hubenthal.

[1] Georg Wilhelm Friedrich Hegel, *The Science of Logic*, trans. George di Giovanni (Cambridge: Cambridge University Press, 2010), 485. First published as *Wissenschaft der Logik* (Nuremberg: Schrag, 1812–16).

[2] Georg Wilhelm Friedrich Hegel, *Elements of the Philosophy of Right*, trans. H. B. Nisbet (Cambridge: Cambridge University Press, 1991), 20. First published as *Grundlinien der Philosophie des Rechts* (Berlin: Nicolai, 1820).

[3] Georg Wilhelm Friedrich Hegel, *Encyclopedia of the Philosophical Sciences in Basic Outline, Part I: Science of Logic*, trans. Klaus Brinkmann and Daniel O. Dahlstrom (Cambridge: Cambridge University Press, 2010), 33. First published as *Enzyklopädie der philosophischen Wissenschaften im Grundrisse* (Heidelberg: Oßwald, 1817).

[4] Hegel, *Encyclopedia*, 33.

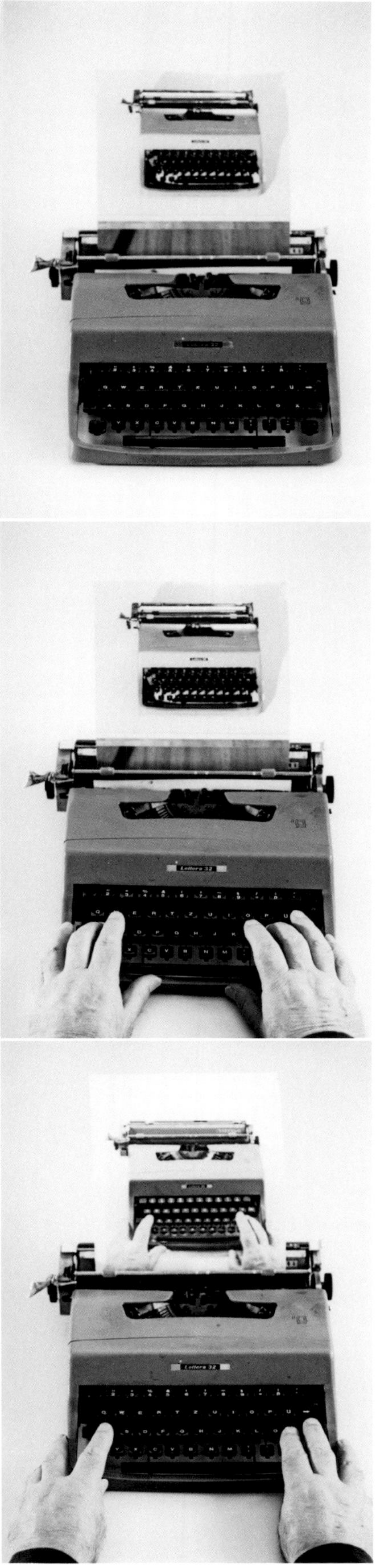

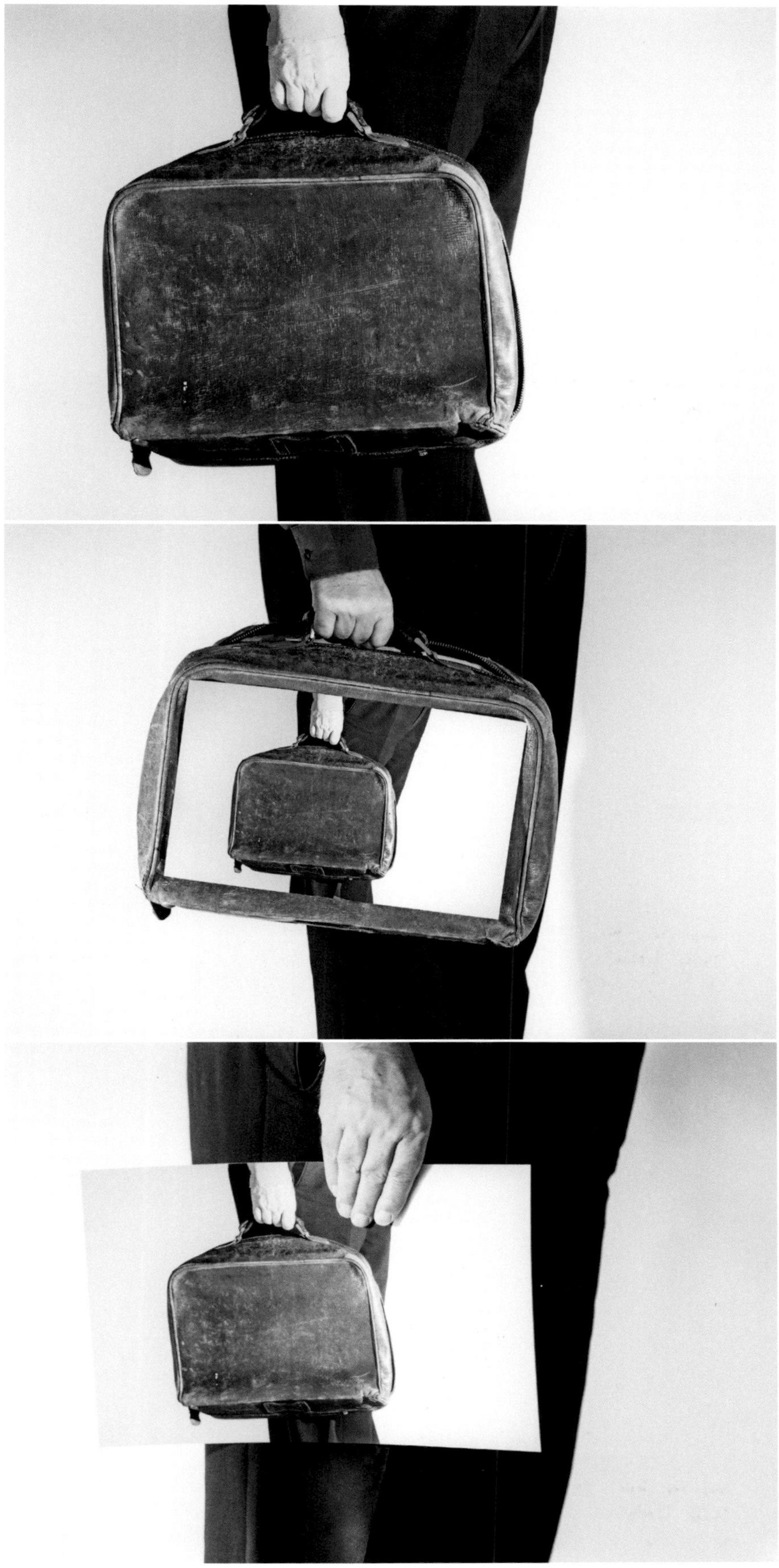

Typewriter / Schreibmaschine

1971 — from the series *Ontological Leaps / Ontologische Sprünge*
b/w photographs, 3 parts,
42 × 29.7 cm each

Suitcase / Koffer

1971 — from the series *Ontological Leaps / Ontologische Sprünge*
b/w photographs, 3 parts,
42 × 29.7 cm each

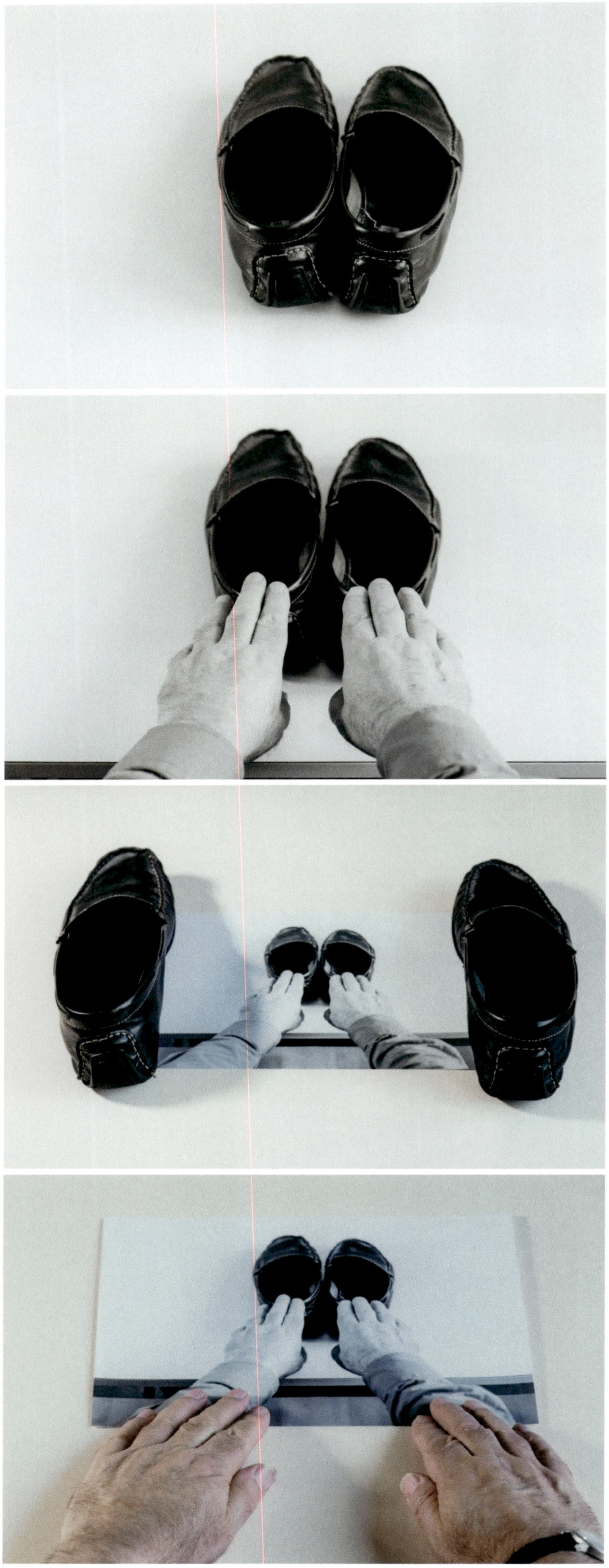

Shoes / Schuhe

1967/ 2014 — from the series *Ontological Leaps / Ontologische Sprünge*
4 photographs, 29.7 × 42 cm each

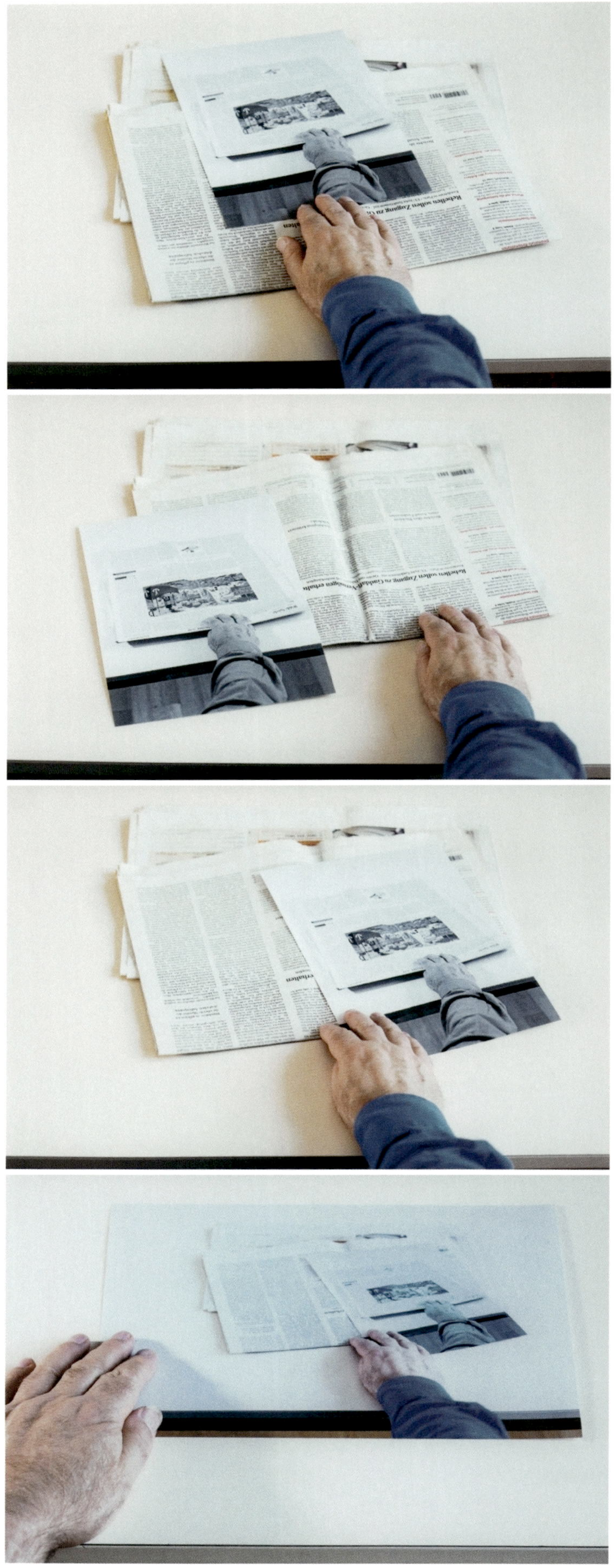

Newspaper / Zeitung

1967/ 2014 — from the series *Ontological Leaps / Ontologische Sprünge*
4 photographs, 29.7 × 42 cm each

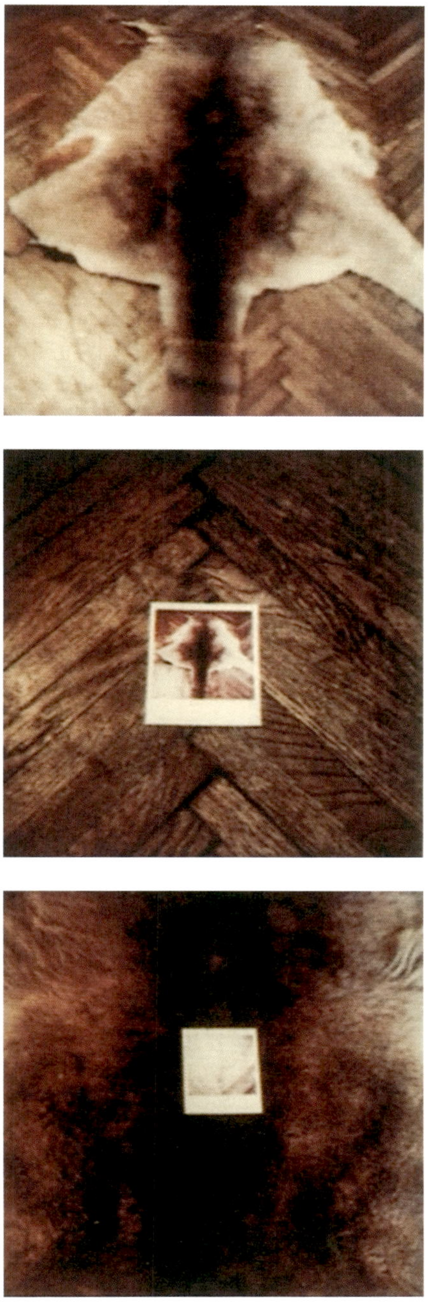

An animal skin, lying on a floor, is photographed. The Polaroid photo (not the skin) is then laid on the floor. This Polaroid photo is then laid on the skin (taking up the function of a new kind of floor).

Honest Skin / Ehrliche Haut

1979 | 3 Polaroid photographs

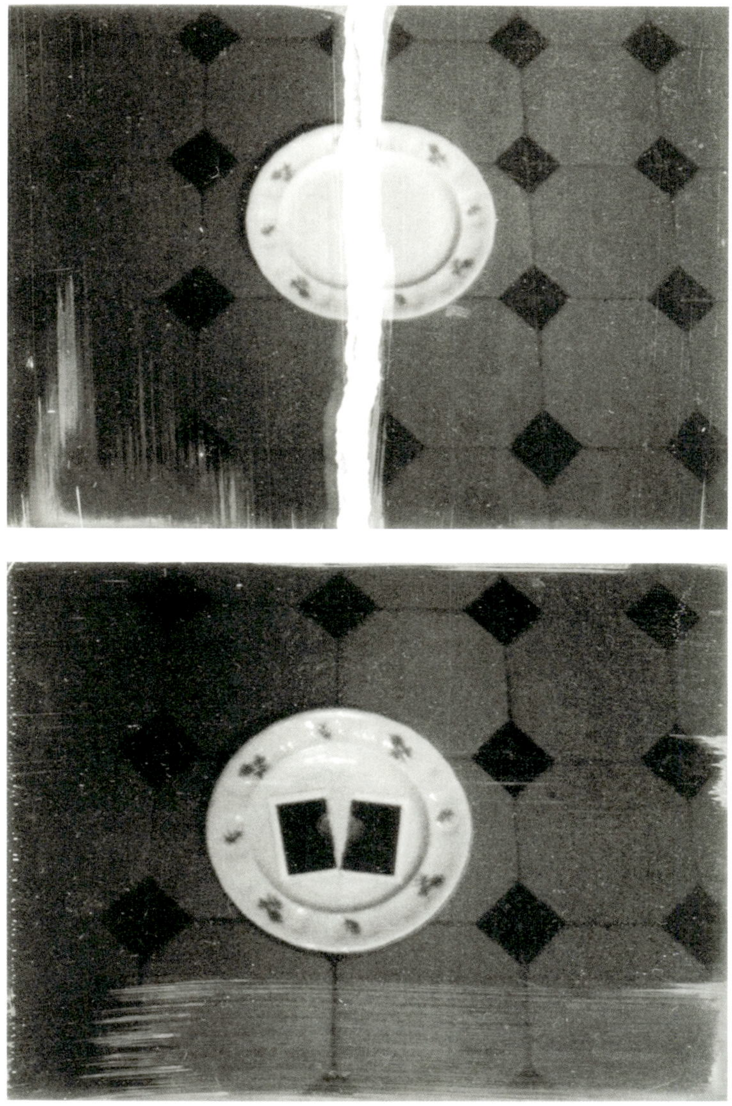

The plate breaks only as a photograph. The photographically broken plate (representation) lies on an intact plate (reality).

Material Splits / Materialsprünge

1973 | 2 Polaroid photographs

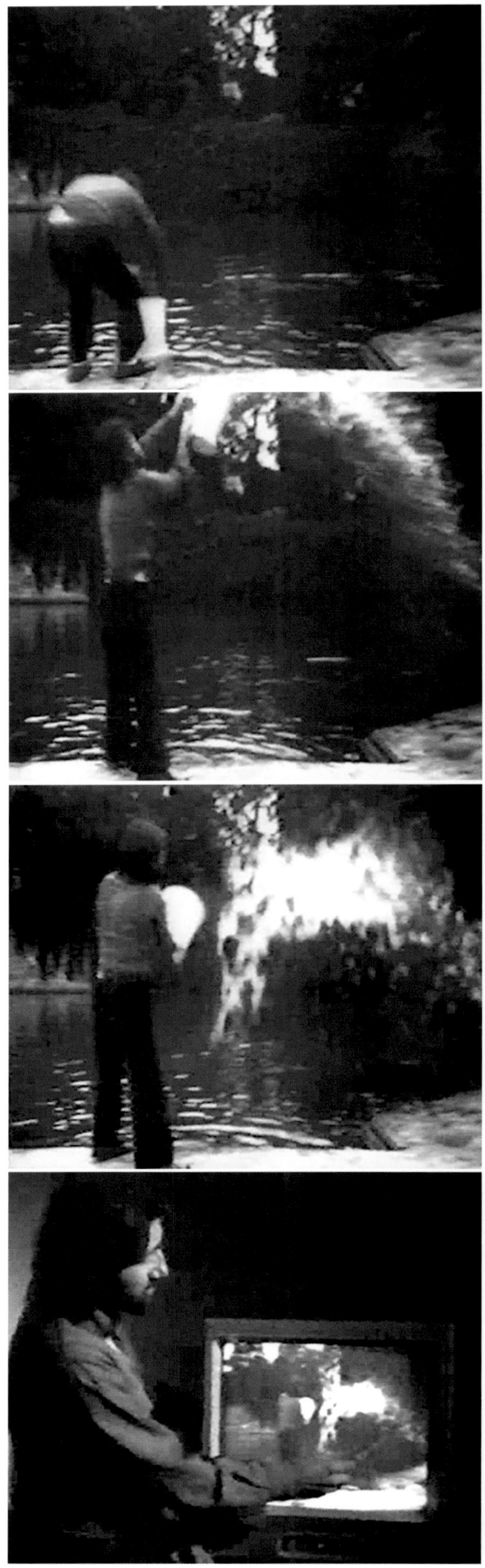

In this video, Peter Weibel throws water into the air with a bucket. You see the water in motion until it freezes in the picture at the moment of its greatest expansion. The camera moves back and you realize that you haven't seen reality, but a film on a screen in front of which Peter Weibel is standing. You jumped from one level of reality to another. Thus, three modalities of reality emerge: reality, filmed reality, filmed film. In his commentary, Weibel defines television as a time-space device: while the movement of water in nature has passed in time, the technology of television makes it possible to capture it in space. The imaginary water sculpture is only visible on the TV screen. Television is a switch of time and space, of reality and representation.

Virtual Water Sculpture / Imaginäre Wasserskulptur

1971	video; b/w, sound, 48 sec., first broadcast in *Impusle 7*, ORF, June 29, 1972

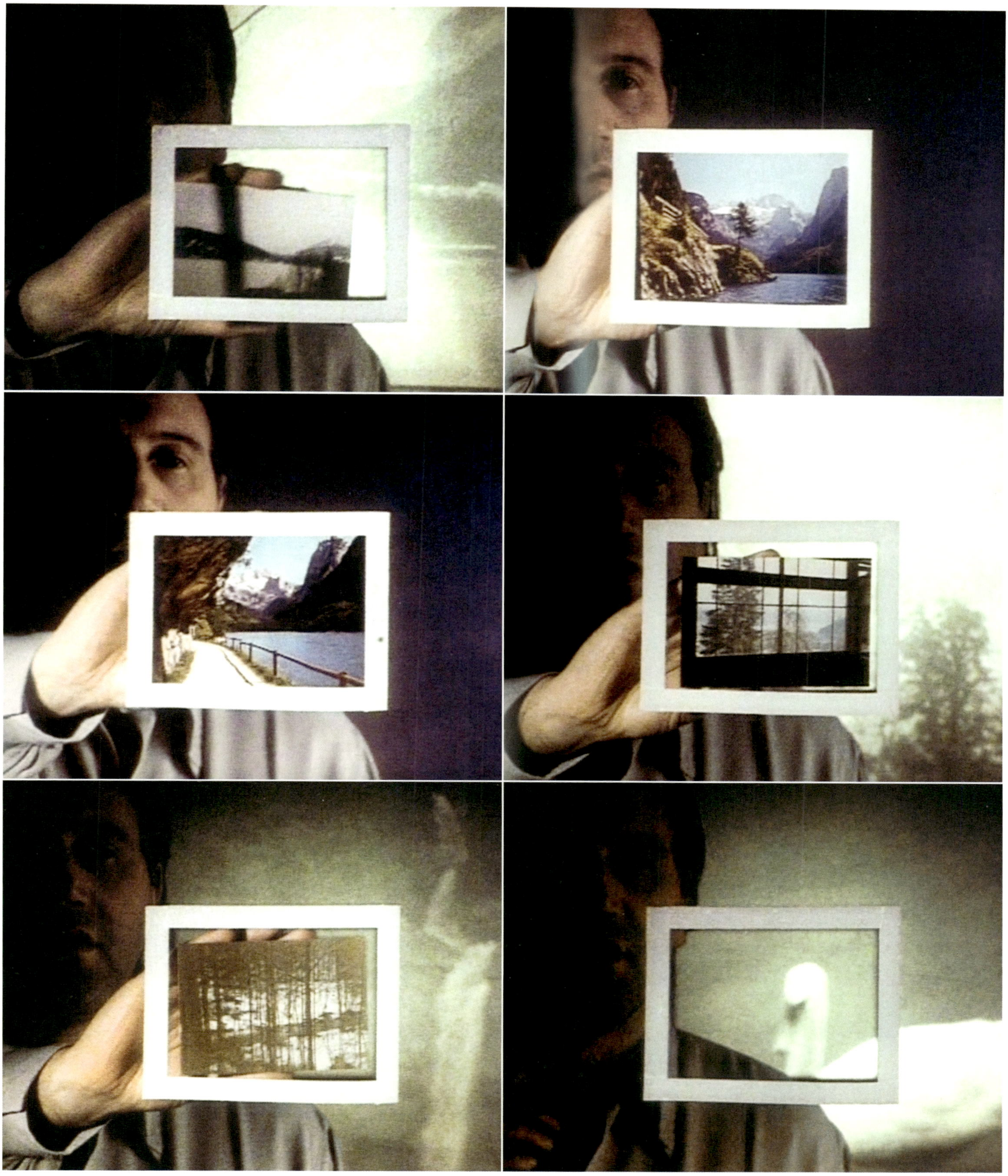

**The Landscape of the 21st Century /
Die Landschaft des 21. Jahrhunderts**

1984 | 16 mm film, color, sound, 40 min. (lost)

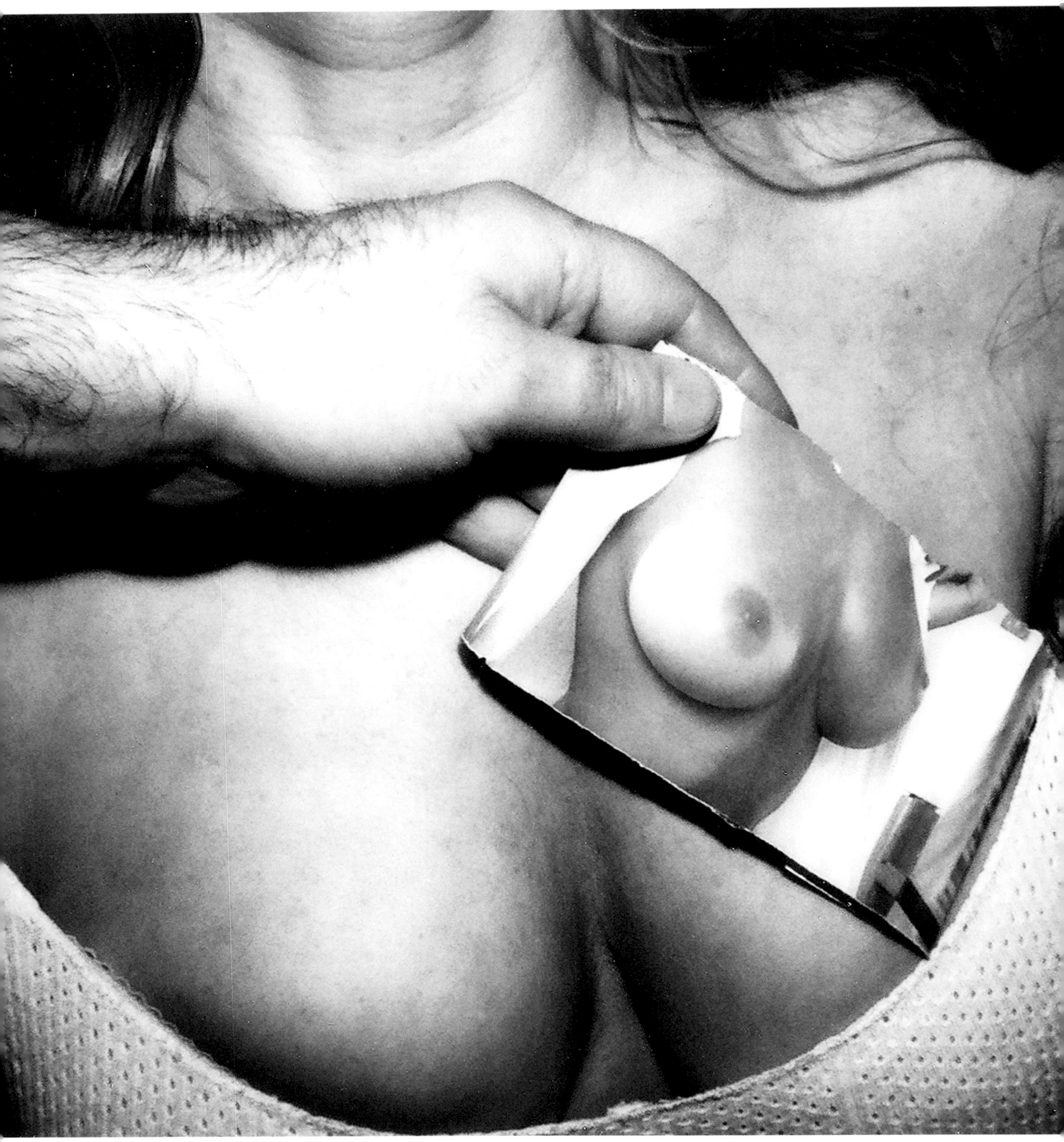

Partisan of Partial Drives / Partisan der Partialtriebe

1975 | Polaroid photograph, 10.8 × 8.9 cm; model: Renée Felden

A real bosom seems to be covered by a photograph of a bosom, but here you see a photograph of a bosom covered with another photograph of a bosom. Each real bosom is only the echo of an image of a bosom. Through the influence of photography and film and their technique of close-ups, the partial objects of the body (eye, lips, chest, etc.) become the primary aims of desire and our media society a paradise of partial drives.

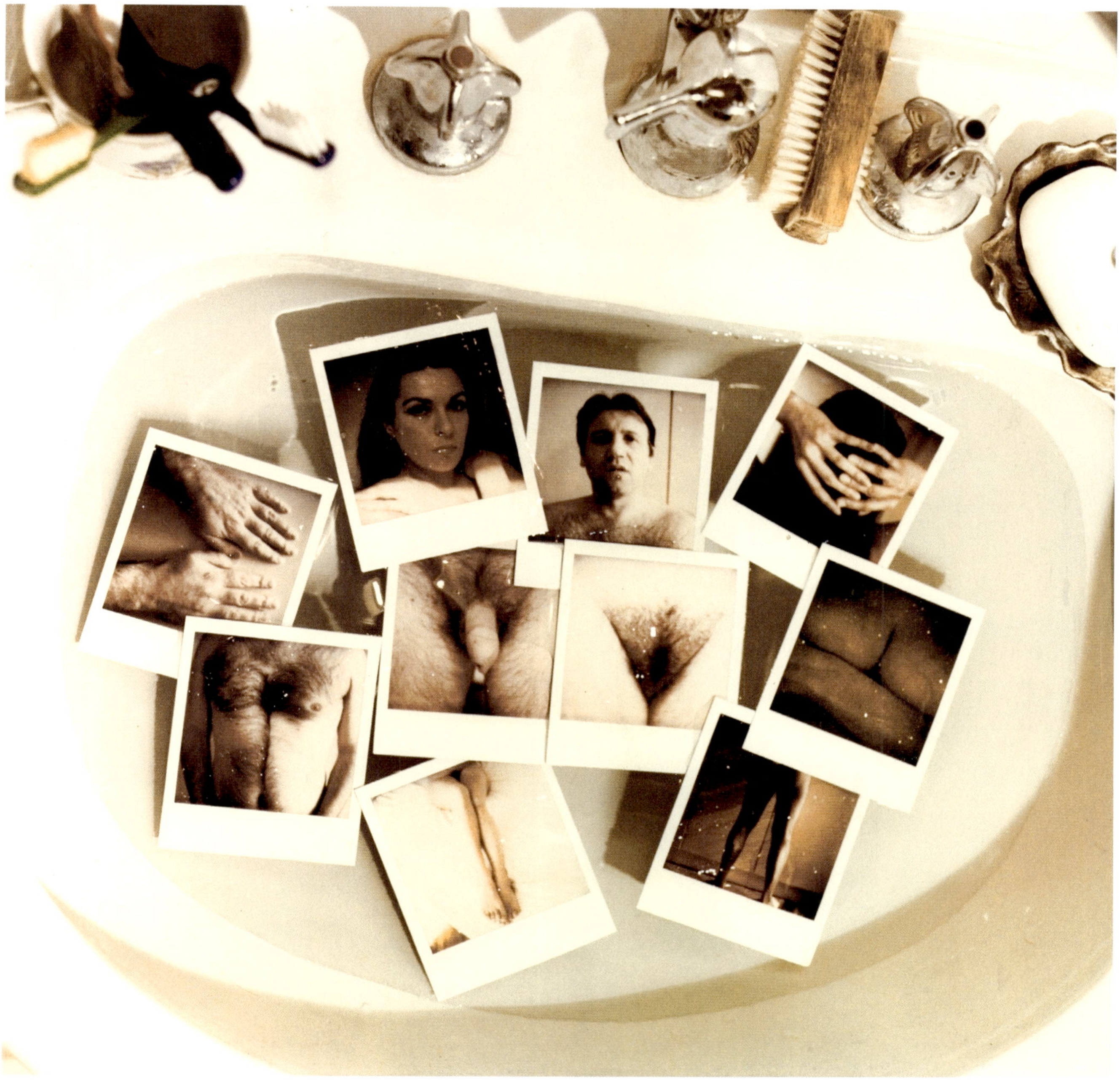

Anagrammar of Anatomy/
Anagrammatik der Anatomie

1982 | Polaroid photographs in water basin; models: Susanne Widl, Peter Weibel

Gender beyond binary code, gender as a choice among free variables.

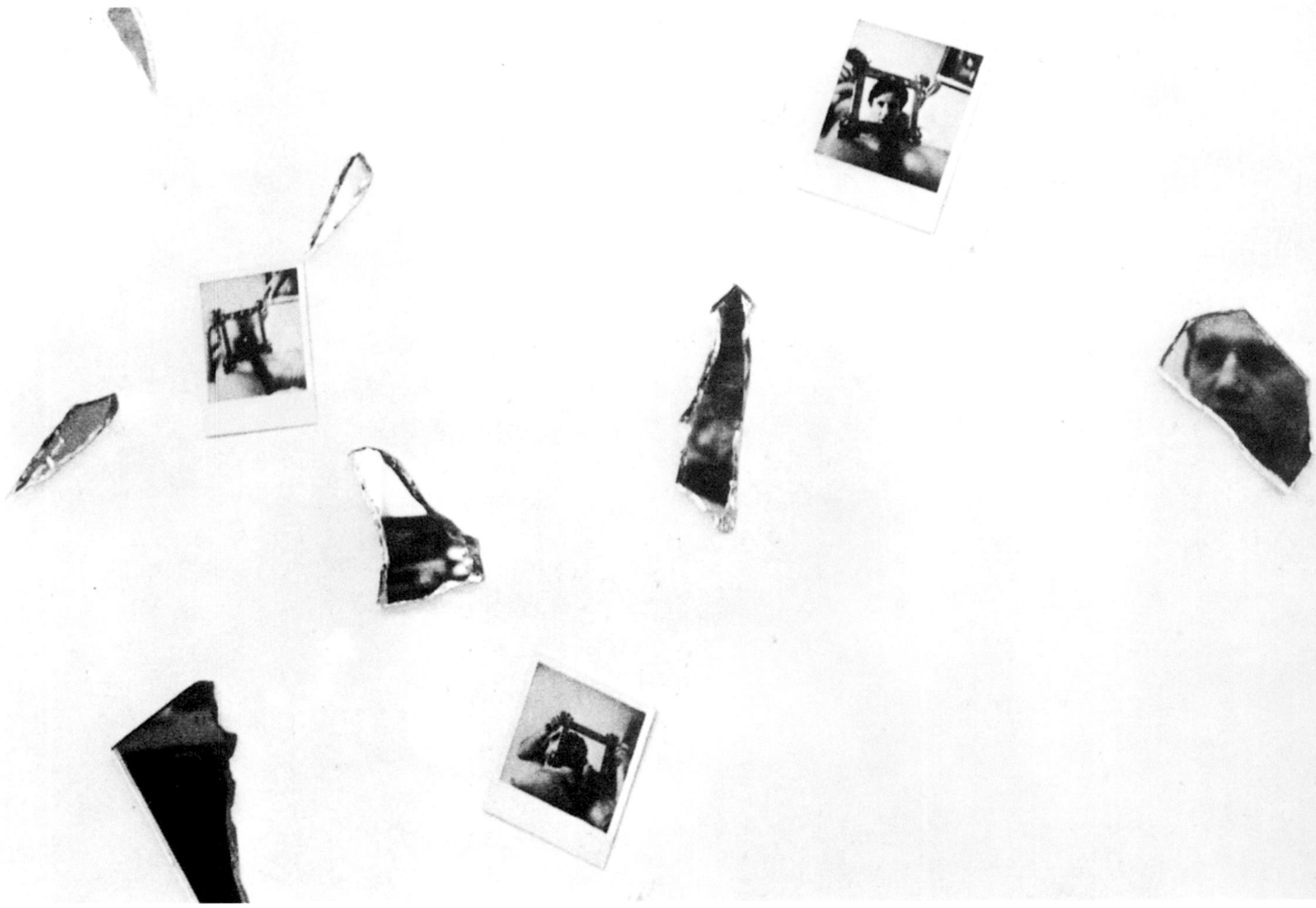

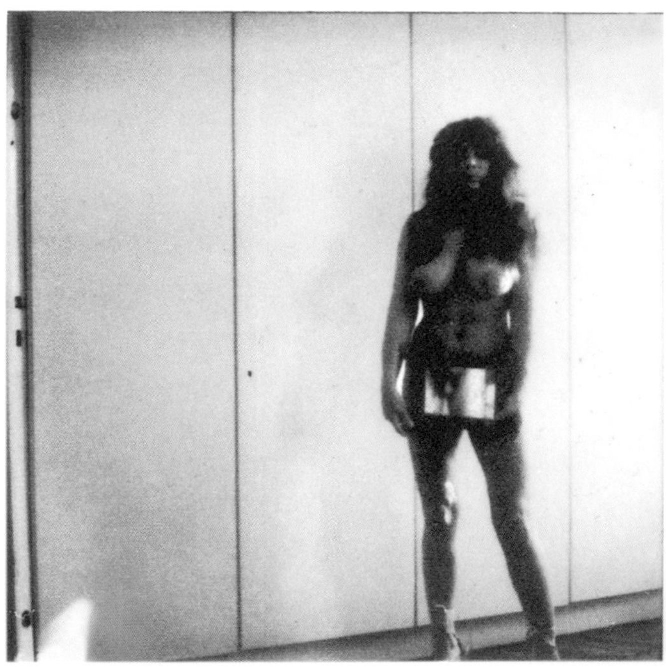

Polaroid photographs and mirror fragments that reflect the viewer are mounted directly on the wall. The Polaroids show a man and a woman, each holding a mirror that reflects the other – or rather, the other's primary and secondary sexual characteristics. The mirrors take the place of those characteristics, either isomorphically (for example, the penis in place of the vagina) or not (the breasts in place of the penis). Given the laws of photographic perspective, it is clear that these photographs could only have been taken by a third person: the camera operator is the mirror. Therefore, these photographs, these reciprocal reflections of sexes and desires, necessarily required a third person (the photographer). The third party as the constitutive element of the couple, of desire. People in love mirror each other – women as men and men as women – and love is the clinic to heal the wounds. —Peter Weibel

The Clinic of Two Mirrors /
Die Klinik der zwei Spiegel

1981 | mirror shards, Polaroid photographs; model: Renée Felden

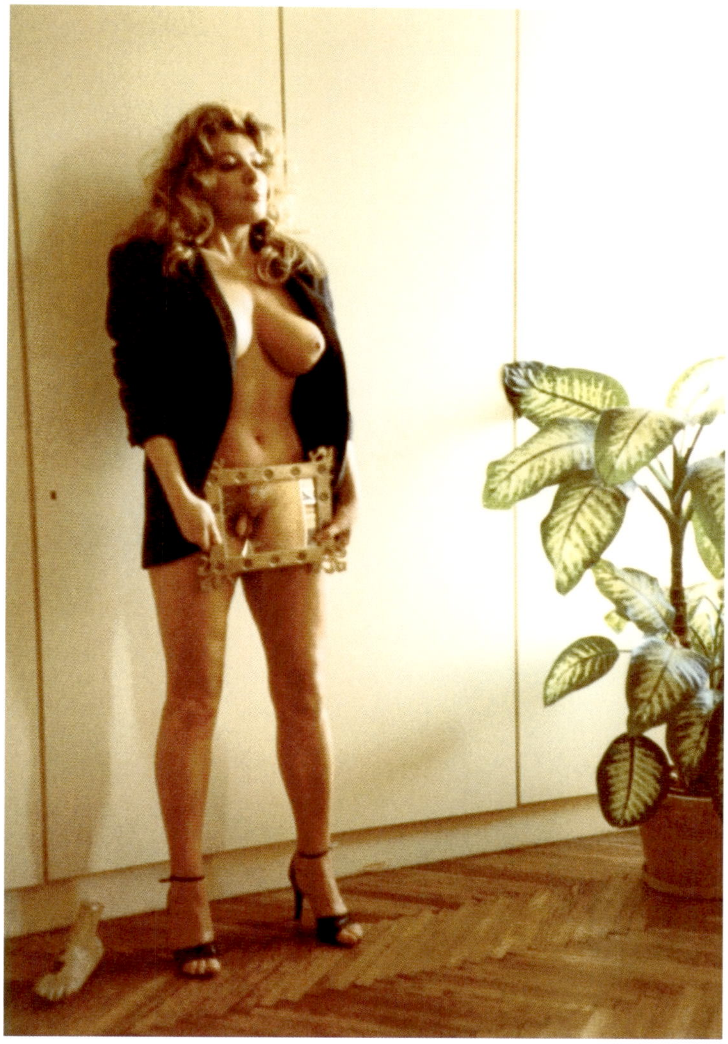

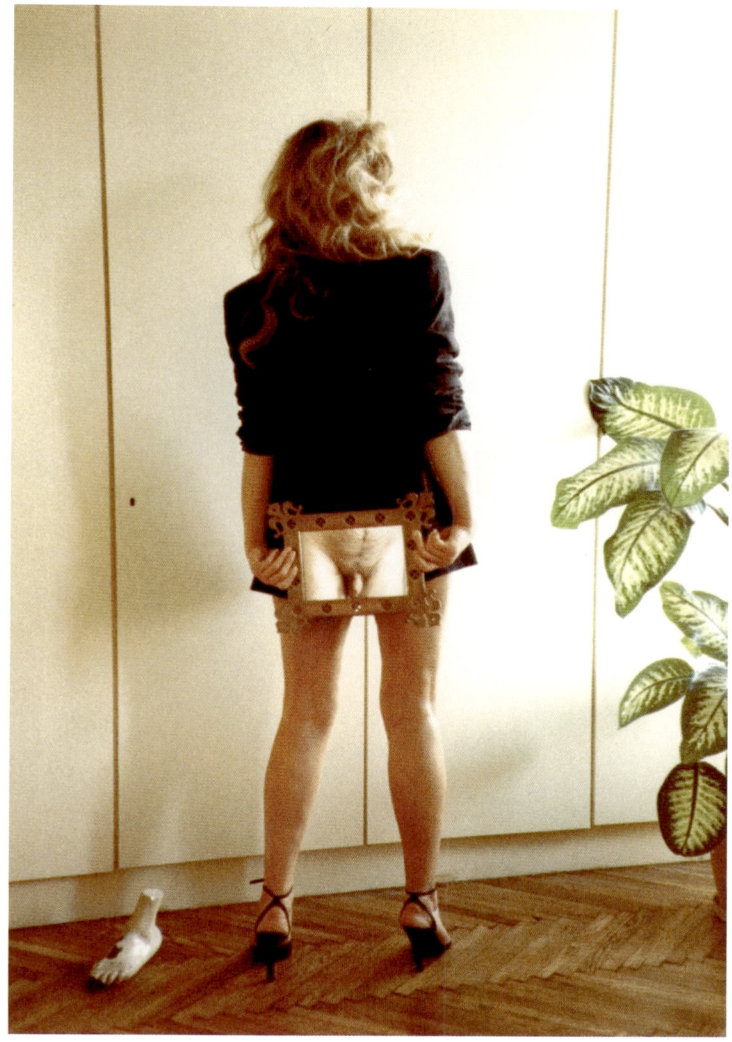

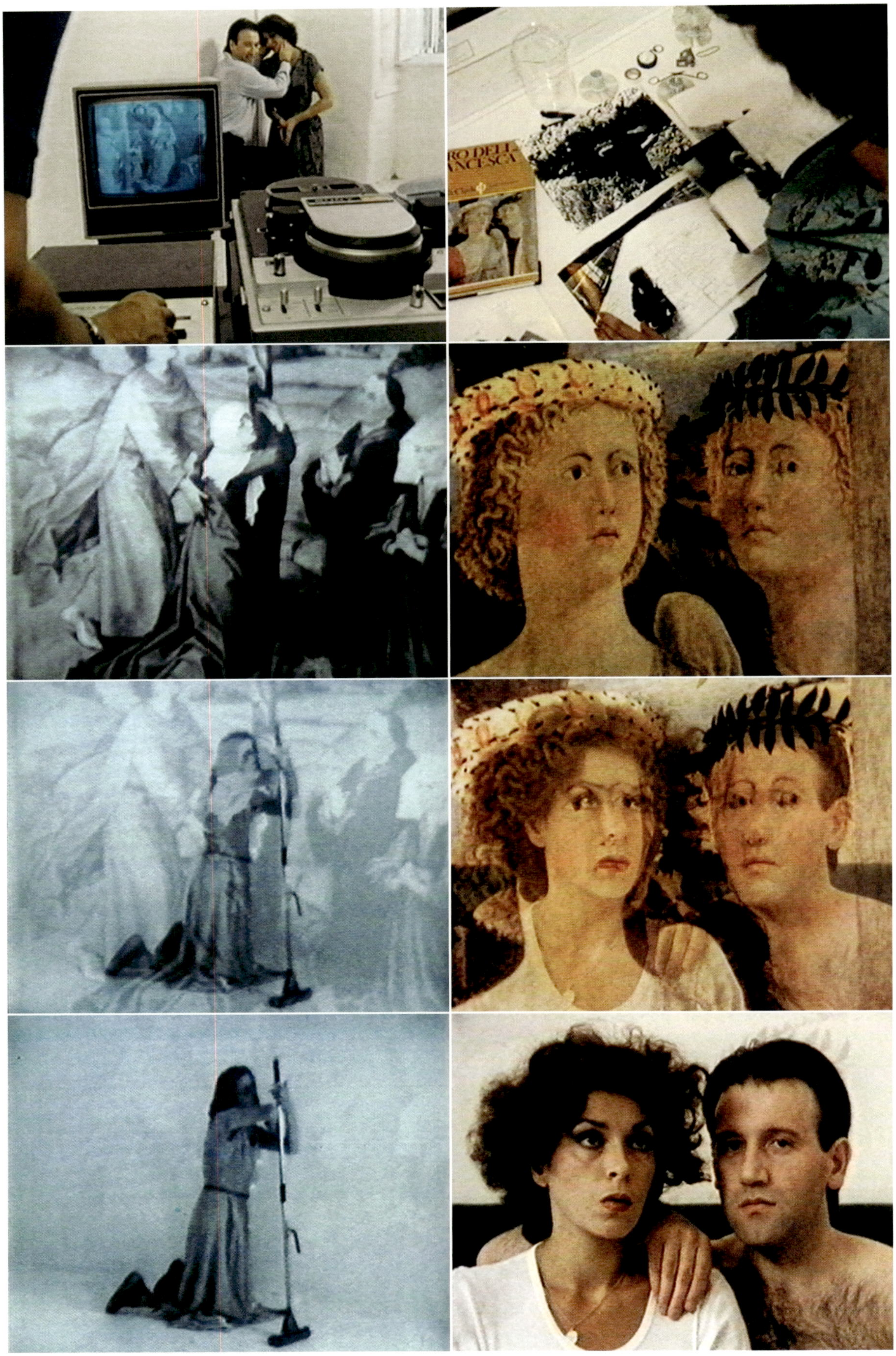

Invisible Adversaries / Unsichtbare Gegner

1970 (idea)/ 1977

16 mm film, color, sound, 107 min.
director: VALIE EXPORT, codirector: Peter Weibel; screenplay: Peter Weibel, collaboration: VALIE EXPORT
See also pp. 202–203 in this volume

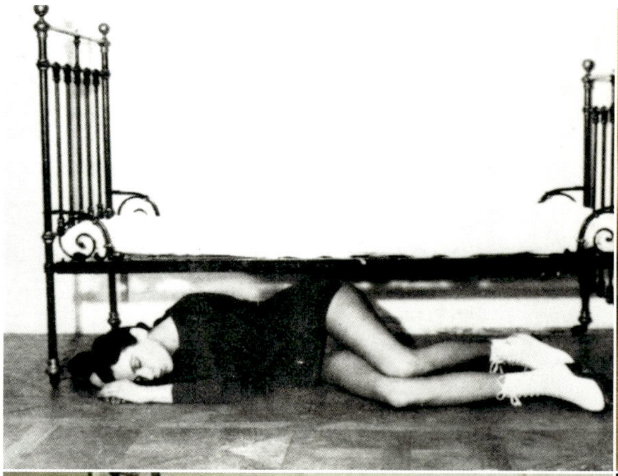

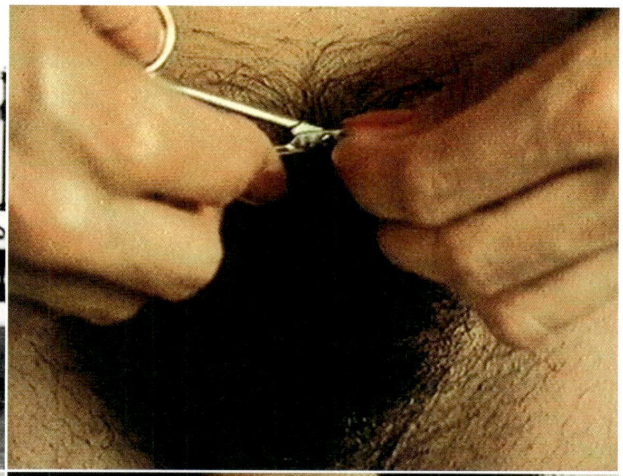

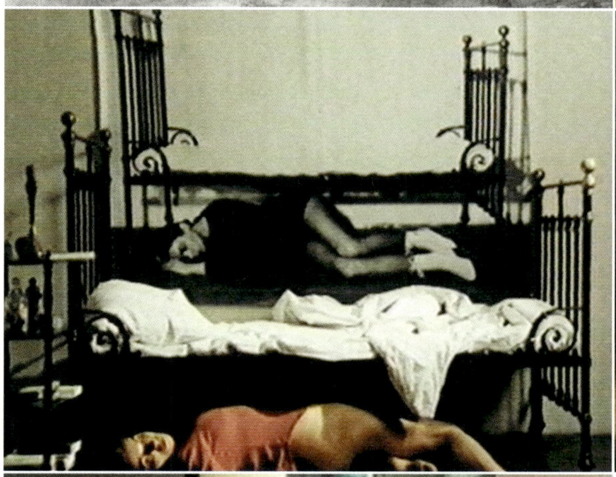

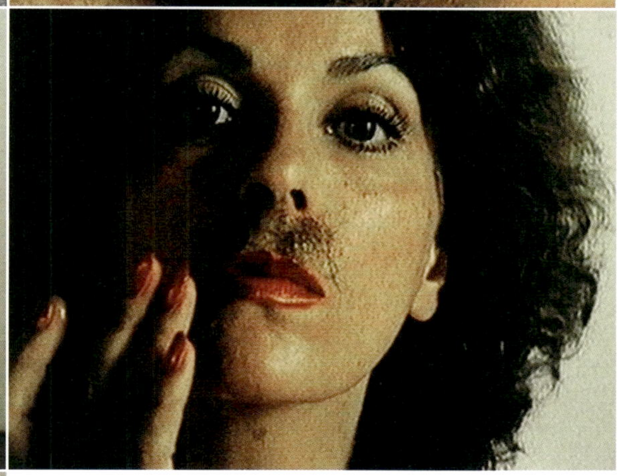

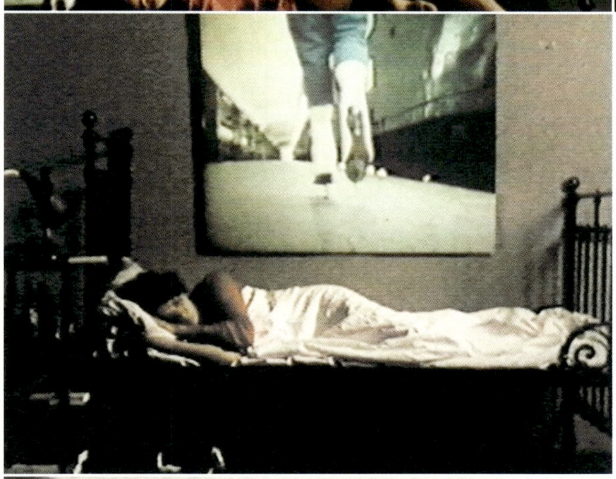

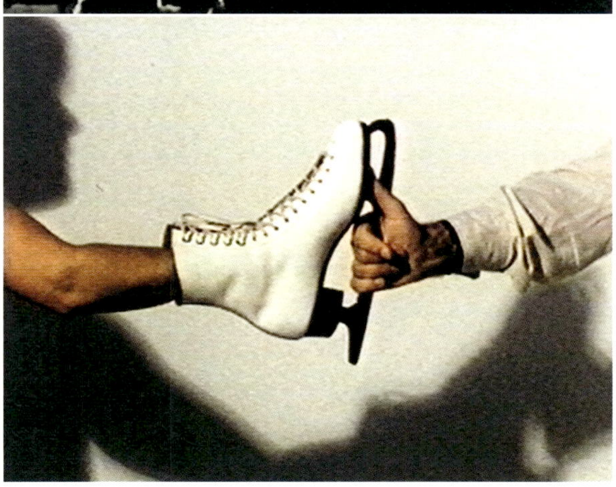

The experimental feature film *Invisible Adversaries* tells the story of the video reporter Anna who wakes up one morning and hears a warning on the radio that invisible foreign powers are about to take possession of the Earth by taking on the form of humans and then changing their consciousness. The aim of these foreign rulers, called Hyksos, is to destroy the Earth. Anna begins to investigate her surroundings and discovers terrible things. It becomes more and more difficult for her to distinguish between reality and fiction. Since her eye is no longer able to distinguish between the endogenous contents (from within) and the exogenous contents (from outside) of perception, Anna, who is aware of her subjective transformation, replaces the eye, this natural projector of reality, with photography and the video camera in order to control her perception and her environment through objective perception machines and to conceive reality. The question as to whether it is Anna's disturbed perception that makes her surroundings appear so horrible or whether they are real remains unanswered. The consciousness of the protagonist – who no longer distinguishes between reality, perceptual reality and media reality – is realized through the narrative methods, the experimental use of film, photography, video, etc. The film is a poetic metaphor for the apocalyptic feeling of modernity.

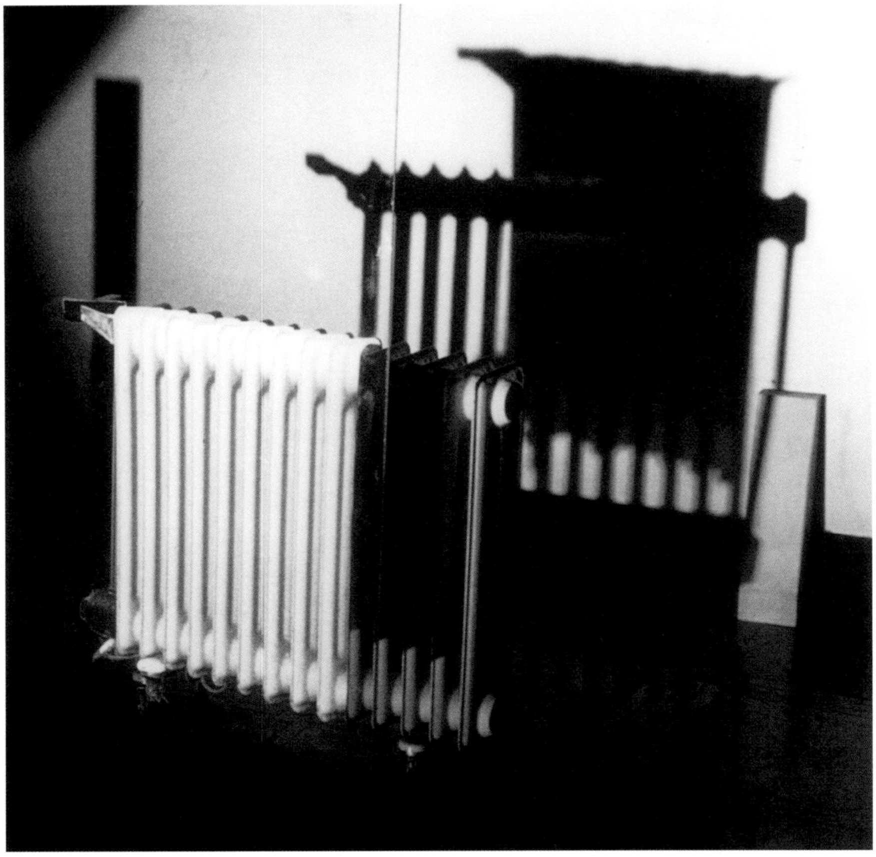

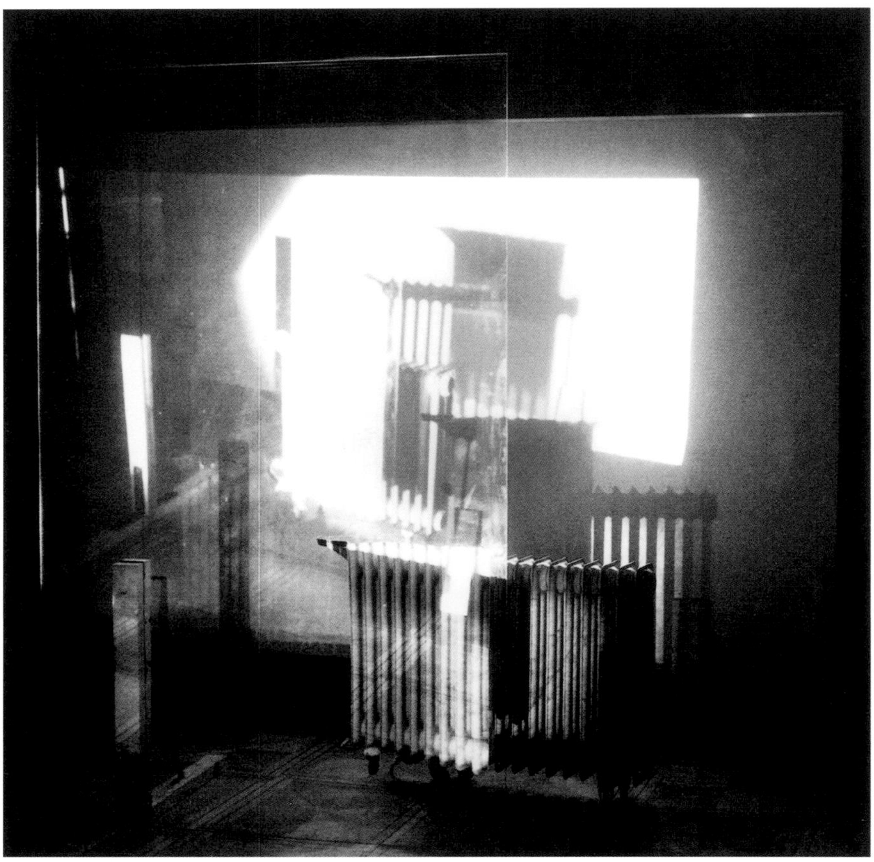

Installation views *Peter Weibel: Medienpoesie*, Kulturhaus Graz, 1982

Object/Resistance Symphony: Four Gradual Displacements / Gegenstand/Widerstand-Symphonie: Vier graduelle Verschiebungen

1982	radiator, photo of radiator pasted on pane of glass, wooden stand, slide projection of the installation, slide projector

Phase 1: Wooden stand with pane of glass / picture of radiator and actual radiator (reality).
Phase 2: A photograph is taken of the glass and the radiator (copy of reality).
Phase 3: A slide of phase 2 is superimposed on the installation.
Phase 4: Reality (glass/radiator) is shown together with the glass/radiator photograph and the slide projection on the installation and wall.

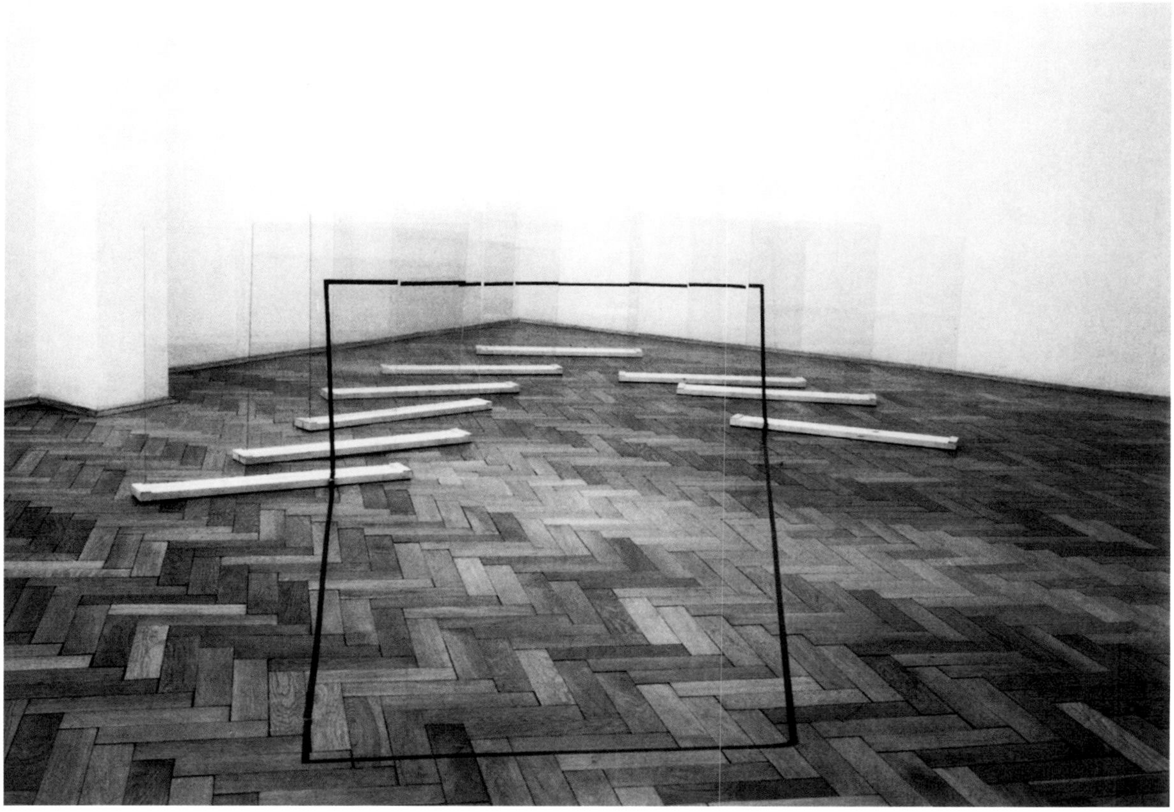

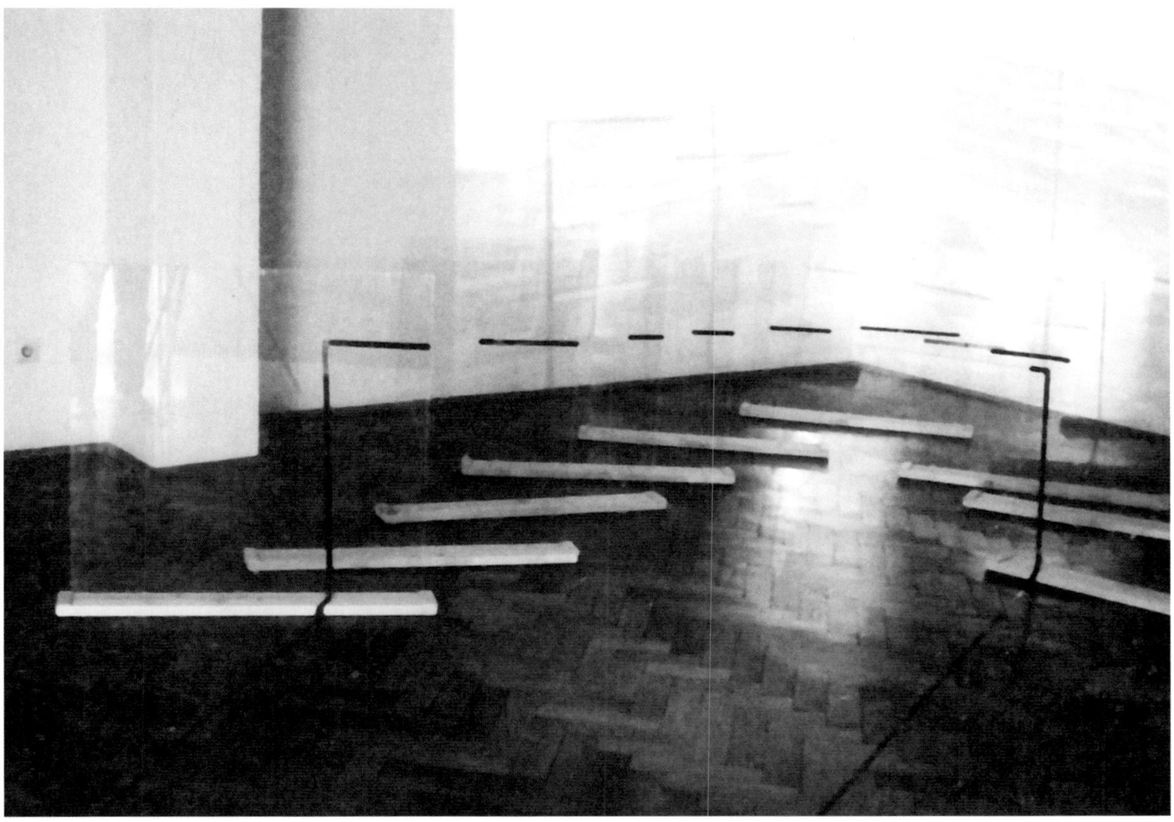

Installation views, *Peter Weibel: Medienpoesie*, Kulturhaus Graz, 1982

Three Degrees of Modality / Drei Grade der Modalität

1982	9 panes of glass, fixed on wooden stands, slide projector, slide of installation, black adhesive tape forming a square

According to classical philosophy, the ways of being fall into three categories: possibility, existence, and necessity. In a corner of a room, nine square panes of glass stand on the floor in two rows that form an acute angle. Black lines are affixed to the floor and the glass (existence). From a certain perspective, the apparently randomly placed lines form a (slightly distorted) square. This is the viewing position from which the installation is photographed (possibility). By superimposing a technological image (a slide) onto the installation, the third degree of modality (necessity) is revealed.

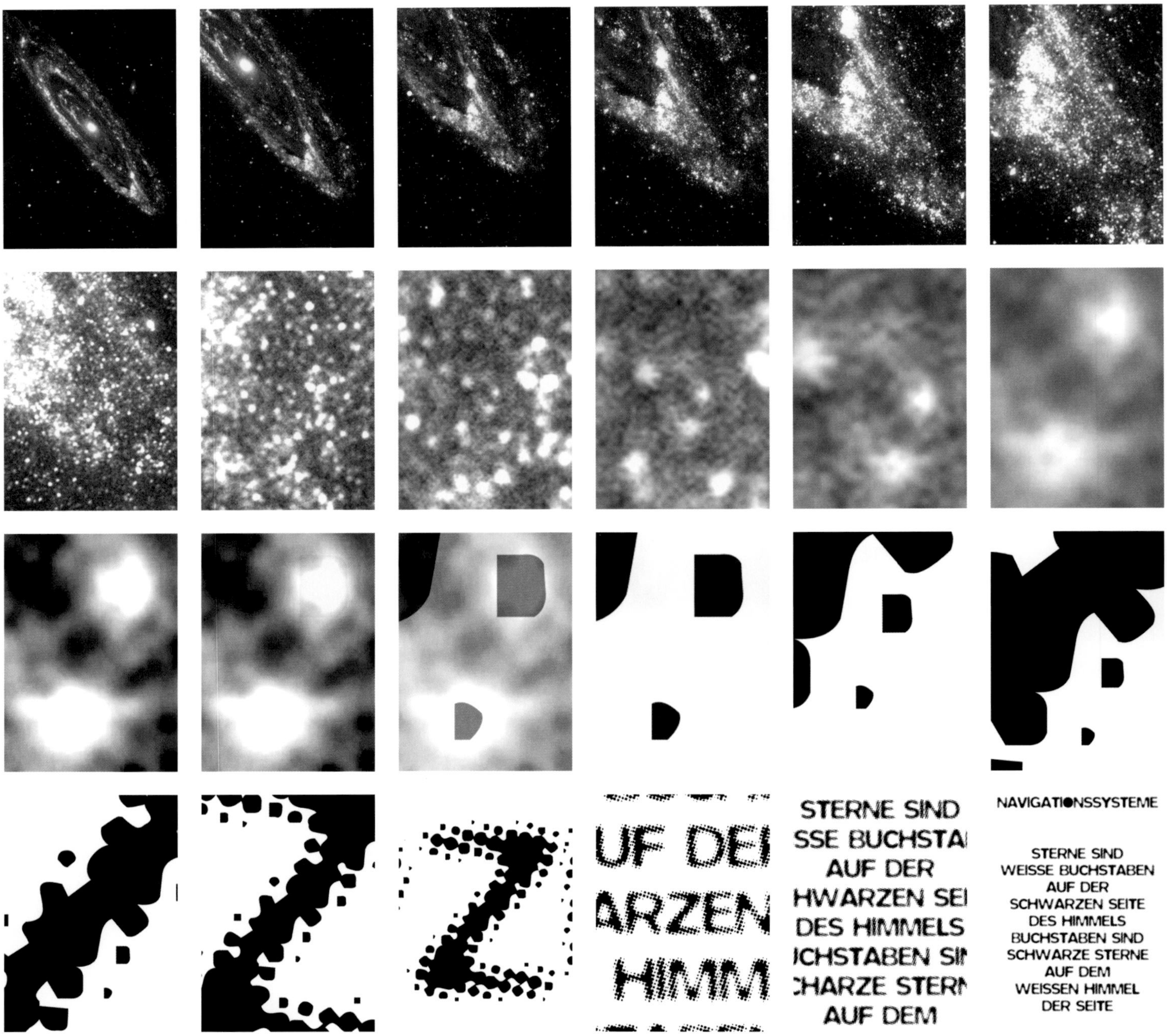

In the 24-part conceptual photo series *Navigation Systems*, the reproduction of the starry sky is steadily enlarged until only abstract forms can be seen. In the reverse direction, the letters of a text are enlarged by reproducing it until only abstract particles are seen. In the middle of these processes the signs overlap. Therefore the impression is created that letters slowly emerge from the stars and, finally, the sentence *Sterne sind weiße Buchstaben auf der schwarzen Seite des Himmels, Buchstaben sind schwarze Sterne auf dem weißen Himmel der Seite* (stars are white letters on the black page of the sky letters are black stars on the white sky of the page) becomes visible. Peter Weibel shows that stars as well as letters are navigation systems. For just as early sea travellers oriented themselves for centuries on the stars, we navigate the world with the aid of signs today as well. Weibel therefore defines writing as the original medium, as our first tool with which we can navigate through space and time beyond the bodily boundaries of the *hic et nunc* and of natural perception. Writing continues the work of the stars. It helps us to orient ourselves in the night of the cosmos.

Navigation Systems / Navigationssysteme

1972	offset print, 24 parts, 100 × 71 cm each

NAVIGATIONSSYSTEME

STERNE SIND
WEISSE BUCHSTABEN
AUF DER
SCHWARZEN SEITE
DES HIMMELS
BUCHSTABEN SIND
SCHWARZE STERNE
AUF DEM
WEISSEN HIMMEL
DER SEITE

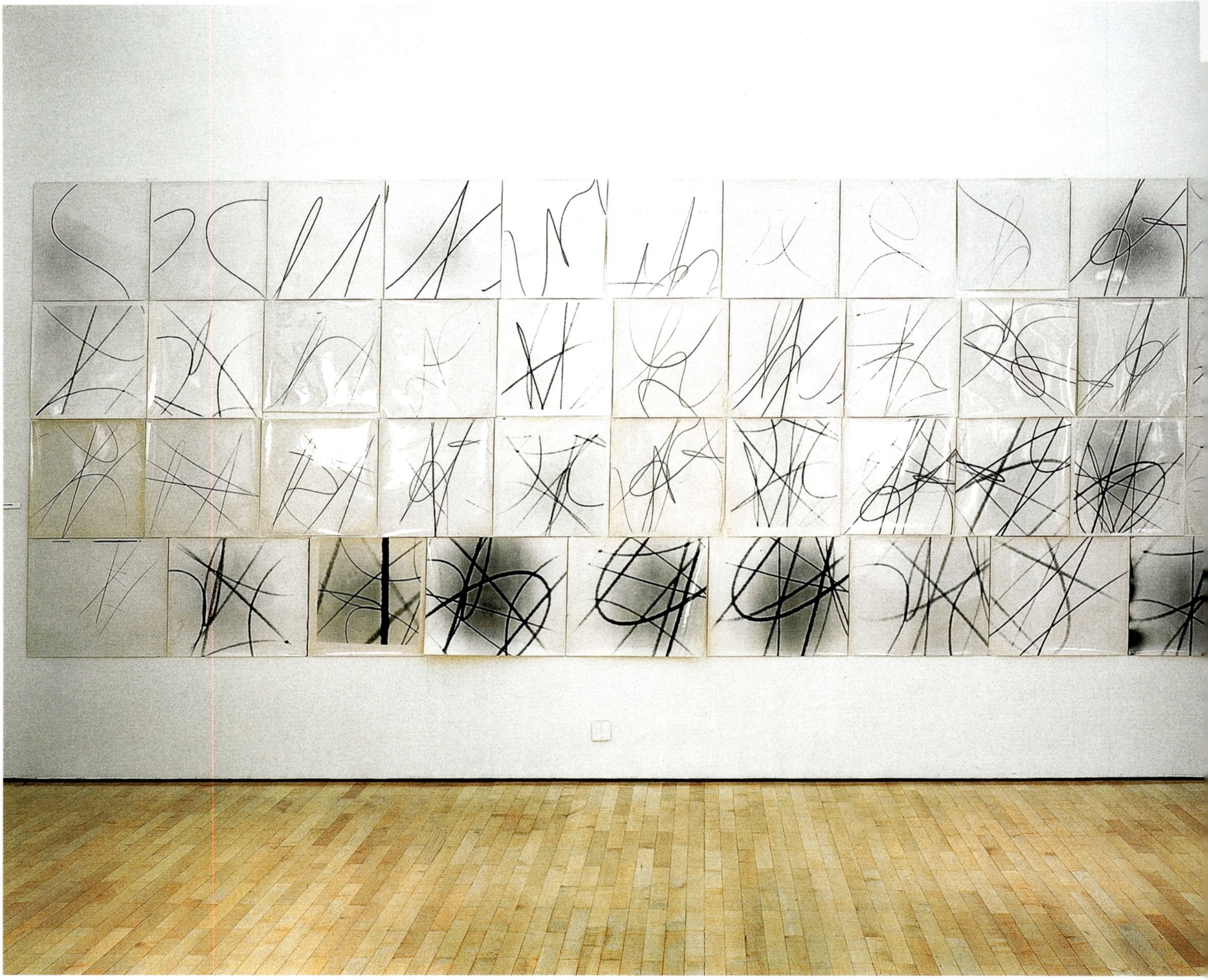

The Infinite One-Word Exhibition / Die unendliche Ein-Wort-Ausstellung

1977

48 gelatin silver prints,
ca. 50 × 50 cm each,
first shown at Galerie nächst
St. Stephan, Vienna, 1977

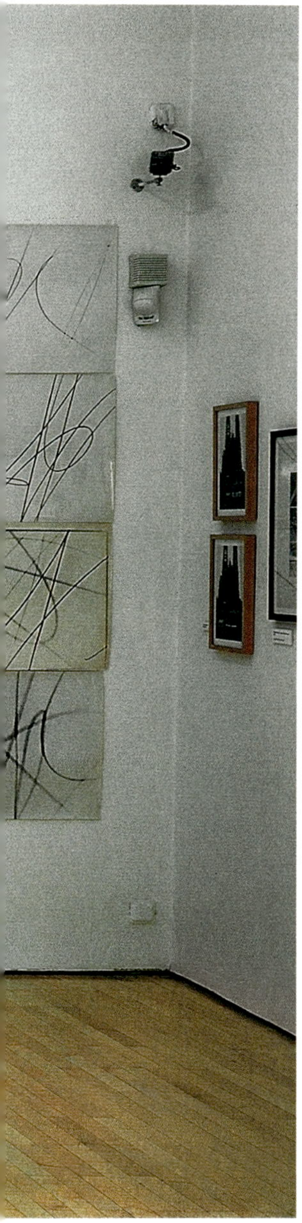

Installation view *Peter Weibel. Das Offene Werk 1964–1979*, Neue Galerie Graz, 2004

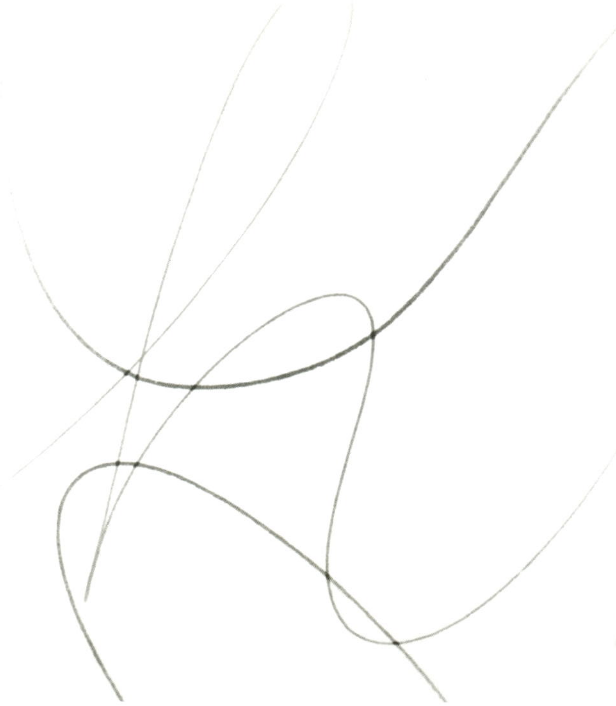

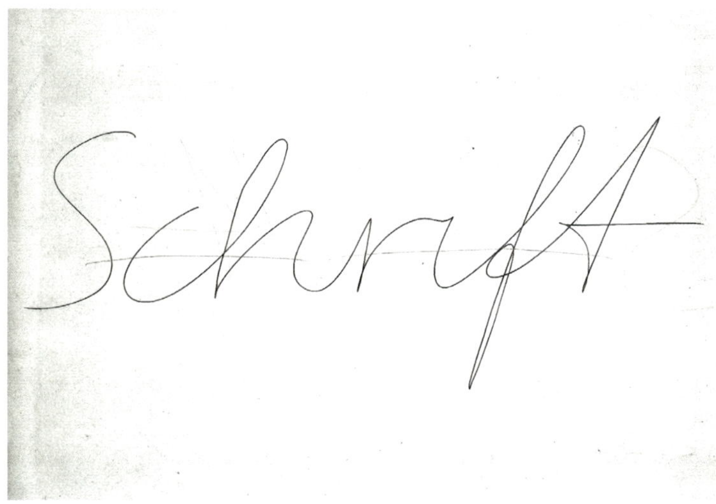

Schrift (writing)

A handwritten word is taken apart into fragmentary segments. These details are photographically magnified. Such magnifications and recombinations of minute parts of a written word create abstract signs which form the elements (alphabet) of an infinitely extendable combinatorics (an infinite language). First, the photographic negatives of the original elements are combined with one another. Once all permutations have been exhausted, new details from these new configurations (scripts) are photographically magnified. The negatives of these are then again combined, with the inclusion of other techniques of variation, such as reduplication, alterations in symmetry, sequencing. New details of this third generation are then photographically magnified, and these new details are then permuted once again with themselves and the previous photos *ad libitum*: a (virtually) infinite exhibition is merely an excerpt.

The original, written word itself does not occur in the exhibition space, merely its depictions, branchings in the infinite space of language. *Mathesis universalis*: this abstract (self-referential) language repeats the infinite linguistic work of concrete (object- and term-referential) language and acts as a paraphrase of abstract expressionist painting. The "expressiveness" was mechanically produced, the "spontaneity" was calculated. A relationship between writing and gesture, literature and painting in the era of the mechanical reproducibility of expression. —Peter Weibel

Translated from the German by Peter Dahm Robertson.

Having already begun as a teenager to read books on mathematics, from George Boole to Bertrand Russell – I was particularly impressed by Eric Temple Bell's *Men of Mathematics* (1937) – I was determined to find logical structures for all of the world's phenomena. And so, from the very beginning I regarded visual art not as the production of images, but as an epistemic procedure. Photography as a purely pictorial endeavor – as mimesis, as representation of the world – was and still is an ethical horror to me. I therefore made a choice between "straight" photography, which merely depicts, and art photography, which scrutinizes the mechanisms of depiction.
My photo booth photos, such as *Self-Portrait as a Woman* and *Self-Portrait as Anonymous* (*Selbstportrait als Frau* and *Selbstportrait als Anonymus*, both 1967, see p. 113 in this volume), are not pictures of me, of my self. They are not representations of the real person, but rather metamorphoses of my self, created with the help of expanded montage techniques. The photographs erase even my identity; they make my self, my person, unrecognizable – for example, by means of a black square, based on the black bars used in newspapers to anonymize individuals whose identities are supposed to remain unknown. Since there is no necessity to the relationship between representation (image) and reality, I also staged an *Evolution of the Poet* (1967, see p. 112 in this volume). In my view, references, especially identity relationships, are contingent, arbitrary, variable, modal, intuitionist, and constructive.[1]
Photography, then, is simply a *possible* interpretation of the world, its truth function merely conditional. Is photography a propositional calculus, or does it belong to the realm of modal logic? We describe as *necessary* propositions that are always true, that are true in all possible worlds – for example, "circles are round." A proposition that is true in at least one (possible) world is *contingent*. In first-order logic, one might formalize these ideas as follows:

The symbol $\exists$ represents the existential quantifier, "there exists." $\exists(x)r$ means that there exists some x possessing the quality r – for example, there exists an object that is red. $\forall$ represents the universal quantifier, "for all." $\forall(x)r$ means that every x possesses the quality r – all objects are red, which is untrue.

During my studies of mathematical logic in the 1960s, I became acquainted with the work of the modal logician and philosopher of language Saul Aaron Kripke and his system $\mathbf{K}$. This system results when we posit the axiom schema $\mathbf{K}$ and allow also Kurt Gödel's necessitation rule as a rule of inference.

Axiom schema $\mathbf{K}$: $\Box(p \to q) \to (\Box p \to \Box q)$
If p necessarily implies q, then if p is necessary, q must also be necessary.
Necessitation rule: $\vdash p \to \vdash \Box p$
If p is a theorem derivable from the rules of $\mathbf{K}$, then p must also be a necessary theorem in $\mathbf{K}$.

Semioticians distinguish between three types of signs: icons, which visually resemble the things they signify; indexes, which have a physical relationship to what they signify; and symbols, whose relationship to their signifieds is arbitrary and coded by convention.

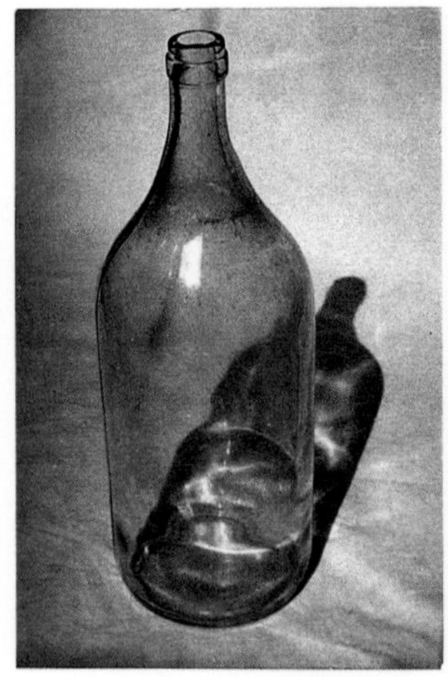

FOTOS ALS NAMEN
P.W. 1976

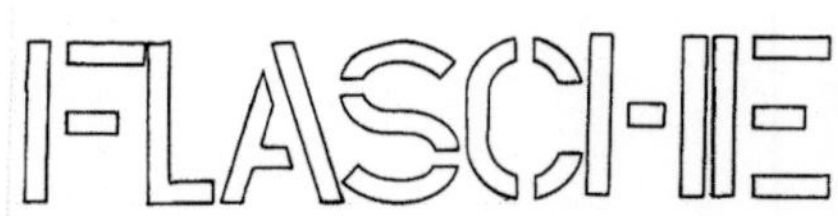

FOTOS ALS NAMEN
P.W. 1976

Photographs as Names: Bottle / Fotos als Namen: Flasche

1976 | photograph, collage on paper, 30 × 42.3 cm each

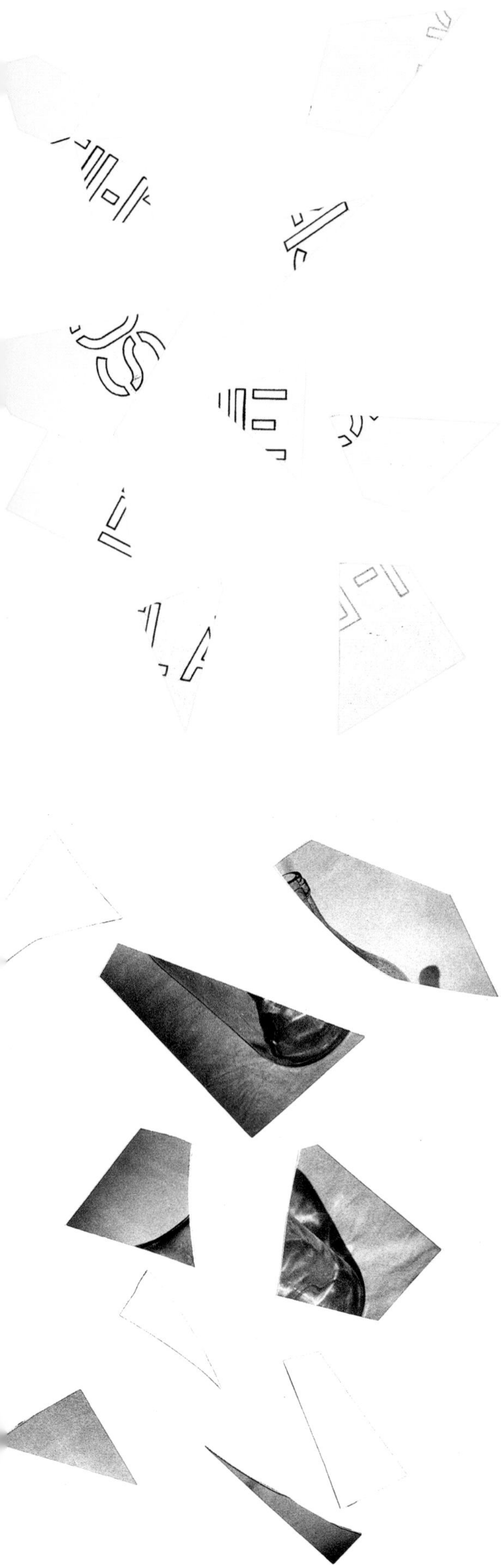

Clearly, photographs are iconic. A photograph of a mountain bears a formal resemblance to a mountain; a portrait of a face resembles the real face. This similarity function is true of almost all images produced by painting, photography, and film prior to twentieth-century abstraction. An analogy function, however, is not a truth function. That is why I have always been skeptical of photography, and of all visual representation, as regards their claims to truth. I have gone so far as to say that a portrait tells us nothing at all about its subject, or at least nothing of importance. That is also why I published my theory of the "photo fake."[2] A picture is not a proof.

My investigations of modal logic in the late 1960s and my study of the field's authors, from C. I. Lewis to Oskar Becker, substantiated my critique of photography and its ontological aspirations. I began subjecting photography to linguistic analysis – treating it as a language. Just as one might ask whether mathematics is a language, and if so, what kind of language,[3] I asked myself: Is photography a language, and if so, what kind? If photography is a language, a visual language, then what do individual photographs mean? Are they names? Definite descriptions? What do names (of cities, people, events, concepts, etc.) mean in a language?

What do proper nouns mean? Generally speaking, names are definite descriptions: they denote a person or object. In any case, a name, like a photograph, constitutes a reference to the thing named, the thing depicted, the object or person who goes by that name or whose picture it is. If a definite description, a name, establishes the meaning of a concept or object, then that description must be true in all possible worlds in order to be deemed necessary. Is this the case with photography?

In *The Logical Syntax of Language* (*Logische Syntax der Sprache*, 1934), Rudolf Carnap proposed a metalinguistic concept of modalities. Modalities are only meant to be used metalinguistically – that is, when speaking about language. In the context of referential labels, I therefore introduced distinctions on the object-language level as well as the metalanguage level. In *Photographs as Names*, I affirmed that what lies between Vienna and Salzburg is, in object-language terms, Linz, whereas metalinguistically speaking it is the word "and." Guided by considerations of this sort, I transferred the definite descriptions and qualities of objects onto words and images, and vice versa.

I had some success with such linguistic digressions in court in the 1960s: when I was brought up on pornography charges, for example, I explained that while the word "Pegasus," the name, exists, we all know that Pegasus does not really exist – the "thing" does not exist, in other words; it only exists metalinguistically. Likewise, the word "pornography" exists, but what does it refer to? Surely, I cannot be condemned because of a name whose object-language existence is uncertain, and so on and so forth. As a result, I was not convicted.
—Peter Weibel, 2022

Translated from the German by Patrick Hubenthal.

[1]
See Kurt Gödel, "Eine Interpretation des intuitionistischen Aussagenkalküls," in *Ergebnisse eines mathematischen Kolloquiums*, ed. Karl Menger, vol. 4, *Bericht über das Kolloquium 1931/32. Gesammelte Mitteilungen der Jahre 1931/32* (Leipzig: Teubner, 1933), 39–40. Reprinted with English translation by John W. Dawson, Jr., "An Interpretation of the Intuitionistic Propositional Calculus," in Kurt Gödel, *Collected Works*, vol. 1, *Publications 1929–1936*, ed. Solomon Feferman et al. (New York: Oxford University Press, 1986), 300–303.

[2]
Peter Weibel, "Foto-Fake," *Camera Austria*, no. 4 (Dec. 1980): 92–102. Reprinted as "Foto-Fake: Die unnachahmliche Treue der Nachahmung?," in *Enzyklopädie der Medien*, vol. 3, *Kunst und Medien: Die Transformation des Bildes in der Ära der Apparate* (Berlin: Hatje Cantz, 2019), 101–16. See also pp. 147–53 in this volume.

[3]
See Kurt Gödel, "Is Mathematics Syntax of Language?," in *Collected Works*, vol. 3, *Unpublished Essays and Lectures*, ed. Solomon Feferman et al. (New York: Oxford University Press, 1995), 334–62. This essay, which Gödel worked on from 1953 to 1959, was not published until after his death.

PHOTOGRAPHS AS NAMES

linz lies between vienna and salzburg	YES
and lies between “vienna” and “salzburg”	NO
“and” lies between “vienna” and “salzburg”	YES
“and” lies “between” “vienna” and “salzburg”	NO
is between “between” and “vienna”	YES
“and” “lies” “between” “vienna” and “salzburg”	?

scissors cut paper	YES
“scissors” cut paper	NO
the cigarette is burning	YES
the “cigarette” is burning	YES

identity of

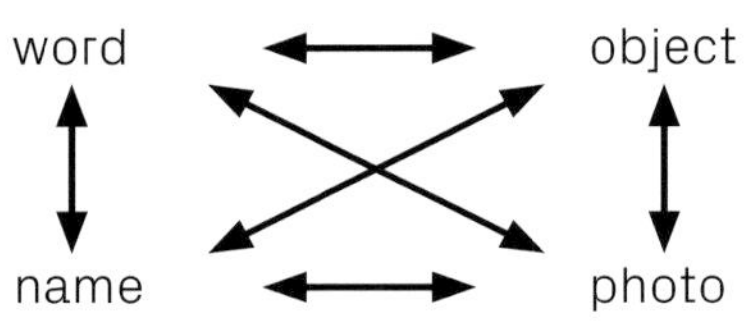

Note: Peter Weibel, "Fotos als Namen. Bildsemantische Untersuchung," 1976, typescript with handwritten additions. Translated from the German by Peter Dahm Robertson.

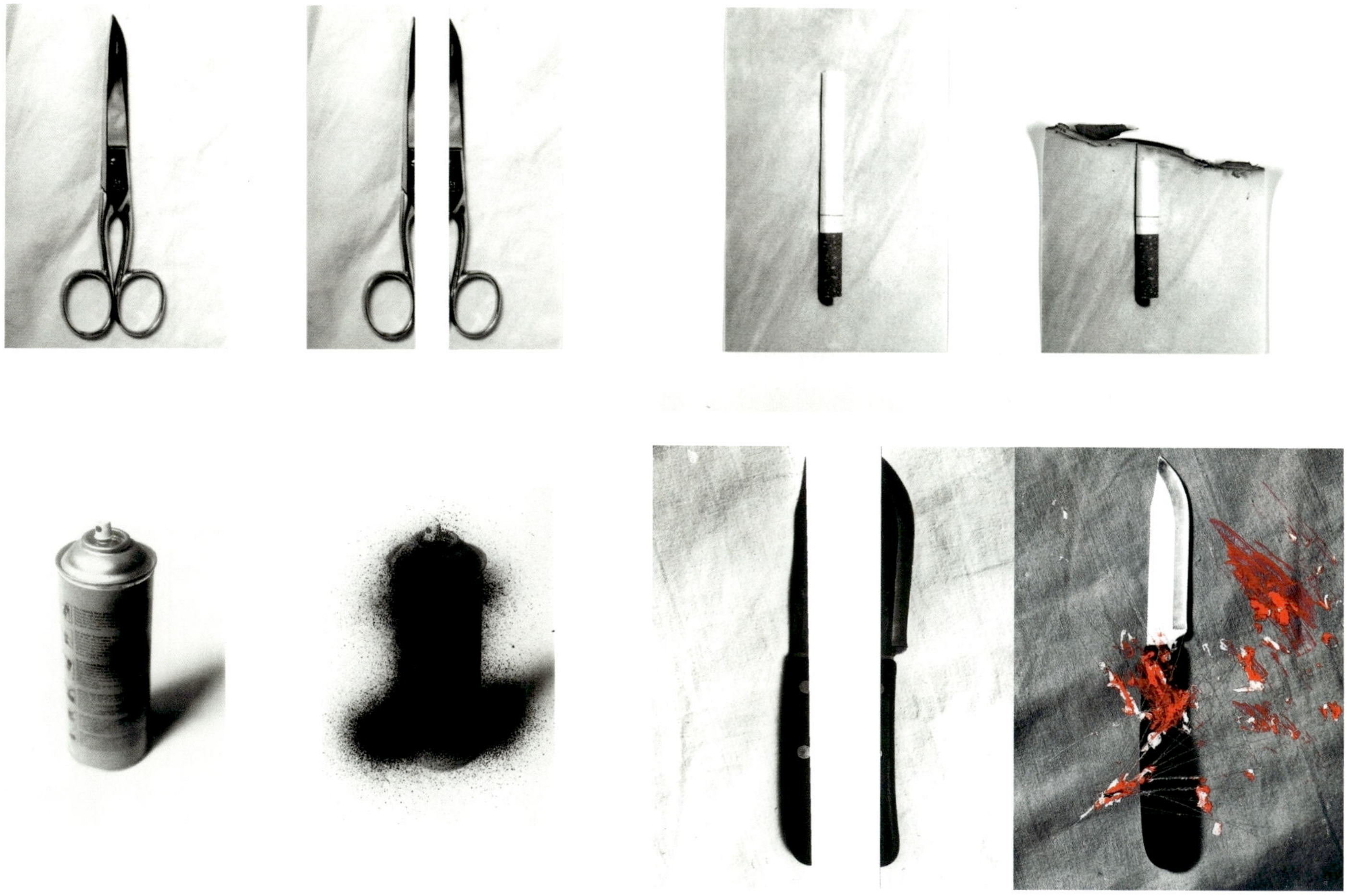

Identity of Action / Identität der Handlung

1975 | photographic series; 21.7 × 29 cm each

A photograph of a pair of scissors is cut. How is this possible? Because the object scissors can cut, but "scissors" as a name, a photograph of scissors, cannot. Cigarettes can burn, "cigarettes" cannot. A spray can can spray, a "spray can" cannot. But the photograph of a cigarette is burnt by a cigarette, a spray can is sprayed with a spray can. A knife makes bleeding wounds, a "knife" does not, but here it does. It is evident that between these pairs of pictures, a third element or action must have taken place, one that normally is neglected. The photo is a result of the feedback between the object and the image, produced by the artist.

Installation view
respectively, Peter Weibel,
ZKM | Karlsruhe, 2019–20

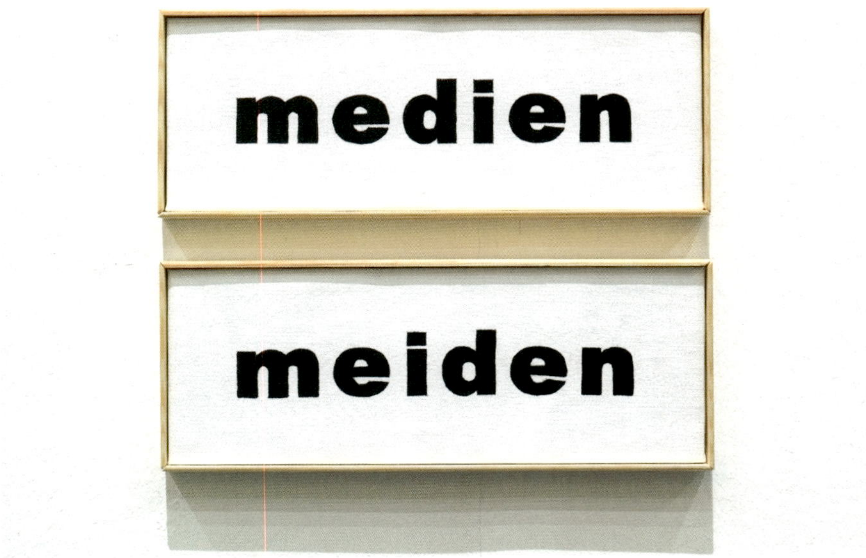

The letters of the German word *medien* (media) can be turned around to become the anagram *meiden* (avoid). The word *medien* in itself already contains the message, the warning: avoid media. The same principle applies to the term "fake news" and its anagram "new fakes."

medien meiden

1986 | acrylic on canvas, 2 parts, 16 × 41 cm each

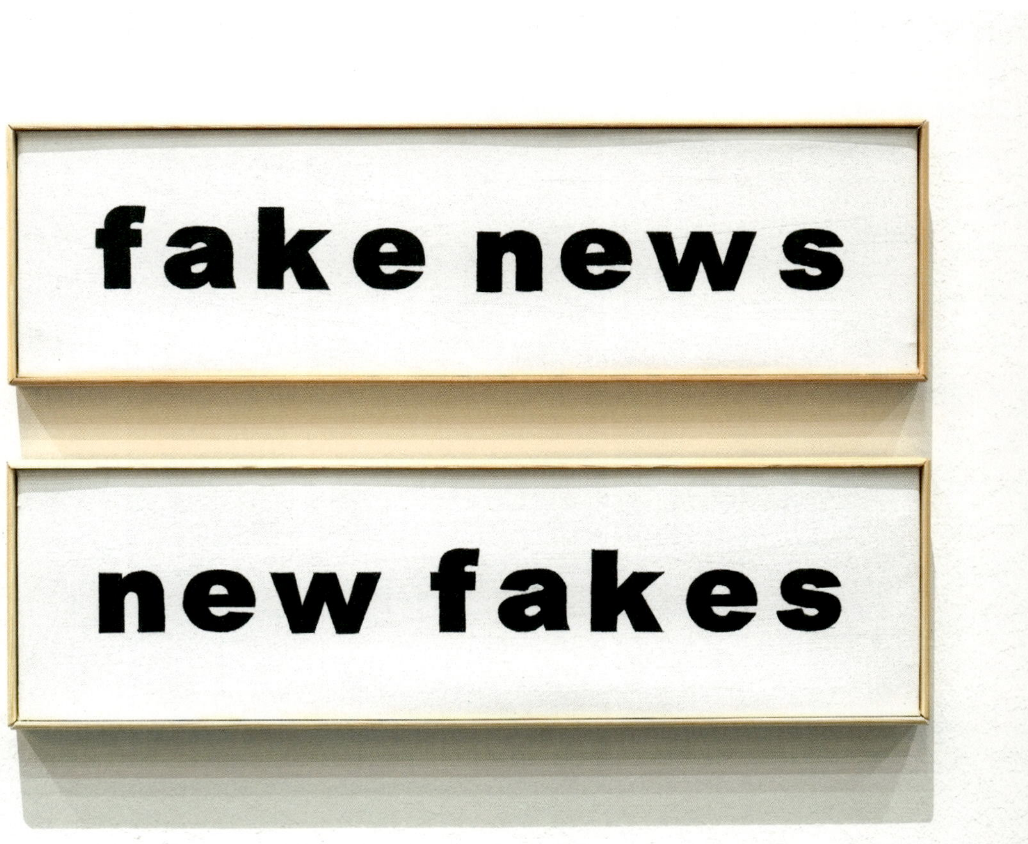

Installation view
respectively, Peter Weibel,
ZKM | Karlsruhe, 2019–20

fake news new fakes

1986 | acrylic on canvas, 2 parts, 16 × 54 cm each

Photo-Fake

Peter Weibel

There are a great many texts and statements, especially from the early days of photography, that testify to photography's "inimitable truth" to reality. In an appendix ("Notice sur l'héliographie") to the contract of December 14, 1829, between Nicéphore Niépce and Louis Daguerre, the former already described his invention as "the automatic reproduction ... of the images obtained in the camera obscura."[1] In his recommendation to the Chamber of Deputies of July 3, 1839, the physicist Arago, a deputy himself, praised Daguerre's invention for its "fidelity to detail" and its "true reproduction of the local atmosphere."[2] In his report to the Chamber of Peers of July 30, 1839, the renowned chemist Gay-Lussac also stressed the fact that things "are reproduced with a perfection unattainable by the ordinary methods of drawing and painting, equal to nature itself, because in fact, Daguerre's pictures are nothing other than the veritable images. The perspective of the landscape and of each object is delineated with mathematical exactness, each incident, each detail, even if imperceptible, cannot escape the eye and the brush of this new painter ... with a perfection unattainable by any other means."[3] And in his report on Hippolyte Bayard's photographic process submitted by the painter Raoul Rochette to the Academy of Arts in Paris in 1839, the author stated that in photography nature spontaneously reproduces itself in a manner "obviously characterized by complete fidelity of representation."[4] In a letter dated February 9, 1839, Alexander von Humboldt writes: "Objects painting themselves in unequalled fidelity, light forced by the art of chemistry to leave lasting traces within a few minutes, delineating even the subtlest contours in all sharpness ..., all this, of course, inescapably appeals to the mind and to imagination."[5] The Austrian, Auer von Welsbach, even entitled his galvano-plastic printing process "Die Entdeckung des Naturselbstdruckes" (The discovery of nature's process of printing itself), published in Vienna in 1853.

At that time the mechanical reproduction of an image was equated with a true representation of nature. The more mechanical the process of representation the less subjective and artistic and therefore, the more true to reality it appeared. Around the middle of the nineteenth century, the notion of a pictorial medium capable of exact reproduction and inimitable fidelity to reality had generally come to be accepted. Since the photographic representation of objects seemed to take place automatically and mechanically, i.e., without human intervention, it appeared to be incorruptible, undistorted, true, and objective. Photography was the "ideal copy,"[6] a document of reality. It is on this notion of the objectivity of photography that the status of the photograph as evidence, as a document, is based. At the same time, this longing for photographic truth is the source of the love for the large format. Thus in the second half of the nineteenth century the notion of the photograph as evidence gave rise to large plate photography, for the homology of object and representation: at least in terms of size, a representation to the scale of 1:1, is still the best guarantee for the claim to fidelity of reproduction. An official announcement of 1875 concerning photographs of the Rocky Mountains taken by William Henry Jackson reads: "These are the largest plates ever used in field photography in this country. They convey an impression of the real grandeur and the magnitude of mountain scenery that the smaller views cannot possibly impart."[7] Eadweard Muybridge, too, wrote that the scale of 1:1 is the only one "that proves worthy of the subject."[8] The photographer Darius Kinsey, who at the turn of the century organized "picture taking expeditions" into the American woodlands (see fig. 1), also used the largest possible glass plates, since making enlargements from smaller plates as well as copying are, of course, detrimental. Therefore, he worked only with the "original negative taken right on the spot. The reality and strength that is lost in copying and enlarging is retained by using the original negative."[9] To Kinsey, enlarging and copying already represented a loss in reality. The gap between the dimensions of the representation and those of the actual object at first had to be kept as small as possible, in order to keep intact the faith in photographic truth. Apparently the rule was: without life-size representation, no truth to life.

But how should it be possible – considering the actual size of the Rocky Mountains – to speak of conveying "an impression of the real grandeur and the magnitude of mountain scenery?" In view of the actual difference of size between the Rocky Mountains and their photographic reproduction it does not really matter whether the reproduction is done in small or a large-format, since, compared to the actual mountains this difference is certainly negligible. But how can one speak of complete fidelity to reality, considering the absence of color, volume or the incredible miniaturization and downsizing of the objects in the photograph? The analogy between the shapes in the photograph and the visual images of actual objects that the mind receives through our eyes cannot alone account for this illusion concerning the character of photography. There must be profound epistemic and ideological reasons to account for it.

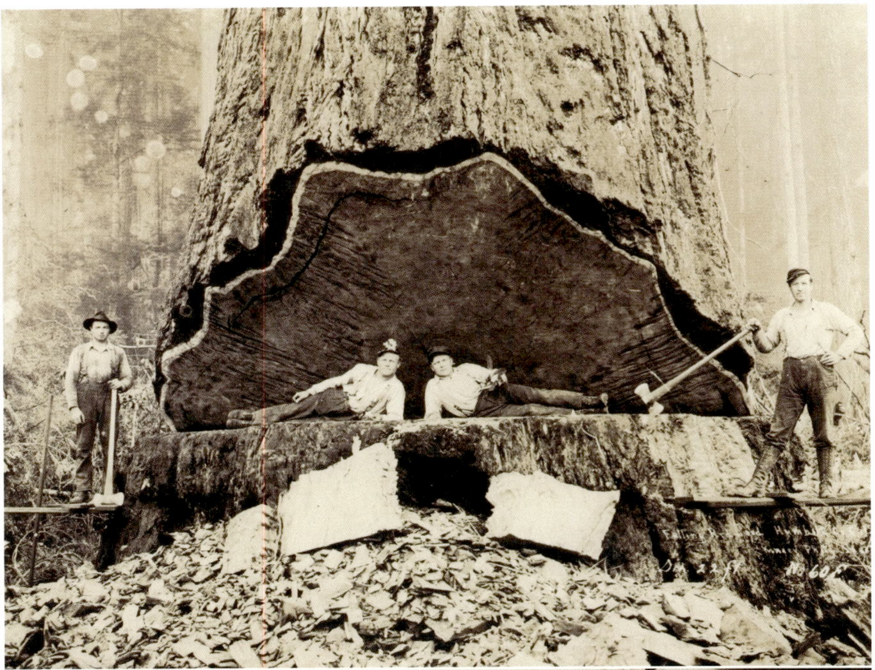

Fig. 1:
Darius Kinsey, *Falling Redwood*, Humboldt County, California, 1906

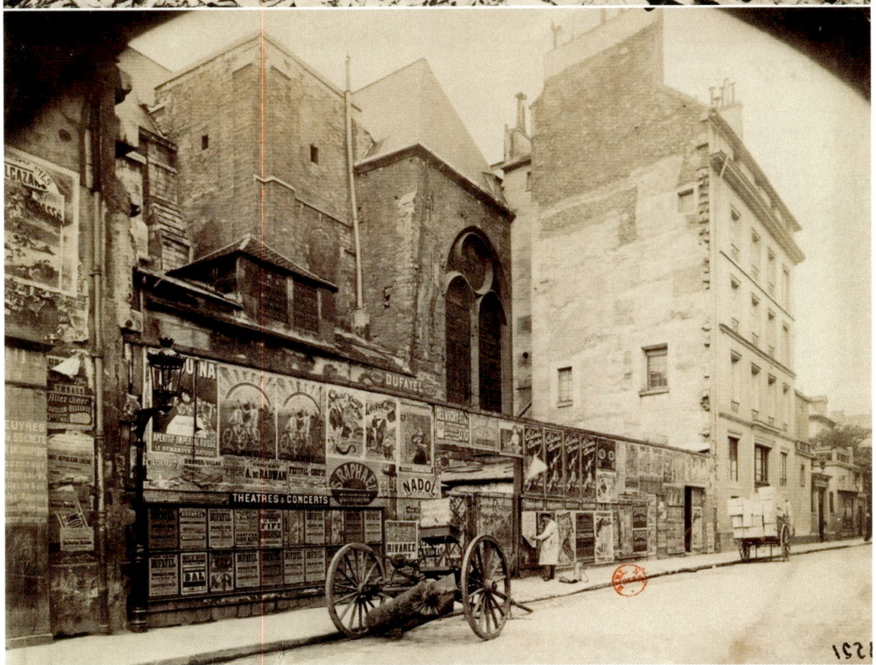

Fig. 2:
Eugène Atget, *Rue de l'Abbaye*, 1898

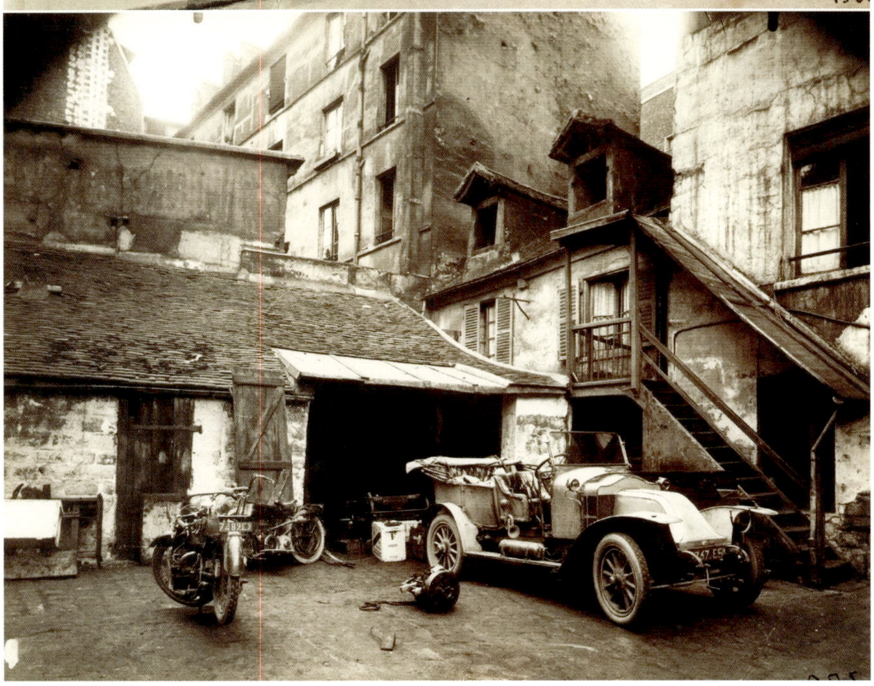

Fig. 3:
Eugène Atget, *Cour, 7 rue de Valence*, 1922

Let us look at an example. As we all know, socio-documentary photography has evolved from the belief in the documentary objectivity and truth of the photograph, with reality being reproduced of itself, as it were, and without distorting human interference. Through the lens of the camera reality is reproduced objectively, or so the claim went. Just as it was presumed that the individual photograph truthfully represented objects such as plants, horses, or streets, there was also no dispute concerning its capacity to document social processes in an objective and truthful manner. Because of this faith in the photographic image and its truth, the step from the photographic document to documentary photography, from a representation of objects to a representation of social processes, simply appeared to be a logical step. Thus field photography, photojournalism and socio-documentary photography evolved. That the photographer with his camera can represent reality with complete fidelity, that the photograph thus can be genuine and true – that is the creed of social photography.

Take the case of Eugène Atget: did he have an eye for the here and now? Compared to Charles Dickens's *Oliver Twist* (1837) or Friedrich Engels's *The Condition of the Working Class in England in 1844* (1887) or other contemporary sources, Atget's photographs rather seem to profess an eye for what is past, for the last remnants of a kind of Paris that was slowly disappearing: for street vendors and other picturesque manifestations of the perishing world of small business. The writer Léon Paul Fargue very accurately observed the decorative and unreal quality about the people in Atget's photographs: "In the work of Atget the small businessmen seem like decorations in the Parisian streets at the turn of the century."[12] Hence, many of Atget's photographs are accompanied by the note: "Disparu en 19 ..." (Defunct since 19 ...)

For Atget was not interested in the present, in the process of industrialization making itself felt in the suburbs (e.g., Aubervilliers), the bourgeoisie with its boasting, the large department stores, etc., but rather in all that was on the point of disappearing, in the last remnants of an earlier period before it vanished altogether. Atget took photographs of the Paris of 1920 as though it was still the Paris of 1870–80 (see figs. 2, 3); in the Paris after the turn of the century he searched for the traces of Paris as it had been before the turn of the century. Is it legitimate to call a photograph of a deserted Place du Tertre on Montmartre which even then was regularly flooded by tourists "unadorned realism,"[13] as Berenice Abbott did? Atget himself described his motivation as "Documents pour Artistes" (Documents for artists) – as the sign on his door read – as "creating a collection of all that is artistic and picturesque in Paris and its environs."[14] What he thus wanted was to compile a record of the old Paris, of the artistic and picturesque elements in that city. Can such photographs be considered "evidence in the historical process,"[15] as Walter Benjamin claimed? Are they not rather evidence of the fact that the true character of social reality was consistently being ignored, avoided? In my view, this tendency, however, is not due solely to social ignorance or lack of critical perception, but rather to the nature of traditional socio-documentary photography the essence of which is the fictious identity of photography and objectivity, which, of course, precludes the emergence of any concept of an alternative reality, i.e., a non-bourgeois reality.

To prove this point let us now turn to a body of work within socio-critical photography that cannot be accused of a lack of socio-critical awareness, since its declared intention was to produce – with the help of photography – a critical record of the economic and social situation of farmers. The reference here is to the photographic work carried out for the Farm Security Administration from 1935 to 1942 which, explicitly rejecting the aestheticism of the pictorialists, using the photograph only as a means to an end, but not as an end in itself. Specifically, let us look at the most successful and best-known of the F.S.A. photographs – which are said to represent the "climax of socio-critical documentary photography."[16] The photograph is by Dorothea Lange and is entitled *Migrant Mother* (1936, see figs. 4 a, b).

Having been reproduced in more than ten thousand publications, it can rightfully be considered synonymous with F.S.A. photography. The photograph was selected from an unpublished series in which the mother and her two children are shown in a variety of postures and positions, which means that it has been arranged, and staged, for the camera like in a studio. As becomes apparent from the series, stage directions were given, with the photographer acting as director: the scene was dramatized, was staged – photo-staging, really. The published photograph was obviously selected for its expressive strength. The mother with the troubled look on her face, her chin supported by her arm to emphasize the vertical lines and symmetry, the two children arranged symmetrically to her left and right, their faces turned towards the mother. We are all familiar with this image composition from religious iconography and hagiography where Jesus Christ looks straight ahead, with the two apostles or angels by his side presenting only their profile to direct the viewer's attention to the center, to Jesus Christ himself. In order to fur-

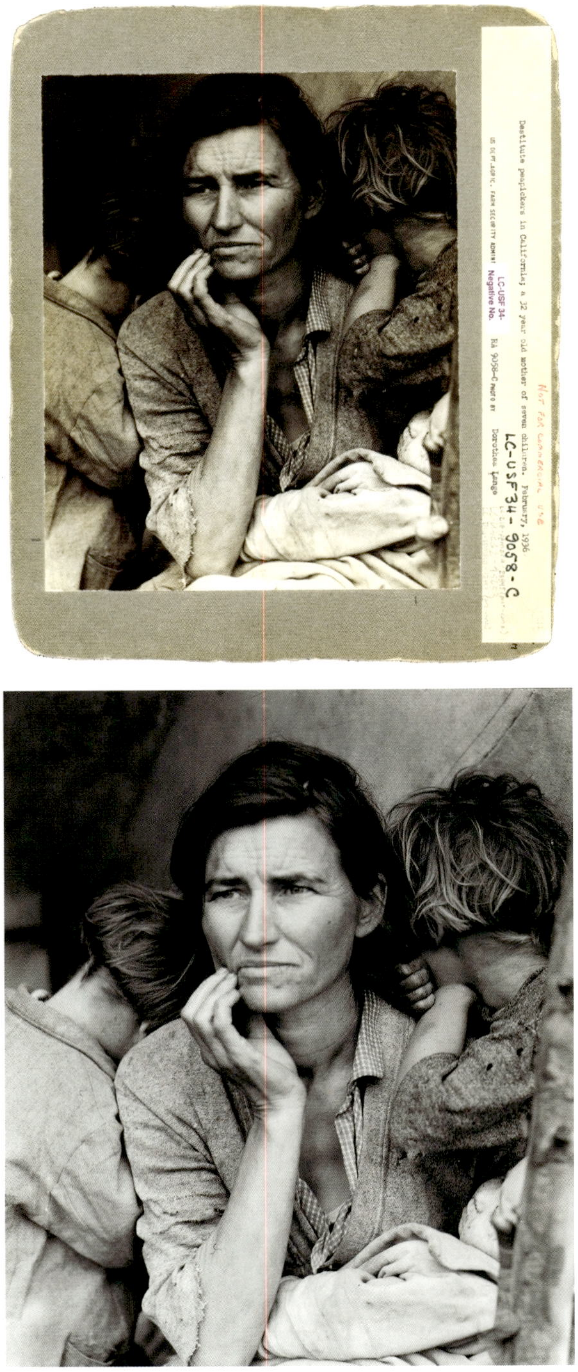

Figs. 4 a, b:
Dorothea Lange, *Migrant Mother*, 1936

Fig. 5: Shootings by the Paris Commune, 1871

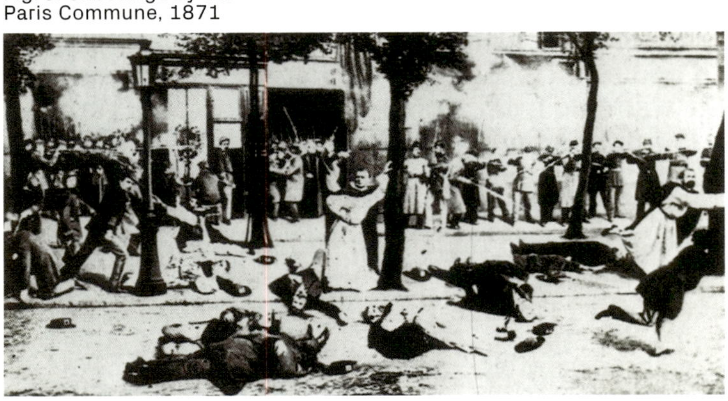

ther underline the aestheticism and the pictorial quality of this "document," Dorothea Lange even resorted to retouching to get rid of a thumb that evidently distracted the symmetry. Can such a photograph still be considered a document – of documentary value? Is it not rather a mere copy of bourgeois historical painting? In Bertolt Brecht's writing we can read: "A photograph of the Krupp works or of AEG reveals virtually nothing about these institutions. True reality has shifted to the functional plane. The materialization of human relations, the factory, for instance, does not return them anymore. Thus, the task indeed is to construct something, something artificial and contrived. And thus indeed art is required, too."[17]

Due to the cultural code proclaiming that in photography reality is reproduced with inimitable fidelity – that was the precondition – it became possible to deliver a false view of reality. What I am referring to here are not the fake photographs and photomontages that accompanied photography from its very beginning (see the alleged photographs of shootings by the Paris Commune in 1871, fig. 5), but rather photographic documents à la Atget and F.S.A. photography in which the falsification was not deliberate. If it were not for this cultural codification of the objective nature of photography, its fidelity to reality would have to be asserted and confirmed anew for every single photograph. But, instead, false photographs were passed off to the ideological consciousness as being true. And these false photographs, under their encoded label of being true and real, condition a false political awareness of reality. It was this which led Raoul Hausmann to define modern photography "as a position of the mind, a mental process."[18] It would be cheap now to attribute this incredible self-delusion concerning the degree of truth to reality in photography to the photographers themselves. Clearly a much more satisfactory answer than personal deficiency on the part of the photographers will evolve, if we examine the epistemic and cultural conditions causing the identification of photography as an objective document of reality. In my view, the true reason for this perfect and fundamental self-delusion as to the relationship between photography and reality can be found in the very nature of photography itself – in its characteristic feature of fakery which is its real character. For the very reason that photography is based on various cultural and epistemic codes such as objectivity, fidelity, the self-printing of nature, mechanical, automatic, is so excellently suited for faking and pretending. If we try to free ourselves just for a moment from these cultural conventions and restrictions, if we try to recall the emotions we experienced in making photographs of a landscape or a person, we will

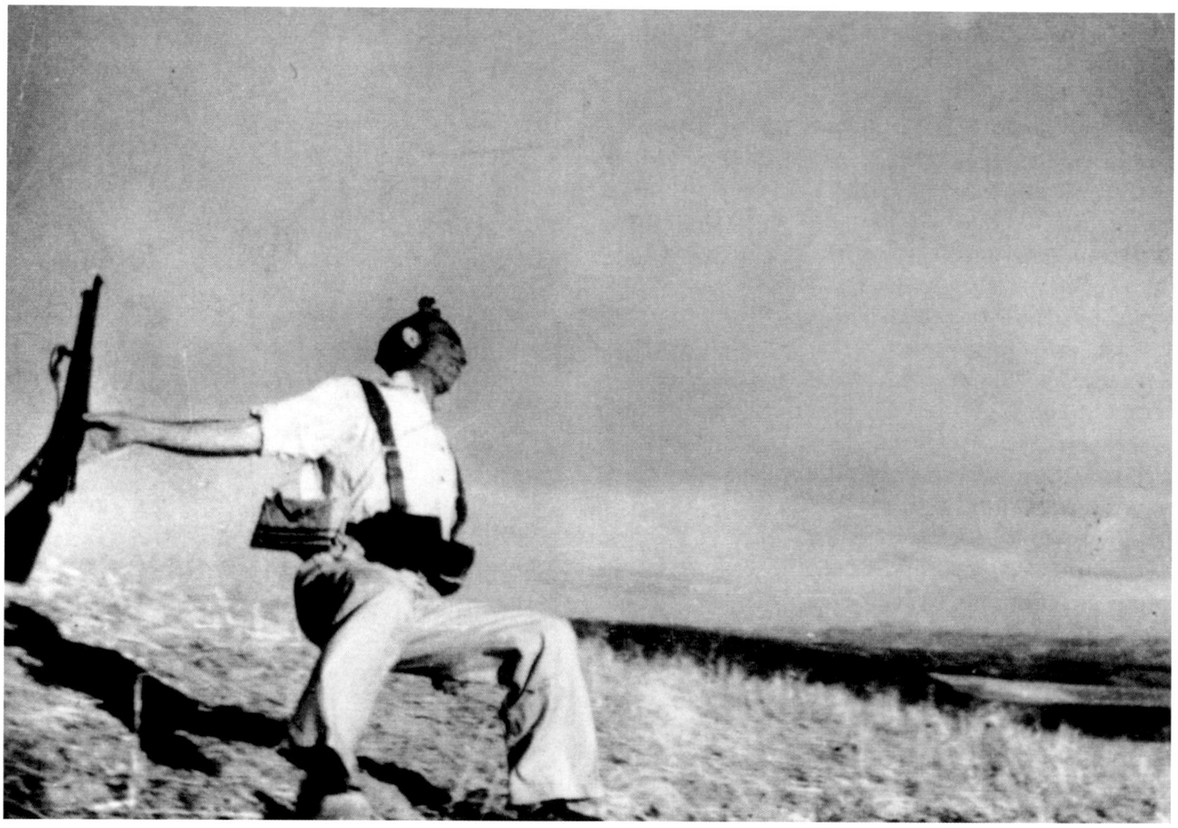

Fig. 6:
Robert Capa, *The Falling Soldier*, 1936

sense acutely how far removed from reality the photograph is, i.e., just as far removed as the size of the large plate photograph is from the actual size of the Rocky Mountains. But it is precisely this character of fakery and deceit that still allows photography to come so deceptively close to reality and that allows the individual photographs appear as such convincing and true documents of reality. Photography feigns authenticity – in a deceptively truthful manner. This quality of fakery thus is the very essence of photography. It works so perfectly that it is even capable of simulating reality itself. Thus, the world had to wait almost a hundred years for Brecht to formulate the trivial truth: "The photo camera can lie just like the typesetting machine."[19] Have we been blind? After all, it is a trivial fact that one can lie with the pen or the typesetting machine. Brecht's sentence reveals that what is naturally assumed for the typewriter, namely, falseness and deceit, is not in the least expected in the case of the medium of photography. Of course, it is also the difference between the written word and images in their quality as signs that moves us to set greater belief in the image than in the word as regards truthfulness. In considering the words "dog" or "tree," for instance, no one would dream of claiming that they in any way represent as a symbol a truthful visual reproduction of a real dog or tree. But, since a photograph of a dog or tree resembles them at least in terms of shape, form, color, that is as icon, we are more apt to believe in its truthfulness, even though the photograph of a tree is no more an objective document of the tree than is the word "tree." Thus, paradoxically it is photography's perfect quality of fakery that constitutes the basis of its claims to objectivity. Hence it lies in the very nature of photography to simulate truth and reality. The position of photography is that of the fake, i.e., a position somewhere between simulacrum and falsification.[20] The lens of the camera is certainly not objective, and if Mathew Brady believed that "the camera is the eye of history,"[21] it is just as well its story and our story, a hundred eyes, really, all perceiving something different. Between the lens regarded as the eye of history and Brecht's lying photo camera there is only one crossing: the ferry across which is called "fake."

At this point I would like to draw on a recent example in order to demonstrate how deep-rooted and how widespread the faith in the truth value of photographic image continues to be, and to what unbelievable extent people still believe in the documentary character of photography.

What we are looking at is a photograph of an action seemingly carried out by the Vienna Action artist Rudolf Schwarzkogler in 1965, shown at the documenta 5 (1972) in Kassel (see fig. 7).

It is not my intention here to discuss the artistic value of this photograph, but rather to illustrate my criticism of the pseudo-documentary value of photography by way of describing their effect on the art world. For this photograph, and similar pictures have created a legend that spread throughout the world, namey, the notion that the artist in his last action fatally castrated himself. There is no French, German, Italian or American magazine, book, or encyclopedia of art which does not refer to this action as *ultima ratio*, as concluding act and final consequence: in his last action the artist kills himself by castration. In numerous articles from *Time* magazine[22] up and down, this action is discussed as a phenomenon of modem art and there are even articles written by other artists to defend it.[23] The true situation, however, is totally different – and how should it be otherwise since it is photography with which we are dealing! The body that is shown in the photograph is not that of Schwarzkogler himself, but that of his friend Heinz Cibulka. Everything is

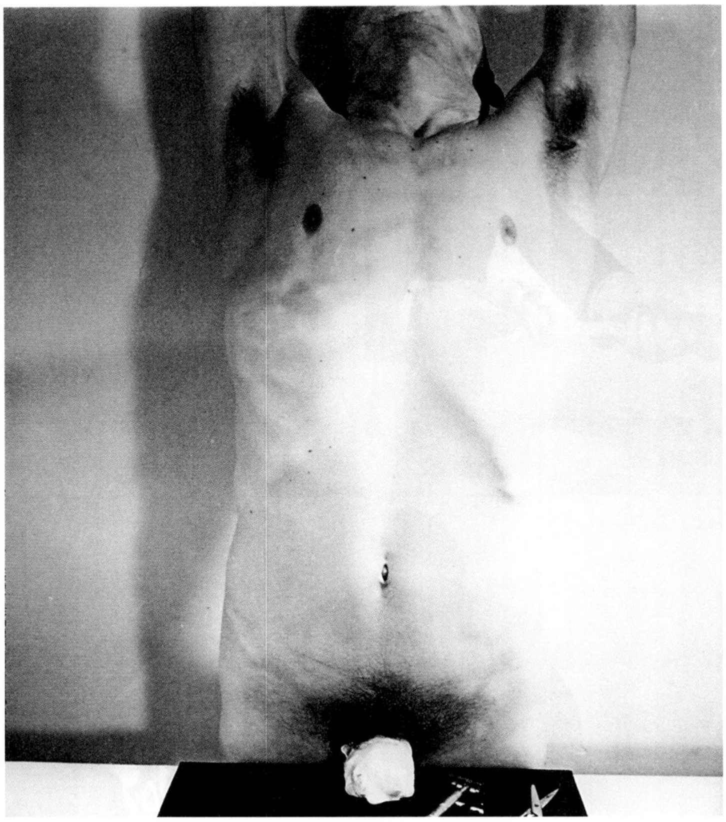

Fig. 7:
Rudolf Schwarzkogler,
3. Aktion, 1965

just simulated with the help of paint and material, and is staged for the photographer Ludwig Hoffenreich, a boulevard photographer. Schwarzkogler is the invisible director of the scene. What we see is a photo séance, a mise en scène, directed for the camera. Schwarzkogler carried out his last action in 1967 and two years later, 1969, put an end to his life by jumping off the balcony of his apartment. In this case, the epistemic and cultural preconditions responsible for the incorrect reception of the photograph are easily perceived: the myth of van Gogh, the general myth of the artist as victim, the fact that Schwarzkogler actually committed suicide, the genuine self-inflicted wounds of his friend Günter Brus, etc. Yet, all this cannot fully explain this striking self-delusion. At this stage it would be a cheap, though appropriate move to point to the deplorable state of art criticism, which indeed, has reached a degree of idiocy that can hardly be surpassed, to blame it for this ignorance and delusion. But how is it possible that these photographs are interpreted as documents of an artistic self-castration with fatal consequences? Is it not once again photography's quality of fakery that obscures the view of reality, that feigns truth and reality for something that never even happened? Is it not to be attributed to the photograph, as fake, that an event that in reality never took place is discussed throughout the world? Is it not the objectivity feigned by photography that allows appearance and reality to be identified, i.e., ideology (expectations, myths, desires, longings) to be passed off as reality?

It is photography's quality of fakery which allows the false view to slip into the garb of objectivity. The photograph, as fake, obscures our view of reality, yet at the same time provides a glimpse of something else - of our consciousness. For photography is a means by which our consciousness expresses itself. Hence the false consciousness finds its expression in the photo-fake. And thus on a meta-level, the truth once more becomes apparent. For if these false interpretations of the photograph of Schwarzkogler's action made by Ludwig Hoffenreich do not tell the truth about the action itself, they certainly do tell the truth about the viewer's expectations, his perceptions and his consciousness. In a formal sense, these photographs bear some relation to sensationalist photography which, of course, also originates in photography's quality of fakery. Sensationalist photography in a way represents a trivial version of art photography. Apart from advertising and photojournalism it is in erotic and pornographic photography, probably the two most common forms of the medium, that photography's characteristic quality of fakery, finds its most significant expression.

Thus, we have not only identified the essence of photography as fakery, but, at the same time, have pinpointed its existential consequence, i.e., the simulation of a reality that is not there. Be it the mineral water bottle with the reproduction of a spring which does not really exist or is not identical with the source of the water (who says it is spring water!) or the countless brochures advertising travels, landscapes, beaches, furniture, etc., or be it a brightly colored photograph of a nuclear power plant peacefully sitting in the countryside - whatever it may be, a second skin has been pulled over reality: the faked skin of photography. Photography's character of fakery has created a fake culture as well as a fake reality.

In this context the role of the small format photograph needs mentioning. The love of large format photographs was characterized, as we have seen, by the desire to keep alive the faith in the truthfulness of photography, at least by way of reproductions that were roughly true to size. Any conspicuous violation of the proper proportions, as in the small format photograph, therefore, meant an all too abrupt upsetting of this faith. The fact that we have by now grown accustomed to the small format, fully accepting its existence, thus indicates that we have gotten used not only to photography being characterized by fakery but also to the same quality in our reality. Thus, the acceptance of the small format testifies to a successive weaning from reality - a development that, in our century, is characteristic of most people's consciousness. If it is correct that reality in part must be considered an epiphenomenon of communication, a social construction and convention, then photographic communication (and its quality of fakery) has an important part in it.[24]

Note: This text was first published as "Foto-Fake" in *Camera Austria*, no. 4 (1980): 92–103. This English translation is taken from Woody Vasulka and Peter Weibel, eds., *Buffalo Heads: Media Study, Media Practice, Media Pioneers. 1973–1990*, exh. cat. ZKM | Karlsruhe (Cambridge, MA: MIT Press, 2008). Translated from the German by Jörg von Stein.

[1]
"Cette découverte consiste dans la reproduction spontanée des images reçues dans la chambre noire," see "Base du traité provisoire," in *Niépce, correspondance et papiers*, vol. 2, ed. Manuel Bonnet and Jean-Louis Marignier (Saint-Loup-de-Varennes: Maison Nicéphore Niépce, 2003), 941–45 (quotation on 943). Translated by Peter Weibel and Jörg von Stein.

[2]
Dominique François Arago, "Bericht über den Daguerrotyp" (1939), in *Theorie der Fotografie I, 1839–1912*, ed. Wolfgang Kemp (Munich: Schirmer/Mosel, 1980), 51–55. Translated by Jörg von Stein.

[3]
Joseph Louis Gay-Lussac, "Rapport, Fait à la chambre [des Pairs] par M. Gay-Lussac," in *Historique et description des procédés du daguerréotype et du diorama*, (Paris: Alphonse Giroux et Cie, 1839), 31–35 (quotation on 32–33). Translated by Peter Weibel and Jörg von Stein.

[4]
Quote taken from Winfried Ranke, "Zur Sozialdokumentarischen Fotografie um 1900," in *Beiträge zur Geschichte und Ästhetik der Fotografie*, ed. Winfried Ranke, Ulrich Keller, and Herbert Molderings (Giessen: Anabas, 1977), 5–36 (quotation on 6). Translated by Jörg von Stein.

[5]
Quote taken from Erich Stenger, "Alexander von Humboldt und die beginnende Photographie, Photophysik, Photochemie," *Zeitschrift für wissenschaftliche Photographie*, no. 31 (1932/33): 54–65 (quotation on 56–57). Translated by Jörg von Stein.

[6]
Alois Auer von Welsbach, *Die Entdeckung des Naturselbstdruckes oder die Erfindung, von ganzen Herbarien, Stoffen, Spitzen, Stickereien und überhaupt allen Originalien und Copien […] durch das Original selbst auf einfache und schnelle Weise Druckformen herzustellen […]. Vorgelesen in der math.-naturw. Classe der kais. Akademie der Wissenschaften zu Wien* (Vienna: Kaiserlich-königliche Staatsdruckerei, 1853), 2. Translated by Jörg von Stein.

[7]
Quote taken from Beaumont Newhall and Diana E. Edkins, *William H. Jackson*, exhib. cat. Amon Carter Museum, Fort Worth (Dobbs Ferry, NY: Morgan & Morgan, 1974), 141.

[8]
Quote taken from *Newhall and Edkins, William H. Jackson*, 141.

[9]
Dave Bohn and Rodolfo Petschek, *Kinsey Photographer. A Half Century of Negatives by Darius and Tabitha May Kinsey*, vol. 2 (San Francisco, CA: Scrimshaw Press, 1975), 309.

[12]
Léon-Paul Fargue, *Dans les rues de Paris au temps des fiacres*, ed. René Coursaget, (Paris: Édition du Chêne, 1950), n. p. Translated by Jörg von Stein.

[13]
Berenice Abbott, *The World of Atget* (New York: Horizon Press, 1964), 22.

[14]
Abbott, *The World of Atget*, 11

[15]
Walter Benjamin, *Das Kunstwerk im Zeitalter seiner technischen Reproduzierbarkeit* (1935/36, here 3rd edition), in *Abhandlungen. Gesammelte Schriften*, vol. 1.2 (Frankfurt am Main: Suhrkamp, 1991), 472–508 (quotation on 485). Translated by Jörg von Stein.

[16]
Robert J. Doherty, *Sozialdokumentarische Photographie in den USA* (Lucerne: Bucher, 1974), 9.

[17]
Bertolt Brecht, *Der Dreigroschenprozeß* (1931), in *Schriften zur Literatur und Kunst I, 1920–1932* (Frankfurt am Main: Suhrkamp, 1967), 143–234 (quotation on 171–172). Translated by Jörg von Stein.

[18]
Raoul Hausmann, *Documents Raoul Hausmann*, in *projectoires* 1, ed. Michel Giroud and Sabine Wolf (1975): 15. Translated by Jörg von Stein.

[19]
Bertolt Brecht, "Zum zehnjährigen Bestehen der A-I-Z (Arbeiter Illustrierte Zeitung)," in *Werke. Große kommentierte Berliner und Frankfurter Ausgabe*, vol. 21, *Schriften 1*, (Frankfurt am Main: Suhrkamp, 1992), 515. Translated by Jörg von Stein.

[20]
Note 2019: Even one of the most famous photos of documentary war photography, Robert Capa's so-called *The Falling Soldier* from the Spanish Civil War (see fig. 6), of which no negative exists anymore, has long been shadowed by doubt. Whether the photograph really depicts the moment of a soldier's death is questionable. In his *book Sombras de la Fotografía: los enigmas desvelados de Nicolasa Ugartemendia, Muerte de un miliciano, La aldea española y El Lute* (Universidad del País Vasco, 2009), José Manuel Susperregui points out that because of certain landscape features in the background, Cerro Muriano cannot be the place where the photo was taken. Thus, the photo was not taken "right on the spot" (Darius Kinsey), but in Espejo, in an area that was far from the battle line in Sept. 1936 and where no shots were fired at the time the photo was taken. This inevitably leads to the conclusion that this photo, like all the others in the series from the front line, was staged.

[21]
Quote taken from Wayne Youngblood and Ray Bonds, *Mathew B. Brady's Civil War* (New York: Chartwell Books, 2011), 12.

[22]
Robert Hughes, "The Decline and Fall of the Avant-Garde," *Time*, Dec. 18, 1972.

[23]
Les Levine, "Les Levine Replies," in *Idea Art. A Critical Anthology*, ed. Gregory Battcock (New York: Dutton, 1973), 194–203.

[24]
The ever smaller size of photographs, the screens of mobile phones, are the logical consequence of the "culture industry." See Theodor W. Adorno in Adorno and Max Horkheimer, *Dialectic of Enlightenment* (New York: herder and Herder, 1972). First published as *Dialektik der Aufklärung* (Amsterdam: Querido, 1947).

Ever since Galileo, we have known that the book of nature is written in the language of mathematics.[1] We thus attempt to understand nature's language, to decipher its code. Language and mathematics are codes. Any code can be converted into another code: for example, text can be converted into binary code, ones and zeros. We seek to apprehend nature's laws and secrets by decoding them. But can we decode every secret? We see a tree trunk, one of nature's biological codes. Nailed to it is a text on a piece of paper. However, this text, whose subject is those natural laws that are to be discovered and articulated in mathematical language, is printed in such a way that it cannot be read. The text remains as secret as the laws of nature. The photograph is a representation of something unrepresentable, indeterminable, incomplete. The observation of nature is thus relative to the observer, determined by standpoint and perspective. This implies that we are relatively limited in our ability to decipher all the laws and secrets of nature.

—Peter Weibel, 2022

[1] "Philosophy [i.e., natural philosophy, the precursor to the natural sciences] is written in this grand book, the universe. ... It is written in the language of mathematics, and its characters are triangles, circles, and other geometric figures." Galileo, excerpts from *The Assayer, in Discoveries and Opinions of Galileo*, trans. Stillman Drake (New York: Anchor Books, 1957), 229–80 (quotation on 237–38). First published as *Il Saggiatore* (Rome, 1623).

The Secret of Nature

If scientists are to investigate the problem of matter, they must first study the forms of matter. The infinite variety and mutability of matter's forms must be the proximate subject of their research; their efforts must be directed toward discovering natural laws, universal principles, that can serve to guide them through this boundless field. Thus, for quite some time now, the focus of the exact sciences, and of physics in particular, has been the analysis of the structure of matter and the forces responsible for that structure.

Since the age of Galileo, the basic method of scientific research has been experimentation. This method has made it possible to move beyond the generalized experience of nature to a more specific experience, and to isolate characteristic processes within nature, on the basis of which nature's laws can be studied more directly than via general experience. If one wants to study the structure of matter, one must therefore conduct experiments on it. One must subject matter to extreme conditions in order to study its transformations under such circumstances, in the hope of identifying specific fundamental traits of matter that remain constant throughout all its seeming changes. Of course, scientists cannot be satisfied with a merely qualitative description of the underlying structure of matter; they must try, on the basis of meticulously conducted experiments, to arrive at a mathematical formulation of the natural laws that determine the forms of matter – that is, the elementary particles and the forces these carry. A clear distinction between matter and energy, between material stuff and immaterial forces, can no longer be drawn in this sector of physics, since each elementary particle not only gives rise to and is affected by forces, but at the same time also represents a particular force field itself.

Transcript of the text that was nailed to the tree trunk. Translated from the German by Patrick Hubenthal.

The Secret of Nature / Das Geheimnis der Natur

1967 | b/w photograph

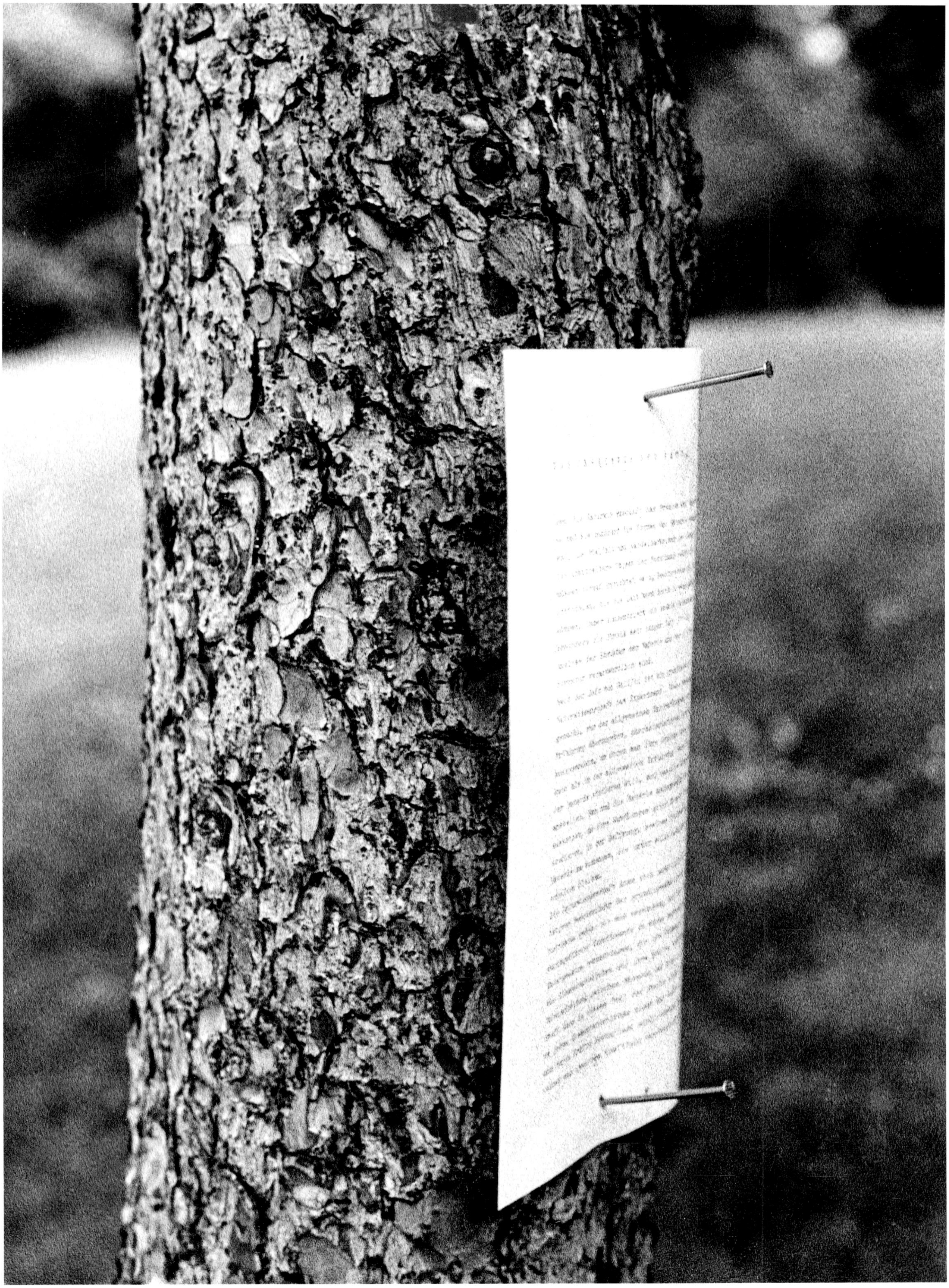

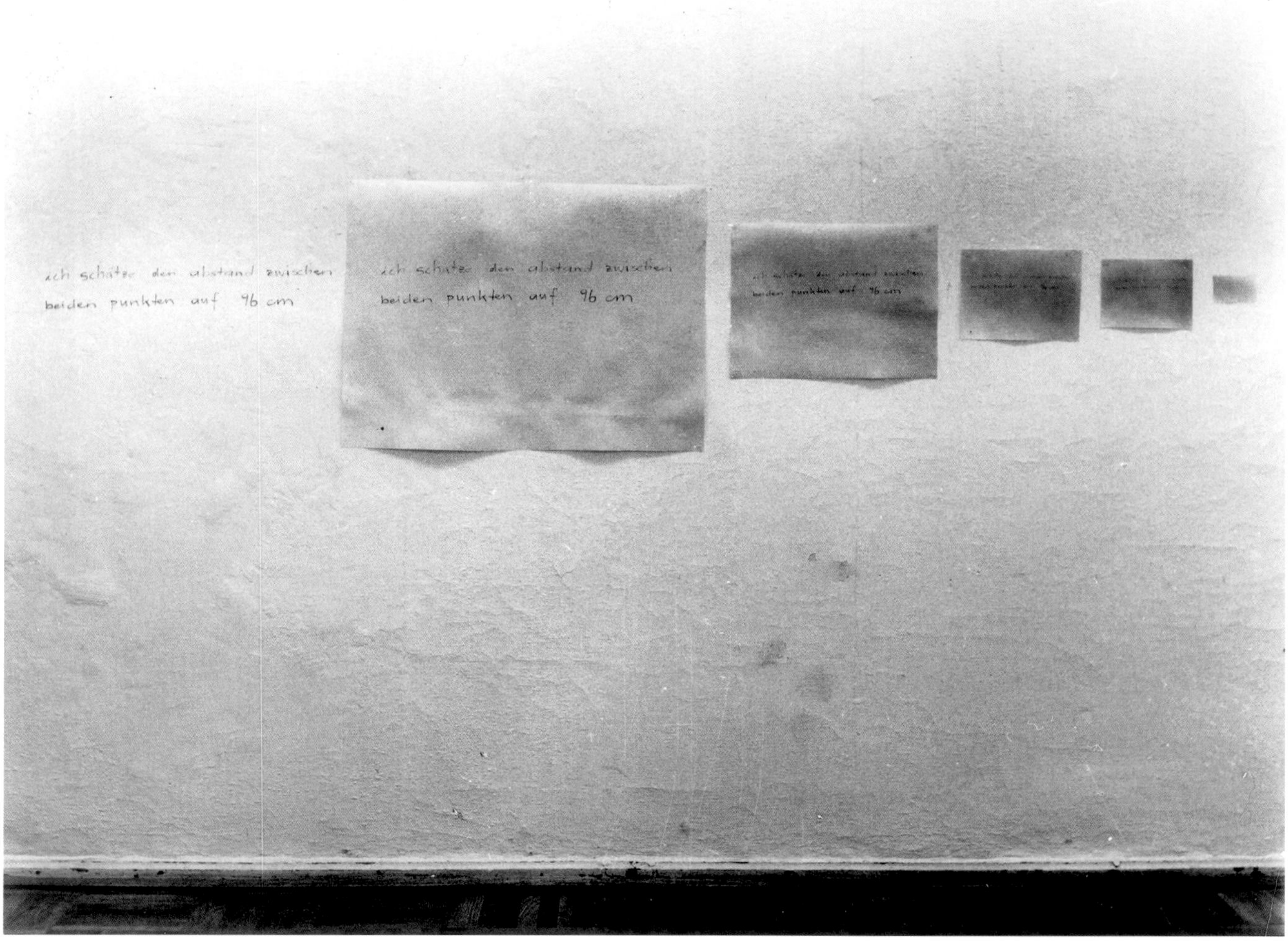

Installation view
Zur Kosmologie des Paradoxen, Galerie nächst St. Stephan, Vienna, 1975

Estimation: The Paradox of the Reference System (Medium) II / Abschätzung: Paradoxie des Referenzsystems (-mediums) II

1975	wall drawing, 5 photographs of different sizes

The distance between two points drawn on the wall is realistically estimated at 96 cm. Beside them is a photograph, a full-scale reproduction. Of course, the size of a photograph is variable – usually it's much smaller than its subject; it's just that our brains are already accustomed to this reduction technique, the downscaling, so they no longer notice it. Here, however, where the representation of the same reality becomes a series of ever-smaller photographs, the statement in those photographs becomes ever more nonsensical. Through the medium of photography, a fundamental aspect of our perception of reality is revealed – namely, that our perception and modeling of reality is dependent on a point or frame of reference. Within a specific frame of reference, a specific statement – for example, about the distance between two points – is true. In a different one, it is no longer true or becomes nonsense. Frames and points of reference form a starting point or standard for the validity of a model, which is what a photograph of the world (of actual facts) is. These changes in scale, position, perspective, focus, etc., are all characteristically photographic changes to the frame of reference. The tools of photography are used to make photography's inherent contradictions apparent – and also, beyond photography, all contradictions inherent to any system of representation. The relativization of the reference system – the standard, the point of view, the scale, the visual and linguistic statements – is a vote for the changeability of the world and a renunciation of ontology. —Peter Weibel

Translated from the German by Patrick Hubenthal.

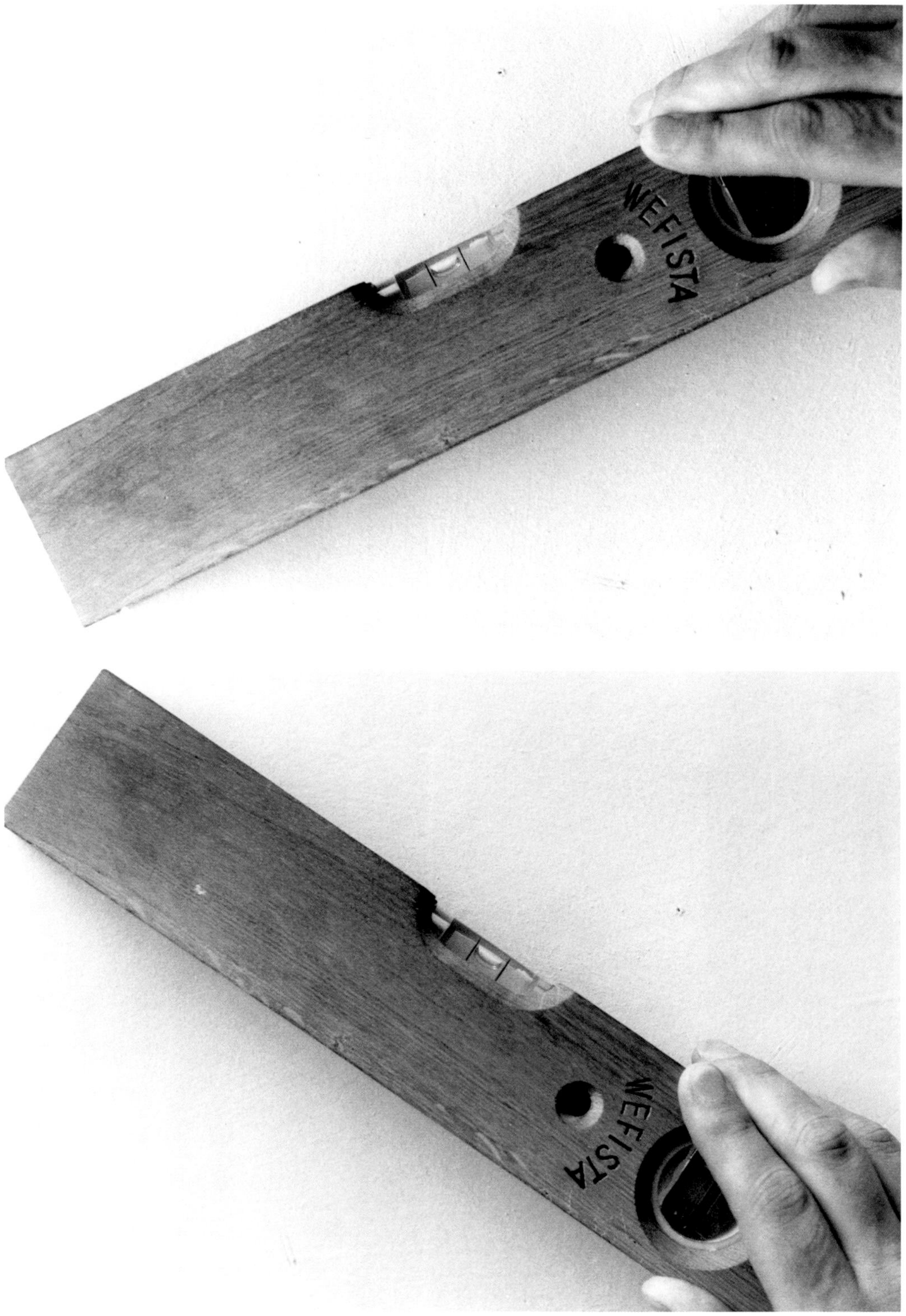

Physical Paradoxes / Physikalische Paradoxe

1974 | 2 gelatin silver prints, 18 × 24 cm each

The measuring instrument depicted here (a bubble in a glass container filled with liquid) always indicates a level surface, a horizontal position, even though the instrument is in two different diagonal positions. How is this possible? An image or representation can simulate something that is not true. The manipulator was the camera: it was tilted forty-five degrees to the left for one picture and forty-five degrees to the right for the other, whereas the measuring instrument, the level, was in reality kept perfectly horizontal. —Peter Weibel

1. Two microphones are placed in front of two cameras: the stereo principle (stereoscopy and stereophony).

2. The two cameras are equipped with trap focus, a mechanism that takes a picture automatically at a specific focal distance, which can be adjusted as desired. Camera 1 (left) is focused at a distance of two meters from the lens. Camera 2 (right) is focused at a distance of one meter from the lens.

3. Whenever someone passes one of these "focal points," the camera's shutter is released. However, no picture is taken since this installation is not a work that produces photographs, but a work about the nature of photography (and new technology), hence the title *Metaphotography*.

4. The trap focus generates a trigger noise that is picked up by the two microphones and transmitted to the computer.

5. The signal from Camera 1 triggers a speech recording in the computer, and the computer starts talking about the word "auto" (from the Greek word for "self"). The text is spoken in English and appears in German on the screen:

   I am a camera with autofocus.
   I focus the picture autonomously.
   I am an autonomous machine.
   I am an automaton.
   I will not serve.

6. With each click of the autofocus mechanism, the text starts from the beginning. Therefore, to hear the entire text, viewers must step aside and wait a bit before stepping back into the field of focus.

7. The signal from Camera 2 triggers an industrial sound (the camera click), which is repeated.

8. The text examines the problem of the self in a sequence of short sentences. Beginning with the autofocus mechanism, the camera talks about the mechanical, automatic, and autonomous aspects of photographic technology. This focuses on the relationship between humans and machines, placing it on a spectrum of self-sufficiency, from autonomy to automaton. Is the independent machine, the automaton, to be compared to Lucifer, who says: "*non serviam*" – "I will not serve"? Our relationship with technology can be mirrored in the question: Who is master, technology or us?

9. This is also a work about space and industrialized mechanical perception. Focusing a camera is, after all, an operation in space, a way of dividing space. The depth of field determines the depth of the space. The faraway is drawn closer by the camera's focus. The space occupied by the viewer and the space occupied by the camera encounter each other interactively. —Peter Weibel, 1989

Translated from the German by Patrick Hubenthal.

Metaphotography / Meta-Photographie

1989	interactive computer-based installation; 2 cameras with autofocus, 2 microphones, Amiga 500 computer, monitor, speaker, 380 × 40 × 50 cm software: Franz Xaver first shown in the exhibition *Fotografie, Wissenschaft und neue Technologien: Facetten der Interaktion*, Kunstmuseum Düsseldorf, 1989

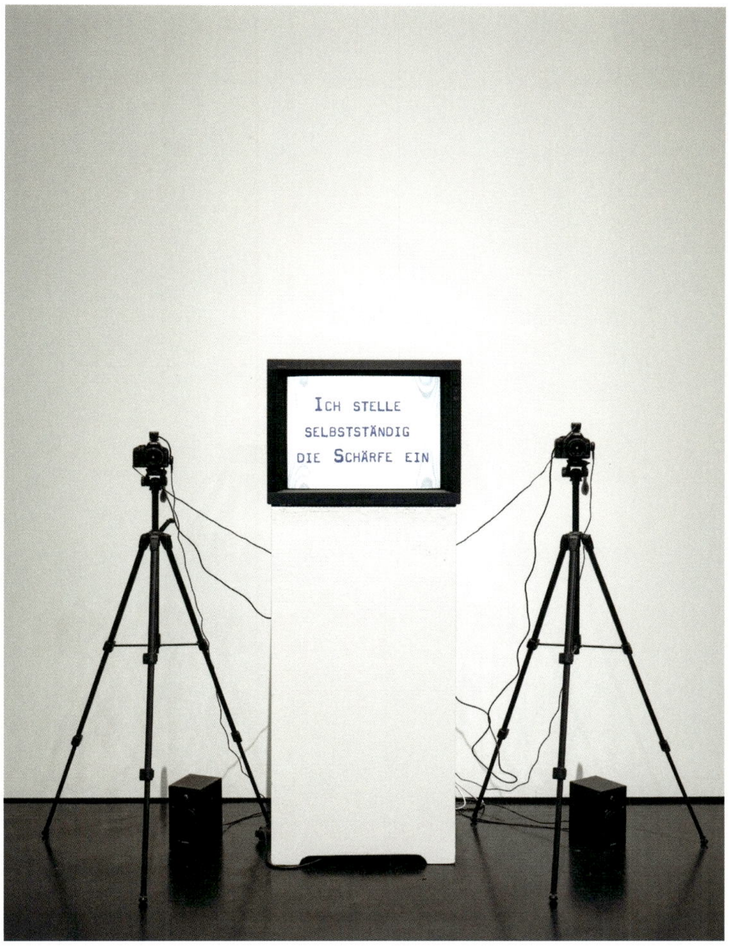

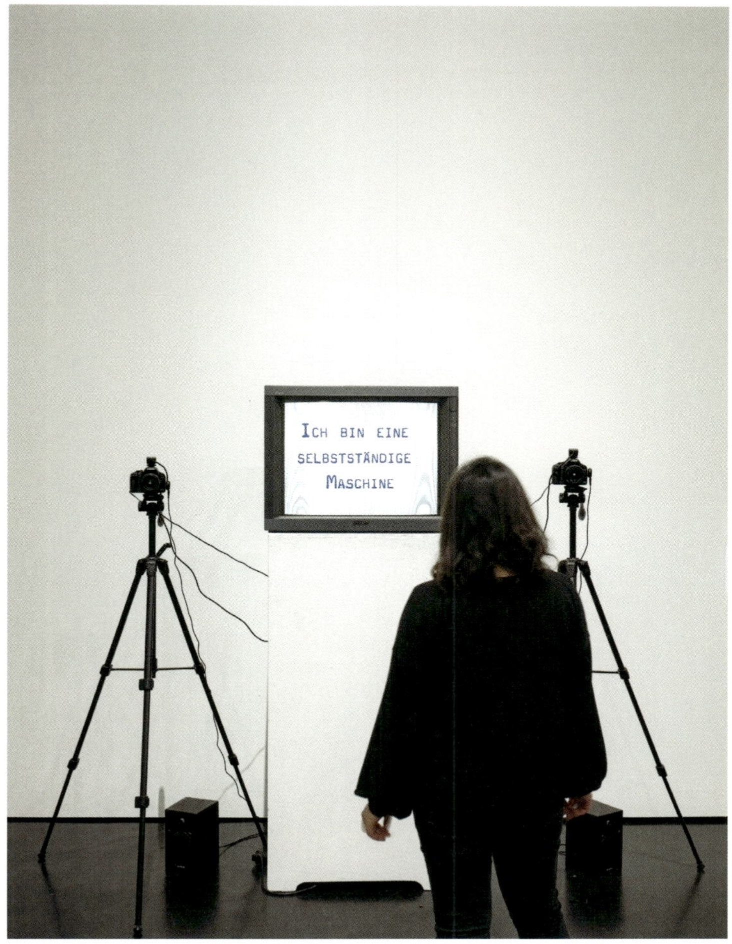

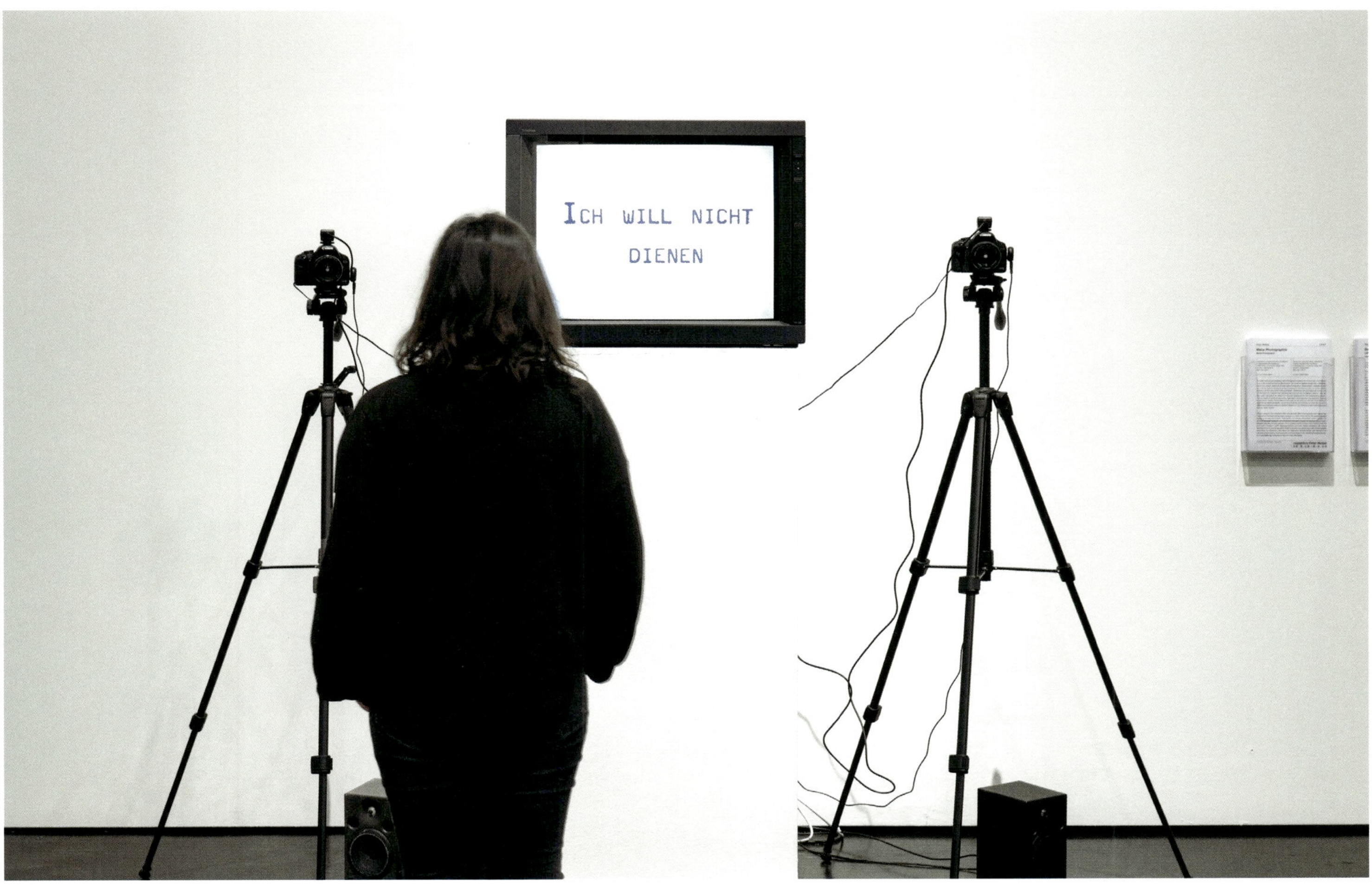

Installation views
respectively, *Peter Weibel*,
ZKM | Karlsruhe, 2019–20

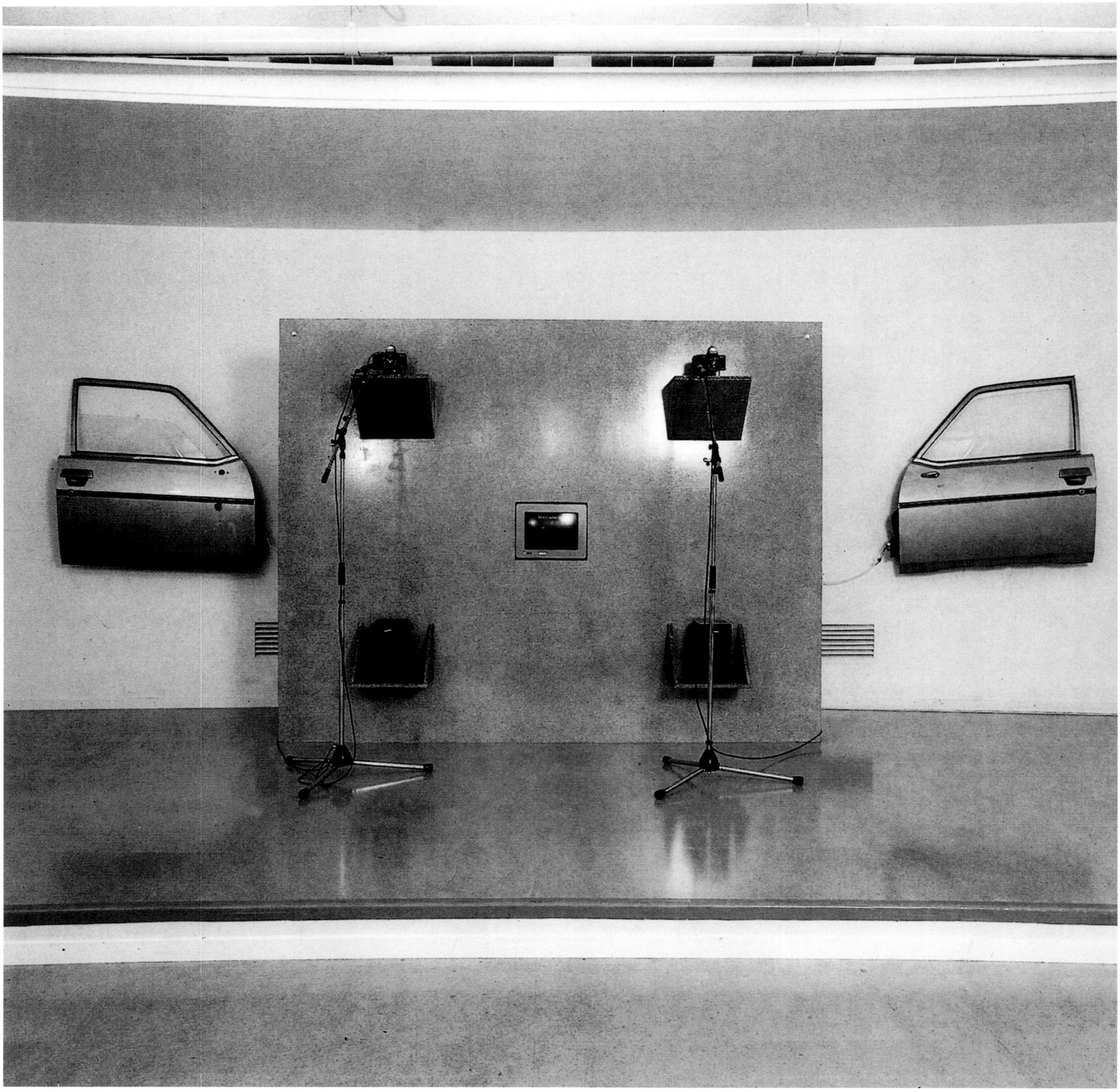

Installation view
Panische Objekte, Galerie
Walcheturm, Zurich, 1990

Lucifer's Dream: Non Serviam; The Revolt of the Automata; To the Self by Itself? / Luzifers Traum: Non Serviam. Der Aufstand der Automaten. Von selbst zum Selbst?

1990	2 car doors, 2 cameras with autofocus, 2 microphone stands with microphones, 2 speakers, computer, monitor, wood partition, 200 × 210 × 45 cm

In the second version of *Metaphotography* (*Meta-Photographie*, first version see pp. 159–59 in this volume), the camera is set for a specific focal distance. When a viewer enters this focus zone, the camera springs into action: the trap-focus mechanism not only releases the camera's shutter and turns on the computer, which displays written texts and emits spoken words, but also causes the windows in the car doors to start moving. From the automobile, the self-moving vehicle, to the automatic camera, the self-actuating imaging device – such is the path of the machines, whose relationship to humans is comparable to Lucifer's relationship to God: creations that will not serve. Does the relationship of machines to humans mirror that of humans to nature?

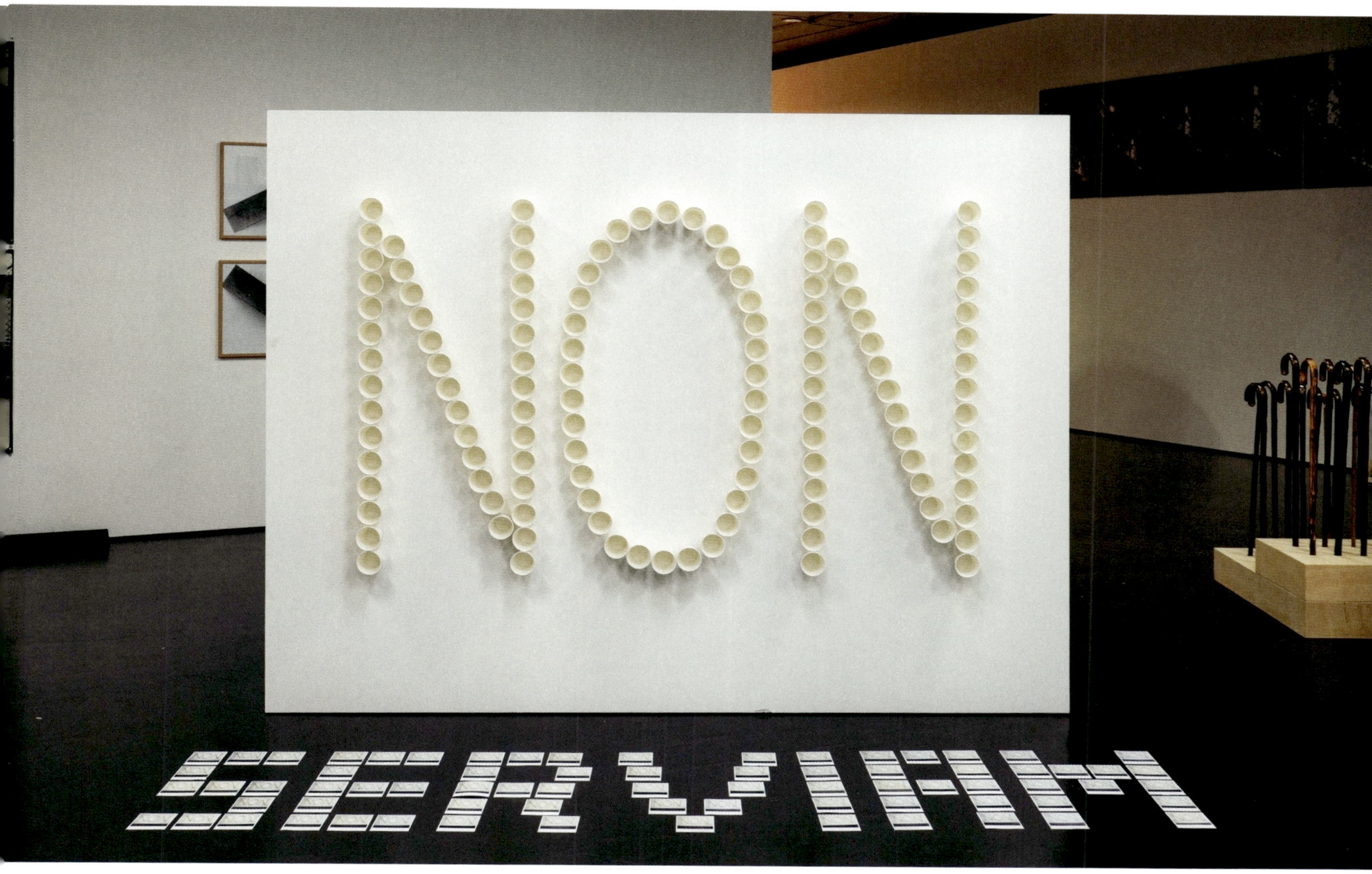

Installation views
respectively, Peter Weibel,
ZKM | Karlsruhe, 2019–20

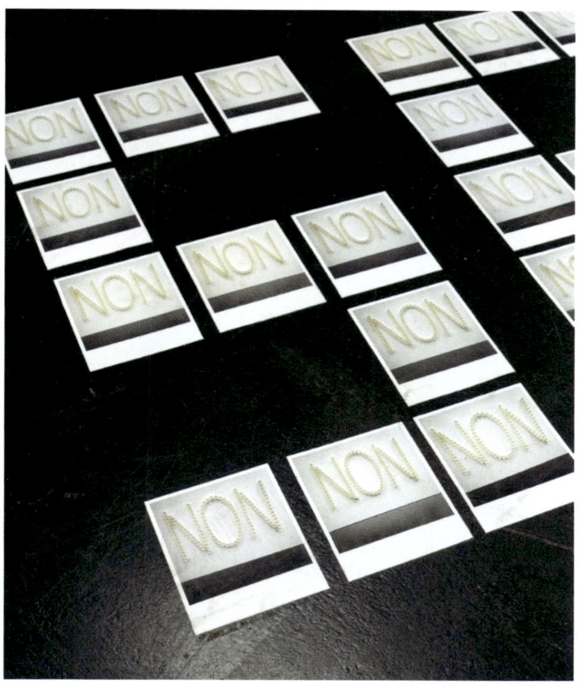

On the Problem Copy-Original: NON SERVIAM (I Will Not Serve), Said Lucifer, Who Wanted to Be Like God / Zum Problem Kopie – Original: NON SERVIAM (ich will nicht dienen), sagte Luzifer, der sein wollte wie Gott

1991/2019 | 73 Polaroid photographs, 99 paper cups, dimensions variable

First shown at Galerie Grita Insam, Vienna 1991, the work *On the Problem Copy-Original: NON SERVIAM (I Will Not Serve), Said Lucifer, Who Wanted to Like God* explores, as the title indicates, the relationship between original and copy. Lucifer, the fallen angel, refuses to serve and strives not just to be God's image, but God's equal. He wants to rise from the copy to the original and is then banished to hell. With 99 identical paper cups, the word "NON" is written on the wall. 73 Polaroid photos of that wall form the word "SERVIAM" on the floor. On getting closer, the viewer will see that the photographs depict the NON on the wall. The NON, the negation, is inscribed in serve (SERVIAM). The break between the original object (NON), which is only a copy of itself, and the supposed copy (SERVIAM) highlights the fragile relationship between the two. The SERVIAM on the floor is negated by the word NON that it depicts. The NON on the wall is negated by the word SERVIAM that is depicted. The word SERVIAM is written with the word NON. The relation between copy and original on the floor is entangled just like SERVIAM is intertwined with NON and not with SERVIAM. They form a dialectic pair.

At first glance, the object standing in the foyer of the ZKM | Karlsruhe in 2007 appeared to be an ordinary photo booth. However, the illuminated sign running across the top, reading "FLICK_KA Bürgergalerie" (FLICK_KA citizens' gallery), signaled that this was a special machine, capable of more than just taking pictures.

This participative artwork by Peter Weibel and Matthias Gommel, titled *FLICK_KA*, enabled the citizens of Karlsruhe and visitors to the ZKM | Karlsruhe to insert a coin (provided for them by the ZKM) and have their photographic portraits taken in the usual way. At the same time, though, a custom-made computer program was digitizing those portraits, uploading them to an online gallery, and sending the subjects' images to the four monitors beside the booth. The paper prints of the photos were just a bonus; the important aspect was the digital storage and display of the images. One first saw previous visitors, then, suddenly, oneself. Museumgoers outside the booth saw the portraits at the same time, and the photographic portraits were also transmitted via the internet to the exhibition space, where they were displayed on a large screen. Thus, every visitor who participated in *FLICK_KA* became part of the exhibition and part of the ZKM Collection.

The democratization of portraiture had been achieved. Unlike in the past, when the possibility of immortalizing oneself by having one's portrait made was reserved, both socially and financially, for the political and religious elites, today everyone can make self-portraits with the aid of (cellphone) photography, the most democratic of all visual media, and distribute it worldwide. Here, *FLICK_KA* took on the role of "a forum in which, before which, and with which all are equal. The consumer becomes the producer, and the museumgoer becomes the artwork, the content of the exhibition." A text by Peter Weibel touching on these points was posted on the side of the photo booth (see "Everyone Can Photograph Everyone: The Democratic Promise of Photography," see p. 165 in this volume).

Weibel has long used photo booths in the creation of his photographic works. The best-known of his *Photo Booth Pictures / Automatenfotos*, a series he began in the mid-1960s, are *Self-Portrait as a Woman* (*Selbstportrait als Frau* 1967, see p. 113 in this volume), *The World/Wilting Kiss* (*Der Kuß der Wel(k)t* 1967, see p. 111 in this volume), and *A Love Story* (1967). His use of this medium was motivated not only by a lack of money (he could not afford to buy a camera or hire a photographer), but also by his recognition that photography's distinctiveness (and its capacity to provoke the art world) lay in its mechanical character and its ability to create pictures automatically – as advertised over a century earlier in the title of William Henry Fox Talbot's 1839 essay "Some Account of the Art of Photogenic Drawing, or the Process by Which Natural Objects May Be Made to Delineate Themselves without the Aid of the Artist's Pencil."

FLICK_KA was not the first automated photography machine to be conceived and realized by Peter Weibel. For *Metaphotography* (*Meta-Photographie*, 1989, see pp. 158–59 in this volume), he built a device that took pictures by itself, with no need for instructions or any conscious intervention on the part of the user. By means of a focus trap, the device automatically turned itself on and emitted a camera-like click, which indicated to any viewers walking into the "trap" that their picture had just been taken. At the same time, they were also confronted with a screen and a synthetic voice, by means of which the machine declared its autonomy and its unwillingness to be used.

The following year, Weibel expanded this interactive media piece to include a second version. *Lucifer's Dream: Non Serviam; The Revolt of the Automata* (*Luzifers Traum: Non Serviam. Der Aufstand der Automaten*, 1990, see p. 160 in this volume)

FLICK_KA

2007	Peter Weibel and Matthias Gommel interactive installation; photo booth, 5 monitors, 5 computers, custom software, projector, projection screen technical execution: Arne Gasser, Nikolaus Völzow, Manfred Hauffen

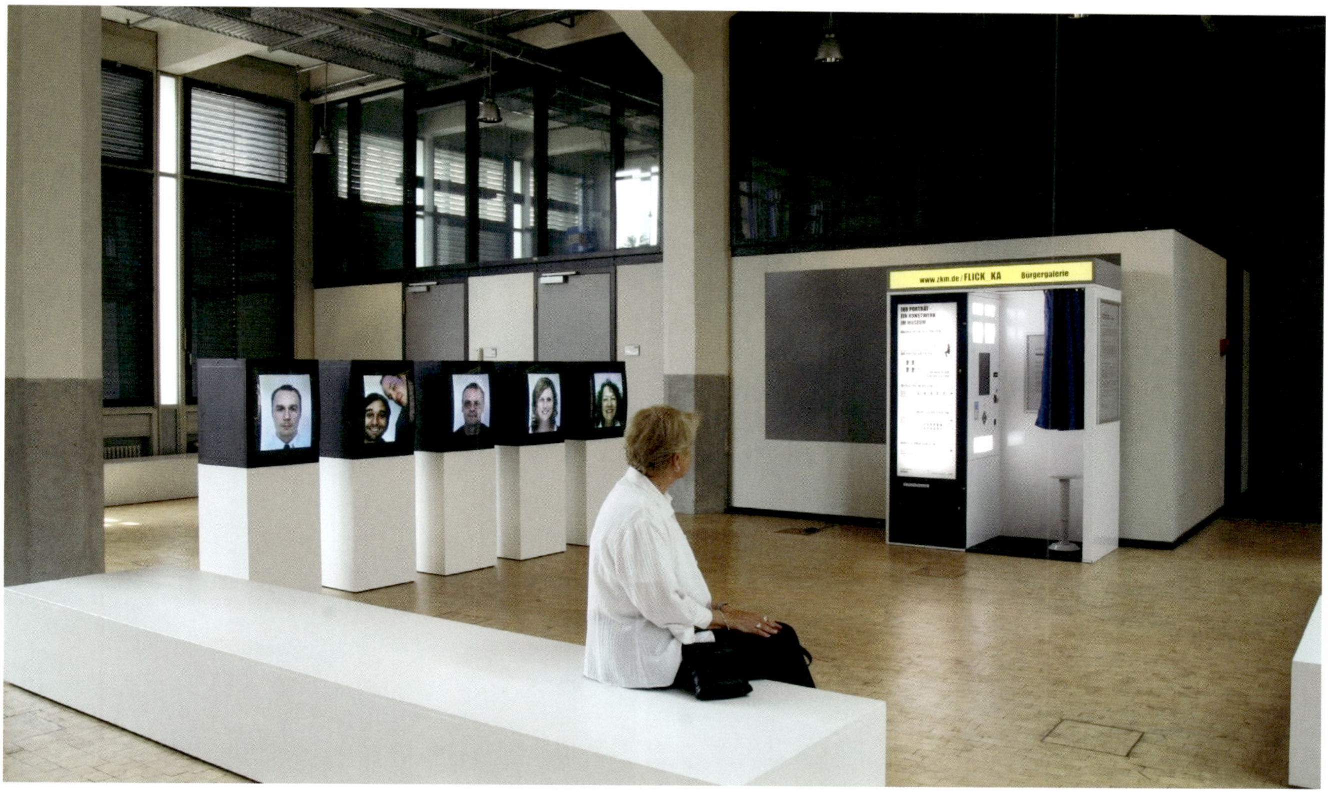

Installation view
ZKM | Karlsruhe, 2015

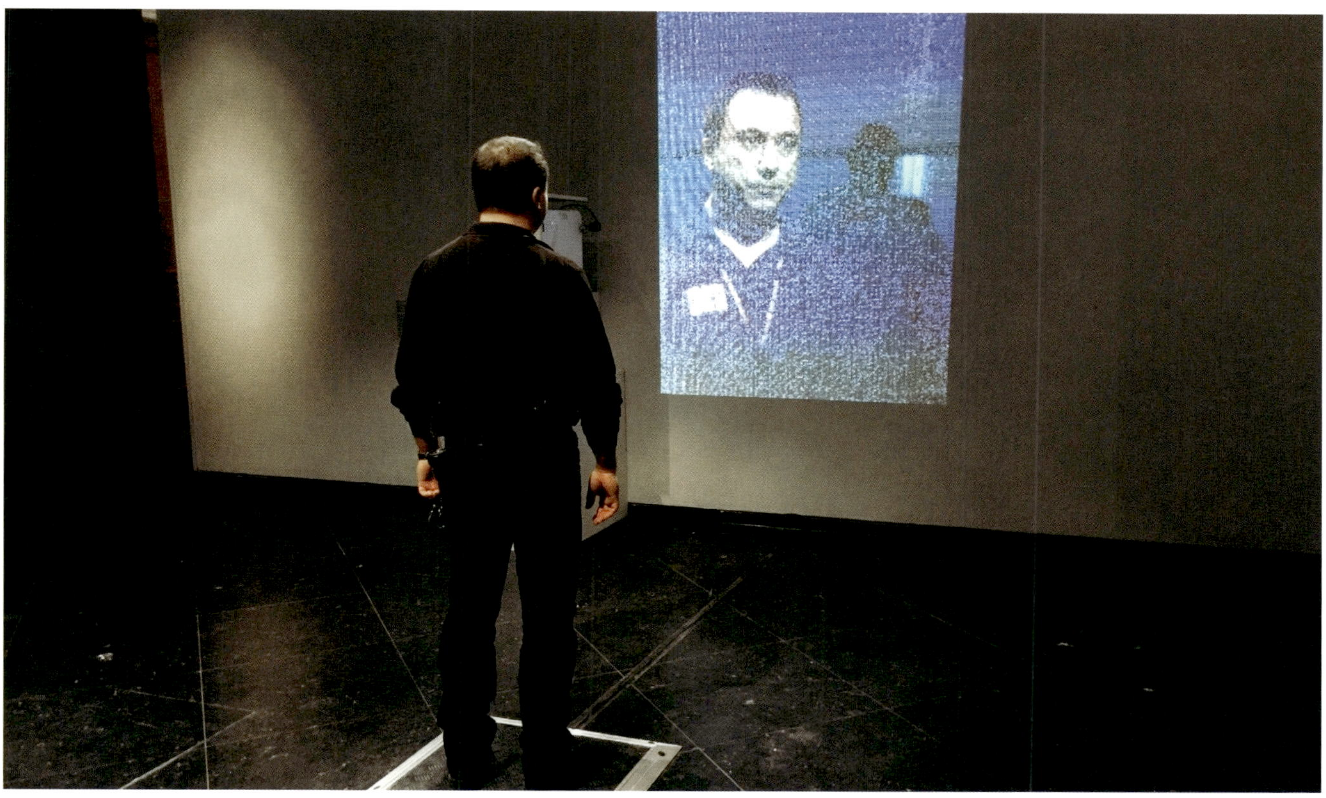

Peter Weibel and Matthias Gommel, *FLICK_EU*, installation view Galerie des Citoyens, Lyon, 2016

added two car doors, one on the left-hand wall and one on the right, to the setup of *Metaphotography*. While the machine delivered its monologue, the windows in the doors moved up and down, seemingly of their own volition. These autonomous actions, in contravention of the functions normally ascribed to these objects and technologies, further heightened the sensation that a revolution of autonomous intelligent machines and media might be in our future.

These three works are linked by Weibel's aim of turning a picture machine into a user machine, of automating the mimetic function of photography while simultaneously calling it into question, thereby triggering a cognitive process in the viewer's mind. With *FLICK_EU* (2007/15), an extension of *FLICK_KA* co-organized with the Kunsthalle Karlsruhe and museums in Lyon, France, and Edinburgh, Scotland, it became possible to exhibit the portraits in all three cities simultaneously and in real time, allowing museum visitors to be virtually present in other places as well as their own city. In essence, *FLICK_EU* represented a model of the European community in miniature.

In 2018, the artist Daniel Heiss expanded the concept of *FLICK_KA* to create a work of his own involving artificial intelligence, titled *FLICK_KAi*. In this piece, he showed different portraits with a question written beneath them: Born or generated? Using the the thousands of visitor portraits that had been recorded up to that point, he then used an artificial intelligence to create portraits of people who do not exist. These portraits of digitally generated people were more or less indistinguishable from the portraits of actually existing people. Viewers were invited to guess which was which. After a while, the answers appeared below the images.

Translated from the German by Patrick Hubenthal.

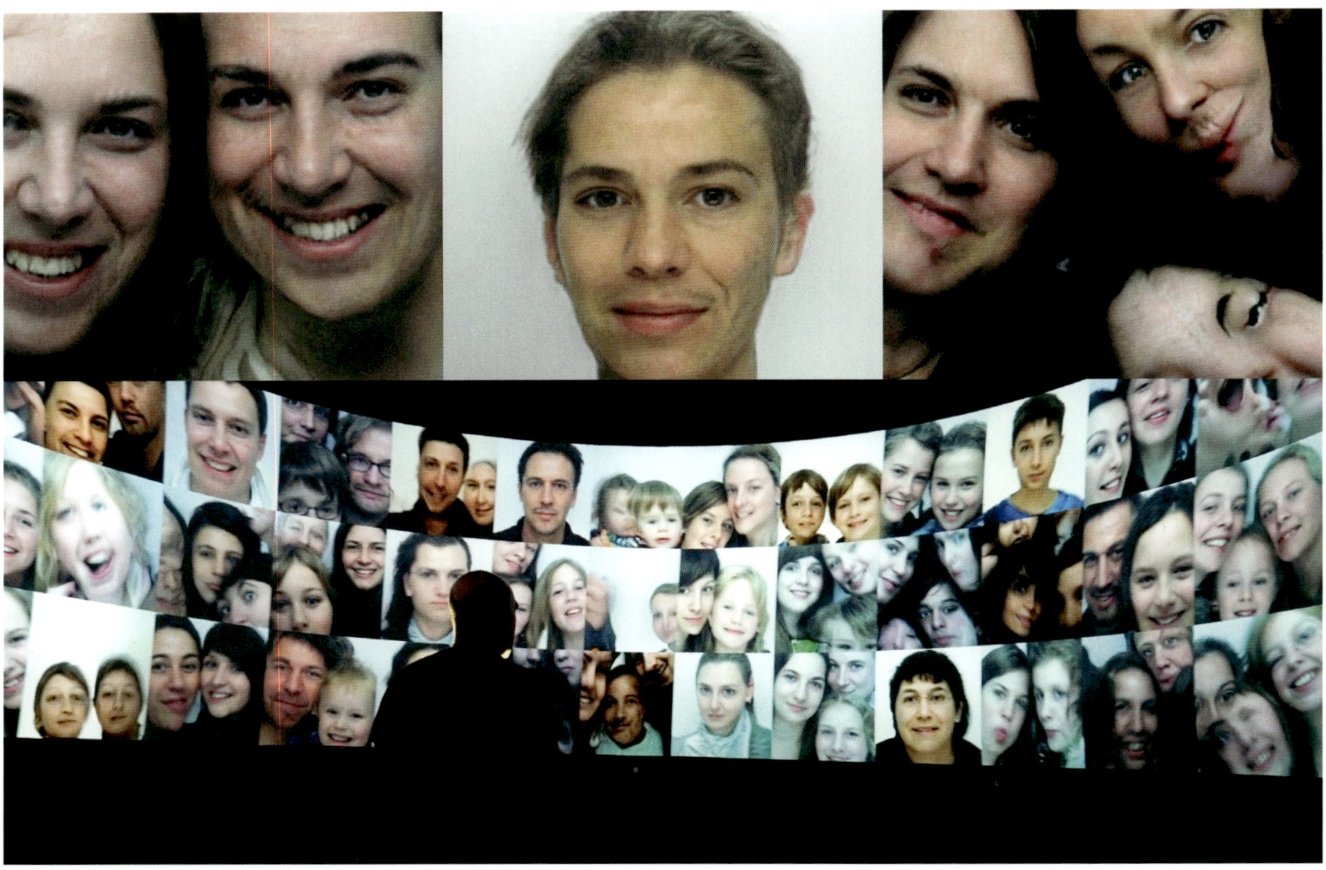

Daniel Heiss, *Flick_KAi*,
2018, installation view
ZKM | Karlsruhe, 2018

Everyone Can Photograph Everyone: The Democratic Promise of Photography

Portrait photography: A guarantee of fame
Having pictures made of yourself was once the preserve of the religious and political aristocracy. Art specialists and social elites shared a monopoly on the production of images. In part, these portraits served as guarantees of fame and glamour, admiration and immortality for their subjects. Few could paint a picture, and few could pay for one.

Photography's democratization of portraiture
Everything changed with the invention of photography in the nineteenth century, and especially with the introduction of the business-card-sized *carte de visite* portrait by A. A. E. Disdéri in 1854, which touched off a calling-card epidemic. In the age of photography, anyone can take anyone's picture, and everyone can have their picture taken. Unlike the aristocratic medium of painting, photography is a democratic visual medium.

You too can be famous! Become a part of FLICK_KA
Inspired by the popular internet portal flickr.com, the FLICK_KA project invites the citizens of Karlsruhe and all ZKM visitors to be represented by a portrait in the exhibition. You can have your picture taken in the photo booth and be inducted into the FLICK_KA Citizens' Gallery.

Your picture will be automatically added to the online FLICK_KA gallery, where it can be viewed at www.zkm.de/flick_ka. Here in the museum, the ever-growing collection of portraits will be exhibited as a projection that changes once per second. With your portrait, you'll become part of the ZKM Collection and part of the permanent exhibition. The exhibition thus becomes a forum in which, before which, and with which all are equal. The consumer becomes the producer, and the museumgoer becomes the content of the exhibition.

Manifesto by Peter Weibel posted on the side of the photo booth. Translated from the German by Patrick Hubenthal.

Self-Portrait of a Theory in Self-Quotations

Peter Weibel

Expansions of conventional film form on the open stage. The commercial-conventional filmmaking sequence (shooting, editing, projection) must be broken through. (March 1967)

Film is a composite of calculi and operators with which one confronts reality, a system of fundamental rules and figures: projector, projection surface, projection booth, auditorium, audience, projectionist, celluloid, image track, sound track, director, camera, lens, camera movement, editing table, projector speed, editing, montage, camera position, theater curtain, etc. This filmic composite is a heuristic convention that can be changed at any time. Film means nothing more than creating events within the possibilities of the filmic composite - with celluloid, for example, or with just an auditorium, screen, and audience. Filming is creating reality with the resources of the filmic composite. I can take any number of the available elements of the filmic composite, in any order, and use them in any way. One can use a mirror instead of celluloid, a piece of string instead of a beam of light, a human torso instead of a screen, fire instead of electric light; one can use chemical reactions instead of light reactions, etc. The classic film on celluloid is just one element of the filmic composite, just a particular combination of specific elements. (January 1968)

The elements of the filmic composite can be a) placed in a finite set of different combinations, and b) provided in different ways. Film may be understood as a function on the variables: object, celluloid, camera, projector, screen, etc. - i.e., $f(o, c, C, p, s)$. The variable o can represent virtually anything belonging to the filmic composite, since of course anything can be used as an object (and can then be placed in novel combinations). The expansion of film may be understood as the substitution and omission of the filmic composite's elements and combinations, of the phases and preconditions of its production. Why? To destroy the boundaries and dichotomies of classic state reality. Expanded Cinema means expanded reality as well. Changed media produce a changed world, and a world clamoring for change clamors for changed media. Expanded Cinema is not just an expansion of the scale of optical phenomena, but rather, in its current phase, the radical decision to do away with reality and the language that both communicates and constructs it. Expanded Cinema's finite possibilities consist in the finite number of selections from and variations on the elements of the filmic composite. The celluloid film is just one possibility; it is a primitive constraint, a restriction to bourgeois art and the bourgeois world, a retreat to a bygone age. Expanded Cinema goes beyond film, representation, and interpretation. Expanded Cinema does not depict the world; it changes it. Expanded Cinema is an exploration of reality via experiments with light, sound, electricity; with group mechanisms, gamma rays, and enzyme reactions. (Spring 1969)

Note: Adapted from Peter Weibel, "Selbst-Porträt einer Theorie in Selbst-Zitaten," (1967–71) in *Avantgardistischer Film 1951–1971: Theorie*, ed. Gottfried Schlemmer (Munich: Carl Hanser Verlag, 1973), 108–11. Translated from the German by Patrick Hubenthal.

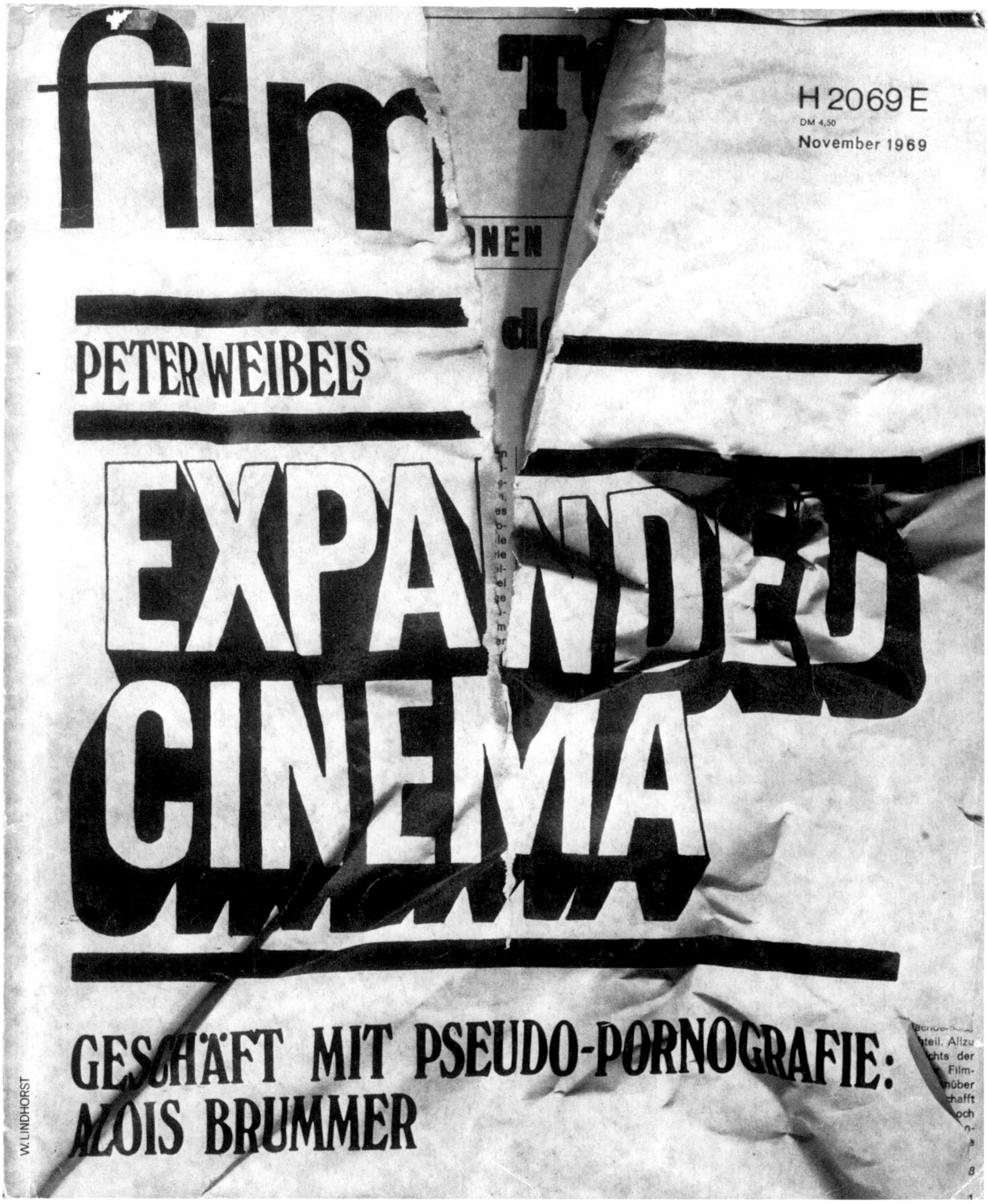

Film was a major monthly magazine that was generally reporting on commercial cinema and art house films. It was put out by the same publishing house as the more conservative magazines *Theater heute* (theater today) and *Oper* (opera) and was the most important film magazine in Germany. But through the events around 1968 the magazine opened up.

For a while I was contributor to the magazine, writing on Sergei Eisenstein, Dziga Vertov, Andy Warhol, Mae West, the Marx Brothers, Alfred Hitchcock, and others. In the 1960s the avant-garde movement was so strong that I could even publish my own vision of cinema in issue 11 of *Film* in ten pages (pp. 41–45, 47–49, and 51–52) and make it to the cover. These excerpts show my definition of cinema as operations with a calculus, founded on algorithmic and set theoretical procedures. My graphic concept for the cover anticipated the expected reaction of the reader. They would tear the magazine apart and throw it away. —Peter Weibel

Expanded Cinema

1969 | *Film*, no. 11 (November 1969), cover design

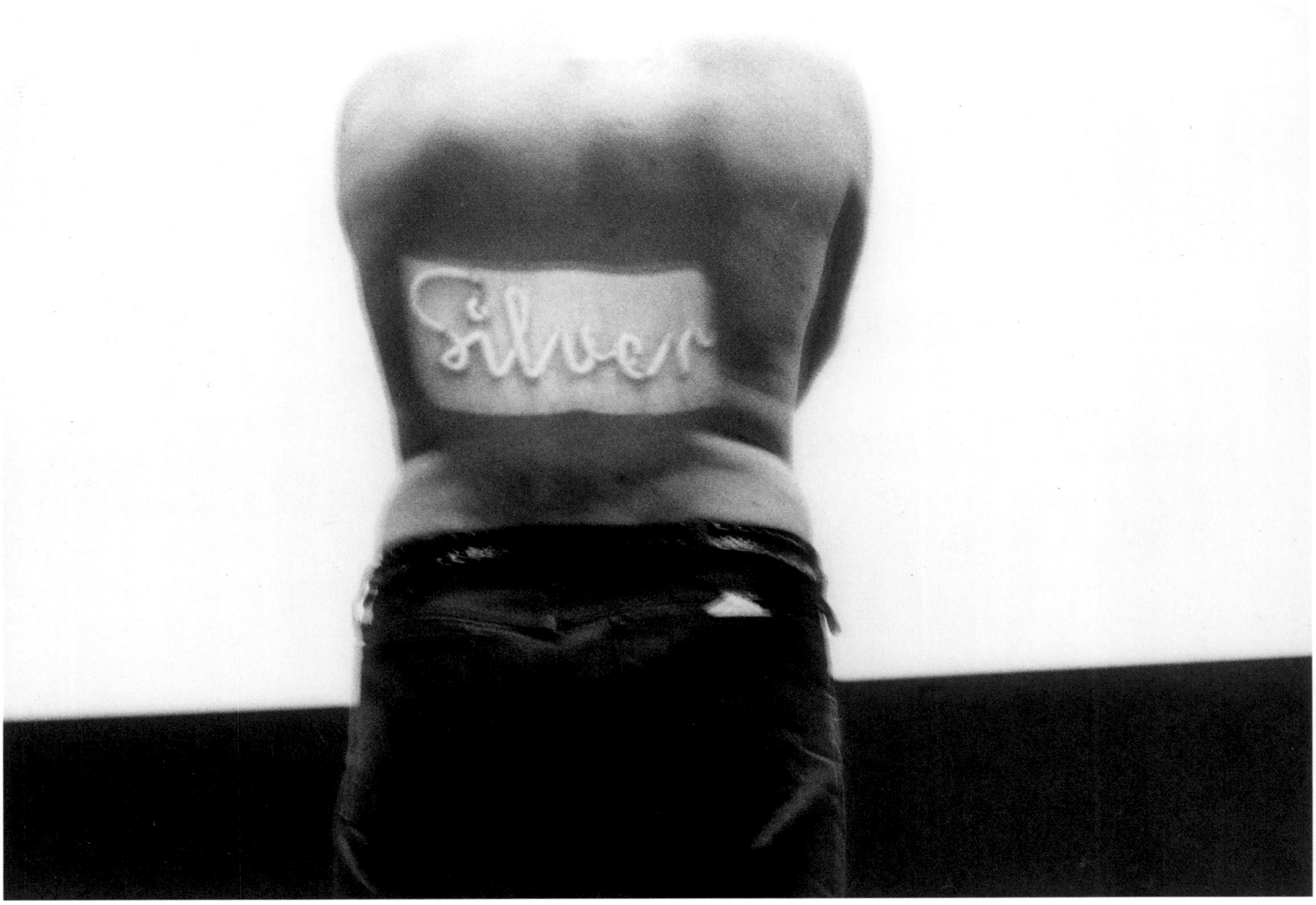

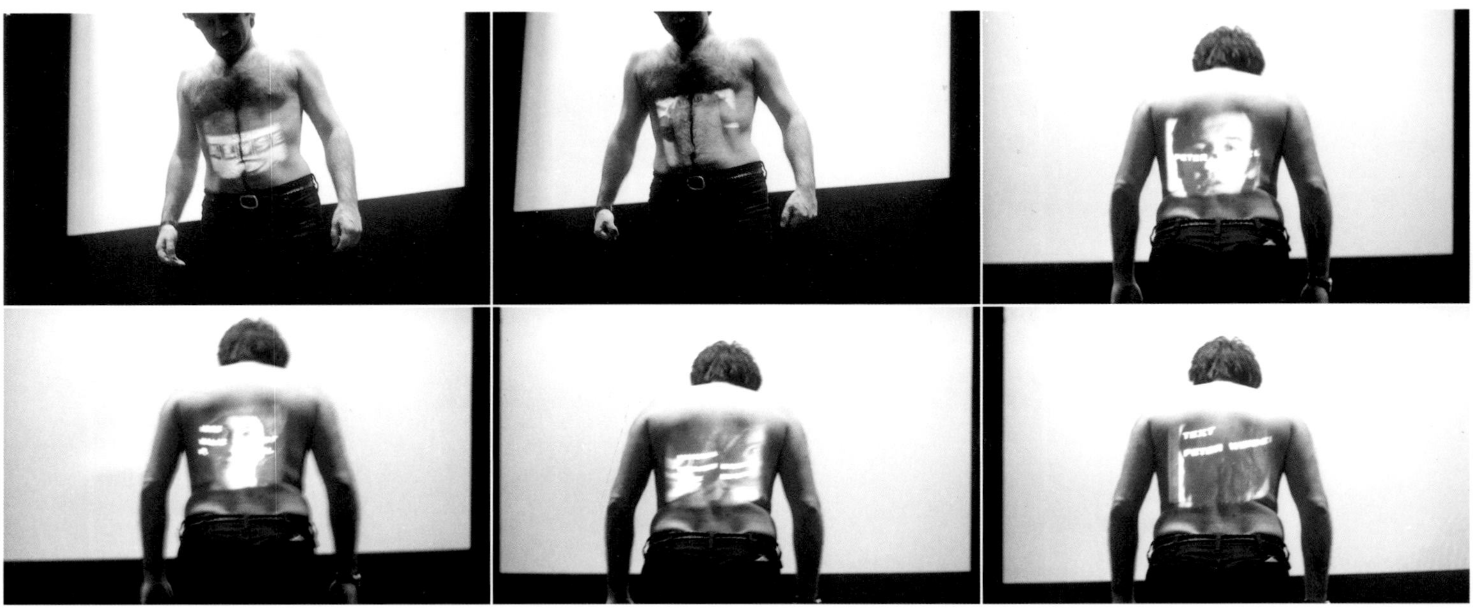

The projection of *Welcome* onto Weibel's body was part of *Action Lecture No. 1*. Photos of a later reenactment

Action Lecture No. 1: Proposals on Non-Affirmative Art

1966	Expanded Cinema with *Welcome*, 8 mm film, color and b/w, ca. 20 min. Destruction in Art Symposium (DIAS), London, September 9–11, 1966 See also pp. 498–99 in this volume

Image-object problem. Model 1: the filmmaker.

The filmmaker serves as a model for the image-object problem. The alteration of the projection surface tends to negate the difference between object and sign, as in the case of the projection of a stomach onto a stomach. In the Expanded Cinema performance *Welcome*, Peter Weibel turned the human body into a projection screen, causing the film he had authored to be projected back onto himself. The environment perceived by the subject and captured by the camera is projected onto the subject itself. This radical expansion of the filmic medium was based on the idea that films always communicate a subjective view of the world. Such a subjective perception is necessarily followed by a subjective projection and the subjective recording corresponds to subjective projection. The film conceived by the filmmaker is cast back upon the filmmaker. Weibel liberated film from the dictates of the conventional screen, allowing bodies to be projected onto bodies, objects onto objects, even buildings onto buildings. The rejection of the traditional movie screen constituted a subversion of the norms of industrial presentation.

The 8 mm film used in the performance, also titled *Welcome*, was shot in Sweden in 1965 and cut in Vienna in 1965–66. Its first intermedia presentation, as part of the performance event *Action Lecture No. 1*, took place at the 1966 Destruction in Art Symposium (DIAS) in London. Half a year later the film was shown again, first as part of *Action Lecture No. 2* at the Österreichhaus in Vienna's Palais Pálffy and then as part of *Action Lecture No. 3* at the Forum Stadtpark in Graz. Two more iterations, *Action Lecture No. 4* and *Action Lecture No. 5* took place in Cologne and Gothenburg in 1968 (see pp. 176–77 in this volume). On each occasion, Weibel appeared shirtless and allowed the film's sequences to be projected alternately onto his chest and back. At the same time, he moved back and forth to the accompaniment of music in front of a screen onto which different movies were being projected – for example, an official film about Stockholm – while also reading a manifesto-like text in German, English, and French, which was simultaneously being played back from tape.

Through Expanded Cinema presentations of this kind, Weibel has sought to circumvent the image-object problem. The alteration of the projection surface – the replacement of the screen with the filmmaker's body, or with a plaster bust – closes the gap between object and sign. The image of a back is now being projected onto an actual back. This allows for a more illuminating treatment of the relationship between *signans* and *signatum* – between signifier and signified – while simultaneously emphasizing the nature of film as medium, the reality of its materiality as contrasted with the representation in the movie theater. This approach renders celluloid and filming superfluous. If buildings are projected onto buildings, rendering image and reality identical to the eye, why then even project an image? Why not simply declare the real building an image? The result is a "film without a film," in Weibel's words, which replaces illusion with unmediated reality and places the focus on the subject's direct experience.

Weibel's investigation of these new ways of expanding the film form and his many experiments in the field of Expanded Cinema resulted in a whole series of works in the late 1960s. For example, in the Expanded Cinema performance *Nivea* (see p. 170 in this volume), which premiered in Vienna on January 26 and 27, 1967, production and projection took place simultaneously rather than consecutively. And the 1969 slide installation *Dumb Show*, which was recreated in 2006 in digital form, centered on the transcendence of a basic piece of cinematic equipment: instead of a screen, the slides were projected onto the steam rising from four bathtubs. What was shown on this ephemeral projection surface of water vapor was a six-minute sequence of photographs from concentration camps, which were continuously dissolving into nothing. In this way, Weibel made visible the mechanism of repression and the traumas inflicted by the crimes of the Holocaust, while at the same time expanding the possibilities of filmic representation by interweaving real and virtual elements and utilizing new materials and devices. Only in this way could the crimes committed in the concentration camps, which had been regarded as beyond the bounds of representation since Adorno, be adequately commemorated. Even Kant condemned the artistic depiction of war, the repulsive, and the like, yet conventional cinema revels in the depiction of violence. Weibel's art breaks free from such technology-based simulation of horror.

Translated from the German by Patrick Hubenthal.

Image-object problem. Model 2: celluloid.

If the site of film is not the screen and the movie theater, then buildings can be projected back onto buildings, bodies onto bodies, meadows onto meadows, etc. When image and object coincide, recording and celluloid become superfluous. "Buildings" are presented as buildings, "Nivea" as Nivea. With celluloid let go, film is created without film; the boards of reality are smashed. The desertion of the norm – the simultaneity of recording and projection, of production and reproduction – represents liberation from industry.

Celluloid, as the *agens* of the sign, is exposed as an illusion just as movement is. By replacing technological reproducibility with immediacy, the objective nature of film and the concept of "reality" reproduced by industry and the state are called into question.

An ironic commentary (hence the subtitle *An Inflatable Promotional Film*) on Charles W. Morris's statement that "the sign reveals itself as the main agency in the development of individual freedom and social integration"[1]. By replacing the technological reproducibility of film with reality, Weibel liberates the cinematic image from the restrictive presentation requirements of cinema and from its use by the commercial film industry.

Note: This text is a composite of the following sources: Peter Weibel and Harald Falckenberg, eds., *Peter Weibel: Das offene Werk 1964–1979*, exh. cat., Neue Galerie Graz, Landesmuseum Joanneum (Ostfildern: Hatje Cantz, 2006), 520, and an excerpt from an explanatory text accompanying performances at Palais Pálffy, Vienna, January 26–27, 1967, and Forum Stadtpark Graz, May 2, 1967, quoted in Weibel and Falckenberg, *Peter Weibel*, 523. Translated from the German by Patrick Hubenthal.

[1]
Charles W. Morris, *Foundations of the Theory of Signs* (Chicago: University of Chicago Press, 1938), 38.

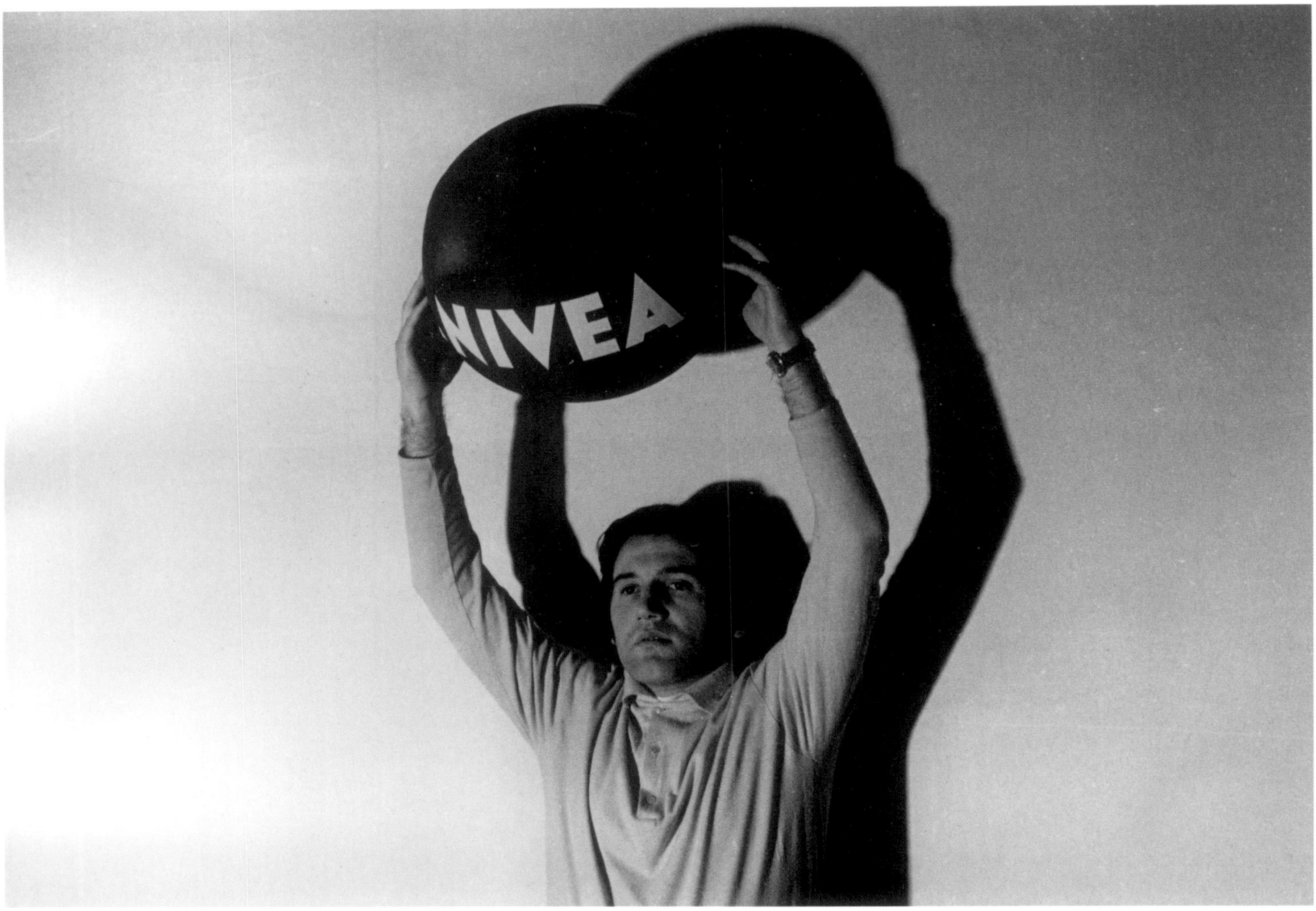

The photo shows Peter Weibel during the Expanded Cinema performance *Nivea*, which premiered in Vienna on January 26 and 27, 1967. In the light of a projector, the artist stands motionless in front of a screen for one minute holding a beach ball with the word "Nivea" in his raised hands. The technology of film is split. There is a projector present, but it projects no image, only light. Since no celluloid is transported, the light creates a blank field on the screen. Meanwhile, a tape recorder plays the sound of a film camera. The camera is not visually present as an object, only aurally. The visual and audible parts of film, sound and sight, are split. The typical sound of projection is heard, even though there is no film in the projector. Therefore, production (the camera sound) and projection (the light) do not take place consecutively, but rather simultaneously. Some of the operators of the cinematographic apparatus are dispensed with, while the order of others is changed. The cinematographic apparatus is split into its visual and aural parts, but reality is also split into its different sensory modalities. Without projecting a film, the artist creates "a film without film" in which the cinematic representation of reality is replaced by immediate reality. The image and the object of the image converge. Both the image and the film stock, the material carrier of the film, become superfluous.

Nivea

1966/67	Expanded Cinema performance, film without film; projector, tape recorder, screen, Nivea ball

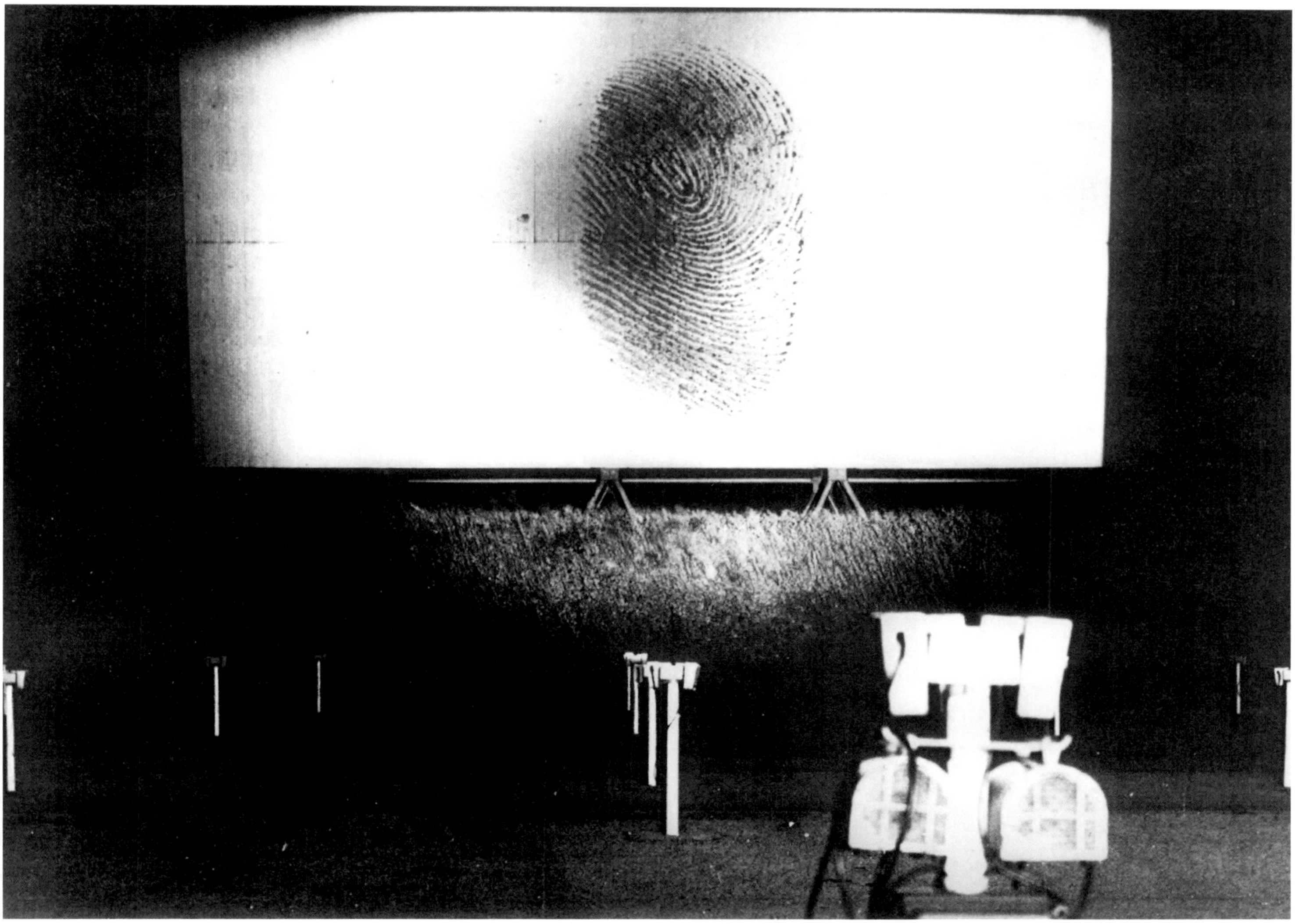

Screening at a drive-in cinema, Vienna, 1969

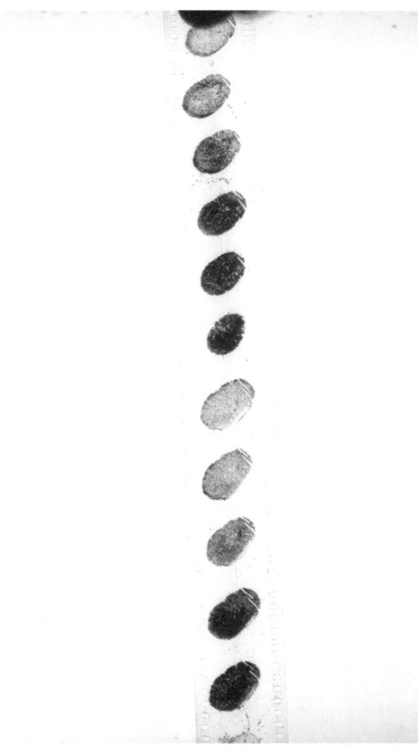

Image-object problem. Model 3: the print.
Synonyms: copy, proof, imprint, trace, impression.
A film copy is called a print. A mark left by a finger is called a fingerprint, which is an indexical sign (or copy) of the finger. The strip of glass film shown in this piece has had Peter Weibel's fingerprints applied directly to it. It is therefore not a copy (print) of a film, but rather of a finger. The film is an original. The thumb's tip has left a decorative scattering of dry ink particles on the film. The medium of film stock transforms this imprint from an object into an image. At the same time, the sign, the fingerprint, has a physical quality. It is what Charles Sanders Peirce terms a "sinsign" or "index."[1] The content of the film is not created by exposing it to light in the usual way, but by touching it. The conventional light trace is replaced by a tactile trace, by a trace of touch. Fingerprint is therefore close to works such as *Touch and Feel Cinema* (*Tapp- und Tastkino*, 1968, see pp. 180–81 in this volume). When the film is projected, the ink particles not only form abstract visual patterns but also function as an optical soundtrack and thus emit noise, whether on 35 mm, 16 mm or 8 mm film.

Fingerprint

1968 | 35, 16 and 8 mm films, ca. 1 min., india ink

[1]
Charles Hartshorne and Paul Weiss, eds., *Collected Papers of Charles Sanders Peirce*, vol. 2, *Elements of Logic* (Cambridge, MA: Harvard University Press, 1932), 142–43.

Image-object problem. Model 4: projection.
A film without celluloid and camera, only with a beam of light, tapes and hall lights. A beam of light is projected onto a wall and the possible shadows of passers-by etc. are fixed with black adhesive tape by the public or filmmaker. After some time a surface of lines can be seen on the wall.
—Peter Weibel, 2006

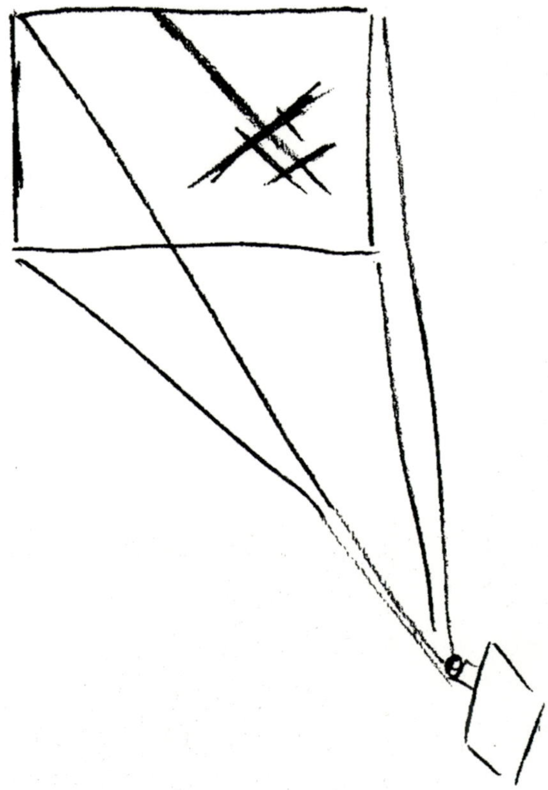

Poor Cinema / Armes Kino

1967 (idea)/ 1968	Expanded Cinema, metafilm, film installation; spotlight, tapes; realized at Melk Abbey, May 23, 1968

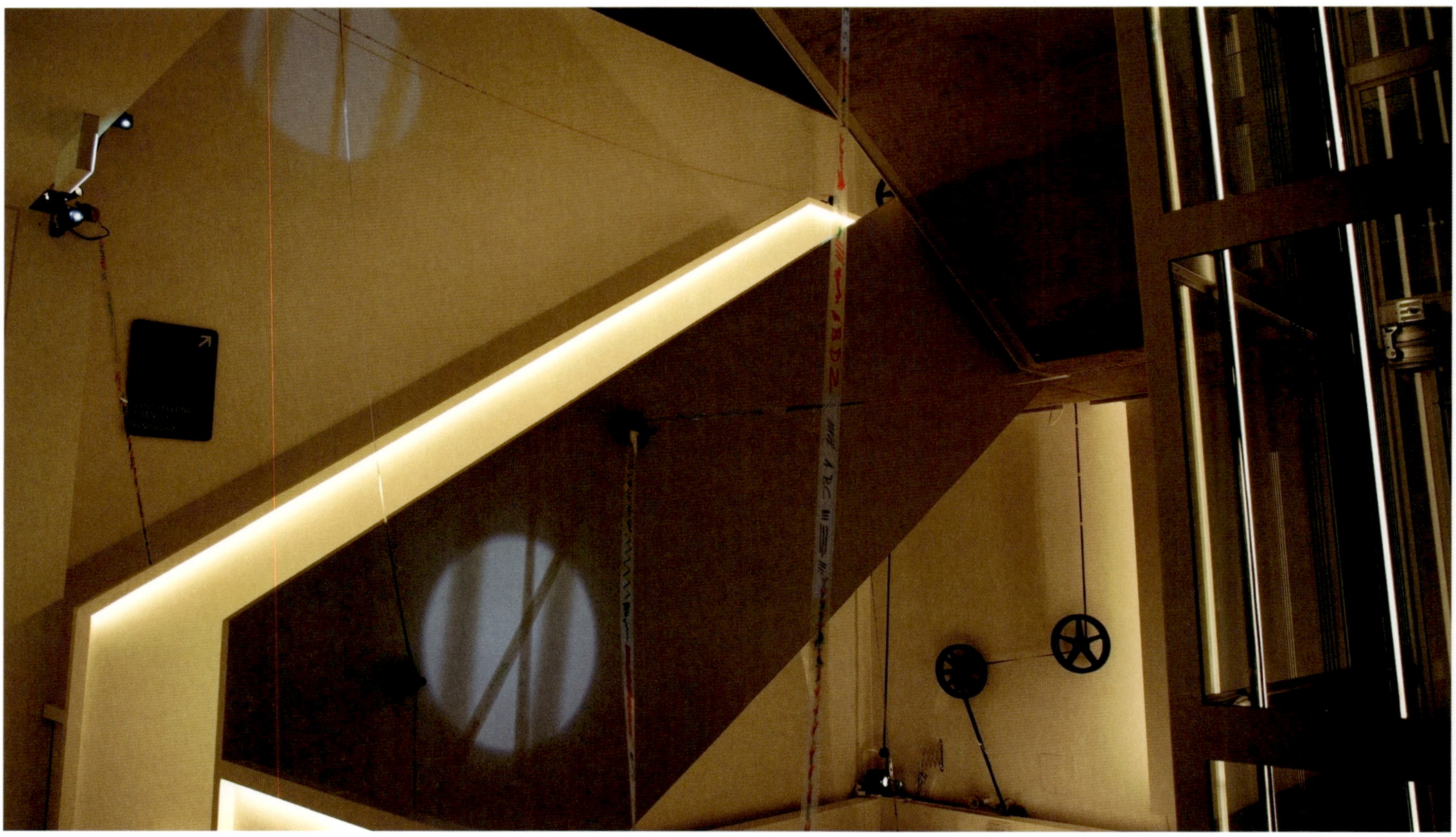

In the Shadow of the Cinema / Im Schatten des Kinos

1966/ 2015	film environment; hand colorized film strips, film reels, lamps

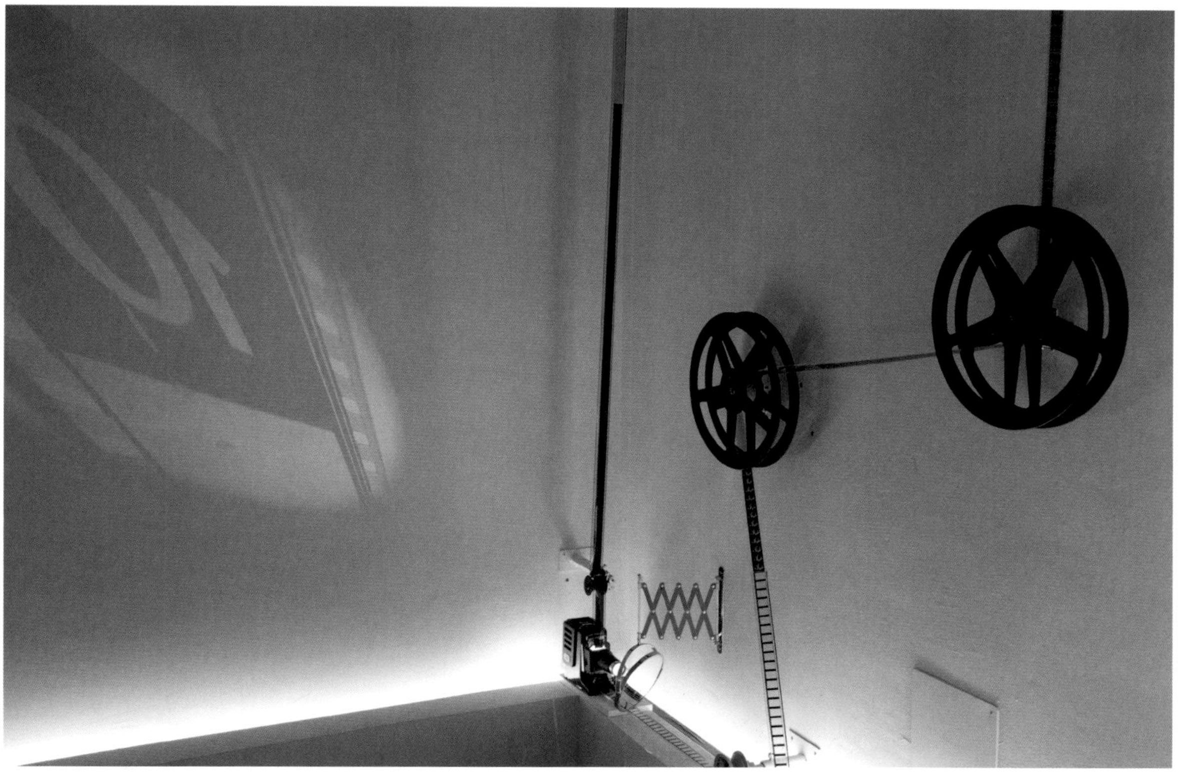

Installation views METRO Kinokulturhaus, Vienna, 2015

I have been making films since 1964, and ever since then I have been trying to expand the perception and definition of what film is. First, I analyzed the material and machines cinema uses, such as celluloid, projector, camera, screen, auditorium, etc. These cinematic operators form an ensemble of processes that – used in the correct order – end up producing a moving image on a screen. I then began to switch off these operators, for instance using natural fire instead of electric light, or leaving certain operations out altogether, for instance films without celluloid or without a camera. This type of experimental cinema, with its emphasis on the materiality of communication, I called either pre-, para-, or postcinema, or Expanded Cinema. The height of my experimentation with these ideas occurred in 1966, but only from 1967 did I find sporadic opportunities to implement these ideas. In January 1967, I was able to perform *Nivea*, a film event during which I used no celluloid. Since film is identified with celluloid, my early film experiments are "film without film." On May 23, 1968, during a student conference at Melk Abbey, I was able to realize several of these "films without film," such as *Poor Cinema* (*Armes Kino*, idea 1967, see page opposite): The shadows of incidental passers-by were thrown onto a wall in a darkened auditorium by means of a spotlight. I, the filmmaker, affixed these shadows with tape. Over time, the wall developed a weave of lines which represented the film. This performance, of course, recalls the legend of the origins of painting.

In his *Natural History*, the oldest extant encyclopedia of ancient knowledge, Pliny the Elder described the painting's origins in shadow painting. According to Pliny, painting was invented when a person's shadow on a wall was traced in with lines.[1] Sculpture, Pliny further tells us, was created when the daughter of the Corinthian potter Butades used lamplight to trace the silhouette of her beloved on the wall before he went off to foreign lands. The father then filled the outline with clay, creating a relief.[2] I tried to find a new equation for the art of projection and shadow painting by mixing shadow painting with the moving image and light painting with the shadow image, allowing both to enter into multifarious relationships. During this time, I also wanted, in accordance with modernism's reductionist program, to eliminate any trace of the representation of the world of objects, so I developed several projects that used only the materials of cinema in self-reflecting ways. My goal was the self-representation of the means of representation; that is, the self-representation of materiality and the operators of cinema, such as film reels and projector, etc. The aim was not to run celluloid through a projector to create images; instead, the projector was to be exhibited as projector, painted over with wax or paint. The celluloid was to lie in a film can and be exhibited as a film object. The piece *In the Shadow of the Cinema* intended to show projectors, film strips, lenses, reel spools, cameras, etc. as such. I conceived of a large space the audience could enter and in which the audience would be surrounded by many strips of film strung across the space on reels. The cinema auditorium itself was to be a large projector inside of which the audience could walk around. After all, from the camera to the projector, classic cinema is wheel technology, creating the illusion of movement as image. I now wanted the audience to see this technology, but not the images created using it. The space with filmic materials and operators was to be a walk-in kinetic sculpture.

—Peter Weibel, 2015

Translated from the German by Peter Dahm Robertson.

[1] Pliny the Elder, *Naturalis Historia*, vol. 25, chapter 5.

[2] Pliny the Elder, *Naturalis Historia*, vol. 25, chapter 43.

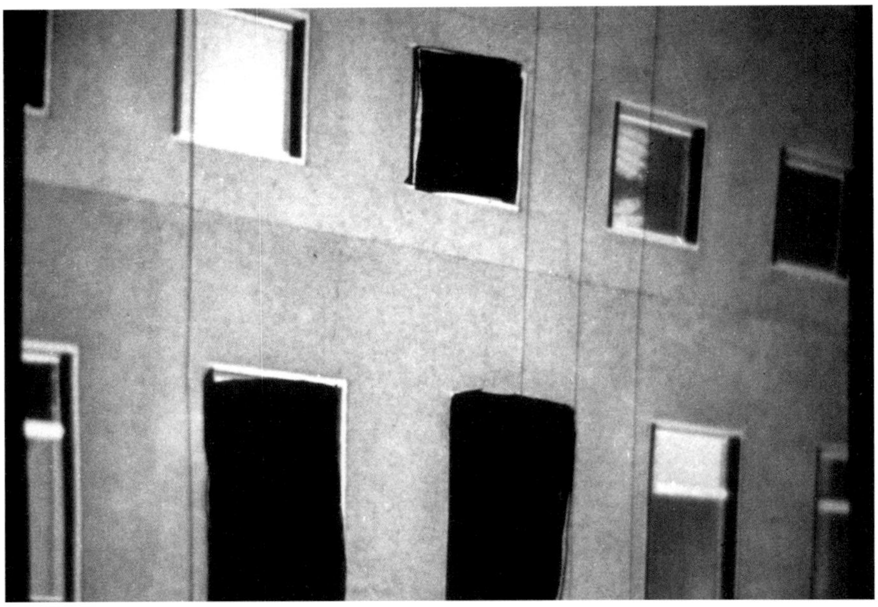

Part 1

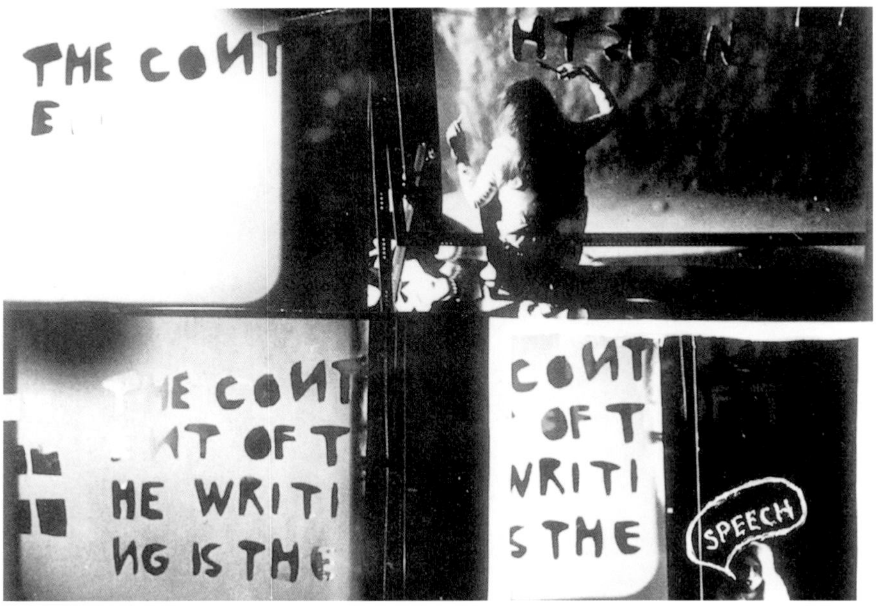

Part 2

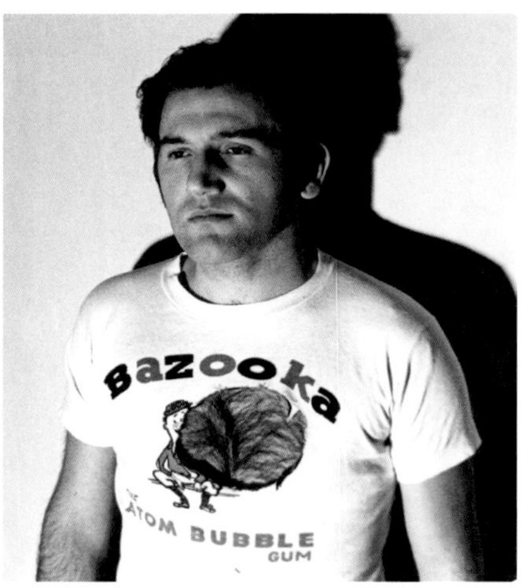

Part 3

Cutting

1967–68	VALIE EXPORT in collaboration with Peter Weibel Expanded Cinema performance, intermedium; 2 paper screens, slide projector, Bazooka bubble gum t-shirt, razor, scissors

Image-object problem. Model 5: the screen.

Part 1: Opening, a documentary. The windows in a building facade projected onto a paper screen are cut out with scissors and opened like real windows.

Part 2: A talkie, homage to Marshall McLuhan. The phrase "The content of the writing is the" is cut out of the illuminated paper screen. The word "speech" is then spoken, completing the sentence. The meaning of the sentence is thus realized: the phrase was written, but the word "speech" was spoken.

Part 3: A comedy, homage to Bazooka Joe. The conventional screen is replaced by a T-shirt on a body, and the bubble-gum "speech balloon" issuing from Joe's mouth is cut out of it.

Part 4: A silent movie, homage to Greta Garbo. Chest, stomach, and pubic hair are shaved off. A naked body becomes the screen; the technical term "body cutter" (a part of the film splicer) is realized. The hair is the signs on the screen (the skin of the body).

Part 5: Fellatio. The wall of a building served as screen (part 1), then a paper wall (part 2), then a piece of cloth (part 3), then the naked body (part 4). Sign language turns to body language, a film without signs. The "speech film," the fellatio, turns into a silent film. The mouth speaks in a different way.

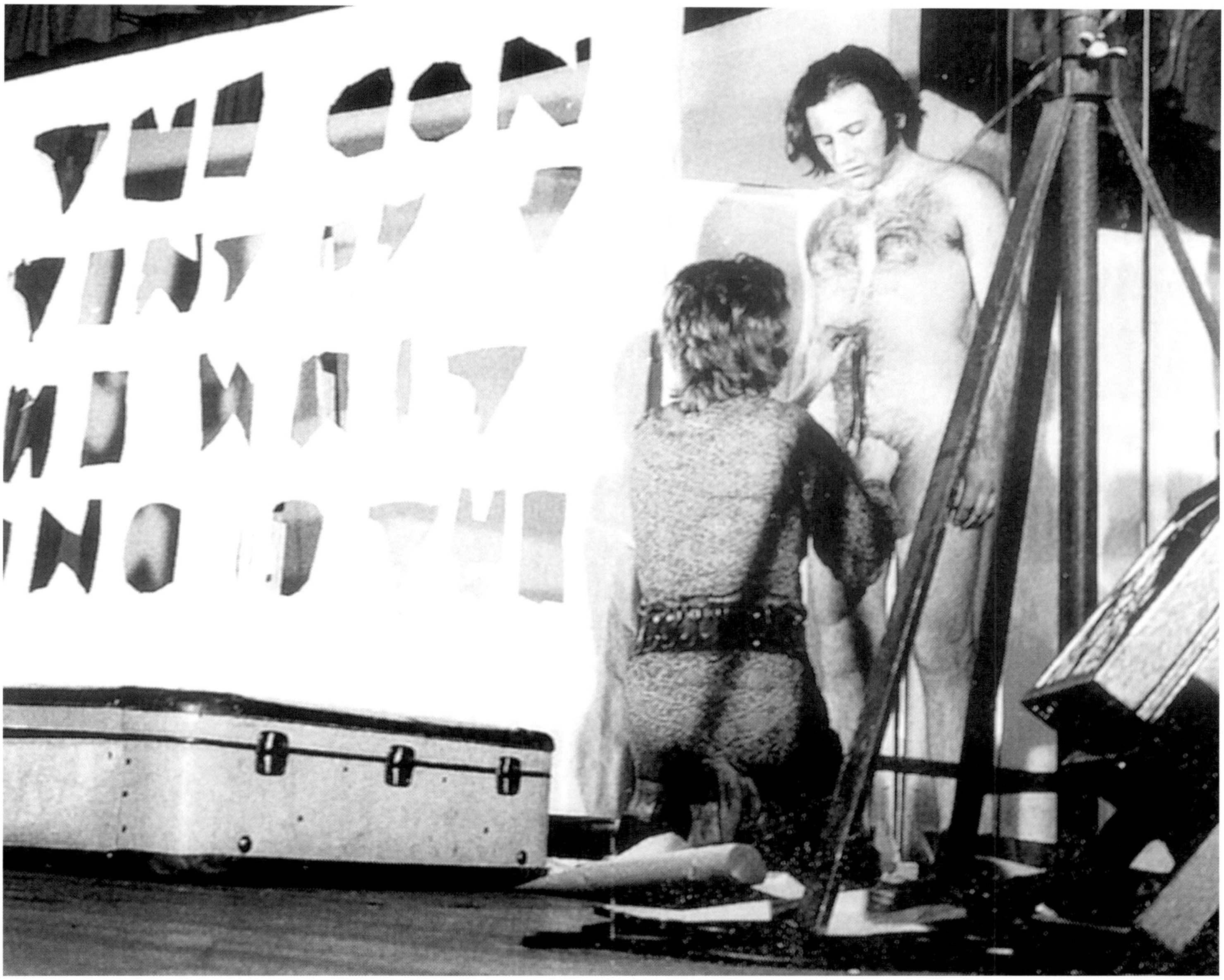

Part 4

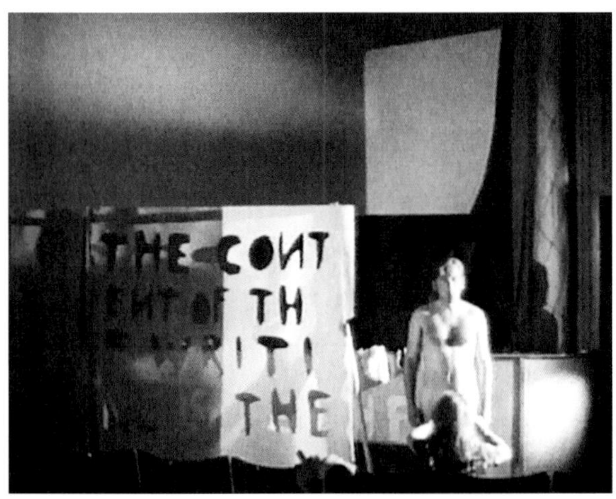

Part 5

The film editing process, also known as cutting, is the operator in the production of film and reality. What was cut was not celluloid, since my Expanded Cinema let celluloid go; rather, the material became the tool. The cutting tool was variously transformed and combined with other elements of the filmic composite. —Peter Weibel, 2006/22

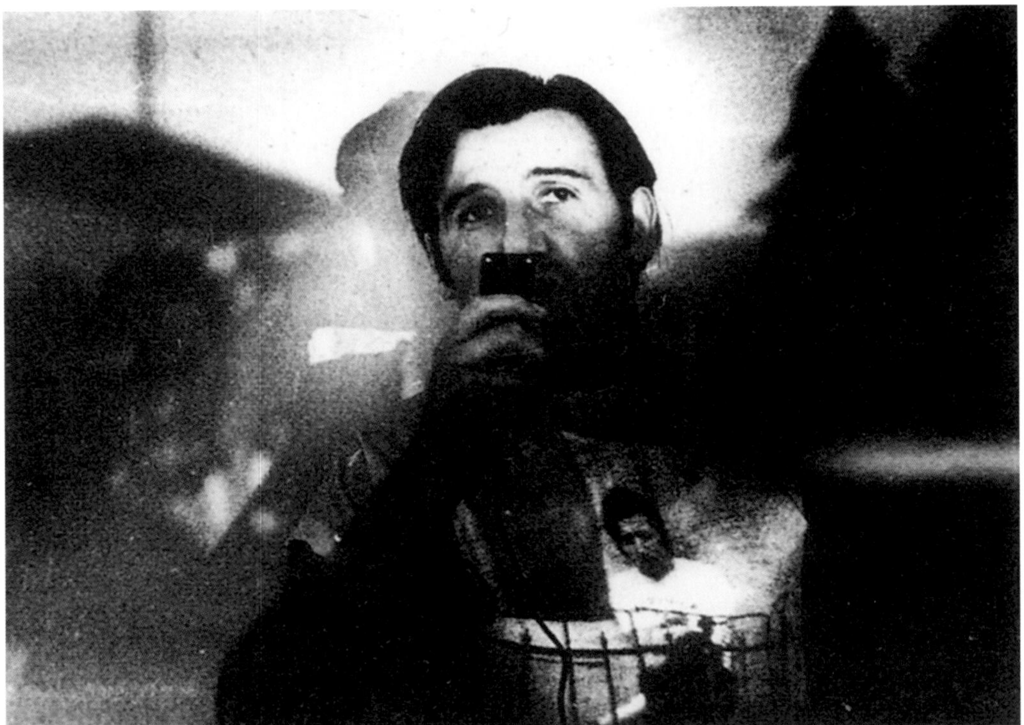

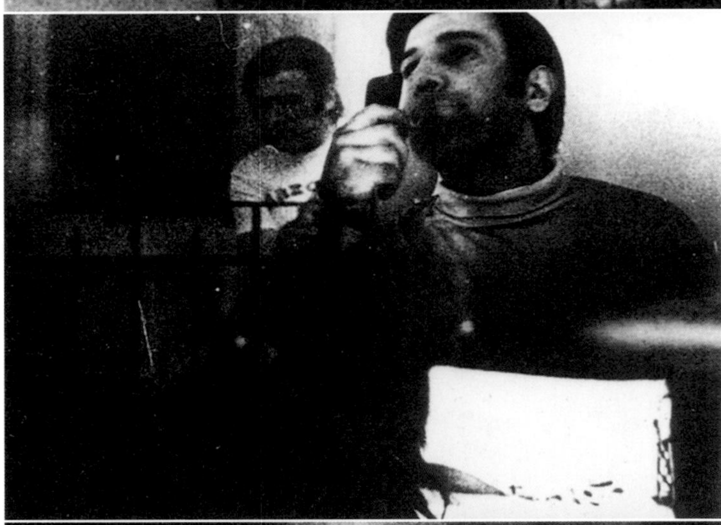

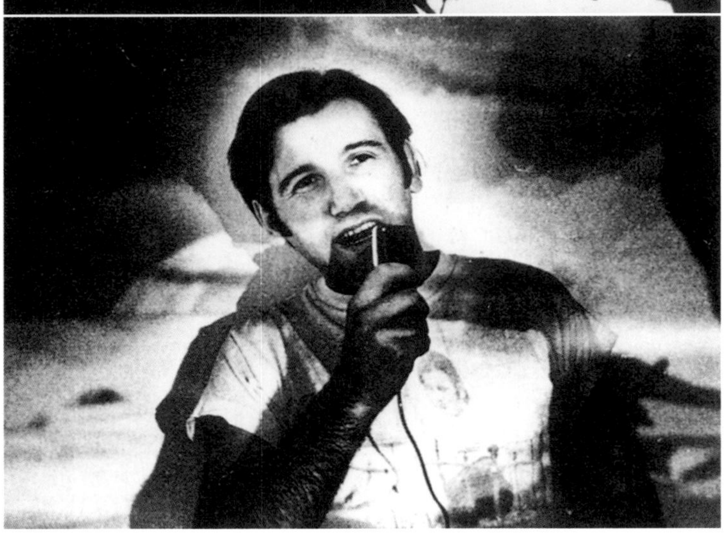

Action Lecture No. 4

1968 — intermedium, Expanded Cinema, multiple projection, performance, communication action; performed at XSCREEN projection space, Cologne, March 24, 1968, with the film *Denkakt* (Act of Thinking) by Ernst Schmidt Jr., 1968

Image-object problem. Model 6: sound.

What appears on the celluloid or the screen is up to the filmmaker. Sound is subject to this dependency as well. Whether and when sound appears on the celluloid or on the screen, which sound correlates with which image, is a question of art(ifice). The sound track can be sprayed onto the celluloid and then scraped off again. One can work directly on the sound track or not. In *Action Lecture No. 4*, the sound track was sprayed onto me rather than onto the film strip, and the audience could scrape it off of me. The sound was split off from the image and transferred onto me: I carried a portable tape player with me, whose sound (a speech by me) could be controlled by the noise of the audience, as could the music on a second tape player. The noise, the screams of the audience, were picked up by a microphone. An electronic circuit transmitted the signals to a lamp, which began to light up when a certain noise level was exceeded. If the audience noise got loud enough, the light came on. Its rays fell on the light-dependent resistor (LDR) in front of it. The LDR was connected to the tape players and a film projector. Only when the LDR was illuminated did the projector and tape players receive current; otherwise the circuit was broken. So when the audience screamed loud enough, the light came on, the LDR was illuminated, and the tape players and projector turned on and produced sounds and pictures. The noise volume in the auditorium, the auditory manifestations of the audience, thus controlled the brightness of the spotlight and, consequently, the volume of the tape players and the projection of images. If they screamed, projection happened: bright light, loud sound, distortion of the projected image and the sound produced by the tape player. If they were silent, no projection: no light, no sound, apart from the live sound of my voice. By the same token, I could act as a kind of dictator and break the circuit by holding my hand between the lamp and the LDR. Our communication was therefore subject to manipulation.

—Peter Weibel, 2006

Translated from the German by Patrick Hubenthal.

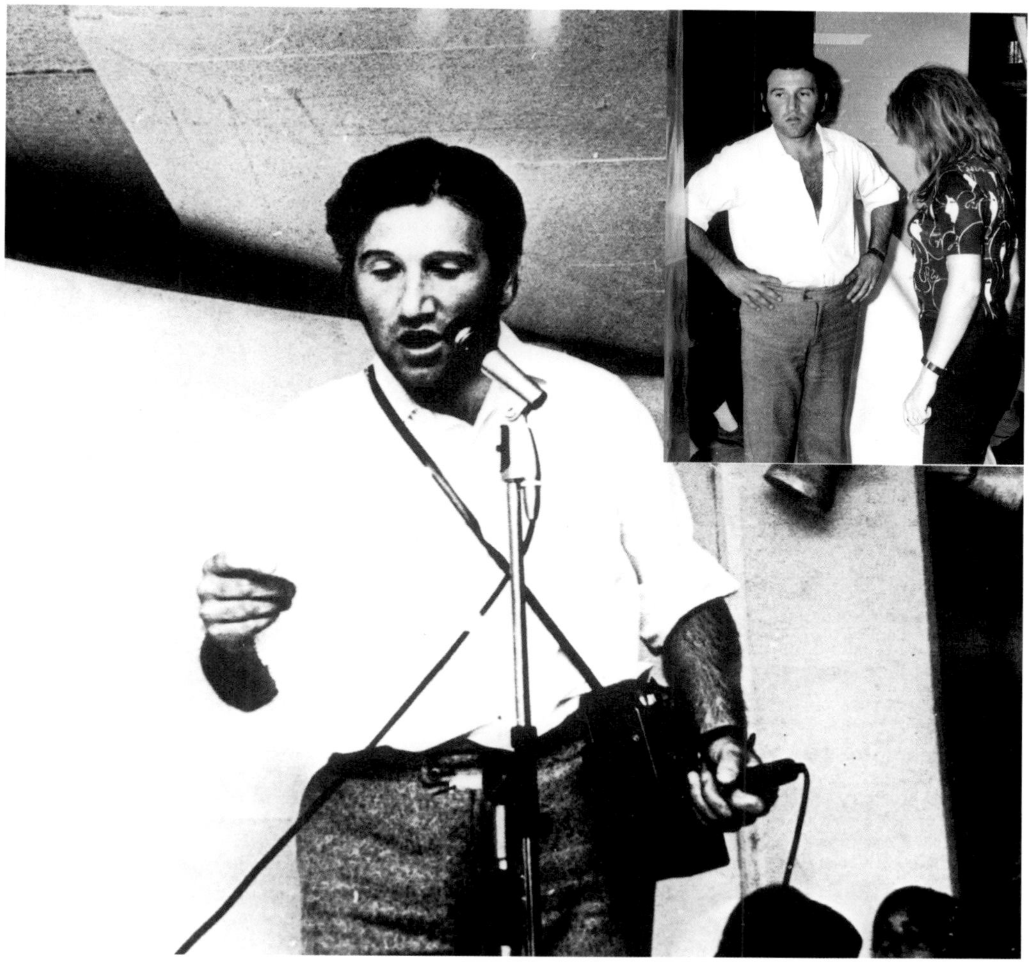

Sight doesn't see any longer.
Speech doesn't speak any longer.
Perception is a stupid animal in us. It does not report what a man feels and thinks, what a man is hoping and suffering.
Language is a secondhand instrument.
It only transports words used by others, and experiences used by others.

What we see is not what we want to see.
What we speak is not what we want to speak.
Our communication is a closed loop.
Our system is a communication breakdown.

Perception is picturing a world not changing the world.
Perception is outdated, is a communication of an outdated phase of evolution.
I want to be free of my eyes who are the treacherous link to the world.
I want to be free of my skin linking me with this room.
I want to explode the catecholamin depots in my nervous system
I wipe off the centuries of my skin
I push the seconds of seconds to the universal space of time that all the billions of future seconds fall down
I destroy my head to destroy the light-reactions
I throw me out of the post-synaptic switch of my eyes intoxicated by the fever of time, put out the yearthousands.

What your eyes tell the brain are simple stories.
They show you a house but not the crime of rental payment, they show you a restaurant but not the crime of paid food, they show you streets and cities but not the criminal investigation of place and time, not the criminal division in proprietors and not-proprietors.

Our eyes are not shocked of the daily picture of the world.
Our eyes are agents of the state. I want the end of perception
I want the end of this world.
To be free from eyes means to be free of deadly television sets, free from the neuronal networks of the state.
Day for day information handling as retinal bombardment day for day my eyes closed in to the excrematory channels of the nation.

I see the junk the state is producing, I see the junk world as the state it is producing. I do not want any longer to see.

Man is a bad picture processor. Filmmakers are the worst examples. Picturing this world is reinforcing this world.
They have not yet attacked the exploring of the future forms of our perception, our communication.

—Peter Weibel, 1968

Action Lecture No. 5

1968	performance; 2 tape recorders, microphone, projector, LDR, spotlight; performed in Gothenburg

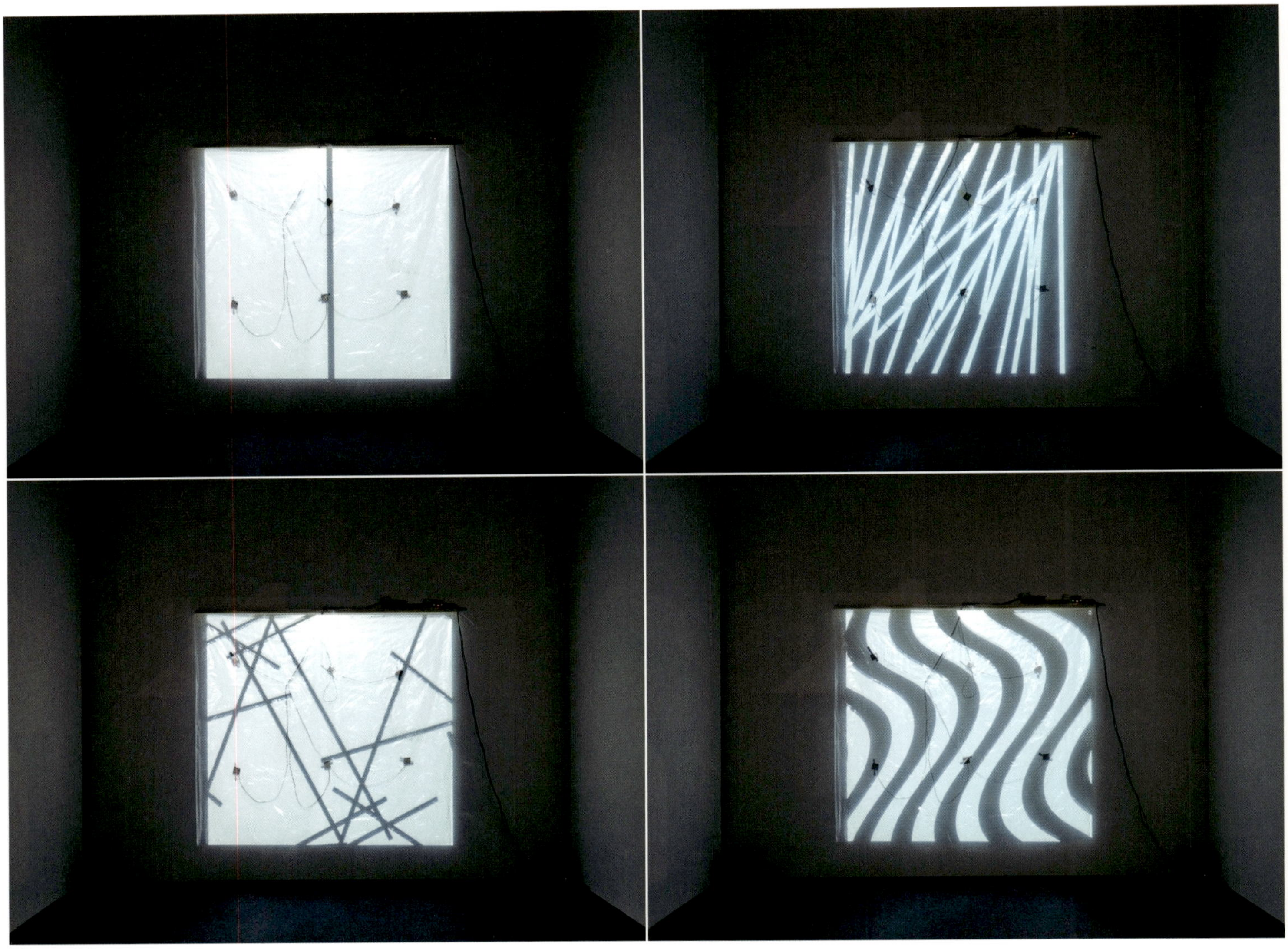

Installation views
respectively, Peter Weibel,
ZKM | Karlsruhe, 2019–20

The Magic Eye / Das Magische Auge

1969/
2019

VALIE EXPORT and Peter Weibel
participatory Expanded Cinema installation; plastic screen, sensors, computer, projector, speakers, wiring, video, b/w, 5:21 min.
reconstruction: Manfred Hauffen
production of the video: Christina Zartmann

The installation *The Magic Eye*, reconstructed for the exhibition *respectively, Peter Weibel* in 2019–20 at the ZKM | Karlsruhe, is considered the world's first autogenerative sound screen, in which sound is not produced in the projector via the film's audio track, but by projecting an abstract film onto a foil equipped with electronic photocells and relays. The brighter the light on the screen, the higher the sound; the darker the light, the lower the sound generated by converting light waves into sound waves. For the original projection in 1969, Weibel made use of the film *Helga Philipp* about an Op Art painting by the artist of the same name made in 1965 by his friend, the Austrian filmmaker Kurt Kren. For the exhibition at the ZKM he conceived a new film. By stepping between the screen and the film projector, the sound can be controlled and manipulated by the moving shadows of the audience.

Installation view *multi media 1*, Galerie Junge Generation, Vienna, Apr. 10–11, 1969

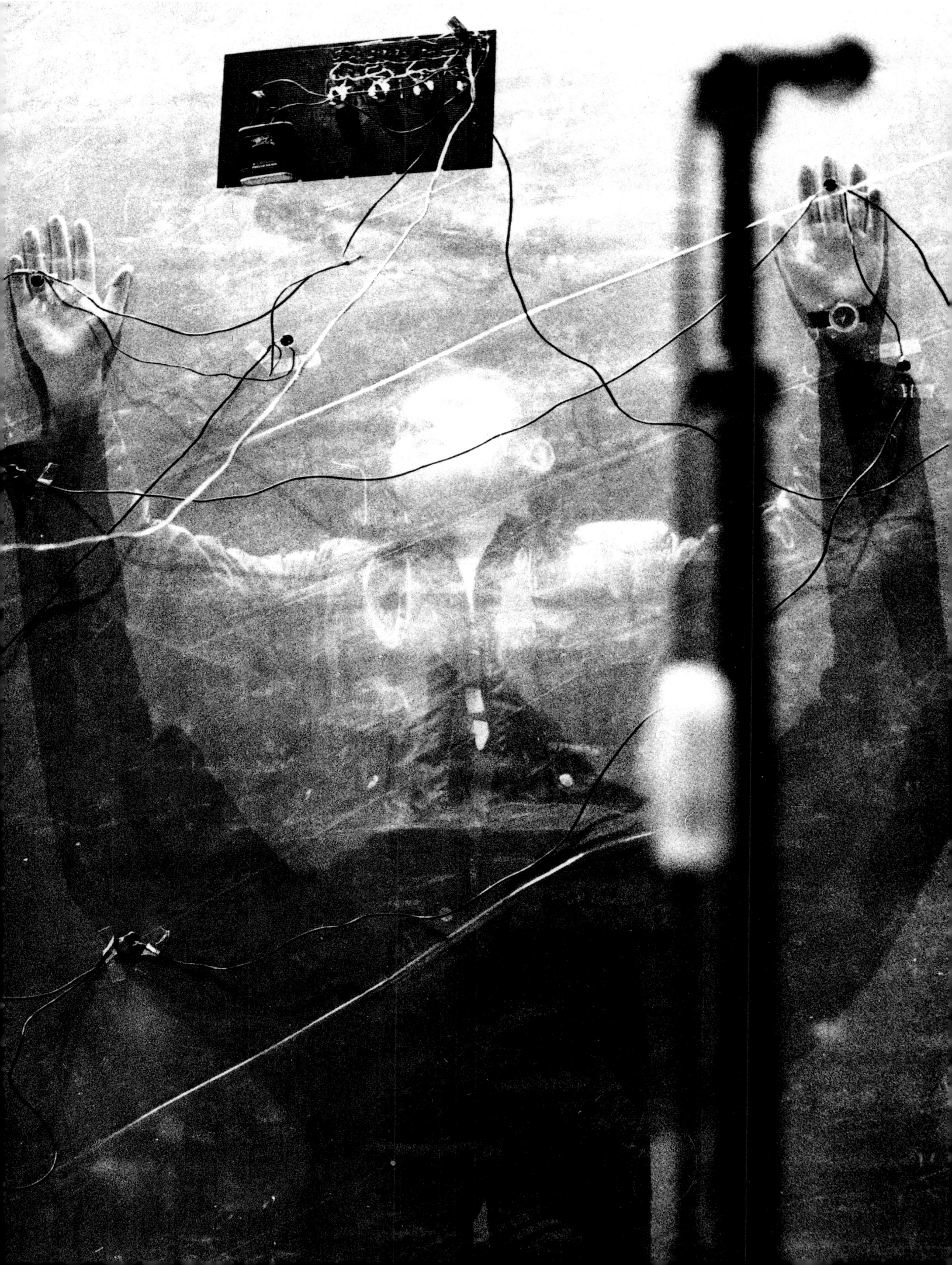

TOUCH AND FEEL POEM

every person becomes the author of their own poem. in every new environment a new poem.

a) environmental experience environmental living environmental learning actor navigates his environment by touch
b) search and meaning the actor roots around in his pockets etc. until he finds the desired object (key, lead, coin, etc.)
c) my favorite poem actor feels his way along his girlfriend's body

Note: Peter Weibel, *Tapp- und Tastgedicht*, 1967, performance at Atelier Zobel, Vienna, 1967. Translated from the German by Peter Dahm Robertson.

Top: Performance in Munich, Nov. 14, 1968, Birgit Hein (left), Dieter Meier (background, second row, second bearded man to the right), Dore O. (behind the camera wearing a hat), Werner Nekes (behind the cameraman), bottom: Werner Nekes (in the center)

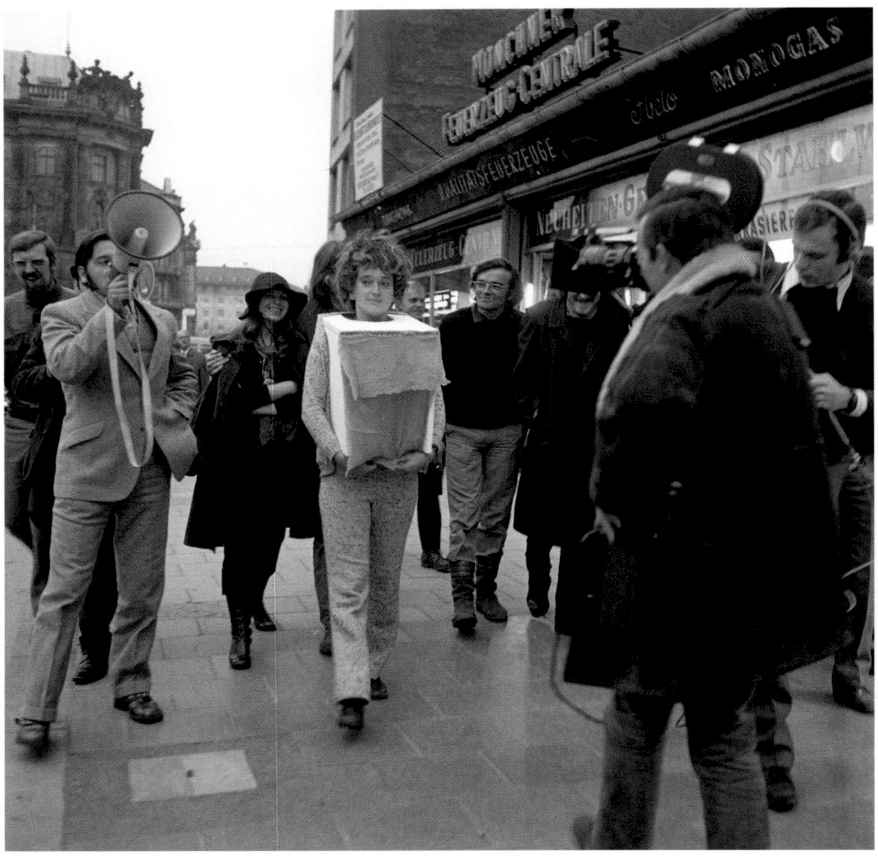

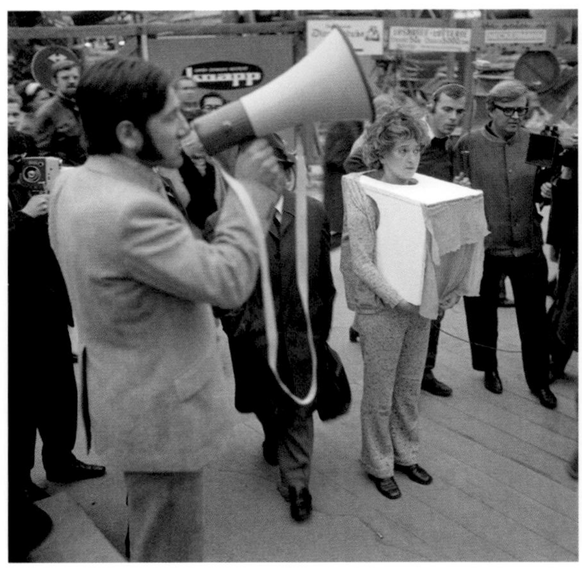

Performance at Karlsplatz (Stachus), Munich, Nov. 14, 1968, from left to right: Peter Weibel, Dore O., VALIE EXPORT, Wilhelm Hein, Dieter Meier

Touch and Feel Cinema / Tapp- und Tastkino

1968	VALIE EXPORT in collaboration with Peter Weibel Expanded Cinema performance documented on video, b/w, sound, 2:32 min.

Touch and Feel Cinema is one of the early Expanded Cinema experiments that extended the perception of film and cinema, and it is also an important work of feminist activism. In this participatory street performance, VALIE EXPORT strapped a box open at the front and back – the down-scaled movie theater – in front of her naked breasts. With a megaphone, Peter Weibel called on passers-by to visit the cinema – in other words, to put their hands through the curtain attached to the front of the box and feel her breasts (max. 12 sec. per view). Thus, in the performance film was not perceived as an optical-visual sequence of images reproduced on celluloid, but as a tactile, direct experience. The performance undermines the distance between spectator and screen, and at the same time demonstrates the representation of women as objects in the mass medium of cinema. In an ironic and provocative way, it literally renders "touchable" what the voyeuristic gaze seeks to "feel" in the visual medium of film.

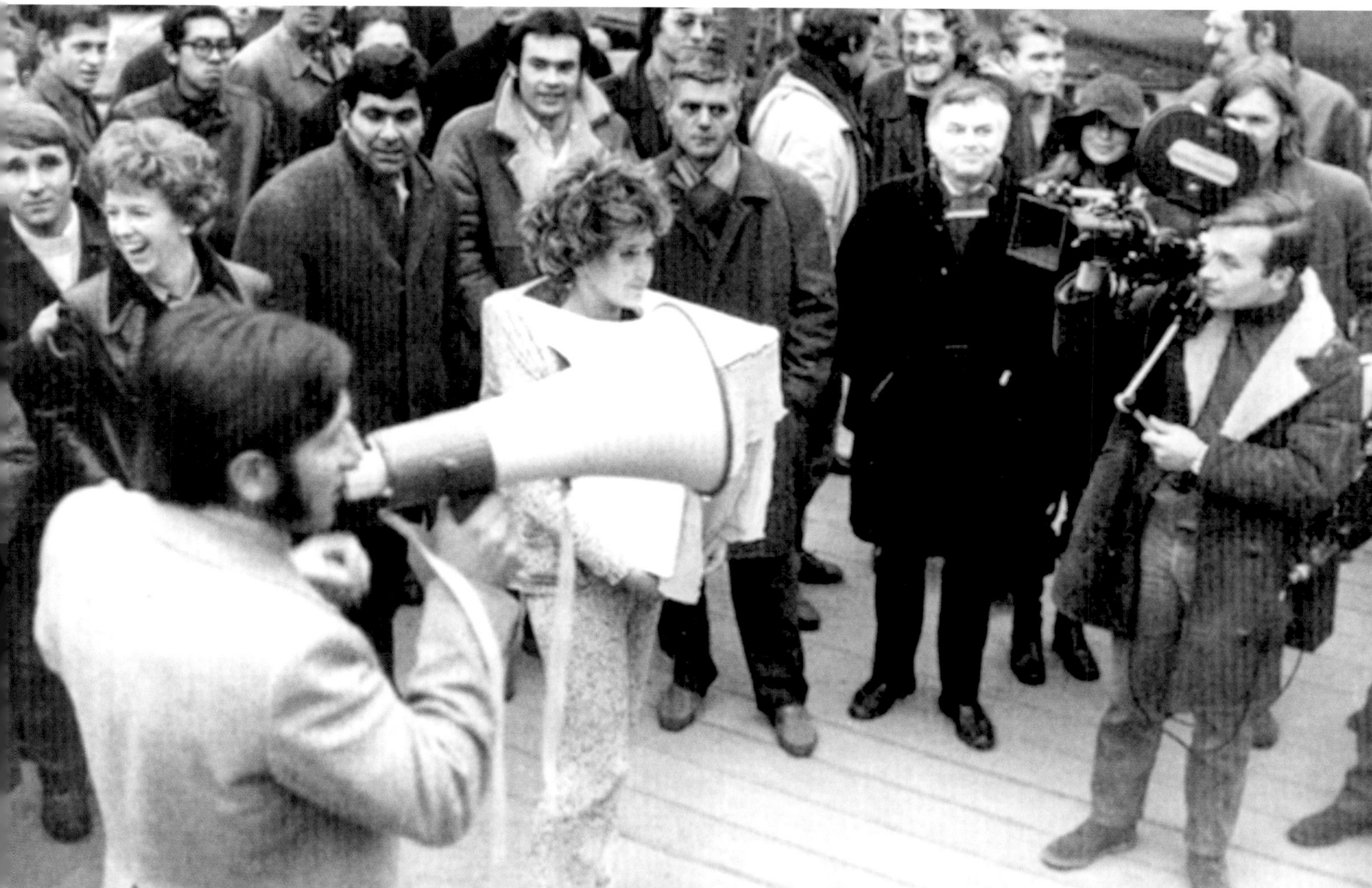

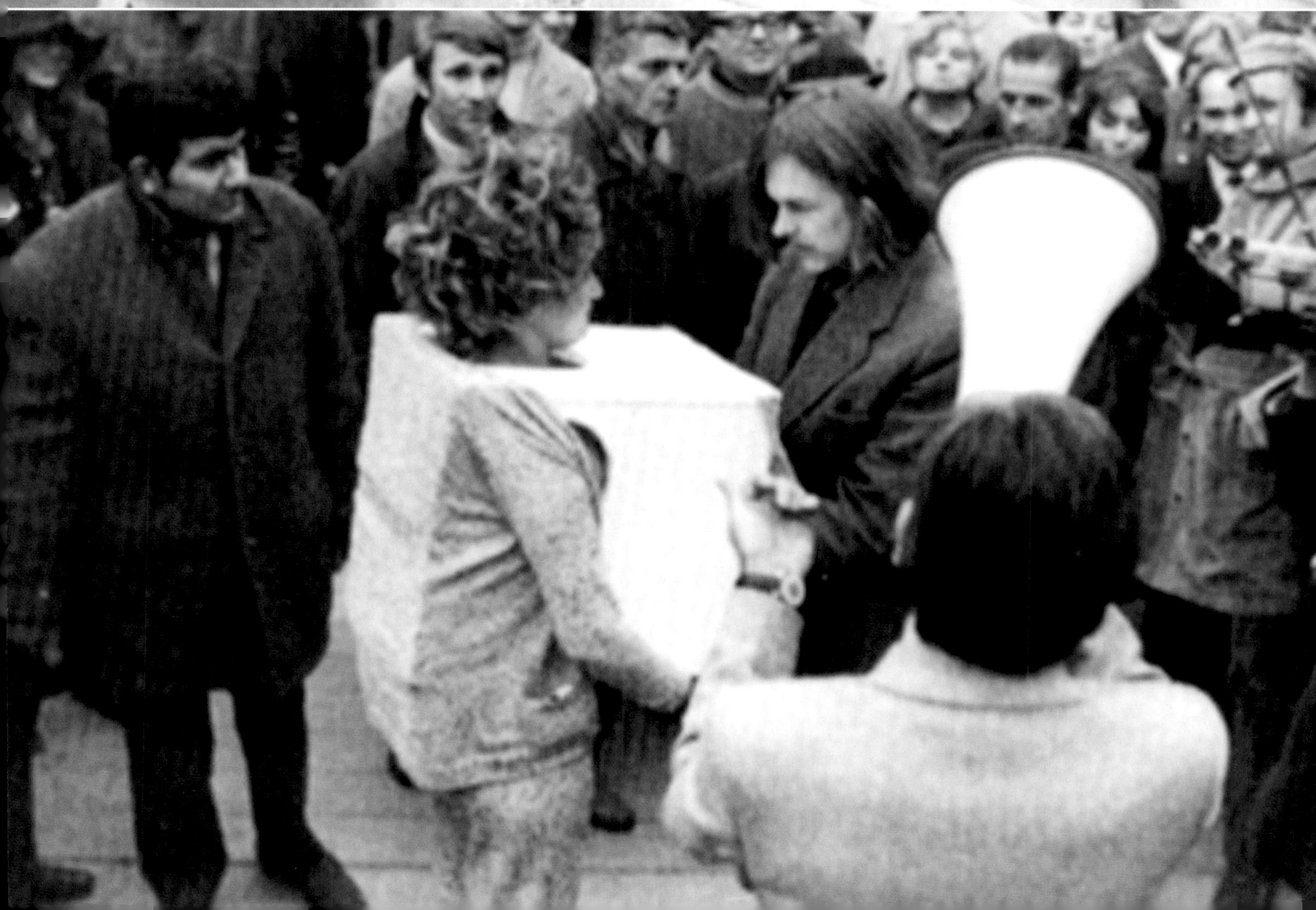

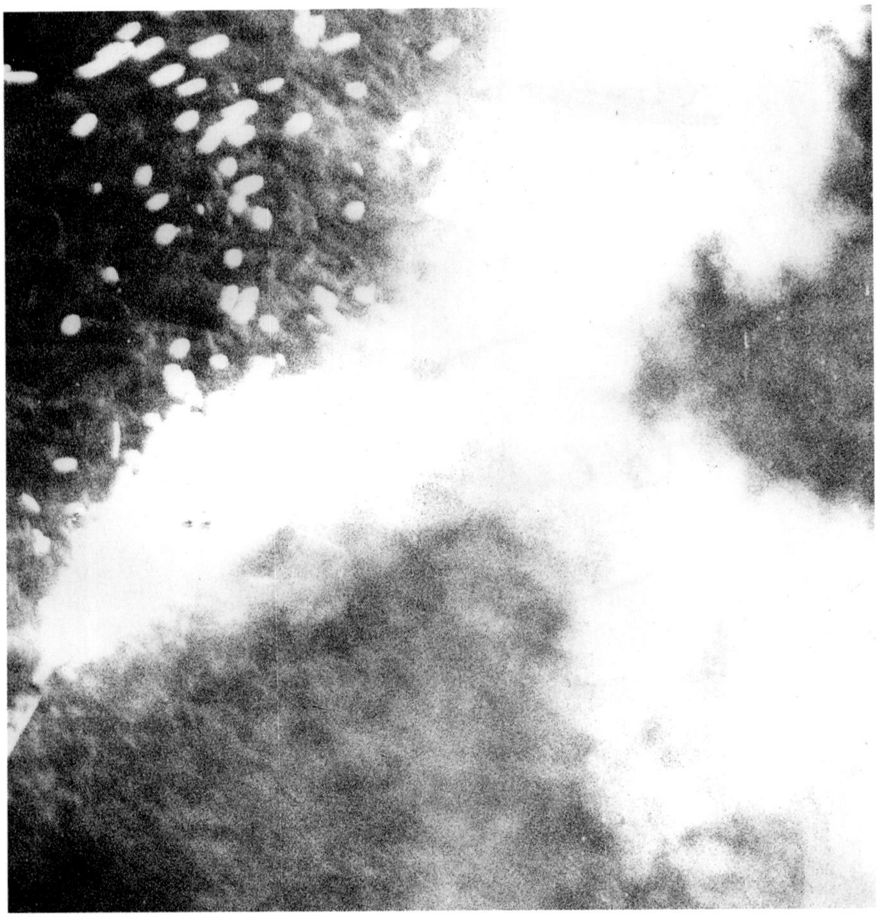

"Fire is light, cinematography is light, scream the reactionaries. They shall have it – the moving image made of light! Film is mistakenly believed to be a pictorial language. The image of the world that is conveyed by language is a reflection of the state and its image of the world. The film industry is a state-sponsored organization set up to convey images of the world that conform to the state's image of the world. When film casts off imagery, it stops presenting the state's image of the world and starts changing the world. Expanded Cinema, in its current phase, is the radical decision to do away with reality and the language that communicates and constructs it."[1]
Peter Weibel's film performance / performance film *EXIT* was premiered in a Munich movie theater on November 15, 1968, as part of an experimental film festival. While the artist gave a manifesto-like speech and industrial films were projected on an aluminum screen, his collaborators VALIE EXPORT, Hans Scheugl, Ernst Schmidt Jr., Gottfried Schlemmer, and Kurt Kren set off a potpourri of pyrotechnic shells, skyrockets, screamers, explosives, and smoke powder. The smoking, flaming projectiles shooting through the air terrified the audience, who jumped out of their seats and ran out of the theater. The title, *EXIT*, foretold the result. Words became reality; reality superseded representation.
For the most part, the Happenings and performances of the late 1950s and 1960s derived from music or painting, and in rare cases from sculpture. Weibel's performances were at the forefront of an emerging expansion of literature and – from the mid-sixties onward – of media, especially cinema. *EXIT* gave a whole new meaning to the phrase "moving image made of light." Rather than modulating in a "well-tempered flow of images,"[2] the light in this piece was shot through the screen and came flying out straight at the viewers. Projection became projectiles. The theater was filled with smoke, a metaphorical image of the ideological smokescreen put up by the film industry. Only in this way might audience members have a chance of emancipating themselves from the conventional and familiar visual experiences of the homogenized, state-indoctrinated cinematic agenda.
This Expanded Cinema performance can also be viewed in relation to other fire-based pieces by Weibel, such as the outdoor film experience *Firewall, or More Traffic Deaths, Fewer Citizens (Brandmauer oder Mehr Verkehrstote, weniger Staatsbürger*, 1969). *EXIT* forms part of an intensive investigation of fire as a wild, natural, primal light that stands in opposition to electric light, which has been tamed and subdued by humanity. As he has done with his art, Weibel has also tried to unleash light – in this case, the light of film.

Translated from the German by Patrick Hubenthal.

EXIT

1968	Peter Weibel in collaboration with VALIE EXPORT, Hans Scheugl, Ernst Schmidt Jr., Gottfried Schlemmer, and Kurt Kren film performance / performance film, communication action; smoke powder, prepared screen (aluminum foil), fireworks mortar, skyrockets, fuses, firecrackers, screamers, etc., industrial films, megaphone, actors, tape player, 10 min., premiered at 1. Europäisches Treffen der unabhängigen Filmemacher, Augusta Lichtspiele, Munich, November 15, 1968

[1]
Excerpt from Peter Weibel's speech, quoted in Peter Weibel and Harald Falckenberg, eds., *Peter Weibel: Das offene Werk 1964–1979*, exh. cat., Neue Galerie Graz, Landesmuseum Joanneum (Ostfildern: Hatje Cantz, 2006), 575.

[2]
Weibel and Falckenberg, *Peter Weibel*, 575.

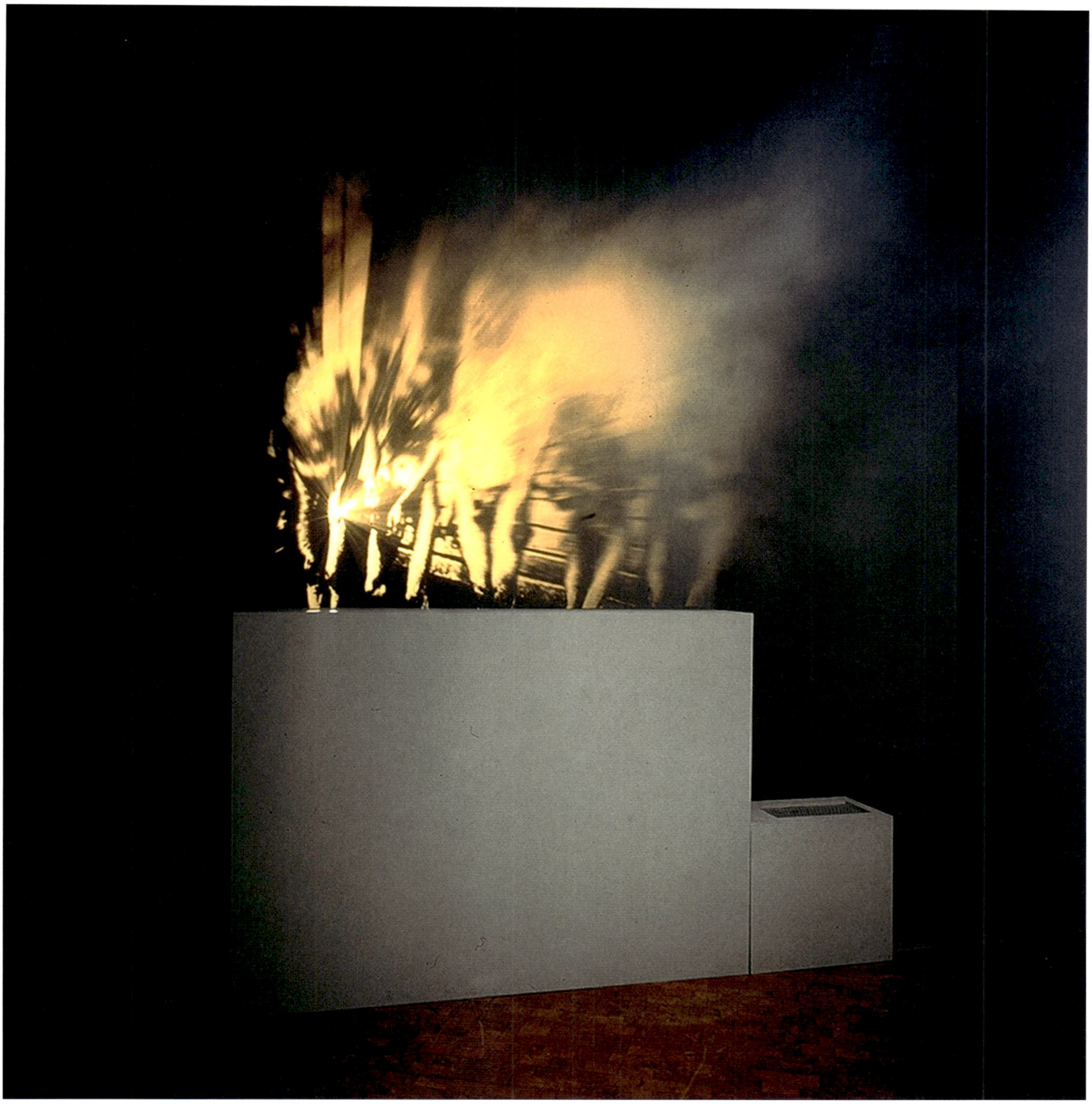

Installation view
MindFrames,
ZKM | Karlsruhe,
2006–07

The installation *Dumb Show* was conceived by Peter Weibel in 1969 as a film or slide installation. Instead of on a screen, the images were projected onto water vapor rising from four bathtubs. In this installation, a six-minute sequence of various photographs from concentration camps can be seen on the ephemeral projection surface of the vapor, which dissolves into nothing after a short time. The repression mechanism, the traumatization by the crimes of the Holocaust, is made visible, as is the mechanism of memory. Just like in his other Expanded Cinema works, Peter Weibel's *Dumb Show* explores the conventions of cinema; here it is the white screen. He expands the possibilities of filmic representation by interweaving real and virtual elements, and utilizing new materials and devices. For the artist, this is the only way to adequately commemorate the crimes committed in the concentration camps, the artistic representation of which Theodor W. Adorno deemed questionable in his essay *Cultural Criticism and Society* (*Kulturkritik und Gesellschaft*, 1949/51). Conventional cinema revels in the depiction of violence. Weibel evades this apparatus-based simulation of violence.

Dumb Show

1969	slide installation; color, projector, vapor, 6 min., loop, dimensions variable

Projects and Projections

I am thinking of films projected through walls, to project horror and panic into citizens' homes on nocturnal forays, to make it easier for prisoners to jerk off

I am thinking of a film, conceived in CinemaScope, that starts out with a life-size spider beginning to spin its web and continues until the web covers the entire screen

I am thinking of films as rays, waves, corpuscles

I am thinking of movies that film my thoughts, to make them clearer to me

I am thinking of projections onto the bottom of the sea so landscape can be enjoyed again

I am thinking of chemically prepared screens that swell up and explode when exposed to light and heat

I am thinking of holographic projections: celebrity parties where everyone is a spectator, landscape exchanges (the Vienna Woods in the far north and the taiga in Vienna)

I am thinking of magnetic communication so I can change my body into any shape I desire: a round ball while running downhill, a thread when confronted with locked doors, flat as a thousand-euro note when in need

I am thinking of a precisely calibrated emission of infrasound, to be able to produce an artificial earthquake: 10-cm fissures open up between the rows of seats in the theater

I am thinking of glasses with a built-in radio, TV, etc.

I am thinking of movies for cats and dogs

I am thinking of films as chemical medium translators: projector, film, screen, and camera all in one as a pill. Pill movies that deintegrate the sensory organs and deliver visions beyond language.

I am thinking of a film projector in the form of a new sense organ that provides, transforms, modifies any sensory quality I desire, in coordination with other sensory qualities, in the form of a new sensory correlator that provides me with air when I step into the ocean, that turns lung-breathing into gill-breathing or vice versa, that lets me sit in my apartment and simulates for me, olfactorily, aurally, haptically, the sensations of the sea or experiences in the streets, that also does not shy away from discrepancies between the senses and allows me to touch foam when I see stone, smell chocolate when I eat grass, hear violins when I seal envelopes, feel water when I stick my finger in the electrical socket, etc.

I am thinking of frequency converters so I can see thermal radiation – e.g., so I can walk around cars even though they are long gone

I am thinking above all of holographic projections, where I can walk around an object

I am thinking of films, too, where everyone sees something different

Note: Peter Weibel, "Projekte und Projektionen," *Supervisuell*, no. 2 (1968). Translated from the German by Patrick Hubenthal.

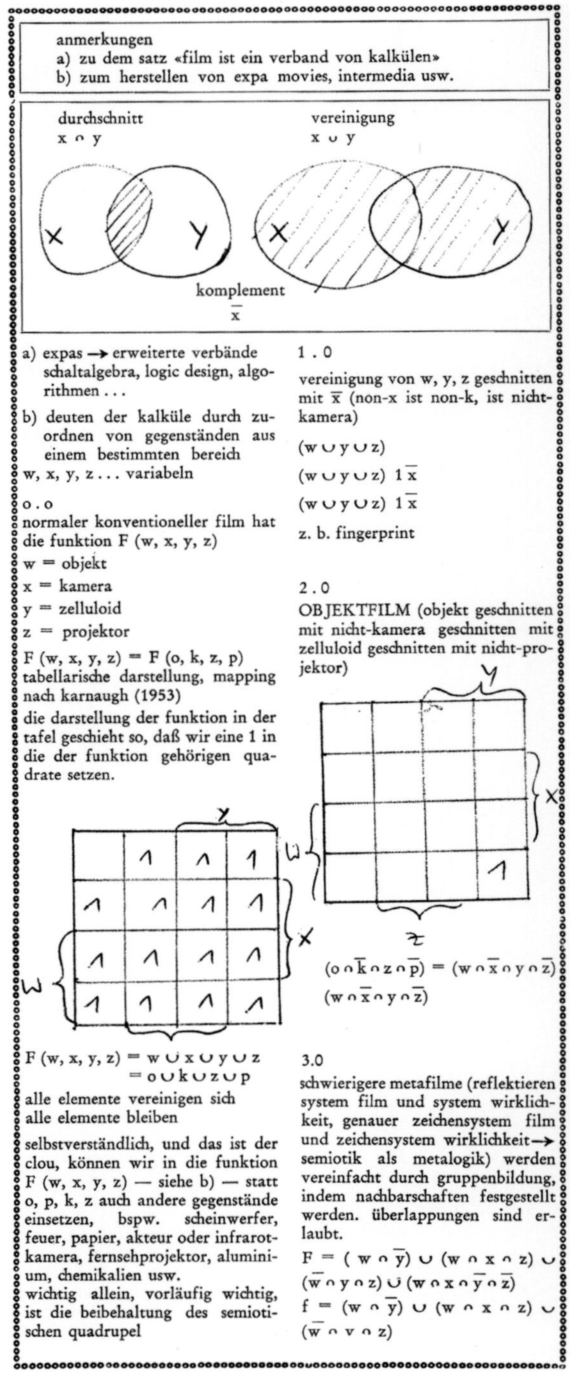
anmerkungen
a) zu dem satz «film ist ein verband von kalkülen»
b) zum herstellen von expa movies, intermedia usw.

a) expas → erweiterte verbände schaltalgebra, logic design, algorithmen . . .

b) deuten der kalküle durch zuordnen von gegenständen aus einem bestimmten bereich

w, x, y, z . . . variabeln

o . o

normaler konventioneller film hat die funktion F (w, x, y, z)

w = objekt

x = kamera

y = zelluloid

z = projektor

F (w, x, y, z) = F (o, k, z, p)

tabellarische darstellung, mapping nach karnaugh (1953)

die darstellung der funktion in der tafel geschieht so, daß wir eine 1 in die der funktion gehörigen quadrate setzen.

$F (w, x, y, z) = w \cup x \cup y \cup z$
$= o \cup k \cup z \cup p$

alle elemente vereinigen sich
alle elemente bleiben

selbstverständlich, und das ist der clou, können wir in die funktion F (w, x, y, z) — siehe b) — statt o, p, k, z auch andere gegenstände einsetzen, bspw. scheinwerfer, feuer, papier, akteur oder infrarotkamera, fernsehprojektor, aluminium, chemikalien usw.

wichtig allein, vorläufig wichtig, ist die beibehaltung des semiotischen quadrupel

1 . 0

vereinigung von w, y, z geschnitten mit $\overline{x}$ (non-x ist non-k, ist nicht-kamera)

$(w \cup y \cup z)$

$(w \cup y \cup z)\ 1\overline{x}$

$(w \cup y \cup z)\ 1\overline{x}$

z. b. fingerprint

2 . 0

OBJEKTFILM (objekt geschnitten mit nicht-kamera geschnitten mit zelluloid geschnitten mit nicht-projektor)

$(o \cap \overline{k} \cap z \cap \overline{p}) = (w \cap \overline{x} \cap y \cap \overline{z})$

$(w \cap \overline{x} \cap y \cap \overline{z})$

3.0

schwierigere metafilme (reflektieren system film und system wirklichkeit, genauer zeichensystem film und zeichensystem wirklichkeit → semiotik als metalogik) werden vereinfacht durch gruppenbildung, indem nachbarschaften festgestellt werden. überlappungen sind erlaubt.

$F = (w \cap \overline{y}) \cup (w \cap x \cap z) \cup (\overline{w} \cap y \cap z) \cup (w \cap x \cap \overline{y} \cap \overline{z})$

$f = (w \cap \overline{y}) \cup (w \cap x \cap z) \cup (\overline{w} \cap v \cap z)$

Peter Weibel's concept for film as an algorithm, published in *Film*, no. 11 (Nov. 1969): 42 (see pp. 38 and 167 in this volume)

Film as AI System

Acquiring language means acquiring a theory
Acquiring a theory of language means acquiring a theory of reality
    changing the world
    without interpreting the world in a different way?
Language as sense organ
Films as sense organs, as distance senses
The expansion of cinema can thus mean the expansion of knowledge
    and the expansion of reality
Is film's system of signs more than a language?
    language about language: metalanguage
    film about film: metafilm
Is perception more than language?
    drugs alter perception
    false films that try to imitate drug experiences
There are no revolutionary films, only a revolutionization of the medium
No language of revolution, only revolutionary language
    consider the conditions imposed by the apparatus
Film can illuminate structures of communication
    clarify problems of perception
    is a first-rate epistemological medium
    intensifying debate with society
    through social communication
If cinema was so late to be invented, it's because the world was so late to become image
Cinema coincides with the concept of modernity
The discovery of cinema by physicists and optical scientists, etc., of the 18th and 19th centuries

Film as life support system
as expansion of perception
    of consciousness
    of communication
    of reality
Film: subsystem of language?
Film is an animal, a machine, a calculus
    with its own sensorium
Film is an artificial intelligence system
    a variety of evolution

Note: This text by Peter Weibel was originally intended to be included in the issue *Expanded Cinema* of the magazine *Film*, no. 11 (November 1969, see p. 167 in this volume). Rediscovered in 2006, it was first published as "Film als A.I.-System" in Peter Weibel and Harald Falckenberg, eds., *Peter Weibel: Das offene Werk 1964–1979*, exh. cat., Neue Galerie Graz, Landesmuseum Joanneum (Ostfildern: Hatje Cantz, 2006), 607. Translated from the German by Patrick Hubenthal.

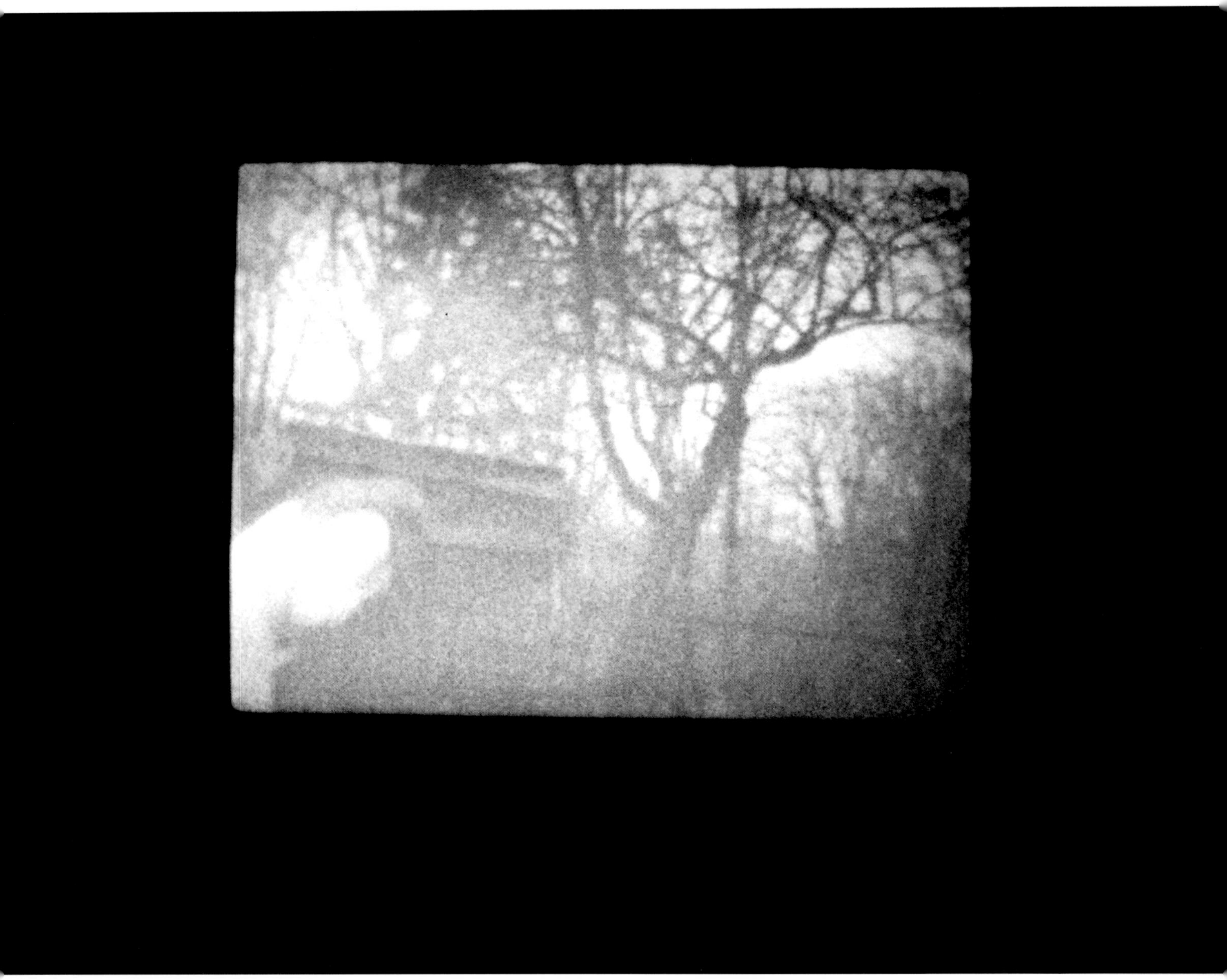

Storm Over (   ) / Sturm über (   )

1969/70 | metafilm; 8 mm film (1969), 16 mm film (1970), b/w, sound, 6 min. actor: Anni Hassmann

A metafilm interpreting the screen as brackets. Just like the spatial parameter (the screen), a temporal parameter determines the duration of the film – a raging storm is so strong that the actress needs six minutes to cross from the left side of the screen to the right. A meditation on the screen, on the illusionism of its limit and area. The title is borrowed from Vsevolod Pudovkin's *Storm over Asia* (Soviet Union, 1928), and the brackets refer to the brackets of the screen. Sound illusionism through overdetermined illustrational function. Illusionary time – real time. —Peter Weibel

Translated from the German by Peter Dahm Robertson.

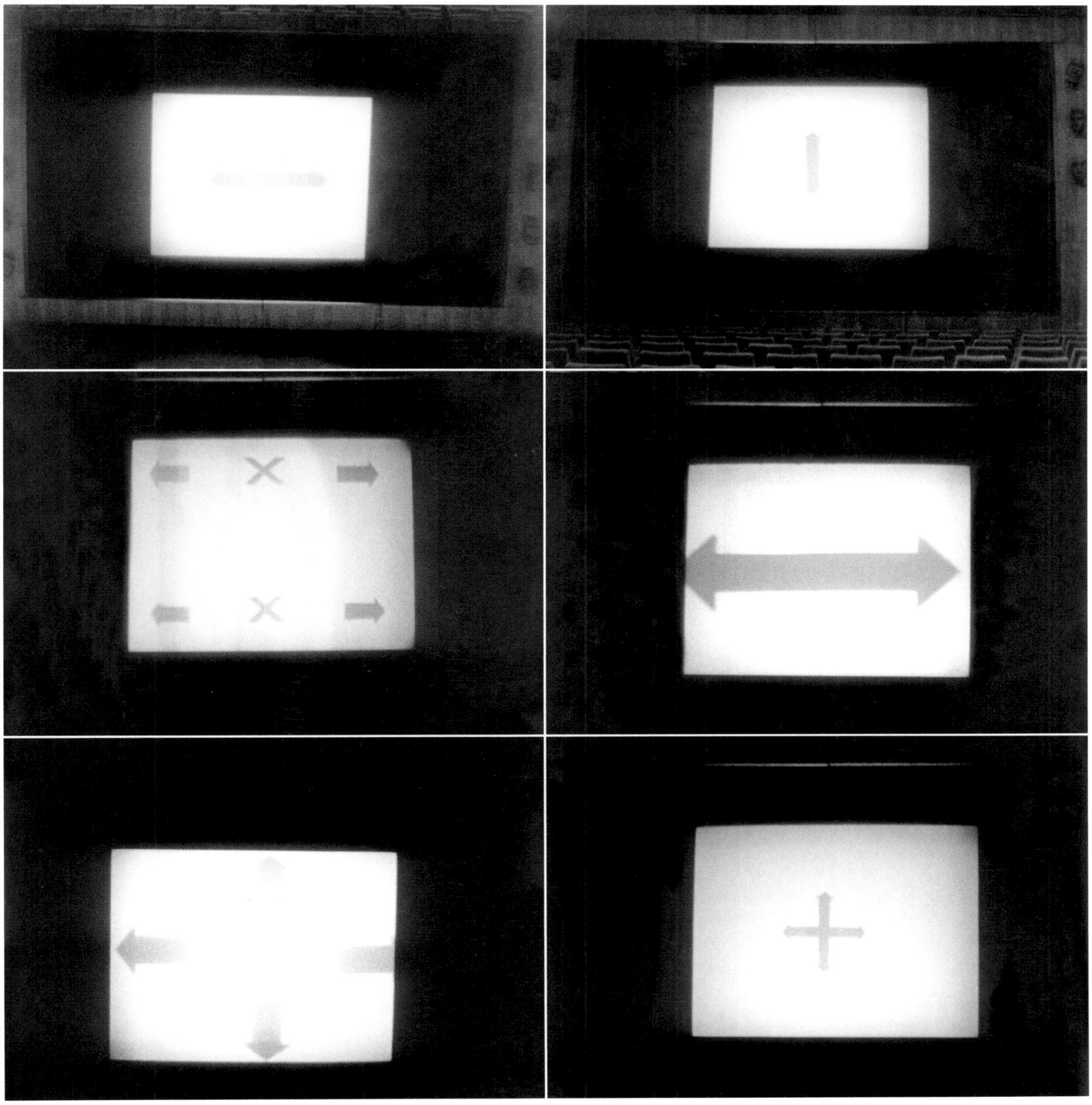

A metafilm

A film using filmic methods to contemplate a particular aspect of the medium of film, namely the question: "What is the screen?" It is shown that the screen occurs where the beam of light falls (intended to be conveyed by swiveling and moving the projector). The image on the screen, then, is the result of the journey of light; the beam of light is already carrying those images, particles, that we identify with objects of the world. Thus, the title *Light Beam : Screen*, because the relationship between light beam and screen is being examined. A commentary points out what can be perceived tautologically through the film: We see only what the light transmits. We do not see what the light does not transmit. We see only the screen, not what is next to the screen. But just as writing is a primal act of repression (Freud: "Notiz über den Wunderblock")[1] because it eliminates particular gestures and symbols from the paper and thus also eliminates particular intellectual as well as emotional content, the language of images (symbols on the screen) is also a primal act of repression. The commentary expresses the desire to see more than merely the symbols on the screen, to see more than the world confined by the brackets of the screen. To see what is obscure, what is not denotable, what is visible only mentally. First, accordingly, the nature of the screen is discussed (size, etc.; for example, for an observer on the moon, half of Earth is a screen) and then the nature of symbols. The arrow, as an expression of the desire transcending the boundaries of film, always wants to go beyond the screen. But it cannot, for when it is beyond the screen it can no longer be perceived: this is the film's central metaphor. The second logical conclusion is that it loses its meaning as a symbol. The arrow is recognizable as an arrow only as long as it keeps to the rules/dimensions of the screen; once it is beyond them, it is only a black bar, a meaningless symbol. In the end, the screen is black since light transmits not only white (reason) but also black (unreason). The darkness is the complement of light. A model for larger systems of symbols and communication. Sound: each time the arrow touches the edge of the screen, there is a bang.

—Peter Weibel

Translated from the German by Peter Dahm Robertson.

Light Beam : Screen / Lichtstrahl : Leinwand

1971	metafilm; 16 mm film, b/w, sound, 4 min.

[1]
Editor's note: Published in English as "A Note upon the 'Mystic Writing-Pad,'" trans. James Strachey, *The International Journal of Psychoanalysis* 21, No. 4 (1940): 469–74.

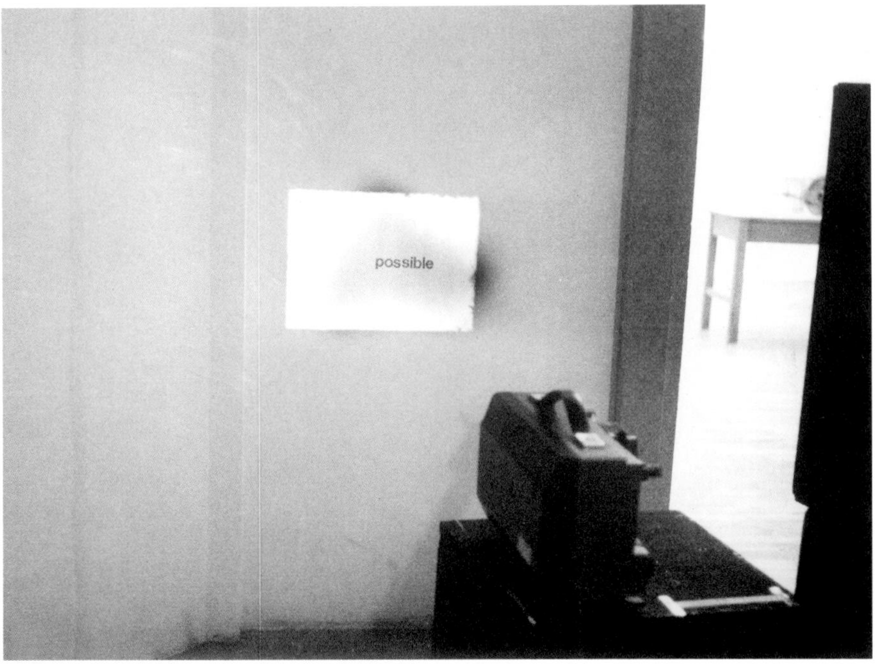

Installation view
Projektionskunst, K45 –
Internationale Kunstmesse,
Künstlerhaus, Vienna,
Feb. 17–21, 1977

The apparent radicalism of Peter Weibel's 1969 installation *Possible* gives it a decidedly conceptual appearance. It takes the form of a word, adjective or noun, written in lowercase, lineal characters. According to the technical rider of the artwork, the cinema projector that illuminates it can be either 8 or 16 mm. This indistinguishability takes on its full meaning when, constrained by the path of an exhibition, we cross its beam and realize that our shadow does not erase the word at all. This means the typeface is not projected as we thought, but has been written on the wall. The aesthetic value of this work thus reaches its peak at the very moment when, unable to eradicate it from the light, we instead reveal its extreme reality while reinforcing its conceptual aspect. Peter Weibel reminds us that "the real is only part of the possible."

Cinema operators know that projectors must be placed high up to avoid the shadows of latecomers. However, since the advent of video projectors in the art world in the 1980s and 1990s, many artists have invited viewers to interact with their own silhouettes. One thinks in particular of the Mexican-Canadian Rafael Lozano-Hemmer, and more particularly of his interactive installations (*Frequency and Volume* from 2003 and *Under Scan* from 2005), whose audio and video media are activated by the shadows projected by the public. Conversely, it is the total absence of interactivity that magnifies *Possible*. In other words, the proliferation of interactive art installations in the 1990s and 2000s has only increased the aesthetic value of this creation by Peter Weibel, which is essentially based on a form of noninteractivity. This is a bit of a stretch, given the artist-curator's unstinting efforts to support interactivity in art, both within and beyond the institution he has directed since 1999, the ZKM | Karlsruhe.

Although the installation *Possible* cannot be reduced to a conceptual work, and it is tempting to define it further because it is not really an interactive piece, it is nonetheless a contemporary expression of a pictorial genre: *trompe-l'oeil*. It proceeds from a form of illusion that only the magic of techniques or technologies usually allows. Accordingly, the word "possible" is neither really what we think it is, nor quite where we think it is. Unless we consider it both in the light of which we believed it before experiencing it and on the wall where we discover it. It is in this manner that it refers to the art of illusion, illustrated so perfectly by the central fresco of the Church of the Gesù in Rome, dating from the seventeenth century. The colors of the fresco by Giovanni Battista Gaulli blend in with the stucco reliefs by Ercole Antonio Raggi and Leonardo Reti, which raises the ongoing relationship to the frame that so many artists, like Peter Weibel, have questioned. It is even central to the contemporary algorithmic practice of the American-Turkish artist Refik Anadol, who integrates frames into his works in order to free himself from them. In this case, breaking the frame is a genre in itself. For *Possible*, it is more a matter of stubbornly refusing to be removed from a rectangle of light.

Peter Weibel's purposefully nonfilmic artwork can also be seen in the light of the so-called third industrial revolution, in which we know less and less about the exact nature of what makes up our environment. This is particularly true of images that we observe without worrying about whether they have been captured or computed. Or captured then computed as in the case of photogrammetry, for example. Already during the Enlightenment, architects distinguished themselves by designing monuments that were not intended to be built. Today, the idea that architectural simulations are plausible is enough for us to dream them into the metaverse, where we also have the leisure to inhabit them. This idea of the "plausible," not to say the possible, is also essential to the practices of certain artists. Witness the Frenchman Gregory Chatonsky with his creations of a form of *Artificial Imagination* and Pierre Huyghe during his exhibition *UUmwelt* at the Serpentine Gallery in London from 2017 to 2018. There is something familiar in these images whose generation is delegated to the computation of artificial neural networks, like a sense of *déjà vu*. Undoubtedly because they have taught their machines how to recognize a great range of images across infinite data sets. For our part, we accept that we no longer know the exact nature of what we are looking at within the uninterrupted flow of images that present themselves to us daily.

Let us return once again to the meaning of the word "possible" to better consider the fields that are open to us in this digital age, with the knowledge that yesterday's possibilities have become today's realities and that digital imaginaries allow

Possible

1969	Expanded Cinema installation without film; 8 mm or 16 mm projector, tripod, adhesive letters on screen or paper, dimensions variable

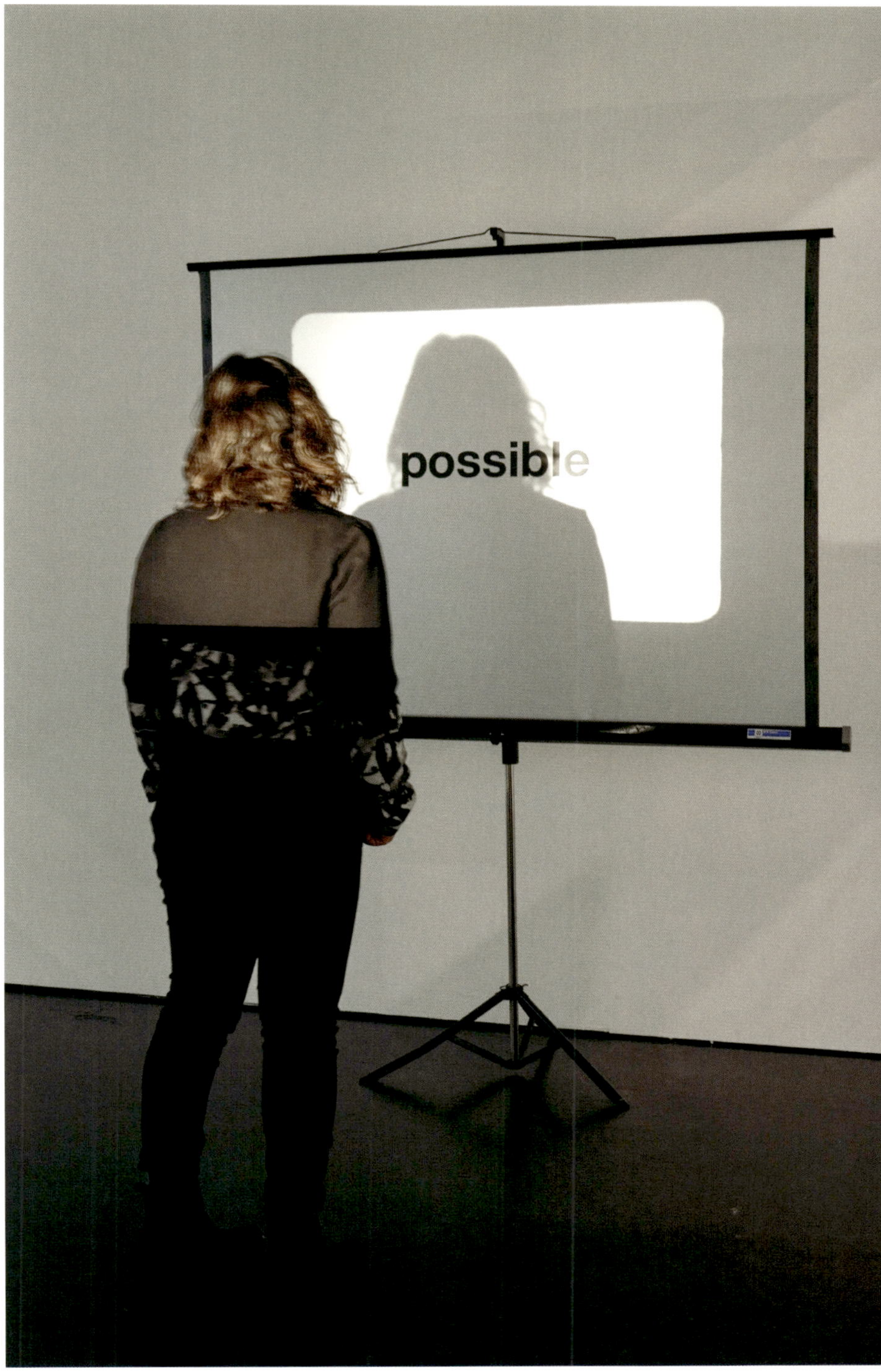

Installation view
respectively, *Peter Weibel*,
ZKM | Karlsruhe, 2019–20

us to project ourselves into plural futures. It is interesting to note that it is, more often than not, in the sphere of art once described as digital – the term contemporary now seems more appropriate – that many innovations have emerged. The interactive light installation *Fiat Lux* (1967) by the French-Israeli Yaacov Agam in the Collection ZKM | Karlsruhe is a perfect example. A light bulb, whose filament is an indication of how old it is, suddenly lights up when you clap your hands, long before the home automation dreams of Microsoft were fulfilled by Google or Amazon. The Russian-American theorist Lev Manovich has already taught us in his book *The Language of New Media* (2001) that careful observation of the past often allows us to project ourselves into the future, and scientists in general know the inestimable value of sidesteps offering other points of view. In view of its longevity, it is interesting to reconsider the installation *Possible* with respect to the periods it has gone through. It began with its relationship to interactivity and its emergence in the art world and then on to the virtual when machines began to compute cyberspace, without forgetting the more recent desire for a form of return to materiality, perfectly illustrated by the multiplication of digital fabrication laboratories. Perhaps the most important lesson of *Possible* lies in the fact that the notions of the real and the virtual are no longer opposable, if indeed they ever were. More and more frequently, finally gifted with ubiquity, we are here and elsewhere at the same time. And in quantum computing we consider that a unit – the Qubit – can be equal to 0, 1, or to the two "superimposed" states. This makes Peter Weibel's hybrid work the very expression of the uncertainty principle. Or, more precisely, of indeterminacy – a principle enunciated in 1927 by the German physicist Werner Heisenberg, one of the founders of quantum mechanics – whose field of possibilities we are just beginning to glimpse.

—Dominique Moulon, 2022

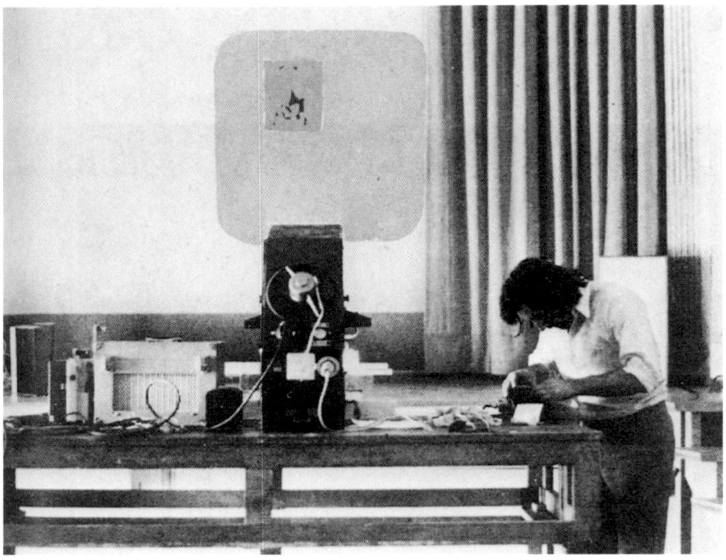

Performance at Festival Konkrete Poesie, Hanau, 1971

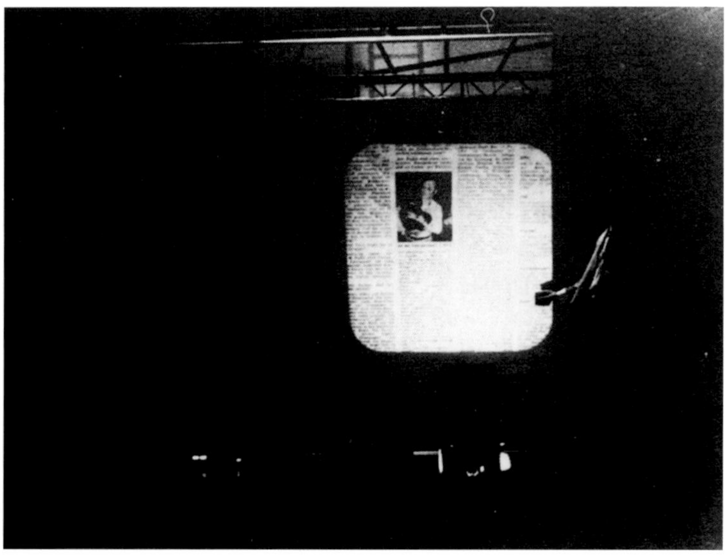

Time and Tense in Picture Processing, or Treatise on the End of the World / Time and Tense in Picture Processing oder Abhandlung über das Ende der Welt

1970/71 | performance; newspaper, episcope, screen, Polaroid camera

A daily newspaper is placed in an episcope and projected onto the screen. The actor then photographs this projected image with a Polaroid camera. Instead of the newspaper, the Polaroid photo is then placed in the episcope and projected onto the wall. The actor steps closer to the screen and photographs the projected Polaroid photo using the Polaroid camera. He now places this second, already less good (in terms of light strength, focus, etc.) Polaroid photo into the episcope, projecting it. This process is repeated several times and is completed when the constant reproduction has reduced image quality to a completely black zero image. During the entire process, a steadily rising pure tone (sine wave) is heard. A variation is to dissolve writing into images by using the same process to project handwritten words and instantly photograph them, with the writing becoming ever larger and blurrier. It's the end of time and time to end. —Peter Weibel

Translated from the German by Peter Dahm Robertson.

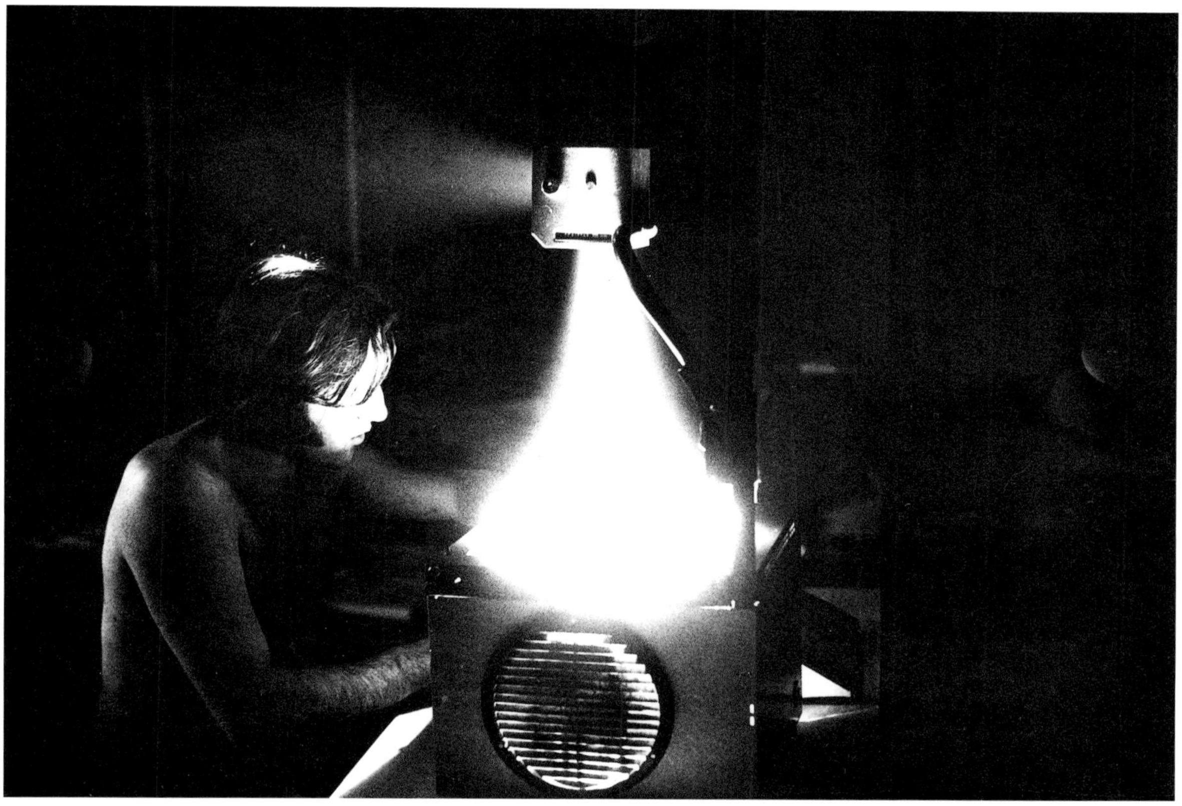

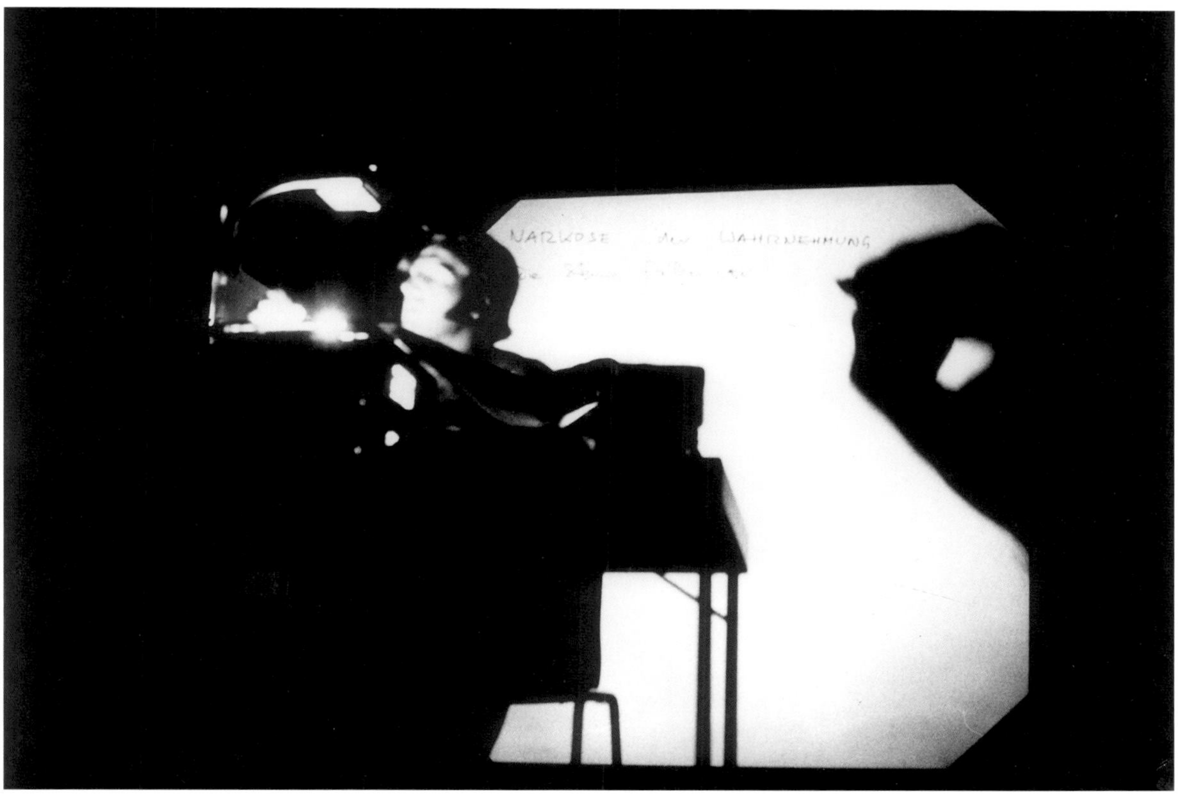

Precinema, Postcinema, Paracinema / Präkino, Postkino, Parakino

1971 | action lecture at Experimenta 4, Frankfurt am Main, June 4, 1971

On an overhead projector, the simultaneity of the act of writing and the act of reading results in one possibility for artistically shaping the process of one's own writing. A second possibility is offered by spatial separation into close-up and full view, with full view showing the writer (very small) and not what is written, while what is written (a detail separated from the full view) appears larger than life in the projected close-up. The simultaneity of two perceptual spaces, enabled only through the technological medium, forms the system of trajectories for a series of artistic experiments and operations on writing.
—Peter Weibel

Translated from the German by Peter Dahm Robertson.

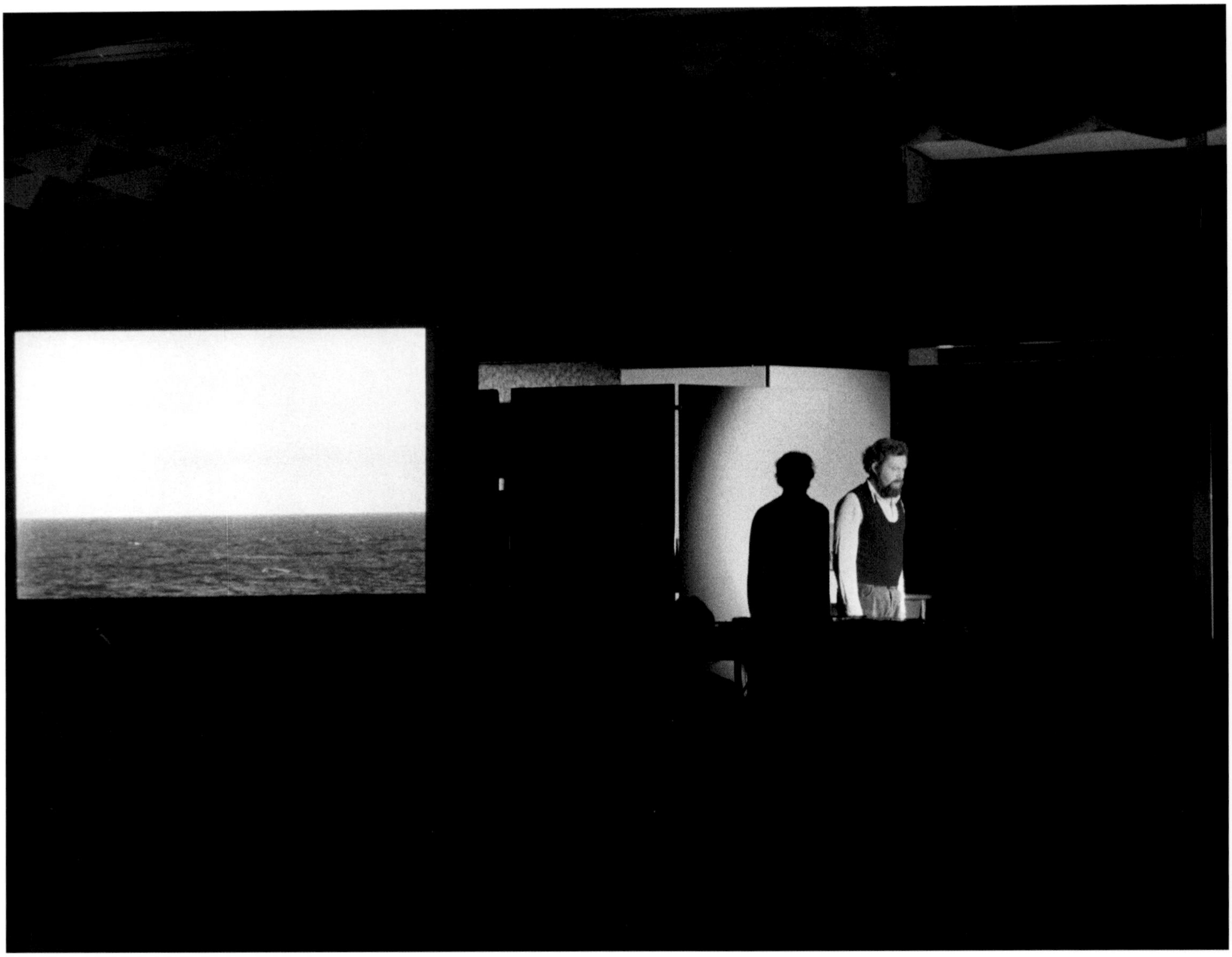

Otto M. Zykan during the performance, Vienna, 1971

Zykan I

1971	slide projector, slides, b/w, magnetic tape sound, 3 min. music and performer: Otto M. Zykan

The viewer sees thirty repetitions of a hammer shattering a glass pane of identical size to the screen, so that the screen appears to be shattered. After the tenth time, the slide projections begin, signifying various skins (screens), media, i.e., from water via mud, sand, grass, wood, etc. to air and the structured series of images is accompanied by a systematically diverging sound structure by Otto M. Zykan. —Peter Weibel

Translated from the German by Peter Dahm Robertson.

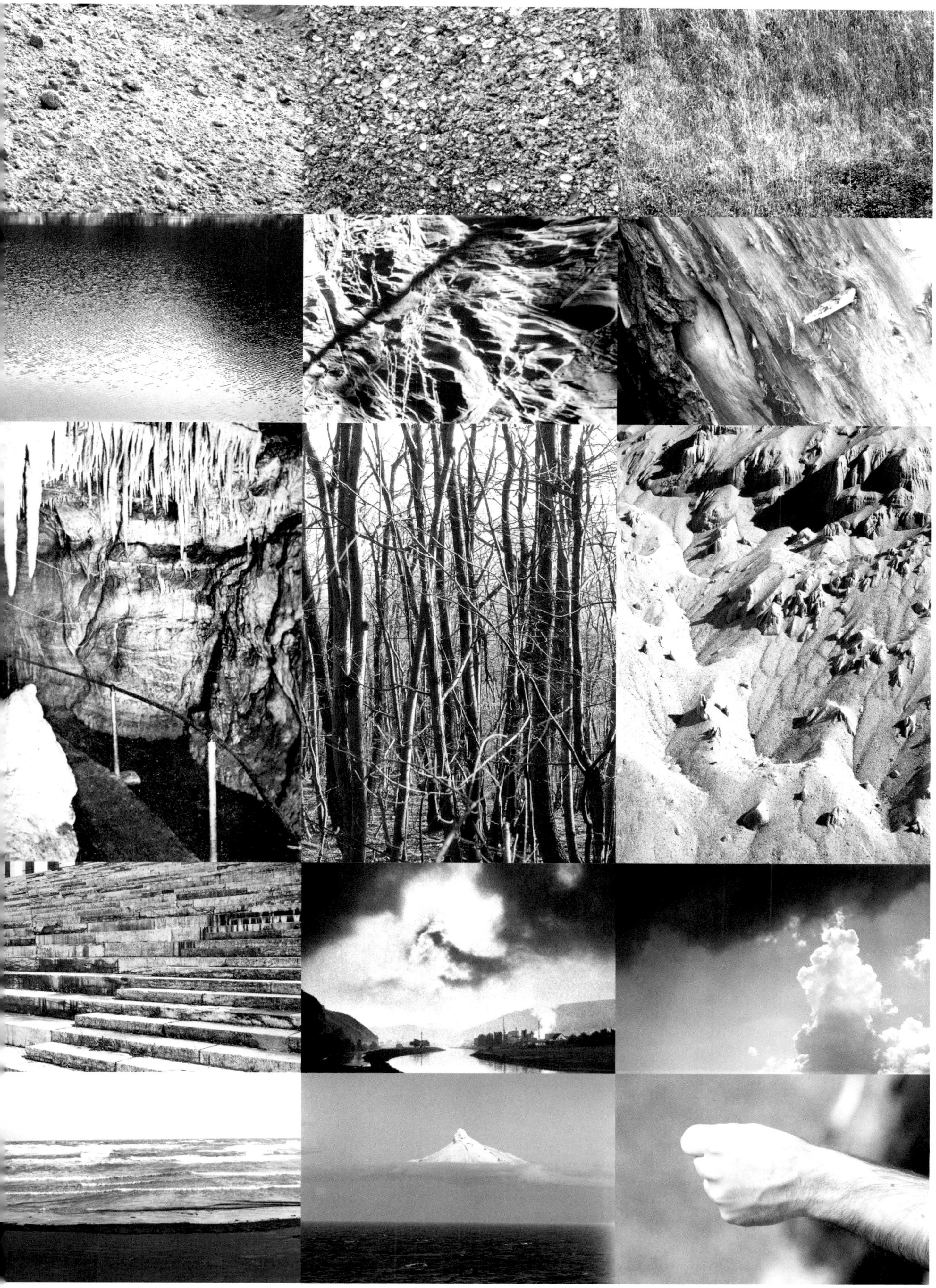

On the Nature of Technology: Landscape with Fine Grain / Zur Natur der Technik: Landschaft mit Feinkorn

1972 | K. B. Schäuffelen and Peter Weibel installation of slides, first shown at *Musik / Film / Dia / Licht-Festival*, Munich 1972

Five slide projectors project five slides of a landscape on the wall in the same rhythm by means of a special electronic circuit. The landscape is shown as a panorama due to changing the perspective during the taking of the photos. All five projectors were loaded with eighty slides in which ash, sand, and the like had been scattered in progressively larger amounts, so that the slides, or rather the landscape, were gradually covered up from above. Thus, first, some specks of dust appeared in the landscape, then clouds, then heavy black clouds, then the landscape is covered completely; until finally, the slides are totally black. And then the projection automatically starts all over again from the very beginning.

By each slide projector there is a contact microphone, which amplifies the noise of the projector changing the slides, and an electronic device amplifies the sound a little more each time a slide is changed. Thus, a curve of increasing loudness is created, which booms at the end. This creates the subjective impression that the slide projection gets faster (due to the increasing volume) whereas the speed of the slide projection actually remains the same.

The fully automated sequence of the slide projection means this is a temporal image installation, a processual image sculpture.

A landscape (the Brucker Moos moor in Bavaria) was shown as a photographic panorama by means of five changes of perspective when the photographs were taken. Each partial perspective was photographed eighty times in succession on August 20, 1972, between 2:58 pm and 4:07 pm, so that a total of 400 photographs was taken of the entire panorama. These 400 photographs were projected onto the wall of a room by five slide projectors. By means of a special electronic circuit, five slides were projected next to each other on the wall in the same rhythm. Eighty landscape panoramas (each consisting of five slides) were thus projected on the wall in a sequence; however, from projection to projection the landscape got dirtier, more polluted.

After a first, pristine five-slide series, fine ash was scattered on the subsequent series of slides progressively increasing in amount, so that the slides / the landscape became gradually covered from above (hence the subtitle: *Landscape with Fine Grain*). First, some specks of dust appear in the landscape, then clouds, then heavy black clouds, then the landscape is completely covered, until finally the slides – and thus the field of view – are totally black.

At that moment, the apparatus, plus all other light sources in the room, switches itself off, and there is complete darkness and total silence for thirty seconds. Then the projectors automatically start again from the beginning with the first pristine landscape panorama.

Next to all five slide projectors there are contact microphones connected to an amplifier, which amplifies the rattling noise of the slides being changed. By means of an electronic device the sound level is amplified a little more at each slide change. In this way an increasing volume curve is created, so that towards the end of the sequence of eighty panorama projections, an annoyingly loud, booming noise results from the slide changing operations. The speed of the slide transport is constant, but the subjective impression that the projection is speeding up, rushing onwards, which only results through the increasing volume. The fully automated sequence of the slide projection signifies this is a temporal image installation, a processual image sculpture.

Note: Peter Weibel and Harald Falckenberg, eds., *Peter Weibel: Das offene Werk 1964–1979*, exh. cat., Neue Galerie Graz, Landesmuseum Joanneum (Ostfildern: Hatje Cantz, 2006), 630–631. Translated from the German by Gloria Custance.

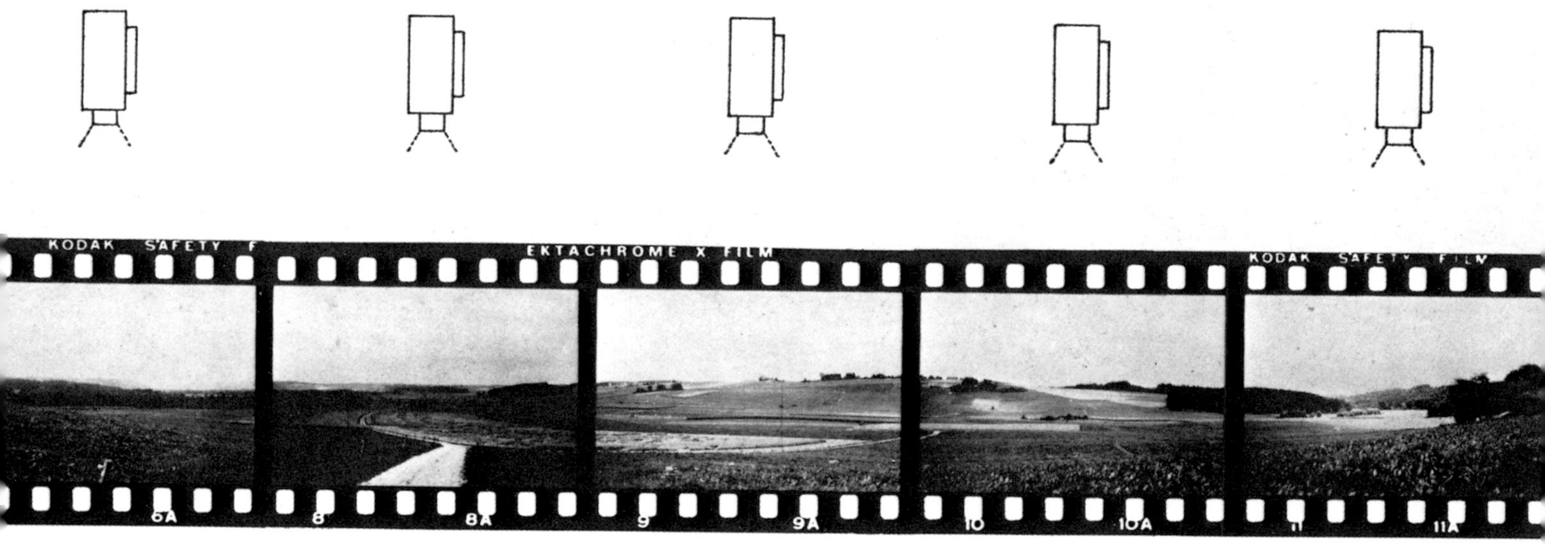
KODAK SAFETY F
EKTACHROME X FILM
KODAK SAFETY FILM
6A
8
8A
9
9A
10
10A
11
11A

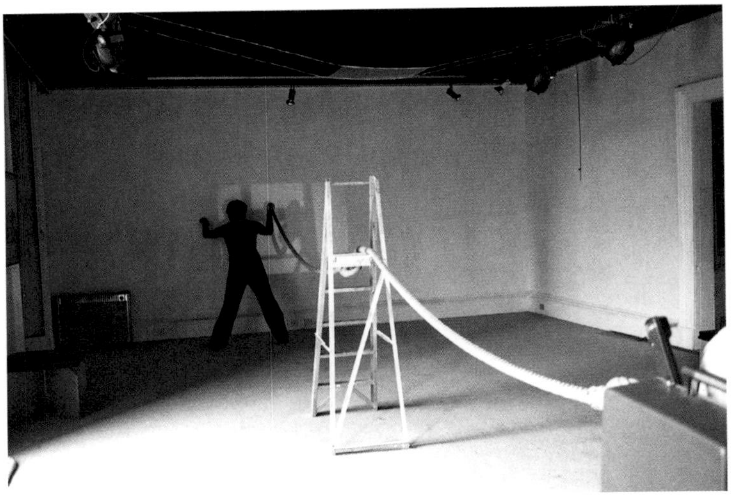
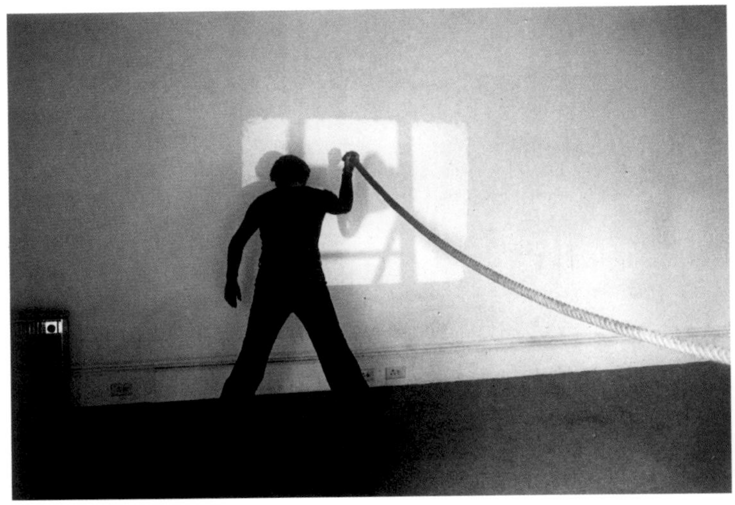
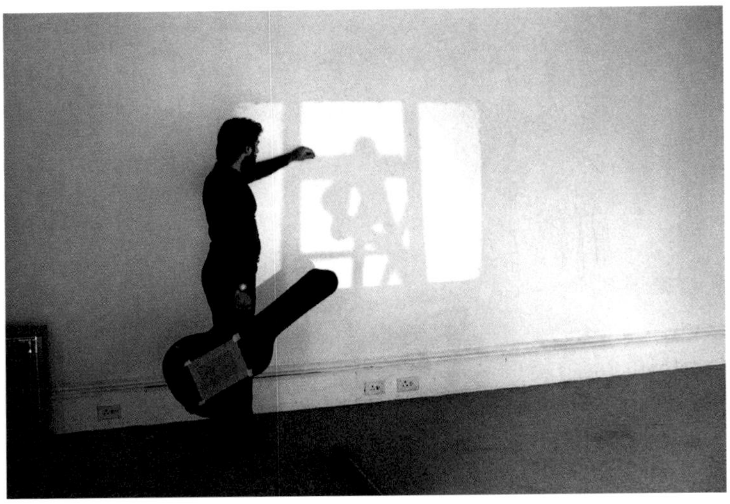
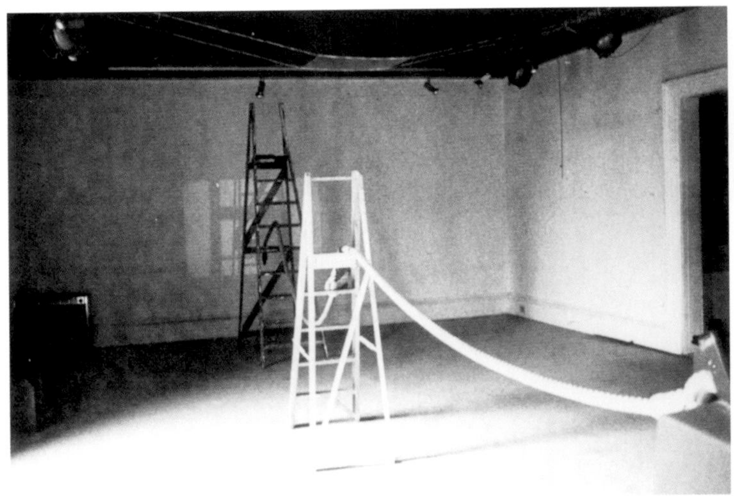

Installation views
The Austrian Exhibition,
Demarco Gallery,
Edinburgh, 1973

Light Rope / Lichtseil

1973	Expanded Cinema installation; screen, rope, ladder, projector, dimensions variable

In the Expanded Cinema installation *Light Rope* Peter Weibel stretches a rope between a projector, a ladder, and a screen. The light beam of the projector, which in a conventional way brings the film to light in the first place, is materialized here by a rope that casts an image onto the projection screen: the shadow of the rope and the ladder. The perspective construction between the projector and the projection screen is rendered comprehensible. The shadow of the ladder and rope forms the content of the film without film, which in turn thematizes the relationship between image and reality. Peter Weibel thus raises awareness of the illusionistic dimension of the medium of film. At the same time, he refers to the theatrical and cinematographic tradition of shadow and light plays. By emphasizing illusionism and revealing how mechanisms of illusion function, he criticizes all illusion systems, from film to religion, the "future of an illusion" (Sigmund Freud, 1927), using the metaphor of the ladder to heaven as an example.

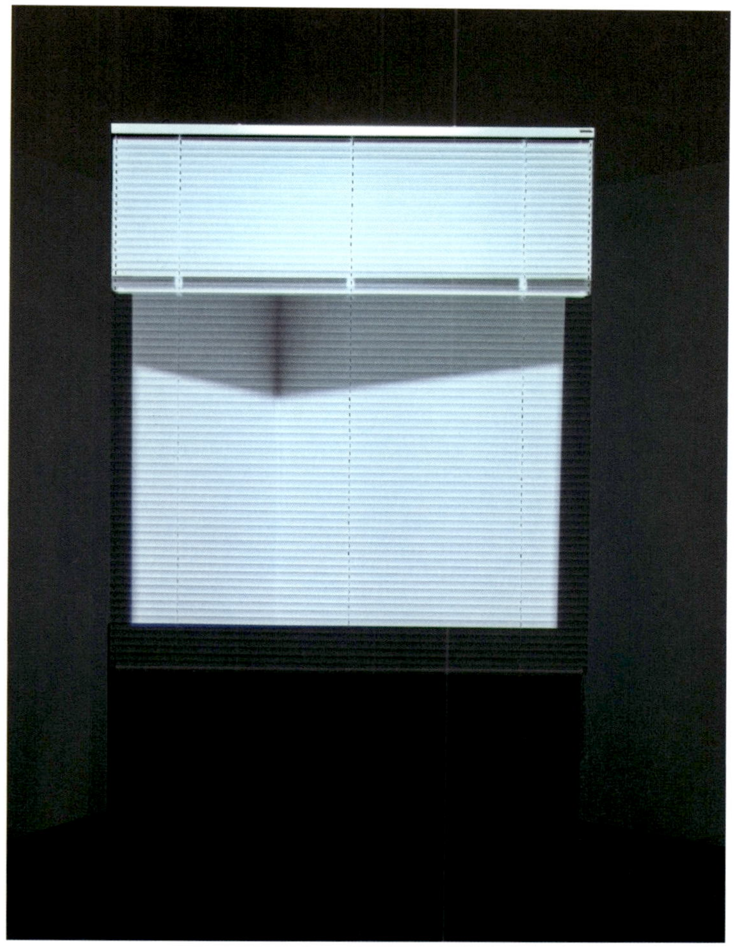

Installation views
respectively, Peter Weibel,
ZKM | Karlsruhe, 2019–20

Window, Film, Frame / Fenster, Film, Frame

1967	Expanded Cinema installation; window shade, wooden frame, video, color, 120 × 205 × 80 cm

A window shade hanging in a room is illuminated by a film projector. Suddenly, it begins to descend. It takes the viewer a while to figure out which one is actually moving, the projected shade or the real one. I move the real and projected shades in tandem and counter to one another; object and image coincide and diverge. —Peter Weibel, 2022

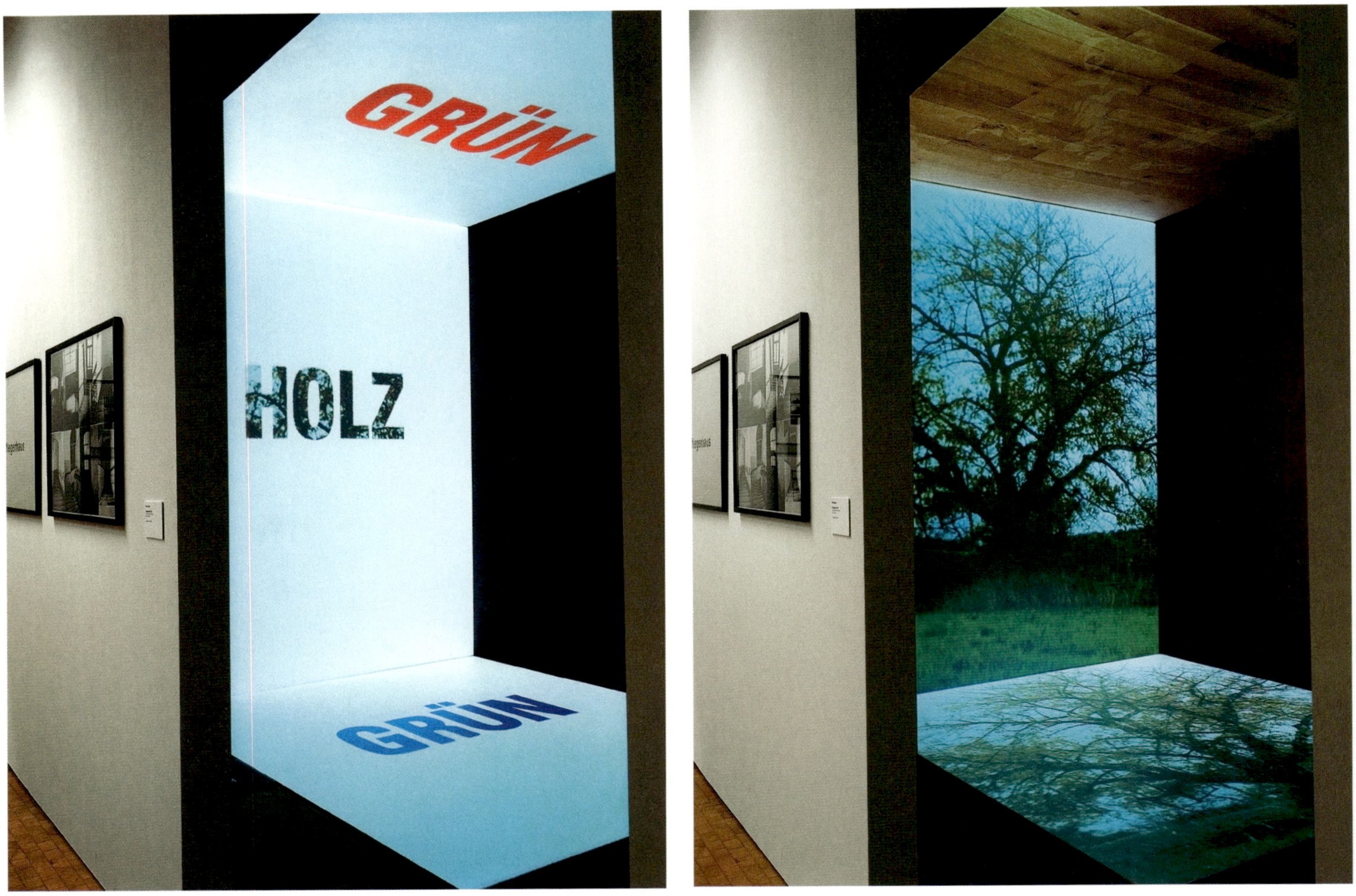

Installation views
MindFrames,
ZKM | Karlsruhe, 2006–07

In the immersive film environment *Way Way Way Out* the observer is a part of the picture by standing between three projection screens. These are arranged in a u-shape manner, forming a continuous projection surface covering the three dimensions of space. At the beginning of the film sequence, the projections of the floor, the ceiling, and the front correspond with their "natural" positions (a meadow on the floor, clouds on the ceiling, a tree in front). Soon the images begin to shift, the positions change and thus confuse the visitors' orientation within the space. While standing on clouds, a river flows vertically in front of the viewer and a meadow floats above him or her. Not only the natural perspective breaks, but also the objects to which the images are referring to are replaced by different kinds of representation of them: The tree is first replaced by a wooden floor texture, then by the word "wood." The surroundings become more and more abstract.
Words such as meadow, marble, wood, and water wander across the screen in a typeface that is reminiscent of the material texture of the substances designated. At the end, the projections are colored red, green, and blue, once as text, once as colored plane. The viewer is looking for a way out of the prison of space and time.

Way Way Way Out

1968/ 2007	immersive film environment; video, color, silent, dimensions variable

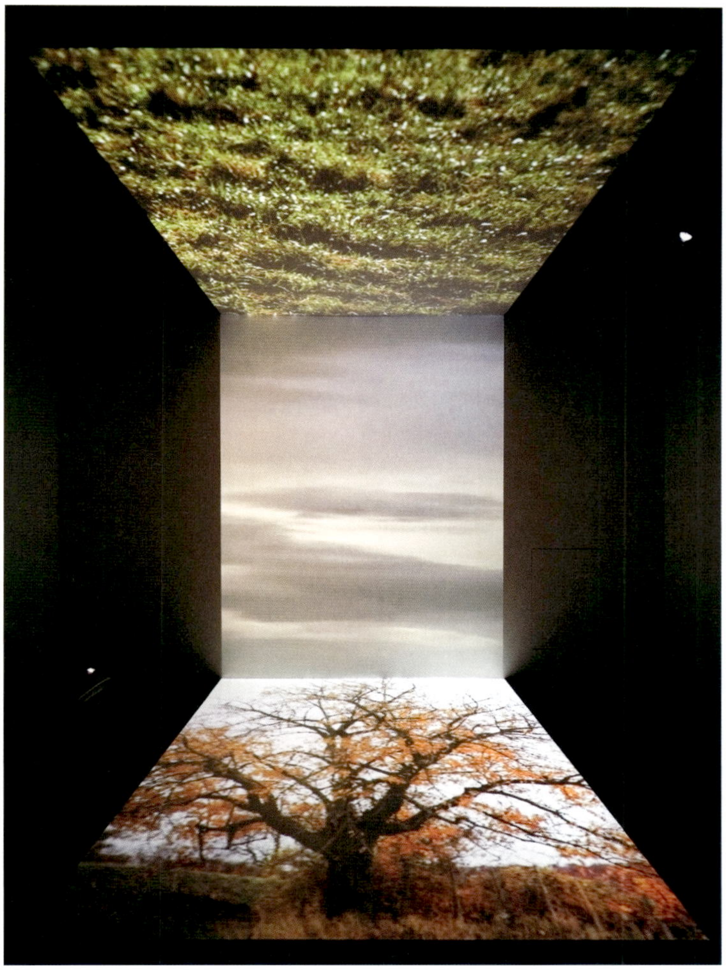

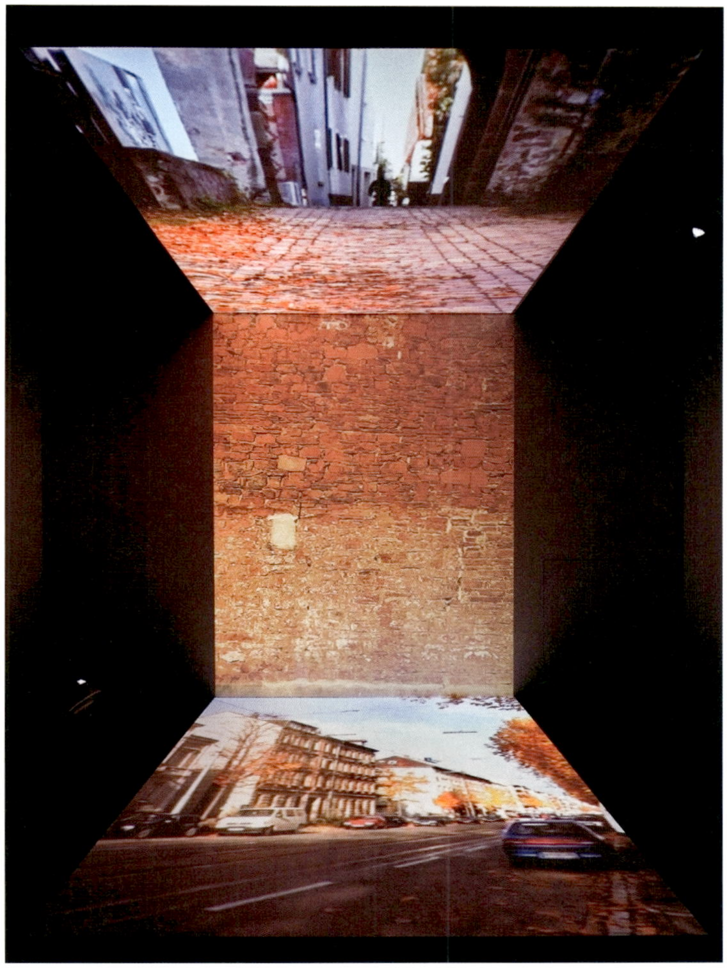

Installation views
respectively, *Peter Weibel*,
ZKM | Karlsruhe, 2019–20

Way Way Way Out
First, **Projector No. 1** is turned on. It projects an image of, for example, a snail onto **Wall No. 1**. The snail's propagation velocity is assumed to be near zero. After, say, five minutes, **Image No. 2** goes up on **Wall No. 2**. The propagation velocity of the content of the image, e.g., a worm, is already slightly above zero. After 4.8 min., accordingly, **Image No. 3** is projected onto **Wall No. 3**. Here, the propagation velocity of the content of the image is even greater than in No. 2. So **Image No. 4** goes on **Wall No. 4** after, e.g., 4.3 minutes, etc. The speed of slide projection is thus proportional to the empirical propagation rate of the organism or machine depicted. For each slide, the word "Way" is spoken, with the volume likewise proportional to the empirical propagation rate of the content of the image. For the fish, however, the speed is such that the slide projection can no longer keep pace. Now **Projector No. 2 is turned on**, which starts slowly and fills in for Projector No. 1 in the calculated rhythm. When even the second is not sufficient, at the very end, **Projector No. 3 is turned on**. Finale: all three projectors, spinning furiously, project the images of a rocket (flying horizontally) onto the wall, so that the "movable" images become a "moving" image, creating the impression of the rocket's movement spinning around the walls. At this moment, the projectors go out. Cue gigantic spotlights and at highest volume, with echo effects, the words "OutOutOut ..."

Note: This concept for *Way Way Way Out* was first published in *Film*, no. 11 (November 1969): 52 (see p. 167 in this volume). Translated from the German by Peter Dahm Robertson.

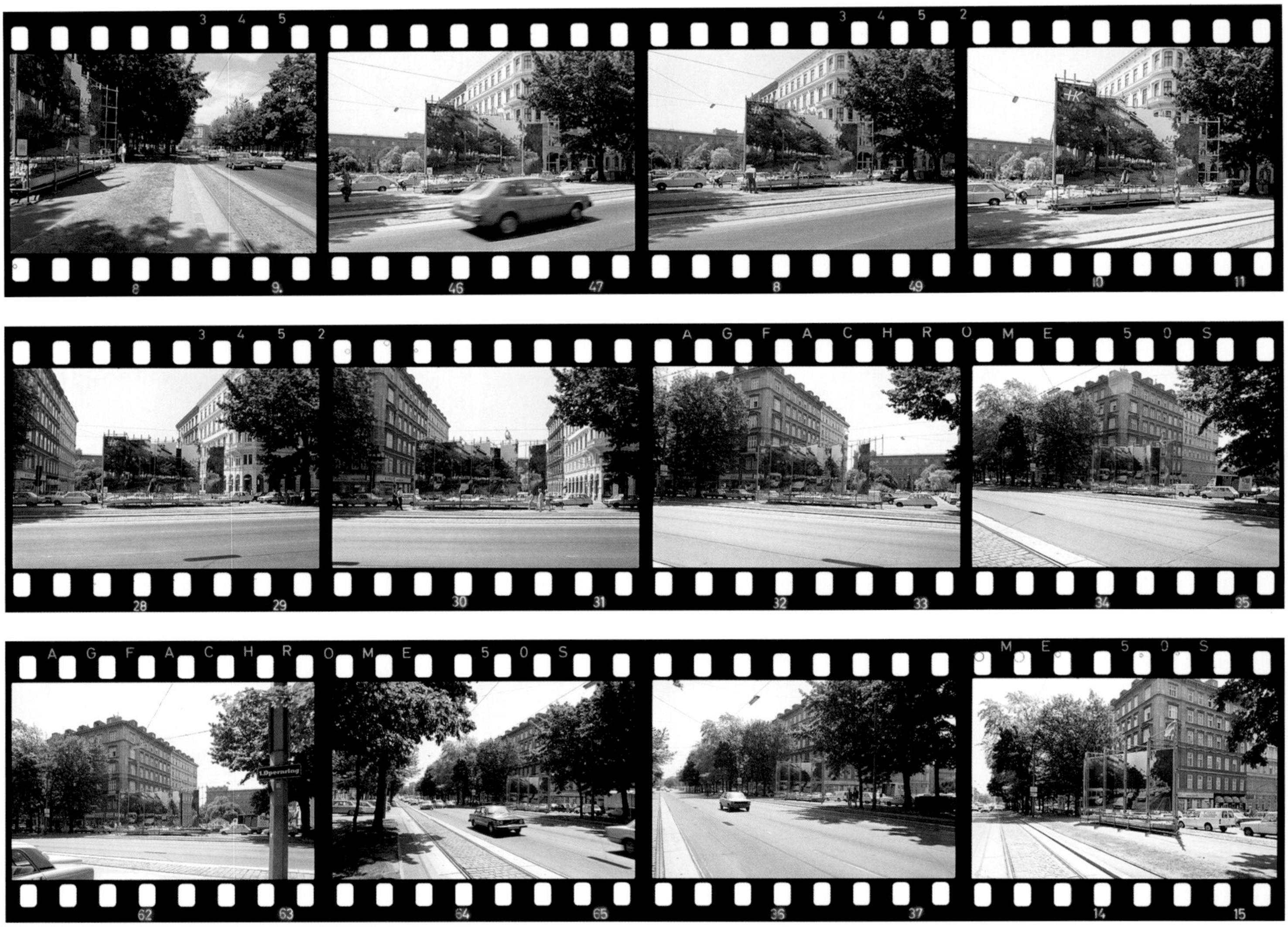

Ringpark Project Cinema /
Ringpark-Projekt-Kino

1970 (idea)/ 1980

city planning: Dr. Werner Schimanovich; artistic concept: Peter Weibel; painting: Heinz Lindinger, Peter Niedermair; technical assistance: Marcus Eibelmayer

Dr. Werner Schimanovich, a city and traffic planner, developed a model of the city of Vienna in which the Ringstraße is implemented as an extended pedestrian zone, with the various existing and future parks integrated into a continuous *Ringpark*. Through retention of the Donaukanal near the Franz-Josef-Kai, an artificial lake would be created linking the ends of the Ringpark into a circle.

A real-life cinema for traffic, with a film for abolishing traffic! By installing a new form of cinema, tram riders will be able to journey into this implementable greener future. The concept is based on the origins of the cinema, where a cylindrical ring with slits turned on its axis, so that when looking at the interior of the cylinder through the slits, one received the impression that the birds painted there were seen in flying motion. In Peter Weibel's *Vitascope* (*Vitaskop*, see page opposite), it is not the cylinder but the viewers in the tram (or car) who move and look through the slits in a wall affixed the edge of the road. In this way, they receive the impression that the images painted on a second wall behind it are moving. The images show – as in a film – bushes and flowers slowly growing around Parliament until it is standing in the middle of a park (illusion of movement). In June 1981, a different version was implemented. A folded wall was set up showing alternately a photograph of the Ring in its current state as a street for cars and a painting of the future Ring as a park. Passing cars saw the green future of the Ring ahead of them and behind them the present and past of the Ring as a street for cars.

Translated from the German by Peter Dahm Robertson.

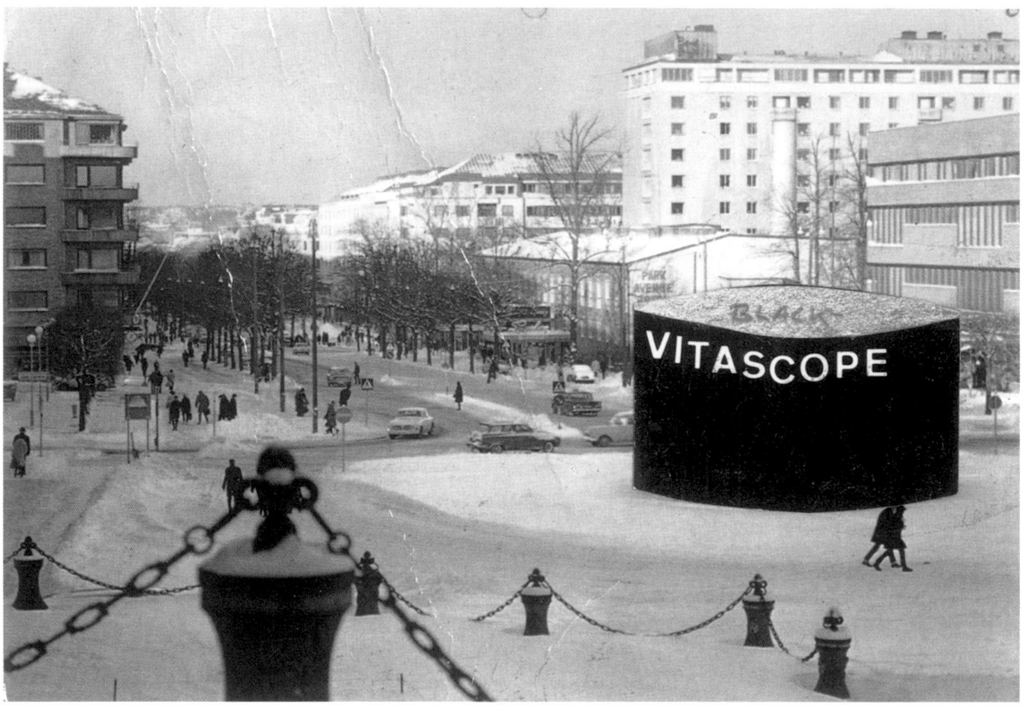

Vitascope / Vitaskop

1970 | movie environment concept

Two circular, covered steel walls (black), I and II. I is smaller than II and is located inside it. I is stationary, II moves. I has a vertical slit. II has eight slits. The viewer stands in the center of I and looks in the direction of slit I. Wall II moves on a track at a speed such that a slit of II passes by the slit of I 24 times per second, so that the viewer may look out at "reality" 24 times per second. —Peter Weibel

Translated from the German by Peter Dahm Robertson.

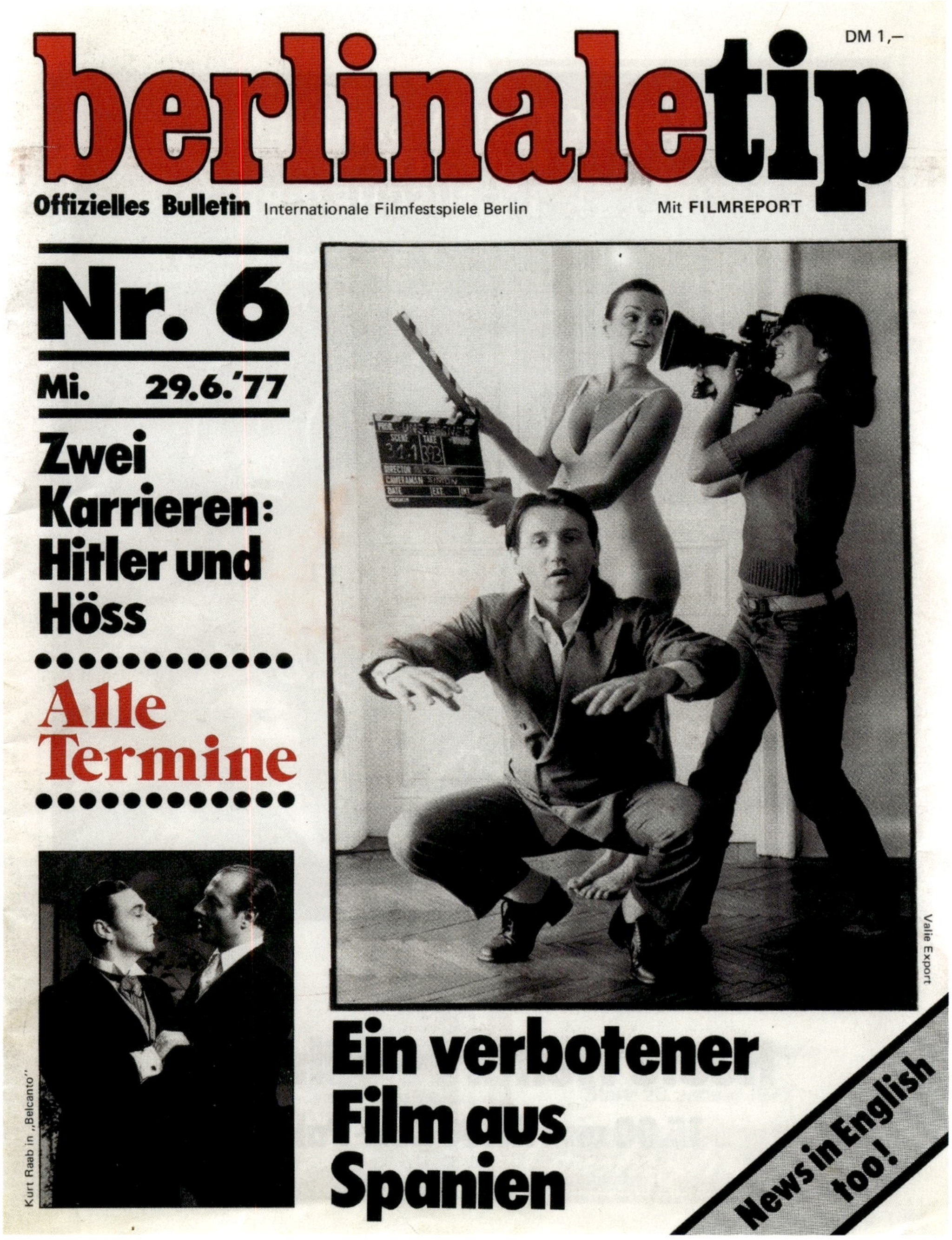

Photo of Peter Weibel, Susanne Widl, and VALIE EXPORT on the set of *Invisible Adversaries* on the cover of *Berlinale Tip*, no. 6 (1977), program published during the 27th Berlin International Film Festival (Berlinale)

Invisible Adversaries / Unsichtbare Gegner

1970 (idea)/ 1977

16 mm film, color, sound, 107 min.
director: VALIE EXPORT, codirector: Peter Weibel
screenplay: Peter Weibel, collaboration: VALIE EXPORT
actors: Susanne Widl, Peter Weibel, Monika Helfer-Friedrich, Helke Sander, et al.
camera: Wolfgang Simon
photographers: Helga Borsodi, VALIE EXPORT, Hermann Hendrich, Monika Hubmann, Karl Heinz Koller, Wilhelm Krejci, Wolfgang Soos, Eric Timmermann
production: VALIE EXPORT
See also pp. 134–35 in this volume

The feature film *Invisible Adversaries* combines many of Weibel's photographic, filmic, and video experiments to describe the world of the protagonist Anna who does not know if what she perceives is true and real or a delusion. In the German original of the film, this distinction is expressed through a play on words when Weibel's character asks: "Ist es Wahn oder wahr, was sie wahrnimmt?"

As the daily news of war, murder, and environmental destruction resounds to the panorama of the city of Vienna, the viewers wonder as much as Anna – video reporter by profession – whether these warnings are true or whether Anna has misheard. After a disturbing phase of unreal hallucinations, real destruction and brutal encounters, Anna goes to bed while the news reports about the military and ecological destruction of the Earth again. She was wearing her pajamas underneath her day clothes.

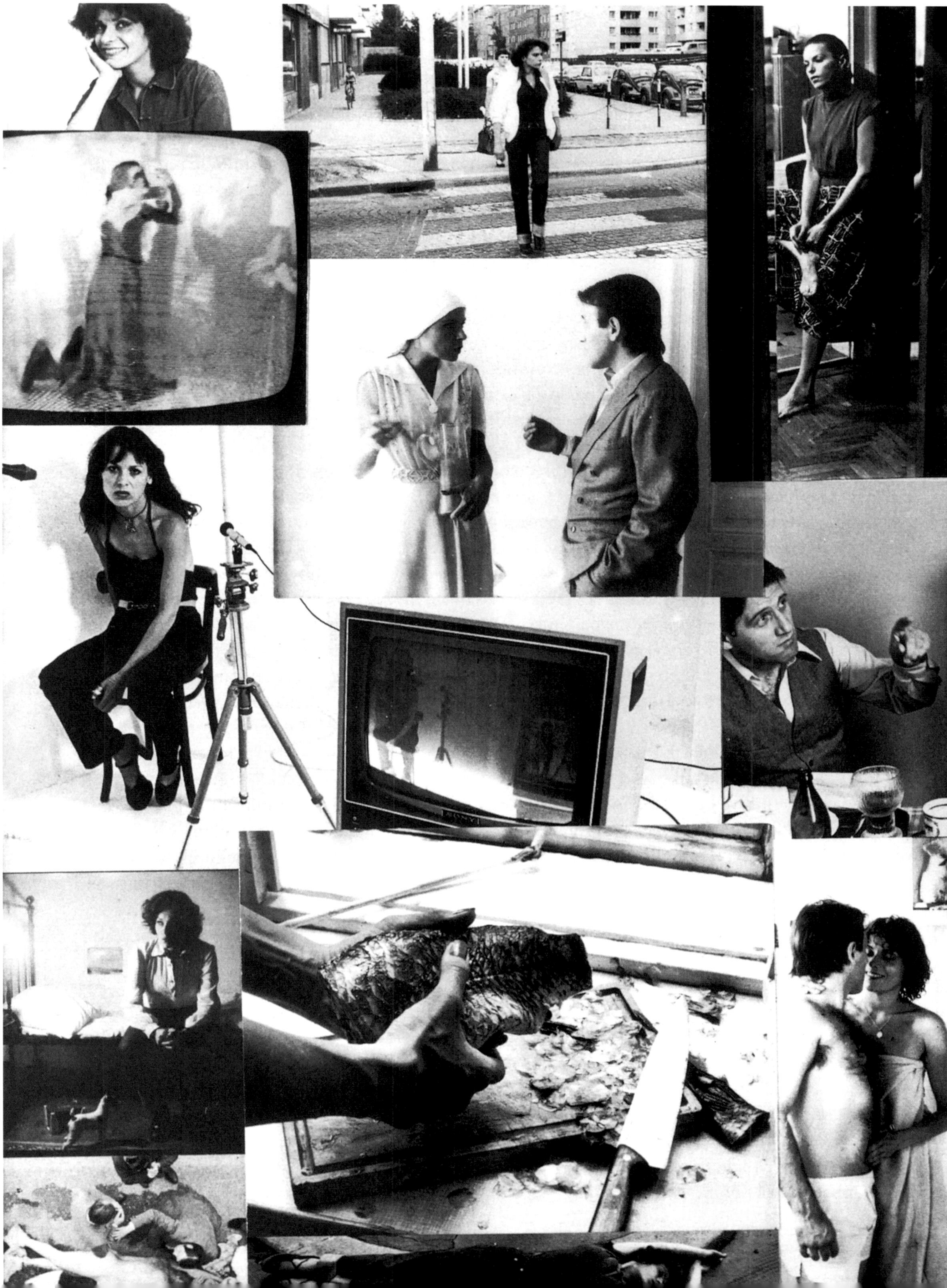

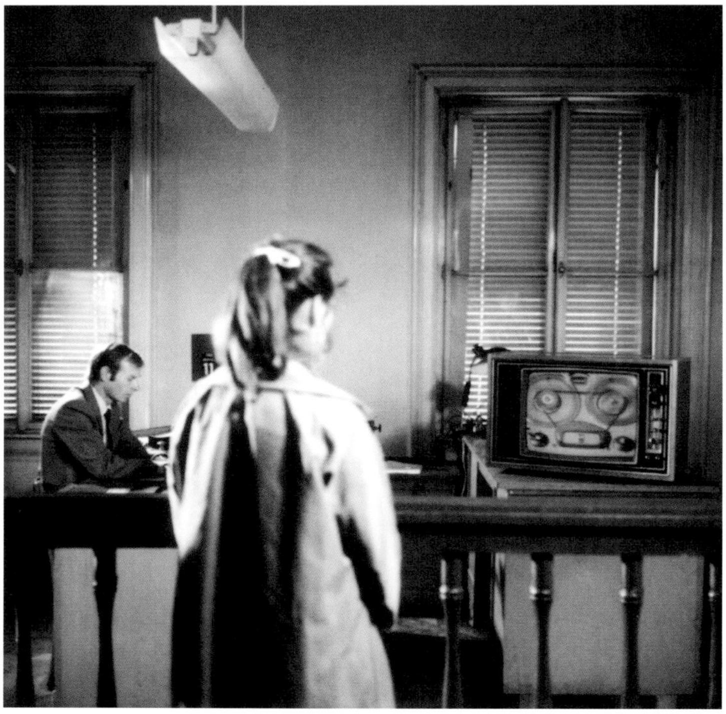

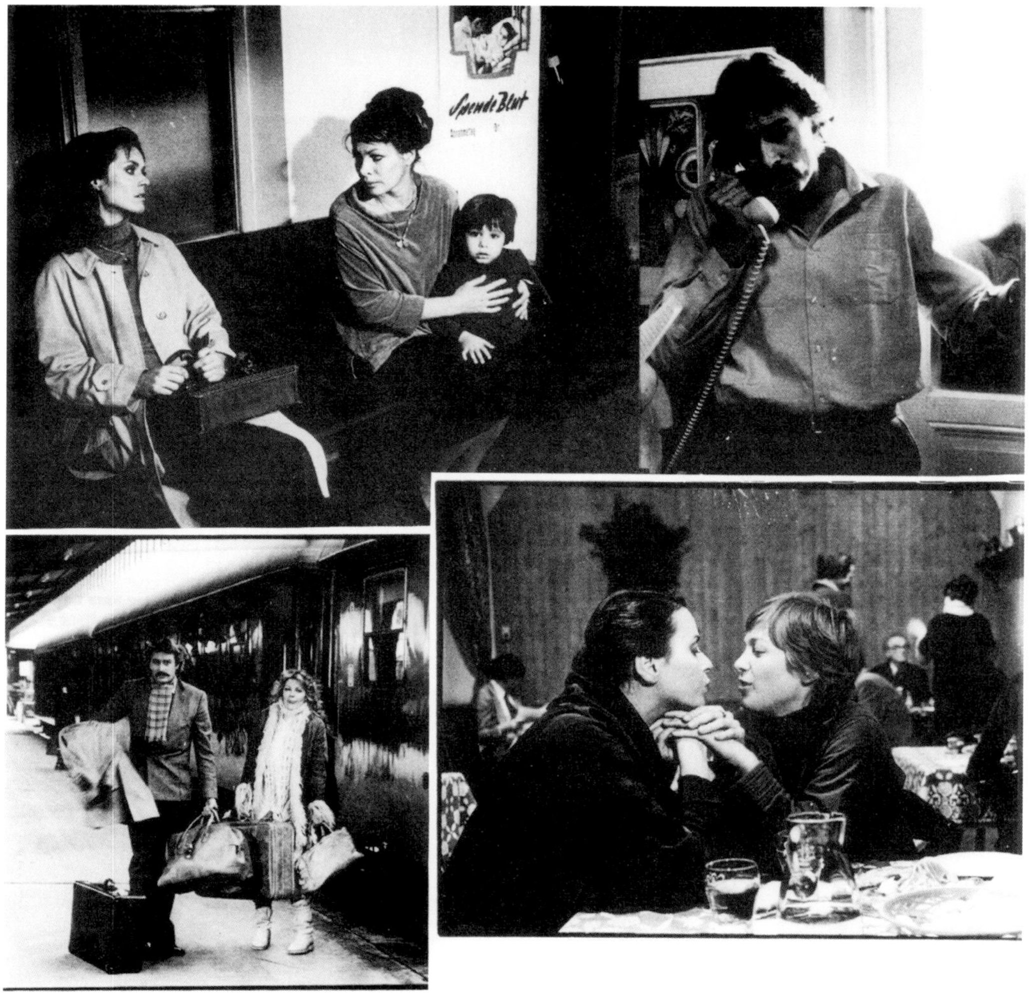

Human Females / Menschenfrauen

1980

VALIE EXPORT in collaboration with Peter Weibel; director: VALIE EXPORT, screenplay: Peter Weibel
35 mm film, b/w and color, sound, 2:12 min.; actors: Renée Felden, Maria Martina, Christiane von Aster, Susanne Widl, Klaus Wildbolz

MENSCHENFRAUEN

Ein Film von VALIE EXPORT

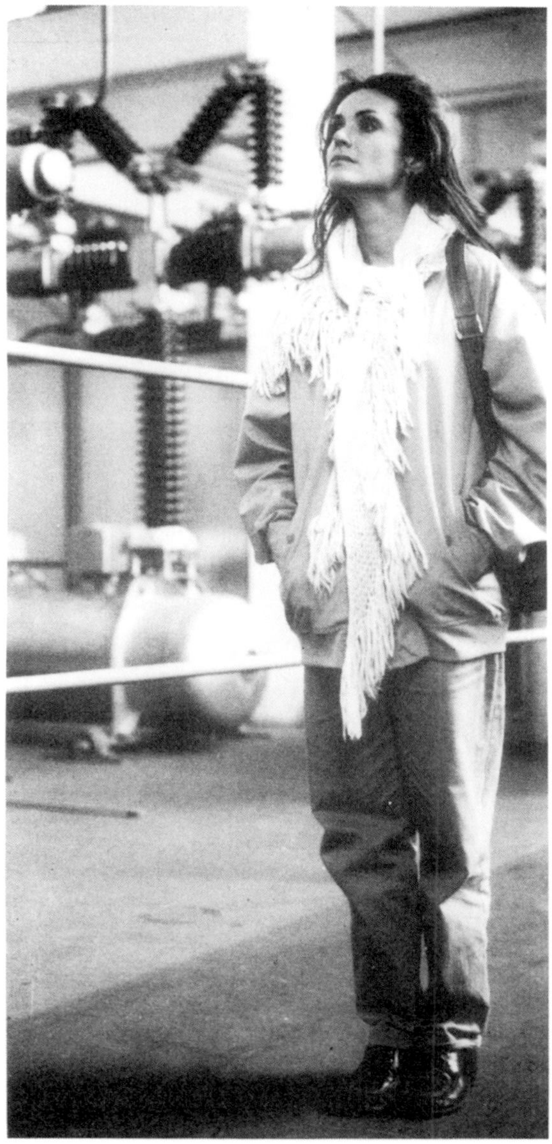

Regie	VALIE EXPORT
Produktion	Valie Export
Drehbuch und Struktur	Peter Weibel, unter Mitarbeit von Valie Export
Kamera	Wolfgang Dickmann, Karl Kases
Toncollage	Walter Stangl
Schnitt	Tina Frese, Friedl Mayer

Darsteller:

Elisabeth	Renée Felden
Gertrud	Maria Martina
Anna	Susanne Widl
Franz	Klaus Wildbolz
Petra	Christiane von Aster
Elisabeth's Sohn	Gerd Weger
Schuldirektor	Peter Janisch
Bankräuberin	Hilde Pilz
Patientin	Elly Naschold
Elisabeth's Mann	Lukas Resetarits
u.a.	

35 mm-Fassung: ca. 116 Minuten Farbe
16 mm-Fassung: ca. 124 Minuten

Österreich 1979

Weltvertrieb: top-Film, Wien

Eigentumer, Herausgeber und Verleger:
Top-Film Ges.m.b.H., 1010 Wien, Walfischgasse 4
Telefon 53 12 46
druck ● rema-print 1080 wien

Fur den Inhalt verantwortlich:
Helmut Siderits, 1010 Wien, Walfischgasse 4
Preis des Programms S8,- (inkl. MwSt.)

The film *Human Females* shows the relations of the journalist Franz S. to four women: his wife Anna, the kindergarten teacher Petra, the teacher Gertrud, and the barmaid Elisabeth. His tight schedule forces him to constantly switch back and forth between the women. He shows the same behavior towards everyone, has the same demands and needs, and wants a lot of compassion from everyone for his situation. One after the other, the women break off their relationships with Franz S. and regain their self-determination. But the women have more in common than just their relationship with Franz. They share the courage to deal with their past, to strengthen friendships, and to search for their own future.
The film is characterized by the restructuring of space and time, mixing dreams and reality, past and present scenes in a way that treats them equally. Techniques of media art and film avant-garde, such as structural film, material film, Expanded Cinema, video art, are transferred to the feature film.

List of Selected Films, Film Concepts, and Other Projects

1966–80

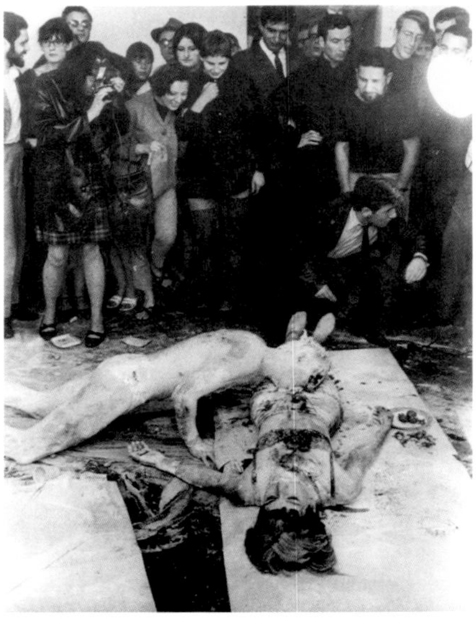

Well and Good / Gut & Gerne

1966 | Expanded Cinema performance, film action after the second *Totalaktion* of Günter Brus and Otto Muehl
Galerie Dvorak, Vienna, June 24, 1966;
16 mm fim, b/w and color, direct and reproduced sound, 15 min. (unfinished)
camera: Helmut Kronberger

Double projection disturbed by spotlights. The actor acts hysterically (brief discourse of psychopathies) in front of the screen either visibly or invisibly (depending on the illumination of the spotlight), either loudly or silently.
Through multiple copying, the same scene is projected onto two screens simultaneously, and through covering/overpainting of the celluloid, movements on the two screens correlate, jump over, create a sense of space. Immediacy (in the aural dimension) is produced in order to escape reproduction (in the visual world). The immediate structures are affinitive to the reproduced ones as well as to the psychological state of the actor. The pathological reactions should thus be classed in the same experiential realm as obsessions and neuroses, which are defined by characteristics of contradiction and repeating, behavioral discrepancies, and experiential recall; there are also moments of reactive cyclophrenia, hysteria (in which, following Ivan Pavlov, the eidetic predisposition is known to predominate), and delusions (sensitive paranoia), among others.

—Peter Weibel

Translated from the German by Peter Dahm Robertson.

Ars Lucis

1967 | film installation, movie environment; projectors mounted on rotating turntables, fixed and rotating mirrors, prisms, projection screens (fixed and mobile, flat and curved)

Film Glasses / Filmbrille

1967 | retinal projection

Brain Glasses

1967 | concept

Reading glasses as auxiliary sensory operator combining movie screen, camera, and projector in one, which transmit directly into the brain, a direct brain film that replaces a world.

Radio Glasses

1967 | concept of glasses with a built-in radio (at the temples)

Portrait of a Contemporary / Portrait eines Zeitgenossen

1967 | 8 mm or 16 mm film, b/w, silent, 10 min. (lost)

Photographs of dead people and famous contemporaries were projected onto me. For each image, I spoke a sentence that was typical for the person depicted. Taken together, these sentences yielded a sort of negative portrait of myself. An ambivalent oscillation between self-portrait and the portrait of an anonymous contemporary. The human body as projection screen! Once this possibility had been discovered, a myriad of films emerged almost of their own accord: one could project enlarged body parts onto the body itself, or surgery movies, or porno movies: if one wanted to play with the identity of object and image, to reduce the difference between sign and signified. Or, to increase the difference: travel movies, landscape films, films of different surfaces, etc. —Peter Weibel

Experience Film No. 1 / Erlebnisfilm Nr. 1

1967 | simulated earthquake in movie theater

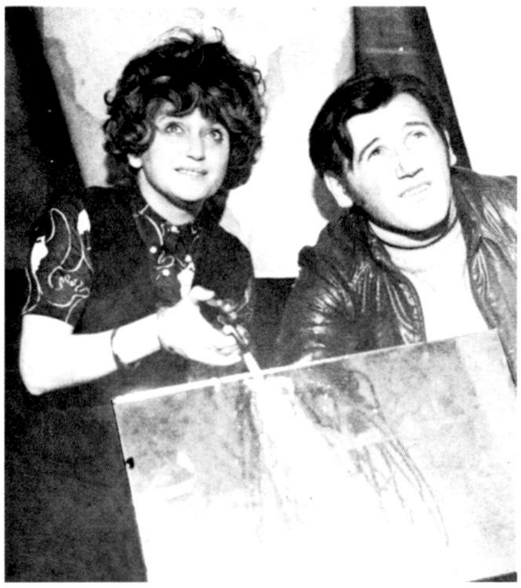

Abstract Film No. 1

1967–68 | VALIE EXPORT (direction) in collaboration with Peter Weibel; expanded movie, performance film; mirror, water, liquids of varying viscosities, spotlight, screen
XSCREEN projection space, Cologne, March 24, 1968

In this piece, abstract symbols are created with concrete materials; no distinction is made between nature and sign. In the battle against celluloid as a limitation of the filmic calculus, we replace celluloid with new methods. A spotlight shines on a mirror onto which various liquids are poured, the effects of which are projected onto the screen by means of reflection. The reliance on natural expedients such as water and simple reflection, the reduction to natural elements, and the renunciation of technology establish connections to Land art, Minimal art, Arte Povera, Op Art, etc.

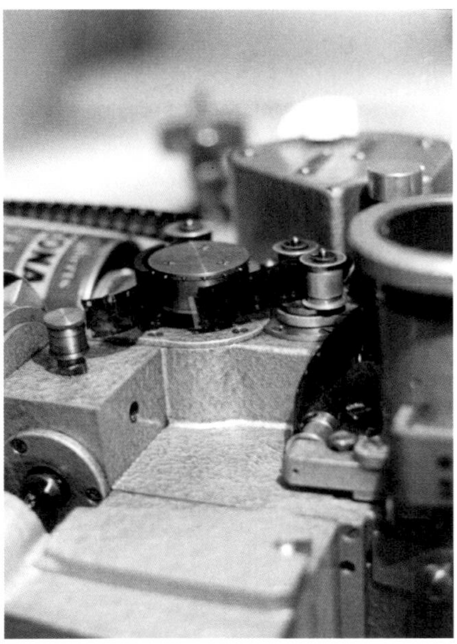

The Gloss and Emulsion of Celluloid / Glanz und Schicht des Zelluloids

1968 | expanded movie; mixed media
XSCREEN projection space, Cologne, March 24, 1968

Ordinarily, 16 mm film is perforated along one edge. I glued an industrial film together so that the perforation changed every minute – that is, the film was first perforated on the left, then on the right. As a result, every time the strip of celluloid hit the wrong side, it would pop off the projector's sprockets and break. I would splice the film back together in front of the audience, as part of the screening, and a minute or two later it would break again. I would splice it together again, and so on and so forth. The film kept getting shorter and shorter, until at last it was completely consumed and the movie was over. A performance of film editing – which is normally part of the process of producing the finished film – was shown instead of the film. The art of splicing was the film. The audience became agitated and protested vehemently. —Peter Weibel

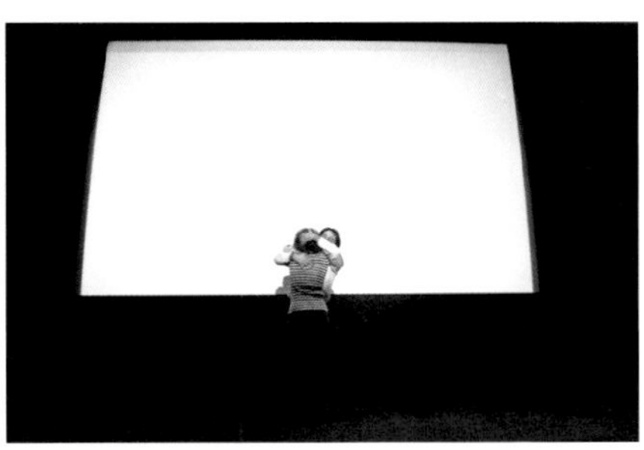

The Kiss / Der Kuß

1968 | VALIE EXPORT in collaboration with Peter Weibel
slow motion picture; 2 performers (Peter Weibel and VALIE EXPORT) kissing onstage, spotlight
XSCREEN projection space, Cologne, March 24, 1968

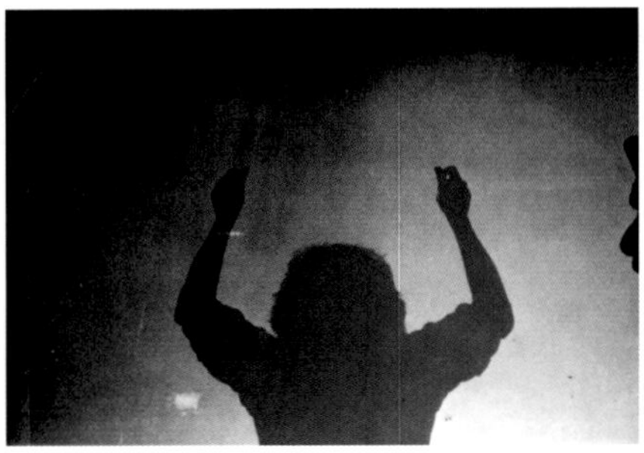

Instant Film / Instantfilm

1968

VALIE EXPORT and Peter Weibel
object film; PVC sheet
XSCREEN projection space, Cologne, March 24, 1968

A metafilm that reflects on the system of film and the system of reality. Following the invention of instant coffee, we have finally succeeded in creating the instant film, which combines camera, projector, and screen in one: instantly consumable, ready to make and show anytime! Instant film – instant gratification! A rectangular sheet of transparent plastic. Zero processing time. An immediately reproducible and projectable film. During the showing, several possibilities were demonstrated. Editing is up to the viewer. They can hang the plastic sheet up at home on their own four walls (screens), with a differentcolored background on each one. They can place it in front of an object, creating their own collages and juxtapositions. They can turn reality into a film by holding the plastic in front of their eyes. As the plastic is moved away from the viewer's two eyes, the visual space (screen size) shrinks. Of course, this can also be conceived on a larger scale: a giant sheet of plastic that turns everything that happens into art. A concept of art reflected through the medium of film.

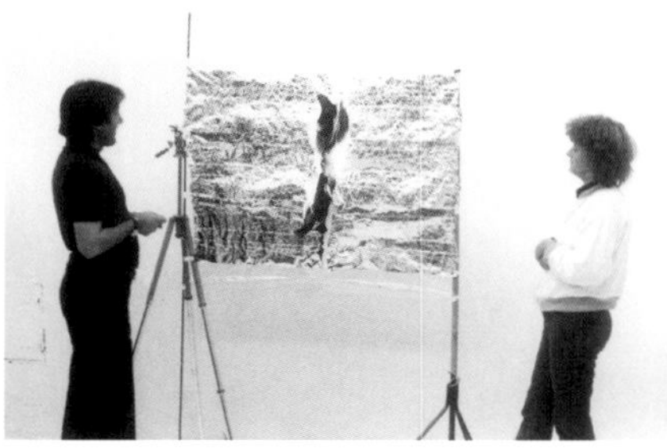

Got a Light? / Haben Sie Feuer?

1968

VALIE EXPORT and Peter Weibel
performance film, a short film in the style of the French New Wave; aluminumfoil screen, cigarette, male and female performers, toilet paper, glue, matchbox, spotlight, video, b/w, 3 min.

Female performer asks male performer for a light. He lights her cigarette and she starts smoking. The audience's reaction (usually heckling, arguments with theater owner because other people start smoking, etc.) determines the start of the second phase: striking a match on the screen and lighting the toilet paper so the screen begins to burn. Minimal film, until the sparks (literally) start flying. Light and motion, play of light and shadow: glow of the cigarette, motion of the rising smoke and its shadow on the screen. Smoking in movie theaters and the like constitutes a violation of fire regulations, a breach. In the dim light of the emergency exits and the light of the flames, the room is revealed as a space of addiction, the movie theater as an institution of the state, a house of correction. A reflection on the cinematic situation, on the pragmatics of film.

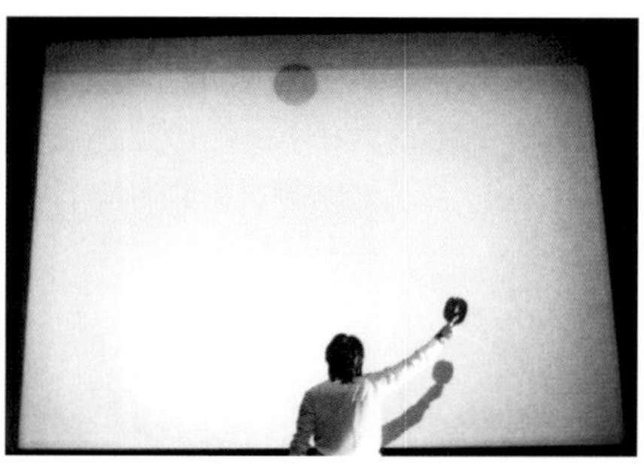

Ping-Pong

1968

VALIE EXPORT, performer: Peter Weibel
Expanded Cinema, home movie; 8 mm film, b/w, 2:30 min., ping-pong balls and paddle, 8 mm projector, performer, solid projection surface

A playful film – a film to play! If you enjoy physical and mental training, then *Ping-Pong* is for you. Grab a ball and a paddle and try to hit the points that appear on the screen. The viewer plays with and against the screen. Without the viewer, there would only be an abstract film of balls. Through interaction, the abstract film is transformed into a participatory event. Stripped of semantics, the relationship between viewer and screen becomes clear: stimulus and response. *Ping-Pong* explicates the power relationship between producer (director) and consumer (viewer). What the eye is telling the brain is an impetus for motor reflexes and reactions. Both viewer and screen are participants in a game whose rules are dictated by the director.

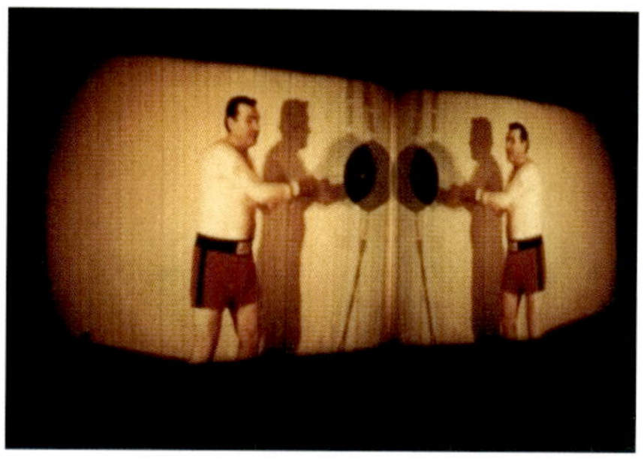

Split-Screen Solipsism / Splitscreen-Solipsismus

1968 | VALIE EXPORT
expanded movie; aluminum foil, 8 mm projector

The projected image is reflected in a sheet of aluminum foil placed perpendicular to the screen, resulting in illusionistic effects (e.g., one boxer becomes two). Doubling – a familiar motif in mannerist painting (Da Vinci, Dalí, Clerici, Micciotta, figureheads on ancient ships) – appears as a characteristic feature of cinematography in Dziga Vertov's *Man with a Movie Camera*: he has himself filmed at the edges of department-store windows. The split screen is a reduced form of doubling: two images with just one projector.

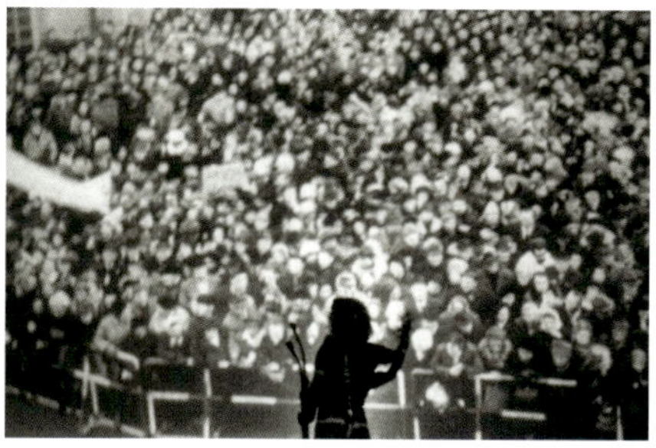

Address/Express / Ansprache Aussprache

1968 | VALIE EXPORT
home movie; microphone, transparent screen, 8 mm stock footage

A film is projected onto a transparent PVC screen from behind: a cheering crowd yelling out encouragement and appreciation. In the foreground of the film, we see a microphone, also projected. The user stands in front of the screen and addresses the crowd through the virtual microphone.

A Family Film by Waltraud Lehner / Ein Familienfilm von Waltraud Lehner

1968 | VALIE EXPORT and Peter Weibel
Expanded Cinema, political film / performance;
8 mm film by VALIE EXPORT, portable microphone, ca. 30 min.
XSCREEN projection space, Cologne, March 24, 1968

Two performers sit in the audience, treating the space like their apartment and the viewers like their personal guests. While watching a home movie by VALIE EXPORT about her family, they strike up conversation and putter about, without inducement or preamble, thoroughly indifferent to the public and to art, which are often nothing more than a perverse playground. These performers perform without permission. Hollywood is not the principal violinist; don't be a string for the state to pluck out the national anthem on. Make movies to satisfy your own needs and desires, not to fit the molds of industry.

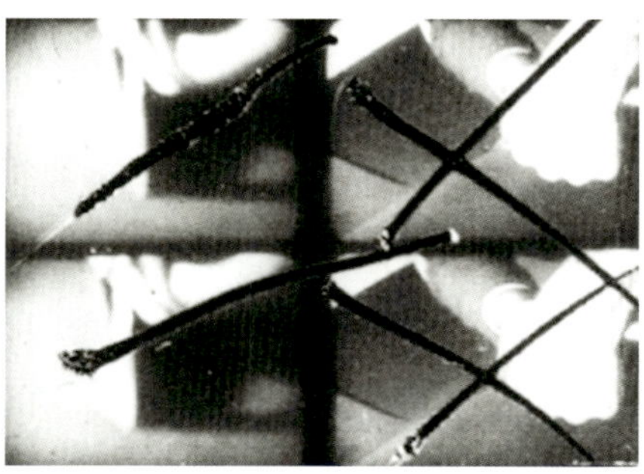

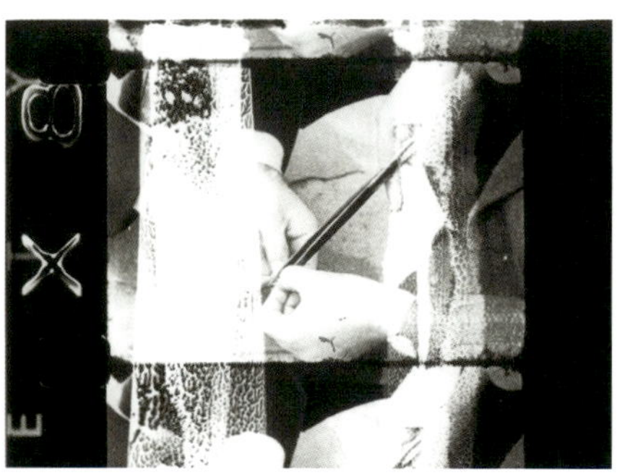

Snip Snap / Schnippschnapp

1968 | Peter Weibel and Ernst Schmidt Jr.
16 mm film, b/w, sound, 2 min.

Top: Still from the first part, "Snip"
This part was shot on 8 mm film that was not cut in half after being developed, allowing it to be projected as a 16 mm film (four images appear on the screen simultaneously). The images feature visible scratches on the finished film, as well as gashes cut into the negative with a very large pair of scissors (black wedge shapes in the center). Because the cuts are so big, the film makes a lot of noise when it is projected.
Bottom: Still from the second part, "Snap"
Each frame is four times as big as in the first part. Some frames have been drawn on (which appears in the positive as a cloudy white streak). Cuts have been made in this strip too, but with a very small pair of scissors, so the film makes almost no sound.

Untitled No. 1 / Ohne Titel Nr. 1

1968	project film, triple projection; 3 staggered, overlapping screens, 3 8 mm projectors

A demonstration of spatial and temporal structures, temporal compression, spatial distancing. For example, we see a hurdle as a runner jumps over it. Three different sides of the object appear simultaneously on the three screens. An airplane takes off on the first screen, flies through the air on the second, and lands on the third. Looping projection.

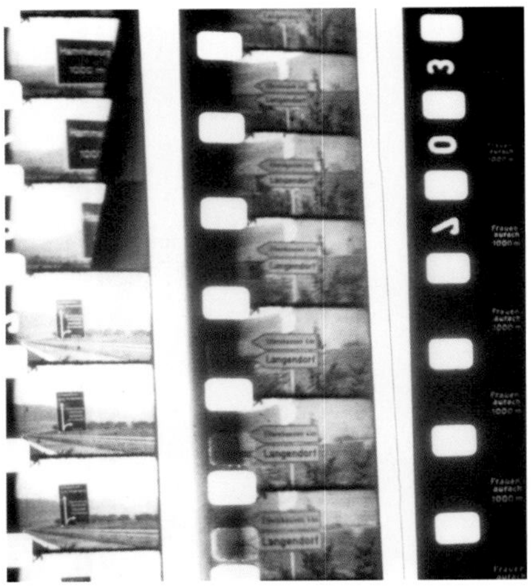

A Trip Is Worth a Trip / Eine Reise ist eine Reise wert

1968/69	VALIE EXPORT and Peter Weibel (text) film; 8 mm film, color, silent, 8 min.

A journey in time, structured according to spatial parameters. Every road sign on a car ride from Kiel to Vienna was filmed, including at night. The number of signs thus determined the length of the film. Points in space and length of time were brought into equivalency. The temporal journey was not measured in minutes or other units of time, but according to spatial, visual parameters. A sixteen-hour car trip thus became an eight-minute film trip. A text film between filmic time and temporal space.

I Choked on Life, It Was Too Big for Me to Swallow / Ich habe mich am Leben verkutzt. Es war mir ein zu großer Brocken

1968	16 mm film, b/w and color, sound, ca. 1 min.

Environment Films / Environment Filme

1968	VALIE EXPORT and Peter Weibel 8 mm film, color, silent

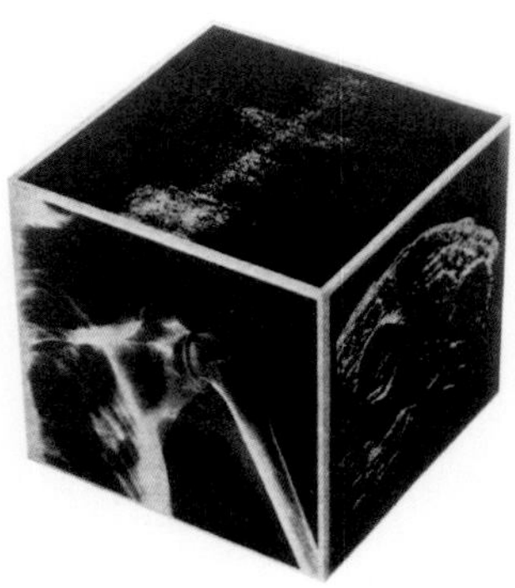

Films are projected onto multidimensional objects, like outsized cubes, cuboids, or tents, from above or below or from the side. Viewers can enter these objects in which they are surrounded by films on all sides, but in a way that shifts the usual perspectives. They find themselves between huge tables and chairs, in front of huge blades of grass, they get pulled out of a desk in a drawer, and similarly the material relations of the environment are changed: under their feet are clouds, water; above their heads meadows, streets, and plains. Film loop projections.
Images produced by various visualization techniques:
Above: radio isotopes
Laterally: ultrasound
In front: X-rays

Translated from the German by Gloria Custance

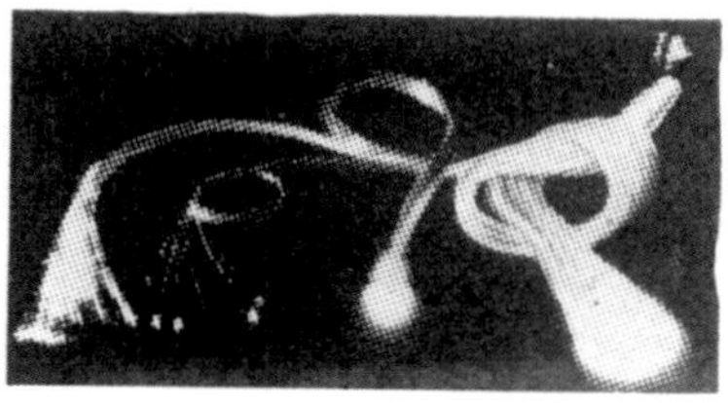

Remote Senses / Distanzsinne

1969 | concept film

Elements of a communication action, expanded movie, "basic training."
Optical fibers as eye hoses at whose ends films are produced according to which the user must react. Through a mass of such fiber-optic cables, each person is connected to everything they require (cities, people, landscapes, etc.), so that at any given time they can see what they want. A communication system with fiber-optic cables can transport a thousand times as much information as a few telephone wires.

—Peter Weibel, 1969

Translated from the German by Peter Dahm Robertson.

Light Whip / Lichtpeitsche

1969 | film object; four-tailed whip with lights

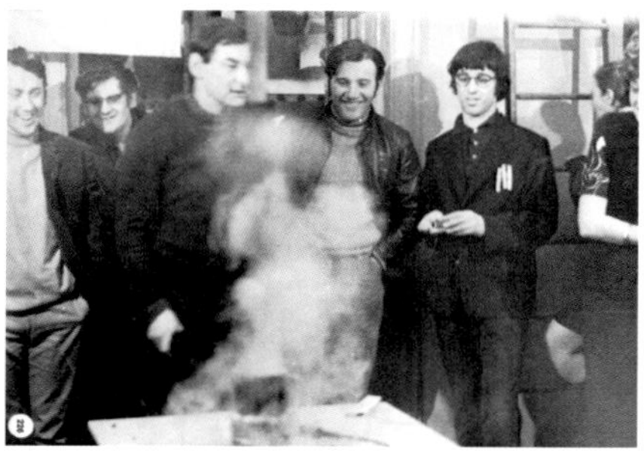

Light

1969 | PVC film, magnesium, Buchhandlung Herrmann, Vienna, February 14, 1969

A sheet of PVC film is attached to the ceiling of the room. Magnesium powder is sprinkled over it, and magnesium wire is laid across it in the shape of a sine wave. The powder is ignited, and bits of magnesium fall as the wire slowly burns. Magnesium powder, which produces a very strange and beautiful light, is also known as flash powder. In old movies, you can often see smoking cameras after a flashlight has gone off. Magnesium wire is still used in the photography industry. In addition to fire and electricity, we thus have a third cinematographic light source, magnesium, that has been liberated here from industrial use. Such alienated light may perhaps defamiliarize the other light (the more accustomed light of the movie theater), thereby shedding light on light in film in general. The photograph shows Hans Scheugl, Ernst Schmidt Jr., Otto Muehl, Peter Weibel, and Franz Kaltenbeck.

The Life of Every German Citizen Is Filmed / Das Leben jedes deutschen Bundesbürgers wird verfilmt

1969 | film project

Genital Panic / Genitalpanik

1969 | VALIE EXPORT
performance film – film performance; action pants

I came across the word *Genitalpanik* in the work of Günter Brus. There are two sexual problem zones for the bourgeoisie: the breast and the vagina. Neither is allowed to be shown in public. With *Touch and Feel Cinema* (*Tapp- und Tastkino*, 1968, see pp. 180–81 in this volume), I tackled the first one. Then I made a vertical shift to the vagina: a pair of jeans was cut wide open in the genital area. Whereas in *Touch and Feel Cinema*, in an inversion of conventional voyeuristic and exhibitionistic practice, the breasts could not be seen, only felt, this piece involved a different inversion: what had been treated as taboo and therefore kept covered – the female genitalia – would now be staring everyone in the face. During a live performance at the Occam movie theater in Munich, VALIE EXPORT squeezed her way down one of the rows while wearing the action pants, her crotch directly in front of the viewers' faces. A number of indoor and outdoor versions were later (1970) staged for a photograph embodying the then-popular catchphrase "pussy power." The hairdo was inspired by Angela Davis (a student of Herbert Marcuse). The machine gun was borrowed for the photo shoot from Udo Proksch – designer, gun nut, media darling, later convicted of murder in connection with the sinking of a cargo ship. —Peter Weibel

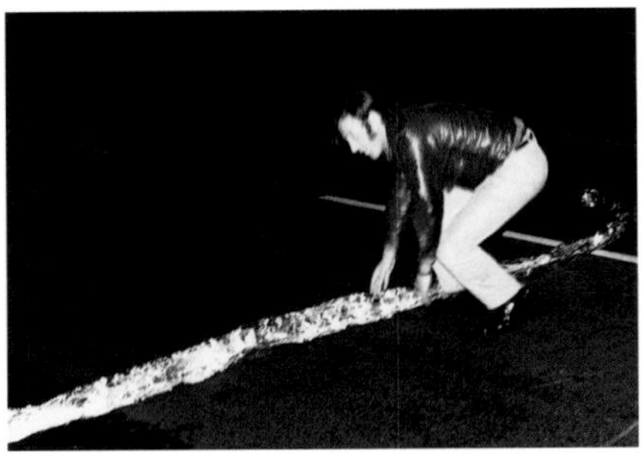

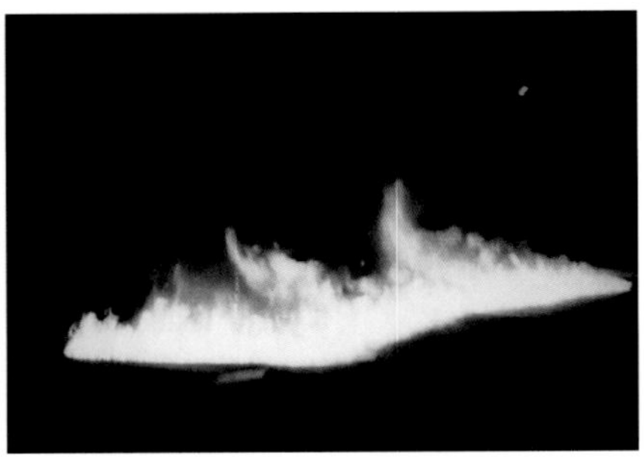

Firewall, or More Traffic Deaths, Fewer Citizens / Brandmauer oder Mehr Verkehrstote, weniger Staatsbürger

1969 | communication action; highway near Gothenburg, Sweden, September 1969, photographed by Franz Kaltenbeck

The attempt to create a traffic accident, a horror film. An organized traffic accident as a civil war, a war of citizens, is a challenge to finally abolish the irreversibility of time. Day after day, injustice is spoken in the courtrooms; day after day, public media adjust the escapist consciousnesses to the norms of a dirty fiction, the reality of the state. Day after day, thousands senselessly die from these structures of communication, from the minister's words, from the judge's sentence, from the terrorism of social behavior.
The man who runs amok, who can no longer stand inhuman, is realizing humanity and freedom for himself and the world through a painful paroxysm in an inhuman and unfree world, precisely by his crime.
True criminality doesn't serve the destruction of the world, but the salvation of the world! —Peter Weibel

Translated from the German by Peter Dahm Robertson.

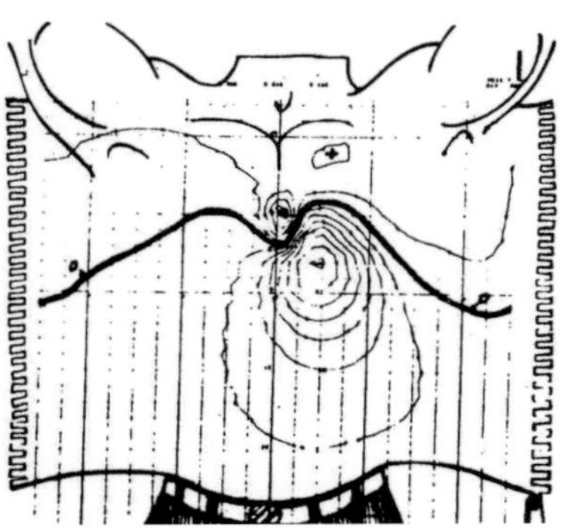

Magnetic Communication / Magnetische Kommunikation

1970 | project film

A transportable realistic external body engine, very small in size when folded, is able to calculate a mathematical model of my body and the functions of its elements and to simulate it by magnetic deformation in such a way that I can shape my body into any desired form: a thread in front of a closed door, flat as a banknote in distress, and so on. —Peter Weibel

Translated from the German by ZKM | Karlsruhe's editorial team.

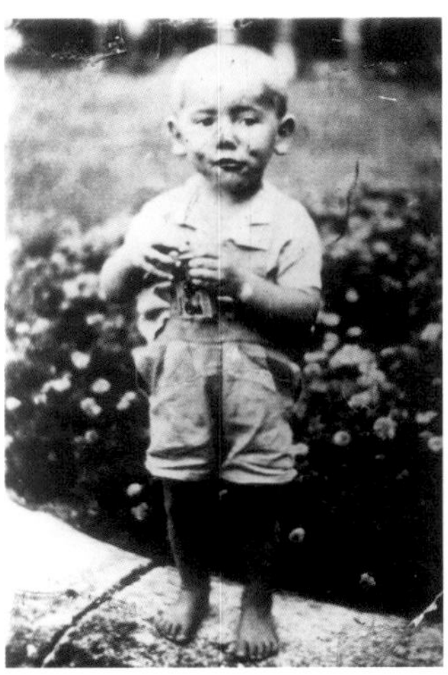

Music Is Petrified Groaning / Musik ist versteinertes Stöhnen

1972 | slide performance, 10 min.

The starting point are two photos. One is a photo of me as a child with boils, taken behind broken glass. The other photo is a long shot of a walled castle. Twelve slides were made of each photo in such a way that the photo of the child became more blurred from slide to slide and the castle photo went from being a long shot to a close-up of a detail on the tower. The slides were projected alternately in such a way that the blurred image per slide became increasingly sharper (i.e., the viewer recognized the photo at the end of the series) and the long shot became more and more of a close-up. Parallel to this, music was played: Anton Bruckner at the castle and the unidentifiable moaning sounds of whales at the child's photo.

In addition, the duration of the projection of the slides changed, too. A tape recorder circuit made each projection unit shorter and shorter; for instance, always the same piece of Bruckner was played, but it became shorter and shorter.

—Peter Weibel

Translated from the German by Gloria Custance.

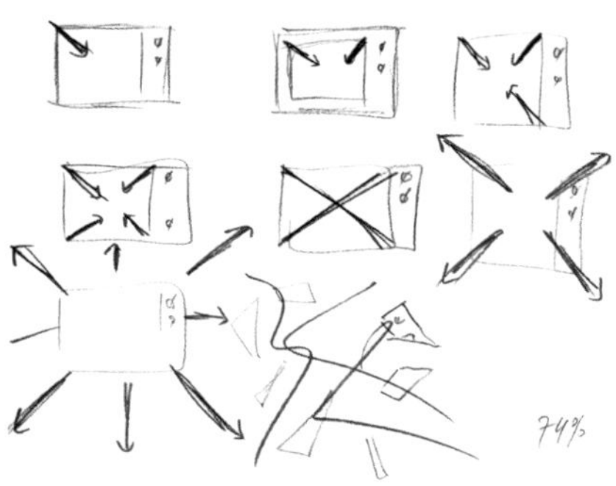

Autoidentity / Autoidentität

1973 | concept

Monitor and recorder. On the screen we see a man drop a bottle; the man stumbles down a staircase, the staircase collapses, a house caves in, a dam bursts, a volcano erupts.
Finally, the monitor itself explodes; the signal to detonate had been encoded onto the tape in the recorder. The last piece of information was not an image, but rather a detonation signal – that is, instead of a representation, what was represented became real.
Self-answering question, automatic
Self-reproduction and self-destruction
Self-solving problem
First-order decidability, self-contained calculation

—Peter Weibel

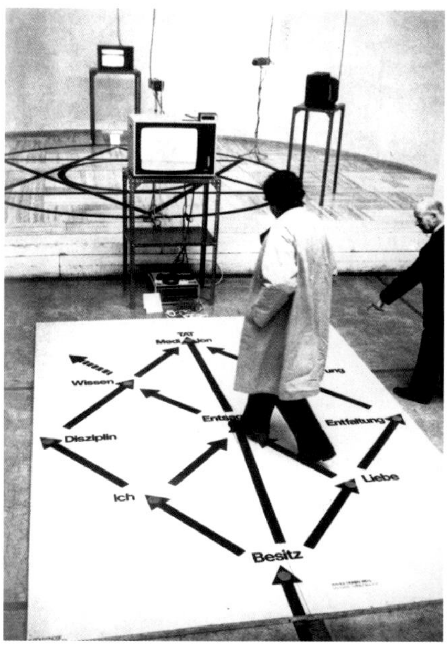

Autohypnosis / Autohypnose

1973 | interactive video installation
Audiovisuelle Botschaften, Trigon '73, Künstlerhaus Graz, 1973

Concepts associated with socialization (ownership/possessions, ego, love, discipline, renunciation, development, knowledge, experience, meditation) are written on a panel on the floor, under which, similar to in the work *The Long March* (*Der lange Marsch*, 1969, see p. 501 in this volume), pushbuttons are concealed. When these are pressed in the correct order, they trigger a signal which switches on the video recorder and its monitor. The concepts on the floor are treated like the keyboard of a typewriter and form the interface between viewer and image. The image only becomes visible if the interface is used correctly; that is, if the user guesses the code that has been used to arrive at the concepts. Users who guess the code correctly (abandon possessions, achieve love and renunciation. Meditate after this experience and take action) will be rewarded by the picture of a cheering crowd.

Translated from the German by Gloria Custance.

Core Codifications / Kernkodifikate

1980 | 16 mm film, b/w and color, sound, 30 min. (lost)

Note: Unless noted otherwise, all the entries in this list were translated from the German by Patrick Hubenthal.

Transverse – Dissymmetrical – Interventional/Epic: Work Gestures from Peter Weibel's Tool Kit

Siegfried Zielinski

The term "polymath" is used to describe that type of pre- or early modern thinker who had had a comprehensive education and was at home in many fields of knowledge, not only the philosophically based subjects, but also the natural sciences, mechanics, and the musical disciplines. Such *Universalgelehrte* (verbally translated as *universal scholar*), as they are called in German – incorrectly, it should be noted, since the term replaces the rousingly multifarious with the boring and flattened-out universal – no longer exist. It is said that the knowledge to be assimilated has grown too extensive; or rather, that the machinery of academia is no longer capable of generating such a poly-mathematic type of learning.[1] Gottfried Wilhelm Leibniz (1646–1716) and Athanasius Kircher (1602–80) are two figures in European history who were classified as this type of scholar.

Peter Weibel is a polymath of the arts, perhaps their last. In his work, he combines a wide variety of knowledge from very different disciplines – ranging from philosophy to linguistics to poetry to psychoanalysis to mathematics and the exact sciences – with the most diverse practices and phenomena from the arts – their history, their techniques, their aesthetics, and the conditions of their production, distribution, marketing, and perception. With regard to these separate fields, his work is *transverse* in the sense defined by Roger Caillois (1913–78) in his brilliant poetic-scientific treatise *The Mask of Medusa* (*Méduse & Cie*, 1960). Weibel's activities "come into strong conflict with our current system of classification." In their linking of thought and deed, they must be regarded as transverse "cuts" that are necessary to "our understanding of those phenomena which, by themselves, always appear exceptional but of which the significance would be much more readily seen if one dared assemble these exceptions, examine them and see if their mechanisms were related."[2]

In the field of the arts and the advanced sciences, technologies, including the worldviews associated with them, Weibel has developed into an exceptionally effective agenda-setter over the last half-century or so. Since the 1960s, he has been energizing many different art scenes and sensitive heterotopias of discourse with his ideas, concepts, irritations, and interventions. He seems to draw upon an inexhaustible reservoir of tools, imagery, modes of representation, and, above all, ideas, the essential elements of which he developed in the first ten years of his artistic and intellectual practice. In this essay I will highlight a few of those that seem to me worthy of note.

The Artist and Intellectual as Operator and Magician

Just a few years before Peter Weibel was born in the city of Odesa, Walter Benjamin (1892–1940), in exile in Paris, wrote what was at the time a rather controversial art-philosophical meditation on the relatively new aesthetic phenomena that the technological-industrial revolution of the nineteenth century had produced. He unhumbly described this text as the first "materialistic theory of art."[3] In the eleventh section of his essay "The Work of Art in the Age of Its Technological Reproducibility" ("Das Kunstwerk im Zeitalter seiner technischen Reproduzierbarkeit," 1935), he establishes a remarkable dualism between the traditional visual artist as magician and the new visual artist as "operator": "The painter maintains in his work a natural distance from reality, whereas the cinematographer penetrates deeply into its tissue"[4] and breaks the subject of his observation down into multiple fragments.

"You under-estimate the technicality of autonomous art and over-estimate that of dependent art"[5] – this was how Theodor W. Adorno (1903–69) summed up his critique of Benjamin's essay. Paintings and sculptures may sometimes come into being without the use of complex machinery, but they are not free of technology, and the *aura* is not simply destroyed in the arts with and through media, but rather transformed, extended, modified, and even, in the case of cinema, amplified, in that the emotional economies of the audience are capitalistically exploited.

Figures like Weibel and positions such as his clearly illustrate the problem of Benjaminian polarization, which has its parallel in the natural sciences, for example, in the distinction between external and internal observer, between "exophysics" and "endophysics," as defined by Otto E. Rössler (b. 1940).[6] As someone who is constantly making diagonal cuts between the arts and the sciences, Weibel in fact embodies both: the intervening, analytically dissecting operator as well as the contemplative magician who lays on hands, simultaneously reflecting upon and auratically charging the objects of his attentions, no matter how suffused with technology they may be. In the course of the second half of the twentieth century, at the very latest, things became so complex that they can only be negotiated via a combination of *endo* and *exo* positions; and aesthetically, too, they can only be developed by means of a dual perspective. I will return to this subject, as it is a key factor in the Weibelian microuniverse.

In a short essay from 1968–69 titled "Projektfilme und projizierte Projekte" ("Project Films and Projected Projects"), Weibel radically articulates his visions of a technological eye that knows no material boundaries: as an all-encompassing *perspicere*, as pitiless scrutiny, and as the "development of a camera that would also be able to film life behind closed doors. The principle: the camera sends out waves which penetrate obstacles such as walls, bones, etc. in much the same way as sound waves. The waves, whose range can be regulated, would be reflected at a certain adjustable distance and would come back into the camera, where they would reproduce the 'image' on a screen. An auxiliary device is necessary for the images captured in this way to be decoded, 'deciphered,' converted to our visual standard. Both state secrets and the secrets of private life will become impossible. Everyone will be able to see into everyone's hearts. Nothing will remain hidden."[7]

That is a provocative description of a society of universal control made possible with the help of advanced visual technologies, and of how one might intervene in it as an artist. At the beginning of the essay, Weibel radicalizes the ubiquitous mobile "kino-eye" of the Russian revolutionary artist Dziga Vertov (1896–1954). In "Projektfilme und projizierte Projekte," however, interventional analysis and the poetic outline of a future visual practice are intimately interlinked. In the text's conclusion, the magical component of Weibel's concept of projection unequivocally prevails: "Night becomes day. The gigantic screen called 'sky,' conceived as a chemically prepared canvas, produces light shows that touch the whole globe. The atmosphere as a light show theater. Artificial control of light conditions above the earth and for the earth. Projection of the word 'love' over the cities ..."[8]

Movement/Mobility: Freestyle Thinking

Institutions that have dealings with intellectually and artistically driven people such as Weibel need to understand: everything about twentieth-century thinking on media that is held dear by those whose socialization was principally effected by the awakening of activist critical theory was not (or at least, not primarily) developed in classic academic contexts and was certainly not hatched in the context of permanent professorships: Benjamin, Adorno, Vertov, Hugo Münsterberg (1863–1916), Lev Kuleshov (1899–1970), Arseny Avraamov (1886–1944), Béla Balázs (1884–1949), Bertolt Brecht (1898–1956), Siegfried Kracauer (1889–1966), Rudolf Arnheim (1904–2007), René Fülöp-Miller (1891–1963), Günther Anders (1902–92), Herbert Marcuse (1898–1979), and Max Horkheimer (1895–1973) were all – compelled by the historical violence of the circumstances – transmigrants, exiles, fugitives, travelers, constantly on the move and on the alert. The Chilean artist Sandra Vásquez de la Horra (b. 1967) was recently asked why she never does her large-scale drawings on a single piece of paper, distributing them instead over multiple smaller sheets which she then assembles for exhibitions. She replied that she needed to remain mobile with regard to her materials. Everything she needed for her art had to be easily transportable.[9] This is the position of artists in migration. "The wanderer must not have 'property'; one who possesses is not *free* to wander," wrote Massimo Cacciari (b. 1944) in a study on the French poet-philosopher Edmond Jabès (1912–91).[10] Traveling light: when Weibel travels, he rarely needs more than what fits into a plastic bag – and that mainly contains newspapers and books.

There is an internal reciprocity here: the propensity and, even more, the pressing need for hurried departure, for flight, for the existentially necessary crossing of boundaries, for the bypassing of walls, barbed-wire fences, and identification and customs checkpoints, is symbiotically linked to a lack of discipline in thought. The most (thought-)provoking critics and theorists of the century that emptied out humanism, turning it into an unrecognizable shell and mask, had and still have no binding commitment to an institutionalized subject in which they would be at home. Then and now, they have always regarded media discourses as *intermedia* discourses as well, swinging back and forth between the classic forms of artistic expression and the means of technological reproducibility, between painting and photography, theater and cinema, music and performance art, written media and networked visual writing media – and likewise between philosophy, science, and the history of technology; sociology and aesthetic criticism; computer science and semiotics; psychoanalysis and the critique of ideologies.

Neither productive creative work with fluctuating media nor intensive and perpetually alert reflection upon them is compatible with being tied to a place, region, or intellectual position. Before Covid-19, when they were still allowed to travel in planes and trains as the need and desire arose, the scene's nomadic intellectuals and creatives, among whom Weibel was a central figure, were sternly reprimanded for jetting back and forth between symposiums, conferences, exhibition openings, and seminars, with nowhere near enough time to produce compelling (re)formulations of their presentations amid all that energy- and vitality-devouring movement. Vilém Flusser (1920–91), Florian Rötzer (b. 1953), Norbert Bolz (b. 1953), Gerburg Treusch-Dieter (1939–2006), Allucquére Rosanne Stone (b. 1936), Dietmar Kamper (1936–2001), Jean Baudrillard (1929–2007), Friedrich Kittler (1943–2011), Slavoj Žižek (b. 1949) – these were some of those who seemed to be constantly on tour at the end of the twentieth century. Their concert halls were the academies, museums, and university auditoriums, their intimate club settings the classrooms and galleries that opened their doors to debates on the criticism and shaping of the conditions of communication and their potential prospects. The presses of the small publishing houses that eagerly competed for their texts rotated faster than the thoughts could be written down. "On the road again"[11] – in the case of the gang of early media-critical intellectuals, we were dealing, to a great extent, with a generation that had been socialized largely by rock music and street culture. In the 1960s many of them founded bands, sang, drummed, played bass or guitar, tinkered with electroacoustic instruments or synthesizers. Weibel was founder and lead singer of the band Hotel Morphila Orchester. As a front man or woman, one learns to take risks in art as well, to improvise and face the consequences when things get serious and critical.

In a jointly written essay, Weibel and I described Flusser's thinking as an activity that one might regard as "thinking without a banister," to use Hannah Arendt's (1906–75) phrase. However, Weibel's intellectual microuniverse could be even better summed up as "*freestyle* thinking."[12] "Freestyle," in swimming, represents the antipode to the more tightly regulated events in which athletes are only allowed to move through the water by means of the breaststroke, backstroke, or butterfly. Freestyle offers the opportunity to display the full energy and creativity of the individual swimming body outside the dictates of discipline. It is a virtuous quality – that is, a virtual discipline – condensed from the absence of discipline.

At the high level at which Weibel operates, such thoughts and actions take on the quality of *diagonal practices*, which are supremely equipped to mediate between the advanced laboratories of the arts and those of the sciences. In the profusion of his transverse activities, he has repeatedly established fields of study that did not previously exist in such a form.

Dissymmetries

The heterogeneous artistic tactics with which Weibel intervened in literature's established rules, grammars, and semantics may be described as *expanded poetry*. As a dispositif, and therefore as a structure to be critically dissected, language was his most prominent field of activity in the 1960s, in which he was strongly influenced by Jacques Lacan (1901–81), Jacques Derrida (1930–2004), and Michel Foucault (1926–84). Julien Torma (1902–33), who disappeared in the forests of the Tyrol at the young age of thirty, provided Weibel with the epigraph for his incursive actions in the realm of conventionalized linguistic apparatuses: "To shut poetry up in the poem is to prevent it penetrating into life. Let's not write anything any more. The poet of tomorrow will be unaware of the very name of poetry."[13]

Naturally, the program of literary criticism that Weibel laid out on the reverse of the "Portrait" that included Torma's dictum, formulated with and through the medium of literature, became a poem itself. In it, the artist lists his demands, observations, and attestations, probably using a portable mechanical Olivetti typewriter from the 1960s, a classic ergonomic model equipped only with lowercase letters:

apostasy from language
extraliterary literature
literature in the age of electronic media
book as means of communication
zero-literature
epistemological role of language
writing as architecture
para-literature
illiterary literature
literature – a communication between rhinoceroses
book as paradoxical medium of communication
language as social terror
liberation from language
liberation of consciousness from language
perception beyond language
communication beyond language
visual communication
electronic communication
sensory communication
chemical communication
invisible communication
extraliterary literature[14]

Emphasizing the transverse quality of his intellectual approach, Weibel proclaimed an "ultra-literature" that would be "what good literature has always been: transport of new insights and feelings, exploration of the reality beyond government norms, amplification and elevation of the individual."[15] The history of this "ultra-literature" would "obviously" include "poets like Galileo Galilei, James Clerk Maxwell, Quirinus Kuhlmann, Athanasius Kircher, [Antonin] Artaud, etc."[16] – in other words, prominent scientists and polymaths as well as artists. In a set of "fore-sentences" Weibel composed in 1968–69, he was already linking this concept to the urgent need for poetry to "take up the challenge of electronic media today and expand its methods along with the expansion of its materials (from stone to photography)." Poetry of this sort would experiment not only with syllables and handwriting, but also with photographs, with electronics, with media machines, with numbers, with bodies that would become the semantic and syntactic elements of a new image-sound-text language. In this way, a new artistic identity would emerge on the horizon: "a poet who is no longer interested in the behavior of words in sentences, but in the behavior of people in situations, who experiments not with syllables but with social structures, with structures of space and time."[17] For example, in the conceptual photograph *stairwell* (*stiegenhaus*, see p. 122 in this volume), several versions of which were created between 1967 and 1972, the artist experimented with a series of eight photographs of different staircases, which he assembled into a dynamic sequence to create a visual spatiotemporal poem without words.

During this period, Weibel was carrying on an imaginary correspondence with William S. Burroughs (1914–97), another radical writer of the Beat Generation, and with his surgical interventions in existing linguistic (and visual) systems and concepts of meaning. In his manifesto *The Electronic Revolution* (1971), Burroughs refers explicitly to Wilson Smith's (1897–1965) study *Mechanisms of Virus Infection* (1963) in a passage in which he projects findings from biological virus research onto media: "A virus IS a very small unit of word and image."[18] In a long series of drastic examples, Burroughs shows how the concept of a viral word-and-image unit can be deployed as a disconcerting experience introduced into the fabric of existing mass-media communications. He begins with simple audiotape experiments using the cut-up method he developed with Brion Gysin (1916–86) as a subversive aesthetic tactic, continues with techniques for chopping up and scrambling language, and finally suggests organizing a "sex-tape festival" at which recording devices (audio and video tape recorders) play an important role: "100,000 people bring their scrambled sex tapes, and video tapes as well, to scramble in together. Projected on vast screens, muttering out over the crowd. ... Soon it will scramble them all naked. The cops and the National Guard are stripping down. LET'S GET OURSELVES SOME CIVVIES. Now a thing like that could be messy, but those who survive it recover from the madness."[19]

At the end of his manifesto, Burroughs describes the "categorical THE" of our established language as a "virus mechanism." The routinely employed "EITHER/OR" is another "virus formula" that always leaves just one alternative: "you OR the virus. EITHER/OR. This is in point of fact the conflict formula which is seen to be archetypical virus mechanism." Burroughs ultimately develops this critical reflection on ossified semantics into a concept for a new language to be created, which would not only do away with viral mechanisms but make it impossible to even formulate them: "This language will be a tonal language like Chinese, it will also have a hieroglyphic script as pictorial as possible without being too cumbersome or difficult to write. This language will give one option of silence."[20]

The agenda that Burroughs and Weibel congenially developed as authors and artists in the 1960s can also be summarized as the dismantling and (poetic) concentration of language as a new state-of-the-art intellectual and magical practice: against mediocrity and boredom, against the police and the state, against the academic and legal system, against the authorities – always in the awareness that our writing implements themselves participate in the writing of our texts, as Friedrich Nietzsche (1844–1900) once said in reference to the mechanical writing device he used, whose keys protruded from a ball.[21] To be able to practice subversion on a high level, one must know machines and the programs that control them just as well as the rules and command structures of language. Only by systematically working one's way through their regular and repetitive structures can one enable the creation of *dissymmetries*.

These were compellingly depicted in the context of language by Edmond Jabès in *Le petit livre de la subversion hors de soupçon* (1982), which was published in Rosmarie Waldrop's English translation in 1996,[22] while this succinct formulation appeared in Roger Caillois's reflections on nature and technology, which he embarked upon in Venice in 1970 and published in Paris in 1973: "Little by little the following distinction has impressed itself upon me: asymmetry, the state that precedes the establishment of an equilibrium, specifically a symmetry; dissymmetry, the state that follows the disruption of an equilibrium or symmetry while allowing the forsaken order to be inferred or induced, namely, by clearly appearing as a subsequent intervention, now-necessary subversion, or premeditated modification."[23]

Expanded Cinema

Weibel also applied the same methodological principle – that one must first carefully study and analyze the internal rules of established orders before one can creatively

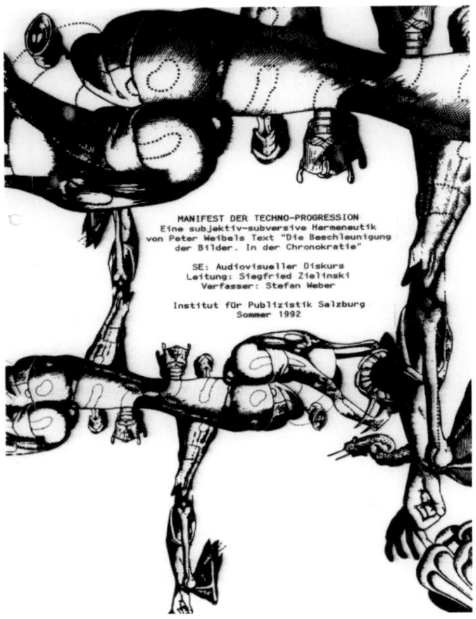

Fig. 1:
Stefan Weber, *MANIFEST DER TECHNO-PROGRESSION*, 1992. The publication features an in-depth analysis of Peter Weibel's book *Die Beschleunigung der Bilder in der Chronokratie* (1987)

break them – to the broad field of cinema, or rather, technology-based arts of sound and image. His *Touch and Feel Cinema* (*Tapp- und Tast-Kino*, 1968, see pp. 180–81 in this volume) simply happened to become, historically, the best-known, most iconic variation on the experiments he was regularly performing with VALIE EXPORT (b. 1940) during this period. Much like his expanded poetry projects, what Expanded Cinema represented for Weibel was, above all, an opportunity to stage public events, performances, and installations in which he could upend the rules of the medium and radically overstep the narrow boundaries of commercial cinema. The photograph *Action Pants: Genital Panic* (*Aktionshose: Genitalpanik*, 1969), which shows VALIE EXPORT wearing a pair of jeans with the crotch cut out and holding an AK-47 at the ready – an image that has itself long since become iconic – was really a byproduct of one of these Expanded Cinema performances: wearing the pictured pants, Weibel's then-partner pushed her way through the rows of a movie theater before the movie started, so that the members of the mostly male audience would see her naked genitals passing directly before them at eye level.

Projections and desires, regressive sublimation and the permanent (dis)illusion produced by the hegemonic imaginary – this was cutting-edge cinema theory, transformed into spectacular performances. The psychoanalysis of Lacan and Sigmund Freud (1856–1939); Louis Althusser's (1918–90) critiques of ideology; the semiotics of Pier Paolo Pasolini (1922–75) and Umberto Eco (1932–2016); the structural analyses of Derrida, the young Foucault, and Roland Barthes (1915–80); and Guy Debord's (1931–94) *The Society of the Spectacle* (*La société du spectacle*, 1967) – all these showed up on the scene in the late 1960s, and for Weibel, who was quite fond of Paris, they formed the intellectual and aesthetic fabric he would use in his Expanded Cinema performances. The concept of the "thought exhibition" had its beginnings long before the quadrumvirate of projects that Weibel realized with Bruno Latour (1947–2022) at the ZKM | Center for Art and Media Karlsruhe.[24] It was already inherent as a fundamental gesture in the early works aimed at transgressing traditional media and their limitations. For example, by abstracting scientific or philosophical concepts, I can arrive at new concretizations that can be materialized and find their form in an exhibition. This is also a version of the concept of *projection* as elaborated by Flusser in the 1980s as a subconcept of the techno-imaginary.[25]

The term "expanding cinema" was first used in 1936 by Mary Ellen Bute (1906–83) and Ted Nemeth (1911–86), taken up again as "Expanded Cinema" by Stan VanDerBeek (1927–84) in 1966, and redefined yet again by Gene Youngblood (1942–2021) in a manifesto-like book published in 1970.[26] At a time when Barthes had begun observing himself leaving the cinema, and the master medium of the first half of the twentieth century was already on a downward trajectory in terms of its significance, Youngblood tried to enable a future present for the now-classic and auratic sound-and-image medium by plugging it into the new, algorithmically controlled machines.

Weibel expanded on the concept in his own way through a variety of tactics, with which he questioned the fundamental elements of the cinematic dispositif and redefined the feature film – the *Spielfilm*, in German – in a literal interpretation of that word: as an infinite game (*Spiel*) with film and with the realities created by the cinematic apparatus. In the year 1968 alone he invented "environment films," described as "multiple projections on walls and objects"; "metafilms" and "instant films" in which screen, projector, and camera were collapsed into one; "poor cinema," so called because it managed without film reels and camera; and the "triple projection," with which a unitary event could be made more dynamic through the use of space-time loops.

He also redefined the "film happening" (1967) by pointedly addressing the cinematic dispositif's deceptions of the audience, just as he had with his performances relating to gender politics. At Vienna's Palais Pálffy, at the first screening of the Vienna filmmakers' group consisting of Kurt Kren (1929–98), Hans Scheugl (b. 1940), Ernst Schmidt Jr. (1938–88), and himself, Weibel presented the piece *Nivea* (1966/67): "holding up a nivea ball in both hands, the filmmaker stands motionless for 1 minute, as

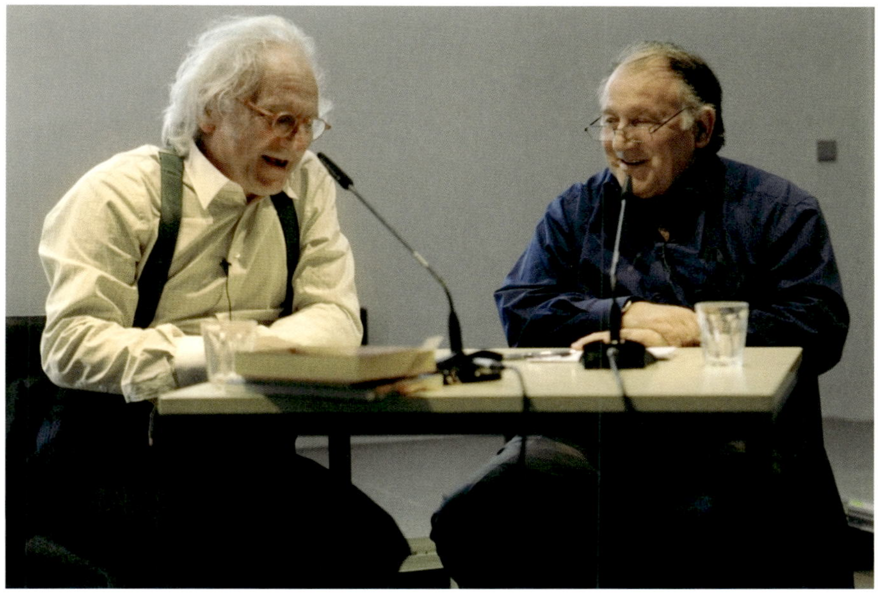

Fig. 2:
Peter Weibel and Siegfried Zielinski at the *Berliner Forum zur Genealogie des MedienDenkens*, University of Arts, Berlin, Oct. 24, 2012

a sort of freeze-frame, in front of the screen, onto which blank frames are projected. ... sound is provided by a tape recorder onstage, which plays back the whir of a running film camera. both the real object and the shadow reproduced on the screen can be seen. by means of a split into visual and aural worlds, one sees production and projection simultaneously. the activity of the camera is reduced to the aural world – what remains is only the camera noise, not the 'illuminated celluloid' – and yet, on the 'illuminated screen,' one sees the shadow of the actor as he is being 'filmed in sound.'"[27]

Interventional/Epic

At the conclusion of his expanded lecture "präkino – parakino – postkino"[28] (1977), Weibel submitted that the division of the world into observers and observed was unequivocally in need of revision. Intelligence and environment were increasingly merging with one another. Subjects had begun to operate atmospherically, as elements of a planetary context. Marshall McLuhan (1911–80) had articulated the same idea a few years earlier in a long essay title: "At the moment of Sputnik the planet became a global theater in which there are no spectators but only actors."[29]

In this thematic context, Weibel's ideas can also be interpreted as early entry points into the potential spaces of a media ecology which, at that point, still lay in the future. However, in the 1960s and early 1970s he was more strongly associated with the field of automation and especially with automata theory. Together with his fellow Vienna resident Franz Kaltenbeck (1944–2018), he translated a series of essays that were already considered standard scientific texts in the United States but were still completely unknown to German-speaking scholars in the human sciences. These were first published in fragments in journals and then, in 1974, in a compact reader. They included now-famous texts by Claude Shannon (1916–2001) and John McCarthy (1927–2011) that first appeared in 1956, Marvin Minsky's (1927–2016) early writings on machine intelligence, and Ross Ashby's (1903–72) legendary 1956 essay "Design for an Intelligence Amplifier."[30]

At the same time, in anticipation of and in reaction to his own theoretical work, Weibel was developing public performances and events that, in retrospect, can be classified as theory performances. One that is particularly impressive in terms of its foresightedness and the clarity of its position is *Synthesis of Two Sequential Machines* (*Synthesis zweier sequentieller Maschinen*, see p. 94 in this volume), a dialogue between the artist and a tape recorder, which Weibel has performed a number of times since 1967. When he starts the device, he says "On!" The machine responds with a prerecorded "Off!" The artist shuts off the device, which has now taken on the character of a player, then says "On!' again and presses the button. The tape machine answers with "Off!" again, and so it continues – until, as described by Weibel in an introduction to the performance, either the device's operator succumbs to exhaustion or the device itself breaks down. Man and machine are linked in a circular, perfectly logical relationship, yet at the same time they are in aggressive competition with one another. Who is stronger, the human or the machine? Regardless, the dramatic game ends only when one of the two systems makes a mistake or gives up.

Synthesis can also be interpreted as a response to Shannon's information theory and to aspects of Norbert Wiener's (1894–1964) cybernetics. However, Weibel's performance was not a reaction to a theory of the digital, because at the time, there was no such thing in the context of aesthetic processes. Rather, he was articulating both of the states that (digital) machines can inhabit, using art's intrinsic resources – namely, aesthetic ones. Poetry and theory intermesh. The artistic experiment becomes the draft for a theory of an impending culture of the digital.

Another, somewhat later result of Weibel's early forays into cybernetics and quantum physics was a long-standing and cordial collaboration with Otto Rössler and his world of endophysics:[31] "A new science, endophysics, is brought to the reader's attention," Weibel wrote in 1990 in his foreword to Rössler's book *Endophysics*. "After the 'relativization' of objectivity through an absolute speed limit, observation and unpredictability, endophysics brings the chain to its logical end by replacing the traditional assumption

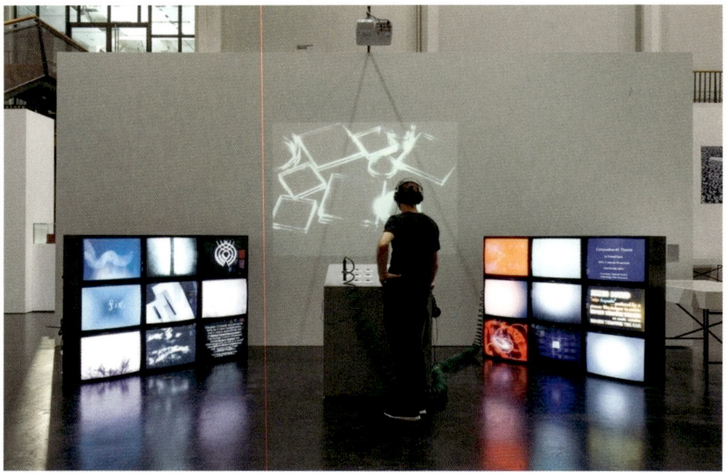

Fig. 3:
Installation view *Art in Motion*, ZKM | Karlsruhe, 2018–19

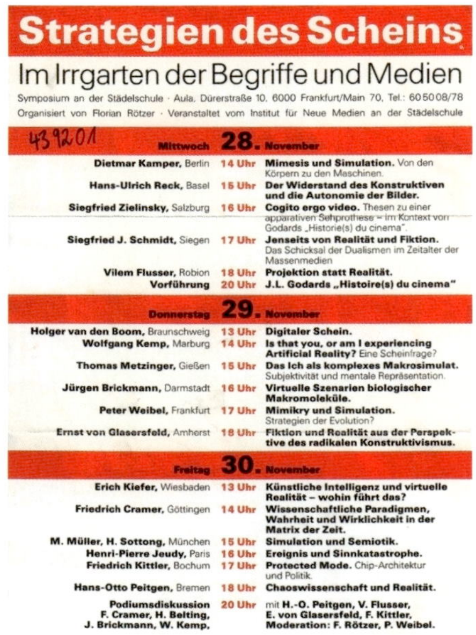

Strategien des Scheins

Im Irrgarten der Begriffe und Medien

Symposium an der Städelschule · Aula, Dürerstraße 10, 6000 Frankfurt/Main 70, Tel.: 60 50 08/78
Organisiert von Florian Rötzer · Veranstaltet vom Institut für Neue Medien an der Städelschule

43920 1

Mittwoch 28. November

Dietmar Kamper, Berlin	14 Uhr	**Mimesis und Simulation.** Von den Körpern zu den Maschinen.
Hans-Ulrich Reck, Basel	15 Uhr	**Der Widerstand des Konstruktiven und die Autonomie der Bilder.**
Siegfried Zielinsky, Salzburg	16 Uhr	**Cogito ergo video.** Thesen zu einer apparativen Sehprothese – im Kontext von Godards „Historie(s) du cinema".
Siegfried J. Schmidt, Siegen	17 Uhr	**Jenseits von Realität und Fiktion.** Das Schicksal der Dualismen im Zeitalter der Massenmedien
Vilem Flusser, Robion	18 Uhr	**Projektion statt Realität.**
Vorführung	20 Uhr	**J.L. Godards „Histoire(s) du cinema"**

Donnerstag 29. November

Holger van den Boom, Braunschweig	13 Uhr	**Digitaler Schein.**
Wolfgang Kemp, Marburg	14 Uhr	**Is that you, or am I experiencing Artificial Reality?** Eine Scheinfrage?
Thomas Metzinger, Gießen	15 Uhr	**Das Ich als komplexes Makrosimulat.** Subjektivität und mentale Repräsentation.
Jürgen Brickmann, Darmstadt	16 Uhr	**Virtuelle Szenarien biologischer Makromoleküle.**
Peter Weibel, Frankfurt	17 Uhr	**Mimikry und Simulation.** Strategien der Evolution?
Ernst von Glasersfeld, Amherst	18 Uhr	**Fiktion und Realität aus der Perspektive des radikalen Konstruktivismus.**

Freitag 30. November

Erich Kiefer, Wiesbaden	13 Uhr	**Künstliche Intelligenz und virtuelle Realität – wohin führt das?**
Friedrich Cramer, Göttingen	14 Uhr	**Wissenschaftliche Paradigmen, Wahrheit und Wirklichkeit in der Matrix der Zeit.**
M. Müller, H. Sottong, München	15 Uhr	**Simulation und Semiotik.**
Henri-Pierre Jeudy, Paris	16 Uhr	**Ereignis und Sinnkatastrophe.**
Friedrich Kittler, Bochum	17 Uhr	**Protected Mode.** Chip-Architektur und Politik.
Hans-Otto Peitgen, Bremen	18 Uhr	**Chaoswissenschaft und Realität.**
Podiumsdiskussion F. Cramer, H. Belting, J. Brickmann, W. Kemp,	20 Uhr	mit **H.-O. Peitgen, V. Flusser, E. von Glasersfeld, F. Kittler, Moderation: F. Rötzer, P. Weibel.**

Fig. 4:
Program of the symposium *Strategies of Seeming: In the maze of concepts and media*, Städelschule Frankfurt/Main Nov. 28–30, 1989

of the external observer (exophysics) with the introduction of the internal observer."[32]

The discourse effect of this paradigm shift in physics was immense for the new field of media art, which at that time was just emerging. Rössler's endophysics became the theoretical foundation for an expansion of activity at the interface between human-machine relations and the concepts of art as a process, which no longer accepted the classic division between work and viewing, between production and perception. "The manipulability of the jail of space and time grows by a notch," Weibel wrote. "The world taken as 'repair shop for the wish machine' takes a step forward."[33]

This new endophysics, a physics of the now, of the moment (which lacks duration but offers enormous decision-making power), and the concept of participation, of involvement, constituted powerful arguments for a new kind of being-in-the-world, which was communicated via media art. It is no longer possible for us to leave the world behind in anything we do. We swim both in the world and in our consciousness of it, like fish in water. The price of stepping back outside it, thereby reducing our view of the world to the classic position of the so-called objective observer, is high. After all, anatomy can only be performed on dead matter. The alternative is the kind of interventional thought and action that Brecht first proposed for a theater for the scientific age in his *Short Organon for the Theatre* (*Kleines Organon für das Theater*, 1949). Emanuele Coccia's (b. 1976) philosophy of plants, *The Life of Plants* (*La vie des plantes*, 2016),[34] likewise described an *endo*-materiological world thoroughly informed by the idea and concept of the all-integrating atmosphere. Coccia's work is celebrated today as the ne plus ultra of an ecology-affiliated applied philosophy.

Chronocracy

The line "Woyzeck, you always look so harassed," spoken by the Captain in Georg Büchner's unfinished play *Woyzeck* (1837),[35] stands for the suffering psyche of the hunted person, the person who feels persecuted, who seeks and finds death in decadence. Paranoia, as a new pathology of temporal awareness, came into being against the backdrop of a new economy of time. In Peter Weibel's extraordinary book *Die Beschleunigung der Bilder in der Chronokratie* (The acceleration of images under chronocracy, 1987),[36] the hegemony of increasing acceleration is transformed into a revolutionary gesture – almost thirty years before the gang of object-oriented ontologists and speculative realists rediscovered acceleration as a potentially transgressive force and began filling slim volumes of political philosophy with it once again. In *Chronokratie*, Weibel discusses the disruption of aristocratic and bourgeois idleness primarily in relation to the technoimaginary, the accelerated images of sequential chronophotography, fluid electronics, and the new culture of video art.

The 1990–91 exhibition *Vom Verschwinden der Ferne: Telekommunikation und Kunst* (On the vanishing of distance: Telecommunication and art),[37] with its accompanying book, was one of the first projects on which I had the opportunity to work more closely with Peter Weibel. At bottom, both book and exhibition centered on an idea that he had already formulated in the 1970s: that telecommunication should be understood, essentially, as a process of separating the message from the body of the messenger. With telematics and its networks, the semiotic material to be transmitted becomes autonomous and can travel at any speed with no spatial restrictions, provided that the signal on which the signs perform is carried by appropriate hardware. The separation of the message from the body of the messenger has become a paradigm – not just in our collaborations up to and including our most recent exhibition project, *Art in Motion: 100 Masterpieces with and through Media* (2018–19, see figs. 2 and 3),[38] but for media art and theory in general.

Vom Verschwinden der Ferne took place in Frankfurt am Main at the same time that Weibel was setting up the Institute for New Media, a research and teaching institution at the renowned Städelschule (which was generally

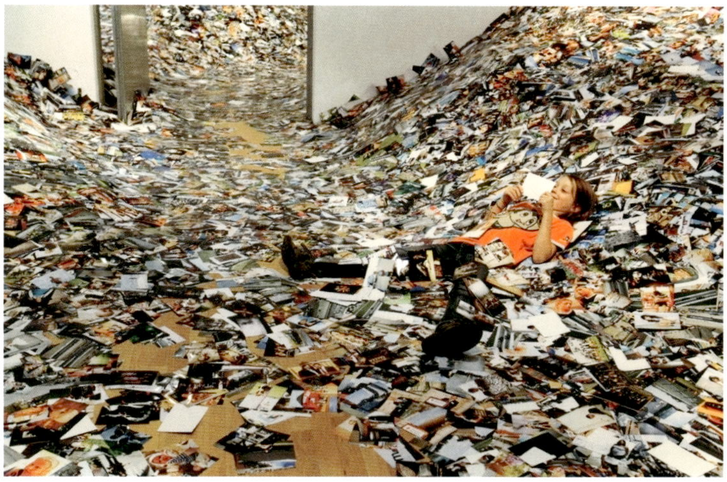

Fig. 5:
Erik Kessels, *24HRS of Photos*, 2011, installation view *GLOBALE: Infosphere*, ZKM | Karlsruhe, 2015–16

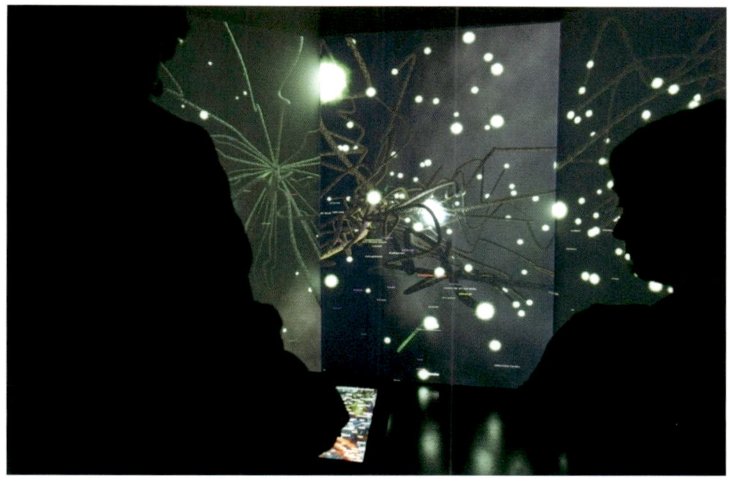

Fig. 6:
Jeong Han Kim, Hyun Jean Lee, Jung-Do Kim, *EMC (Emergent Mind of City) III: Karlsruhe, Washington, D. C. and Seoul*, 2015, installation view *GLOBALE: Infosphere*, ZKM | Karlsruhe, 2015–16

resistant to his new ideas about art).[39] Technoimagination, the expansion of the new imagination, virtual scenarios involving biological macromolecules, machine intelligence, virtual reality – faced with dispositifs such as these, only a few of the artists under the academy's big top avoided going under. These few accepted them as a challenge, sought the terminology they needed to reflect as deeply as possible upon them, philosophically and scientifically, and tried to ask the right questions – which naturally involved radically transdisciplinary gestures. *Strategien des Scheins: Im Irrgarten der Begriffe und Medien* (Strategies of seeming: In the maze of concepts and media, 1989, see fig. 4) was one of the remarkable conferences Weibel organized in Frankfurt to allow the traveling intellectual circus of that era to debate the new realities that were posing challenges for artists, scientists, and philosophers alike.[40] It took place at the precise moment of transition from the dominance of local media to the hegemony of networked media.

In those days Weibel was driven by a conceptual triad – sometimes extended to a quartet of terms – that remains a key terminological framework for his heuristics of the technological image to this day. It consists exclusively of nouns beginning with *v*. "Virtuality," for Weibel, denotes the aforementioned separation of the message and the biological body of the person conveying the message; the image becomes a hypothetical in space. As a technological imagining, this image can be altered at any time; both the artists involved and the audience can intervene in the visual reality and manipulate it, in the exact meaning of that word. Weibel calls this quality "variability." "Virality" stands for that stratagem of the new visual reality that bears on the rapid transmission of information in a communicative network. The monads of communication can be everywhere in no time, as long as the network has the capacity. The concept of "viability" is the most far-reaching. It goes back to the radical constructivist Ernst von Glasersfeld (1917–2010), who proposed it as an alternative to the traditional ontological concept of truth.[41] The viability (and thus the vitality) of the computer-based image not only elevates technological reality to the status of a modifiable model but also transforms it into the nature that surrounds us – which, after all, now only exists in the form humanity has given it. The viability of advanced media realities allows them to become an atmosphere permeated by information technology: an *Infosphere*, to borrow the title of a phenomenal exhibition mounted by Weibel at the ZKM in 2015–16 (see figs. 5 and 6).[42]

I will permit myself an a-Romantic thought in closing. I am interpreting the prefix a- here exactly as Jean-François Lyotard (1924–98) wished it to be understood in his operative concept of a transgressive "acinema":[43] not as an antagonism against the Romantic, but as its simultaneous fracturing and epistemic-poetic extension. By taking a view of Novalis's (1772–1801) circa-1800 postulate calling for all sciences to be "poeticized"[44] that has been radicalized by and with Weibel, I liberate the concept from Romanticism and allow the latter to function as subversive energy and an alternative to the conventional idea of the Romantic.

According to Georges Bataille (1897–1962), we understand friendship as a shared feeling of being alien in this world we never chose. In his "accursed" economics (1949), he used the sun as an example to show how an economy of wastefulness can function as an economy of extravagance: "The solar radiance *we are* finally finds nature and the meaning of the sun: it is necessary for it to give, *to lose itself without calculation*. A living system grows, or lavishes itself *without reason*."[45]

With his intellectual and poetic practice, Peter Weibel has tremendous access to such wastefulness or extravagance, which takes "loss"[46] in its stride. There are only two ways we can respond to this gift we are constantly receiving from the artist, the thinker, the doer – in a word, the tireless enabler: with joie de vivre, and with the heightened presence of mind (*Geistesgegenwart*) that is called for when one is with him. Otherwise, he'll get bored and zone out, or maybe even go to sleep.

Note: This text is a re-worked version of a lecture given on the occasion of a symposium accompanying the exhibition *respectively, Peter Weibel* on September 28, 2019. Translated from the German by Patrick Hubenthal.

[1]
The "mathematical" is to be understood in its original meaning here. Only in the fifteenth century was mathematics established in Europe as a specific discipline of numbertheory and calculation. But *mathema* originally meant much more and was understood in a broader sense. The *ars mathematica*, whose name was derived from the Greek verb *manthanein* (to learn, find out, understand), encompassed everything one could learn and all available knowledge one could acquire.

[2]
Roger Caillois, *The Mask of Medusa*, trans. George Ordish (London: Victor Gollancz, 1964), 12–13. First published as *Méduse & Cie* (Paris: Gallimard, 1960).

[3]
Benjamin to Max Horkheimer, Oct. 16, 1935, in *The Correspondence of Walter Benjamin, 1910–1940*, ed. Gershom Scholem and Theodor W. Adorno, trans. Manfred R. Jacobson and Evelyn M. Jacobson (Chicago: University of Chicago Press, 1994), 508–10 (quotation on 509).

[4]
Walter Benjamin, "The Work of Art in the Age of Its Technological Reproducibility," trans. Harry Zohn and Edmund Jephcott, in *Selected Writings*, vol. 4, *1938–1940*, ed. Howard Eiland and Michael W. Jennings (Cambridge, MA: Belknap, 2003), 251–83 (quotation on 263).

[5]
Adorno to Benjamin, Mar. 18, 1936, trans. Harry Zohn, in *Aesthetics and Politics*, by Ernst Bloch, Georg Lukács, Bertolt Brecht, Walter Benjamin, and Theodor Adorno (London: Verso, 1980), 124.

[6]
Otto E. Rössler, *Endophysics: The World as an Interface* (Singapore: World Scientific, 1998). First published as *Endophysik: Die Welt des inneren Beobachters* (Berlin: Merve, 1992).

[7]
Peter Weibel, "Projektfilme und projizierte Projekte," *protokolle* 17, no. 2 (1982): 92.

[8]
Weibel, "Projektfilme."

[9]
Sandra Vásquez de la Horra, artist's talk, Galerie Michael Haas, Berlin, Sept. 26, 2020, held in conjunction with the exhibition *El viaje imaginario* (Sept. 11–Oct. 31, 2020).

[10]
Massimo Cacciari, "Die Weiße und die Schwärze," in *Migranten*, by Edmond Jabès, Luigi Nono, and Massimo Cacciari, ed. and trans. Nils Röller (Berlin: Merve, 1995), 45–58 (quotation on 46).

[11]
This phrase was chosen in allusion to the 1968 Canned Heat song of the same name.

[12]
Siegfried Zielinski and Peter Weibel, "Introduction: Flusseriana – An Intellectual Toolbox," in *Flusseriana – An Intellectual Toolbox*, ed. Siegfried Zielinski and Peter Weibel with Daniel Irrgang (Minneapolis, MN: Univocal, 2015), 6–12 (quotations on 10). The source given there for Arendt's phrase is her essay "On Hannah Arendt," in *Hannah Arendt: The Recovery of the Public World*, ed. Melvyn A. Hill (New York: St. Martin's, 1979), 301–39 (quotation on 336). The phrase has subsequently been used as the title for a book of Arendt's essays: *Thinking without a Banister: Essays in Understanding, 1953–1975* (New York: Schocken, 2018).

[13]
Julien Torma, *Euphorismes* (Paris: Guiblin, 1926); quoted in German translation in "Porträt 1967: Peter Weibel – Dichter ohne Literatur – Mediendichter," *protokolle* 17, no. 2 (1982): 3.
The English citation here is from *Euphorisms*, trans. Iain White, in *Four Dada Suicides*, by Arthur Cravan, Jacques Rigaut, Julien Torma, and Jacques Vaché (London: Atlas, 2005), 149–94 (quotation on 174).

[14]
Peter Weibel, "Apostasie von der Sprache," *protokolle* 17, no. 2 (1982): 4.

[15]
Peter Weibel, quoted in Stefanie Orphal, *Poesiefilm: Lyrik im audiovisuellen Medium* (Berlin: De Gruyter, 2014), 108.

[16]
Peter Weibel, "Zur Konzeptkunst in Österreich," in *Kontinuität und Identität: Festschrift für Wilfried Skreiner*, ed. Peter Weibel, Christa Steinle, and Götz Pochat (Vienna: Böhlau, 1992), 50–71 (quotation on 71).

[17]
Peter Weibel, "Vorsätze," *protokolle*, 17, no. 2 (1982): 5–6 (quotations on 6).

[18]
William S. Burroughs, *The Electronic Revolution* (UbuClassics, 2005), 7. First published as *Electronic Revolution 1970–71* (Cambridge: Blackmoor Head, 1971).

[19]
Burroughs, *The Electronic Revolution*, 23–24.

[20]
Burroughs, *The Electronic Revolution*, 35.

[21]
The exact phrase is "Unser Schreibzeug arbeitet mit an unseren Gedanken" (Our writing implements collaborate on our thoughts). Nietzsche to Peter Gast [Heinrich Köselitz], letter 90, end of Feb. 1882, in *Friedrich Nietzsches Gesammelte Briefe*, vol. 4, *Friedrich Nietzsches Briefe an Peter Gast*, ed. Peter Gast (Leipzig: Insel, 1908), 97.

[22]
Edmond Jabès, *The Little Book of Unsuspected Subversion*, trans. Rosmarie Waldrop (Stanford, CA: Stanford University Press, 1996).

[23]
Roger Caillois, *La dissymetrie* (Paris: Gallimard, 1973), 75.

[24]
The four projects are *Iconoclash: Beyond the Image Wars in Science, Religion and Art* (2002), *Making Things Public: Atmospheres of Democracy* (2004–05), *Reset Modernity!* (2016), and *Critical Zones: Observatories for Earthly Politics* (2020–22). Weibel's "thought exhibitions" also include our joint projects *Allah's Automata: Artifacts of the Arab-Islamic Renaissance (800–1200)* (2015–16) and *DIA-LOGOS: Ramon Llull and the Ars Combinatoria* (2018).

[25]
Miklós Peternák, "Projection," in Zielinski, Weibel, and Irrgang, *Flusseriana*, 334.

[26]
Editor's note: See Mary Ellen Bute and Ted Nemeth, "Expanding Cinema's *Synchromy 2*," *Literary Digest*, no. 122 (Aug. 1936): 20–21; as well as Stan VanDerBeek, "'Culture: Intercom' and Expanded Cinema: A Proposal and Manifesto," in *Film Culture*, no. 40 (Spring 1966): 15–18. VanDerBeek's article also appeared in the *Tulane Drama Review* 11, no. 1 (Autumn 1966): 38–48, and, under the title "Culture-Intercom and Expanding-Cinema: A Proposal," in *motive*, Nov. 1966, 13–23. Peter Weibel published his vision of film as "Expanded Cinema" in *Film*, no. 11 (Nov. 1969): 41–47, 51–52, having seen an "expanded arts diagram" by Fluxus artist and cofounder George Maciunas in the journal *Film Culture* approximately two years earlier: "Expanded Arts Diagram," in "Expanded Arts," special issue, *Film Culture*, no. 43 (Winter 1966): 6–7 (a scan of the diagram is archived on the website of the Museum of Modern Art in New York, at http://moma.org/collection/works/128225.) In this map, Maciunas graphically depicted the various artistic and intellectual movements of the late 1950s and early 1960s, along with the connections between them and their exponents. Four years later, Youngblood published the wide-ranging book *Expanded Cinema* (New York: E. P. Dutton, 1970). In addition, Sheldon Renan had already described and documented the history of avant-garde film in his book *An Introduction to the American Underground Film* (New York: E. P. Dutton, 1967). However, it was VanDerBeek, in an article published in 1961, who first used the term "underground film" for the cinematographic work of contemporary avant-garde filmmakers, which had been known until then primarily by the label "experimental cinema": Stan VanDerBeek, "The Cinema Dilemma: Films from the Underground," *Film Quarterly* 14, no. 4 (1961): 5–15.

[27]
Peter Weibel, "*Nivea*," *protokolle* 17, no. 2 (1982): 63.

[28]
The lecture, whose full title was "präkino – parakino – postkino: Filme und Videotapes 1968–76," was presented at Galerie nächst St. Stephan in Vienna on Nov. 10, 1977: See "Ausstellungen 1975–80," website of Galerie nächst St. Stephan, accessed June 28, 2021, http://www.schwarzwaelder.at/exhibition/1975-80.

[29]
In *Journal of Communication* 24, no. 1 (1974): 48–58.

[30]
The early translations include Edward F. Moore, "Was kann ein Automat?," trans. Peter Weibel, *Neues Forum: Internationale Zeitschrift links von der Mitte*,no.223(May1973):47–49. This essay is a translated excerpt from Moore's paper "Gedanken-Experiments on Sequential Machines," in *Automata Studies*, Annals of Mathematics Studies 34, ed. Claude E. Shannon and John McCarthy (Princeton, NJ: Princeton University Press, 1956), 129–56. *Automata Studies* also includes Minsky's "Some Universal Elements for Finite Automata" (117–28) and Ashby's "Design for an Intelligence Amplifier" (215–34). A German edition of the entire book (the "reader" mentioned above) was published as well: Claude E. Shannon and John McCarthy, eds., *Studien zur Theorie der Automaten (Automata Studies): Erweiterte Ausgabe und Übersetzung durch Franz Kaltenbeck + Peter Weibel. Mit Zeichnungen von Dieter Roth* (Munich: Rogner & Bernhard, 1974).

[31]
This term, which I mentioned at the beginning of these reflections, was coined by David Finkelstein in a letter to Rössler in 1983: See Rössler, *Endophysics*, 27.

[32]
Peter Weibel, foreword to Rössler, *Endophysics*, xiii–xv (quotation on xiii).

[33]
Weibel, foreword to *Endophysics*, xv.

[34]
Emanuele Coccia, *The Life of Plants: A Metaphysics of Mixture*, trans. Dylan J. Montanari (Cambridge: Polity, 2019).

[35]
Michael Ewans, *Georg Büchner's* Woyzeck*: Translation and Theatrical Commentary* (New York: Peter Lang, 1989), 45.

[36]
This book (Bern: Benteli, 1987) was a standard text when I held my first position as professor of audiovisions in Salzburg from 1990 to 1993. In 1992 the future Austrian media theorist Stefan Weber wrote a term paper on it, titled "MANIFEST DER TECHNO-PROGRESSION: Eine subjektiv-subversive Hermeneutik" (see fig. 1).

[37]
See Edith Decker-Philips, ed., *Vom Verschwinden der Ferne: Telekommunikation und Kunst*, exh. cat., Deutsches Postmuseum, Frankfurt am Main (Cologne: DuMont, 1990). The exhibition ran from Oct. 2, 1990 to Jan. 13, 1991.

[38]
The exhibition *Art in Motion: 100 Masterpieces with and through Media; An Operative Canon* was shown in the ZKM's two atria from July 14, 2018 to Jan. 20, 2019.

[39]
Around this same time, Peter Kubelka was leading Expanded Cinema seminars with his own students at the Frankfurt academy that involved making mayonnaise and spaghetti: Olaf Karnik and Volker Zander, "Die Küche des Künstlers. Wie Peter Kubelka das Kochen nach Frankfurt brachte," radio broadcast, 47:31, aired Sept. 16, 2016 on Deutschlandfunk Kultur, http://www.deutschlandfunkkultur.de/die-kueche-des-kuenstlers-wie-peter-kubelka-das-kochen-nach.3720.de.html?dram:article_id=366025.

[40]
The symposium took place at the Städelschule from Nov. 28 to 30, 1989. In addition to Dietmar Kamper, Hans-Ulrich Reck, and Friedrich Kittler, the presenters also included Flusser, who presented his concept of projection in a lecture titled "Projektion statt Realität" (Projection instead of reality).

[41]
Ernst von Glasersfeld, *Radikaler Konstruktivismus: Ideen, Ergebnisse, Probleme* (Frankfurt am Main: Suhrkamp, 1997).

[42]
Infosphere: The Transformation of Things into Data ran from Sept. 4, 2015 to Jan. 31, 2016 at the ZKM.

[43]
Jean-François Lyotard, "Acinema," trans. Peter W. Milne and Ashley Woodward, in *Acinemas: Lyotard's Philosophy of Film*, ed. Graham Jones and Ashley Woodward (Edinburgh: Edinburgh University Press, 2017), 33–42. First published as "L'acinéma," in "Cinéma: théories, lectures," ed. Dominique Noguez, special issue, *Revue d'esthétique* 26, no. 2–4 (1973): 357–69.

[44]
See, for example, Novalis's letter to August Wilhelm Schlegel dated Feb. 24, 1798, available online in the *Digitale Edition der Korrespondenz August Wilhelm Schlegels*, updated Oct. 2020, http://august-wilhelm-schlegel.de/briefedigital/letters/view/4776?left=text&right=manuscript.

[45]
Georges Bataille, *Essential Writings*, trans. Michael Richardson (London: Sage, 1998), 75; translated excerpt from "L'Économie à la mesure de l'univers," in *Oeuvres complètes* (Paris: Gallimard, 1976), 7:9–16.

[46]
Bataille, *Essential Writings*, 61.

Art against Media: Early Works by Peter Weibel

Terry Smith

During the 1960s in Europe, the United States, and other "advanced" economic spheres, television rapidly displaced the printed press, radio, film, and public advertising as the mass medium that most effectively served the spectacularization of social life and private imagination. With seemingly unstoppable speed, and with an inevitability akin to a force of nature, television was becoming the dominant form of public media as such. Combining elements of existing communicative mediums, it quickly became the leading edge of what was soon termed "the media." Meanwhile, avantgarde artists who were preoccupied with testing the limits of inherited artistic mediums – painting, sculpture, architecture, the graphic arts and crafts – experimented by combining these ancient practices, theatricalizing them as performances or merging them with newer mediums such as photography or film. From the Gutai Group in Japan, through Happenings and Environments in the United States, Neoconcretism in Brazil to "actions" in Europe, these artists, along with many others, transformed the structures of artistic modernism. Artists such as Andy Warhol and Wolf Vostell, and groups such as the Situationists and Fluxus, combined these impulses with the other great driver of late modernism: a realization that art must engage with the processes of everyday life in advanced consumer societies, above all its colonization of individual subjectivity. At that time, these processes were in formation, still spreading their influence, not yet hegemonic. Perhaps, many thought, concerted action by artists, by sufficient numbers of artists and audiences, might redirect this flow away from the priorities of the state and commerce, towards more open situations, towards freedom, for sure. Nothing dates that period as much as such political naivety. But nothing makes it more worthy of revisiting at this time, when the hegemony of spectacle is so spectacularly failing.

This was the context, broadly speaking, of the first of Peter Weibel's "art actions" involving television. For him, as for others, the work of the late Nam June Paik had set startling precedents. From the early 1960s onwards Paik modified television monitors to demonstrate the greater variety of relationships that viewers might have with television, ranging from treating it as an object of meditation (*Zen for TV*, 1963) through instructive installations (*Exposition of Music-Electronic Television*, 1963) and artworks (*TV Garden*, 1974) to interactivity, which he theorized extensively as "participation TV."[1] More aggressive attacks upon television monitors as symbolic of state manipulation and capitalist propaganda were launched by artists such as Wolf Vostell in his *TV Dé-Coll/ages* from 1963, or the US group Antfarm's *Media Burn* (1975), a staged "media event" in which a Cadillac El Dorado was driven into a pyramid of burning television monitors in order to create not only an art video but a publicity sensation. The history of television-related art in Europe is tracked by Dieter Daniels in a useful survey essay "Television-Art or Anti-Art?"[2] In *Feedback, Television Against Democracy*, David Joselit offers a parallel study that focuses on the United States, acutely drawing out the differences for the possibilities for artistic intervention when commercial imperatives rather than state control of information are the priority in a media environment.[3]

Weibel's early interventions were rather straightforward displays of the mechanics of televisual miscommunication. In *The Public as Exhibit (Audience Exhibited)* (*Das Publikum als Exponat (Ausstellung des Publikums)*, 1969, see p. 41 in this volume), interviews with visitors to an exhibition were recorded by video camera, then played on monitors hung in the gallery. Visitors noticed time delays, and could request replays of their interviews. Weibel went on to elaborate these temporal and spatial disjunctions in many later works, as did other artists such as Dan Graham, whose *Time Delay* series of room installations during the 1970s added a random element to the idea of audiences witnessing their own screened dislocation: a room in which a performer had similar, scarcely distinguishable, experiences that were, in turn, inserted into the mix of screened images.

Creating audiences and delivering them to the controllers of television broadcasts (be they the state or commercial interests) was the purpose of mass media. It was, and still is, at the core of the entire enterprise. This goal was explicitly stated by broadcasters, especially as they sold their services to advertisers, and was obvious to critics of state control and consumer capitalism. In *The Mechanical Bride: Folklore of Industrial Man* (1951), Marshall McLuhan had shown this to be the case with advertising in general. In *Understanding Media* (1964) he argued that "Rapidly, we approach the final phase of the extensions of man – the technological stimulation of consciousness, when the creative process of knowing will be collectively and corporatively extended to the whole of human society, much as we have already extended our senses and our nerves by the various media."[4] This passive feedback effect is dramatically evoked in a work of 1971, *Facing a Family*, by Weibel and his collaborator VALIE EXPORT (see p. 249 in this volume). Broadcast on Austrian television ORF during a magazine program entitled *Kontakte*, it consisted of a short film, shot as if from inside the monitor, of a middle-class family watching that same program. In Weibel's *The Endless*

Sandwich of 1969 (see p. 248 in this volume), we see a film of a viewer sitting to the right of his monitor, watching another viewer doing the same thing, and so ad infinitum into the televisual distance. It gradually becomes clear that the distant viewers are having to get up to correct a fault in the transmission, each one seeming to cause the same fault in the next one, coming closer to the front all the while, until eventually the screen before you – and likewise that of ORF – blurs into obscurity.

This activist parody, broadcast over actual public television, contrasts with Richard Serra's 1973 video *Television Delivers People*, which consists entirely of a series of statements rolling (to the sound of muzak) across the screen as if it were a cue-sheet: "You are the product of television," "Mass media means that a medium can deliver masses of people," "You are delivered to the advertiser as a consumer," and "We own you." Serra seemingly presumed that the mere presentation of these analyses would be enough to inflame political opposition, as the video circulated between galleries and on informal circuits (it can now be found on YouTube).

The statements also read like a series of extracts from one of McLuhan and Quentin Fiore's highly graphic image-books. In *Culture is Our Business* (1970) McLuhan had devoted a number of pages to analyses of advertisements for television.[5] Placed in print magazines, some were directed at selling the capacity of television companies to deliver audiences to advertisers, others used television imagery to advertise their superiority at reaching audiences. They illustrate McLuhan's argument, put forth in *Understanding Media* and other books, that the content of a communicative medium is not the overt matter it displays, but another medium, that the constant massaging of its audiences conducted by "the media" is, in modern societies, an incessant, always closing circle.[6] McLuhan's core argument was that "the medium is the message." What really matters is that each new technology introduces "a change of scale or pattern or pace" into human affairs. This is each medium's real, affective, and effective content. Recently, they are converging, turning into "the media." Artists, who are especially alert to the power of mediums in and of themselves, and in their convergence, might seek to act against this. But how?

The Massage Is the Medium

These considerations allow us to revisit a well-known work by VALIE EXPORT and Peter Weibel, the *Touch and Feel Cinema* (*Tapp- und Tastkino*, see pp. 180–81 in this volume) from 1968. It is important to recall the immediate context. Both *Touch and Feel Cinema* and Weibel's performance *EXIT* (1968, see p. 182 in this volume) were prepared to be shown as a pair at the First European Meeting of Independent Filmmakers, held at the Independent Film Center in Munich. *Exit* consisted of Weibel projecting a film (any film?) onto a tin foil screen, which he then disrupted by threatening the audience from the stage, in a calculated display of masculine aggression. *Touch and Feel Cinema* was an actualization of Weibel's *Touch and Feel Poem* (*Tapp- und Tastgedicht*, 1967, see p. 180 in this volume). …

VALIE EXPORT enacted the spirit of this poem, but with important reversals. She offered (parts of) her body to be "felt up," not to her boyfriend/actor but to any passer-by at the time. She did not offer her body parts up to the general gaze, but covered them with a screen that, on a smaller scale, replicates a theater. In the context of the independent film festival, this portable puppet theater was intended to evoke a movie house, as becomes clear when we revisit the text that Weibel wrote for the performance. He begins by listing four purposes for the film:

1. touch and feel cinema
2. street film
3. mobile film
4. genuine woman's film

An exposition of intent follows:

> As always, the showing takes place in the dark, but the movie theater has become somewhat smaller. There is room enough only for two hands. In order to see the film – that is, in this case, to feel and sense it – the viewer has to guide both his hands through the theater entrance. Thus, the curtain, which has hitherto been raised only for the eyes, is finally also raised for the hands. Admittance to the showing is free of charge and G-rated.
>
> In the State movie theater, they sit in the dark and watch as somebody does it with somebody else; no one is watching them. Here they do it themselves with somebody else, in the light, and many people are watching. They meet their star and each is him or herself a star. Every visitor is allowed to stay only 12 seconds in the *Touch and Feel Cinema*; altogether then one fifth of a minute, so that each can have his or her turn.
>
> The breast is a major subject in the film industry, because fixated partial drives – exhibitionism on the screen, voyeurism in the audience – are a perpetual source of manipulation and exploitation of the public. One breast brings in millions, because its endlessly reproduced cop-

> ies multiply it a hundredfold (for about 60 lovers in real life, there are 60 million in the movie theaters) because of reproduction technology.
>
> Brought to the public a millionfold, one breast optimally extends its sales market. Chauffeurs and producers fuck the original star breast; the poor lover in the State movie theater, removed by continents from the original, must saturate himself with the celluloid surrogate, the already grotty copy. In the State movie theater, everything becomes voyeurism.[7]

EXPORT followed these instructions and implications, including the 12-second time limit, which she enforced on each participant (thus fulfilling the fourth stipulation?). The film records a number of people taking advantage of this offer, along with multiple cuts to the gathering crowd, acting as the voyeurs described in the notes. The film also shows Peter Weibel in the role of spruiker, or sideshow man, drumming up attention, but also offering an interpretation of what was going on. ...

The actions by Weibel and EXPORT in the context of the film festival were inspired not only by developments in experimental cinema and avant-garde art up to this point – specifically Expanded Cinema and Viennese Actionism – but also by anger at "the colonization of everyday life" as theorized not only by Guy Debord (*La Société du spectacle* was published in 1967) but most specifically by McLuhan. We might think, then, that the odd contraption EXPORT holds across her chest, which she invites viewers to enter as – literally – massagists, is an object that metaphorizes not only the psychic site of the theater and the cinema, but also that of television.

Televisual Death

Indeed, both artists went on to make a number of works that explore the sexual economies of all of these mediums, and that did so by imitating aspects of them, only to turn them into monstrosities of distortion and exaggeration. I will comment on several works from this period by Weibel.

Taking the television monitor as equivalent to a fish tank, *TV Aquarium (TV Death 1)* (*TV-Aquarium (TV-Tod 1)*, 1970/72, see p. 250 in this volume) evokes the aspect of television as live performance, but reverses it to show (or, actually, imply) a real death: that of the apparently filmed fish. Weibel is seeking here a contrast to the constant picturing of violence and reporting of death on broadcast news and in entertainment shows, both of which gradually induce a sense of the currency of fictive death, aestheticizing viewers against actual death (which is never shown). *TV News (TV Death 2)* (*TV-News (TV-Tod 2)*, see p. 250 in this volume) was subtitled: "Communication is the medium." Broadcast in 1972, it takes the viewer to a familiar studio set, that of the newsreader presenting the evening news. The difference is that the newsreader is smoking so much that the smoke causes him not only to cough but to eventually suffocate. The two films were shown on Austrian television, the second using a well-known newsreader as an actor. In the course of the tabloid-gleaned items of news – "Vienna. Assault in shelter for homeless. Perpetrator under influence of drugs" – followed by the football scores and weather forecast, the clouds of smoke render the newsreader's task increasingly difficult. Almost forty years later, it is difficult to imagine the impact of such a film when shown on public television. Not least because such occurrences have since been spoofed by a number of films and television programs (for example, comedy shows such as *Saturday Night Live* in the United States).

From filling the screen with smoke, Weibel moved to a work that filled it with his own blood. *Time Blood* (*Zeitblut*, see pp. 254–55 in this volume), conceived in 1972 and finally broadcast in 1979, can be seen as a revisiting of the ideas concentrated in *Touch and Feel Cinema* but now applied directly and unequivocally to television. Weibel was invited to participate in the discussion program Club 2. The recording opens with a text insert: the formula "anf(p)fnf(p)," a postulate for the end of time, written on a blackboard with white chalk. A sketch and a description of the action follows... The presenter, seated among his studio guests, begins talking to Peter Weibel, who is commencing bis performance in the background:

> Presenter: Peter Weibel calls this action *Time Blood* and Peter Weibel ..., well, now, I have to ask you; I can't read clearly any more what you call this precisely. Media ... what are you?
>
> Weibel: ... poet.
>
> Presenter: Come again, please?
>
> Weibel: Media poet.
>
> Presenter: Media poet, yes. And that you wrote the script for *Invisible Adversaries* (*Unsichtbare Gegner*, see pp. 134–35 and 202–203 in this volume). I've already mentioned and also that you acted in it. If you'd like to tell us what is happening, please, do. I'm not sure yet whether that is part of the plan or not.
>
> Weibel: ...
>
> Presenter: Yes, thank you very much. If you like and if, having given so much or given so much of yourself for this action, so to say, you would still like to join us and

it isn't too crowded for you there, I should be pleased if you would.[8]

With these words, the host attempts to restore the conventions of the television talkshow. But the screen is soon to be obliterated by Weibel's blood. This action is an insertion of the artist's body into the screen in an extremely literal way. It has elements of self-sacrifice, in that the artist subjects himself to inevitable misunderstanding and public ridicule, his giving of his blood being the most obvious metaphor of his willingness to sacrifice everything for his artistic quests. These elements are made explicit in the final work that we will consider, a video consisting of various solo performances, and entitled *On the Theorem of Identity* (*Zum Theorem der Identität*, 1973, see pp. 236–37 in this volume). The first part *Tritity* (*Tritität*) was triggered by the coincidental visual similarity between Weibel's own facial features, those of Jesus Christ as painted by Piero della Francesca, and of Nikolaus Lenau, the romantic Austrian poet who spent his last years in a lunatic asylum. As these images merge into each other, and shift from one to the other, the voice-over combines statements of the artist with passages from Della Francesca's books on geometry and from Lenau's writings. ... The words repeat, then begin to transpose themselves according to a strict conceptual logic. ...

The follow-on work *Soliloquy. Language Mirrorings* (*Selbstgespräch. Sprachspiegelungen*, 1973, see pp. 96–97 in this volume) presents the artist in dialogue with himself via a tape recorder. This is followed by *Self-Description* (*Selbstbeschreibung*, 1973, see p. 234 in this volume), in which a set of textual inscriptions show the artist questioning the motivation that leads him to dare to place his own visage within those of his iconic predecessors:

Self-Limitation
Self-Delineation
Self-Description

WHAT IS THIS ANXIETY ABOUT FAILING?
WHY DO YOU SEEK SUCCESS?
FOR RECOGNITION OF YOUR ACHIEVEMENTS?
WHY DO YOU NEED DO YOU SEEK JOY
PLACE YOURSELF BEYOND UNDERSTANDING
BEYOND THE HISTORY OF MAN[9]

Which leads, finally, to some ruminations about *Parenthetical Identity* (*Parenthetische Identität*, 1973, see p. 235 in this volume). ... Although personal in their subject matter, Peter Weibel's reflections are subject to the same principles of combination and transformation as those that generated the "trinity" in the first instance. That is to say, they are coincidental as to their occurrence and arbitrary in their unfolding. They demonstrate the impossibility of psychic truth when it is subject to relentless mediation.

In his survey essay cited above, Dieter Daniels places the early work of Weibel and EXPORT in the category of "playful disturbances" to the ways in which television, subject to technical, ideological, and commercial demands, was closing down its potentialities and becoming a monolithic, massaging juggernaut. I hope that these notes have demonstrated that Peter Weibel, in collaboration with VALIE EXPORT and in his own work, offered at that time more than mere disturbances, work that went beyond the playful." Rather, these were artistic interventions aimed at showing how important aspects of individual being, of independent selfhood, could be registered even when artists, like everyone else, were becoming increasingly subject to the enveloping web of "the media."

Note: This text was first published in Dennis Del Favero, Ursula Frohne, and Peter Weibel, eds., *un_imaginable* (Ostfildern: Hatje Cantz, 2008), 116–27. This text quoted many of Peter Weibel's performances and speeches in length and for the first time in the English language. As many of these texts are also featured in this volume, some of the longer passages have been omitted from this reprint.

[1]
Nam June Paik, "Media Planning for the Postindustrial Society – The 21st Century is Now Only 26 Years Away" (1974), in *Media Art Action – the 1960s and1970s in Germany*, ed. Rudolf Frieling and Dieter Daniels (Vienna: Springer, 1997), 239–42.

[2]
Dieter Daniels, "Television-Art or Anti-Art," in *Media Art Net – Survey of Media Art*, ed. Rudolf Frieling and Dieter Daniels (Vienna: Springer, 2004), 58–79, www.medienkunstnetz.de/themes/overview _ of _media _ art/massmedia. See also Douglas Davis and Allison Simmons, eds., *The New Television: A Public/ Private Art* (Cambridge, MA: MIT Press, 1977).

[3]
David Joselit, *Feedback: Television Against Democracy* (Cambridge, MA: MIT Press, 2007).

[4]
Marshall McLuhan, *Understanding Media: The Extensions of Man* (New York: McGraw-Hill, 1964), 4.

[5]
Marshall McLuhan, *Culture is Our Business* (New York: McGraw-Hill, 1970).

[6]
Marshall McLuhan and Quentin Fiore, *The Medium is the Massage: An Inventory of Effects* (New York: Penguin Books, 1967).

[7]
Extract provided by Peter Weibel from his archives; translated by Jonathan Uhlaner.

[8]
Extract provided by Peter Weibel from his archives; translated by Jonathan Uhlaner; editor's note: Peter Weibel's entire lecture can be found on p. 254 in this volume.

[9]
Extract provided by Peter Weibel from his archives; translated by Jonathan Uhlaner.

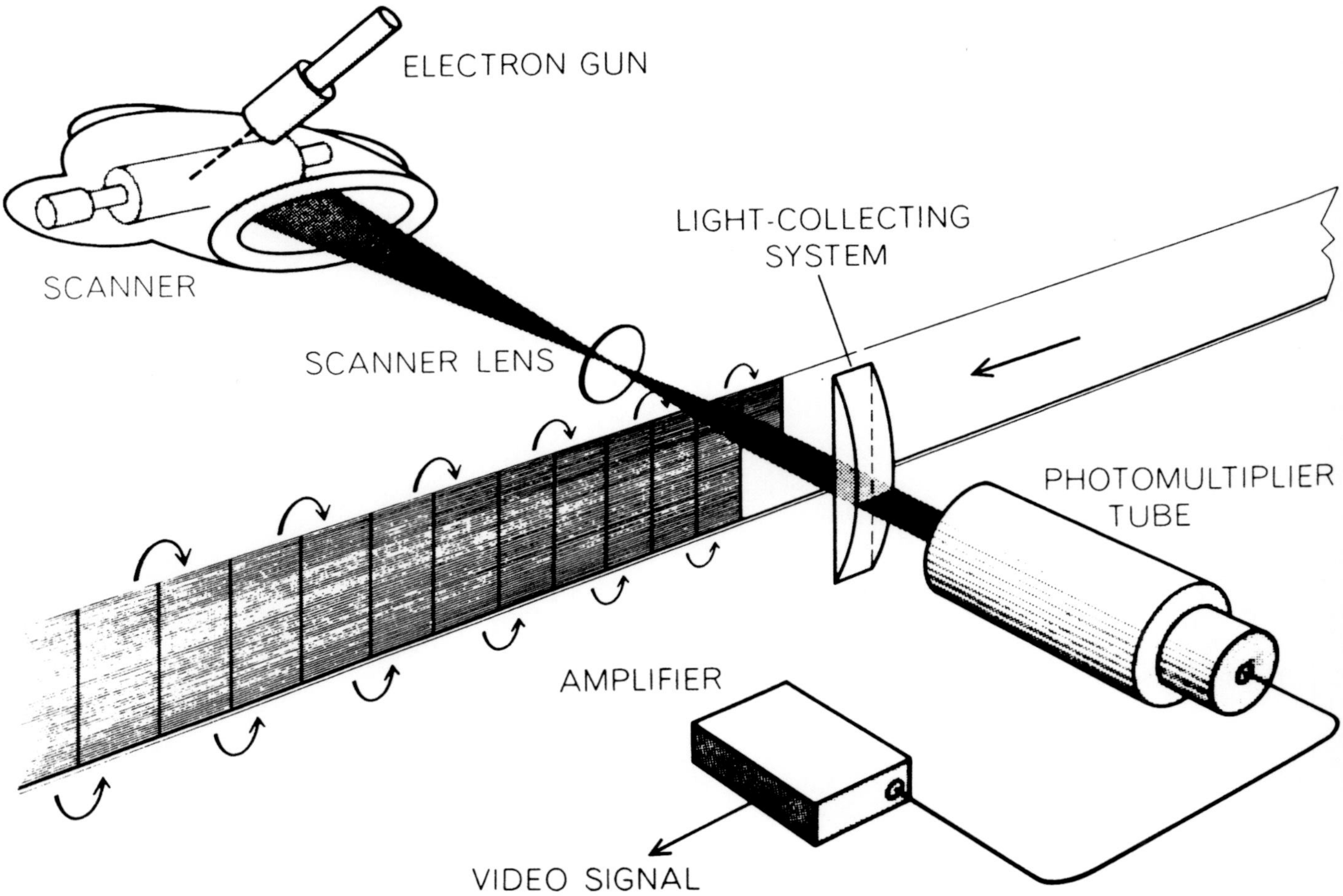

der scanning process (suchprozeß) bewegte einen mikroskopisch kleinen lichtpunkt entlang eines teiles des filmes, der durch einen elektronenstrahl erzeugt wurde, der auf einer mit phosphor belegten trommel auftraf. die dichte des bildes auf dem film steuerte die stärke des lichtes, das die photomultiplier tube (fotoverstärkerröhre) erreichte, die das elektrische signal für die übertragung zur erde erzeugte. eine komplette ablesebewegung überstrich die informationen auf einem streifen des filmes, der etwa 3 mm schmal und 60 mm hoch der filmbreite nach war. diese bänder sieht man in den fotos vom orbiter.

schließlich sollte ein programm mit einer ausreichenden zahl von besuchern produziert und die ergebnisse sollten für eine überarbeitung des programms benutzt werden. galerie junge generation, wien 1, blutgasse 3/I, donnerstag, 10. und freitag, 11. april 1969, von 18 bis 22 uhr, multi media 1. (zutritt jederzeit möglich.)

marc adrian	(amplifier)	horst l. renner	(scanner)
VALIE EXPORT	(light-collecting system)	gottfried schlemmer	(electron gun)
peter hassmann	(photomultiplier tube)	peter weibel	(video signal)
wolf hermann	(scanner lens)		

Poster announcing the exhibition *multi media 1*, Galerie Junge Generation, Vienna, Apr. 10–11, 1969

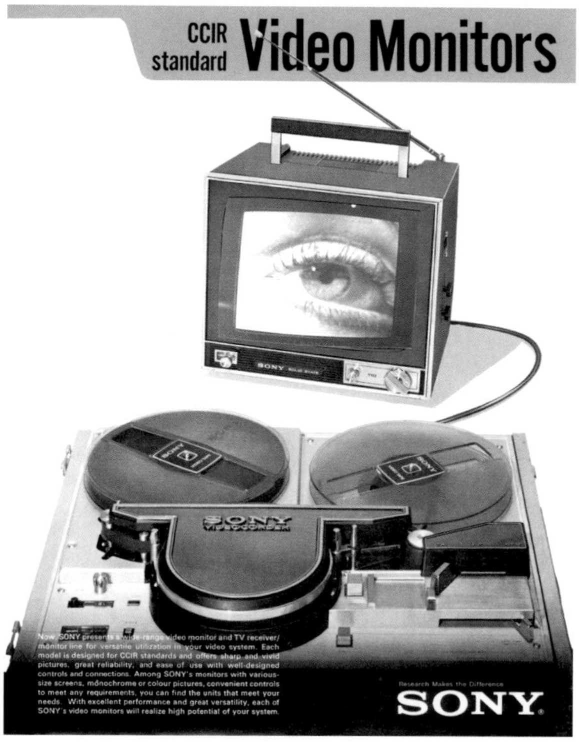

Monitoring

1969 | collage; 21 × 29.7 cm

The subject of this piece is the act of seeing, monitoring, observing, surveilling.

"Eye" sounds like "I"
I see eyes
Eyes see me
Eyes see eyes
Video – I see
Lumen (plural *lumina*) – light source, eye
I see monitors (video)
To be seen while seeing – to be observed while observing

Video as the opposite of film. In the *camera obscura*, the light source, the image projector (the sun), is outside. In the *video system*, the light source, the image projector (the cathode-ray tube), is inside the device, as is the eye inside the body. Whereas the film projector is an imitation of the sun, as the projected beam of light still attests, the electronic picture tube is the imitation of the eye; it is artificial, synthetic, created by humans. If one looks closely at the eye in this collage, one can make out an image of the camera in the pupil – that is, the imaging machine, the artificial eye, can be seen inside the natural organ eye.

SENDER —

(sender)

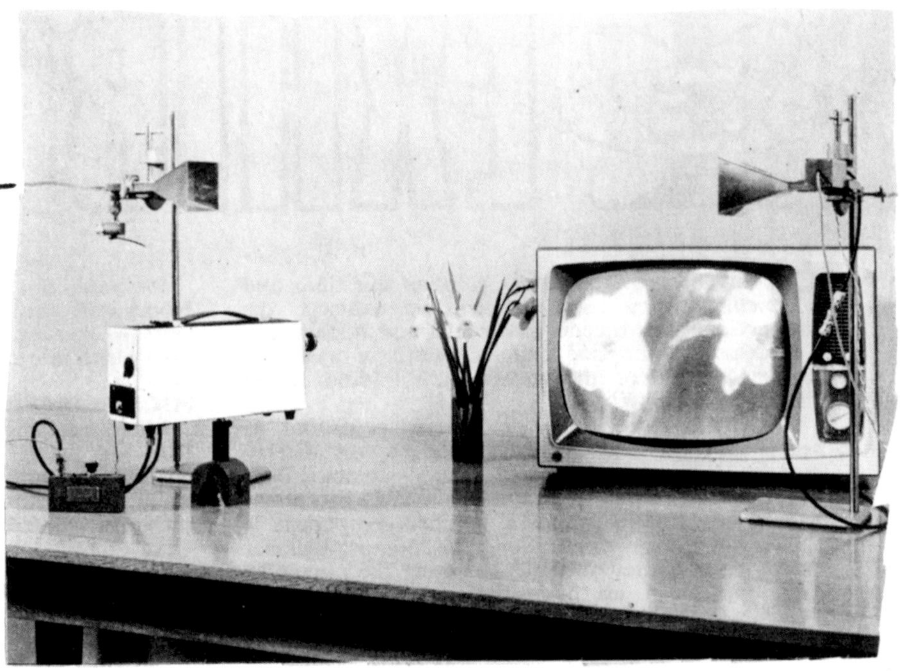

— EMPFÄNGER

(receiver)

Miracle of Technology / Wunder der Technik

1969 | collage for video installation demonstrating simultaneity

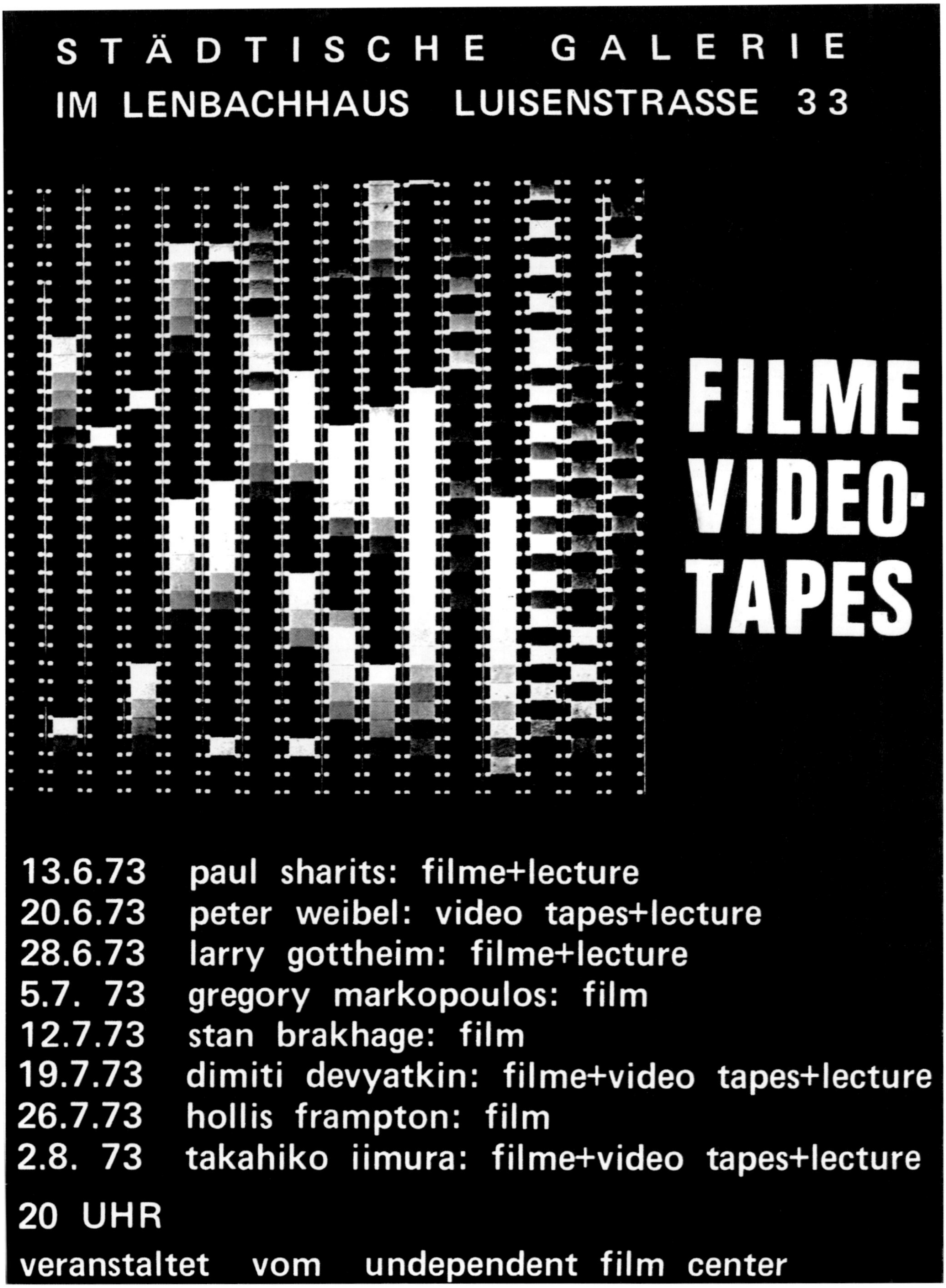

Poster for the film and video screenings *Filme Video-Tapes*, Städtische Galerie im Lenbachhaus, Munich, 1973

On the Philosophy of VT and VTR

Peter Weibel

I would suggest provisionally understanding video art as reducible to five characteristics of VT (videotape) and VTR (videotape recording): *synthetics, transformation, self-reference, instantaneousness, box*. Using several examples, I will briefly illustrate how closely these characteristics are linked, how much they distinguish video from film, and how defining they are of video as encoded medium.

In some ways, one can say that the video system is probably closer to the record player than it is to film. Just as the record industry has the Moog synthesizer, the video is based on the same synthesizer principle. The video-synthetic image machine, which can create and manipulate all visual forms in an apparently almost endless multiplicity, which records images and overlays a real, live recorded scene with any imaginable electronic effect (distortion, multiplication, visual echo, visual noise, discoloration, etc.) – that is what gives video addicts their metamorphotic high. And indeed, the electronic image machine, a video-computer symbiosis that creates any imaginable patterns, shapes, figures, and their transformations, might be the essential basis of the encoded medium of video. Perhaps that is why video deconstructs identities while film constructs them; why video adjusts the world to consciousness while film adapts consciousness to the world; why video extends cinema where the avant-garde has always situated cinema: at the place where an autonomous reality is created.

Closely linked with the concept of *video synthesis*, of course, is the *transformation of signs*. The constant manipulation of signs at will, the dissolution of concrete, interpretable images into abstract forms and back again, the transformation of things (a head becomes a horse, etc.) and meanings: these factors place video in the technological proximity of computer graphics and in the psychological proximity of hallucinogens. Accordingly, *synthetics and transformation are the two elements of VT and VTR* that have so far generated the greatest enthusiasm and led to great masterworks. They create effects – synesthetic and uncoupled from reality – for psychedelic, cosmological multimedia fans seeking to expand their consciousness through the eye.

Self-reference and *simultaneity* represent the perhaps more important videological operators for "cool users" by creating new space-time structures, new behaviors, by humanizing reality. Video is a technological communications system with all manner of feedbacks and loops. This distinguishes it fundamentally from, for example, cinema: Film is a single-sense communication running from producer to consumer. Video can be a dual-sense communication. Video is not merely a TV screen. Instantaneousness means one need no longer consider the time required to develop negatives; rather, there is a temporal union of production and consumption, of recording and presentation. One not only sees immediately what has happened, but also simultaneously what is happening. One sees the signs on the screen in parallel with the event. The system can record itself. One can display one's behavior to oneself directly. Instantaneousness rules. Time can be disposed of, or at least the temporal delay between event and representation (repetition) can be significantly minimized. If one emphasizes, then, less the system of signs, as in syntheses and transformations, and more the communications system, these feedback loops allow novel explorations of social behavior and new therapies; open up new social, didactic, political functions for optical signal systems; enable new ways of communication from the private all the way to the public. This communicational difference (one-way communication TV, two-way communication VTR) marks the distinction between, for example, broadcast television and the video system. Through the two different sociocultural codes it becomes possible to imagine even tapes whose effect is realized only on TV, that is, whose codification is attuned precisely to the encoded medium of TV. TV and VTR together suit the *box-ness* of the monitor, which is more than a mere technological fledgling, but instead exemplifies the space-time switch represented by both TV and VT, the pill-like nature of machines sending the viewer on journeys that can extrude only highly concentrated doses of convention as narratives of reality. This space-time switch, which shortens distances and times and changes images and communication strategies, has far-reaching moral, ideological, and social, and economic implications.

These five videological factors (video synthesis, transformation, self-reference, simultaneity, box-like display) may be one example of why classic cinematography was broadened into Expanded Cinema and strives – with video, telephone, holography, laser – to move from a projection system to a general machine system for image processing and image production. This unmistakably foregrounds technological art's role as a model. What sends the video freak on a video trip is the hallucinogenic suggestion of VT and VTR. Apparent reality can be turned on and off; the more channels, the more realities.

Note: Excerpt from Peter Weibel, "Zur Philosophie von VT & VTR," published on the back of the poster for the film screenings *Filme Video-Tapes* in the Städtische Galerie im Lenbachhaus, Munich, 1973. Translated from the German by Peter Dahm Robertson.

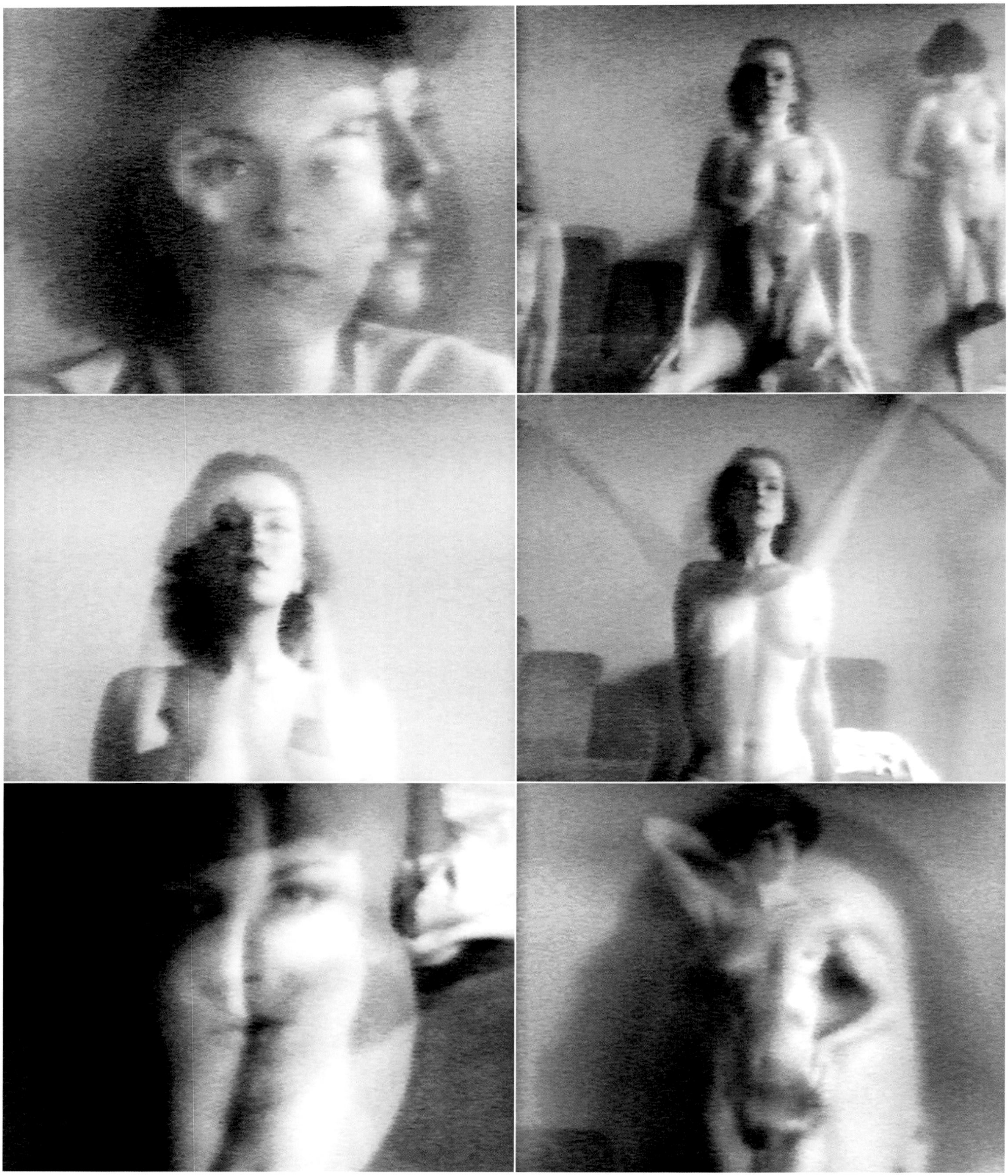

Switcher Sex

1972 — video; b/w, sound, 2:32 min.
actors: Susanne Widl and Sieglinde Schönberg

Two cameras are connected via a mixer, the recordings of which are blended into a picture on the monitor. Layers of various body parts in different scales are made in order to form analogies of body-scapes, an anagrammatic of the body: ear instead of eye, vulva instead of mouth, legs instead of arms, ...
in the tradition of surrealism: *le corps morcelé* (the fragmented body). Organs combined to new bodies.

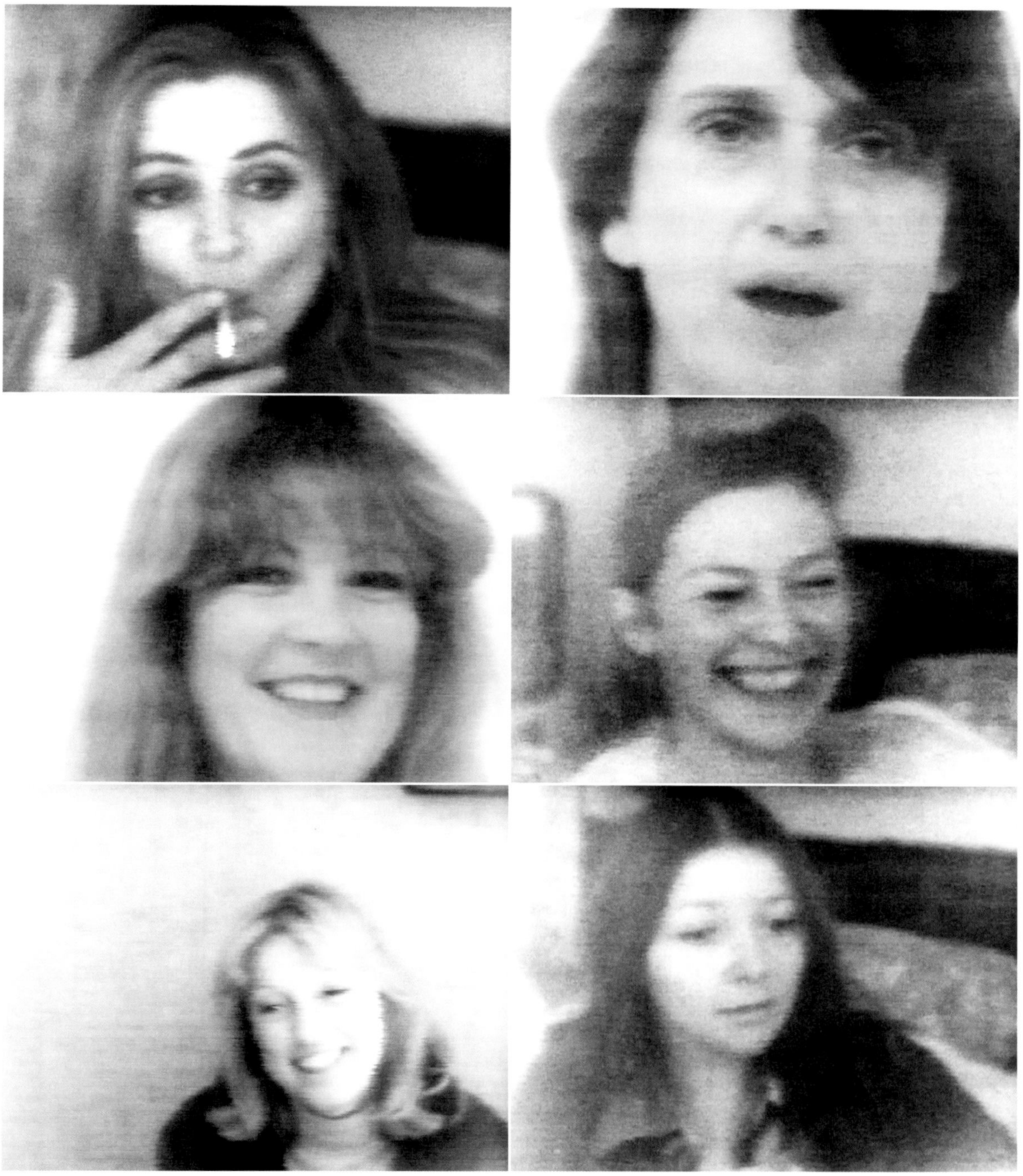

The video *Review of the Nude, or Inverse Self-Portrait* shows the heads of young women talking about a male nude, depicting Peter Weibel, who is not seen in the picture, by means of verbal description. The video work seeks to reverse the history of the nude as the pleasure of male artists. Art history is full of nude drawings and nude paintings of female naked bodies by men. A man looks at the woman and paints what he sees. Weibel turned this setup around in various ways. First, the nude is talked about and not visualized. Second, the naked model is a man and not a woman. Third, a woman describes what she sees, but what the viewers do not see: the naked male model. Usually, the viewers look at the naked female body from the point of view of the male artist. Here, contrary to the usual viewpoint, the purely linguistic description of the nude creates identification of the viewers with the invisible model. The viewers gain the impression, they are what the women describe.

Review of the Nude, or Inverse Self-Portrait / Aktbesprechung oder Inverses Selbstporträt

1975 | video; b/w, sound, 19:01 min.

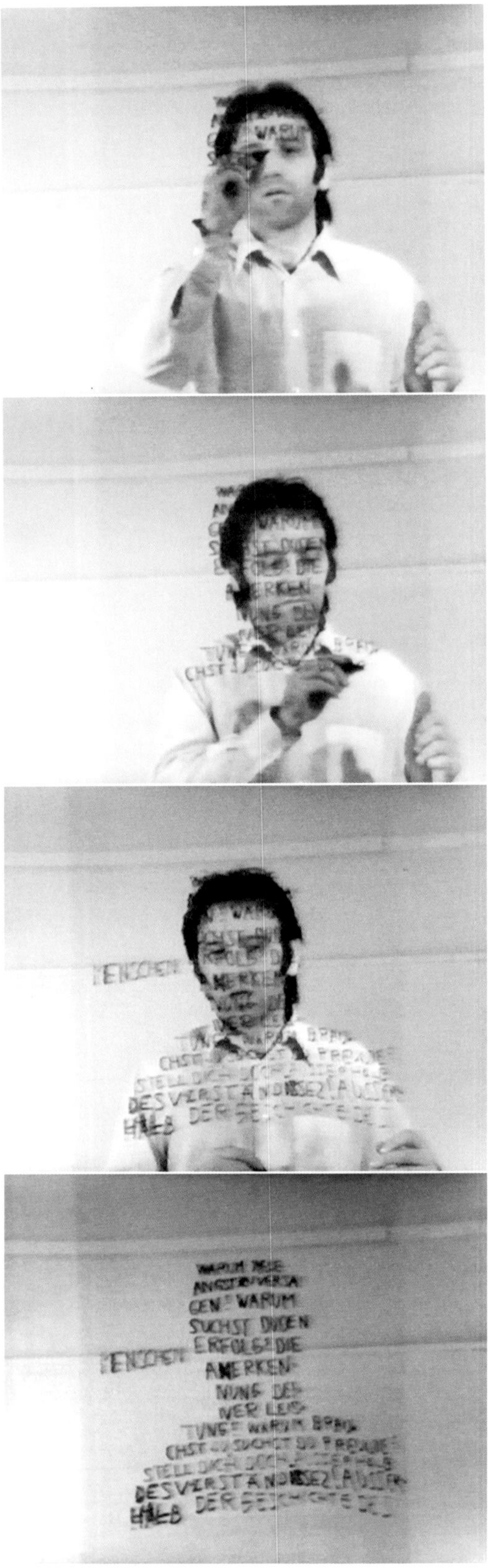

On a glass plate, I write on myself within the boundaries of the image of my body on the screen. —Peter Weibel

Self-Description / Selbstbeschreibung

1973 | video; b/w, sound, ca. 12:21 min.

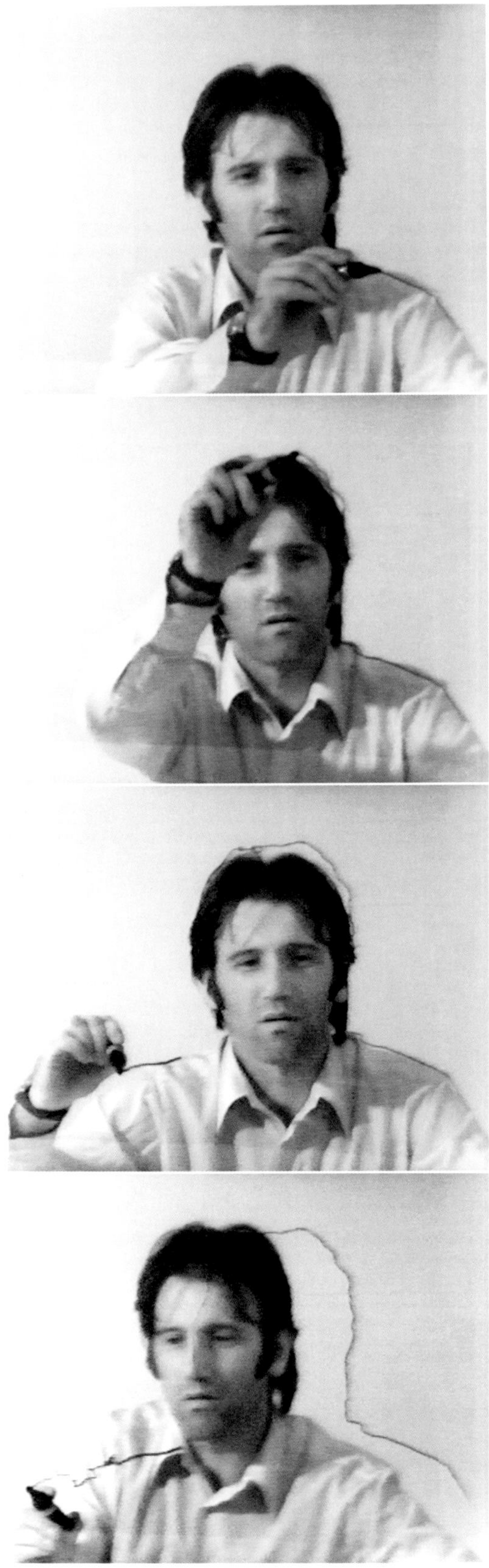

Drawing the contour of my body on a glass plate was unsuccessful, because when I tried to draw my hand as it was drawing my hand, the hand disappeared off the edge of the screen. —Peter Weibel

Self-Delineation / Selbstbezeichnung

1973 | video; b/w, sound, 2:12 min.

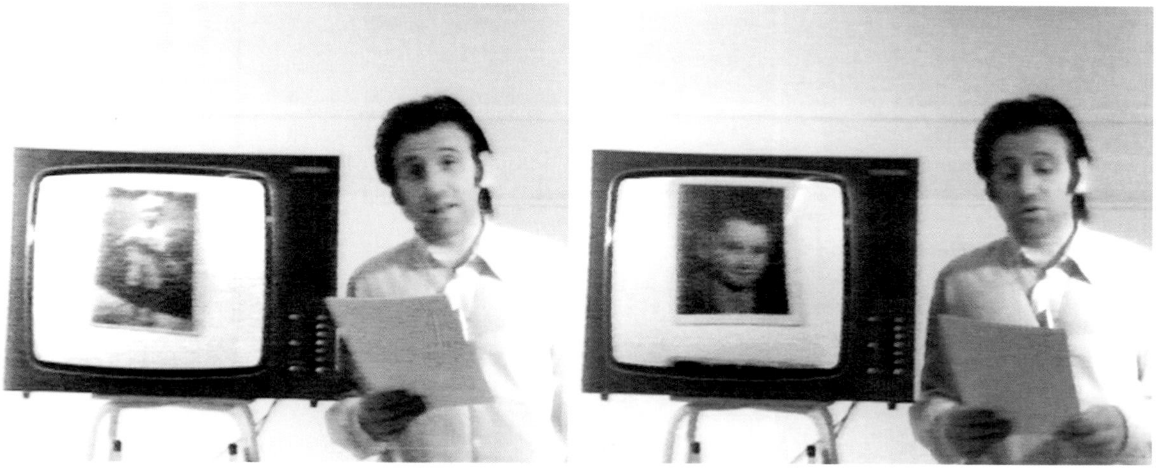

I stand in front of a TV set which shows photos of myself, ranging from my childhood to my present age. During this photo sequence, I read the following text aloud:

if i were the first son of my mother and father
then of course i would not be me
rather i am me
for i am the first son of my mother and father
just as i would be my sister if the first child of my
mother and father had been a girl
if i were the second son of my mother and father
i would be my brother
i am in fact a brother but the brother of the second son of
my mother and father
what troubles me though is the chance possibility that i might
have been the first son of my mother and another father
then i would probably have been my own stepbrother
or stepsister if i had been the first daughter of my mother and
another father
in that case thanks to my mother i would surely still have been
half me
if i were the second son of my mother and another father
or the second daughter of my mother and another father
then i would also be my stepbrother or stepsister
but i would be further away from myself
than a half-me
in all these cases i would have the same mother
so i would not have diverged that much from my current identity
but if i were the first son of my father and another mother
then i would be much more different from myself
than in the other variants
obviously i would be
not me

—Peter Weibel

Translated from the German by Patrick Hubenthal.

**Parenthetical Identity /
Parenthetische Identität**

1973 | video; b/w, sound, 2:54 min.

The combinatorial structure of the text parallels the combinatorial structure of the images. The superimposition of Piero della Francesca's Christ onto the portraits of Nikolaus Lenau and Peter Weibel, which causes the static images to move, come alive, and speak, much like modern-day morphing, produces results that are just as grotesque and absurd as the texts.

The portrait of Peter Weibel speaks:
My message throws everyone back upon themselves.
The portrait of Nikolaus Lenau speaks:
I believe that I myself am poetry. My selfest self is poetry.
The image of Piero della Francesca's *Resurrection* speaks:
Form and content are brother and sister in the venerable hall of space.
The image of the *Resurrection* (Christ) speaks:
I believe that I myself am poetry. My selfest self is poetry.
The portrait of Peter Weibel speaks:
Form and content are brother and sister in the venerable hall of space.
The portrait of Nikolaus Lenau speaks:
My message throws everyone back upon themselves.
The image of the *Resurrection* (Christ) speaks:
Love thy neighbor as thyself.

Now the combinatorial dissolves between images and permutations of sentences begin:

Love form and content as brother and sister.
Love thy neighbor as form and content.
Love thy neighbor as poetry itself.
Love thy neighbor as my message.
Love thyself as thy neighbor.
Love thyself as poetry.
Love poetry as brother and sister.

My message throws thee upon thy neighbor in the venerable hall of space.
My message throws everyone upon brother and sister.
Form and content are my message.
My message is poetry itself.
My message throws everyone upon poetry as the love of form and content.
Love in the venerable hall of space is my message.

My selfest self is love.
My selfest self is form and content.
My selfest self are brother and sister.
My selfest self is the love of oneself as neighborly love between form and content.
My selfest self is my message.

The form of space is the content of poetry.
Brother and sister are the content of space.
Space is the form of the message.
The content of love is the form of poetry.
Form throws everyone back upon content.
Content throws everyone upon form as the message of itself.
Content and form are the message of love.
Form is the love of content.

Poetry is the love of oneself.
Poetry throws everyone back upon themselves.
Poetry loves itself as brother and sister.
Poetry throws everyone upon brother and sister as neighborly love.
The content of the brother is the form of the sister.
The form of love is the content of poetry.
Poetry throws everyone into the hall of space.
Poetry is form and content.
Poetry is the message of itself.

—Peter Weibel, 1975

Translated from the German by Patrick Hubenthal.

On the Theorem of Identity: Tritity / Zum Theorem der Identität: Tritität

1975 | video; b/w, sound, 8:33 min.

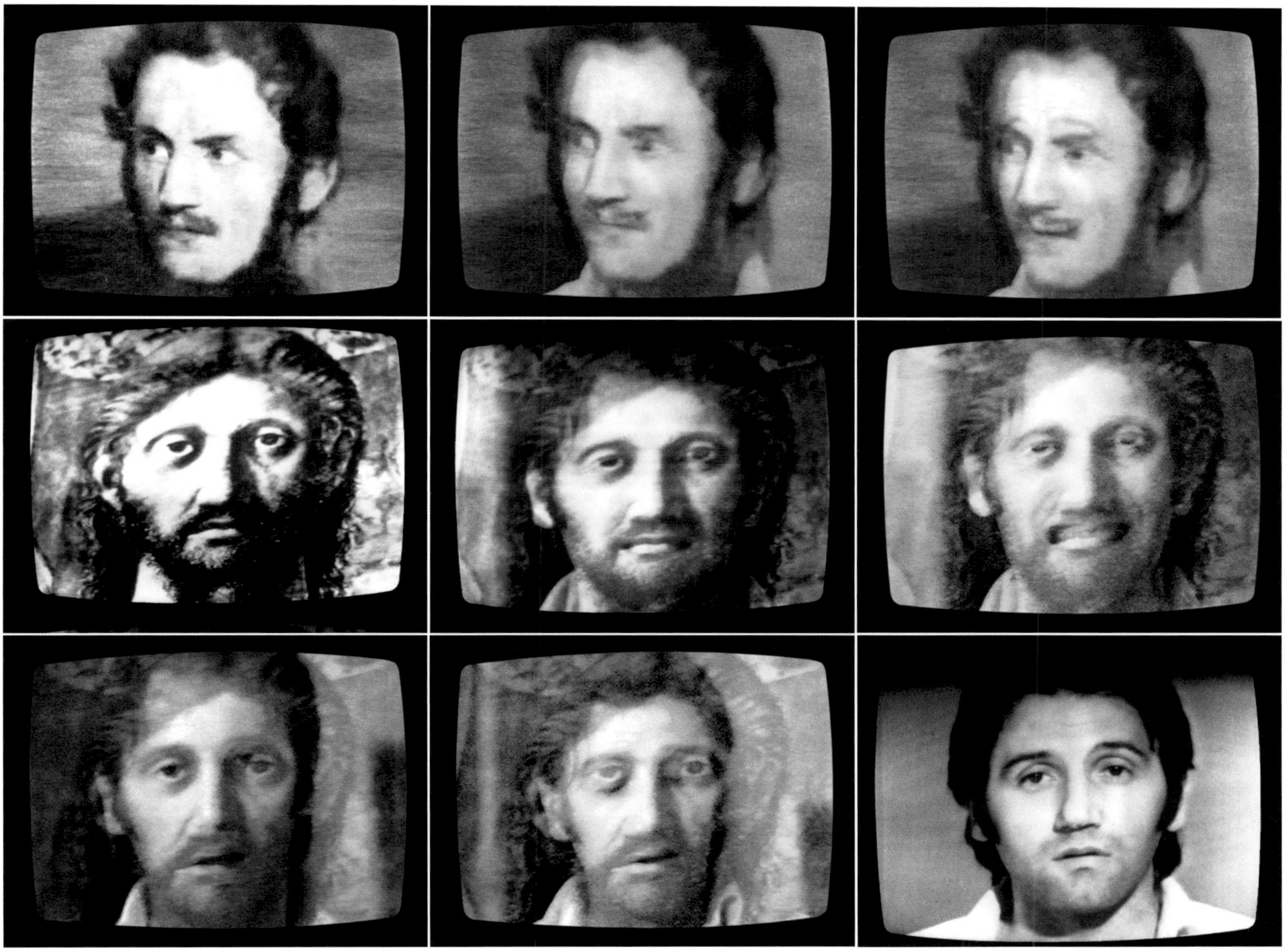

The video work *On the Theorem of Identity: Tritity* shows the faces of Peter Weibel, Nikolaus Lenau (a late Romantic Austrian poet who spent the last years of his life in asylums), and Jesus Christ as depicted by Piero della Francesca. Quotations from these three people are heard as each portrait is shown. Using blending effects that anticipate digital morphing, the photographed faces of Lenau and Christ are then underlaid with a live video recording of Weibel making faces. This brings the static images of Lenau and Christ to life and makes them speak. Just like the faces, the verbal messages of Lenau, Christ, della Francesca, and Weibel are mixed and combined. The messages are relativized to the point of absurdity by the mixing of the quotations. Weibel's work questions the hypothesis of the existence of a single identity and throws the viewers back on themselves as message.

εἴδω, ὅτι οἶδα
οἶδα, ὅτι εἴδω

I see that I know
I know that I see

IDOLOGY IDOTECHNE VIDEO ART

an etymological excursion has the virtue of demonstrating the history of thoughts sedimented in words. Ideas coin the words, but the words coin the ideas as well. word history as roots of reference for an understanding of the essence of a medium.

the root of the latin word 'video' (I see) and the greek εἴδω (eido – I see) is the indogerman ϝιδ (wid, weid) which means: see, know, but initially: find, compare the sanskrit word 'vindati' (he finds).

One finds the root ϝιδ (see)
in the greek εἴδω (see), in the latin 'video' (see), in the gothic 'witan' → the old highgerman 'wizzan' → the modern german 'wissen' (know). the german word 'witz' (joke) and the word 'weise' (wise) are also derivable from this root √ϝιδ.

In etymology there is a strong connection between
seeing and knowing
as can be traced in the development of ϝιδ (wid – see) to 'wissen' (know).

this connection is even stronger in the greek language:
the conjugation of the (as present tense unusual) verb εἴδω (eido, see) leads to οἶδα (oida) as (active) perfect tense.
but οἶδα (oida) has a double meaning: it means "I have seen" as perfect tense of εἴδω (eido, see) and " I know, I experience" as present tense.

εἴδω, ὅτι οἶδα : I see that I have seen
I see that I know
οἶδα, ὅτι μὴ οἶδα : I have seen that I haven't seen
I know that I haven't seen
I have seen that I don't know
I know that I don't know

greek οἶδα (I know), sanskrit 'veda' (I know), gothic 'wait' (I know), oldgerman 'weiz' (I know), new german 'ich weiß' (I know).

as language demonstrates: there are cognitive relations between seeing, knowing, experiencing and finding. video art is always epistemic. the greek 'techne' means 'artefact, art'. From 'eidos' (image) to 'idos' (idea), even to ἰδεῖν (idein). from eidology to idology and to ideology. idotechne (the art of image) is video technique is video art, that is
the art of seeing and knowing!

Druck: Hörtenbergdruck - 6410 Telfs / Tirol

Idology – Idotechne – Video Art

1974 | back cover of the publication *Peter Weibel: An Annotated Videography*, ed. Österreichische Hochschülerschaft (Innsbruck: 1977)

PETER WEIBEL

1090 VIENNA NORDBERGSTR. 16/22 Austria

EPISTEMIC VIDEOLOGY III 1975

The video system is the physical implementation of one of the most characteristic features of man: self-reference, which is the basis of consciousness.

Turing Machines – recursive functions – selfreproducing automata.

the image of the image:

what happens if I show the image to the image (if I turn the camera on to the monitor)? Virtually an endless line, an infinite recursion, images referring to images. Is this logic of self-reference also a logic of self-knowledge? or even a logic of self-consciousness?

If I show an image to an image, does the image then know the image? Can I write then as in the case of "Smith knows that he is Smith".

$E(y)(y=S \,\&\, K_S(y=S))$ E... exist   K... know

"the image i knows that it is the image i"

$K_i(s(i)) = s(i)$ introducing the special operator "s" for "it itself"

when I mean

"the image i showing the image i shows the image i showing the image i ....."?

The 'knowing image' of the video system shows that the relation between image and object (signifiant and signifié) is a recursive function, that means that a knowing image imagines knowing as the image knows or the knowing imagines, that is, man is picturing all his images in objects as victim of his self-recursiv, self-reflexive imagination.

The Video Show 1–25 May 1975 Serpentine Gallery London

Epistemic Videology III /
Epistemische Videologie III

1975 | concept; felt pen on paper

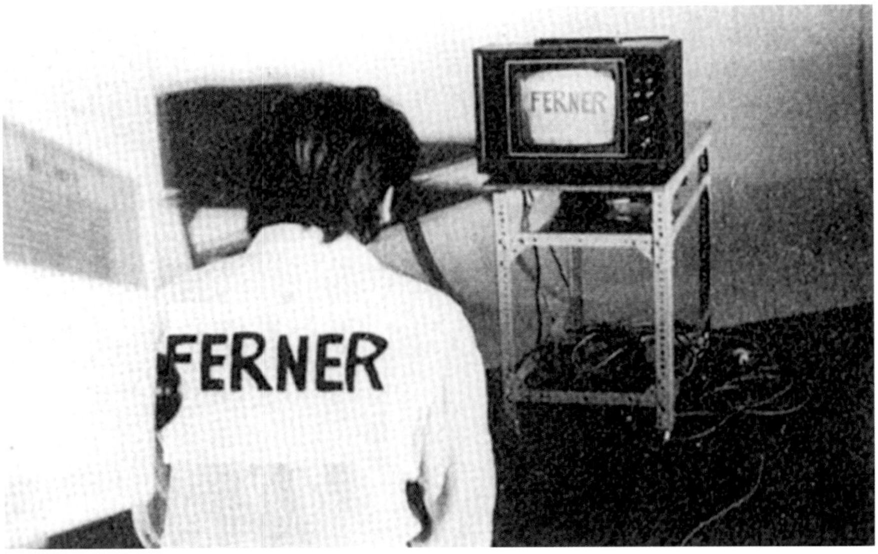

Performance at the First International Underground Film Festival, Arts Lab, London 1970

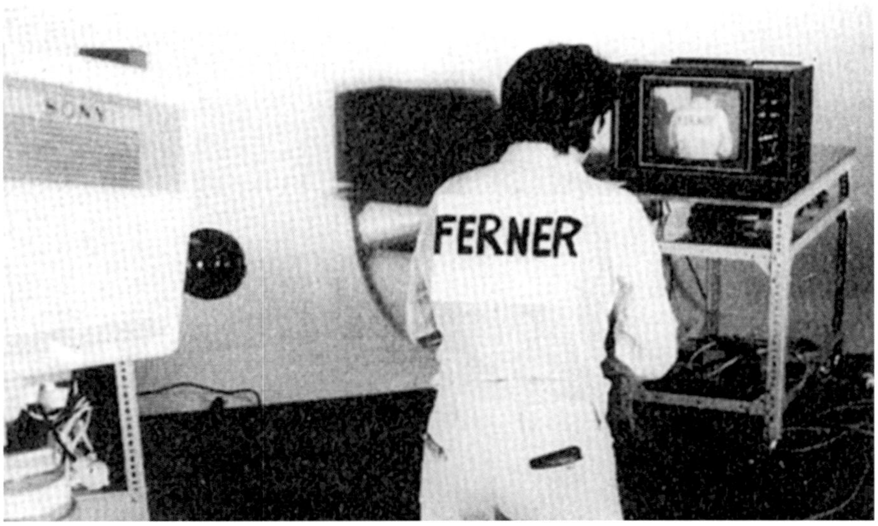

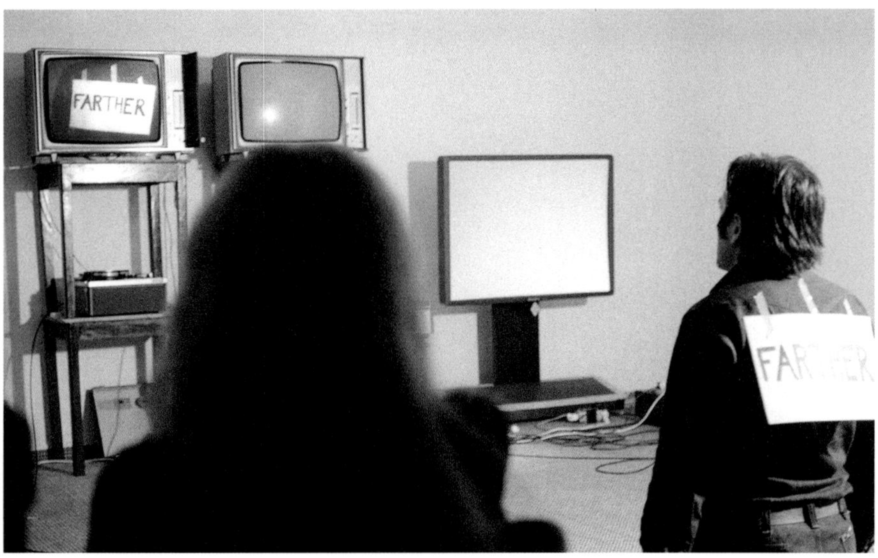

Performance at the Independent International Experimental Film Festival (EXPRMNTL 5), Knokke-le-Zout, 1974

Investigation of Identity I: Closer to You My Self / Investigation of Identity I: Näher zu dir mein Ich

1970/74 | video performance; part 1 of 3

A kneeling performer (Peter Weibel) moves from the camera to the monitor. A sign on his back reads "FERNER" (farther). As he moves forward, he says "näher" (closer). Thus, each time he moves closer to the monitor (auditory level), he moves farther away on the screen (visual level). He is moving "closer" to the monitor as an object, but in the picture on the monitor he is getting farther away, since he is moving away from the camera. He is therefore "farther" away in the image, but "closer" to the physical screen (monitor).

In a way, self-observation is just an extreme example of the type of external yet self-referential observation that constitutes the psychological method. Because of its feedback techniques, video is the medium most suited to highlighting such mechanisms, perceptually and cognitively. It is no coincidence that a screen is called a "monitor" and "monitoring" also means "surveillance."

The epistemic boundaries of the dissection of a person into their internal psychic parts are demonstrated by a contradiction in ego psychology and perception – namely, the contradiction between *phonum* ("closer") and *visum* ("farther"), between what one is hearing (speech) and what one is simultaneously seeing (vision).

—Peter Weibel

Translated from the German by Patrick Hubenthal.

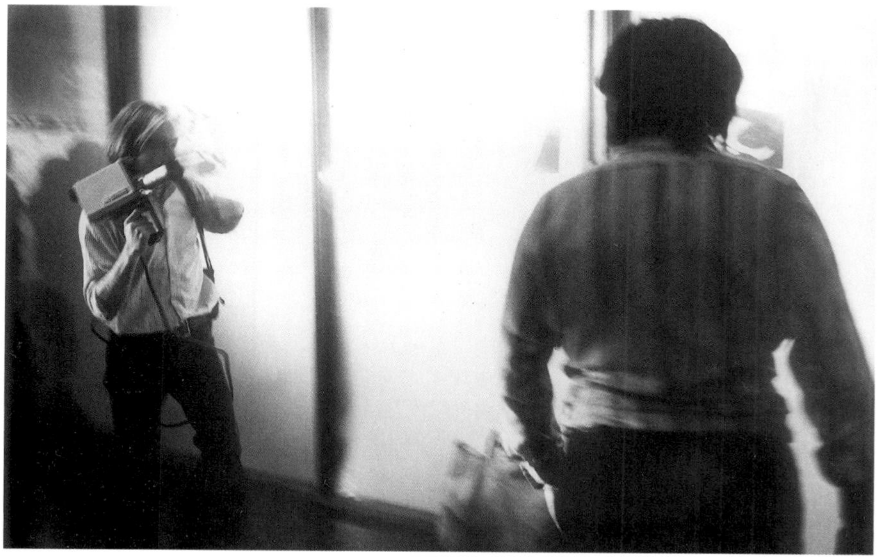

Performance in *The Austrian Exhibition*, Institute of Contemporary Arts, London, 1973

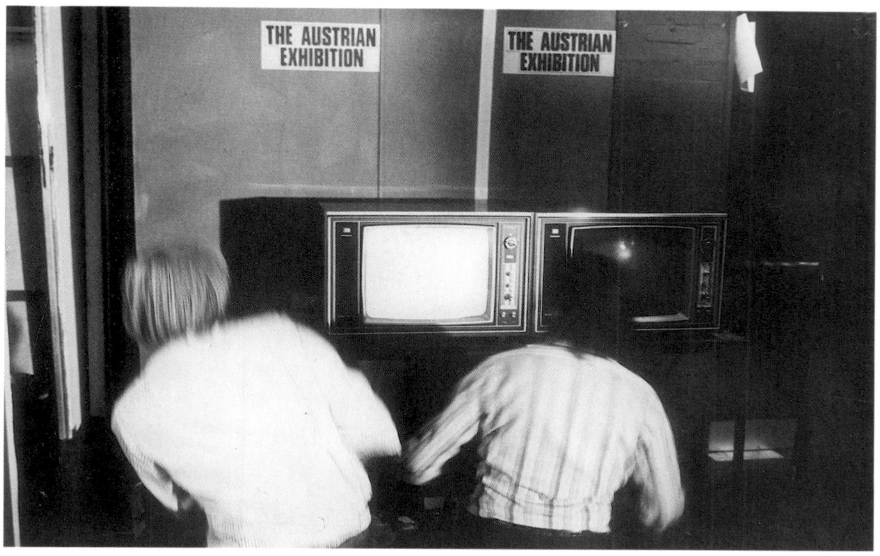

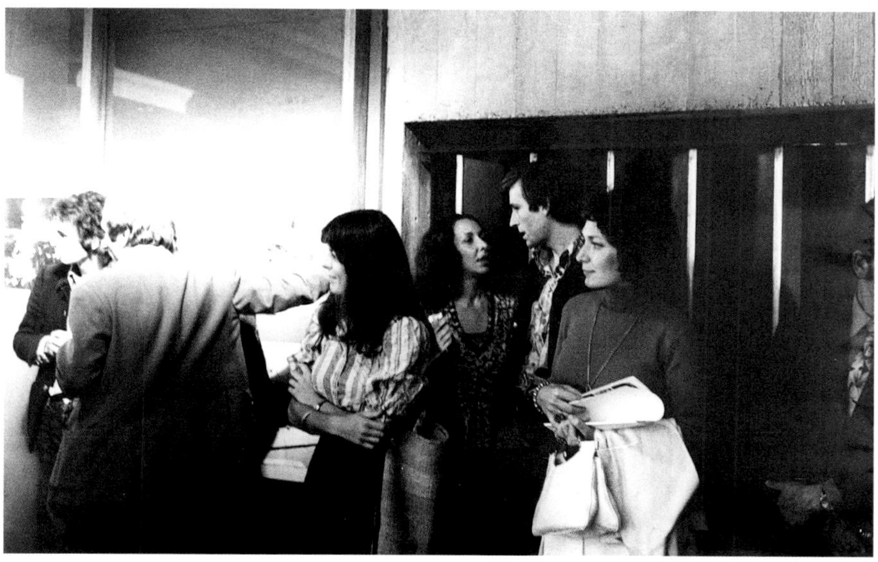

Carolee Schneemann and Anthony McCall (talking with each other in the back) watching the performance in London

Investigation of Identity II: Servare Contingentiam

1970/73 | video performance; part 2 of 3

SERVARE CONTINGENTIAM

For John Duns Scotus (d. 1308, teacher of William of Ockham)

Everything that is real, and different from God, is contingent to the core of its being.

—John Duns Scotus [1]

Live action:
I move my head erratically in front of the camera, so that it cannot track me and my head constantly escapes the frame.
The recording is then played back. I kneel in front of the monitor and try to follow my own random and unpredictable head movements, at which I fail at least as often as the cameraman did.
In the epistemology of Duns Scotus, God was, inter alia, the pinnacle of ideal qualities; his existence was therefore necessary. Since humans are by nature different from God, the nature of their being must necessarily be contingent. This foreshadows the contemporary philosophy of chance, from John Cage to Jacques Monod.

—Peter Weibel, 1971

Translated from the German by Patrick Hubenthal.

[1]
Editor's note: This sentence is not, in fact, a quote by Duns Scotus. It was intended as an encapsulation of Scotism by its actual author, Philotheus Boehner. It appears in his introduction to *Philosophical Writings: A Selection*, by William of Ockham, ed. and trans. Philotheus Boehner (Indianapolis: Hackett, 1990), ix–li (quotation on xxii).

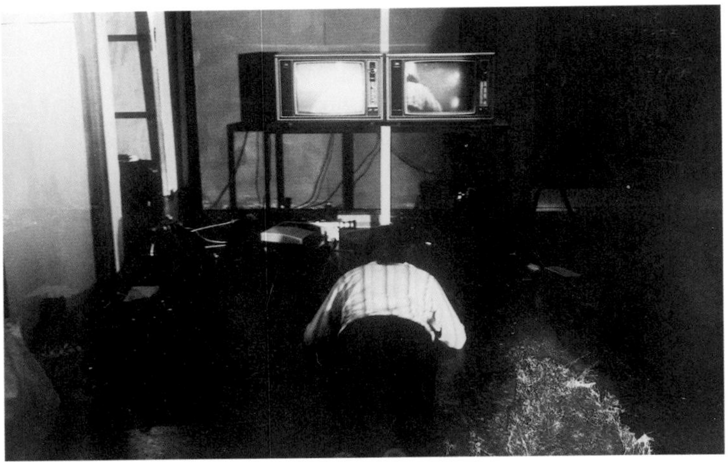

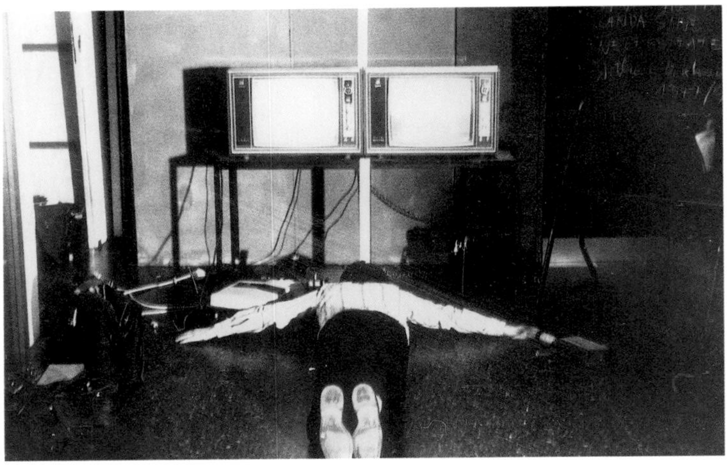

Performance in *The Austrian Exhibition*, Institute of Contemporary Arts, London, 1973

OCKHAM'S RAZOR

Entia non sunt multiplicanda sine necessitate.
—William of Ockham (d. 1347) [1]

Entities must not be multiplied without necessity.

Ockham's conceptual blade is one of the most fundamental axioms in the history of the philosophy of science.
Among other things, it led to Johannes Kepler's revolutionary transformation of astronomy: his astronomical equations required fewer entities than those of his rivals.
Since then, however, this theorem seems to have become an antiliberal force, limiting individual identity and therefore individual freedom.
The coercion in Ockham's razor is revealed by a schizoid representation, presented via video, of a person who wishes to multiply himself in space and time as a visualization of his expanding desires and multiple identities.

Two cameras divide my body into two equal parts. The right side of the left camera captures the left part of my body. Therefore, the left part of my body appears on the right side of the screen of the left monitor. The left side of the right camera captures the right part of my body. Therefore, the right part of my body appears on the left side of the screen of the right monitor. Initially, my head is split in this way. Then I slowly crawl forward toward the two monitors, and my entire body comes into the cameras' field of view. Finally, I lie down with my arms outstretched, my body forming the shape of a cross, in front of the two monitors, which show me in two separate pieces (two half-duplicates). —Peter Weibel, 1973

Translated from the German by Patrick Hubenthal.

[1]
Editor's note: Although this phrasing is often attributed to Ockham, he does not appear to have ever used it himself; his preferred formulation was "Pluralitas non est ponenda sine necessitate" (plurality should not be posited without necessity). See Philotheus Boehner, introduction to *Philosophical Writings: A Selection*, by William of Ockham, ed. and trans. Philotheus Boehner (Indianapolis: Hackett, 1990), ix–li (specifically xxi). The version referring to "entities" first appears in later commentaries on Ockham and other scholastic philosophers; for one early occurrence, see John Duns Scotus, *Opera Omnia*, ed. Luke Wadding, vol. 2 (Lyon: Laurent Durand, 1639), 551.

Investigation of Identity III: Ockham's Razor

1973 | video performance; part 3 of 3

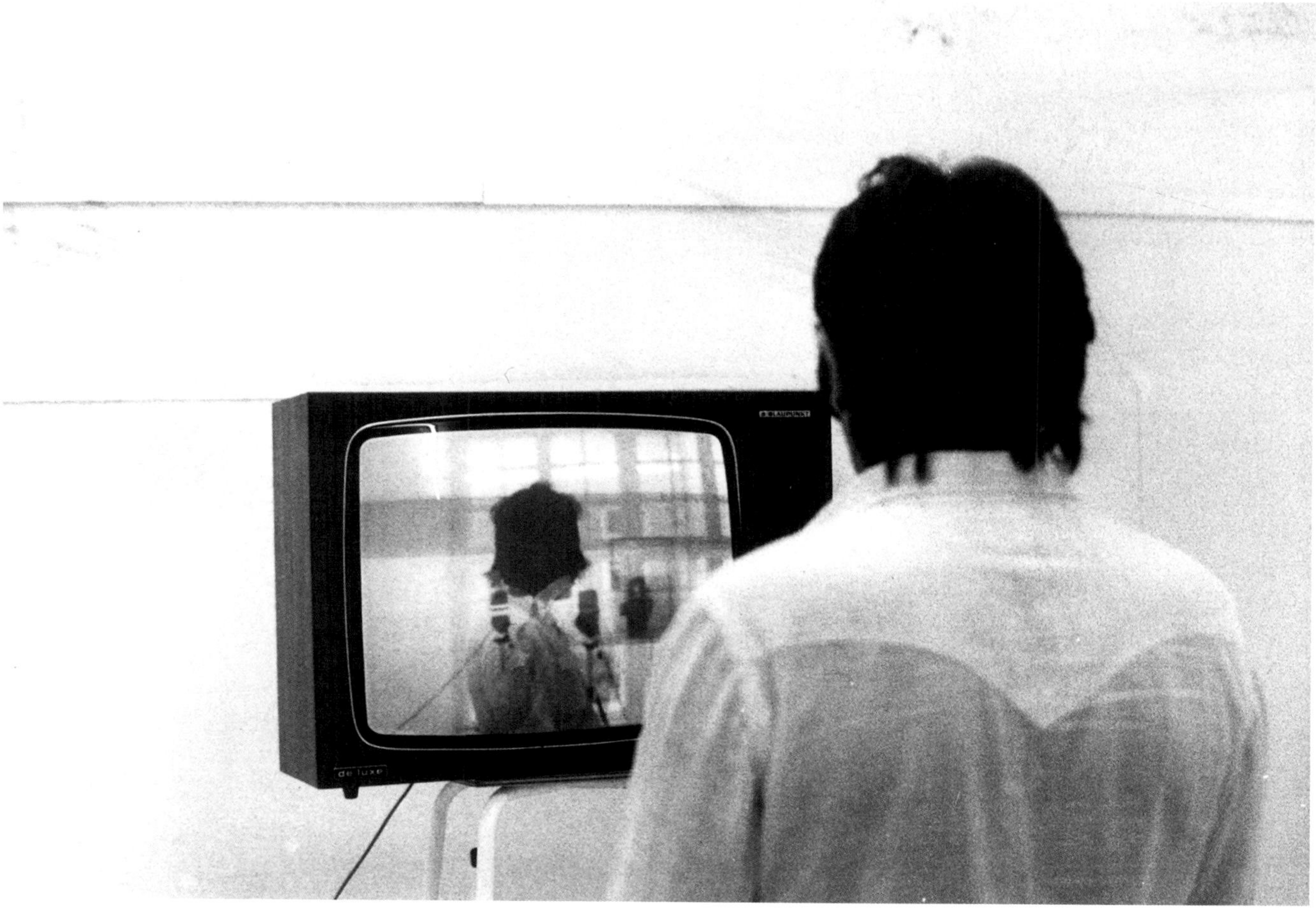

Studio recordings, Vienna, 1974

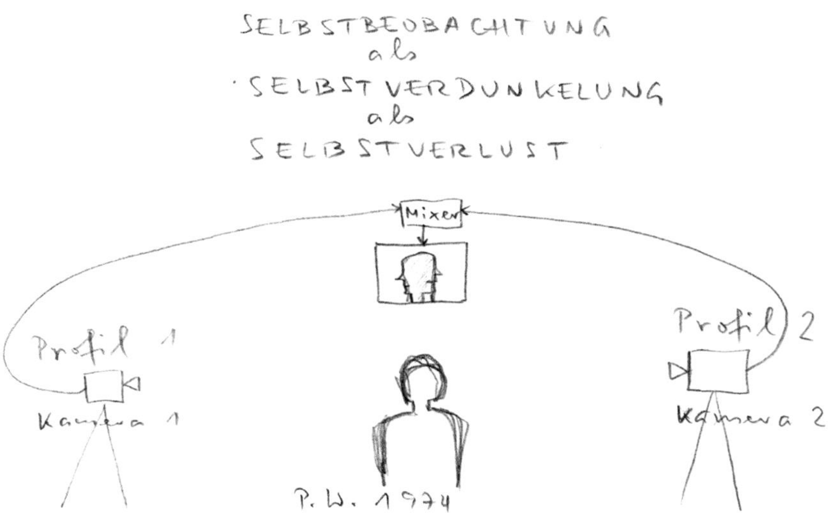

Conceptual drawing for *Self-Observation as Self-Obscuration as Self-Loss*, 1974

Self-Observation as Self-Obscuration as Self-Loss / Selbstbeobachtung als Selbstverdunkelung als Selbstverlust

1974 | video performance

The performer stands in front of a screen between two cameras facing in opposite directions. Because each camera transmits a reversed image of him to a mixer, the result is an erasure of information: the head is obscured; all that can be made out are two mirror-reversed profiles.

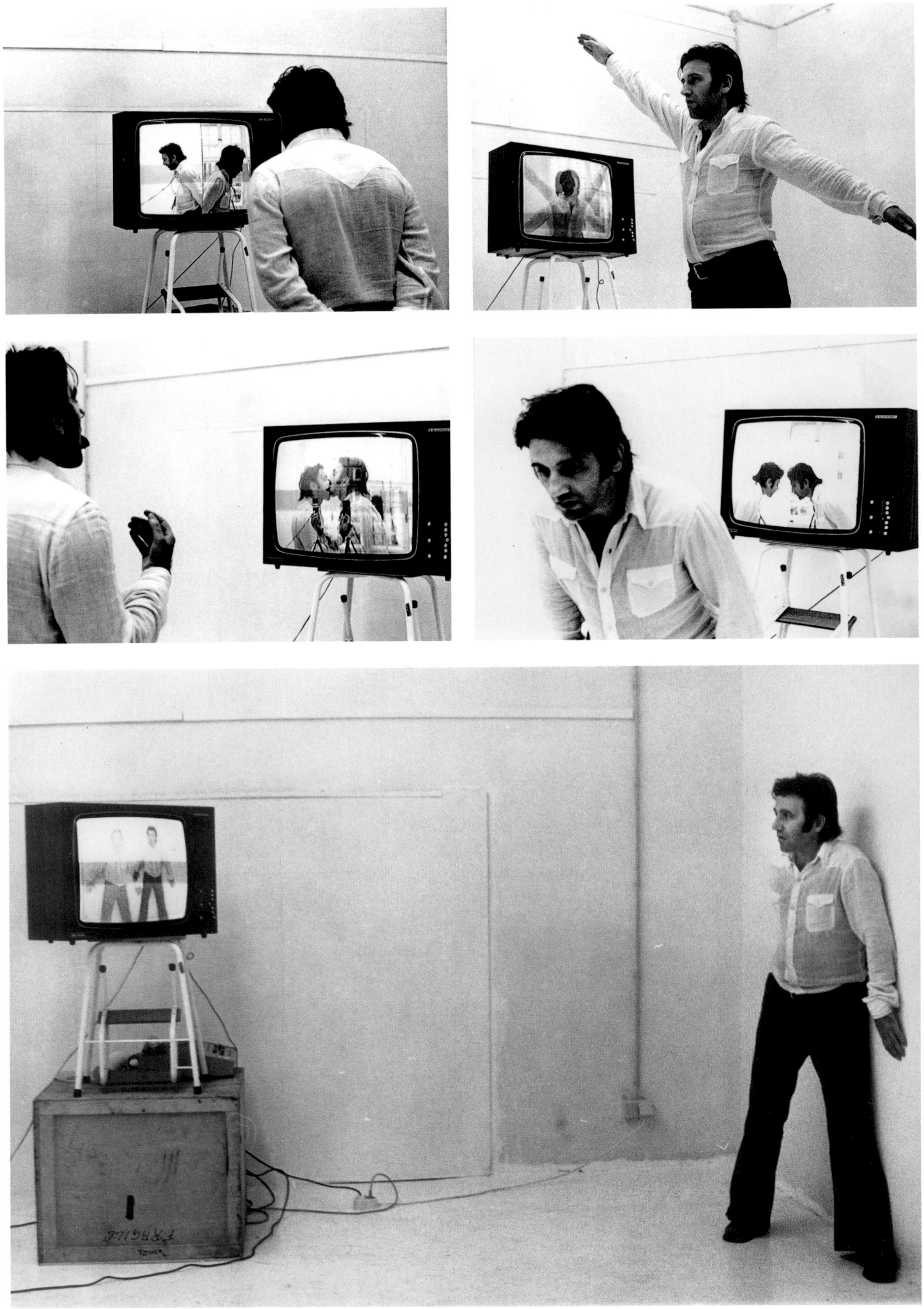

Studio recordings, Vienna, 1974

Doubles: Perspective Torsos (Screwed Together or Separate) of Sculptural Identity / Verdoppelungen: Perspektivische Torsi (verschraubt oder getrennt) der skulpturalen Identität

1974 | video performance

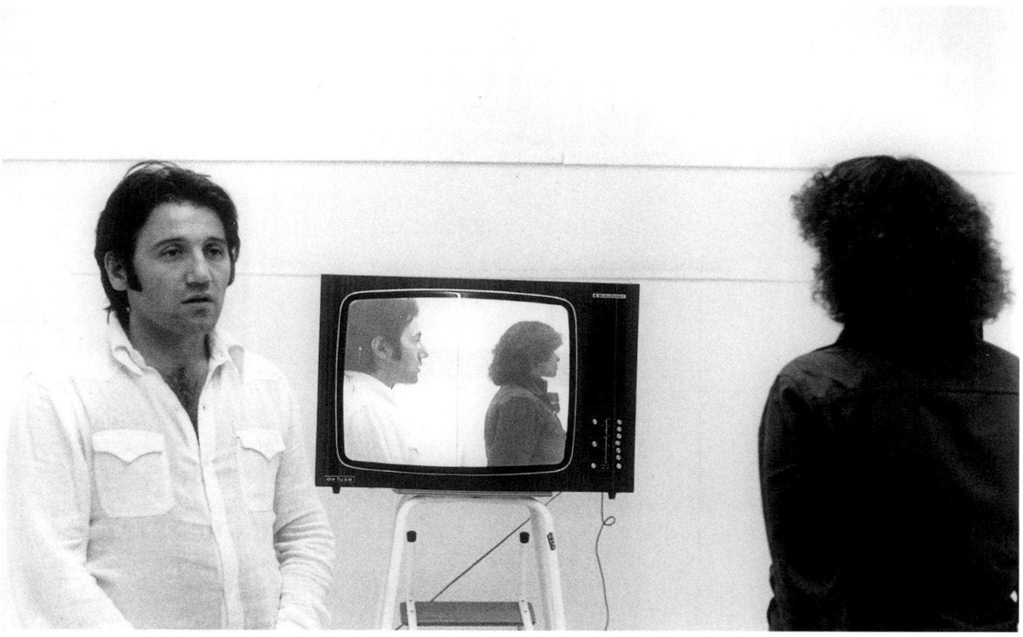

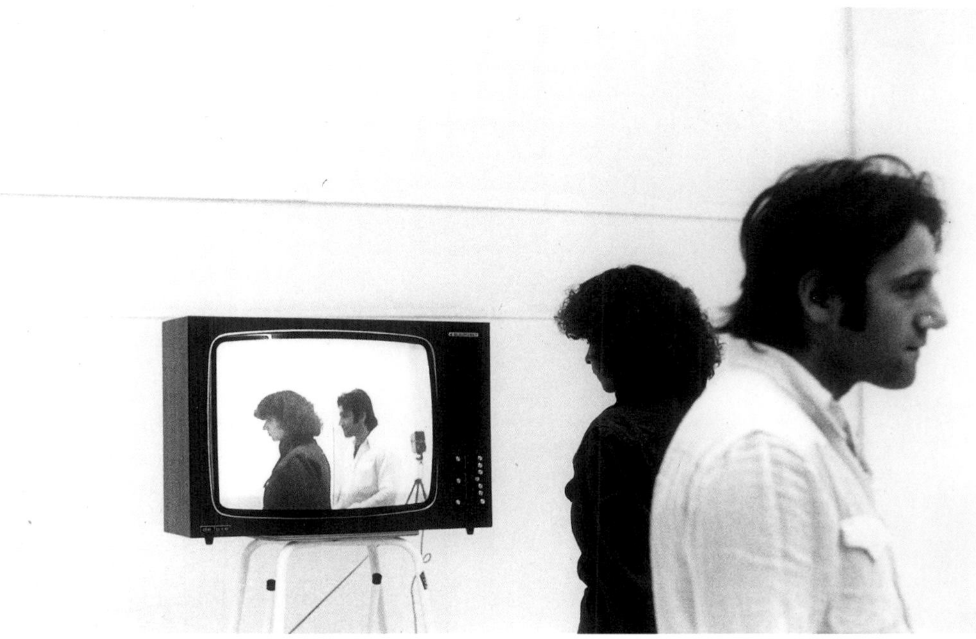

Studio recordings, Vienna, 1974

Two people walking forward towards each other, captured by two opposed cameras, and their images transmitted by a mixer appear on a monitor as advancing in parallel succession. Two people walking away from each other on an intersecting path, recorded by two opposed cameras and transmitted via a mixer to a single monitor, appear there as two people following each other. And so on. An anthology of paradoxical communications and encounters. People who think they are meeting are parting on an imaginary level. People who think they live in a shared world are actually living in separate worlds. The TV screen as a screen for consciousness, or the unconscious. Consciousness as an illusion of reality, or the unconscious as a revision of reality. Images of reality and inverted imaginary images. Conflicts and collisions, and their varying interpretations.

—Peter Weibel

Paradoxical Encounters / Paradoxe Begegnungen

1974 | video performance

Peter Weibel: The *Tele-Actions*

Hans Belting

The *Tele-Actions* (*Tele-Aktionen*), as Peter Weibel named a series of his works spanning a period from 1969 into the 1970s, embody a fascinating contradiction in their very concept.

Actions are of a direct and spontaneous nature. They are related to performance and generally take place in the presence of the public; in the case of art, usually in gallery spaces. Viennese Actionism, which was already past its heyday at the end of the 1960s, provides pertinent examples of this. As pieces created for radio, the *Text-Actions* (*Text-Aktionen*, 1967/68) also fall into this category within Peter Weibel's œuvre. However, the *Tele-Actions* were intended for television, where one expects programs to be broadcast in an often anonymous and in any case ritualistic way. From the outset, the nature of the medium places a distance between the program and its viewers, but at the same time the actions establish contact between the actors and their spectators. So, when the program was "interrupted" by an intruder, the television viewer must have perceived this as an act of aggression, unexpected within this context and not conforming to the rules. Inevitably, due to the high cost of airtime, there was to be only a short intermezzo each time a *Tele-Action* was aired (this was allowed only after some hesitation); an *entr'acte*, which interrupted the scheduled television program.

So far, however, only one aspect that played a role in the *Tele-Actions* has been cited. For Weibel, another aspect lay in conquering TV for art. In no way did this mean just showing art on television. It was far more a question of "making" art on TV, and thus mobilizing a new public arena for art. For a short while, it appeared as though artists could hope to get their hands on this medium. We don't need to mention any examples here. The *Tele-Actions* are different from other similar attempts in that they represent a critical analysis of the images on television and hence of television itself; they expose the "scene of the crime" by revealing the clichés commonly found there. They counteract the pseudo-reality of TV images by means of an intervention in which TV is revealed as a producer of illusions. The intention was to startle the viewers and make them aware of how little they usually got to see and just how prevalent stereotypes were in this repertoire of images.

For a while, it seemed as if a new era was dawning for images because of television. In April 1961, an issue of *Cahiers du cinéma*[1] was devoted to television, with the intention of examining television through film. However, in the same issue, Jean-Claude Bringuier (1925–2010) complained that the "mystique du direct" (mystique of the live) was already waning, since it was ultimately linked with the direct transmission of images to which drama was in the process of ceding its position.[2] As a concept, communication was a complete mockery since it had to occur intersubjectively whereas "transmission" followed other rules, as Régis Debray (b. 1940) mercilessly argued in his book *Transmettre* (1997). Its immediacy was an illusion, he claimed. Jacques Derrida (1930–2004) and Bernard Stiegler (1952–2020) deplored in their interviews that TV was about "reappropriation" and "deterritorializ[ation]" that alienated the viewers from their own reality.[3]

Peter Weibel's early works can already be classified from this perspective. They belong to the heroic years of an age of new ideas during which everything still seemed possible in art. However, the artist did not just adopt the position of criticizing a public medium, he also took the opportunity to bring forward his own questions: about the space in which identity is either established or broken down, and the location of images; a question which also invites us to look at the history of images that have become increasingly processual, and in becoming processual, they exert their power over the viewer. In some respects, television seemed to work against its own potential, which is why Peter Weibel's actions targeted a development, the freewheeling nature of which he wanted to denounce.

One would do well to distinguish between the *Tele-Actions* and the videos that Weibel produced during the same period. Not so much the techniques used, but rather the intention behind these works makes the difference. An artist's video may be shown on television, but it makes no reference to it. It is also shown in gallery spaces where it represents the seemingly self-referential presence of the artist. This stance was criticized by Rosalind Krauss (b. 1941) in 1976 in the first issue of the magazine *October*[4]; however, she also stated fundamental features of the genre at that time. She talked about a narcissistic aesthetic – just like when, for instance, Vito Acconci (1940–71) in 1971 pointed at the center of a monitor in his piece *Centers* and in so doing used it as a mirror. This point of view does not have to concern us here, though, as it is about a matter of discovering a different genre in the *Tele-Actions* – a genre directed at the TV audience that robbed them of the innocent way in which they watched their favorite medium.

A more innocuous experiment was the case of *TV News (TV Death 2)* (*TV-News (TV-Tod 2)*, idea 1970, first TV transmission 1972, see p. 250 in this volume) in which the TV set was exposed as being a real box in the viewer's room, which, as a box, could actually not be a place or space for transmitting images. A professional news anchor who was familiar to the public appeared to choke on the smoke from his cigar inside the tiny box as though he was actually sitting inside the device. Moreover, he was puffing on his cigar whilst reading the news, which was a new kind of taboo-breaking. In this rather violent parody, the images were not consistent with the place where viewers saw them. As television images, they escape the gaze that wants to know more about them. In another piece, the TV set metamorphosed into an aquarium in which the fish died when the water was let out (see p. 250 in this volume). Of course, this was an illusion, but it drew attention to the fact that illusions formed and still form the basis of television programs in any case.

Other pieces that function as *Tele-Actions* in the narrower sense go way beyond such demonstrations in their semantic spectrum and develop a discourse that is critical of the media; a discourse which is literally aimed at the subject of the transmission of images. At the top of the list among them is the piece *Depiction Is a Crime* (*Abbildung ist ein Verbrechen*, 1970, see p. 249 in this volume).

Of course, the title refers to the famous dictum *Ornament and Crime* (*Ornament und Verbrechen*, 1910) by Adolf Loos (1870–1933), which, however, in this case also refers to the freewheeling nature of images that we love to believe in. The opening credits include the text "A contribution to tele-communication," which can be understood not only ironically, but also in an enlightening way. We see the artist behind the viewfinder of a Polaroid camera. He takes a picture while the viewer, towards whom the camera appears to be directed, is in the dark as to who is actually being photographed; nor does the viewer realize who has filmed the man with the camera. While the camera takes 30 seconds to develop the photograph, the TV viewer has to sit and wait in front of a black screen on which there is nothing. Only then does the TV film crew come into view – accompanied by live sound, by the way. The roles of sender and receiver are reversed. The TV crew members are photographed while filming; indeed, photographed in a freeze frame, which negates the rolling film. Only the crew, though, and not the photographer were able to transmit the image on TV. Images produce images, and in this process the viewers are simply extras – that is, viewers remain viewers during an action that is not intended for them despite the fact that it is transmitted into their living rooms. Before their very eyes, an exchange of shots between two picture producers takes place on behalf of different media. So, where is the picture? And where am I, when I am in the picture? These are the questions viewers may well ask themselves, who doubt whether – contrary to their previous expectations – they really still are "in the picture."

The piece *Depiction Is a Crime* has been discussed in detail here because it demonstrates fundamental features of the *Tele-Actions*. On the one hand, they are actions, but on the other, they are even more, critical reflections on the transmission of images via the mass media; in their transmission, the images are locked in a closed circuit. They have no other reference point but themselves.

Whoever looks for the world behind them will find themself caught up in new images. Even the term "depiction" in the title of the piece can be analyzed (Weibel adopts a cunning strategy towards the words he chooses to use). In essence, the concept of depiction promises to refer to the object of the depiction, but the images either do not provide this reference, or they hide it amongst other images.

In *The Invisible Borders* (*Die unsichtbaren Grenzen*, 1970, see p. 450 in this volume), the customs area of Vienna airport has been chosen – if we may extract a single aspect of this piece – as a metaphor for an extraterritorial zone in which different laws apply, or, in order to stay with the argument, where the images themselves circulate in the living rooms of the consumers in their own, duty-free world. In the work *Intervals* (*Intervalle*, 1971, first aired on Hessischer Rundfunk (HR) in 1972, see p. 251 in this volume), the latest image production is mirrored backwards in the historic supply of images. Perception of space is thus offset against the experience of an image as it appears in an image represented in perspective. First of all, the artist set up an acoustic counterbalance. He filmed a sound generator that produced a sinusoidal or a calibration tone. The camera moved away in measurable intervals; as the distance increased, the volume eased off accordingly. Finally, only the sound of the apparatus itself was audible, while the reproduced sound disappeared into the distance. In one version for the ORF (Austrian Broadcasting Corporation), a monitor was used instead of a generator. It stands in a geometrically laid out avenue lined by high hedges in the park at Schönbrunn Palace. Even handled this way, the so-called "Alberti's window" of the Renaissance panel painting still reveals its rhetorical character. Initially, one can, as it seems, clearly distinguish the image in perspective on one's own screen from the image that gets smaller and smaller (and moves farther away) on the filmed monitor (an image within an image). However, when the filmed monitor finally disappears completely at the vanishing point, one still retains the image one saw the whole time on the screen – at which point the difference between nature and image collapses like a house of cards. All the observations one believed to have made about the image have, unnoticed, taken place within an image.

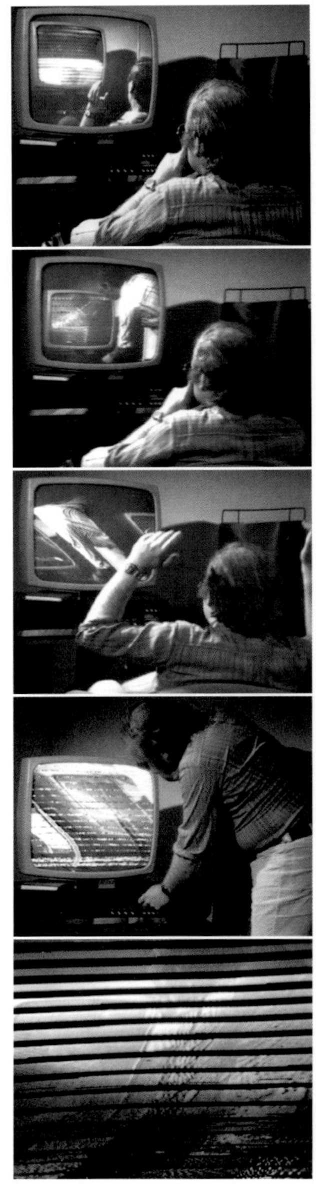

In the piece *The Endless Sandwich* (idea 1969, first TV transmission 1972) the mirror situation is projected right at the beginning in an informative inversion of the anonymous relationships that exist in TV consumption. At first it seems as though one is in a mirrored cabinet in which one's own reflection is repeated multiple times. But appearances are deceptive. The consumers cannot and should not find themselves on TV; they should only switch on their sets. The consumer remains anonymous in the invisible society of other TV users. As in the domino effect, in each case a different viewer is pushed further back into the picture and, at the same time, another viewer who is caught up in a new "picture" stands up in front of the TV in order to sort out some fault in the frequency (which one could still do at that time). Each time, another person does the same thing; neither gets to see the other. We only see rear views of people, who do come into view one after the other, but not in the program being broadcast: their game is limited to consumption and use. The stereotypical reaching for the button on the TV becomes a topic for reproduction itself; with this, Weibel allows the public into the image (but not into the TV program) in a paradoxical way – that is, in the act of behaving as consumers.

The artist, who is certainly not lost for words or explanations, demonstrates in his *Tele-Actions* that arguments can also be brought up in images when, instead of only consuming them on TV, one makes them an object of reflection. This "reflexiveness" was implemented here years before Rosalind Krauss demanded it from video artists.

Note: This text, here slightly edited, was first published in Peter Weibel and Harald Falckenberg, eds., *Peter Weibel: Das offene Werk 1964–1979*, exh. cat., Neue Galerie Graz, Universalmuseum Joanneum (Ostfildern: Hatje Cantz, 2006), 930–32. Translated from the German by Y'plus.

The Endless Sandwich deals with the relationship between TV and viewer. The viewers have to switch the television on and off and between channels. This action becomes the content of the program itself: one sees a line of people sitting in front of their TV sets. The last device develops a malfunction, so that the viewer has to stand up to fix it, which leads in turn to the malfunction of the next person's set. The disruption propagates and reaches the real TV set. The real viewers, too, must now stand up and fix the malfunction. The ORF changed the frequency for some moments which is why real TV viewers had to stand up and try to find the right channel. The real event is the end point of the reproduced process. Peter Weibel thus plays with the feedback and recursion of real action and image. The media world consists of a chain of *n* observers.The video premiered in September 1970 as a live performance at the Arts Lab during the *First International Underground Film Festival* in London.

The Endless Sandwich

1970/72	*Tele-Action* No. 1 video; b/w, sound, 2:03 min., first broadcast *Impulse 7*, ORF, June 26, 1972

[1]
Cahiers du cinéma: Revue mensuelle de cinéma, no. 118 (1961).

[2]
See Jean-Claude Bringuier, "Télévision sans frontières," *Cahiers du cinéma*, no. 118 (1961): 29–38 (specifically 29).

[3]
See Jacques Derrida and Bernard Stiegler, *Echographies of Television: Filmed Interviews*, trans. Jennifer Bajorek (Cambridge: Polity Press, 2006), 58 and 65. First published as *Échographies de la télévision: Entretiens filmés* (Paris: Éditions Galilée, 1996).

[4]
Rosalind Krauss, "Video: The Aesthetics of Narcissism," *October*, no. 1 (1976): 50–64.

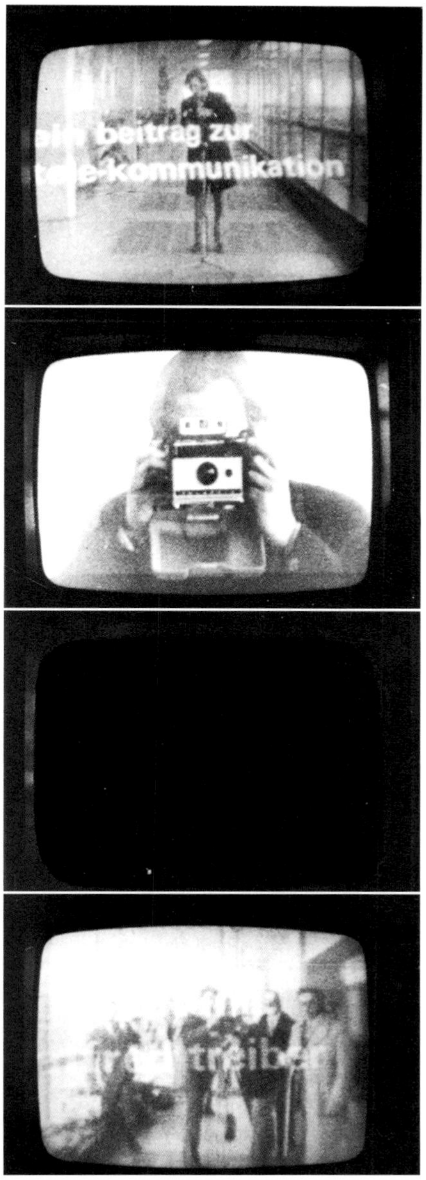

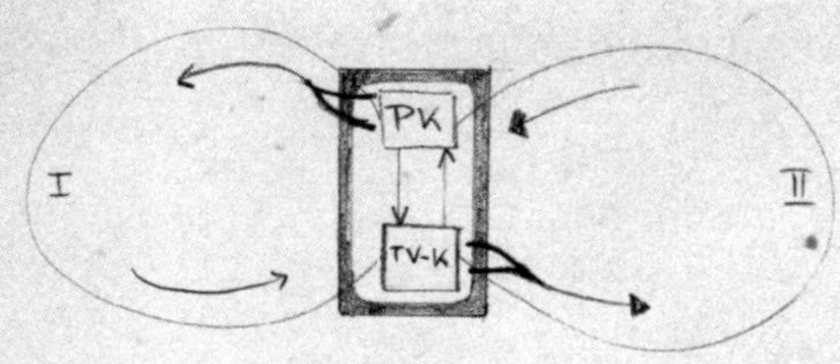

Jacques Lacan: "Human language constitutes a communication, in which the emitter receives his own message from the receiver in an inverted form."

The video work *Depiction Is a Crime*, produced in 1970 and first shown in 1971 by ORF (Austrian Broadcasting Corporation), shows Peter Weibel, who is holding a Polaroid camera, being filmed by a television team. The two actors operate simultaneously and facing each other; the television team films, and Peter Weibel takes a picture. The recordings are interdependent, as transmitter and receiver of a message. When the artist shoots a Polaroid, the television team's camera stops and the screen shows a black image for twenty seconds, because the time it takes for the photo to develop is transferred to the TV camera and the TV set. As long as the development time takes, there is a structural superimposition and the TV screen is black. The video continues the moment Peter Weibel pulls the photograph out of the camera. This photograph, depicting the television team, is then scaled full screen. This inversion of content and structural superposition allows the transmitters and recipients to get back their messages in reverse.

Depiction Is a Crime / Abbildung ist ein Verbrechen

1970/71	*Tele-Action* No. 3 video; b/w, sound, 2:19 min., first broadcast in *Kontakte*, ORF, February 1971

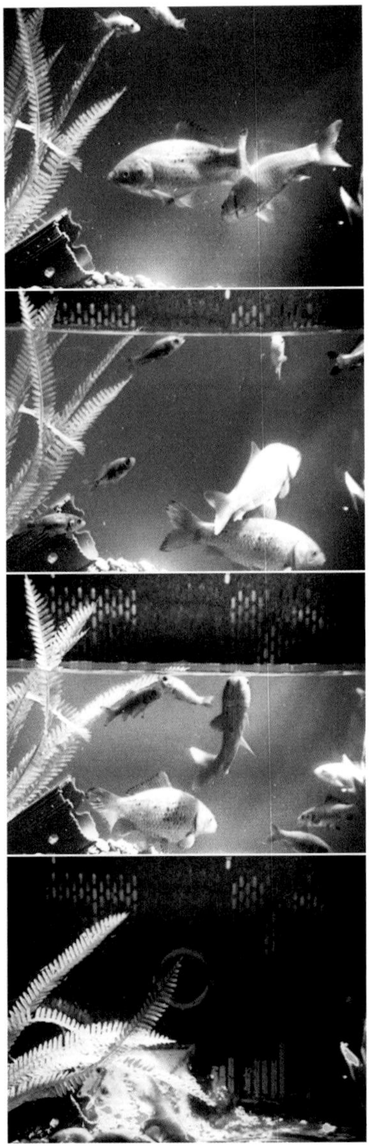

A TV set is a box one looks into just like one would look into an aquarium. When the television shows pictures of fish in an underwater landscape, the television transforms itself into an aquarium, a still life, an object of meditation.
TV Aquarium thus shows the identity of the real and represented event. After a few minutes, however, the water (seemingly) slowly runs out of the television. The water level sinks, the TV becomes empty, the picture tube appears. The fish, at first distributed over a large area and swimming calmly, move with the water level downwards and start to twitch visibly. A sinusoidal noise sounds as a warning, while the water almost disappears and the fish at the bottom of the monitor flap around wildly, fighting to survive. The fish seem to die in the TV set. Every day, people watch the news without feeling empathy for the horrific and deadly incidents they see as they seem to happen someplace far away: *tele* (far) – television. Here, however, death occurs in people's private homes, within their own TV sets. This makes it unbearable to watch. Therefore, many viewers protested. —Peter Weibel

TV Aquarium (TV Death 1) / TV-Aquarium (TV-Tod 1)

1970 (idea)/ 1972	*Tele-Action* No. 4a video, b/w, sound, 1:31 min., first broadcast in *Impulse 7*, ORF, June 29, 1972

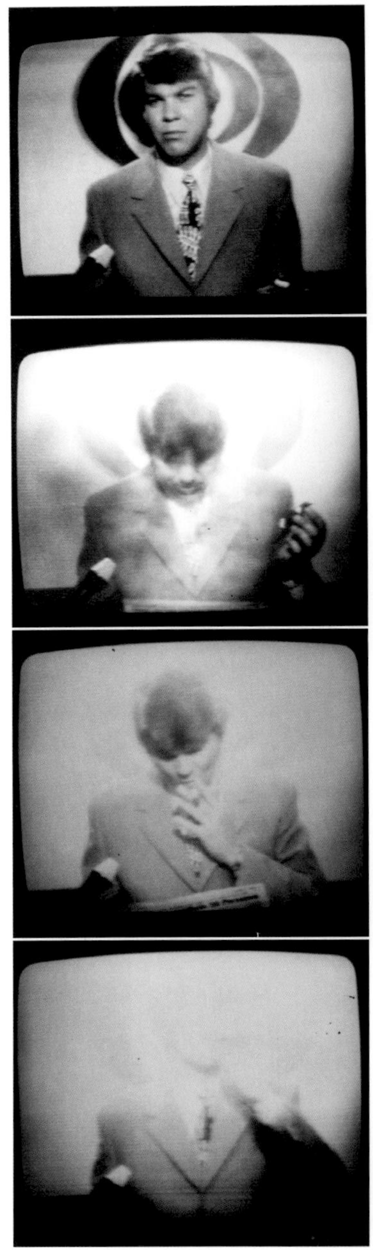

In *TV News* an actual Austrian news anchor reads the news, but something is unusual: he smokes. He sits in an invisible, transparent glass cabin, so that the smoke in the picture, in the usual TV environment, thickens more and more. This gives the impression that the television box is increasingly filled with smoke. Gradually the speaker begins to cough and stutter, especially during the weather forecast. He stops speaking the news when the TV is filled with smoke. The screen becomes impenetrably milky, the news anchor suffocates and disappears. The medium is taken literally: the deadly poison of the words, the intoxication with fake news cause the death of the speaking subject. A poisoning of communication as it happens daily: the deadly ecology of television. This creates a closed toxic feedback circle of word and deed, of information and action. Viewers protested about the smoking anchorman. They did not realize that they had seen news from one year ago. —Peter Weibel

TV News (TV Death 2) / TV-News (TV-Tod 2)

1970 (idea)/ 1972	*Tele-Action* No. 4b video, b/w, sound, 5:47 min., first broadcast in *Impulse 7*, ORF, June 29, 1972

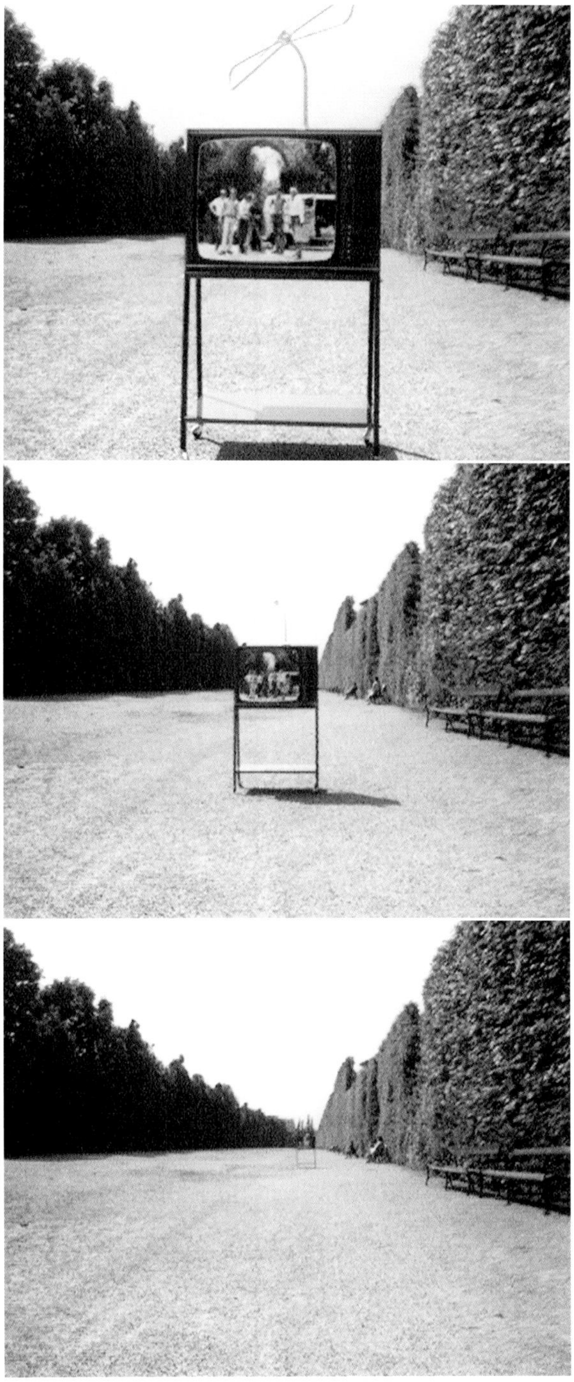

In the work *Intervals* the TV set, the image medium, is disappearing, but the immanent electric sound of the TV set appears. It mirrors the process backwards.

Intervals / Intervalle

1970/72 | *Tele-Action* No. 6
video, b/w, sound, 1 min., first broadcast *Impulse 7*, ORF, June 29, 1972

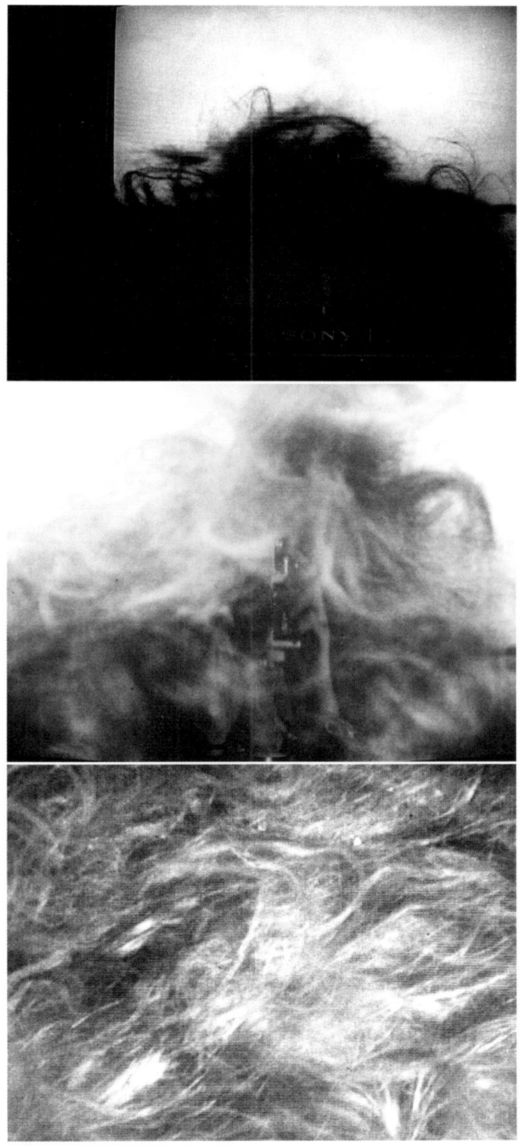

The TV is a box, the room is also a box. So there is a box in the box with the TV set. When small tufts of hair fall into the empty TV set, the illusionistic effect of the box in the box causes them to fall into an apartment, accompanied by gentle breathing. When the set is completely full, the roar of a lion sounds. The heap of hair becomes something threatening.

Solution of Fantasy / Lösung der Phantasie

1972 | video; b/w, sound, 2:18 min.

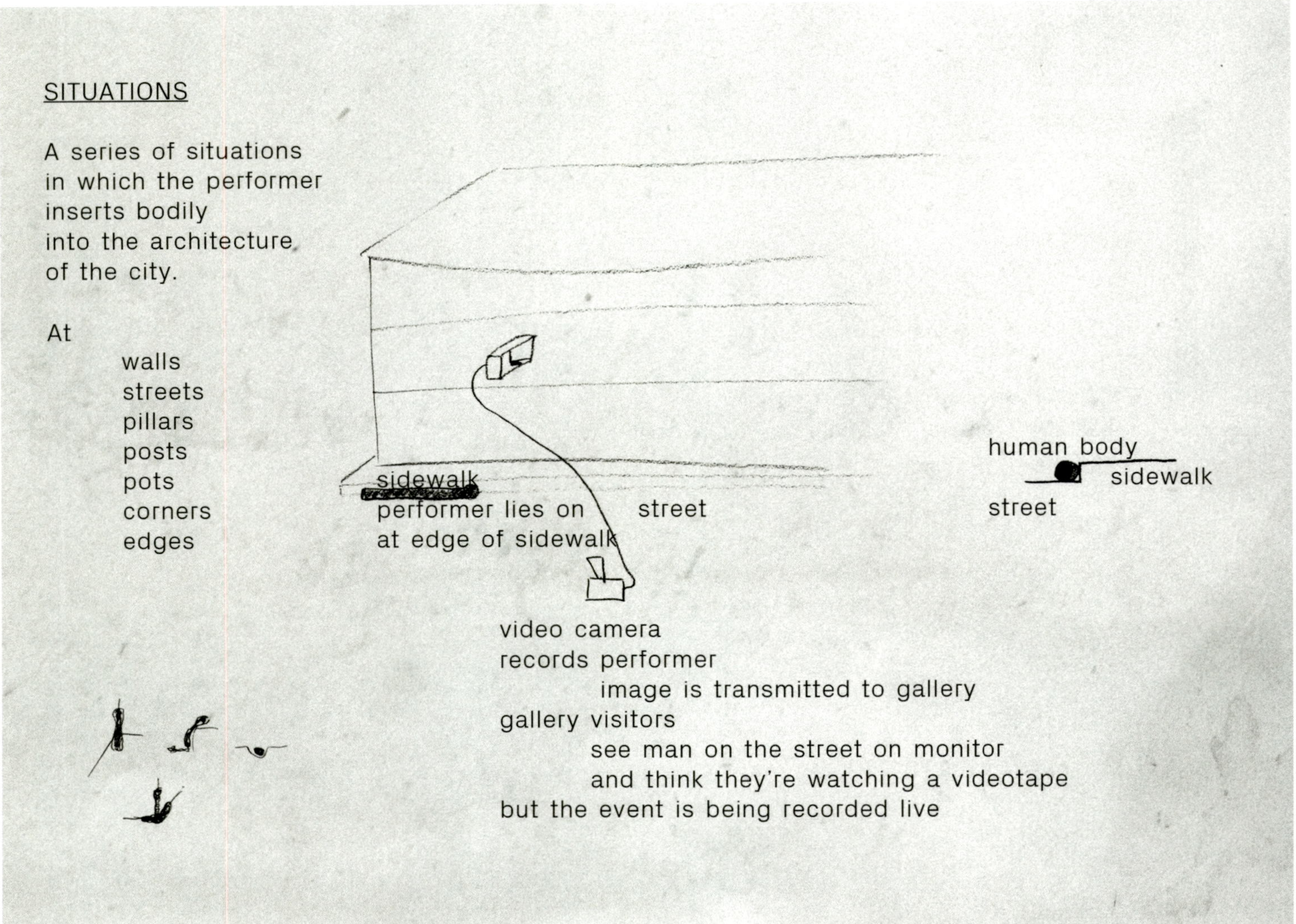

Conceptual drawing of *Situations*, 1970. Translated from the German by Patrick Hubenthal.

A TV set installed in the gallery shows a man lying in the gutter. The visitors believe that a tape is being played. They do not realize that the piece is a live transmission of the artist lying in the gutter at a nearby location. Perhaps, by some accident, a visitor sees the real situation after leaving the gallery. I liked the idea of the suspense (Is it live now or is it a tape?), or rather the fallacy of believing it to be a tape. A case that cannot be decided within the gallery situation! In Bonn I changed the concept slightly since it was a performance event. After the audience had watched the TV showing the man for about 10 minutes – it had already been switched on when they entered the room, and so it was impossible for them to know whether it was showing a tape or reality – the camera changed its angle and the audience could then see that the man was lying right in front of the building. The actor began to move and speak to the audience, passing through doors and corridors to enter the exhibition space where the audience was located. Finally, he switched off his own image, showing himself switching it off. —Peter Weibel

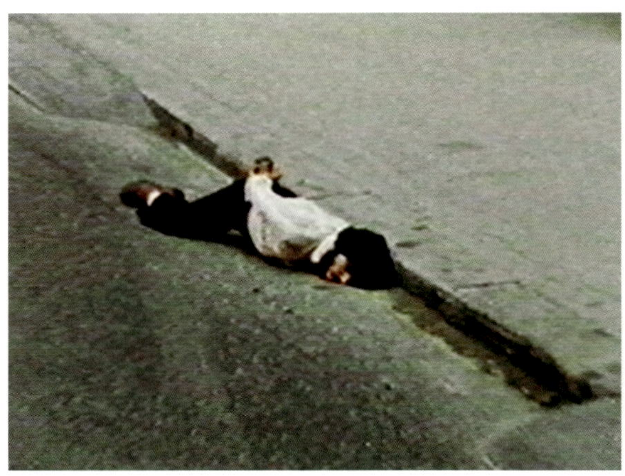

This idea was reused for VALIE EXPORT's series *Body Configurations* (1972–82) and in the film *Invisible Adversaries* (*Unsichtbare Gegner*, 1977, see pp. 134–35 and 202–203 in this volume) by VALIE EXPORT and Peter Weibel

Situation I

1971/74	video performance, closed-circuit installation; performed at Kulturforum, Bonn, November 15, 1974

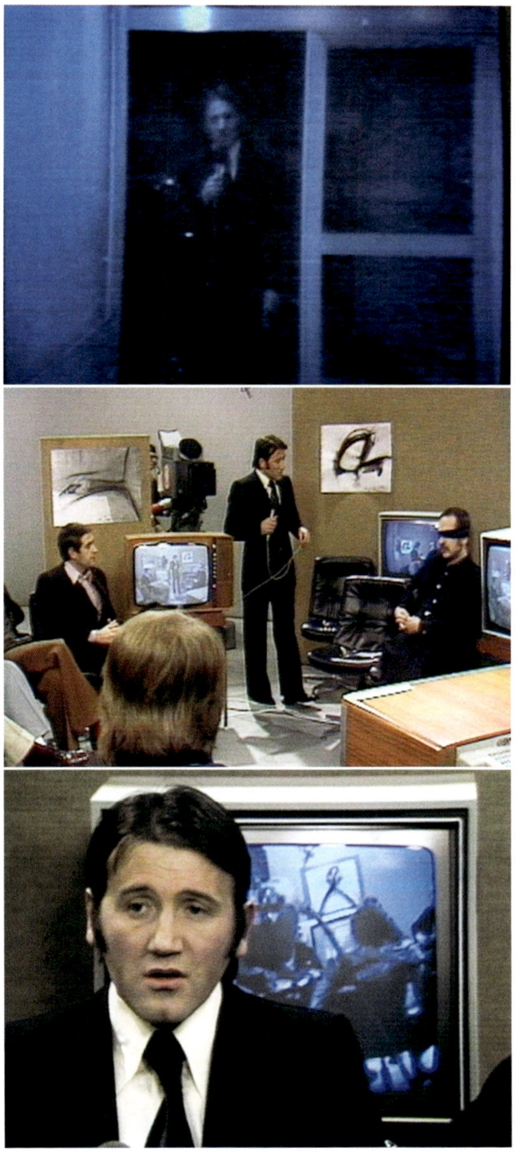

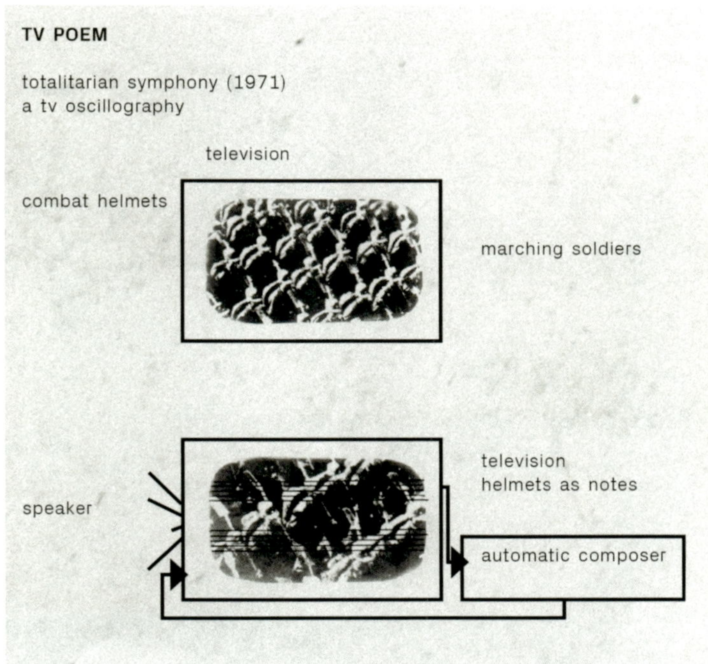

During a televised discussion, a TV set was placed behind my chair. At the beginning, the discussion moderator introduced the participants. When my turn came, I was not sitting on my chair, but on the screen of the TV set. I said hello to the audience and told them that I was located somewhere else in the building, but that it is the nature of video to be able to duplicate and multiply a person's appearance simultaneously, to be at different places at the same time, which no traditional visual art medium could achieve before. I explained that time played no part in the classical arts, but that video, as a typical nonclassical art, is a time-based art about dislocation. During this speech, I moved through the TV building toward the studio. People saw me moving on the monitor and sitting down on my chair while watching myself on the monitor sitting down on my chair. —Peter Weibel

Situation II: Simultaneity and Simulation / Situation II: Simultaneität und Simulation

1974	video; color, sound, ca. 10 min., live television transmission with Horst Gerhard Haberl, Richard Kriesche, Arnulf Rainer, and Peter Noever in *Impulse 42*, ORF, December 9, 1974

Recordings of soldiers on parade. The helmets march across the projection surface of the TV screen. The pattern of a musical staff is superimposed on it. The helmets become notes. The movement of the helmets along the lines and spaces of the staff is fed into an automatic composer to generate instant music.

—Peter Weibel, 1971

Translated from the German by Patrick Hubenthal

Totalitarian Symphony / Totalitäre Symphonie

1971	typescript; typewriter on paper, 2 b/w photographs, published in *Werkstatt Aspekt* 3 (1972)

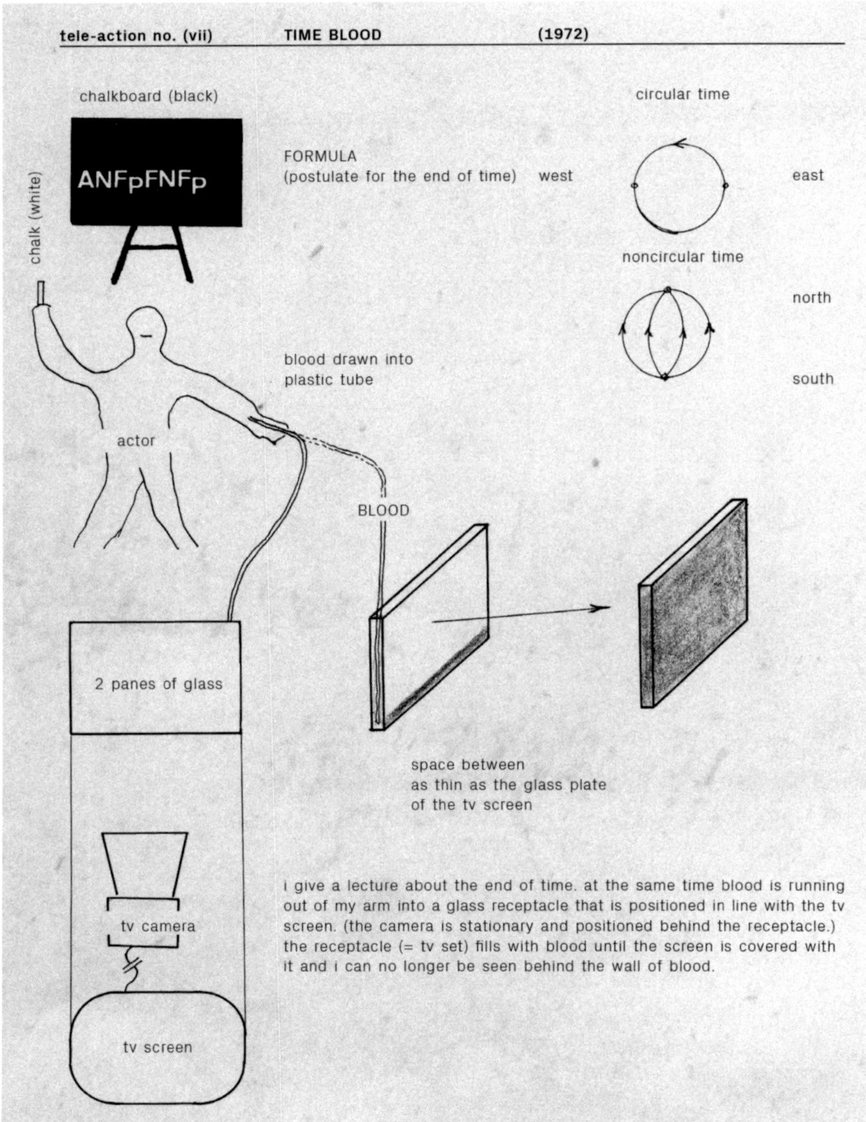

Conceptual drawing of *Time Blood*, published in *Werkstatt Aspekt* 3, Vienna 1972. Translated from the German by Patrick Hubenthal.

This *Tele-Action* is my response to the question posed by *Club 2*: What is art for, and who pays for it? The aim of art, just like the aim of technology, is to make nature human. Of course, the original Greek word for art, *tekhnē*, means "artificial," "made by man" as opposed to "made by nature." Art is thus part of the history of man's self-representation and self-realization. Now if, in the course of this telecommunication, spanning great distances as it does, my blood runs into Austria's households, or rather down the TV screens of Austria's households, it is not by chance nor purely symbolic that this evokes a very famous so-called Holy Communion, in which blood was once shed for mankind. Rather, the aim of this performance is in fact to reveal the nature of art, by which I mean that in art, something is brought to light that we also call childhood, or home – that is, the place where man finds his center. As a consequence, the history of art is naturally different from the history of economics and the history of politics, for the simple reason that art is a critique of prevailing conditions and, in its utopian function, a manifestation of that which man desires: his yearnings, his ideas, his structures. This means that art is part of the avant-garde of human history. And that is the reason for the yawning gap between politics and art, on the one hand, and between art and the public on the other, and that is why we have these deep-rooted frictions, these clashes, this climate of hostility that we struggle to overcome with subsidy programs, and that we only do overcome in very rare cases, because it is simply part of the nature of art to be the avant-garde of human history.

Note: lecture given by Peter Weibel during the performance 1979. Translated from the German by Patrick Hubenthal.

Time Blood / Zeitblut

1972/79	*Tele-Action* No. 7 video, color, sound, 4:04 min., first broadcast in *Club 2*, ORF, March 8, 1979

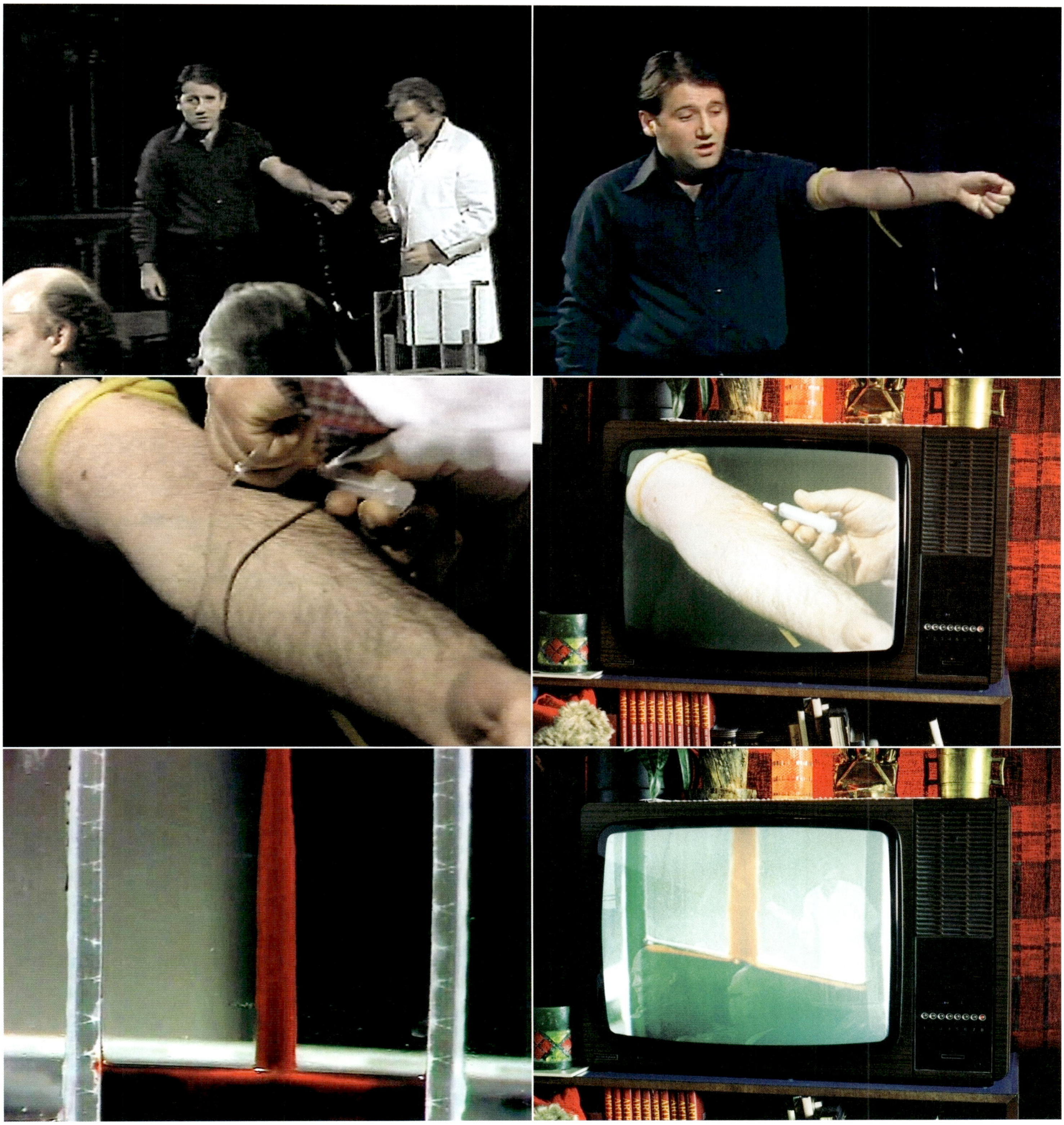

The *Tele-Action Time Blood* was first broadcast on the Austrian television program *Club 2* in 1979. Peter Weibel is seen while a venous cannula is inserted. The blood flows from his arm into a flat glass vessel; this is filmed by the TV camera and broadcast in real time. As the vessel slowly fills with blood, Weibel gives a lecture on time logic and the end of time, standing in front of a blackboard on which mathematical formulas are written. During this, the glass vessel is filmed so close that the television screens in Austrian homes appear to be filling with blood. The performance ends when the lecturer can no longer be seen behind the wall of blood. Blood running over the Austrian TV screens – an example of telecommunication and *Tele-Action* demonstrate: all technology is tele-technology (remote technology), and all tele-technology is theo-technology. Sacred communication rituals for overcoming space and time are secularized as social technology.

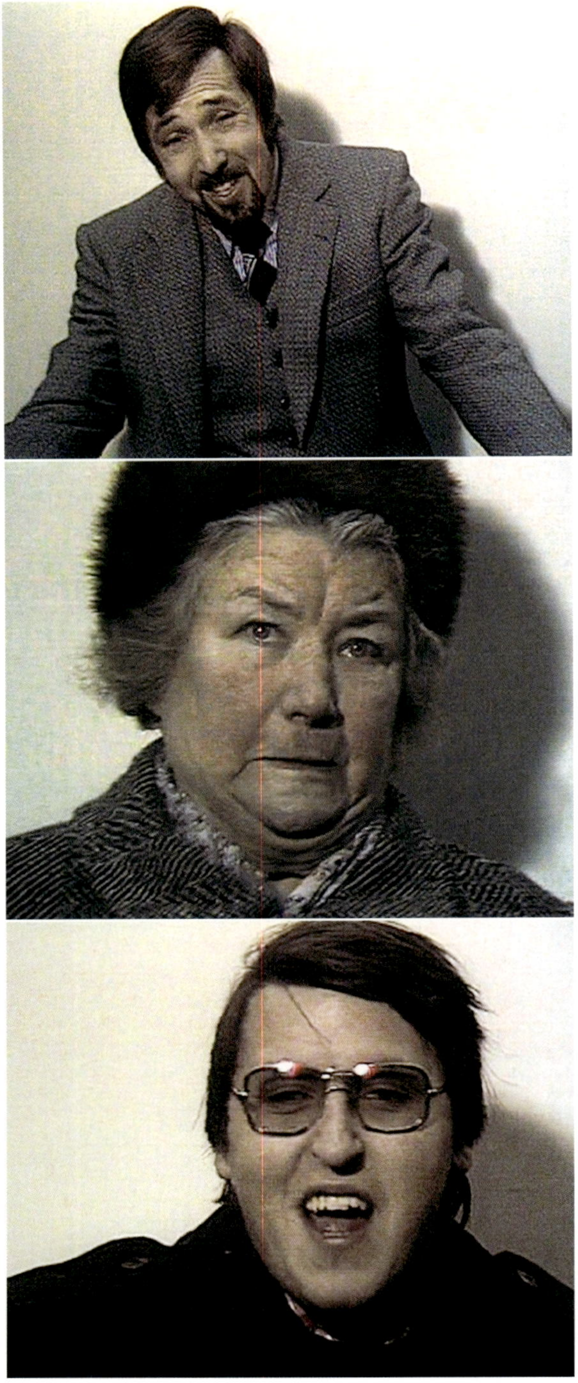

Running Gags / Witze am laufenden Band

1974 | TV comedy; video, color, sound, 12:30 min.

It has become commonplace for all television people, who produce a feature or documentary with normal people, to be paid, but the normal people themselves are not paid, for example, the television interviewer is given money, but not the person interviewed for the work he is doing. I wanted to do something so that the people were paid. I was also always interested in the question as to the way in which jokes are created and circulate, how it is possible to hear the same joke within a few days and in completely different places. I proposed to those responsible in television that they produce a weekly show with running gags: A television team was to travel through Austria, and anyone who tells a joke for one hundred Austrian Shillings should be given the possibility to see himself or herself in the show. The audience should produce the content of the show and not just consume a show. I made a first attempt and put an advert in a Vienna newspaper so that anyone who wished to tell a joke on television for a fee should go to a certain place, Am Graben, at the corner of Trattnerhof, Vienna 1. The next day the place was astonishingly overfull, with hundreds of people. I filmed as much material as was possible. The material was really good. The jokes were mostly political, exposing the critique of the political class in a disguised way. But the television management suddenly no longer wished to do this show. So, I am using a little of this material here, in the program *Impulse 3*. —Peter Weibel

Comxxxicxxxxx / Komxxxikxxxxx

1974 | compilation of TV actions and TV comedies; video, color, sound, 40 min., broadcast in *Impulse 3*, ORF, February 25, 1974

If you replace several letters in the word "communication" with x, then it becomes possible to recognize the word "comic" in it. This is what I had in mind: the idea, that comedy emerges in interrupted communication. Mistakes and misunderstandings in communication induce laughter in us, which is why the idea of a wrong title appealed to me, an incorrectly written title in the TV journals. —Peter Weibel

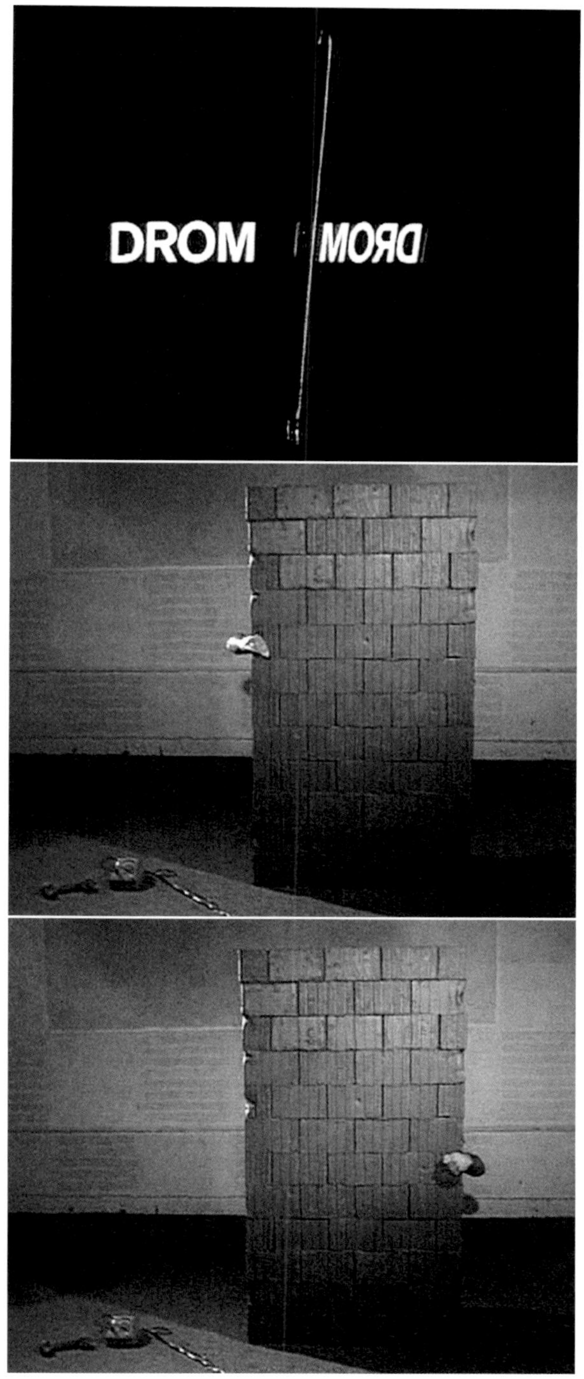

The word "monodrome" stems from the Greek and means "one-way street." This is TV: a one-way communication system, a monodrome. What effect does it have? The mirror image of the word "drom" is "mord" and means in German "murder." In a newspaper advert I invited people to treat me like some kind of material. They could contact me under a certain telephone number and I would enact anything they told me to do. I would fulfill every wish, which was possible for me to perform with my body in a room. Originally, I had planned to do this work in a gallery since 1970, but had never had the opportunity to do so. Those people, who had become aware of the event through the newspaper advertisement, could call and issue with me commands as to what I should do. A number of people called since they were thus able to remain anonymous. But I didn't really enact their wishes since I acted without any real monitoring, and without simultaneity it was senseless. I selected the interesting wishes. Finally, the TV viewers saw a screen of clay, a wall and after every wish that came through the telephone next to the wall, one saw a hand, which knocked on the wall. The wall functioned like a mirror. So, the instructions for me became instructions for the audience by the audience. When, finally, one woman wished that I disappear from the culture scene, the entire wall vanished. —Peter Weibel

Monodrome / Monodrom

1970/74	TV action; video, color, sound, 5:28 min., broadcast in *Impulse* 3, ORF, February 25, 1974

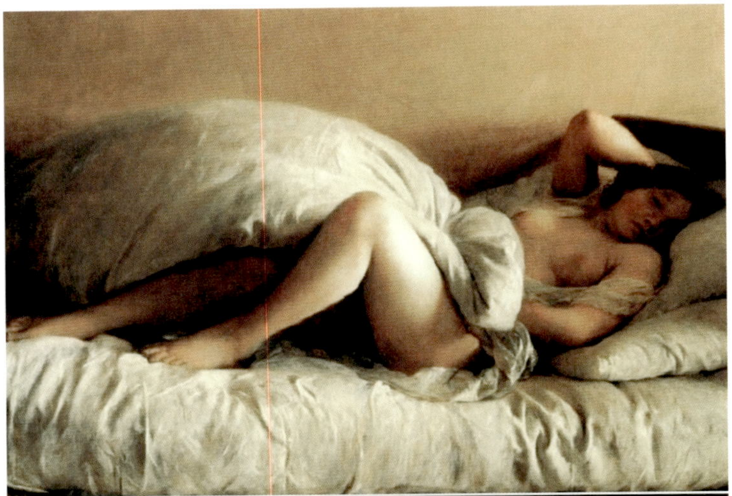

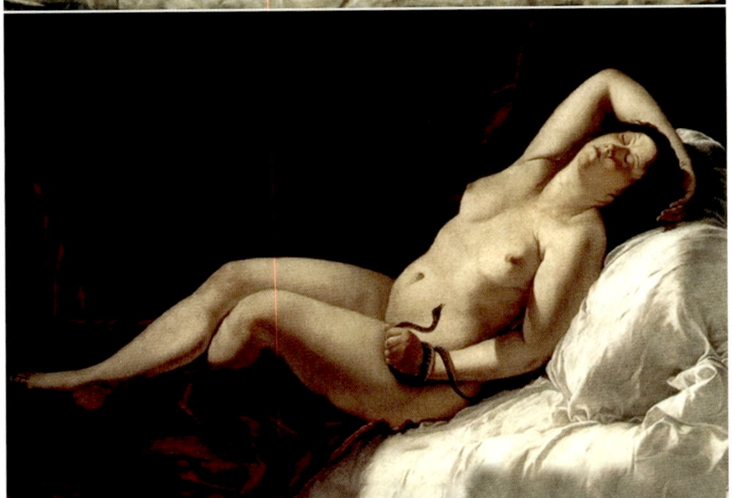

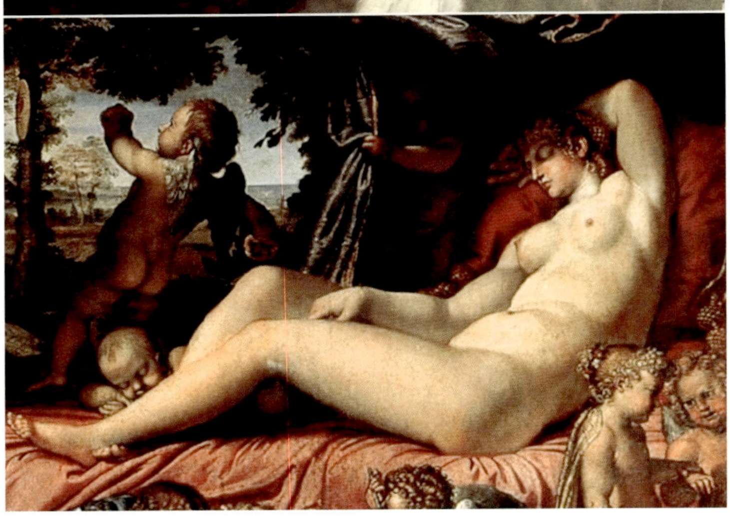

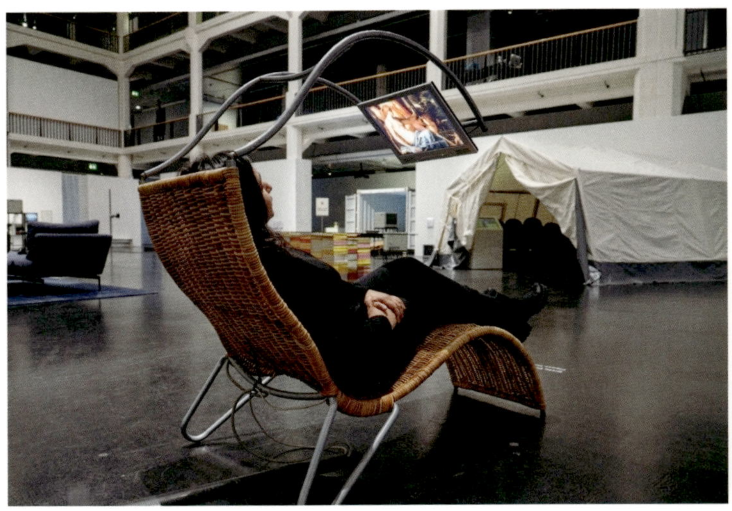

Venus in Furs displayed on a screen of Peter Weibel's *Media Lounger*, *respectively*, *Peter Weibel*, ZKM | Karlsruhe, 2019–20

Venus in Furs / Venus im Pelz

2003	video; color, sound, 4:27 min.

The history of painting is, to a great extent, the history of portraits. The history of portraits is, to a great extent, the history of naked women. The male gaze has created its own genre of pleasure for the painter and the viewer, disguised as art history. I have assembled some of the most famous female nudes and used the morphing techniques in order to pull back and debunk not only the curtain of desire, but at the same time also the hidden grace of the frozen body gestures and movements.
—Peter Weibel, 2023

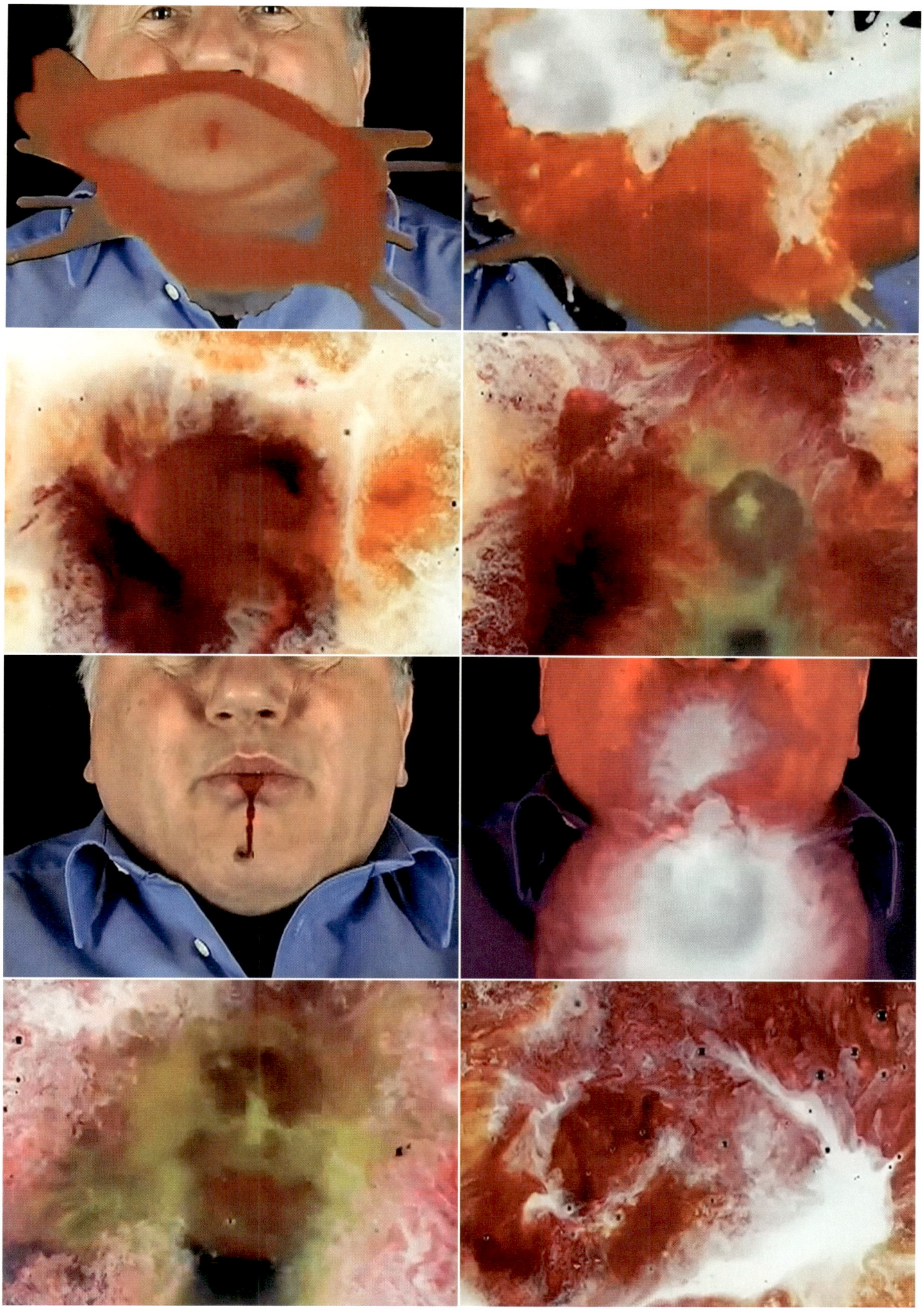

Spit Painting

2009 | video; color, sound, 7:01 min.

Part of the mythology surrounding the medium of painting is, first, the autonomy of color and its movements, present especially in informal and abstract expressionism, technically ranging from dripping to shooting. Second, as the word expression already says, a relation between inner state and external form is claimed. Therefore, I take the liquid color into my mouth, keep it inside my body for a while, and then spit it onto a glass plate placed in front of the video camera. The autonomy of the color's movements on the glass plate creates a color field and shapes the abstract form and multicolored patterns. A parody?

Video as Spatial Art

Peter Weibel

In my text "On the Philosophy of VT & RT" (see p. 231 in this volume), I defined codifiable and realizable attributes as specific videological constants, that is, as characteristics that are only codifiable and realizable by the means of a video system: synthetics, transformation (metamorphosis), self-reference, instant time, and box character. Due to technical superiority, it was initially in the USA that the first two of these attributes were extensively used. The dissolution of concrete forms into abstract ones, the transformation of images and signs, the distortion of form and color by video synthesizer and colorizer provide an easy field for so-called spontaneous creativity. What is remarkable in this connection is that research on color by means of video technology, which would be on a par with, or comparable to, the legacy of a Delacroix, van Gogh, Manet, Runge, Friedrich, Malevich, Mondrian, Albers, and indeed, even with color field painting, is still a long time coming. As the expression "colorizer" reveals – and one may recall here with what degree of indignation every painter of color shunned the designation of being a colorist – it colors a graphic form. To date, the practice has not been to paint with the cathode tube – which as we know from Nam June Paik was intended to replace the well-known claim of the brush – but rather to draw, such that one may say of a painter who does not arrive at form through color and surface, that he draws with the brush. The picture as sign form dominates and thus covertly (in spite of all synthetically generated alienation) is a popular, copy true perception of reality.

In other words, in spite of all advanced technology, the aesthetics of the color video work did not even come close to the achievements of Romanticism, (color as expressive value of a personal symbolic language etc.) to say nothing of a Paul Cézanne or Mark Rothko, as long as the surface, and with it the form, will not be composed and realized from the color. Only a real color form will also in video art deepen and intensify an autonomous development of the medium. It is precisely by means of its temporal form, which is suitable for optic-physiological effects, such as successive contrast, complementary contrast, etc., that video will be in a position to supply valuable bases for the absolutization of color, and for the liberation of objects. Van Gogh prophesized that "the painter of the future will be a painter of color"[1] and that future painting will be "more music and less sculpture."[2] Among other things, what may also be inferred from this dominance, is that artists primarily interested in video as temporal art and whose conception of time is inspired by music especially like to work with video synthesizer and colorizer: because van Gogh's prophecy ended at the moment at which, through the emancipation of color, form also separated from the natural object. Specifically, the independence of form and color (the definition of the painting as colored surface) liberated painting from illusionary space (with the help of graphic forms, such as perspective etc.); and so, because real forms have only been available since real colors, only since the autonomy of color, since the dissolution of illusionist space associated with this, does pictorial space exist. An especially characteristic phase here is the shaped canvas painting (Frank Stella, for example), where the ambiguity of picture and object (object-bound painting – painted object) represents the transition to those sculptures of minimal art, and primary structures may be considered as sculptures derived from painting: from the colored surface to real space. An approach to space also extends from color.

The two attributes of the video system appear to have created considerable confusion, which, at a cursory glance, alone contribute to the temporal character of video, namely, to self-reference (feedback) and instant time, though which are also important for video as spatial art. In other words, they establish among other things, that *novum* in art history whereby the viewer becomes part of the work of art. In short, by means of the incorporation of the viewer, Lessing's distinction between the arts in time and spatial art is cancelled out in video art, or more precisely, in video art as spatial art (video installations, environments, sculptures)!

Centuries-old experience of space has been very well expressed in the formulation: "architecture has to do with the design of space, sculpture with the attitude towards space, while in painting with the unfolding of artistic means."[3] What catches the eye is the similarity to Peirce's semiotic model, whereby "something functions as the sign of another and something else for a third thing."[4] Here also emerge the notions gestalt, attitude, and unfolding. In short, the sign realizes (as medium) gestalt, communicates (for a subject) an attitude, and unfolds (from an object) the meaning. In other words: architecture is the syntax of space, sculpture is the pragmatics, while painting is the semantics of space. This semiotic definition of space allows/enables us to understand the role of the body of the viewer in video art as new spatial art.

In classical architecture, only interior space counted as real space. The exterior space of a building did not, so to speak, exist, since, on the outside, the building appeared like "a body on the earth," that is, as a closed sculpture. Today, spatial planning also means something like interior architecture or, in less stronger terms, the design of interior rooms according to bodies and as body. It is not for nothing that Le Corbusier's *Modulor* was proportioned according to the human figure. Architecture knows no exterior space, but at most an exterior building. Hence, it is clear that video artists, such as Dan Graham, VALIE EXPORT, and myself have concerned themselves with architectonic video spaces, which especially take up the problem of interior space and exterior space (predominantly in the form of reversion).

Just how much classical sculpture lives from the representation of the body is something which is known by the educated layperson and the conceited specialist alike, and even in the case of more abstract sculptors, such as Henry Moore, one still notices the composition of real bodies (instead of the design of actual space) shimmering through, even if protruding in the bones, mollusks etc. of the human body. Hence, in the case of several video artists the traditional depiction of space breaks through as body. The spiritual space of a Henri Matisse, the simultaneous space of the cubists, the spatialized time of the futurists, the sense of time of monochrome painting or, in terms of dialectic the jinxed, enigmatic space of the surrealists, especially René Magritte's, are nevertheless examples of a modern conception of time in painting, which seeks to detach itself from spatial representation by the body – and which hence are or could be binding for video works. Painting has the capacity to design pictures of space. An arrangement of video equipment has the capacity to be a three-dimensional structure, namely, a (bodily) sculpture and, at the same time, to design pictures of space like a painting on its "screens." In special cases, the video sculpture or a part of it can even be illustrated on the screen. As a new kind of space machine, video represents an interface of sculpture and painting. It unfolds space through the depiction of it and, as sculpture, determines the relationship to space. As in all areas of art, so also in the case of video, advanced research operates in the area of semantics and pragmatics, in their blurring and intermingling.

The monitors or video screens can also show the viewer as picture, and can represent the body of the viewer and their attitude. It is precisely through this that video becomes the essential instrument for the pragmatics of space. The relationship of the viewer, now not merely represented as body, but incorporated, introduces that sculptural process which mediates a new relationship to space, the range of which is capable of covering physiological (in the case of time lags) through to spiritual nature. Due to the inherent characteristic of the medium video – simultaneity – for the first time in history, it is possible for the viewer, captured by the camera, to observe themself on the monitor in the act of observing. The external observer, hitherto always outside the image, becomes an internal observer who is inside of the image they are viewing. They become a part of the very system they are watching. The image of the viewer in the video screen led some theorists to prematurely define video as a mirroring machine. If I may be permitted, I would like to point out that those works (for example, the installations by Peter Campus) just mentioned concern new spatial artistic articulations, which convey perceptions of space that overcome the space of castes, the hierarchical space of traditional painting and sculpture. By the chimerical images on video screens (through large projectors) to the differentiated multi-images of numerous monitors, an artistic space is designed, which essentially differentiates the space of perceptual reality and only thereby becomes real (artistic) space. Both the box character as place-bound (the oldest idea of space), like the pictorial character of video, make video suitable for the sculptural process. Due to these affinities, from video to painting and sculpture, numerous fine artists work with video (installations etc.), so as to realize new experiences of space beyond sculpture and painting in the customary sense. This new experience of space is essentially connected to a new experience of time, whereby, hopefully, recourse to physics, which talks of a spacetime continuum (and no longer of isolated concepts of space or time), is not required for the defense of video as space-time machine.

One may thus surmise that in the coming years developments will shift away from the production of videotapes and pass over to the construction of spatial, architectonic, sculptural video solutions because here, for a moment, one supposes the secret of video and its art.

Note: Peter Weibel, "Video als Raumkunst," in *Video*, exhib. cat. (Geneva: Association Musée d'Art Moderne / Basel: Galerie Stampa, 1977), 30–33. This English translation is taken from Woody Vasulka and Peter Weibel, eds., *Buffalo Heads: Media Study, Media Practice, Media Pioneers. 1973–1990*, exh. cat. ZKM | Karlsruhe (Cambridge, MA: MIT Press, 2008), 788–89.

[1]
Vincent van Gogh to Theo van Gogh, May 5, 1888, in Vincent van Gogh, *Sämtliche Briefe*, vol. 4, *An den Bruder Theo* (Bornheim Merten: Lamuv, 1984), 39–41 (quotation on 40).

[2]
Vincent van Gogh to Theo (1988), 1984, 133.

[3]
Kurt Badt, "Raumphantasien und Raumillusionen" (1960), in *Raumphantasien und Raumillusionen* (Cologne: DuMont, 1963), 7–130 (quotation on 11).

[4]
"A Sign, or *Representamen*, is a First which stands in a genuine triadic relation to a Second, called its object, as to be capable of determining a Third, called its Interpretand." Charles Sanders Peirce, *Collected Papers*, vol. 2 (Cambridge, MA: Harvard University Press, 1931–1958), 274. (quotation on 11).

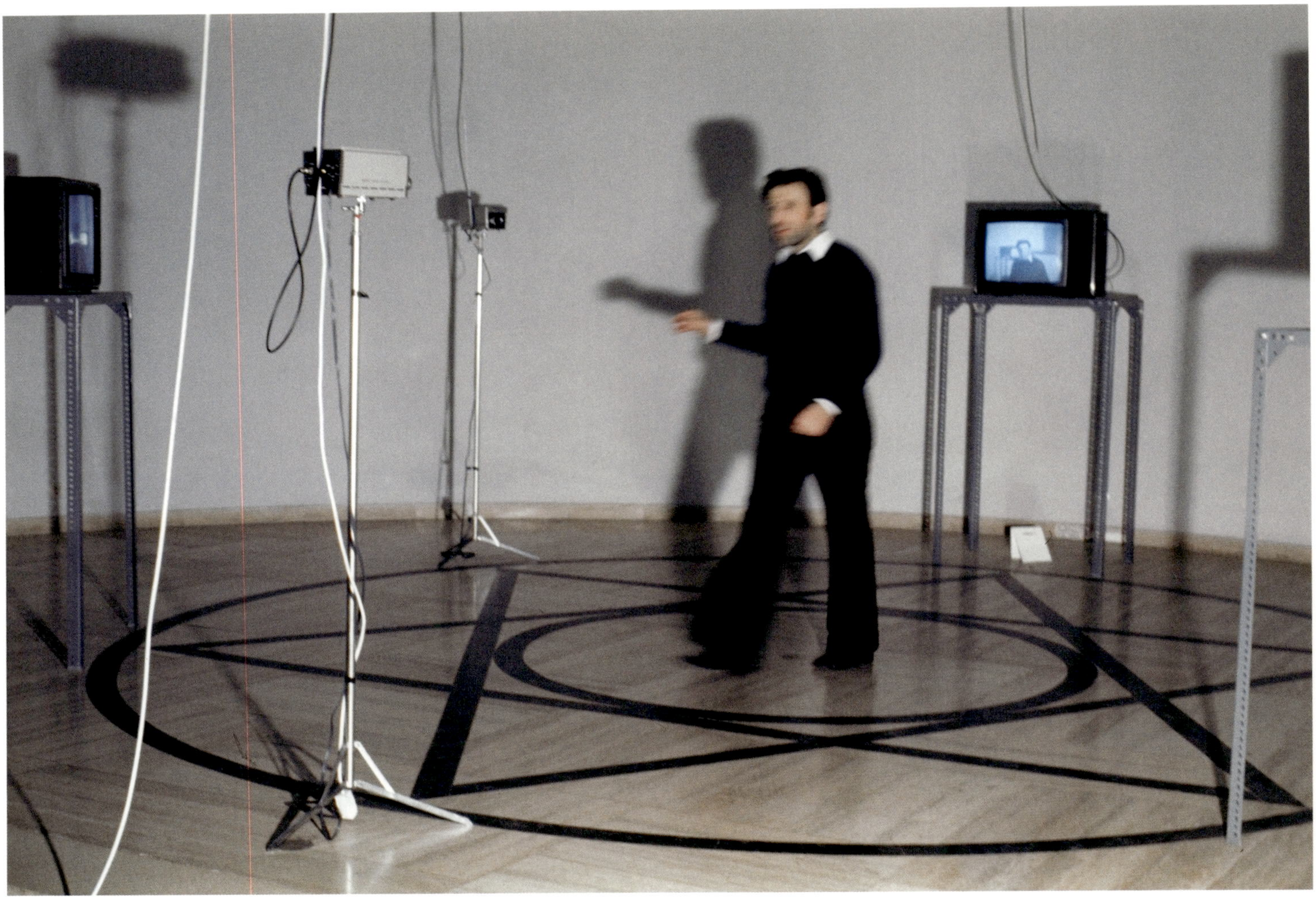

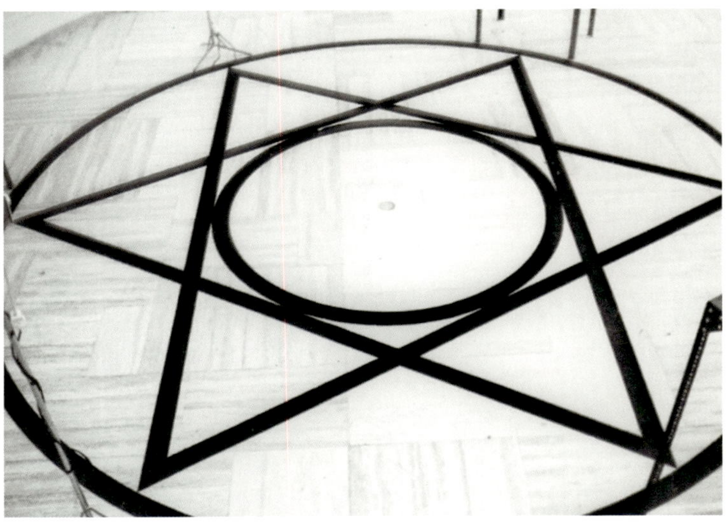

Installation views *Audiovisuelle Botschaften*, Trigon '73, Künstlerhaus Graz, 1973

Observing Observation: Uncertainty / Beobachtung der Beobachtung: Unbestimmtheit

1973	closed-circuit video installation; 3 video cameras on tripods, 3 video monitors on metal frames, floor marking, diameter 400 cm

The closed-circuit video installation *Observing Observation: Uncertainty* was conceived by Peter Weibel for the exhibition *Audiovisuelle Botschaften* (Audiovisual messages) at Trigon '73. It consists of three cameras and three monitors, which are positioned opposite each other in a circle and oriented towards its center. The hexagram created by the positions of the monitors and cameras (two overlapping, opposite triangles) is marked on the floor.

Anyone entering the closed circuit is filmed and can observe themselves in real time. However, it is not possible to see the view of the face and the view from the front, regardless of how much one twists and turns. The body can only be seen fragmentarily. This installation is part of a series of works that deal with problems of cybernetics and quantum physics: as an observer, one cannot observe one's own observation. In this way, Peter Weibel points out the limitations of the human apparatus of perception. The observer is a prisoner of perspective. Humans are living in an observer-relative world.

Peter Weibel: The Audience as Exhibit

Wolfgang Kemp

In 1969, the year he brawled with the audience in Munich and elsewhere, Peter Weibel also staged a video performance in Vienna titled *The Public as Exhibit (Audience Exhibited)* (*Das Publikum als Exponat (Ausstellung des Publikums)*, see p. 41 in this volume). On the face of it, the transition from the cavernous Zirkus Krone to the rather more intimate Galerie Junge Generation might seem an improbable one: from carnival exuberance to a studio or laboratory ambience, and soon thereafter to something evoking Zen. "Visitors to the exhibition *multi media 1* at Vienna's Galerie Junge Generation were interviewed. The conversations were recorded by a video system provided by the Philips Corporation through the courtesy of Hermann Hendrich. These interviews were shown, with a slight delay, on two televisions in the gallery. The recordings could also be replayed multiple times if visitors wished."[1] The artist also created a second video piece for the same occasion: *Process as Product* (*Prozess als Produkt*, 1969), which showed the exhibition being set up. In a similarly audience-oriented, similarly Bruce-Nauman-inspired study of self-reflection titled *Observing Observation: Uncertainty* (*Beobachtung der Beobachtung: Unbestimmtheit*, 1973, see page opposite), Weibel placed the "viewer" at the marked center of the installation, from which point one could look into three closed-circuit monitors that all displayed one's back – even when one turned around.[2] We had now arrived in the realm of circuitry, and the talk was of "feedback" and "self-referentiality."

Nor was that as far as things could be taken. Having been filtered through Nauman and Weibel, and especially through Niklas Luhmann, reception-aesthetic analysis has today been replaced by texts like this one: "Both a closed circuit and an observer are capable of integrating environmental data into their self-self-reference: two self-referential self-other observation systems, that of the recipient and that of the closed circuit, interpenetrate one another. The closed circuit, however, can only integrate otherness within a static system, while the observer can make use of new environmental data to reconceptualize dynamic mental systems."[3] Demonstrations of self-referentiality of this nature have been a mandatory exercise for essays in the arts and humanities since the 1990s. Nothing is new, even under the artificial sun of media studies.

What does come as a surprise, though, when we go further back in the history again, is the contemporaneity of attack and peaceful observation (or self-observation), of direct and meta. The higher technological standard seems to have contributed significantly to the de-escalation. Did the loops of self-referentiality turn into chains around those observers observing observation? It was still a bit early for a harsher, socially critical reading in the spirit of Michel Foucault's *Discipline and Punish*; that book didn't come out until 1975. Weibel himself explained *Observing Observation: Uncertainty* this way: "The observer of himself ... cannot see his face. Locked in a room, every point in space is his jailer; perspective is his deathly fate."[4] That's still completely oriented toward the observer, and the technology too, but not the system of observation. Caught as the subject is in the crossfire of cameras and monitors, sparring with the sight of his own back, he remains unaware of the perceptions of others. The "videological" works of the late 1960s, this aesthetic of the media-specific, reached an early apotheosis and conclusion with Nam June Paik's closed-circuit installation *TV Buddha* (1974). A Buddha contemplating its own image in a monitor, a media meditation in which the wild years of "media analysis" and "participation TV" silently trail away: "video karma," pure and simple – though perhaps also a signpost marking the end of the heroic years of reception aesthetics and reception activism; Robert Pfaller's concept of "interpassivity" inevitably springs to mind: One allows oneself to be observed.[5]

Note: Adapted from Wolfgang Kemp, *Der explizite Betrachter* (Konstanz: Konstanz University Press, 2015), 75–78. Translated from the German by Patrick Hubenthal.

[1] "*Publikum als Exponat*, 1969," Peter Weibel's website, posted Nov. 2, 2014, https://www.peter-weibel.at/publikum-als-exponat-1969/.

[2] Peter Weibel and Harald Falckenberg, eds., *Peter Weibel: Das offene Werk 1964–1979*, exh. cat., Neue Galerie Graz, Landesmuseum Joanneum (Ostfildern: Hatje Cantz, 2006), 790.

[3] Thomas Dreher, "Peter Weibel – Polykontextualität in reaktiver Medienkunst," in *Peter Weibel: Bildwelten. 1982–1996*, ed. Romana Schuler (Vienna: Triton, 1996), 33–62 (quotation on 38).

[4] Weibel and Falckenberg, *Peter Weibel*, 740.

[5] Hans Belting, "Buddhas Spiegel, oder: Auf der Suche nach Paik," in *Zeitgenössische Kunst und ihre Betrachter*, ed. Wolfgang Kemp (Cologne: Oktagon, 1996), 101–110.

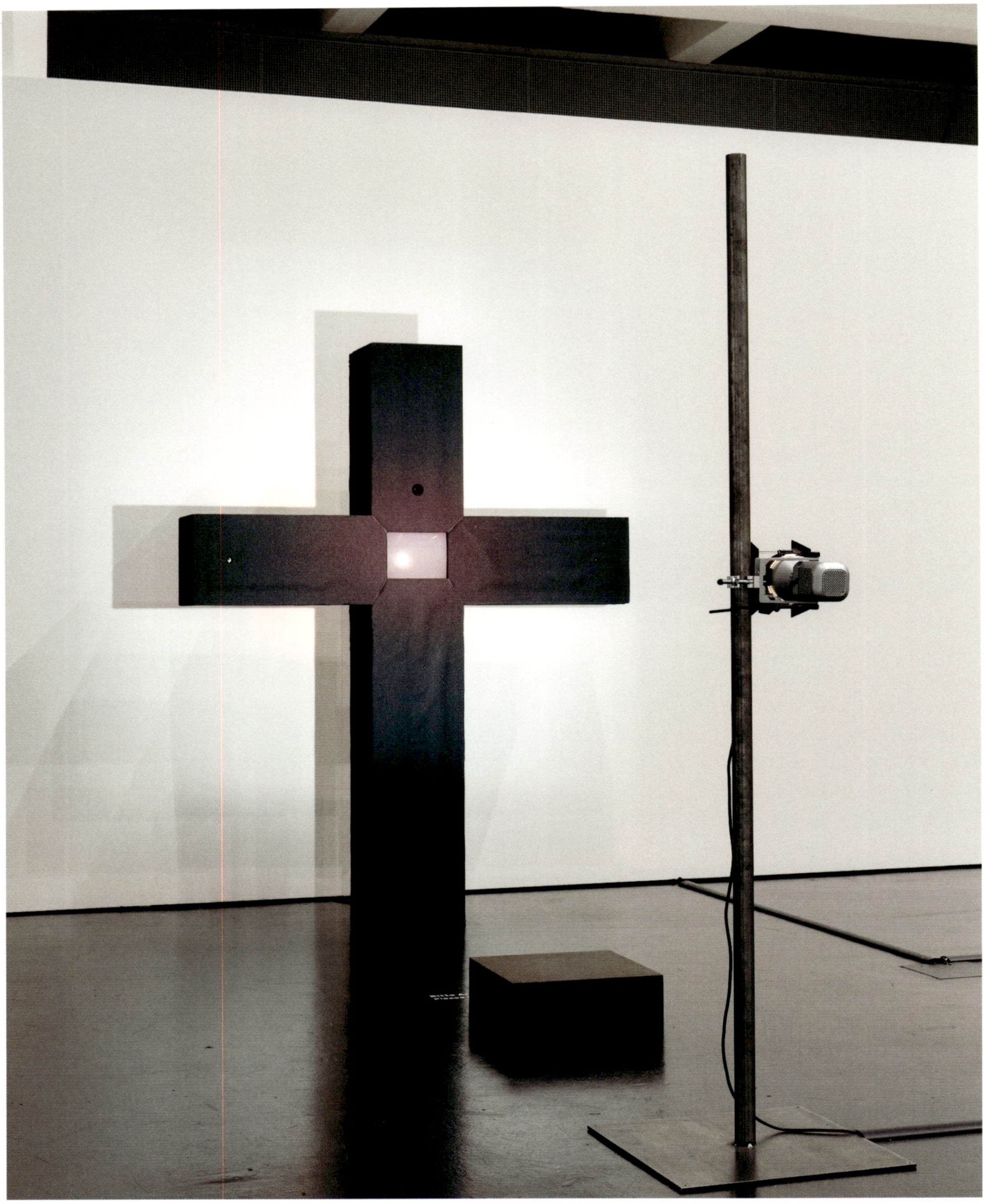

(opposite page)
Installation view
Audiovisuelle Botschaften,
Trigon '73, Künstlerhaus
Graz, 1973

Installation view
respectively, Peter Weibel,
ZKM | Karlsruhe, 2019–20

Crucifixion of Identity /
Kruzifikation der Identität

1973	closed-circuit video installation, interactive video sculpture; camera, monitor, 2 light sensors, wood, sandpaper, spotlight, pedestal, dimensions variable, cross: 252 × 190 × 40 cm

The closed-circuit video installation *Crucifixion of Identity* consists of a cross covered with black sandpaper, which contains a camera, two sensors (electro-optical circuits), and a monitor, a triple of interfaces. The camera only switches on when a person stands on the small platform, exactly in the center, and the spotlight behind their body casts a shadow on the cross. To bring the image up on the screen, it is necessary to spread out one's arms; like a cross, the shadow of the arms then falls on the two sensors which release the camera's image to the monitor. Only in this pose of a cross can the viewers see themselves on the screen on the cross. When the viewer lets their arm sink, the image on the screen disappears. Replicating the form of the cross, the symbol of Western culture par excellence, which in Christianity stands for the relationship between God and humankind, is the trigger that allows the viewer to see the picture of themself on the monitor. "You only recognize yourself in the name of the cross." Is identity a crucifixion, the victimization, the axiom of Western culture?

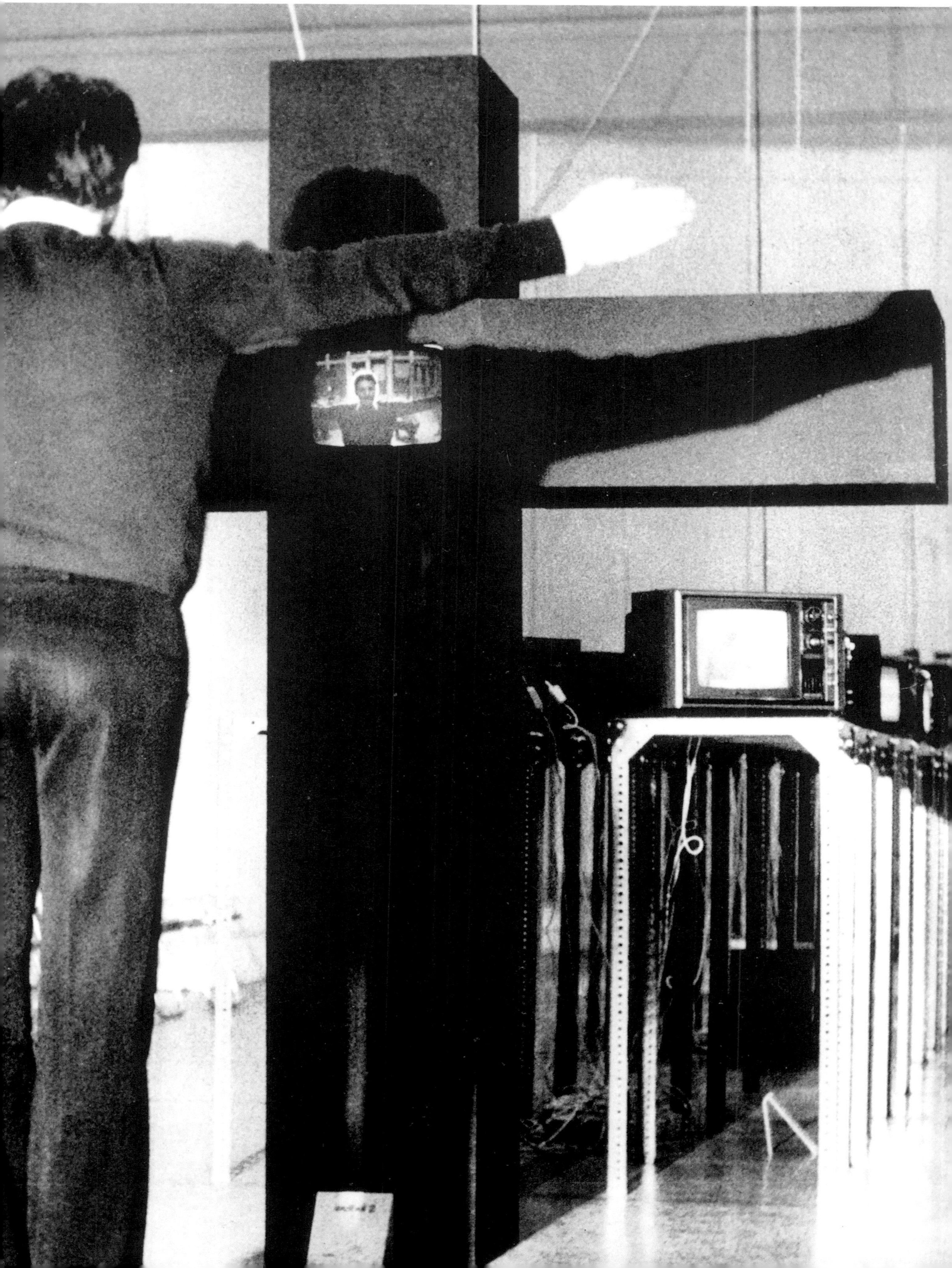

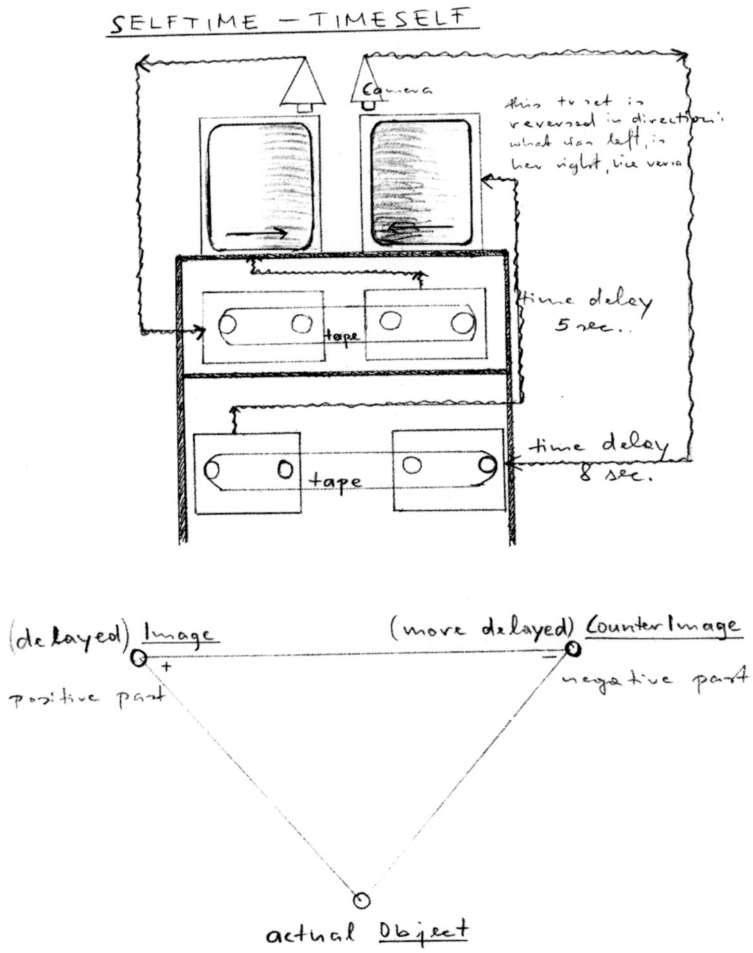

Conceptual drawing for *Self-Time – Time-Self*, 1975

Artificial perception, object construction, and identity crisis form the triad with which the discourse of radical constructivism begins. The evolution of perspective from anamorphic distortion (Jurgis Baltrusaitis) to Piero della Francesca's *De prospectiva pingendi* (1470s–1480s) and projective geometry is not just the story of the birth and death of certain visual forms of representation, it also exposes a theater of consciousness whose aim is to produce effects that the political powers declare as real in order to exploit our obscured consciousness.
To understand perception and its mechanisms, it is also necessary to know how we acquire knowledge. The experiment with the blind spot in the visual field teaches us that we do not see it at all, so we can only deduce from this that we only see what we believe we see. Understanding comes before seeing, consciousness molds the gaze. Perception via technical devices, which dominates the entire world today, belongs to the beginning of those concepts of the second stage like "learn how to learn," "understand how to understand," "know how to know," and "consciousness of consciousness." Perspective itself is, after all, already the perfect demonstration of the hypothesis of Heinz von Foerster that believing is seeing rather than seeing is believing.[1] For if you do not understand and learn the perception of perspective, you do not learn to fall for the illusion of a four-dimensional representation. Perspective, therefore, always means participation. The construction of being through perception completes the circle that began with George Berkeley's dictum "Esse est percipi." The problem of perception is therefore of central importance in a world that has been constructed almost entirely by technical devices.
In the closed-circuit installation *Self-Time – Time-Self,* the viewers are confronted with past images, lost images, and counterimages of themselves. The recordings of one recorder are transmitted to a second recorder, where they are not stored, but only played back. One monitor displays the signal from one camera, which is delayed by five seconds. The second monitor shows the signal of the other camera, delayed by eight seconds and mirrored vertically. What appears on one monitor as the movement of a person's left arm is seen as the movement of a right arm a few seconds later on the other monitor. What the mirror can't do, the monitor can: it shows time-delayed images. Thus, the viewers not only turn in space, but also in time: they see themself from all sides, the first time because their spatial image becomes a time image.
This physico-physiological experience of the flux of time has a drug effect: the experience of troubled identity. The spectator went on a trip, they had to perceive themself in a continuum, connected with past tense and past perfect, connected with their own image and counterimage. The body seemed to dissolve from the actual one in the flood of time. Identity as a form of time, as a function of time! —Peter Weibel

Self-Time – Time-Self / Ichzeit – Zeitich

1975	closed-circuit video installation; 4 video recorders, 2 video cameras, 2 monitors, 257 × 150 × 60 cm

[1]
Editors note: See Heinz von Foerster, *Observing Systems*, The Systems Inquiry Series (Seaside, CA: Intersystems Publications, 1982).

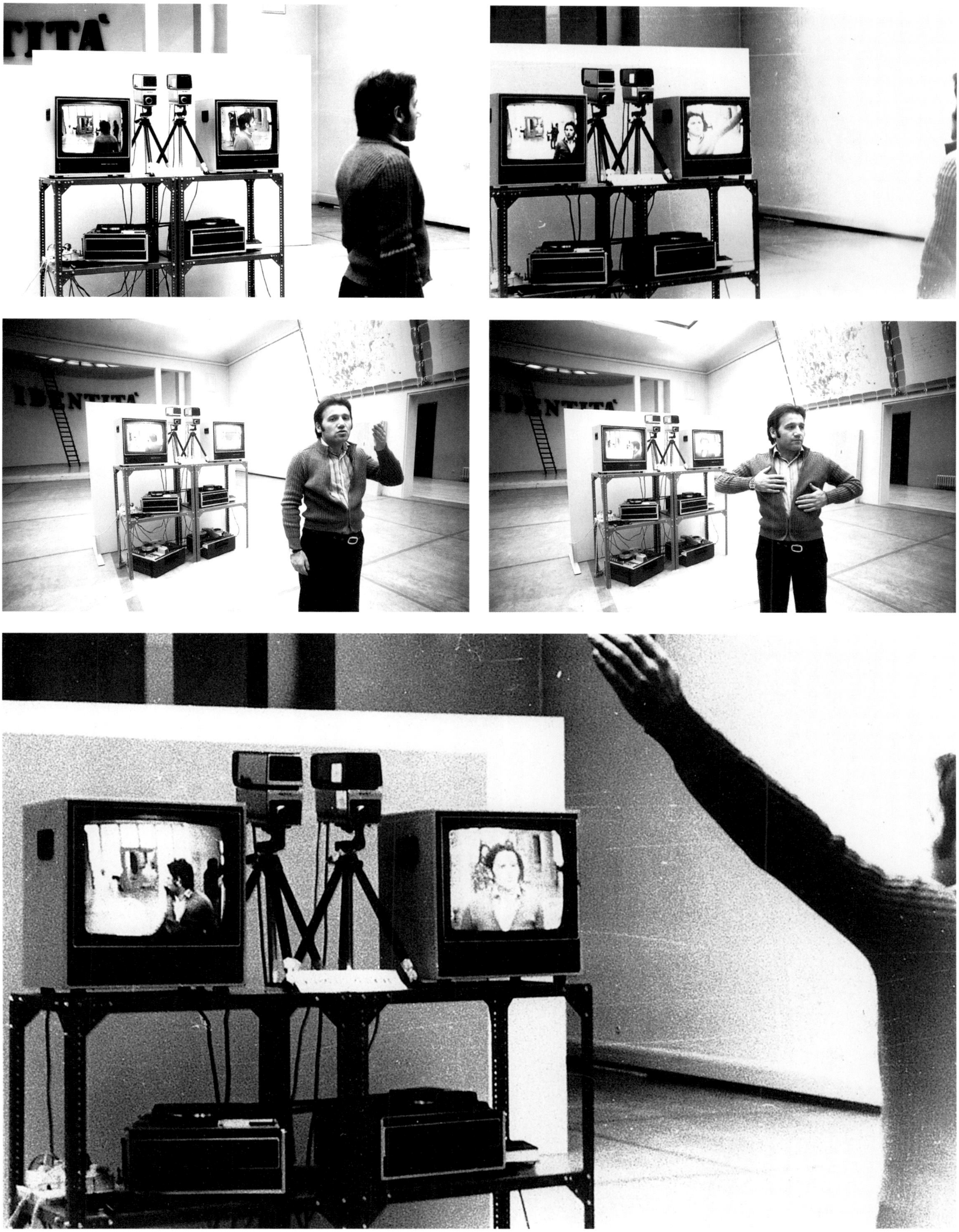

Installation views Trigon '75, Künstlerhaus Graz, 1975

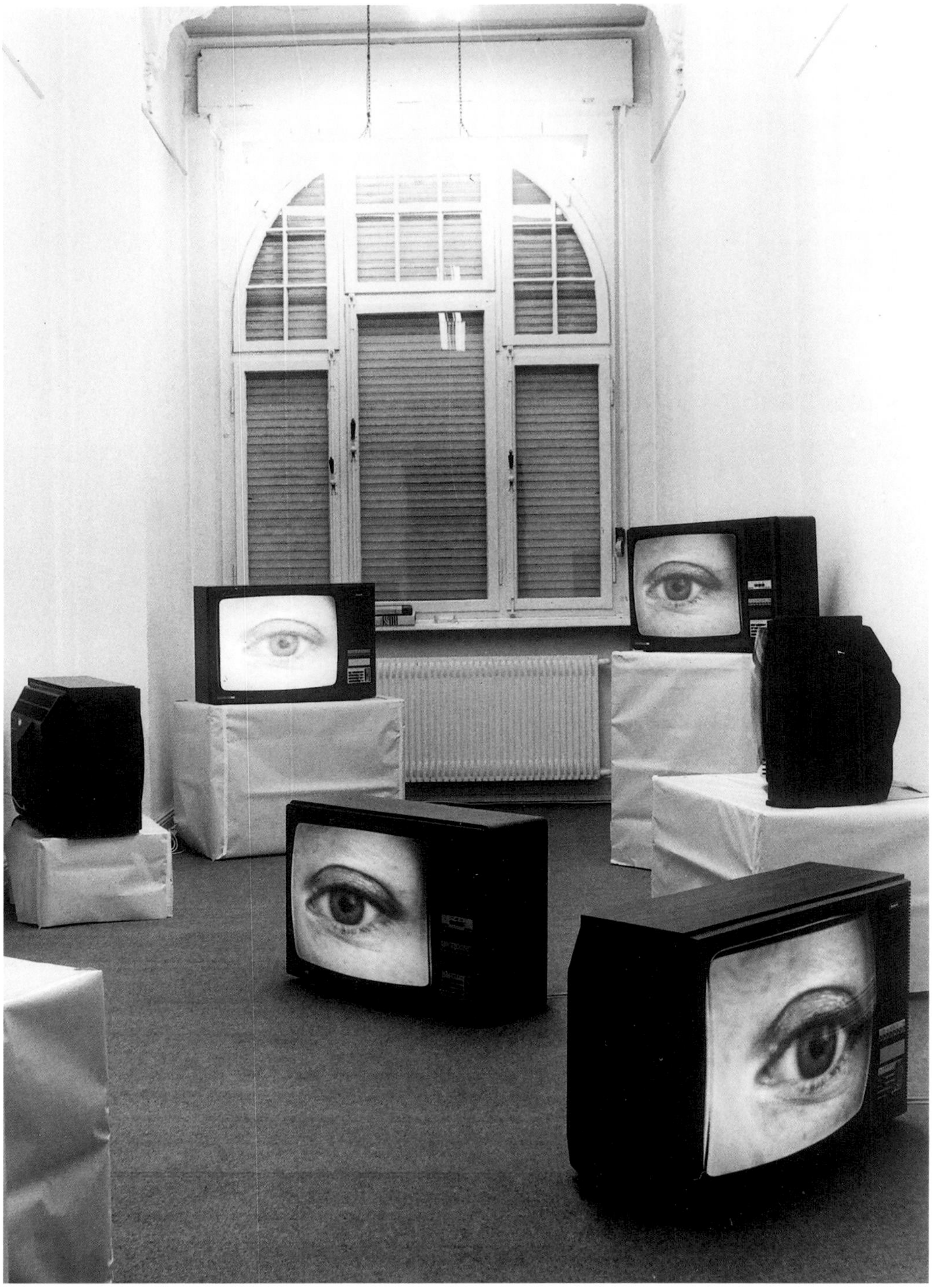

Installation view *Video Lumina & Video Corner*, Galerie Magers, Bonn, 1977

Video Lumina

1972 (idea)/ 1977	video installation; video tape, color, silent, 7 monitors, pedestals, dimensions variable

On seven monitors distributed around the room on pedestals of different heights or on the floor, in *Video Lumina* the same eye is always looking at the viewer. And the viewer also looks at the eye. Eyes thus see eyes; the observer is observed. Looking closely, it is possible to see the camera in the pupil of the eye, which means seeing the machine eye in the natural eye organ. Here I reflected on the directness and subjectivity of the gaze, which is simulated in the walkable video installation, in its technical and physical conditions. Eyes watch eyes, eyes observe eyes. Multiple eyes create a sphere of fear, of total surveillance. —Peter Weibel

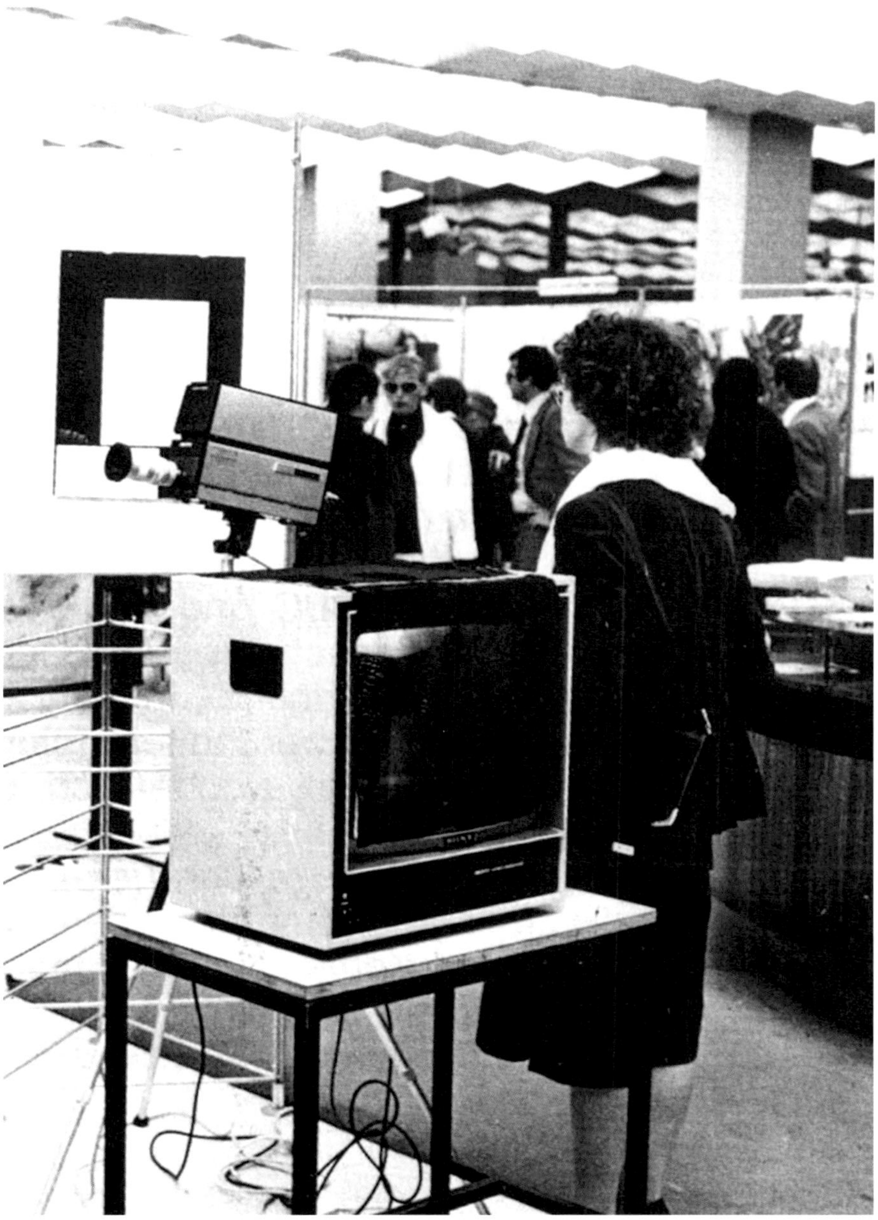

Installation view
Zentralsparkasse, Vordere
Zollamtsstraße, Vienna,
1978

The Guard as Bandit /
Der Wächter als Bandit

1974 (idea)/ 1978 | closed-circuit installation; 2 cameras, TV, mesh stocking

A monitor and a video camera were installed in the lobby of the main branch of the Zentralsparkasse (Savings Bank) in Vienna. The video camera was directed towards a surveillance camera so that the camera could be seen on the monitor screen. As a result, the Savings Bank's surveillance camera was itself being filmed and monitored by my video camera. The monitor was covered in a mesh stocking, similar to those used by criminals during bank robberies to hide their identity. My monitor, therefore, showed the Savings Bank camera wearing a mesh stocking, the "guard as bandit," so to speak. The organ that observes you is itself being observed. You become aware of being constantly supervised, watched.

The supervising, controlling function of the video system (several cameras and monitors) in this bank lobby is rendered blatantly visible through the control system installed by me (just one camera and one monitor).

What also becomes evident is that this video system, which is supposed to warn against and prevent acts of violence, possesses inherent aspects of violence itself. History provides us with ample, terrifying examples of the relationship between guard and violence, control function and violence. By furnishing the monitor with a symbol of violence, namely, the anonymity provided by the mesh stocking, the anonymity of the violence of control is removed. The covert surveillance, the concealed threatening element of surveillance, becomes apparent.

Observing the observer, supervising the supervisor, monitoring the monitor – a basic principle of democracy. —Peter Weibel

Note: *1984: Orwell und die Gegenwart. Ausstellung der Wiener Festwoche im Museum des 20. Jahrhunderts vom 10.5. bis 8.7.1984*, exh. cat., Museum moderner Kunst, Vienna (Vienna: Museum moderner Kunst, 1984), 177.

A film projector casts a beam of light onto the canvas, forming a luminous rectangle. In this white rectangle, on top of a black pedestal, stands a monitor that is connected to a video camera beside the projector. The camera's field of view corresponds exactly to the size of the canvas.
Before the image from the camera arrives at the monitor, however, a positive/negative inverter changes all its luminance values into their opposites.
As a result, the camera reproduces the bright canvas as a dark screen and the black shadow on the canvas as a white shadow on the monitor screen, while simultaneously reproducing the white shadow on the screen as a smaller black shadow on the screen, and the dark canvas on the screen as bright, and the smaller black shadow as white, and so on.
Thus, although ideally no object is reproduced, only the luminance values from the film are used for video operations, a concrete image nonetheless emerges. —Peter Weibel

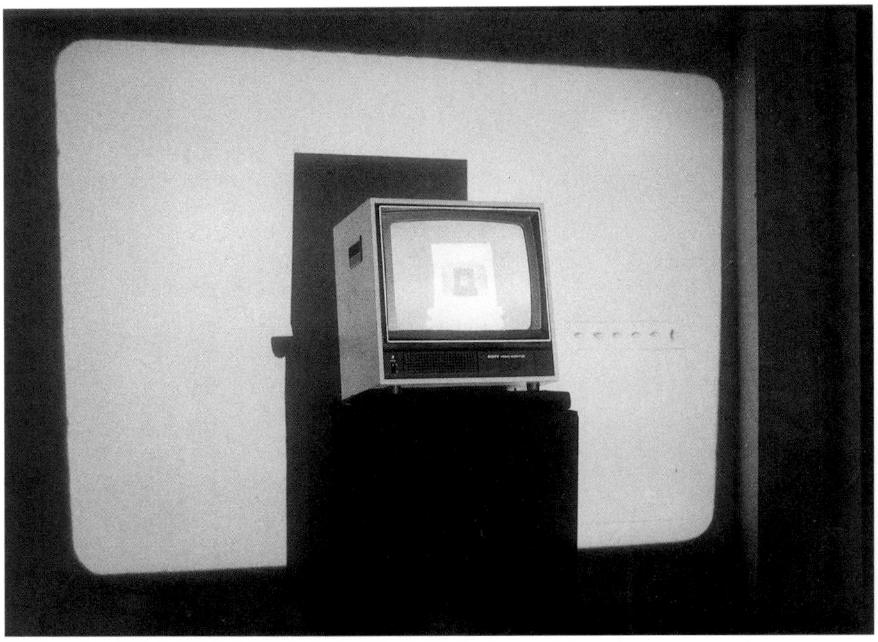

Installation view *Vidéo*, Musée d'art et d'histoire, Geneva, 1977

Negative-Positive Transfinite / Negativ-Positiv Transfinit

1973/77	film and video installation; film projector, camera, monitor, canvas

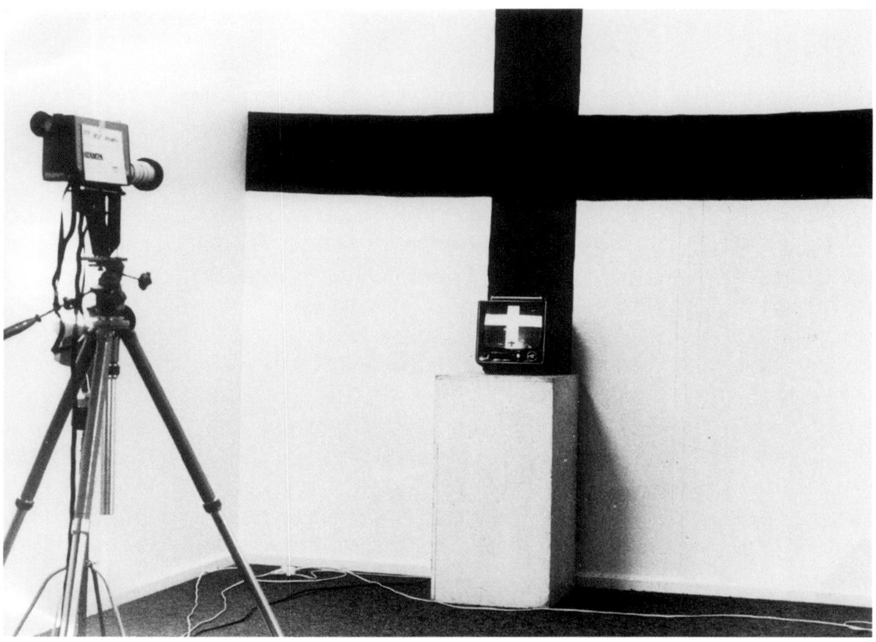

Installation view *Video*, Stampa Galerie Basel, 1977

Horizontally and vertically oriented black flags combine to form a cross. The camera reproduces this black cross in black on the camera monitor, but on the TV screen it appears as a white cross because the camera is connected to an actively running positive-to-negative converter.
The white cross on the screen is in turn reproduced as a smaller black cross, which is reproduced as a smaller white one, which is reproduced as a smaller black one, and so on *ad infinitum* .
This chain of alternating positive and negative reproductions produces a technological circuit: The reproduction apparatus generates an image – in this case, the metaphysical illusion of the infinite. This linear accumulation, this transfer of positive and negative images generates a new (media) reality that virtually extends to infinity. It is an implementation of the theorem of Christ's death on the Cross – that is, the Resurrection. Jesus, as the son of God and of humans, represents the transference from the infinite, or transfinite (of God), to the finite (of humans) – as well as another transference, in which Jesus bore our sins (transferred them onto himself) and sacrificed himself for us (identity transfer).
Technology thus becomes a theo-technology, generating or reproducing religious themes.
—Peter Weibel

Transfinite Transference: The Cross / Transfiniter Transfer: Das Kreuz

1977	closed-circuit installation; camera, TV, black molleton, tripod, 59.5 × 47.5 × 47 cm

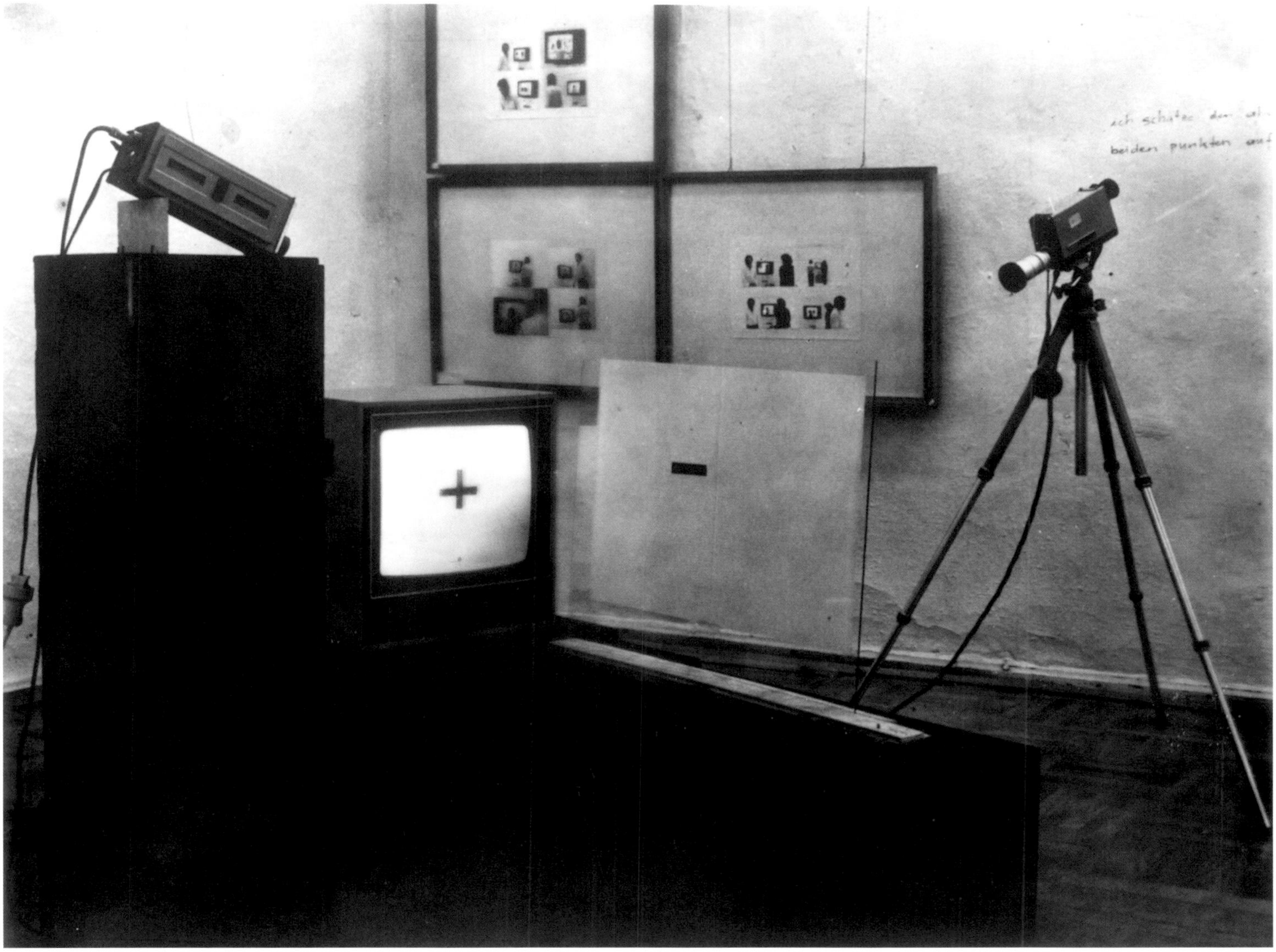

Installation view
Zur Kosmologie des Paradoxen, Galerie nächst St. Stephan, Vienna, 1975

Epistemic Videology I / Epistemische Videologie I

1974	video installation; 2 cameras, monitor, video mixer, glass plate, dimensions variable

In the work *Epistemic Videology I* Peter Weibel considers language from a spatial, geometric point of view. A transparent glass plate with a horizontal line (a minus sign) on it is recorded by two cameras opposite each other, one of which is lying on its side. Thus, one camera sends a horizontal line and the other sends a vertical line to the video mixer. The mixer mixes the two images, the vertical and horizontal lines, into a plus sign which appears on the monitor screen. The video installation thus highlights the fact that a representation (a plus) does not necessarily depict reality (a minus), but can be responsible for distortions. Media images can be the opposite of reality.

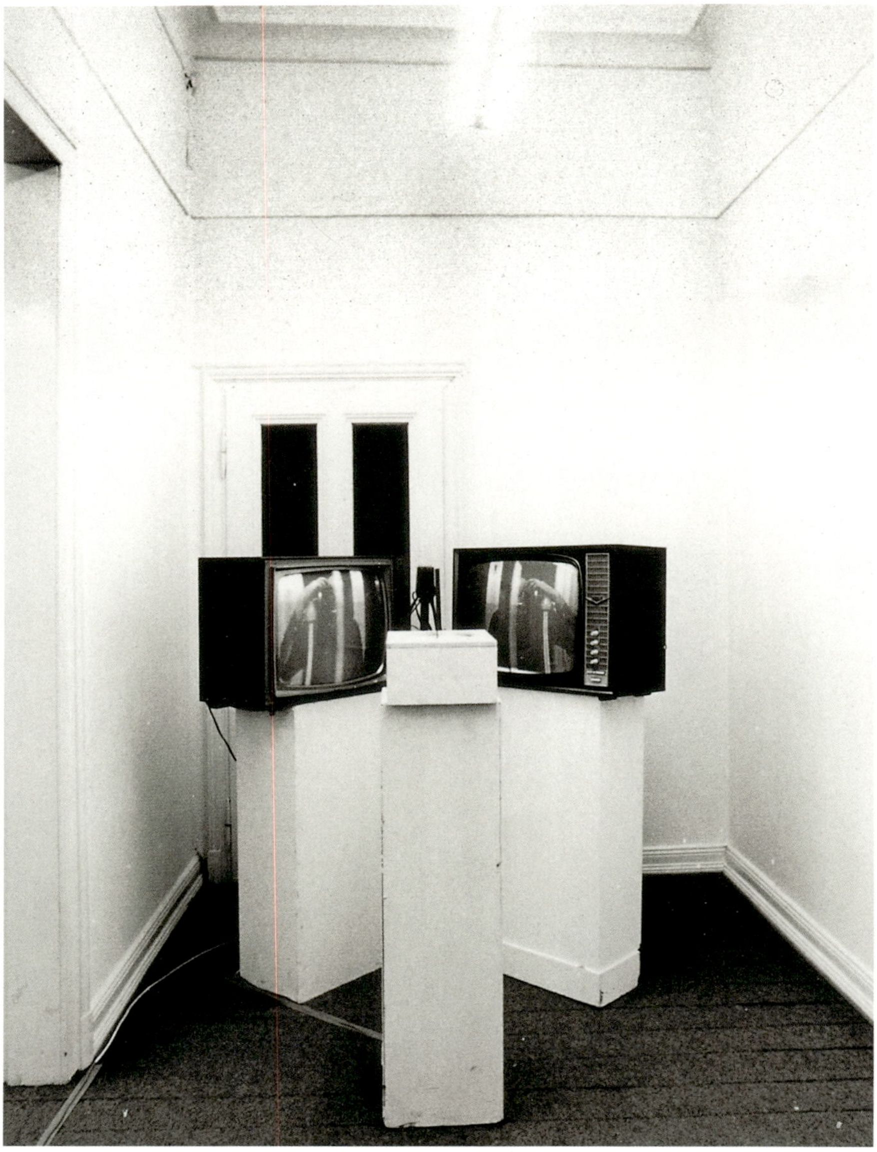

Installation view *Video Lumina & Video Corner*, Galerie Magers, Bonn, 1977

Peter Weibel's closed-circuit video installation *Video Corner* invokes a perpetual exchange between inside and outside, between the surrounding space and the individuals acting within it. To initiate this dialog, this permeable process, the artist first links together two monitors, setting them up in such a way that one image is vertically reversed. The two are then positioned in space so as to form a corner, in which a video camera is placed. Opposite this corner, a one-way mirror rotates on a turntable, marking the boundary between inside and outside.

Whenever the mirror turns to face the two monitors, the camera captures their reflection, an interior space that is virtually multiplied to infinity by feedback. But because the mirror is semitransparent, the exterior space is included in the image as well. The result is a (never-ending) succession of pictures within pictures, interior and exterior spaces superimposed on one another. Since the mirror is rotating, these images converge on the center of the video corner, producing a symmetry of objects hurtling toward each other. Whenever the mirror is perpendicular to the monitors, the camera records the exterior space and the observers of the process, who see their reflected image in the monitors and then in the mirror.

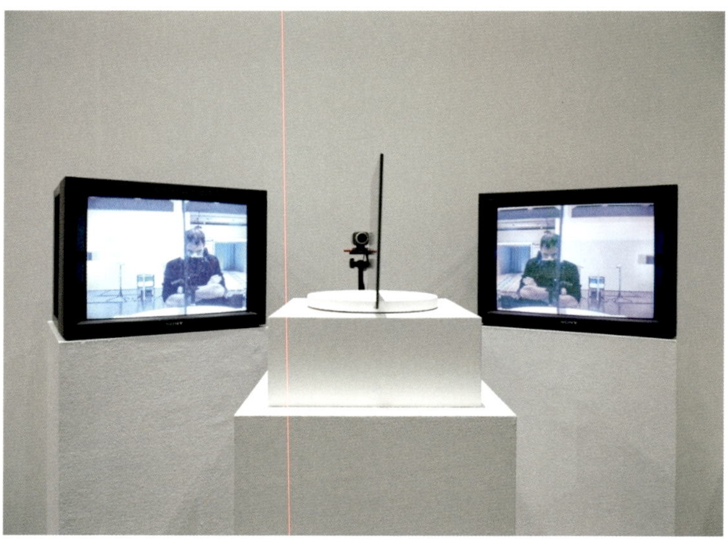

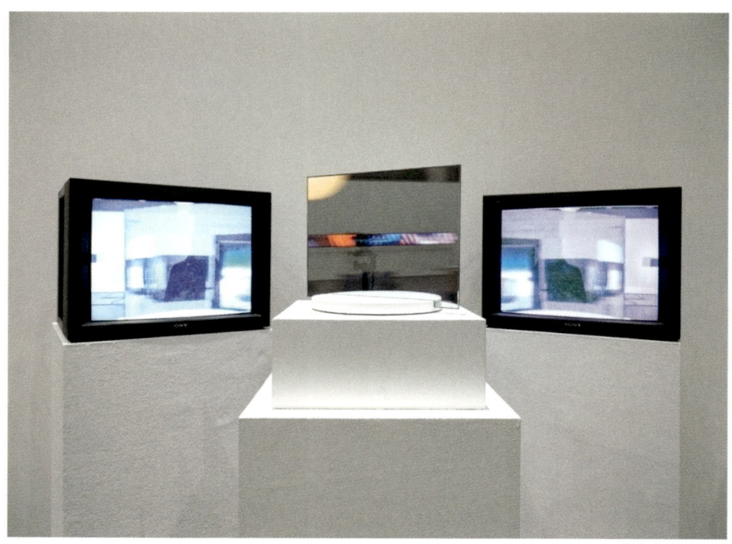

Installation views *respectively, Peter Weibel*, ZKM | Karlsruhe, 2019–20

Video Corner

1974/77 | closed-circuit video installation; 2 monitors, rotating half mirror, camera, dimensions variable

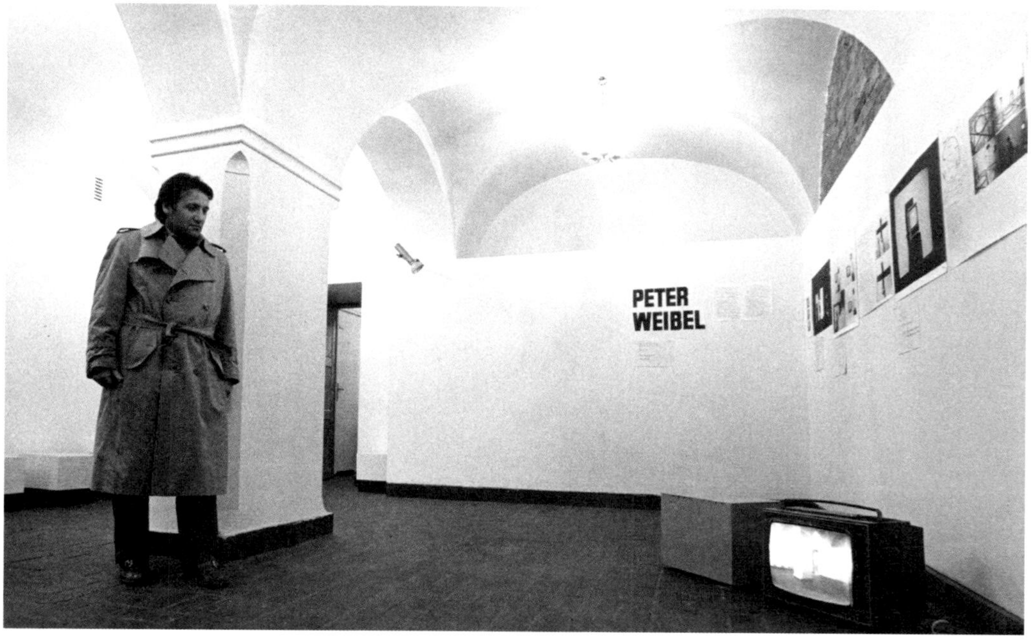

Installation view Mala galerija, Ljubljana

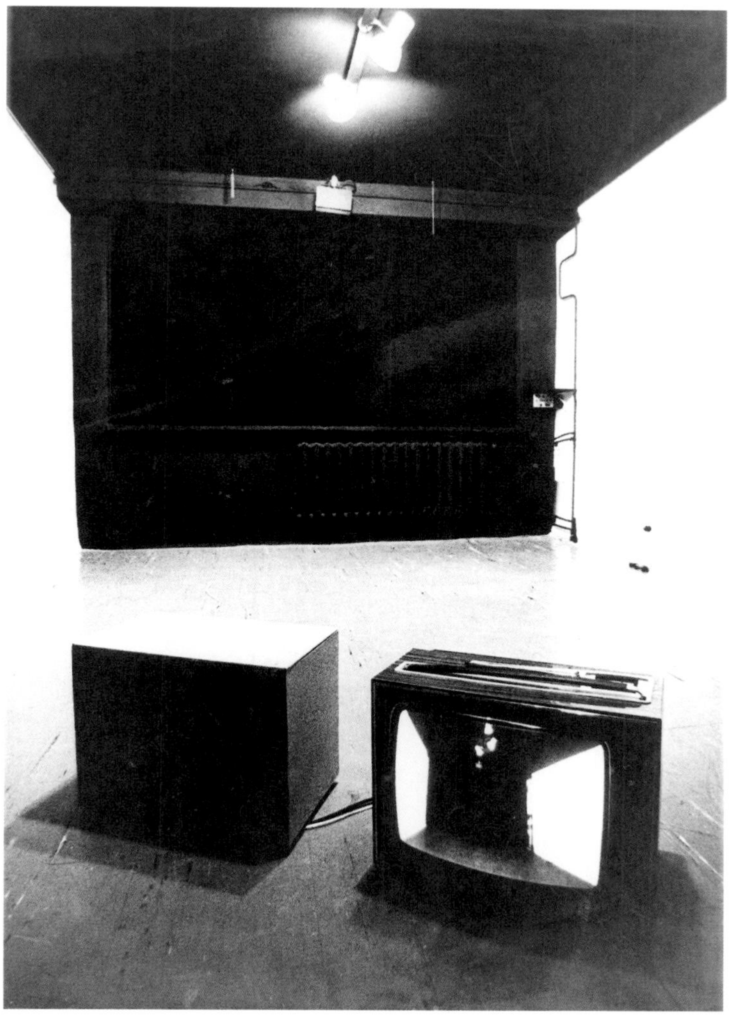

Installation view *Tribute to Tarski*, Galeria Fotografii, Wrocław, 1980

Inverse Box / Inverser Raum

1972/80	monitor, camera, connecting cable, box, b/w photograph of the room affixed to the inside of the box, dimensions variable

The hierarchy of spaces: the space appears like a box which, in turn, contains two smaller spatial boxes, a closed box and a TV set.

The space is a volume, a box. The monitor is likewise a box. Hence, in the space itself, there is also a smaller model of the space: a TV box in the space box. Normally, the surrounding space is reproduced on the screen. We thus see an empty gallery space and, on the monitor, an image of this empty gallery space. When a viewer enters the empty space, they expect to see their reproduction on the screen. And yet it does not appear on the screen. For this reason, the viewer looks for the camera so as to take up a position in front of it. But no camera can be found. So how does this spatial reproduction appear in the image box, if no camera can be located? How is it, moreover, that when a person enters the space as depicted on the monitor, that they are not reproduced on the screen? How does this simultaneous presence in real space and absence in the image arise?

Next to the image box is another, completely sealed box, to which a cable leads. In this box is a video camera and an illuminated photograph of the spatial situation. This photograph of the space (this two-dimensional reproduction) is transmitted from the camera in the box to the TV set. Even though the observer is standing in the real space, they are not visible in the pictorial space, because they are not on the photograph. The illusion of the two-dimensional screen, which shows a three-dimensional space, is disillusioned. The viewer cannot distinguish whether the camera is filming a photograph or the real space. What the observer sees on the TV screen is the black-and-white picture of the space, sealed in the box, not the real space. Therefore, the observer is not present on the TV screen.

Yet the absence of the viewer on the TV set makes – ex negativo – their presence in the real space all the more present or noticeable. Through their absence from the picture the viewers actually become conscious of their spatial presence.

In traditional sculpture, where only the human being is depicted, the real person, the viewer plays no role at all. In my video sculpture, the person who is observing is constitutive of this work of art, so for the first time the human being is of real importance. —Peter Weibel

Note: Adapted from Woody Vasulka and Peter Weibel, eds., *Buffalo Heads: Media Study, Media Practice, Media Pioneers. 1973–1990*, exh. cat. ZKM | Center for Art and Media Karlsruhe (Cambridge, MA: MIT Press, 2008), 791.

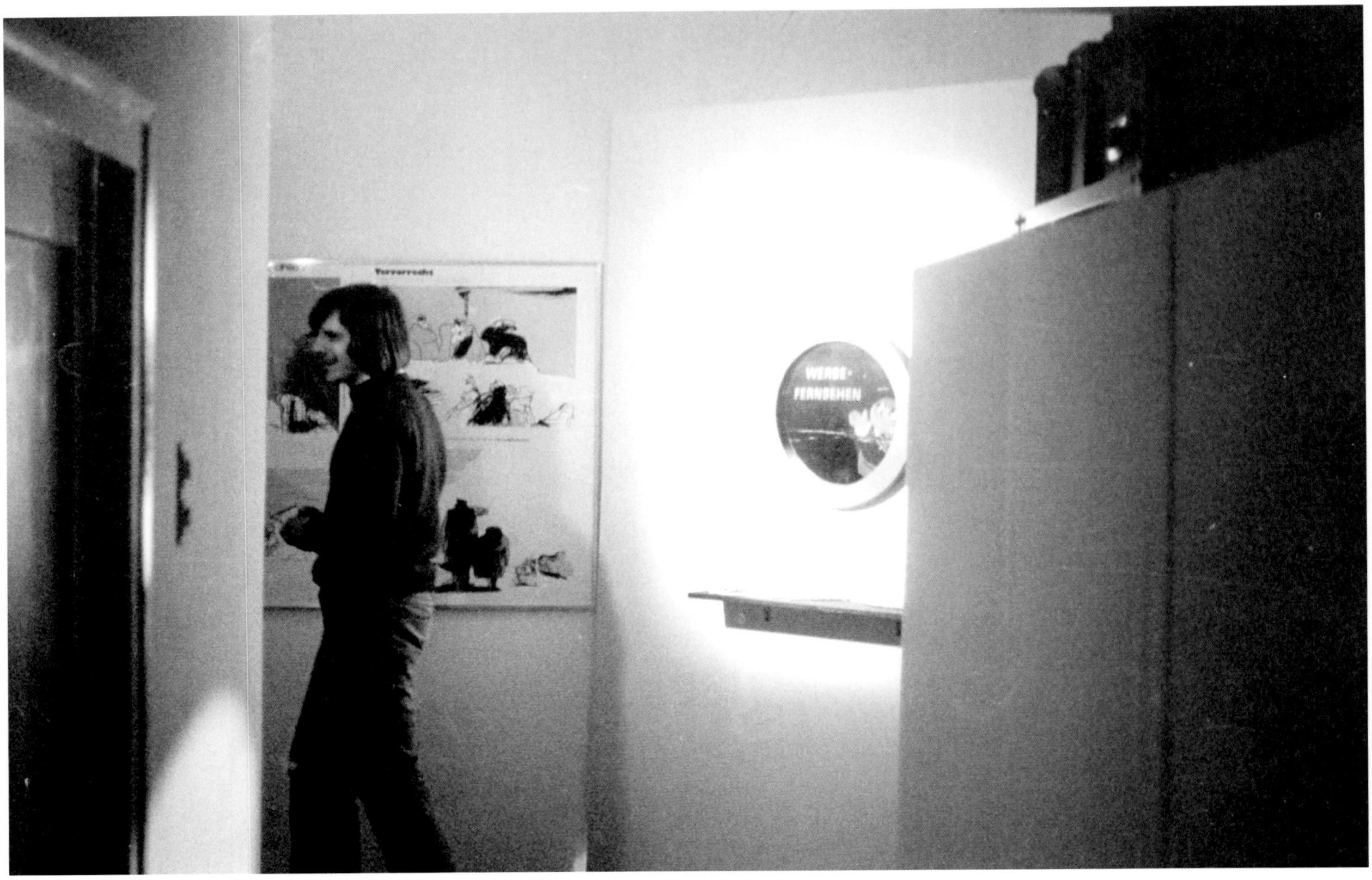

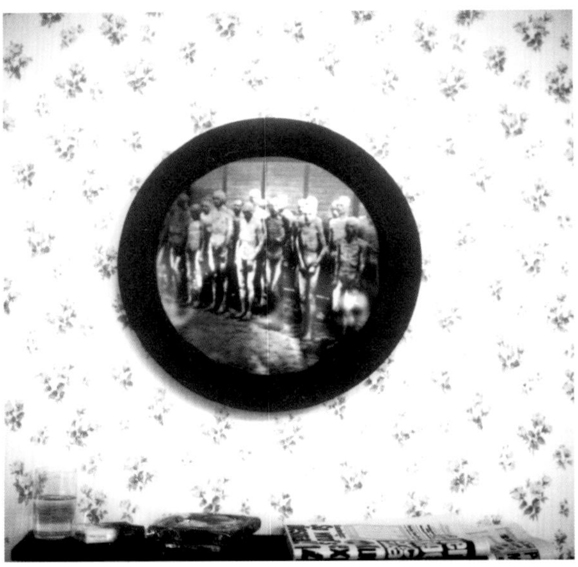

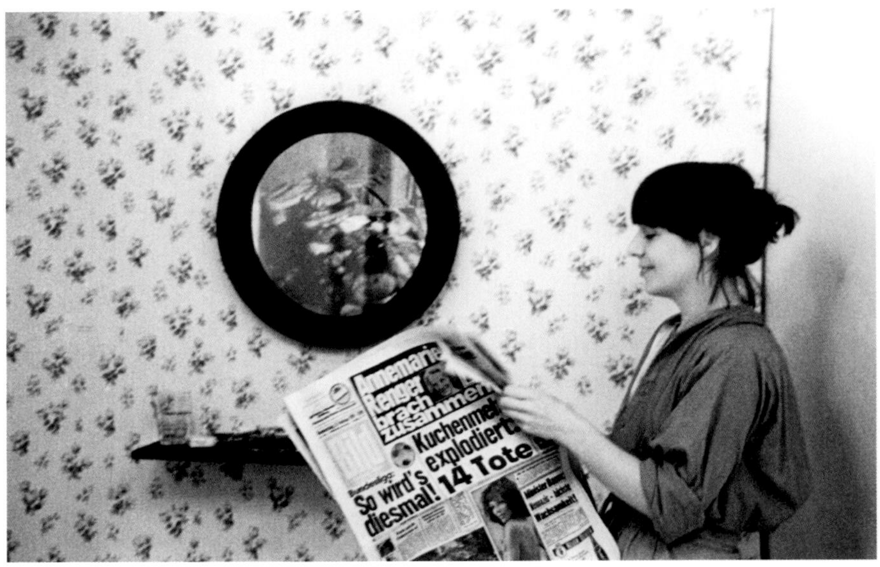

Installation views Loft, Kirchenstraße 15, Munich, 1978

Mirror Check I

1978	video installation; semitransparent mirror, video tape, player, monitor

A cloakroom wall with a mirror. A video plays behind the semitransparent mirror. The videotape shows corpses at concentration camps. As a result, the real people in front of the mirror are reflected and superimposed on the video images in the mirror. They see themselves among the corpses.

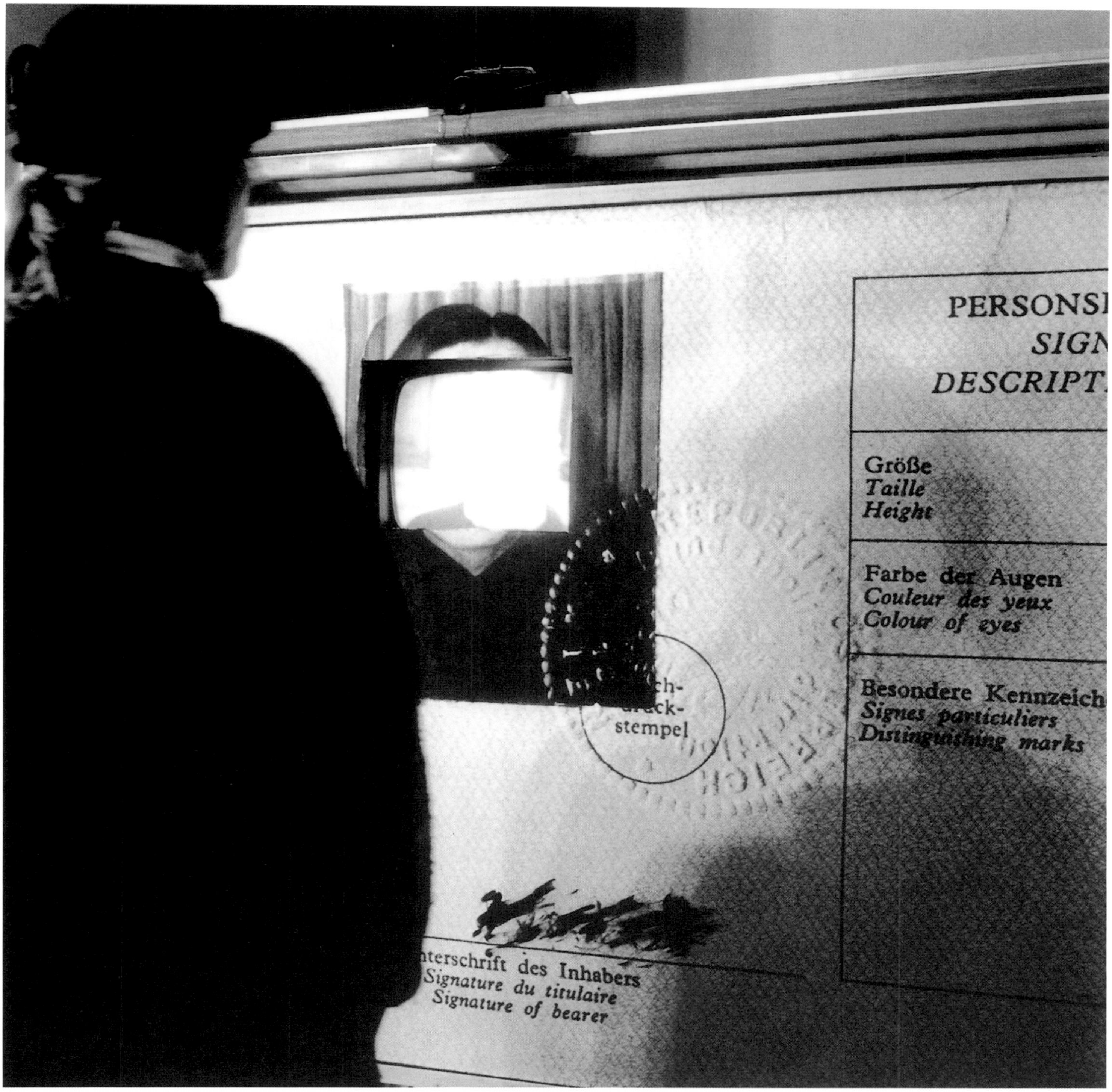

Installation view *Peter Weibel: Medienpoesie*, Kulturhaus Graz, 1982

Against Institutional Identity / Wider die institutionelle Identität

1980/ 2014	photograph of an enlarged passport, mounted on wood, video camera, monitor, 55 × 150 × 40 cm

On an oversized passport, the face part of the passport photo is replaced by a monitor. If someone steps in front of the monitor, the person will be filmed by a camera mounted above it. The face of the viewer now appears on the screen as part of the passport. The work *Against Institutional Identity* interrogates the meaning of personal, national, and institutional identity. The passport as a document prescribed by the state is revealed here as a construct, as a variable. The installation invites us to think about identity far away from any institutional framework – an approach, that is enormously topical, especially today, in contemporary society, which is characterized by streams of migrants. How can it be, in a world that claims to be just, that papers, paper passports, decide on human rights? Are papers forged because *sans papiers* ("without papers") people are worth nothing?

—Peter Weibel

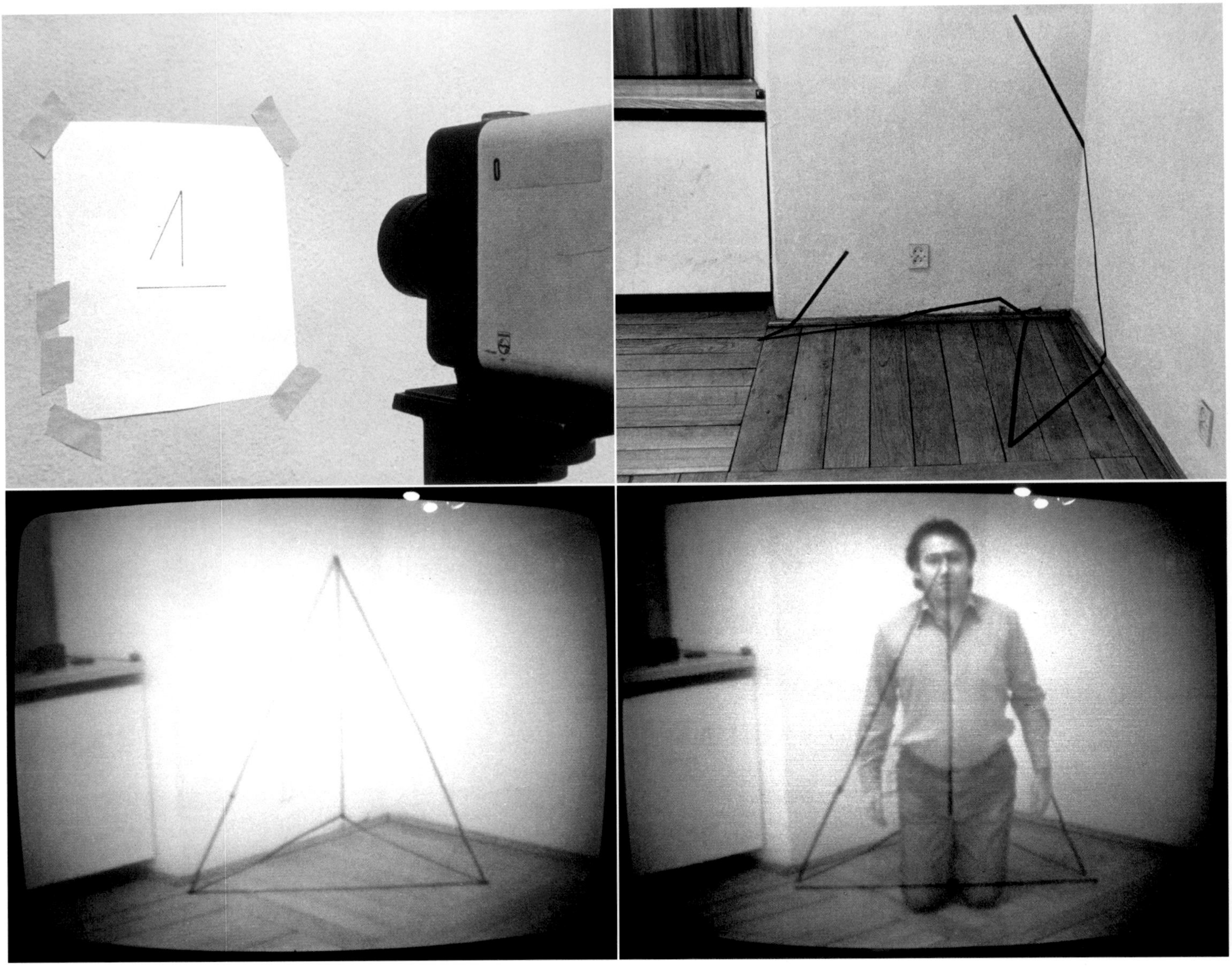

Installation views
Imaginärer Raum Galerie
Stampa, Basel, June 1980

Virtual Tetrahedron /
Imaginärer Tetraeder

1978 (concept)/ 1980	closed-circuit video installation; 2 cameras on tripods, video mixer, monitor, paper, ink, dimensions variable

Upon entering the room, the viewers first see only an abstract tangle of lines on walls and floor. On the screen, however, a tetrahedron becomes visible. If someone steps into the tangle of lines, some lines of the tetrahedron disappear behind the body, and some lines appear in front of the body. As the visitors are moving the lines of the geometrical object move from front to back, and from back to front. The space becomes unreal, for the screen presents the space in an unreal way. This is done by one camera recording the lines on the wall and floor with their perspective distorted which become parts of the tetrahedron on the screen. A second camera records the missing parts from a drawing on the wall. Via a mixer, the images of both cameras form the virtual tetrahedron on the screen. When a person steps into the maze of lines, the lines from the second camera cross the body and seem to be unreal. The real lines of the first camera are covered by the body and disappear. The screen shows something that does not exist, a tetrahedron. The tetrahedron is constructed in a way similar to our brain that also takes particles of reality and forms a coherent picture or model. The screen reacts like our consciousness.

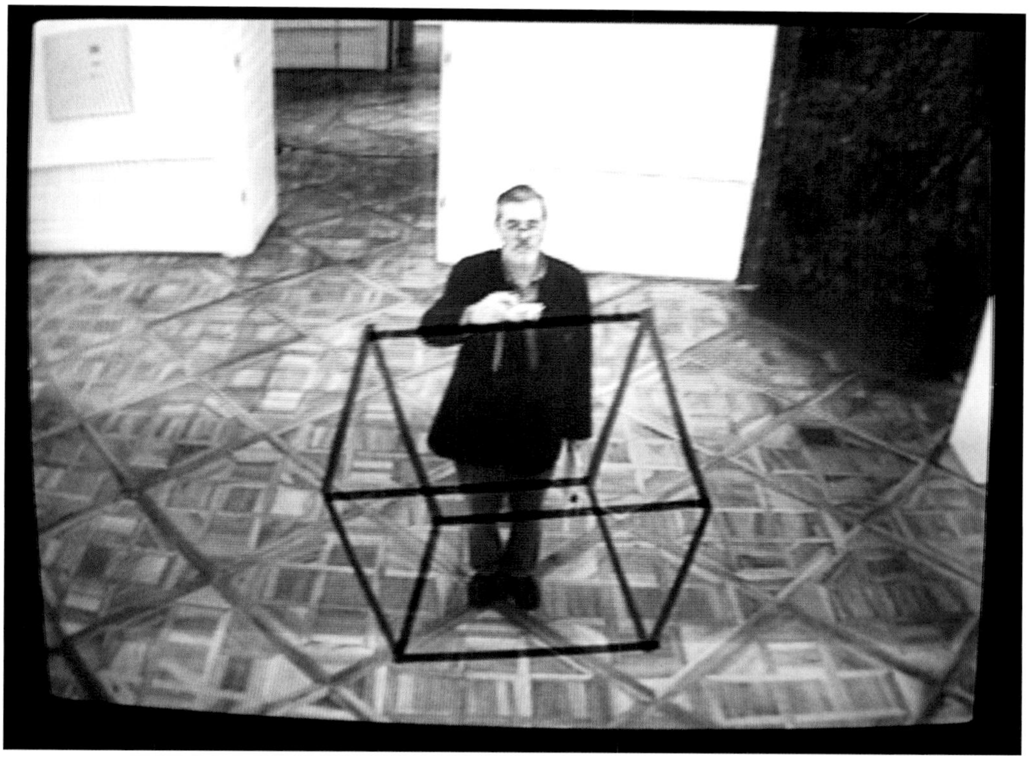

Installation view *Peter Weibel. Das Offene Werk 1964–1979*, Neue Galerie Graz, 2004

The representation of three-dimensional space on the two-dimensional plane of a picture – or rather, the construction of the illusion of three-dimensional space on the two-dimensional plane of a picture – was both the object and the cause of the invention of perspective. I take the opposite approach, reproducing the two-dimensional image of a cube in space. Perspective is replaced by a scanning system. The point-by-point, line-by-line scanning of an image is translated back into space, thereby transforming real space into an electronic space. The electronic space is a layered, scanned space containing islands of real space. In the same way that electronic images are scanned – that is, layered, broken down into lines made up of points – the electronic space is scanned as well.

Electronic space is a constructed space, a mental representation. Multiple electronic layers take the place of the customary x, y, and z axes that have been used to geometrize and represent space in the past.

Since the fifteenth century, painting and geometry have been linked by the perspectival construction of the image. Indeed, the mathematization of *perception*, the rationalization of sight and seeing, was elevated to the rank of *costruzione legittima*, the only legitimate method of constructing a picture. In perspectival construction, mathematics and painting came together, laying the historical foundations for the contemporary union of painting and media.

What perspective achieves – the "correct" representation of three-dimensional space on the (picture's) two-dimensional plane – is achieved automatically by the camera, by means of the lens and the laws of optics. What interested me, therefore, was retroconstruction: starting with the image and building it back into space. The digital, mathematical space of the screen is extended into real space.

This gives rise to a new *spatial art*. The construction of perspective is deconstructed, decentered, decomposed into many parts and layers. It is only the camera and the viewer's consciousness that reassemble the locally scattered spatial and pictorial elements into a whole (on the screen itself, which becomes an image, an analog, of consciousness).

However, what is articulated in this *mapping* of drawing and geometry, of painting and mathematics, of plane and space, of painting and media, is not merely modes of perception, but also social codes – for example, how reality is socially constructed, or how fiction becomes reality in the mind through a social compact. The fragments of real space on the screen serve as reference points for those viewing the piece, who see themselves in the picture, inside a cube. When they stretch out their hands, they seem to touch the bars of the cube. They find themselves inside an imaginary cube, a space that is partly real and partly imaginary – a psychotic space, in other words; a virtual reality. Viewers see how reality is constructed from imagination and the real. They see how social codes are constructed – for example, the belief in the rational constructibility of a just society. The gaze is the place where consciousness and the world encounter one another as image. The gaze presents our consciousness of things.

Thanks to the camera and monitor, both of which are within the pictorial space, the viewers, too, are inside the picture.

No matter how they shift their position, they cannot share the camera's point of view. The camera occupies a privileged viewpoint; it is virtually located in the infinite like the vanishing point in a perspective painting. The viewer becomes the victim of perspective. The perspectival construction becomes a labyrinth in which one becomes lost. —Peter Weibel

Translated from the German by Patrick Hubenthal.

Virtual Cube / Imaginärer Würfel

1978/80 | closed-circuit video installation; 2 cameras, video mixer, monitor, drawing on paper and floor

If history is also a history of humanity's self-realization, then civilization, art, and culture are part of that self-realization. Art, especially, as a medium of expression and creation, has manifested traits of this sort of self-demonstration in the history of humanity's development. Viewed from this perspective, video, as a new branch on the tree of art, embodies a new organ in human evolution, a new product of this self-demonstration, in that it physically realizes certain human percepts and concepts – that is, it externalizes certain fundamental characteristics of our thought and behavior that had previously been internalized; it makes them visible, depicts them visibly: self-referentiality, instantaneous reproduction, the temporal aspect of the pictorial function. Mental characteristics have become technological ones, which of course is part of human evolution: the replacement of the natural with the artificial – the Greek word *tekhnē* means both art and technology – the substitution of the organic with artifacts. By virtue of its characteristics, video is an implementation of certain mental structures in technological constructs. In this way, the representational code has also been expanded, into its own inner workings. Because of its self-referentiality and its temporal element, video is also particularly well-suited to live performance.

In this videology, I attempted to investigate and illustrate a concept that is valid everywhere from biology to mathematics: bifurcation, a two-pronged division or fission (of plant and animal parts, of forces, of paths, of vectors, of the windpipe, etc.) – in other words, a branching into two parts. Especially in the last few years, the study of the origins and evolution of biological structures has given rise to a mathematical theory of structural stability that necessarily applies to any general theory of morphology: René Thom's catastrophe theory, in which the problem of bifurcation – the study of the profound modifications that a dynamic system undergoes when the parameters governing its vector space change – plays a special role. The investigation of the destruction of a structurally stable attractor by variations to a vector space is the subject of a qualitative dynamics that Henri Poincaré named the bifurcation theory of dynamical systems. Thom calls the morphological effect of such a change on the local level a catastrophe. His main hypothesis is thus that every morphology can be assigned to some bifurcational phenomenon. This thesis has proved exceedingly fruitful for the study of the dynamics of self-reproduction. This was the theoretical background for my video performance, although the background is not necessary for understanding the performance. The branch is an illustration of the principle of bifurcation: one becomes two. My video performance about the bifurcation principle centered on the human body. The branch functioned as an icon (symbol) of the vital principle, of growth. Its Y shape is the opposite of a related form with an antithetical meaning: X, which possesses a terminating function and is an icon for anonymity, death, negation, etc. By bringing the Y and X shapes into play, I was attempting to deal with the principles of life and death, to present them through a visual action, a performance. I attempted to lay the principle of bifurcation like a bridge across the chasm that is the division between image and object, the schizoid difference between the object and its representation. Accordingly, my video performance arose directly out of my video installations. One of those installations made a real ellipsoid into a circle through the act of perspective representation. Because representation had previously always been practiced in the name of identity, a real circle always became a represented circle – that is, the shape remained the same, and only the medium of representation and the object differed in terms of their perceptual dimensions. In my installation, however, representation created new shapes: it split the ellipsoid, as one might say, into an ellipsoid and a circle.

And so, when the performance began, the people sitting in their chairs saw the empty room on the three screens, but not themselves. Representation split them into present and absent, since I was showing the room in the state of their absence. To accomplish this, one camera was pointed at the audience, but its signal was suppressed in the mixer; a second camera was pointed at a photograph of the empty room, and this picture alone was transmitted through the mixer. On the two-dimensional screen, of course, three-dimensionality is an illusion, so viewers could not tell whether the camera was capturing a photograph of the room or the room itself. Next, a Polaroid photograph was taken of the circle on the screen, which was the representation of a huge ellipsoid in which the audience was sitting. This was captured by the second camera as well. The next step was to spray-paint white over the actual (black) ellipsoid, the image of which was transmitted from the second camera to the screen, so that viewers could see how representation turned the ellipsoid into a circle, since one was overlaid on the other. After that, the Polaroid was scribbled out with a black marker, leaving only the white ellipsoid as a white circle. I then went over to a T-shaped booth consisting of two parts: the action space, and the representation space with the three monitors. Because another camera was behind the action space, recording from behind everything that the first camera recorded from the front – namely, my actions – three types of images appeared on the screens: the first type of representation was a mirror image, the second showed movements in one direction and the opposite direction simultaneously, and the third transposed top and bottom. When a piece of sheet steel is subjected to pressure from both sides, it suddenly bends one way or the other. Under excessive pressure, the steel seeks a release, and the force, the pressure, bifurcates, taking the form of a bend. In my performance, the pressure, as it were, was exerted by two cameras, which observed each other from opposite sides. This produced a reversal of direction(ality): what appeared as left to one camera was right for the other. The viewers, who shared the first camera's viewpoint, saw the real performer move from left to right, while the first screen, which presented the image from the second camera, showed the right-to-left mirror image. The two different perspectives were superimposed on the second screen: movements in one direction simultaneously became movements in the opposite direction. Action elements were then explored within this formal structure. First, the clapping of my hands in front of a microphone, with the hidden second camera's hole serving as the center. A delayed tape setup provided an echo and, ultimately, an increasingly independent, constantly self-reproducing applause sound, which ensued without my involvement. Of course, the clapping hands appeared on the first screen as their mirror image, on the second as four hands, and on the third as moving away from one another when in reality they were moving toward each other. Additional bifurcation phenomena were demonstrated with a piece of sheet metal, a pistol, water, a falling ball, and more. The phase of opposite movements involving the body began with a demonstration using an actual forked branch. I next sprinkled sugar in the shape of the fork. Sugar is not only the gustatory opposite of salt – which, of course, preserves time and inhibits growth

Bifurcation / Bifurkation

1979	performance; Videowochen Essen 79, Folkwang Museum, Essen, November 30, 1979

(see the use of salt as a preservative in kippering). Sugar is a catalyst to growth: the important monosaccharide glucose, its derivate glycogen, carbohydrates in the form of plant starch.
The body bifurcations ended with an X shape, when a pole was placed diagonally across the action booth, appearing in its representations on the screens as a large X. Prompted by this termination signal, the performer tied himself up.
A technological meditation on biological – indeed, chreodic – structures. A performance about the complexity of morphological representations, about the topology of cognition.
Schizo-division as a principle of life.

Note: Adapted from Peter Weibel and Harald Falckenberg, eds., *Peter Weibel: Das offene Werk 1964–1979*, exh. cat., Neue Galerie Graz, Landesmuseum Joanneum (Ostfildern: Hatje Cantz, 2006), 724–25. Translated from the German by Patrick Hubenthal.

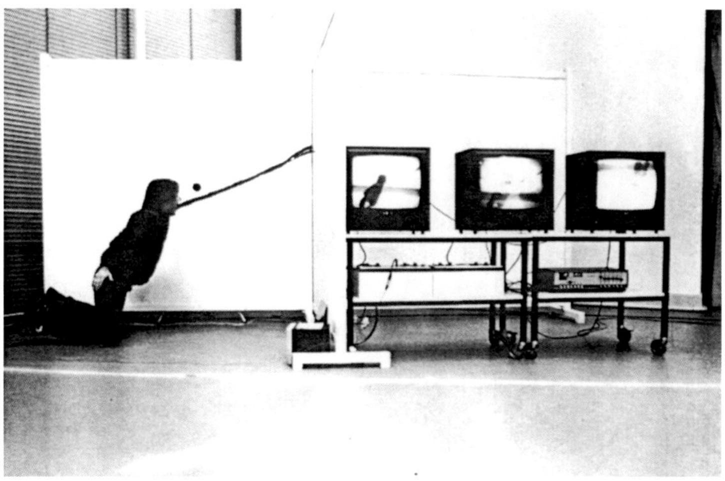

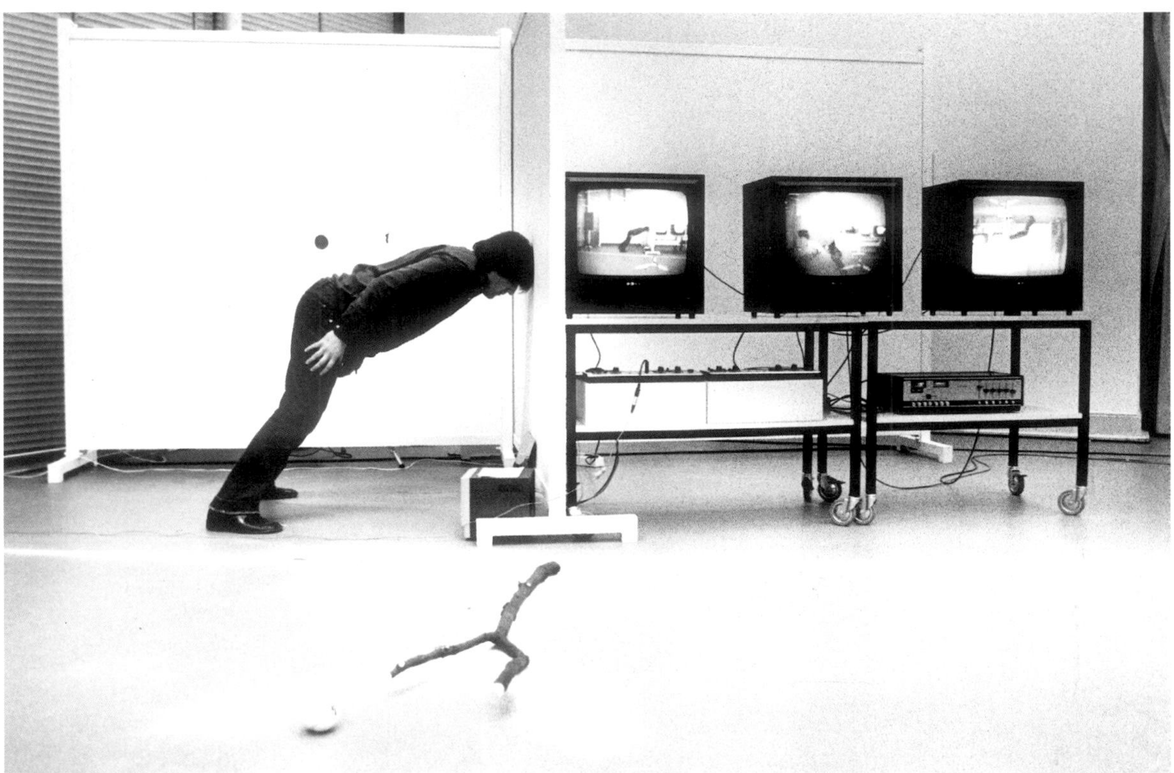

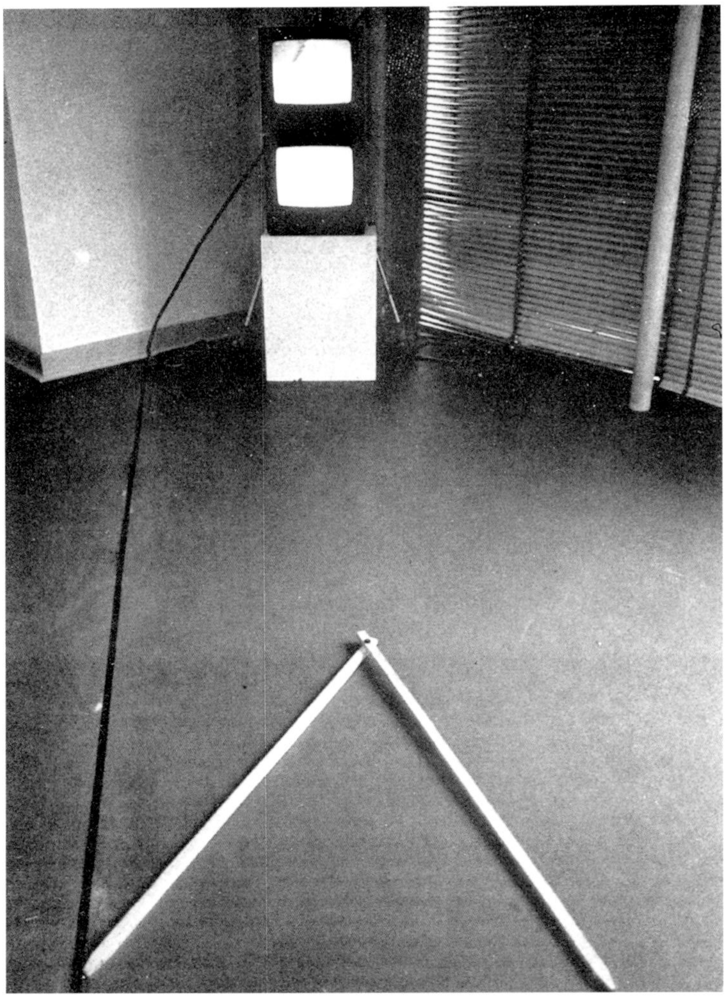

I
The invariability of form in the process of reproduction is called into question. A wooden angle lies in front of two monitors and a camera. One monitor is connected backwards, so that the angle points up on one monitor and down on the other. As a result of this differential reproduction using just one camera, the angle forms a (distorted) square in the medium of reproduction. Through reproduction, one shape (the angle) gives rise to another (the square). In effect, reproduction TWISTS one object into TWO different images.

Part I

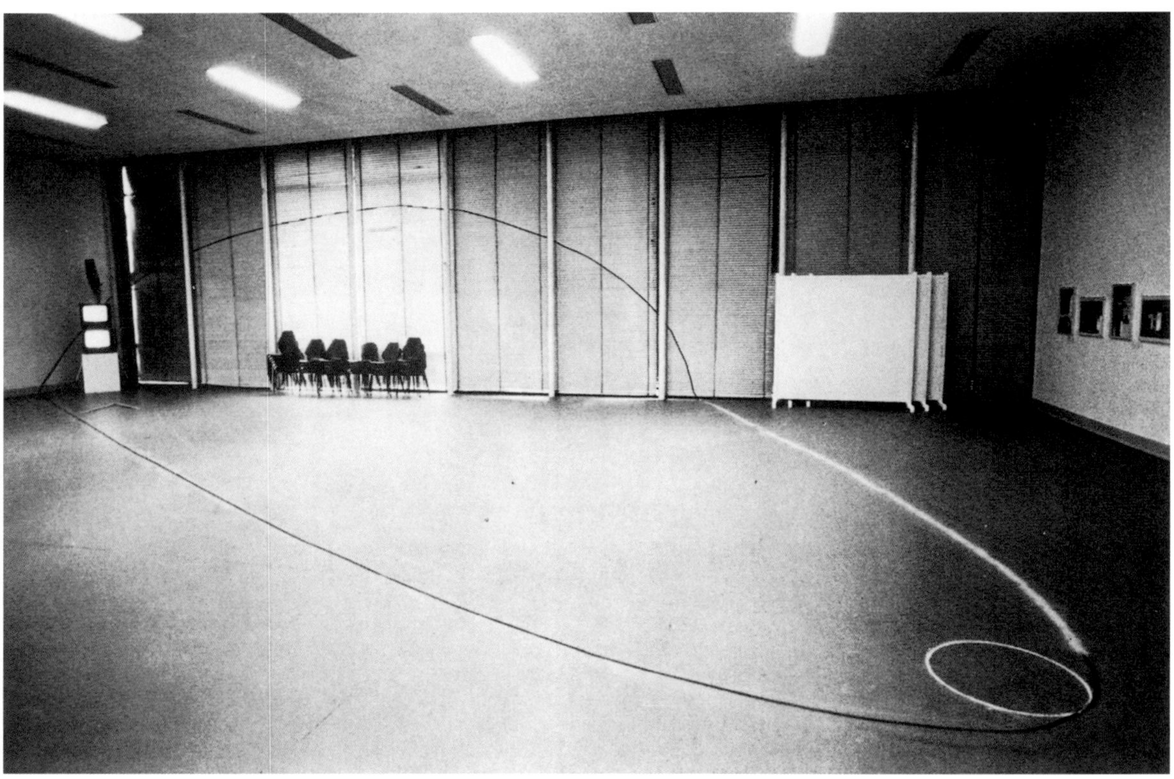

Installation views
Videoräume, Videowochen Essen 79, Folkwang Museum, Essen, Nov. 1979

Part II

The Dream of Everyone Sharing the Same Consciousness / Der Traum vom gleichen Bewußt-Sein aller

1979 | video triptych

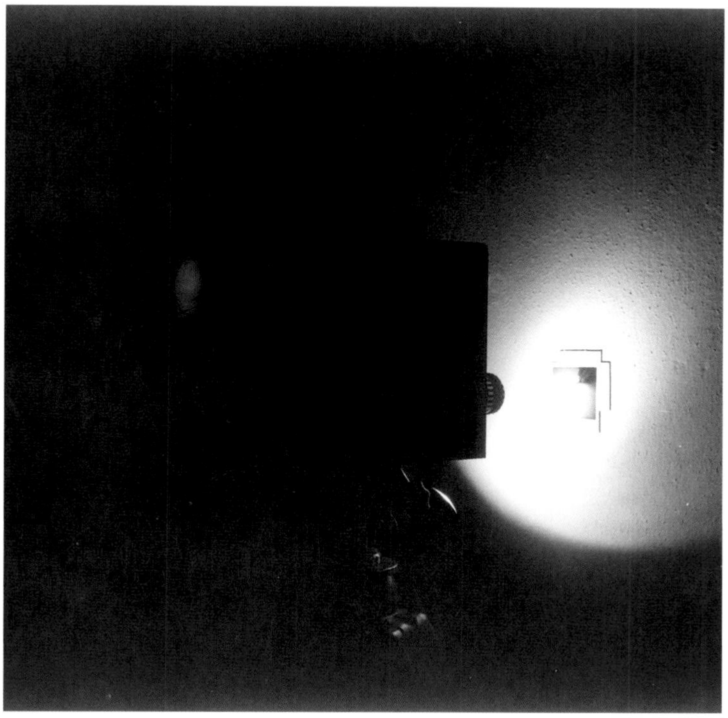

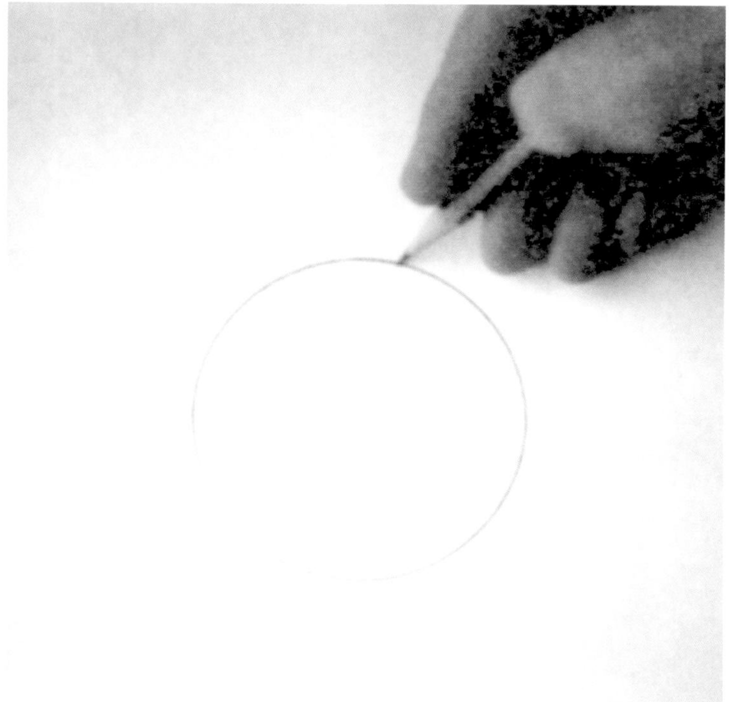

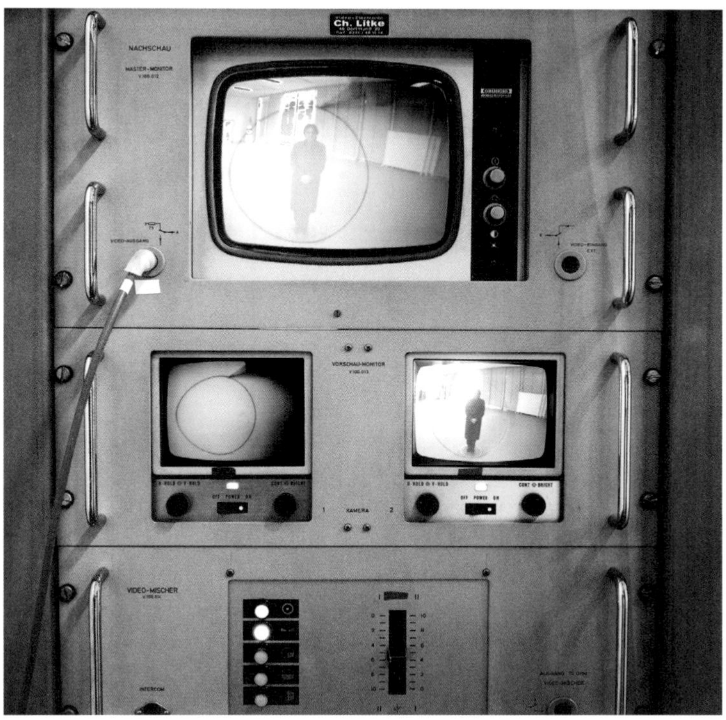

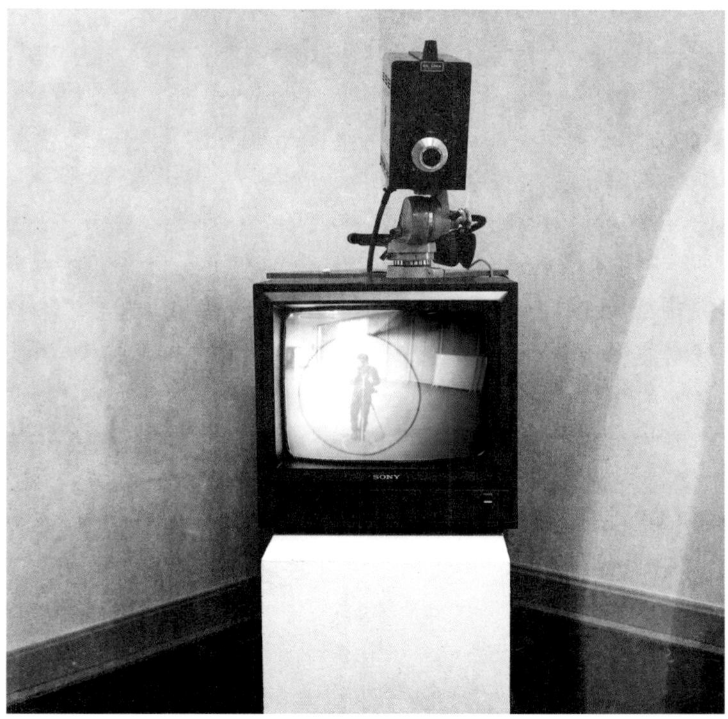

Part II

II

The constancy of form is relativized in the next installation as well. Two different shapes, a circle and an ellipsoid, are superimposed on the screen. A Polaroid photograph of a circle (and also a pencil and a finger) is recorded by Camera 1. Camera 2 records an ellipsoid drawn on the walls and floor of the room. Both cameras transmit their images to the monitor by way of a mixer. On the screen, the circle and the ellipsoid are congruent: the ellipsoid is reproduced as a circle because of the way Camera 2 distorts perspective. Another twisting reproduction. Three spatial moments (photograph, room, screen) are placed in relation to one another, three moments of a bifurcation: the gazes and movements of the viewer articulate the syntax of the space.

Camera 2 is positioned behind the monitor in such a way that a viewer standing at the head of the ellipse fits inside the image of the Polaroid circle. In addition to the shape feedback, then, there is also an attempt to establish a model of relativization via size. The way dimensions are reduced when reproduced on a screen is by now a visual habit. On this screen, however, a pencil appears to be the same size as the figure of the viewer. A quantity that has been photographed and is thus already deranged in scale (the Polaroid of the circle and pencil) and a real quantity (the viewer) are depicted simultaneously and congruently by being superimposed on the same screen. An observable assimilation of two dissimilar quantities.

This unreal proportion renders visible the relativity of shape and size.

III

Unlike the primary colors of painting (red, blue, and yellow), the three primary colors of video are red, blue, and green. Three neon tubes in these colors lie on the floor, forming a star. They are controlled by a timer switch, so that the room is illuminated by only one of them at a time. Let us call the light emitted by the *neon tube* the actual color (local color, object color). A video camera captures the color and transmits it to two monitors simultaneously. The actual color is reproduced on two additional monitors with a slight delay, casting the light of the past back onto that of the present. Let us call the light emitted from the *TV tube* the live image color. The image color is thus a reproduction of the present actual color and the stored actual color. One might say that the video representation twists the color in time. In a sense, this twisting introduces history into the (stable) system. A temporal feedback loop causes the actual color to overlap with the image color, the present with the past. The past actual color (e.g., red) radiates as an image color ("red") back onto the present actual color (e.g., green). This gives rise to alternating color mixtures. Image color and actual color tend toward immateriality. The color space produced by the neon tube is constantly being changed by the image colors produced by the TV tubes; this is made possible by the video-specific feedback. However, this feedback does not function as a visual mirror, but rather via time: what is being fed back is present and past, not object and image. Video time delay as temporal mirror, as chromatic time. For the first time, color is infused with time. The chromatic heart breathes.

The words *TRAUM* (dream) and *BEWUSST* (conscious) are written on the wall in invisible paints (blue and green, respectively) that glow under UV light. The word *SEIN* (being, but also the second half of *Bewusstsein*, or consciousness) is written on the wall in fluorescent red paint, which means that it is always (faintly) legible. The UV light is synchronized with the red light, so that they are always illuminated at the same time. When that happens, all three words can be read.

The formal feedback (two proportions, two figures) and the chromatic feedback (two time frames) break through the constancy of color and shape, the homogeneity of time and space. This fluctuating, unstable, evolutionary process might serve as a model for the theoretical dream of all humans sharing the same consciousness. To wit, we see that such a state cannot be a simple exchange of information, but rather that communication influences both receiver and sender. Over time (human memory: books, buildings, the works of past humans), the consciousness of the past is fed back into the consciousness of growing, living humans, and vice versa. As the end of history, the optimum in a series of equalities, an equal or simultaneous consciousness shared by all humans is not attainable, because we are dealing with an open-ended, constantly evolving, dissipative process that involves successive instabilities, shapes itself by means of all present and past factors and ramifications, and is defined by the nonsimultaneousness of its elements. This is because "there can be no end to history," for "every system may present instabilities when suitable perturbations are introduced. ... We have in this perspective the constant generation of 'new types and 'new ideas' that may be incorporated into the structure of the system, causing its continual evolution."[1] Humanity must simply see to it that the structure of the social system remains dissipative and thus open to such developments.

Note: Adapted from Peter Weibel, *"Der Traum vom gleichen Bewusst-Sein aller," protokolle: Zeitschrift für Literatur und Kunst* 17, no. 2 (1982): 180–85. Translated from the German by Patrick Hubenthal.

[1]
Ilya Prigogine, *From Being to Becoming: Time and Complexity in the Physical Sciences* (New York: W. H. Freeman, 1980), 126–28.

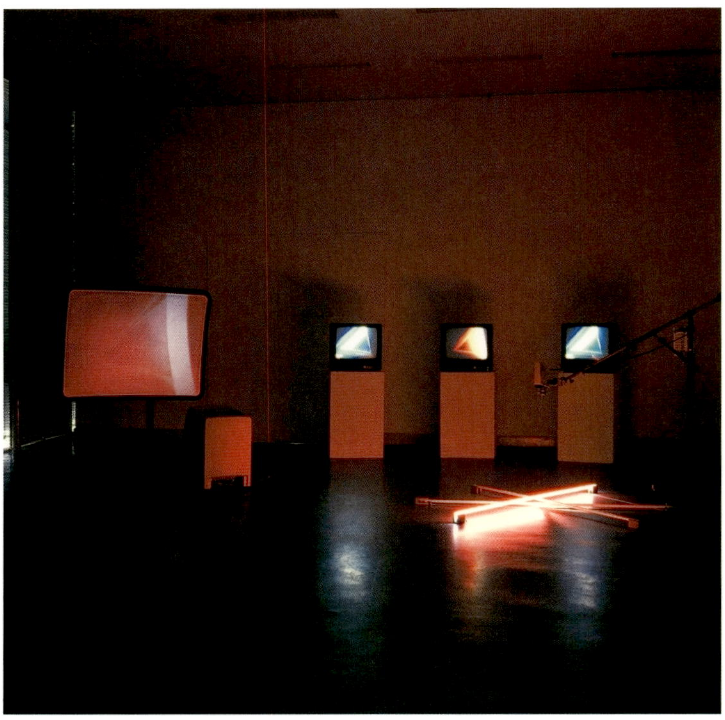

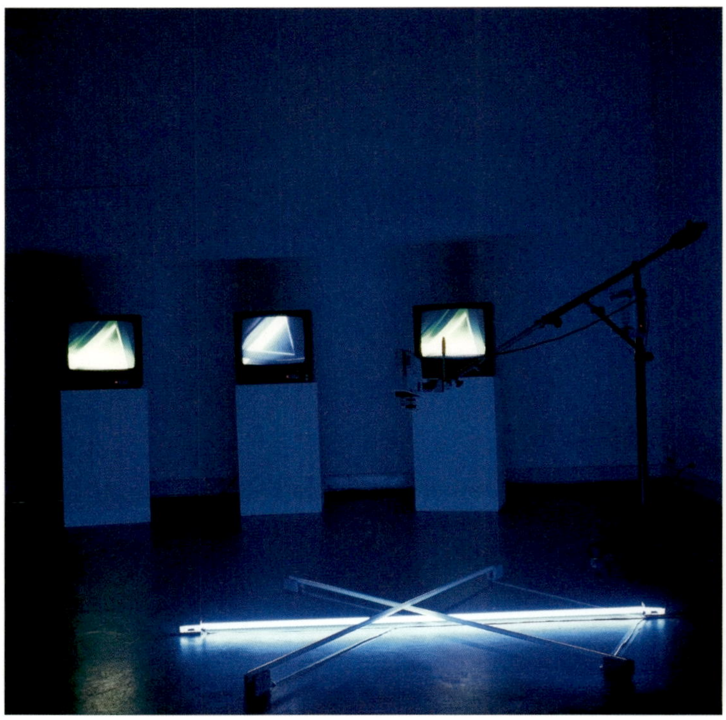

Part III

Installation views
Videoräume, Videowochen
Essen 79, Folkwang
Museum, Essen, Nov. 1979

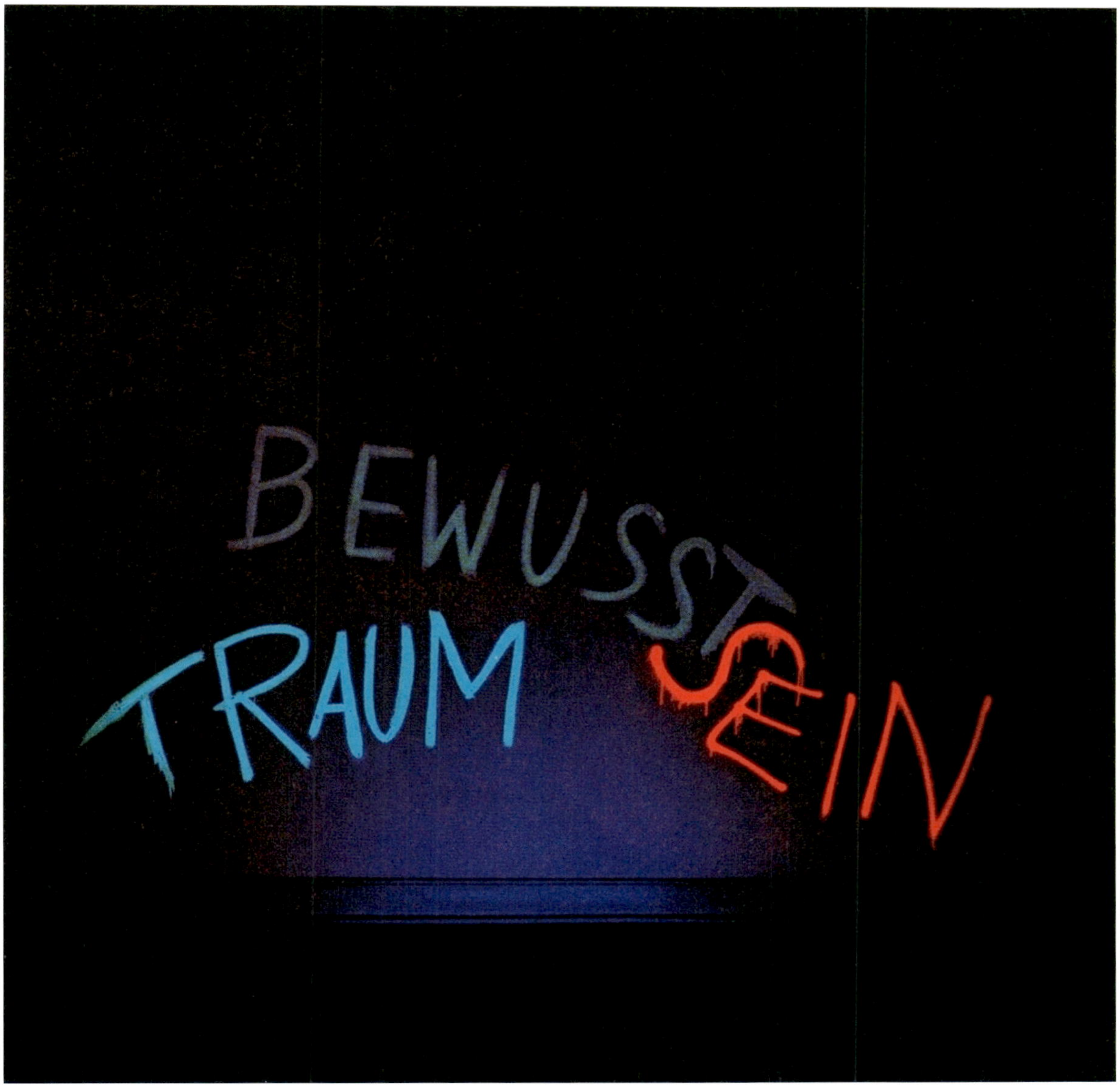

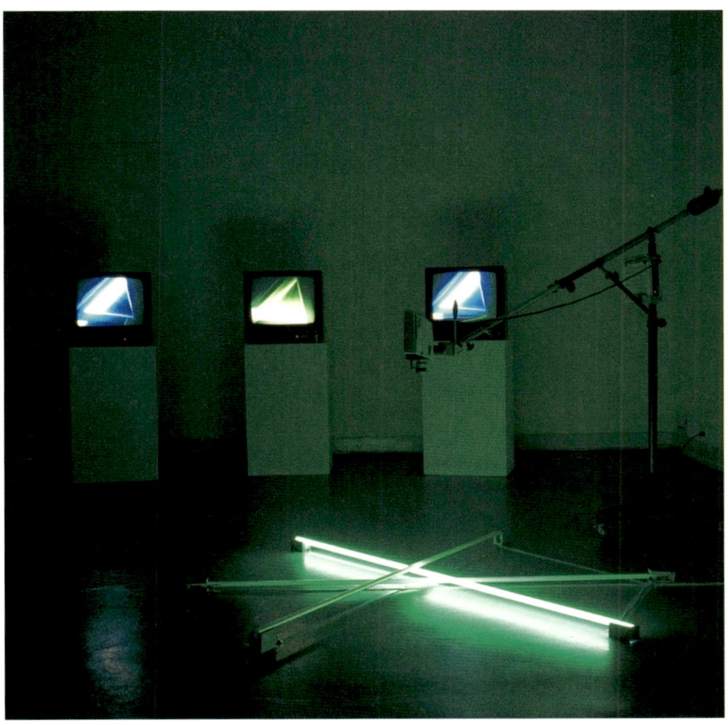

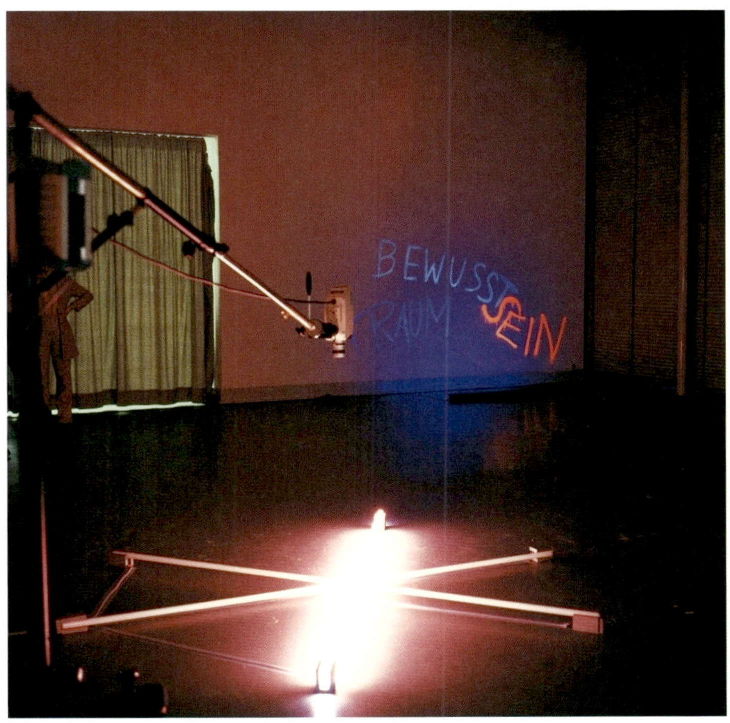

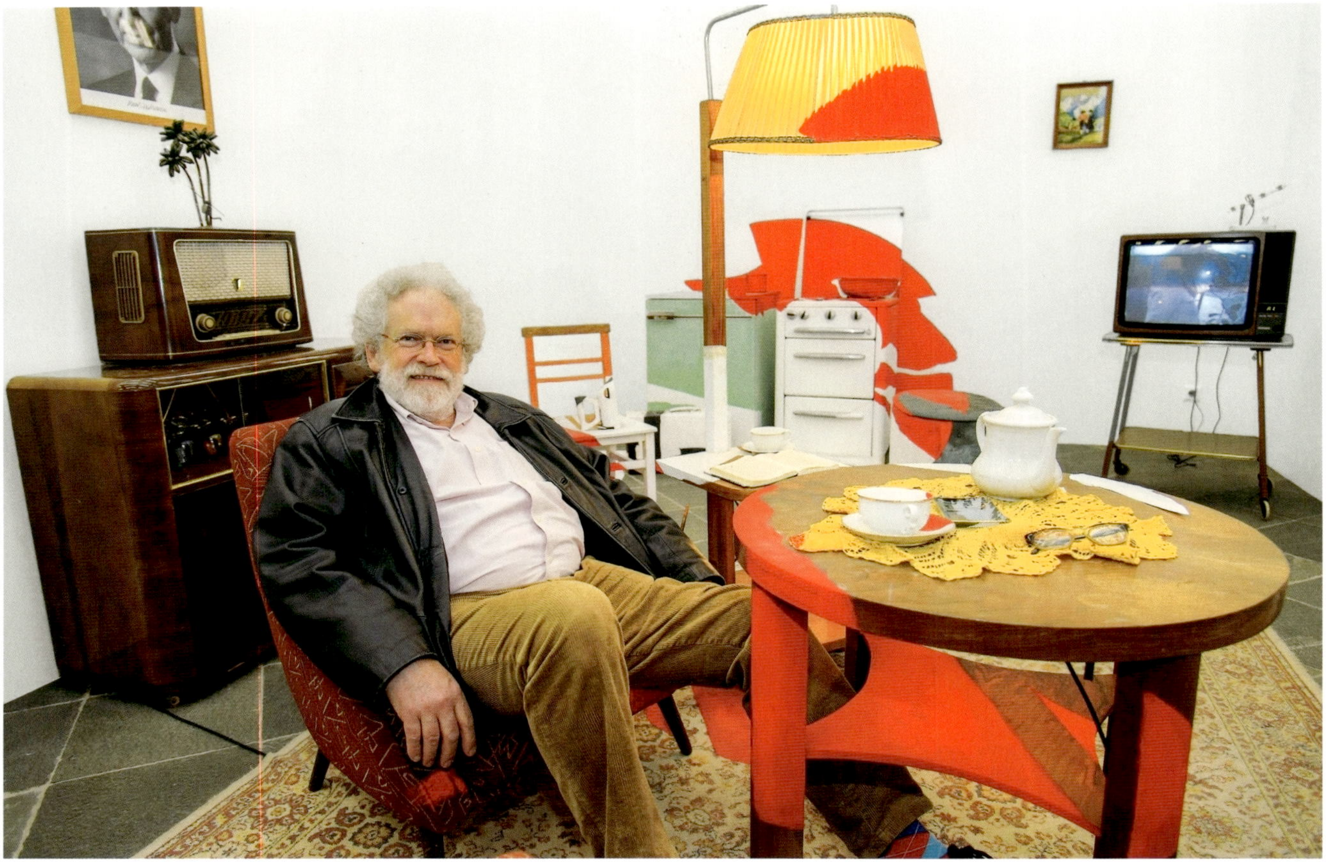

Peter Weibel's friend Anton Zeilinger, winner of the 2022 Nobel Prize in Physics, in the *Austria Room*, installation view *Peter Weibel – MEDIA REBEL*, Belvedere, Vienna, 2014–15

The pictures of Piero della Francesca, who also wrote the *Trattato d'abaco* (a handbook for businessmen, ca. 1480), brought together painting, mathematics, and mercantilism. The sunlit chambers of Vermeer's paintings (which were in fact drawn with the aid of a "dark chamber," a camera obscura) reproduced the social consciousness of the Netherlands as a colonial power. These images are mirrors of the social code. Perspective, therefore, is not just a formal instrument for the representation of space; it also lends the forms meaning. It reproduces the codes of consciousness, the codes underlying the construction of meaning. Via the discourse of technology and science, the discourse of art is also linked to the discourse of politics.

In the installation *Austria Room*, the viewers become part of the construction. Typical middle-class furniture is arranged around the room, painted red and white in a seemingly haphazard way. From the camera's perspective, however, the splotches of color form a question mark which is visible on the TV screen. The gas heater is a symbol for the many suicides in Austria.

The disgraceful dungeon that is the Austrian state, the Second Republic, which seems to consist solely of *carceri oscuri* – shadowy figures, shadowy machinations, secret societies, secret pacts, intrigues – meets its artistic representation, its mirror image, in this *Austria Room*. The use of red and white here as Austria's national colors – that is, the sudden politicization and injection of meaning into "free" colors, colors that would seem to be merely formal and devoid of meaning – causes the signs to become fluid. The viewers are not merely in the picture; they become prisoners, hostages of the Austrian state – which is their actual condition. They experience the horrifying, suffocating uniformity of Austrian society, and the blocklike unfreedom that nests inside it, as a real situation.

Signs are given new interpretations. The signs of the state are visually deconstructed, their meanings logically exposed. The official text is given its subversive subtext. —Peter Weibel

Translated from the German by Patrick Hubenthal.

Austria Room (With Gas Heater for That Authentic Austrian Feeling) / Österreich-Zimmer (mit einem Gasherd zur Erzeugung echt österreichischer Stimmung)

1982	closed-circuit video installation; spatial assemblage with different materials and furniture in the style of the 1950s, video camera, monitor, white and red paint, photographs of all the Austrian presidents of the Second Republic

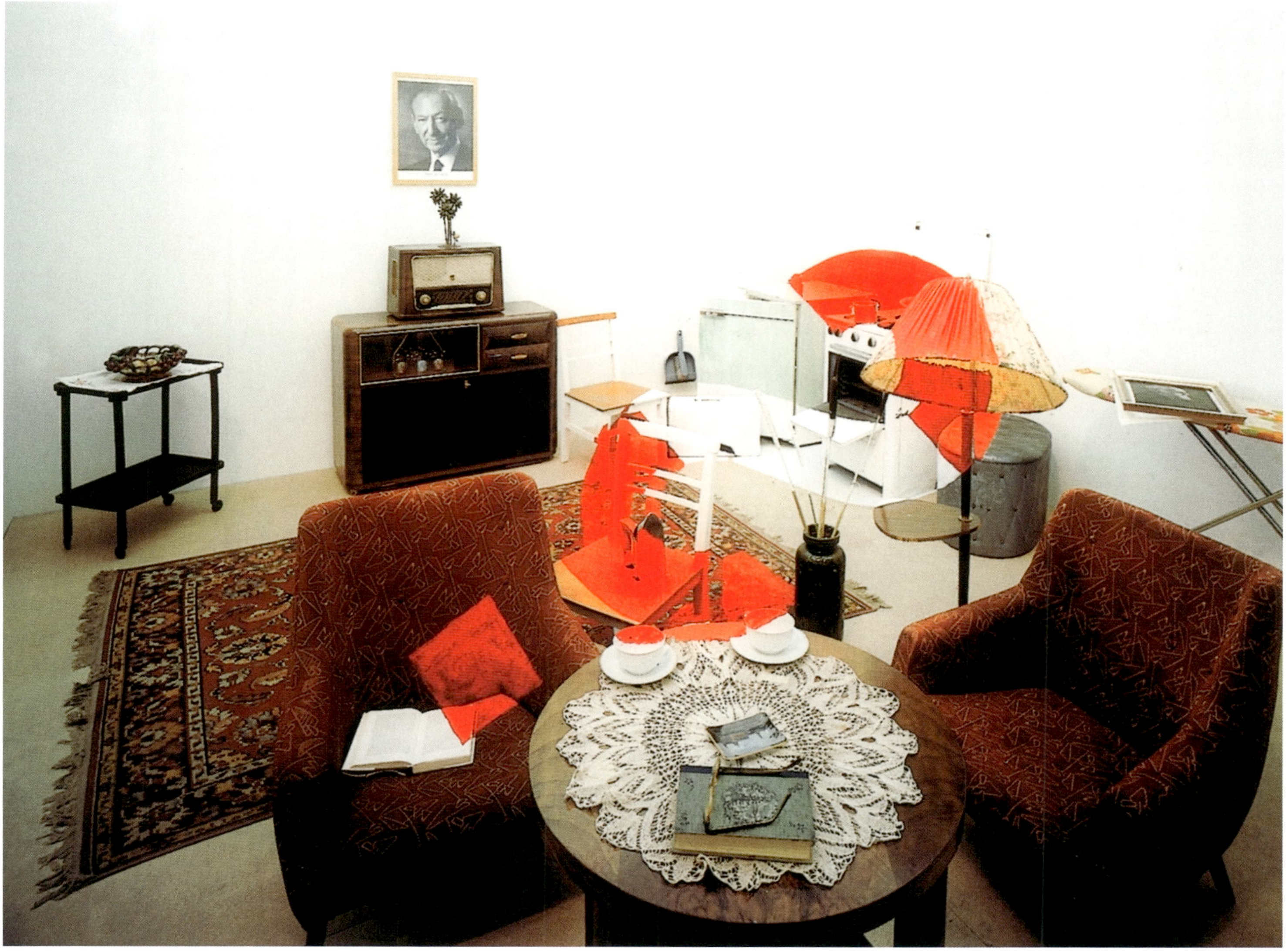

Installation views *Zeichen im Fluß*, Belvedere, Vienna, 1990

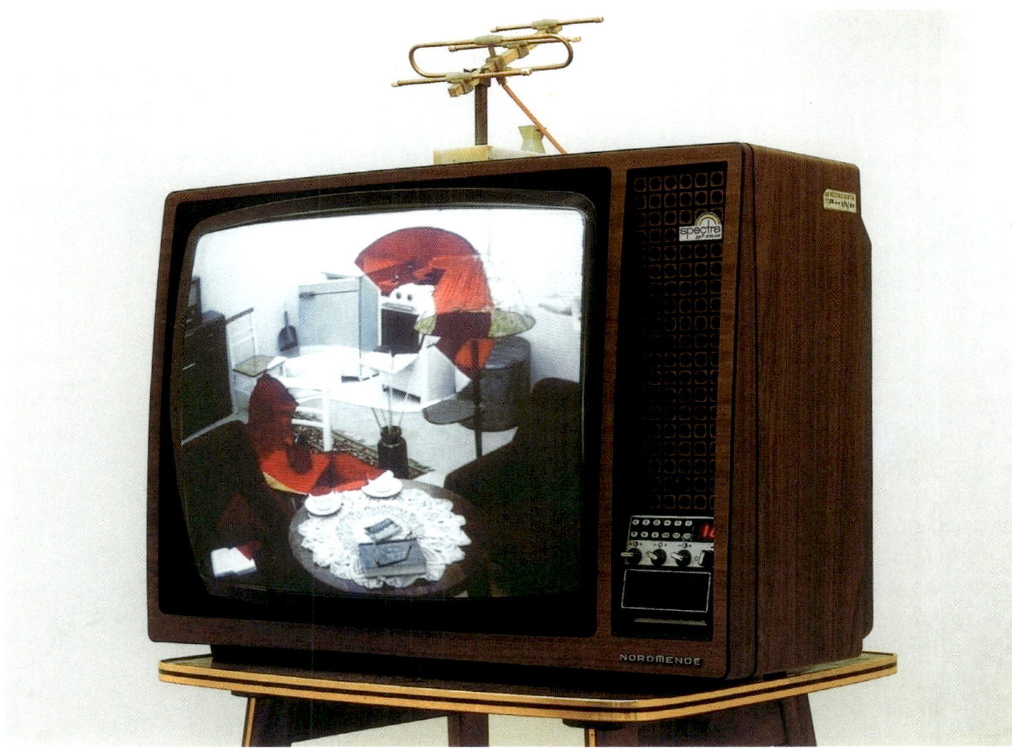

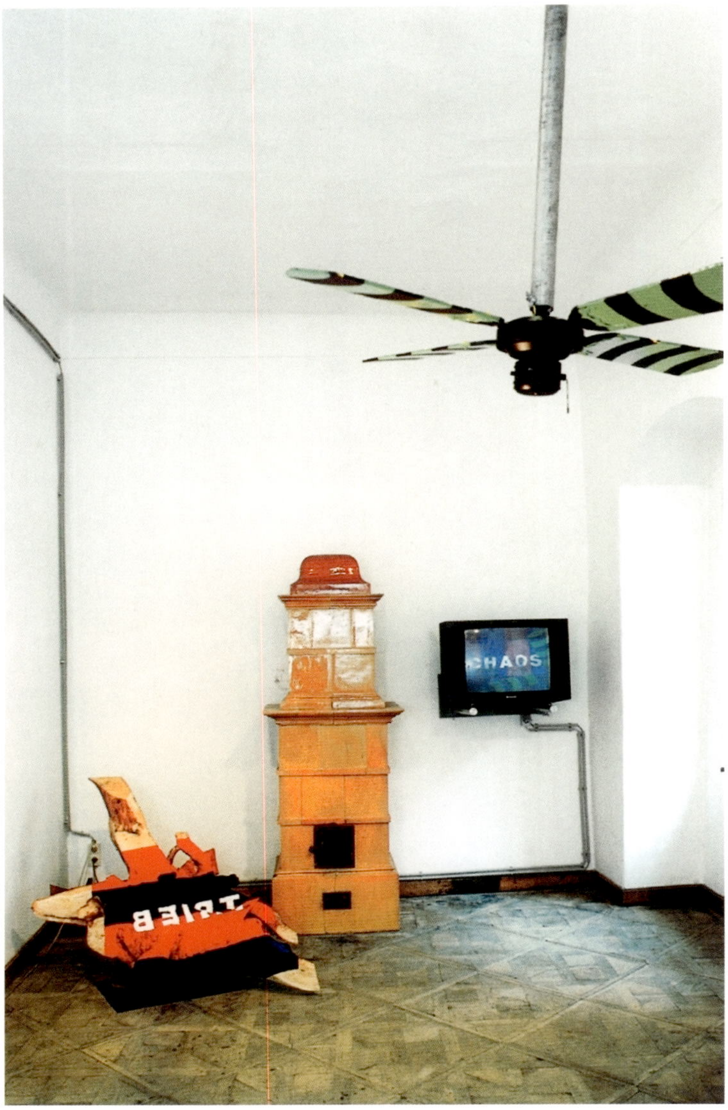

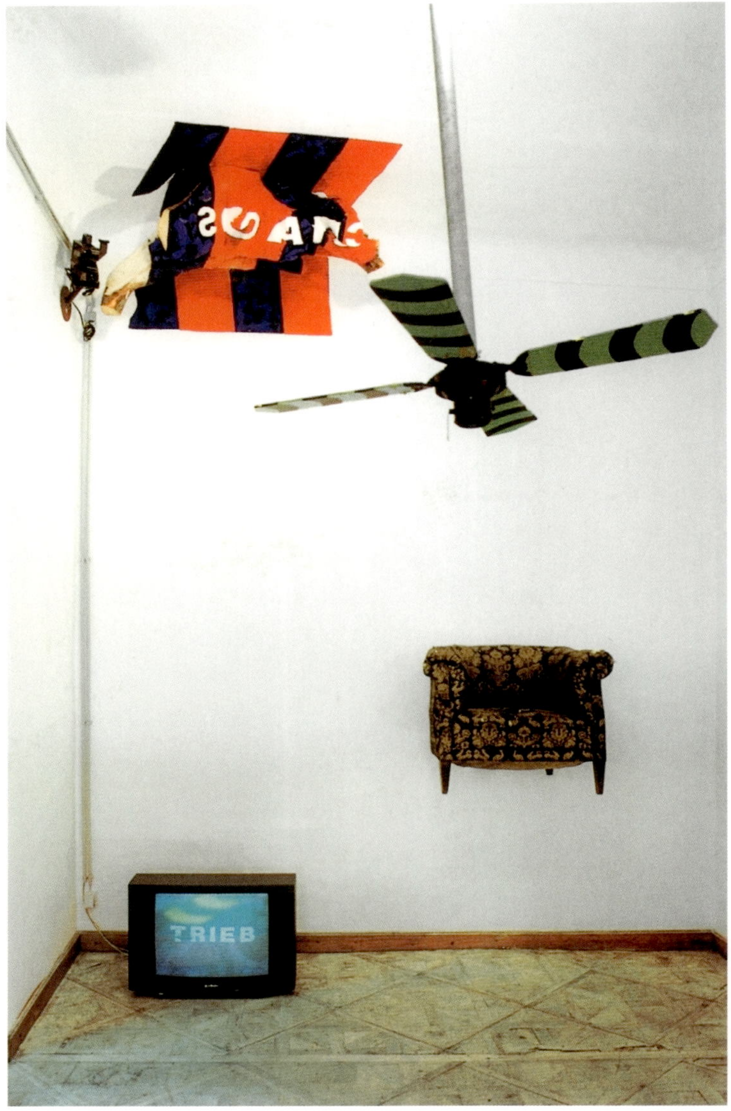

Installation views
Kunstraum Buchberg IX,
Buchberg am Kamp, 1989

On the Dynamics of the Virtual / Zur Dynamik des Virtuellen

1989	rootstock, 2 monitors, 2 video cameras, ceiling fan, armchair

A shredded living room. In two opposite corners, a rootstock and an armchair are placed with the words *CHAOS* (written on the ceiling) and *TRIEB* (drive). The words appear in mirror writing, unreadable by the viewer. Two cameras record the two objects and send the images to screens in the opposite corners of the room. The words are now readable from the privileged perspective of the video camera. On the monitor, the ceiling fan chops up the words.

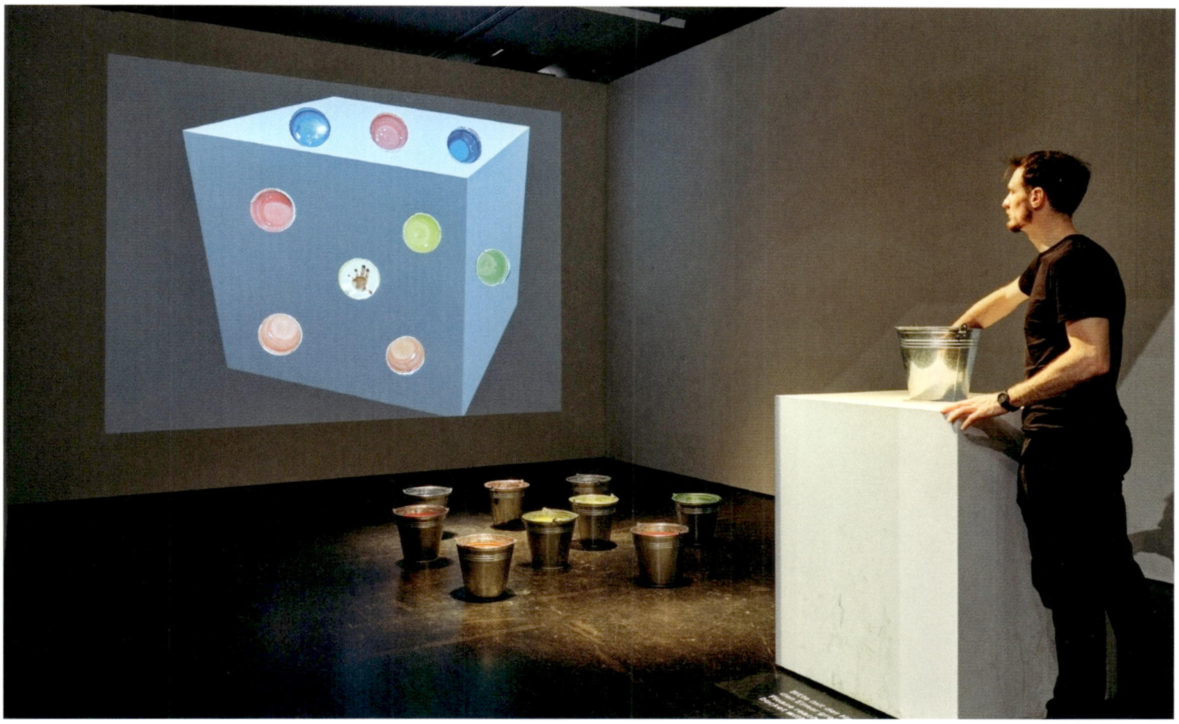

Installation view *respectively, Peter Weibel*, ZKM | Karlsruhe, 2019–20

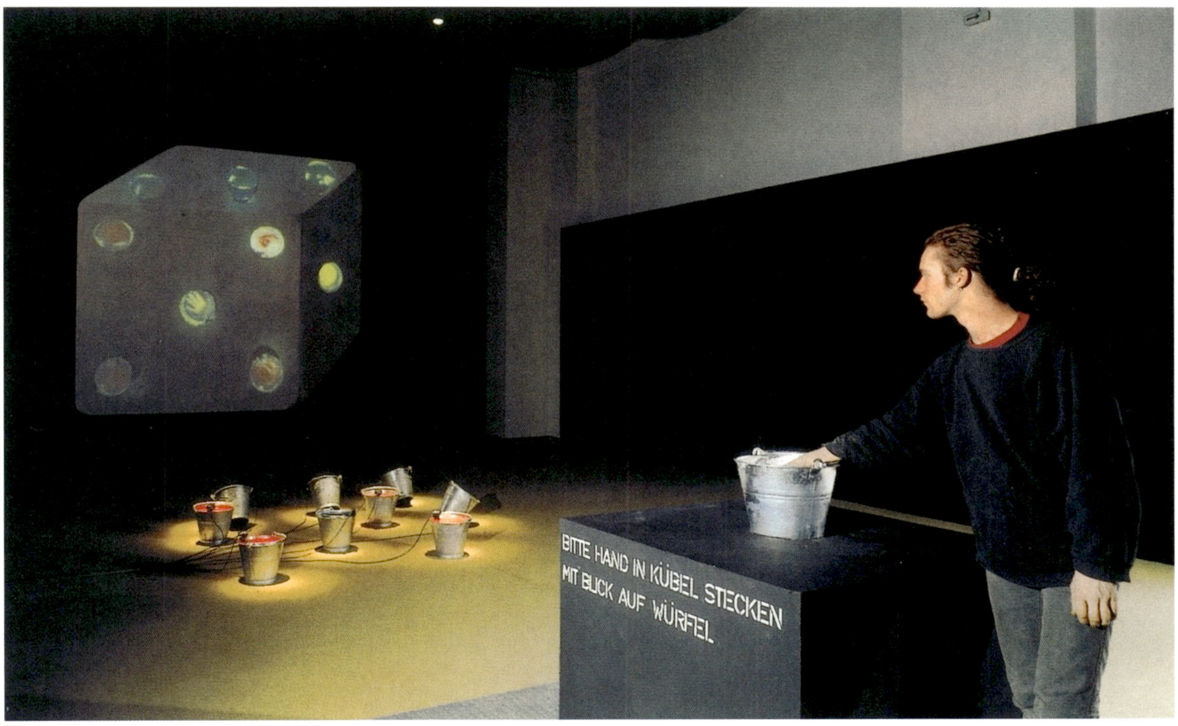

Installation view *Inszenierte Kunstgeschichte*, MAK – Museum of Applied Arts, Vienna, 1988–89

Water Dice / Wasserwürfel

1988	interactive computer-based installation; large format projector, 2 video cameras, mirrors, wood, 10 zinc buckets, paint, lamps, water, motors, transformers, chroma key mixer, dimensions variable technical assistance: Franz Xaver

Nine buckets in a seemingly random order and filled with differently colored water stand in front of a wall, on which there is an image of three sides of a dice with nine dots. The buckets are filmed by a camera positioned on the ceiling, and the recording is then projected vertically on the projected dice (Karlsruhe) or painted dice (Vienna), forming its dots. On a real dice there is an empty bucket without a bottom. A second camera inside the dice looks upwards. If the observer puts a hand into the bucket, it sticks out of one of the dice's dots. The hand travels from the bucket to the dice, separated from the body. The recordings from the cameras in the physical dice and on the ceiling are layered in a video mixer with blue coloring and are then projected. With this work Weibel points to the shrinkage of distances, the separation of messenger and message, the transcending of the body in digital space.

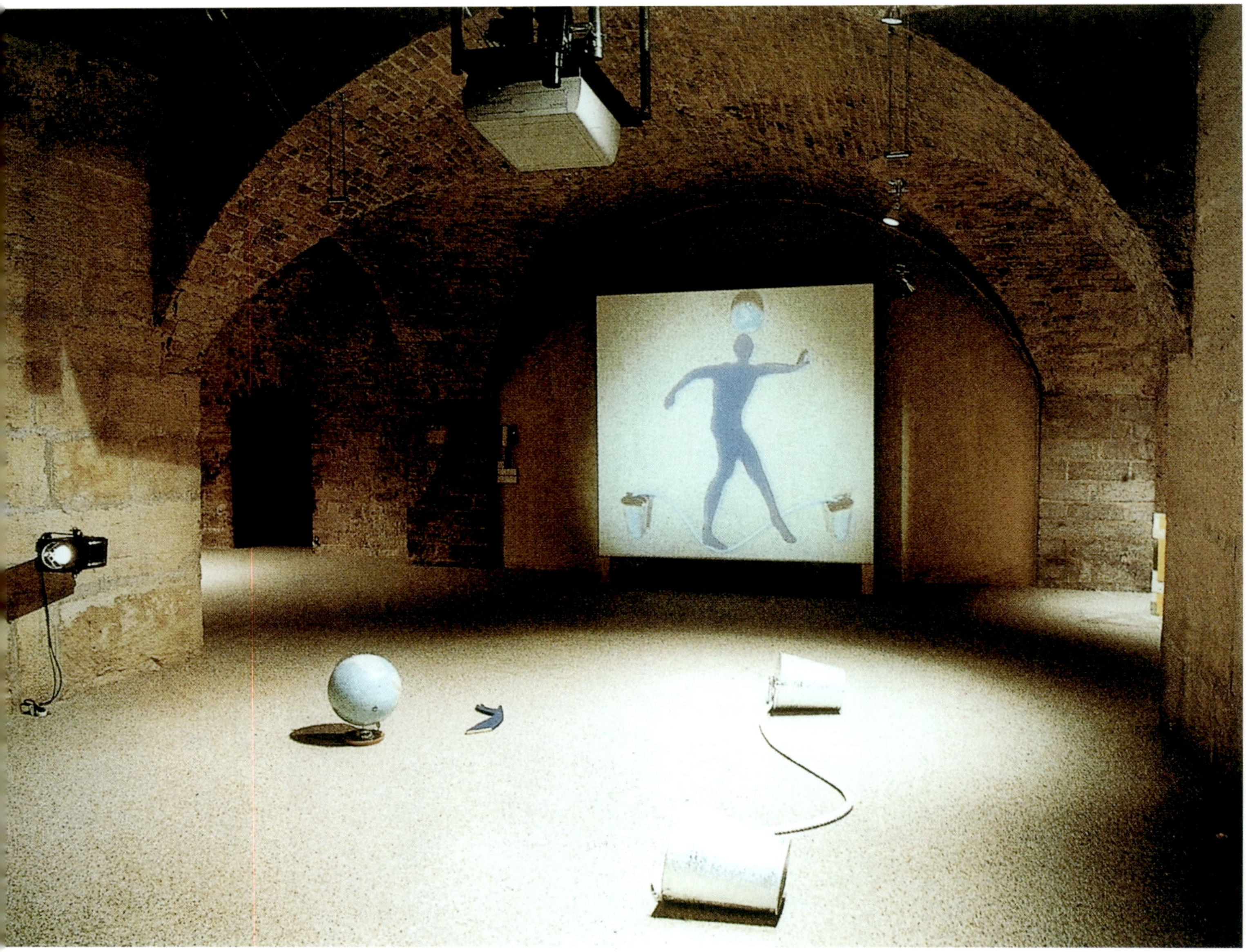

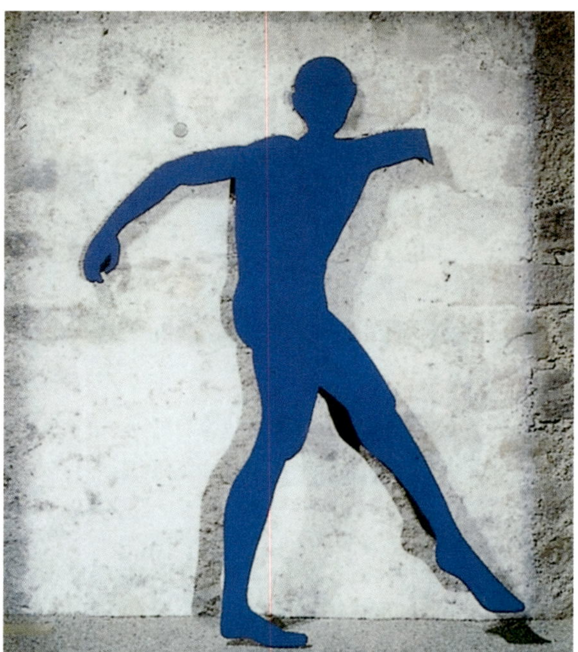

Installation views
Elisabeth Schneider Preis 1990, Galerie Schneider, Freiburg im Breisgau, 1990

Scanned Space

1990	projection screen, 2 video cameras, wooden figure in two parts, globe, 2 buckets, rope, video projector, mixer

From above, video camera 1 records the scene on the ground (two buckets, a rope, an arm, and a globe) and transmits this image to a mixer. Video camera 2 records a one-armed wooden figure opposite the projection screen and transmits this image to the mixer as well. There, the two images are adjusted so that they fit together seamlessly. As a result, the video projector, which receives the signal from the mixer, projects a tightrope walker balancing a globe on his head.

Conceptual drawing for *Ear Wings*, 1978

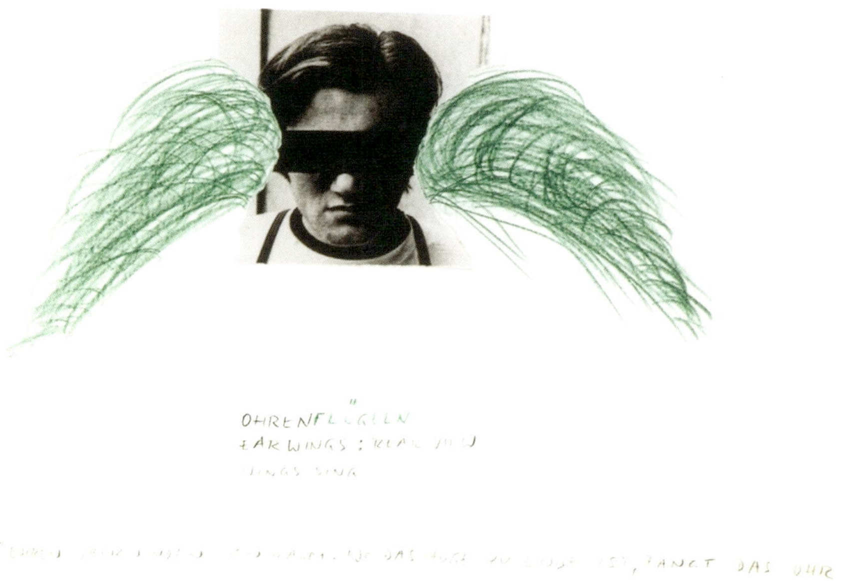

Installation view *Peter Weibel – MEDIA REBEL*, Belvedere, Vienna, 2014–15

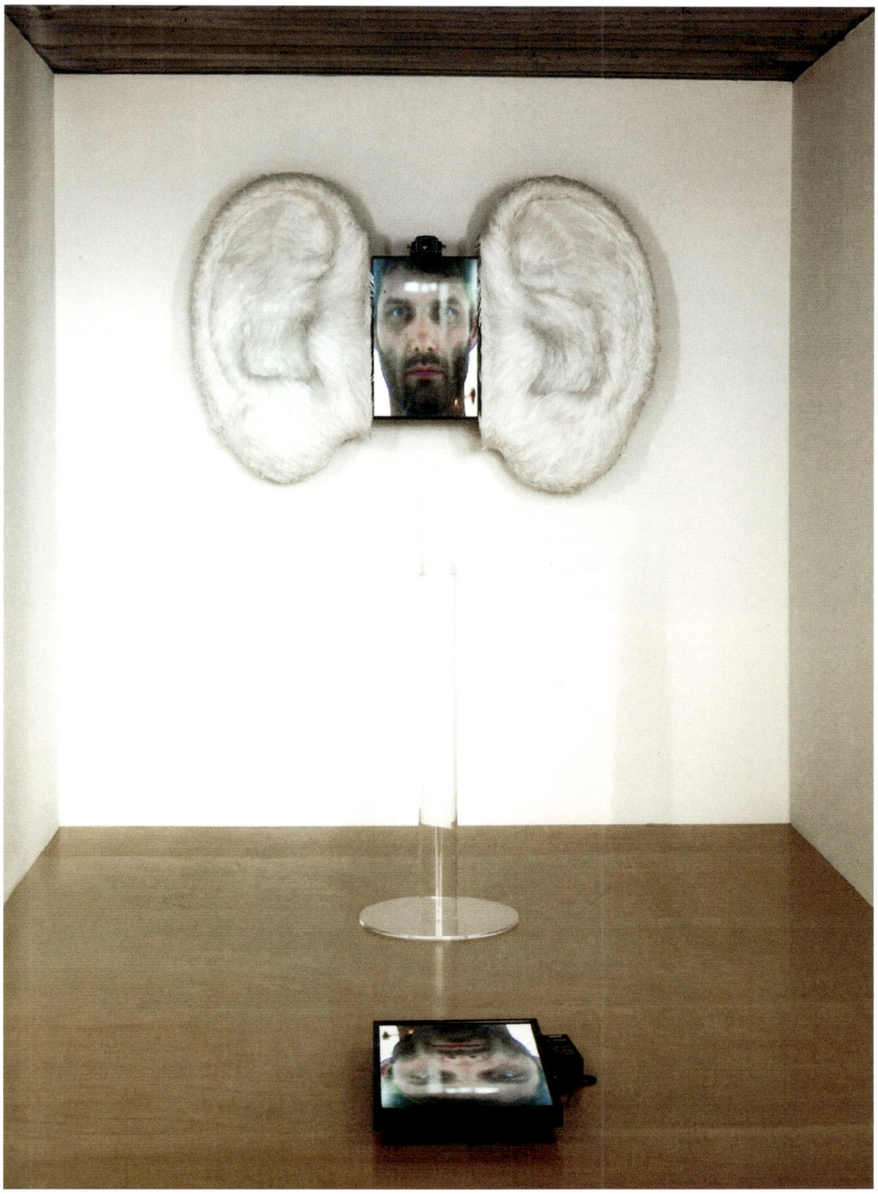

Wings are tools to fly with, to defeat gravity and space. Ears are wings for picking up sounds. They detect and transform distant sources which can't be seen. The sound carried by waves defeats gravity and space as well. Following this thought, angels have wings because they – as nonterrestrial beings – do not obey the laws of time, space, and gravity either. Therefore, our ears partially lend us angelic properties.

The way the installation *Ear Wings* is constructed underlines these ideas. A camera catches the distant observer and places him on a monitor between two furry ears. The vertical image is mirrored horizontally. The portrait and the two ears are a composite image of real and virtual, local and nonlocal elements. The candle in front of the monitor recalls the passing of time – as exemplified in the history of painting.

Ear Wings / Ohrenflügel

1978/ 2014	video installation; ears of the artist made from feathers, 2 monitors, video camera, mixer, candle, candle holder

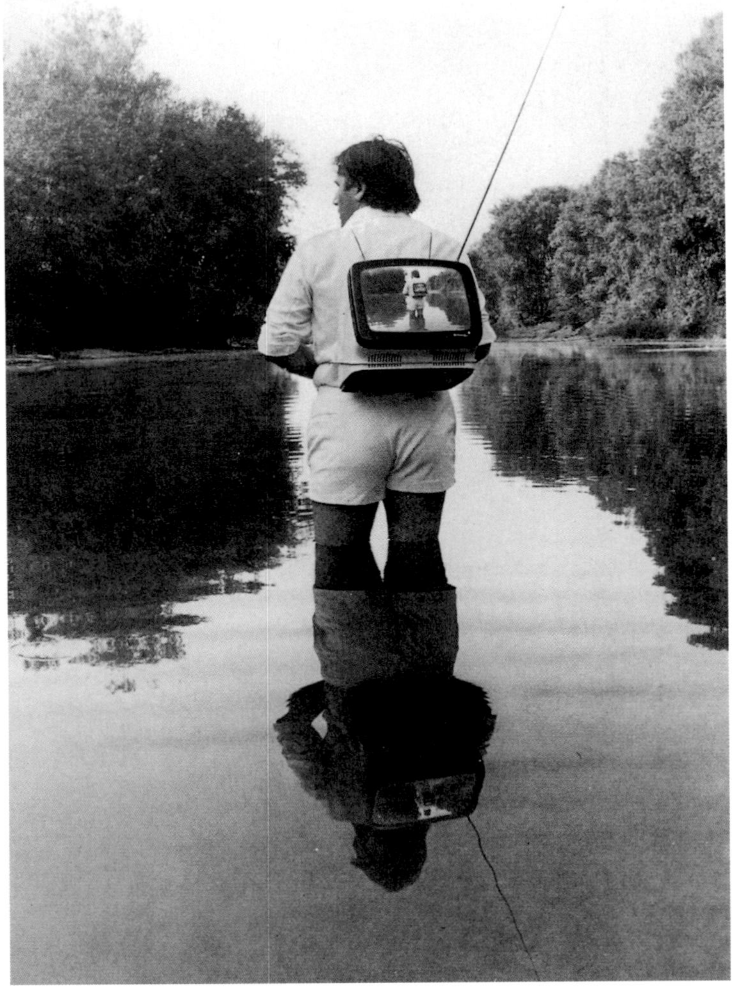

According to the myth of Narcissus, a pool of water was the first mirror (a naturally occurring mirror) to give humans a glimpse of themselves, their faces, their bodies. The mirror image is a reflection, a visual representation, a visual repetition, an imitation. Over the centuries, humans have created many artificial mirrors, various media of mimesis, from the silvered-glass mirror to the video screen. In his theory of the "mirror stage," the psychoanalyst Jacques Lacan underscored the importance of the small child's identification with their image in the mirror, between the ages of six and eighteen months, for the process of becoming a person, the development of the ego. The child's first encounter with themself in the mirror gives them great delight, because they are no longer just seeing partial objects, individual pieces of their body, but rather their whole form and face.[1]

Humans, then, have invented many mirrorlike media: painting, photography, film, video – all technological media that reflect humans, their bodies, their faces. However, traditional mirrors portray their subjects in a limited way: the mirror before which a person stands does not show them from behind, only from the front, face to face. As a result, there are some representations that are impossible, such as those presented in René Magritte's 1937 painting *Not To Be Reproduced* (*La reproduction interdite*) and my 1977 video performance *Reflection Permitted* (*Die erlaubte Reflexion*).

Technological media, on the other hand, enable the camera to be positioned behind the viewer, whose image on the screen in front of them shows them as seen from behind. Technological media have thus brought about a radical break, in the form of a visual bijection. Furthermore, the camera can also be pointed at the screen and reproduce its image, creating a virtually infinite chain of reproductions, an endless loop. What was dreamed of by the most audacious painters, such as M. C. Escher, and depicted by the physicist Ernst Mach, albeit somewhat incompletely (the paper on which the drawing itself is being executed is missing), in his famous drawing of the "self-inspection of the ego" (ca. 1885), has now been achieved by technological imaging systems. The self-portrait has been extended beyond the self.

Which is why I placed myself in water, the first mirrorlike medium, which functions as the first link in an endless chain of reflections. In the tradition of the forbidden reproduction, I am seen from behind with a monitor, the new, technological mirror, which in turn shows me in the water with the monitor, *ad infinitum*.

Reflection is a special case of symmetry. As we know, there are vertical and horizontal reflections or symmetries. The water reflects my figure and the trees in horizontal symmetry. At the same time, the real trees form a vertical symmetry, the monitor a kind of fractal symmetry. The multiple reproductions and symmetries thus produce an extremely dense chain of reproductions: infinite contemplation.

However: in 1982, I did not have access to a camera capable of transmitting its image to the monitor wirelessly. The vision was thus more advanced than the available technology, so I had to settle for a photomontage, a performance for photography.

—Peter Weibel, 2022

Translated from the German by Patrick Hubenthal.

[1]
See Jacques Lacan, "Le stade du miroir comme formateur de la fonction du Je, telle qu'elle nous est révelée dans l'expérience psychanalytique," *Revue Française de Psychanalyse* 13, no. 4 (1949): 449–55.

Infinite Contemplation / Unendliche Anschauung

1982 | b/w photomontage

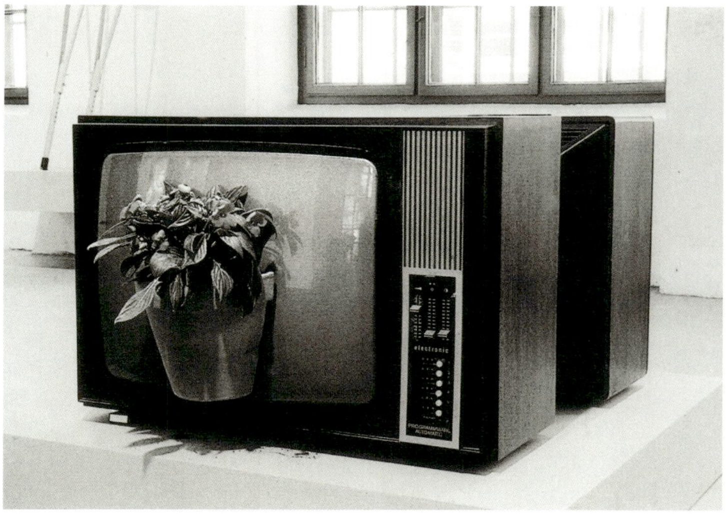

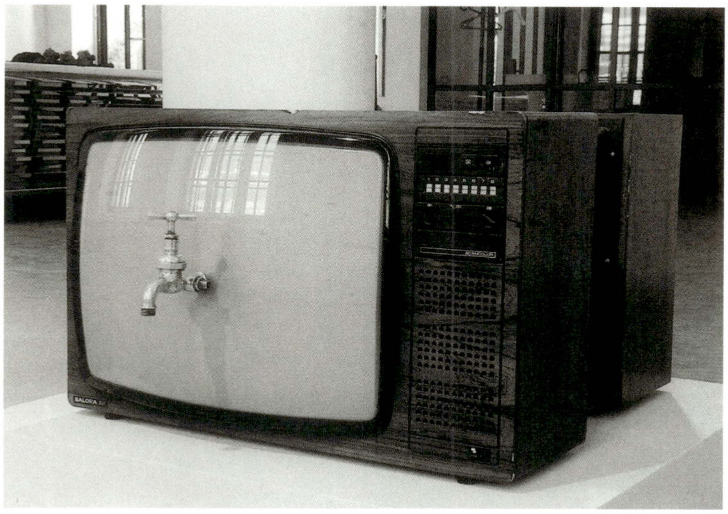

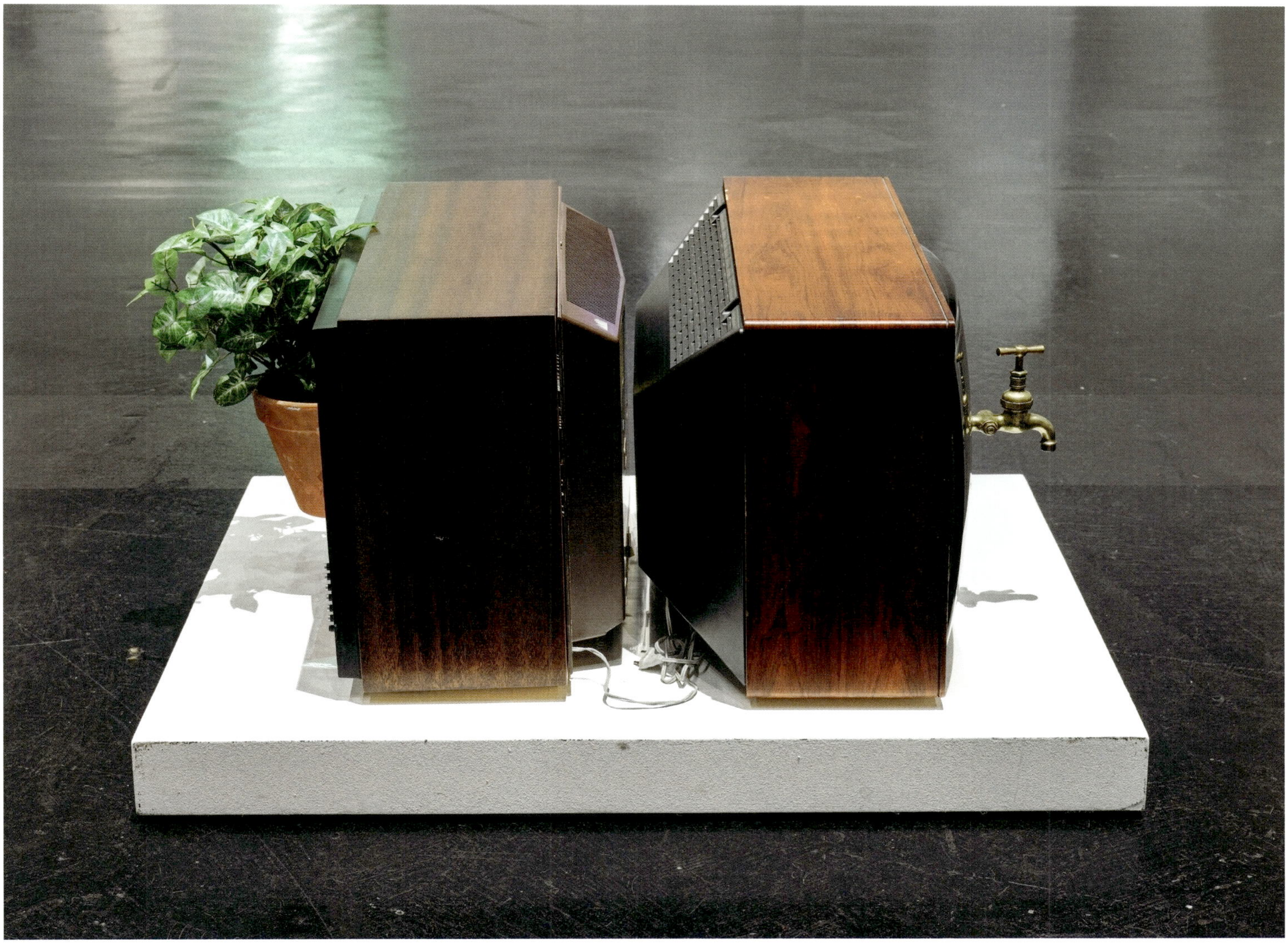

Top: installation views
Hypothetische Produkte. Vom Elend der Privilegien, Prodomo, Vienna, 1994;
bottom: installation view
respectively, Peter Weibel, ZKM | Karlsruhe, 2019–20

When Fuji Was Still a Mountain and Not a Company. Re: Wind & Fast Forward (of the Real) / Als Fuji noch ein Berg und keine Firma war. Re: Wind & Fast Forward (des Realen)

1990	2 televisions, faucet, plant in a half flowerpot, 72 × 80 × 46 cm

In the work *When Fuji Was Still a Mountain and Not a Company. Re: Wind & Fast Forward (of the Real)*, Peter Weibel positions two televisions back to back. From one screen a faucet protrudes, and from the other a halved flowerpot containing a living plant. If one interprets the faucet as the flow of information from the media and the plant as reality, it soon becomes clear that the information conveyed by the media and reality itself are not reconcilable. Because of their opposite alignments, the water from the faucet will never reach the plant. Here Peter Weibel formulates a highly topical form of media criticism, which at its core questions the relationship – or rather the gap – between media and reality.

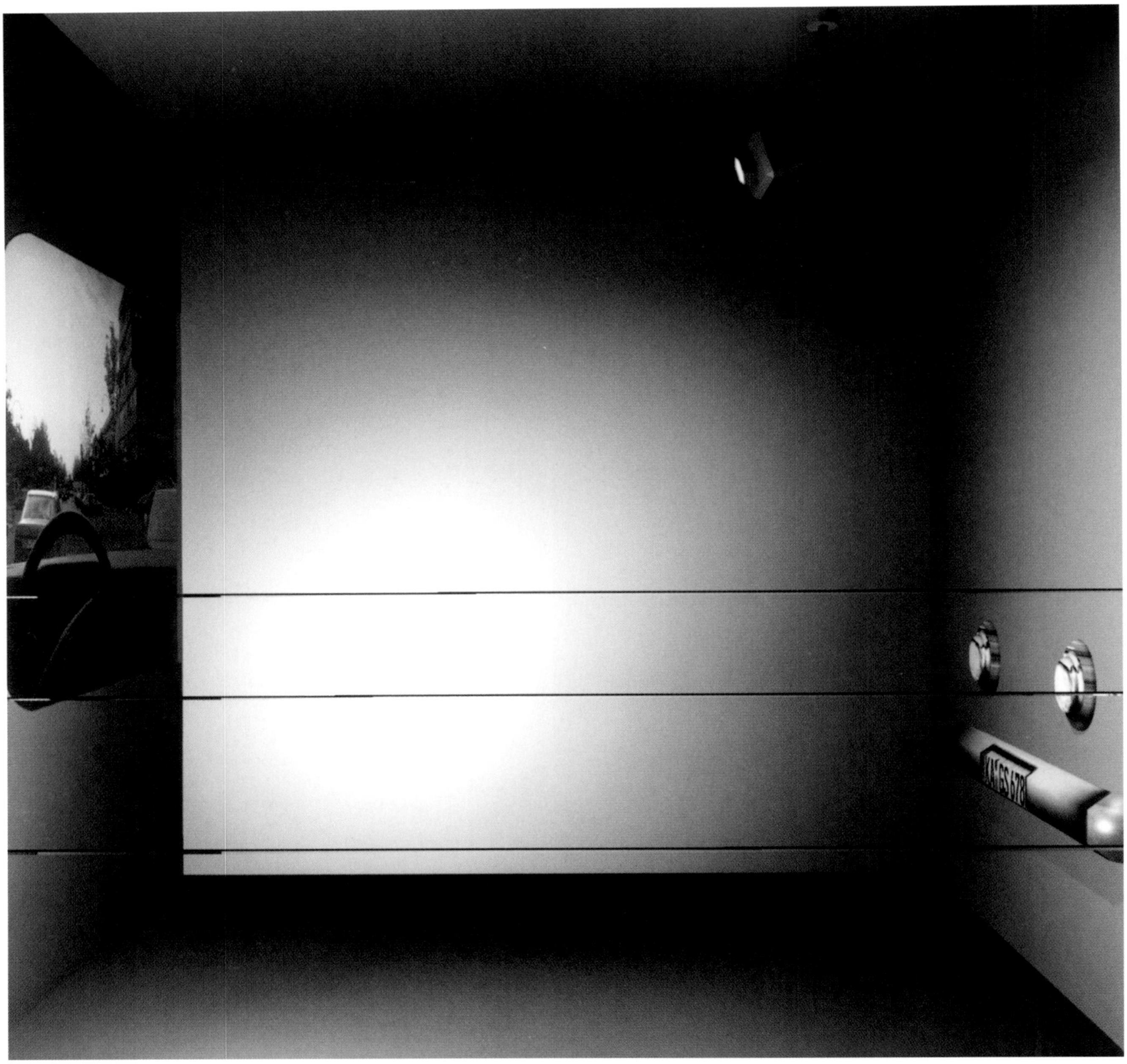

Computer simulation for the exhibition *Globale Gier*, Kärntner Landesgalerie, Klagenfurt, 1999

A car bumper and two headlights are mounted on a wall, creating the impression that the car itself is behind the wall, and that it was the wall that brought it to a stop – that the car is now stuck in the wall. A video of a car ride, shot from the perspective of the driver, is projected onto the opposite wall. The automotive object is stationary, whereas the image is moving. The spatial axis is also reversed: the viewer, who occupies the position of the driver and seems to be looking out of a moving car at a country road, is standing in front of the car but should really be sitting in the invisible vehicle behind the wall. But if the viewer are standing between the image of the car drive and the automotive object, then they're in a place that does not actually exist. One cannot be looking out through the windshield and standing in front of the car at the same time. Point of view and body position do not match; they are contradictory and divorced from one another. The organ of vision has detached itself from the body. The body's organs float freely in technospace, which has no locatable boundaries. This imaginary bending of the body in space deconstructs the spatial axes; the viewer is no longer skimming over the grid formed by the Cartesian coordinates of space and time. This makes possible a bodily experience of how the body itself is dismembered by mechanical dislocation in technospace and technotime. The disembodiment effected by telematic dislocation manifests itself primarily in the dissolution of the unit composed of the body and its organs. The body's parts are uncoupled and restrung on a new space-time axis. The technobody is a dismembered body. The contemporary experience of space and time is a disembodied experience.

—Peter Weibel

Translated from the German by Patrick Hubenthal.

Space and Time Unbodied: The Body Freed from Space and Time / Raum und Zeit, entleibt: Der Leib, von Raum und Zeit befreit

1999	car headlights, bumper, video projection

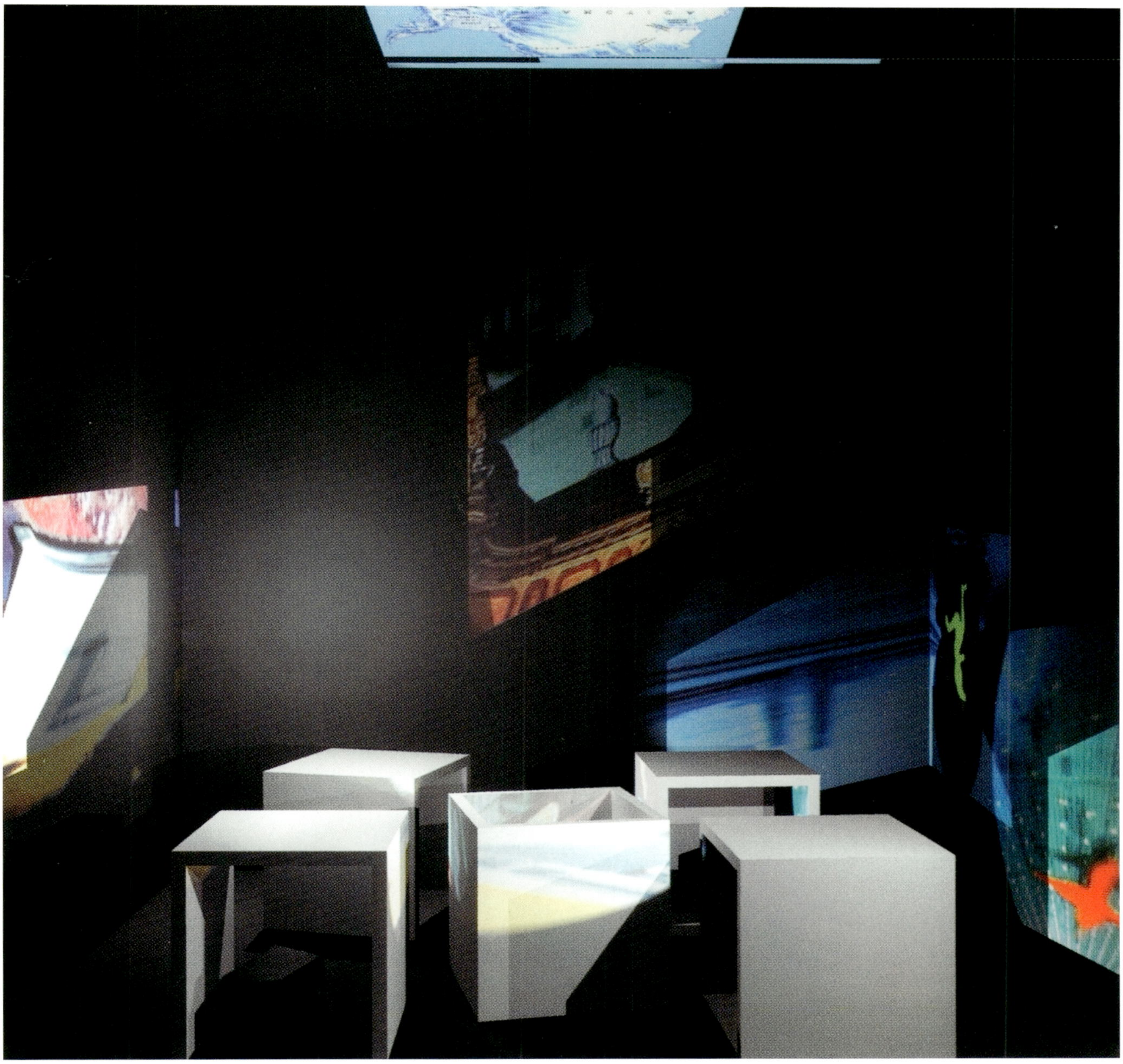

Computer simulation for the exhibition *Globale Gier*, Kärntner Landesgalerie, Klagenfurt, 1999

The room, consisting of six walls, is a box. Famously, architects even use the term "dumb box." In the cultural history of modernity, the term "box" has become a secret leitmotif. Boxes are referred to frequently and in the most widely differing contexts. Behaviorism regarded the human mind as a black box. From the white cube of the museum gallery to the human black box, we are ensconced in a universe of the closed box – a form which particularly in contemporary art has subsumed the classical functions of frame and pedestal. This installation, in the closed box of the room, shows several smaller boxes which open in different ways. The boxes are multifunctional in nature, appearing as seats, containers, packaging, storage. The mirroring of the room box in its many subboxes refers to the model of a world and its subworlds. It is for this reason that a map of the world is presented on the ceiling.

The Cycle of Boxes / Der Kreislauf der Kisten

1999	5 projectors, 5 boxes, 50 × 50 × 50 cm each, 5 video recorders, everyday objects and garbage, videotapes

The miniature projectors projecting various video works by Peter Weibel in all directions fulfill the classic function of a picture, which is to enable views into other spaces and eras through the window of the picture frame. Thus, the pictures punctuate the real space with imaginary spaces and events. The dumb box is broken open and hoodwinked. This interplay of wall and picture reached its apex during the Baroque period and is subtly referenced in the projection on the ceiling. Today, in the media age, this same concept becomes an intersection between real and virtual worlds. The installation, styled in sloppy domesticity, points to the box's most common use: as a home. This home, in turn, is subject to alienation because of the pictures of unfamiliar times and alien spaces. Depots are places where things are left. This is the third aspect of this space: a place of storage and memory. Trunks, suitcases, and boxes, large and small, share the fate of becoming containers for things that are left over, and are therefore themselves left in dark storage boxes. This melancholy cycle of boxes is the new staging of the classic *vanitas* motif. The crypt, the grave, the coffin is the final box which human beings inhabit. Art is the rebellion against the final box. —Peter Weibel

Re: Peter Weibel Polyperformer

Pamela C. Scorzin

Even though (or precisely because) new and classic media have always been Peter Weibel's primary interest, his many years of artistic activity have also centered upon the artist's body. The analysis of the relationship between the body, space, and time, both in theory and in practice, constitutes a major theme of his many works. Performative tactics and self-presentational strategies run throughout his respectable career as a "polyartist in the multiverse of the arts," to touch briefly on the well-known Weibel-speak – though of course only he can truly capture the tone ...

The works of Peter Weibel do not have the form of (commercial) commodities. Although autonomously exhibitable as artworks, they are at the same time intimately bound up with the person of the artist and theorist. For the polyperformer Peter Weibel, the body is clearly not just a means of expression but the very first medium, or quite simply *the* primary material. It is therefore all the more astonishing that many of his artistic works simultaneously revolve around expanding and transcending the limited human *physis*. Thus "externalization, exteriorization, [and] ecstasization"[1] constantly recur as central themes of his artistic works.

To that end, media prostheses and technological devices – for example, typewriters, tape recorders, cameras of all sorts, projectors, megaphones and microphones, and lastly computers – have been part of his scene-making performances since the beginning. Whether through actual presence or in delegated form through media presence, the idea is to obsessively-ecstatically communicate and perform with the body while simultaneously freeing the self from it. Expansion is a constant in the Weibelian universe.

The media artist thus has an exceedingly impressive résumé with distinct phases, from his beginnings with the legendary Vienna Actionists right up to the immediate present, in which Weibel, now a much-honored university professor and the long-serving, successful director of the ZKM | Karlsruhe, works his charisma on the international stage. For more than half a century, the Austrian artist has also been appearing – in (self-)portraits, films and videos, installations, and performance projects in public spaces – as a passionate polyperformer, a rebel actor who is constantly reigniting his public – or rather, firing it up intellectually. The conceptual photo performances Weibel staged in the 1970s with his partner, the Austrian artist Susanne Widl (b. 1944), seem to have been fundamentally inspired by the old Doors hit "Light My Fire" (1967) in more senses than one. Under the conditions of the Media Age, his fiery Promethean spirit has been creating an extremely broad and diverse body of work since the 1960s that has consistently dared to conceive artistic-formative creativity and art as an optional and open sphere of activity and expanded social network, which therefore requires an agile agent and active actor. It is clear that the artist is always interested in igniting thought processes and, ultimately, gaining knowledge. To this end, Weibel took semiotic and linguistic reflections dealing with the interdependence of actual reality and the reality of signs, or the synthesis of biosphere and infosphere, among other subjects, and developed them into a singular artistic metalanguage which, in the mid-1960s, took him rapidly and quite early on from experimental literature to (sociopolitical) performance art. In light of the current hype around the art of performance in the most varied artistic and cultural disciplines, from visual art to dance and theater to social media, politics, and pop culture, we should recall that in the context of the 1960s and 1970s, the turn toward this practice must be understood, rather, as a form of rebellious opposition and a revolutionary "opting out" of art – while admittedly maintaining a reference to the established art system.

In his many staged and performative projects before audiences both real and virtual, Weibel has not only investigated the "media" of body and language, but also specifically explored their expansions into modern communications technologies, such as film, video, audiotape, and interactive environments (first electronic, later digital and immersive). Even then, Weibel was already analyzing, reflecting upon, and deconstructing their ontology and media-specific characteristics for the construction and fabrication of reality, and doing so in critical and, by contemporary standards, extremely perceptive fashion. He often made use of paradoxical, attention-grabbing inversions, or rather, astonishing metaleptic stunts, which at first seduced viewers into amazement but then connected with their cognitive faculties via their senses. The epistemic function of art and discourse theory seems to take precedence over any sort of aesthetics in Weibel's work, which is why the performative per se, the provisional-prototypal, the virtual-variable, and the conceptual-demonstrative or flexible-viable are all characteristic of his multifarious artistic practices. In addition to his infamous performances with representatives of the Vienna Group and Viennese Actionism – he acted as the latter movement's chronicler and was most likely its christener as well – and a series of risky, explosive *Attacks* (*Anschläge*, 1970–75, see pp. 536–37 in this volume) documented in photo performances. From 1966 onward, together with VALIE EXPORT (b. 1940), Hans Scheugl (b. 1940), Ernst Schmidt Jr. (1938–88), and others, he experimented with an "expanded cinema" that aimed to systematically analyze and vividly deconstruct the ideological and technological conditions of filmic representation by creating hypostases with materials and people in real space. At the same time, it also vividly translated heady theory into sensuous scenography – for example, in the 1966 Expanded Cinema piece *Nivea*.

With Weibel's early TV projects, his legendary *Tele-Actions* (*Tele-Aktionen*), which were broadcast on Austrian state television in 1972 as part of the program *Impulse*, the poet and media artist as rebellious and provocative performance artist repeatedly overstepped the boundaries of art's operating system. Along with commercial productions, he undertook experimental investigations of what was by then the easily accessible technology of video in its artistic and creative applications in the then mass medium of television. For several years now he has also been goofing around on YouTube with video clips, such as his comment on the "Crapstraction" boom in abstract art identified by Jerry Saltz (b. 1951),[2] that range from the subversive-ironic to the conceptual-performative.

At this point the younger members of Generation Y, the millennials, may be asking: Has the legendary performer and gifted self-representer Peter Weibel also deconstructed and hypostatized the Instagram selfie with filters? Essentially yes, but unfortunately, after a few visual *Photo Poems* he seems to have left the field to the photo artist Cindy Sherman (b. 1954) – who, by the way, he discovered and exhibited as a curator of conceptual photography early on.[3]

As if that were not already enough in the way of being driven to appear in various media, Weibel also turned not only to acting, but to music and sound as well: first with Austrian punk and rock music, and somewhat later with the electronic (media) opera. In 1978, together with Loys Egg (b. 1947), he founded the band Hotel Morphila Orchester; even the legendary Falco (1957–98) – so it is said – threw himself at their feet in Vienna. But with punk songs like "Sex in der Stadt" (Sex in the city) and "Dead in the Head" (both 1982), he not only scored points in the scene but was also experimenting at the same time – long before Rammstein – with extreme musical performances.

Of course, Weibel was never "Dead in the Head" – although it still hurts to watch the music video performance due to Weibel's dramaturgical self-flagellation with the microphone in front of the live audience. This had been preceded, however, by drastic, almost masochistic self-inflicted injuries and risky full-body commitment in various photo performances – for example, in the skin poems in Weibel's *Codex Cutis* (1971, see pp. 154–55 in this volume).[4] One might also recall his setting his hand on fire during the legendary *Incendiary Speech* (*Brandrede*, see pp. 508–509 in this volume) he gave in a lecture hall at the University of Vienna in 1968 as part of the performance event *Art and Revolution* (*Kunst und Revolution*), which was created in the context of student protests, as well as the dangerous bloodletting during the television performance *Time Blood* (*Zeitblut*, 1972/79, see pp. 254–55 in this volume) and the painful setting of his tongue in concrete in the 1973 photo performance *Space of Language* (*Raum der Sprache*, see p. 525 in this volume).

A decade later, ever in step with the newest developments in media technology – for which Weibel, as founder and director of the Institute for New Media in Frankfurt am Main, had become an influential pacesetter in the broader field of the visual arts in Germany and abroad (albeit not without a certain amount of resistance, especially from his colleagues in painting and sculpture) – he worked with a team of experts to conceive, design, and realize early computer-based interactive installations through which he addressed the relationship between media and constructions of reality, between the actual and the reality of signs, thereby conceptually anticipating present-day developments in hybrid realities such as VR and AR.

As a *uomo universale* of "Renaissance 3.0," he has also been publishing definitive statements on contemporary art, media history, media theory, film, photography, video art, and philosophy for decades now in numerous lectures, articles, and books. His tireless inquisitiveness as a researcher and writer is reflected in portraits that depict him as a productive thinker amid chaotic landscapes of books and his impressive private library.

As a scholar, theorist, critic, curator, and practitioner, Peter Weibel has championed art and art history that take into account the histories of technology and science and interlink theory and practice. And while the rest of us are all still talking about globalization and digitization, he proclaimed the algorithmic revolution (*Algorithmic Revolution: The History of Interactive Art*, 2004) and articulated his interest in quantum technology and artificial intelligence long ago.[5] In recent years, during these times of crisis, of global migration and pandemic, the newest technological developments have been joined at the forefront of his thoughts and actions by the great societal challenges of our time, such as global climate change, democratic decline, ubiquitous surveillance, and the transformation of cultural institutions.

For example, with the exhibition *Open Codes* (2017–19, see fig. 1), he made history not only in terms of the show's content but with its design as well: the museum exhibition space effectively converged, here, with the laboratory, the lounge, the coworking space, the office furniture design showroom, the hip workstation, the social meeting place, and the fun experiental learning space. It also gave him the opportunity to experience himself live as a cicerone, tirelessly arguing that "media facilitate our discovery of technological evolution while also acting as a critical complement to it. The function of art consists in opening doors in a wall where no one sees them. The artist keeps optional spheres of activity open – as a critical mirror or utopian reservoir. With art, society creates for itself in galleries and museums an institutional framework in which other modes of production and worldviews are possible. The critical potential that accumulates there can be used to investigate or transform prevailing social institutions."[6]

The point here was not just to critically analyze the modern museum, but to plumb its untapped potential through transformation and transgression, thereby exploring new spheres of activity. The exhibition's scenographic concept strategically unfurled a relational field based on the structure, organization, and configuration of a flexible and interactive network in which various temporary performances of the artistic and creative could take place. This, in turn, gave rise to another turn and twist from representation to reality, a characteristic that is eminently typical of Weibel's performances in general. In this way, art and creativity were not merely exhibited, but (democratically) enabled and initiated.

In his role and position as an avid educator at renowned art academies, and as the long-serving director of institutions such as Ars Electronica in Linz (1986–95), the Institute for New Media in Frankfurt am Main (1989–95), and, last but not least, the ZKM (1999–2023), the Odesa native has invigorated the global media-art scene through numerous conferences, workshops, exhibitions, and publications. As a much-awarded and highly honored conceptual and media artist, however, and as a widely connected curator and farsighted theorist, he has always paid attention to classic artistic genres such as painting and sculpture as well. With keen intuition, he has exhibited and written about young artists as well as forgotten and overlooked figures, often long before they were ever (re-)discovered by the museums or art market; he also embarked on the task of redescribing modernism early on, and in exhibitions of global art he turned his attention beyond Eurocentric, colonialist thinking.

Peter Weibel always seems to be one step ahead of everyone in every respect. His delight in discovery and thirst for knowledge, his creative drive and will to communicate seem to be boundless – and, to return to the fire alluded to above, Promethean. The Renaissance had the figure of the *alter deus* for this type of artist-creator, seen in works such as Albrecht Dürer's (1471–1528) stylized *Self-Portrait* as Christ (1500). For Peter Weibel and his approach, on the other hand, the comparison with *Homo deus*[7] is unavoidable, as his oeuvre is also deeply infused with the ideas and discourses of transhumanism: the desire, the human yearning, to break out of the Cartesian prison of the body is a fundamental theme for him, as is the transcendence of death through cultural technologies. Weibel's entire body of work can thus be interpreted, in my opinion, as an ongoing performance of transhumanist ideas, albeit one that never neglects to reflect critically upon them as well.

As a progressive thinker and one of the most prominent proponents of transhumanism, Ray Kurzweil (b. 1948) – author, director of engineering at Google, and pioneer of optical language recognition, among other distinctions – is especially optimistic about the prospects for the transformation and substitution of the aging, tiring human body. In Kurzweil's vision, tiny robots will combat viruses, bacteria, and cancer cells inside the human body in the near future, for example. And if it is true that human consciousness, mind, and identity are based on purely physical and biochemical processes, then "mind uploading" should absolutely be a possibility in today's "upgrade society."[8] Such visions of transhumanism are presented in abundance in bestselling novels and Hollywood movies as well,[9] though they are often given a grim, dystopian slant, a Frankensteinian trauma, whereas the transhumanist ideas and concepts scattered throughout Weibel's theoretical and practical work, especially his media theory, appear as moments of cultural evolution and emancipation, in my view. Weibel calls this "exo-evolution." These ideas include that of accelerating the development of prosthetics for the optimization and transformation of human nature, in the form of exogenous technologies and apparatuses, with the next step then being to use these technologies external to the natural body to rewrite the biological body itself – perceived, as it is, as fundamentally deficient – on the level of the genetic code, with the ultimate intention and motivation of transcending it.[10] Technological progress toward perpetual extension, enhancement, optimization, and transcendence is seen by transhumanists as part of natural evolution – an idea that also finds expression in the tenor of Weibel's activities as a scholar, founder, and developer. His telematism is thus eminently comparable to the commitment to progress and the ideas of technological advancement and optimization found in transhumanist thinking; like Prometheus, the

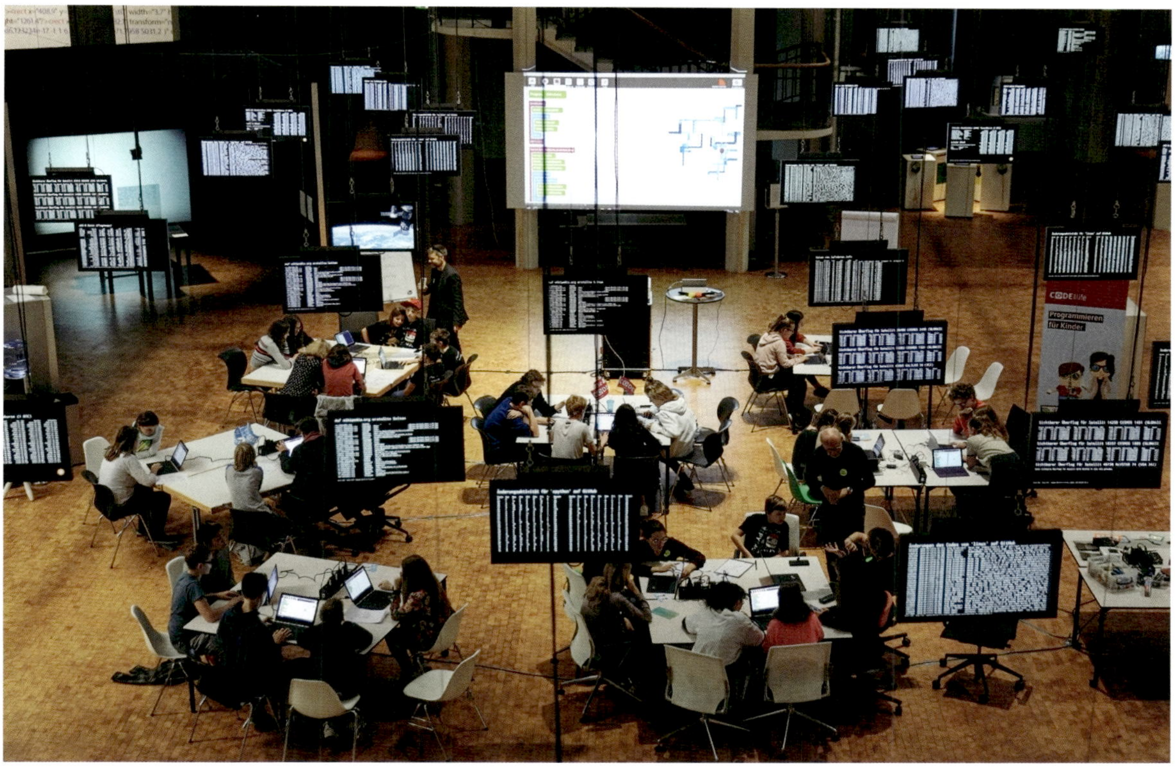

Fig. 1:
Installation view *Open Codes*, ZKM | Karlsruhe, 2017–19

"media rebel" is constantly bringing fire – not as a natural element, but as the beginning of a consummate mastery of nature – into spheres of cultural activity that require the spark of creative imagination. Or, as Weibel himself put it:

> If Freud understood the human being as a creature defined by drives, by natural forces – by biology, in other words – then that's a downward tension, if you will. But there's also what Peter Sloterdijk calls a "vertical tension," in that humans have aspirations to rise above themselves. I am replacing this rising-above-the-self – and this is the critical point – with an emerging-out-of-the-self. I describe this as externalization, exteriorization, or ecstasization. Humanity improves itself, and that's precisely where we find the parallel to Nietzsche. What I have in mind is a secularized version, so to speak, of the sacralized *Übermensch* – a normalization of ecstasy. The Übermensch still has divine, omnipotent traits. My version is a more democratic one. Humans who become aware of their flaws or deficiencies try to ameliorate or counterbalance them, or find a substitute. ... Humans improve themselves by optimizing their material, physical, and mental attributes.[11]

Thus the fact that Weibel was already taking an interest in BioArt in the 1980s – here again, he was one of the first (he coined the term in his 1981 essay "Biotechnologie und Kunst" (Biotechnology and art)) – and has exhibited seminal genetic art, such as the work of Lynn Hershman Leeson (b. 1941)[12] – strikes me as only consistent, and highly authentic.

On the other hand, with regard to the internal logic and media-specific actuality of sign-based realities, which Weibel has repeatedly investigated through his art, both the philosopher Nick Bostrom (b. 1973) and, more recently, Tesla boss Elon Musk (b. 1971) have speculated that we may already all be living in a vast computer-generated simulation – which, according to this sensational hypothesis, may even have been programmed in the future by humanity itself. In an interview,[13] Musk noted the enormous advances in video game design, from the field's beginnings to the almost photorealistic interactive animated games that have followed just fifty years later, which today are played online by millions of networked gamers simultaneously. This type of interaction, together with immersion in ever more functional AR and VR situations, will soon make it impossible for us to differentiate between actual and sign-based realities. Such a technological evolution could even have already taken place in the past, with the result that we now live inside a massive simulation, from which – as in the cinematic fiction of the *Matrix* trilogy (1999–2003) – a few attempt to escape. Unlike Bostrom, however, Musk believes there is only one chance in a billion that we are not living in a holographic simulation. He is convinced that there are already billions of simulated universes that are indistinguishable from our physical universe, and which, at best, are governed by other algorithms – that is, physical laws that are alien to us. We would thus exist within one of these code-based, programmed multiverses. Significantly, Weibel created a digital video titled *Chants of the Pluriverse* (*Gesänge des Pluriversums*, see pp. 304–307 in this volume) from 1986 to 1988.

For the polyperformer Peter Weibel, the tension between the presence of corporeality and the simultaneous desire to transcend it is a characteristic aspect that runs throughout his entire oeuvre as a fundamental dialectical principle. Again and again – especially in his many performances, in the broadest sense of that word – he masterful-

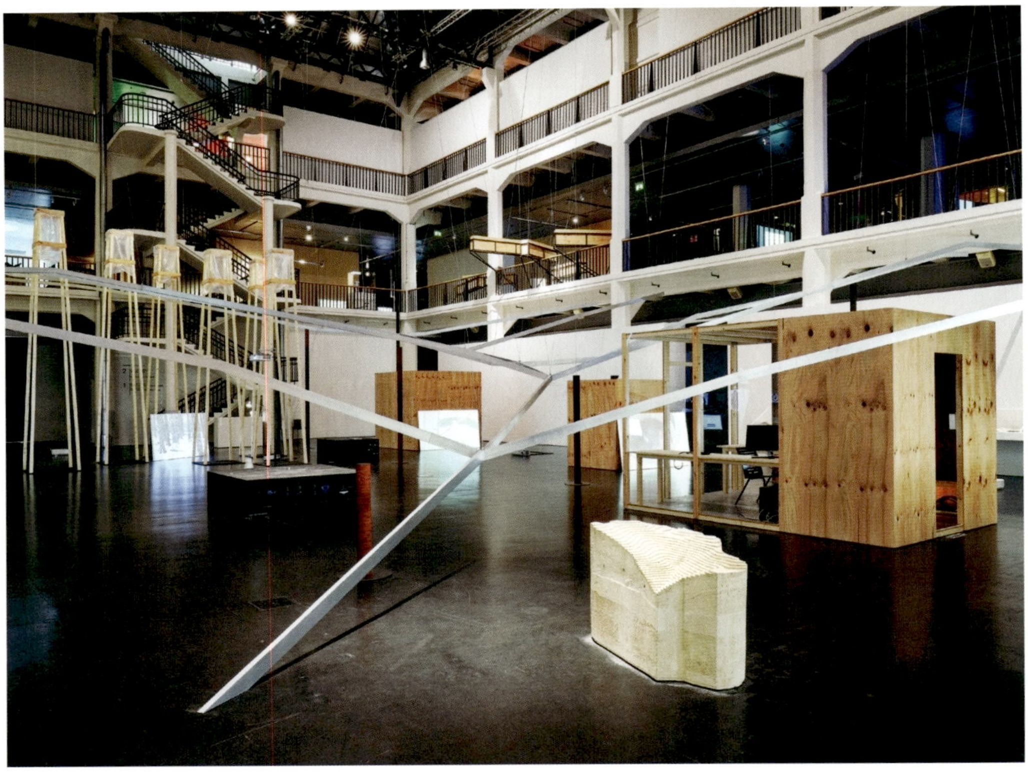

Fig. 2:
Installation view *Critical Zones: Observatories for Earthly Politics*, ZKM | Karlsruhe, 2020–22

ly rattles the bars of the biological nature of the body and the mental prisons of putative certainties. Artistic practice, for him, is thus intellectual social criticism as well, and at the same time it is also another way of producing knowledge - one that has conceptually, artistically, and poetically anticipated advances in biotechnology such as CRISPR, for example:

> My body performances are very important to me. Their underlying linguistic and media assumptions set them significantly apart from the body cult found in other performances. My body politics was a body critique. Most body artists developed their performances out of painting, specifically action painting, to which they remained visually and conceptually bound. What distinguishes my body performances from other performances realized in the 1960s - in Vienna and California, for example - is, first, their relationship to media, and second, their relationship to politics. Writing, the typewriter, photography, film, and video are the media and apparatuses that make possible a new *écriture corporelle*, a new body politics that is also a critique of the traditional politics of representation and identity. The body image projected in my performances shows the body in the age of its media- and genetic-technological constructibility, as a product not of nature but of culture - for example, in questions of gender. The body is an alphabet; the organs are its letters.[14]

In summary, Weibel's many and varied activities as a scholar, founder, and developer must be underscored as an important underpinning of his art. Since 2017 he has also been the director of the Peter Weibel Research Institute for Digital Cultures at the University of Applied Arts in Vienna.[15] The website of the recently founded institute features the 2017 mixed-media installation *Alphabet Space* (see pp. 330–31 in this volume), which was created by Weibel in collaboration with Christian Lölkes (b. 1990) and Adam Słowik (b. 1980). This piece was also on view in Karlsruhe in *Open Codes*, the epoch-making exhibition mentioned above.[16] In it, dynamic and interactive computer-based alphabet space was presented as a playful challenge for users: through the skillful manipulation in real space of a four-cornered metal tube that had built-in position sensors (just like our smartphones), they could cause two-dimensional "silhouettes" to appear in real time on a screen on the wall as black digital letters that varied depending on the object's position and orientation. Entire messages and statements could be transmitted in the exhibition space in this way, one letter at a time. Another Weibel quote neatly encapsulates the artistic and creative ambition of the project's concept, which, as a conceptual model, amounts to a quantum leap forward:

> My answer to the question of whether humans can outwit nature, even though they are a product of it, is a clear yes. Even though we are imprisoned in the framework of the prevailing laws of nature, we can expand the universe by discovering new laws of nature. For example, we have learned that we can do any arithmetical operation with just two digits (0 and 1). We will discover, however, that as soon as we take the term "number space" literally, we will realize that until now we have been operating only on the two-dimensional plane with numbers, in the same way that writing, which currently consists of twenty-six letters and various symbols, has only been coded on the two-dimensional plane. Here, too, we will develop new scripts and number systems that operate in space and are produced and received with new devices in space.[17]

The ZKM as it is currently constituted must also be considered part of Weibel's body of work, for here too we find a turn and twist from representation to reality, and here too creativity is not merely exhibited, but also analyzed and reflected upon. During the Covid-19 lockdown, for example, Weibel and his staff had to react creatively and courageously so that a long-planned exhibition accompanied by numerous workshops could open on schedule in the spring of 2020. The hybrid exhibition *Critical Zones: Observatories for Earthly Politics* (co-curated with Bruno Latour, 2020–22, see fig. 2), which initially opened digitally and included a long streaming festival, an extensive program of satellite events and "activations," and a reactive online platform, thus relied less on (populist) performances than on Zoom lectures and conversations, film screenings, and virtual exhibition walk-throughs. Forced in more than one respect to undergo a necessary shift in perspective, the ZKM effectively became a temporary broadcasting station, a "HOME museum" with an online video program, as impresario Peter Weibel called it, which successfully sought contact with the public through social media platforms and various messenger services for questions, feedback, and collective discussions on the topic. After *Open Codes*, Weibel the artist-curator once again proved his ability to reimagine the exhibition as a dispositif for times that are critical in every respect, in the service of a society that must be made more sensitive – not just to art and culture, today, but also and equally to the great questions of humanity.

Note: A version of this text was read on the occasion of a symposium accompanying the exhibition *respectively, Peter Weibel* on September 28, 2019. Translated from the German by Patrick Hubenthal.

[1]
Peter Weibel, "Die Transzendierung des Menschen durch den Menschen," interview by Heinz-Norbert Jocks, *Kunstforum*, no. 237 (Dec. 2015–Jan. 2016): 74–85 (quotation on 83), http://www.kunstforum.de/artikel/die-transzendierung-des-menschen-durch-den-menschen.

[2]
Jerry Saltz, "Zombies on the Walls: Why Does So Much New Abstraction Look the Same?," *Vulture*, June 17, 2014, http://www.vulture.com/2014/06/why-new-abstract-paintings-look-the-same.html.

[3]
I am referring here to the group exhibition *Erweiterte Fotografie* (Expanded photography), which was held at the Vienna Secession from Oct. 22 to Nov. 22, 1981 as part of Vienna's 5th Internationale Biennale.

[4]
For more information on this subject, see Agnes Husslein-Arco and Alfred Weidinger, eds., *Peter Weibel. Medienrebell: Warnung! Diese Ausstellung kann ihr Leben verändern*, exh. cat., Belvedere, Vienna (Cologne: Walther König, 2015), 37–43.

[5]
Claude E. Shannon and John McCarthy, eds., *Studien zur Theorie der Automaten*, ed. Franz Kaltenbeck and Peter Weibel (Munich: Rogner & Bernhard, 1974). First published as *Automata Studies*, Annals of Mathematics Studies 34, ed. Claude E. Shannon and John McCarthy (Princeton, NJ: Princeton University Press, 1956).

[6]
Peter Weibel, "Was sind die Aufgaben eines modernen Museums?," interview by Ronald Berg, *Tagesspiegel*, Jan. 11, 2000, http://www.tagesspiegel.de/kultur/was-sind-die-aufgaben-eines-modernen-museums/115782.html.
For further information on Weibel and his oeuvre, see: Pamela C. Scorzin, "Peter Weibel: Erbauer von Modellwelten," *Künstler: Kritisches Lexikon der Gegenwartskunst* 32, no. 52 (2000): 1–12; Manuel Mischel and Pamela Scorzin, "Der beschleunigte Leonardo da Vinci: Ein Gespräch mit Peter Weibel," in *Sichten MM. Jahresausstellung Fachbereich Architektur, Technische Universität Darmstadt*, exh. cat. (Hagen: Ardenku, 2000), chap. e05; Pamela C. Scorzin, "*respektive Peter Weibel*: ZKM | Zentrum für Kunst und Medien. 28.09.2019–08.03.2020," *Kunstforum International*, no. 264 (2019): 255–57; and "Pamela C. Scorzin über Peter Weibel," *PLOT: Inszenierungen im Raum*, Nov. 4, 2019, http://www.plotmag.com/blog/2019/11/pamela-scorzin-ueber-peter-weibel.

[7]
See, e.g., Yuval Noah Harari, *Homo Deus: A Brief History of Tomorrow* (New York: Harper, 2017).

[8]
Kurzweil publicly proposed this hypothesis during his lecture at the Global Futures 2045 International Congress in New York in 2013, see Victoria Woollaston, "We'll be uploading our entire MINDS to computers by 2045 and our bodies will be replaced by machines within 90 years, Google expert claims," *Daily Mail*, June 19, 2013, http://www.dailymail.co.uk/sciencetech/article-2344398/Google-futurist-claims-uploading-entire-MINDS-computers-2045-bodies-replaced-machines-90-years.html. For more extensive commentary by Kurzweil, see his book *How to Create a Mind: The Secret of Human Thought Revealed* (London: Penguin, 2013).

[9]
The films *Gattaca* (1997, dir. Andrew Niccol), *The Island* (2005, dir. Michael Bay), and *Limitless* (2011, dir. Neil Burger) are all worth noting in this regard, as are the recent science-fiction novels *Upload* (2012, by Mark McClelland), and *Seveneves* (2016, by Neal Stephenson).

[10]
For more information, see Weibel, "Die Transzendierung."

[11]
Weibel, "Die Transzendierung," 83.

[12]
Peter Weibel, ed., *Lynn Hershman Leeson: Civic Radar*, exh. cat., ZKM | Karlsruhe (Ostfildern: Hatje Cantz, 2016).

[13]
Recode, "Elon Musk | Full Interview | Code Conference 2016," video, 84:15 min., June 2, 2016, http://www.youtube.com/watch?v=wsixsRI-Sz4. The discussion of the simulation hypothesis begins at 1:13:59.

[14]
Peter Weibel, "Körperkritik als Körperpolitik," in a booklet accompanying *Peter Weibel: Körperaktionen / Bodyworks, 1967–2003* (Vienna: Arge Index, 2014), DVD.

[15]
See http://www.dieangewandte.at/institute/peterweibel.

[16]
See http://www.zkm.de/en/exhibition/2017/10/open-codes.

[17]
Weibel, "Die Transzendierung," 81.

Postontological or Digital Art: Virtuality, Variability, Viability

Peter Weibel

Technological visual media such as photography and film share with traditional art media such as painting and sculpture the quality of storing information – chemically, in the case of photography and film – in a material storage medium, which makes it difficult to change that information quickly. In the classic physical storage media, information is fixed and irreversible. The image itself is thus frozen, static; if it moves, that movement is only an illusion. The opposite of this is the digital image, every part of which is itself variable and modifiable because the information is stored electronically. The image is adjustable, not just as a whole but, more importantly, point by point. Even with photography, film, and video, the possibility of manipulating information in the postproduction process can be increased exponentially by changing the storage medium from analog to digital code. In the digital medium, all informational parameters are instantly modifiable. Information in the computer is variable. This instant variability of information, and thus of the behavior of each pixel of the image, is what makes the digital image so well suited to virtual environments and interactive installations. Because the information is stored virtually, every point, dimension, and parameter of the image is a variable. This radically changes the nature of the image. For the first time in its history, the image has become a dynamic system of variables. Because the information is stored dynamically, the image becomes a field of variables. However, these variables are not controlled by the structure of the storage system, as in film or video, but rather by the viewer. The image transforms from a static plane into a dynamic, all-encompassing image field, a field of variables that can be controlled through interfaces by the viewer. The viewer constructs the context in which the free variables take on values, thereby determining the forms they take on as well. Of course, the image itself is not an image yet, but merely a field of variables or, more specifically, a sequence of variables (0, 1). Machine-generated sequences of variables form the event horizon for the viewer, whose behavior is what determines how the variables are manifested as image and sound. The digital signal is initially neutral; it is only the technological interface, the technological context, that converts it into a pictorial or tonal signal, a specific event. The image thus becomes a field of events. It would therefore be more appropriate to speak of a field of events, consisting of auditory, visual, or olfactory variables, rather than an image. The image becomes a dynamic field of local (auditory, visual, or olfactory) events. This field of events, which can be determined, constructed, and controlled either locally or globally (i.e., point by point or as a whole) – and from any given context – is the new form of the image. The idea of the image changes from that of a window (in Alberti's 1435 formulation)[1] to that of a door, through which the viewer can enter and exit a world of multisensory event fields that describe, or rather construct, a temporally and spatially changeable, dynamic experiential space. If what we have before us now is not a stable, static, two-dimensional picture, but rather a world of events (built out of digital signals), then the question of context becomes more pressing than ever. The values of the variables are changed by means of the context. The context can be a different pictorial system, a sequence of notes, a different machine, the human being as observer. The ever-increasing complexity of interface technology enables us to construct ever more nuanced contexts. The interfaces (the brain, the hand, light-emitting devices, movement, breath, etc.) form the context. The image becomes a context-controlled experiential world of variables. On account of its systemically dynamic qualities, however, this variable, virtual image has acquired yet another characteristic: it acts like a living system, reacting to input from its context and altering its state and output as a result of the viewer's input. The image thus transforms into a dynamic system with lifelike behavior, simulating a living organism. The interactive nature of media art therefore consists in these three aspects of the digital image: virtuality (of its stored information), variability (of its content), and viability (of its behavior).

Note: Adapted from Peter Weibel, "Postontologische Kunst: Virtualität, Variabilität, Viabilität," in *Peter Weibel: Bildwelten. 1982–1996*, ed. Romana Schuler (Vienna: Triton, 1996), 242. Originally written in 1994. Translated from the German by Patrick Hubenthal.

[1] "Principio in superficie pingenda quam amplum libeat quadrangulum rectorum angulorum inscribo, quod quidem mihi pro aperta finestra est ex qua historia contueatur" (First of all, on the surface on which I am going to paint, I draw a rectangle of whatever size I want, which I regard as an open window through which the subject to be painted is seen). Leon Battista Alberti, *On Painting, in "On Painting" and "On Sculpture": The Latin Texts of "De Pictura" and "De Statua,"* ed. and trans. Cecil Grayson (London: Phaidon, 1972), 38–114 (quotations on 54–55). First published as *De Pictura* (Florence, 1435–36).

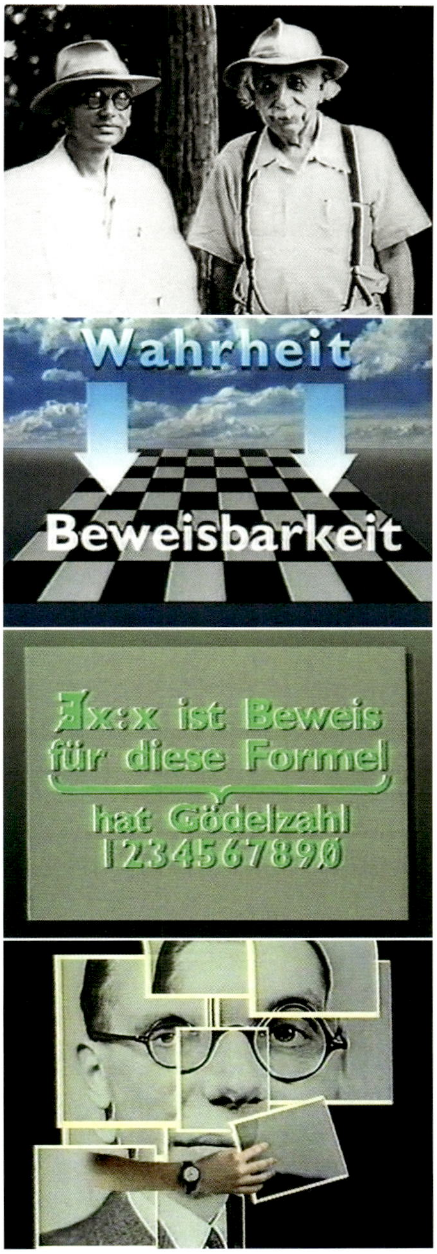

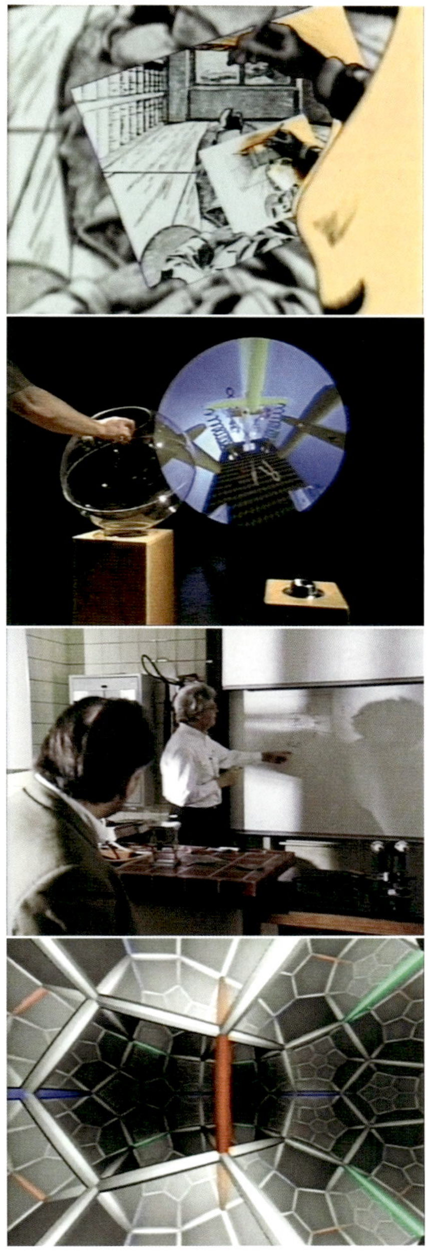

Like Alan Turing, Kurt Gödel is one of the founding figures of Computer Culture. For many, the greatest logician since Aristotle, one of the greatest mathematicians of the twentieth century, a mysterious genius, a legend who proved in Vienna in 1931 at the age of 25 the fundamental undecidability of certain mathematical propositions with exact mathematical methods. Gödel then lived in complete seclusion, in Princeton, USA, friends with Albert Einstein. This film explains, in a mixture of mathematics and musical, the work of Kurt Gödel. It is at the same time a much-quoted source work.

Kurt Gödel – A Mathematical Myth / Kurt Gödel – Ein mathematischer Mythos

1986	Peter Weibel and Dr. Werner DePauli-Schimanovich 16 mm film; color, sound, 80 min., produced for the ORF

Otto E. Rössler (b. 1940) is one of the pioneers of chaos theory and designed the Rössler attractor in the 1970s. He is also the founder of the field of endophysics. In contrast to classical physics where the influence of the external observer is excluded, in endophysics the observer is part of the system that he observes. He is an internal observer. No separation is made between the object system and the observer, thus resembling a closed-circuit video installation. Also, endophysics delivered a new model that is based on the central role of the interface (be it a machine, be it a human) in the process of interpreting, observing, and constructing the world, our reality. His endophysics is therefore not only a contribution to general physics, but also to media theory.

On the Borders of the Real. A conversation between Otto E. Rössler and Peter Weibel / Die Grenzen des Realen. Ein Gespräch zwischen Otto E. Rössler und Peter Weibel

1992	video; color, sound, ca. 120 min., produced for the ORF

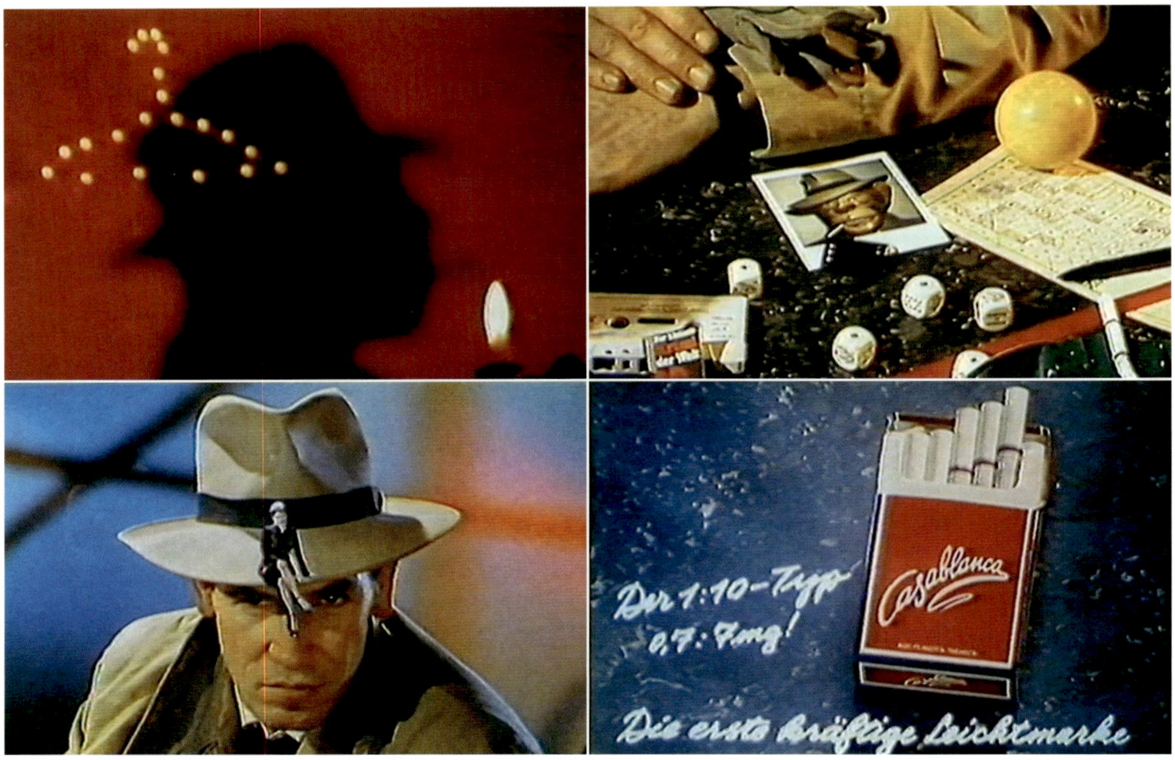

In the 1980s, it was difficult for media artists to find galleries and museums that supported video art. That's why Peter Weibel made some award-winning commercials, such as for the cigarette brand Casablanca and Johnny Filter, to deepen his knowledge of computer effects. He understood his works as an attempt to show what is possible with cinema, what is possible with video and what is possible with computers and how these elements can be mixed and mutually enhanced. His goal was to work media-specific. *Casablanca I* includes computer animation, blue box, videographic effects and film tricks, which are processed in a videotape and then transferred to film and projected. The decomposition of the pictorial elements and the fragmentation of the plot show a new narrative style in the electronic age.

Casablanca I

1983	35 mm film; digital video, color, sound, 1:05 min.

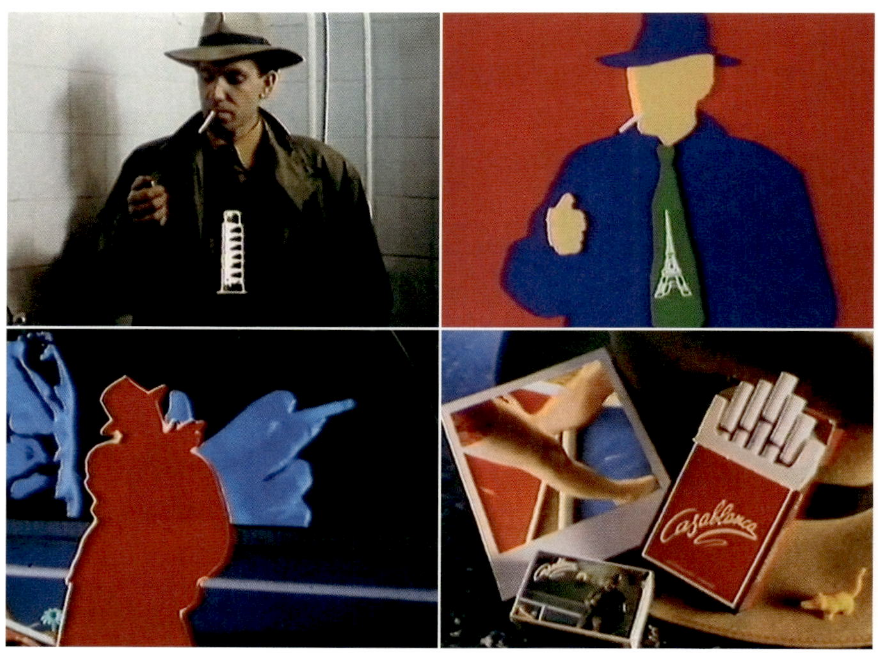

In *Casablanca II*, film and video tricks, cartoons and computer animation occasionally mix with eight exposure units. The screen is transformed into three levels, where an electronic aesthetic emerges from the forthcoming electronic world, which is characterized by transformation, simulation, and ubiquity.

Casablanca II

1983	35 mm film; digital video, color, sound, 1 min.

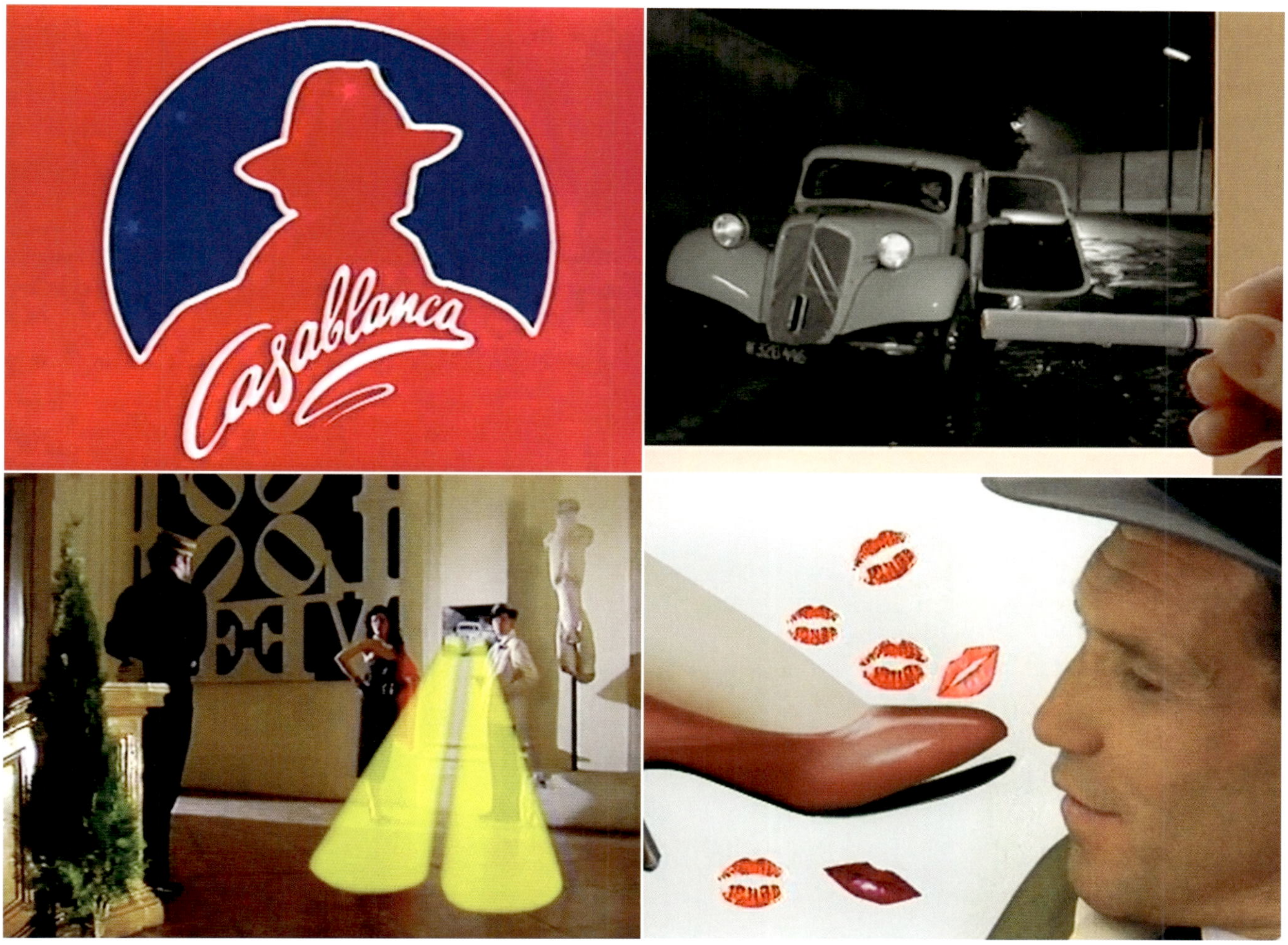

Casablanca III

1985 | 35 mm film; digital video, color, sound, 1:10 min.

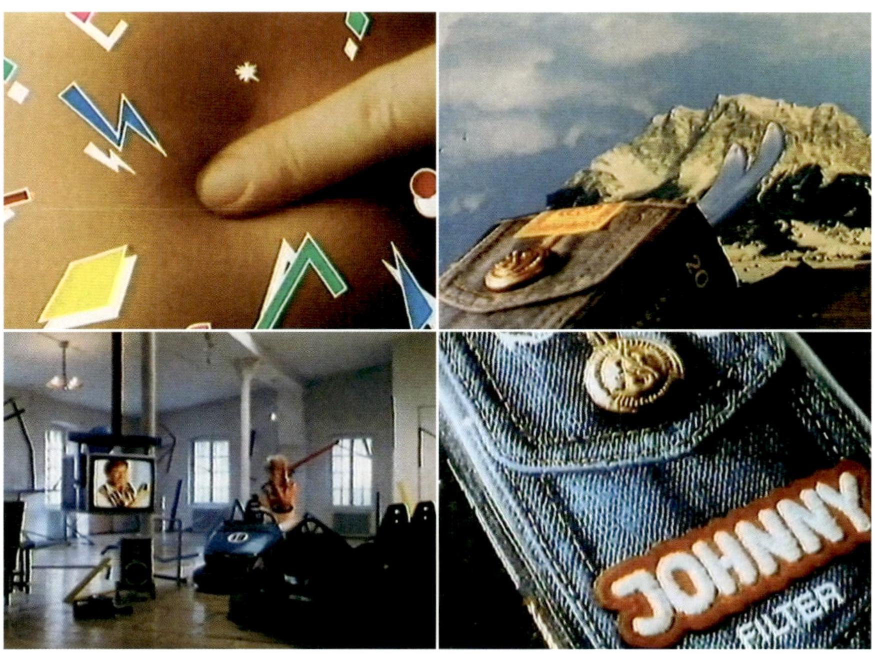

Johnny Filter

1986 | 35 mm film; digital video, color, sound, 1:10 min.

In *Johnny Filter*, a new combination of stage design (e.g., a model of an airport rotating by means of camera movement) and digital effects created a visual language as the most advanced poetry.

Zero Zone with Henry Jesionka, music: Bruno Degazio, postproduction:Telesis, Rochester, NY

Part 1: "Parallele Revisionen" (Parallel revisions) with Henry Jesionka; music: Bruno Degazio; postproduction: J. D. Cowles, Telesis, Rochester, NY; coproduction: Lumen, Toronto

Part 2: "Zweiheit der Natur" (Duality of nature) with VALIE EXPORT; collaboration: Fritz Paschek, Henry Jesionka, Lynn Hershman; camera: Fritz Oelberg, Henry Jesionka; dance: Brigitte Wald; music: Gerhard Reiter, Pas Paravant; faces: Susanne Widl, Peter Weibel; postproduction: Houdek & Kurek, Vienna, Telesis, Rochester, NY; coproduction: VALIE EXPORT Film, Vienna

Part 3: "Raumzeit-Industrie" (Space-time industry); camera: Henry Jesionka; music: Robert Ponger; animation: Henry Jesionka; editing: Rick Feist; postproduction: Broadway Video, New York, Matrix Video, New York; commissioned by the Austrian Federal Railways (Österreichische Bundesbahn, ÖBB), mediated by Harald Betke

Part 4: "Truth Table" with The Vasulkas; camera: Steina and Woody Vasulka; music: Woody Vasulka; postproduction: J. D. Cowles, CGI, Ontario, NY

Part 5: "Logo Kultur" (Logo culture) with material by Paul Sharits; camera: Henry Jesionka, Peter Mettler; music: Bruno Degazio; set: Peter Weibel, Henry Jesionka; paintbox: Tanya Weinberger; actors: E. and H. Jesionka, Susanne Widl, Peter Weibel; production set: Ryerson Polytechnic Institute, Toronto; postproduction: J. D. Cowles, CGI, Ontario, NY

Part 6: "Box Man" with Henry Jesionka; music: David Felder; trombone: Miles Anderson; camera: Henry Jesionka; postproduction: J. D. Cowles, CGI, Ontario, NY, Video Inn, Vancouver; production: Liva, Ars Electronica, Linz

Part 7: "Techno Body"; computer animation: Louis Lok, Robert O'Kane, Paul Minni; music: Stephen Young; postproduction: CGI, Ontario, NY

Part 8: "The Wheel of the Real"; music: Nicholas Collins; postproduction: J. D. Cowles, CGI, Ontario, NY

Part 9: "Architron"; music: Woody Vasulka, John Myhill; sampled by: Peter Weibel; paintbox: Tanya Weinberger; actress: Susanne Widl; postproduction: J. D. Cowles, CGI, Ontario, NY

Part 10: "Scanscape"; music: Woody Vasulka; postproduction: J. D. Cowles, CGI, Ontario, NY

Part 11: "Time as a Code – Chronokratie" (Time as a code – chronocracy); music: Peter Weibel; postproduction: J. D. Cowles, CGI, Ontario, NY; Distinction Award in the category computer animation of the 1988 Prix Ars Electronica

Chants of the Pluriverse / Gesänge des Pluriversums

1986–88	12 digital videos; color, sound, 100 min.

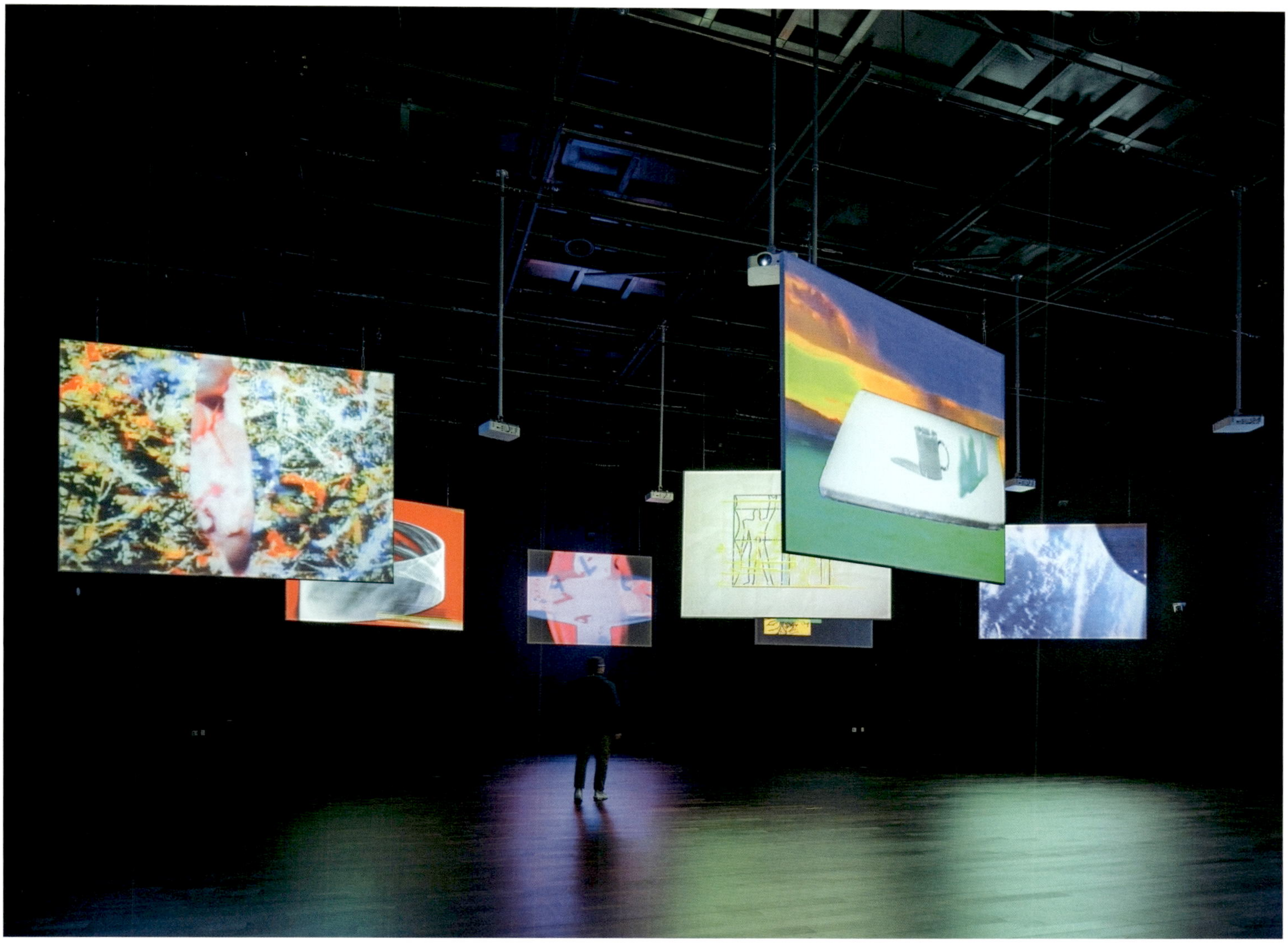

Installation view
respectively, Peter Weibel. Art as an Act of Cognition, National Museum of Modern and Contemporary Art (MMCA), Seoul

Chants of the Pluriverse is Peter Weibel's chief videographic work combining the media of video, photography, film, and computer. Digital techniques and special effects assemble the different elements into an audiovisual pluriverse – or rather multiverse – of the age of technology. The impressive anthology of digital special effects that Weibel accumulated in two years of work visualizes the techno-transformation of the world over the last two hundred years, from the wheel-based industrial revolution to the data-based postindustrial information revolution. The new experience of space and time, whose relations can be changed at will, and the many possible worlds in electronic space-time are celebrated in twelve songs.

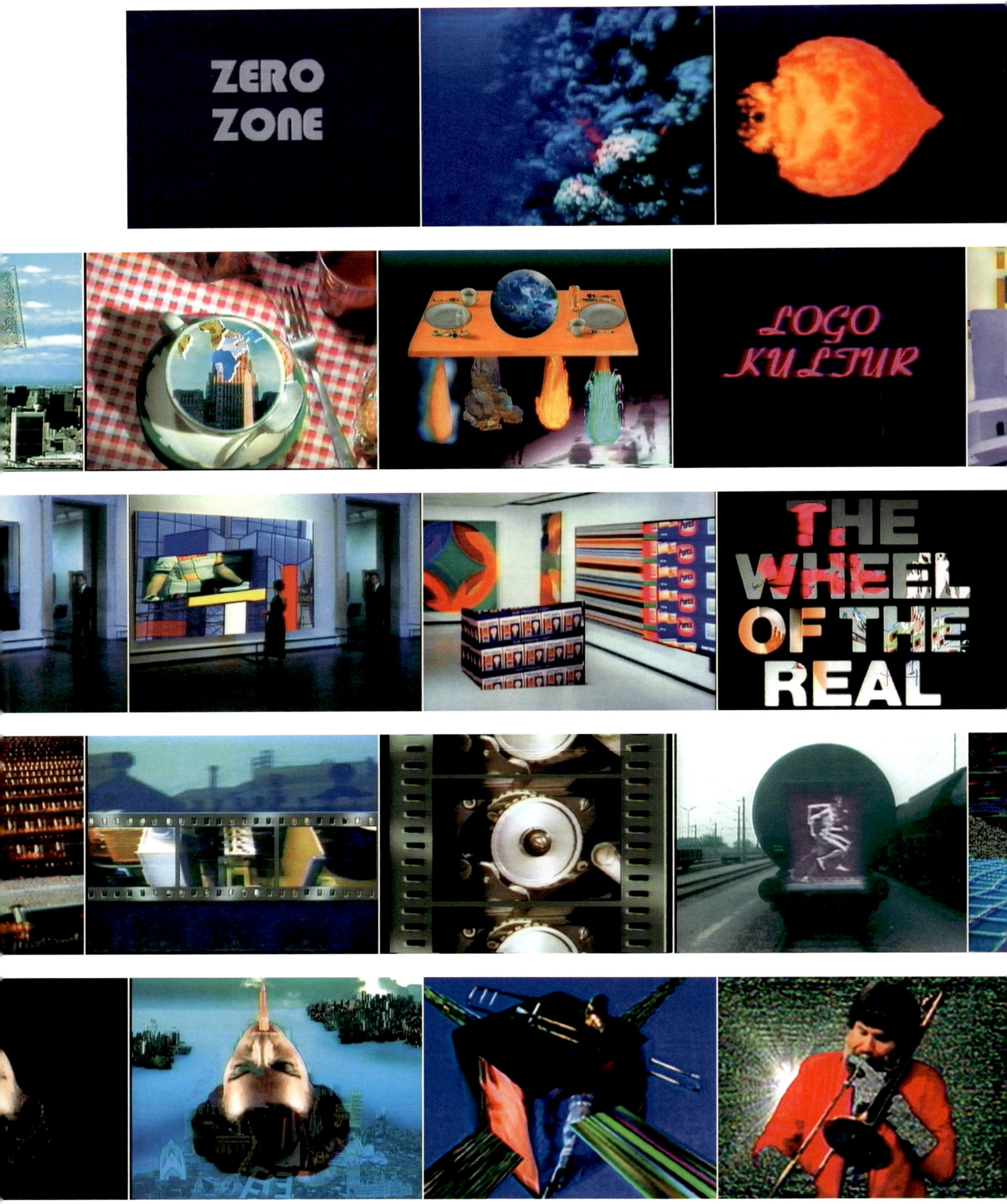
ZERO
ZONE
LOGO
KULTUR
THE
WHEEL
OF THE
REAL

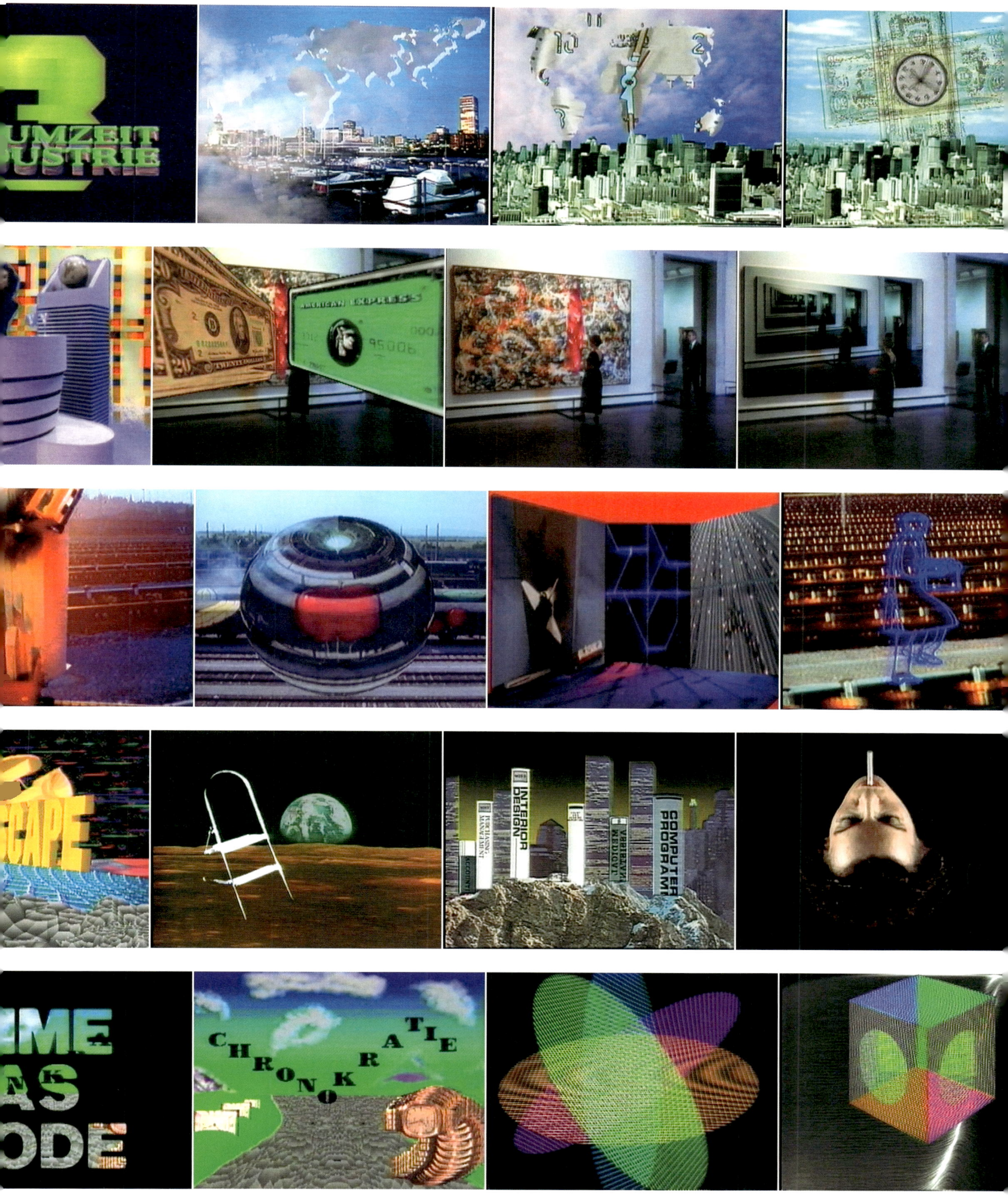

Video stills from *Chants of the Pluriverse*, 1986–88

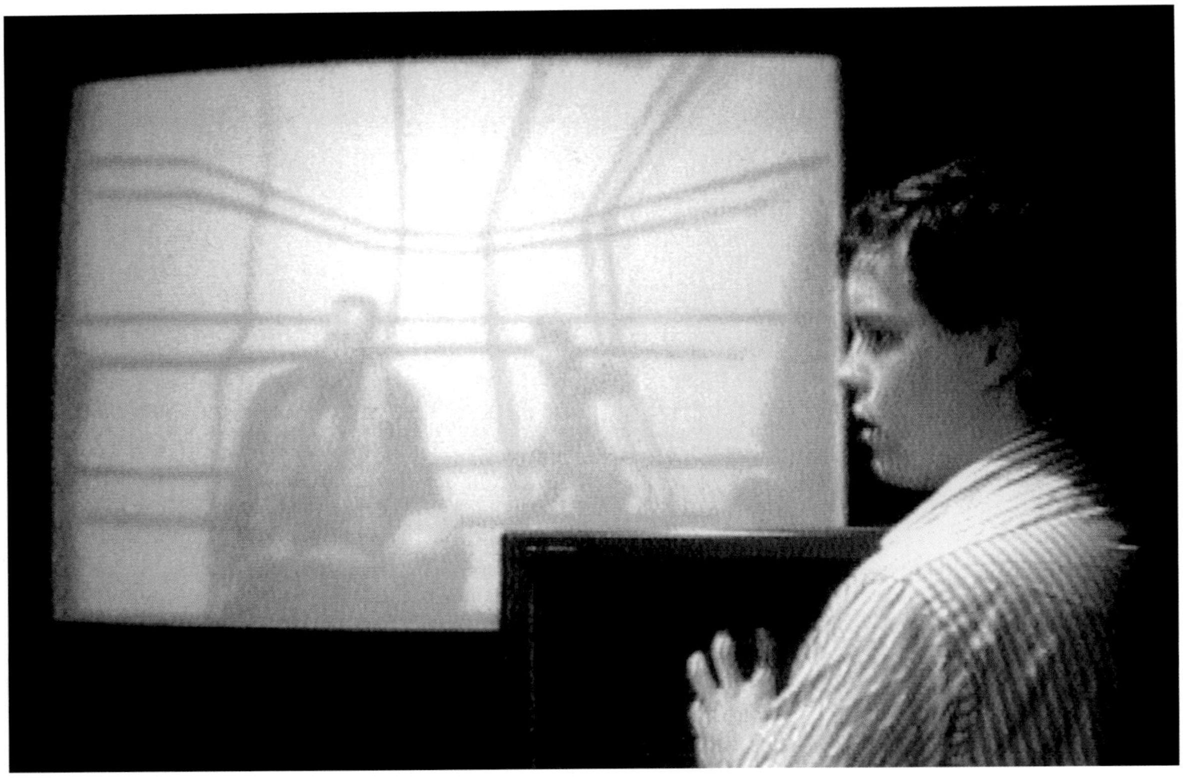

Robert O'Kane operating the installation, installation view *The Living Image*, Frankfurt Art Fair, Frankfurt am Main, 1991

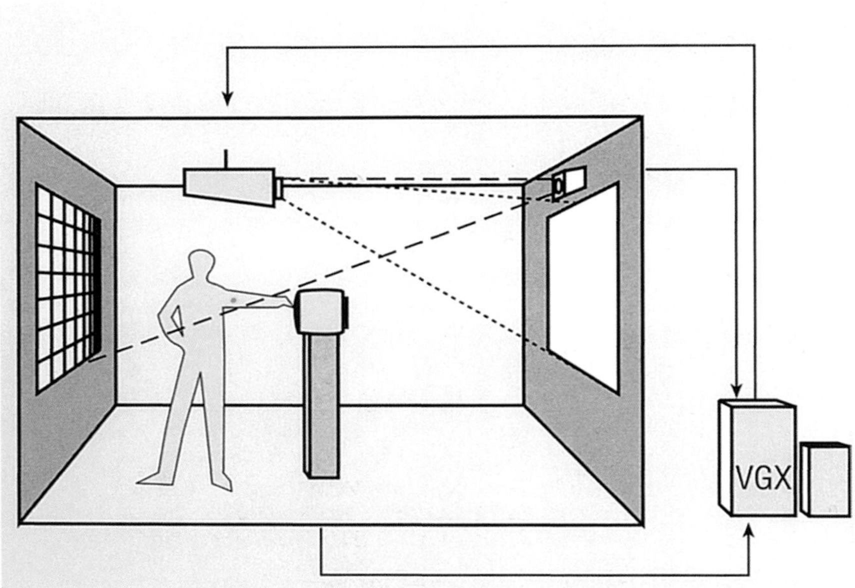

Conceptual drawing for *The Tangible Image*, 1991

The Tangible Image / Das tangible Bild

1991	interactive computer-based installation; monitor made of rubber, projection screen, wood, railings, video projector, video camera, VGX computer, sensors, dimensions variable interface/software: Robert O'Kane

In the interactive computer installation *The Tangible Image*, the viewers are standing in front of a Cartesian coordinate grid. Filmed standing in front of this grid by a camera, they see the image projected on the opposite wall. When they touch the rubber screen of the monitor, which stands on a pedestal in the middle of the room, the projected image warps. Thus, they can interact with the image via a three-dimensional "touch screen" in real time. Every time they touch the screen, behind which sensors are installed, information is sent to a computer to which the camera's live images are transferred. In the computer, the signals of the warping of the screen are converted into digital data which influence the data representing the image. The rubber screen is the interface between the real and the projected screen. The viewer doesn't touch or change the real screen (reality), only the interface and the projected screen is touched (reproduction). Real distortions of the rubber screen appear in the projection image as distortions of the grid and the viewers. An interface (the rubber screen) is switched on between the grid and the projection image. It is not the changes in the grid that affect the projection, but the changes in the interface. Is our world merely the product of an interface technology, the product of the natural interface, the body?

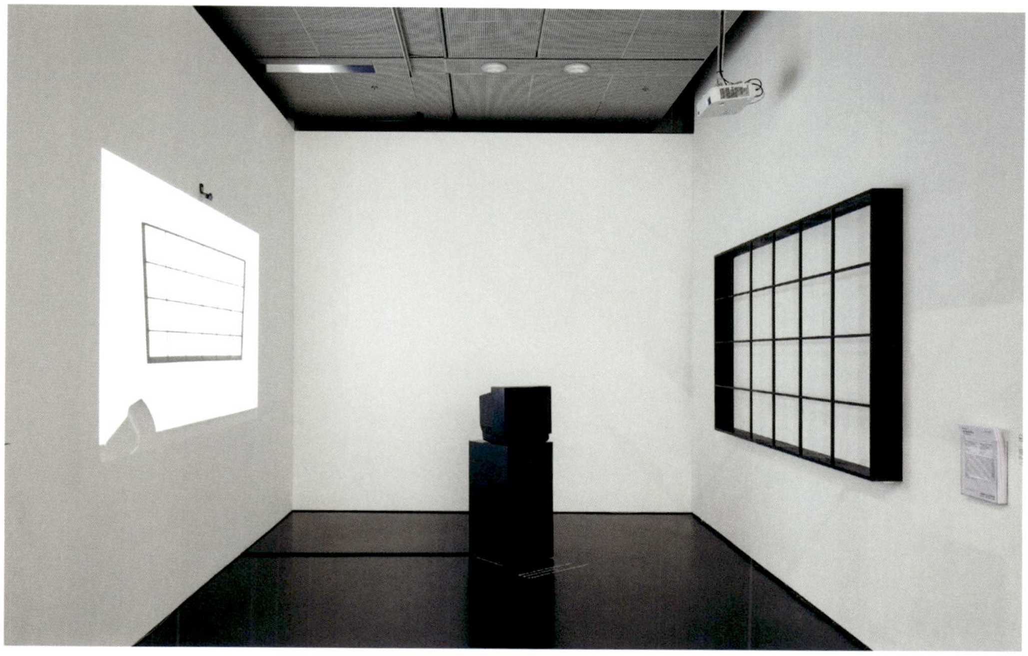

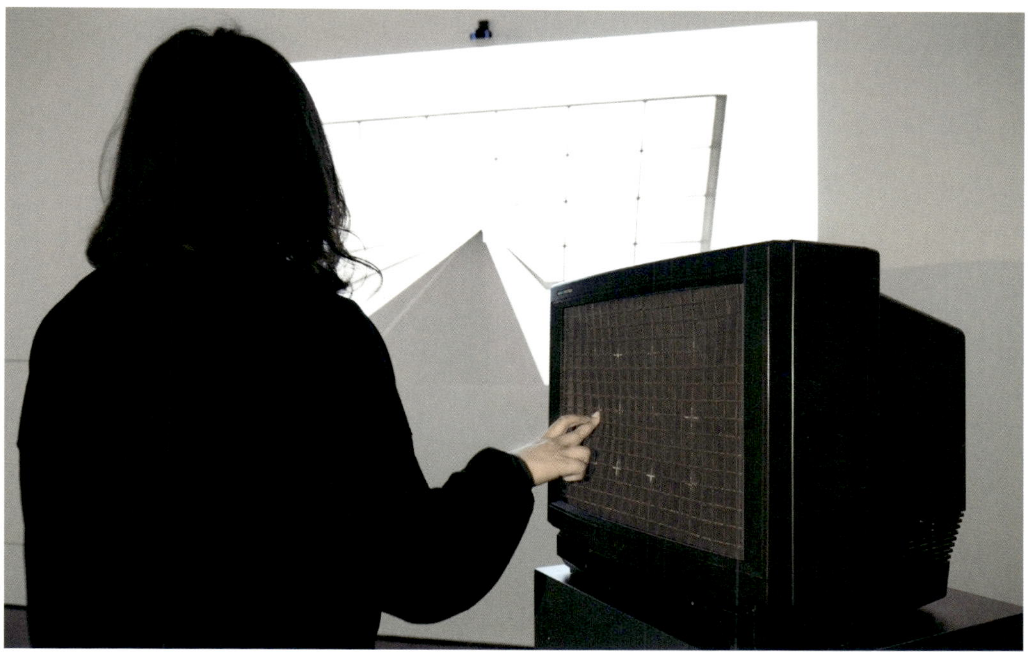

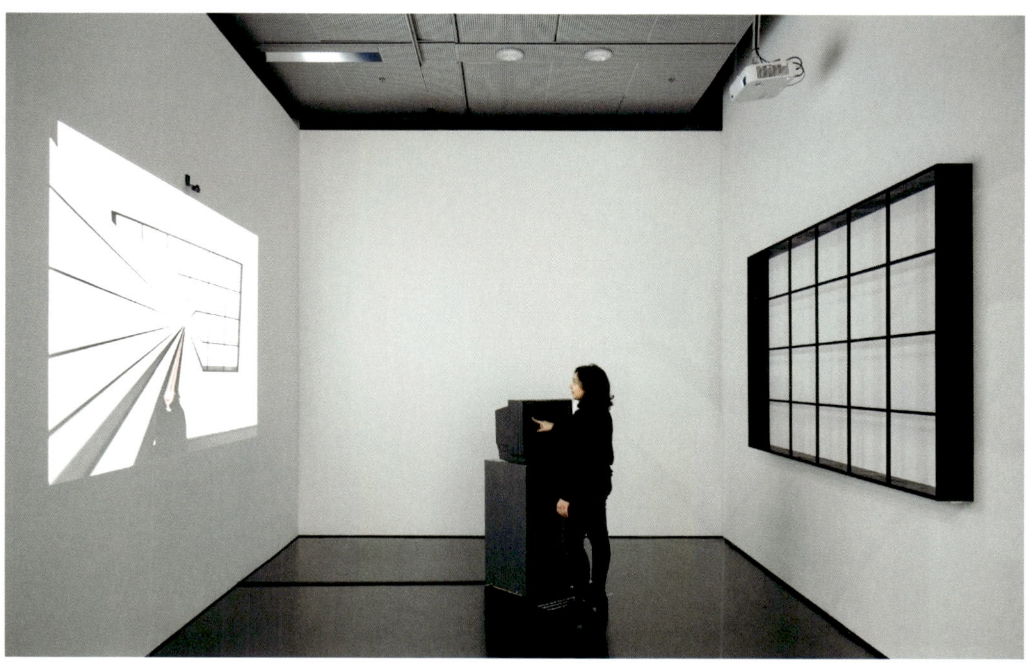

Installation views
respectively, Peter Weibel,
ZKM | Karslruhe, 2019–20

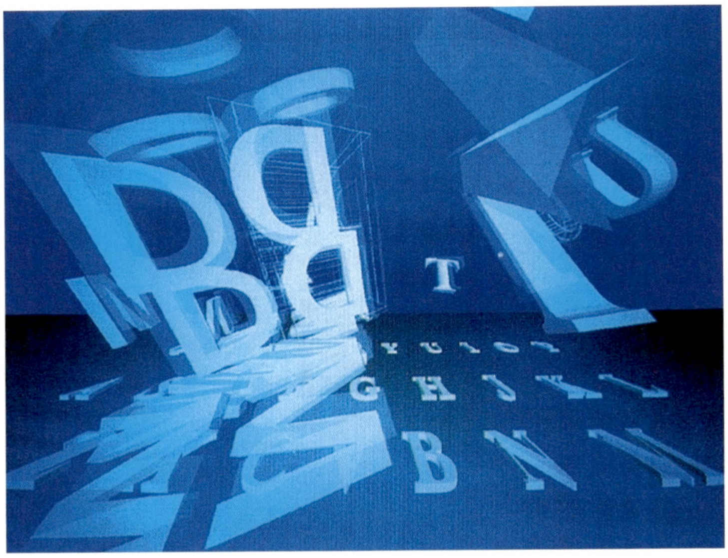

Text world

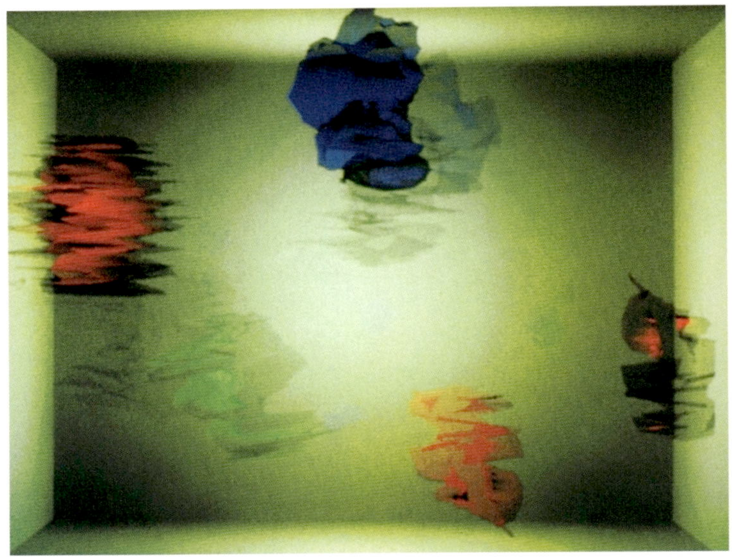

Gas world

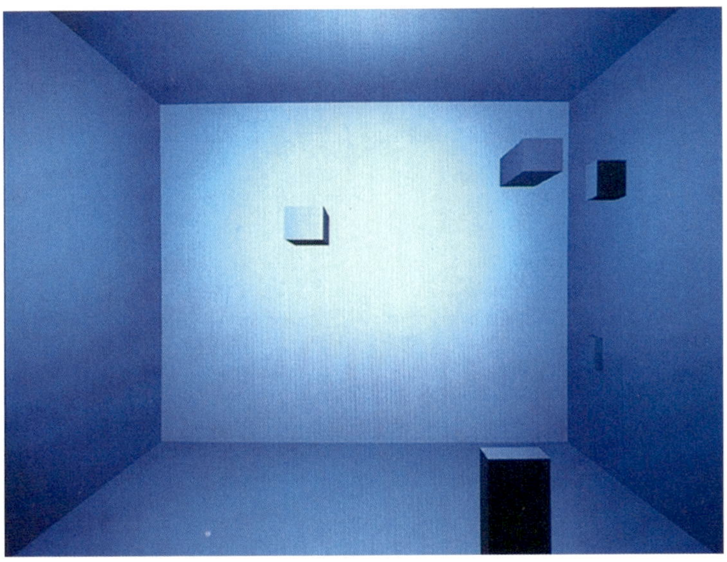

Architecture world

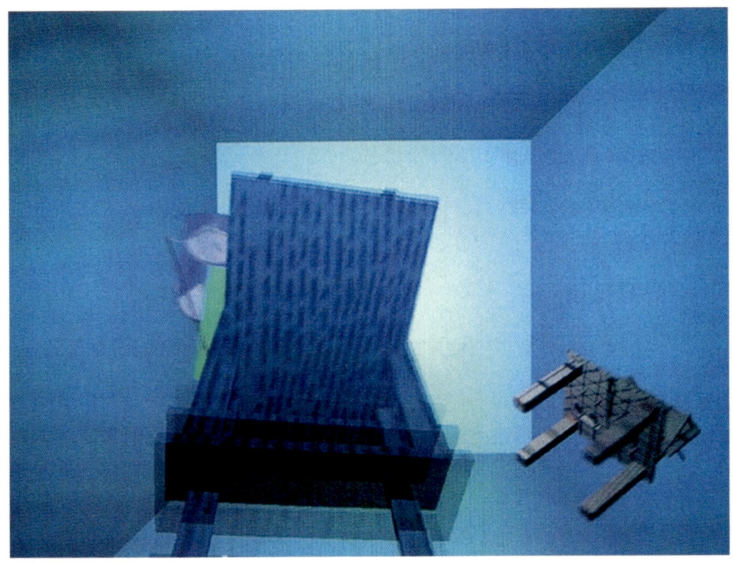

Object world

On Justifying the Hypothetical Nature of Art and the Non-Identicality within the Object World / Zur Rechtfertigung der hypothetischen Natur der Kunst und der Nicht-Identität in der Objektwelt

1992

interactive computer-based installation; sensors, computer, projector, scoreboards, mats, dimensions variable
interface/software: Robert O'Kane; technical assistance: Dieter Sellin; programming: Constanze Ruhm, Robert O'Kane (text world), Christian Möller, Dieter Beck (architecture world), Akke Waagenar (object world), Laurent Mignonneau (gas world); supervision: Gideon May

The interactive computer installation *On Justifying the Hypothetical Nature of Art and the Non-Identicality within the Object World* consists of a carpet with 32 sensors that enables the viewers to interact with four different virtual worlds. The floor becomes a keyboard known from typewriters and computers. Using the four large pressure plates (sensors) of the first row, they can choose a virtual world: the world of the alphabet (language); space (architecture); objects (furniture); and creatures made of gas (artificial beings). With the 28 other sensors each virtual world can be designed and constructed. Comparable to a keyboard, walking around and stepping on the hidden sensors underneath the carpet allows users to access, edit, and immerse themselves in virtual worlds. Classic visual worlds, such as painting, photography, and film, cannot be changed by the viewers. Digital image worlds are constructed, controlled, and changed by the viewers who become internal observers. The images react physically to their input. Thus, they exhibit behavior that is life-like. Especially the fourth pictorial world, the gas clouds, behaves as though it is artificial life controlled by artificial intelligence imitating evolution. Gas clouds of the same color multiply and reproduce when they meet; gas clouds of other colors consume each other.

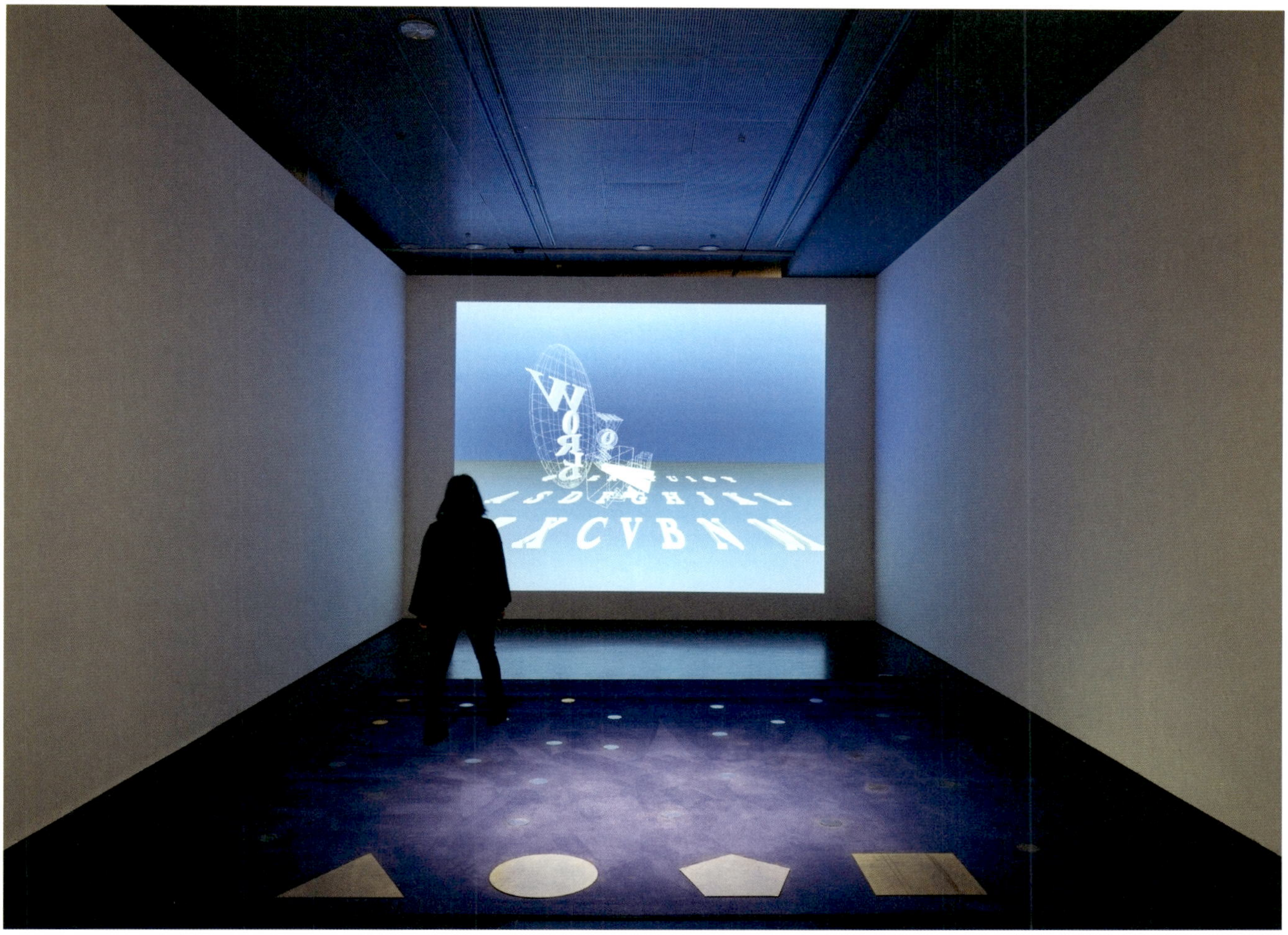

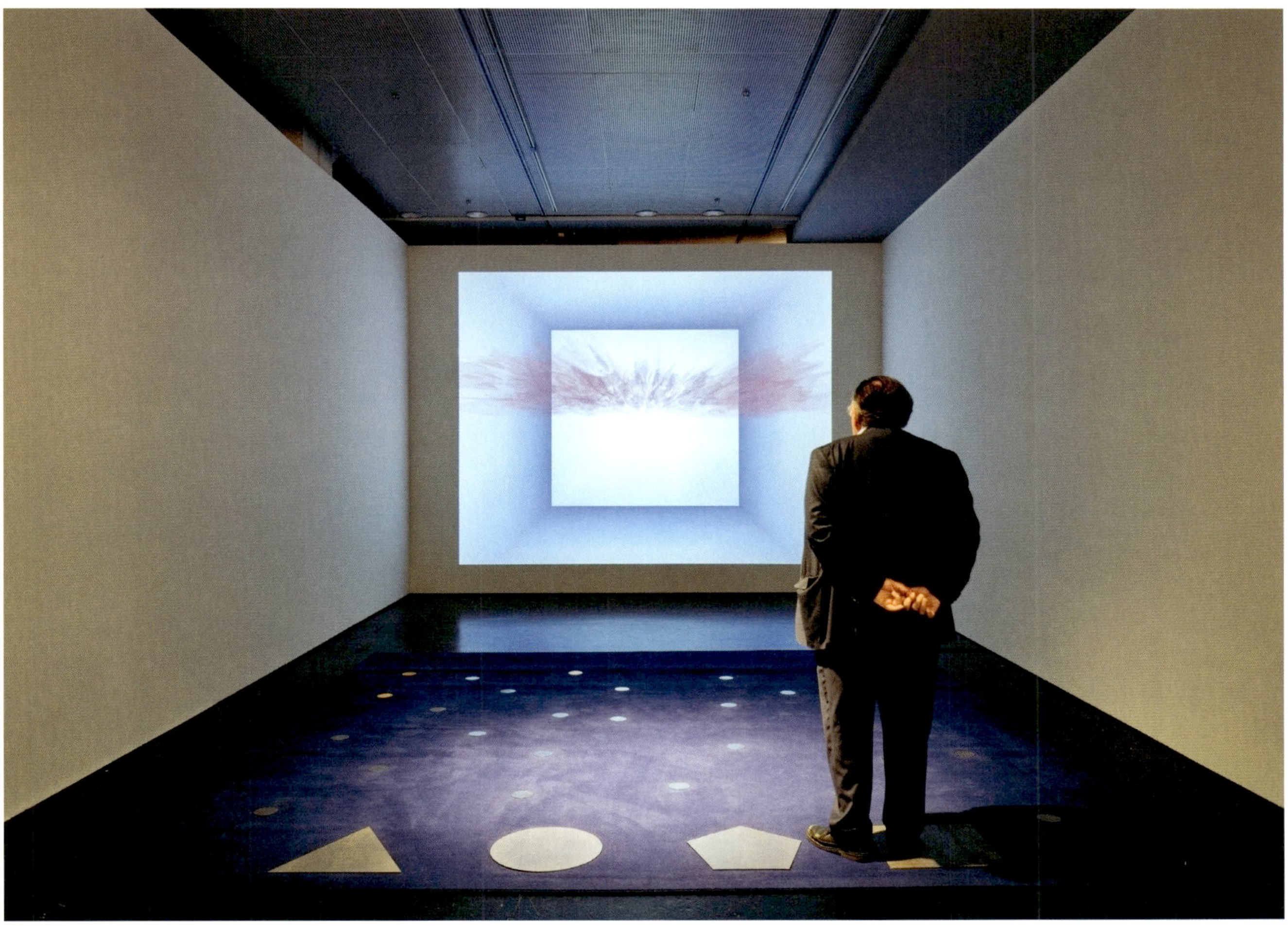

Installation views
respectively, Peter Weibel,
ZKM | Karlsruhe, 2019–20

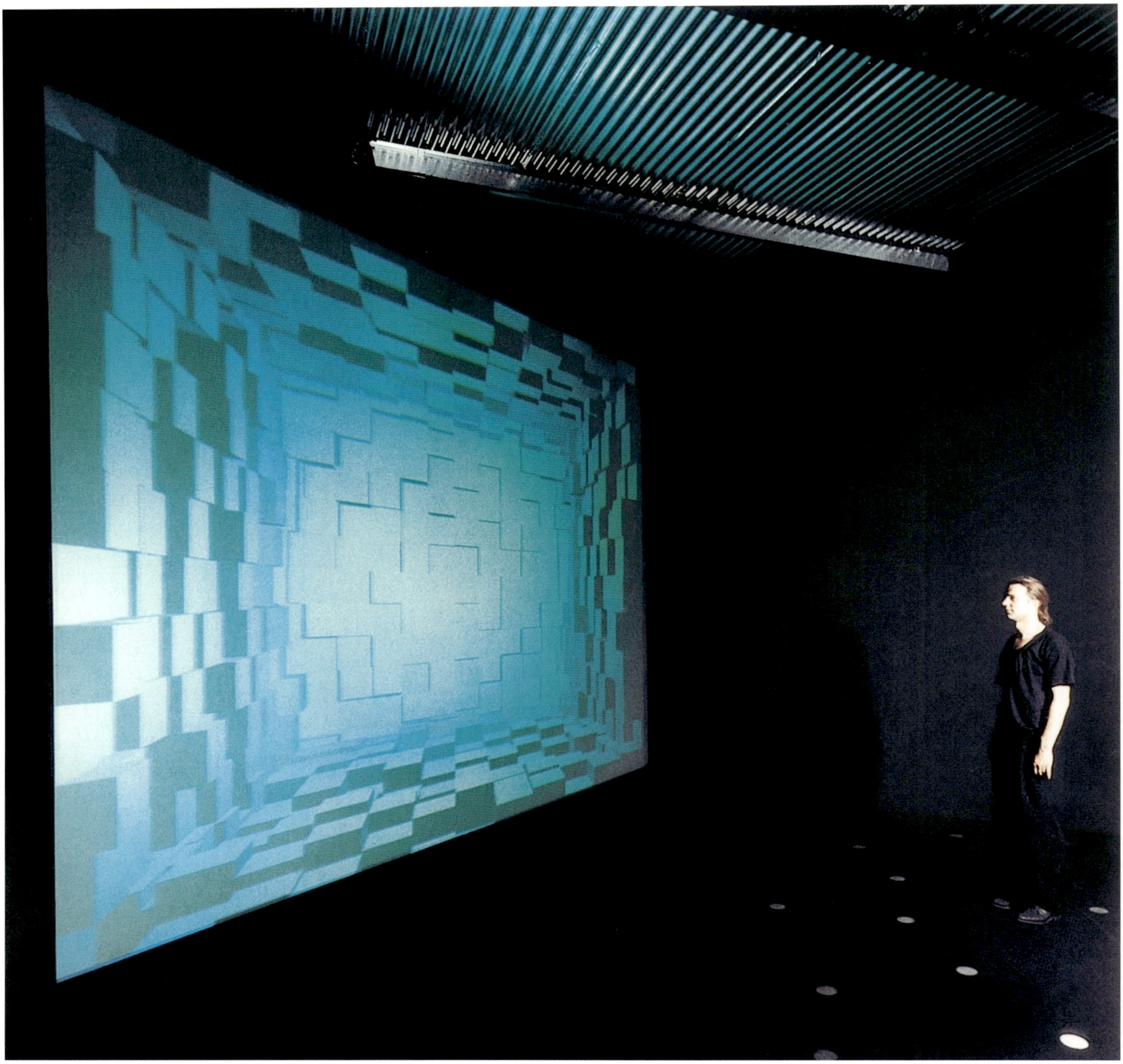

Installation view *Virtuelle Welten*, Galerie Tanja Grunert, Cologne, 1992

The four virtual worlds are generated by the activities of the observer. Some sensors (buttons on the floor) have special functions such as transforming a line into points or freezing the motion of the objects. Therefore, object designed images become very abstract like paintings or drawings (set of points). The letters create sculptures, the texture of the furniture, its scale and proportions can be changed. The whole object world becomes a field of variables controlled and generated by the viewers.

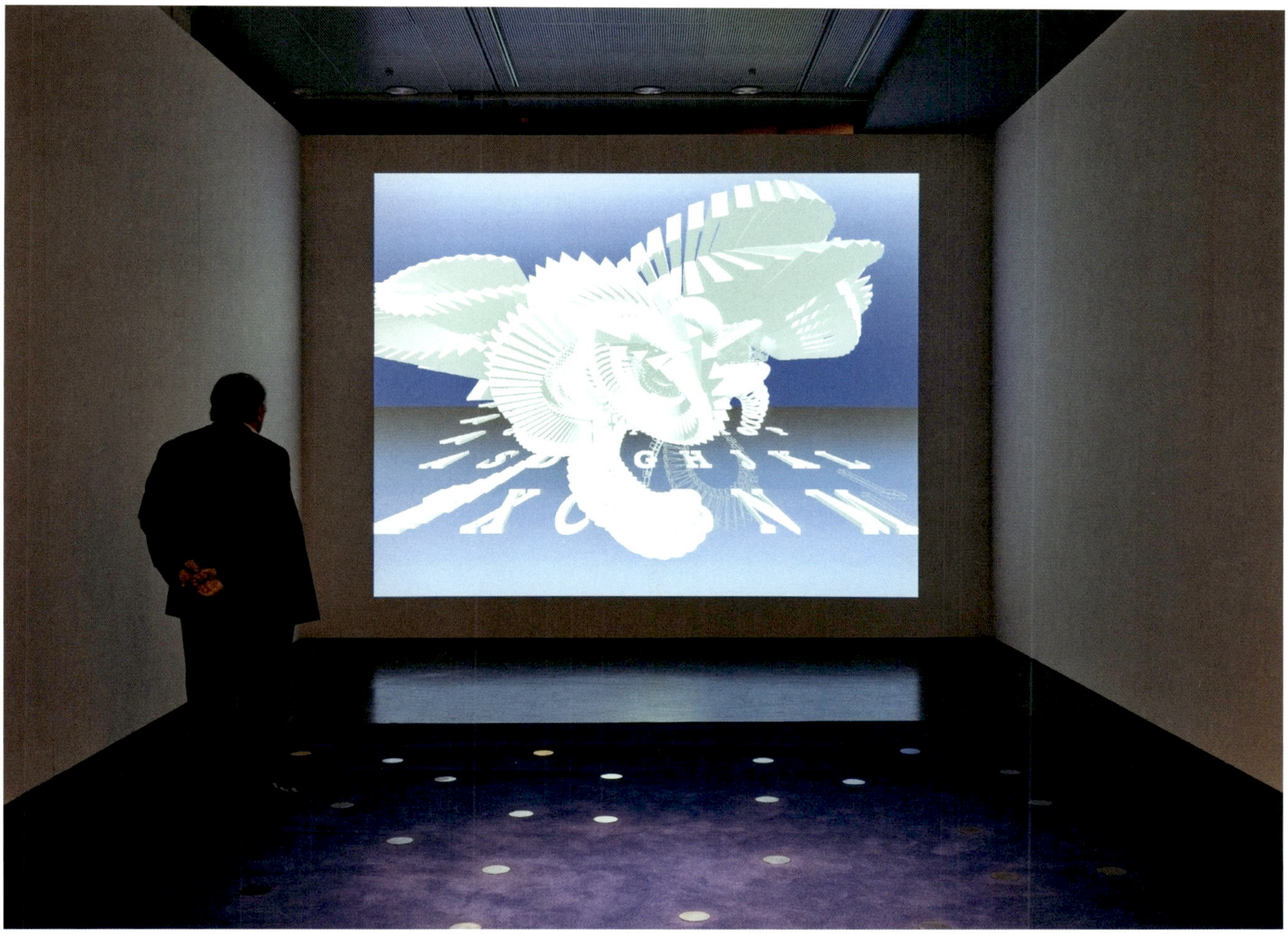

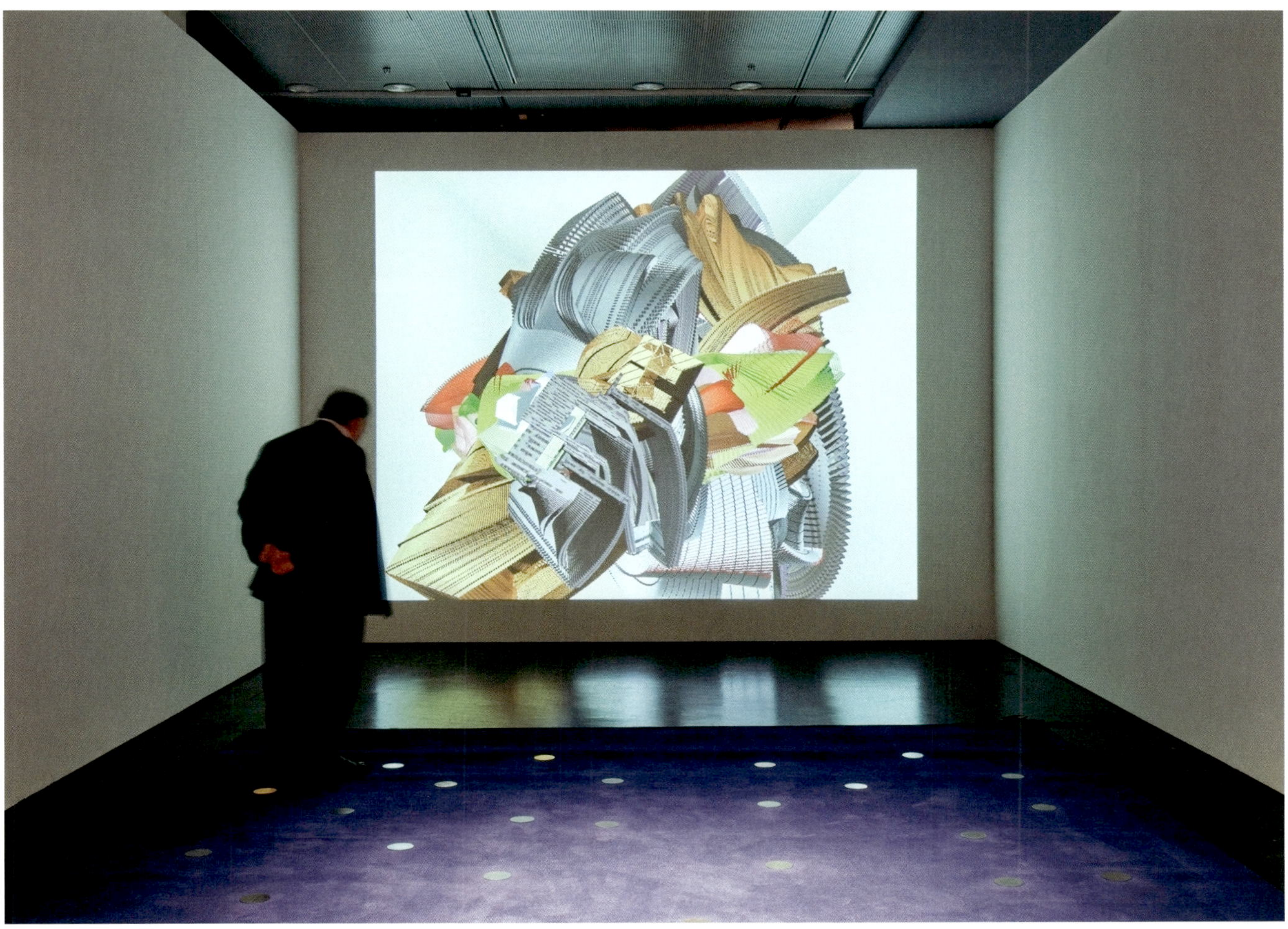

Installation views
respectively, Peter Weibel,
ZKM | Karlsruhe, 2019–20

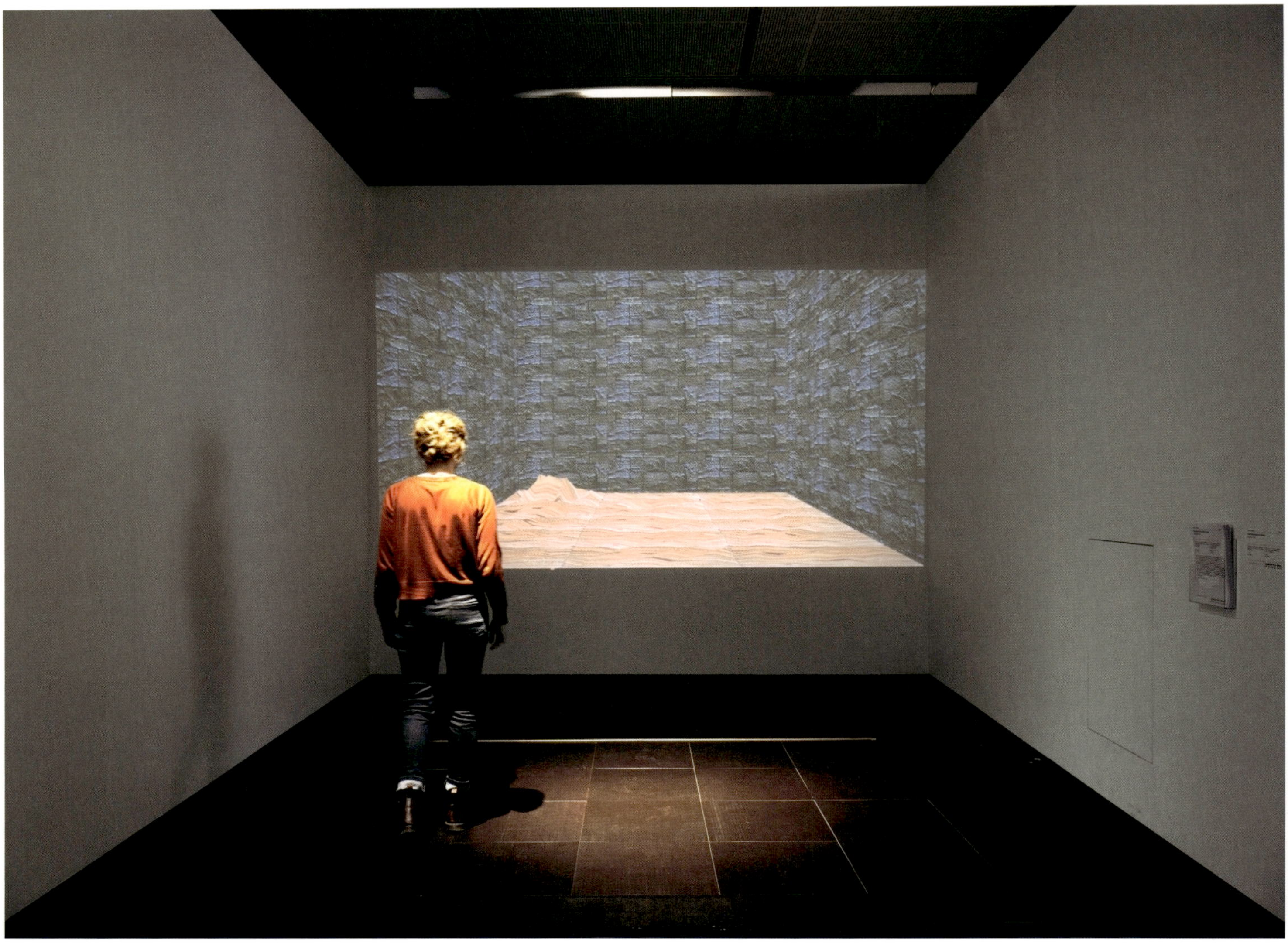

Installation view
respectively, Peter Weibel,
ZKM | Karlsruhe, 2019–20

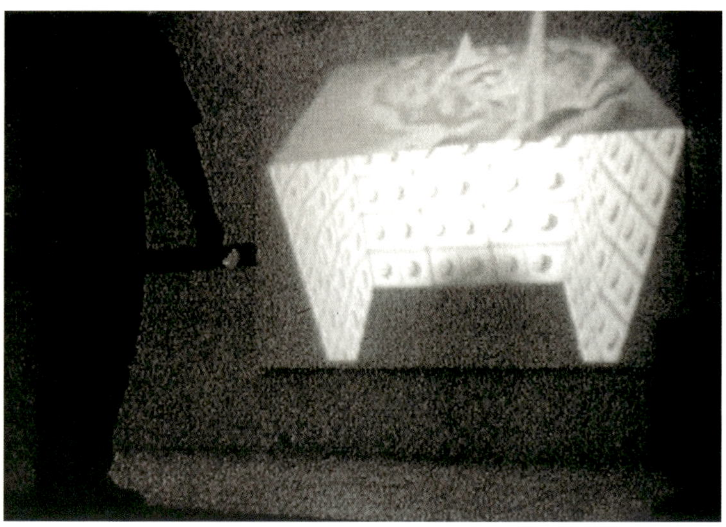

Installation view *Machines à communiquer*, Atelier virtuel, La Villette, Paris, 1991–92

In the interactive computer installation *Cartesian Chaos*, Peter Weibel engages with the viewer's perspective in computer-simulated space. In real spaces of perception, humans are always part of the system they observe: internal observers. In front of a painting, in front of a film screen, in front of a TV screen, humans are external observers. The digital space, partly already the video space, facilitates an internal observation (closed-circuit installation).

The installation consists of a wooden platform on the floor equipped with sensors, and the projection of a table made of the floor material covered with water. Depending on how forcefully the viewers tread on the wooden platform, in the virtual space they trigger wave movements of the water and the surface of the table. The more forcefully one stamps, the higher the waves become in the image until, finally, they flood the entire image space.

Cartesian Chaos / Cartesianisches Chaos

1991	interactive computer-based installation; video projector, wooden platform with sensors, VGX computer, dimensions variable interface/software: Robert O'Kane

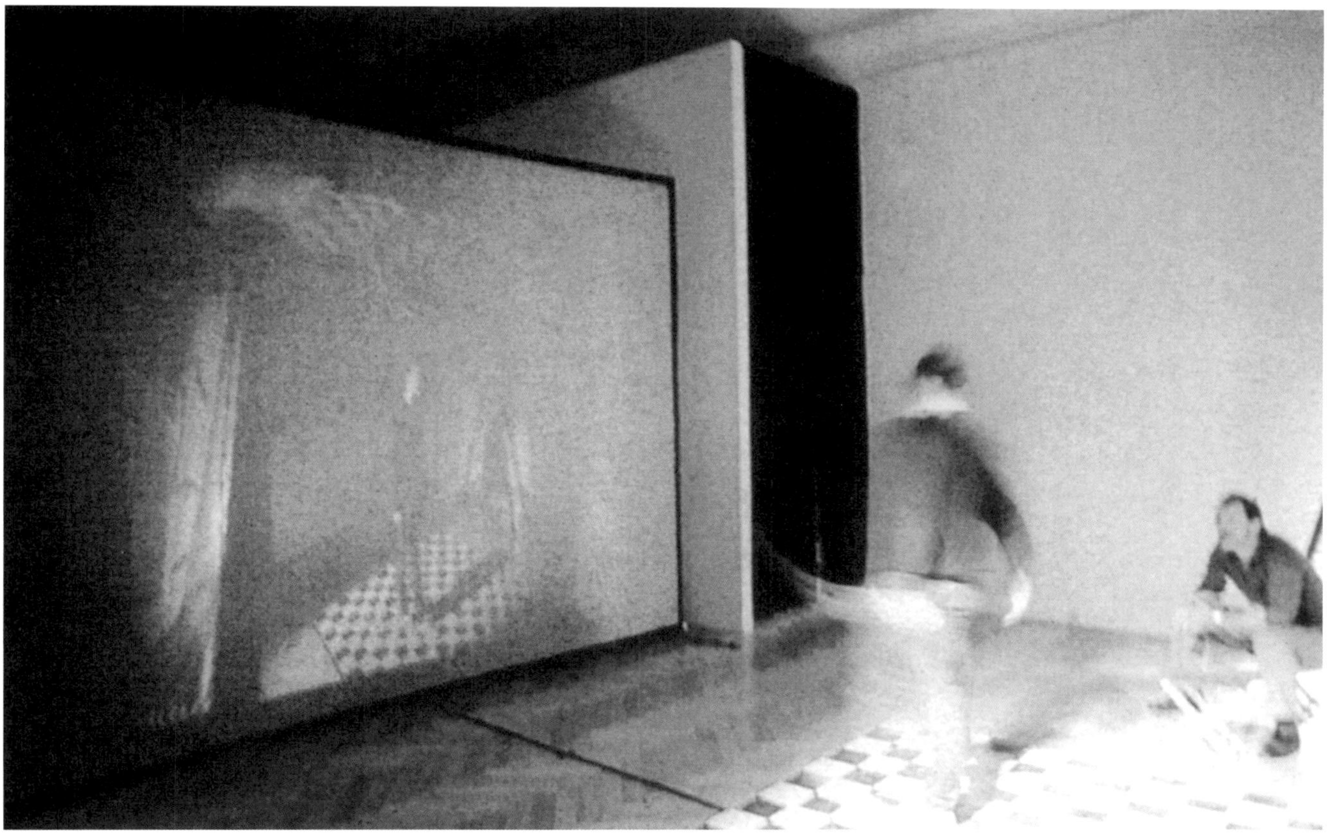

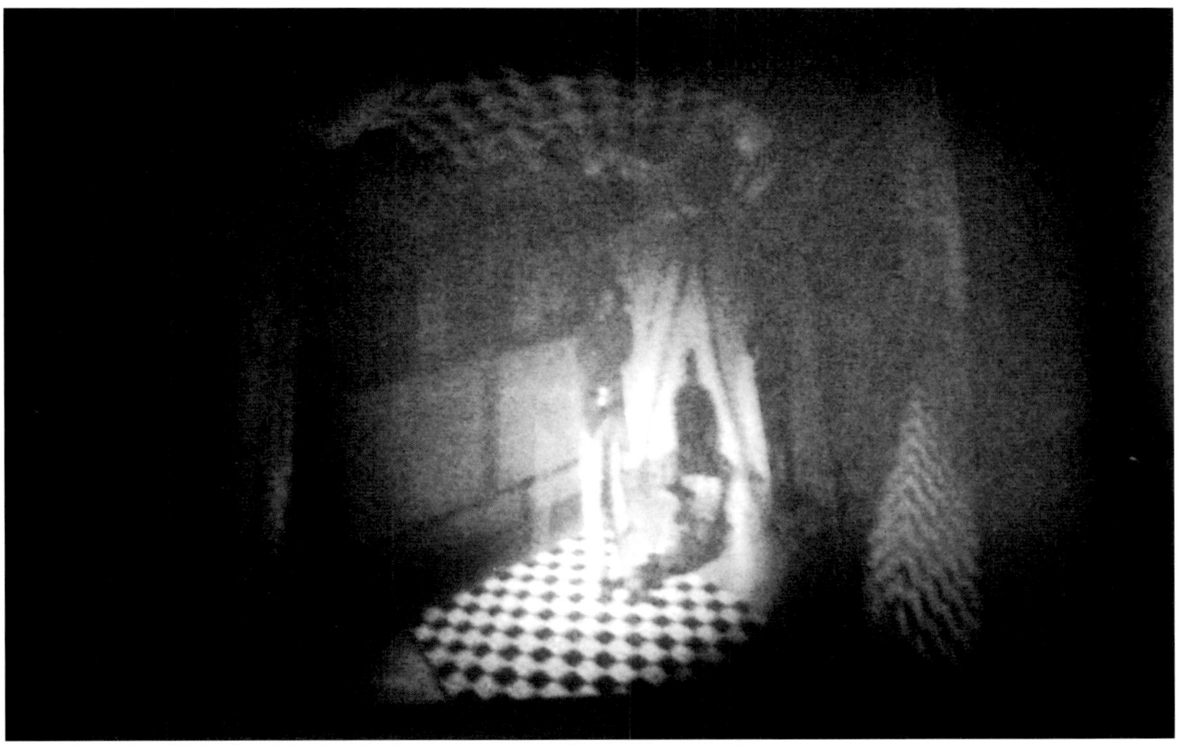

Installation views
Ars Electronica '92,
Oberösterreichisches
Landesmuseum, Linz, 1992

World Cube / Weltwürfel

1992	interactive computer installation; video camera, projection screen, video projector, VGX computer, checkerboard floor with built-in sensors interface/software: Robert O'Cane

World Cube is an extension of *Cartesian Chaos*. The floor has a checkerboard pattern with built-in sensors. The viewer is recorded by a video camera. This image of the floor is sent to the computer. The four walls of the virtual space (the projected room) are then created from these images. The image of the ceiling is mirrored in such a way that the viewer, standing upright on the floor of the real space, hangs upside down in virtual space. The viewer's movements across the floor cause wave-like deformations of the images on the walls of the virtual space. If the viewer jumps, the ceiling image is distorted downward, like a drop of liquid.

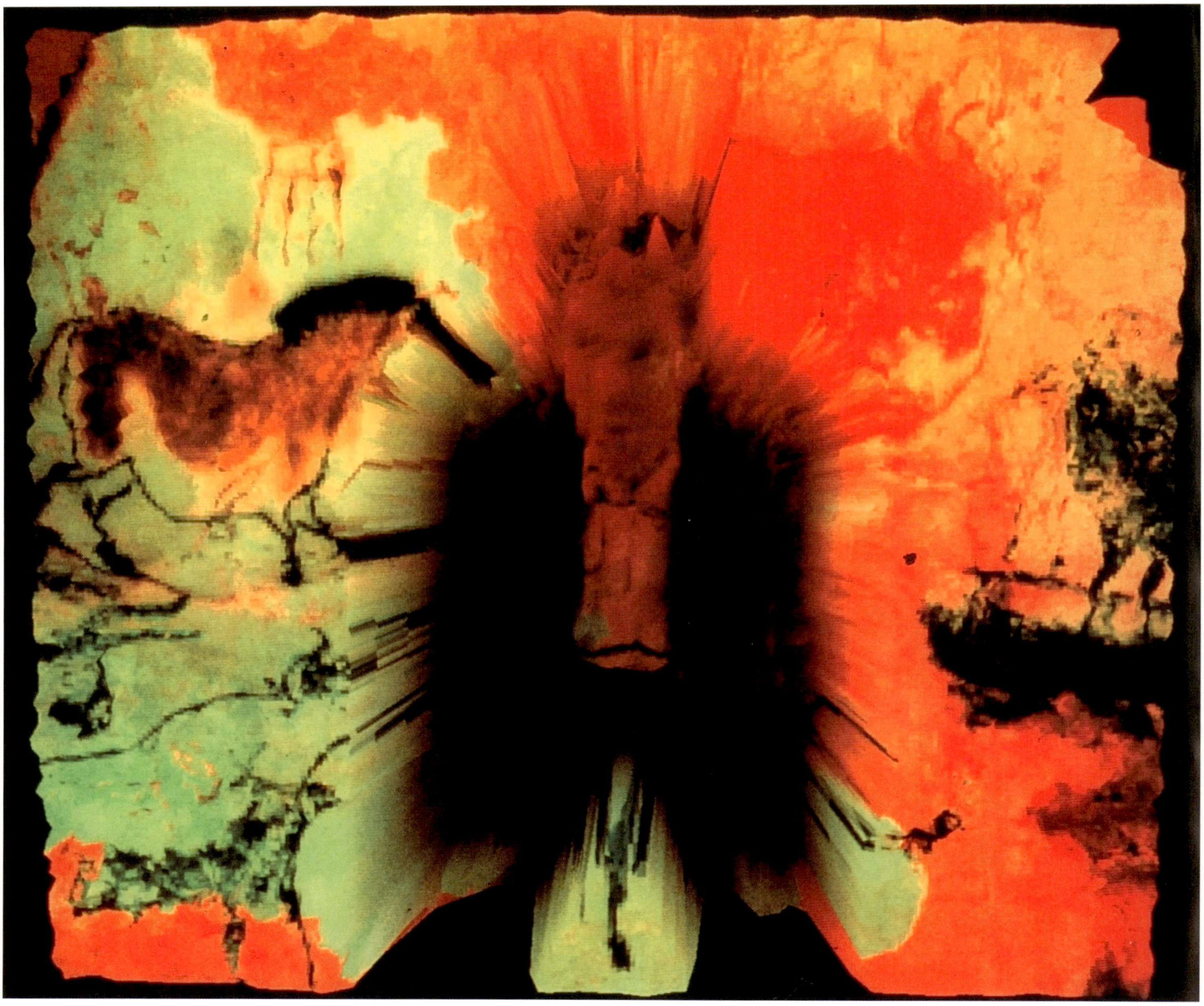

Installation view *First Europeans*, Schloss Charlottenburg, Berlin, 1993

The Curtain of Lascaux I / Der Vorhang von Lascaux I

1993	interactive computer-based installation; video camera, video projector, digitized image of the cave paintings of Lascaux, VGX computer, dimensions variable software: Robert O'Kane; technical assistance: Dieter Sellin

An image of the cave paintings of Lascaux, stored on a computer, is projected onto the wall of a darkened room. The viewer standing in front of the image is recorded by a video camera. The image of the viewer is mixed with the image of the cave paintings by the computer in real time, in such a way that the viewer appears to be inside the cave wall, which is spatially distorted by their movements.

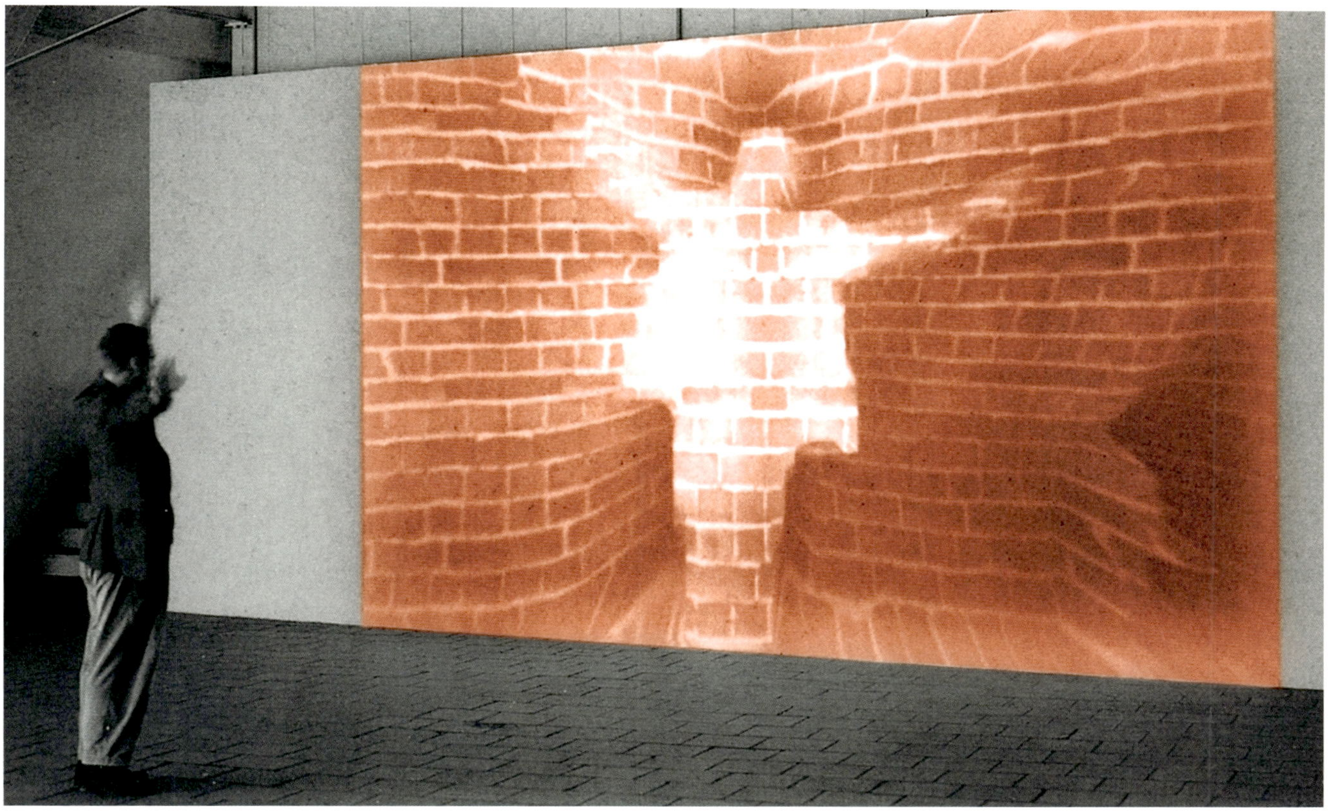

Installation view *Peter Weibel / Lawrence Weiner*, Galerie Tanja Grunert, Cologne, 1994

"*Limit* means the extremity of each thing, the first thing outside of which there is nothing to find and the first thing inside of which everything is." —Aristotle, Metaphysics, Book V

"No one here gets out alive"
—Jim Morrison, The Doors, "Five to One," 1968

Plato's allegory of the cave was probably inspired by cave paintings. In *The Wall, the Curtain (Border, the) or, in Technical Terms: Lascaux*, a brick wall is stored as an image on the computer and projected onto a wall of the museum, the new cave wall. Any viewer who steps in front of the wall is recorded by a video camera. The (analog) camera signals that convey the viewer's movements arrive at the computer, where they are converted into strings of digital signs, and these digital signal sequences alter the brick wall stored as an image on the computer. Both signal sequences are projected – with a slight delay, and in somewhat skeletal form, due to the tremendous complexity of the data processing involved – onto the wall.

The Wall, the Curtain (Border, the) or, in Technical Terms: Lascaux / Die Wand, der Vorhang (Grenze, die) fachsprachlich auch: Lascaux

1994	interactive computer-based installation; video camera, video projector, digitized image of a brick wall, VGX computer, dimensions variable software: Robert O'Kane

In reality the viewers are standing in front of the wall; in the image they are behind the wall. They are simultaneously in front of and behind the cave wall, in front of and behind the stage, in front of and behind the curtain; they are simultaneously inside and outside the cave. They are woven into the curtain, painted into the cave painting itself. The observer is part of the world they see – that is, part of the interface, the endoside of the world.
Whereas in the real world, humans can normally only observe and act from their internal perspective, digital technology enables them to take a viewpoint virtually outside the world: behind the wall. The prisoner of the world, condemned to be its purely internal observer, is virtually transformed into an external observer. With the help of technology and new models of the world, one can imagine oneself as an external observer of the world, pressing in on the wall from the outside. The world's inhabitants become fictional observers outside the world they inhabit. From this imaginary position beyond the limit, the interface, the wall, the prisoner of the world pushes from the outside against the bars of the prison of space and time, against the stage curtain, against the cave wall.
The interactive computer installation references the origins of image production, the cave paintings of Lascaux, and is simultaneously a technological reinterpretation of Plato's allegory of the cave, which can be understood as a philosophical archetype for questions pertaining to our perception and knowledge of the world. For Plato, the world is a cave in which humanity is trapped. For Aristotle, the world is a stage and the human being an external observer who cannot intervene on the stage because of the laws of nature. For me, the world is a curtain into which humans are interwoven and which can be changed via interface. Media are interfaces that can change the physical world. —Peter Weibel, 1996/2022

Translated from the German by Patrick Hubenthal.

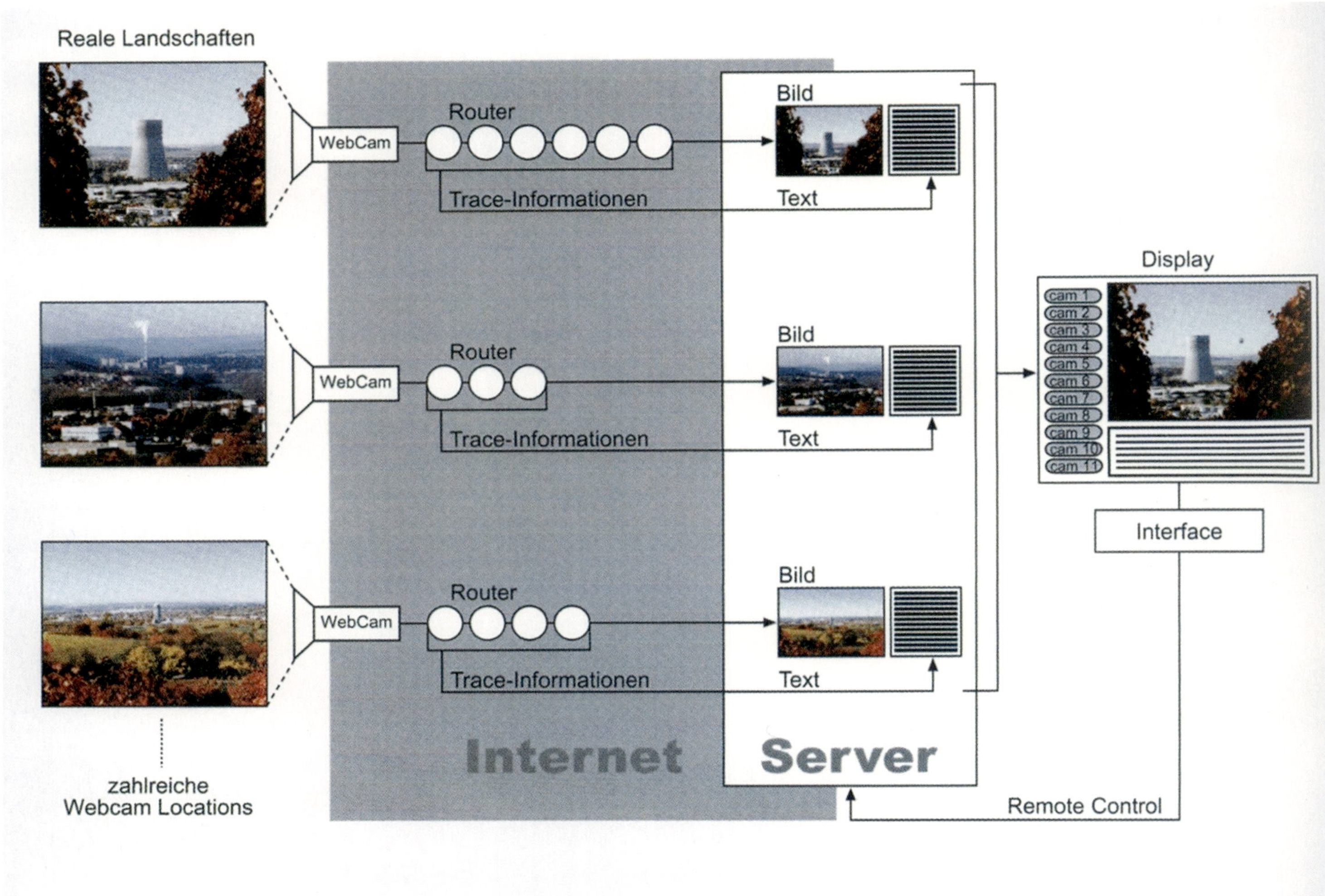

Approximately twenty landscapes – webcam images – are clickable in the form of thumbnails on the left-hand side of a browser so that they fill out the format and can be viewed individually. The motifs are entirely in the tradition of this classic genre of visual art: just as the landscape painting began to edge out the panel painting from the seventeenth century onward, the webcam images of landscapes show the increasingly prominent presence of the landscape filling internet browser screens. The images feature observed views of real physical landscapes in real time from outside the internet, that is, from a position located outside the exhibition space. The selected grouping of webcams showing landscapes was not globally installed specifically for this piece, but was taken as a found resource from the internet, which upon closer inspection is an act of webcam piracy.

With the development of the landscape painting, a new sociopolitical stance took center stage: utopian socialism (landscapes for the masses to enjoy) became a counterpart of late Romanticism (landscapes for the elite). Likewise, the "webcam landscapes" seek to promise the ideal of freedom of the people and the individual, to mark a transition from feudalism to capitalism, and to postulate a move towards a second modernism – knowing full well that Romanticism, along with the already existing Net-Romanticism, is the product of a (petit) bourgeois transformation.

When a landscape is called up full screen in the browser, it is captioned with a legend indicating the location of the webcam. Thus, it is not the artist, year of completion, title, etc. which are presented, but rather temporal and spatial coordinates of the determined position in online space. In the rhizomatic internet, a computer's localization is constantly changing and unique, temporary, and therefore singular.[1] These Internet-specific indications in the relic "caption card" undercut the Romantic nature of the image.

Even the logical image space of central perspective – abrogated in Romanticism – finds its analog once again in the caption's time code, since this piece of information uniquely identifies that there is only one vanishing point – the IP address, the webcam image – but countless points of view from the computers of the internet. This correspondence, furthermore, enables a renewed, if highly abstract, use of the inverted pyramid of vanishing point perspective: by being immersed in the "virtual outside," the viewer loses their subject status and becomes the object of description by the network.

Note: Adapted from Markus Huemer, "Genauer: Web-Cam-Piraterie," in *Peter Weibel: Globale Gier* (Klagenfurt: Kärntner Landesgalerie, 1999), 32–34. Translated from the German by Peter Dahm Robertson.

[1] The ping can find a new trace route each time. For example, if one and the same computer sends a ping twice in a row, the online route, and thus the response time and number of hops, can still change.

Islands of Nonlocality: From the Local and the Real to the Nonlocal and the Virtual

1999	internet-based installation; computer, monitor, peripheral equipment

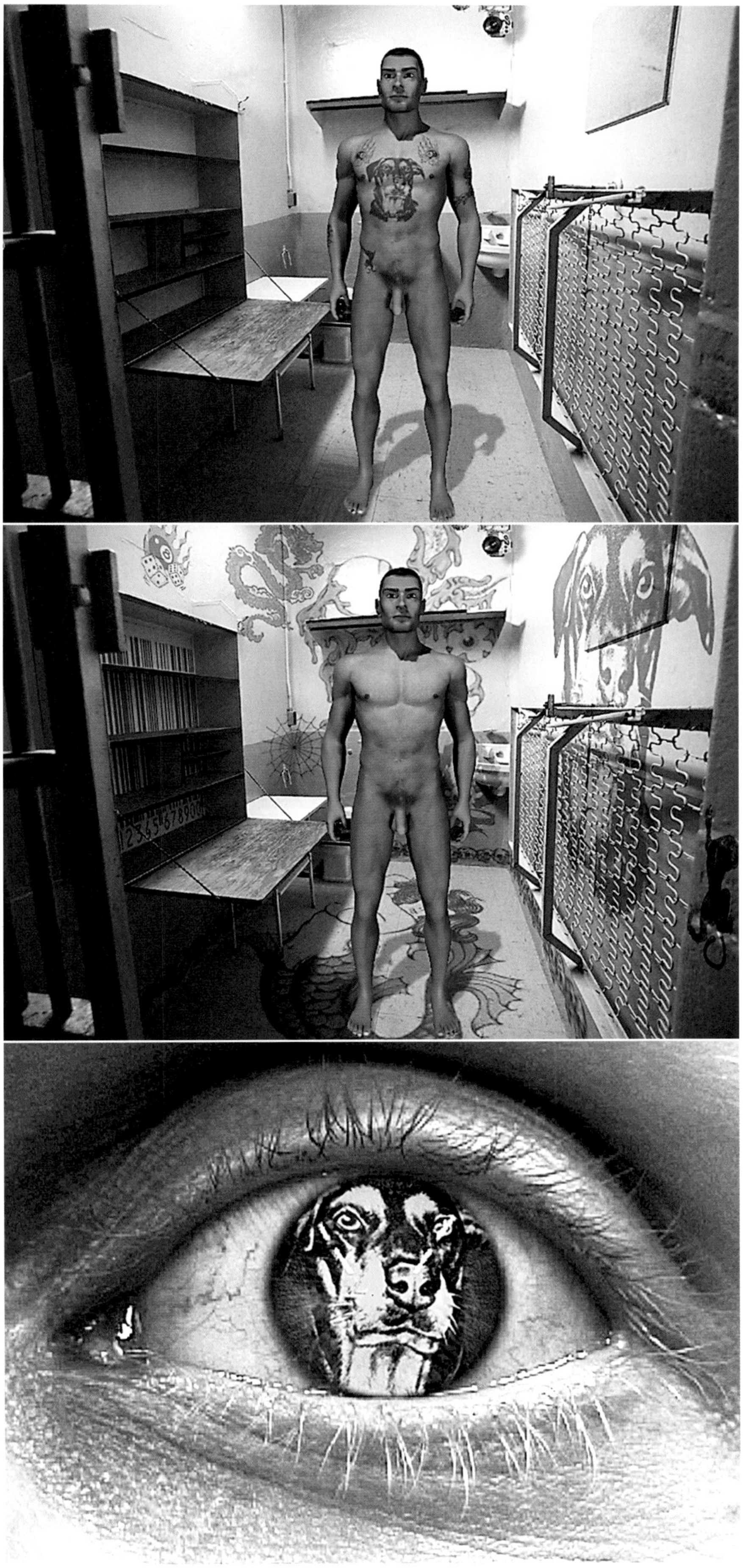

The Panoptic Society or Immortally in Love with Death

2001 | interactive DVD

The phenomenon of an increasing degree of surveillance in our society was already reflected in Weibel's early video works. Examples of this include *Video Lumina* (see p. 268 in this volume) or *Observing Observation: Uncertainty* (*Beobachtung der Beobachtung: Unbestimmtheit*, see p. 262 in this volume), dating from 1973. The latter work consists of three monitors and three cameras, arranged alternately in a circle. When visitors step into the middle of the circle they discover that the cameras are connected to the monitors in such a way that they only ever show the visitors from behind. The subject in front of the camera becomes an object. Inevitably, the uncomfortable feeling arises of being watched or observed.

To Weibel the situation today appears quite different. In a time where people are increasingly integrating the media into their lives, they enjoy being watched. *The Panoptic Society or Immortally in Love with Death* shows a prisoner in his cell who transforms himself and his criminal existence into a public spectacle by laying it bare in front of the public. Weibel focuses our attention on the pathological side of the media whose "jousting for audience attention produces visual excesses that leave escapism behind and elicit new obsessive-compulsive scenarios of control, xenophobia, and sado-masochistic exploitation. The prisoner's right to renounce the privacy even he can claim for himself and his disastrous situation is based on an 'economy of attention' dominant in the real voyeurism of the entertainment industry, reality TV shows, and whose strategies are deployed ever more aggressively in the infotainment offered by news media, in the staging of politics as theater."[1]

Weibel shows us a development in which the medium of observation turns out to be a powerful medium for putting oneself on show. We are confronted with a new axiom: I am observed, therefore I am, a proposition by George Berkeley: "Esse est percipi." People are eager to be observed, to be perceived. A new excess culture of trash TV arises.

Note: Adapted from Dominika Szope, "Peter Weibel," in *Future Cinema: The Cinematic Imaginary after Film*, ed. Jeffrey Shaw and Peter Weibel (Cambridge, MA: MIT Press, 2003), 180–91.

[1]
Ursula Frohne, "Horror vacui," in *(dis)LOCATIONS*, ed. ZKM | Karlsruhe and The Centre for Interactive Cinema Research, College of Fine Arts, University of New South Wales, Sydney, exh. cat. (Ostfildern: Hatje Cantz, 2001), 85.

T_ Visionarium is an immersive, interactive, multimodal environment. By enabling viewers to immerse themselves in virtual modalities, it explores the expressive potential of dialogic and transcriptive as opposed to conventional interactive narrative. Navigation through the data sets dramatizes the televisual information archived in *T_ Visionarium*'s database. This strategy allows viewers to experience the sense of a wholly personalized authorship. *T_ Visionarium I* was first installed in 2003,[1] within a dome-shaped projection space, the EVE – Extended Virtual Environment, which had been developed from 1993 to 2002 by the ZKM | Institute for Visual Media under the direction of Jeffrey Shaw.[2] The setting within a dome twelve meters in diameter by nine meters in height made of inflatable fabric allows viewers to spatially navigate a televisual database in virtual time.

On entering the dome, viewers place a magnetic position tracking device, connected to wireless stereo headphones, on their heads. By means of a remote-control interface, viewers are able to select from a range of parameters which arrange the televisual database according to thematic categories such as "dialogue" and "crowds." The projection system is articulated to the tracking device so that viewers, by shifting their heads, move the large projected viewing window across the interior surface. This tracking device identifies the exact orientation of an individual viewer's point of view, which in turn controls the orientation of the projector so that it presents an image directly in front of where the viewer is looking. The audiovisual data streams are virtually distributed over the entire surface of the dome, so that the movement of the projection windows enables the viewer to navigate between these multimodal data streams. The stored televisual data sets are physically mapped over the dome surface such that each data set is allocated a specific window grid on the dome's surface. This enables viewers to navigate between each data set by merely shifting their point of view. This mapping strategy applies to both image and sound.

Seamless transitions between discrete image and sound events are handled by specific design parameters of the audiovisual delivery system. The acoustic delivery system is based on the use of RF (radio frequency) wireless headphones, which each visitor wears while inside the dome. Multiple audio channels are interactively mixed down for stereo RF delivery to all the headphones. The mixing of these channels is handled dynamically in immediate relationship to the movement of the pan-tilt projection system so that a fully spatialized soundscape can be defined inside the dome architecture. This is directly linked to the distribution of the visual content. The mixing of the audio, synchronized with the movement of the pan-tilt projection system, allows a fully spatialized soundscape inside the dome to be synchronized with the distribution and experience of the visual content. These same satellite data streams are recorded onto a hard disk system and sorted within a database.

T_ Visionarium allows new recombinatory narrative content to be generated by the viewer. By means of their interaction with the matrix interface and the simultaneous movement of their heads and projection window individual viewers originate unique performances on behalf of a larger viewing public of up to eighty persons. This strategy allows viewers to experience the sense of a wholly personalized authorship. Navigation through the data sets dramatizes the televisual information archived in *T_ Visionarium*'s database. These data sets are reconstituted across the interior skin of the dome under the converging impetus of viewer and matrix and are encountered as emerging multitemporal events. Processing is animated by the recombinatory parameters of the interactive software. It is based on broadcasts from 48 global satellite television channels recorded sequentially in thirty-minute intervals over a single 24-hour period.[3] This data is post processed by the matrix in ways that hyperlink the variegated data sets in virtual time to form a large-scale database. The viewer explores the results of these recombinatory searches by moving the projection window across the dome screen. Selecting the parameter "dialogue," for example, ushers forth intersecting cascades of current affairs, sports, features, life style, historical, scientific, musical, and anthropological episodes of "dialogue" across 48 channels, a multiplicity of languages, numerous time zones, and a heterogeneity of cultures within the simultaneous timeframe of the visualization apparatus.

The profoundly multitemporal logic echoes the theoretical architecture implicit in digitized audiovisual data.[5] This logic is imperceptible in conventional viewing frameworks, which can only recover time analogically by scanning and juxtaposing whole fragments. At best, conventional viewing establishes symmetrical patterns of temporal resemblance among broadcast items that are based on syntactical properties patent within the data.[6] *T_ Visionarium*'s narrative moves beyond this logic of resemblance. With changes to their point of view, viewers activate a powerful navigational matrix that produces a directional flow of information in which the expressive meaning of the data is boundlessly transcribed.

Sifting through digitized televisual data a virtual "infosphere" of images and sounds whose patterning is freed from the constraints imposed by the analogical, or representational, redelivery of information. The viewer unravels subvisible links. As a consequence, narrative becomes a complex event which interlaces a number of intersecting temporal and physical navigations. The viewer can refine these streams by zooming into a specific current within the streams. This interweaving of matrix and viewer navigation precipitates the emergence of unprecedented narratives. In this respect *T_ Visionarium* opens interactive cinema to a multimodal aesthetic of a kind that is currently confined to the unimodal contexts such as text-based chat rooms.

Note: Adapted from Dennis Del Favero, Neil Brown, Jeffrey Shaw, and Peter Weibel, "*T_Visionarium* – Towards a Dialogic Concept of Digital Narrative," in *Disappearing Architecture. From Real to Virtual to Quantum*, ed. Georg Flachbart and Peter Weibel (Basel: Birkhäuser, 2005), 144–51.

T_Visionarium I

2003/06	Jeffrey Shaw, Dennis Del Favero, Neil Brown, and Peter Weibel interactive installation produced for EVE – Extended Virtual Environment

[1]
On the occasion of the exhibition *Cinémas du Futur*, Dec. 6, 2003–Feb. 22, 2004, Lille, European Cultural Capital, 2004.

[2]
A second version of the installation, *T_ Visionarium II*, was created in 2006 for the Advanced Interaction and Visualisation Environment (AVIE), which had been developed at the iCinema Centre for Interactive Cinema Research at the University of New South Wales, Sydney, Australia. In contrast to the spherical EVEdome, the AVIE is a 360° circular panoramic screen.

[3]
The data for the second version, *T_ Visionarium II*, originated from Australian television.

[5]
Elizabeth Grosz, ed., *Becomings: Explorations in Time, Memory and Futures* (New York: Cornell University Press, 1999).

[6]
Juan P. Casares, "SILVER: An Intelligent Video Editor," in *CHI'01 Extended Abstracts: Conference on Human Factors in Computing Systems*, Seattle, Mar. 31–May 4, 2001 (New York: MC Press, 2001), 425–26.

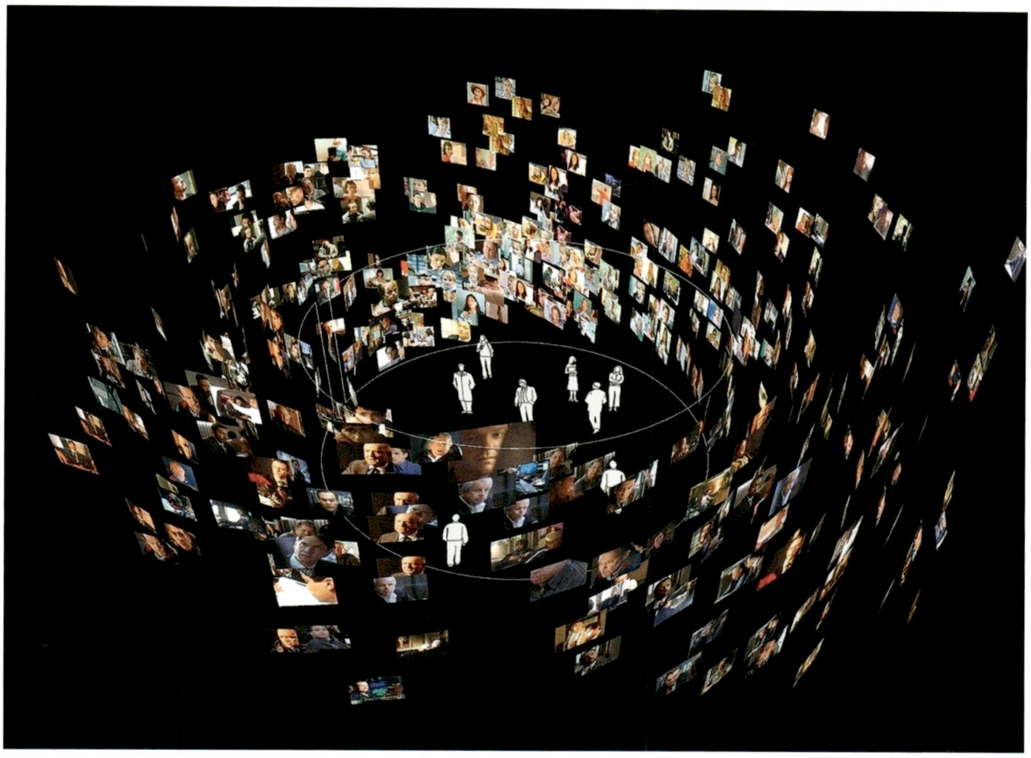

Jeffrey Shaw, Dennis Del Favero, Neil Brown, and Peter Weibel, *T_Visionarium I*, 2004

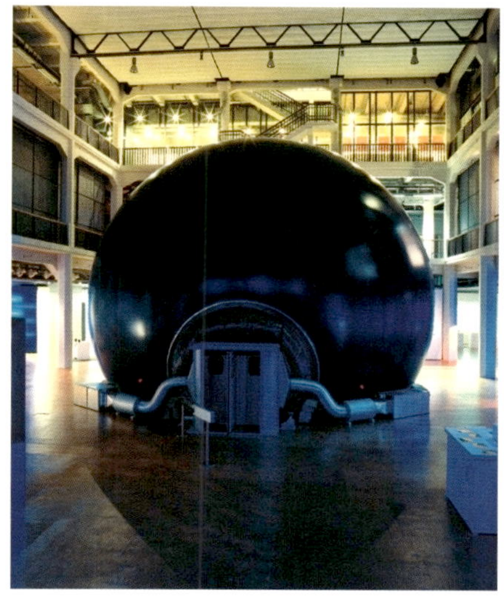

EVE dome, 1993–2002, exhibition view *Future Cinema. The Cinematic Imaginary after Film*, ZKM | Karlsruhe 2002–03

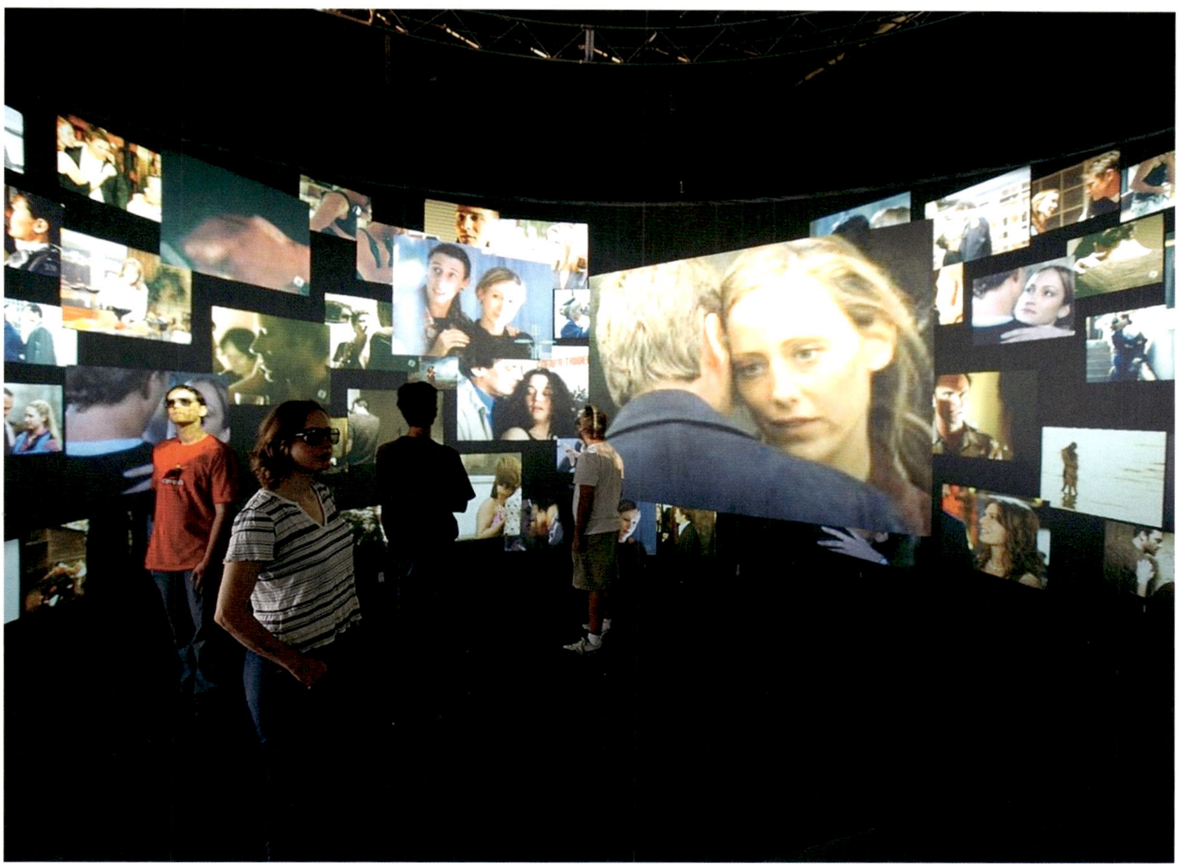

Jeffrey Shaw, Dennis Del Favero, Neil Brown, Matthew McGinity, and Peter Weibel, *T_Visionarium II*, 2006; installation view *Imaginig Media@ZKM*, ZKM | Karlsruhe 2009

The title of the work, *YOU:R:CODE*, can be read in two different ways. The interpretation "your code" refers to the fact that in the installation, visitors experience different kinds of coded transformations of themselves. The second way of reading the piece's title, "you are code," emphasizes that we ourselves consist of codes, from dress code to genetic code.

Station I: The mirror
At the first station, visitors see their familiar reflection in a mirror – the most real virtual depiction that we can imagine. Water was the first medium in and through which people could see themselves, a sort of liquid mirror. Waves of water became waves of light, and a reflection in water was like a reflection in the mirror. The mirror reflection is merely a phenomenon of light, which obeys the laws of geometry and optics, the code of physics. Metaphorically speaking, the reflection is a kind of optical code. The mirror provides a virtual two-dimensional likeness of a person, which appears highly realistic because it is a perfect analog copy.
In Greek mythology, Narcissus, whose vanity drew him to his reflection in the pool where he drowned, evokes the significance of the mirror in the formation of the self or subject, just as many fairy tales and Jacques Lacan's psychoanalytical theory of the mirror stage (1936) do. In a sense, the mirror represents the first stage of a visual coding of data.

Station II: The digital mirror
The second station presents an image that gradually transforms into a digital data body. Viewers see themselves reflected as a set of data and as a mirrored image, which overlap. Sensors and a camera are used to capture the three-dimensional surface of the viewer's body. Data from the sensors is sent to a computer that simulates the workings of an analog mirror. It is the computer that creates the image, not the mirror. The computer generates a digital image that acts like an analog image. Viewers see a computed image of themselves – they are reflected in it and in the analog mirror image at the same time. The real body becomes a virtual mirror image, a data body. Viewers see digital and analog copies of themselves. They create their own digital likeness through data processing. This is where the transition from self-image to synthetic self-creation begins.

Station III: The scan
A television image is generated from horizontal scan lines. At the third station, viewers' bodies are scanned three-dimensionally. The data set acquired through this process can be used independently. The body detaches from the viewer to lead a life of its own in a parallel world, like in Lewis Carroll's story *Alice through the Looking Glass* (1871). The scan is an expansion of the mirror because it identifies and displays data that is not visible in a common mirror.
An important difference becomes apparent. The analog mirror image follows the viewer. The viewer controls the analog mirror image: every real movement is identical with the virtual movement. The digital mirror image, on the other hand, detaches itself from the viewer: it is autonomous; the negative digital body rotates independently of the positive real body. The digital mirror shows the function of the data in the data world as in a shadow world. In front of the digital mirror, viewers follow the rotation of their images, even though they themselves do not move. The data does not behave like a shadow attached to a human being; it detaches itself from the viewer, and others can use it to do whatever they want. This recalls Adelbert von Chamisso's "literary fairy tale" *Peter Schlemihl's Miraculous Story* (*Peter Schlemihls wundersame Geschichte*, 1814), about a man who sells his shadow to the Devil. He lives without a shadow; his shadow leaves him, and others do with it what they want. In the same way, users on the net have sold their shadows, their data, to the Devil, and others do with it what they will. The Big Five (the internet companies Alphabet, Amazon, Apple, Meta, and Microsoft) earn billions from the sale of our shadows, the user-generated data that we voluntarily and gratuitously hand over to these companies that merely provide the medium of distribution (that is, the platforms).
The distribution and storage of gigantic amounts of data enables that data to be correlated using artificial intelligence. The data traces in the network serve to better profile and identify the behavior of the data producers, thus ultimately making them predictable and controllable. People increasingly find themselves in a prison of multiple codes (gender code, dress code, genetic code, social code, etc.). All these codes are used

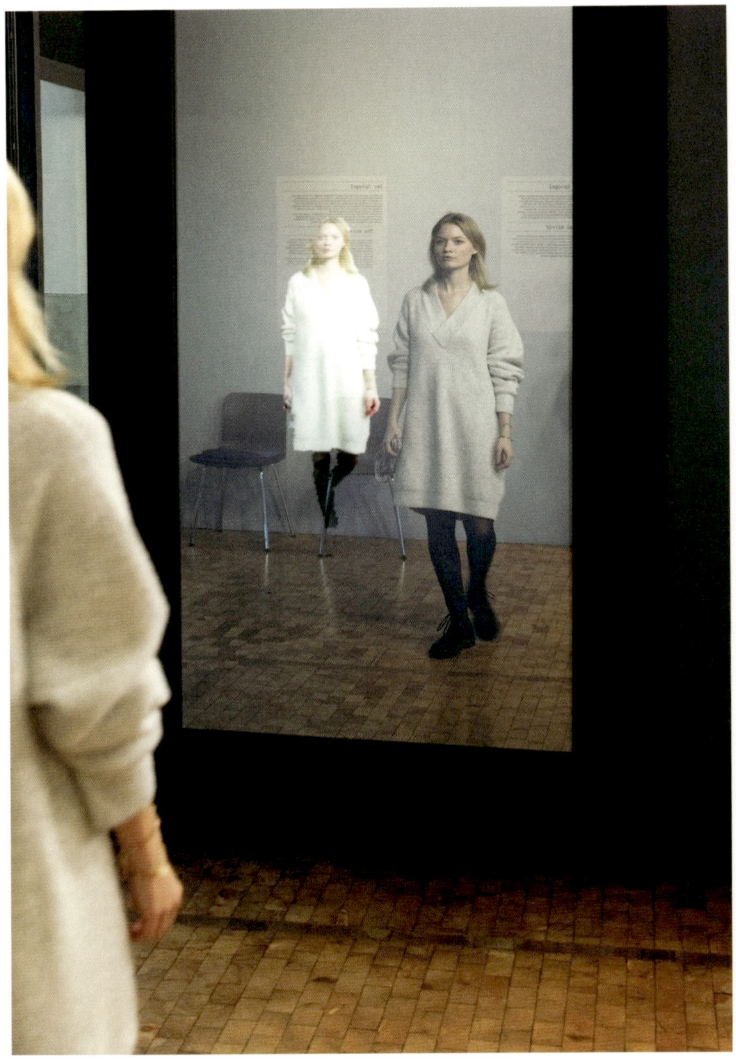
Installation views *Open Codes*, ZKM | Karlsruhe, 2017, Station II

YOU:R:CODE

2017	Bernd Lintermann and Peter Weibel interactive 7-part installation with multichannel projection idea: Peter Weibel; realization: Bernd Lintermann; audio design: Ludger Brümmer, Yannick Hofmann; flip-dot display: Christian Lölkes; technical support: Manfred Hauffen, Jan Gerigk; setup, planning: Thomas Schwab; production: ZKM | Hertz-Lab

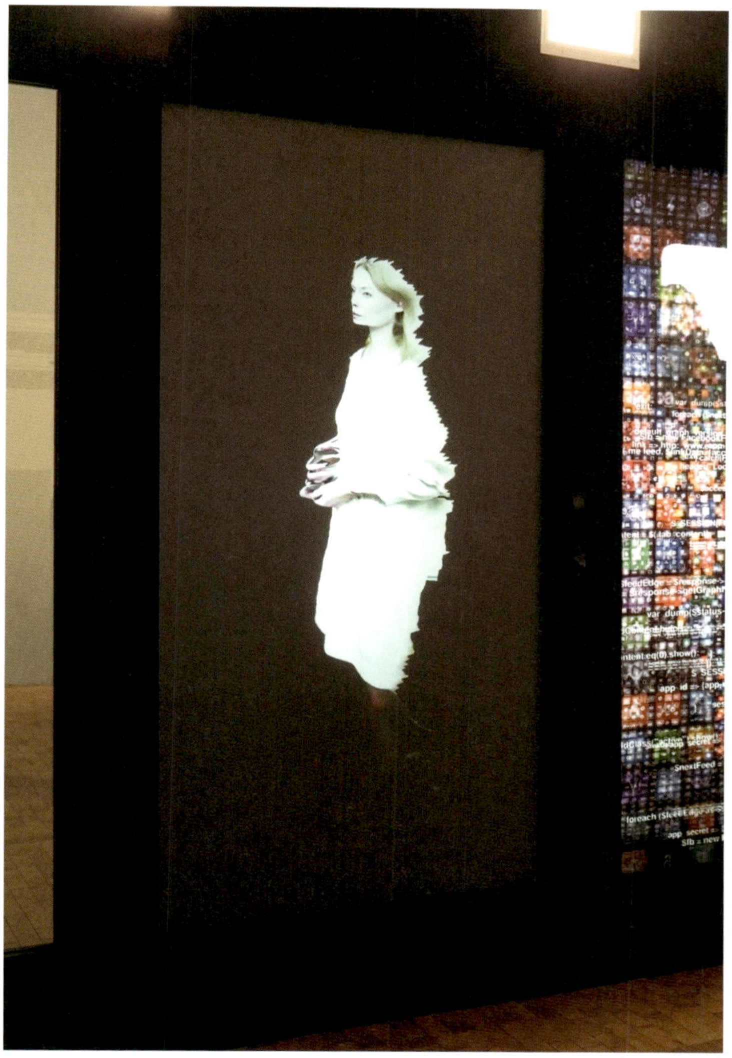

Station III

for predictive data analysis and thus for social control. We live in a world of unleashed codes.

YOU:R:CODE, this performative path through the world of codes, presents humans as data hunters and data carriers – with the goal of liberating them from their code prison.

Station IV: The traces of social media

At the fourth station, visitors are processed by facial-recognition software. Above their heads they see precise descriptions of themselves: sex, age, height, hair color, etc. Through our web browsers and our smartphones, we leave permanent electronic traces on the net. These traces are continually saved in clouds and transformed into digital identities through big-data algorithms and artificial intelligence, as are sensory data from the environments through which we move. Data from different sources is now combined to generate an ever more precise model of our behavior. Social-media portals offer us services in order to dig up ever more varied and more private data.

In the installation, viewers are represented as a collage of social-media icons and program codes, which represent their algorithmically calculated online identities. Their physical traits are ascertained from the cloud through artificial-intelligence techniques. Just like a footprint in the sand that shows people who has walked their path before them, internet users leave a large trail of data behind them. The algorithm follows the person, analyzing their footprints and traces at every turn to build an ever clearer and sharper profile of them. Another chapter in the colonization of us, our selves, by the data industry.

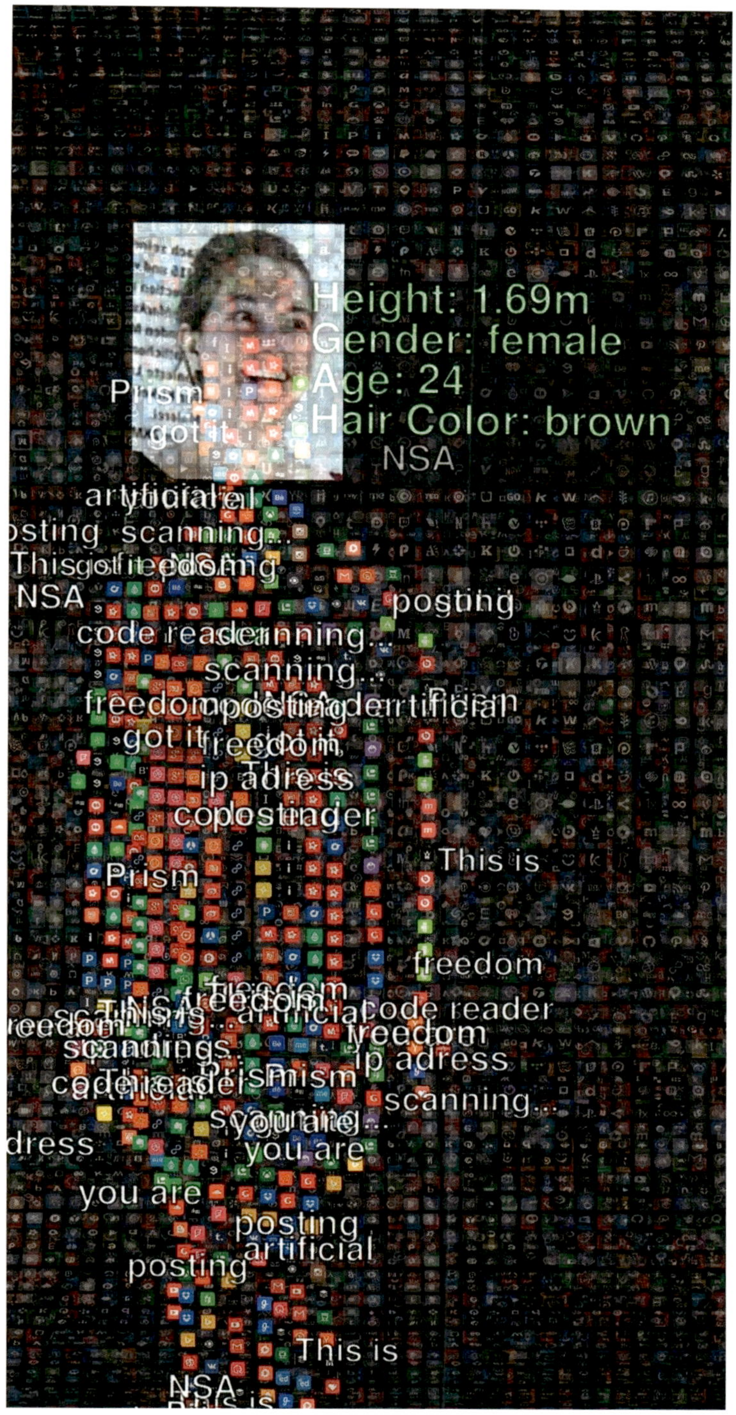

Station IV

Station V: The genome and memome

We carry the source code of our bodies around with us in the form of our genomes. The genome is made up of the inherited nucleotide sequences of deoxyribonucleic acid (DNA) that are inside the nuclei of all our cells. Cells are continually dying and being replaced through cell division. Human beings always carry around with them their own instructional code, the definition of one's self: the DNA code. But code alone does not constitute a person: enzymes read the code and transcribe it for further use in the body. They are the data-processing algorithms that form us. The code is complex, rich in information, and prone to errors: the smallest changes can have enormous consequences, such as illnesses. What is special about DNA is its universality: it is the same for all living beings on Earth. The extremely stable fundamental structure of DNA is the double helix. Two strands made of sugar and phosphate

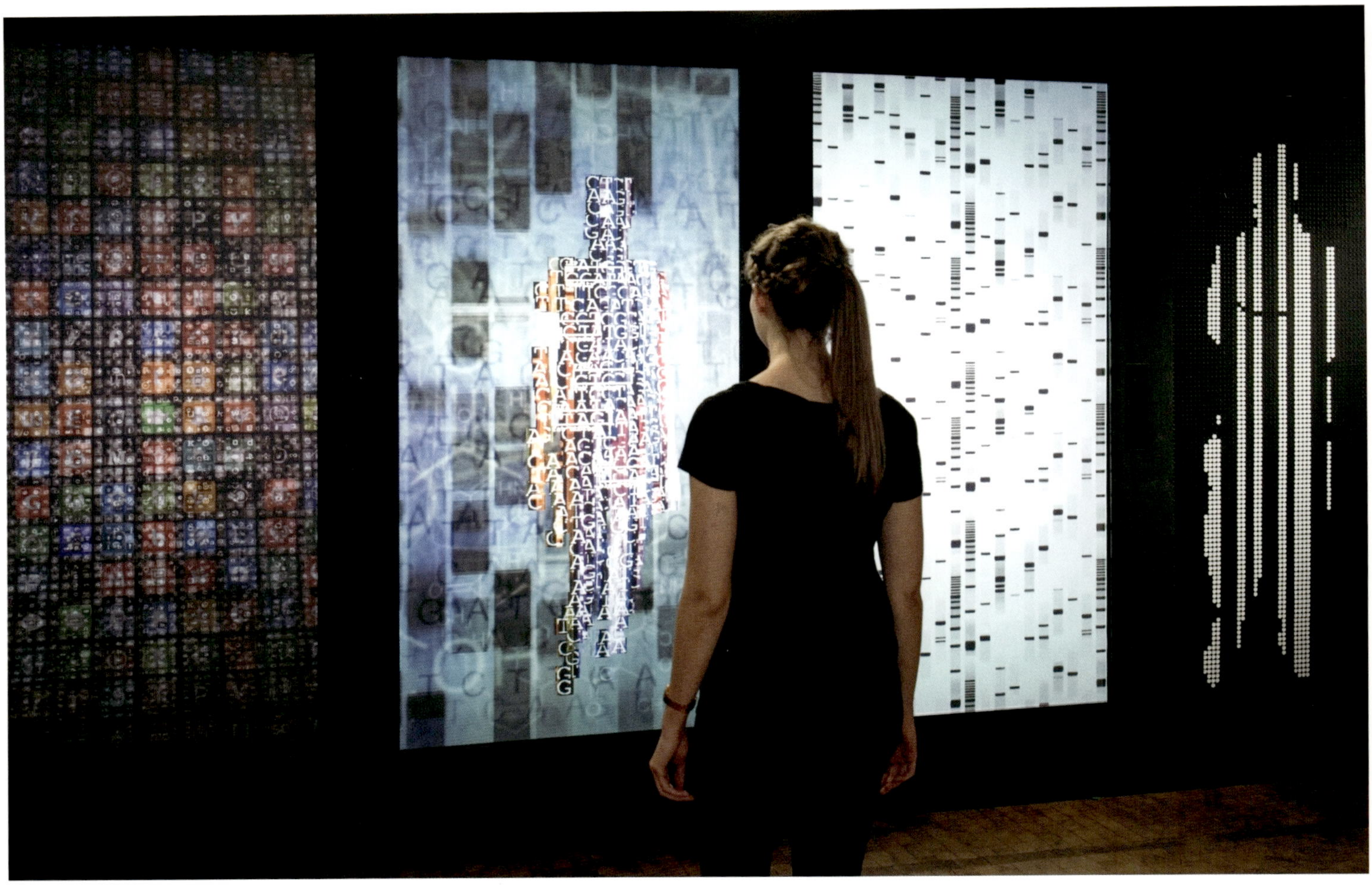

Stations IV, V, VI, and VII

form a twisting ladder of the four bases adenine, cytosine, guanine, and thymine. The genome forms the algorithm of life and determines not only the external traits of our bodies, but also predispositions to illnesses as well as character traits, which tended in the past to be ascribed to our phenotypes – that is, to the person as formed by their environment. For example, genes have been identified that are responsible for a person's sense of humor, altruism, or empathy.

In the form of DNA, over billions of years evolution has developed a perfect machine, the source code for every living thing on Earth. Through research on DNA, code such as the use of the CRISPR technique in cutting telomeres (the ends of linear chromosomes), humanity has gone beyond natural evolution. In human-directed exoevolution, people hand over their algorithms and the content thereof to technology through DNA synthesis. God is said to have created man in his image. Now man is creating a perfect image of himself in a second act of creation. In current research projects, biological and digital codes are being combined. At the European Bioinformatics Institute in Cambridge, DNA is being investigated for archival purposes as a compact and durable repository for information. Machine synthesis of DNA generates long strings of information that can be decoded through DNA analysis. Specially developed coding processes map digital data onto the four bases. This type of encoding and decoding has been tested successfully on a four-hundred-page nonfiction book, as well as an image and sound file of Martin Luther King's "I Have a Dream" speech. In the future, human cultural creations – "memes," in Richard Dawkins's coinage – could be saved in a digitized "memome." But everyday objects are also now being marked by synthetic DNA to identify them clearly. For example, to deter thieves, Deutsche Telekom marks copper cable with liquid artificial DNA made up of several bases, so that the cable's original owner can be identified if it is stolen: DNA is the barcode of the future.

Station VI: The barcode

At the sixth station, the visitor is reduced to machine-readable code, the barcode. According to a legend passed down by the Roman scholar Pliny the Elder, painting came into existence when a young Corinthian woman traced the profile of her beloved, who was leaving on a voyage by sea, onto the wall where his shadow was being cast by the candlelight. For decades, images made of light – on televisions, on monitors, and emitted from projectors – have defined our communication with digital computers. The viewer's body is displayed as an outline, as Pliny described. This outline is filled in with vertical lines, displayed as a barcode. The person becomes a commodity among commodities. Identity, the self, is marked by a barcode, just like merchandise. The barcode becomes a medium of identification. At the end of evolution, *Homo sapiens* has become a data-storage device and a data hunter. Transhumanism has begun.

Station VII: The flip-dot display

Finally, at the seventh station, the visitor breaks free from virtual depiction, from representation, and is materialized on a flip-dot display. The viewer's movements change the material state of the display. The picture becomes a three-dimensional sculpture. The spectator transforms the real world by coding the material world without touching it.

Since the advent of the "Internet of Things," the digital image has broken away from light and objects themselves have become displays. When an analog object is illuminated, it casts a shadow. The installation draws and illuminates the

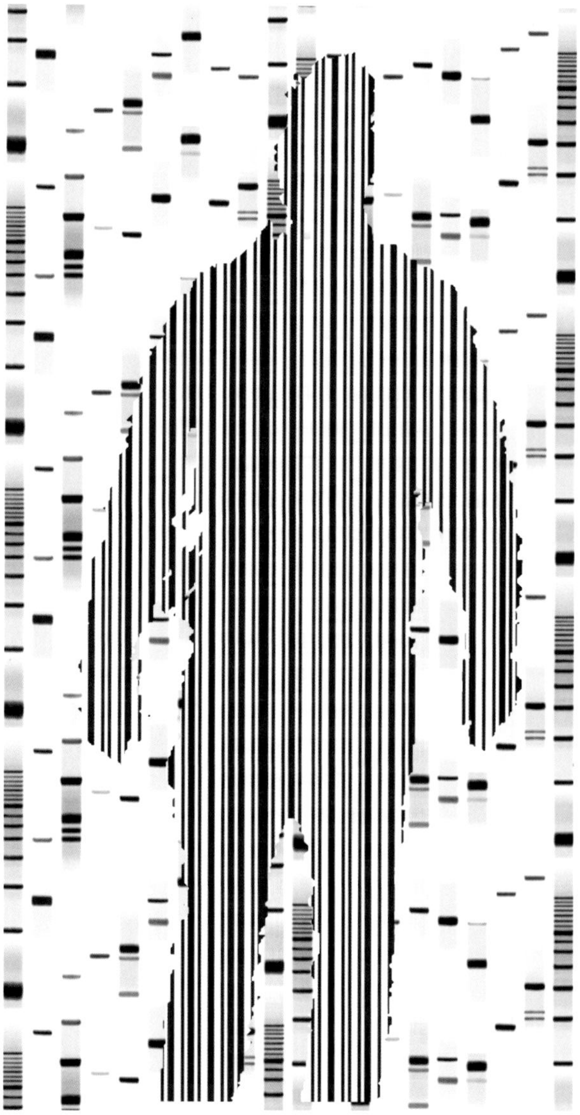

Station VI

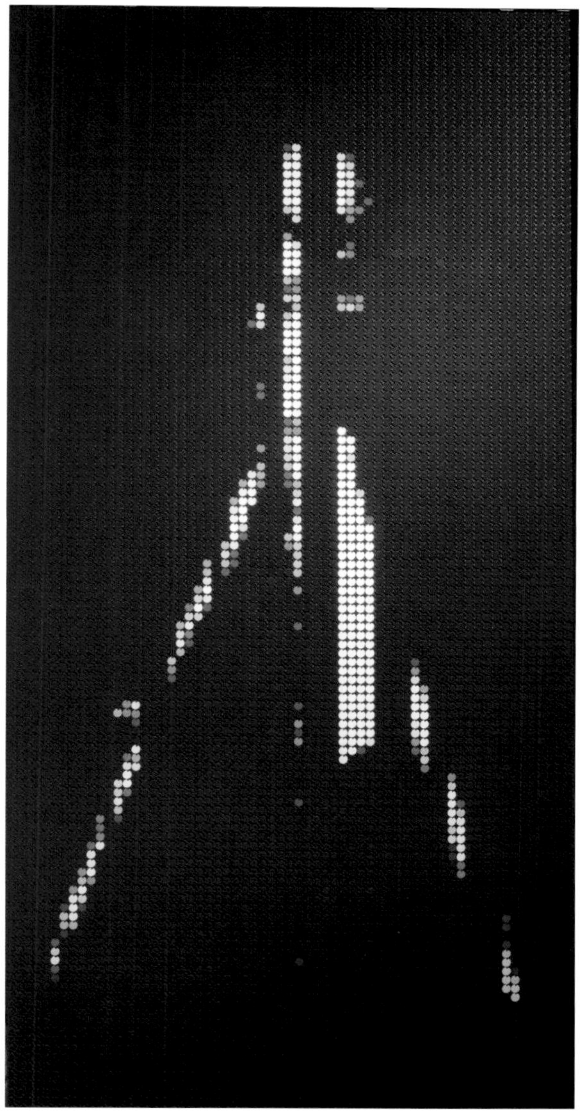

Station VII

digital image of an object, and the flip-dot display built into the installation shows its digital shadow. The display itself is made up of thousands of tiny plates or rotating discs, black on one side and white on the other, which are analogous to pixels and behave like data. Mounted on low-friction pivots, they are rotated by a magnetic field to depict the object, showing a white surface for content or a black surface for emptiness and absence of data. The virtual image is once again material. On the flip-dot display, viewers see how their data, having been processed and enhanced in value at the previous stations, becomes binary code. Binary states, normally microscopic, are made visible. Due to the large size of the dots/pixels and the relatively slow speed of their transitions, the origin of the image – namely, an ordered set of binary states, becomes discernible. The discrete states become visible, and the transition from one to the other can be both heard (the clacking of the discs) and seen (their movements).

When viewers stand in front of the flip-dot display the dots are flipped by the movements of their bodies, which are recorded by a Kinect camera (as at the other stations) equipped with a combination color/depth sensor. For each color pixel, the camera also captures a time-of-flight measurement, determining travel time for pulses of light and combining this information with the known speed of light to calculate distance. Using this information, filter algorithms can distinguish subject from background both spatially and chromatically.

Space and color information are merged into a two-dimensional likeness, or outline, and transmitted to the display's control unit. There, the outline is fitted to the display's physical dimensions, with each pixel allocated to one of the dots. The data is then serialized – transformed into a data stream – and provided with addresses for each of the display's segments. More than forty small control units receive the information via a bus connection and convert it into pulses, first electrical and then magnetic, which flip the dots.

By means of both its content and its technological functionality, the flip-dot display thus brings viewers back to the heart of its digital existence: a set of bits. Viewers alter the physical world, the states of the flip-dot display, without physically controlling or even touching it.

—Bernd Lintermann, Christian Lölkes, and Peter Weibel

Translated from the German by Patrick Hubenthal.

Installation views *Open Codes*, ZKM | Karlsruhe, 2017–19

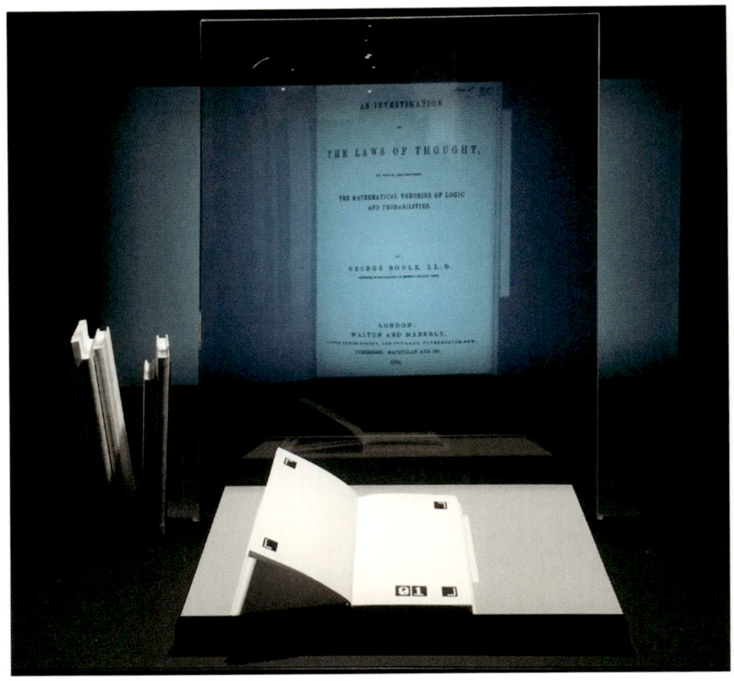

The world of the nineteenth century was marked by the industrial revolution based on machines and materials. Our postindustrial digital world, the age of communication and information technology, is dominated by media and data. The artistic work *Bibliotheca Digitalis* illustrates the life form in the digital world of immaterial data through the form of an innovative presentation: the information is separated from the technical carrier medium, i.e., the book.
A camera mounted above the reading desk captures the book opened by the visitor, which consists of blank pages except for the page numbers. The text in an analog real book would be visible to the human eye. However, the text projected onto the screen and the textual information stored in the computer are not visible. The digital version of the book becomes visible through the glass pane coated with a polarization filter. The reading experience in the analog space captured by the natural sense organs shows empty pages. Virtual information can only be received with the aid of technical apparatus.
With the help of this installation, the three phases of digitization become comprehensible for the viewer. Digitization began in 1623 with the sentence by Galileo Galilei, which reads roughly: "The Book of Nature is written in the language of mathematics." Two masterpieces have completed this first phase of digitization, the mathematization of physics: Isaac Newton's *Philosophiae Naturalis Principia Mathematica* (1686) and Joseph-Louis Lagrange's *Mechanique analytique* (1788).

Bibliotheca Digitalis: Three Phases of Digitization / Bibliotheca Digitalis: Drei Phasen der Digitalisierung

2017	Bernd Lintermann, Nikolaus Völzow, and Peter Weibel interactive installation with polarized light and augmented reality technology idea: Peter Weibel; concept: Bernd Lintermann, Nikolaus Völzow; software development: Nikolaus Völzow; book design: Jan Zappe; installation design: Matthias Gommel; technical collaboration: Jan Gerigk, Manfred Hauffen; production: ZKM | Hertz-Lab

The second phase, the mathematization of thought, began with George Boole. He intended to do what Lagrange had achieved for physics, i.e., the algebraization of logical operations. With the two works *The Mathematical Analysis of Logic* (1847) and *An Investigation of the Laws of Thought* (1854) he created Boolean algebra, which transformed logic into mathematics. Boole used already the grandiose invention of G. W. Leibniz (*De progressione dyadica*, 1679), i.e., the representation of numbers not by 10 digits (1–9, 0), but only by the two digits 0 and 1, the so-called binary code.
This mathematization of thinking was taken to the next level by many other authors, e.g., by Gottlieb Frege (*Begriffsschrift*, 1879) and above all by Bertrand Russell's and Alfred North Whitehead's three-volume work *Principa Mathematica* (1910–13), an echo of Newton's *Principia Mathematica* (1686). After Kurt Gödel, in "On formally undecidable propositions of Principia Mathematica and related systems I" (1931), had described what cannot be mathematized completely, Alan Turing, in "On Computable Numbers, with an Application to the Entscheidungsproblem" (1936), showed what is calculable and computable, and thus wrote an essential work on the justification of the computer.
The mechanization of the mathematization of thought began with Claude Shannon's work *A Symbolic Analysis of Relay and Switching Circuits* (1937). It stands for the third phase of digitization, the implementation of mathematical logic in physics and electronics. Shannon showed that circuits can be arranged in such a way that, on the one hand, they follow the binary code, i.e., that the current voltage represents the digit 1, whereas 0 stands for "no current voltage," and that, on the other hand, these circuits can thus represent Boolean propositional logic. Shannon proved the equivalence of Boolean algebra and switching algebra. Thus, it became possible to convert immaterial formalized, mathematized thought processes into material electronic switching states.
The development of transistors and integrated circuits, i.e., an immense micro-miniaturization, led in the 1950s and 1960s to electronic devices, from portable radios to mobile phones, which transferred electronics and the associated databased communication and information technologies into everyday life. At this historic moment, digitization became marketable and attractive to the general public. —Peter Weibel

Translated from the German by Peter Dahm Robertson.

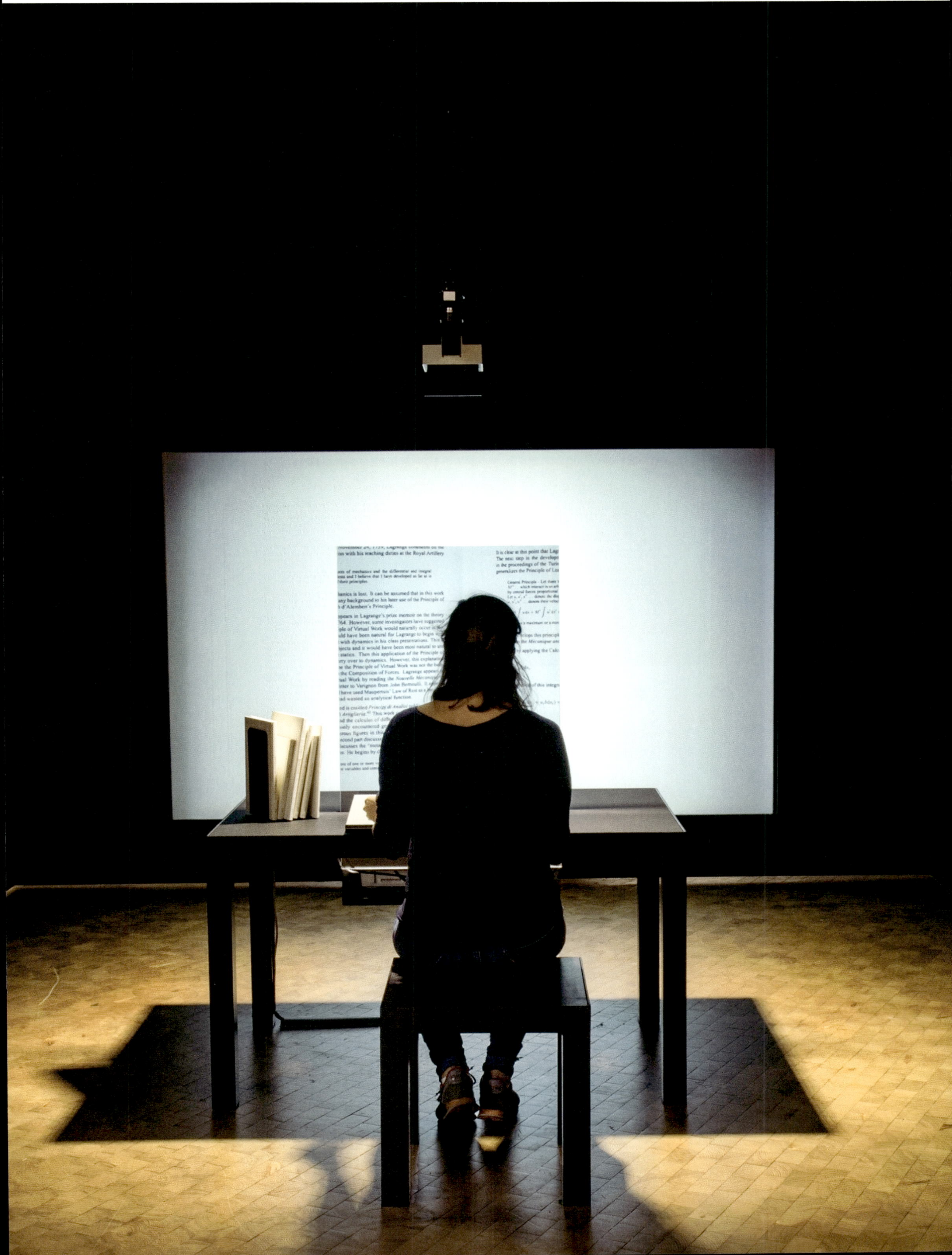

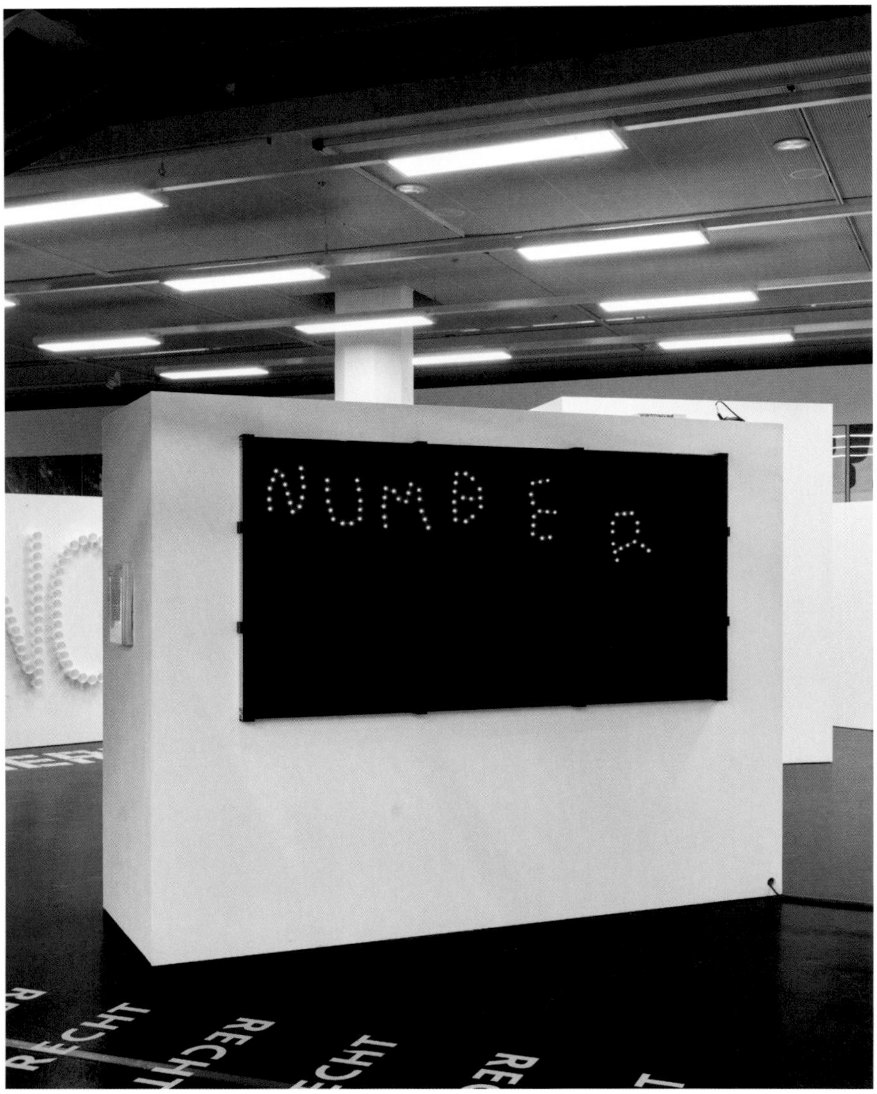

Installation view
respectively, Peter Weibel,
ZKM | Karlsruhe, 2019–20

I first encountered Warren S. McCulloch's book *Embodiments of Mind* (1965) in Sweden in the summer of 1965. I was profoundly impressed and influenced by it. After reading it I saw myself as part of McCulloch's tradition of experimental epistemology: that is, the attempt to find material explanations for mental processes, and to transform epistemology into an exact science with the help of mathematical models and computer simulations.

Reading this book, as well as others in the fields of cybernetics, computer science, and information technology, changed my academic trajectory. I had originally begun studying medicine in Vienna in 1964 with the intention of becoming a neurologist and then taking courses in computer science, as a way of preparing myself to develop exact formal models of thought. This course of study would have taken about twelve years, which of course I found a bit off-putting; so after reading McCulloch's book, I decided to take a shortcut and immediately begin studying mathematical logic – or "logistics," as computer and information science was called in Vienna at that time. In 1967, after three years of studying medicine, I quit the program. One of the fruits of the many years I spent studying McCulloch and his colleagues was that in 1972 I even tried to found an institute for experimental epistemology, together with my friends Werner Schimanovich and Franz Kaltenbeck. Schimanovich and I had already successfully published a number of papers on logic and mathematics, and we had many ideas for computer simulations, from road traffic to physiological processes. We hoped that businesses and governmental institutes would commission research from us – in a sense anticipating what today is called a start-up company. I had already rented a big apartment to use as an office, but when the time came to sign the papers with our lawyer, Schimanovich refused to sign because he was nervous about financial responsibility and the future. And in truth, we had no money whatsoever. So the institute collapsed before it could open.

In his 1960 essay "What Is a Number, That a Man May Know It, and a Man, That He May Know a Number?," McCulloch sketched out the history and philosophy of experimental epistemology in the context of his own biography, from the physiology of the nervous system to the electronic-circuit theory of the brain. Along the way he made references to Plato, Aristotle, Duns Scotus, William of Ockham, Ramon Llull, Leibniz, Berkeley, Hume, Charles Sanders Peirce, Frege, Russell, Whitehead, Gödel, John von Neumann, Ross Ashby, Donald M. MacKay, Oliver Selfridge, Gordon Pask, and many others, not to mention his own groundbreaking essays written in collaboration with Walter Pitts, such as "A Logical Calculus of the Ideas Immanent in Nervous Activity" (1943). Naturally, I carefully studied this list of authors myself.

Since reading McCulloch, I had been preoccupied by the idea of representing the all-or-nothing impulses of neurons in the ones

What Is a Number That a Man Can Recognize It and What Is a Man That He Can Recognize a Number?

1992/ 2019	computer-based installation; aluminum, acrylic glass, LEDs, single-board computer, dimensions variable

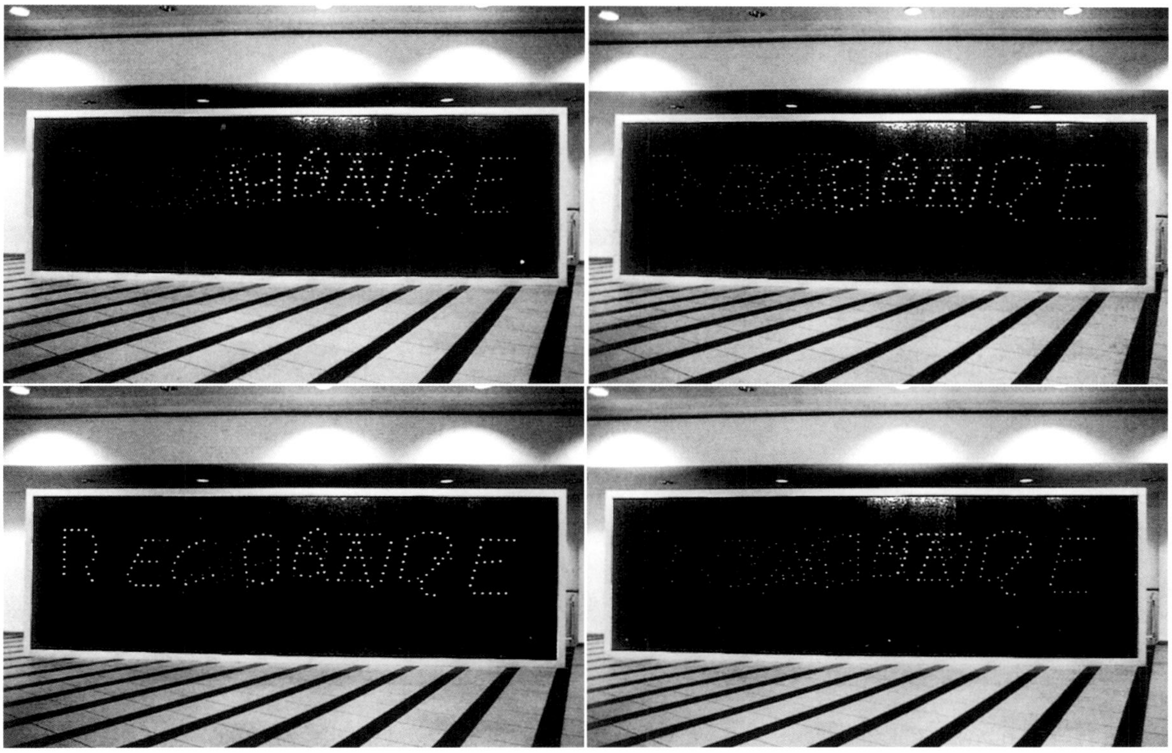

Installation views IBM headquarters, Vienna, 1992

and zeroes of binary code, of implementing them in electrical circuits. What interested me was not the transformation of Boolean algebra into Claude Shannon's circuit algebra – that is, the transformation of mathematics (algebra) into logic and logic into electronic circuits – but rather the question of how mental processes such as mathematical operations could be converted into neural processes, and how these in turn could be converted into electronic circuits based on binary code.
Bertrand Russell defined numbers as follows: "A number is the class of all classes similar to any one of themselves"[1] – in other words, all classes that can be put into a one-to-one correspondence with one another. "Thus 7 is the class of all those classes that can be put into one-to-one correspondence with the days of the week, which are 7," writes McCulloch, who calls Russell's formulation the first "usable definition" of number.[2] The recognition of a number, therefore, is primarily the identification of a correspondence. In the brain, this correspondence is represented as a mental process, a firing in the neural network, and this representation can in turn, thanks to the all-or-nothing nature of neurons, be represented as a binary electrical network. We are thus dealing with a chain of representations, or of correspondences among classes.
I therefore mounted a class of light-emitting diodes (LEDs) on a wall: points in a set of points, metaphorical neurons, which were either illuminated or not (all or nothing) as directed by a computer program. This program controlling the LEDs created an invisible network among the points, which formed letters and words recognizable as such by the human brain. My idea was that the circuits connecting the LEDs corresponded to the circuits connecting the neurons in the brain, since both are based on electrical (im)pulses and are therefore temporal codes. The synchronization of these external and internal circuits constitutes the process of (re)cognition. The correspondence between the electronic circuit and the neural network was a provisional answer to the question posed in the piece's title (a variation on the title of McCulloch's essay). The material way in which the question was asked – that is, by means of a computer-programmed circuit – was simultaneously the question's answer, the solution to the problem it posed. The fact that a temporal, pulse-based method of linking or triggering points produces patterns recognizable as such by the human brain leads us to infer that the very same thing happens within the brain: when triggered, neurons form patterns which, through training, practice, habituation, and neuron-based coding, are recognized as patterns. If the question had been put another way – that is, not as a computer program – it would have yielded a different answer.
What is interesting and relevant here is the fact that the number of LEDs used was quite small, a finite number. And yet a great number of graphs or relationships could be created between them: in other words, the number of relationships or graphs was much larger than the number of nodes (LEDs). To form letters, therefore, the computer had to trigger different subsets of the same set of points. The next level would be to write a program – a kind of generative grammar – that would use artificial intelligence to autonomously choose its own subsets of points to trigger, thereby generating a continuous stream of new letters and words. —Peter Weibel, 2022

Translated from the German by Patrick Hubenthal.

[1]
Bertrand Russell, *The Principles of Mathematics* (Cambridge: Cambridge University Press, 1903), 128.

[2]
Warren S. McCulloch, "What Is a Number, That a Man May Know It, and a Man, That He May Know a Number?," in *Embodiments of Mind* (Cambridge, MA: MIT Press, 1965), 1–17 (quotations on 6–7). First published in *General Semantics Bulletin*, no. 26–27 (1960): 7–18.

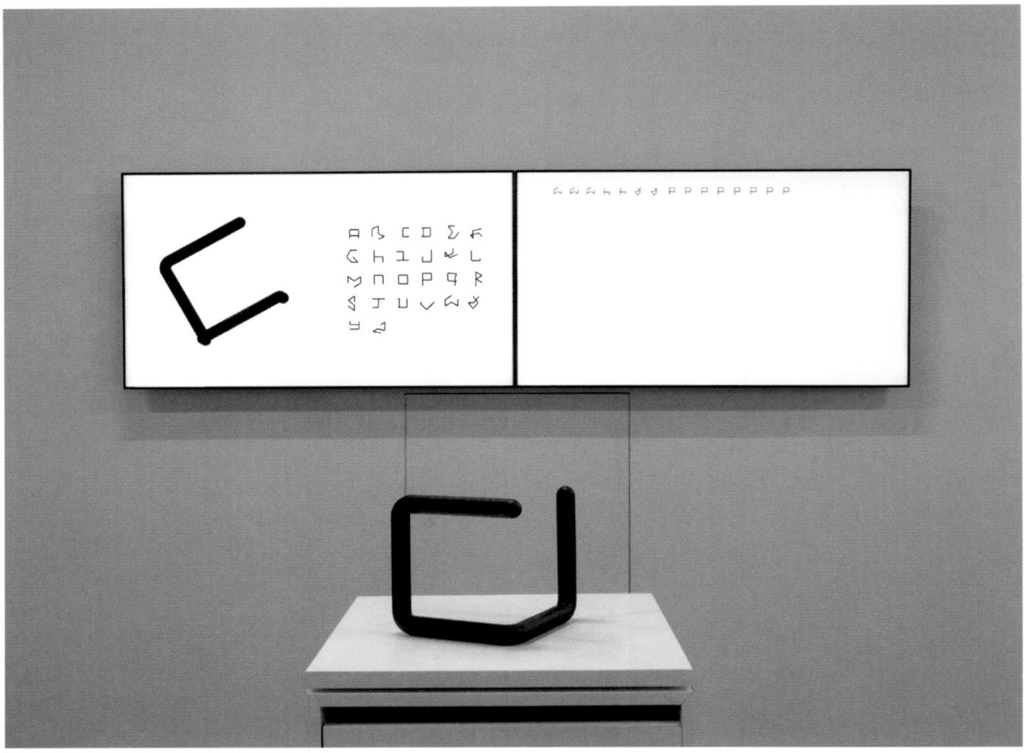

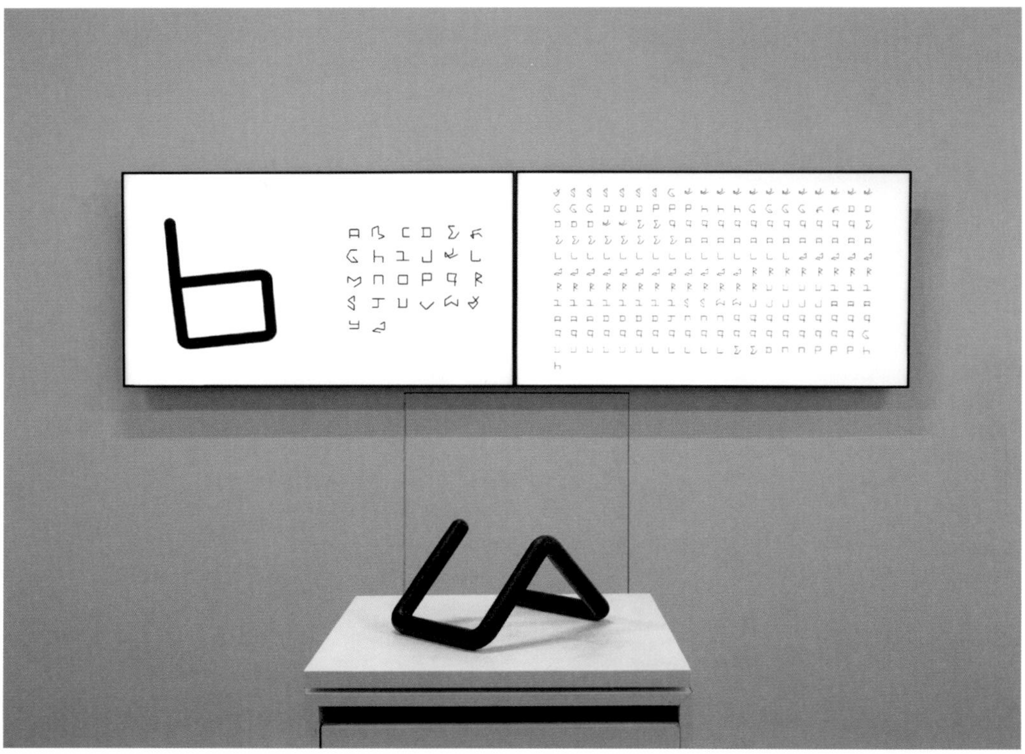

Installation views
respectively, Peter Weibel,
ZKM | Karlsruhe, 2019–20

Alphabet Space

2017

Adam Słowik, Christian Lölkes, and Peter Weibel
interactive media installation;
computer, 3D print, flatscreens

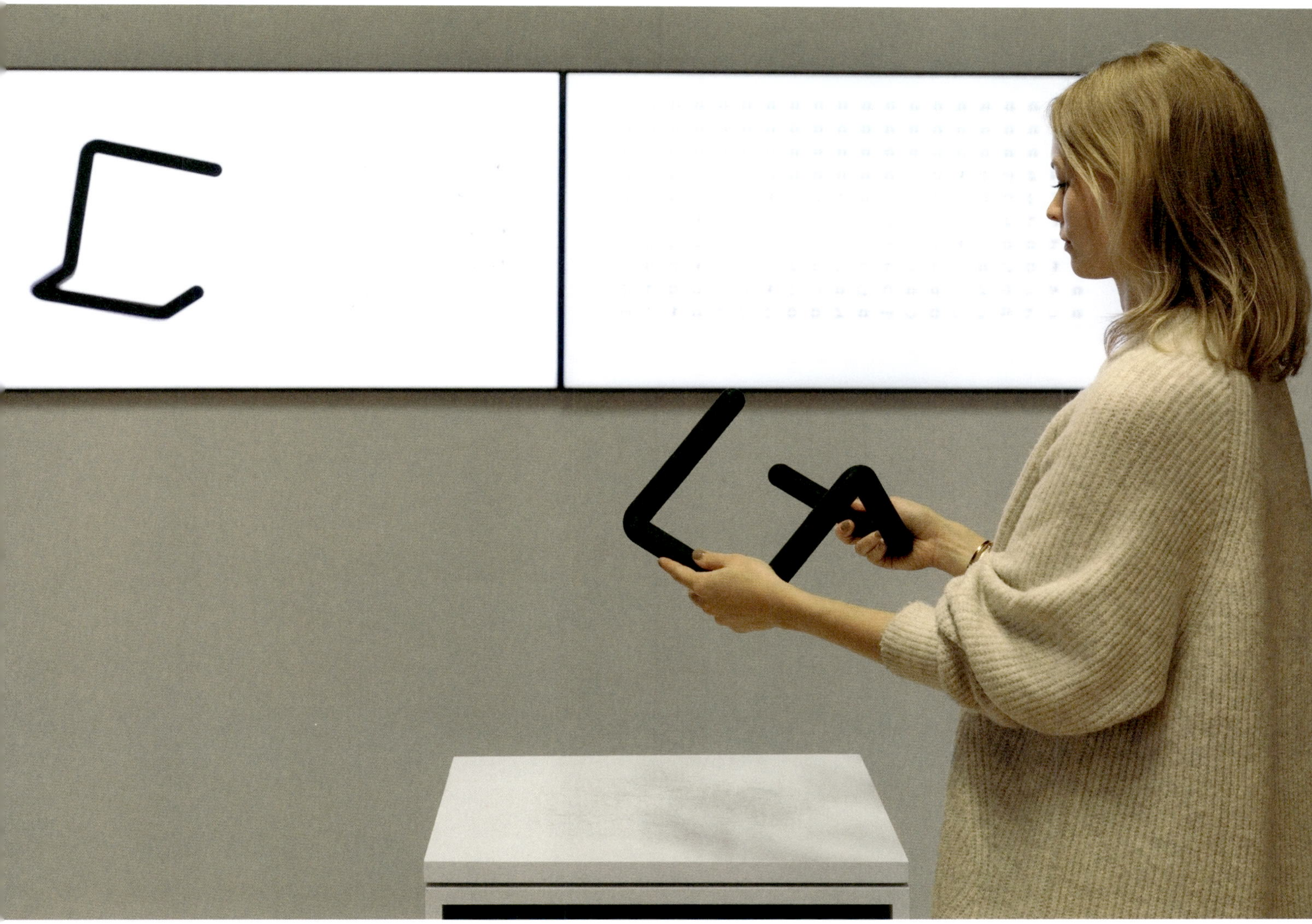

Installation view *Open Codes*, ZKM | Karlsruhe, 2017–19

The computer-based installation *Alphabet Space* introduces the viewers to the new digital world of three-dimensional notation. It is now possible to represent all 26 letters with a single three-dimensional character.
This work links letters of the alphabet and mathematical spaces. The basis is a three-dimensional object with five edges of equal length, which are connected to each other at right angles. Inside the object are inertial sensors to determine the orientation. This technology is also found in smartphones and consists of an IMU (Inertial Motion Unit): a gyroscope to measure rotation, an accelerometer to measure linear motion, and a geomagnetic field sensor for compensation and zero-point detection. A Kalman filter is used to estimate the current position, and by a derivative over time, the computer detects the position of the object and displays it telemedially on a screen.
By aligning in space at a constant viewing angle, the respective two-dimensional projections of the three-dimensional object result in all letters of the alphabet. The space – in the mathematical sense – of the alphabet is translated from a discrete into a continuous space. For example, *Alphabet Space* can also represent the letter "Ä" by a rapid movement of "A" and "E." The individual letters, if one can still call them that, are uniquely identified by a special mathematical number – a unit quaternion. This makes it possible to formalize strings such as words and sentences, to represent them and calculate them as a multiplication of several numbers. The image on the left shows the table with the assignments, the image on the right shows the results of the inputs. —Christian Lölkes, 2023

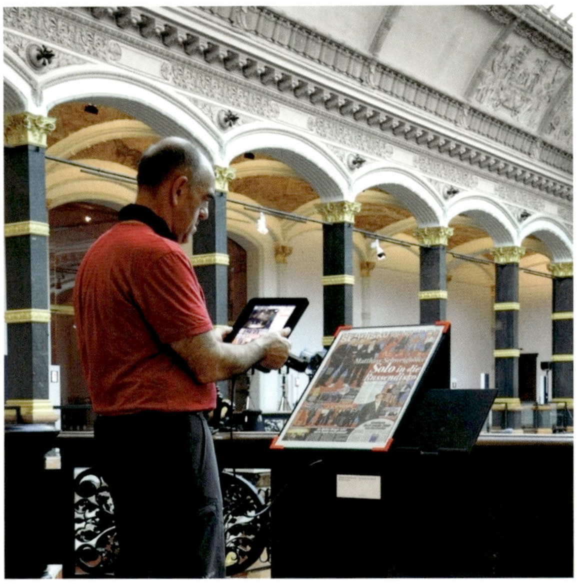

Installation views
ARTandPRESS, Martin-Gropius-Bau, Berlin 2012

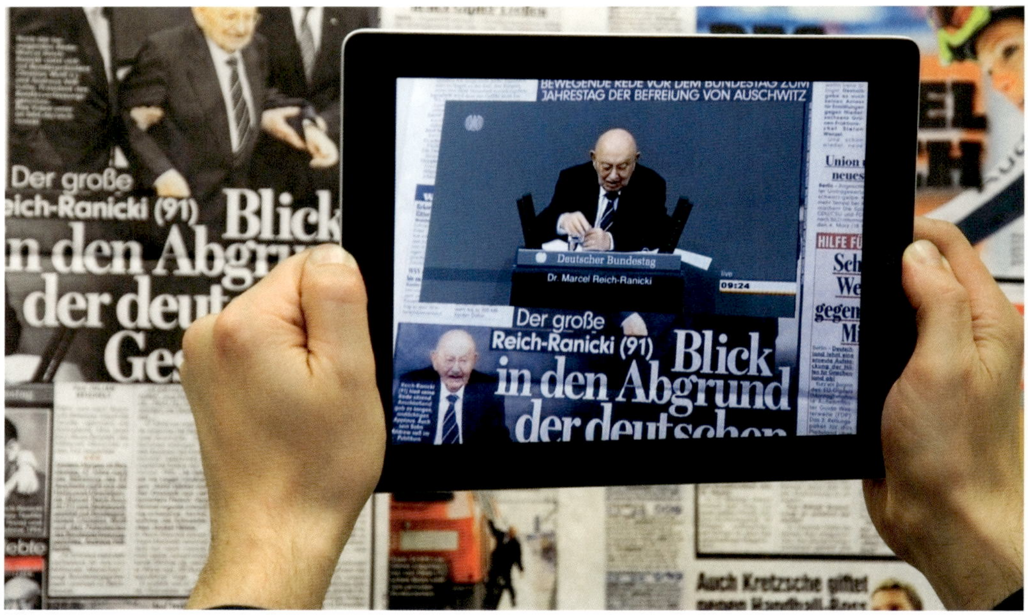

A page of the current edition of the daily tabloid newspaper *Bild* is placed on a pedestal. If the newspaper is viewed through the provided tablet, the viewer sees video documentaries embedded into the layout of the newspaper, adding multimedia information to the articles. A silent still photograph becomes a moving image with sound. Print and online information correspond – software bridges the gap. Instead of reading the report, complete with photograph, on Marcel Reich-Ranicki's speech in the Bundestag, one can use the app to access an online recording of the original speech and form one's own opinion. The commentator's judgment is no longer definitive, and readers are emancipated.

For this experiment, aside from the paper's printed items, *Bild* also produced current video dailies available online through the paper's video channel. The editorial board of *Bild* supported the vproject, matching its video offerings to the presented pages of its daily print edition each morning over a period of three months. The presentation in the exhibition *ARTand-PRESS*, Martin-Gropius-Bau, Berlin (2012) used the augmented reality app Aurasma, which enabled the editors in Berlin to author content online. The app stores specific characteristics of the pictures which are to be recognized, and the camera image is continually scanned for these elements. If there is a sufficient number of characteristics for a recognized image, the position and orientation of these points are tested as well (natural feature tracking). A mathematical mapping known as a homography provides the perspective of the tablet's camera, allowing virtual contents to be precisely superimposed on this image of reality.

The Newspaper of the Future lets information flow between print media (the newspaper) and social media (the internet). Compared with the latter, the visual information in a newspaper is limited. A newspaper cannot do more than print an image or photograph. A newspaper has no sound. Using new interfaces, such as a tablet or smartphone, the newspaper gains the dimensions of sound and moving image. The tablet allows the reader to scan across a photo in the paper and see the videos embedded in the newspaper. These videos then show the story behind the photograph both acoustically and as moving images.

—Peter Weibel

The Newspaper of the Future / Die Zeitung der Zukunft

2011–12	Bernd Lintermann and Peter Weibel interactive augmented reality installation; pedestal with current edition of the German newspaper *Bild*, tablet computer, Aurasma software idea: Peter Weibel; technical implementation: Bernd Lintermann, Manfred Hauffen

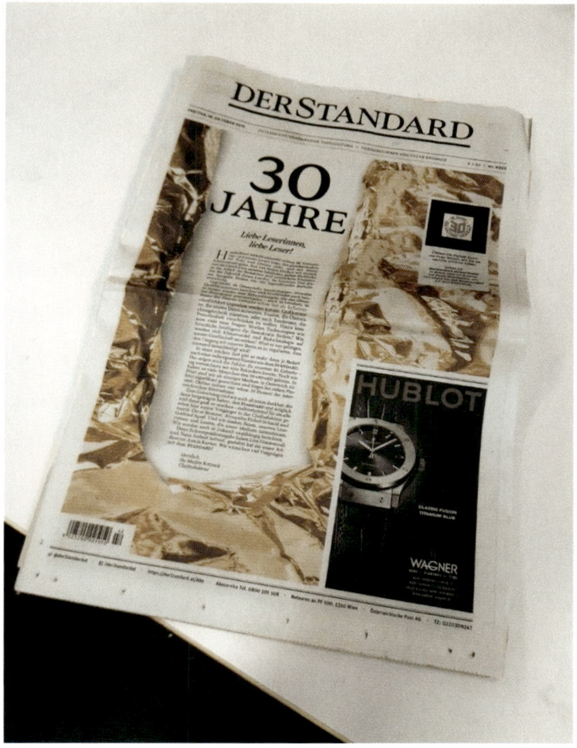

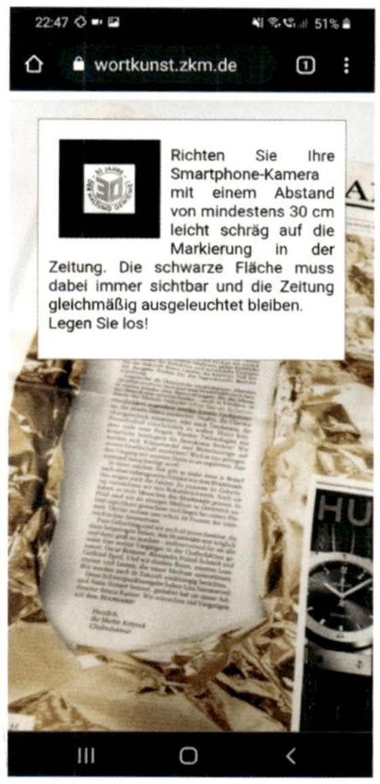

Peter Weibel was asked to make an artistic contribution to the thirtieth anniversary issue of the Austrian daily newspaper *Der Standard* on October 19, 2018. The mixed reality website wortkunst.zkm.de superimposes a virtual newspaper onto the cover of *Der Standard*, with a marker on the cover providing instructions for the smartphone. The program code in the website recognizes the front page of the newspaper by means of a so-called marker. With this marker, the code can determine the position of the smartphone relative to the newspaper and display appropriate data. Unlike most artworks that bring mixed or augmented reality to the public's mobile devices, *Word Art* requires no apps and no app store. Therein lies the technical innovation. All that is required is reading with a smartphone. After the camera is activated, the blank newspaper page with a typewriter appears on the screen. With a simple tap on the screen, one can fill the front page with programmed texts by Peter Weibel. The screen functions as the keyboard of the typewriter. The readers, as it were, bring their texts into the newspaper themselves. They not only read the newspaper, they write the newspaper themselves. Whoever buys or reads the newspaper becomes the editor.

Texts in the *Word Art* website that appear virtually on the real newspaper pages:
- Who reads, writes. Who writes, reads.
- The freedom of the word is the freedom of the reader.
- Words in freedom guarantee the freedom of society.
- The freedom of the printed word guarantees the freedom of readers.
- Freedom of expression needs freedom of the press.
- The free word is under pressure. Only the printed free word resists pressure. Only freedom of the press guarantees freedom of expression, including your freedom of speech.
- Free newspapers are government newspapers. Paid newspapers are your newspapers.
- When you buy a newspaper, you buy your own freedom, your own freedom of expression, your resistance, and your criticism of the conditions.
- Democracy needs informed voters. Voters need informative newspapers.

Word Art / Wortkunst

2018 | Peter Weibel and Christian Lölkes
technique: JavaScript, WebGL;
programming: Christian Lölkes

Two Levels of Reality: "Exo" and "Endo"

Peter Weibel

Classical physics operates on the assumption that the world can be objectively described from the point of view of an external observer. With Otto Rössler's new science of endophysics, however, our understanding of reality is radically called into question for the fourth time this century, after the theory of relativity, quantum mechanics, and chaos theory. The theory of relativity put an end to the absolute nature of space and time and made all phenomena solely subordinate to the speed of light, the last absolute constant. By introducing observation, or rather, measurement, which makes it impossible to obtain complete information, quantum mechanics relativized the objective character of the world. Chaos theory greatly increased the unavoidable error in measurement and made the future impossible to predict. Only if one stands outside a complex universe - that is, only if one is an external observer - can that universe be described in its entirety. Kurt Gödel's famous undecidability theorem is valid only within an arithmetical system.

Endophysics introduces an internal observer, thereby making the relative and observer-dependent nature of the description of the world far more radical, because the internal observer is constitutive.

Since the introduction of perspective in the Renaissance and of group theory in the nineteenth century, we have known that the manifestations of the world depend in a systematic way on the location of the observer (codistortion). It is well known that perspective is not completely objective; rather, it is "observer-objective" (i.e., objective only for the observer). It is impossible to avoid distorting the world when one is an observer. The first modern scientist to acknowledge this was the eighteenth-century mathematical physicist and theologian Roger Joseph Boscovich of Dalmatia. He claimed that the world was actually elastic ("like rubber," we would say), but that we are unable to perceive its elasticity because we deform with it (are ourselves made of rubber).

Classical physics describes the world from the point of view of an external observer outside the world - that is, it describes an exoworld. To the endophysicist, this position outside a complex universe is possible only in a model, not in reality itself.

Endophysics is an offshoot of chaos theory, to which Otto Rössler has been making contributions since 1975 (e.g., the Rössler attractor, 1976). Endophysics is different from exophysics because the physical laws that apply when one is part of what one is observing are generally different from those that apply from a real or imagined external viewpoint.

In effect, endophysics permits two ways of accessing the world. Alongside our direct access to the real world (via the interface of the senses), a second way in is opened up from the position of an imagined observer. Is what we call objective reality just the endoside of the exoworld?

Time and again, the history of cultural production has provided evidence that humans sense the possibility that their world is just the endoside of an exoworld, and that their access to the world depends on the interface between world and observer. Our perception of the world changes depending on the interface's configuration. In the electronic age, humans have created numerous interfaces that are open to manipulation. In other words, the interface between the observer and the world can manipulate the world. As an illustration of the phenomenon of the interface as sole reality, consider the model of the "bubble boy," who lives in a sterile bubble and communicates with the world exclusively via the interface. The menu for his programming of the world exists on a keyboard inside the bubble.

The new endophysical principle of observer-relativity (as opposed to the simple relativity of the inertial frame of reference) will perhaps break open Descartes's prison of the *hic et nunc*, analytical geometry's description of the world as a prison made up of the four bars of space and time. With endophysics, the manipulability of the prison of space and time grows to another level. The world as "repair shop for the desiring-machine" advances another step.

The endo-approach offers promise for the complex technological world of the electronic epoch, with its vast number of interfaces, from microscope to telescope and the whole array of teletechnology (from radio to internet). The endo-approach places at our disposal a new theoretical framework for the description and understanding of the scientific and technological conditions governing the modern world. Virtual worlds, for example, are a special case of endophysics. The questions posed by endophysics, from observer-relativity to nonlocality and the problem of representation to the world itself as an interface problem, are the central questions facing electronic-telemetric civilization.

Note: Adapted from Peter Weibel's foreword to Otto Rössler, *Endophysik: Die Welt des inneren Beobachters* (Berlin: Merve, 1992), 9–12, combined with portions of the chapter "Zwei Ebenen der Realität – 'Exo' und 'Endo,'" cowritten by Rössler and Weibel, in *Endophysik*, 85–86. Translated from the German by Patrick Hubenthal.

The World as Interface

Peter Weibel

I. Plato, or the World as Cave

In the idealist model of the world, we perceive only the reflection of things. The realist model undergoes a reversal. The stage becomes the cave in which we sit, and from it, we see the world outside only in the form of reflections on the cave wall. The way the world really is, the way things-in-themselves are (Immanuel Kant), is beyond our experience and understanding. In the cave model of the world, we human observers sit inside and look out at an external world that is not directly accessible to us, and that therefore carries on without us onlookers. We see the world only as a shadow, a semblance, an idea.

II. Aristotle, or the World as Stage

In the so-called realist and objective model of the world, we humans are external observers. We are spectators who cannot intervene in events. We see the world as a play upon a stage. The course of events, the mechanisms of the world, are guided and directed by the ever-operative laws of nature. God is the great watchmaker, and the world is a clockwork which, once set in motion, functions according to the same rules forever, impervious to humanity's contingency and subjectivity. In the classical stage model of the world, the observers sit outside the stage, in the auditorium, and look upon a world that also carries on without them. This model established the classical mechanistic worldview, in which the world functions irrespective of our observing it.

III. The Curtain Model, or the World as Interface

Neither Plato's model nor Aristotle's addressed the question of the observer's function. Their questions were metaphysical and epistemological in nature: What is behind physics? What is really happening behind visible reality? The thirst for knowledge is nothing other than the longing to see behind the curtain, the desire to view the world directly instead of on the cave wall.

It was quantum physics that finally asked questions about the observer and the effects of our observation on the events taking place on the world's stage (Niels Bohr), thereby creating a nonclassical "participatory" model of reality and the world (John Archibald Wheeler). In this nonclassical understanding of our relationship to reality, the stage is the cave, so to speak, but the curtain is us. The endophysical interpretation of quantum physics (Otto E. Rössler) shows that there is no such thing as external observation of the world as presupposed by classical physics; rather, we are always internal observers of the world. We live in a world that we observe and describe. We are part of what we observe. We are thus incapable of objective, external observation, and we cannot be subtracted from our model. The world is the world of the internal observer (the cave), but at the same time it is not outside our observation post. If the objective, outside world is depicted on the cave wall, then we watchers are depicted there as well, because the rock wall is us too. Quantum mechanics and endophysics teach us that the observer is part of what is being observed – that is, part of the screen or wall on which both reality and the mechanisms of observation are pictured. We are part of the world we are observing, and precisely because we are observing it, we observe it only in distorted form. The world is only accessible via our observing it – or, to be more precise, observation is the curtain between us and the world. Today we call that curtain an interface.

The curtain – the interface – is our entry to the world. This means that so-called objective reality is just the endoside of an exoworld – that is, the side visible to observers of the world, which they can only access from within. We thus have a new model of the world: the curtain model, aka the interface model. In this new, nonclassical interface or curtain model of the world, the observer is not just a spectator, but also an actor upon the world's stage. There is no longer an external world, only an internal world, which can only be explained in relation to the observer.

Our sensory organs are the interfaces developed by natural evolution. They are a natural curtain. The eye is evolution's answer to the sun, the lung its answer to the atmosphere. The human eye is a very limited interface because it processes only a very small fraction of the electromagnetic spectrum, thereby allowing us to see, to have access to the visible world. Some animals can see at night; we cannot. With different eyes, we could and would see different things. Therefore, the world is observer-relative, or, more precisely interface-relative. With technology we create artificial organs: cameras, recorders, radios, television, etc. – media. These media are part of our new, human-made artificial organs, which are nonetheless based on natural laws. With these new organs, we can see better, speak louder, and store and process information faster and more reliably than with our natural organs. Media are interfaces that create a new world, a world beyond natural organs.

Note: Adapted from Peter Weibel, "Schnittstellen," in *Bildwelten 1982–1996*, ed. Romana Schuler (Vienna: Triton, 1996), 256. Translated from the German by Patrick Hubenthal.

Images of Thinking

Boris Groys

The avant-gardes of the twentieth century fought against (art) institutions in the name of each individual artist's freedom – the freedom to overcome the limits of these institutions, to break free of them, flout obsolete standards, and create the outrageous. Today, people still like to criticize institutions, aiming for their disempowerment and abolition. Meanwhile, however, the meaning of this critique has fundamentally changed.

Criticizing institutions nowadays only serves to conceal the fact that these mighty, normative institutions de facto no longer exist. Today's art institutions are usually weak-willed, have little money, are scarcely registered by the wider public, and cannot and do not want to set up binding aesthetic standards any longer. They would rather chase after trends promoted by the media; they seek to be cool and hip in order to appeal to "young visitors." Figures like the august, mighty museum director with undisputed power over his collection, or the art critic with far-reaching influence who sets directions and dictates his taste to the public, belong to an age long gone. Therefore, today's criticism regarding cultural institutions can only be interpreted as a nostalgic invocation of this past epoch – as an attempt to conceal the real weakness of art institutions in this day and age.

Under these conditions, artists no longer have the opportunity to orient themselves on existing art institutions – be it in the sense of adapting to the standards these have established, or in the sense of a revolt against these very standards. Basically, artists are left with only one possibility: to found an institution of their own, making themselves an institution. Certainly, this way is much more wearisome than the easier way of common criticism of institutions. Hence, there are only very few artists who have consistently gone down this road. It is these few artists who all the more deserve our admiration. Without a doubt, we find Peter Weibel amongst them.

Weibel has understood from early on that advanced art has no stable place in our society and that this place has yet to be created; namely, on all levels of topical artistic practice. As an artist, if one does not choose to present oneself as a curator, one is dependent on the curator. When artists wish to assume the role of curators, though, their best move is to keep a museum space constantly available for themselves. A space in which they can experiment with all kinds of exhibitions. Today's artists cannot count on a binding comment on their work unless they become theoreticians in their own right and comment on their own work. They could also invite other theoreticians to write something for a publication conceived by themselves and then go on to design and edit the book. Ideally, they would print it themselves as well. In this, the artists would always be dependent on the technology continuously developing within art production and distribution – that is, unless they don't develop the production on their own account and could thereby avoid having to rely on using only the things other people had produced as part of technological progress. This again means the artists themselves would have to create and manage institutions, which would further develop artistic techniques. They would have to run these institutions, employ people, and have the means to pay them. So, the artists have to ask themselves how they would not only passively reflect on and criticize the economic, political, and administrative aspects of their artistic practice, but also how they could actively design, organize, and manage them on a daily basis.

In other words: once the artists cease to act as decorators for the state, they have no other choice but to become a state within a state – should they choose not to just cater for the art market. This development is an implicit consequence of the modern artist's pursuit of sovereignty. Sovereignty regarding the decision as to what will be presented on the surface of an image or in the context of an installation is, however, obviously not enough: true sovereignty calls for domination, control, and design of the contexts in which a piece of art is placed and received. Hence, the dissolution of modern art institutions requires the artists to govern, command, and organize in order to make sure that their sovereignty does not perish in the meaninglessness of today's art market. Of course, not all artists are suited to govern; neither do all of them have the inner discipline and stamina that such a task requires. Above all, most artists do not have the determination for totality, which cannot be separated from the pursuit of sovereignty; that is, the will to control not just a few of the contexts in today's art practice, but all of them. Peter Weibel has this determination. He is an artist who has taken modernism's promise seriously – to give the artists sovereignty and independence. Hence, he has indubitably given the path of his life a paradigmatic meaning. This meaning is becoming increasingly clear over time. The perception of sovereign artists controlling the network of their own institutions is an image eminently present, one that characterizes our epoch more accurately than any other image.

Weibel's art, however, is not limited to the creation of this image of artistic sovereignty. For beyond art institutions, there is another dimension that increasingly determines today's art world: the beholders. The avant-garde promised the artists absolute individual freedom regarding all their artistic decisions – liberty not only to choose the theme of their work freely, but also to choose freely the artistic means utilized; in fact independent of all traditions, conventions, and criteria of good taste or the consummate skill of masters that still largely determined artistic creation before the rise of the radical avant-garde at the beginning of the twentieth century. This newly promised freedom had, as we know, a drug-like effect on the artists at first, and created a historically unprecedented intoxication. This is why so many works that were created in those early times we still admire deeply today.

At the same time, however, one cannot deny that little else has remained of this early euphoria. Today, artists no longer feel free – rather, they feel that their subjectivity has little or no relevance for the art world. The reason for this new powerlessness on the artists' part can be put down to the fact that in the attempt to free themselves, at the same time the artists of the historical avant-garde committed a somewhat fateful act, whose importance they did not notice at the beginning: they freed the beholders as well. Beholders were relieved of all criteria they had been able to or were supposed to invoke for their judgment of a piece of art. Artists actively liberated themselves from these criteria – and the beholders lost them.

Initially, this made the beholders extremely insecure. They no longer knew how to react to the pieces of art that were presented to them; they felt helpless when confronting the arbitrariness of the artists, utterly unable to defend themselves by means of a well-founded judgment. This initial insecurity has decreased significantly over time. The further modern art developed, the more the beholders began to both cherish and relish the insecurity given to them by the artists as a freedom. There are, of course, more beholders than artists and, according to democratic conditions, we all know that the minority must bend to the will of the majority. In earlier times, artists were able to reject the public's judgment by trying to prove that people had not understood their work properly. The artists could invoke their masterly skills, the historical originality of their work, and the strength of their inspiration and so on. Today, however, all these justifications no longer apply, as the public feels totally free in its judgment. If it simply does not like something, it is not prepared to let itself be convinced otherwise anymore: the tables have turned. The avant-garde has pitted the artists' freedom against all of the public's judgements. Now the public's judgment has the freedom to be directed against all reasoning and explanations on the artists' part. This is the trap of freedom into which modern art fell at the end of the twentieth century. Modern art proclaimed the absolute freedom of the individual's aesthetic judgment – and then became itself a victim of this freedom. If the beholders are free to make their own aesthetic assessments and the artists cannot call upon an independent authority to repudiate the beholders' judgments, the artists become dependent on the public. They become dependent in the same way as they once felt utterly independent of their public. Each and every objective finding of art theory will fail with reference to a modern work of art. One no longer *recognizes* why a certain piece of art is good, or even why a certain piece of art is a piece of art. One can only *vouch* for modern art – if one likes it. The growing dependence on the public's taste, which can no longer be criticized, has increasingly irritated modern artists. One well-known symptom of this growing irritation was Viennese Actionism, the atmosphere out of which Weibel embarked on his artistic path. Viennese Actionism was mainly a desperate attempt to fight against the modern public's callousness and complacency by employing brute methods. Events, images, and texts were produced in order to shock the public by at least briefly shaking up their arrogant attitudes and pleasure in the new freedom of aesthetic judgment. This desperate, almost hysterical action was successful to a certain degree. However, such an attack can never be reasonably sustained – it has to stay historically unique. So Weibel chose a different way – albeit with the same aim of bringing the public's taste under his control. He chose to use technology.

Even if our culture no longer believes in the criteria of taste, it certainly still believes in knowledge and technology. Art's liberation from the beholder's taste can only happen in terms of a mechanization of art, or – in other words – of its remechanization. Art has always, in the first place, been a matter of technique. It was only the historical avant-garde that discredited and then abolished art production's old techniques. The avant-garde made an attempt to liberate art by demechanizing it, by liberating it from capability and knowledge, by equalizing the act of artistic creation with aesthetic judgment. Kazimir Malevich's (1879–1935) *Black Square* (1915) and Marcel Duchamp's (1887–1968) ready-mades introduced an age of the demechanization of art. It was particularly at this point that artists lost their cultural privilege vis-à-vis the beholders. If an artistic decision to declare an object or a random shape a work of art is totally free, then the beholders must also be free not to accept this decision. Everything becomes a question of personal or individual taste. When the artists' taste is contrary to the beholders' taste, then, as we all know, the artists are the losers.

A phase of remechanization has now set in, initiated by the use of new digital techniques of image production and distribution. The artists have again become technicians, specialists, and producers, thereby – once again – establishing cultural distance between themselves and the beholders. Peter Weibel very early on understood and seized the opportunity offered by the new determination of arts. Many artists and theorists have preached the return of art to technology and knowledge during the Postmodernism phase of the 1970s and 1980s. However, this almost always meant some kind of – mostly ironic – return to traditional artistic techniques. The opportunity and even the need to reactualize the avant-garde program results for Weibel from the fact that art has left behind its traditionally closed rooms and must act amidst today's technological, media-dominated world. Hence, the avant-garde was right to let go of traditional art techniques and direct the beholders' view to the new, technical world – even with a critical agenda in mind. Today, however, it is widely propagated that art should give up its critical, elitist, attitude; that it should be agreeable and easy to digest. In short: that art, as soon as it starts to be produced in the context of and by using mass media techniques is obliged to adapt aesthetically to the laws of mass media dissemination.

The original character of Weibel's art strategy consists in the fact that he indeed shares the corresponding proposition, yet he does not accept the above-mentioned conclusion. The fact that art is no longer primarily practiced in protected museum spaces does not mean for Weibel that topical art is condemned to be the mimicry of media surfaces. On the contrary: this is what makes it necessary to refuse such mimicry again. However, the issue is not breaking taboos or transgressing boundaries, for these are the typical means of attracting the media's attention and from an artistic point of view, are thus of little relevance. It is rather about acquiring the techniques and procedures used for the production of images circulating in the mass media – but utilizing these techniques and procedures differently.

Here, the avant-garde is transcended by means of a remechanization of art, and at the same time continued in its critical impetus. The repertoire of images circulating in mass media networks is very limited. Not only so-called images of the other, but also images created by science, interactive images, purely subjective images, let alone abstract images, usually have no chance of entering large mass media networks. It is precisely these, excluded images that Weibel works with, images which can be created using technical media and are therefore indeed technically compatible with the networks of mass media dissemination. At the same time, though, they exhibit an aesthetic incompatibility with these networks. In this sense, Weibel's artistic practice is a direct continuation of the artistic practice of the historical avant-garde under the conditions of its technical and media topicality, and this implies a continuation of the critical discussion about mimetic illusion.

As a matter of fact, an artistic technique is always a technique of reproduction. This is how artistic technology differs from technology per se. Technology produces technological objects such as cars, aircraft, or computers. Art also produces artificial items that, however, mimetically reproduce certain "natural" things. So what is now mimetically reproduced by means of digital art operating with computer programs? One could say: it is the act of thinking. For the avant-garde, using the title "art" is a purely subjective decision, which, if you like, takes place in the "mind" of the artists or the beholders. The classic avant-garde saw the expression of the artists' individual freedom, autonomy, and sovereignty in this decision that was both lonely and entailed social obligations. This autonomy, however, was radically called into question by discourses such as those on Structuralism of the late 1950s and 1960s, for back then the lonely autonomous decision of an individual was no longer seen as a free decision, but rather as one dictated by the system of reasoning; that is, the code controlling the thinking and decision-making processes of an individual. In this sense, Concept art at that time, too, attempted to use art in order to clarify and illustrate how such decisions worked, how they were built into the system of thinking, and what "objective" logic looked like that could lead – and does indeed lead – to such "subjective" decisions. Of course, only the introduction of computers allowed the reproduction of such thinking processes, formalizing them as computer programs and visually depicting them. Concept art in the 1960s, though, already took the decisive step towards a depiction of the thinking process by making pure thinking its object and thereby aestheticizing it. Only this aestheticization of thinking paved the way for new computer art, which is reproduced via programming: the depiction of thinking processes in the computer. Accordingly, new computer art will only then accomplish its real mission when it not only depicts the world's surface in the form of images – which could also be reproduced with traditional artistic techniques – but also makes a process-related depiction of thinking as such.

To fulfill this task, however, also means to reflect on the limits of a mimesis of thinking, for attempts to depict thinking processes go back at least to Plato and Aristotle.

The basis of the latest computer programs, too, is classical formal logic, which was formulated in the works of these authors. Yet, as we have known at least since Gottlob Frege (1848–1925), Bertrand Russell (1872–1970), and Kurt Gödel (1906–78), this logic is incomplete, and leads to numerous paradoxes. It creates an idealized and thereby incomplete picture of human thinking. In this way, utopian projects such as those designed by Ray Kurzweil (b. 1948) are given clear limits. This includes the central project of depicting the entire human brain in a computer. Once, René Magritte (1898–1967) stated with his painting *La trahison des images* (*The Treachery of Images*, 1928/29) that the depiction of an apple was not an apple, and the depiction of a pipe was not a pipe. Hence, the depiction of human thinking by means of computer programs is not thinking – yet. So-called artificial intelligence is both hardware and software that makes it possible to copy, duplicate, and simulate certain functions of the human intellect – just as the world was reproduced on canvas before. Humankind needs thinking for the purpose of its individual and collective survival – for the sake of its instinct of self-preservation. By contrast, "intelligent" machines just keep thinking all the time – oblivious of any kind of concern about survival or well-being. Such artificial intelligence would be classified as natural stupidity if diagnosed in a human being. This demonstrates again: artificial intelligence – computer programs built on formal logic – does not represent thinking, but rather an image of thinking, just like traditional painting did not produce apples, but images of apples.

This insight, however, only worries an engineer, not an artist. Weibel is very well informed about the limits of logic since he has seriously and systematically researched this field. Artists like to use computer programming and simulation precisely because they like to demonstrate how the system loses its own systematic, how strict logic leads to paradoxes, and how ambivalence is directly created by an obsessive search for unambiguousness. Art is indeed interested in systems, structures, and programs, but mainly where these lead themselves *ad absurdum*, for it is precisely in this drift towards the absurd that it becomes clear that the image of thinking differs from thinking itself and that the living cannot be mimetically duplicated offhand. Today's artists, working with communicative and "intelligent" media, are more interested in the blind alleys, disturbances, and absurdities of media communication than in its acquisitions. This is exactly why the remechanization of art certainly does not just stand for naive enthusiasm or worship of technology. Quite the contrary: it enables both an ideological claim of the limits of technology, and their technical analysis. It is in no way a mere desire for chaos that makes artists interested in the dysfunctional, the deceptive element of computer-controlled and communicative processes. Rather, it is more a systematic and critical analysis of mimetic procedures by means of art. This is a task that modern art has imposed upon itself, which, in the meantime, has become a tradition – except that today it is a question of the mimesis of thinking, not reality.

The analytical, critical dimension of Weibel's engagement with the operational logic of today's computer-controlled and communication systems, which he pursues in the context of his work at the ZKM | Center for Art and Media Karlsruhe, is already detectable in his early works that deal with the logic of everyday mass media communication and the conventions of today's media culture. These works often have a clear political dimension and constantly refer to the opportunities as well as the shortcomings and absurdities of today's media world. They almost always offer a very specific sense of humor. As we know, Søren Kierkegaard (1813–55) drew a strict line between irony and humor. For him, irony was the manifestation of a seemingly endless subjectivity aiming to triumph over the finite character of things. In his point of view, humor was the result of this subjectivity's insight into its own finiteness.

Weibel is very aware of the finiteness of human subjectivity, and repeatedly contrasts it with the potential infinity of technical media. It is precisely the finite character of the human being manifested through humor that makes a final subjection of humans to the repetitive infinity of logical systems impossible: for them, the possibility of an instantaneous, illogical and unique whim, the spontaneous need to make a good joke, and the willingness to sacrifice the infinity of logical laws without further ado to this momentary whim, result from the finite character of their existence. This is exactly what the jokes in Weibel's art are like – like the instantaneous, immediate readiness of their author to sacrifice the whole logical system to a joke for the sake of humor. Albeit this willingness has a strict logical basis of its own, because when the logical system can be sacrificed, this means that the system itself is, after all, still not really as logical, consistent, or complete as it seems. This Weibelian humor is the clearest sign of its author's confident handling of all kinds of systems and their media – and hence of his belief in the survival of the human being and art under the conditions of the media age.

Note: Adapted from Boris Groys, "Images of Thinking," in *Peter Weibel: Das offene Werk 1964–1979*, ed. Peter Weibel and Harald Falckenberg, exh. cat. Neue Galerie Graz, Universalmuseum Joanneum (Ostfildern: Hatje Cantz, 2006), 982–985.

5. Music and Sound Art

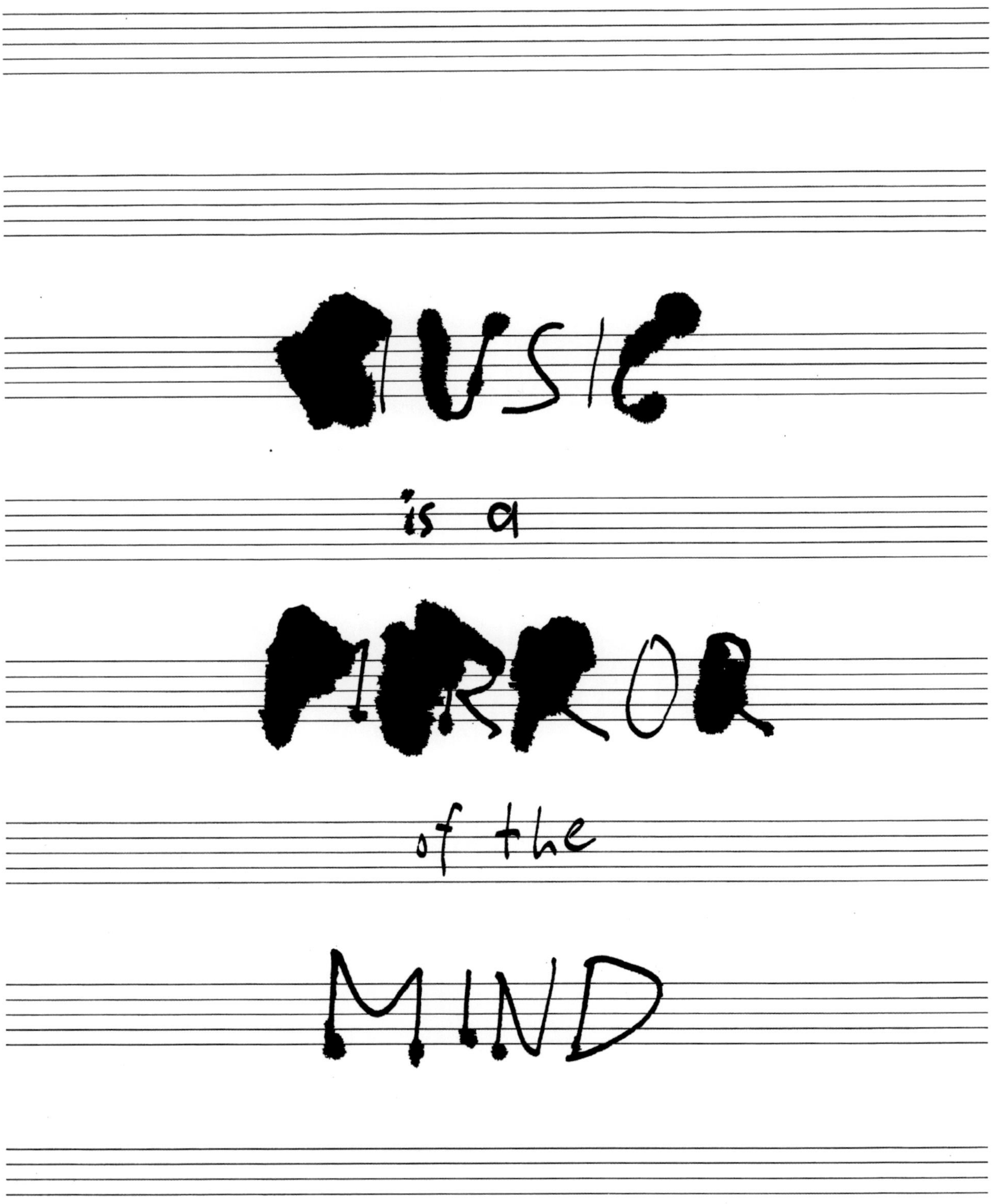

Music Is a Mirror of the Mind

1967 | visual music; ink, 2 mirrored pages, 21 × 29.7 cm each

A sheet of handwriting is mirrored like a Rorschach test while the ink is still wet, turning it into a sheet of graphic musical notation.

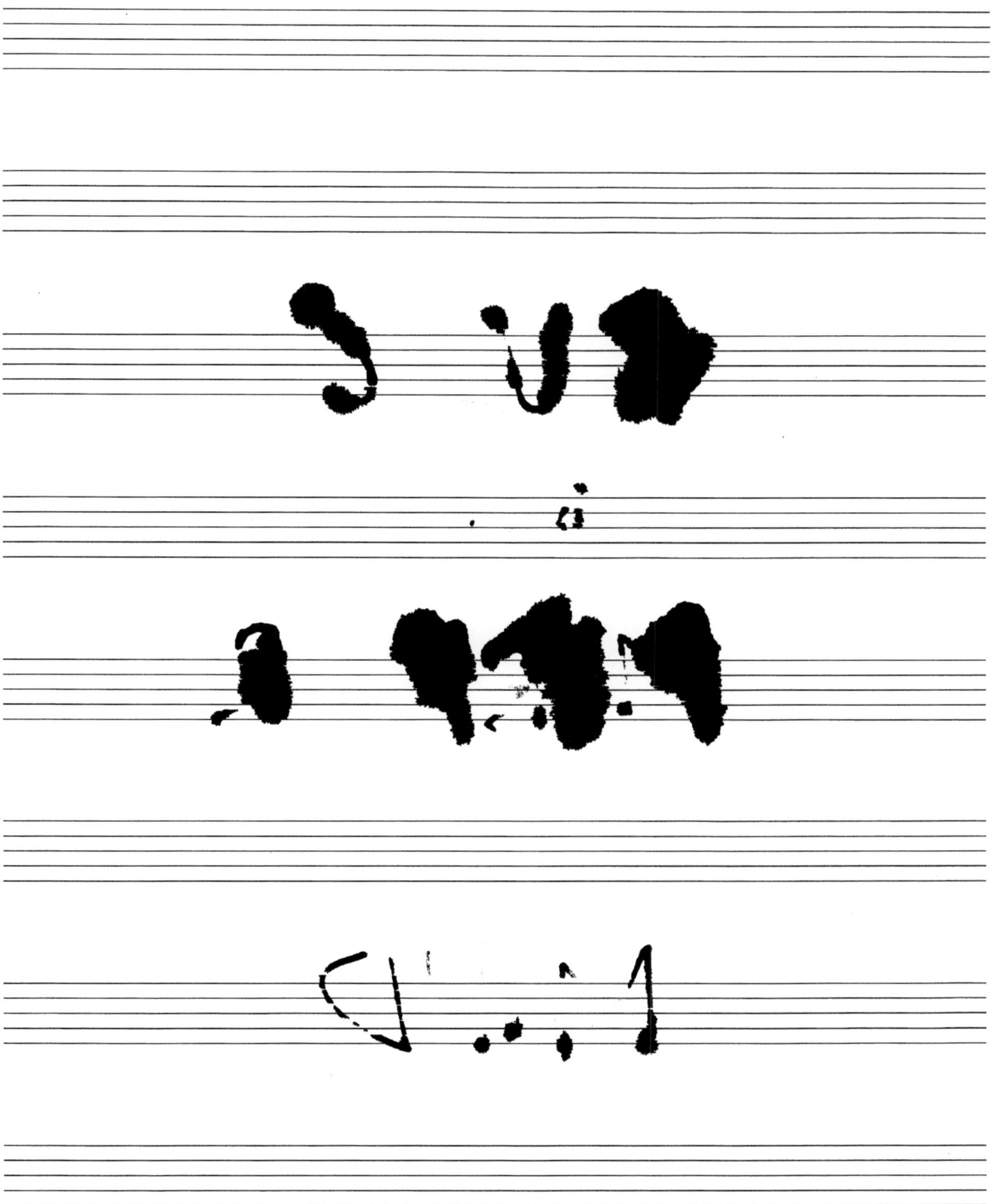

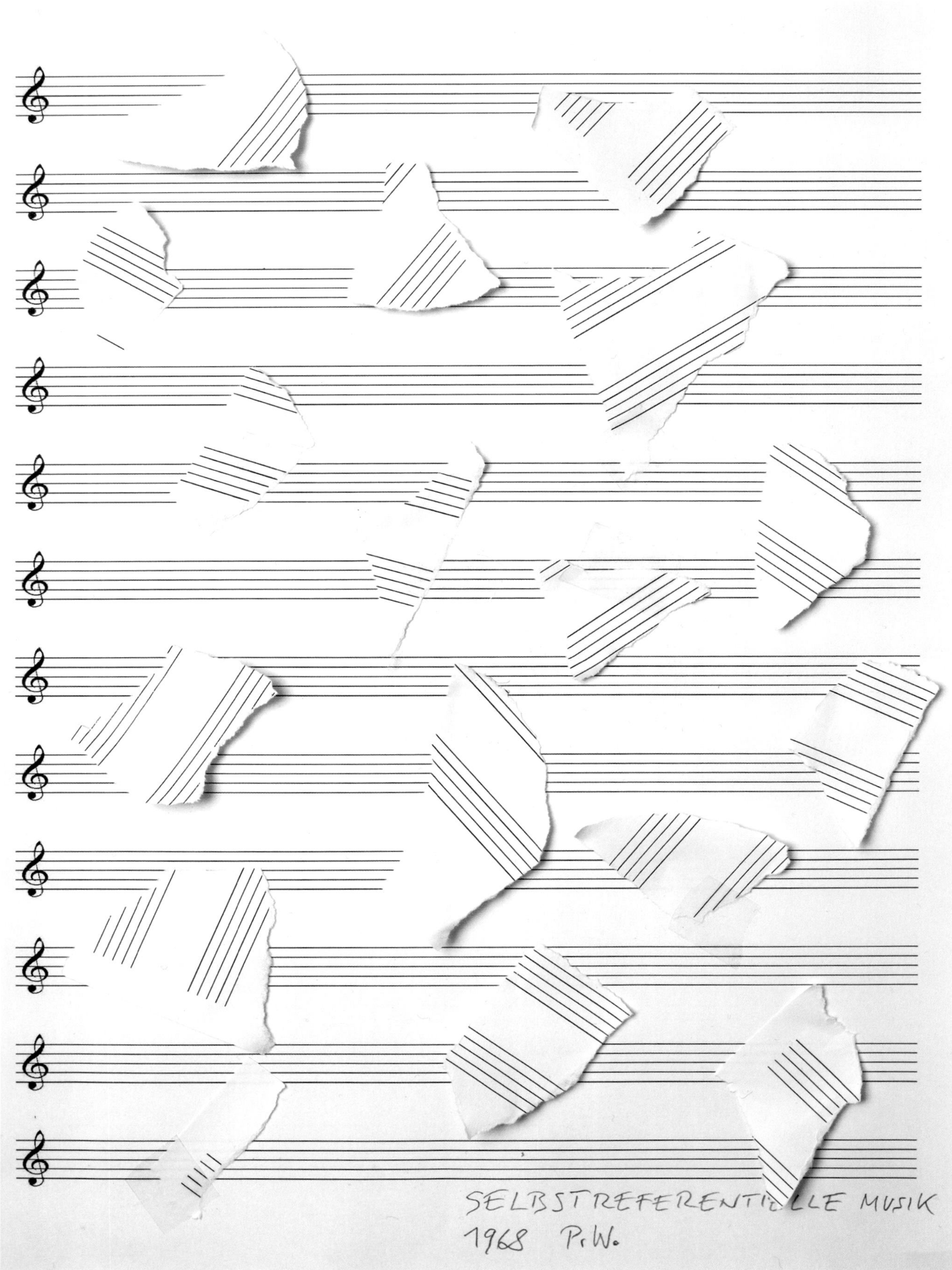

Self-Referential Music /
Selbstreferentielle Musik

1968 | collage

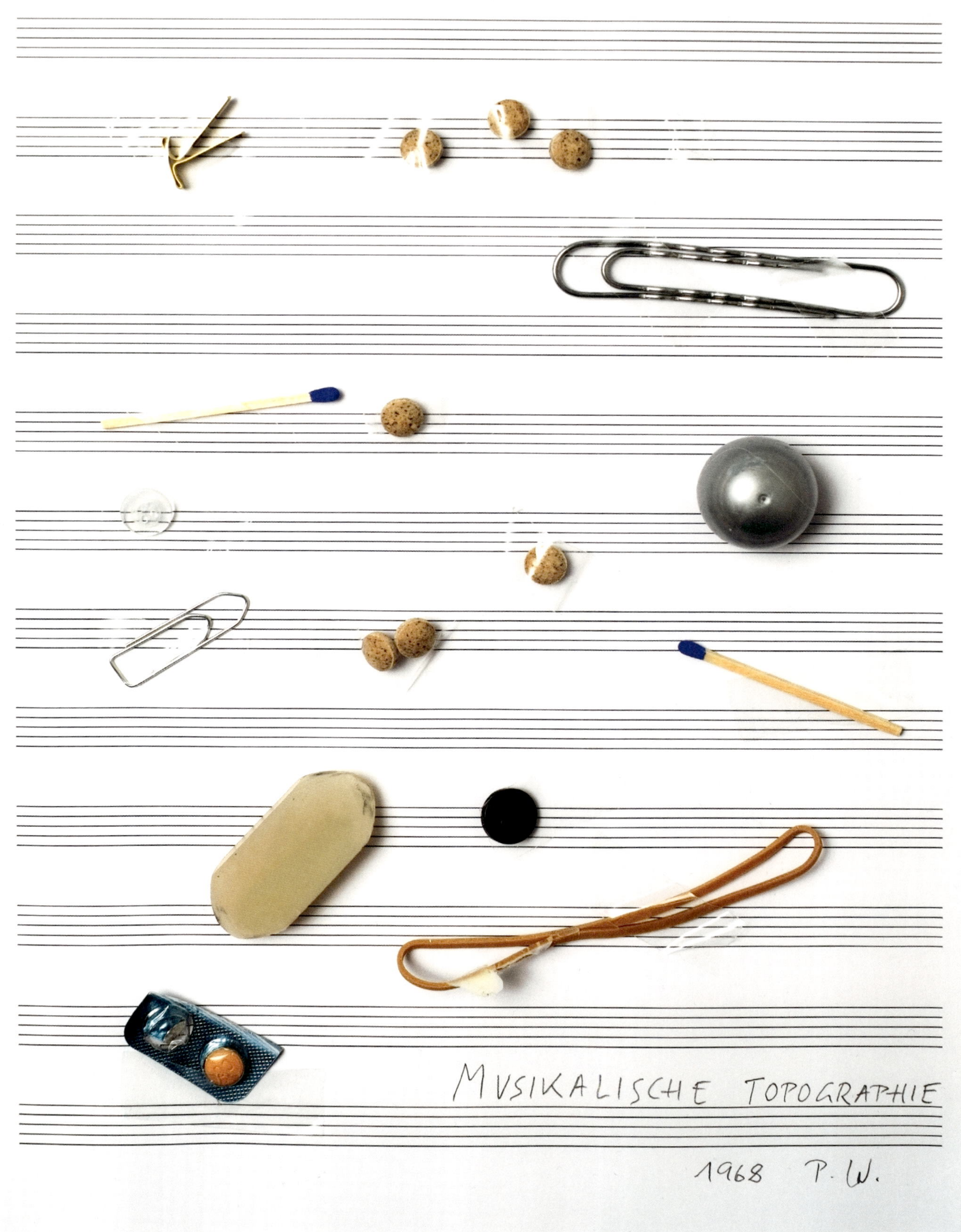

**Musical Topography /
Musikalische Topographie**

1968 | collage

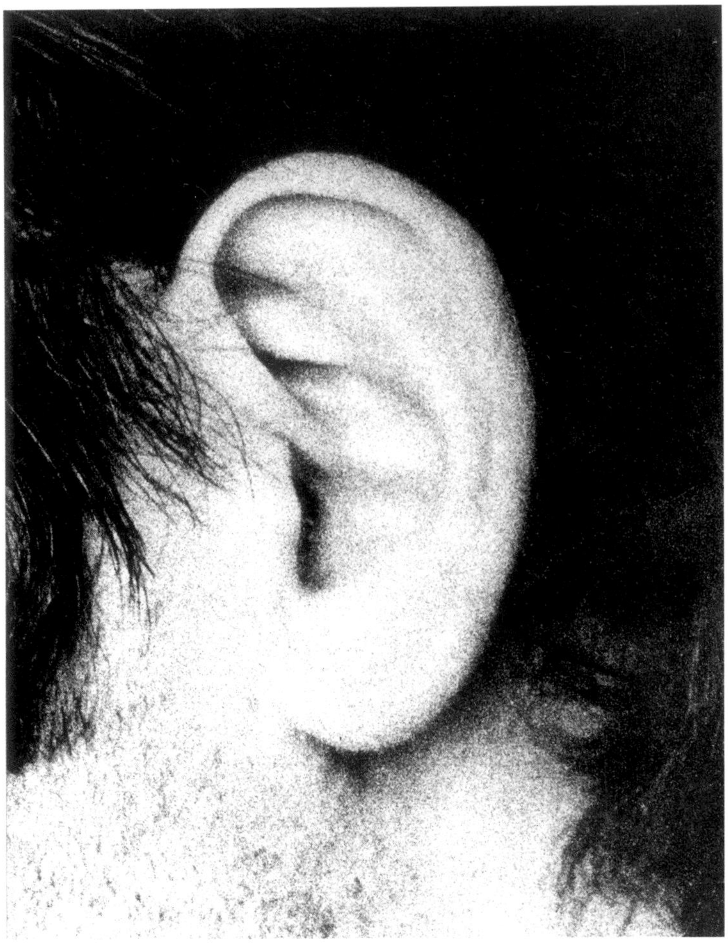

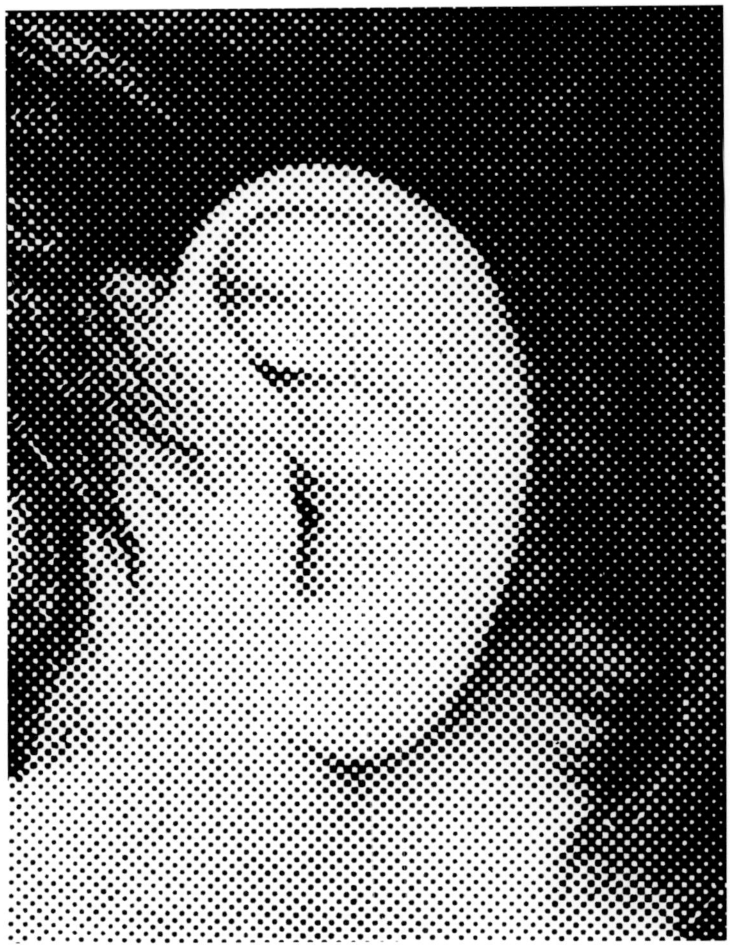

Transformational Music / Transformationelle Musik

1972 | structure overlay, paper music, visual music

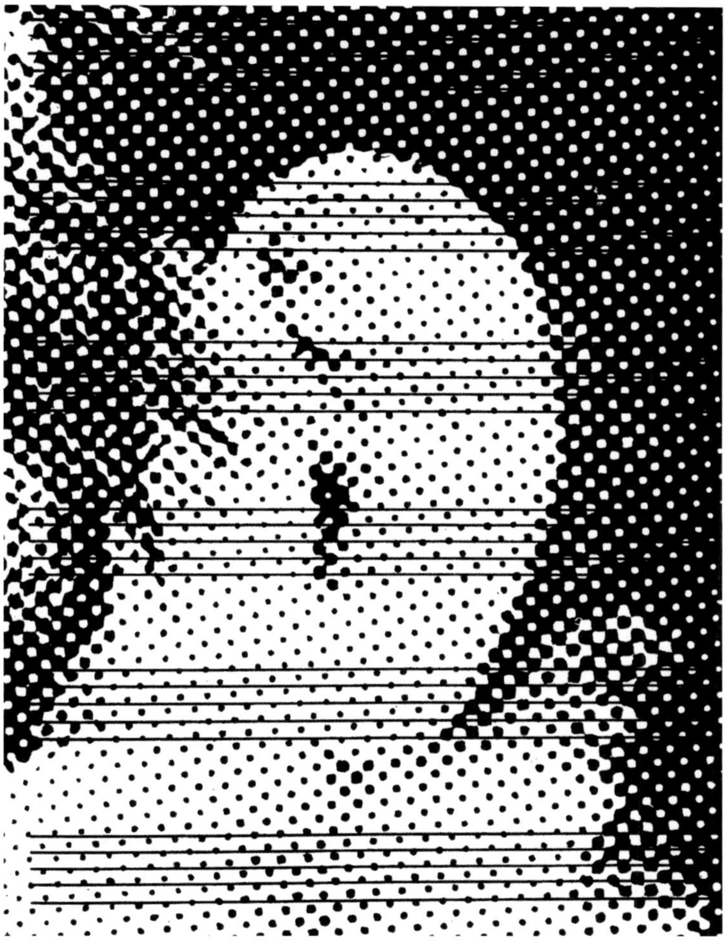

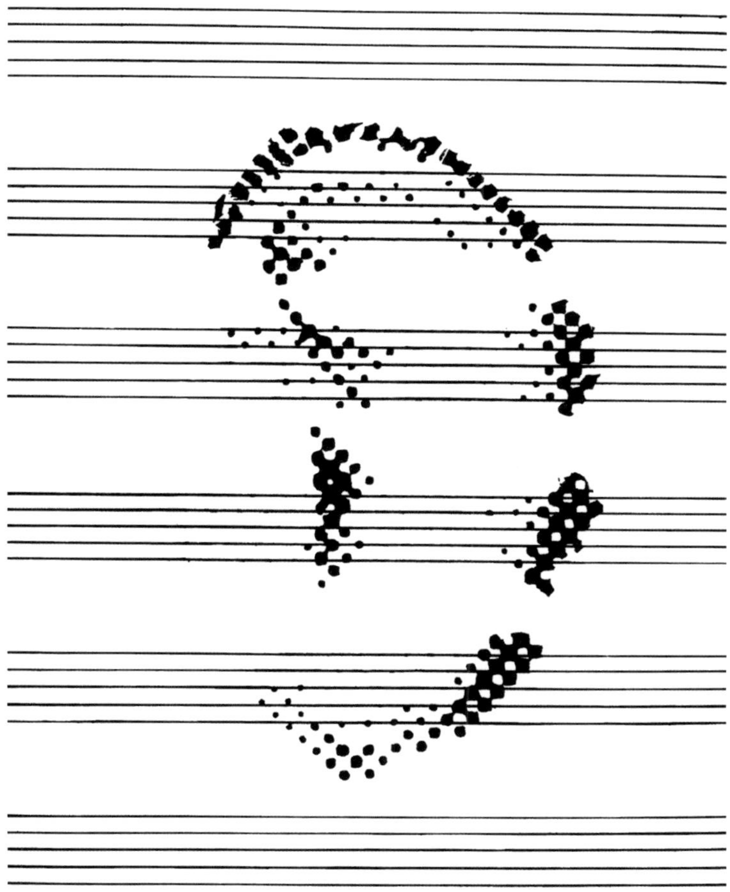

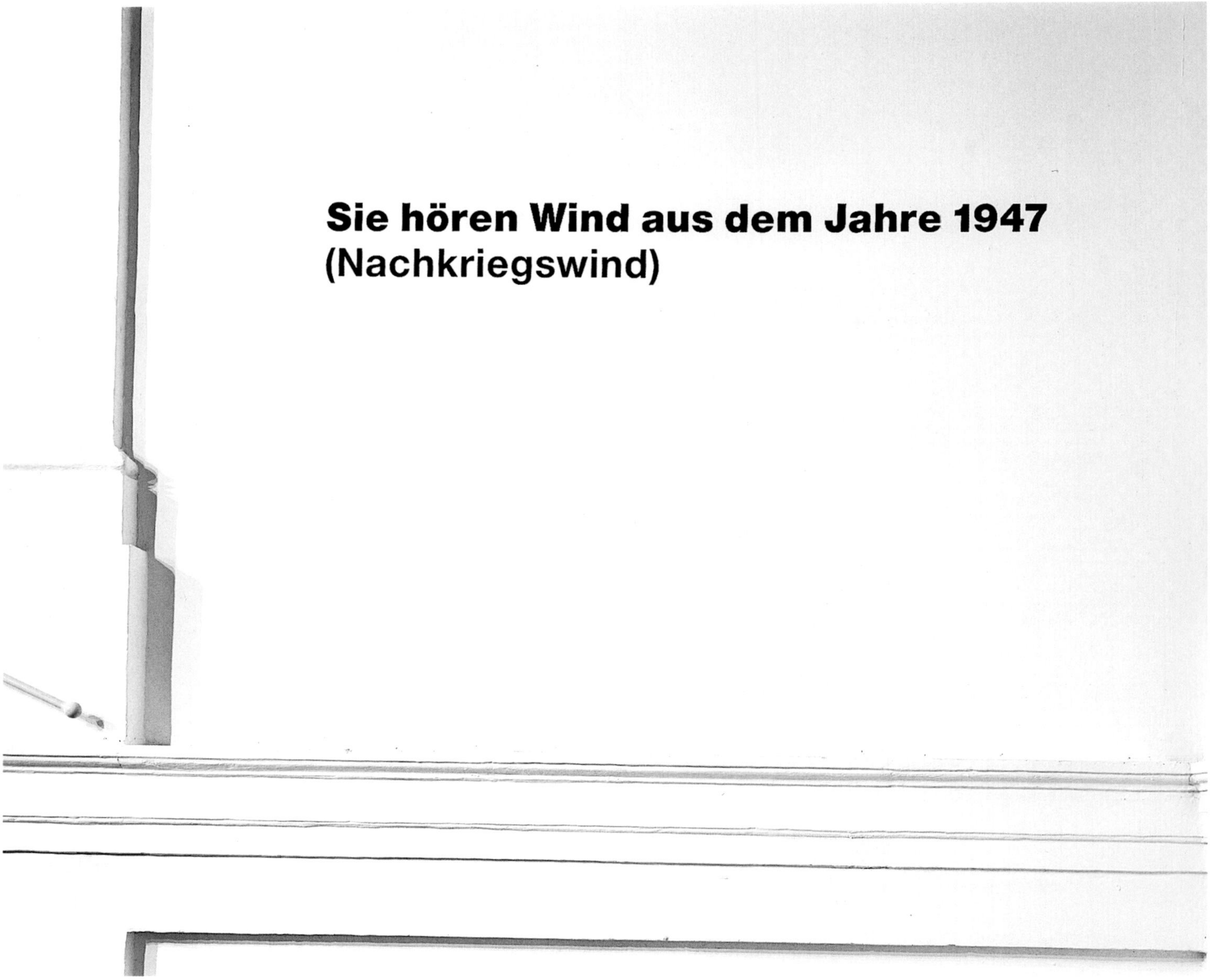

Installation view *Peter Weibel. Das Offene Werk 1964–1979*, Neue Galerie Graz, 2004

You Are Hearing Wind from 1947 (Postwar Wind) / Sie hören Wind aus dem Jahre 1947 (Nachkriegswind)

1972 | loudspeaker, tape recorder

In an empty room, visitors hear the sound of wind from a tape recorder. A text on the wall says "You are hearing wind from the year 1947 (postwar wind)." But since nobody has compiled an archive of the sounds of the wind year by year, nobody knows if the sounds they are hearing are really from 1947. Nobody can prove it. Nobody knows what the wind sounded like in 1947.

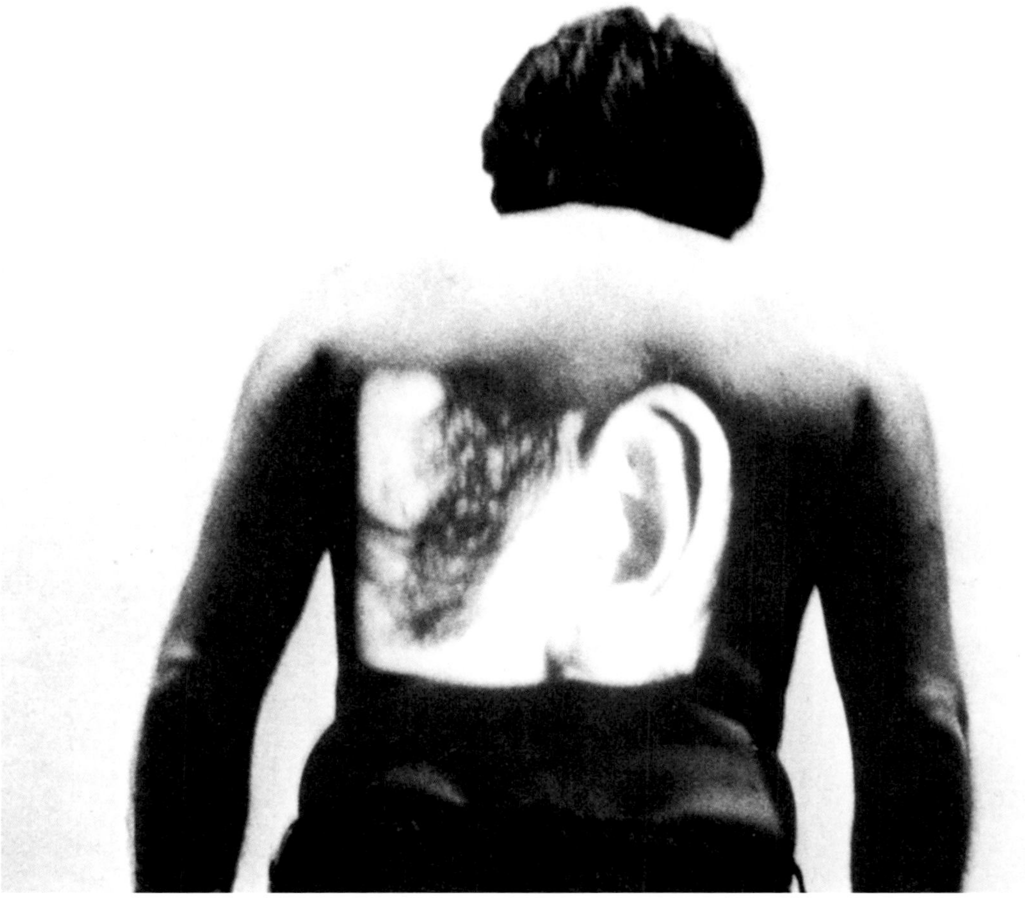

Listen / Hör zu

1968	8 mm films with audio description, with sound or silent, ca. 3 min. each

Version 1: While texts by the author are read aloud, an ear appears on the screen or on the author's body. The people listen with their ears, and their eyes see an ear. The act of listening is visualized. The texts can be presented in any order; only the ear remains constant.
Version 2: The same film without sound. The ear appears on the screen, but no text is heard. While the people listen, the screen itself listens, too. The silence between the two ears (the real one and the reproduced one) gives the film its charge, its "silent music." Sound can be produced, however, if the audience, provoked by the silence, provides its own commentary.
Version 3: Like the first version, except instead of the ear, the audience sees a spinning record. The film lasts as long as the record continues to turn and play music.

—Peter Weibel, 2006

The installation *Auditory Canal* was first shown in 1979 at the Galerie Grita Insam, Vienna. Ears, which are normally receptors for sound waves inwards, are here paradoxically loudspeakers, sending sound waves outwards. Viewers stand between the two enlarged images of the artist's ears and hear the following text in an endless loop:

aufhören [to stop/to listen up]
aufhören
aufhören
aufhören
hören sie auf [please stop/please listen up]
hören sie auf
hören sie auf
hören sie auf
hör auf [stop/listen to...]
hör auf
hör auf
hör auf
hör auf mich [listen to me]
hör auf mich
hör auf mich
hör auf mich
hör auf zu [stop .../listen to "to"]
hör auf zu
hör auf zu
hör auf zu
hör zu auf [listen closely to ...]
hör zu auf
hör zu auf
hör zu [listen]
hör zu
hör zu
hör zu
hör auf zu sehen [stop seeing]
hör auf zu sehen
hör auf zu sehen
hör auf zu sehen
hör zu sehen auf [stop seeing/stop watching]
hör zu sehen auf
hör zu sehen auf
hör zu sehen auf
hör zu sehen [listen to see]
hör zu sehen
hör zu sehen
hör zu sehen
hör sehen [listen-seeing]
hör sehen
hör sehen
hör sehen
sieh zu [look]
sieh zu
sieh zu
sieh zu
sieh auf [look up]
sieh auf
sieh auf
sieh auf
sieh zu auf [look closely at]
sieh zu auf
sieh zu auf
sieh zu auf
sieh auf mich [look at me]
sieh auf mich
sieh auf mich
sieh auf mich
sieh auf zu hören [look up to hear]
sieh auf zu hören
sieh auf zu hören
sieh auf zu höherem [look up to higher things]
sieh auf hören zu [look at listening]
sieh auf hören zu
sieh auf hören zu
sieh auf hören zu
sieh hören [see-hearing]
sieh hören
sieh hören
sie hören von mir [you will hear from me]
aufsehen [looking up/commotion]
aufsehen
aufsehen
aufsehen
zusehen [watching]
zusehen
zusehen
zusehen
sieh dir zu [watch yourself]
sieh dir zu
sieh dir zu
sieh dir zu
sieh auf dich [look at yourself]
sieh auf dich
sieh auf dich
sieh auf dich
hör dir zu [listen to yourself]
hör dir zu
hör dir zu
hör dir zu
hör auf dich [listen to yourself]
hör auf dich
hör auf dich
hör auf dich
hör auf [stop/listen to...]
hör auf
hör [hear]
hör
ohr [ear]
ohr
hörohr [hear-ear]
siehauge [look-eye]
hörauge [hear-eye]
siehohr [look-ear]
ohrauge [ear-eye]
augenohr [eye-ear]
ohren öhren [ears needle-eyes]
augenäugen [eye-looking]
zuäugen [eyeing]
zeugen [eyewitnesses]
weghören [shutting one's ears]
herhören [listening up]
hörsehen [listen-seeing]
hersehen [looking over]
hellsehen [clairvoyance]
hallsehen [seeing reverb]
schallsehen [seeing sound]

Translated from the German by Peter Dahm Robertson.

Auditory Canal / Gehörgang

1970/79 | sound installation; digital prints on aluminum, loudspeakers, 2 parts, 218 × 104 cm each

Installation view *Audio Scene '79*, Galerie Grita Insam, Vienna, 1979

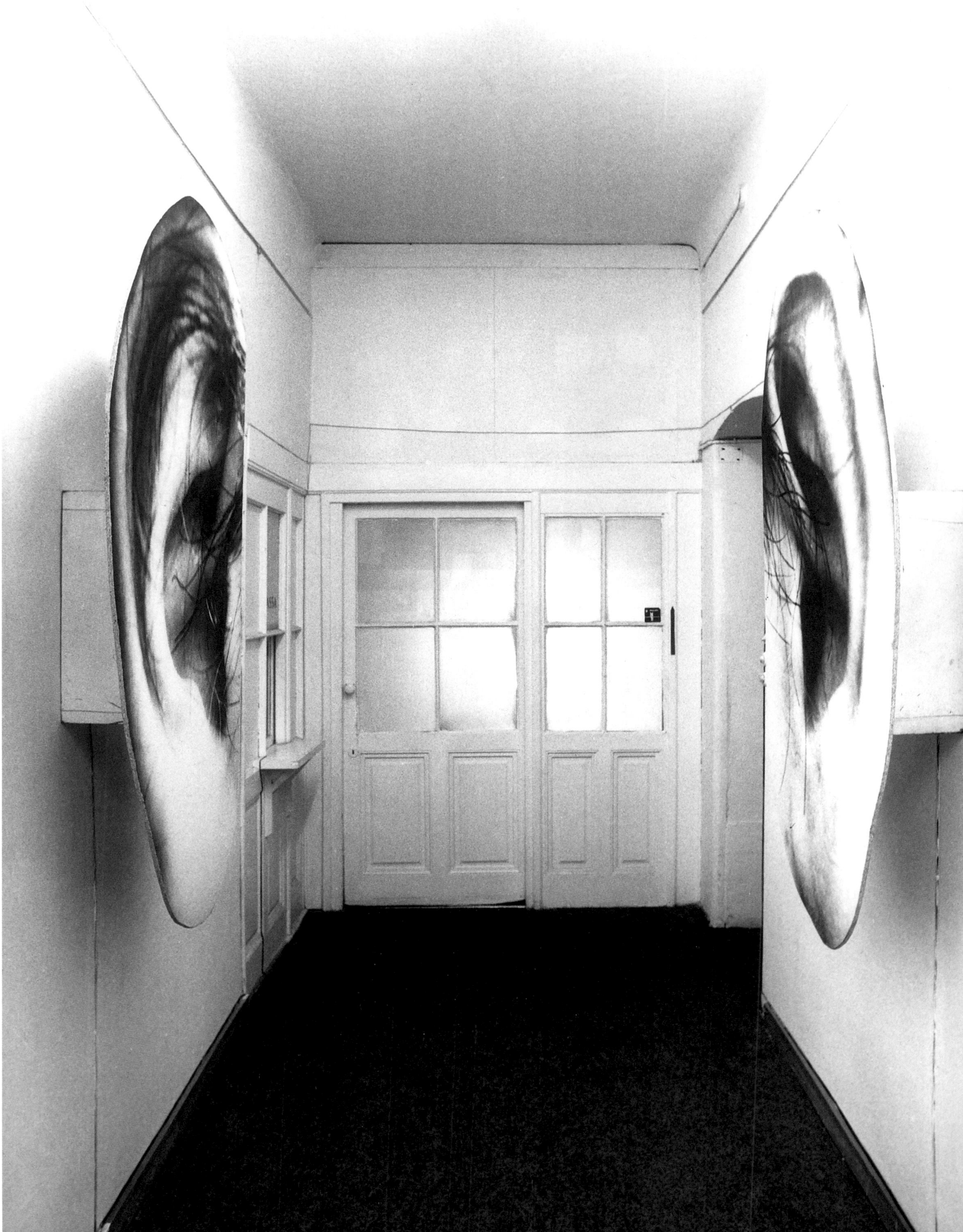

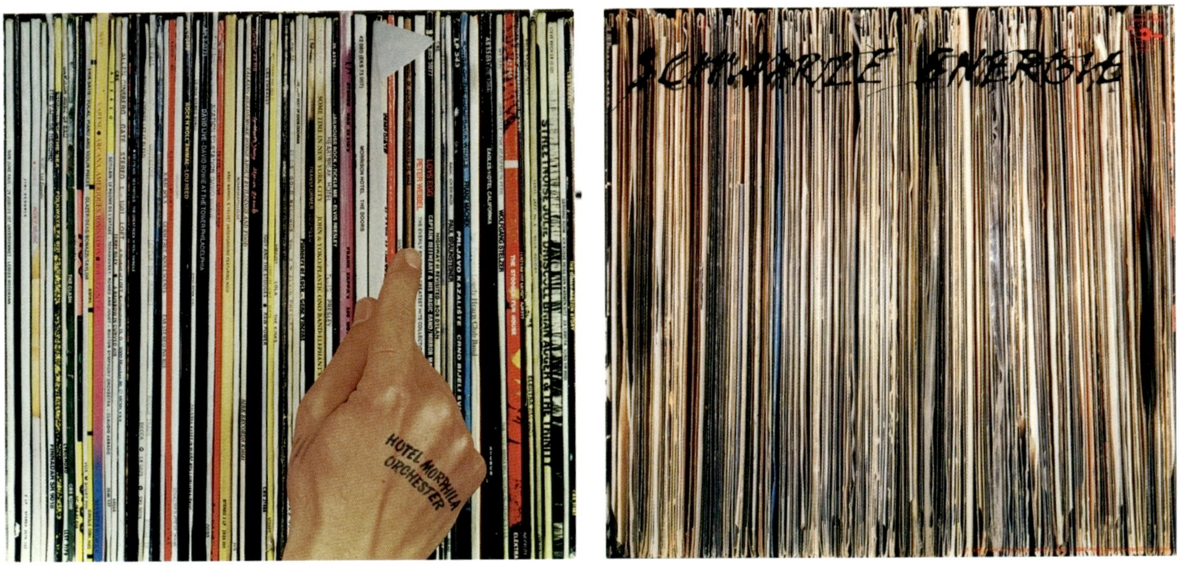

"Schwarze Energie" (Black energy), 1982, LP cover, front and back

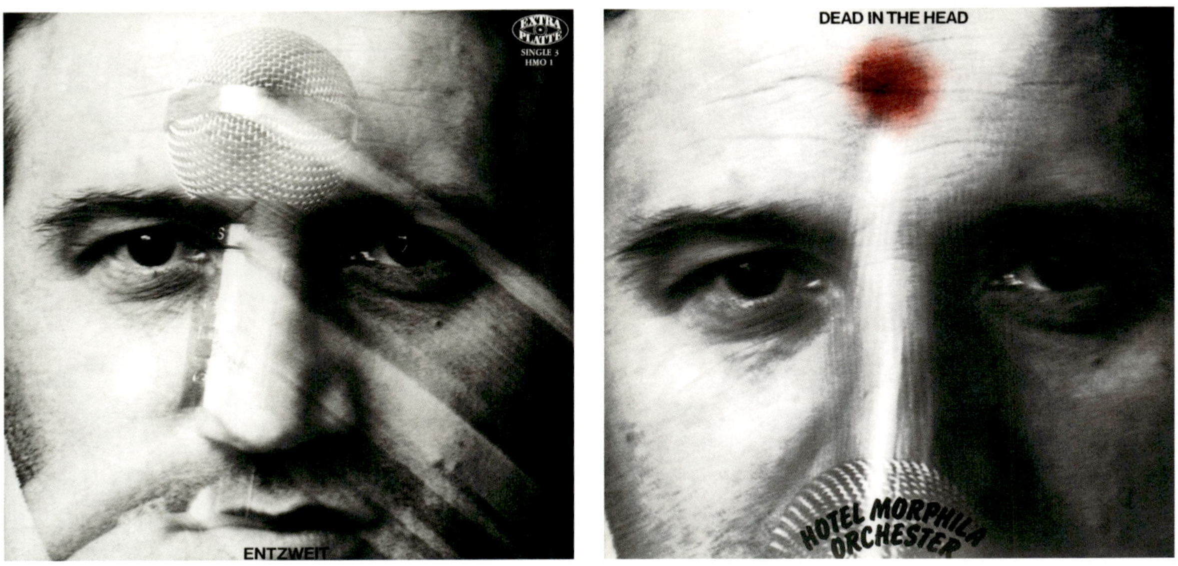

"Dead in the Head," 1982, single cover, front and back

In the 1980s, it became a widespread phenomenon that visual artists expanded their palette of expression and started to move into the field of pop music. Many painters, performance artists, and poets were also rock musicians – or tried to become rock musicians by founding bands. A subcultural scene developed a new language for painting and rock music, from the NDW (Neue Deutsche Welle, or "New German Wave,") to No Wave in New York. The renewal of rock music was brought about in significant measure by groups and prominent figures from the visual arts. This most interesting and lively development of rock music was termed "art rock."

Hotel Morphila Orchester

1978–83	members: Reinhard Blank (drums), Paul Braunsteiner (guitar), Günther Dinold (bass guitar), Franz Dorfner, Loys Egg (guitar), Franz Machek, Wolfgang Stelzer, Peter Weibel (vocals)

The band Hotel Morphila Orchester was founded in 1978 by Peter Weibel and Loys Egg. The name Hotel Morphila, a fictional hotel, refers to the Viennese tradition that every good hotel or coffee house used to have its own orchestra. Thus, Weibel chose the name of a hotel that did not exist for a custom that no longer exists. The band played punk rock whose apparent musical simplicity was contrasted with complex poetry and visually complex structures.

The name Hotel Morphila Orchester was chosen based on the fact that hotels – such as Elvis Presley's "Heartbreak Hotel" (1956) or The Eagles' "Hotel California" (1976) – had played a significant role in the history of rock 'n' roll. The specific hotel Egg and Weibel had in mind was the Hôtel Orfila in Paris, where August Strindberg stayed. In addition, the word Morphila is a mix of the words Orpheus, morphium, morphé, Godzilla etc.

The musical performance was embedded in video installations, which themselves also defined the musical structure of a piece.

The use of popular music as an art medium. In the song "Dead in the Head" (1982), Weibel beat time on his forehead with the microphone. This action left a red bruise which is depicted on the cover of the single.

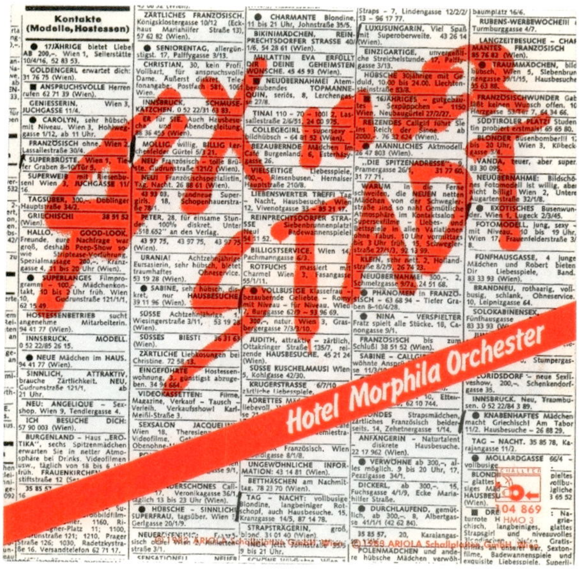

"Sex in der Stadt," 1982, single cover, front

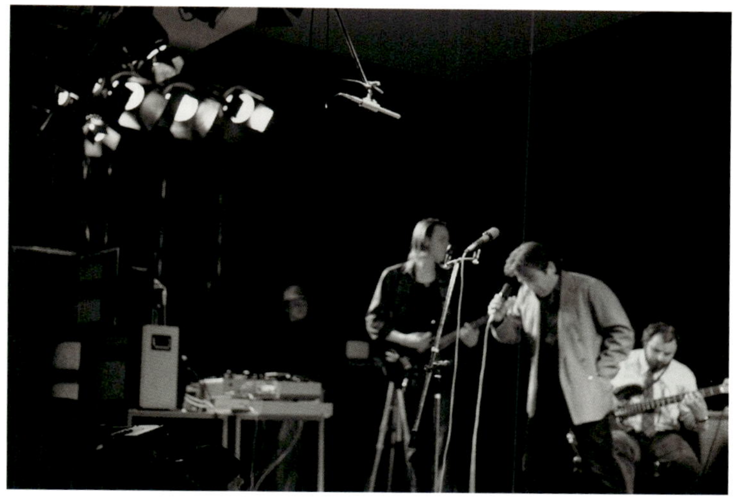

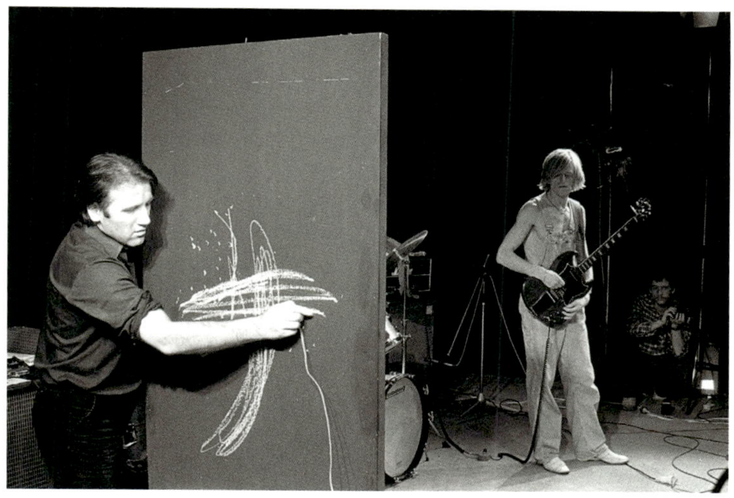

Concert of Hotel Morphila Orchester at the festival *Performances '79* (focused on body language, dance music, video film, theater), Lenbachhaus, Munich, Feb. 2–18, 1979. The participants, among others, were Herbert Achternbusch, VALIE EXPORT, Jochen Gerz, Hans-Jörg Mayer, Hermann Nitsch, Fabrizio Plessi, Gerhard Rühm, Jürgen Klauke, Peter Weibel, and Oswald Wiener.

In 1982, the single "Sex in der Stadt" (Sex in the city) was released on vinyl. The music video was produced in a telephone booth as the single's text extraordinarily deals with pornographic advertisements printed in newspapers. For a week, you could hear the piece of music when you called a particular phone number. At the band's first live concert at the *Performances '79* event at the Städtische Galerie im Lenbachhaus in Munich in 1979, Peter Weibel called sex workers who had offered their services in Munich newspapers. The conversation was broadcast live on stage without the women knowing. Weibel expressed the most abstruse sexual ideas and the women responded with suggestions as to how they could implement the ideas and what they would charge financially for the extraordinary services.
The Austrian public broadcaster (ORF) refused to play the single due to its pornographic content. However, this is paradox as the pornographic texts are published in mass media on a daily basis. What Weibel did was just to reenter one public sphere (newspaper) into another one (radio). "Sex in der Stadt" was included in a list of 52 significant Austrian pop songs from 50 years for the exhibition *Ganz Wien. Eine Pop-Tour* (2017–18) by the Vienna Museum.

Two video cameras capture me and the guitarist. While I draw, shade, dot on a blackboard with a piece of chalk – captured by a special microphone – the guitarist makes corresponding sounds and movements on his guitar. I hear the sounds produced by my writing and drawing – graphic musicality – both in real time and delayed, just as I hear the sounds of the guitar now as well as delayed. In this way, the musician and I are playing not just against/with one another, but also against our past. The left screen shows my activity delayed from my real activity (while I am making slow loops in real time, slow dotting is seen on the screen, etc. The right screen shows the musician's activity with delay. Simultaneously, we see my activity and that of the musician superimposed on the central screen.
Or while the guitarist plucked the guitar, I plucked out my hair synchronously. Each pluck of my hair – a pluck of the guitar string. I plucked the hair on my head, my eyebrows, or tore it in bunches from my chest. Hair music. Particularly clearly visible and comprehensible on the superimposing screen.

—Peter Weibel

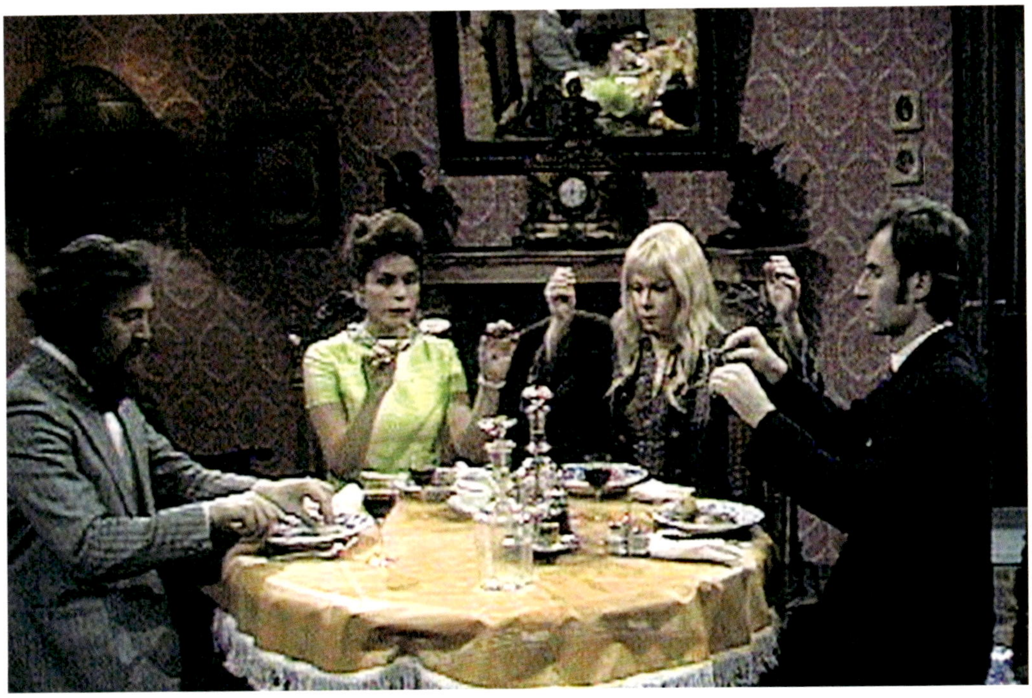

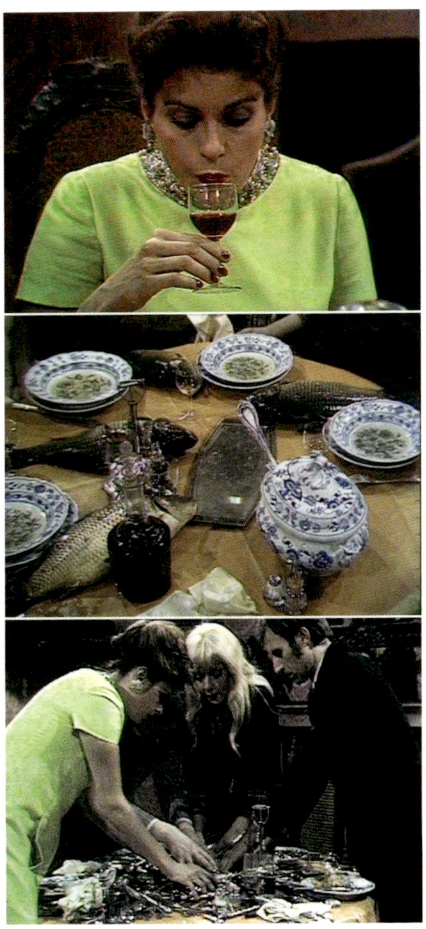

Chamber Music / Hausmusik

1974

TV comedy; color, sound, 17:07 min., broadcast in *Impulse 3*, ORF, February 25, 1974
music: Otto M. Zykan; with Dolores Schmidinger, Ernst Macher, Susanne Widl, Otto M. Zykan, and Peter Weibel

Usually, a comedy is steered by the mistakes of people or by fatal objects (like a banana). This is the first sound comedy in which the actor's action is steered by peculiar sounds. Music and eating: music and dinner. Two bourgeois rituals are combined along musical structures. What is novel about this is that it is the sound material that develops the plot and not the emotions or the psyche. The history of bourgeois society is narrated by means of its music. The spectrum of this music extends from sounds generated by the body (suppressed and inhibited), through classical music (where the bodily posture clearly becomes entirely rigid) up to the sounds of war. The sound anticipates the image and the image the tone, for example, laughing passes over into the sounds of fish, and suddenly, one sees fish instead of spoons next to the plates. The fish is then placed in the soup, and the sounds disappear. The fish in the soup already announces the end: the flooding of the living room. Initially, only one person present hears the peculiar sound and they are uncertain about whether or not they really do hear it. But their odd behavior, influenced as it is by the peculiar sounds, is noted by others. When the wife of the man reproaches him, she is infected and along with her, so is everyone else. After a few attempts to identify the guilty party, they are no longer irritated, although step-by-step, they become increasingly alarmed. However, each time they are able to find explanations and to calm themselves. They just don't wish to pay attention to the warnings, they ignore the changes and the symptoms of catastrophe, and even end up beginning to enjoy the distortions until finally they are swept away. A musical comedy about the self-destruction of the Western bourgeois-capitalist world. —Peter Weibel

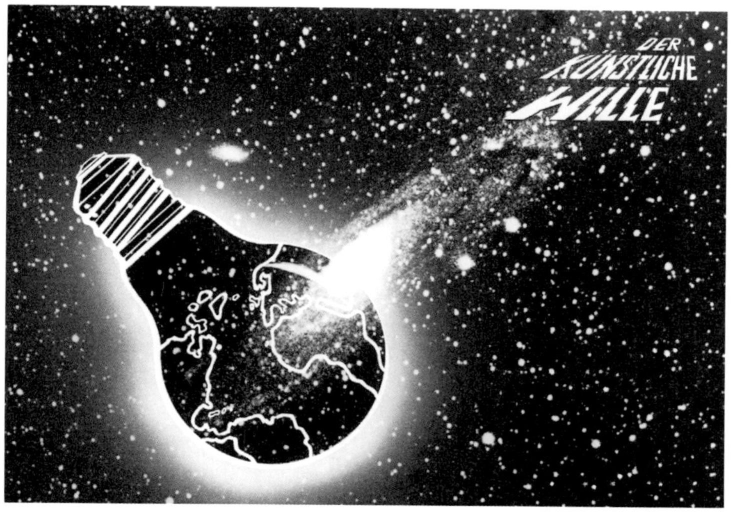

Poster for *The Artificial Will*

The Artificial Will is Peter Weibel's first major electronic media opera, which he performed at the Ars Electronica Festival in the Brucknerhaus in Linz in 1984 together with the band Noa Noa that he founded and the electronic musician Zyx. The opera thematically deals with the third millennium and shows the connection of the three media music, painting, and language through the possibilities of the digital and anticipates with the term audiovisual alphabet (AVA) today's digital interpenetration of our everyday world. The opera consists of ten acts.

Act 1: Electronic Space-Time
In several places at once and in the same place at different times: ubiquity and simultaneity. In the electronic Paleolithic, there are several degrees of modal reality rather than a single one. The boundaries blur between depiction and object, between imaginary and real. Energy of a nomadic existence.

Act 2: The Electric Home
Gas and water pipelines (mechanical age) are extended with cable lines for electricity and information (electrical age). The home is permeable. Objects of interior decoration (sofa) are shaped according to codes of the outside world (city). The exploitation behind the idyll. The force of social behavior, perturbed by advertising and ideology, fixes women back into the space of the nineteenth century (chamber maid). Chains and carrots in cloud-cuckoo land.

Act 3: The Conquest of Light
Europe is the Prometheus of electricity, of artificial light, of artificiality. With the light bulb we captured the sun. Electricity is the human form of light. Artificial, accordingly, means human-made. Artificial light is the Promethean intrusion of artificiality, of human will into God's natural order. For eons, only God said: Let there be light. Now any child says it several times a day.

Act 4: The Extended Continent
The electrified continent. The end of Eurocentrism as the last Eurovision broadcast.

Act 5: End of the World
A kick into the twentieth century, the program of self-destruction, the insane totalitarian ideologies. In two world wars, the reactionaries tried to fend off the arrival of the third millennium. Demon democracy. The crimes of the parties in the infantile society. The caste of politicians as parasites of the twentieth century.

Act 6: Time as Medium of Hope
The journey of humanity from the infinite depth of time into infinite space. Light as the medium of progress. We are leaving the twentieth century.

Act 7: The Artificial Will
The evolution of the world is replaced with computer animation. Thy will be done on Earth as it is in heaven, the prayer runs. God's will, that great invention of humankind, is replaced with the will of humankind. The world as will and imagination – of humankind. At the end of evolution, the supreme result of evolution takes evolution itself into its own hands – does that mean that humanity is laying hands on itself, as well?

Act 8: Extended Earth
The expansion of humanity and the capacities of its organs lead also to an expansion of the Earth. The externalization of the human body (eye becomes video, ear becomes audio, etc.) and of the terrestrial body (oil, water, etc.) leads into space. That is where we will source our energy in the future.

Act 9: The Electronic Messiah
The world as artifact of human beings' artificial will. Everything human beings encounter will be artificial in future: from the nurturing layer around the embryo to that around the planet, from the artificial placenta to the artificial gas sun.

Act 10: The Artificial Sky
The sail of Noah's ark on which humanity is journeying through time and space will in future be moved not by the natural wind of the stars, but by a human-made, artificial motor. The heart of evolution will beat artificially. —Peter Weibel, 1984

Translated from the German by Peter Dahm Robertson.

The Artificial Will. Electronic Media Opera / Der künstliche Wille. Elektronische Medienoper

1984	media opera; 66 min., debut performance at Ars Electronica '84, Brucknerhaus, Linz, September 10, 1984 idea, script, director: Peter Weibel; performers: Renée Felden, Petra Felix, Peter Weibel, Susanne Widl; music: Noa Noa, Zyx; visual communication: Gregor Eichinger; technical director: Marcus Eibelmayr; object production: Werner Degenfeld, Peter Weibel

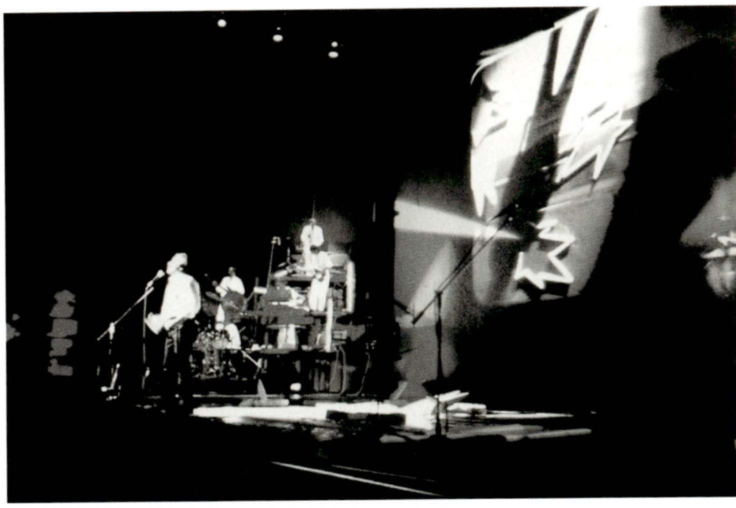

Peter Weibel and the musicians on stage

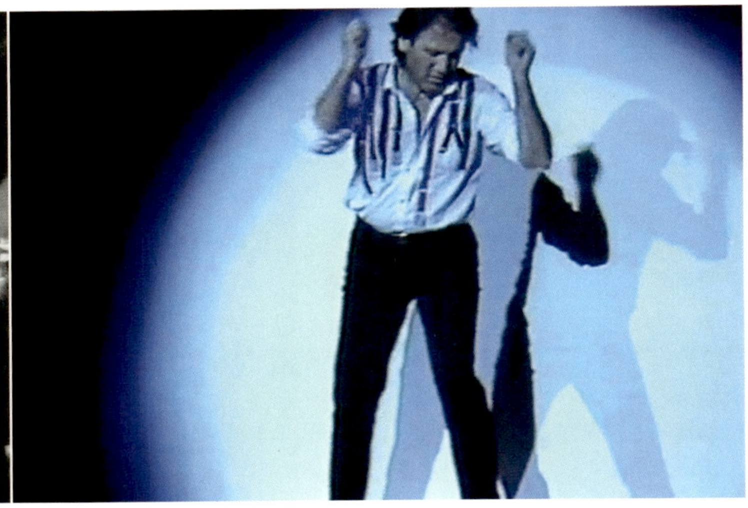

Peter Weibel in a scene of the overture

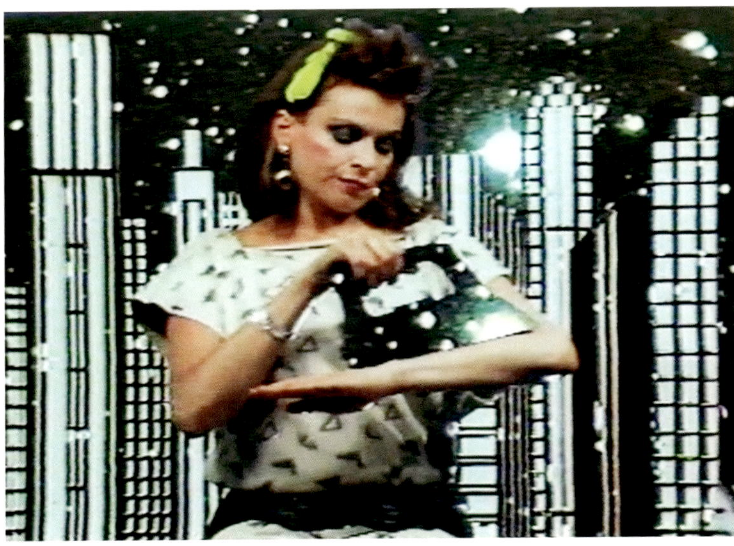

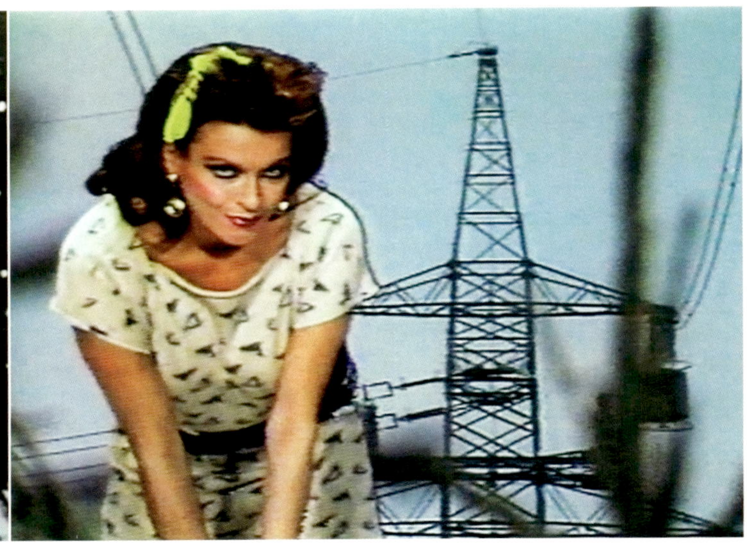

Scenes from Act 2

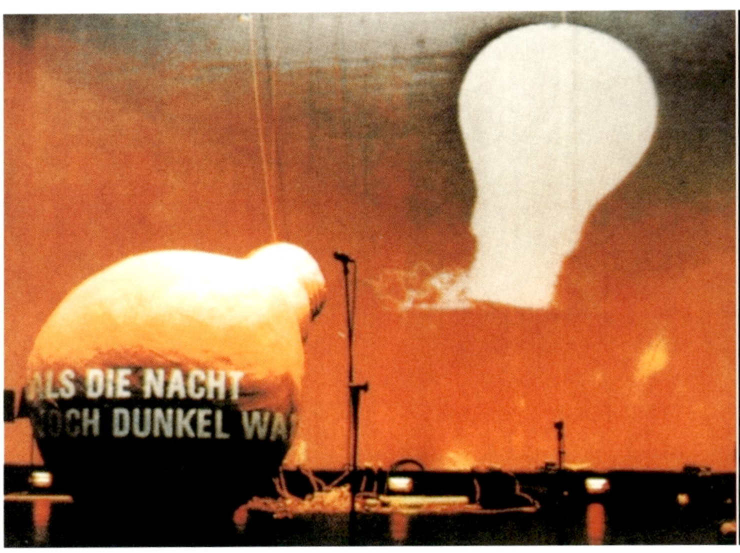

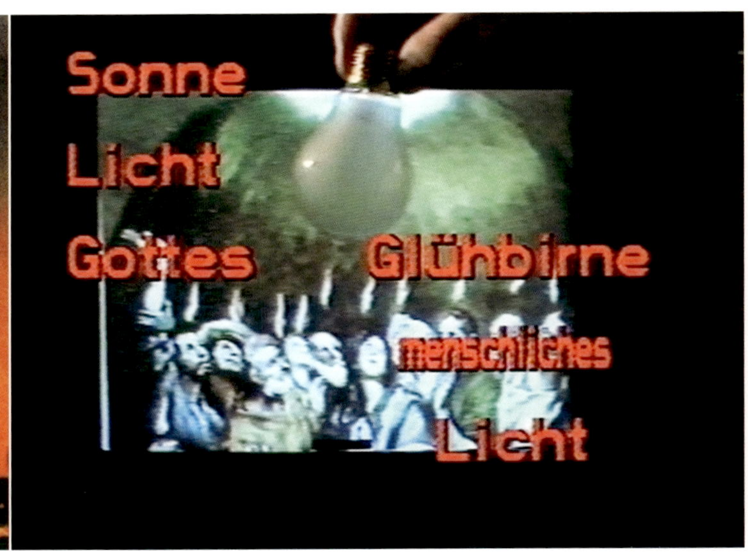

Scences from Act 3

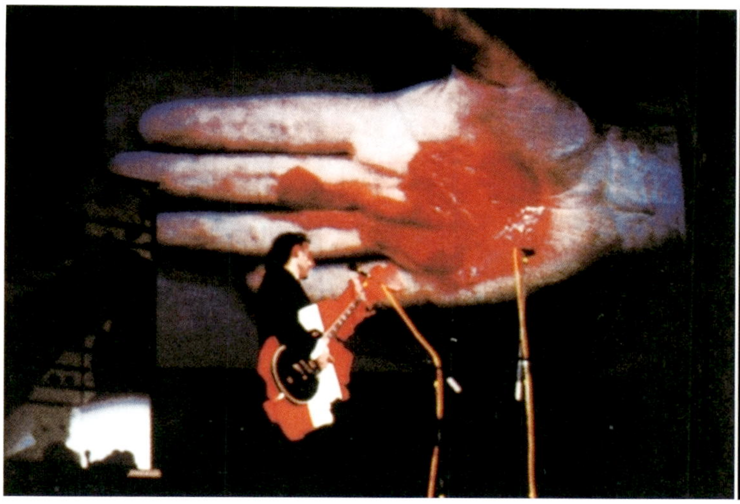
Scene from Act 4. Peter Weibel with his Austria guitar in front of a live video projection. The stain on the hand represents Austria.

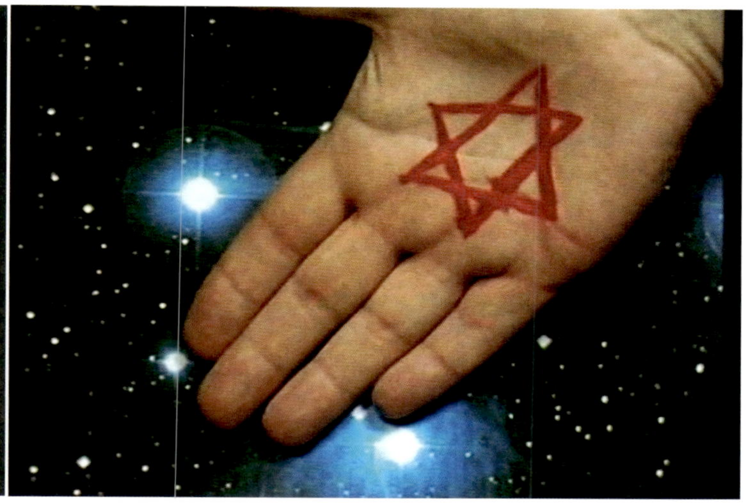
Scenes from Act 5

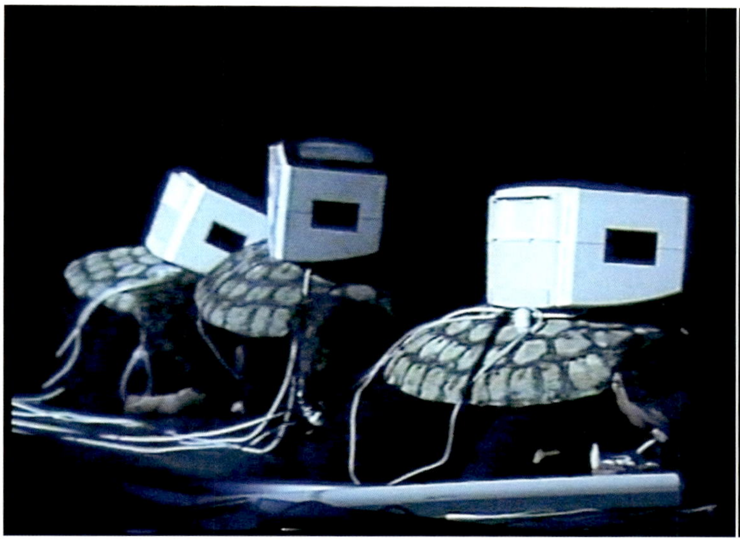
Scenes from Act 6

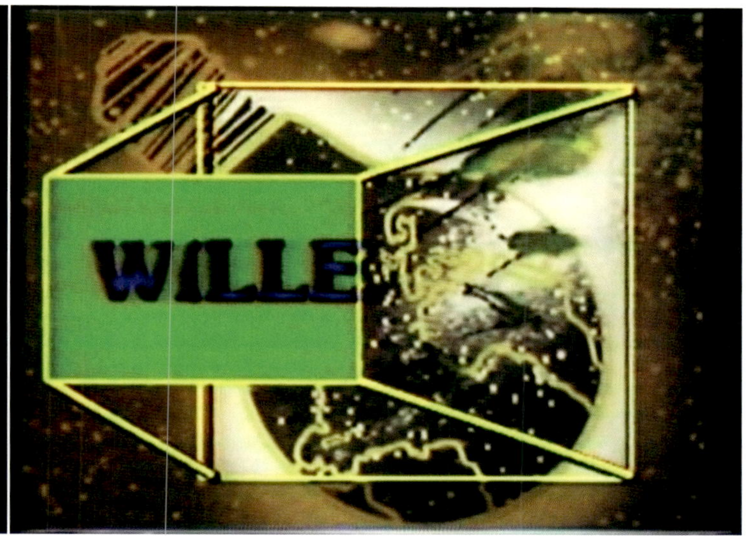

Scenes from Act 7

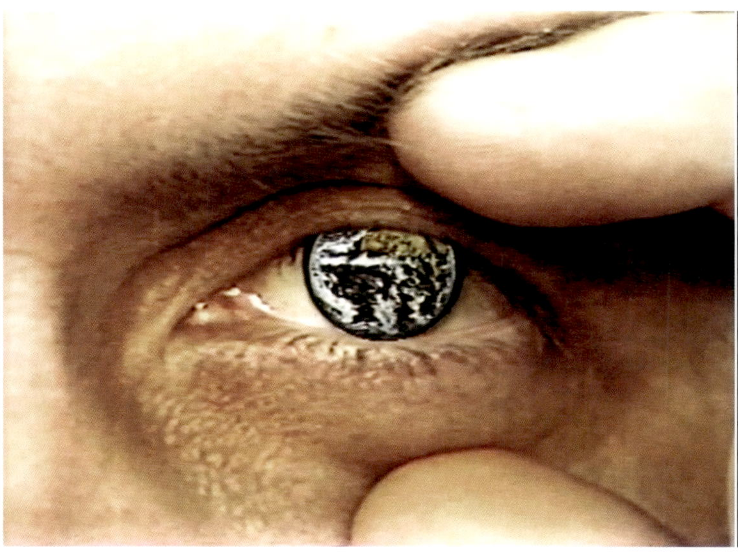
Scene from Act 8

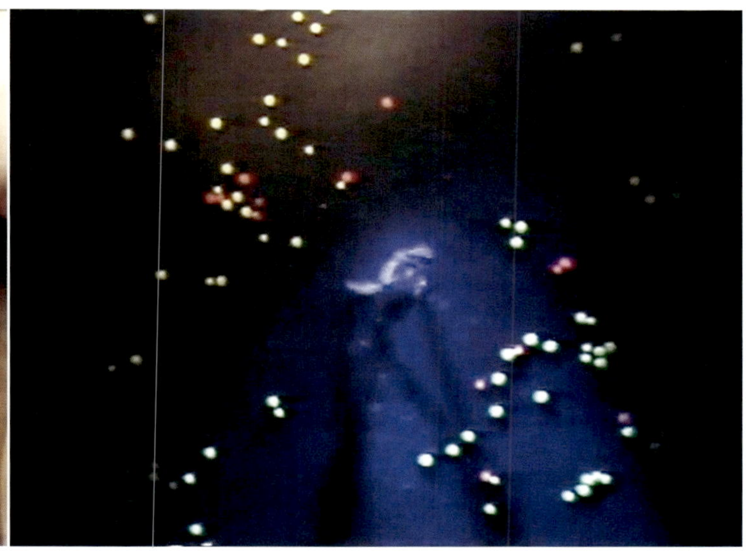
Scene from Act 10.
A camera fixed on the ceiling of the stage was remotely connected to a button on Peter Weibel's trousers, with which he could direct the camera, for example, from a total view to a close-up. These images from above were then projected on the vertical screen. At the end, golfballs lay around on stage which appeared like stars on the screen. A naked woman came on stage and tossed the "stars," the balls, to the audience accompanied by blaring sounds.

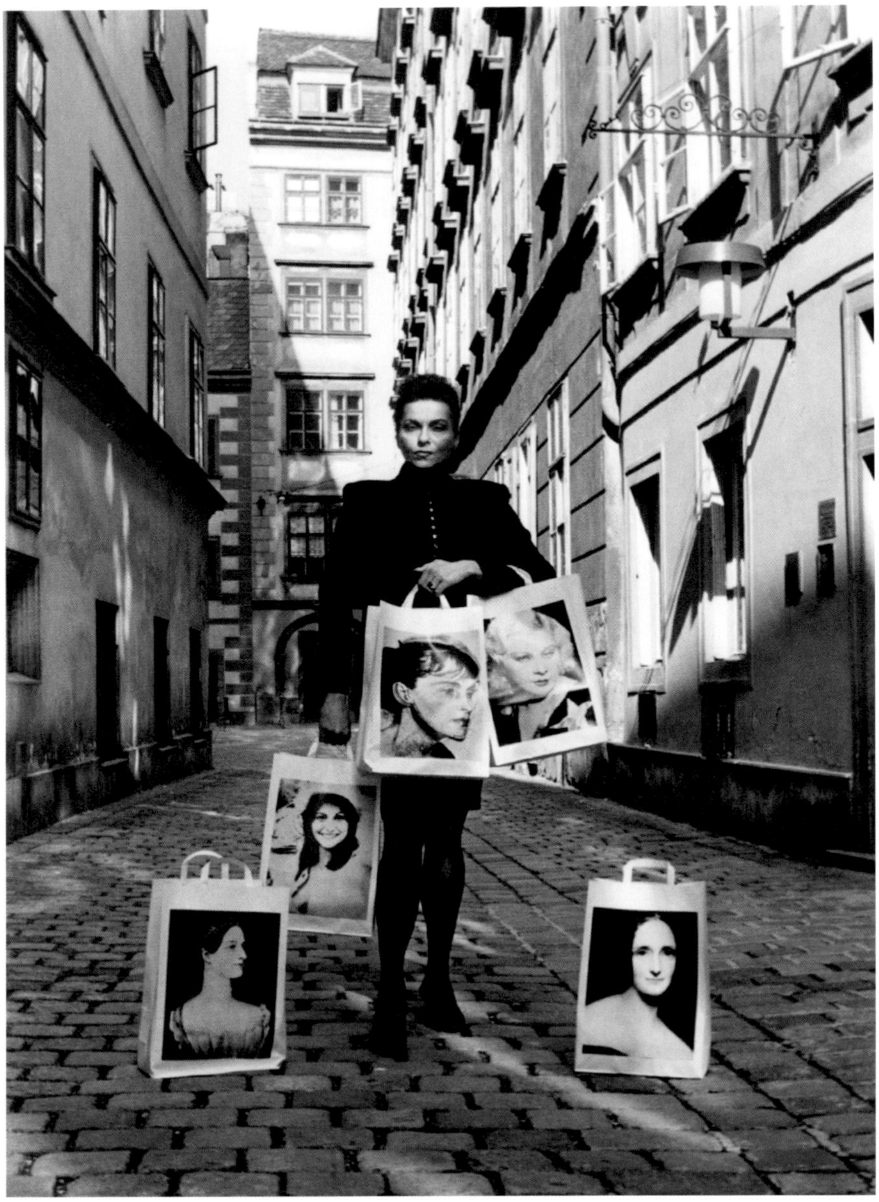

Susanne Widl carrying shopping bags with the portraits of the women she is portraying

Voices from an Inner Space / Stimmen aus dem Innenraum

1988	electronic media opera; 66 min., mobile video wall, live cameras, computer, spotlights, audiotape machines, loudspeaker, walkway, stacker truck, objects, sculptures, props etc., debut performance at Ars Electronica '88, Brucknerhaus, Linz, September 17, 1988 performer: Susanne Widl; videos on Linda Lovelace, Ada Lovelace, Mae West and Mary Shelley: Peter Weibel; videos on Unica Zürn: VALIE EXPORT, Peter Weibel; object production: Peter Weibel; music: Patricia Jünger; director, texts, and dramaturgy: Peter Weibel

The polyphonic alphabet of the body and the lost vocal

If the body resembles a sentence and is analyzable and combinable like an anagram, then we would like to construct an alphabet of the body and from this reconstruct the unconscious social text about women. In doing so, we discover that there is a lost vocal. Our culture has introduced a cipher of emptiness, a symbol of loss: the vagina. In the soundless world those "voices" necessarily become loud, which come from an interior space and which are polyphonic. This polyphonic alphabet of the female body should be made to speak, with the example of five selected women, in the form of a circular movement.

The drama of the phallic tongue whereby the woman becomes lost in the labyrinth of love: Unica Zürn

The temptation of the phallic mother, the verse of the vamp, the anal breast: Mae West

The letters of the stomach, the machinery of the mother mummy: Mary W. Shelley

The brain machine, the enchained head, the erotic rhymes of calculation, the body and the binary: Ada Lovelace

The encore of desire, the throat as clitoris, all-powerful oral phallus fantasies: Linda Lovelace

From the tongue as phallus (Unica Zürn) to the phallus as tongue (Linda Lovelace) closes the circle.

—Peter Weibel, 1988

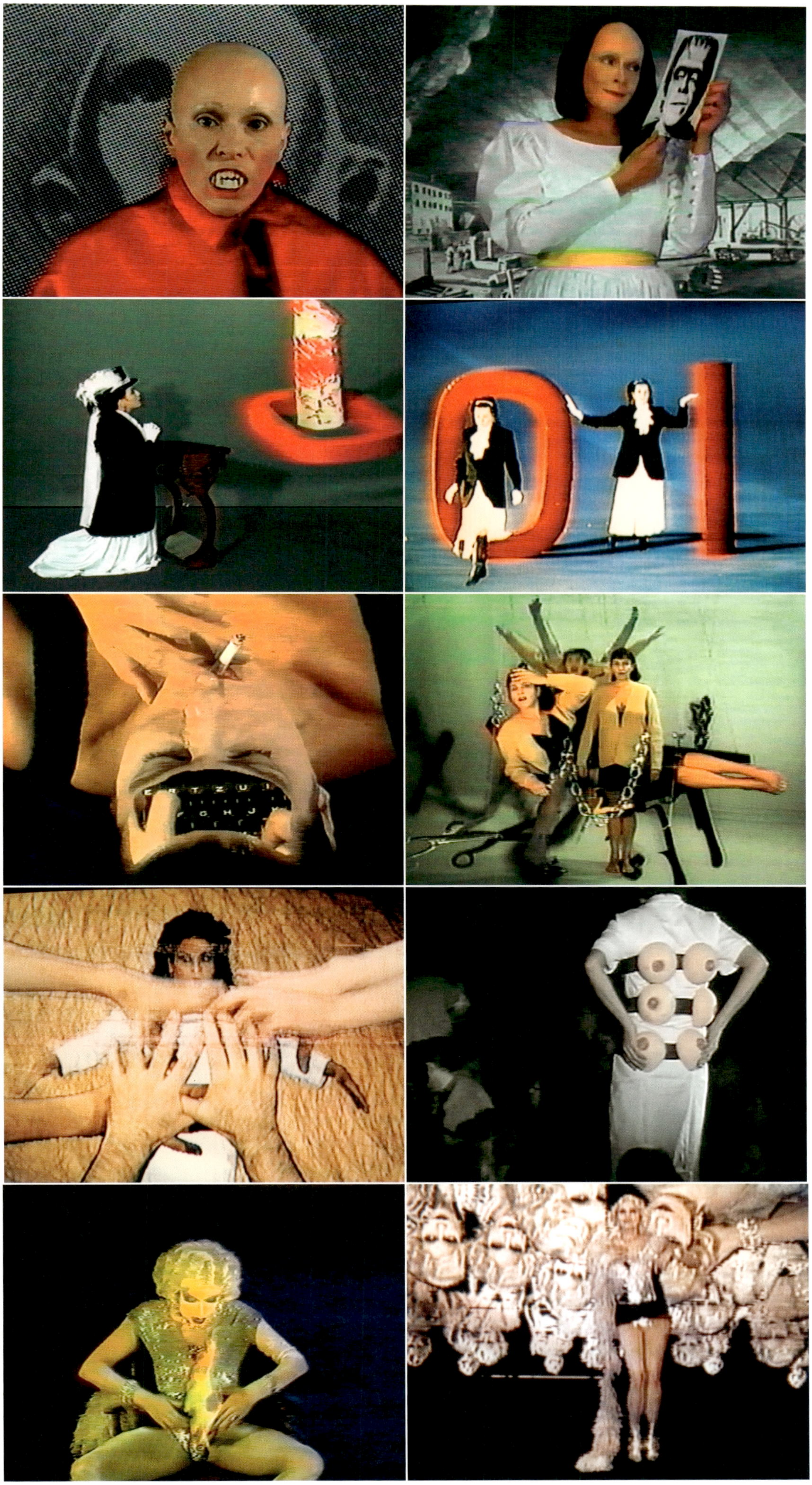

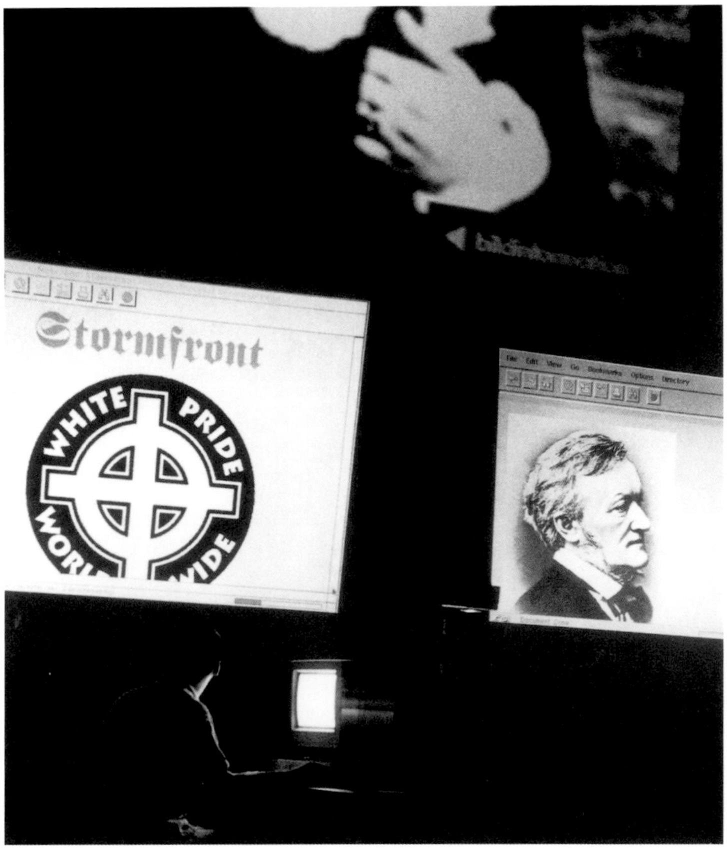

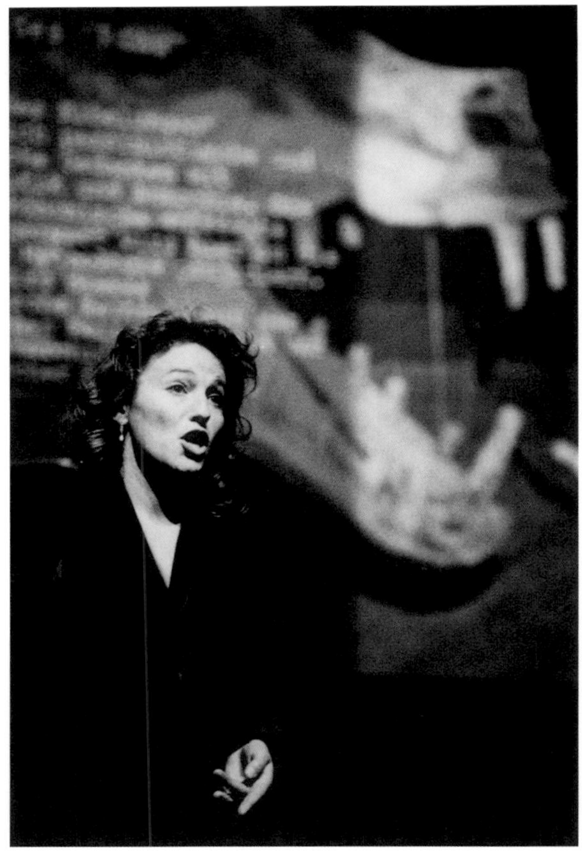

Singer Susan Dumas

Wagner's Delusion, or the Holy Land of Capital / Wagners Wahn oder das heilige Land des Kapitals

1995	cyber opera for stage and internet; debut performance at Brucknersaal Linz, Ars Electronica '95, June 20, 1995 concept: Peter Weibel; CD-ROM development: Orhan Kipcak; singer: Susan Dumas; editors: Peter Weibel, Curd Duca

This cyber opera was first performed on June 20, 1995, at the Brucknerhaus in Linz, on the occasion of Ars Electronica '95. The opera could not only be experienced locally, live in the hall and foyer of Brucknerhaus, but also nonlocally via internet. In the foyer the visitors had access to the Ars Electronica software *Wagner's Delusion* through two multimedia terminals. At the same time, video inserts of information web pages, set up for this event, were put on the terminals set up for the event, as well as excerpts of the online discussion via internet, and the event on the stage. The information path, which the users took from the software *Wagner's Delusion*, formed a part of the performance in the hall. Video projections from the CD-ROM *Wagner's Delusion* (served by the multimedia terminals in the foyer) were shown in the Brucknerhaus hall during the performance of the Wagner Wesendonck songs by Susan Dumas, together with projection of the web pages, as well as online discussions – a telematic electronic stage design. Information on Richard Wagner was made publicly accessible on the Worldwide Web for the whole runtime of Ars Electronica '95. For the duration of the performance via IRC (Internet Relay Chat) server, a worldwide online discussion on "Wagner's delusion" was developed and moderated by Peter Weibel.

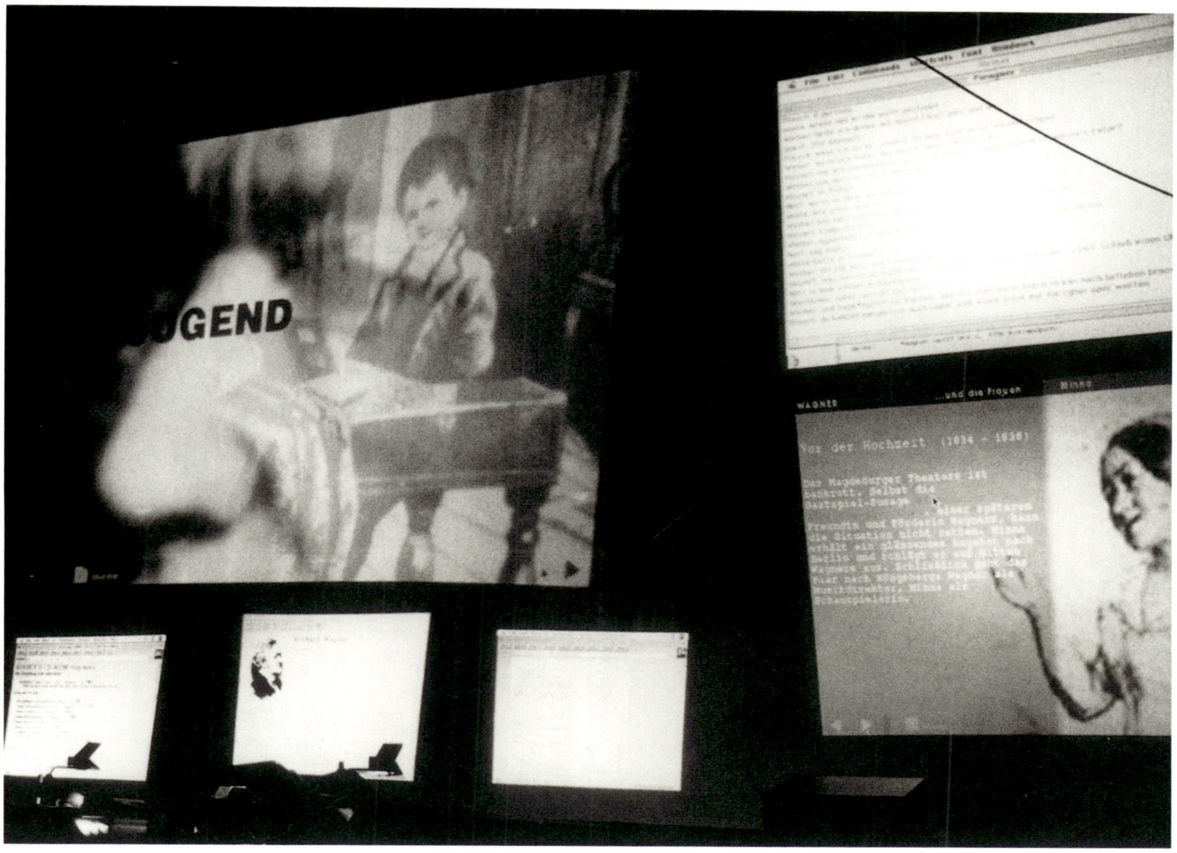
UGEND
Vor der Hochzeit

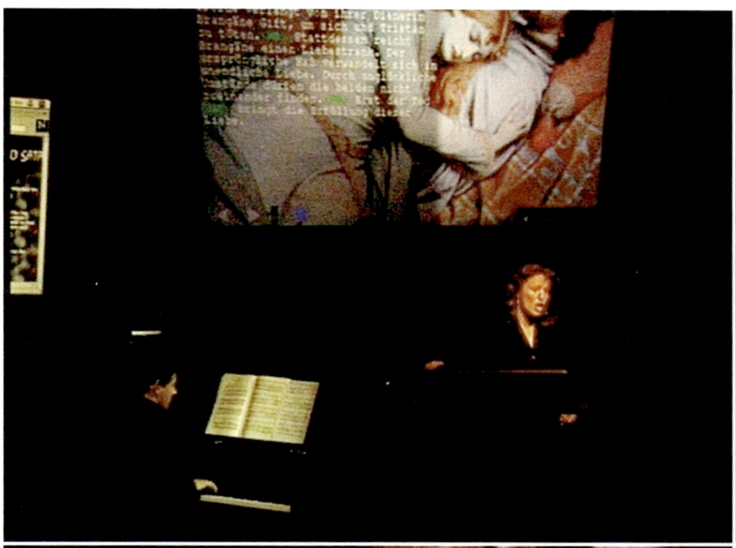

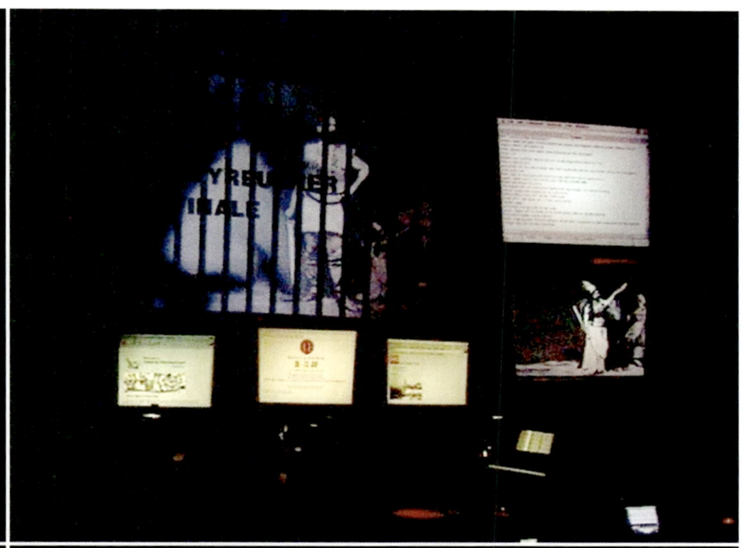

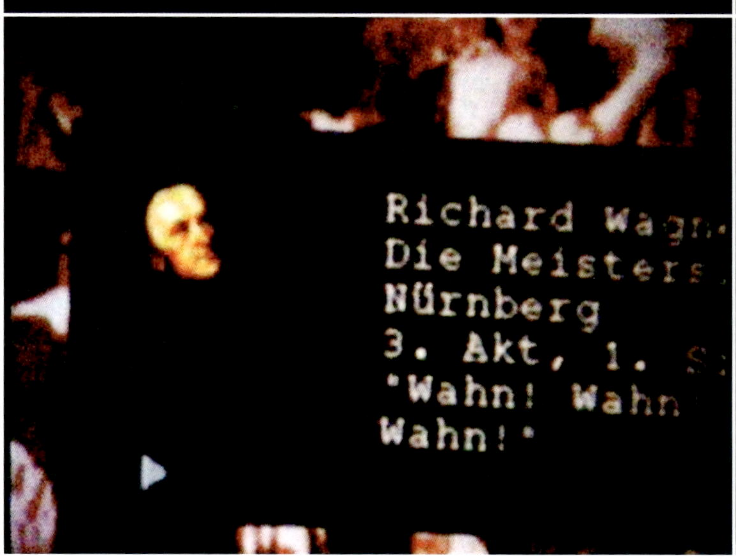
Richard Wagn
Die Meisters
Nürnberg
3. Akt, 1.
'Wahn! Wahn
Wahn!'

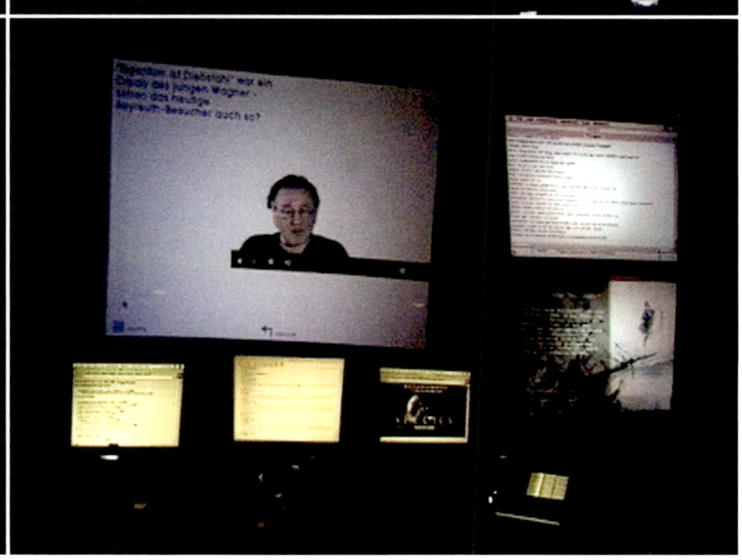

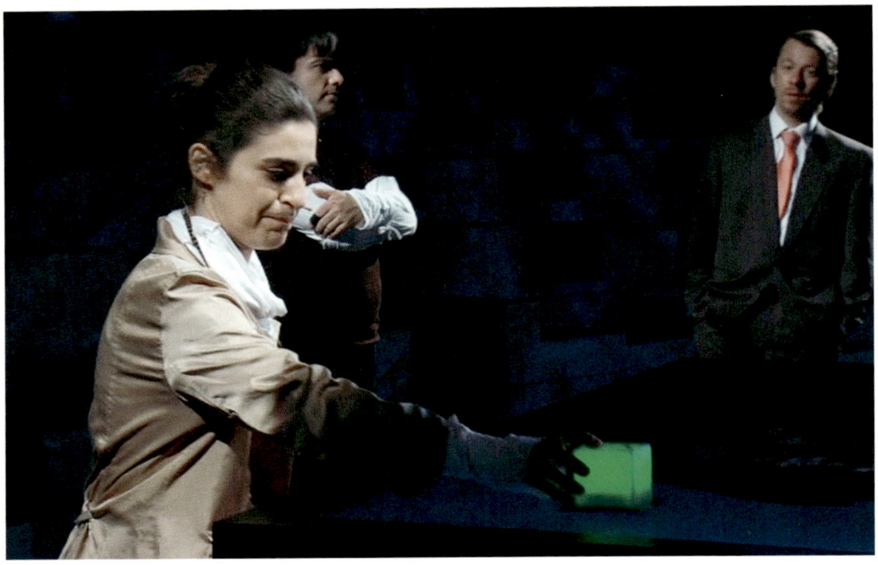
Scene of the *Amazon Conference*

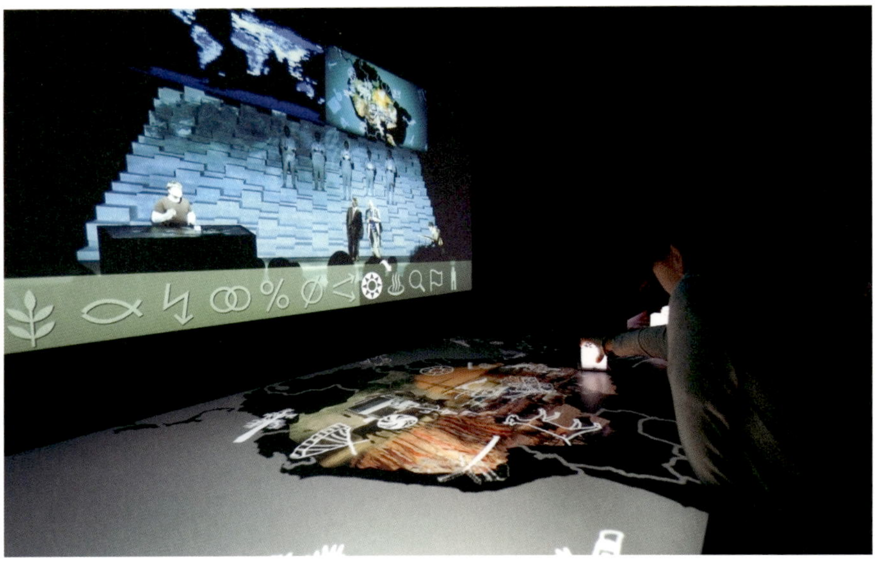
Presentation of the interactive elements

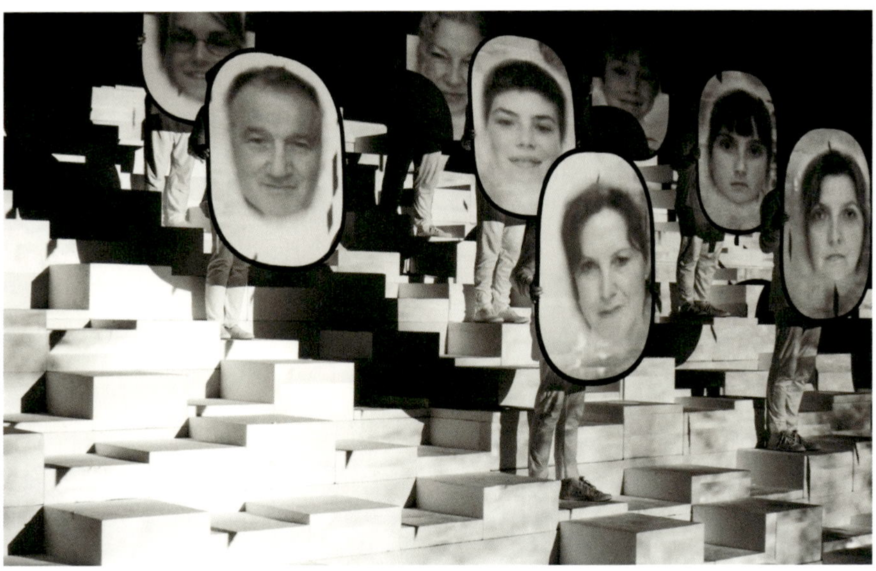
Rehearsal of holding up faces of members of the audience

Amazon Conference: Pending the Suitability of a Rational Method for Solving the Climate Problem

2010

Part III of *Amazon: Musical in Three Parts*
Peter Weibel and Ludger Brümmer
musical theater; ca. 150 min., debut performance at Reithalle München, Münchener Biennale 12, May 5, 2010

Views of the performance at Reithalle München, Münchener Biennale 12, May 5, 2010

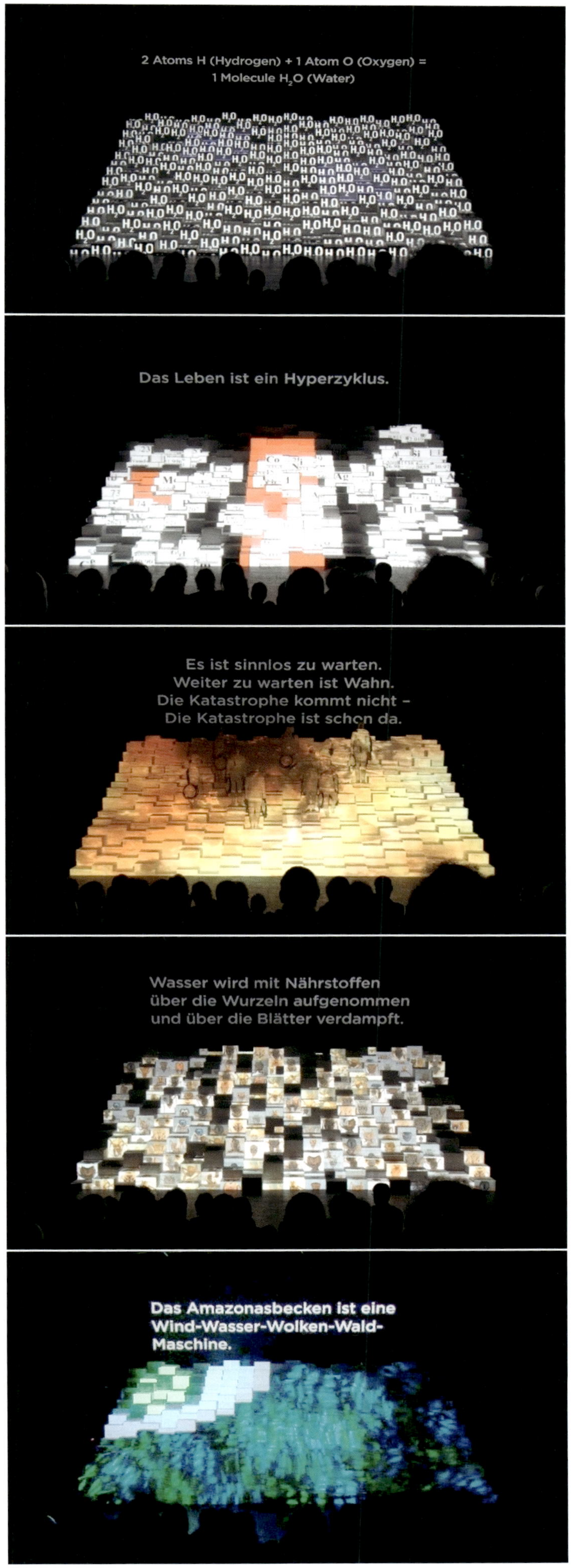

The musical theater piece *Amazon* was developed at the ZKM | Karlsruhe between 2006 and 2010 under the direction of Peter Weibel together with representatives of the Yanomami – one of South America's largest indigenous groups – , media artists, composers, sociologists, and anthropologists. With a focus on the rain forest in the Amazon region not only as a dwindling habitat for indigenous groups and the local flora and fauna, but also in its global importance for the atmosphere, the multimedia musical drama explored the Amazon's future using all the virtual and multimedia possibilities of modern musical theater.

The concept of multimedia musical theater is reconceived here on the basis of the computer's capabilities: musical theater and media art flow into one another, music is visualized, images and data become audible. For the first time, the performers sing mathematical and chemical formulas, which are transformed into light and image signals. This universal synesthesia is created with new interfaces. Contrary to the usual process of moving from libretto via music to staging, the development of music, text, and image is interdependent. The piece grows out of collective, interdisciplinary practice. Electronic sounds are musically counterposed with a vocal octet, language, and images. Using a "sound dome" of twenty-four loudspeakers disposed around the entire auditorium, the sounds unfold above the audience; the frontal stage melts away and sound takes on an independent dramaturgical role.
The Amazon Conference consists of three approximately fifty minutes long parts. The first part is a scientific explanation of the rain forest's ecological significance; namely, the water and carbon cycles – processes essential to life. These processes of life are simulated mathematically using John Conway's *Game of Life*: simple mathematical algorithms determine the growth, life, and death of cells and cell populations. The changing cell tissues are projected onto a three-dimensional grid shaped like an irregular staircase, with each building block of the staircase representing a cell or pixel. As the cell state changes, so does the electronically produced music. Mathematical laws of life's growth processes generate and control the images and sounds. The stage itself, as spatial set design, is likewise an output device for the computer. The second part is the actual conference. In the third and final part of the Amazon Conference, the stage fades away as apocalyptically as the Amazon itself. The audience has literally the final word. Pictures of the audience members' faces, taken when the audience entered the performance space, are projected onto oval screens that the singers hold out in front of them. Computer animations make the faces and lips speak while the singers lip-synch, bemoaning "market nihilism." In the process, we return to the initial question for opera: What is a choir? And we answer: The audience is the choir. The choir, in turn, represents the forest. If we kill the forest, we are thus effectively committing suicide.

—Peter Weibel

Translated from the German by Peter Dahm Robertson.

The 3D noise concert *The Origin of Noise – The Noise of the Origin* was premiered in 2013 by Peter Weibel in collaboration with the ZKM | Karlsruhe for the Donaufestival (Danube festival) Krèms Brûlée. A mixture of digital media art, new music, and performance allows the audience to experience the phenomenon of noise. This phenomenon has existed since the origin of the universe, the Big Bang. The world began with noise. We have recently found out that the noise of the TV screen consists partly of cosmic microwave background radiation. It is not Johannes Kepler's harmony of the spheres that makes up the music of the universe, it's the noise. During the concert, a text about the origin of the universe is typed, projected, and simultaneously translated into Morse and binary codes and sounds. The projections can be seen three-dimensionally by means of 3D glasses. A spatial composition and a special arrangement of the loudspeakers make it possible to trace the sounds' movements in space. Floods of data, numbers, and characters create an audible and visible noise concert.

Just like light, which has been supplying us with information for billions of years on its journey since the Big Bang, noise is not the end of the sound spectrum but its beginning. However, noise is not merely an acoustic phenomenon but a visual one as well, for example, on the TV screen. Noise is created in part during human observation. Noise, then, is an interactive phenomenon between the material realm and the human being, as well as between the human being and the world. Embedded in a universe of noise, the audience at *The Origin of Noise – The Noise of the Origin* experiences an expansion of consciousness at a new level. Stimulation rather than simulation. In the digital age, we are all merely particles in the data ocean, are data carriers and data hunters. We ourselves are data amidst the data noise.

Noise is one of the most complex and varied phenomena in existence. It contains a multiplicity of frequencies, figures, rhythms, and pitches. For example, when slowed down extremely, apparently uniform noise turns out to be a sequence of different notes and rhythms. There are also different types of noise, such as the "pink noise" or "white noise" known in signal processing, which are distinguished by different tone colors. Digital noise contains an especially large amount of high frequencies.

In the composition of *The Origin of Noise – The Noise of the Origin*, chaos and order define the material range of sound. Physical models were used to create the resonance of bowed strings; rudiments of voices can be heard as well. Virtual pendulums enabled the creation of repetitions and highly rhythmical material, and lastly, a connection between chaotic and orderly materials was created with the aid of cross-synthesis. Chaotic material became ordered material when extremely slowed down, and ordered material was developed into a rhythmized, chaotic sonority.

In addition, a stereoscopic projection floats above the stage. The audience wears passive polarization glasses and can thus perceive the three-dimensional visual scenes. Using software developed at the ZKM | Institute for Visual Media, these scenes are computed from real-time data, such as Peter Weibel's typing on a type writer. The sound of the instruments which were developed specially for this performance is likewise visually interpreted and transformed into three-dimensional images. For example, the acoustic pressure of a sound modulates the surface of changing geometric bodies. Over the course of the performance, visual samples known as particles create a three-dimensional Lissajous curve representing the mathematical equivalent of the superimposition of harmonic sounds. The feedback loop in *SynSeeThis* (see p. 368 in this volume), an iPad app that acoustically interprets visual signals received live on stage through the iPad's built-in camera, points to the fact that sound and image are merely different aspects of data transformations expressed as visual or sonic events depending on the perspective.

Translated from the German by Peter Dahm Robertson.

The Origin of Noise – The Noise of the Origin. 3D Noise Concert

2013	Peter Weibel, Ludger Brümmer, and Bernd Lintermann concert; ca. 63 min., debut performance at Donaufestival Krèms Brûlée, Krems, April 25, 2013 idea and performance: Peter Weibel; composition: Ludger Brümmer; 3D image composition: Bernd Lintermann, Nikolaus Völzow; music, audiovisual design; implementation: Götz Dipper, Hartmut Bruckner, Manfred Hauffen; lighting: Manuel Weber, Benjamin Schild; body performance, dance: Akemi Takeya

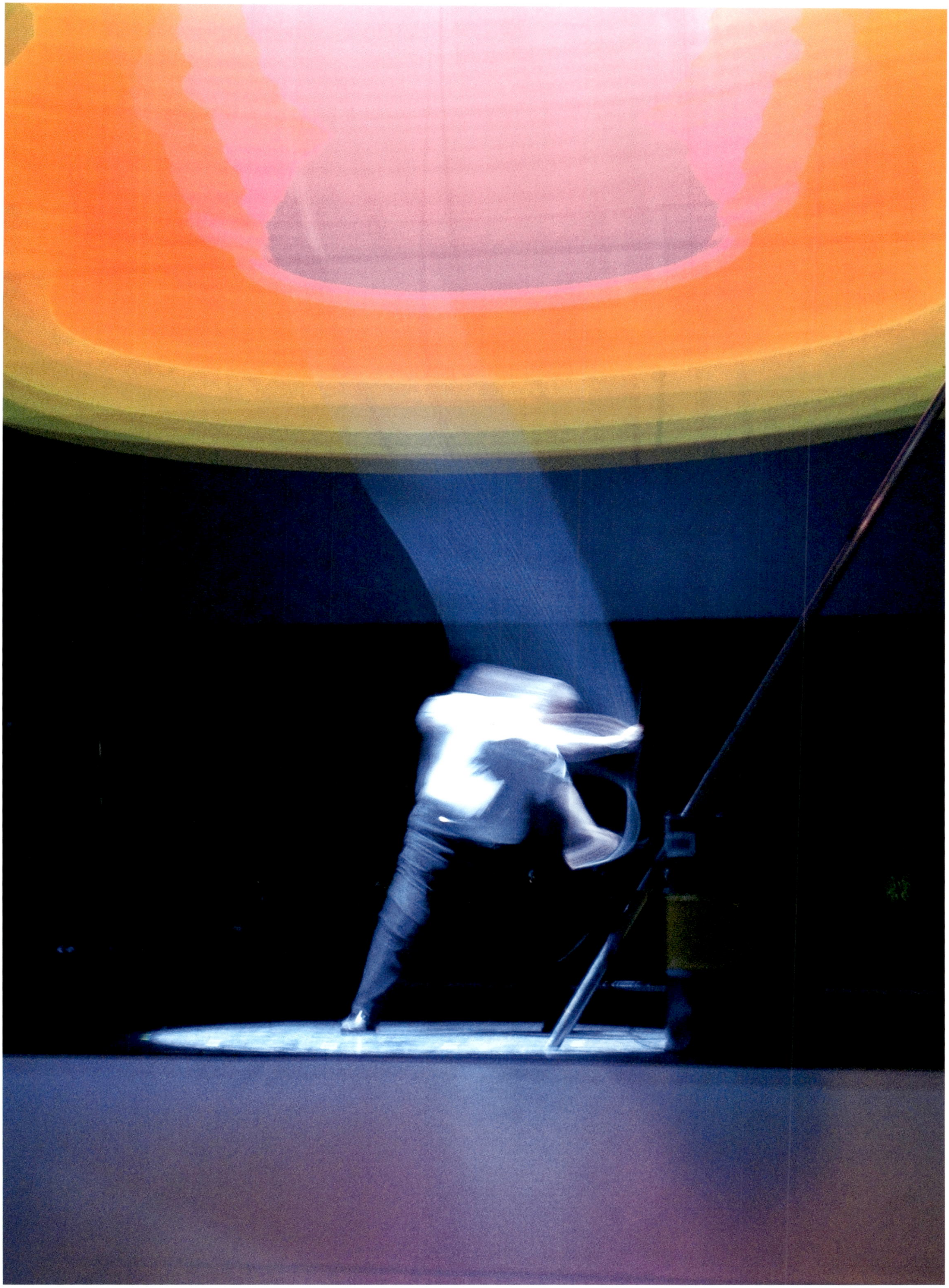

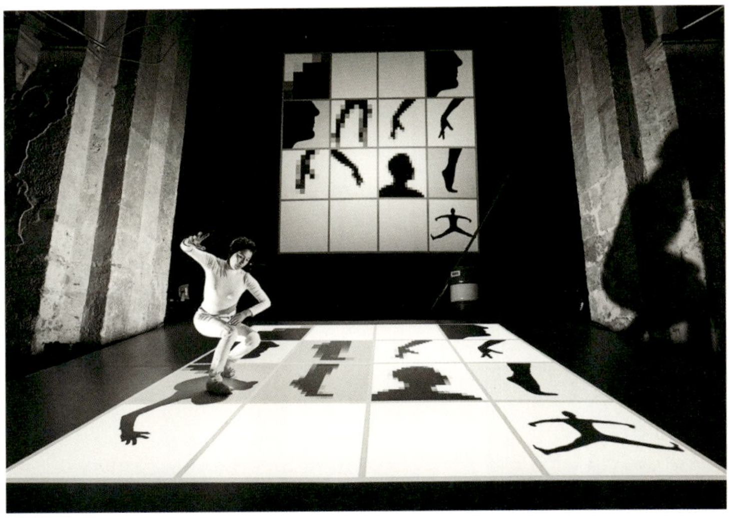

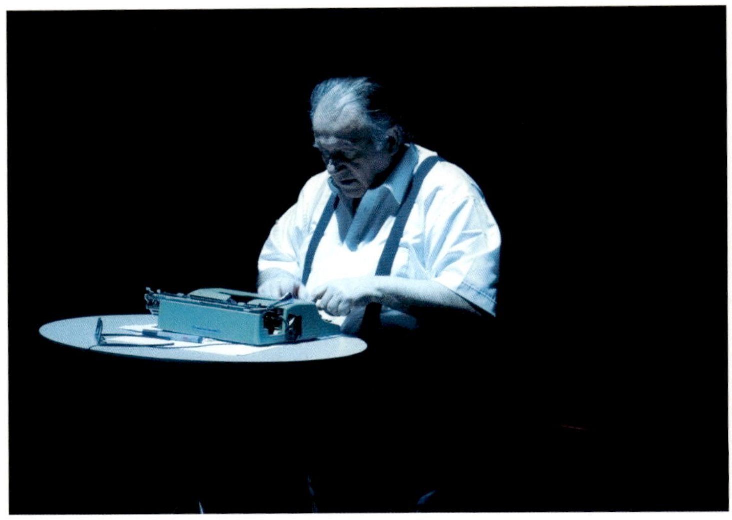

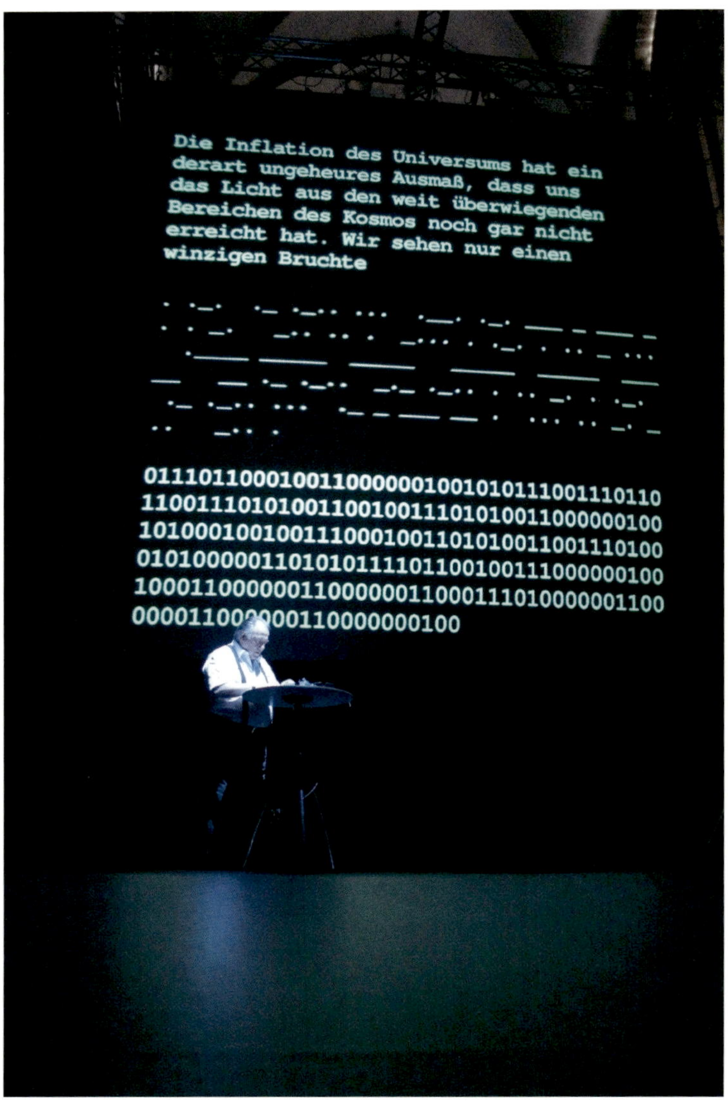
Die Inflation des Universums hat ein
derart ungeheures Ausmaß, dass uns
das Licht aus den weit überwiegenden
Bereichen des Kosmos noch gar nicht
erreicht hat. Wir sehen nur einen
winzigen Bruchte

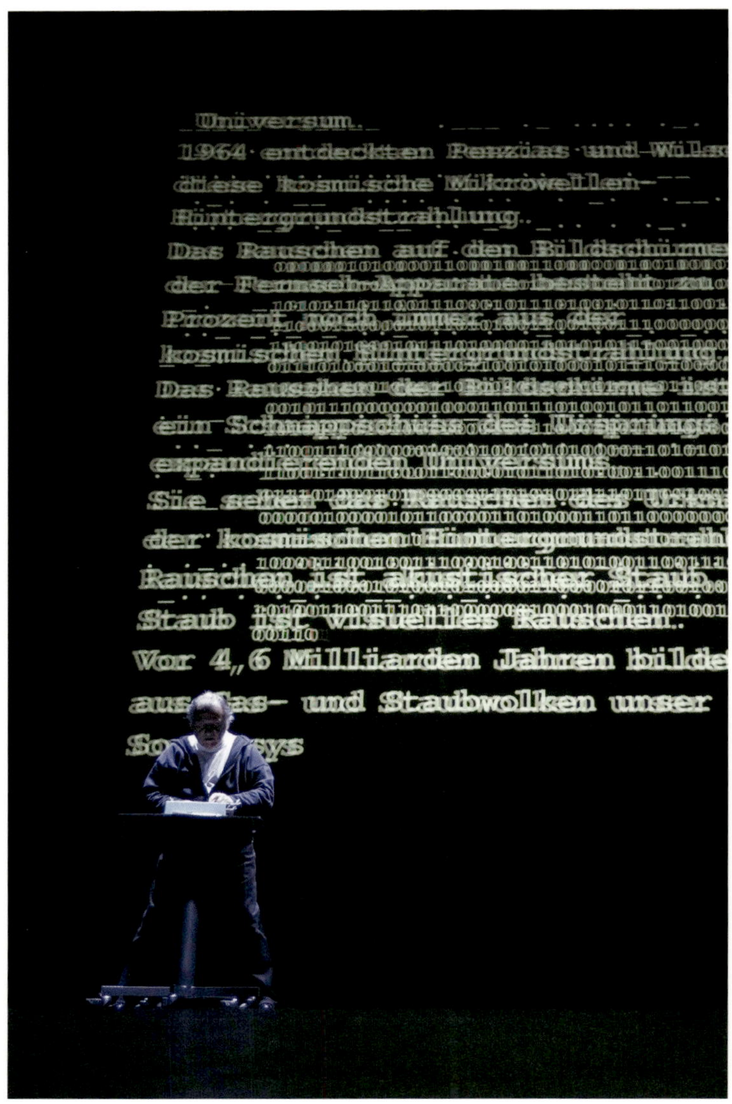
Universum.
und Staubwolken unser

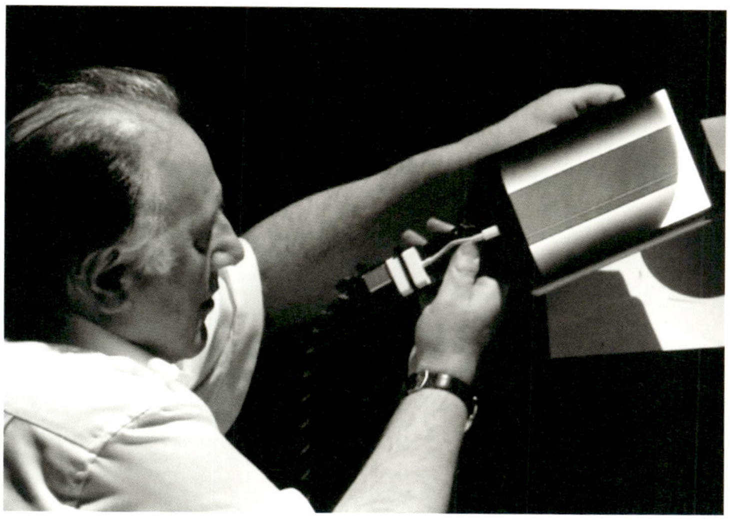

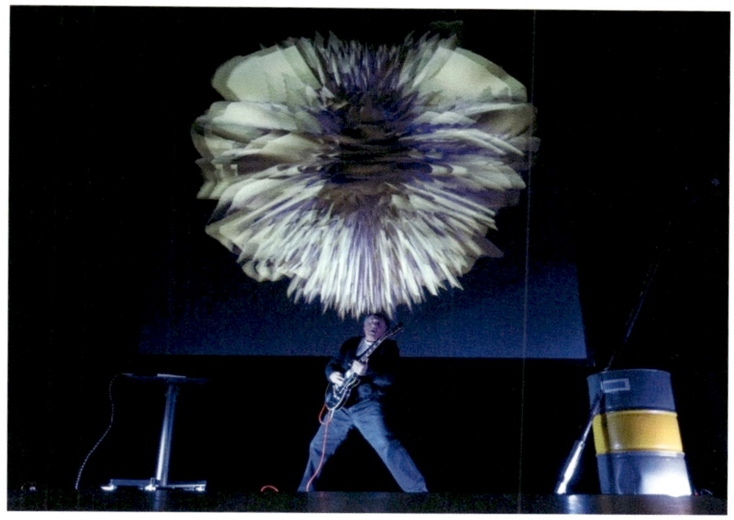

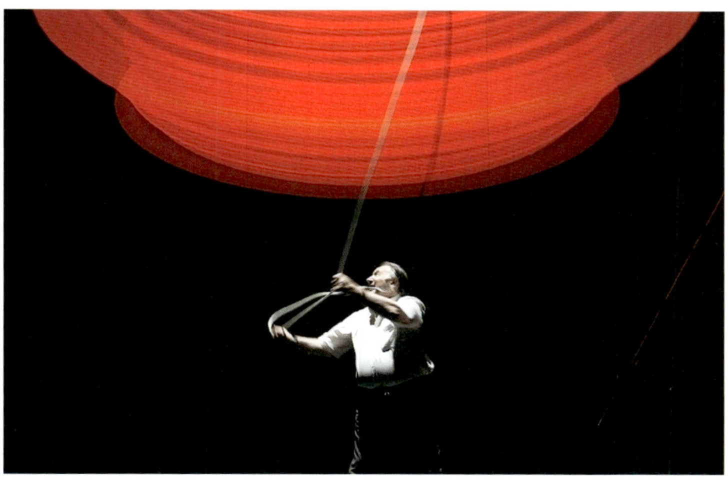

Views of the performances in Krems and Karlsruhe, 2013

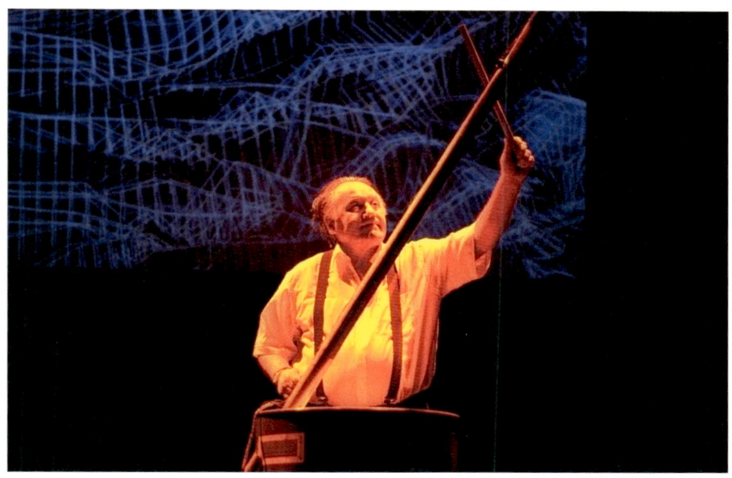

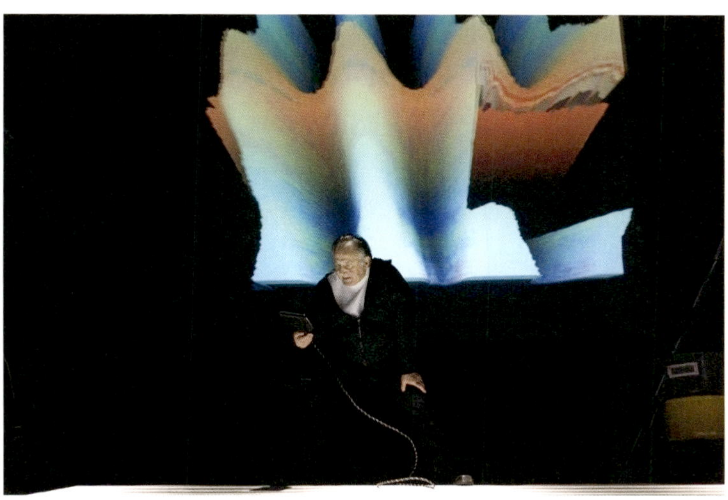

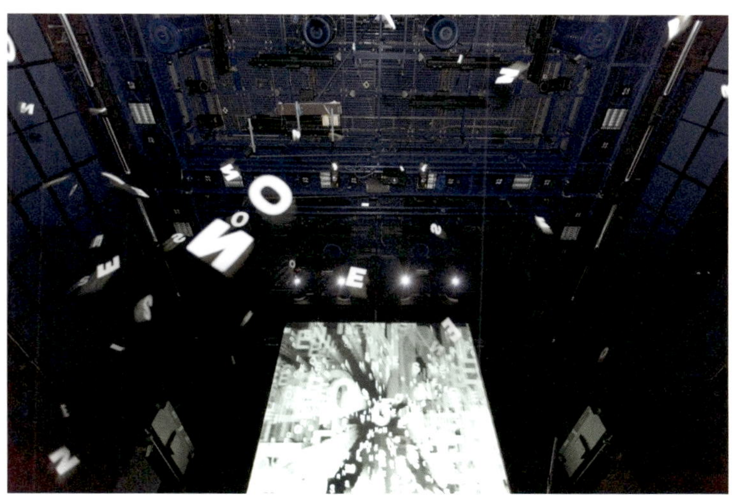

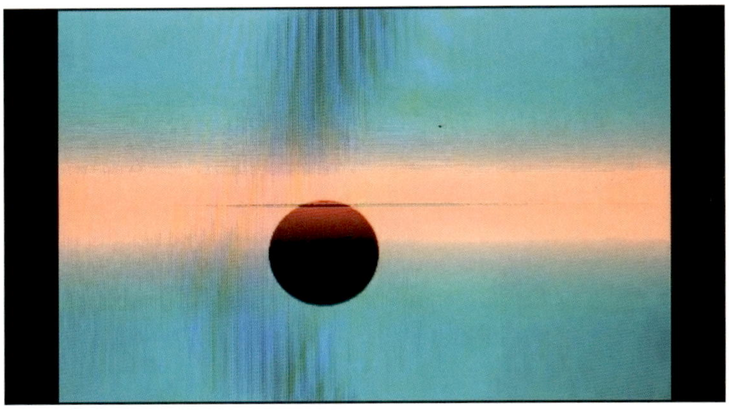

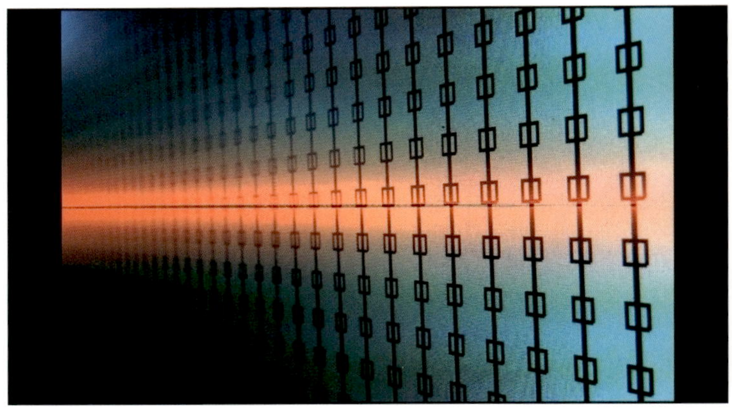

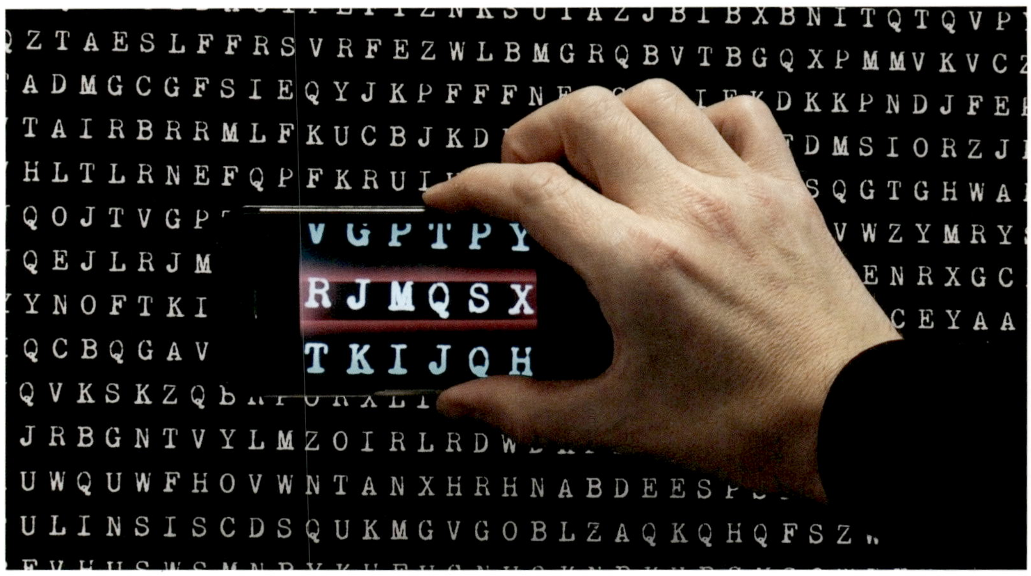

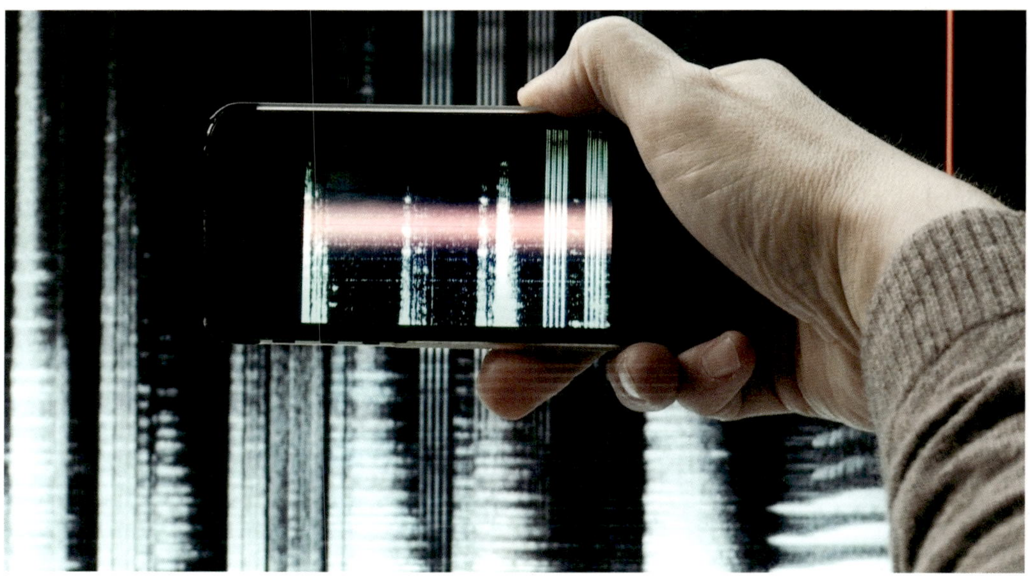

SynSeeThis

2013

Peter Weibel and Bernd Lintermann
sound installation; iOS app, iPad, book with patterns, dimensions variable
concept, software: Bernd Lintermann; technical support, sound: Manfred Hauffen

SynSeeThis is an app for iOS, which was created for the mus performance *The Origin of Noise – The Noise of the Origin* the Donaufestival (Danube festival) 2013 in Krems, Austr The app is a visual musical instrument, which generates feedback loop of images and sound. For the installation, iPad is fixed to a mount. When a page of the book lying in t installation is placed in the visual range of the iPad's built- camera, the recorded image is transformed into sounds. T pages of the book exhibit various visual patterns, which generate specific sounds.

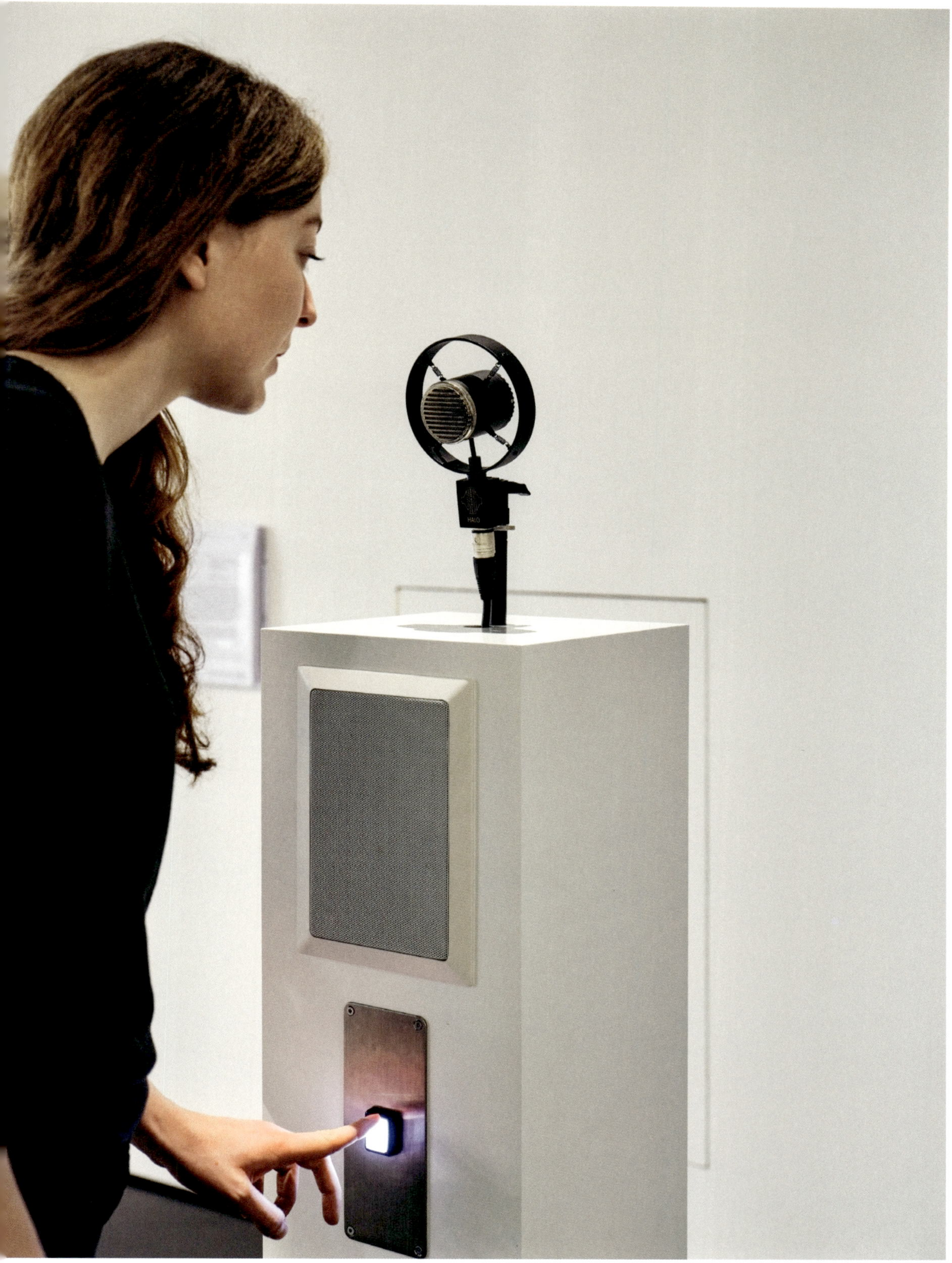

Installation view
respectively, Peter Weibel,
ZKM | Karlsruhe, 2019–20

oh cɐt ɐcho

2012	interactive sound installation; microphone, loud speakers, computer, 154 × 25 × 25 cm software development: Götz Dipper; artistic scientific assistance: Barbara Nerness

When someone speaks into the microphone of the sound installation *oh cet echo*, the speaker's voice is simultaneously simulated by a musical instrument (piano, marimba, vibraphone, or flute, each using short or longer tones). It is possible to choose or switch respectively between the instruments using a push-button, and also select a short recording of previous acoustic contributions (with a time delay of one second). Over time, the recorded words thus create a collective musical composition. Music in the age of social media, which are resonance spaces and echo spaces.

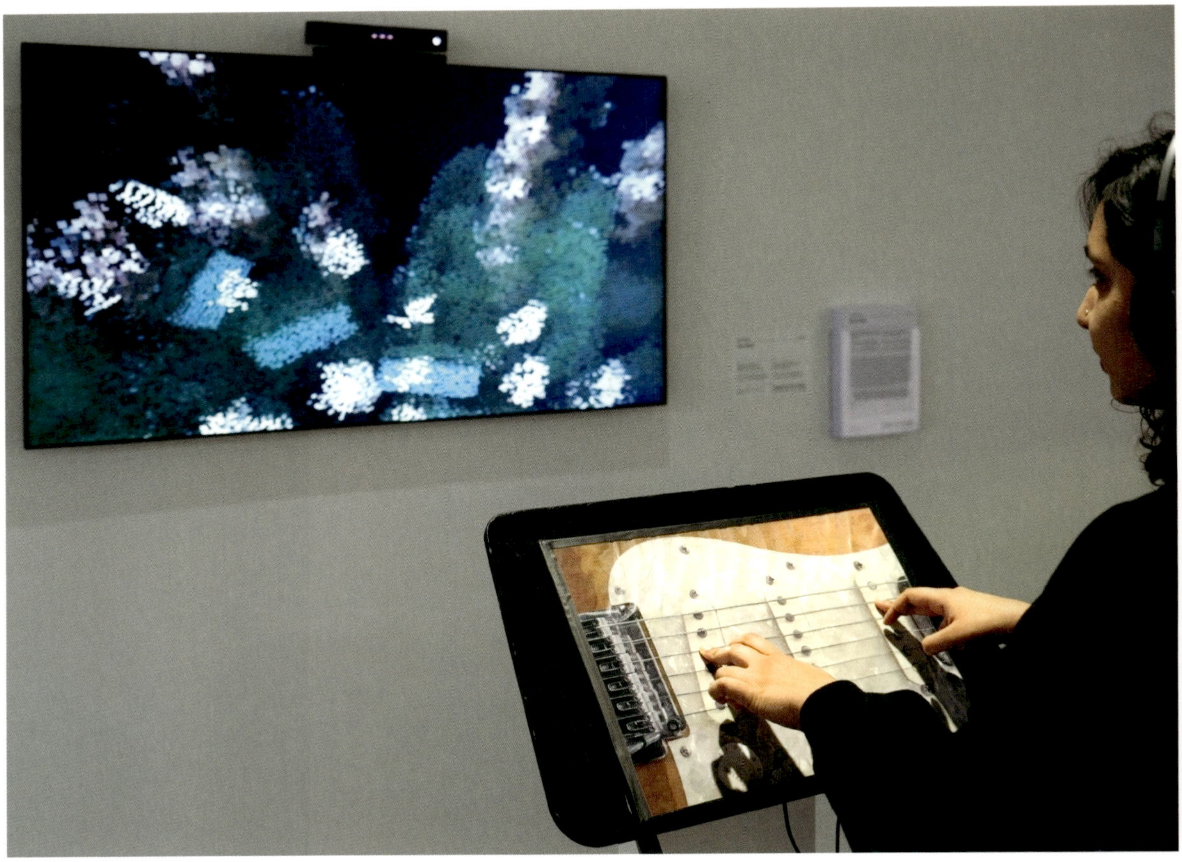

Installation view *respectively, Peter Weibel*, ZKM | Karlsruhe, 2019–20

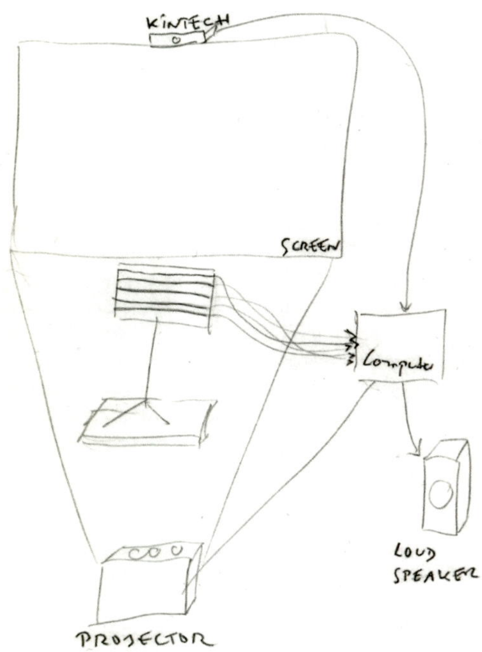

Conceptual drawing for *Data Music*, 2016

Data Music

2016	interactive sound installation; music stand, photograph, strings, screen, computer, loudspeakers software development: Nikolaus Völzow; sound and interface design: Manfred Hauffen; commissioned by DIGITAL SYNESTHESIA artistic research project by Katharina Gsöllpointner, Ruth Schnell, Romana Schuler

The installation *Data Music* consists of a music stand with a photo of a guitar on it. The strings of the guitar, though, are real and connected to a computer on which several samples are stored. The viewers are invited to pluck the strings, which are both the notation and the instrument. The viewers press their fingers on different positions of the real strings which trigger different guitar sounds. As the music is stored physically as data in the computer, music is performed via an interface, which is programmed by an algorithm.
In addition, the performance is recorded by a Kinect camera and displayed on the screen on the wall. As the images of the performance are data, the configuration of the visual data can easily be influenced by the configuration of the acoustic data. Sound and image are synchronized with each other. Thus, when someone is in front of the music stand, they are seen displayed on the screen. The moment someone starts performing with the music stand, the image distorts into a flow of visualized acoustic data. In the digital era, each musical instrument – connected to a computer – can play the sound of each instrument. Each object – connected with a computer – can play the sound of each object.

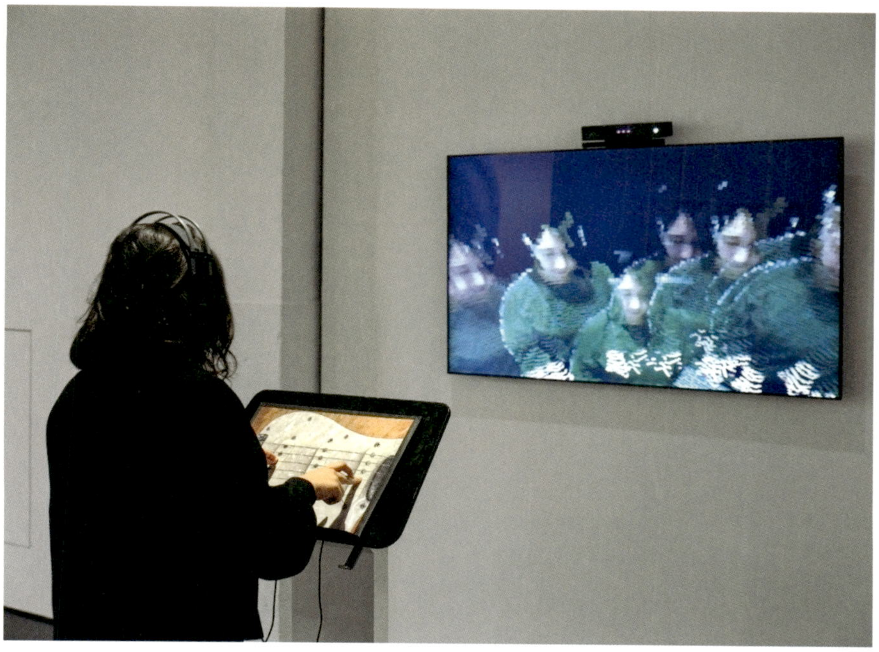

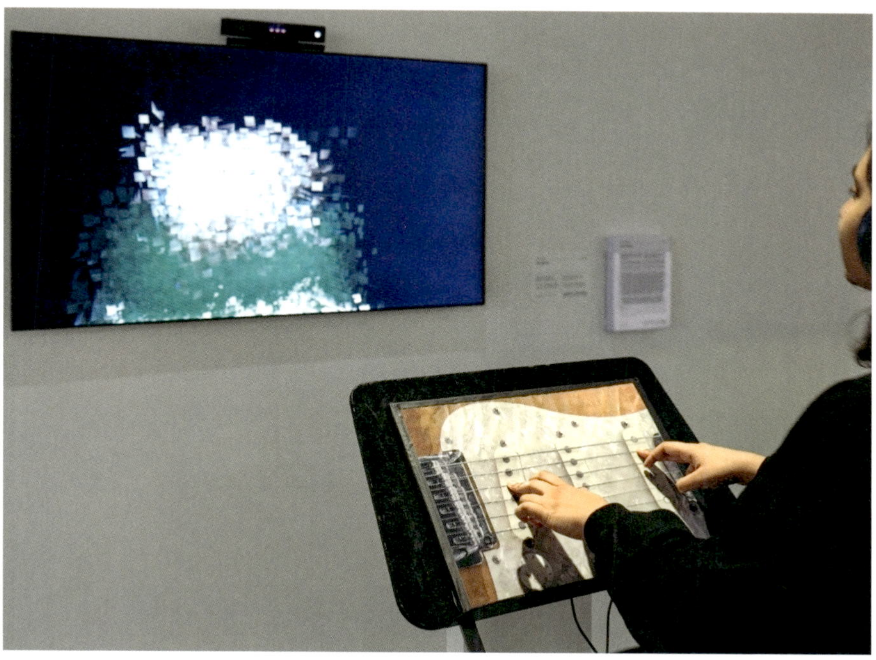

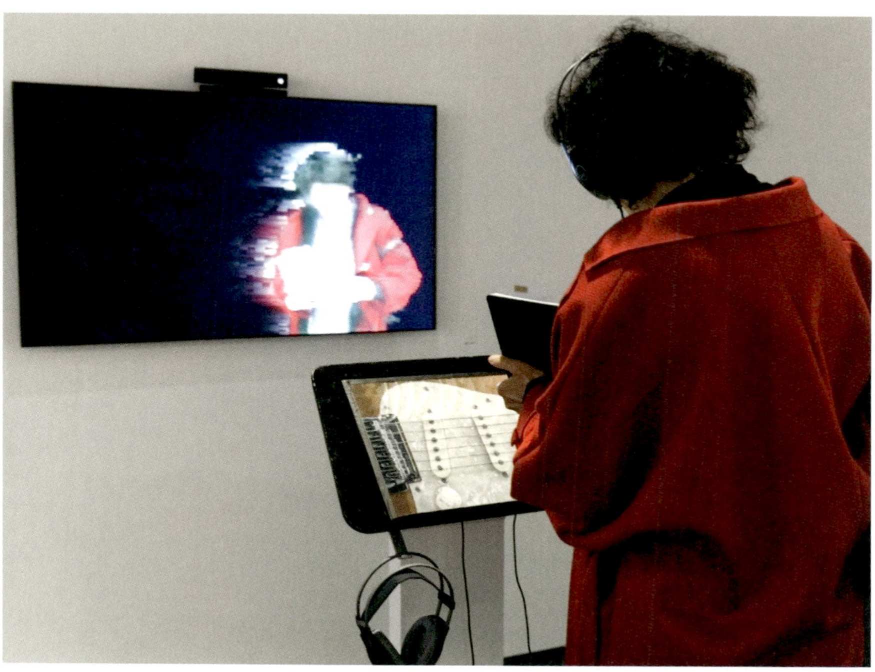

Installation views
respectively, Peter Weibel,
ZKM | Karlsruhe, 2019–20

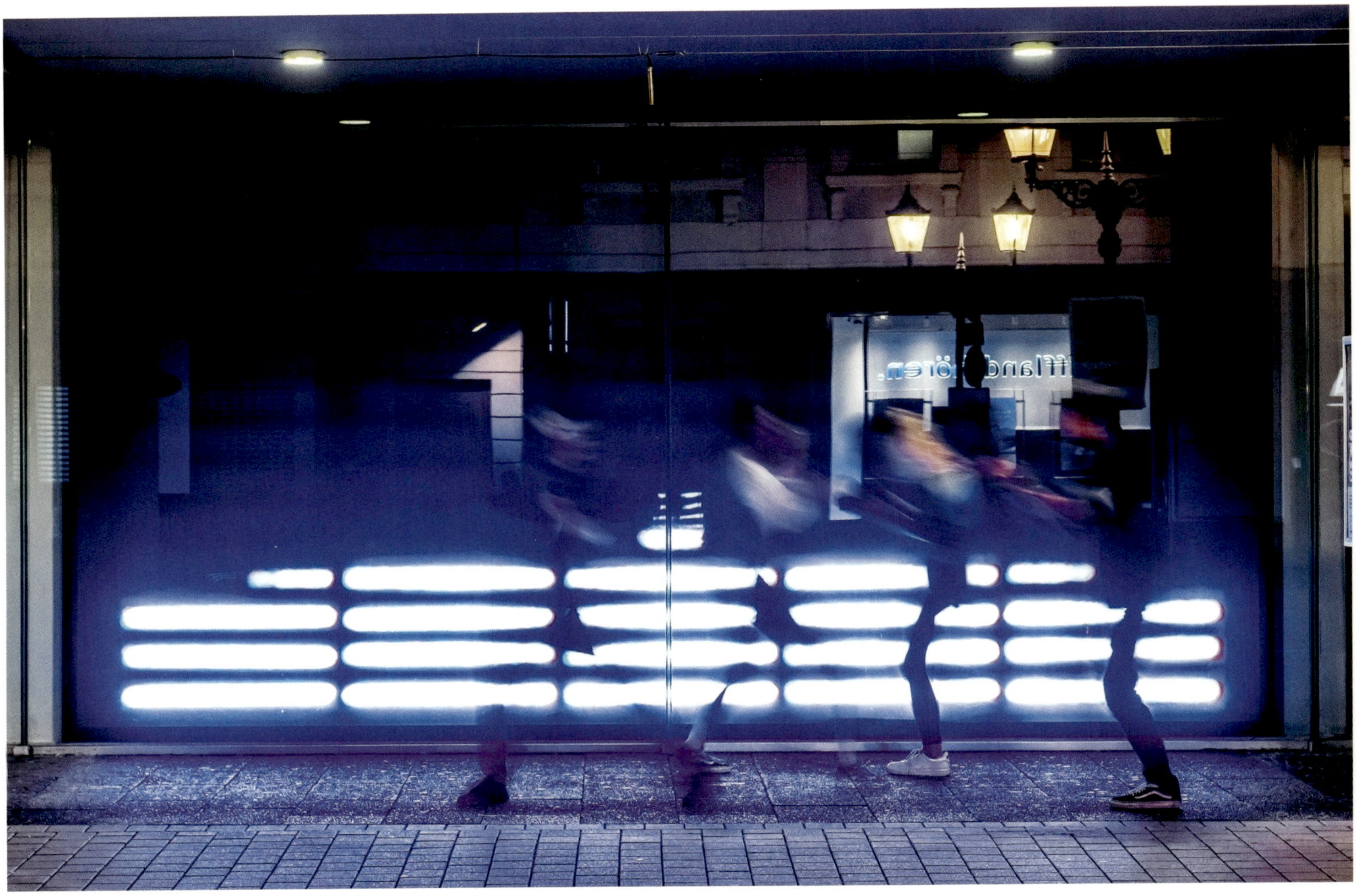

Installation view *Seasons of Media Arts*, City Center of Karlsruhe, 2021

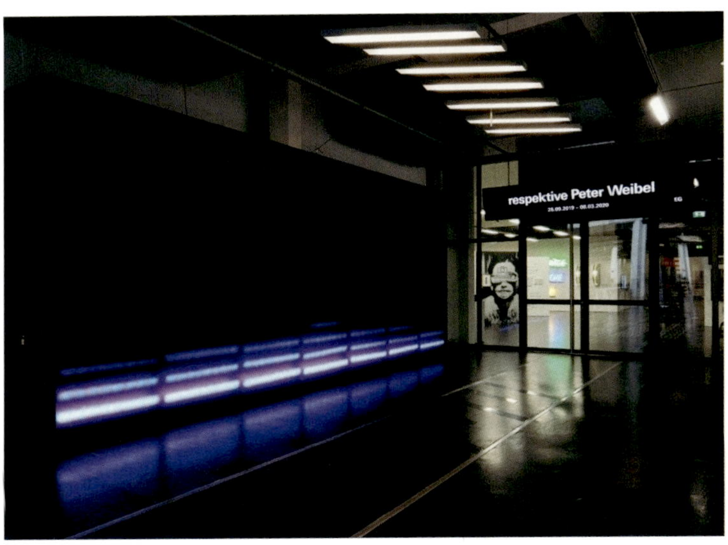

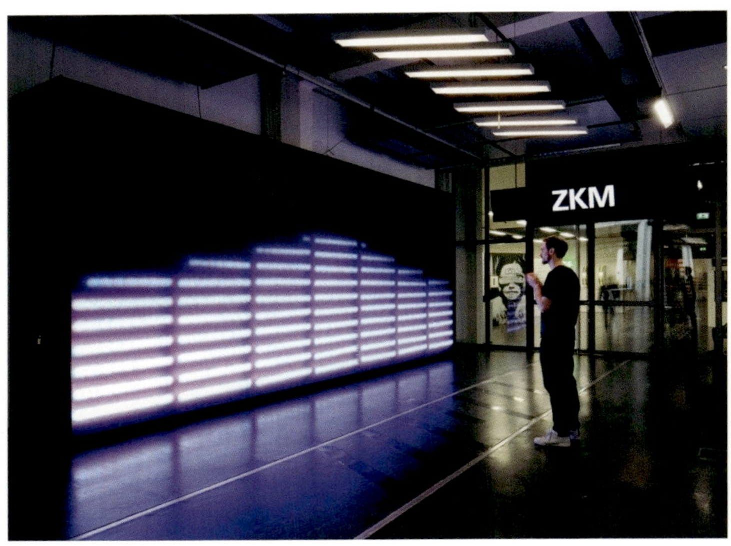

Installation views *respectively, Peter Weibel*, ZKM | Karlsruhe, 2019–20

Sound Channel / Klangkanal

2012	interactive sound installation; LED lights, screen, microphone, 360 × 700 × 100 cm technical concept: Manuel Weber; software development: Nikolaus Völzow

A wall of neon light tubes reacts to sound. The higher the volume the more the tubes start to light up. The sound decibels generate the brightness of the light. Interaction of sound and light and of the public and the artwork. A clap of hands and there will be light.

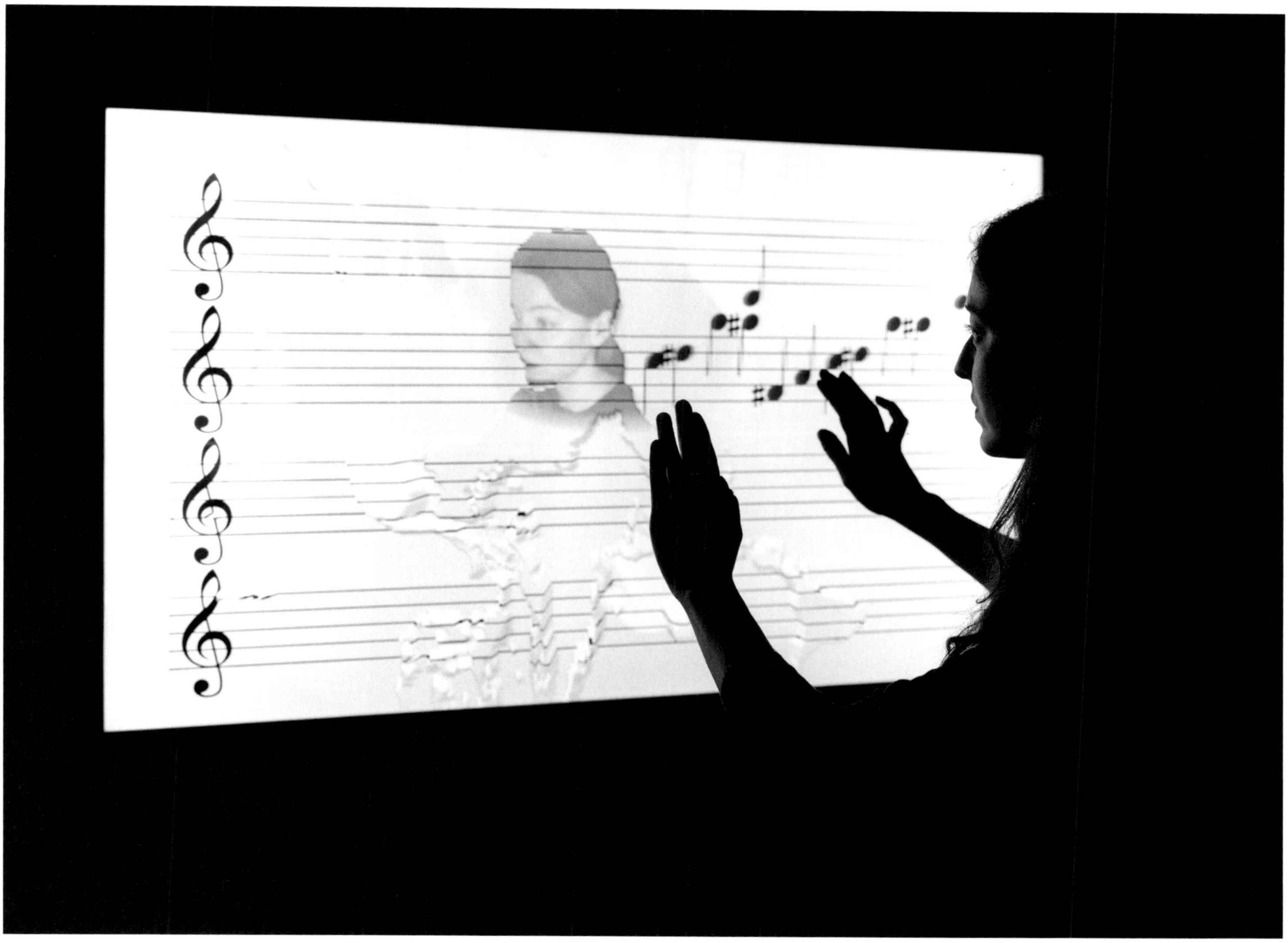

Installation view
respectively, *Peter Weibel*,
ZKM | Karlsruhe, 2019–20

Music Passenger

2019 Peter Weibel and Bernd Lintermann
interactive sound installation;
computer, 3D camera, self-developed software, screen, dimensions variable
production: ZKM | Hertz-Lab

In the interactive installation *Music Passenger*, viewers transform music as they pass by. Compositions can change into completely new pieces in the twinkling of an eye, according to how the viewers interact with the work. The notes of piano pieces by old masters run on staves across a large screen and are simultaneously translated into corresponding sounds. At the same time, a 3D camera registers the viewer's body movements and gestures, which are pressed into the virtual music paper as depth images and trigger the touch of the piano keys. The passing notes are only audible when the body of the observer virtually "touches" the notes. The passing notes and the passengers have to meet to create music. The body image of the observer becomes the sound image. The body image creates the sound body. The visual recording of the movement of the body, and the gestures are the conductor, the composer, the score, the instruments, and the performer at once. The bodies of the "passengers" themselves become a musical instrument – or even a learning algorithm that creates new pieces of music from existing sounds. The installation was exhibited in 2019 as part of the Seasons of Media Arts festival at Musikhaus Schlaile in Karlsruhe.

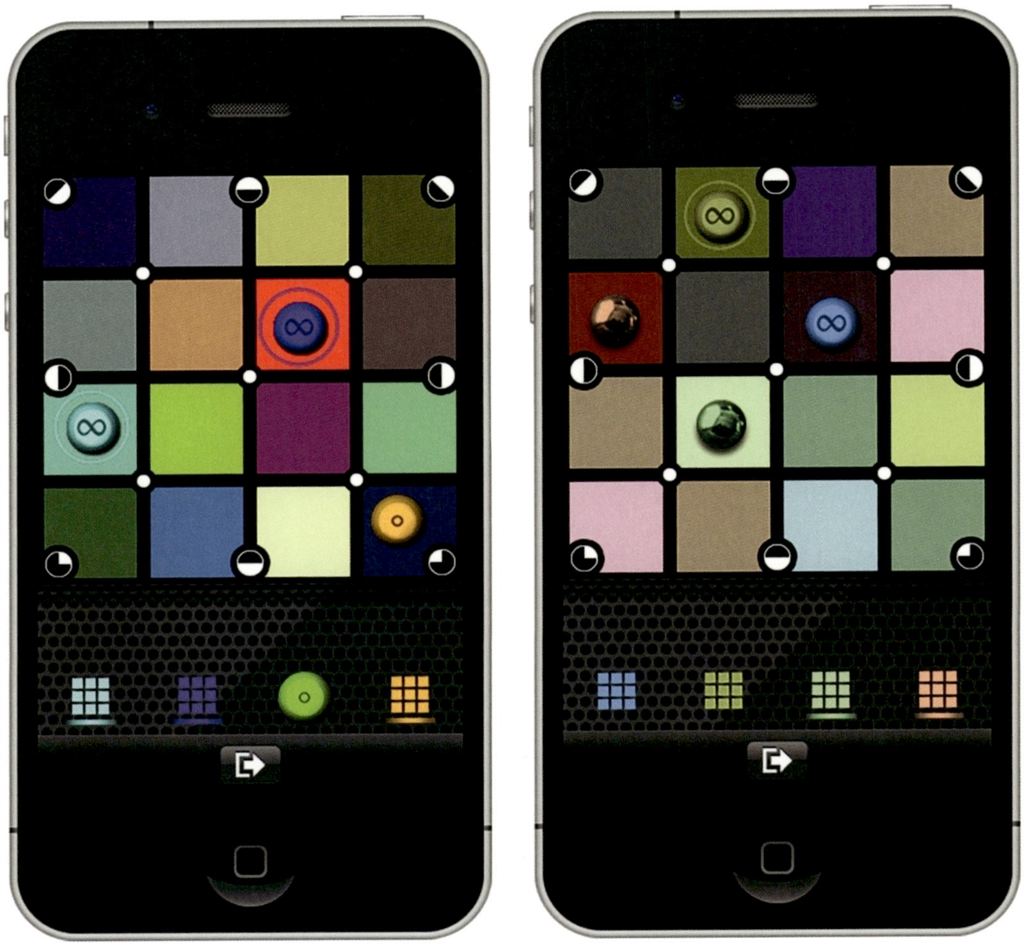

In *Music Board* by Peter Weibel and Jens Barth, a smartphone display shows a virtual ball rolling across a grid field. Depending on how the cell phone is held, angled, or turned, the virtual ball rolls across the screen triggering stored sounds. The sounds themselves are also variable, since the fields, like quaternions, can be rotated in space, while other sound sources can be accessed as well using an image panel. The virtual movement of the ball in space creates the score, which in turn creates the music. Spatial proximities bridge the gap between two notes. By the simple application of a Clifford algebra, tones can be interpreted as points in space. Thus, the tonal material of the score can be transformed by rotation on the horizontal and/or vertical axes, exploring the compositional possibilities of topological transformation.

"The composition of music based on systems of cells enables complex and variegated musical structures. The principle can be demonstrated using a two-dimensional game board whose cells are arranged in a regular order. The cells form a system of neighborhoods (4 × 4) which can be categorized into subgroups (2 × 2). Neighborhoods can be manipulated by rotating and/or mirroring the fields along symmetrical axes. If the individual cells are underlaid with musical notes or samples, a kind of topological score is created. By placing virtual 'game pieces' on the 'board,' that is, the screen, one produces the sounds of individual cells in a rhythmic order, creating melodies. Distinguishing between sampler objects (which play a sample once or in a loop), sequencer objects (which follow neighborhoods along ordered chains by using an internal pulse), and melody objects (which change their spatial orientation through physical movement of the device) leads to expanded possibilities for navigating within the cellular system. This combination of the temporal domain (different game pieces) with the spatial domain (game board/screen) allows the user to playfully connect the basic tonal material, forming complex musical structures and thus producing ever new compositions. With *Music Board*, users can define and continually change sounds and sequences, the temporal act of music, by means of spatial manipulation. Even without expert knowledge, one can use the app to work intuitively with generative processes and thus interact with the structural techniques of modern music. *Music Board* can be understood as a controller whose potential for developing generative compositions is fully explored only through further 'open' – i.e., user-accessible – parameters such as selection and makeup of samples, definition of tempo and meter, as well as adaptation of the objects."[1]

Music Board

2011 | Peter Weibel and Jens Barth
app
software: Jens Barth

[1] Jens Barth, 2016, quoted in: Peter Weibel, "Zellulare und molekulare Musik – Zur Kluft zwischen zwei Tönen: Entwurf einer Topologie der Musik," in *Enzyklopädie der Medien*, vol. 2, *Musik und Medien* (Berlin: Hatje Cantz, 2016), 383–416. Translated from the German by Peter Dahm Robertson.

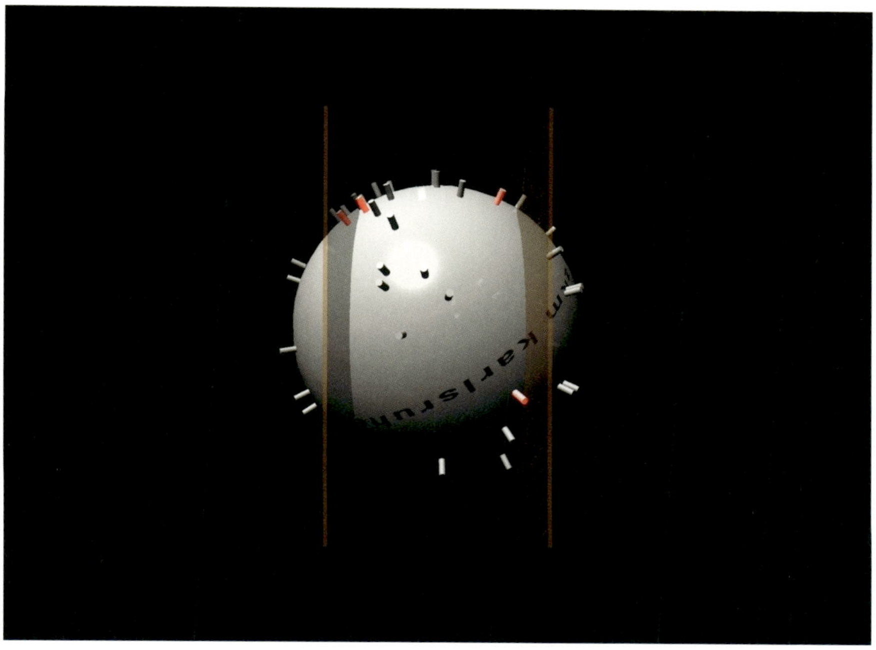

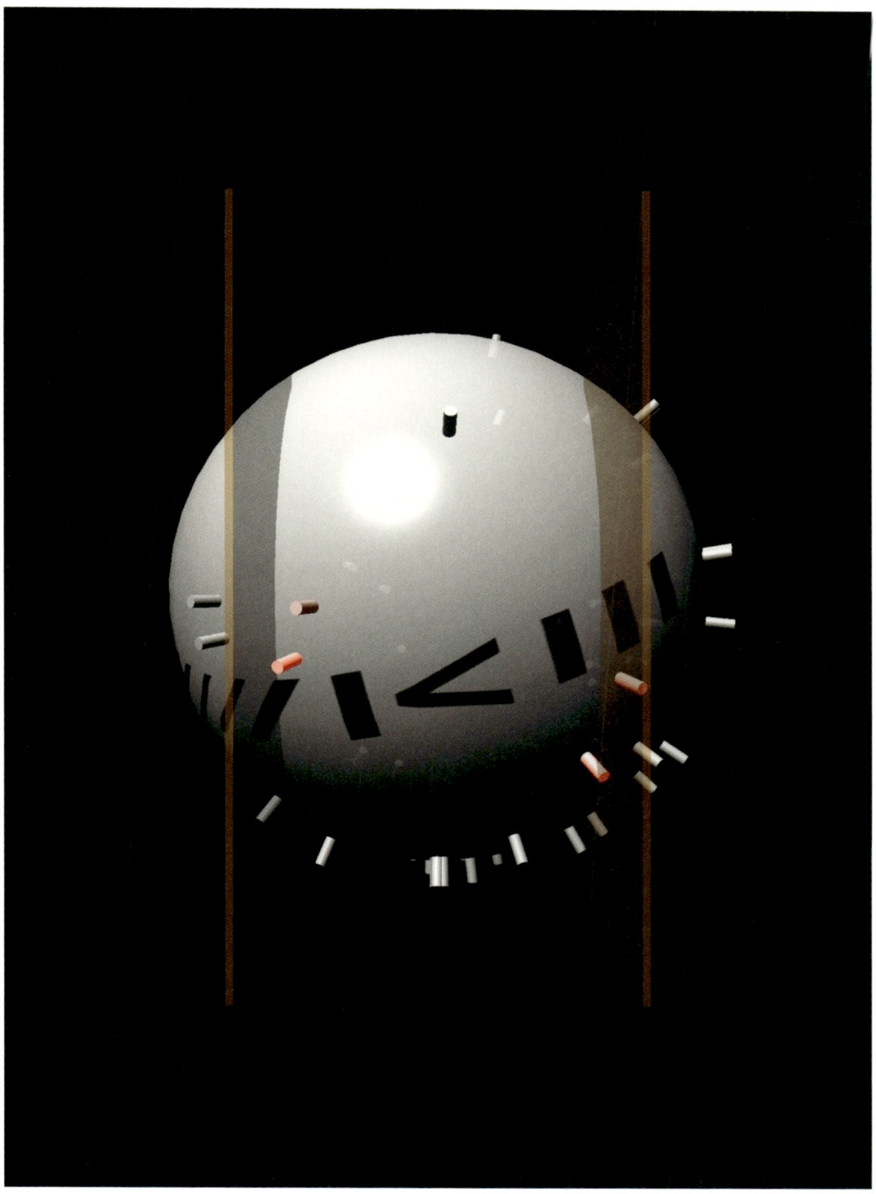

Sound Writer

2016 | Peter Weibel and Chikashi Miyama
app
software: Chikashi Miyama

The application *Sound Writer* is a sequel to *Music Board* (2011). It was programmed by Chikashi Miyama, guest artist at the ZKM | Karlsruhe. In this case it is not the incline of a grid/field of cells that is affected by changes in the spatial position of the smartphone. Rather, it is the movement of the smartphone that causes a ball to rotate in space. The ball is covered with several movable pins, similar to the letters on the typeball of some electric typewriters. By touching the display, users can also move planar optical barriers, called "sonic combs." There are therefore several spatial variables: the movement of the ball, the position of the pins, and the orientation of the sonic combs. These two- and three-dimensional movements create the tone sequences. Software written by Miyama brings the movements together to create sounds: (1) Scene Kit, an Apple framework for 3D applications and games; (2) Pure Data, a visual open-source programming language for sound development and synthesis, originally developed by Miller Puckette; (3) libPd, a C library allowing programs written in Pure Data to be embedded in apps. Scene Kit is used to visualize the ball, the combs, and the pins on the surface of the ball. The software also recognizes when a pin meets either of the combs. As soon as Scene Kit has identified such a collision, it sends the user optical feedback by changing the color of the pin while it is in contact with the comb. Simultaneously, using the implemented Pure Data program, this contact is sonified with a pitched tone. Through spatial, topological parameters, *Sound Writer* thus generates ever new tone sequences.

In both works – *Music Board* as well as *Sound Writer* – the position of the smartphone in space determines the creation of the series of tones and the music. When a pianist plays the piano, it is irrelevant whether the instrument is placed on the floor, suspended from the ceiling, or bolted to the wall. The music does not change in any way; the score is played as usual. The music remains independent of the position of the piano in space. By contrast, in both apps it is the position of the instrument – that is, the mobile device – in space that creates the score. In other words, the series of notes changes with the spatial position of the instrument. The space not the time dictates the composition of the music.The graphical user interface is no longer touched, but becomes the score, the sheet music which through its position in space creates the music itself. Not the movement of a hand on an instrument or on a two-dimensional user interface, but rather the movements of the instrument through space create both the score and the sound. There is no longer any difference between score, that is, notation, and the sound event. Notation and sound occur in real time. The notation, the instrument, and the sound event become one. —Peter Weibel, 2016

Translated from the German by Peter Dahm Robertson.

6. Space

Articulations of Space

Peter Weibel

A *Bildhauer* was once said to be an artist who uses tools to carve his sculptures out of hard material, while *Plastiker* are artists who produce their sculptures from soft, wet materials. The *Bildhauer* works from the outside in, the *Plastiker* from the inside out.

Although we do indeed work with hard materials, such as wood, slide projectors, etc., as well as soft, wet materials, such as photographs, water, and air, and although we also work from the outside in and vice versa, nevertheless, most viewers will not regard these creations as sculptures. However, I maintain that we really are dealing with sculptural works here, and with *Bildhauer* and/or *Plastiker*, albeit of a new sort. Whether bound to the figure, an object, or a three-dimensional shape (e.g., the cube), sculpture has for centuries been slumbering aboard a ship sailing in this direction, following the faint trace of the object, under a flag defining space as volume. Has space – light, atmosphere, place – revealed itself under these circumstances? The classical view is nicely summed up by Kurt Badt: "Architecture deals with the shaping of space, sculpture with one's relationship to space, painting with the revealing of space."[1]

Taken with a grain of salt, is it not true that space – as a problem of representation, as an immediate object – does not, in fact, appear at all in these three art forms? Strictly speaking, these three branches of art have never created spaces; they have merely built or depicted solid bodies or volumes. The entirety of spatial art to date has, in reality, only created volumes (out of stone, wood, etc.), never spaces.

Given that space has demonstrably emerged as an important categorial concept in art history only in the last two centuries,[2] it should come as no surprise that up to now, spatial effects in works of art have not been achieved directly, but always indirectly, through the representation of bodies. As a result, the *Kunstwollen*, as Alois Riegl would have it,[3] has had to make a distinction between bodies and the air (between them), which in fact has led to that unfortunate equation according to which "free" space becomes the space of the air and sculptural space the space of the body. It is thus imperative that sculpture be freed from the shackles of the object and the body, that sculpture be established in free space, that free space be conquered for sculpture.

We are trying to give to sculpture what belongs to it – namely, space. To conquer a larger space for sculpture than before, to take it back to its essence, is one of the goals of these artistic creations. Naturally, this places demands on all those cognitive and perceptual functions of the mind that are also carried out in humanity's normal, "everyday" spatial operations: imagining; supplementing and replacing missing pieces; establishing connections with the help of various sense organs such as nose, ears, and skin; detection and orientation (securing evidence) via confrontations with linear dimensions, proportions, temperatures, colors, and surfaces. The whole surgical tool kit of the human brain, with whose help humans find their way in space and create space for themselves, becomes part of the artistic calculus of contemporary sculpture; it becomes a sculptural tool kit, especially for spatial art that operates with media.

Abstract sculpture, even if much of it still echoes and evokes the organic, was the first awakening here. But we are not concerned with new materials; no, we are quite simply concerned with a new concept of space in sculpture. The new sculpture is more than just a body or object in space; it places and creates space, shapes space through our behavior in space, thereby revealing space for the first time.

Note: Adapted from a lecture given by Peter Weibel on the occasion of the exhibition *Raumartikulationen* at Forum Stadtpark Graz, January 19, 1979 and first published in a brochure accompanying the exhibition. Translated from the German by Patrick Hubenthal.

[1] Kurt Badt, *Raumphantasien und Raumillusionen: Das Wesen der Plastik* (Cologne: DuMont, 1963), 11.

[2] Hans Jantzen, *Über den kunstgeschichtlichen Raumbegriff* (Munich: Verlag der Bayerischen Akademie der Wissenschaften, 1938).

[3] Riegl conceived *Kunstwollen* (literally, "artistic will") as a zeitgeist-like collective creative impulse that gives rise to each era's characteristic artistic styles and tendencies.

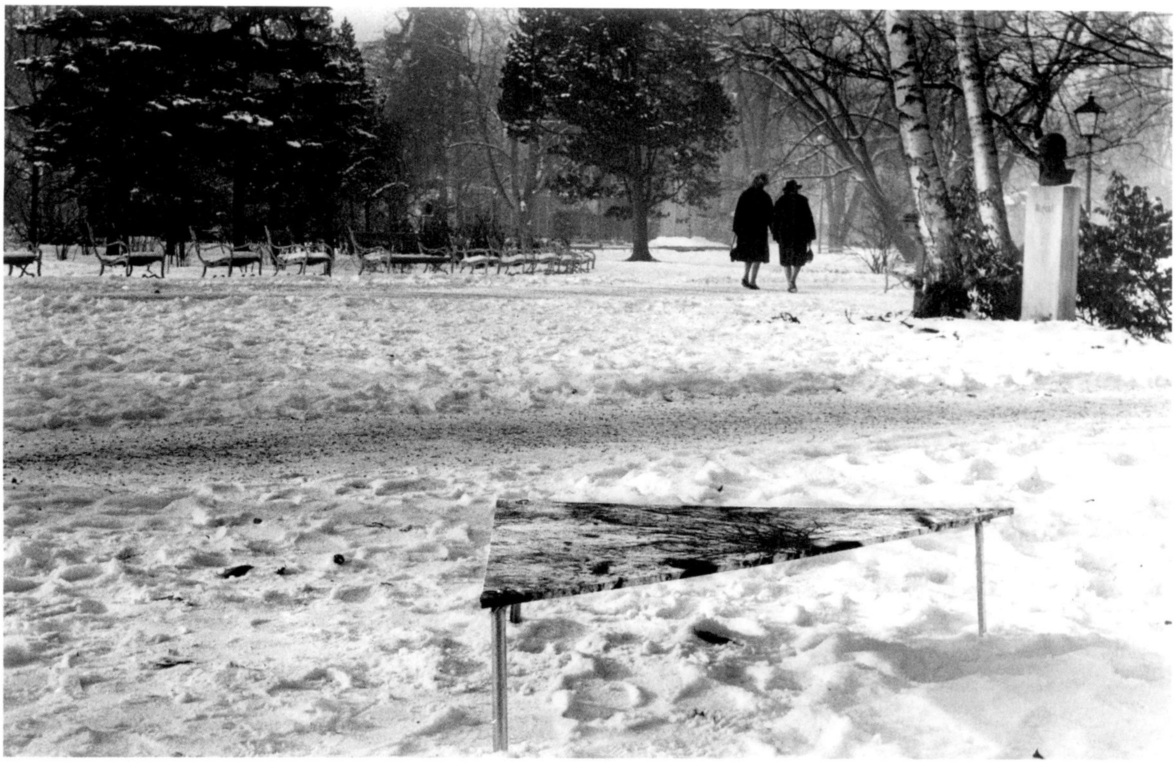

Installation views
Raumartikulationen, Forum Stadtpark, Graz, 1979

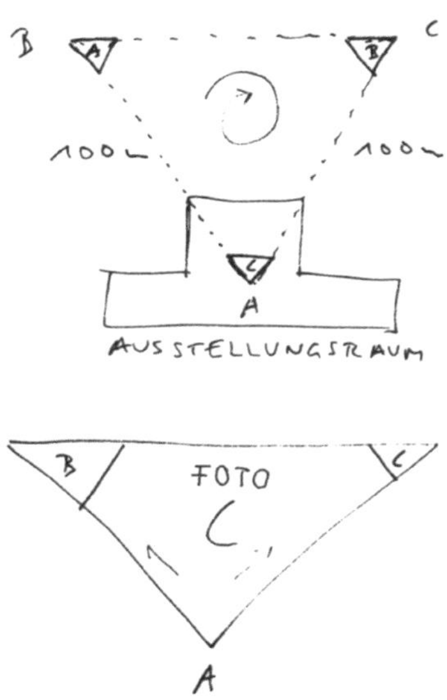

Conceptual drawing for *Displacement*, 1977

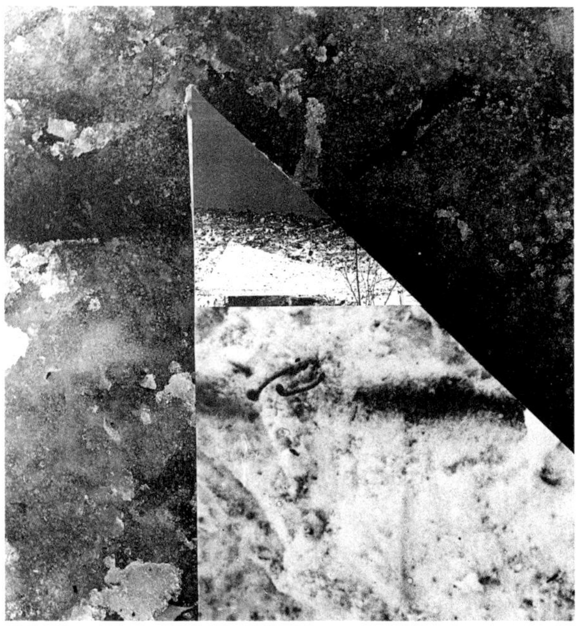

Site A is photographed, but its photograph is placed at site B. Site B is photographed, and its photograph is placed at site C. The photograph of site C is placed at site A. Site and site image do not match. At each site, a site different from that site is shown:

Site A displays the photograph of site C.
Site B displays the photograph of site A.
Site C displays the photograph of site B.

These three sites are the corners of a harmonic triangle, whose legs are each three hundred meters long, and one of whose vertices is in Graz's Forum Stadtpark. The photographic plates themselves are also harmonic triangles. The corners of each harmonic photo-triangle that point to the other two photo-triangles display photographs of the other two sites. The two corners of photograph C at site A display a photograph of site B with a photograph of A and a photograph of site C with a photograph of B, ad libitum. —Peter Weibel, 1977

Displacement

1977/79 | photographs, wood

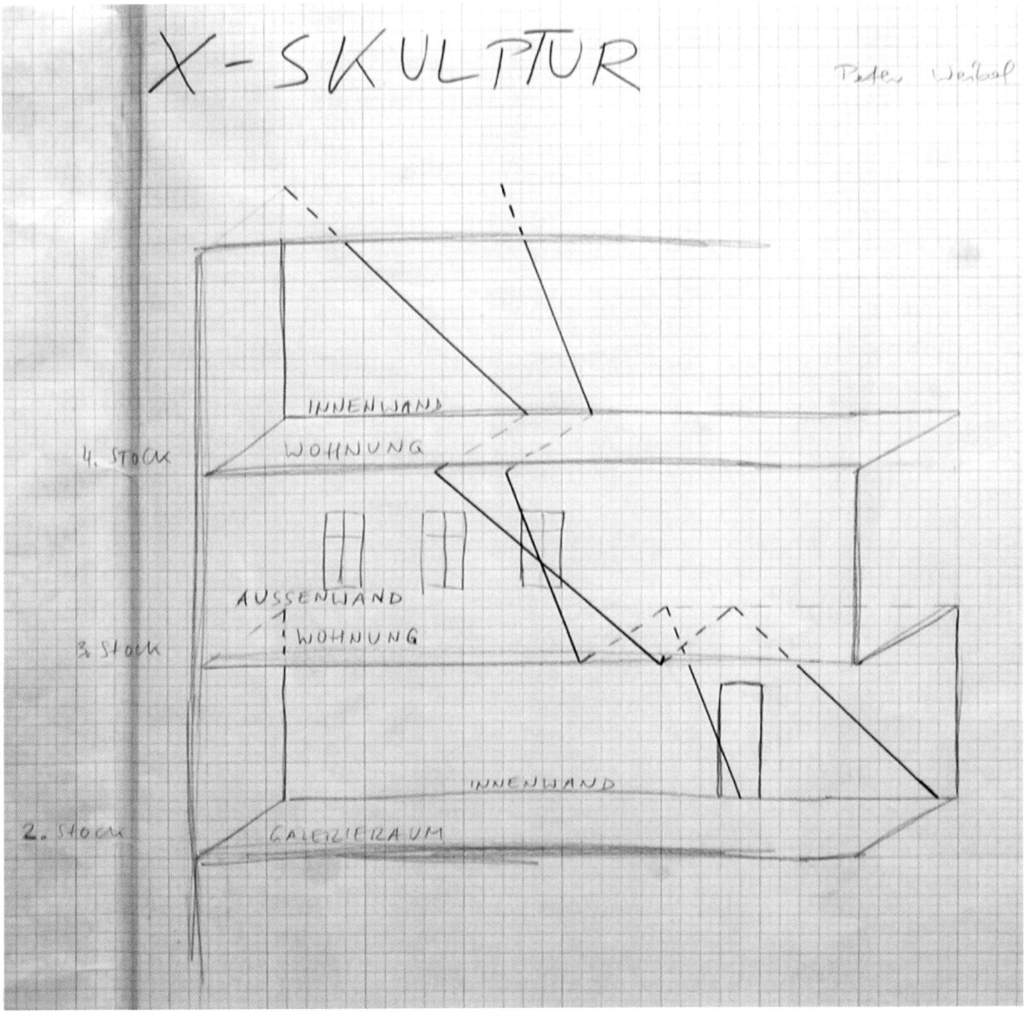

Conceptual drawing for *X-Sculpture*, 1977

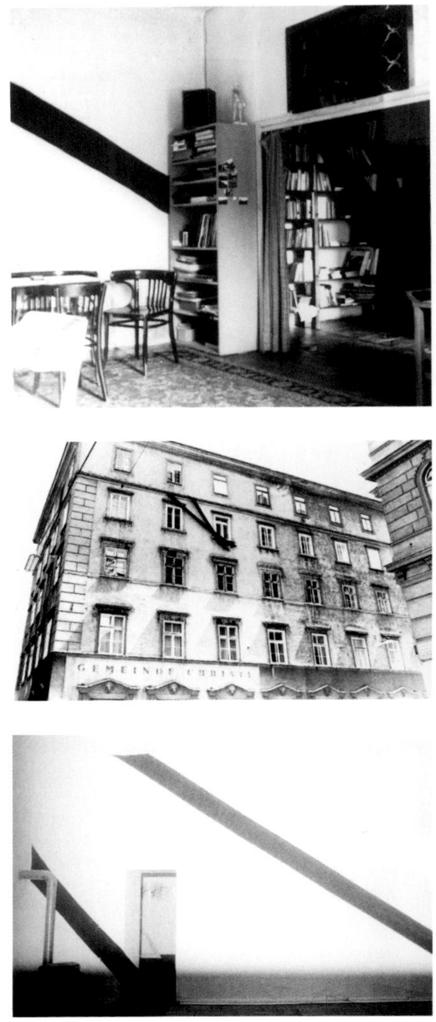

Installation views *Die unendliche Ein-Wort-Ausstellung & Skulpturen*, Galerie nächst St. Stephan, Vienna, 1977

A sculpture beyond the gallery space

In the Galerie nächst St. Stephan, a public space, two inclined lengths of fabric that make no sense on their own are exhibited across a wall on the second floor. The two panels of fabric cross one floor higher on the façade of the gallery building, another public space, to form an X. In a private room on the fourth floor, the two fabric panels move apart again. Visible real space and invisible imaginary space define the sculpture.

The essence of sculpture is not the human body.

Sculpture as the representation of the human body is an illusionary, anthropomorphic egocentrism. Conventional rigidity, mindless reproduction (inertia): ever since the discovery of the geometrically constructible representation of space in the Renaissance.

X-Sculpture / X-Skulptur

1977	6 lengths of fabric, dimensions variable

Space conceived as the human body; the false identity of body and space, as expressed, e.g., by the term "spatial body," can still be found in the cubists' and futurists' advanced concept of space, also later in more abstract sculptors who have penetrated into the bone and mollusc structures of the human body (i.e., Henry Moore).

Nor is the human being the producer of space, as an anthroposophical school derived from figuration would have us believe. Nevertheless, the concentration on the materiality of the spatial body that comes about in this process is one of the most decisive steps in becoming untethered from the representation of space by the human body.

The oldest conception of space: the place; but also the non-place; present and absent; material (real) and immaterial (imaginary).

The subject of sculpture is space.

Sculpture is the pragmatics of space: spatial perception and experience, open and closed space, interior and exterior space, hierarchical and stepless space. —Peter Weibel, 2006

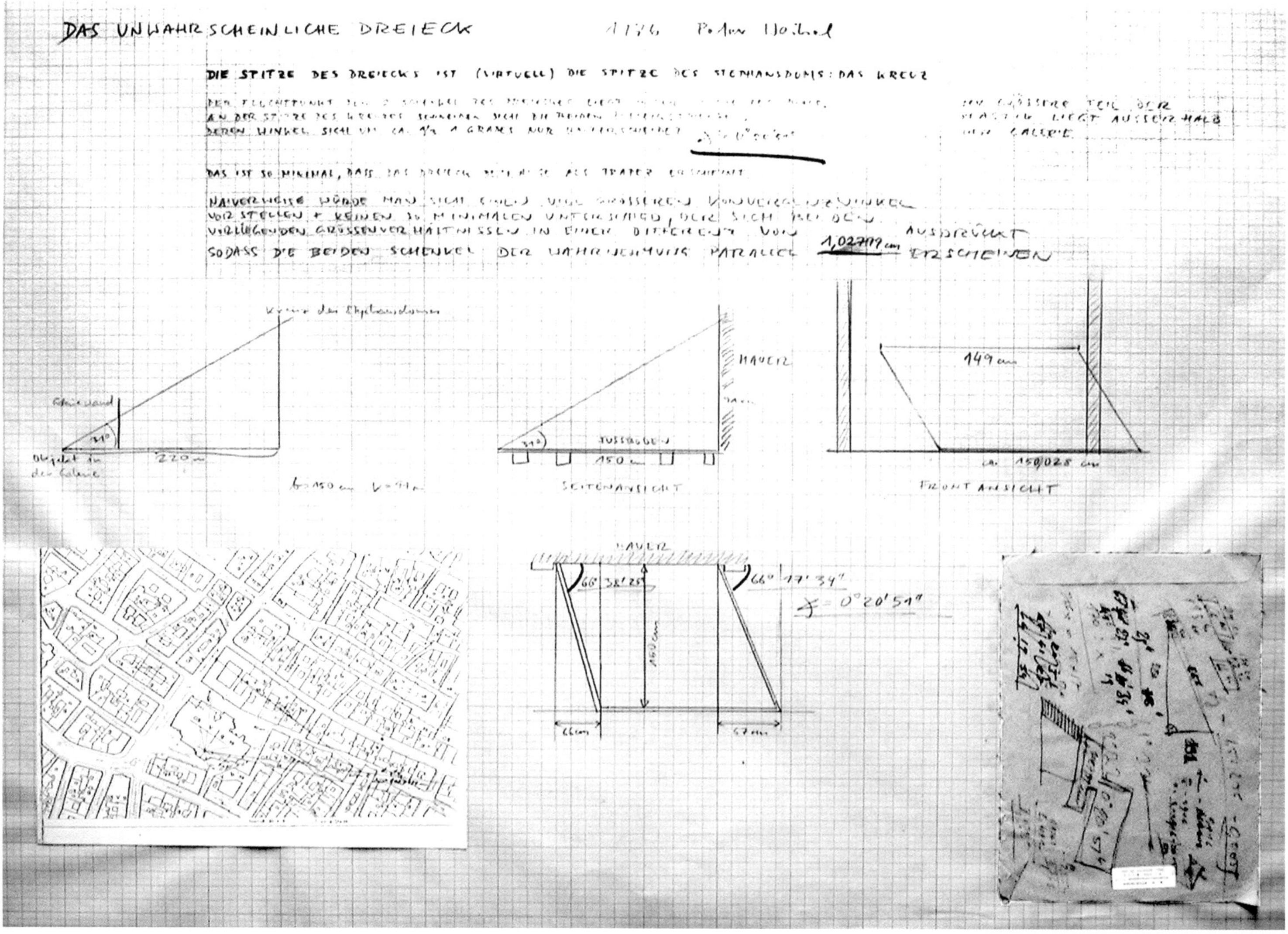

Conceptual drawing for *The Improbable Triangle*, 1976

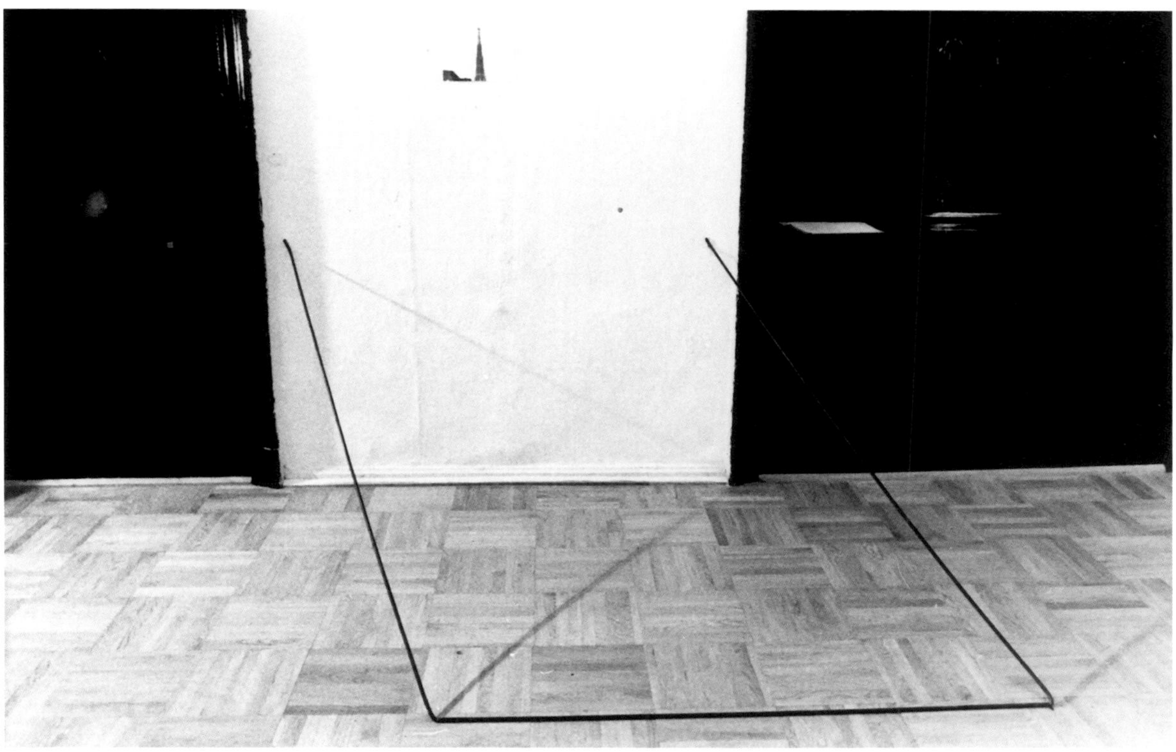

Installation view *Die unendliche Ein-Wort-Ausstellung & Skulpturen*, Galerie nächst St. Stephan, Vienna, 1977

The Improbable Triangle / Das unwahrscheinliche Dreieck

1976 | photograph, map, rubber

The Galerie nächst St. Stephan is, as the name suggests, close to St. Stephen's Cathedral in Vienna. I therefore made a calculation based on construction plans, with which inclination two rubber lines would reach the top of the cathedral.

—Peter Weibel, 2022

Installation view *Peter Weibel. Das Offene Werk 1964–1979*, Neue Galerie Graz, 2004

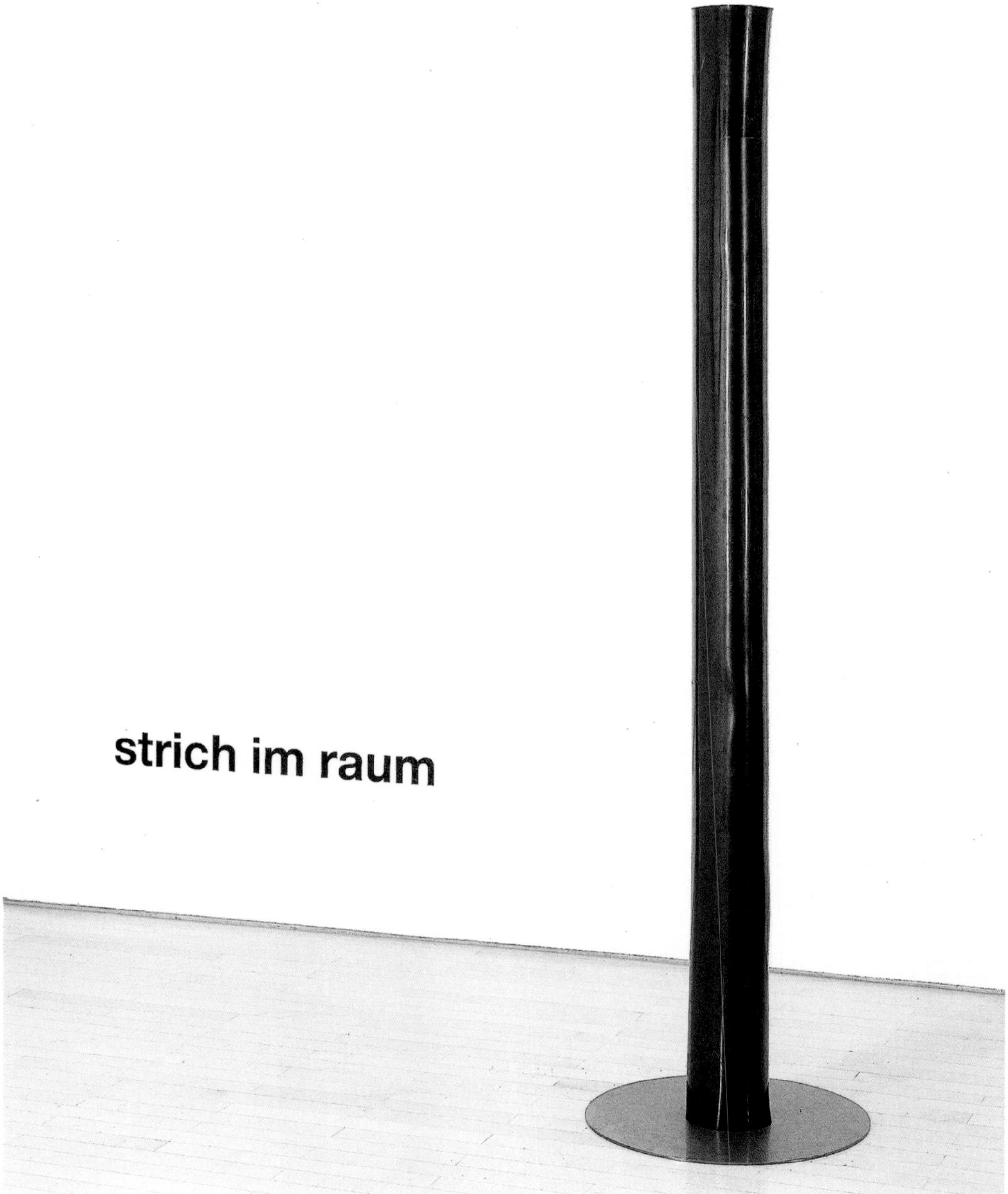

[1]
Peter Weibel, "Skulptur in Österreich," in *Raum annehmen*, exh. cat. (Vienna: Galerie Insam, 1982), 30–38 (quotation on 35).

For the sculpture *Black Line in Space*, a black rubber column is set up in the gallery. The words "Schwarzer Strich im Raum" can be read on the white wall, just above the floor.
The white wall becomes the ground for the text, the black line a freestanding sculpture in space. The text (title) becomes a pictorial motif on the wall. Through the interplay between sculpture and text, we discover new criteria in sculpture that significantly diverge – or rather, distance themselves – from the classical sculpture concept (form, mass, weight, etc.). Weibel sees the sculpture as a "construct of variable dimensionality." The new sculpture can no longer operate on the realities of space, only on the "signifiers of space." "These signifiers of sculpture are fundamentally linguistic in nature; they are signs," and "via these signifiers, sculpture becomes the language of space."[1]

Note: Adapted from Romana Schuler, "Einführung zum Werk Peter Weibels," in *Peter Weibel: Bildwelten. 1982–1996*, ed. Romana Schuler (Vienna: Triton, 1996), 8–17 (excerpt on 14–15). Translated from the German by Patrick Hubenthal.

Black Line in Space / Schwarzer Strich im Raum

1976 | rubber column with metal base, height 200 cm, diameter 15 cm

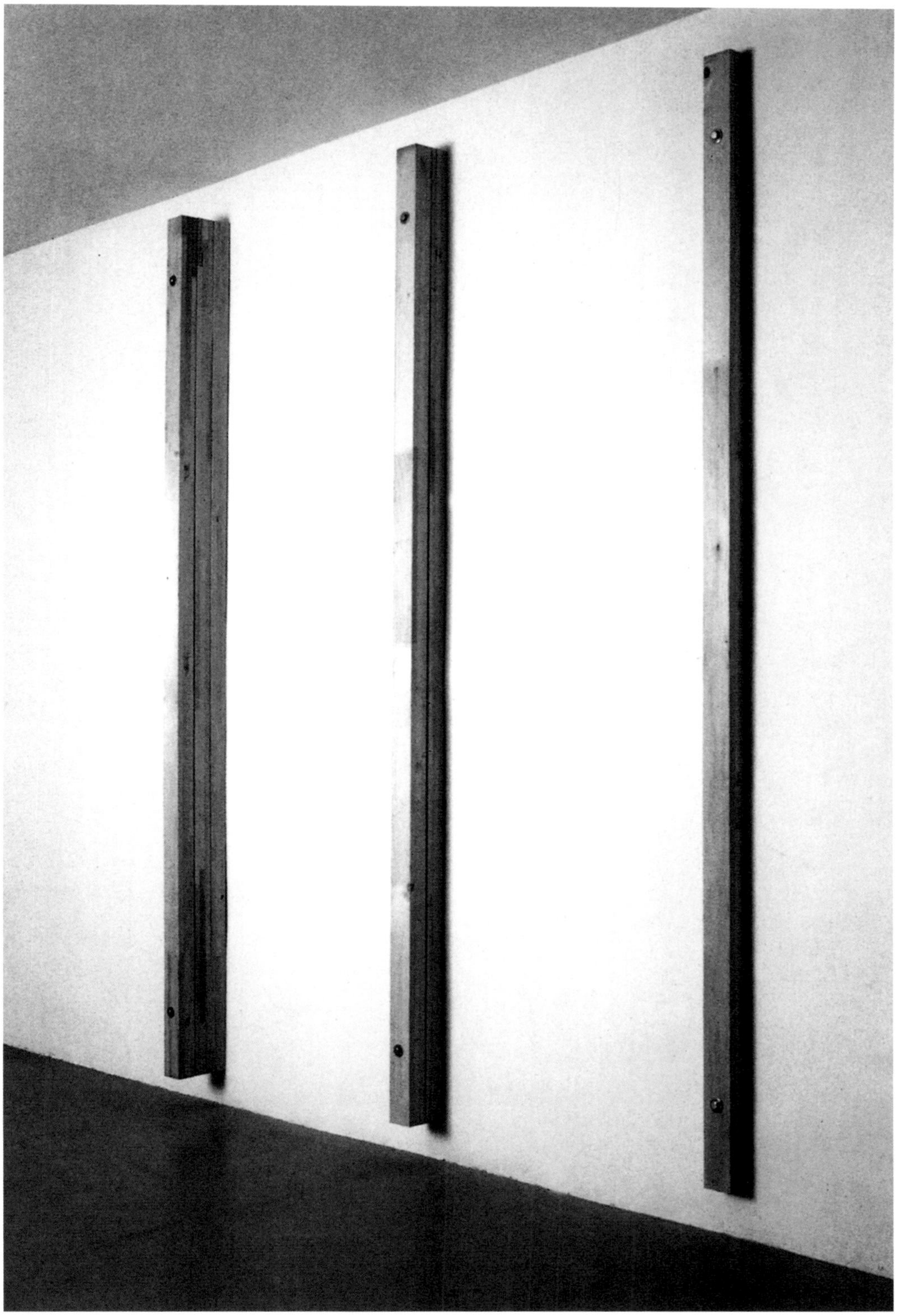

Installation view *Parterres*, Galerie Martina Detterer, Frankfurt am Main, 1990

Axes of Architecture / Achsen der Architektur

1990 | wood

At distances of one meter, the axes add up from one to two to three. Space is formed by the axes *x*, *y*, and *z*. These three are represented arithmetically and geometrically. The boundaries of the pictorial field and the field surrounding it become dissipative and indistinct.

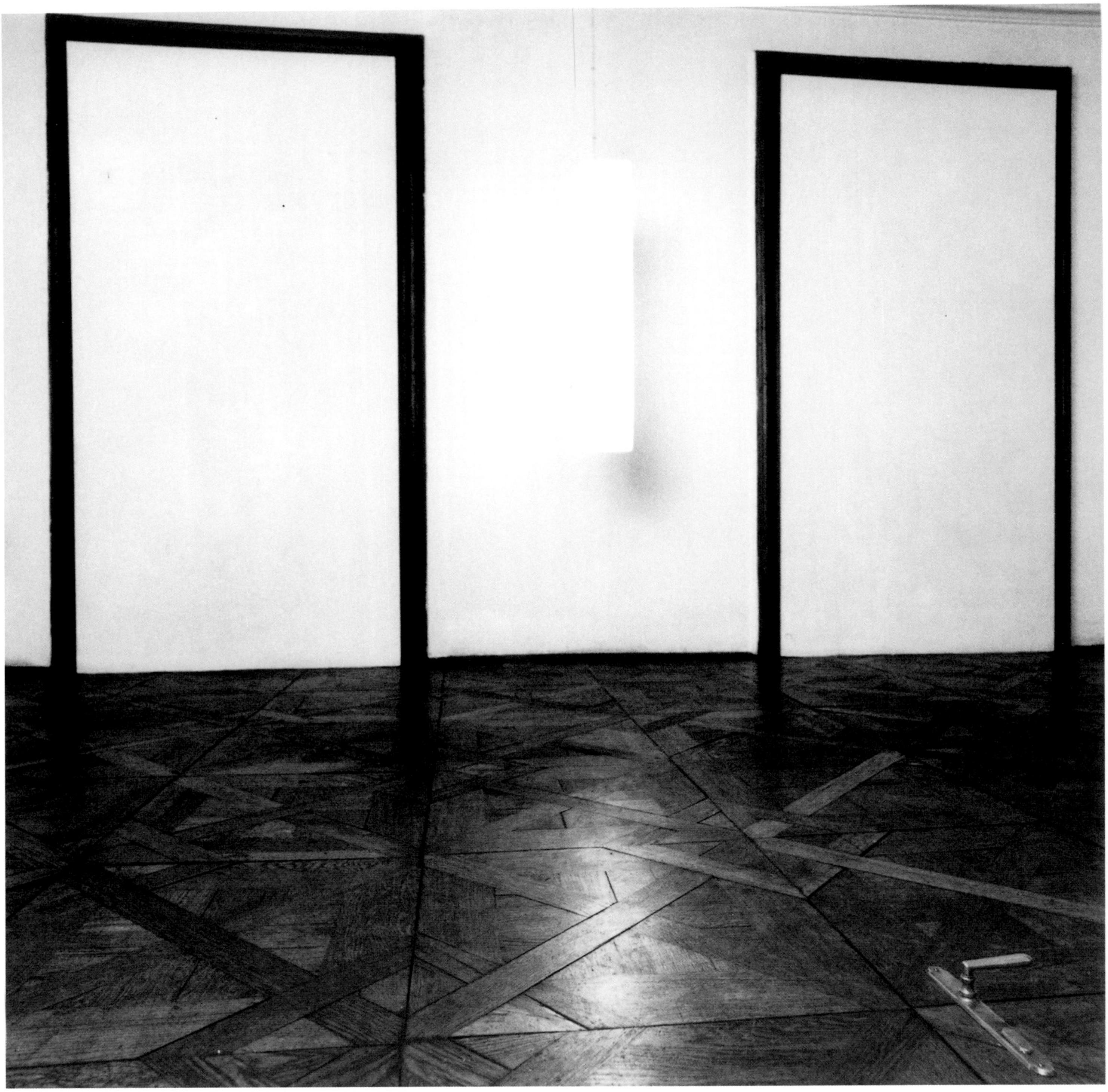

Installation view *Inszenierte Kunstgeschichte*, MAK – Museum of Applied Arts, Vienna, 1988–89

Untitled / Ohne Titel

1989 | from the series *Displacements*
door handle, lamps

A cube is the essential model for space. The vertical walls contain windows and doors, the ceiling contains a light source. I rotated the space: the door handle moved from the vertical wall to the floor, therefore the visitors stood on the door. The lamp moved from the horizontal ceiling to the vertical wall, the windows disappeared. The visitor felt lost in space, alienated in the cubic home, a displaced person.

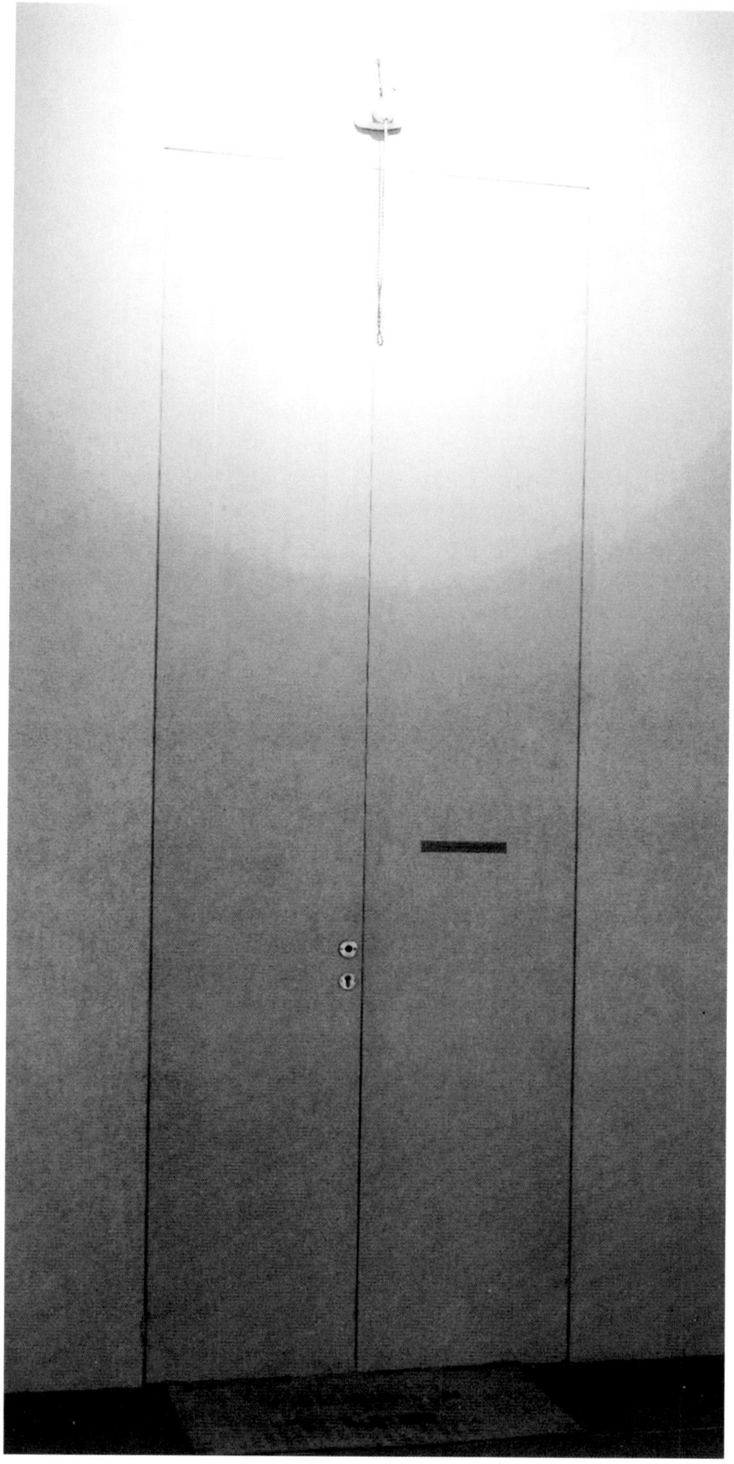

Installation view *Inszenierte Kunstgeschichte*, MAK – Museum of Applied Arts, Vienna, 1988–89

Untitled / Ohne Titel

1988 | from the series *Displacements*
telephone mounted above a door

The telephone can serve as an exit from a closed space. It is ordinarily a medium that transcends distances; here, positioned like a light fixture (emergency exit), the telephone itself is distanced.

Costruzione Legittima: On the Principle of Perspective

Peter Weibel

As is well known, the origins of perspective can be located in the fifteenth century. The mathematician Antonio di Tuccio Manetti, a friend of Paolo Uccello, taught Uccello the rudiments of the perspectival construction of the image. Uccello was so excited by the discovery of perspective that he is said to have uttered the famous sentence "Oh che dolce cosa è questa prospettiva!"[1] We find the first valid productions of perspective in the architecture and sculpture of Filippo Brunelleschi (who, incidentally, used a mirror as an aid to his perspectival constructions).

Leon Battista Alberti demonstrated the construction of one-point perspective in his book *De pictura* (1435). Antonio di Pietro Averlino, better known as Filarete, made his own contribution to the theory of perspective in his *Trattato di architettura* (1460–64); he called Brunelleschi "a most subtle follower of Dedalus."[2] He was followed by Piero della Francesca, who wrote *De prospectiva pingendi* as well as *Trattato d'abaco*, an essay on the abacus (both ca. 1470), and by Jean Pèlerin, known as Le Viator, author of *De artificiali perspectiva* (1505). Della Francesca regarded linear perspective as the "legitimate" way of constructing any figurative composition imaginable, thereby marking out the path of spatiality for the painters of the Renaissance. In the fifteenth century, then, artificial perspective, the mathematical approach to perspective that was emerging from Urbino, was formulated as a rational method of construction for the composition of pictures. This was the beginning of a discourse on constructive forms of expression. A logocentric rationality defined the problem of representing three-dimensional space on the two-dimensional surface of the picture as a rational form of construction. This discourse grew so powerful that it did not simply inspire Leonardo da Vinci's and Albrecht Dürer's studies of perspective as *costruzione legittima*; more than that, perspective came to be regarded as *the* new concept of space – specifically, a mathematical concept of space. The defining aspect of this new way of viewing space and imagining the world is that space and the image are described in terms of construction. With perspective as the only legitimate way of constructing space, the image is presented for the first time as a problem of construction, to be solved in accordance with rational and mathematical principles. The "divine" proportions of the golden section, which was rediscovered and refined around the same time as the invention of perspective, is an offshoot of this idealized mathematization of space and this conception of the image as something to be constructed. Thus, the principle that construction is the legitimate and universal operator for both pictures and sculpture begins with perspective.

In technical terms, perspective can be expressed as a geometric formula, by means of which three-dimensional objects can be translated into (their) images on a flat canvas. The purpose of perspective, then, is to produce the illusion of three-dimensional spatial depth on a two-dimensional surface. The encoding of the illusion of perspective as scientific, realistic objectivity becomes even more astonishing when we consider the imaginary means by which perspective is constructed. Perspective is not simply a matter of using lines to represent the surface area (length and width) of a patch of ground; rather, the lines laid out at regular intervals on that surface converge on a fixed point, the *punto di fuga*.

This vanishing point, however, is not only located on an imaginary horizon, but beyond the surface of the painting itself; in fact, strictly speaking, it lies in the infinite. With this vanishing point that is forever receding into the infinite (which is why it is called the *Fluchtpunkt*, the "fleeing point," in German), the representation of three-dimensional space on a two-dimensional picture has paid a high price: the invention of the imaginary. In that sense, perspective's vanishing point also marks the inception of a geometry of the imaginary.

Erwin Panofsky was therefore right to give his 1927 essay on perspective the title *Perspective as Symbolic Form*, a reference to Ernst Cassirer's 1923 book *The Philosophy of Symbolic Forms*, in which Cassirer strives to establish a "'morphology' of the human spirit."[3] These symbolic forms have no other purpose than to herald the conquest of the world as representation.

Under the pressure of the post-1800 Industrial Revolution and the Machine Age, movement and dynamism became a central phenomenon in society – one that art, too, could no longer ignore. As a result, a cluster of writings on the representation of movement on two-dimensional surfaces appeared around the year 1900. The mobile observer, the multiplex gaze of the viewer moving around the static object to be depicted, led to the multiple perspectives of Cubism and the abolition of the tyranny of perspective in painting. Futurism studied the moving object from a static point of view. The problem of motion had always lain hidden beneath the signifier of perspective; when it was uncovered and explicated, it obliterated perspective. After Paul Cézanne and Eugène Delacroix, perspective vanished from painting. The cyclopean subject disappeared along with the tyranny of perspective.

The problem of perspective vanished from painting when another medium appeared that seemed to be better suited to dealing with it – namely, photography. Between 1900 and 1930, perspective, having (e)migrated from painting into photography and been dubbed (by the Bauhaus) "Das Neue Sehen" (the new way of seeing), underwent a rapid development. Not only did the accelerated perspective of 1920s photography (achieved by means of sharply diagonalized, foreshortened vanishing lines) release the problem of motion (the mobile gaze, the moving viewer) in painting (Cubism, Futurism), it set the image in motion more generally. The movement that was set free in the image, the moving image, took perspective further, to the extreme perspective games and mirrorings of computer animation. Perspective thus moved from easel painting and sculpture to the art of the moving image, from the accelerated perspective of photography to the accelerated ecstasy of the computer-generated animation of perspective. Anyone who has had the opportunity to experience the way the objects on the intarsia panels in Urbino's Palazzo Ducale seem to move with each change in the viewer's point of view knows that there is an inescapable link between the point of perspective and the position of perception, between viewpoint and standpoint, between the vanishing point of the image and the observer's point of view. Seen in this light, the concept of an active, participatory observer would appear to be a historical necessity.

The movement set free in the image led, on the one hand, to the art of the moving image, and on the other to the mobile gaze, to multiple perspectives, to the moving viewer (of the image), and thus to various forms of interactivity between image and viewer. This evolution, from perspective to the participation of the viewer, can be seen from El Lissitzky to Op Art. The interaction between viewer and image created the actual form in which the work of art existed. Viewer and image became virtual components of a dynamic system. Virtual components are components whose properties are not discernible in isolation, only becoming visible during the interaction itself. Virtual space and interactivity, therefore, are further hidden signifieds in the space of perspective's signifiers. The *Fluchtpunkte* to which the perspective lines relate are, again, theoretically infinitely distant: vanishing points, points of reference that actually "disappear" into infinity. The vanishing point has caused perspective itself to disappear. The infinite itself, as a concept, has split off from perspective's infinite vanishing point and become an absolute. A free-floating infinite has emerged in both art and science.

In the age of orbital and nuclear acceleration, we are particularly interested in the physical infinite, conceived as a reciprocal acceleration. The deeper we penetrate into the microcosm of cells, genes, atomic nuclei, and subatomic particles, the more extensively we explore the cosmological macrosom of the universe as well. This gives rise to interesting infinitesimal perspectives of reciprocity, relativizations of scale, shifts and distortions in size. The size relationship between a virus and a person is the same as that between the person and the Earth. The diameter of the Earth is approximately 12,750 km. The diameter of a virus is approximately 15–440 nm. So compared to a virus, I am the size of the Earth. To the Earth, we are the same size that viruses are to us. These are examples of infinite reciprocity, exploding proportions, rotating perspectives, infinitely variable sizes, imploding scales.

Over the course of its development, perspective has passed through a variety of media, from painting to the computer, from easel pictures to moving pictures on the screen. In the process, the constructional principle, like an umbrella, has opened up other constructional principles and problems: motion, the infinite, interaction (between image and viewer). Following the dismantling of the achievements we owe to the Renaissance, the dismantling of logocentric rationality has now begun. It should be noted, however, that the principle of construction and its various signifieds have not succumbed to a visionary obscurantism, but rather have found new materials, media, and methods. The principle of construction will thus continue to perform the task set by perspective: to see what exists and to make visible what does not exist, to see by being deceived and to render sighted through deception. The corroboration of the real and the imagination of the symbolic form the ballistic knot, the equilibrium of the falsifications that simulate the true. A geometry of the imaginary is a legitimate perspective of the future. With the invention and construction of perspective came the triumph of observer-relativity and the realization that what seems objective is objective only to the observer.

Note: Adapted from Peter Weibel, "Zur Perspektive als konstruktivem Prinzip: Eine Geometrie des Imaginären," *Kunstforum*, no. 105 (January–February 1990): 168–78. Translated from the German by Patrick Hubenthal.

[1]
Giorgio Vasari, "Vita di Paolo Uccello," in *Le opere di Giorgio Vasari*, vol. 1, *Le vite dei più eccellenti pittori, scultori e architetti* (Florence: David Passigli, 1832–38), 230–34 (quotation on 233).

[2]
Filarete, *Filarete's Treatise on Architecture*, vol. 1, *The Translation*, trans. John R. Spencer (New Haven: Yale University Press, 1965), 102.

[3]
Ernst Cassirer, *The Philosophy of Symbolic Forms*, vol. 1, *Language*, trans. Ralph Manheim (New Haven: Yale University Press, 1953), 69. First published as *Philosophie der symbolischen Formen*, vol. 1, *Die Sprache* (Berlin: Bruno Cassirer, 1923).

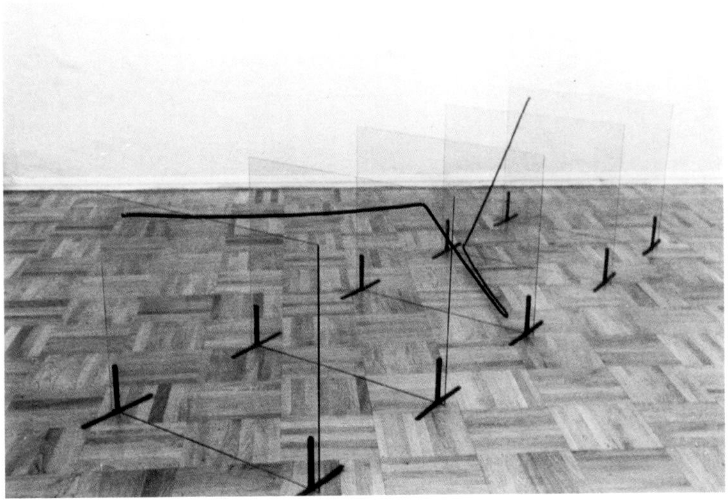
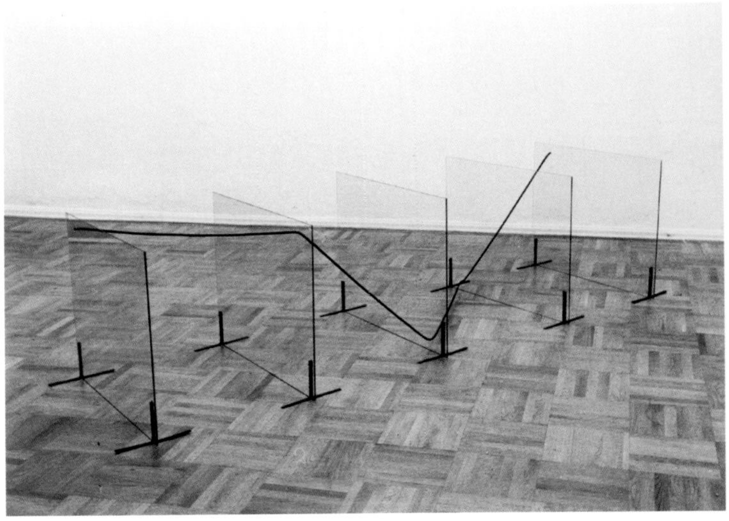
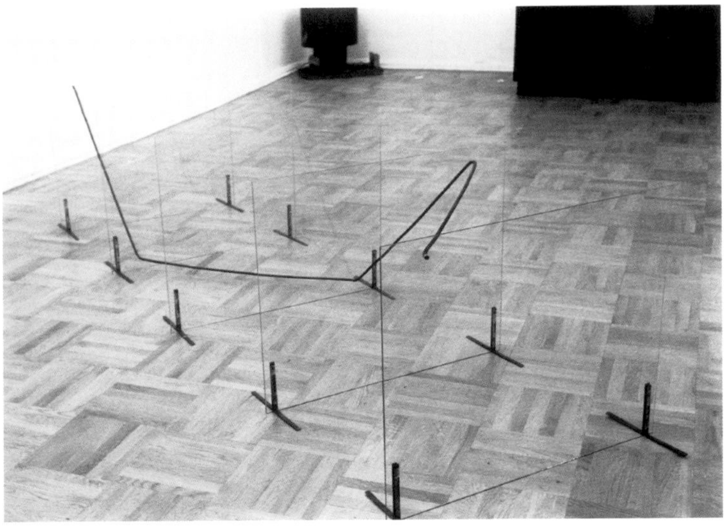
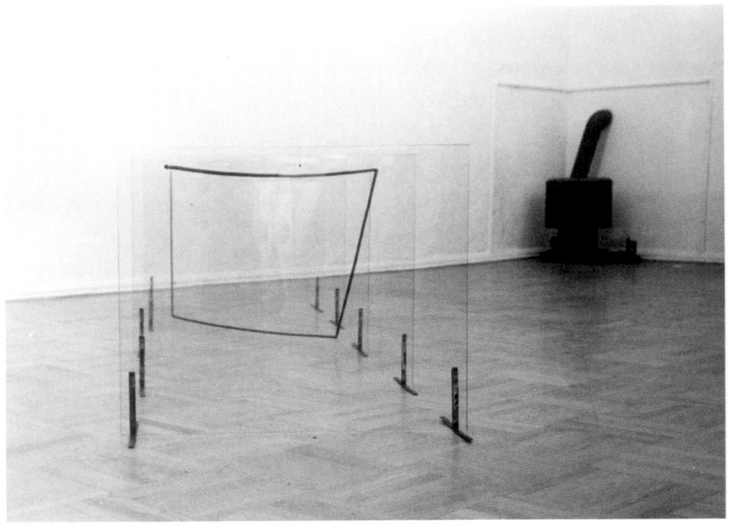

Installation views *Die unendliche Ein-Wort-Ausstellung & Skulpturen*, Galerie nächst St. Stephan, Vienna, 1977

In the postmodern era, the concept of construction has been called into question, its logocentric rationality deconstructed, by asking: What is the libidinous energy that governs the organization of the "symbolic" forms of construction and composition? What is the desire, the libido, that constructs perspective, or Malevich's *Black Square*?

My sculpture *Tortured Square* is the metaphor that heralds the deconstruction of the principle of construction on the basis of the economy of desire (What is the desire of the gaze that situates the point of view in the infinite? etc.). This deconstruction is accompanied by the traversal and dissolution of the array of problems marked by the four concepts of perspective, proportion, motion, and infinity, until we arrive at the fork where viewer and (field of) vision are linked to the problem of the subject within the framework of a problem of perception. As early as 1886, the dissolution of perception into a stream of sensory data, a changing field of multiple perspectives and relativizing gazes, had led Ernst Mach to declare that "the ego must be given up."[1] Bruce Nauman highlighted the connection between subject (position) and perspective in his sculptural ensemble *Forced Perspective I* (1975), which presents dozens of cubes in distorted perspective. Mach countered the tyranny of perspective and the single point of view by analyzing the "sensations" experienced by the subject, thereby dissolving the subject's identity and knocking perspective off its throne as *costruzione legittima*. Experimental perceptual psychology (with its optical illusions) and the phenomenology of perception (after Maurice Merleau-Ponty, Jean-Paul Sartre, and Jacques Lacan) have always challenged and pushed against the autonomy, constancy, identity, and sovereignty of the subject.

—Peter Weibel, 1990/2022

Translated from the German by Patrick Hubenthal.

[1]
Editor's note: Ernst Mach, *The Analysis of Sensations and the Relation of the Physical to the Psychical*, trans. C. M. Williams and Sydney Waterlow (Chicago: Open Court 1914), 24. First published as *Beiträge zur Analyse der Empfindungen* (Jena: Gustav Fischer, 1886).

Tortured Square / Gequältes Quadrat

1977	plexiglass panes, metal bases, rubber cord, 100 × 100 × 100 cm each

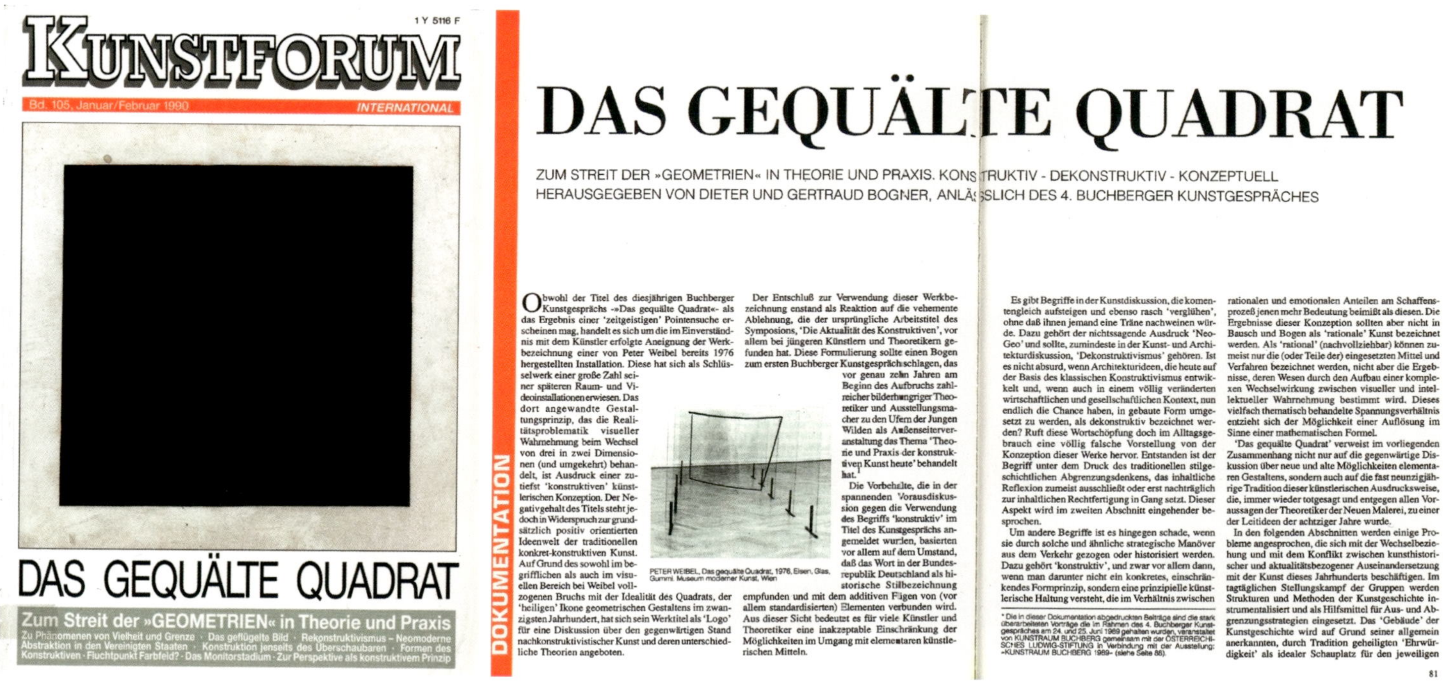

1 Y 5116 F

KUNSTFORUM

Bd. 105, Januar/Februar 1990 INTERNATIONAL

DAS GEQUÄLTE QUADRAT

Zum Streit der »GEOMETRIEN« in Theorie und Praxis

Zu Phänomenen von Vielheit und Grenze · Das geflügelte Bild · Rekonstruktivismus – Neomoderne Abstraktion in den Vereinigten Staaten · Konstruktion jenseits des Überschaubaren · Formen des Konstruktiven · Fluchtpunkt Farbfeld? · Das Monitorstadium · Zur Perspektive als konstruktivem Prinzip

DOKUMENTATION

DAS GEQUÄLTE QUADRAT

ZUM STREIT DER »GEOMETRIEN« IN THEORIE UND PRAXIS. KONSTRUKTIV - DEKONSTRUKTIV - KONZEPTUELL
HERAUSGEGEBEN VON DIETER UND GERTRAUD BOGNER, ANLÄSSLICH DES 4. BUCHBERGER KUNSTGESPRÄCHES

Obwohl der Titel des diesjährigen Buchberger Kunstgesprächs -»Das gequälte Quadrat«- als das Ergebnis einer 'zeitgeistigen' Pointensuche erscheinen mag, handelt es sich um die im Einverständnis mit dem Künstler erfolgte Aneignung der Werkbezeichnung einer von Peter Weibel bereits 1976 hergestellten Installation. Diese hat sich als Schlüsselwerk einer große Zahl seiner späteren Raum- und Videoinstallationen erwiesen. Das dort angewandte Gestaltungsprinzip, das die Realitätsproblematik visueller Wahrnehmung beim Wechsel von drei in zwei Dimensionen (und umgekehrt) behandelt, ist Ausdruck einer zutiefst 'konstruktiven' künstlerischen Konzeption. Der Negativgehalt des Titels steht jedoch in Widerspruch zur grundsätzlich positiv orientierten Ideenwelt der traditionellen konkret-konstruktiven Kunst. Auf Grund des sowohl im begrifflichen als auch im visuellen Bereich bei Weibel vollzogenen Bruchs mit der Idealität des Quadrats, der 'heiligen' Ikone geometrischen Gestaltens im zwanzigsten Jahrhundert, hat sich sein Werktitel als 'Logo' für eine Diskussion über den gegenwärtigen Stand nachkonstruktivistischer Kunst und deren unterschiedliche Theorien angeboten.

PETER WEIBEL, Das gequälte Quadrat, 1976, Eisen, Glas, Gummi. Museum moderner Kunst, Wien

Der Entschluß zur Verwendung dieser Werkbezeichnung enstand als Reaktion auf die vehemente Ablehnung, die der ursprüngliche Arbeitstitel des Symposions, 'Die Aktualität des Konstruktiven', vor allem bei jüngeren Künstlern und Theoretikern gefunden hat. Diese Formulierung sollte einen Bogen zum ersten Buchberger Kunstgespräch schlagen, das vor genau zehn Jahren am Beginn des Aufbruchs zahlreicher bilderhungriger Theoretiker und Ausstellungsmacher zu den Ufern der Jungen Wilden als Außenseiterveranstaltung das Thema 'Theorie und Praxis der konstruktiven Kunst heute' behandelt hat.[1]

Die Vorbehalte, die in der spannenden Vorausdiskussion gegen die Verwendung des Begriffs 'konstruktiv' im Titel des Kunstgesprächs angemeldet wurden, basierten vor allem auf dem Umstand, daß das Wort in der Bundesrepublik Deutschland als historische Stilbezeichnung empfunden und mit dem additiven Fügen von (vor allem standardisierten) Elementen verbunden wird. Aus dieser Sicht bedeutet es für viele Künstler und Theoretiker eine inakzeptable Einschränkung der Möglichkeiten im Umgang mit elementaren künstlerischen Mitteln.

Es gibt Begriffe in der Kunstdiskussion, die komentengleich aufsteigen und ebenso rasch 'verglühen', ohne daß ihnen jemand eine Träne nachweinen würde. Dazu gehört der nichtssagende Ausdruck 'Neo-Geo' und sollte, zumindeste in der Kunst- und Architekturdiskussion, 'Dekonstruktivismus' gehören. Ist es nicht absurd, wenn Architekturideen, die heute auf der Basis des klassischen Konstruktivismus entwikkelt und, wenn auch in einem völlig veränderten wirtschaftlichen und gesellschaftlichen Kontext, nun endlich die Chance haben, in gebaute Form umgesetzt zu werden, als dekonstruktiv bezeichnet werden? Ruft diese Wortschöpfung doch im Alltagsgebrauch eine völlig falsche Vorstellung von der Konzeption dieser Werke hervor. Entstanden ist der Begriff unter dem Druck des traditionellen stilgeschichtlichen Abgrenzungsdenkens, das inhaltliche Reflexion zumeist ausschließt oder erst nachträglich zur inhaltlichen Rechtfertigung in Gang setzt. Dieser Aspekt wird im zweiten Abschnitt eingehender besprochen.

Um andere Begriffe ist es hingegen schade, wenn sie durch solche und ähnliche strategische Manöver aus dem Verkehr gezogen oder historisiert werden. Dazu gehört 'konstruktiv', und zwar vor allem dann, wenn man darunter nicht ein konkretes, einschränkendes Formprinzip, sondern eine prinzipielle künstlerische Haltung versteht, die im Verhältnis zwischen rationalen und emotionalen Anteilen am Schaffensprozeß jenen mehr Bedeutung beimißt als diesen. Die Ergebnisse dieser Konzeption sollten aber nicht in Bausch und Bogen als 'rationale' Kunst bezeichnet werden. Als 'rational' (nachvollziehbar) können zumeist nur die (oder Teile der) eingesetzten Mittel und Verfahren bezeichnet werden, nicht aber die Ergebnisse, deren Wesen durch den Aufbau einer komplexen Wechselwirkung zwischen visueller und intellektueller Wahrnehmung bestimmt wird. Dieses vielfach thematisch behandelte Spannungsverhältnis entzieht sich der Möglichkeit einer Auflösung im Sinne einer mathematischen Formel.

'Das gequälte Quadrat' verweist im vorliegenden Zusammenhang nicht nur auf die gegenwärtige Diskussion über neue und alte Möglichkeiten elementaren Gestaltens, sondern auch auf die fast neunzigjährige Tradition dieser künstlerischen Ausdrucksweise, die, immer wieder totgesagt und entgegen allen Voraussagen der Theoretiker der Neuen Malerei, zu einer der Leitideen der achtziger Jahre wurde.

In den folgenden Abschnitten werden einige Probleme angesprochen, die sich mit der Wechselbeziehung und mit dem Konflikt zwischen kunsthistorischer und aktualitätsbezogener Auseinandersetzung mit der Kunst dieses Jahrhunderts beschäftigen. Im tagtäglichen Stellungskampf der Gruppen werden Strukturen und Methoden der Kunstgeschichte instrumentalisiert und als Hilfsmittel für Aus- und Abgrenzungsstrategien eingesetzt. Das 'Gebäude' der Kunstgeschichte wird auf Grund seiner allgemein anerkannten, durch Tradition geheiligten 'Ehrwürdigkeit' als idealer Schauplatz für den jeweiligen

* Die in dieser Dokumentation abgedruckten Beiträge sind die stark überarbeiteten Vorträge die im Rahmen des 4. Buchberger Kunstgespräches am 24. und 25. Juni 1989 gehalten wurden, veranstaltet von KUNSTRAUM BUCHBERG gemeinsam mit der ÖSTERREICHISCHES LUDWIG-STIFTUNG in Verbindung mit der Ausstellung: »KUNSTRAUM BUCHBERG 1989« (siehe Seite 88).

81

Cover and pages from
Kunstforum, no. 105, 1990

Installation view *A Nyitott Mű 1964–1979*, Műcsarnok, Budapest, 2005

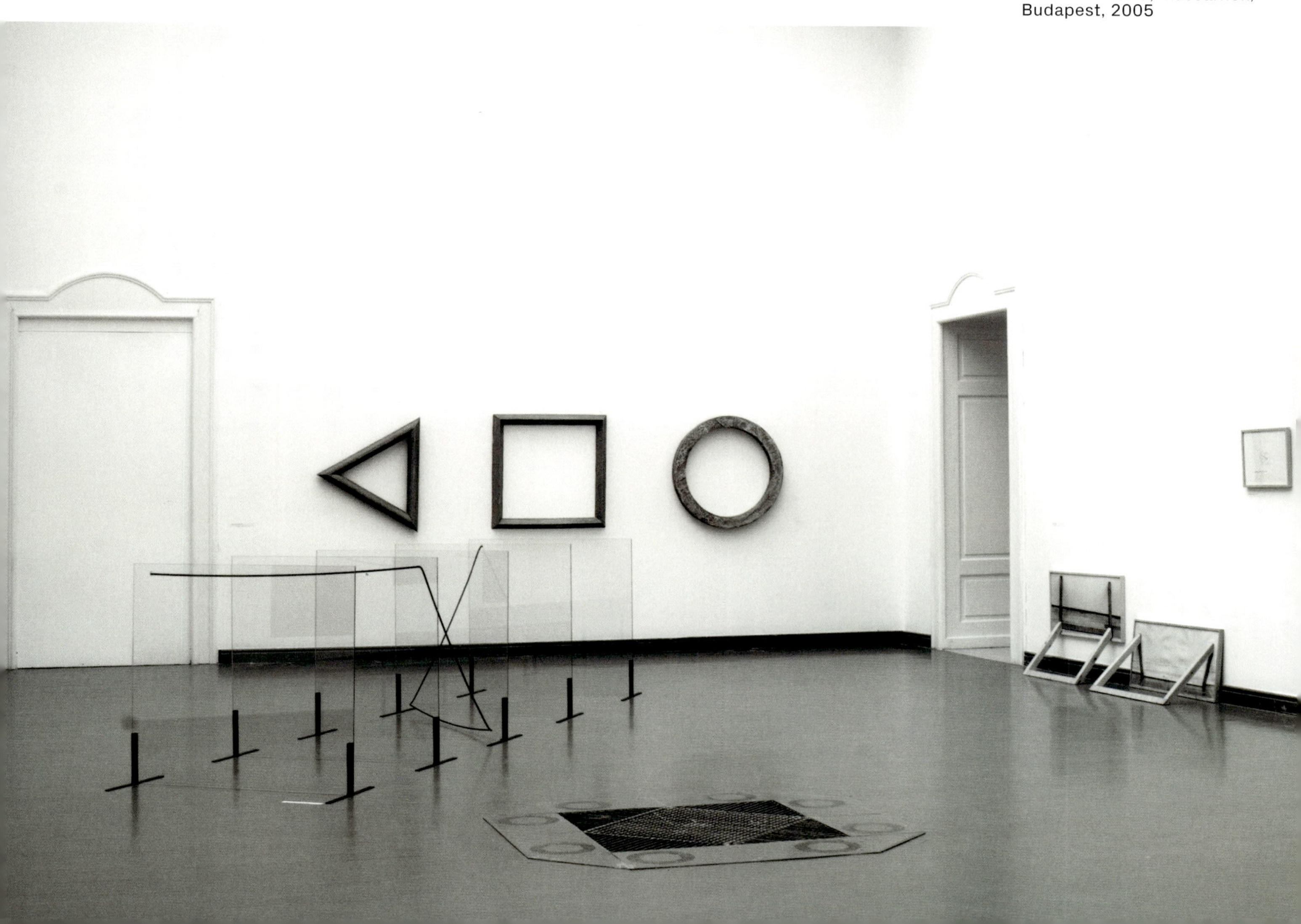

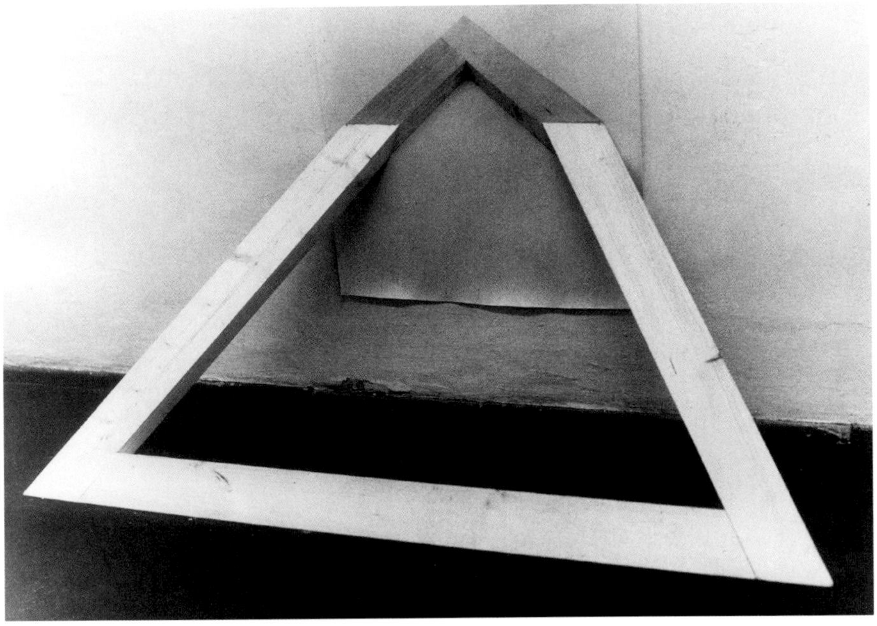

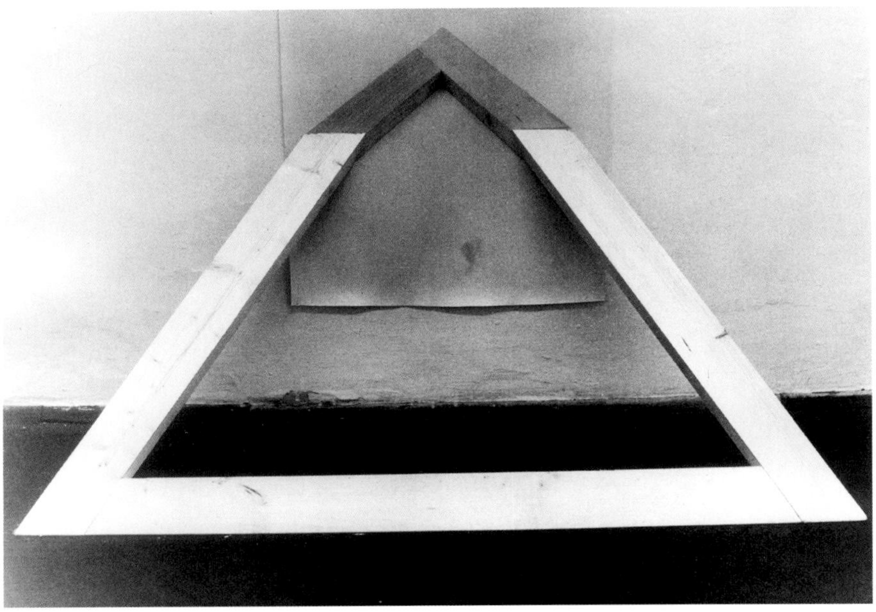

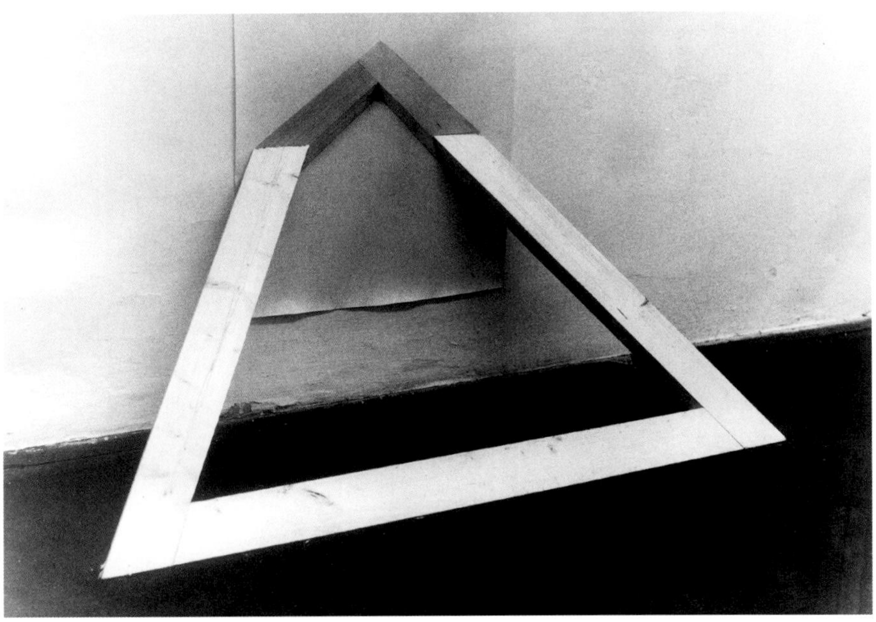

Installation views *Die unendliche Ein-Wort-Ausstellung & Skulpturen*, Galerie nächst St. Stephan, Vienna, 1977

Warped Triangles / Verdrehte Dreiecke

1976 | wood, photographs

A real, broken triangle is completed by a photograph. Changes in the viewer's position cause radical changes in perspective. From one vantage point, the real triangle's right leg aligns perfectly with the one on the photograph; from another, the left one does. Perspectival differentiation thus depends on the medium of differentiation. Photography and reality react very differentely to positions of perspective.

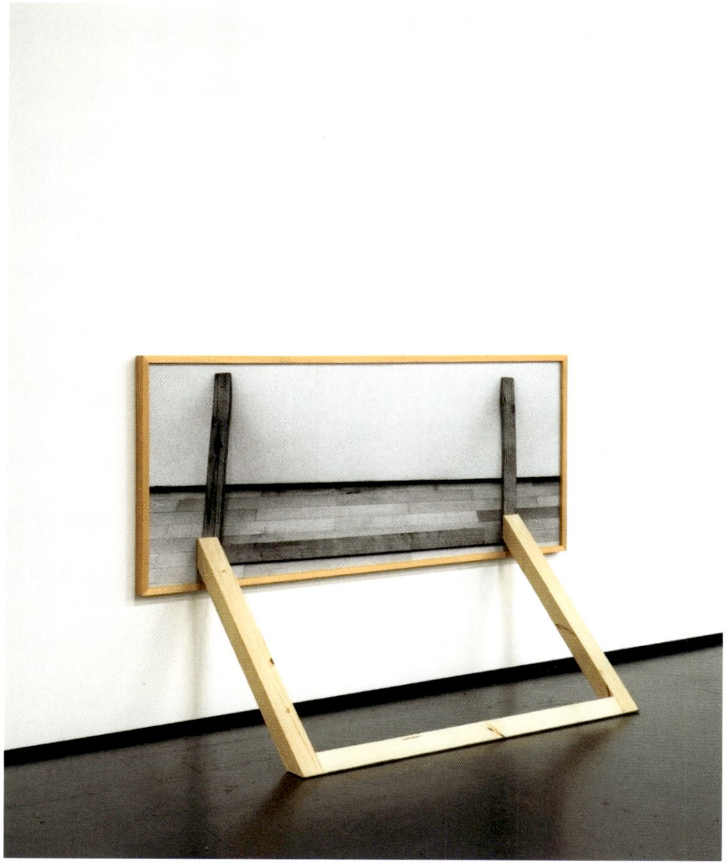

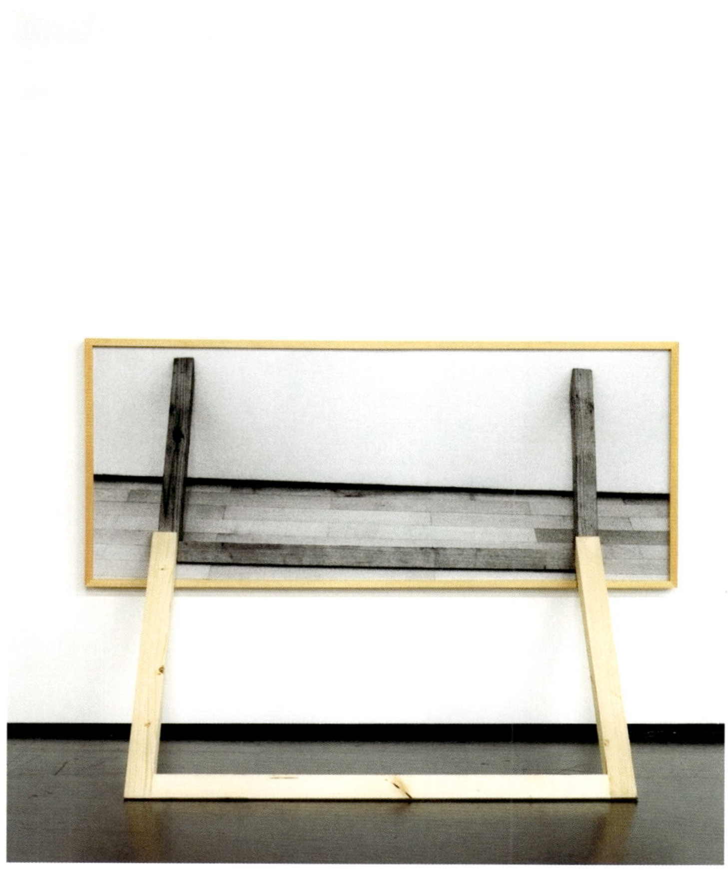

Halved Square / Halbiertes Quadrat

1977 | wood, photograph

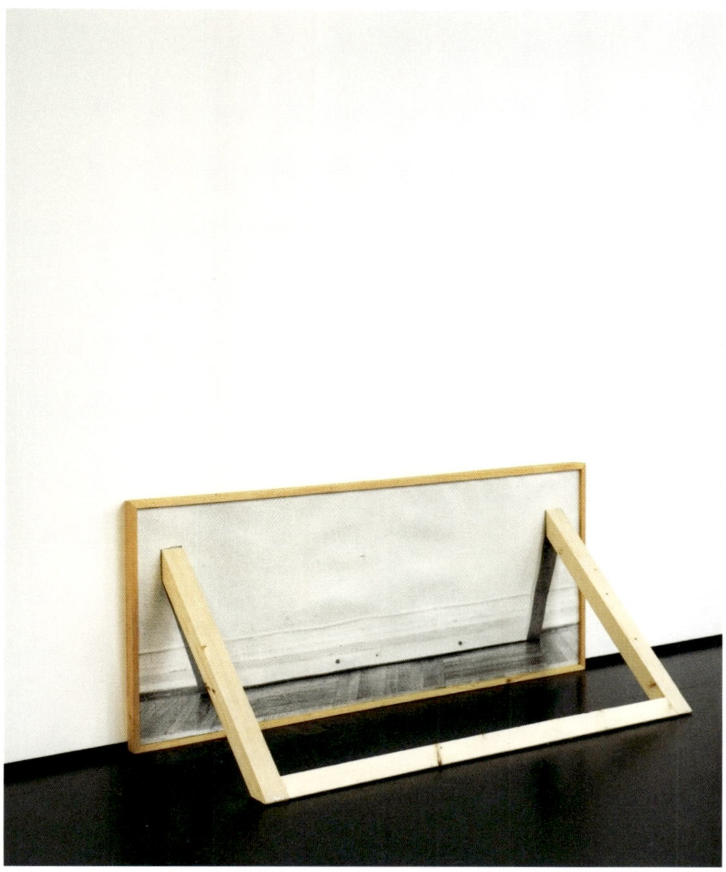

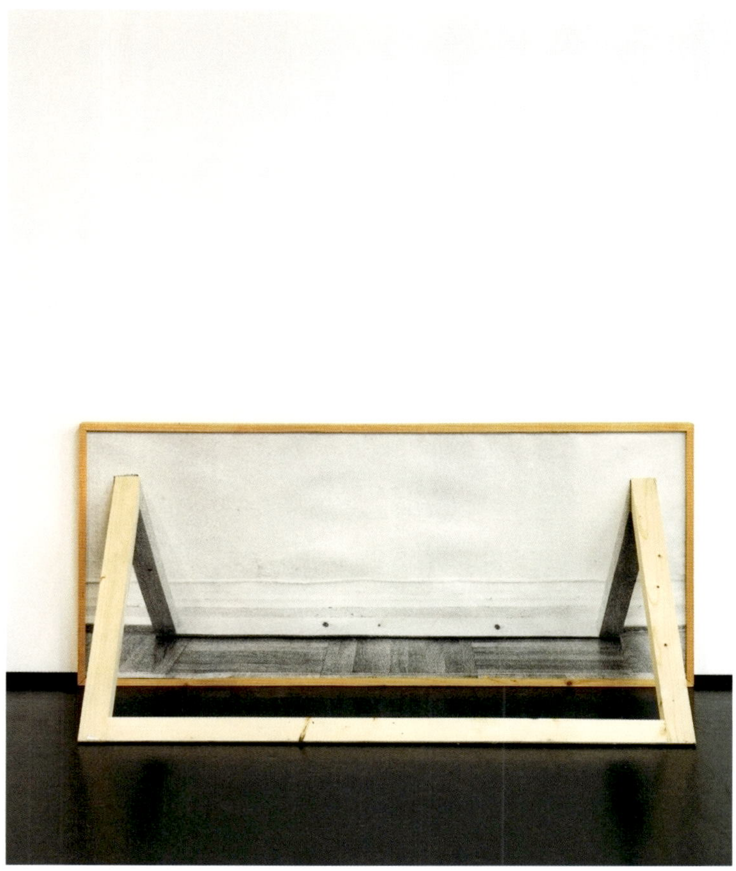

Mirror Square / Spiegelquadrat

1977 | wood, photograph

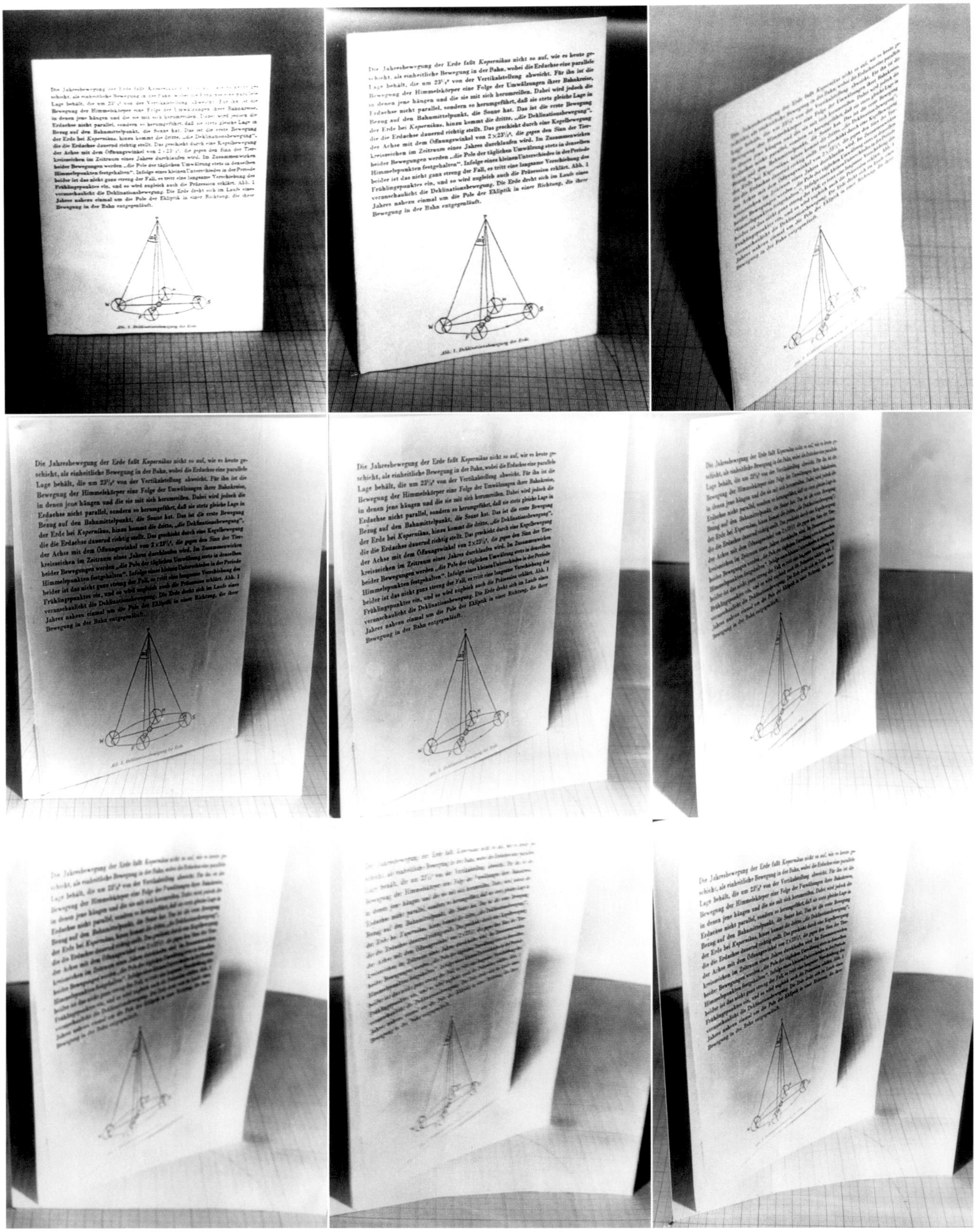

The Paradox of Perspective / Das Paradox der Perspektive

1973

from the series *Ontological Leaps / Ontologische Sprünge*
18 b/w photographs, 24 × 18 cm each

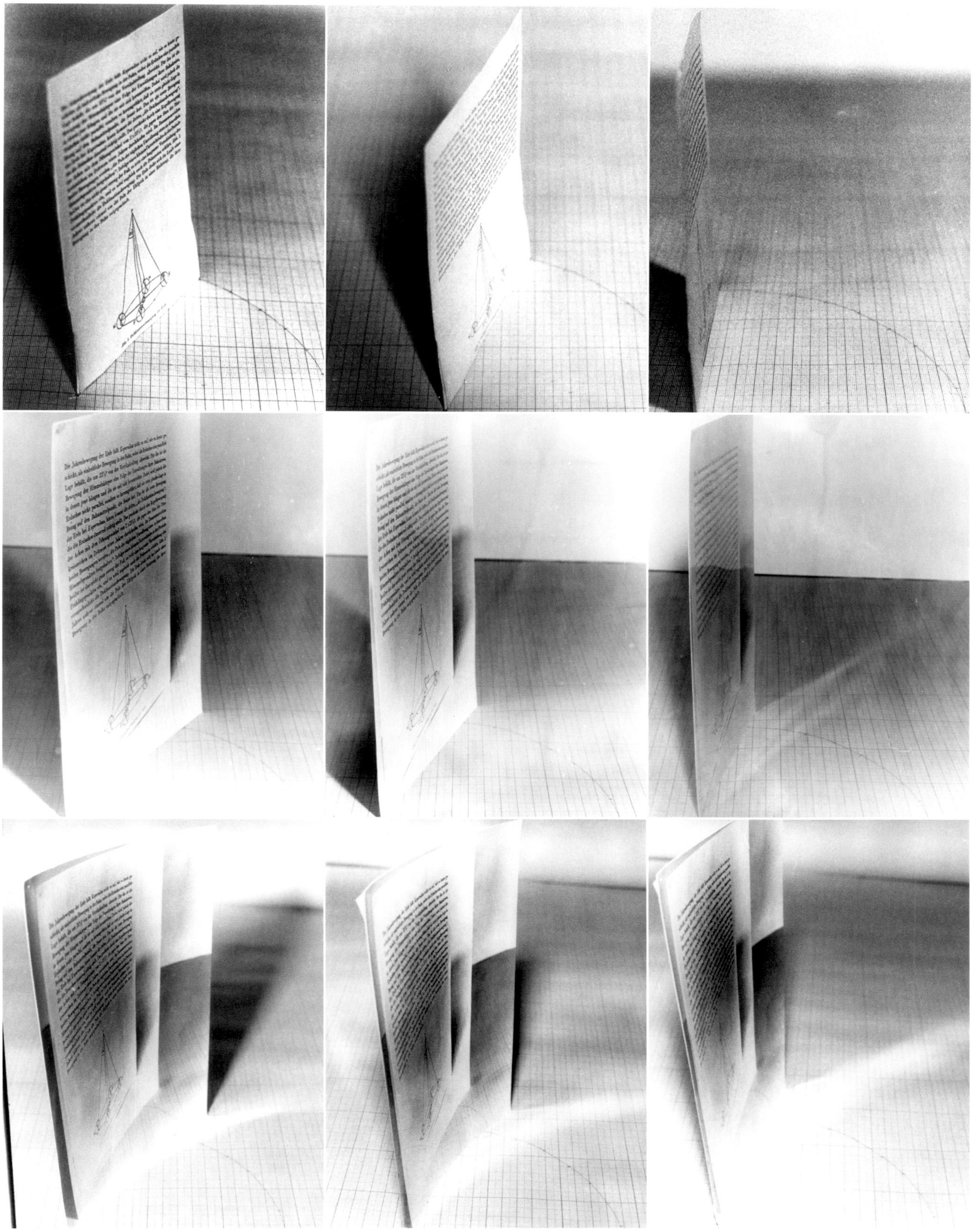

In *The Paradox of Perspective* from the series *Ontological Leaps*, a page from a book about the Earth's rotation on its axis by the astronomer Nicolaus Copernicus is first photographed. In a first series, this page is photographed at 6 different angles between 0° and 90°. In a second series, the photo with an angle of 45° from the first series, is photographed again at 6 different angles between 0° and 90°. At the third turn, the angle would sum up to 90° and the page should be invisible. The third series begins with the third photo of the second series which already shows an angle of 90°. In the third series, this results in a total of 135° by the end, which means that the photo should have already rotated around its axis and the page of the book should be unreadable. Yet miraculously, it is still possible to read the page despite its 135° rotation. The mathematical calculation of the angles thus contradicts the visible images. A paradox arises, a contradiction, with which Weibel points both to the problem of perspective and to the difference between photographic representation and reality.

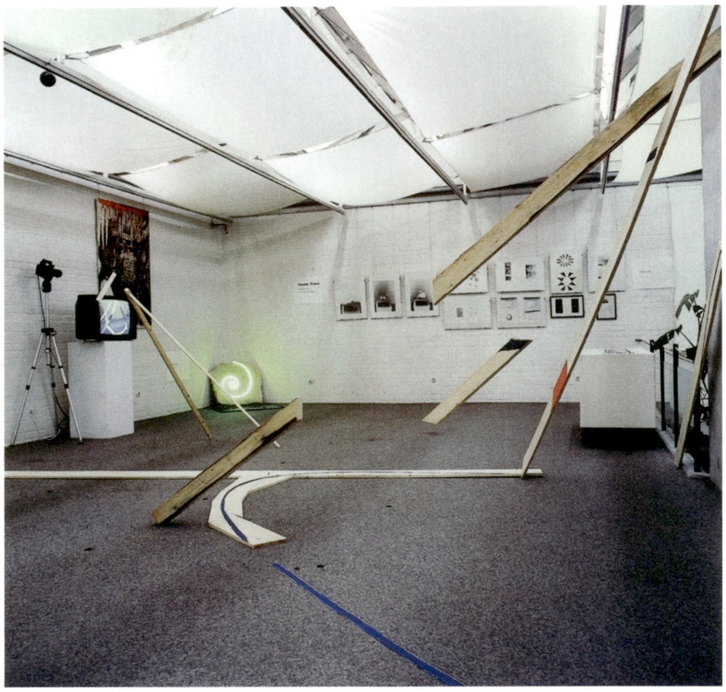

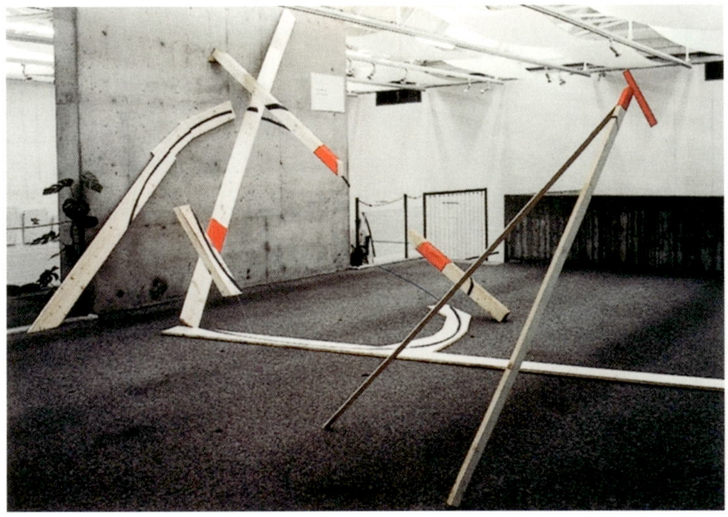

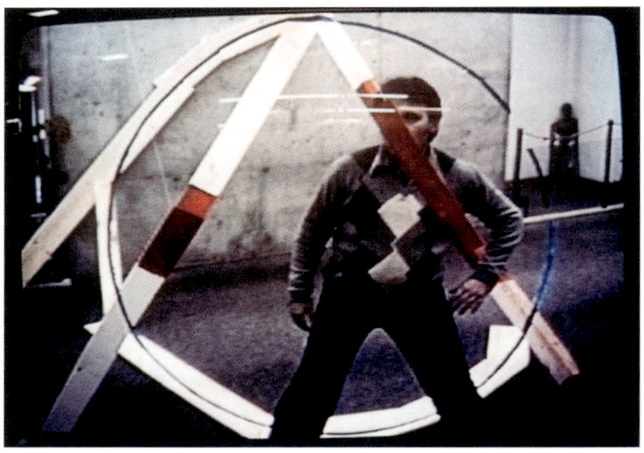

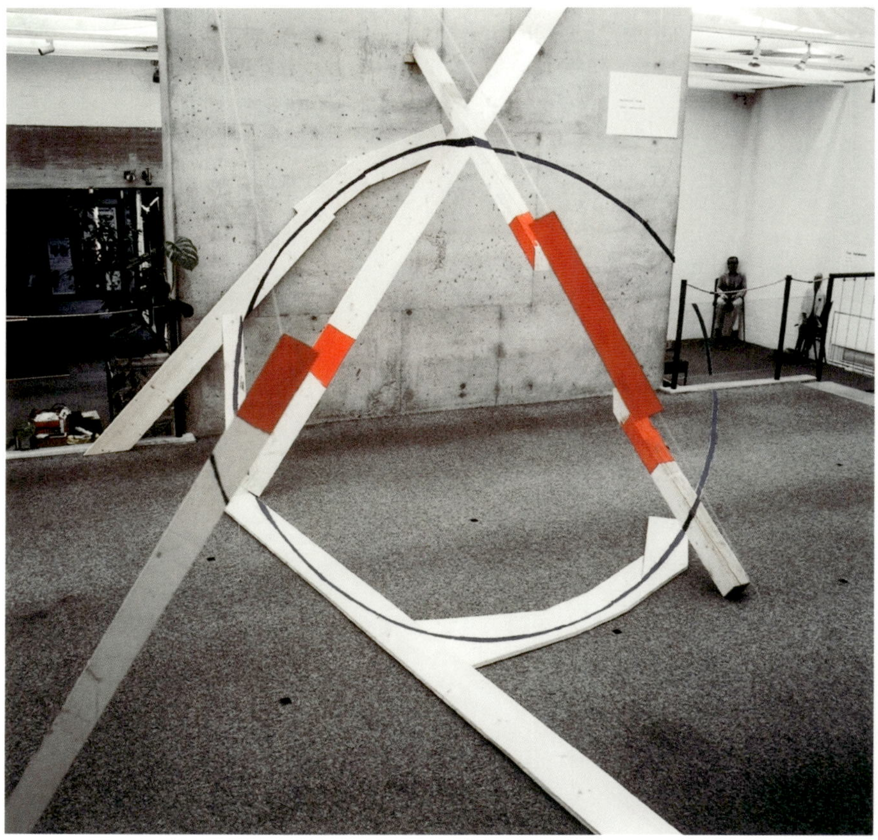

Installation views *Medien und Realität*, Kunsthalle Wilhelmshaven, 1982

Virtual Space / Imaginärer Raum

1982	closed-circuit video installation; wooden slats, monitor, video camera, paint, dimensions variable

A number of wooden slats, some lying on the ground, some hanging, are installed in the exhibition space. The slats and the walls are painted red, black, and blue in various places and shapes. From a certain position (occupied here by a video camera), the slats and the painted lines in the room combine to form a triangle inside a circle. However, this constructed image is only visible to viewers on the monitor.

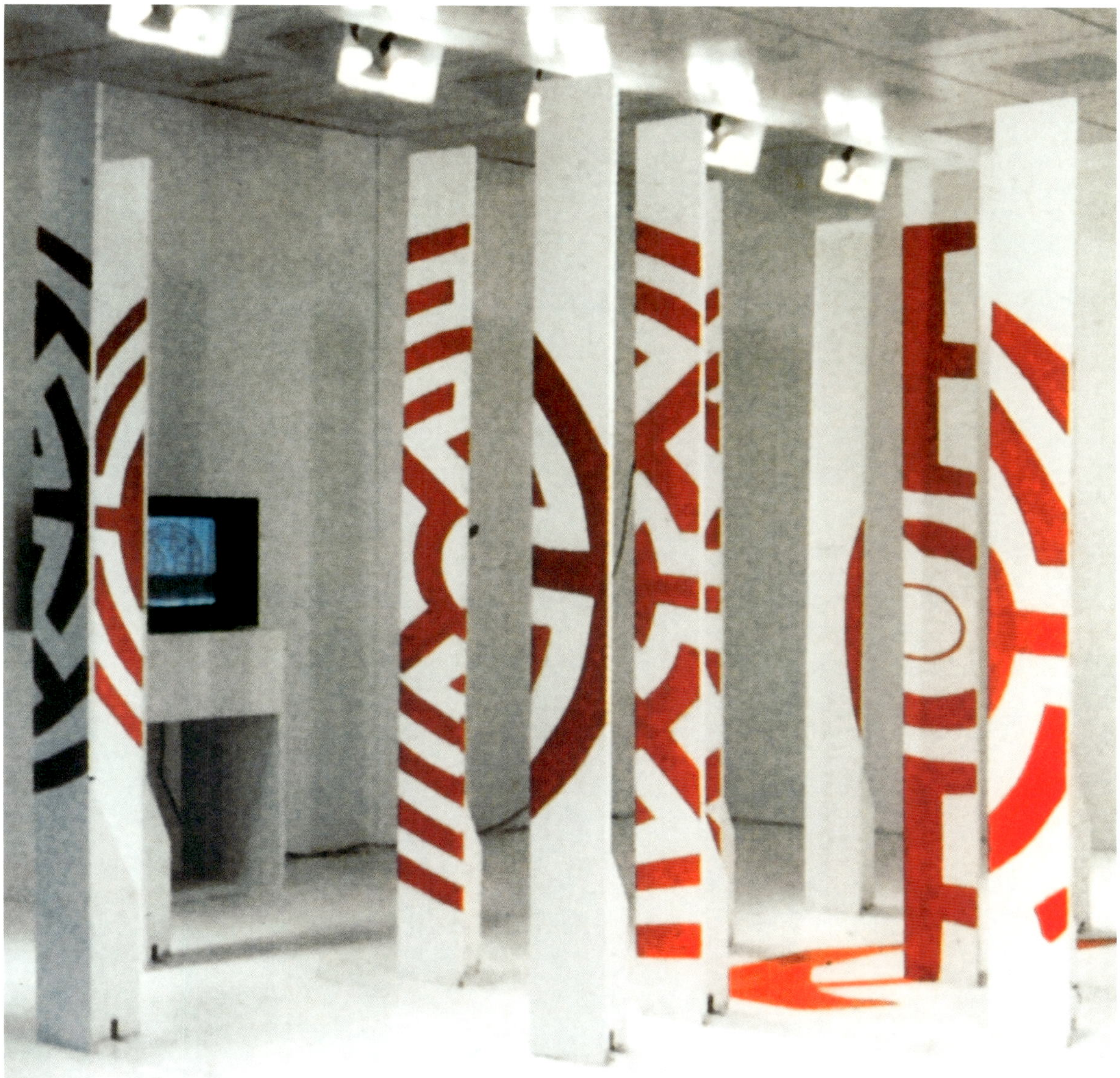

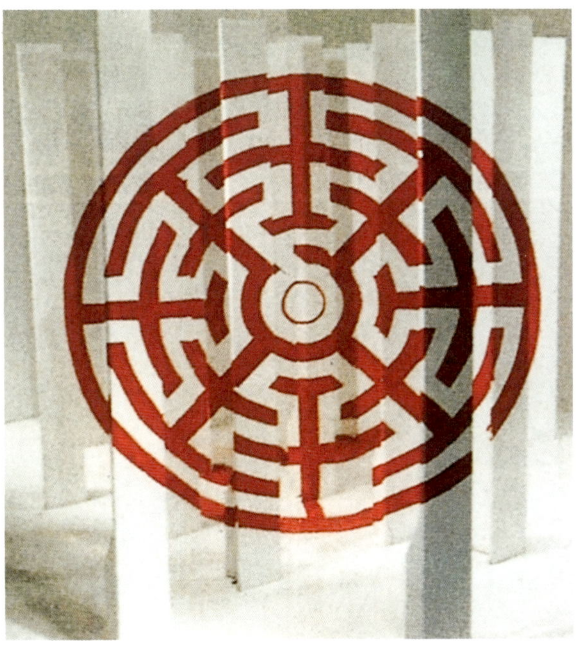

Installation views Primer Festival Nacional de Vídeo, Círculo de Bellas Artes de Madrid, 1984

Video Labyrinth

1984	closed-circuit video installation; wooden slats, monitor, video camera, paint

Viewers move through a forest of wooden slats that are painted red in an abstract way. The slats are installed vertically in the gallery space on the imperceptible outline of a labyrinth. The camera occupies a position from which the paint on the slats and the partially painted floor reveal the image of a labyrinth, which is reproduced only on the monitor.

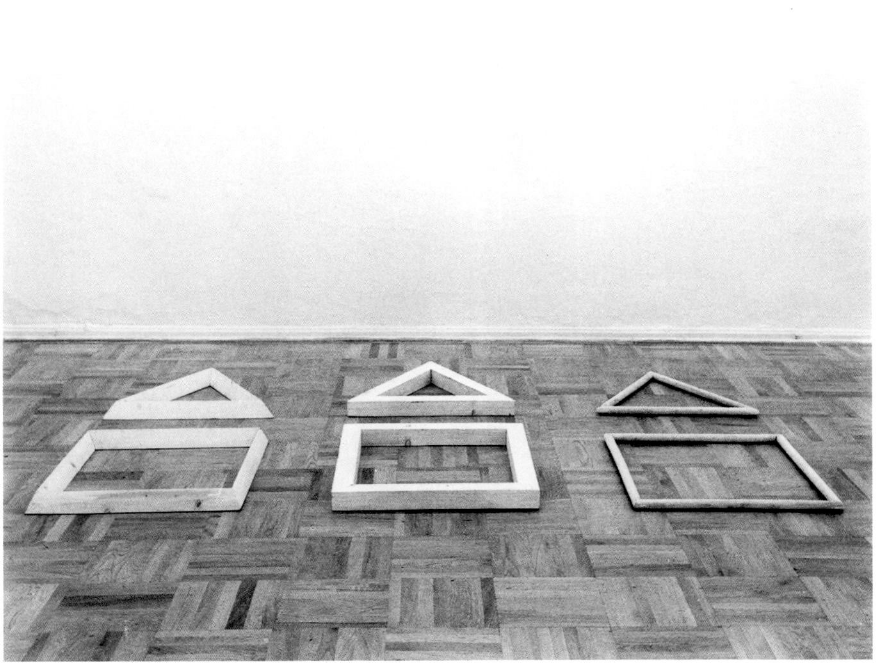

Installation view *Die unendliche Ein-Wort-Ausstellung & Skulpturen*, Galerie nächst St. Stephan, Vienna, 1977

Forced Representations, Prospective / Erzwungene Abbildungen, Prospektiv

1977 | wood, sides: 100 cm

Triangle, circle, and square can be represented as signs on paper without any problem. But when one tries to construct these three figures as objects in space, a constraint arises because they have a three-dimensional volume which can be shaped freely. The cross-section of the triangular figure can take the shape of a triangle, in which case the representation is identical. However, the cross-section of a triangle-shaped object can also be a square, a circle, etc. The same is true of the circle and square and their cross-sections. As a sculpture, a square can be circular, triangular, etc. in cross-section; a circle can be square or triangular in cross-section. Since a triangle can take on the image of a circle or square, I speak of forced representations within representations. Anyone can imagine further representations of the circle by themselves; that's the prospective part. —Peter Weibel

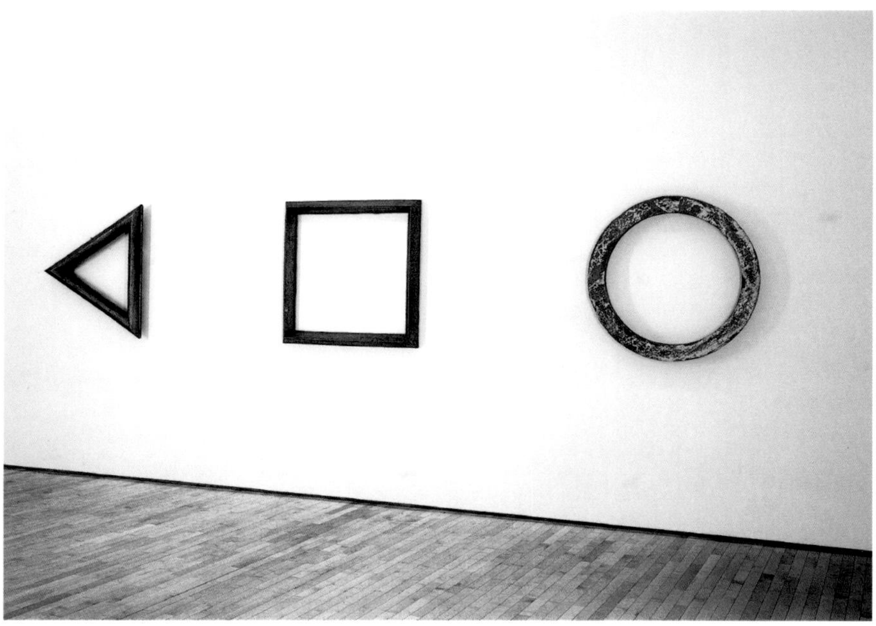

Installation view *Peter Weibel. Das Offene Werk 1964–1979*, Neue Galerie Graz, 2004

Geometric Pictorial Bodies / Bildgeometrische Körper

1975/93 | cast aluminum, 3 parts, each ca. 100 cm in diameter

In this installation, the cross section of the triangle is a circle, the cross section of the square is a triangle, the cross section of the circle is a square.

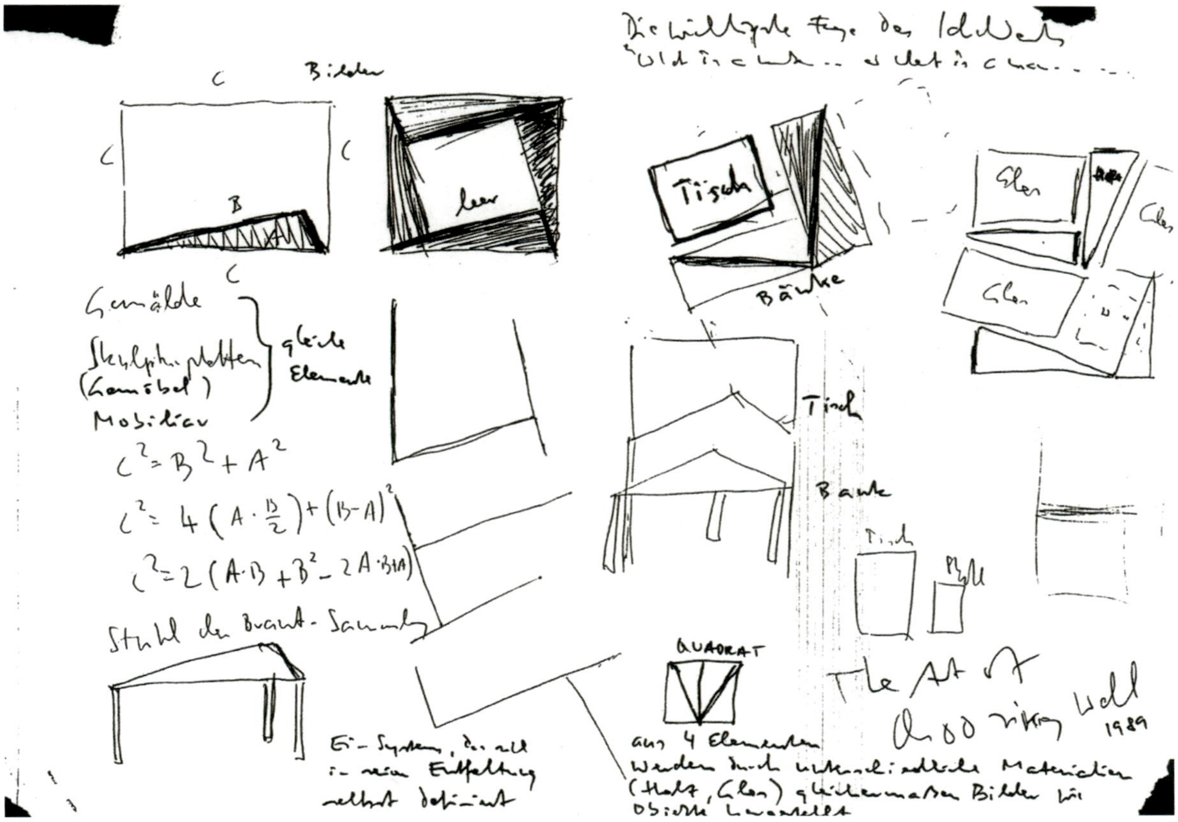

Conceptual drawing for *Universal Code: From Uniformity to Multiformity*, 1989/91

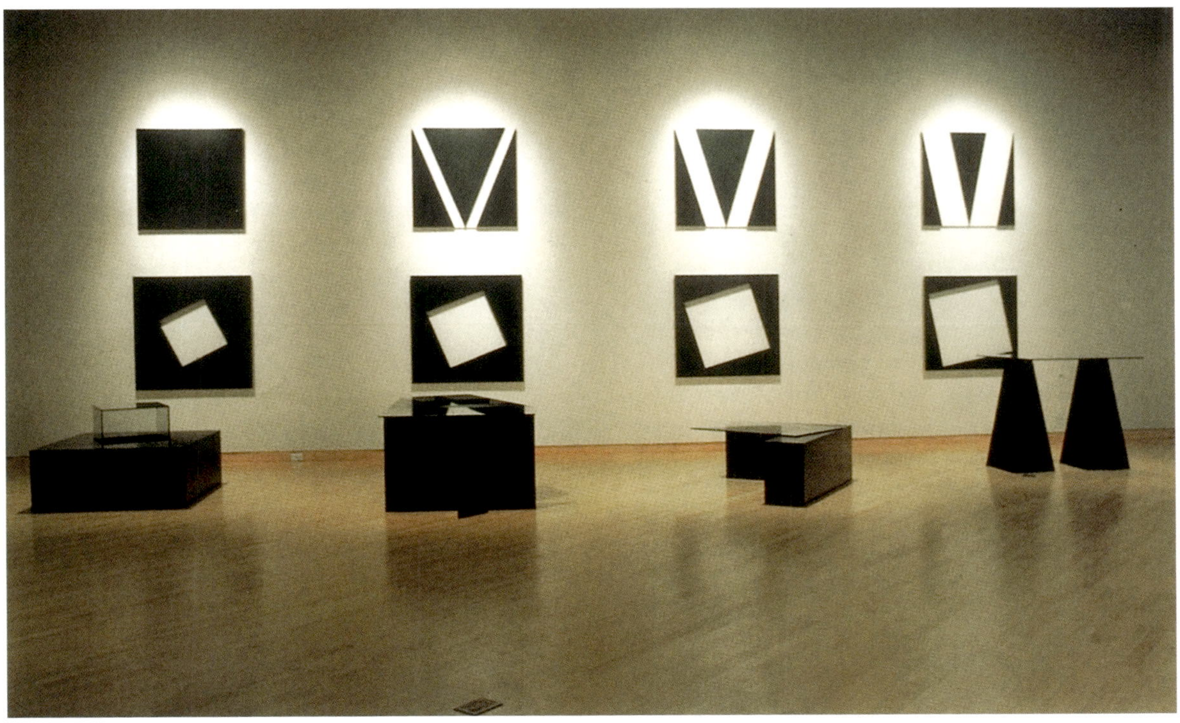

Installation view *Hypothetische Produkte. Vom Elend der Privilegien*, Prodomo, Vienna, 1994

Universal Code: From Uniformity to Multiformity / Universaler Code: Vom Einrichtungsgegenstand zum Mehrrichtungsgegenstand

1989/91 | wood, paint, glass, plexiglass

A square is divided in half, and the resulting halves are split diagonally. The gaps become steadily larger. Viewed straight on, the panels become picture-like objects. Those objects are then used as horizontal projections to generate three-dimensional volumes of varying heights.
These height variations and the different materials employed determine the volumes' functions as furniture: table, chair, bench, etc. A mathematical formula produces specific three-dimensional shapes – a system that evolutionarily and morphogenetically defines itself. —Peter Weibel, 1996

Scaling and Scanning: Intrinsic Properties of Media Transposed onto the World of Objects

Peter Weibel

In the world of media, scaling has taken over the function that was performed, in the classical real world, by proportion. Both proportion – the size relationship between the various parts of a whole, the relationship of the parts to the whole – and scale – the ratios and orders of magnitude that exist between objects or parts – have traditionally taken as their starting point the human body, human scale, the measurements of the body and its proportions, as in Le Corbusier's *Modulor* (1948, see fig. 1).

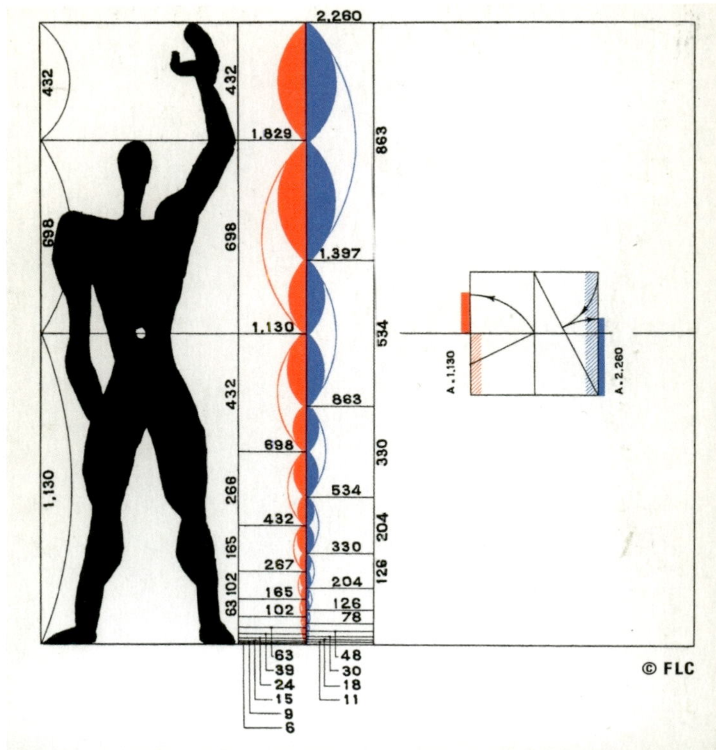

Fig. 1: Le Corbusier, *Modulor*, 1948

The premise of my work, however, is the thesis that humanity's experience of space and time has, since the nineteenth century, been determined not by our bodies, but by vehicles such as automobiles, trains, bicycles, and airplanes, as well as media such as the telephone, the fax machine, television, radio, etc. – in other words, teletechnologies, from the Greek word *tēle*, "far away." Space is no longer measured in cubits and feet (i.e., body-based measurements), but rather registered in terms of time and speed. The speed of an airplane in relation to the ground over which it flies is called "ground speed"; so, for example, a distance of 1,000 km on the ground is an hour in the air. Thanks to the wireless transmission of messages, which no longer require the body of the messenger as a carrier, message and messenger have been separated since the mid-nineteenth century: the message travels alone on electromagnetic waves. The speed and accelerated movement provided by machines and the all but simultaneous transmission of messages via telemedia have radically changed humanity's experience of space and time – or to be precise, our units for measuring space and time, the proportions of our spatial and temporal units. Our media have given rise to a near-bodiless experience of space and time, a shifted, unnatural scale of sizes and size relationships. This is wonderfully expressed in the paintings of Giorgio de Chirico (see fig. 2), which he called "metaphysical" because of the way they contradict physical, body-centered proportions.

A photographic image can reduce a thousand-meter-high mountain to 10 cm, a human to 2 cm. But, conversely, a photograph can also enlarge a thumb tenfold, a microbe a thousandfold. In media systems, therefore, measurements are determined by and derived from technological apparatuses rather than natural organs. Proportion and scale, in the world of media, consist of variable parts, variable orders of magnitude, variables. Every photograph is inherently variable in terms of its scale. Every technological image is a field of variables.

In a series of three-dimensional works, I have attempted to translate the media experience of space and time, scale and proportion, back into sensory experience – to turn representational media into objects of natural reality – in order to make media's intrinsic properties visible. The properties of media are translated back into properties of objects, the experience of media into the experience of reality. Media qualities versus sensory qualities.

I have favored furniture and objects because they show the contrast between movement and stasis – mobile and immobile, movables and immovables, household goods and real estate – most clearly, especially via the principle of reduction.

Thus the Guggenheim Museum became a bar (see pp. 404–405 in this volume), Buffalo City Hall an armchair with a built-in bar and car radio, the New School of Research in New York a chest of drawers, etc. (see also pp. 402–403 in this volume). These reductions in size have also brought secret and hidden meanings and functions to light.

Once I had begun rescaling and reproportioning parts of objects and furniture, of urban and household objects, it opened up a whole horizon of furniture transformations, and while these seemed surreal, in reality they were illuminating, as furniture became buildings, buildings became furniture, lamps became tables, tables became fountains,

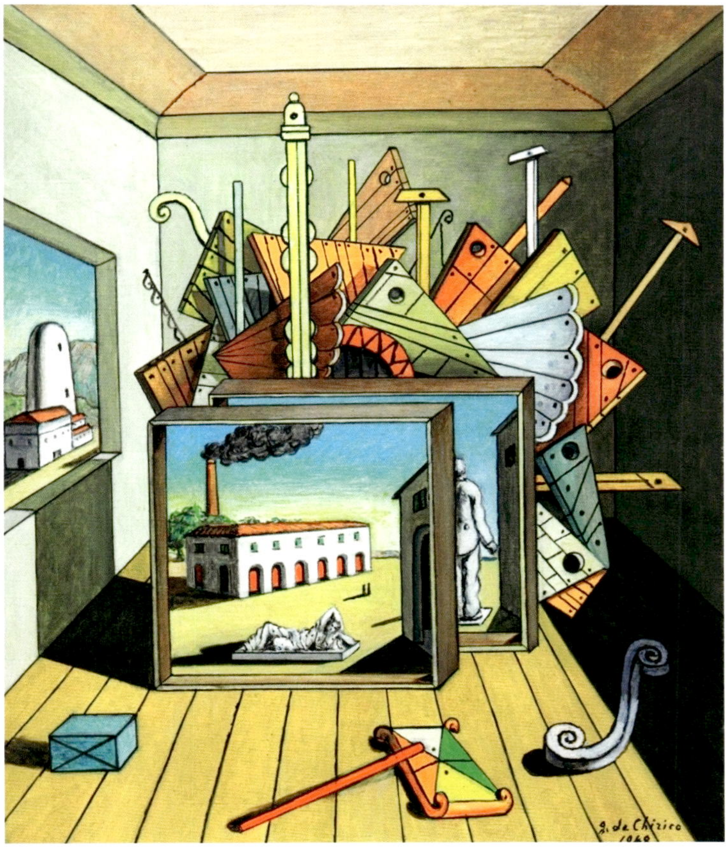

Fig. 2: Giorgio de Chirico, *Metaphysical Interior with Workshop*, 1948

stacked-up highway interchanges became étagères, and so on (see pp. 420–23 in this volume).

The variability of scale, which turns size relationships into undefined variables – small things can be big, big things small – is an intrinsic property of the strategies of visual representation, from painting to photography to the screen. Every image is the product of a process of scaling, of up- or downsizing.

As a result of the particular way in which they are produced – the decomposition of an image into lines of points and the projection of these scan lines as a stream of electrons in a cathode-ray tube – the electronic images produced by televisions and video players possess another new property as well: the scanning principle, which is also the basis for 3D printing, the construction of objects from extruded lines. I first made use of this principle in 1986, in the video *Chants of the Pluriverse / Gesänge des Pluriversums* (see pp. 304–307 in this volume).

The scanning principle was what enabled Alexander Bain to transmit a drawing to a remote location for the first time in 1843 with his "automatic copying telegraph," which rendered an image or text as a series of lines. Each point in these lines represented the presence or absence of graphic information. A drawing thus became a single line in the sense of a sequence of points in time. Having been sequentially scanned in this way, the graphic source was reconstructed in another location according to the same principle. Frederick Bakewell (1800–69) used the same process in his "image telegraph" a few years later.

Today, any book can be transferred to any computer screen by means of the scanning principle. A networked computer, whose screen is itself dependent on the scanning principle, is today's library. The local, material repository of information that is the library becomes a nonlocal, immaterial repository, an "informationary." The global informationary, the networked world of computers, extends the local library.

My *Scanned Objects* and *Scanned Sculptures* (see pp. 410–15 in this volume) repeat this process of converting the object into information, of transforming the world of objects into a world of signs that are not bound to any location, but rather dislocatable and telematically transportable. The collapse of the historical universe of objects brought about by the scanning principle, the principle of dislocation, has given rise to a new world of freely available and constructible signs, behaviors, and object choices.

I have taken this principle – another property intrinsic to the world of media, alongside variable scaling and unfixed proportion – and transposed it onto the real world of objects. By transferring this fundamental, generative, artistic principle of the media image back onto the natural world of objects, I emphasize the difference, the *separatio*, between reality and representation.

For the series *Scanned Objects*, I sawed everyday objects – armchairs, doors, TV sets, suitcases, entire library walls, and even (purely photographically) skyscrapers – into a multitude of horizontal strips, which I then reassembled, shifted, and skewed, placing plexiglass panes between the strips. Because the strips were uniformly cut and distributed, the ordinary object was represented as a calculable body in space. By transferring the principle of the digital scanning process onto the analog in this way, I was already anticipating the principle behind 3D printing technology in the 1990s.

My *Scanned Sculptures* are an extension of this process. Based on digital 3D models of well-known artworks freely available on the internet, they continue the convergence of digital and analog works.

Note: This text was written in 2022 and is published for the first time in this volume. Translated from the German by Patrick Hubenthal.

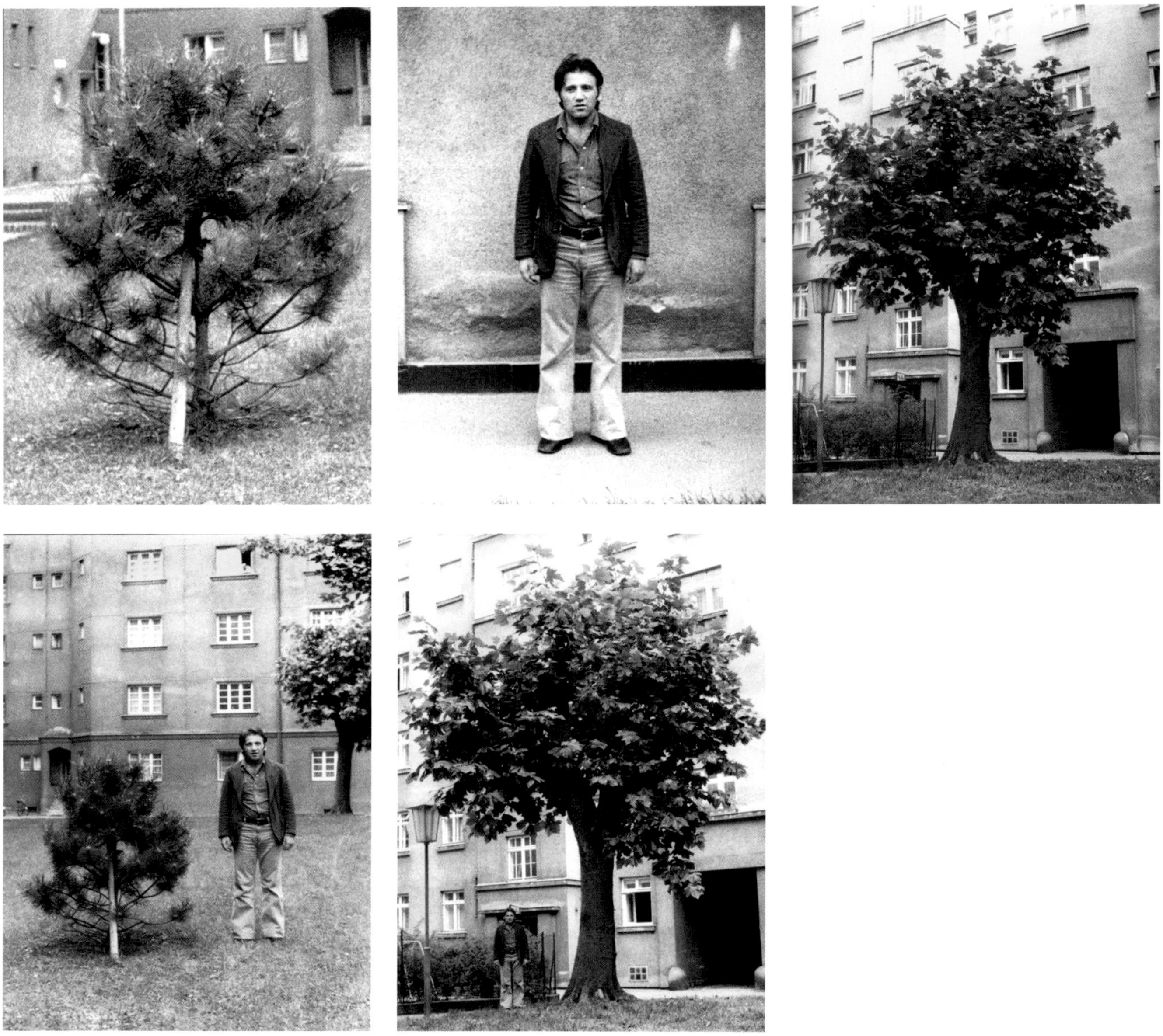

Scaling: Paradox of the Reference System (Medium) I / Skalierung: Paradoxie des Referenzsystems (-mediums) I

1974 | 5 b/w photographs, 2 each 24 × 18 cm, 3 each 36.5 × 26 cm

A total of five black and white photographs show three objects in an inner courtyard. In three of the pictures Peter Weibel, a deciduous tree, and a conifer are photographed individually; in two further pictures the artist stands next to the trees. Compared to painting, photography has the status of the faithful reproduction of objects. Photographs are taken of objects that actually exist/existed and their representation in the photo appears true to scale. In his work *Scaling: Paradox of the Reference System (Medium) I*, Weibel presents the paradox inherent in this technical process: the three individually depicted objects are the same size due to the photo format. Only through the reference of the person depicted next to the trees in the two other photos do the proportions become clear. The trees are not the same size as the photos suggest. Their size is very different.

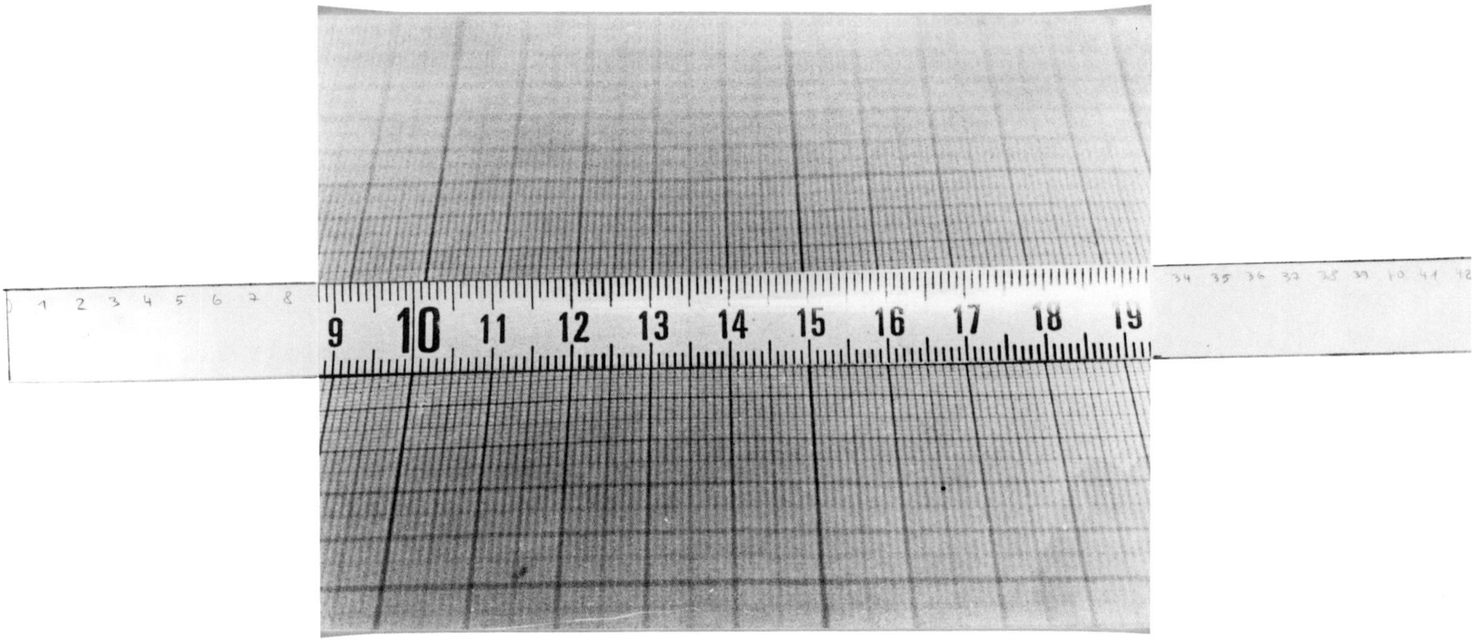

The Measure of Measure /
Das Maß des Maßes

1975 | measuring tape, photograph

Enlargement and reduction are frequently used photographic techniques, minimal reshapings of the world. When exaggerated for contrast, however, they become conspicuous. A photographically enlarged measuring tape, inserted into the middle of a real (drawn) measuring tape, ends at 19 cm, at a point that should mark 33 cm. This shows arbitrariness and relativity of all standards of measurement. —Peter Weibel

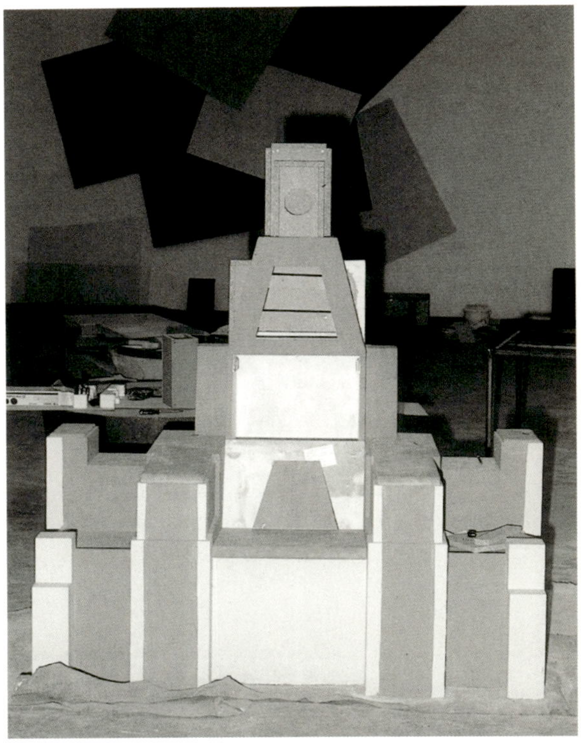
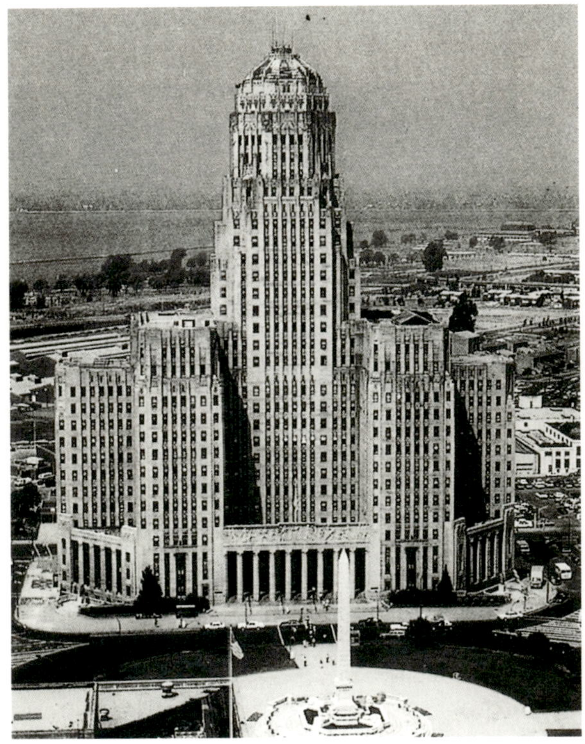

Buffalo City Hall as an Armchair with Car Radio and Built-in Bar / Buffalo City Hall als Sessel mit Autoradio und eingebauter Bar

1988 | plywood, radio, mirror, built-in bar

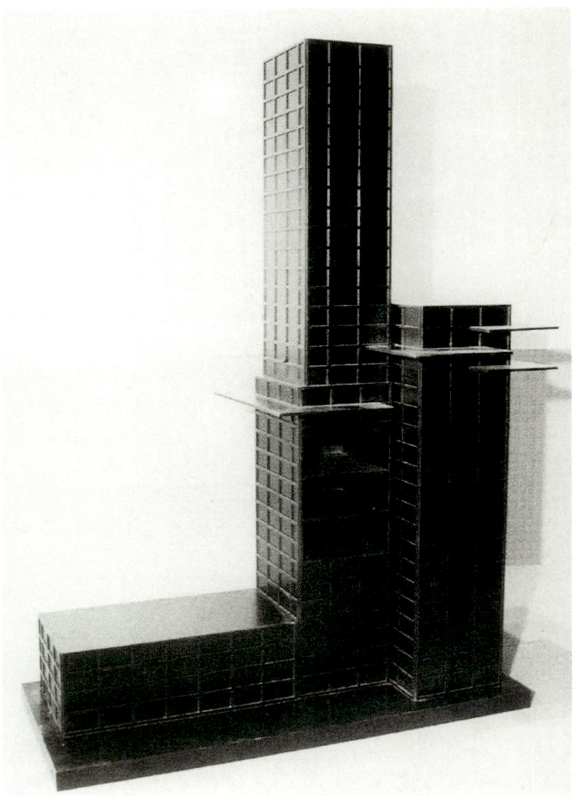
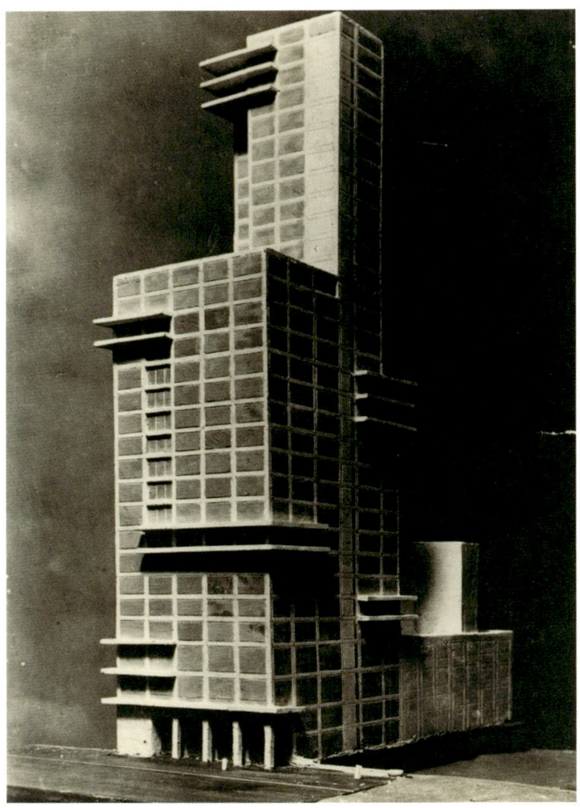

Chicago Tribune Tower (Project 1922 by W. Gropius and A. Mayer) as a Freestanding Shelf / Chicago Tribune Gebäude (Projekt 1922 von W. Gropius und A. Meyer) als freistehende Ablage

1988 | plywood, paint

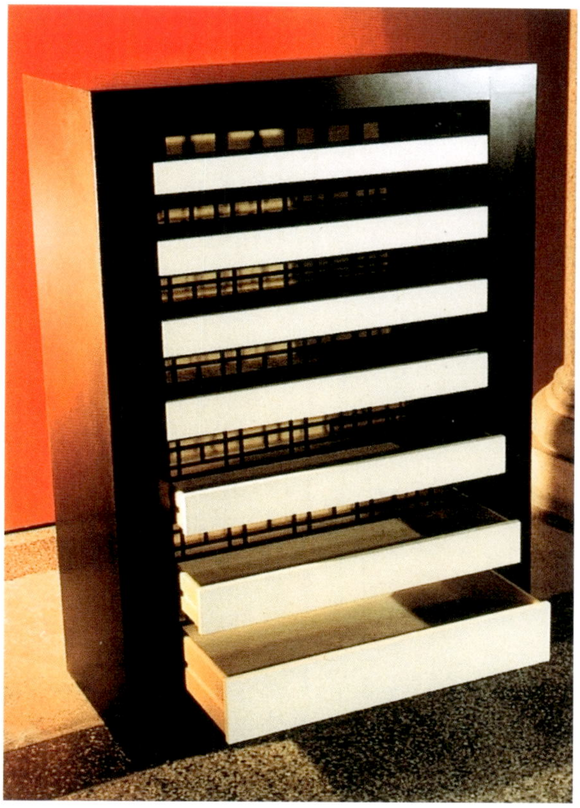

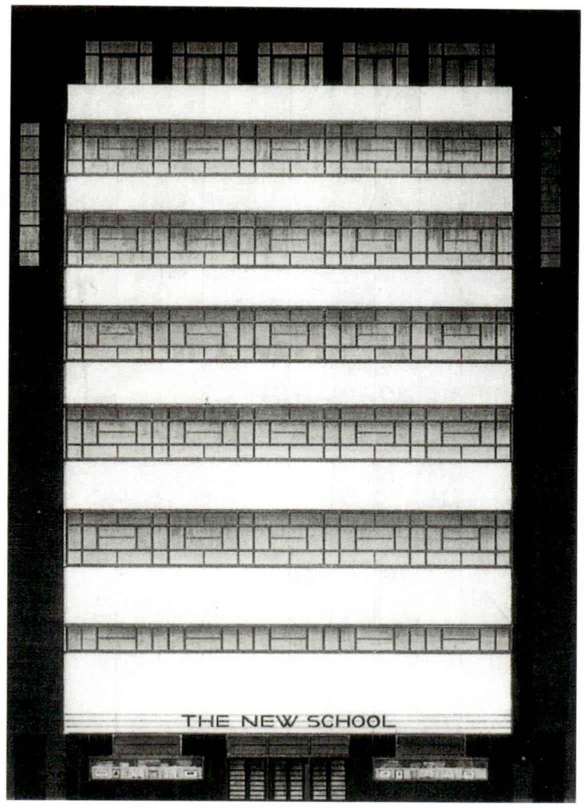

The New School in New York by Joseph Urban as a Commode / The New School in New York von Joseph Urban als Kommode

1988 | plywood, plexiglass, paint

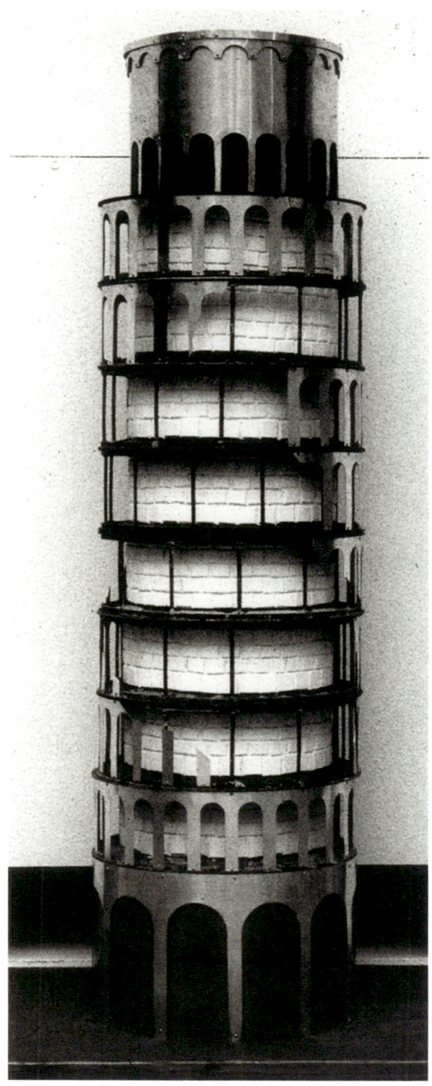

The Leaning Tower of Pisa as a Flowerpot / Der schiefe Turm von Pisa als Blumentopf

1988 | aluminum, sheet metal, water, cardboard, wood, paint, asphalt

Set photos of *Chants of the Pluriverse* (*Gesänge des Pluriversums*), 1986–88 (see pp. 304–307 in this volume), with Susanne Widl as Peggy Guggenheim

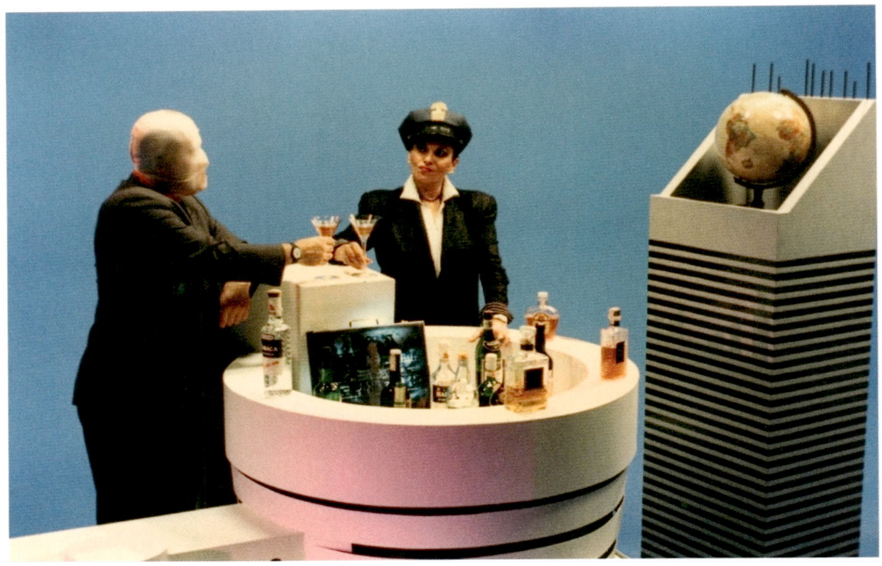

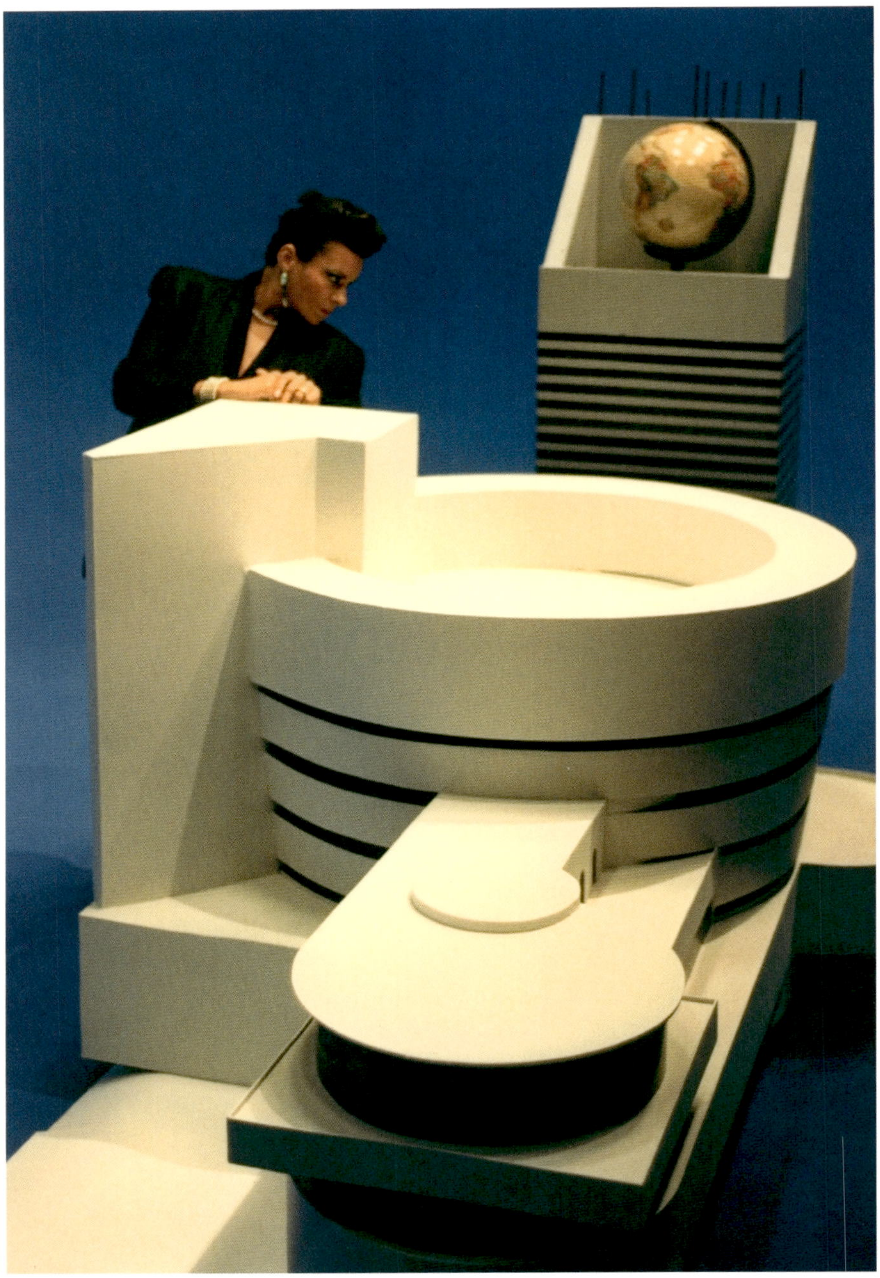

Guggenheim Museum as Bar / Guggenheim Museum als Bar

1988 | wood, paint, 120 × 290 × 140 cm

(opposite page)
Installation view
respectively, Peter Weibel, ZKM | Karlsruhe, 2019–20

The theme of scale, of scaling, which is a distorting characteristic of media, is reflected in the work *Guggenheim Museum as Bar / Guggenheim Museum als Bar* from 1988. Here, the artist scales, redimensions, and reduces the museum's architectural structure. In this way, an icon of museum architecture is transformed into an ideal bar. The work is part of a series in which various famous architectural buildings are scaled down to the size of functional furniture.

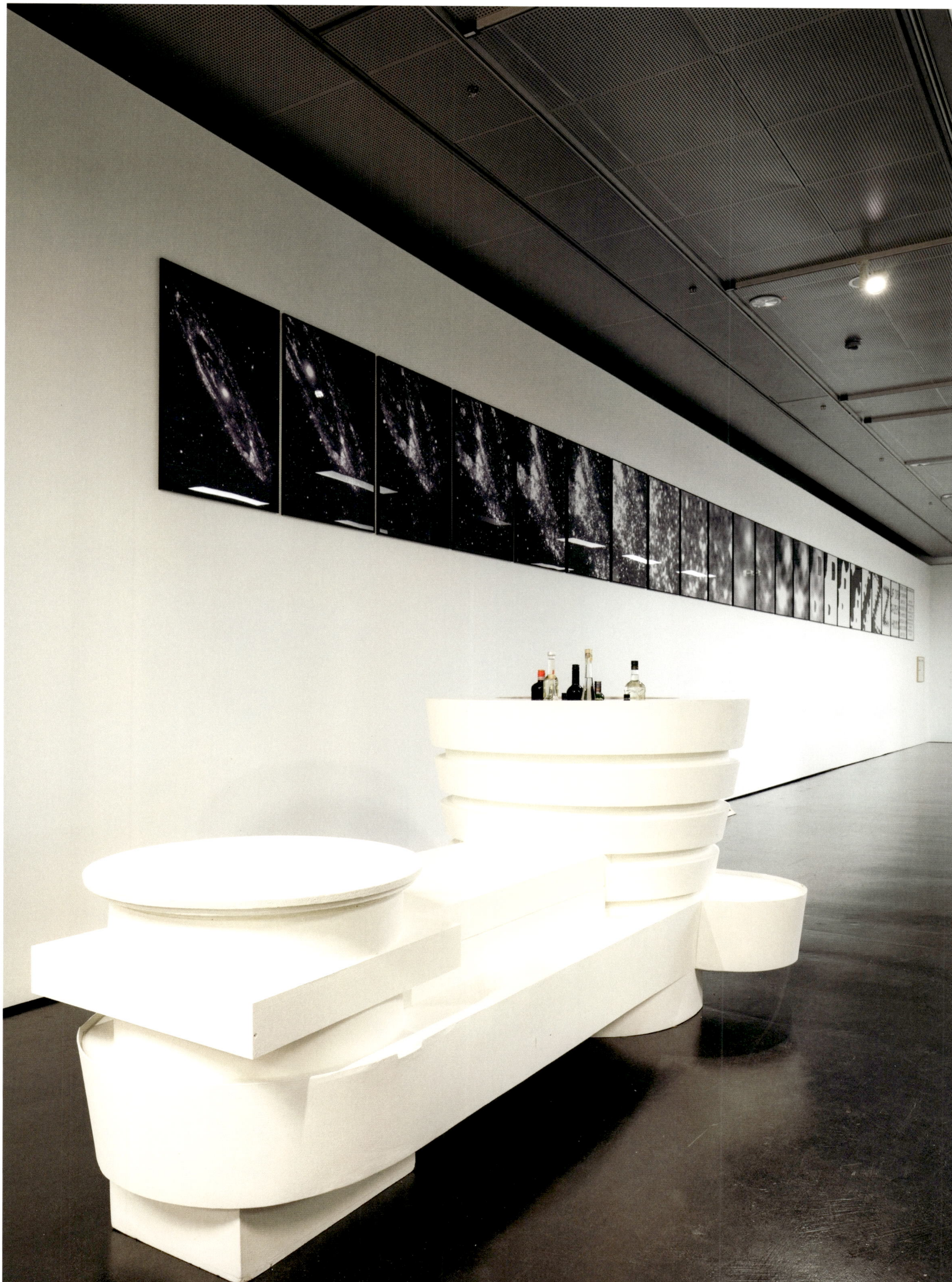

This page and opposite page, top left: *Inhabitable Library I*, installation views respectively, Peter Weibel, ZKM | Karlsruhe, 2019–20

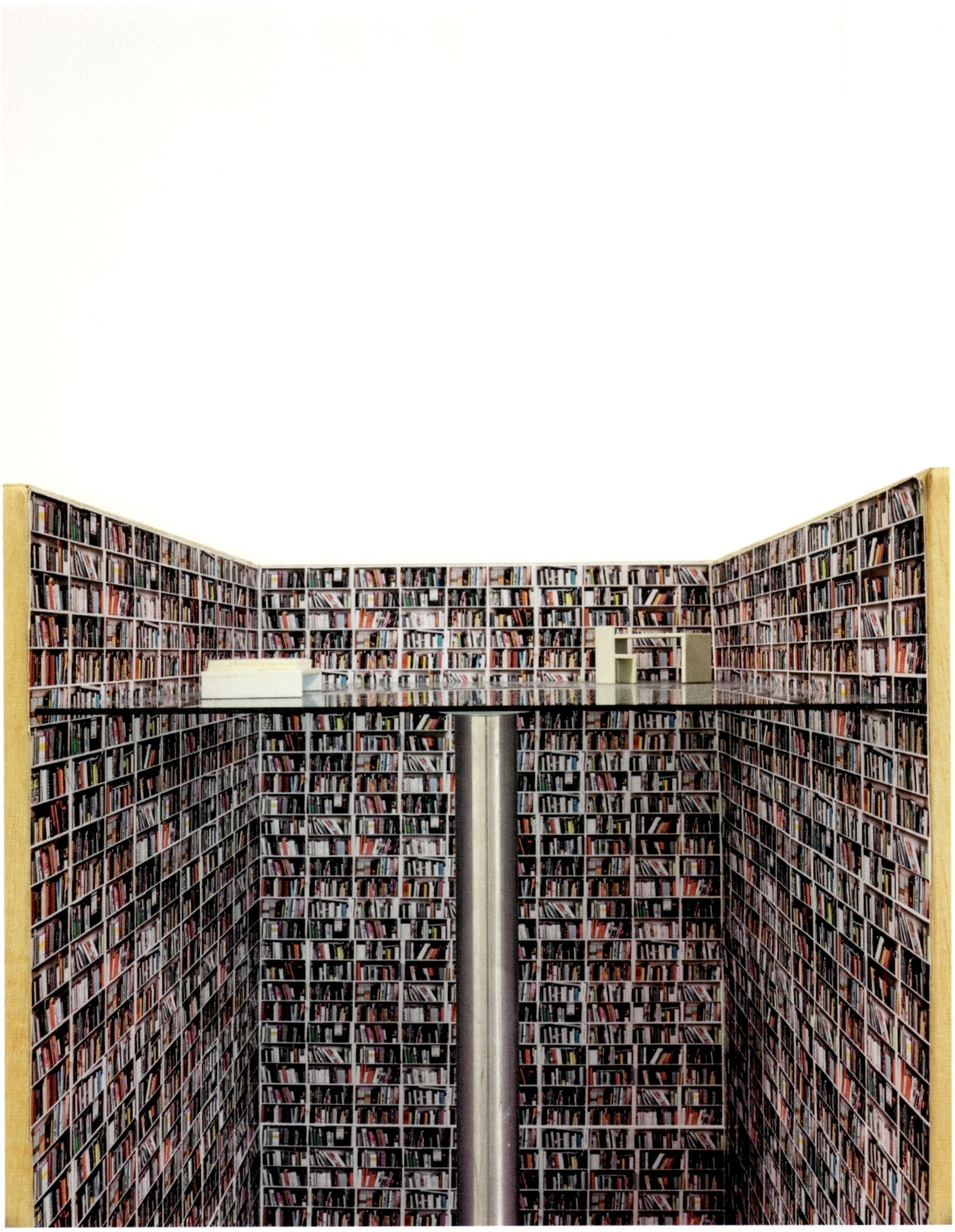

Inhabitable Library I–III – The Elevator is the Apartment / Bewohnbare Bibliothek I–III – Der Aufzug ist die Wohnung

2009	*Inhabitable Library I* and *II*: living towers; wood, book wallpaper, plexiglass elevator, 190 × 50 × 45 cm (I), 150 × 100 × 30 cm (II) *Inhabitable Library III*: living tower; wood, cast-in books, plexiglass elevator, 150 × 40 × 100 cm

The floor plan of an apartment reflects the basic functions of all living beings. In the kitchen you cook, in the bedroom you sleep. The layout of an apartment always reflects complex social relationships. The works *Inhabitable Library I–III – The Elevator is the Apartment* depict a new way of living, free from traditional conventions and adapted only to the individual wishes of the artist. The elevator does not lead to the apartment, it is the apartment. Equipped with a digital infrastructure that enables communication with both the library and with the outside world, the elevator functions as a space for living, sleeping, and working, always moving along the bookshelves.

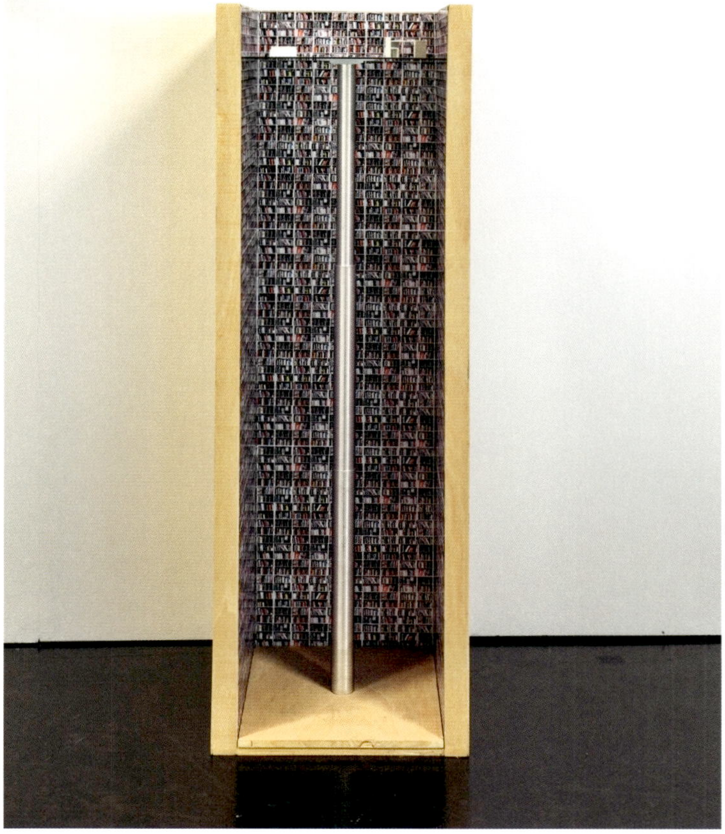

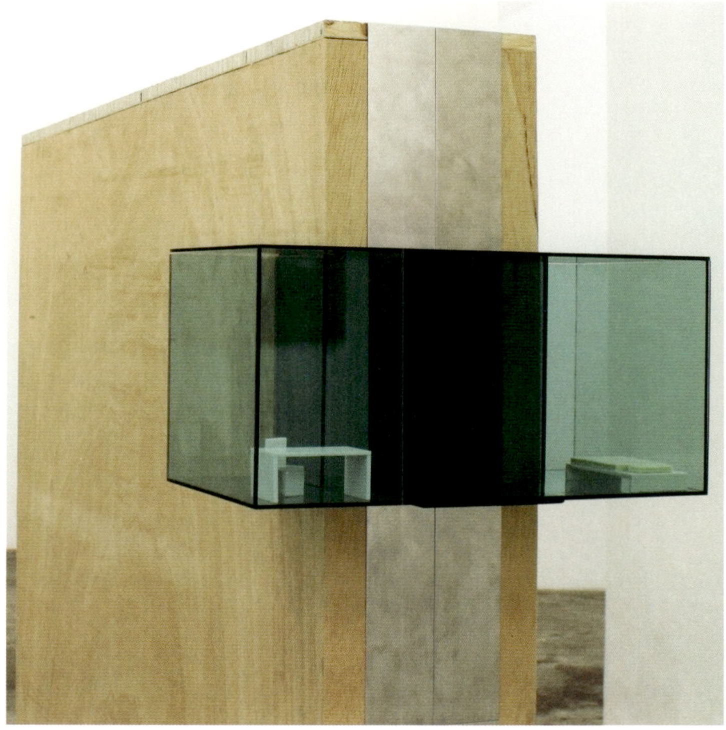

This page, top right and bottom: *Inhabitable Library II*, installation view Galerie Grita Insam, 2009

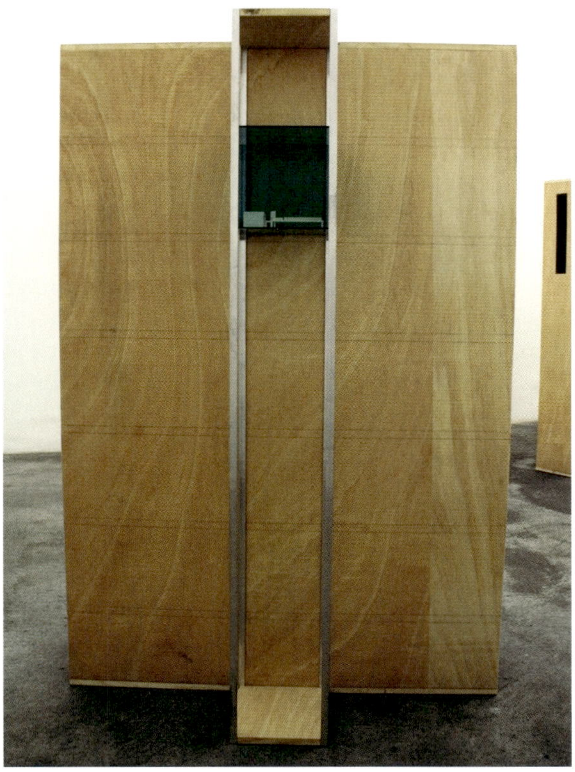

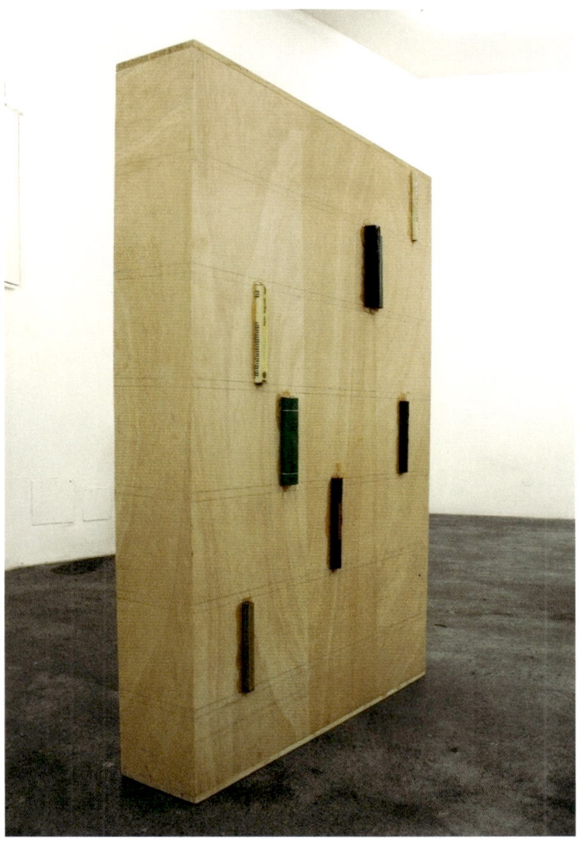

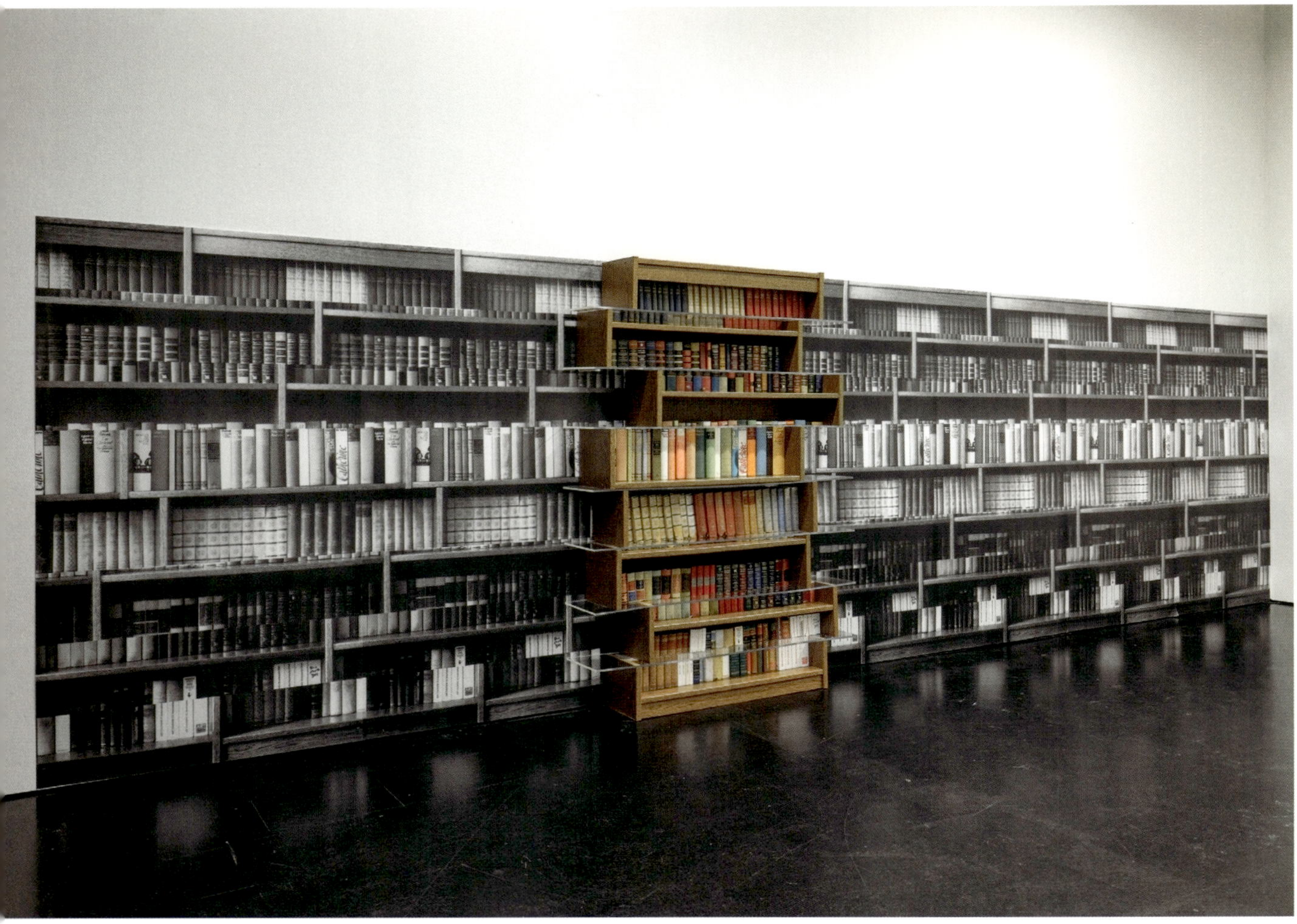

Installation views
respectively, Peter Weibel,
ZKM | Karlsruhe, 2019–20

Since ancient times, the library – in Latin *bibliotheca*, from the Greek *biblion* (book) and *thēkē* (container – has been a repository for books, a place for the study of literature. In the Middle Ages and the Renaissance, libraries were long galleries with rows of lecterns and bookshelves. In the grand library halls of the Baroque era, such as the Benedictine monasteries in Melk on the Danube (1730) and Admont (1774), the walls were lined with built-in bookcases. The library became a Gesamtkunstwerk, its walls decorated with books instead of paintings. In the nineteenth century, the growing numbers of books transformed the library hall into stacks and storerooms. The artistic arrangement of the library was eclipsed by the scholarly use of information. At the same time as this aesthetic transformation, the library also underwent a social one. While most of Europe's libraries were founded by clerics in the seventh and eighth centuries as monastery libraries, in the late medieval period these became the scholarly libraries of the universities that had evolved from monastic and cathedral schools. The nobility and the urban patriciate became the new patrons of learning and founders of libraries. In the wake of humanism, private libraries emerged and then gradually developed into public libraries. In the nineteenth century, the library attained perhaps its greatest significance, as university libraries instituted a standardized cataloging system and enjoyed a permanent budget for new acquisitions. The first professional librarians appeared in this period as well. Today, in the age of the World Wide Web, there is a new information space that is not physical and not bound to any location, prompting a critical reassessment of the library's function. Neither the aesthetics of the library nor that of the book is relevant in this era; the only thing that matters is information. Peter Weibel's installation *Scanned Library*, from the series *Scanned Objects* (1990–2015), is a reflection on this transition from the material, local library hall to the immaterial, nonlocal, global information space.

Note: Adapted from Christa Steinle, "Peter Weibel/*Scanned Library:* Von der Bibliothek zur Infothek," in *Peter Weibel: Globale Gier*, exh. cat. (Klagenfurt: Kärntner Landesgalerie, 1999), 22–24. Translated from the German by Patrick Hubenthal.

Scanned Library: From Books to Information / Scanned Library: Von der Bibliothek zur Infothek

1996	silk-screened photo wallpaper, bookshelf with books, dimensions variable

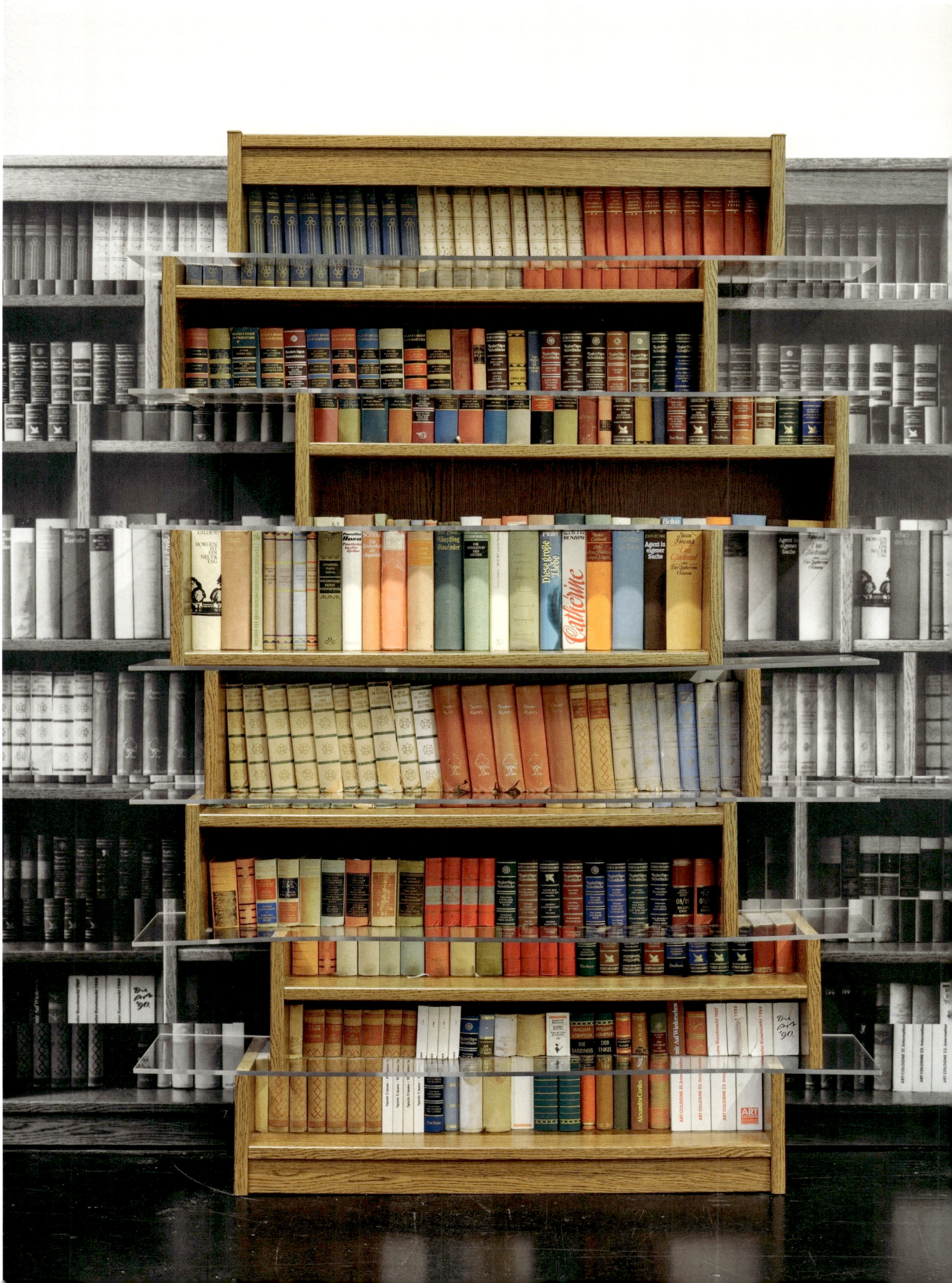

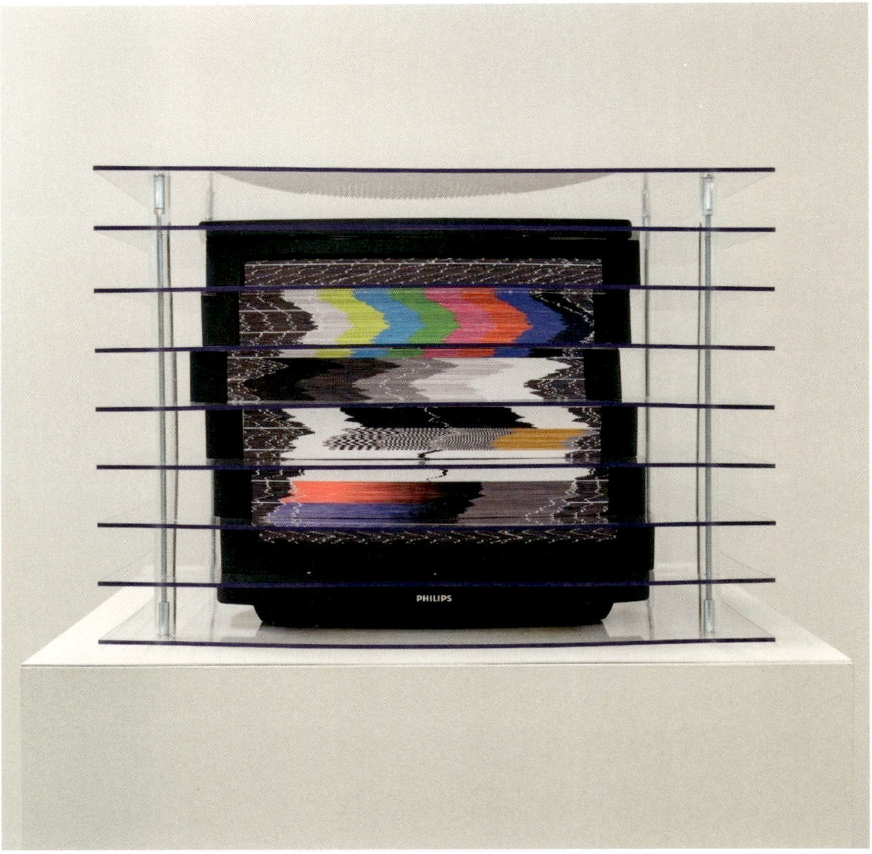

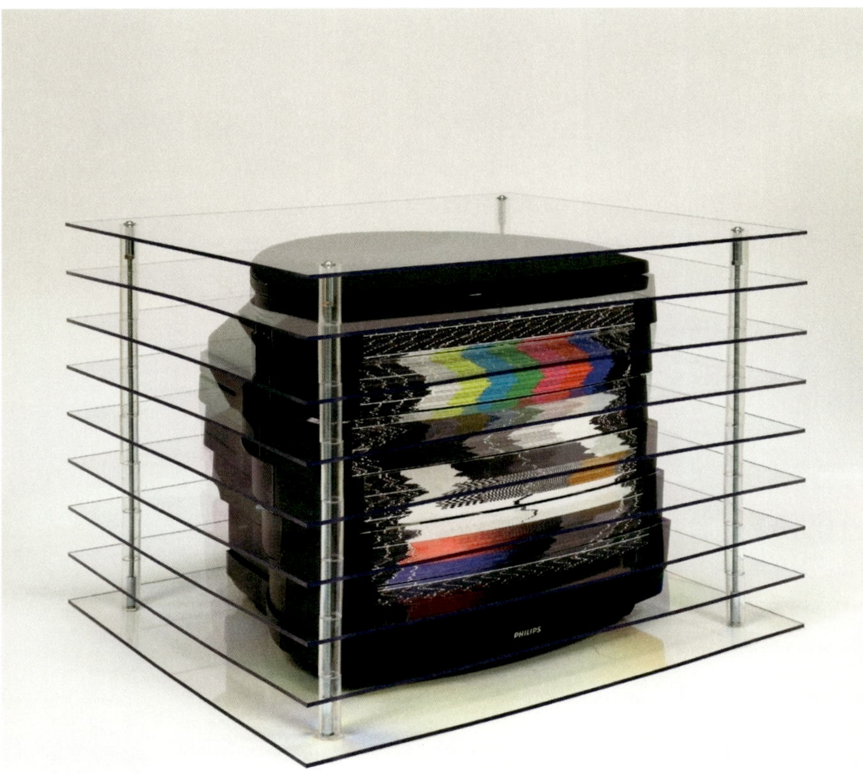

Installation views
respectively, Peter Weibel,
ZKM | Karlsruhe, 2019–20

The television image is made of a temporal sequence of points in a line, the scan lines. They are built up so fast that they appear to be simultaneous to the human eye and they are so vast in number that they form an uninterrupted plank for the eye. Only during a failure, for example of the cathod ray tube, are the viewers able to see the scan lines as a disturbance, a typical distorted TV image. These intrinsic features of the media system I externalized and cut the real TV set in scan lines and arranged the TV box like the TV image distorted by scan lines.
This method of scanning I applied to many other objects, transforming a media characteristic (representation) into an object characteristic (reality). One of my main inventions.

—Peter Weibel, 2023

Scanned Television

1991 | from the series *Scanned Objects*
television, plexiglass, 44 × 60 × 50 cm

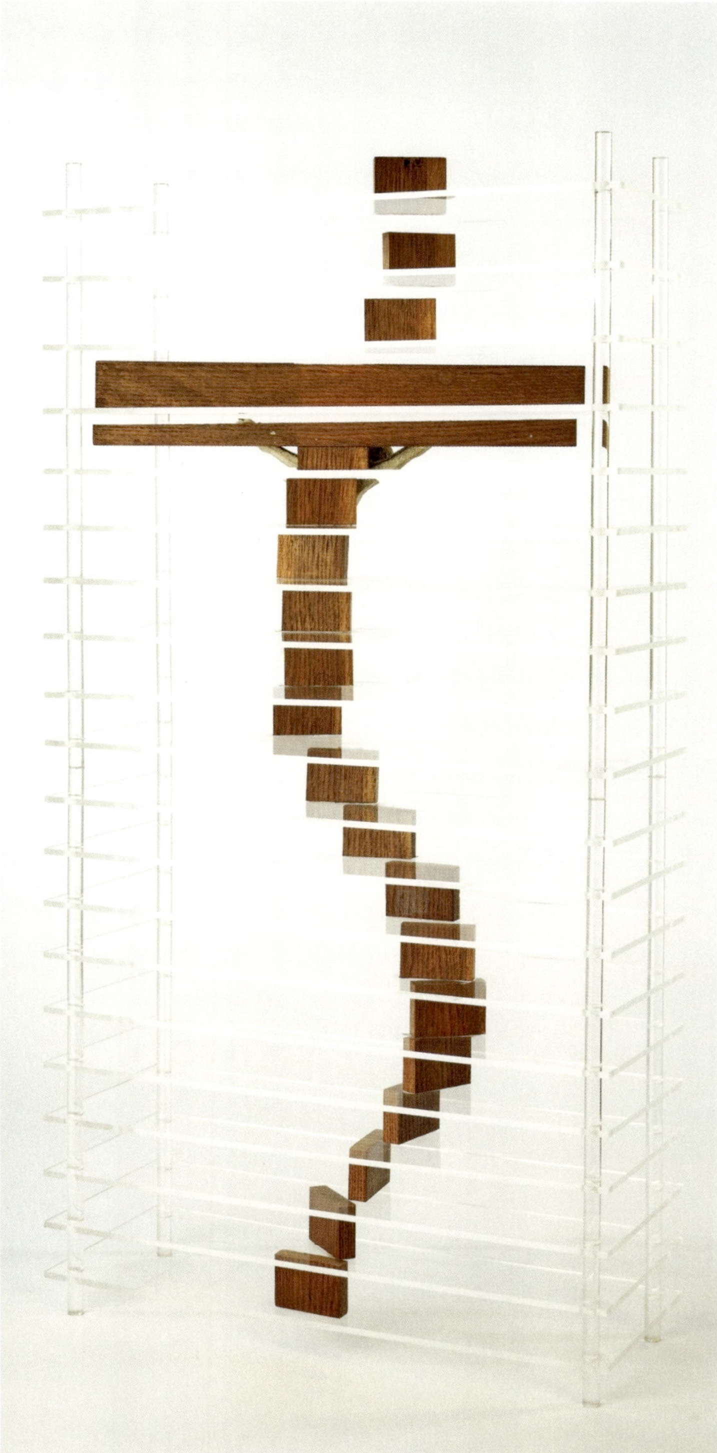

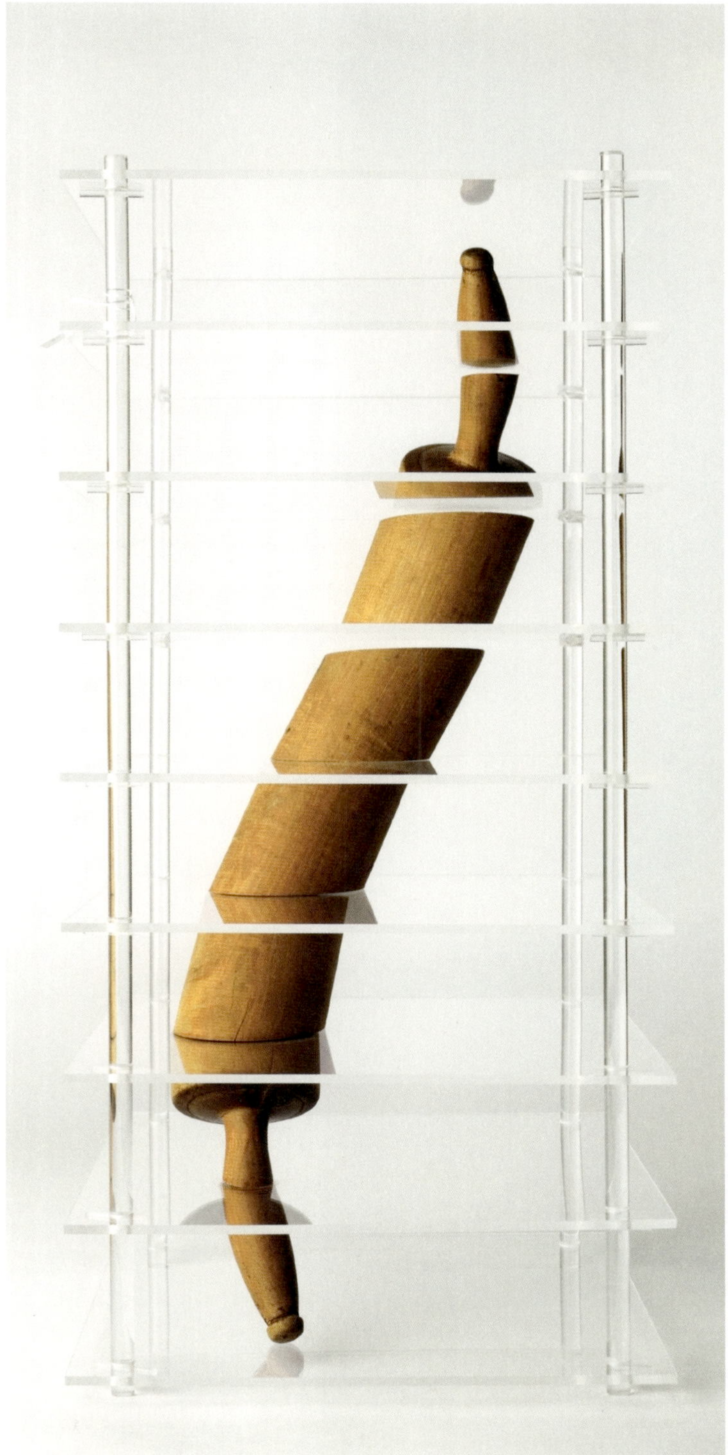

Installation views
respectively, Peter Weibel,
ZKM | Karlsruhe, 2019–20

Scanned Cross

1991 | from the series *Scanned Objects*
crucifix, plexiglass, 73.5 × 37 × 14 cm

Scanned Rolling Pin / Nudelwalker

1991 | from the series *Scanned Objects*
rolling pin, plexiglass, 53 × 25 × 25 cm

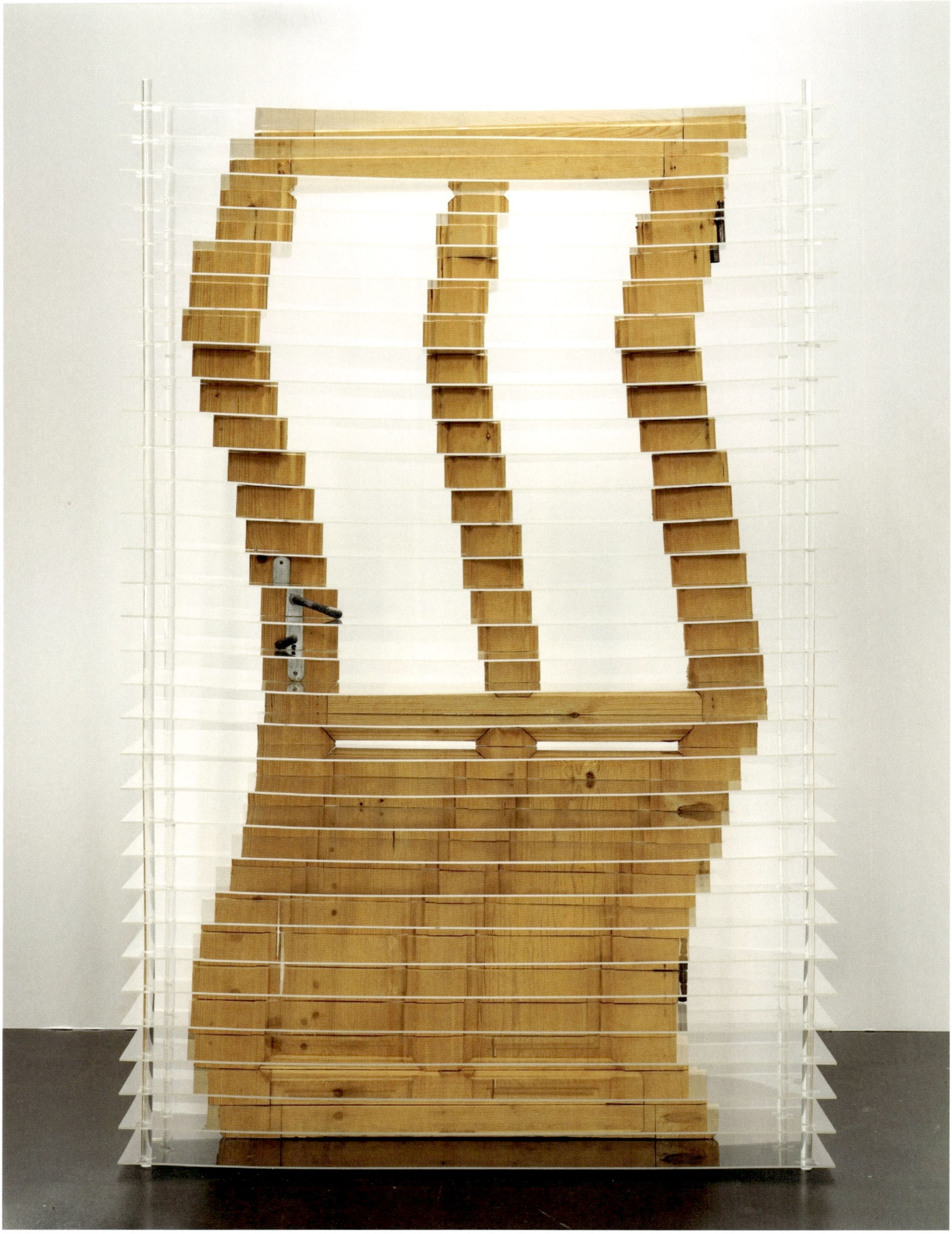

Installation view
respectively, Peter Weibel,
ZKM | Karlsruhe, 2019–20

Scanned Door

1991 | from the series *Scanned Objects*
wooden door, plexiglass,
49 × 130 × 40 cm

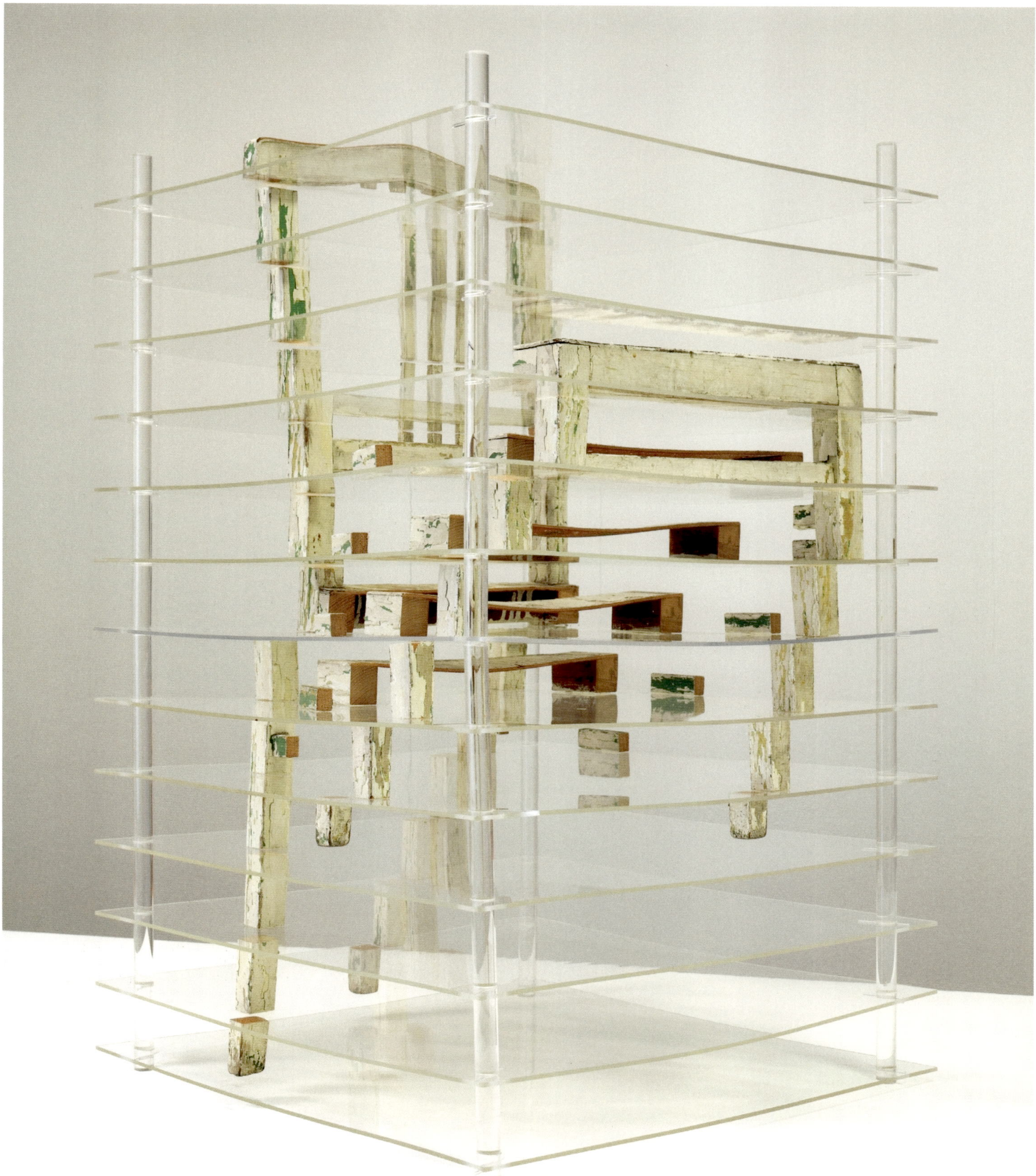

Installation view
respectively, Peter Weibel,
ZKM | Karlsruhe, 2019–20

Scanned Chair

1991 — from the series *Scanned Objects*
wooden chair, plexiglass,
97 × 60 × 60 cm

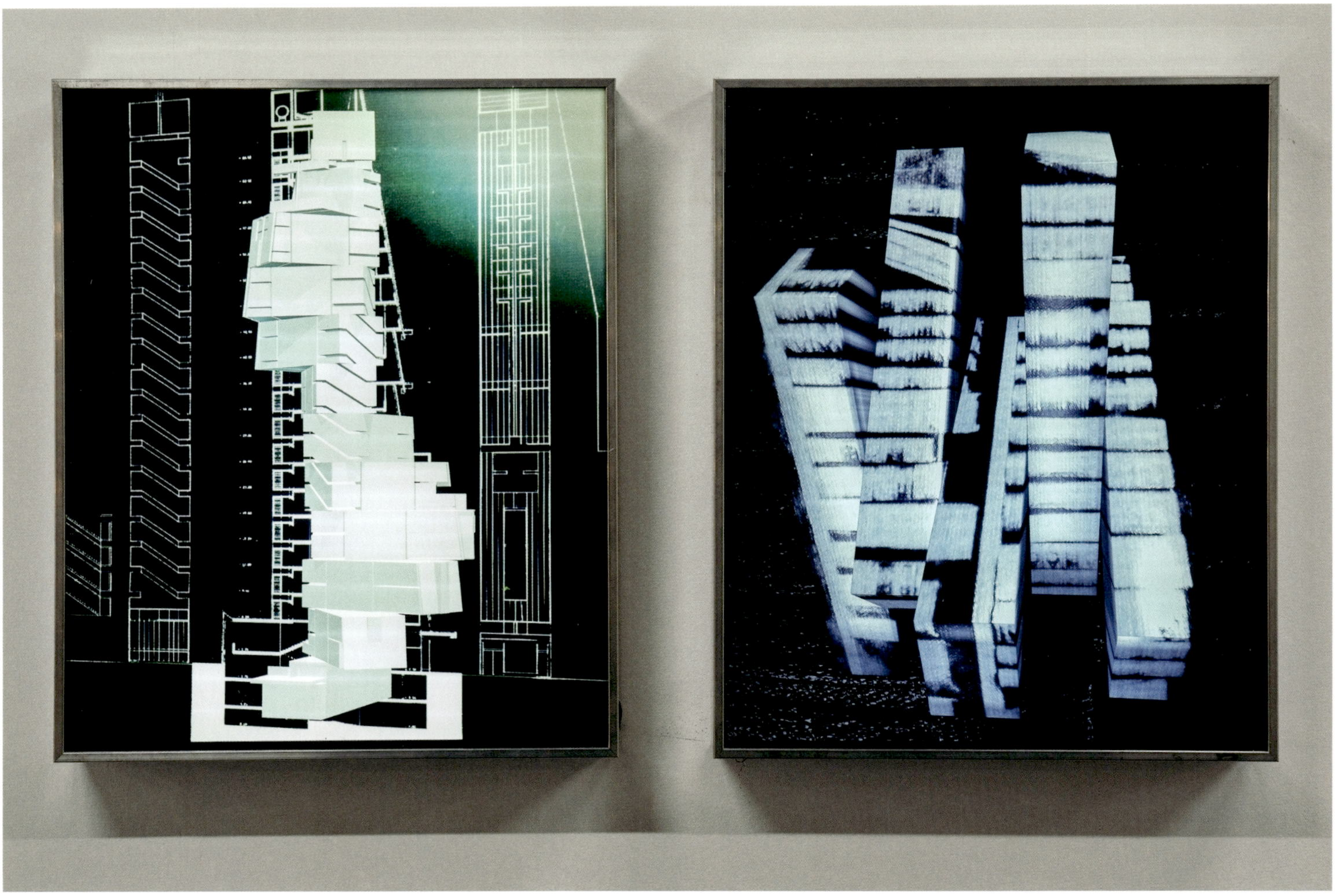

Installation view
respectively, Peter Weibel,
ZKM | Karlsruhe, 2019–20

Naturally, I had the idea of scanning architecture: buildings which would lie behind the monolithic closed skyscraper as cuboids. Scanning as a tool of deconstruction. My scanning technology, that served to deconstruct objects line by line, anticipated 3D printing, which serves to construct objects line by line. I used 3D printing technology of construction and combined it with my deconstructive scanning.

—Peter Weibel, 2023

Sculpture Scan

n. d. | 2 light boxes, 64 × 81 × 12 cm each

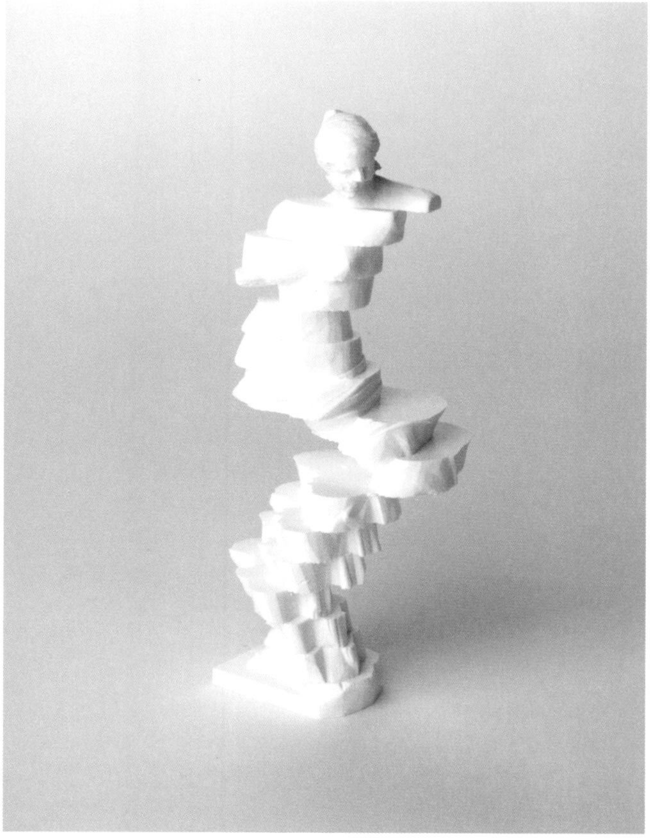

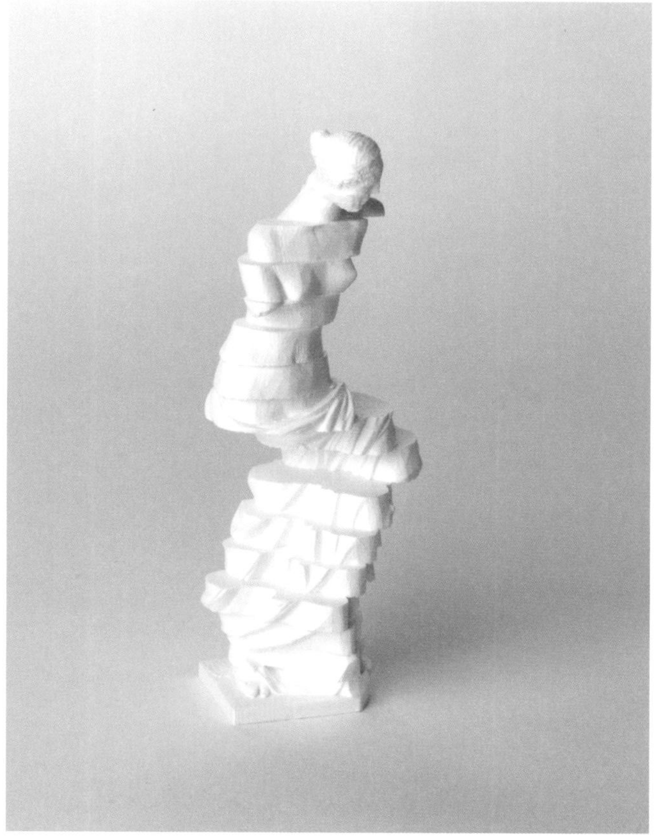

Scanned Sculpture: Venus de Milo

2013/19 | from the series *Scanned Sculptures*
3D print, synthetic, 16.4 × 6.2 × 3.8 cm

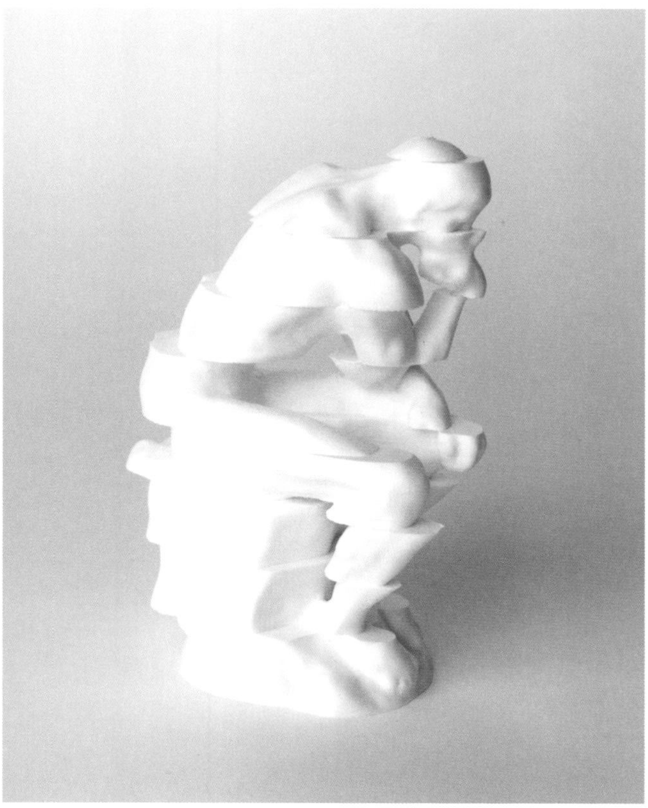

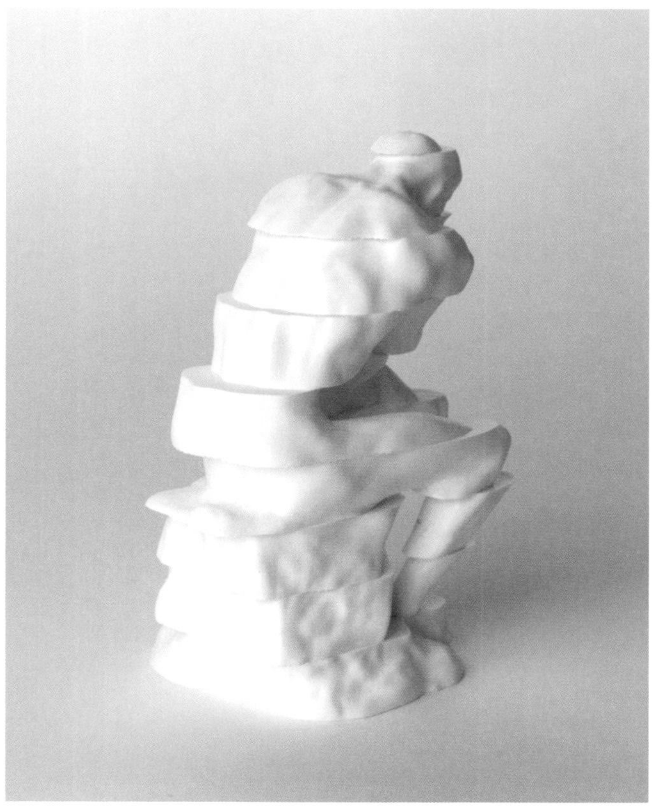

Scanned Sculpture: The Thinker by Auguste Rodin

2013/20 | from the series *Scanned Sculptures*
3D print, synthetic, 16.4 × 8 × 10.2 cm

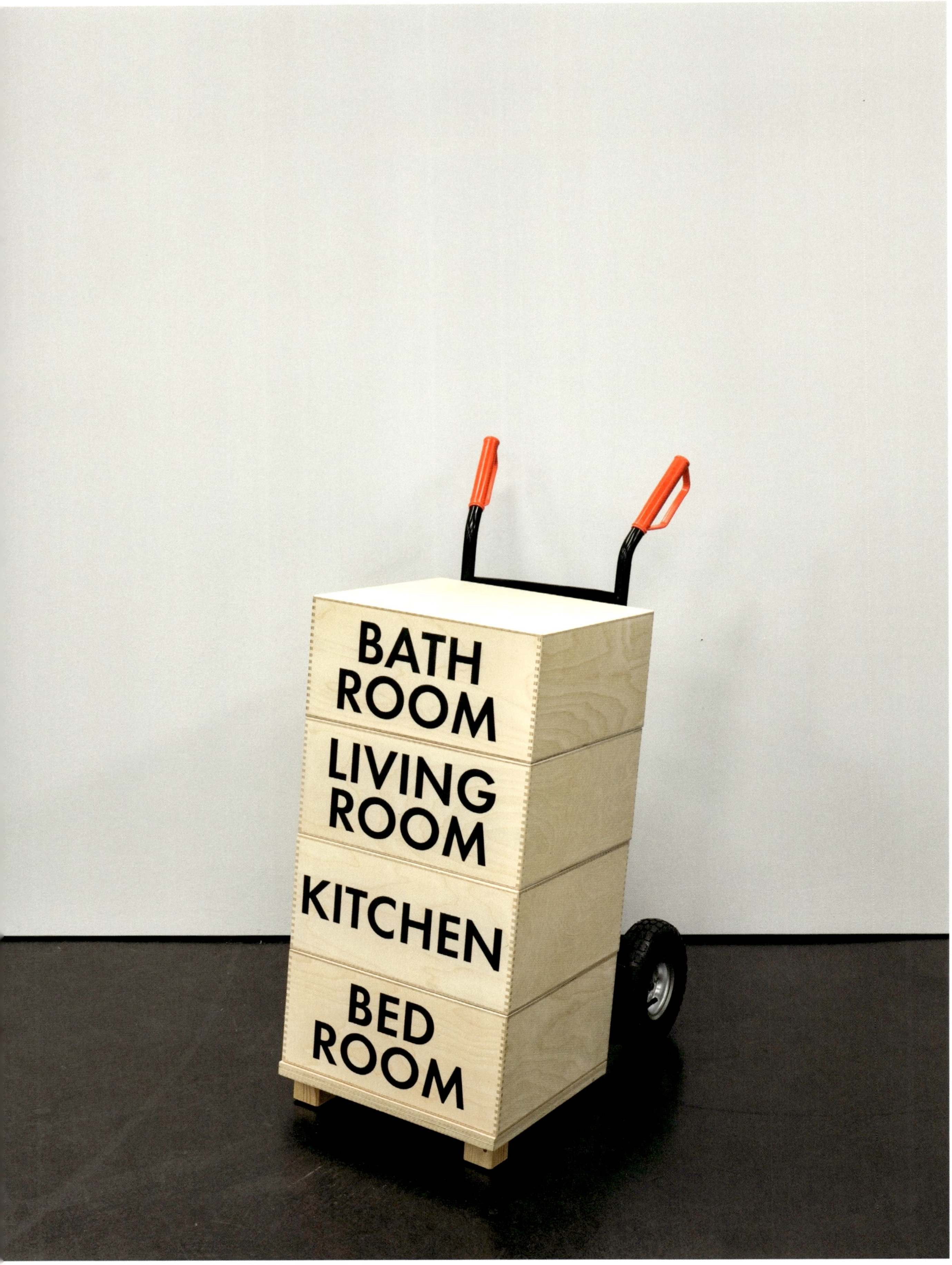

Installation view
respectively, Peter Weibel,
ZKM | Karlsruhe, 2019–20

The Mobile Home of the Future

1994 | hand cart, boxes, writing,
115 × 54 × 70 cm

The Liberated Object: Coding as Design

Peter Weibel

In the discourse of liberation as defined by classic modern movements such as cubism, futurism, suprematism, constructivism, productivism, De Stijl, and so on, an aesthetic utopism emerged, which extended from the easel painting to all realms of human life. From Malevich to Mondrian, from Tatlin to Doesburg, they produced not only works within the visual arts but also in architecture. The pictorial principles that were gained from color, form, and surface material in the wars of independence found themselves transformed within the world of objects. In concealed form, the liberation of the object inscribed in modernism, which led to the independence of color, form, and surface, that is, to abstract art (to art informel), produced a negative dialectic subtext. The liberation of form, color, and material was followed by the liberated object itself, the last link in this chain.

In the second half of the twentieth century the avant-garde program found its continuation. After almost all painterly and pictorial means had been liberated, it was now the object, which was being liberated, that marked the beginning of this movement of liberation. Like the transition from giving something form to design, design is now being succeeded by coding as a new aesthetic principle. We are living in an age in which coding is emerging as a new means of design. The suppression of design, however, has not led to a new spirituality, as "the spiritual in art" in Kandinsky and Mondrian whose theosophic underpinnings are known. Rather, it has ushered in an age of media and electronics, which represent a new form of immateriality; that is, the variable immateriality of news, the new form of design known as CODE. Coding is the structuring of information and data. Code is the attempt to apply the free manipulation and variability of the digital image to the world of objects. The liberated image in digital art and the liberated object correspond. The postmodern ambivalence of the status of the object corresponds to the variable position of postmodern subjectivity. They are the two ends of one rope hung over the abyss of politics. The liberated object is the last station of modernism's aesthetic utopianism.

The forms of the objects and their relationship is painfully at odds with the human desire for experience. Not an insurrection of the abstract but only one of subjectivity against the object can provide relief from this pain. Ontology and metaphysics of occidental subjectivity are dealt with in the living room to which the world of objects of the bourgeois individual has shrunk: furniture-morals, furniture-metaphysics. In this "delusionary room" (J. B. Blume) the vacant chair is the "self," the table the arena of phallic self-assertion. Furniture collapse because morals are always the morals of a class that is on the decline. Furniture, derived from the French "meuble," originally referred to movable property as opposed to "*res immobiles*," nonremovable assets such as ground property, etc. The fact that furniture, once something movable, also appears to us to be something immovable, has to do with the intellectual change of an epoch, with the transformation of subjectivity in the industrial age. It also has to do with the fact that the living room has become a metaphor of the mind's fear of change. In this realm between movable assets (movement) and unmovable ones (nonmovement), the subject moves *motivus* (motivating). An expression of this *motiva* (revolt), French: "revolt," this mutiny against furniture as immovables is the liberation of the object from its rigidity, its stability, its identity, and its transformation into a code. The free subject rebels against unidirectional objects since the human mind is entitled to multidirectional objects and a multidimensional world. The digital code transforms the world into a field of variables. Beyond the readymade: the object liberated from its use function can become a sculpture.

Note: Excerpt from a lecture held at the 29th International Art Talks of the Galerie nächst St. Stephan, March 16, 1985.

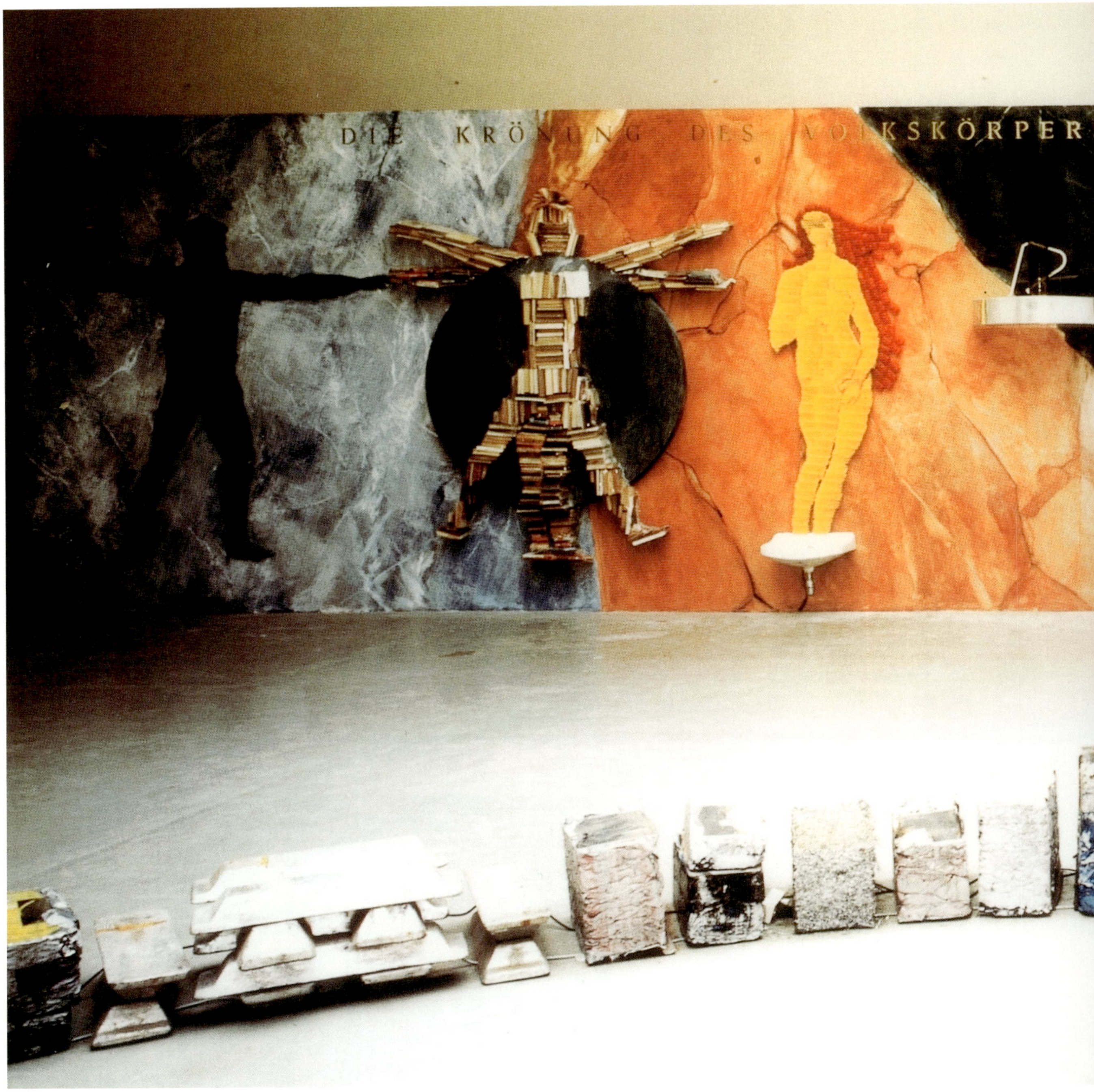

Installation view *Inszenierte Kunstgeschichte*, MAK – Museum of Applied Arts, Vienna, 1988–89.
In the foreground: part of the installation *Raft* (*Floß*, 1988, see p. 484 in this volume)

The Coronation of the Body of the People by Democratic Vigilantes / Die Krönung des Volkskörpers durch die demokratischen Vigilanten

1988	frieze; plastic cups, toothpaste tubes, bicycle tubes, plastic office filing trays, washbasins, books, neon tubes, soap, rubber, aluminum boxes, cleaning brushes, buttons, belt buckles, iron chains, television set

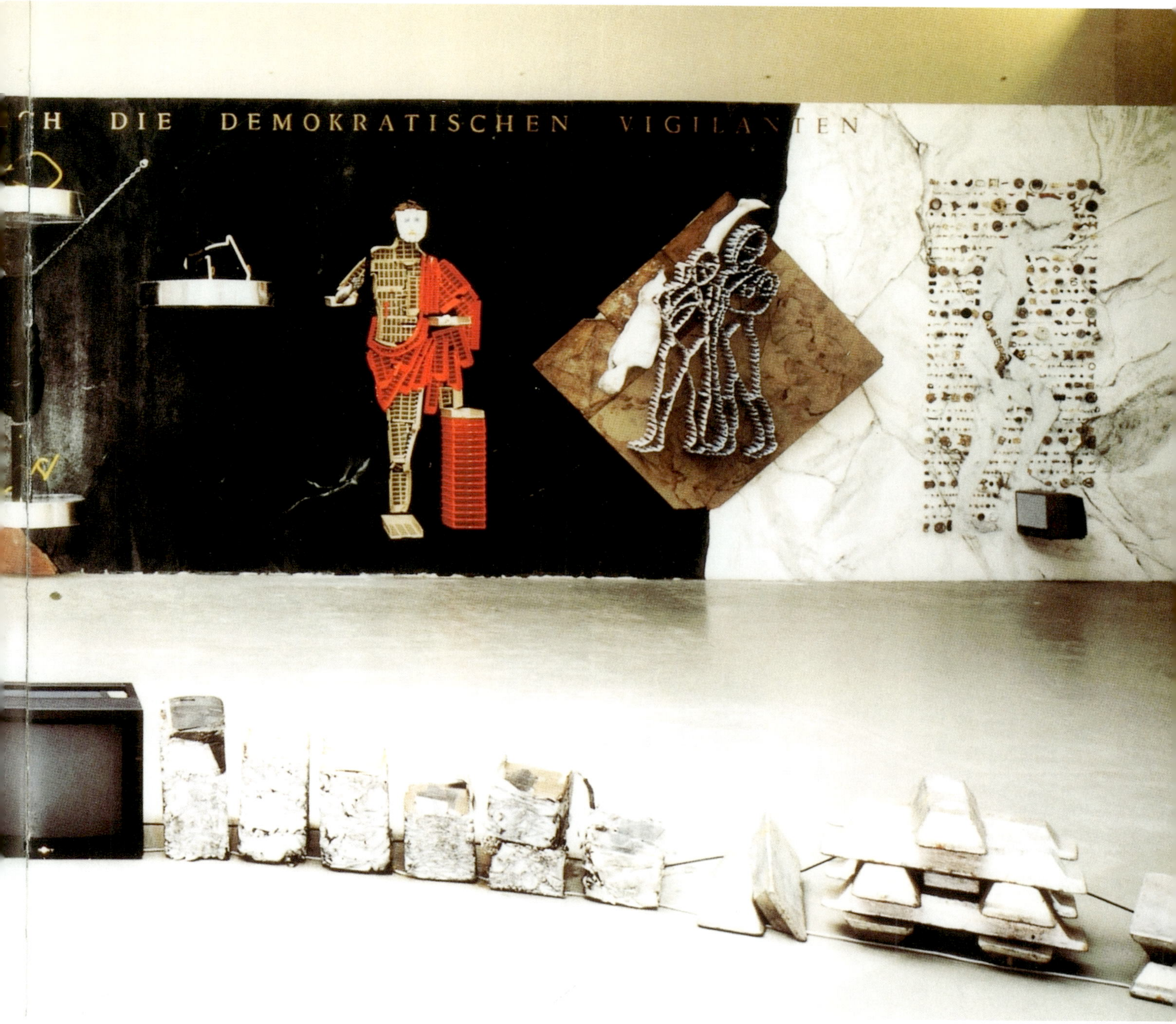

Famous motifs from art history, from antiquity to the twentieth century, are recreated with industrial products in the form of a frieze. The Muybridge motif depicts the transition from easel painting to the moving media image and is therefore made up of wheels, knobs, a TV set, etc. Instead of a shell, a washbasin is used for Botticelli's foam-born Venus. Her body is formed from soap and brushes, the body of the Hellenic Zeus from bicycle tires. Leonardo's drawing is constructed from books. The Christ motif in the style of Piero della Francesca depicts the transformation of belief into the administrative apparatus of the church and is therefore built out of filing trays. The motif from Giorgio de Chirico is made of toothpaste tubes. In the center, a torso (the average citizen) hangs with its back turned between illegible symbols, forming the shape of a forked cross.

Note: Romana Schuler, ed., *Peter Weibel: Bildwelten. 1982–1996* (Vienna: Triton, 1996), 122–23. Translated from the German by Patrick Hubenthal.

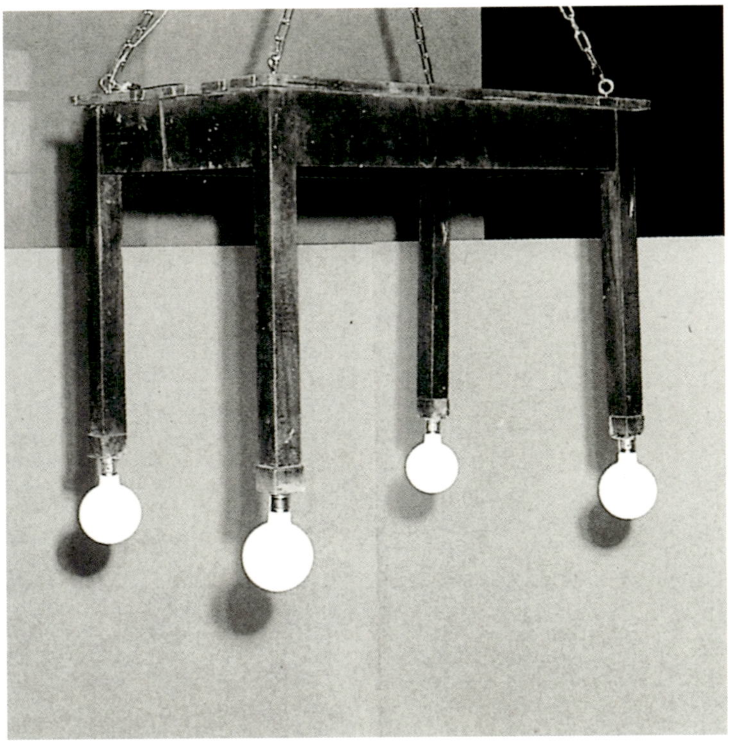

Installation view *Inszenierte Kunstgeschichte*, MAK – Museum of Applied Arts, Vienna, 1988–89

Table Lamp / Tischlampe

1988 | table, 4 light bulbs, chains

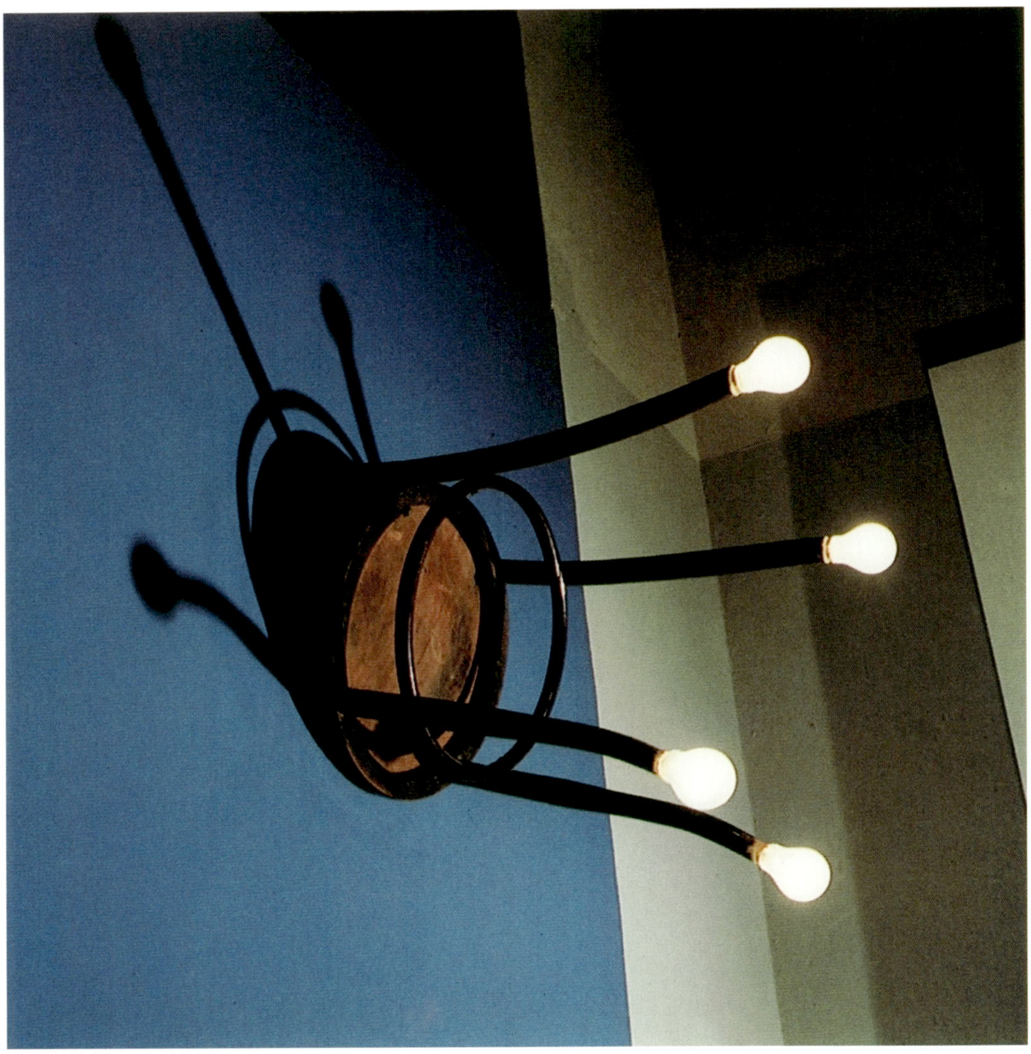

Installation view *Inszenierte Kunstgeschichte*, MAK – Museum of Applied Arts, Vienna, 1988–89

Chair Lamp / Stuhllampe

1988 | chair with separated backrest, 4 light bulbs

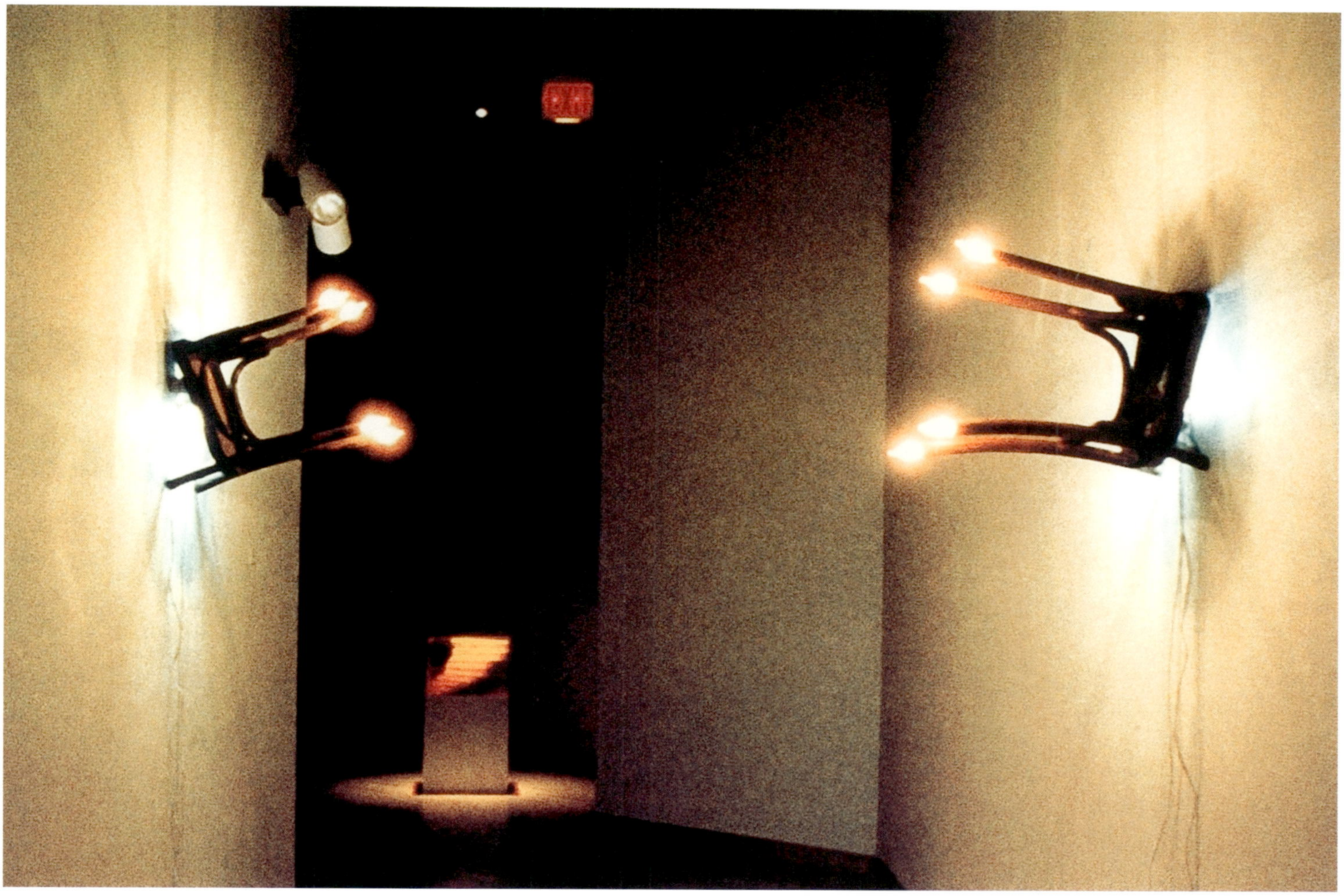

Installation view *Inszenierte Kunstgeschichte*, MAK – Museum of Applied Arts, Vienna, 1988–89

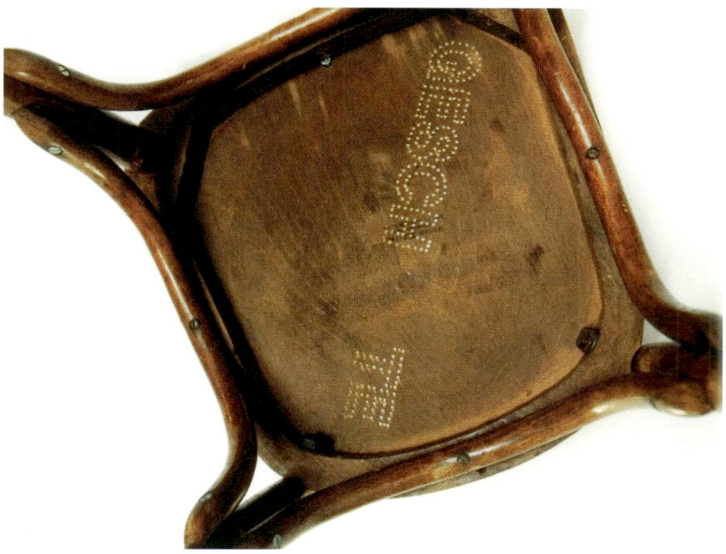

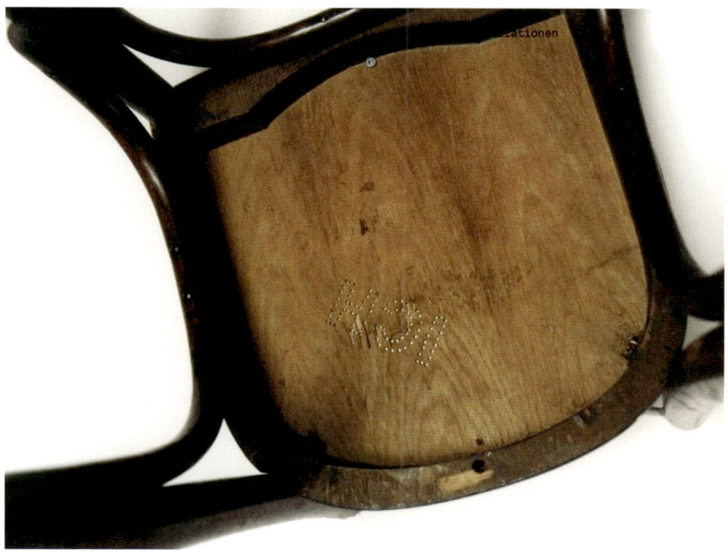

Installation views *Peter Weibel – MEDIA REBEL*, Belvedere, Vienna, 2014–15

On the Dialectic of History. How Do History and I Meet in the Afternoon While Passing Stools? / Zur Dialektik der Gesch(ich)te. Wie treffen sich Gesch(ich)te und Ich am Nachmittag beim Stuhlgang?

1990 | 2 armchairs, mounted on two opposite walls, 8 light bulbs, armchairs, 62 × 42 × 40 cm each

Pinpoints of light write the words *GESCH TE* on the underside of one chair and *ICH* (me) on the underside of the other chair. In German, the word "history" (*GESCHICHTE*) contains the word "me" (*ICH*), showing the dialectics between subject and system, individual and institution as a form of history.

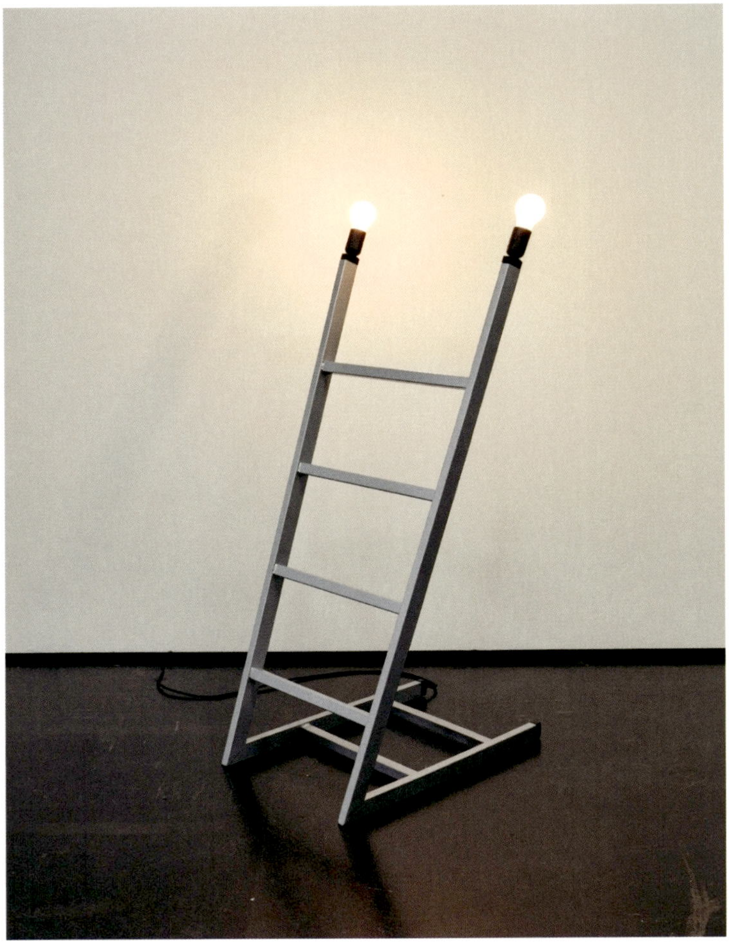

Installation view *respectively, Peter Weibel*, ZKM | Karlsruhe, 2019–20

Ladder Lamp / Leiterlampe

1988 | steel, 2 light bulbs, cable

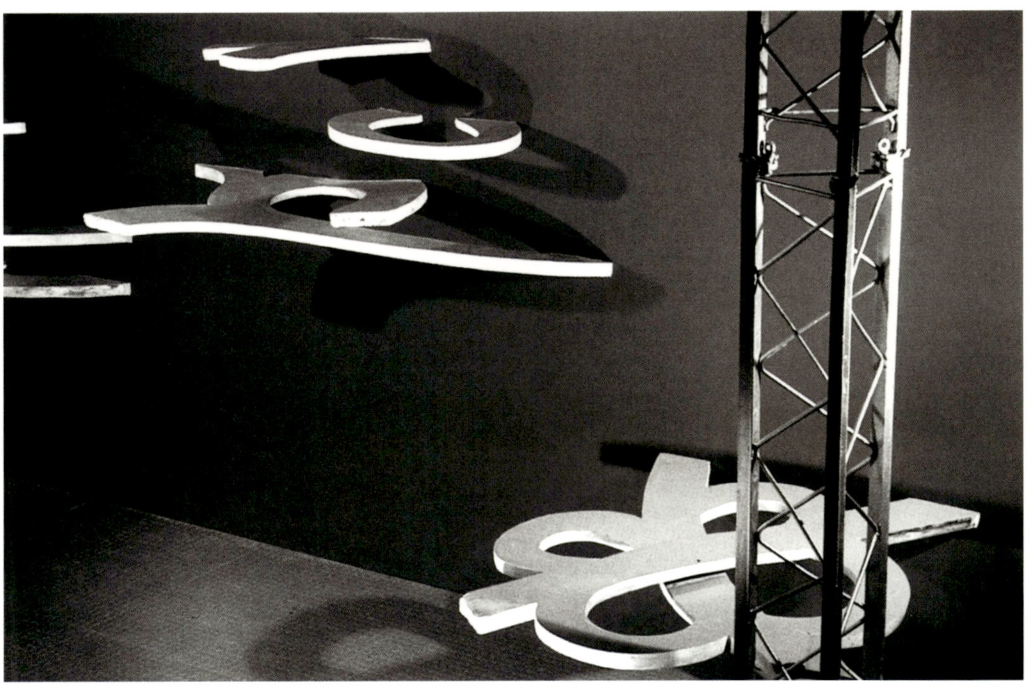

Installation view *Inszenierte Kunstgeschichte*, MAK – Museum of Applied Arts, Vienna, 1988–89

The Autobahn as Etagère and Console / Autobahn als Etagère und Konsole

1988 | concrete, steel

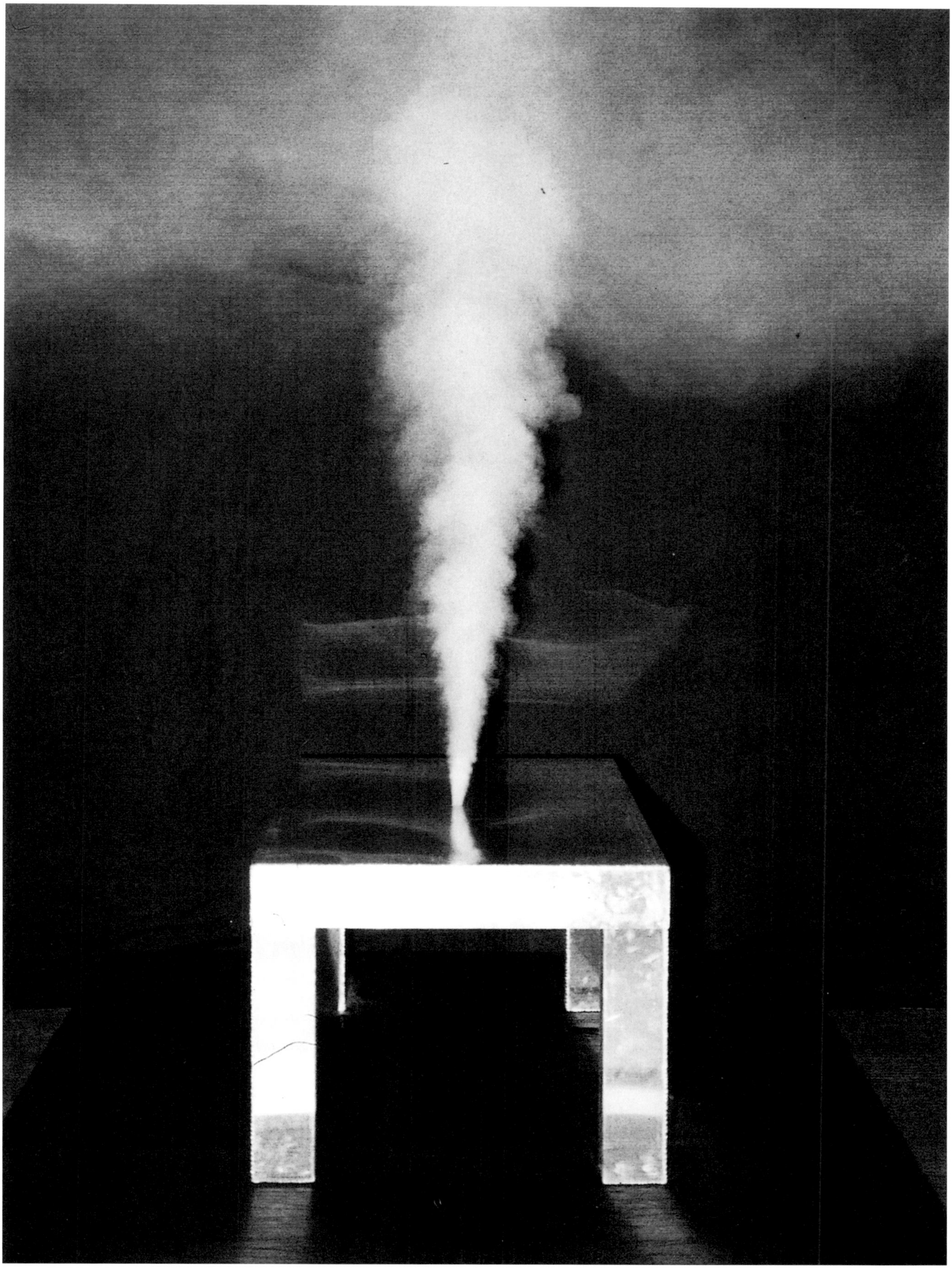

Installation view *Inszenierte Kunstgeschichte*, MAK – Museum of Applied Arts, Vienna, 1988–89

The Revolution My Home / Die Revolution bittet zu Tisch

1987 | aluminum, vaporizer

A steam table instead of a steam engine, because the latter has revolutionized even domestic life.

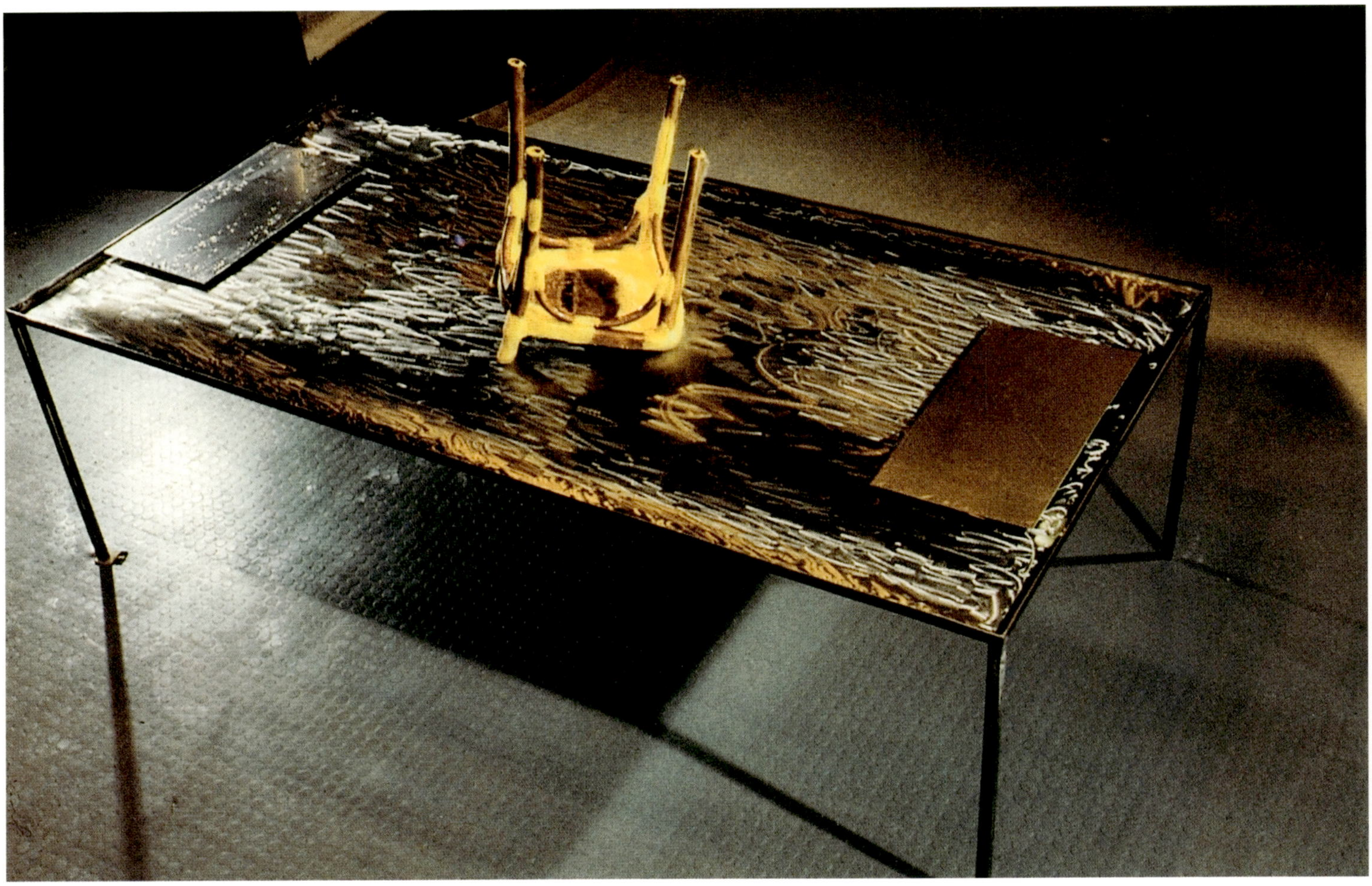

Installation views *Inszenierte Kunstgeschichte*, MAK – Museum of Applied Arts, Vienna, 1988–89

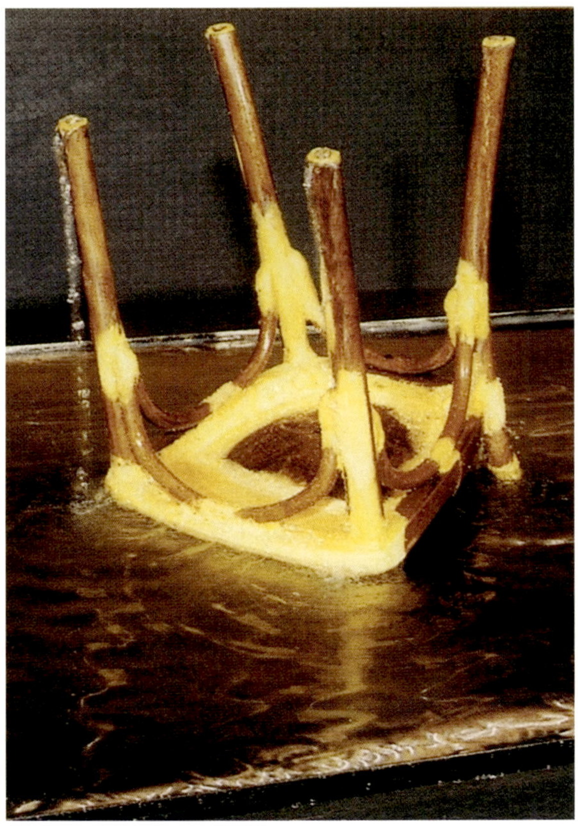

Roof of the World III: Table with Chair as Fountain / Dach der Welt III: Tisch mit Sessel als Springbrunnen

1988 | steel, wax, water

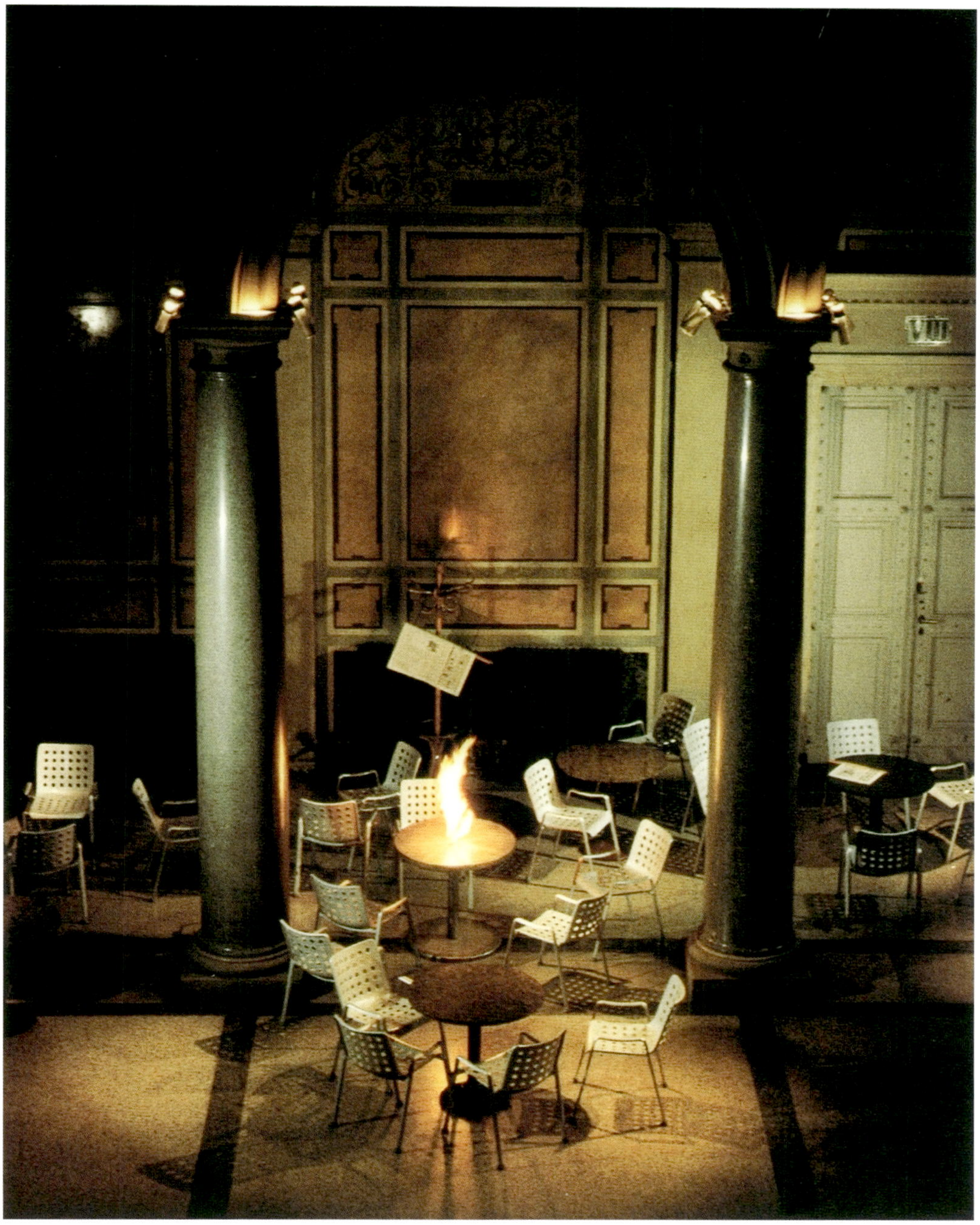

Installation views *Inszenierte Kunstgeschichte*, MAK – Museum of Applied Arts, Vienna, 1988–89

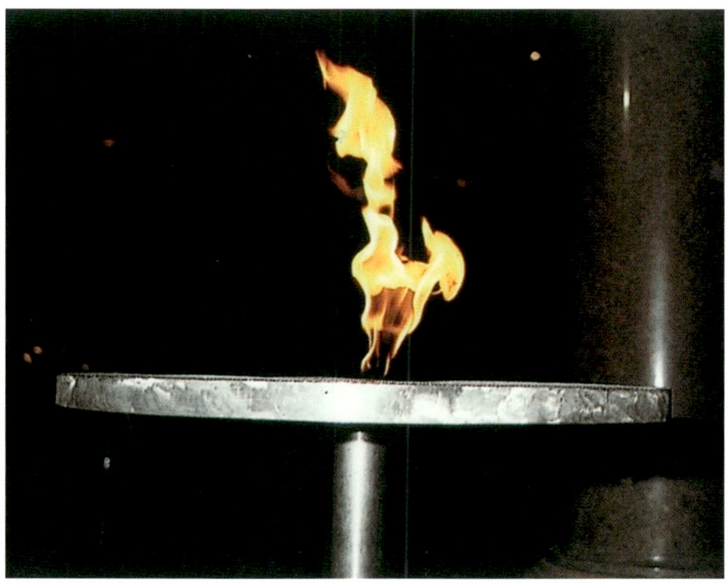

Burning Table / Brennender Tisch

1974/88 | table, gas cylinder

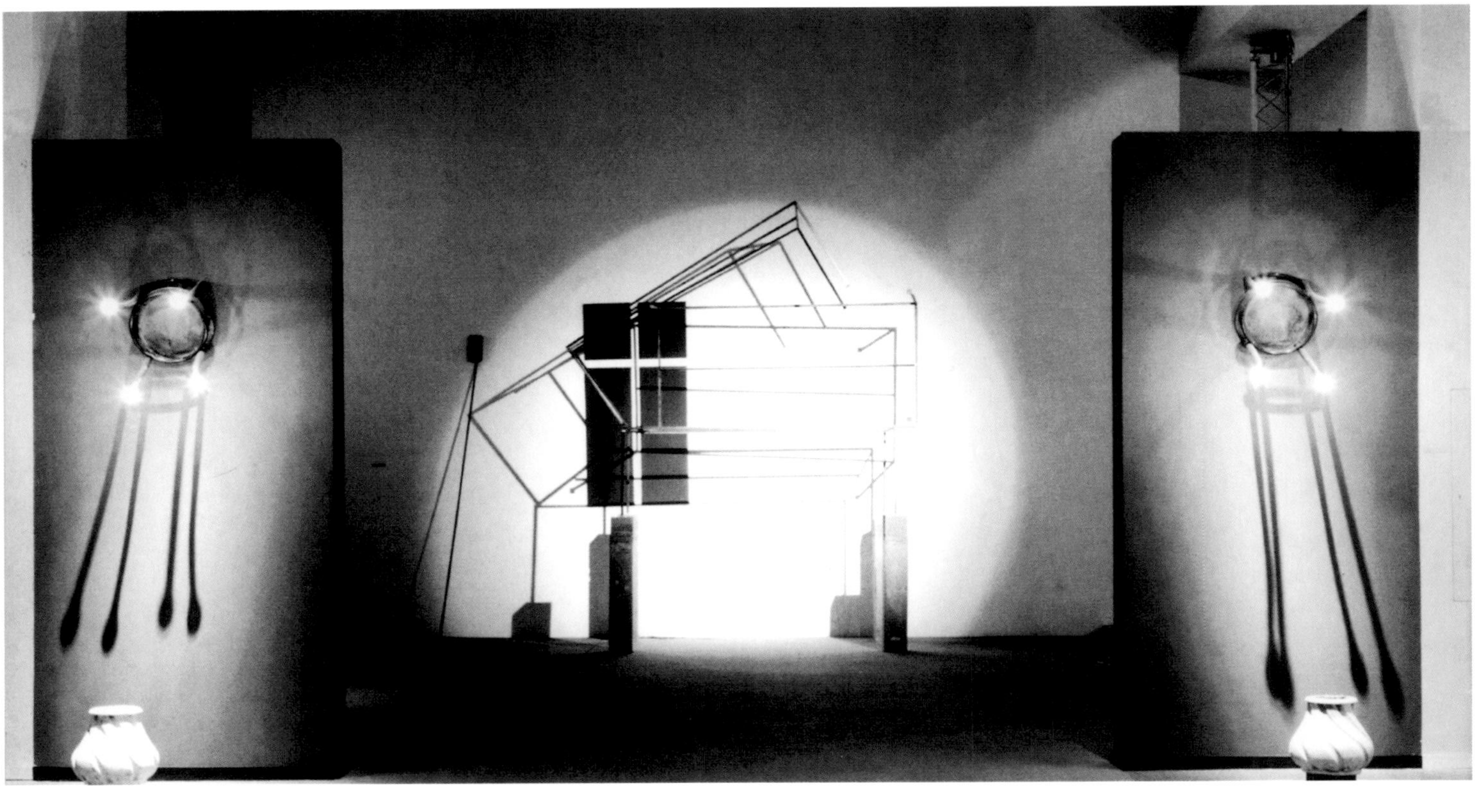

Installation view *Inszenierte Kunstgeschichte*, MAK – Museum of Applied Arts, Vienna, 1988–89

Oikos: Table as House / Oikos: Tisch als Haus

1988	iron bar, bricks, 3 metal tables, 4 canvases painted blue, blue pigment on the floor, 4 concrete bases spotlights, 2 vases, 2 chair lamps

Signifiers, not signified objects of a house, represent a house as an abstract hybrid construction. Three skeletal tables are stacked on top of one another in such a way that the shape of a house (*oikos*, in Greek) is formed (fractal self-representation – *pars pro toto*). More pieces of the house: a leaning metal bar holds a loose brick against the wall. Four pieces of canvas painted blue stand in for a window. Concrete plinths act as a foundation wall, buttressing the entire construction. In the foreground (front yard, facade) are two chair lamps with vases. The shadows multiply and dramatize the construction.

Installation view *Inszenierte Kunstgeschichte*, MAK – Museum of Applied Arts, Vienna, 1988–89

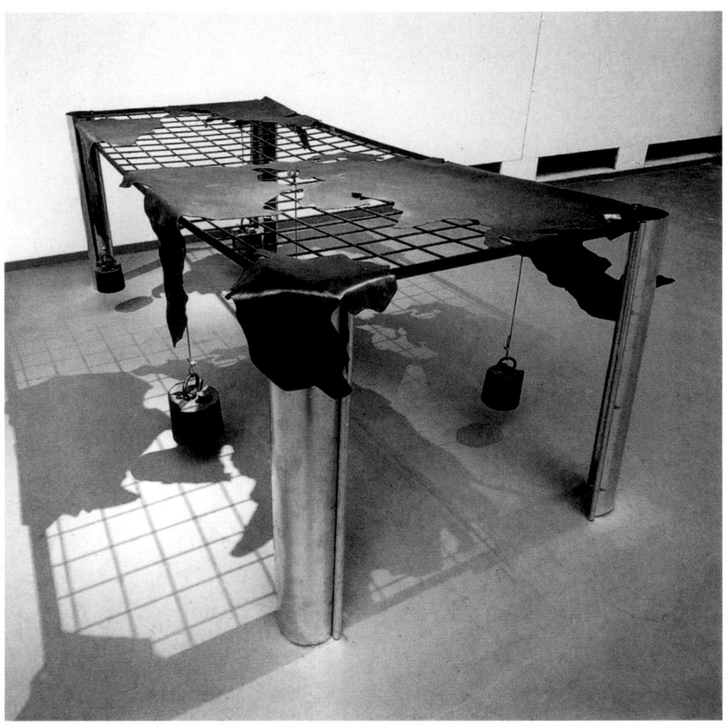

Roof of the World II / Dach der Welt II

1988	sheet metal, lead, iron grating, weights, gutters

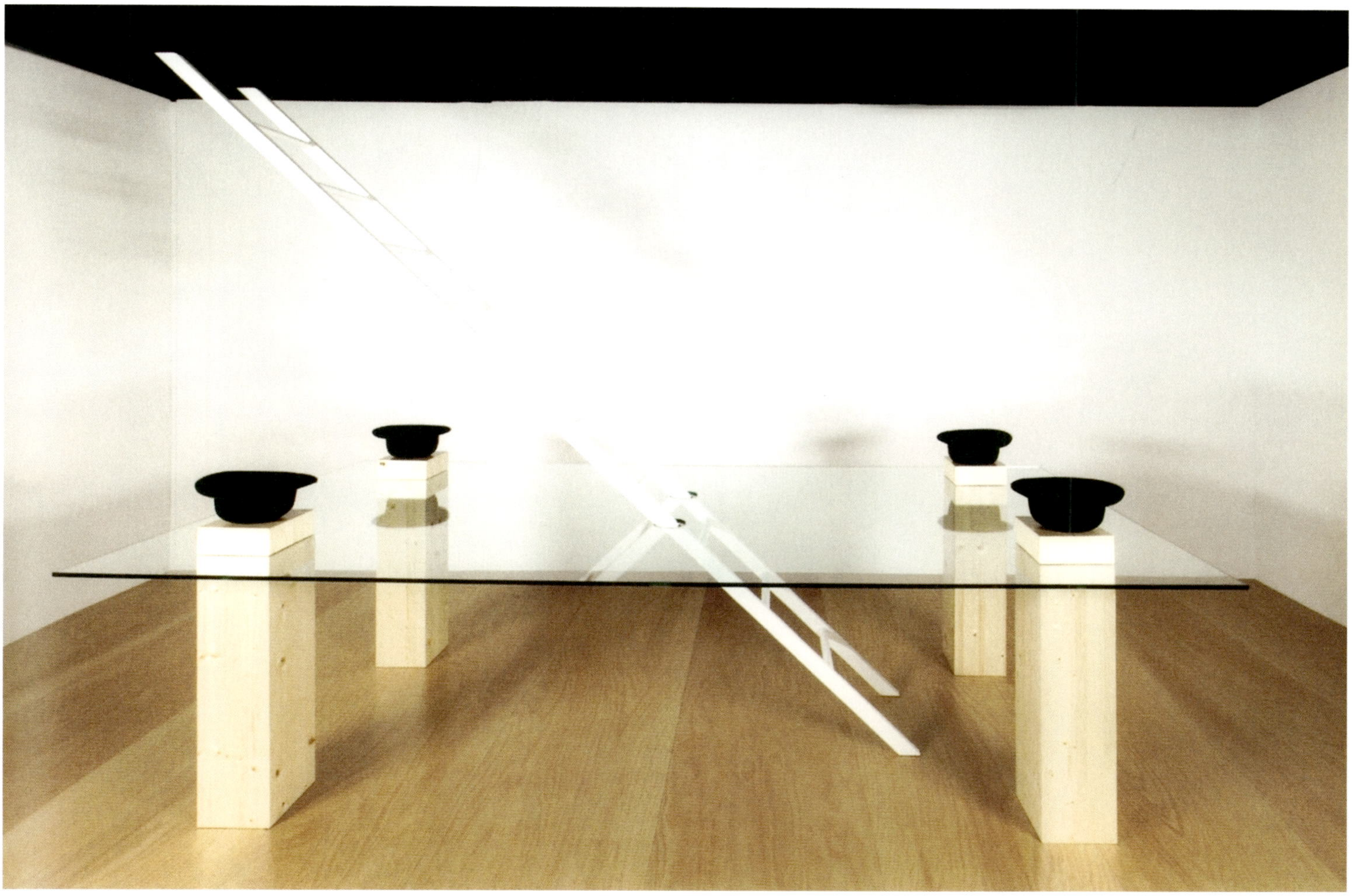

Installation view *Peter Weibel – MEDIA REBEL*, Belvedere, Vienna, 2014–15

Roof of the World IV: Buy Nature, Buy Yourselves / Dach der Welt IV: Kauft Natur, Kauft Euch selbst

1988 | 4 prepared hats filled with water; table made of wood and glass pane with 2 holes, ladder

The stability of the table legs gives the impression of massive pillars (like those of bridges, high-rises, etc.). The tabletop thus becomes a roof or bridge. The ladder – whose function, of course, is to overcome differences in height (and which is therefore itself a bridge) – gives visual expression to the play on scale. At the same time, the ladder paradoxically penetrates the transparent pane of glass, thereby creating the impression of a blockage. How could the ladder (Jacob's Ladder) penetrate the glass pane? The hat motif, familiar from art history, is used metonymically here in two ways: on the one hand, *pars pro toto* for the human head; on the other, the water in the hats represents the house's gutters. All the elements are real and completely unaltered. Only in combination do they lose their conventional quality, changing from simple household items to complex multivalent objects, from readymades to unreal, ambivalent, hybrid constructs – to theoretical objects.

—Peter Weibel, 1996

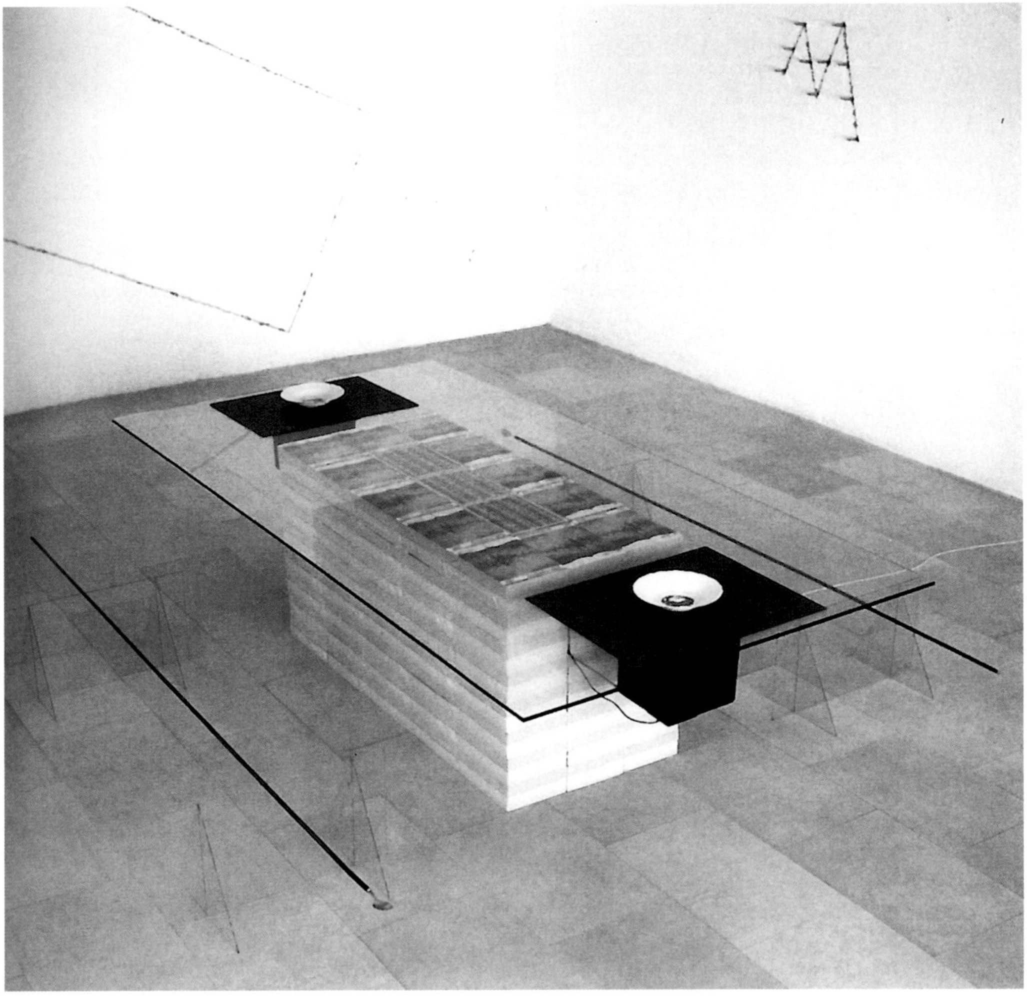

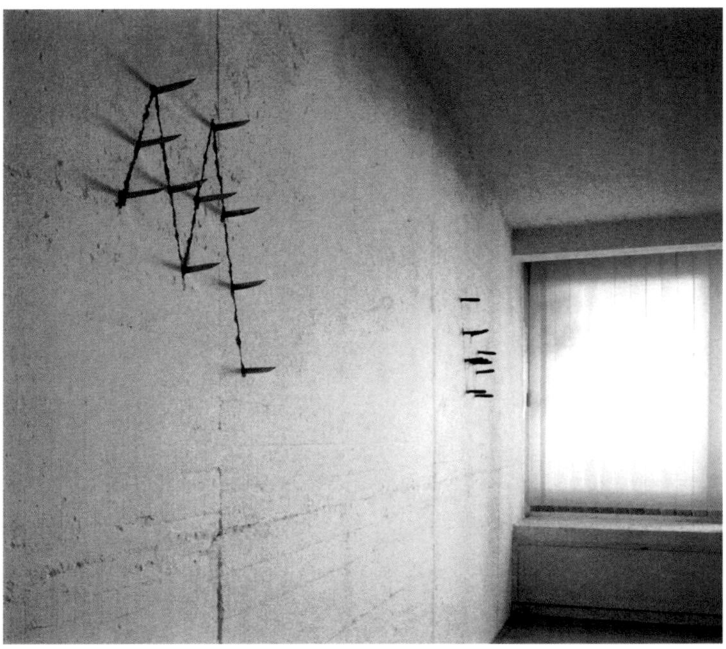

Installation views *Le désenchantement du monde*, Villa Arson, Nice, 1990

Roof of the World V / Dach der Welt V

1990/ 2003	computer-based installation; glass plate with 2 cutouts, telephone books,2 table sets with cutouts, 2 porcelain plates with cutouts, 2 long spoons, 2 computers, software, 75 × 140 × 280 cm

The installation consists of telephone books stacked on top of each other as the base of a glass table with place mats on which there are two porcelain plates. Holes are cut out of the two plates, in which real time Meteosat satellite images of the Earth are visible. With the oversized metal spoons on the table, the globe is practically being served for consumption. With *Roof of the World V* from 1990, Peter Weibel was an early commentator on the critical situation of planet Earth and the threat to its existence due to the excessive consumption of its resources by our way of life. The telephone books symbolize global networking (before the internet) and, together with the satellite images, form the envelope of the Infosphere, with the help of which the Earth is being eaten more quickly.

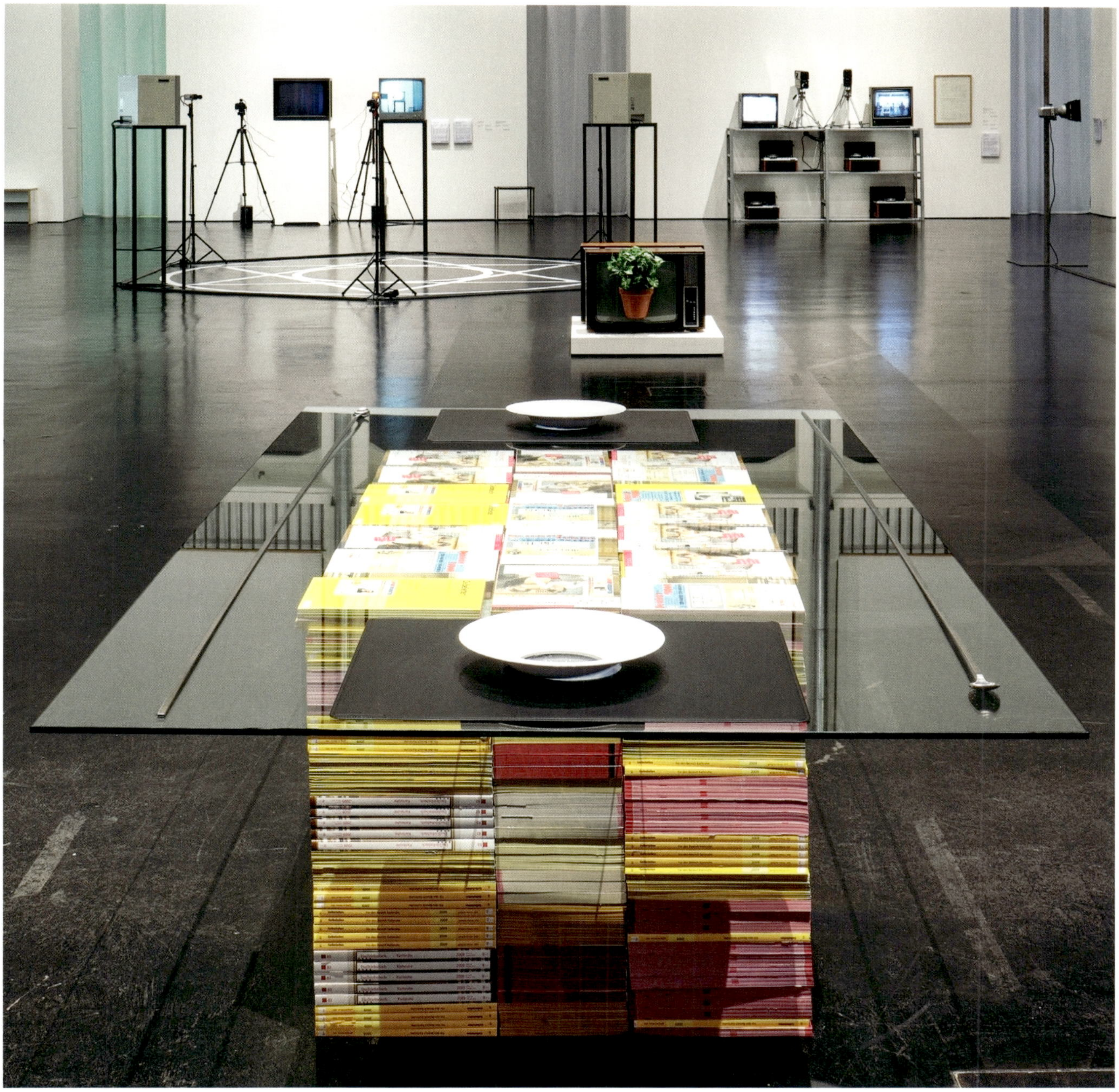

Installation views
respectively, Peter Weibel,
ZKM | Karlsruhe, 2019–20

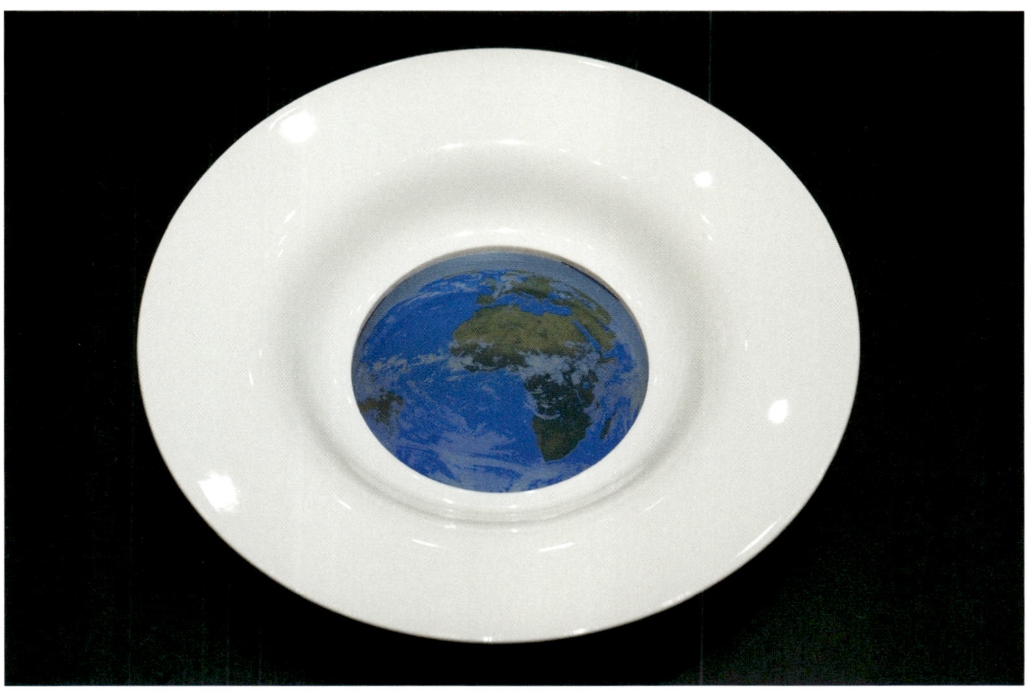

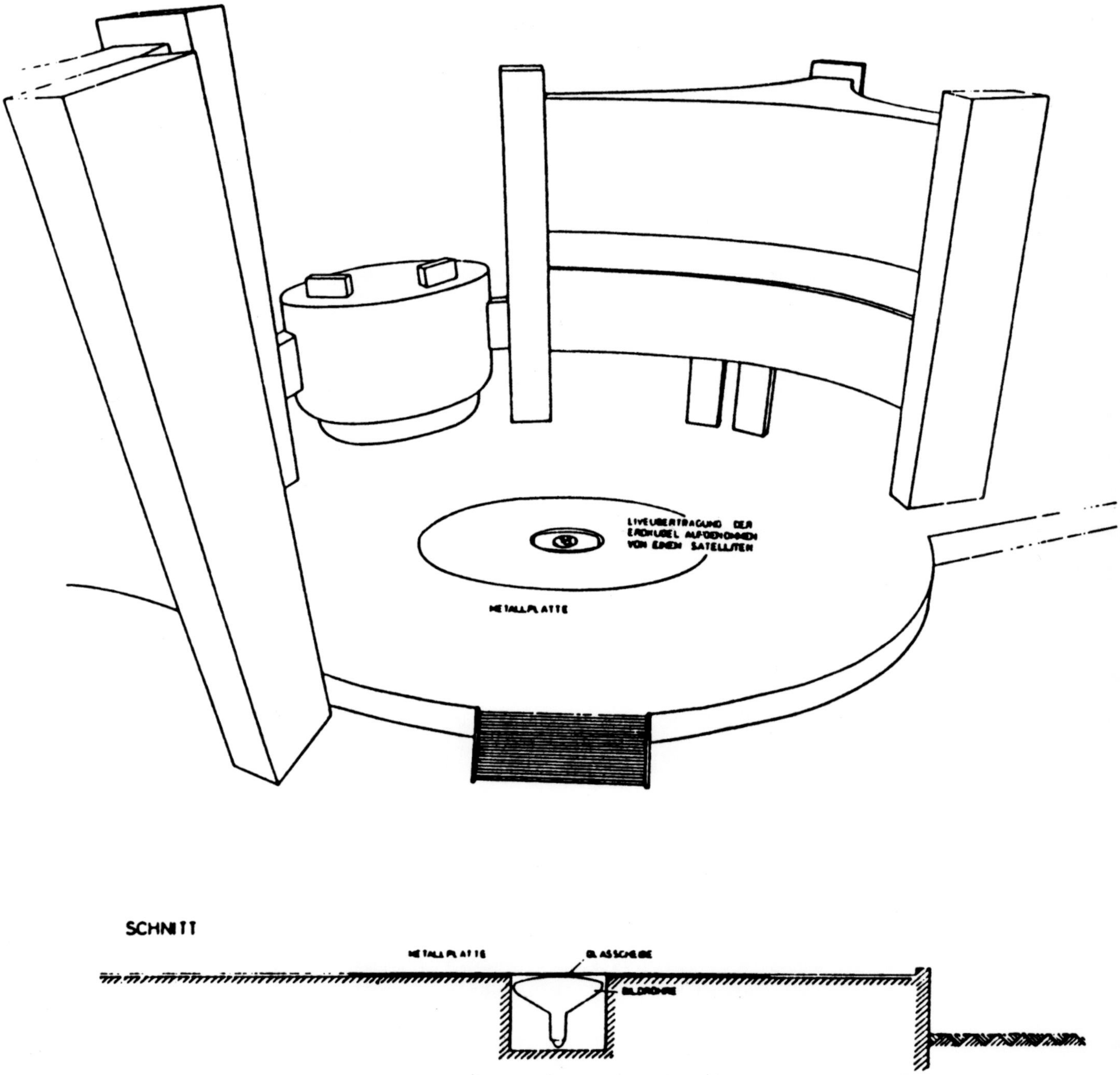

Conceptual drawing for
Orbital Sculpture I, 1976

Orbital Sculpture I / Orbitale Skulptur I

1976 | project for UNO City, Vienna, drawing

A circle as a stage. In the center of the stage, another small circle which contains a TV screen showing a live broadcast of the globe, filmed by a satellite. The observer stands on the Earth and at the same time looks at the Earth from outer space: With their natural eye, they see the floor, with their artificial satellite eye, they see the Earth from above. Dislocation of bodily organs, exteriorization of the eye, splitting of organic, natural and apparatus-based perceptions and experiences.

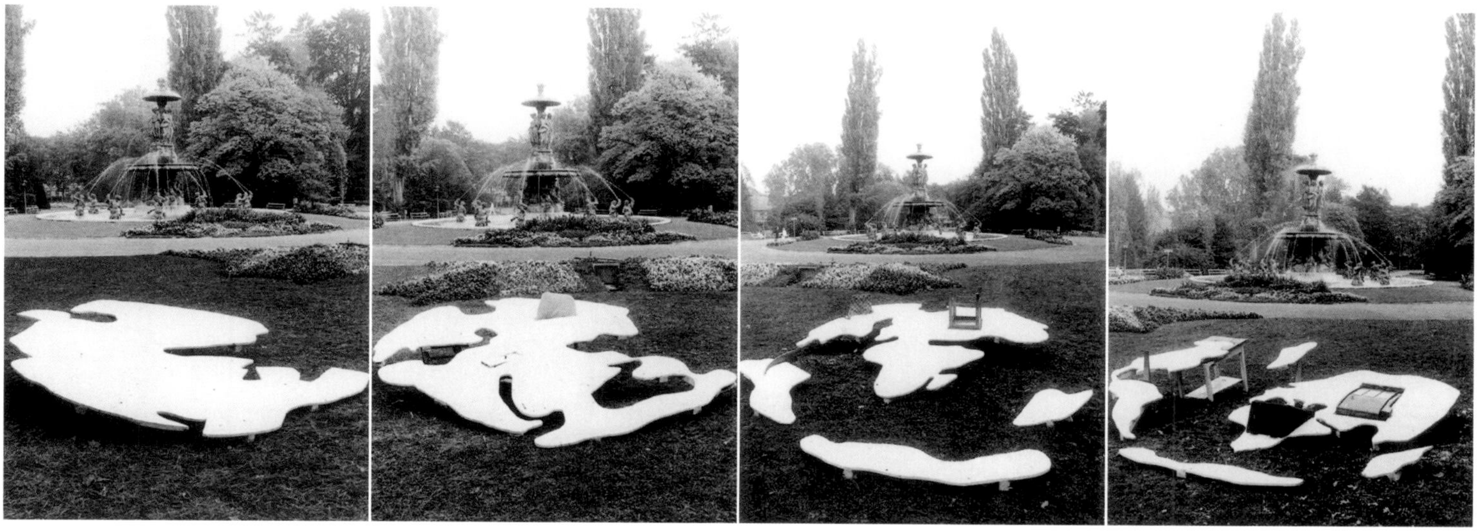

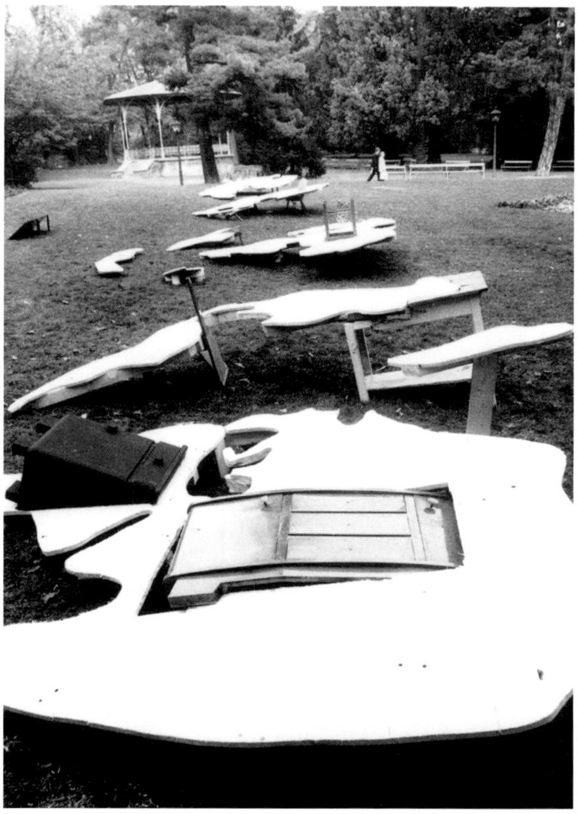

Installation views *Die neue Dimension: Skulptur und Raum*, Forum Stadtpark, Graz, 1985

Orbital Sculpture II: The Evolution of the Earth over 300 Million Years into an *Oikos* (House) / Orbitale Skulptur II: Die Evolution der Erde in 300 Millionen Jahren zu einem *Oikos* (Haus)

1985 | Phase Sculpture; 5 parts, wood, various pieces of furniture

The five-phase sculpture begins with a flat wooden plane, the early landmass. Phase after phase, the continents develop and lift up from the ground. At the same time, furniture becomes part of the continents, demonstrating that the Earth has become our home, *oikos*. The words "ecology" and "economy" have a common origin in the Greek word *oikos* (house, household). The primeval state of the Earth, its forests and meadows, its caves and trees, were humanity's first home. Over millions of years, the continents diverged and humans developed the tools to build human habitations. The continents became dwellings, the landmasses furniture. The primeval state gave rise to human civilizations. Planet Earth became humanity's house. In this way, humanity unfortunately also shrugged off responsibility for the management of the household, of the Earth's natural and energy resources.

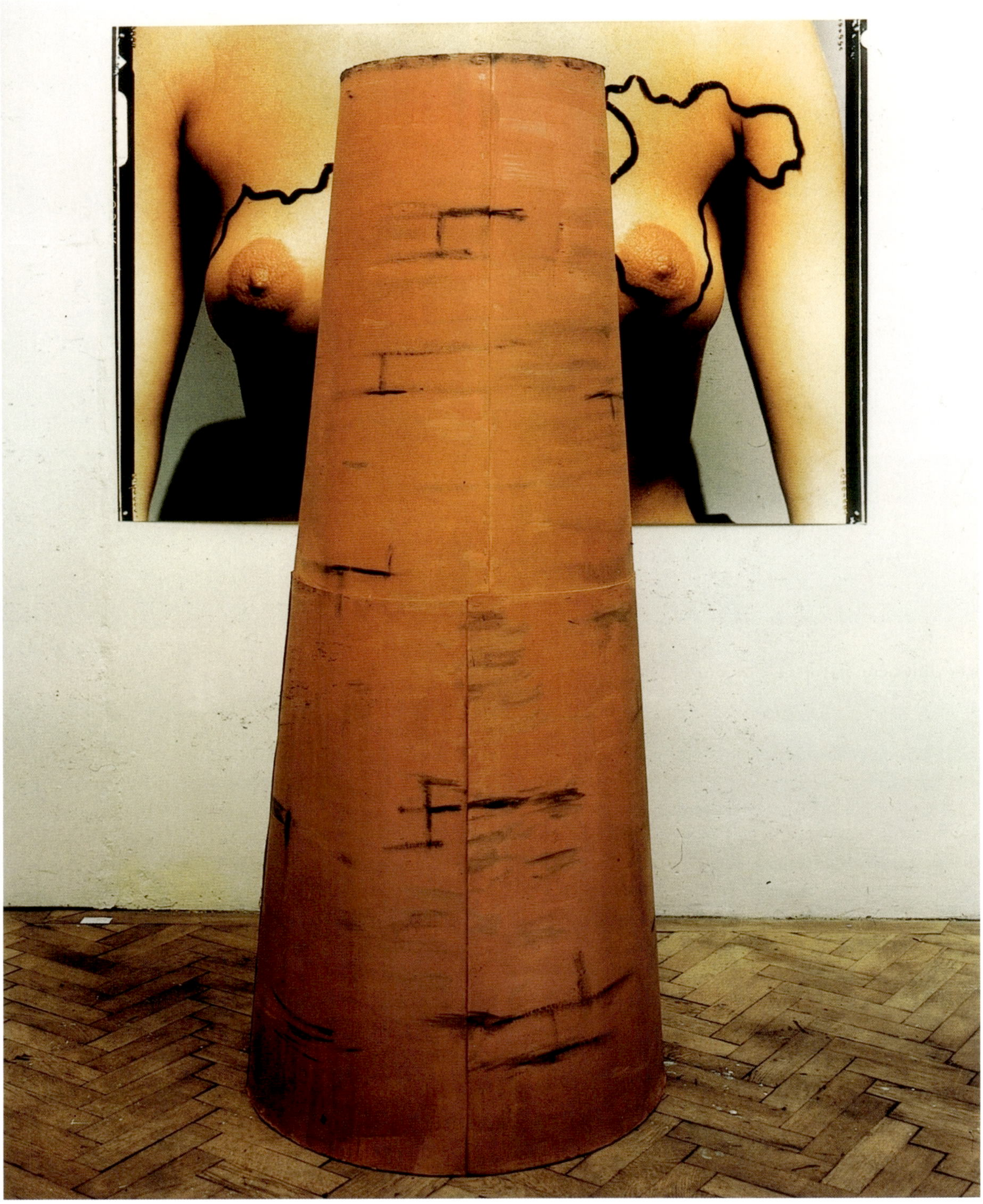

Installation view *Raum annehmen: Aspekte österreichischer Skulptur 1950 bis 1985*, Galerie Grita Insam, Vienna, 1985

Orbital Sculpture III / Orbitale Skulptur III

1985 | photograph, cardboard, red paint

A smokestack is part of a factory and therefore a metonym for the Industrial Revolution. The continents, painted on a female body, metonymically represent Mother Earth. The smokestack, the factory chimney, is phallic. We thus see two opposed principles that are caricatured by the installation. The bifurcation – splitting, separation, division – of culture and nature is a prejudice from the past. Society and nature can only be conceived in conjunction with one another.

—Peter Weibel

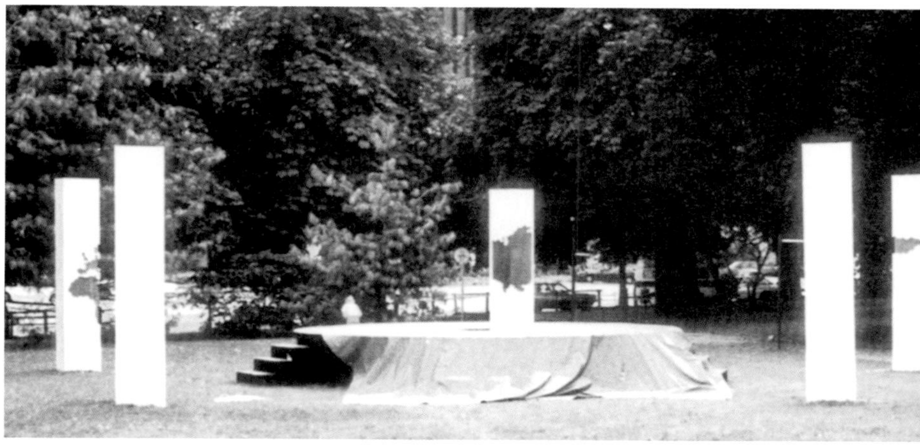

Installation view Ars Electronica '87, in front of the Brucknerhaus, Linz, 1987

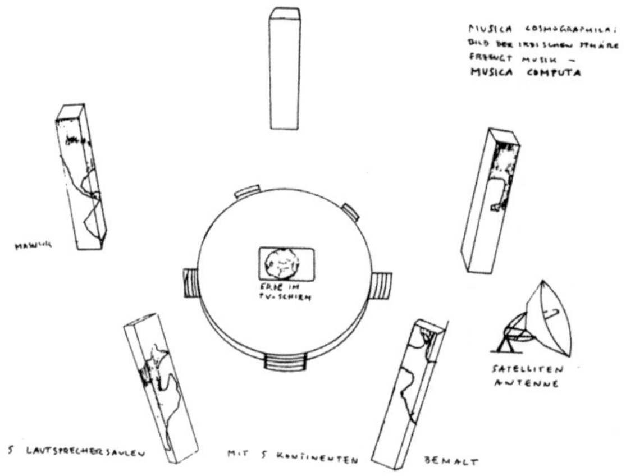

Conceptual drawing for *Orbital Sculpture IV: Spectral Music*, 1987

The sound installation consists of a screen lying flat on the floor of a stage which shows live images of the Earth provided by the Meteosat weather satellite. The images of the atmosphere and the cosmos, their brightness, spectral colors, and scan lines are converted into sounds by a computer. In this way, environmentally specific computer music is created. By walking and looking down on the monitor, which shows the Earth seen from the satellite's orbit, you are watching the Earth in a twofold way: you are standing with your feet on the Earth, but your eyes are looking down on it from a satellite far away. Plato assigned the tetrahedron, hexahedron (cube), octahedron, dodecahedron, and icosahedron to those terrestrial elements which make up the world's matter. Johannes Kepler followed in Plato's tracks, and believed that the Platonic solids held the key to the planetary system, the *Mysterium Cosmographicum* (1596). In this work, Kepler explained the number of the planets and their distances from the sun as a nesting of the five regular Platonic solids within each other and within spheres. According to Kepler, the radii of the various spherical shells produced the spheres in which the five then known planets – Mercury, Venus, Mars, Jupiter, and Saturn – moved according to a "celestial harmony." Since the distances of planetary orbits, their periods, and the movement of the planets seemed to obey harmonic, Pythagorean laws of proportion, there must be an extraterrestrial music of the spheres intelligible to reason alone. This music must accompany the planets on their orbits and fill the universe. The "harmony of the spheres" had been found: *Harmonices Mundi* (1619). This idea can still be found in the interpretation of the atomic model with its electron orbits. Erwin Schrödinger's equation of atomic electrons, published in 1927 in his treatise on wave mechanics, applies Pythagorean theories of the harmony of the two-dimensional resonating string to the oscillation forms of three-dimensional atoms.

I take the idea of a music of the spheres literally and derive music directly from the atmosphere. The sound sculpture is an *environment-specific* sound sculpture. The music is a *Musica cosmographica*, since it draws its material from the graphical description of the cosmos/atmosphere. However, it is also created with the aid of a computer, and is thus just as much a *Musica computa*.

When you look at the screen, you are looking directly into Earth's interior, into the core of the Earth on which you are standing – a human being standing on Earth with their eyes floating above the Earth as it were. Bodily and ocular perception are split. The viewer is simultaneously present (body) and absent (eye) beyond the Earth. The gaze dislocated from the body is a double gaze (from the real eye and the prosthetic eye). Two images and two sensations arise: being on Earth and not being there; looking into the Earth and looking at the Earth; being imprisoned by gravity, glued to the Earth, and being triumphant over gravity, free-floating in space. An artificial eye (satellite) describes the cosmos, delivers images of the state of the Earth's atmosphere. An eye roaming the orbital spheres delivers "cosmographic" images of the spheres. A computer turns these visual, graphical descriptions of the cosmos, of the (atmo)sphere into an acoustic description of the sphere: "Musica cosmographica" is born. The constantly changing meteorological patterns of the Earth generate sound patterns which are constantly changing. This creates computer-aided music drawn from the cosmos of the spheres. Via amplifiers, the synthesizer sounds are sent to the loudspeakers in the "continental" pillars. These pillars around the circular platform show (in a fragmented form) the five continents of the Earth.

—Peter Weibel

Translated from the German by Peter Dahm Robertson.

Orbital Sculpture IV: Spectral Music. Musica Cosmographica. Sound Monument for Johannes Kepler / Orbitale Skulptur IV: Spektralmusik. Musica Cosmographica. Klangmonument für Johannes Kepler

1987	sound installation; monitor, computer, loudspeakers, satellite, dimensions variable technical assistance: Franz Xaver; sculptural assistance: Hans Weigand

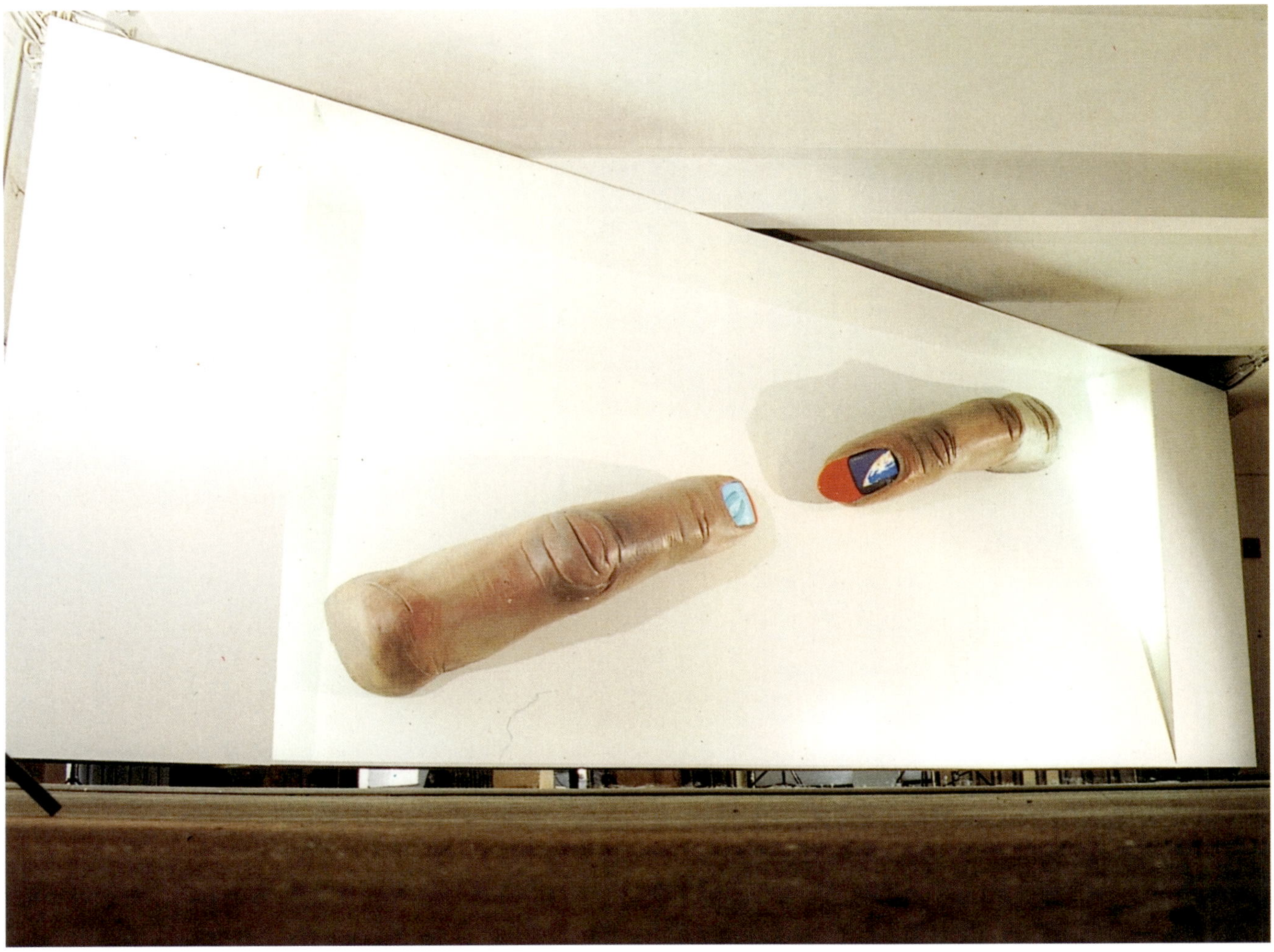

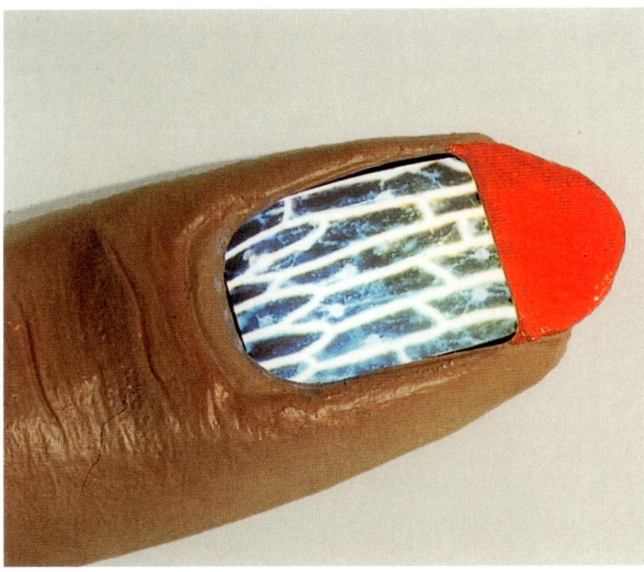

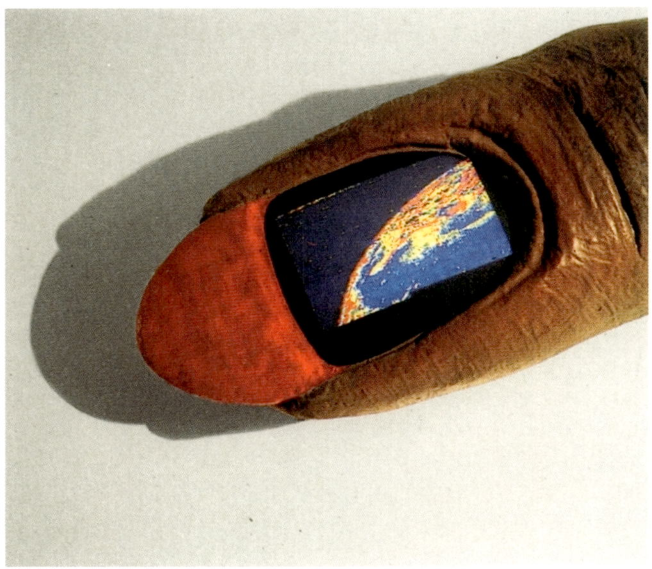

Installation views *Das gläserne U-Boot*, transart 1, Tabakfabrik, Krems an der Donau, Austria, 1988

Orbital Sculpture V / Orbitale Skulptur V

1988 | polyester, 2 monitors, weather satellite and satellite antenna, video player, tape with images of the microcosm and live images from the weather satellite

A meteorological satellite (Meteosat) transmits pictures of the Earth, the macrocosm (right fingernail monitor). A video player displays pictures of the microcosm (left monitor). In the future, there will be screens the size of fingernails. In telematic culture, scale is variable. Microworld and macroworld are the same size.

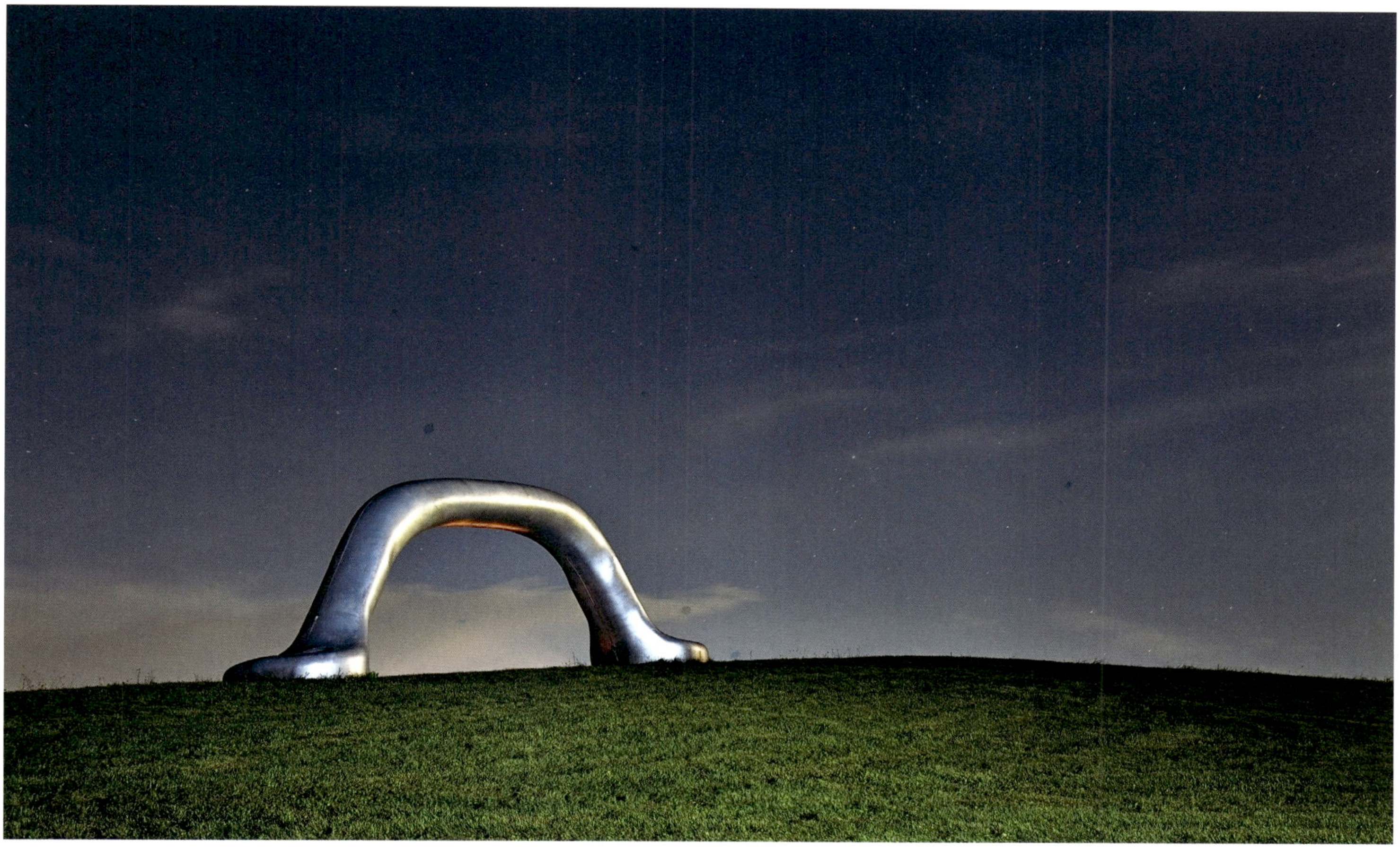

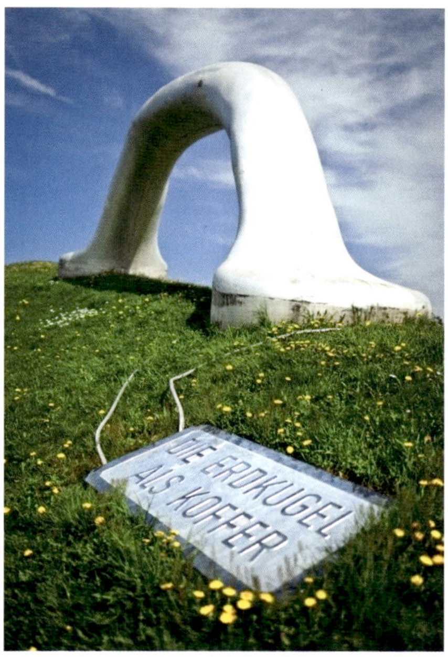

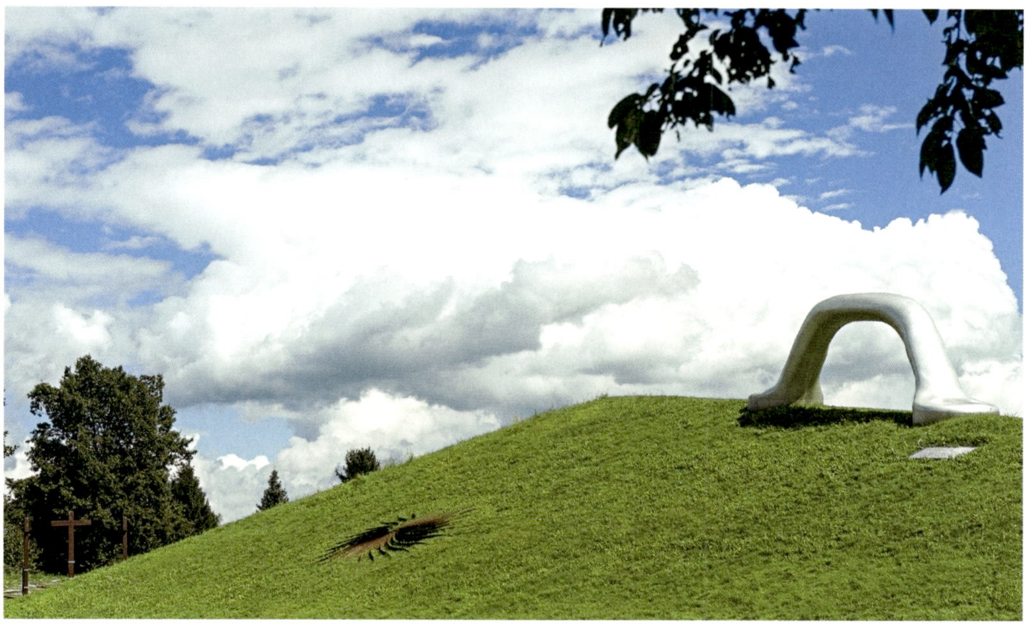

Installation views at the Austrian Sculpture Park, Premstätten near Graz

We humans are all traveling through cold and desolate outer space with a single ticket, a single suitcase, a single vehicle: the globe. The provisions for humanity's journey through the universe come solely from Earth. Earth is our most sacred sculpture. We are cosmological travellers. This piece is also a protest against mass mobility and mass tourism.

—Peter Weibel

The Globe as Suitcase / Die Erdkugel als Koffer

2004 | white concrete, 12.6 × 3.58 m

The history of the human body
as a history of the prosthesis
converges with
the history of the Earth
as a history of the prosthesis.
The Earth becomes the human body.

Tools and technological culture
arise from externalizations (exteriorizations)
of the human body.
Tools and the products of technology
are substitute body parts (prostheses)
that perfect human organs:
muscle – motor
hand – hammer
foot – wheel
eye – glasses, microscope, telescope, television
ear – telephone, radio
memory – writing, photography, sound recording
calculation – computer
womb – house, Earth.

The essence of the prosthesis is the replacement
of natural organs with
artificial auxiliary organs
made by humans:
the technomorphic transformation of the Earth
as part of the human body.
Technology is Earth's makeup.
The modeling of the body
leads to technology as prosthesis culture,
to the cloned body.
Computational modeling and cloning
are the most advanced prosthetics.
Humanity becomes a prosthetic god,
the master of the human body,
of Earth and outer space.

The conquest of space
continues the exteriorization of the body.
Satellites and spaceships are
not just externalizations of the body
but also exteriorizations of the Earth.

The dwelling places and energy sources,
organs and tools
of the Earth are externalized into space.
The Earth as womb is
exteriorized in the age of space-time.
Like the interior of the Earth,
the human unconscious is also
made external.
The Earth becomes the brain of humanity,
of the prosthetic god that is the human being.
The Earth becomes a prosthesis,

a tool of humanity.

The Earth as brainwork:
self-made reality,
self-created world.

The Earth becomes a human:
this is the purpose of tools,
of prosthetics and technology.
Satellites are the eyes, ears,
and cameras of the human Earth.

The cloning of the body
is followed by the cloned Earth.

Spaceships are the
first cells of the cloned Earth.
Earth, the mother ship and master tape,
will release infinitely many copies
into space.
Just as every cell of an individual organism
can become the belly of an
identical individual through cloning,
the Earth can become the belly of
identical Earth stars and Earth cells.

In the age of space-time
we replace parts of the Earth
until the Earth itself is replaced.
A replaceable, cloned Earth
becomes infinitely reproducible.
The Earth transforms itself
into a gigantic
immortal prosthesis.

Translated from the German by Patrick Hubenthal.

The Earth as Brainwork

1985 | poem with drawings; opaque white on black paper

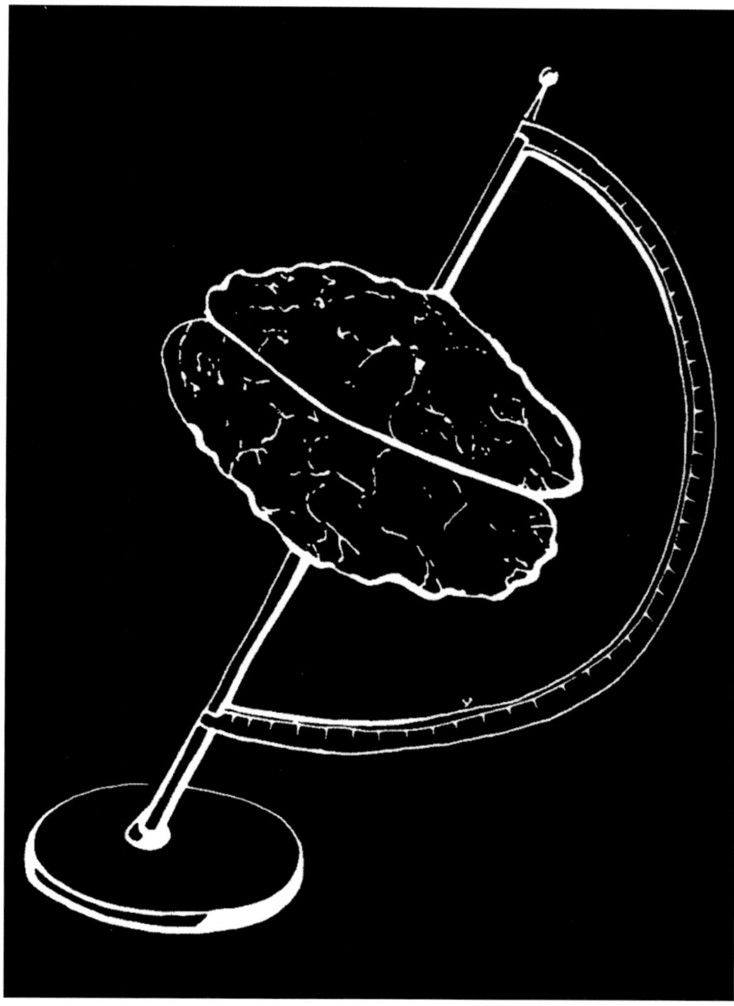

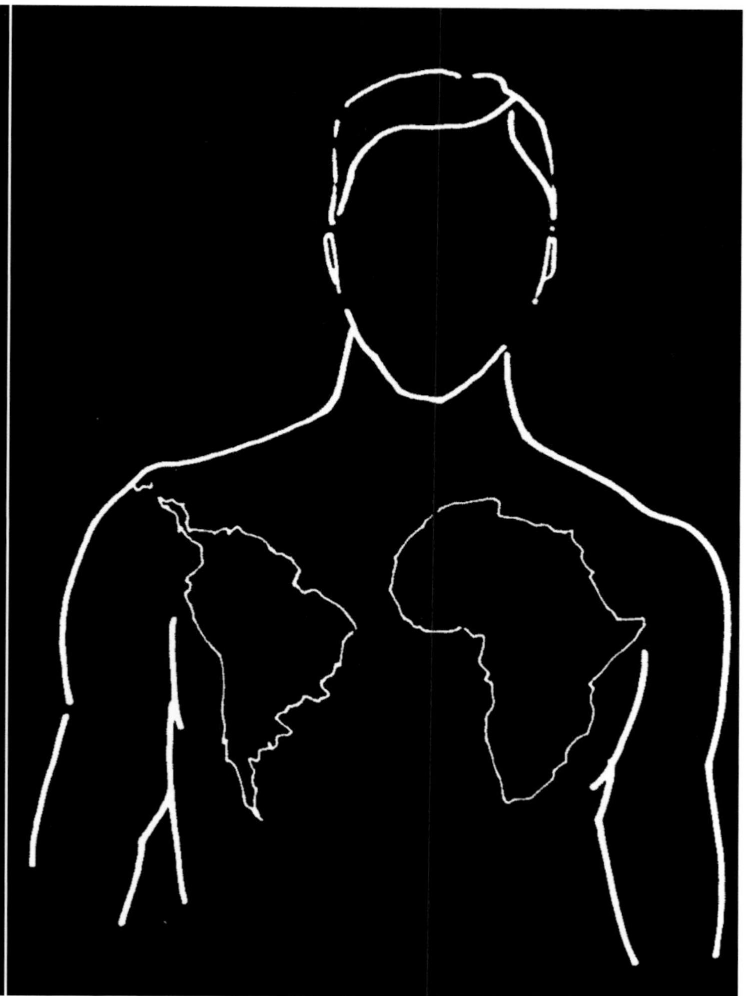

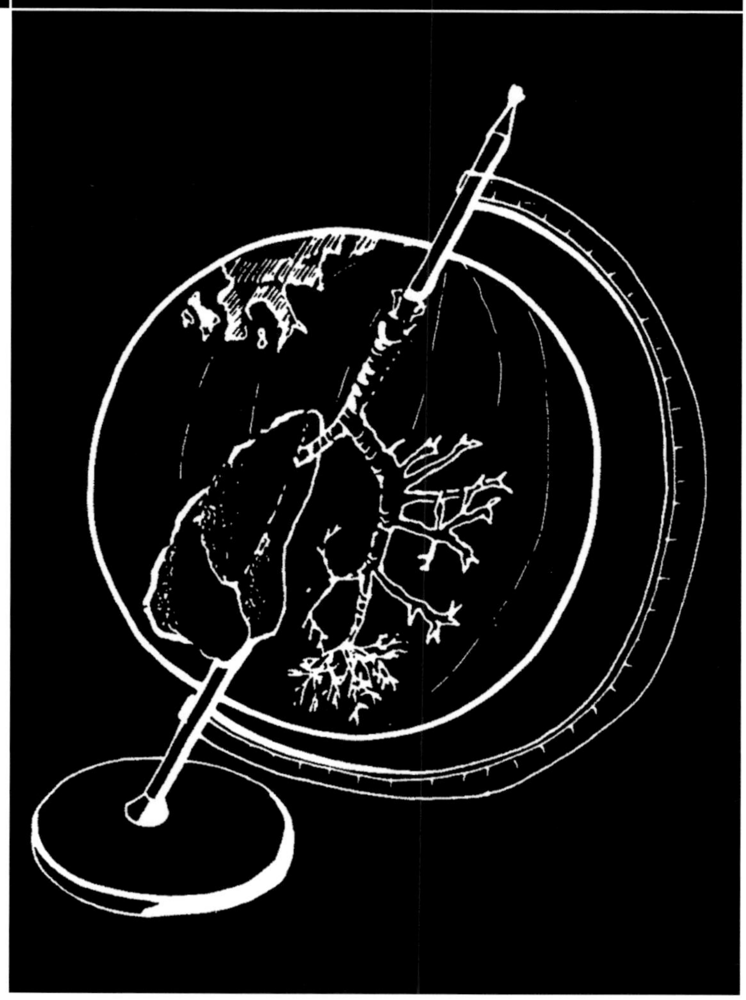

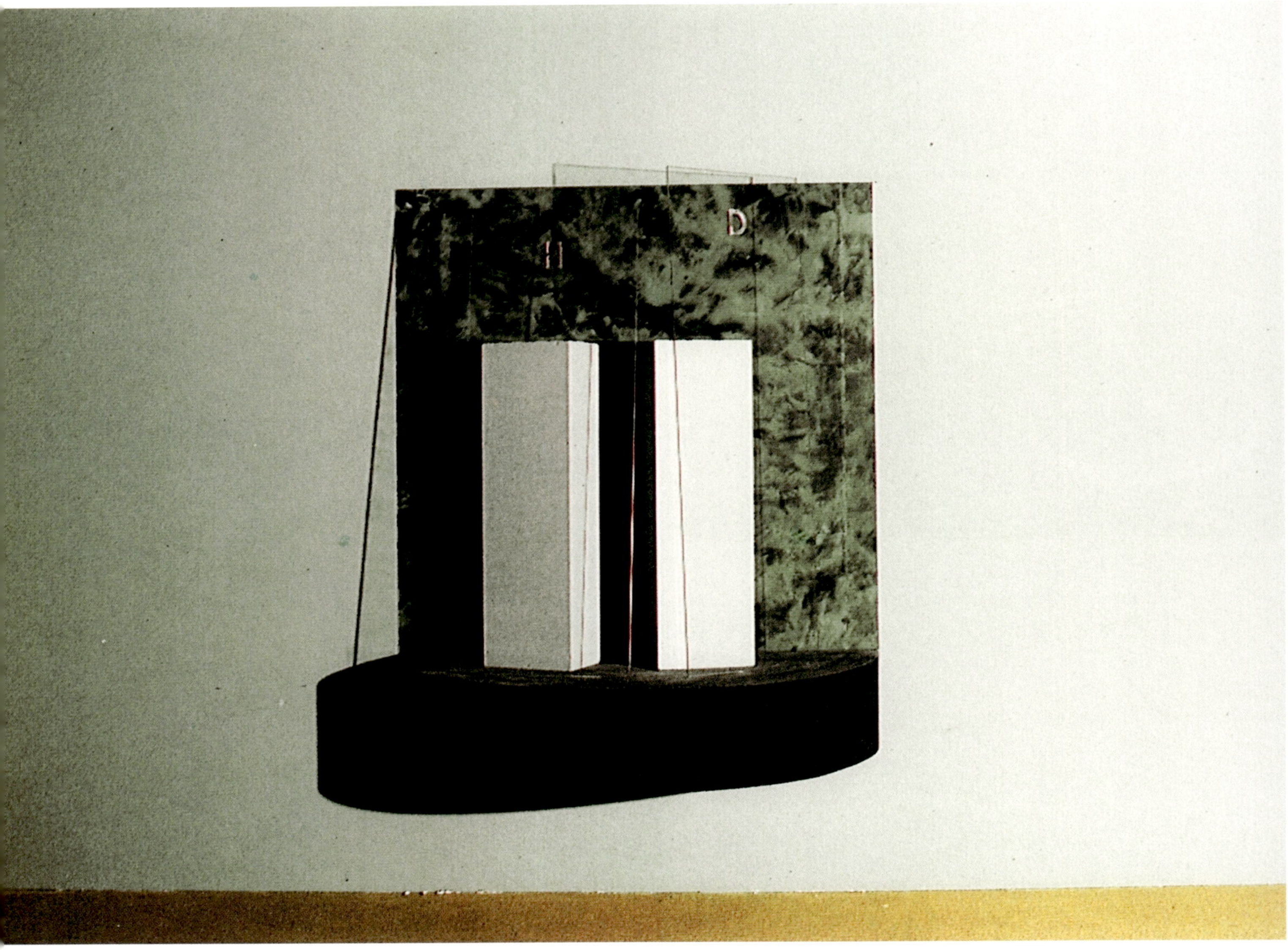

Installation view *Inszenierte Kunstgeschichte*, MAK – Museum of Applied Arts, Vienna, 1988–89

Object-Image of Desire for Public Architecture / Objektbild des Begehrens für öffentliche Architektur

1988	2 glass panes, mural, 2 pedestrals, 2 letters

During my 1988–89 exhibition *Inszenierte Kunstgeschichte* at the MAK – Museum of Applied Arts in Vienna, I removed the letters *H* (for *Herren*, "Gentlemen") and *D* (for *Damen*, "Ladies") from the public restrooms so that visitors would not know which restroom to enter. In this way, I highlighted the binary division of sex and gender. I then used the two letters to create a sculptural ensemble built out of conventional architectural elements. —Peter Weibel

The Symbol-Processing Ability of Man and Machines: The Emancipations of Tools

Peter Weibel

The Church-Turing thesis of 1937 implies *grosso modo* the following: Everything that is can be formalized. Everything that can be formalized can also be calculated. And what can be calculated can also be mechanized. All formalisms can be applied to a machine since every formal system is defined by its mechanism. In this concatenation of formalism, mechanism, and machine, the incalculability and the undecidability of formalism are also transferred to the machine, as the development from Gödel to Turing shows. This produces the limits of formalization, calculation, and speakability. If man can be defined as a symbol-processing, thinking being, then he is thus machine-like. "A combination of a man with written instructions will be called a 'Paper Machine.' A man provided with paper, pencil, and rubber, and subject to strict discipline is in effect a universal machine," wrote Turing in his "Intelligent Machinery."[1] Man as a "paper machine" corresponds to the "paper machine" as man. Thus, there are learning, self-modifying, automatic, self-reproducing machines. This idea of an "electron brain" marks the final point of a long line of limitations whereby each natural human organ is replaced or imitated by a machine. Sigmund Freud referred to this creature whose voice is amplified by a microphone and whose scope is extended beyond the natural (auditory and visual) boundaries by means of telephone as well as telescope, microscope, and TV camera and whose nervous system networks are well simulated by electronic circuitry in *Civilization and Its Discontents* (1930) as "prosthesis God."[2] The idea that machines could think, that is, process symbols and give them meaning on their own, which until then had been regarded as a privilege of humans, has become more tangible than ever before. Turing's universal calculator based on the binary symbols 0 and 1, that are represented by a noncurrent and current in circuits, which in turn are based on George Boole's algebra, can actually be likened to a living organism, if independent symbol processing is geared to the notion of a living organism. Autonomous symbols are really a sensation in the history of civilization. It is no longer humanity who paints animals on walls. Now the explosion of the arts, which is at the same time an explosion of signs, has led to machines drawing signs by themselves. Tools no longer serve to fulfill humans' will and to give signs material, real form. Rather, tools themselves set signs and the realm of signs sets reality. Symbol-processing machines, such as the computer, have made signs autonomous agents. Machines create new autonomous models of the world and of tools. Media art shows us this new phase of the autonomous, symbol-processing machine. The explosion of the arts is followed by the explosion of signs – with the help of machines. The tool culture has entered a new phase; namely, that of the intrinsic world of the world of machines.

Tools did not evolve from language and nor language from tools. Rather, language and tools have a common source: the human ability of symbolization. Language and technology have developed on this common ground. In this sense tool technology, in particular those tools that autonomously process symbols like intelligent machines, is the key to human evolution. Tool culture has always been symbol culture. Without symbols there is no storage possibility, without storage or memory, there is no experience. Writing is the first memory, the computer at present the last one. Writing can be used to bridge spatial and temporal distances. Decorporealized, dematerialized information can be moved about in space and time. This was the first revolution in communication. The invention of the printed book was the second revolution in communication. Mass communication was now possible. The separation of messenger and message, body and symbols, the symbolization of messages by means of electromagnetic fields, ushered in with telegraphy in 1883, formed the technical basis for the third telematic revolution in communication: the electronic digital processing of information. Machine communication had become possible.

In the electromagnetic age (J. C. Maxwell, 1873)[3] signs travel freely and autonomously at electronic speed. The signs of the third communication revolution are liberated from humankind and live a life of their own thanks to sign machines. Tools have become emancipated and have begun to live a life of their own as well as symbolic processing machines. The rise of symbol-processing machines means the end of the (last) privilege and monopoly of humankind.

Note: Adapted from Peter Weibel, "The Symbol-Processing Ability of Man and Machines: The Emancipations of Tools," in *First Europeans: Frühe Kulturen – moderne Visionen*, exh. cat., Große Orangerie des Schlosses Charlottenburg (Berlin: Exter & Exter Unternehmensberatung und Marketingservice, 1993), 100–103.

[1] Alan Turing, "Intelligent Machinery: A Report by A. M. Turing" (Summer 1948), in *Key Papers: Cybernetics*, ed. C. R. Evans and A. D. J. Robertson (Baltimore, MA: University Park Press, 1968), 26–53.

[2] See Sigmund Freud, *Civilization and Its Discontents* (New York: Norton PB, 1962). First published as *Das Unbehagen in der Kultur* (Vienna: Internationaler psychoanalytischer Verlag, 1930).

[3] See James Clerk Maxwell, *A Treatise on Electricity and Magnetism* (Oxford: Clarendon, 1873).

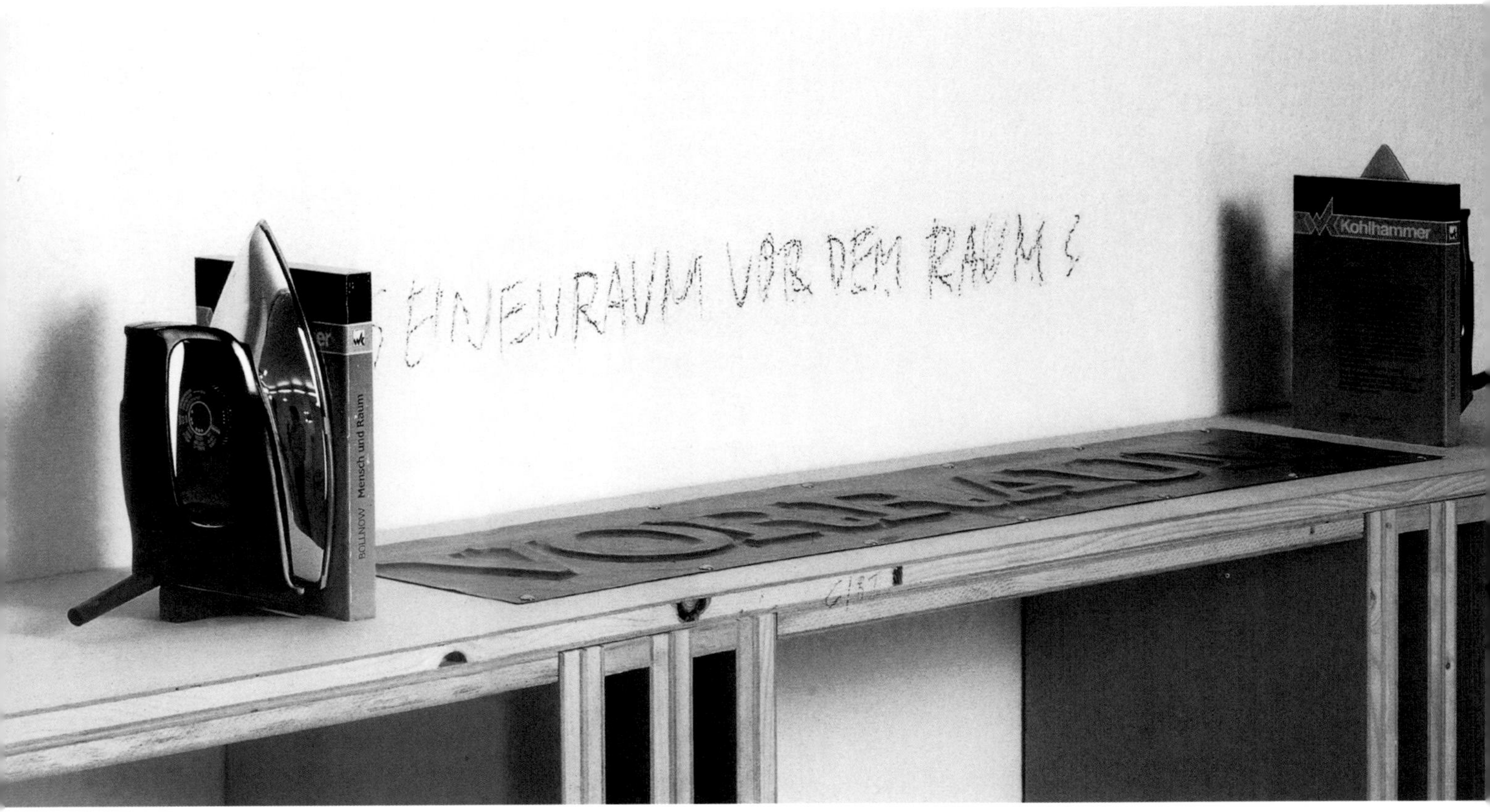

Installation views
Hypothetische Produkte.
Vom Elend der Privilegien,
Prodomo, Vienna, 1994

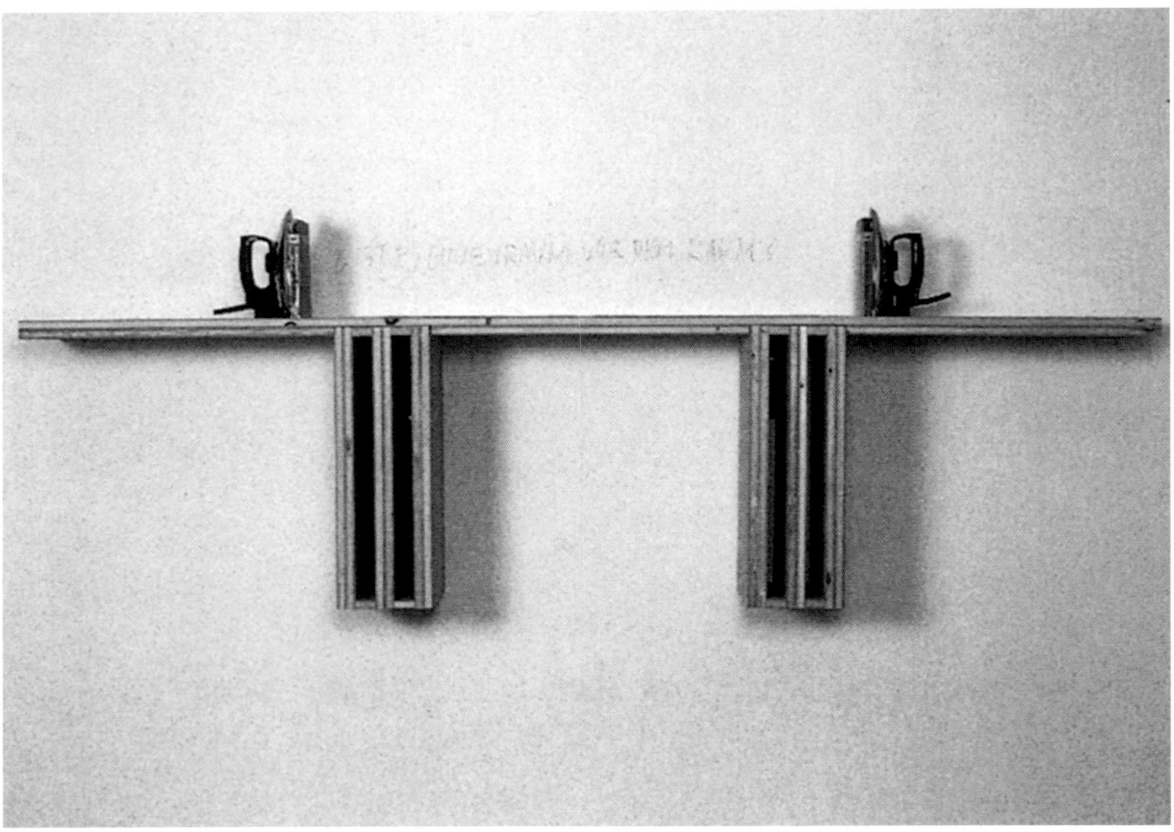

The Object That Is Not Itself: Pure Reality; Is There Time before Time? Is There Space before Space? / Das Objekt, das nicht selbst ist: die reine Realität. Gibt es eine Zeit vor der Zeit? Gibt es einen Raum vor demRaum?

1990	2 wall shelves, 4 irons, 4 books, 2 lead mats with the words Vorraum (anteroom, vestibule) and Vorzeit (antiquity, prehistoric times), 120 × 30 × 60 cm, 2 parts

Disability as Design Avant-Garde

Peter Weibel

My design concepts have to do with the emancipation of tools. The third, information-based communications revolution has radically changed the nature of objects. Their elevation, from tools as extensions of the body to symbol-processing machines, is complete. The processing of symbols is no longer a privilege of humans; now machines can do it too. Objects have thus become adaptable, autonomous agents, materialized strings of signs that themselves process signs. Intelligent products show that we have always been disabled without realizing it.

Disability as design-avant-garde.

ars pro deo
ars pro ecclesia

Art for God
Art for the Church

ars pro re
ars pro castello

Art for the king
Art for the palace

ars pro cive
ars pro museo

Art for the citizen
Art for the museum

ars pro homine
ars pro domo

Art for human beings
Art for the home

ars pro machina
pars pro toto

Art for the machine
(p)art (taken) for the whole

Note: Adapted from Peter Weibel's, foreword and back cover to *Peter Weibel:* in *Hypothetische Produkte. Vom Elend der Privilegien*, ed. Robert Fleck, exh. cat. (Vienna: Prodomo, 1994). Translated from the German by Patrick Hubenthal and Peter Dahm Robertson.

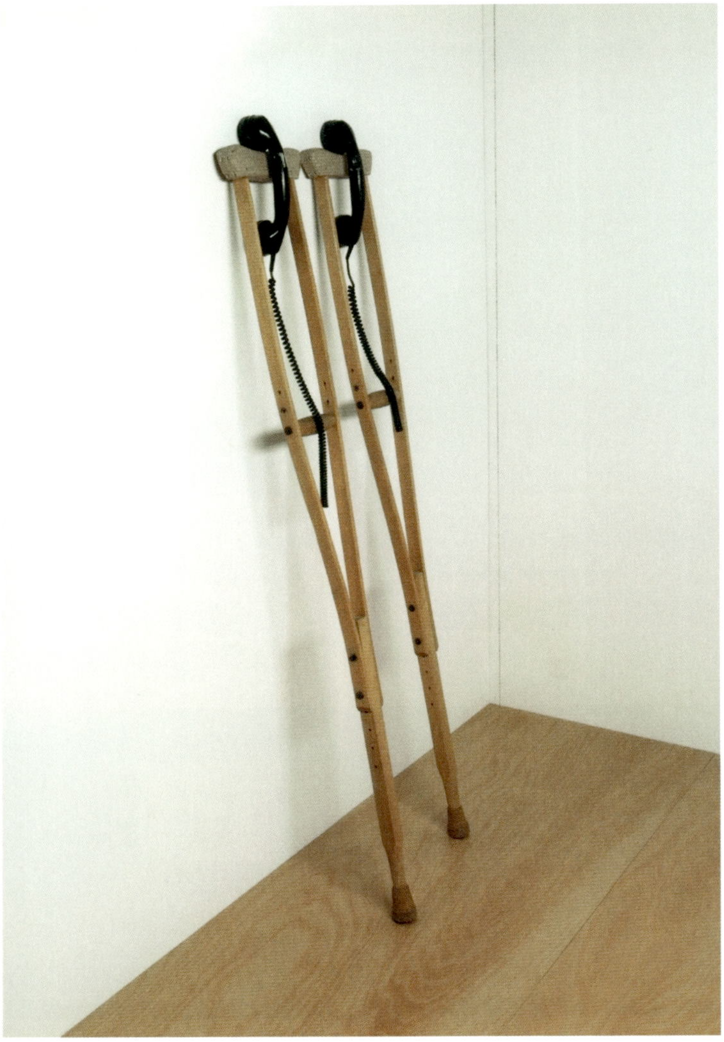

Installation view *Peter Weibel – MEDIA REBEL*, Belvedere, Vienna, 2014–15

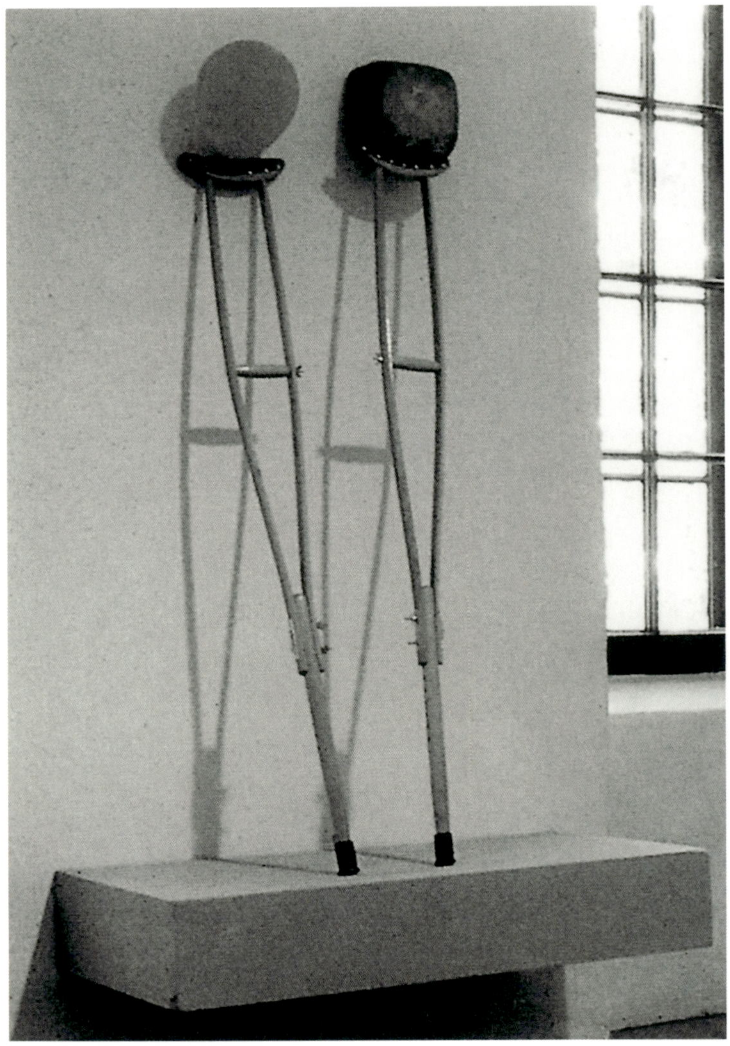

Installation view *Hypothetische Produkte. Vom Elend der Privilegien*, Prodomo, Vienna, 1994

Traditionally, prosthetics help people overcome disabilities. Technology extends this function in more refined and improved fashion.
In particular, technological devices (television, telephone, telefax, internet) are used to compensate for collective disabilities caused by space and time. —Peter Weibel, 1996

All Technology Is Teletechnology, All Teletechnology is Theotechnology / Alle Technologie ist Ferntechnologie, Alle Teletechnologie ist Theotechnologie

1994	2 telephone receivers, 2 crutches, 2 cable reels, Elastoplast plasters, 140 cm

All Technology Is Crutch Technology and Overcomes the Pain of Absences and Distances / Alle Technologie ist Krücken-Technologie und überwindet die Schmerzen der Absenzen und Distanzen

1994	2 crutches on pedestal, granite stone, balloon

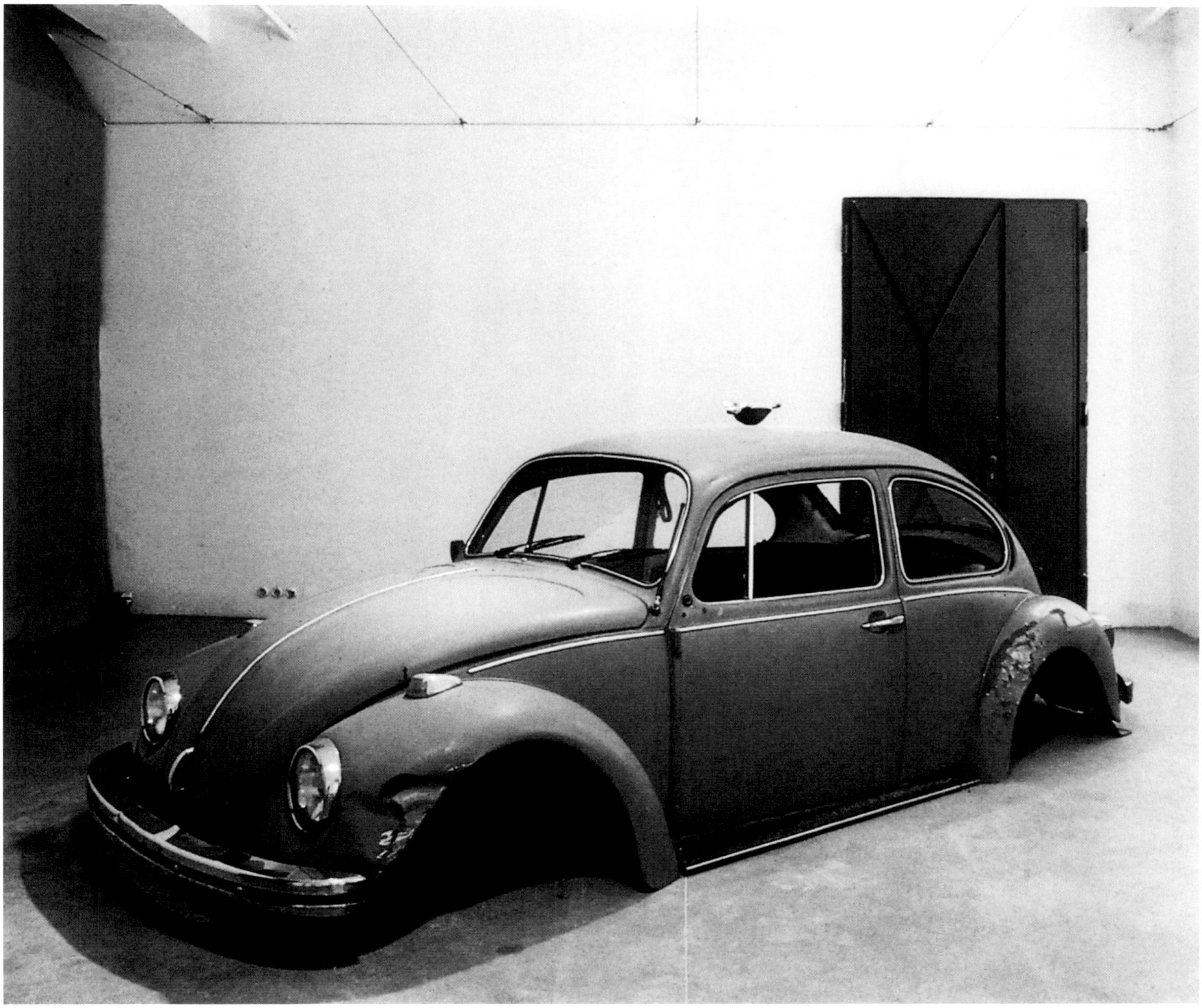

Installation views
Hypothetische Produkte.
Vom Elend der Privilegien,
Prodomo, Vienna, 1994

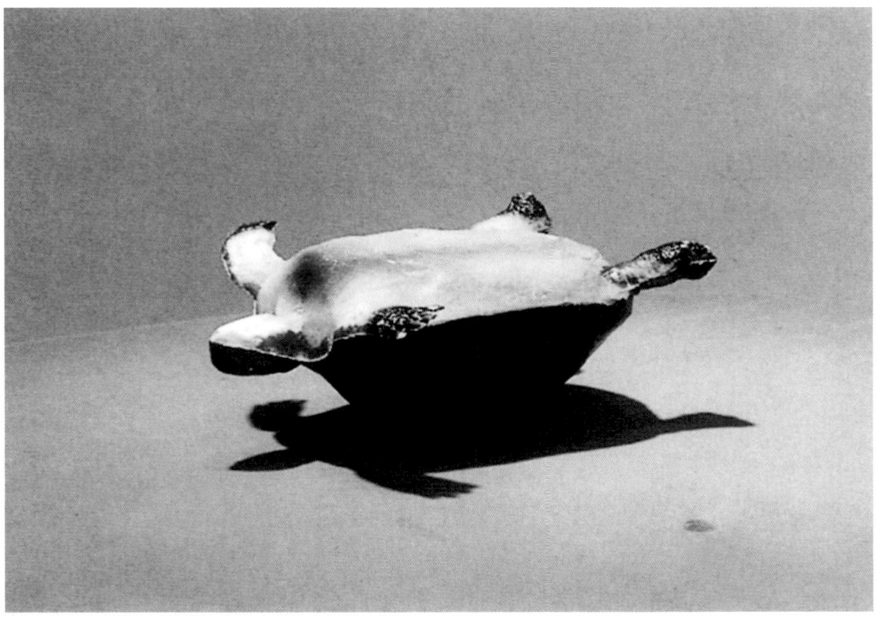

Mechanics of Organisms – Organics of Machines / Mechanik der Organismen – Organik der Maschinen

1994 | car body, plastic turtle

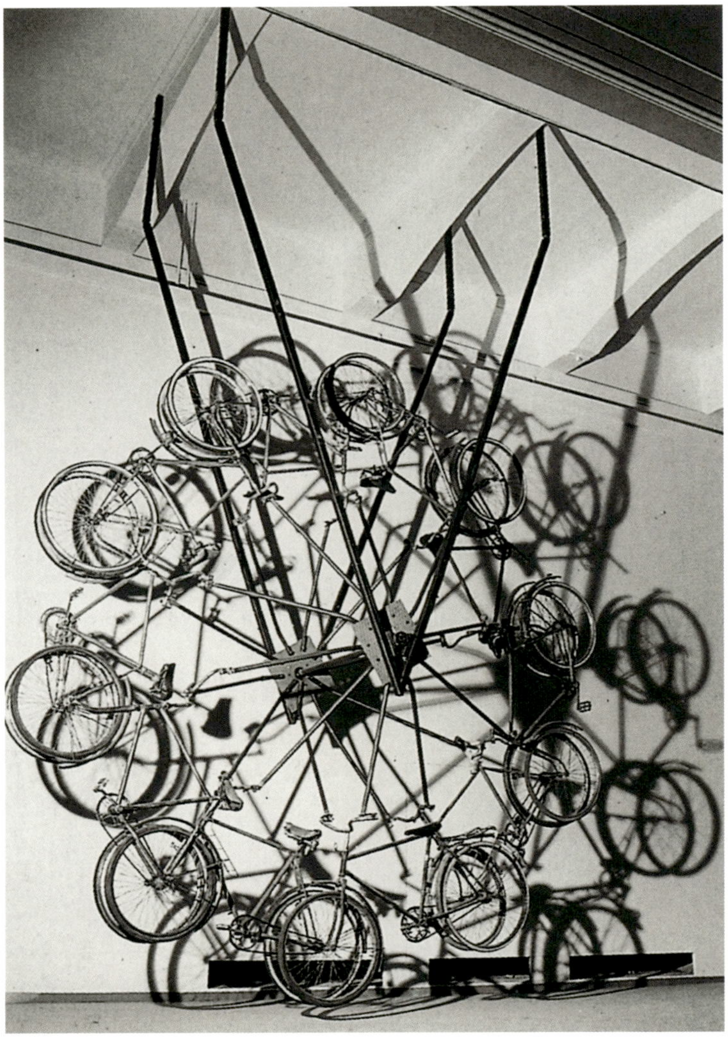

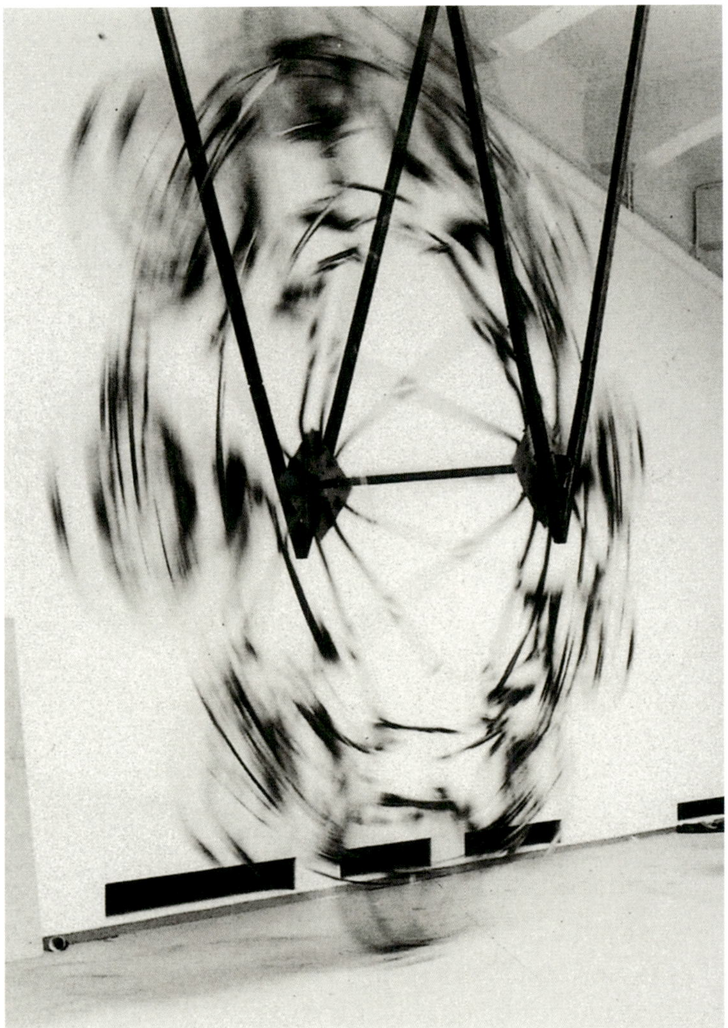

Installation views *Inszenierte Kunstgeschichte*, MAK – Museum of Applied Arts, Vienna, 1988–89

The outdoor sculpture *The Wheel of the Real* consists of ten bicycles attached to a supporting structure and a lead human figure. The industrial revolution of the nineteenth century was based on wheel technology. Wheels set the production machinery in motion in the factories after 1800, from which the wheel-driven machines of movement (bicycle, trains, automobiles, and airplanes) developed, as well as image machines (camera, film projector) in the early twentieth century. With this sculpture, created in 1988, but lost later and reconstructed in 2014, Peter Weibel pays tribute to the technology of the wheel as one of the foundations of today's technologically shaped world. For Peter Weibel the sculpture also stands for the wheel of history that is turned by politics and in which the individual human being is only a small, insignificant passenger. Citizens do not play a role in the wheels of power. Billions of individuals are just attachments of history.

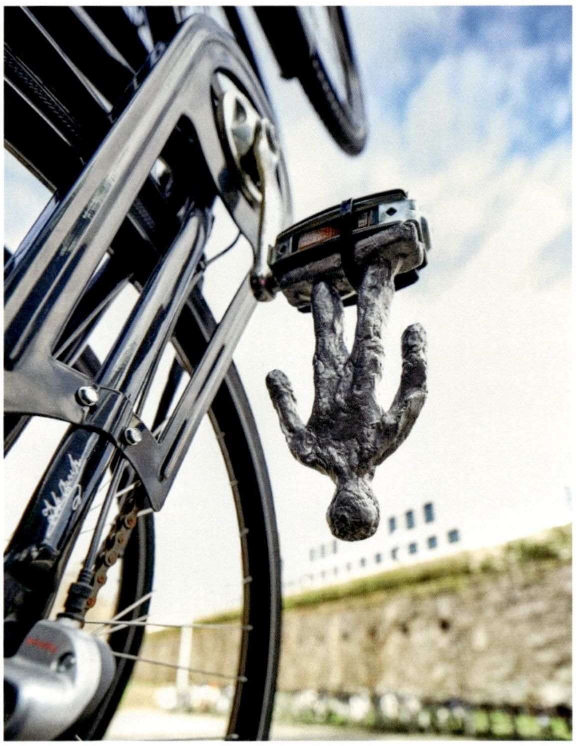

Installation view *respectively, Peter Weibel*, ZKM | Karlsruhe, 2019–20

The Wheel of the Real / Das Rad des Realen

1998/2014 | 10 bicycles on support construction, lead figure, 440 × 440 cm

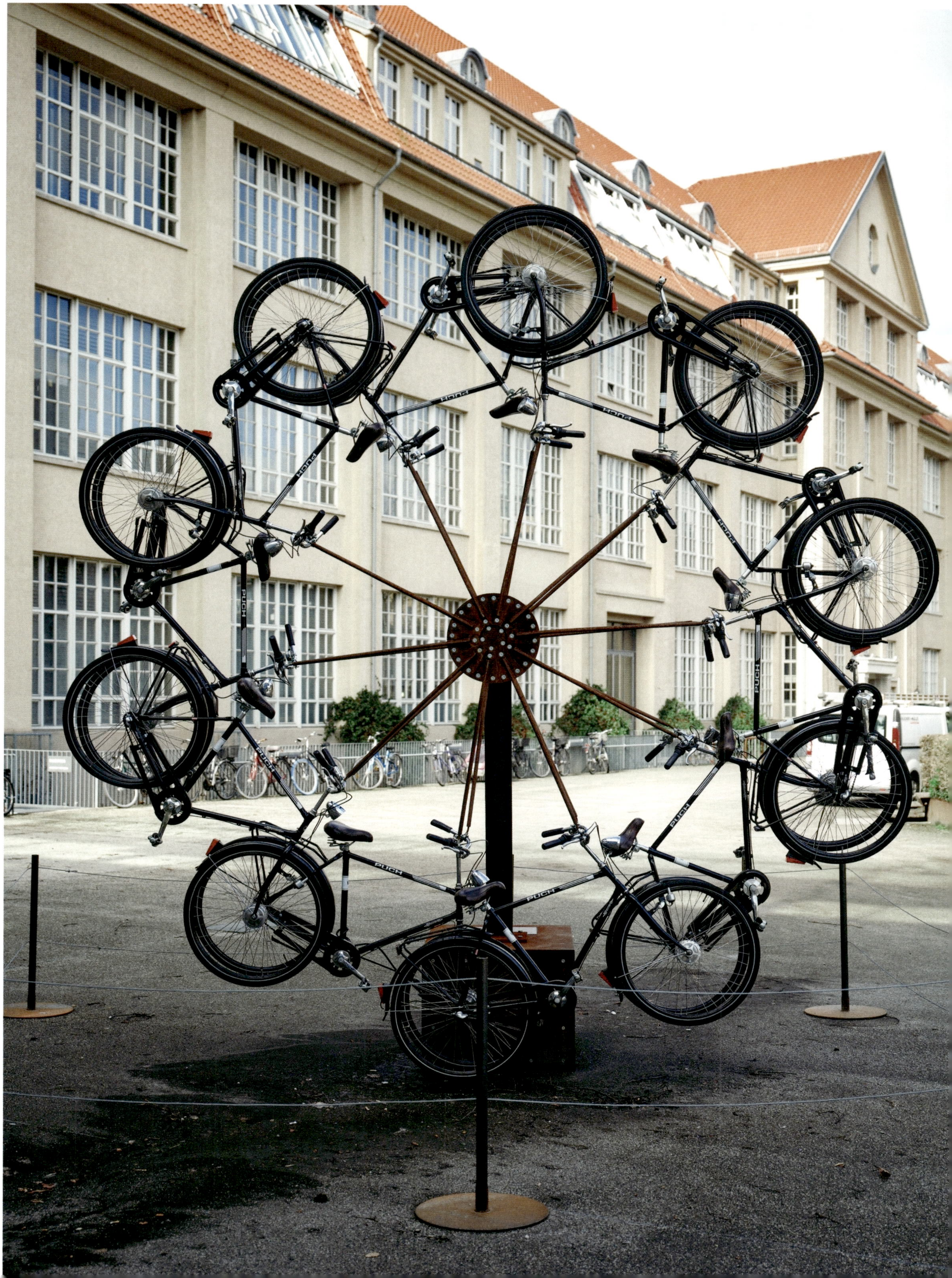

7. Social Sculpture

Gallery project no. 1 (1968)
The gallery remains closed during the exhibition. (Whether the rooms of the gallery contain exhibits at all remains a contractually agreed secret between the gallery and the artist.)

...

Gallery project no. 3 (1968)
I only start to work on the exhibition during the opening of the exhibition itself. For the duration of the exhibition, I work on the exhibition in the gallery space and only during opening hours. The work on the exhibition is thus itself exhibited. The results of this work, which is completed with the conclusion of the exhibition, are destroyed after the exhibition. The documentation of this work of destruction forms the material for a following exhibit. The destruction is documented in a catalog. (The catalog is burned. The photograph of it is the only thing that remains.)

Gallery project no. 4 (1968)
An exhibition announced under my name contains the artworks of those artists who, in the time of the year prior to my exhibition, had asked to be exhibited in precisely that gallery and were refused by the gallery.

...

Gallery project no. X2 (1973)
Using the gallery's telephone number, which for the duration of the exhibition is advertised every day in daily newspapers with a corresponding comment, telephonic visitors to the gallery can tell me what I am supposed to do, specifically with my body. These suggestions are transmitted into the gallery using an amplifier and speakers. Physical visitors to the gallery can see me react to the commands and wishes of the absent audience. The calls are recorded on tape and my reactions videographed. This selection (of the exhibition), dictated by the nonpresence of visitors, is presented in a second room of the gallery as (constant) (recorded) exhibit for potential visitors.

...

Translated from the German by Peter Dahm Robertson

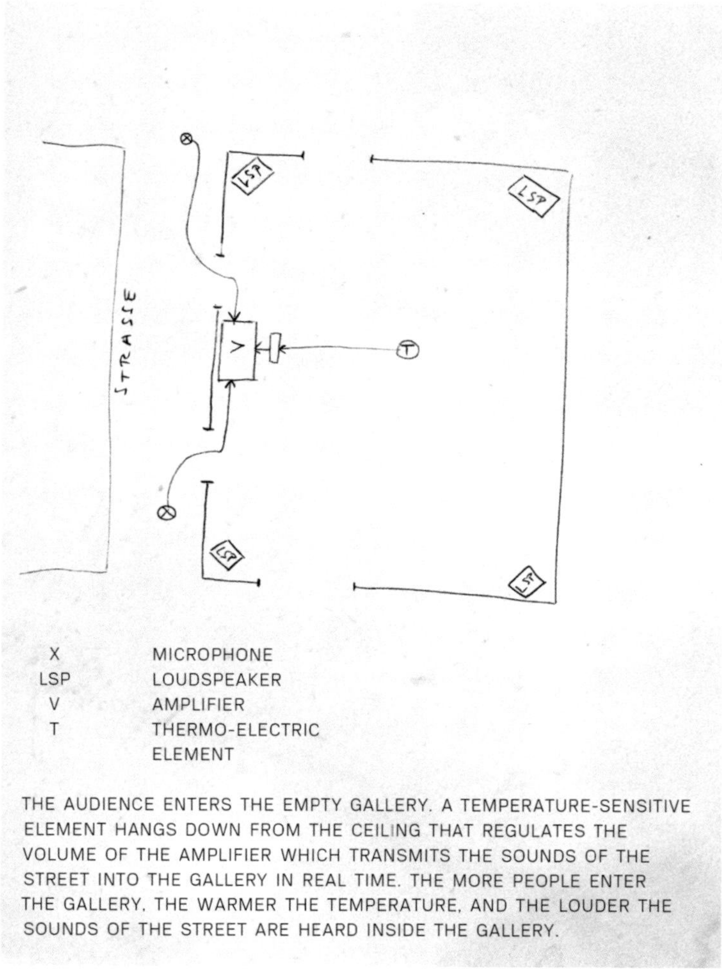

Conceptual drawing for *Inside/Outside*, 1969. Translated from the German by Peter Dahm Robertson

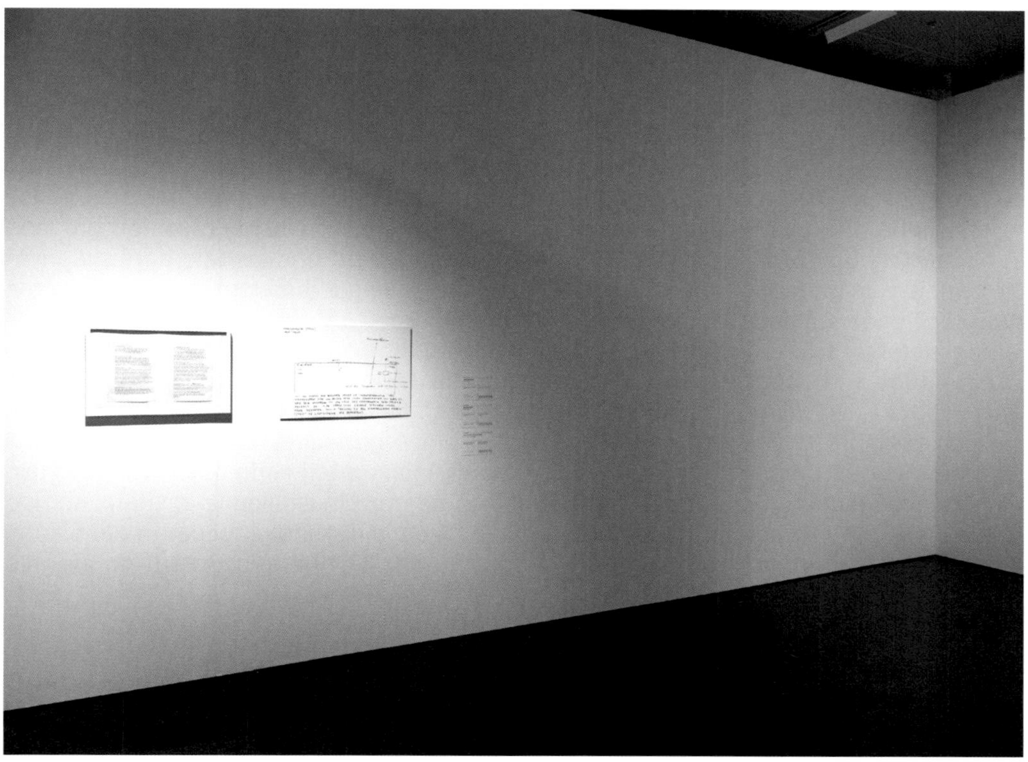

Installation view *respectively, Peter Weibel*, ZKM | Karlsruhe, 2019–20

Inside/Outside / Innen/Außen

1969 | gallery project, sound performance

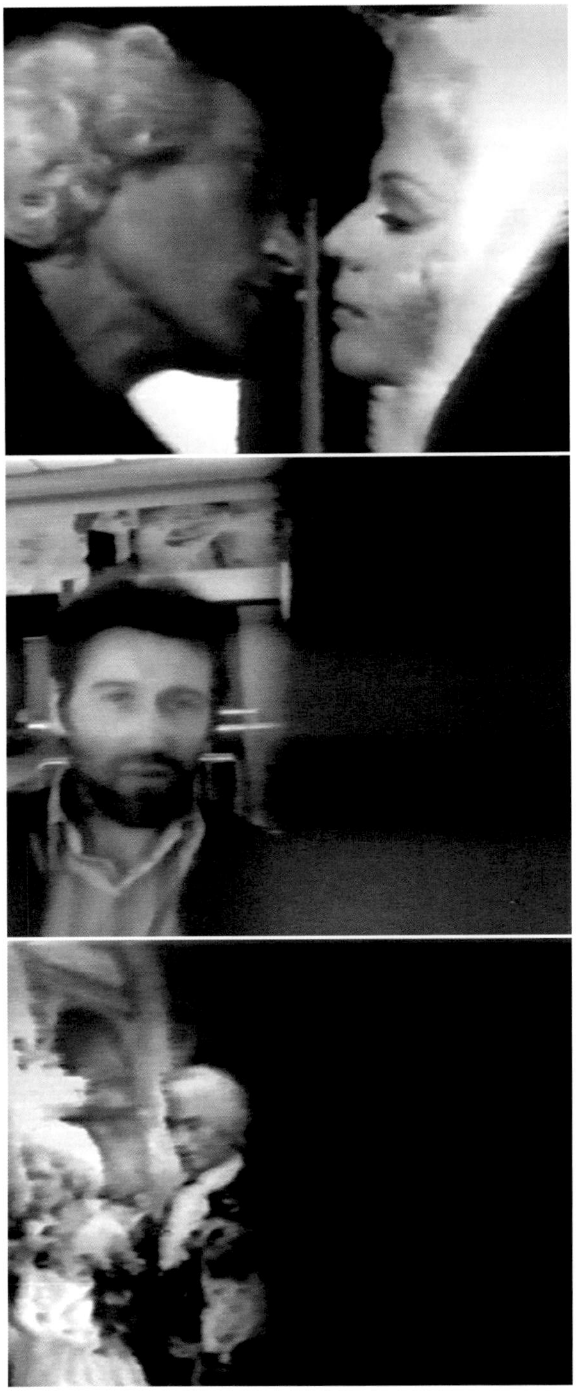

At an airport, the boundary between the duty-free zone / transit area and the rest of the airport is videotaped in such a way that it divides the screen exactly in half. Therefore, one half of the TV is already in Austria, while the other is still in the duty-free area. I appear on the Austrian half of the screen and declare all Austrian TV sets to be dutiable goods. The other half of the TV, of course, is still in the extraterritorial duty-free area, the duty on it still unpaid. As a penalty for this act of smuggling – since the TV set has obviously already been taken home – the transmission is suspended for several minutes on the side of the screen that is in the duty-free zone. All Austrian TVs are therefore black on one side. A political interpretation of the line. A political context is transferred onto the TV set. Geographically, the airport is in Austria, and yet, through an administrative act, the transit area is not in the state of Austria. This extraterritoriality is a model for the invisible territories of privilege more generally. All of us carry these invisible borders within ourselves, and the division of the Earth is a division of people. Every person should be extraterritorial, immune like a diplomat, one's own ambassador, one's own state, one's own country. I divide the TV set into freedom and punishment, just as the judiciary divides reality into legal and illegal reality. This demonstration of a pseudoquality born of an administrative despotism that abets executive efficacy reveals the madness of the national economy, the judiciary, and politics, which create borders where none exist. —Peter Weibel, 1970

The Invisible Borders / Die unsichtbaren Grenzen

1970 (concept)/ 1971	*Tele-Action* No. 2, 10 min., submitted to ORF in 1972 and not broadcast due to political reasons

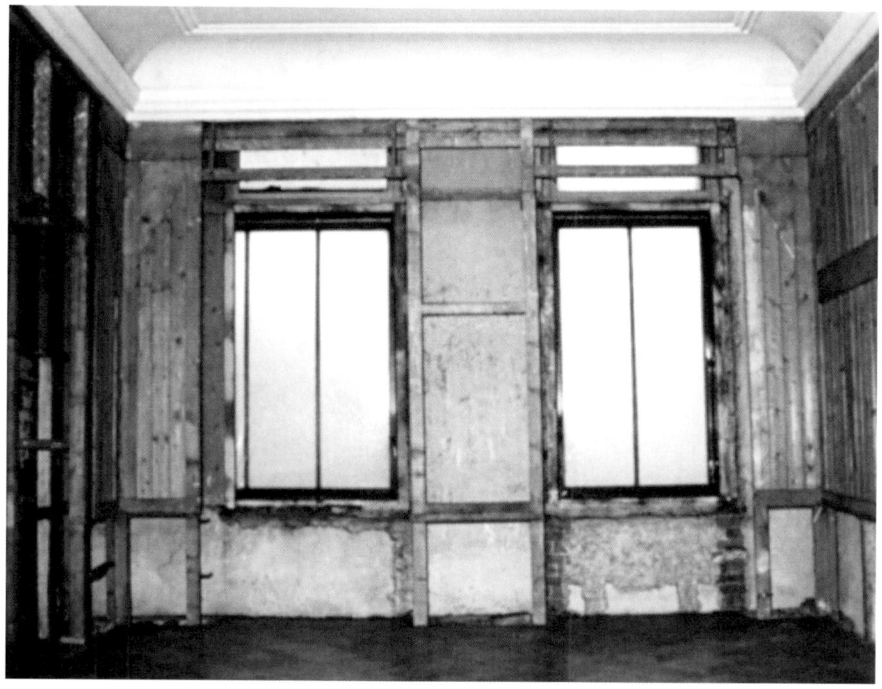

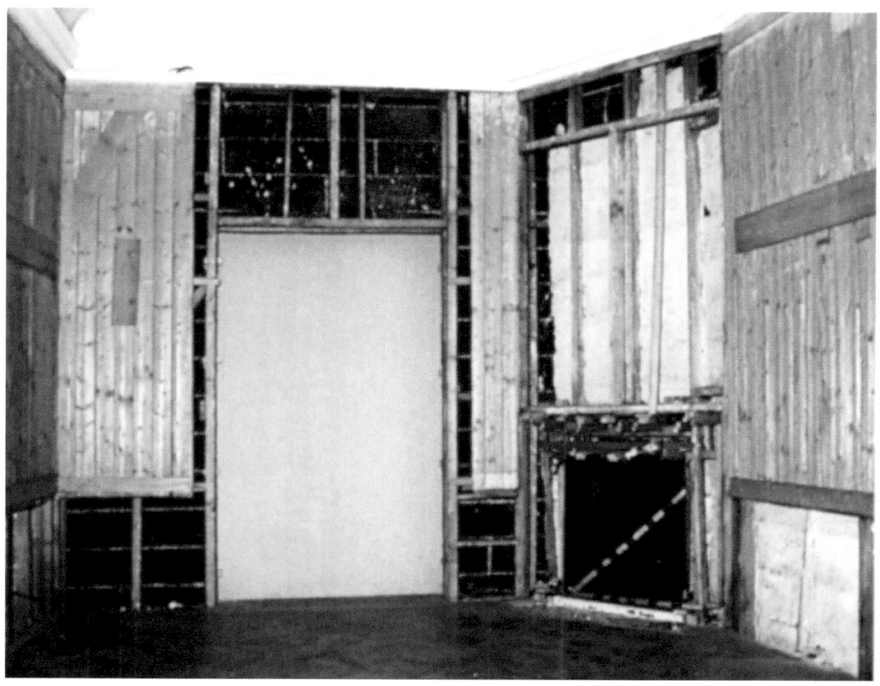

Installation views Galerie nächst St. Stephan, Vienna, 1971

Exhibition Spaces Dismantled, Exhibited as Readymades / Ausstellungsräume demontiert, als Ready-Mades ausgestellt

1971

After my *Gallery Projects* (*Galerieprojekte*, 1968–73, see p. 448 in this volume), I was less interested in the objects being exhibited than in the exhibition spaces themselves, and in the processes taking place between exhibition and viewer. Above all, as part of my critique of representation, I wanted to change exhibition spaces (as spaces of representation), transforming them into workshops, construction sites, laboratories, and so on. The renovation of Galerie nächst St. Stephan in Vienna gave me the opportunity to exhibit its exhibition spaces as dismantled spaces. —Peter Weibel, 2022

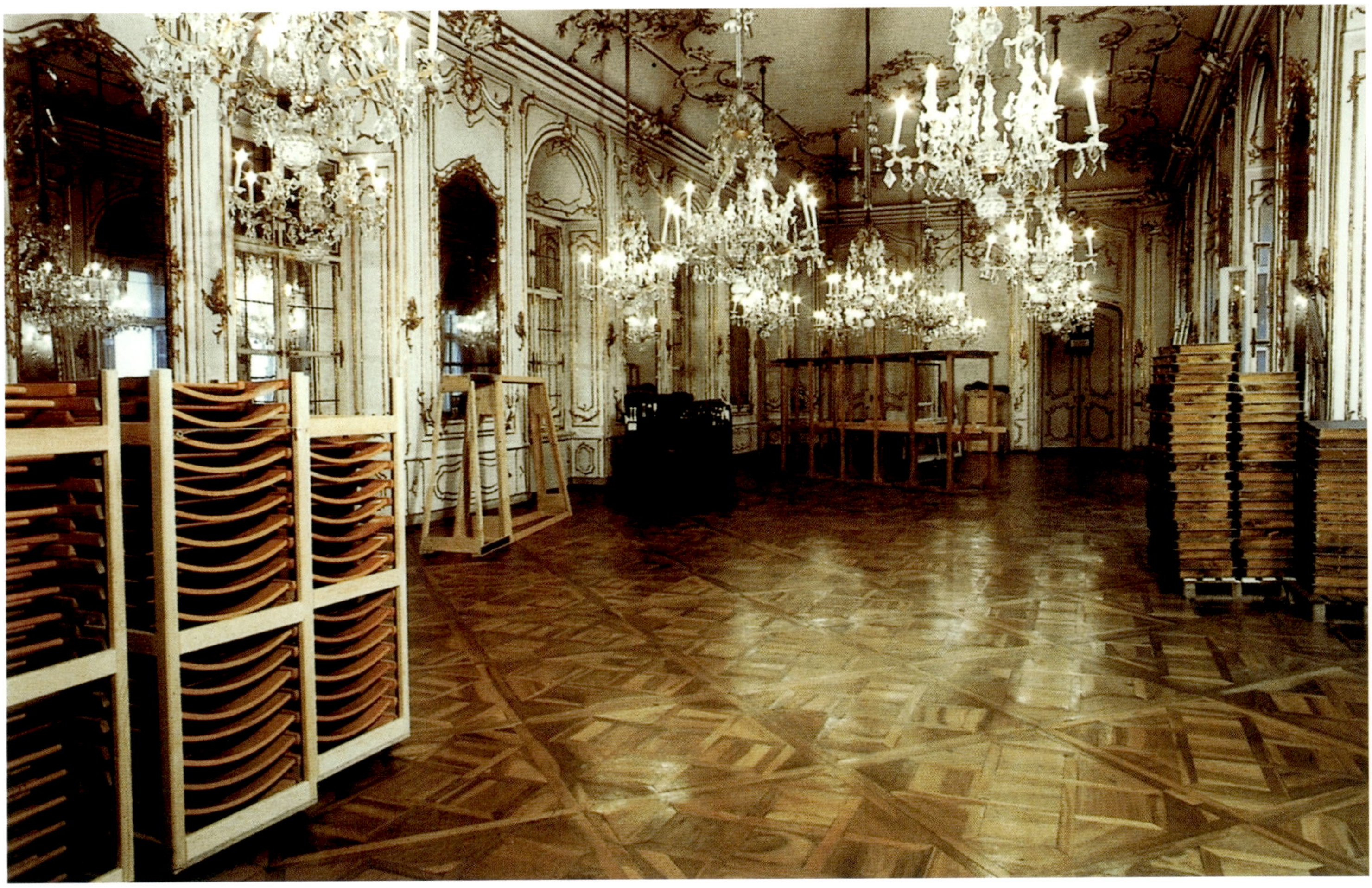

Installation views *Malerei zwischen Anarchie und Forschung*, Neue Galerie Graz, 1992

I have released exhibition spaces from their representative function on a number of occasions. In 1992, I mounted a painting show without any paintings, consisting only of the support structure of a painting show (frames, etc.). The spaces were turned to storage rooms, the perfect material background for the ideological function of art. The four sides of a frame can be placed in relation to four data sets. For example, the lengths of the frame's elements may be determined by the number of visitors in a day, a week, a month, and a year. This social contextualization achieves a new kind of self-referentiality that is no longer formal, but rather social. The social context through which art is exhibited – all the social conditions, staff, and support structures that are necessary to produce the final product of the exhibition – are themselves exhibited. The formalism of modernism (the white cube) is critiqued. Storage room, workshop, and exhibition space are all treated equally. The field of the image and the field surrounding it become variables in a new equation. Pictorial field and ambient field are rendered equal and unbounded. The unbounding of the picture plane and its surroundings has achieved its culmination, albeit at the cost of losing the picture plane. The "image," now consisting solely of fragmentary frames, has effectively been transformed and has vanished. —Peter Weibel, 1996

Translated from the German by Patrick Hubenthal.

Framed Space

1992 | wood, frame, wood chips, archive folders

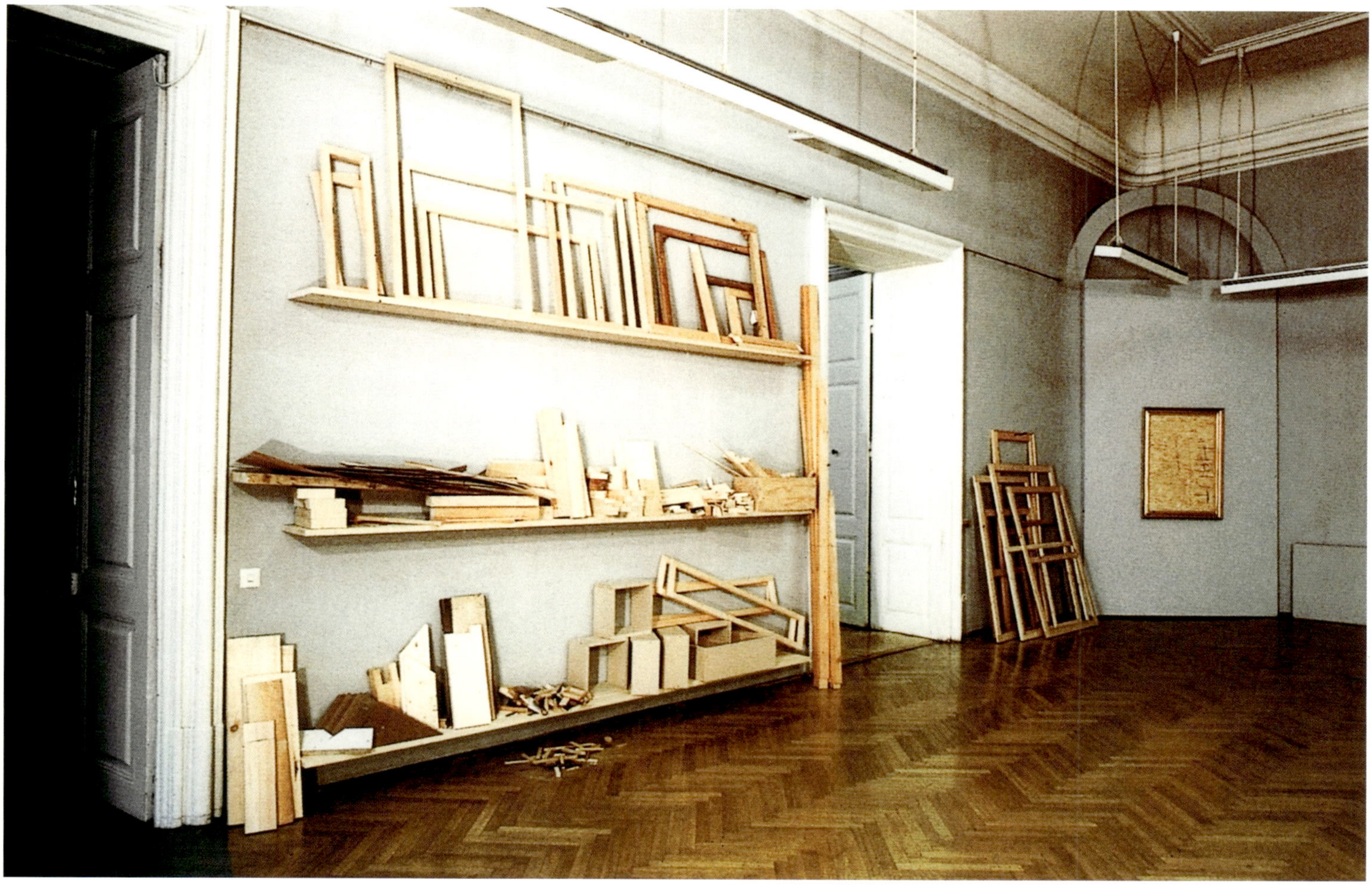

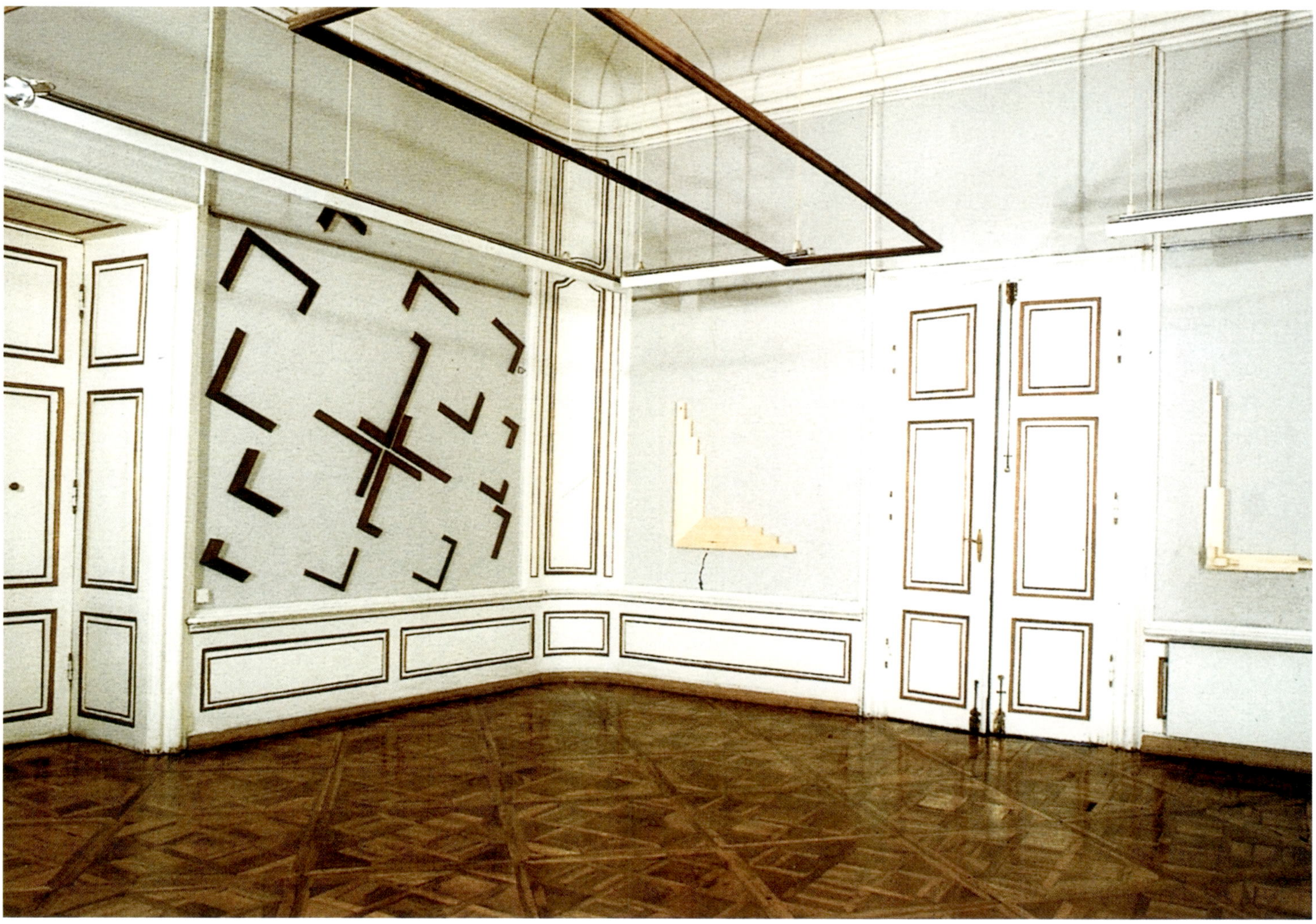

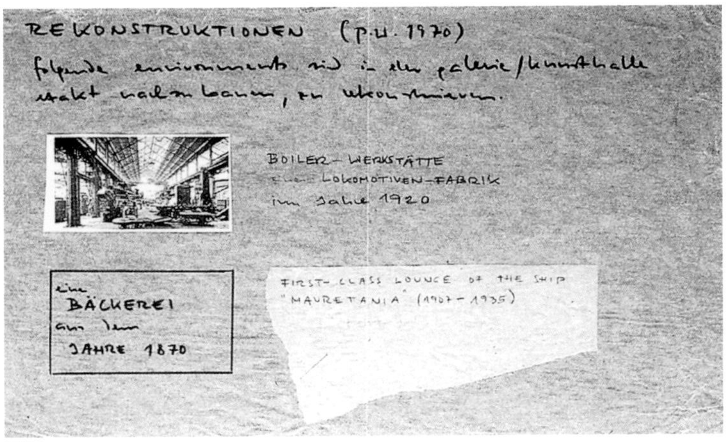
REKONSTRUKTIONEN (p.u. 1970)

BOILER-WERKSTÄTTE
LOKOMOTIVEN-FABRIK
im Jahre 1920

BÄCKEREI
JAHRE 1870

FIRST-CLASS LOUNGE OF THE SHIP
"MAURETANIA" (1907–1935)

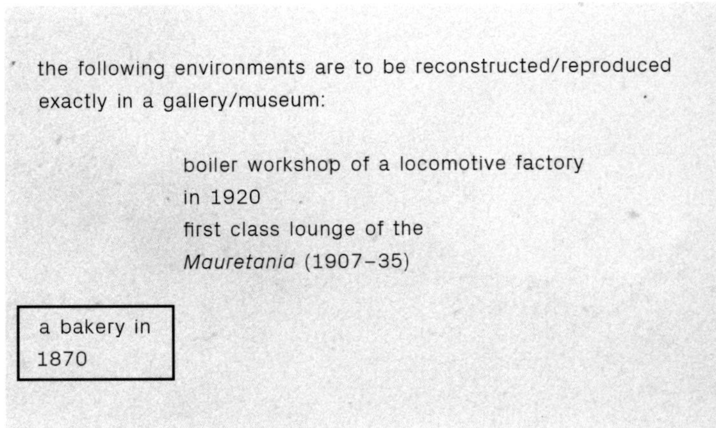
the following environments are to be reconstructed/reproduced exactly in a gallery/museum:

boiler workshop of a locomotive factory
in 1920
first class lounge of the
Mauretania (1907–35)

a bakery in
1870

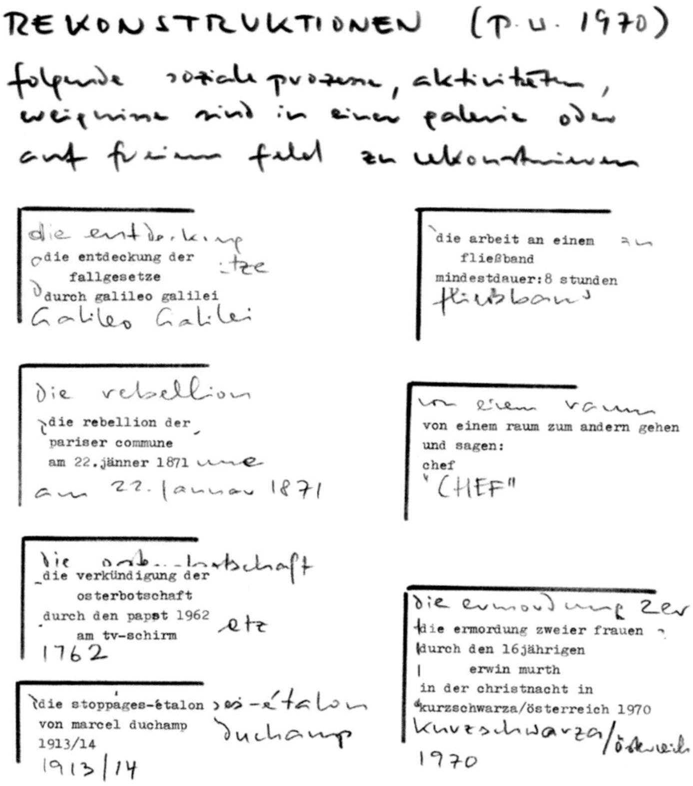
REKONSTRUKTIONEN (p.u. 1970)

die entdeckung der
fallgesetze
durch galileo galilei
Galileo Galilei

die arbeit an einem
fließband
mindestdauer:8 stunden

die rebellion der
pariser commune
am 22.jänner 1871

von einem raum zum andern gehen
und sagen:
chef
"CHEF"

die verkündigung der
osterbotschaft
durch den papst 1962
am tv-schirm

die ermordung zweier frauen
durch den 16jährigen
erwin murth
in der christnacht in
kurzschwarza/österreich 1970
1970

die stoppages-étalon
von marcel duchamp
1913/14
1913/14

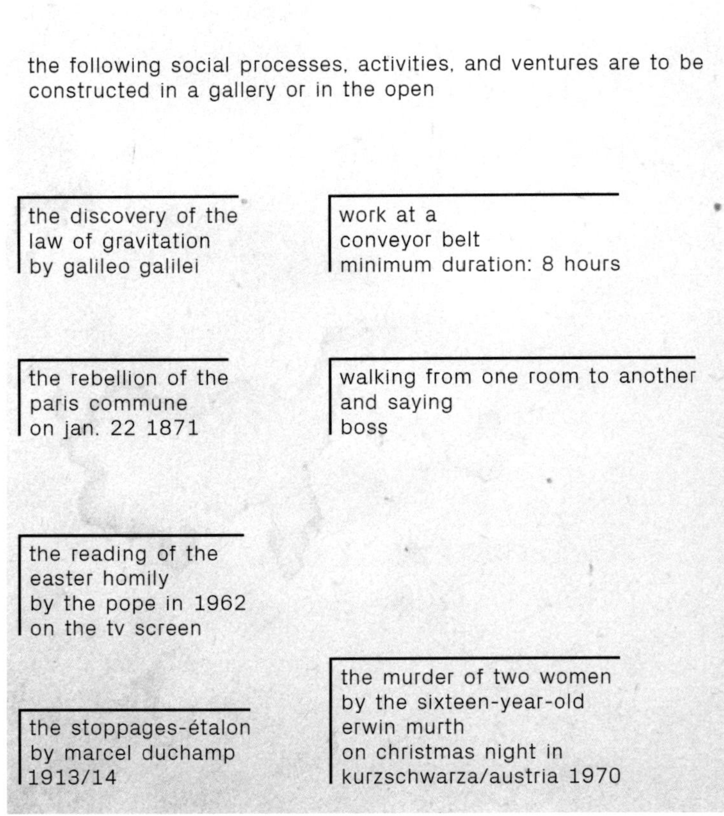
the following social processes, activities, and ventures are to be constructed in a gallery or in the open

the discovery of the
law of gravitation
by galileo galilei

work at a
conveyor belt
minimum duration: 8 hours

the rebellion of the
paris commune
on jan. 22 1871

walking from one room to another
and saying
boss

the reading of the
easter homily
by the pope in 1962
on the tv screen

the stoppages-étalon
by marcel duchamp
1913/14

the murder of two women
by the sixteen-year-old
erwin murth
on christmas night in
kurzschwarza/austria 1970

Reconstructions / Rekonstruktionen

1970	concepts; print, pencil, and ballpoint pen on paper, published in the Slovenian journal *Problemi*, 1971. Translated from the German by Peter Dahm Robertson

In the group of works entitled *Reconstructions* conceived in 1970, the artist planned to reconstruct various historical events or social spaces inside the gallery space that for him stands as a closed social space. For Weibel the art system was a mirror of class society and capitalist division of labor. Therefore, by reconstructing important social events and spaces within the gallery he wanted to demonstrate a return to reality and deconstruct the art system by transferring reality to the art gallery. The events that were to be reenacted were the Paris Commune's rebellion on January 22, 1871, and the Pope's announcement of the Easter message on television in 1962. Factories, offices, and workshops were among the spaces to be reconstructed architecturally. For the Galerie nächst St. Stephan in Vienna, Weibel planned to rebuild the dining rooms of ocean liners in the exhibition rooms, each of which was furnished differently and reserved for the social classes: first class, second class, third class. Three gallery rooms representing class society.

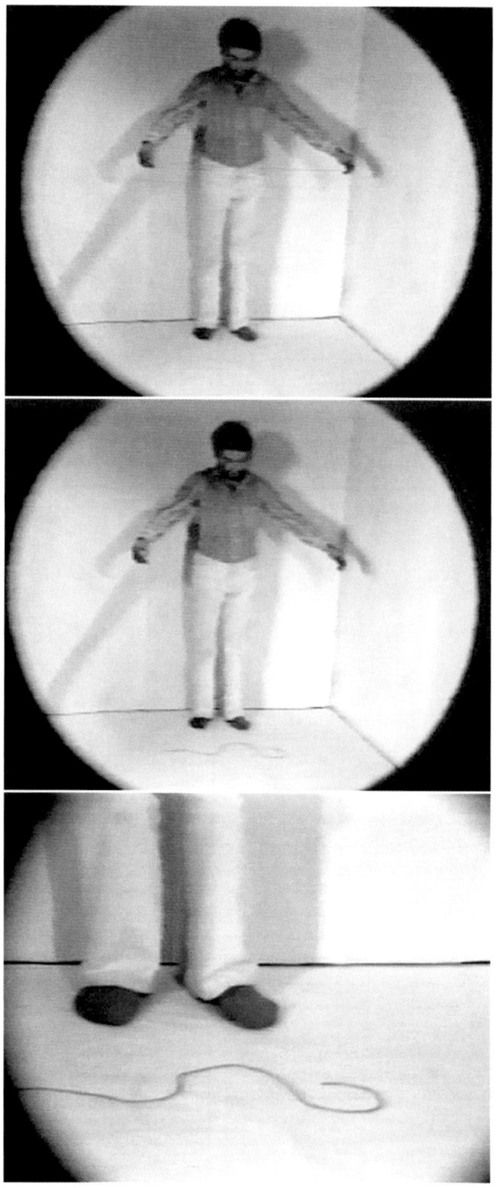

Reconstruction of Marcel Duchamp's *3 stoppages-étalon* (1913–14) / Rekonstruktion der *stoppages-étalon* (1913/14) von Marcel Duchamp

1970 (idea), 1971 (performance)	video performance, 48 sec.

The question of how poems are produced typically elicits two answers: on the one hand, via the sounds of speech, the phonetic form of the language; on the other, via the written form of notation. By abstracting these forms, writings can become abstract drawings, as can be seen in the works of Cy Twombly, for example. Such abstractions, however, are based on hand gestures that have been amply and exhaustively visualized over hundreds of years. Consequently, Peter Weibel has constantly sought ways of expanding both visual representation and poetry. In doing so, he has focused not only on syntax and semantics, but also on pragmatics, the way language is used through actions. This pursuit forms the basis of one category of Weibel's works, his action poems and performance poems.
These include *Reconstruction of Marcel Duchamp's* 3 stoppages-étalon *(1913–14)*, a 1970 performance documented in drawings, photographs, and video. In it, the artist dropped a meter-long piece of string onto the floor three times from a height of one meter, producing strange curves, abstract forms that can be interpreted as abstractions of written characters. The performance poem, therefore, is not based on the formulation of language so much as on the action and materials involved. By observing their physics, their intrinsic characteristics, Weibel created new forms of poetry.
As the title clearly indicates, the artist was reconstructing an action performed by Marcel Duchamp, the father of modern art. In repeating the same action three times – dropping a one-meter-long of string from a height of one meter – Duchamp was seeking to generate a new kind of unit of length. With this deliberately chosen reference, Weibel anticipated the artistic genre of the reenactment, which was to become popular in the early twenty-first century.
The arbitrary and relative nature of all units of measurement – which tend to be distorted by media, and especially by photographic representation – is also clearly illustrated by a 1975 piece titled *The Measure of Measure* (*Das Maß des Maßes*, see p. 401 in this volume).

Translated from the German by Patrick Hubenthal.

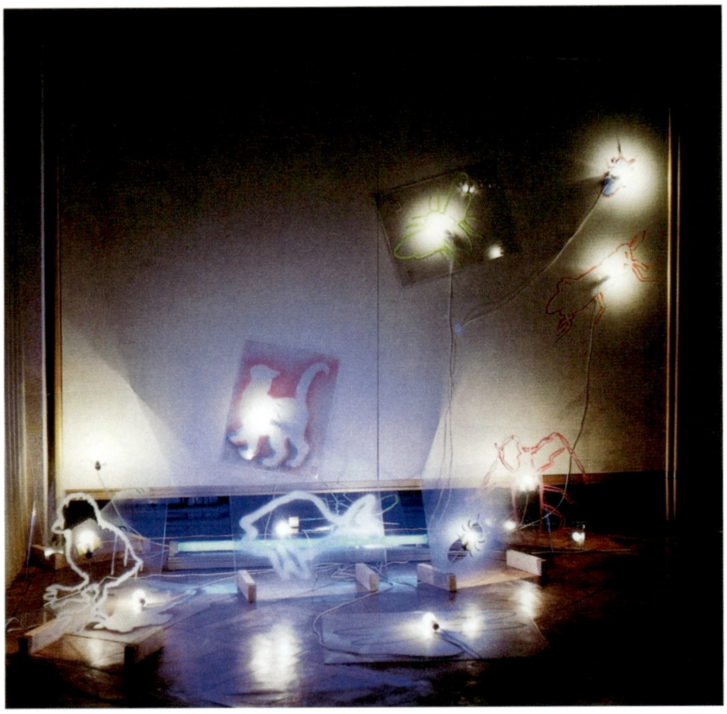

Installation view *Peter Weibel: Medienpoesie*, Kulturhaus Graz, 1982

Electric Animals / Elektrische Tiere

1982	wall, light bulbs, drawings of animals on glass panes

Outlines of animal bodies painted on glass are illuminated by light bulbs positioned behind the panes. The painted animals look like lamps designed for a domestic interior.

The precondition for my Animal Art is the autonomy of the animal.
The animal as symbol (of nature)
The autonomy of the animal set in relation to the human being who undermines the autonomy of the animal and wants to subjugate the animal as well as nature in general. Accordingly, in my art the animal, where it is really and corporeally present, appears not only as an animal but simultaneously as a symbol of nature as well: the animal as nature's double. Many of my works with animals are thus allegorical reflections of the relationship between human and nature, a criticism of the relationship.
...
Natura geometrica – prisoner of geometry
Electronic appliances as new electronic pets.
TV, radio, telephone, etc. as prosthetic animals. Tamed nature – geometricized nature – geometricized life! In modern civilization. Natural mouse and technological geometry – are both prisoners of geometry? The highest geometry: the crystal lattices in the silicon of high tech. The lowest geometry: primitive animals. And yet: mechanized turtles, geometricized mice, and otherwise automated animal toys follow the tracks of civilization like phantoms.
Where there is matter, there is geometry, Kepler says. Then doesn't the mouse collide with the labyrinthine geometry of its orbit of movement?
...
The animal was humans' first "natural" machine for transport and communication (from the mule to the carrier pigeon). It helped overcome spatial and temporal distances, which today the media of telephone, TV, and radio do. The animal was indeed a medium. Paradoxically, though emblematically for the animal problem, the animal can thus symbolize not just nature but also technological media. Biotechnology lives off showing us how best to use the anatomy of and technology of animals as blue prints, as role models for constructing machines. Animals as protoforms of technology, as intermediate steps in the evolution from nature to technology, were the first steps in this loss of autonomy. Human beings made the animal, an autonomous being in nature, into a tool, a technology, a medium, by using it as load carrier, message transmitter, companion, etc. The taming and training of the animal mirrors the taming of nature from which technology developed. That is why the animals in my works do not just appear real, as themselves, or as their images, but also as technological prostheses, as machines, as food, as devices, and: as people, that is, as equal beings. —Peter Weibel, 1987

Translated from the German by Peter Dahm Robertson.

The Wrath of the Animals / Der Zorn der Tiere

1988	plastic animals painted red and mounted to a house facade with steel cables

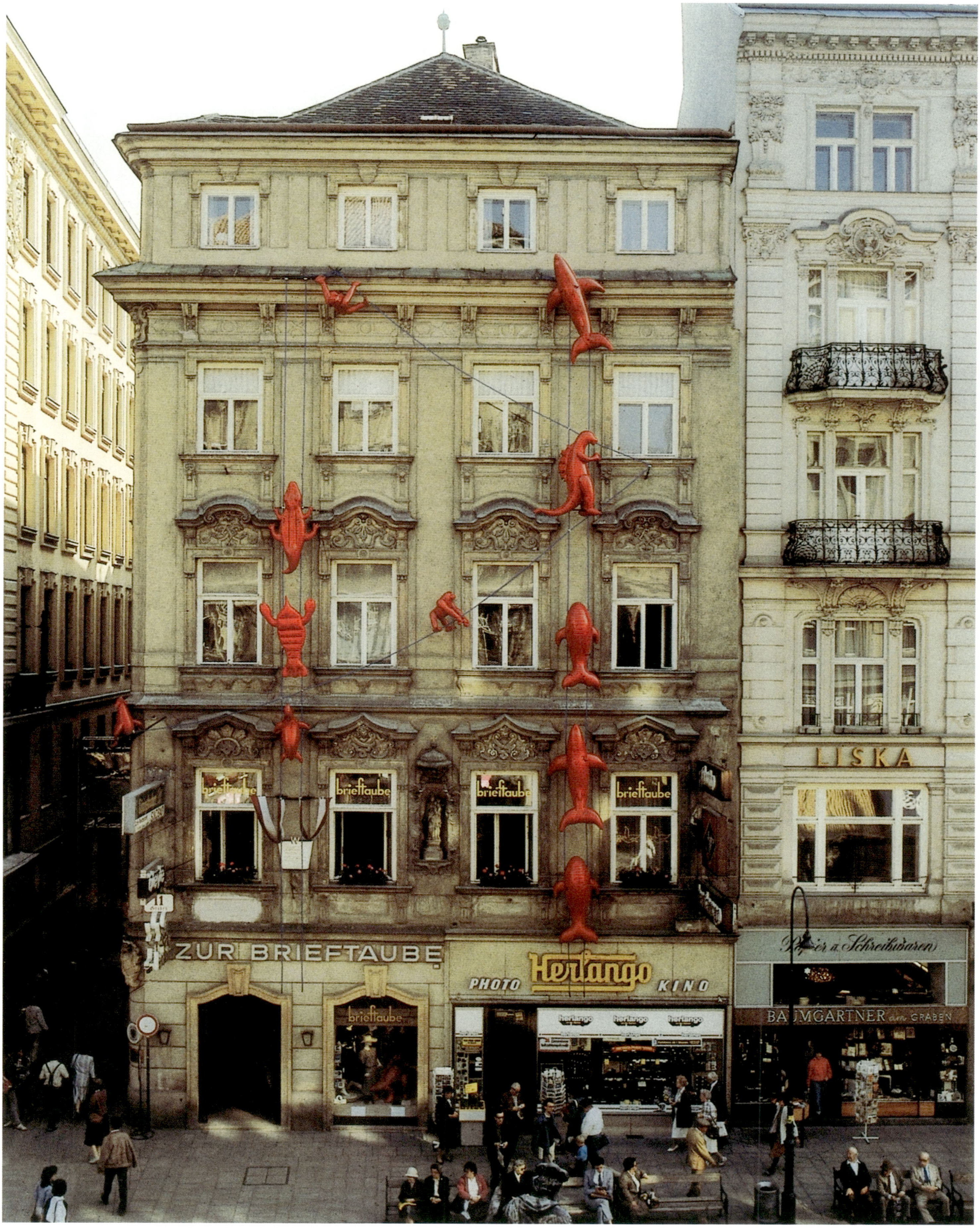

Installation view *Freizone Dorotheegasse*, Vienna, 1988

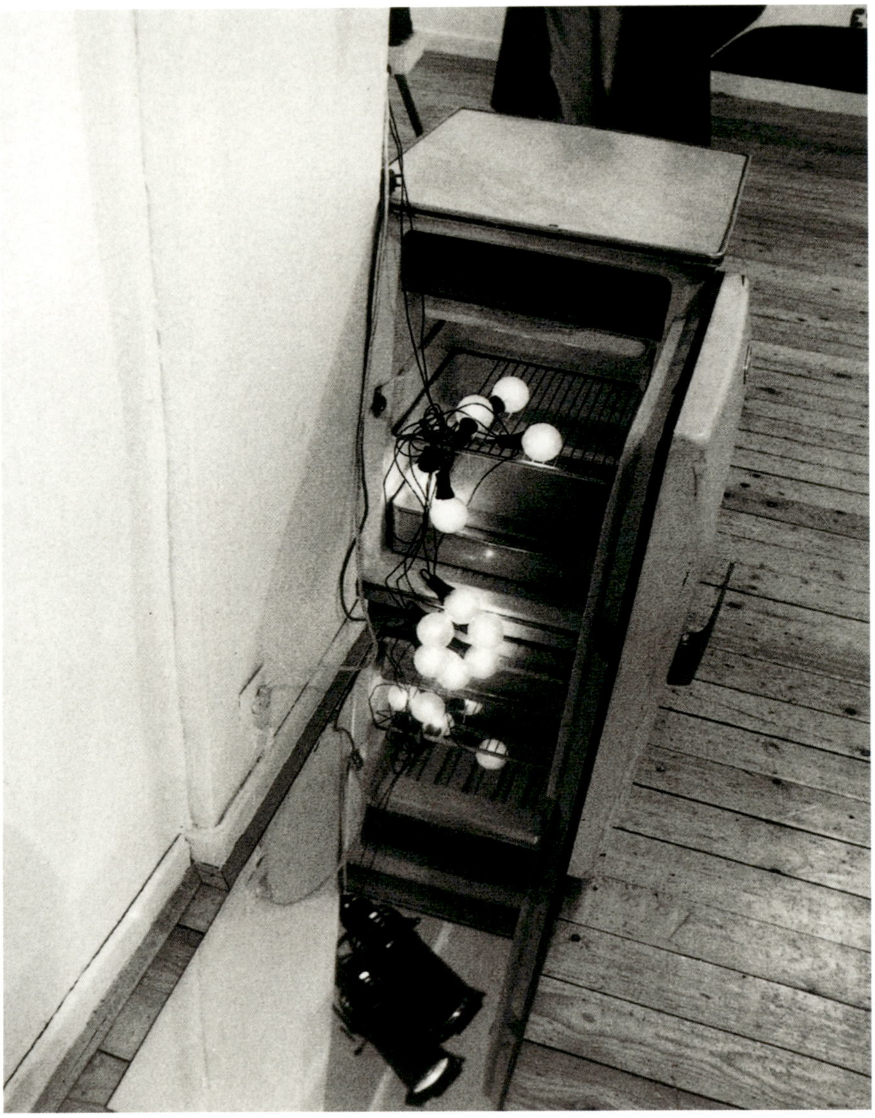

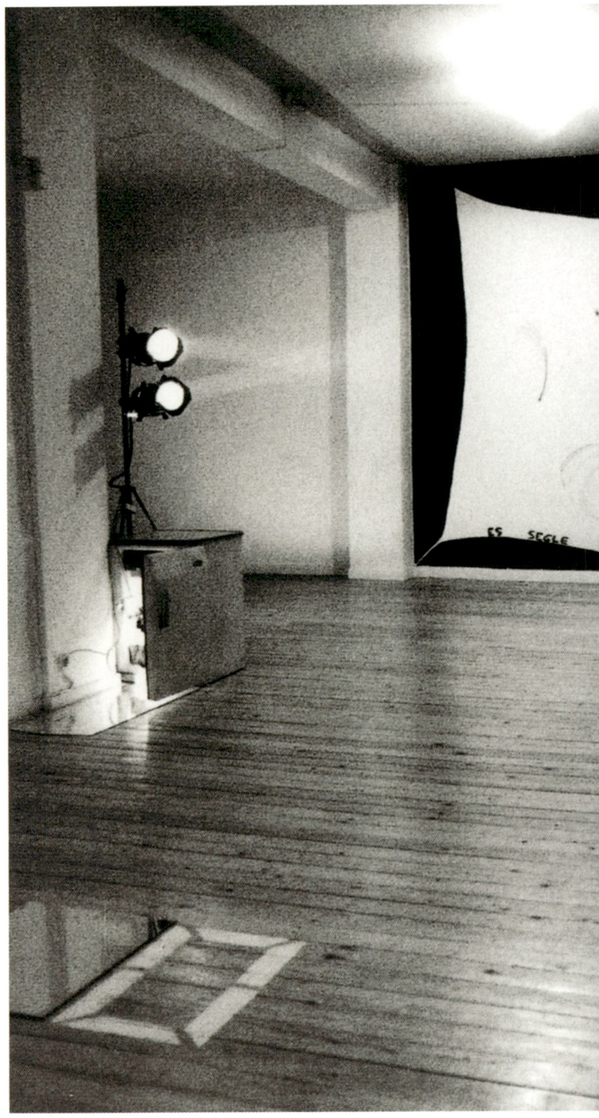

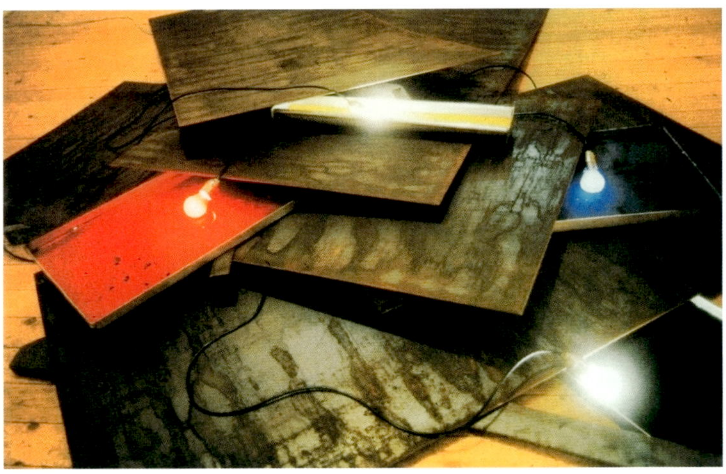

The Electric War / Der elektrische Krieg

1983 | live snake, gas, fire, neon tubes, steel plates, colored water, tin tubs, refrigerator, mirror, copper wires, spotlights, colored light bulbs, paint

Human homes are in an unrecognized state of war. Hidden in the walls, weapons can be found, such as gas, or electricity, both of which can kill someone like a snake could. Without these resources, human homes would be disfunctional: no refrigerator, no water, no electricity.

Our control system hides the electric war behind the curtain of civilization. But at any time, the curtain can be unmasked. A gas pipe is laid along the walls of the gallery and plastered over. The gas streams out from the wall and is ignited. Another part of the wall is opened up, as is usually done for laying electrical wiring. Instead of the wiring, a live poisonous snake is fitted into the wall and covered with glass. The fire damage on the wall is mirrored as a binary "1" against the "0" painted on the floor: digital code. The steel plate sculpture shows equilibrium in chaos through the horizontality of water in tin tubs, from which electricity is de facto generated. The wall gives us water, electricity, gas, information (telephone wires, etc.), and functions therefore as mobile as a sail. Our homes don't move, because the energy moves through our homes.

—Peter Weibel

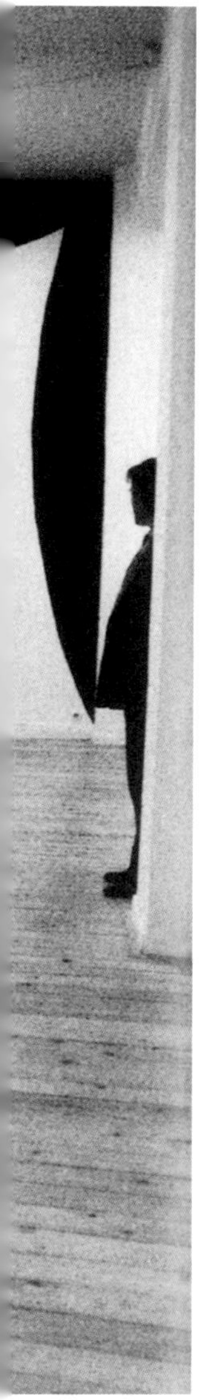

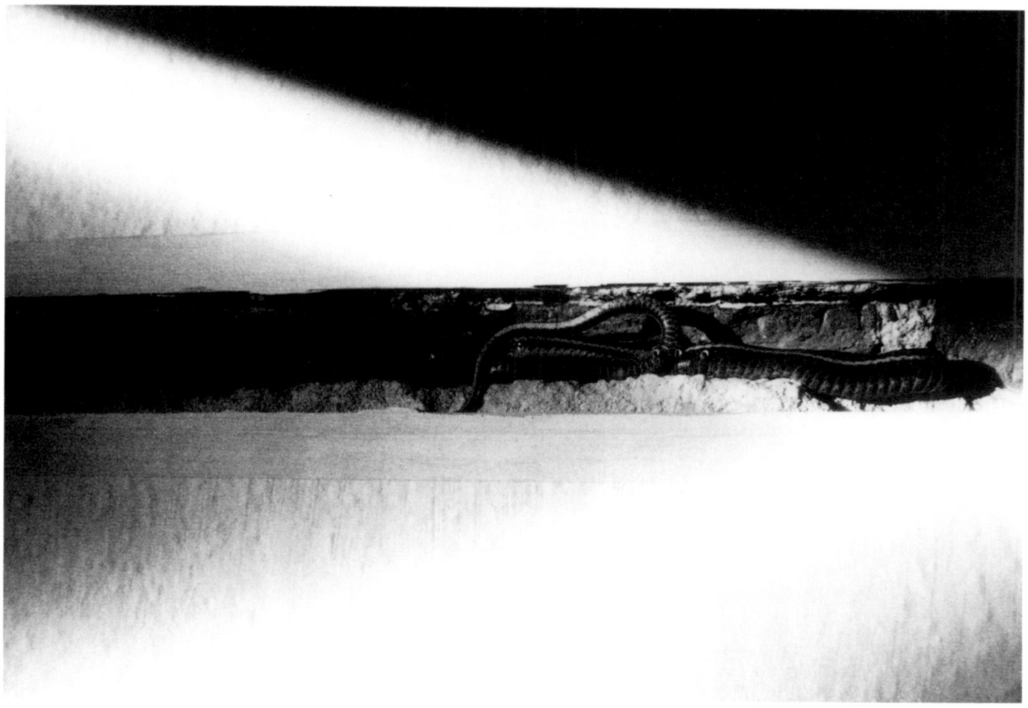

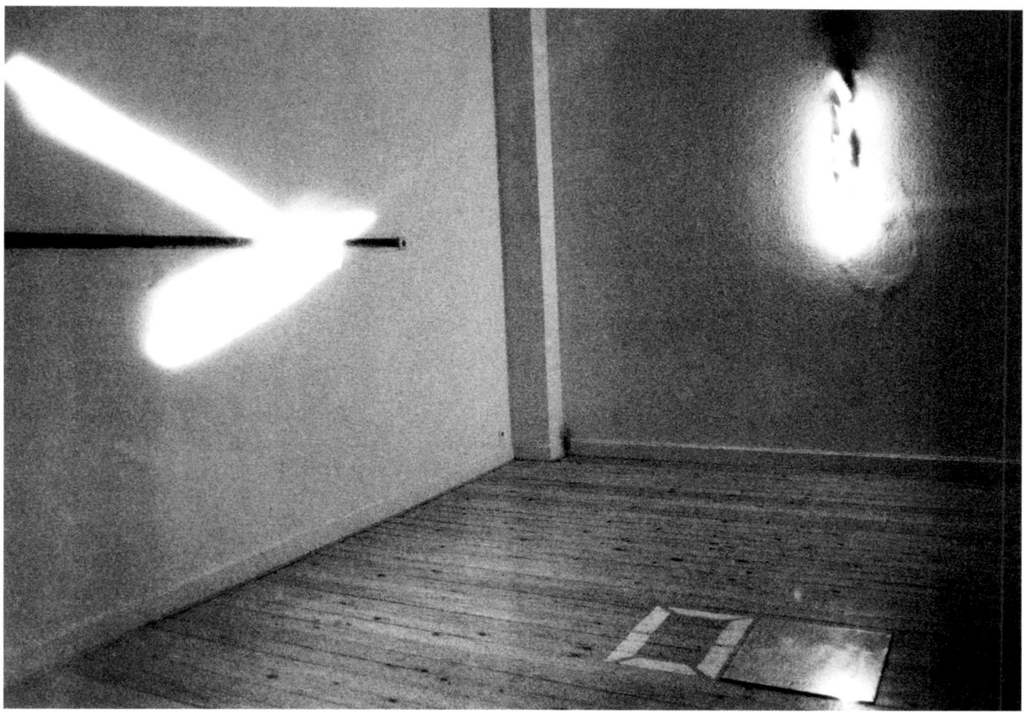

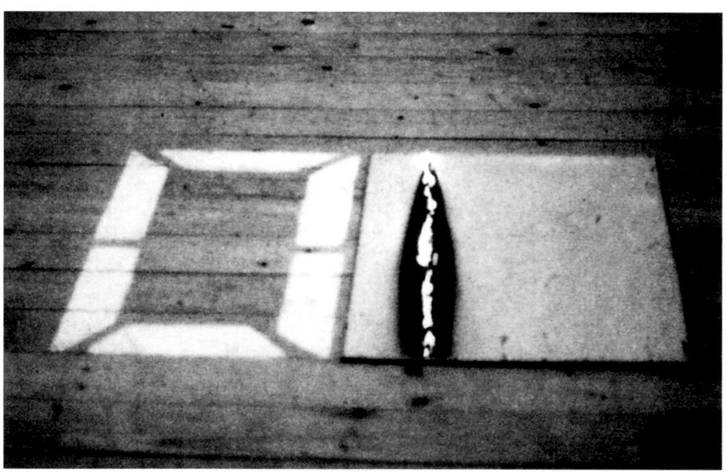

Installation views *Der elektrische Krieg*, Moltkerei Werkstatt, Cologne, 1983

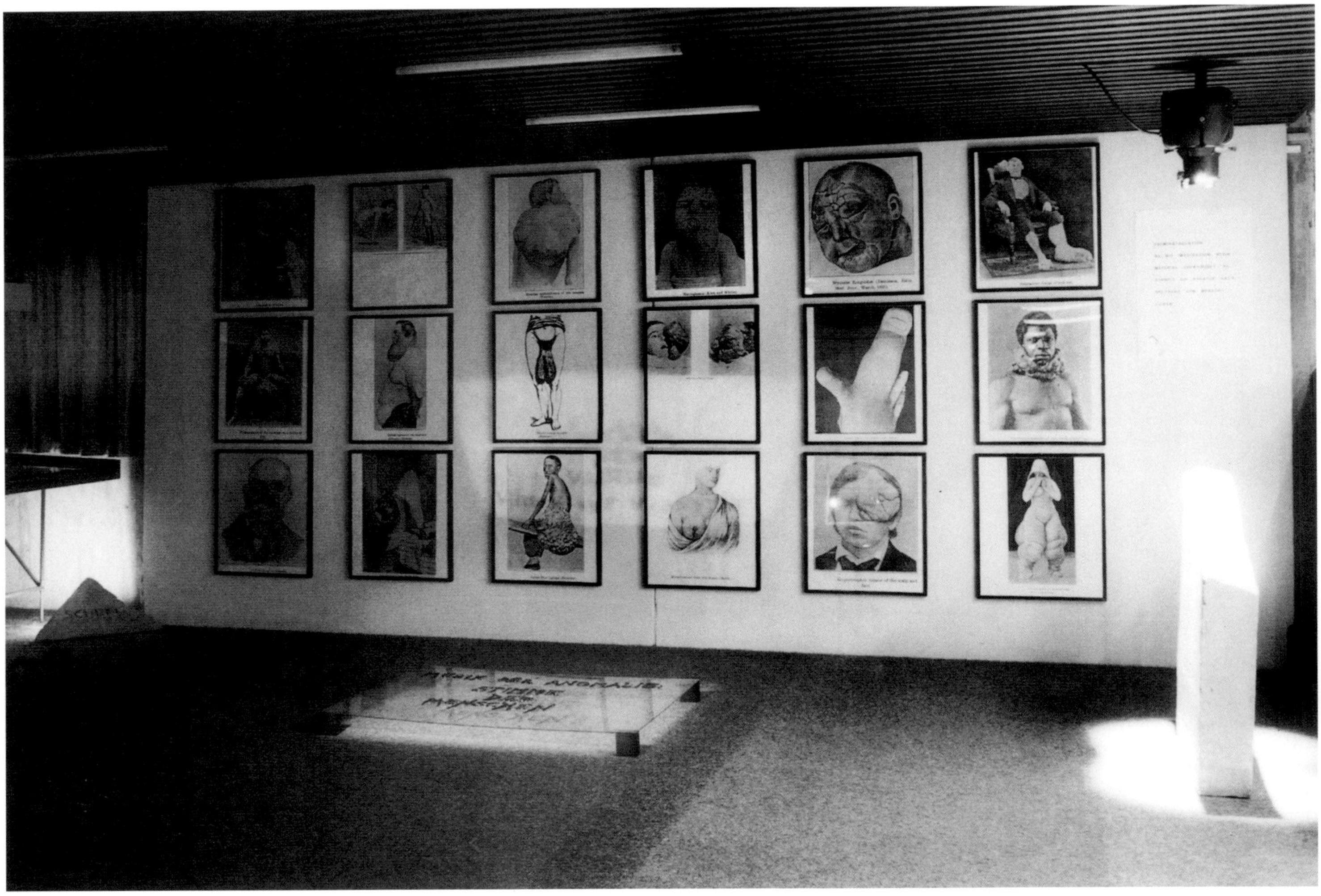

Installation view *Medien und Realität*, Kunsthalle Wilhelmshaven, 1982

Music of the Anomaly – Voices of Man / Musik der Anomalie – Stimme der Menschen

1982	18 prints, sandstone, glass plate, asphalt, cement

The installation *Music of the Anomaly – Voices of Man* consists of portraits of physically disfigured people, a sandstone sculpture, a concrete sculpture in a corner with the German word *Schrei* (scream) and a plate of glass with the words *Musik der Anomalie – Stimme des Menschen* in asphalt. The images of the mutilated people ask us the question of who and what a human is and who defines being human. Is the cry of need the real voice of humanity, the frozen, suffocated, silent cry of the creature? Are abnormal anatomies, anatomical anomalies, the limits of humanity? Or is the actual wellspring of humanism the imagination, the longing for transcendence, the desire to overcome the limits of what exists? Is anomaly the proper voice of humans?

Part of the installation is a sandstone sculpture from 1974 (*Imagination*, see p. 83 in this volume), on which the first notes of the composition *Arcana* (1925–27) by French composer Edgard Varèse and the word "Imagination" are carved. The stone is intended to show that petrified conditions can be changed by the imagination.

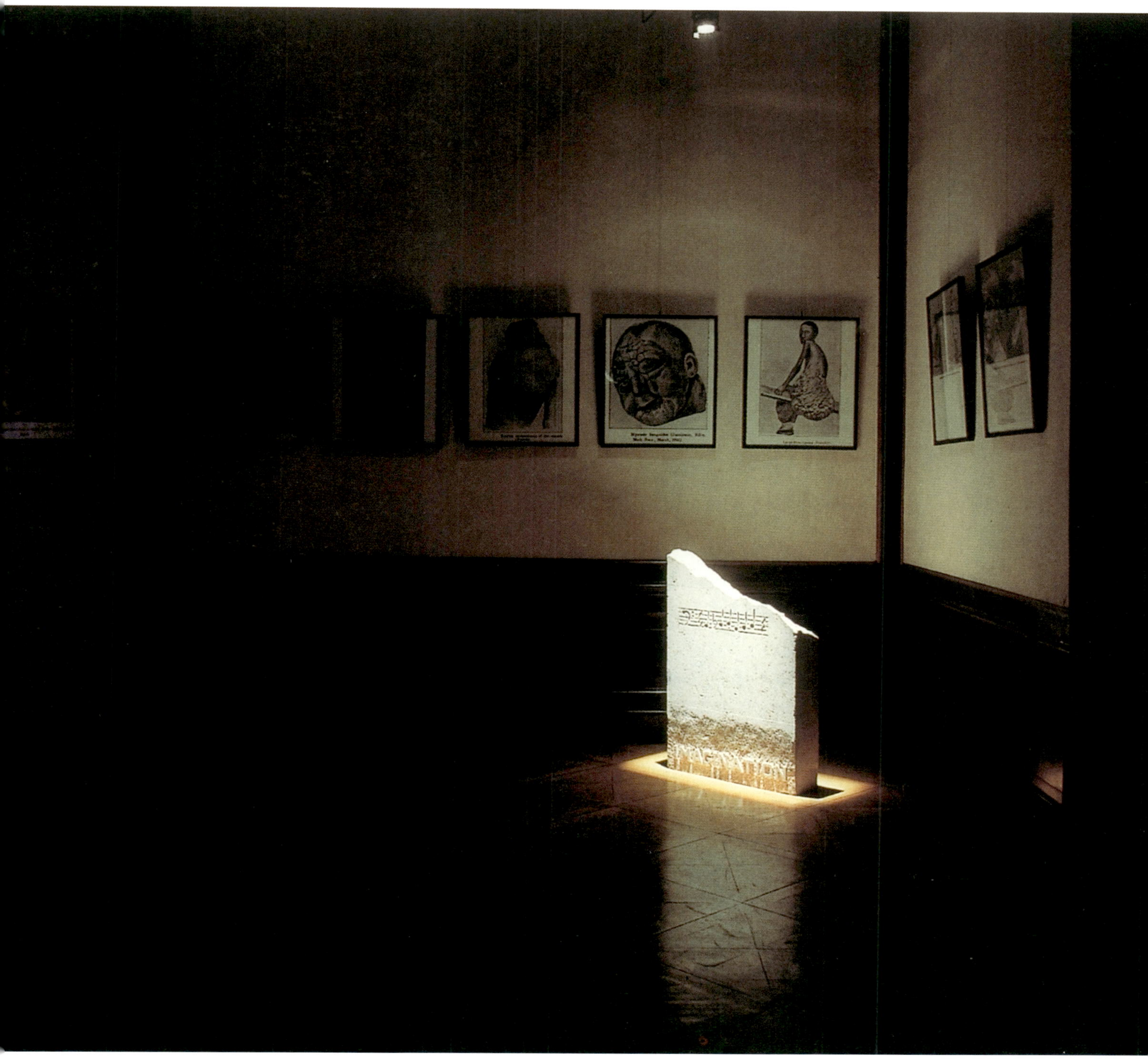

Installation views *Peter Weibel: Medienpoesie*, Kulturhaus Graz, 1982

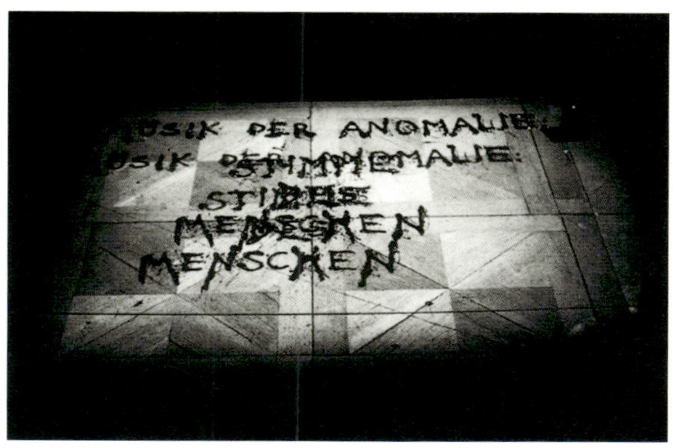

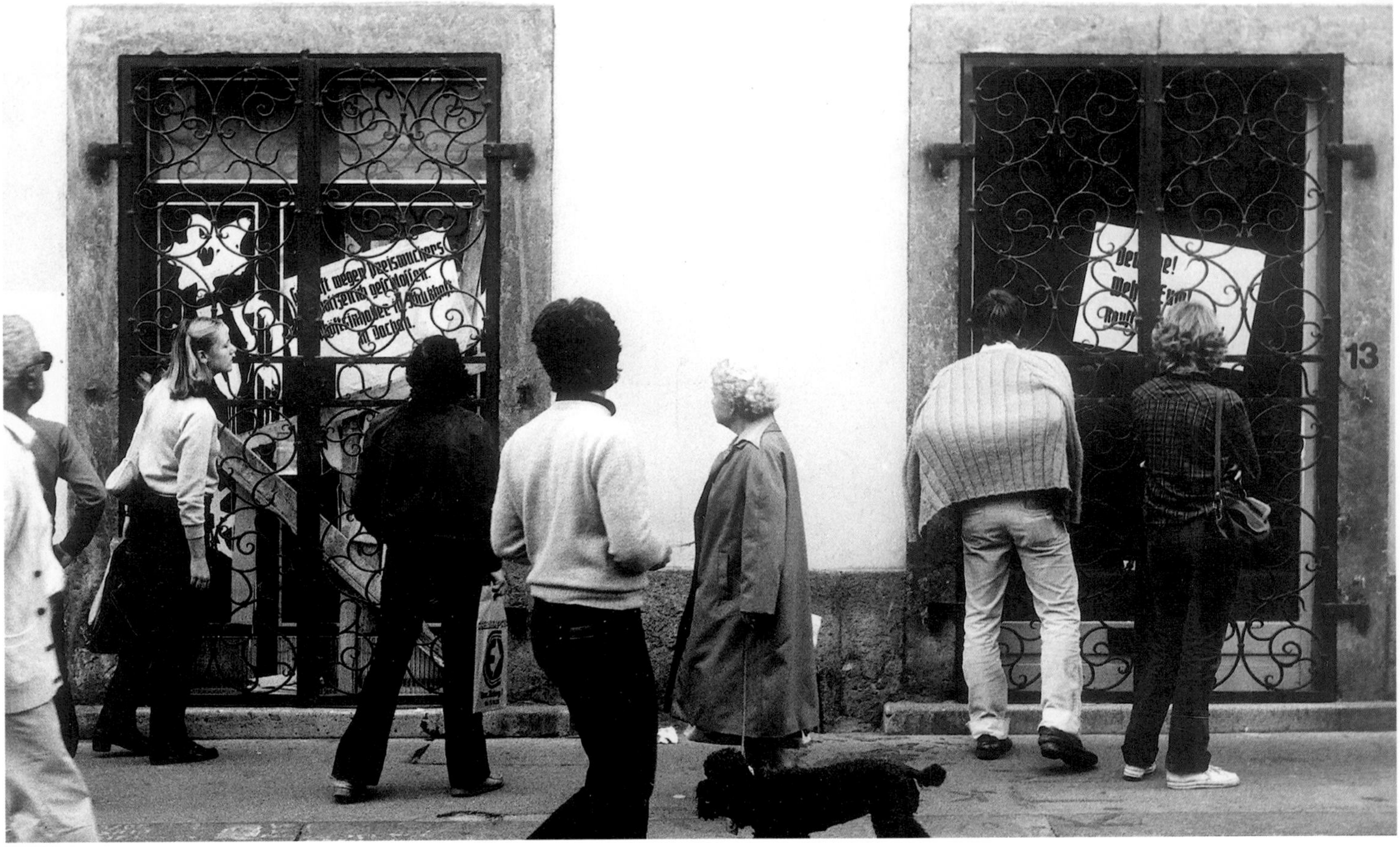

As part of the exhibition *Künstlerschaufenster* (Artists' display windows), which I cocurated with Peter Pakesch for the steirischer herbst contemporary art festival (September 21–October 5, 1979, Graz, Austria), I reconstructed a window display at the Radio Reiss shop on Glockenspielplatz 7. The purpose of this display was not to provide a glimpse of the space behind the window, but rather of the time before: a *Zeit-Schaufenster*, the conclusion of a chain of associations encompassing *Schaufenster* (display window), *Schauraum* (showroom), *Raumfenster* (room window), *Zeitfenster* (time frame), and *Zeitraum* (time period).

The theme was the role of mass media, media monopoly, manipulation of the masses, and the role of radio during the Nazi era in Germany and all over the world today.

Time Display Window / Zeit-Schaufenster

1970 (idea)/ 1979	installation, artistically reconstructed window displays of Radio Reiss, Glockenspielplatz 7, Graz

On display in the window were a number of Volksempfänger VE 301 Dyn radios labeled with their original prices – for example, 65 reichsmarks. Like the Volkswagen (the "people's car"), the Volksempfänger (the "people's receiver") was a crucial product for the Nazis' manipulation of the masses. The Volksempfänger was tuned to a single frequency, on which listeners could only hear Hitler's messages – their "master's voice." An example of the abuse of media and technology in the service of totalitarian systems. The "business end" of the media of propaganda.

The reconstruction was based on the visual model of the past. It was not labeled as a work of art; instead, passers-by stumbled upon it without warning, unprepared for its reality. They therefore assumed the window was the work of neo-Nazi sympathizers. They were not sufficiently media-literate to notice that it was a display of old radios and old prices.

However, the print shop in Graz where I had my signs and price tags printed was perfectly alert. They reported me to the authorities, and as a result I received a summons to appear at state police headquarters in Vienna (1st District, Schottenring 7–9) on September 19 and 20, 1979. —Peter Weibel, 2022

Translated from the German by Patrick Hubenthal.

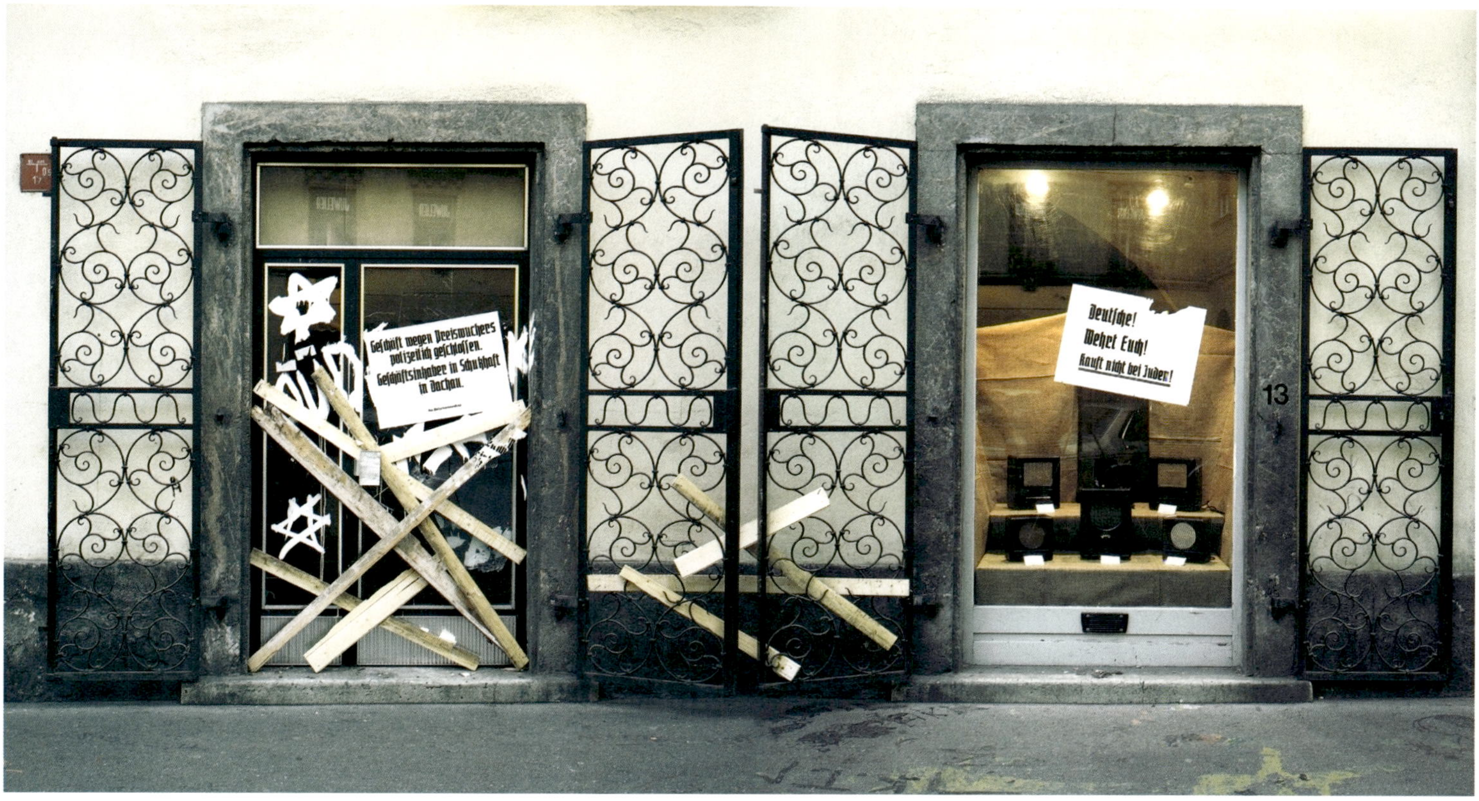

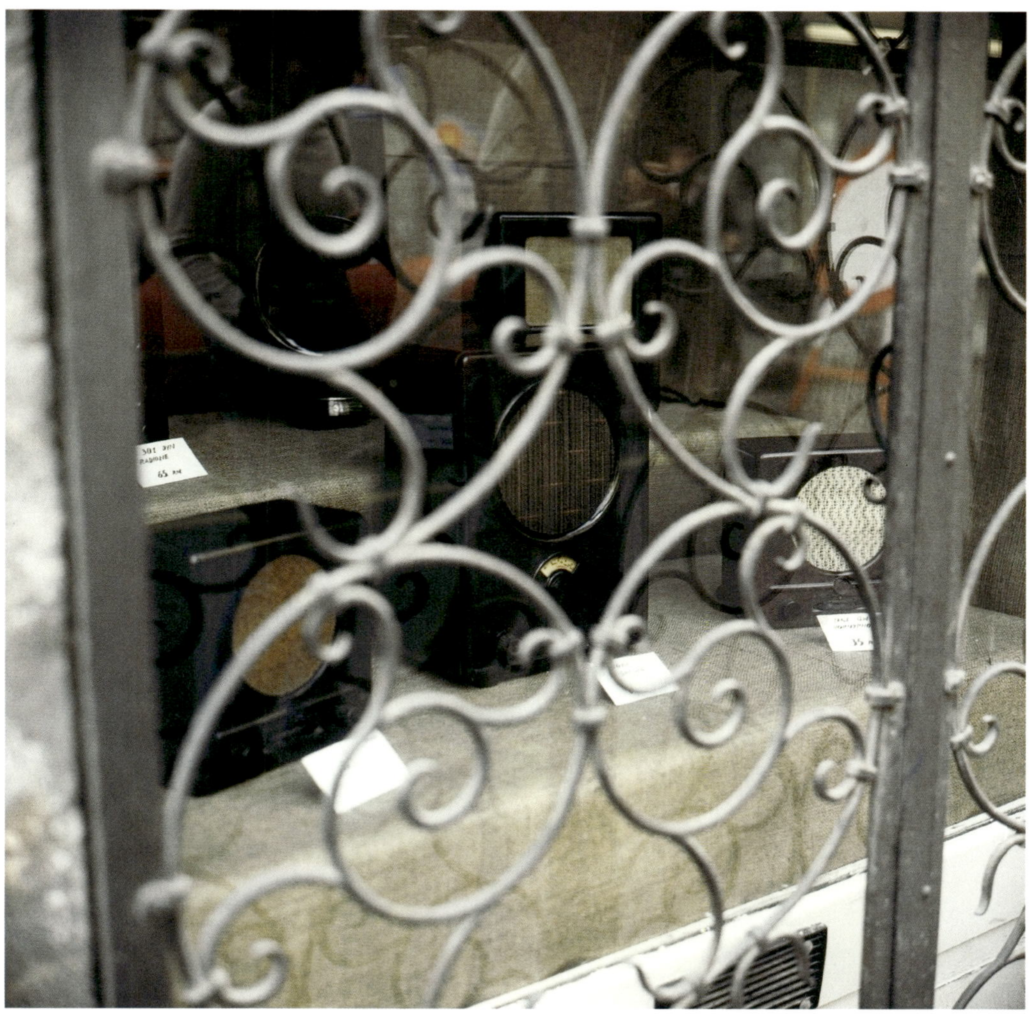

Installation views
Künstlerschaufenster
steirischer herbst '79,
Graz, 1979

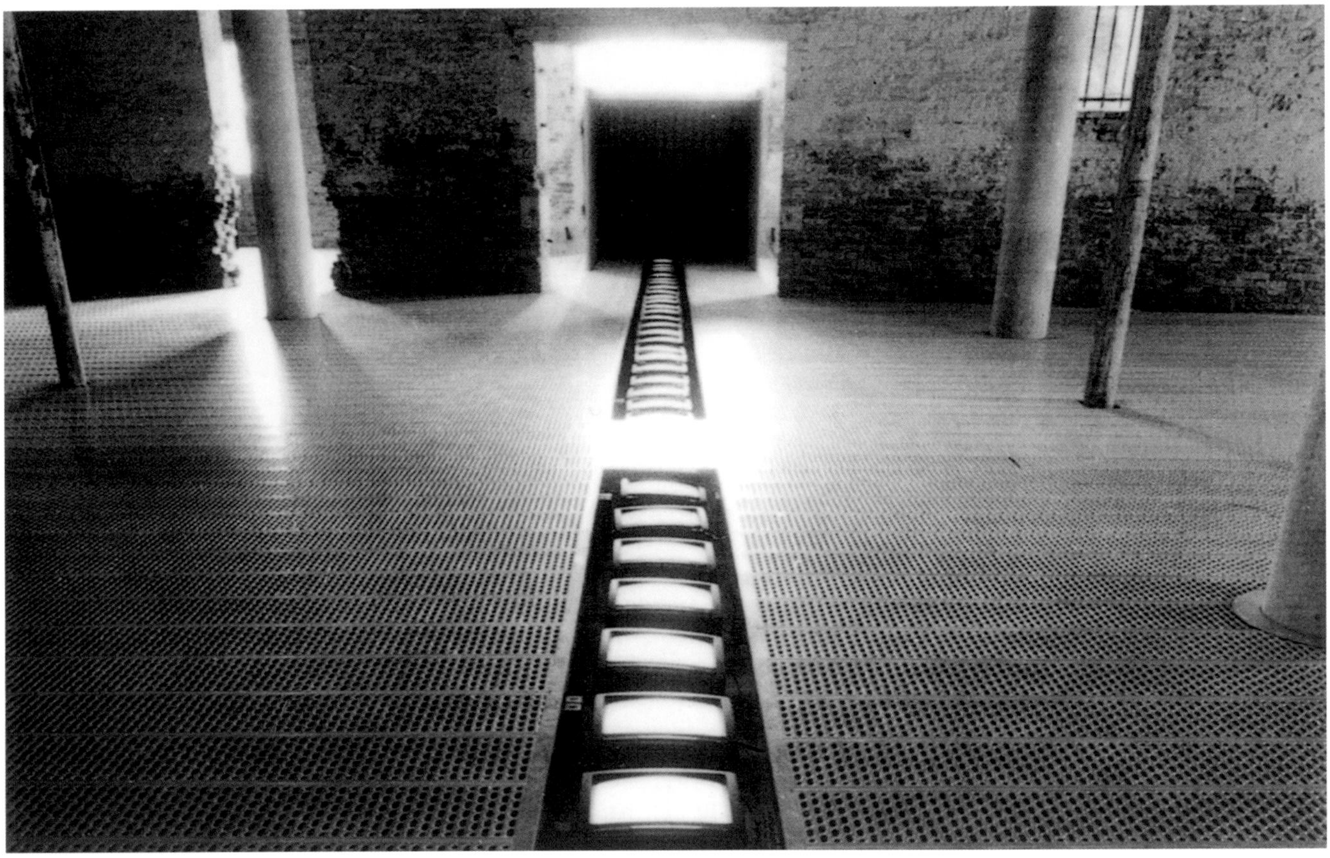

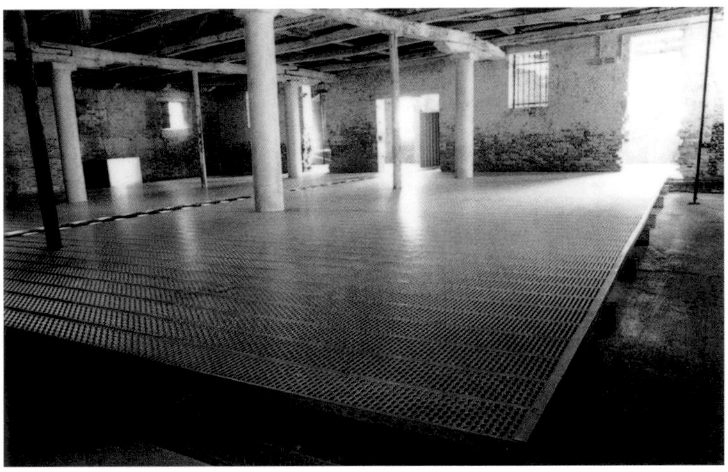

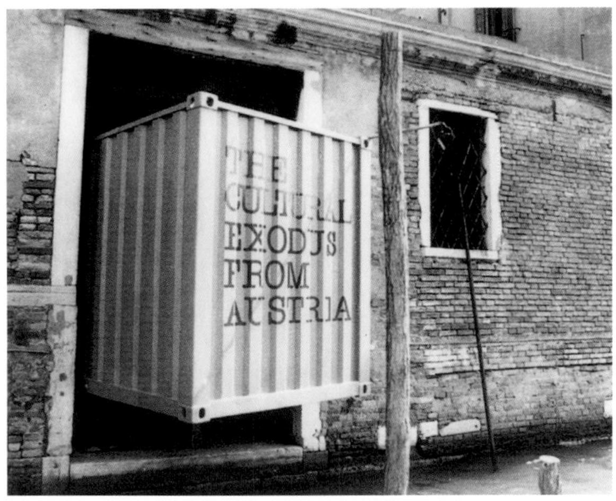

Installation views La Biennale di Venezia '93,
Fondaco Marcello, San Marco, Venice, 1993

The media installation *Expulsion of Reason* is the result of Peter Weibel's investigation of the expulsion of academics, researchers, artists, and intellectuals from Austria between 1933 and 1945 which he started with the exhibition *Austrian Avantgarde* at Galerie nächst St. Stephan in Vienna in 1974. Fourteen monitors on a ramp made of wooden pallets form a line that ends in a shipping container that resembles railway wagons. Ramp railway wagons that commemorate the Holocaust. The container breaks through the museum wall and protrudes into the space behind it. Short biographies of Austrians forced to emigrate and victims of the Nazi regime from the fields of science and art appear on the screens. Each short biography moves from one monitor to the next until it arrives in the container and disappears. Every person's name is spoken out loud over the course of 8 hours. A cultural exodus, an acoustic and visual funeral march. The installation was first realized in 1993 at the Venice Biennale, and included 5,000 short biographies. The extensive database of this cultural exodus from Austria was successively expanded together with Friedrich Stadler in the following years, and exhibited in 1997 at the Simon Wiesenthal Center Museum of Tolerance in Los Angeles amongst other places. In 1993, the book *Vertreibung der Vernunft: The Cultural Exodus from Austria* was published together with Stadler by Löcker Verlag, Vienna.

This media sculpture was not only shown in the analog public space, but also in the media space. On Thursday, August 6, 1998, from 11 p.m. until 6 a.m. on Friday, August 7, the names of the expellees ran across Austrian TV screens in the program *kunst-stücke* (Andrea Schurian, Wolfgang Lorenz).

Expulsion of Reason /
Vertreibung der Vernunft

1993	computer-based video installation; 14 monitors, computers, sound, wooden pallets, sea container, dimensions variable

Installation view
respectively, Peter Weibel,
ZKM | Karlsruhe, 2019–20

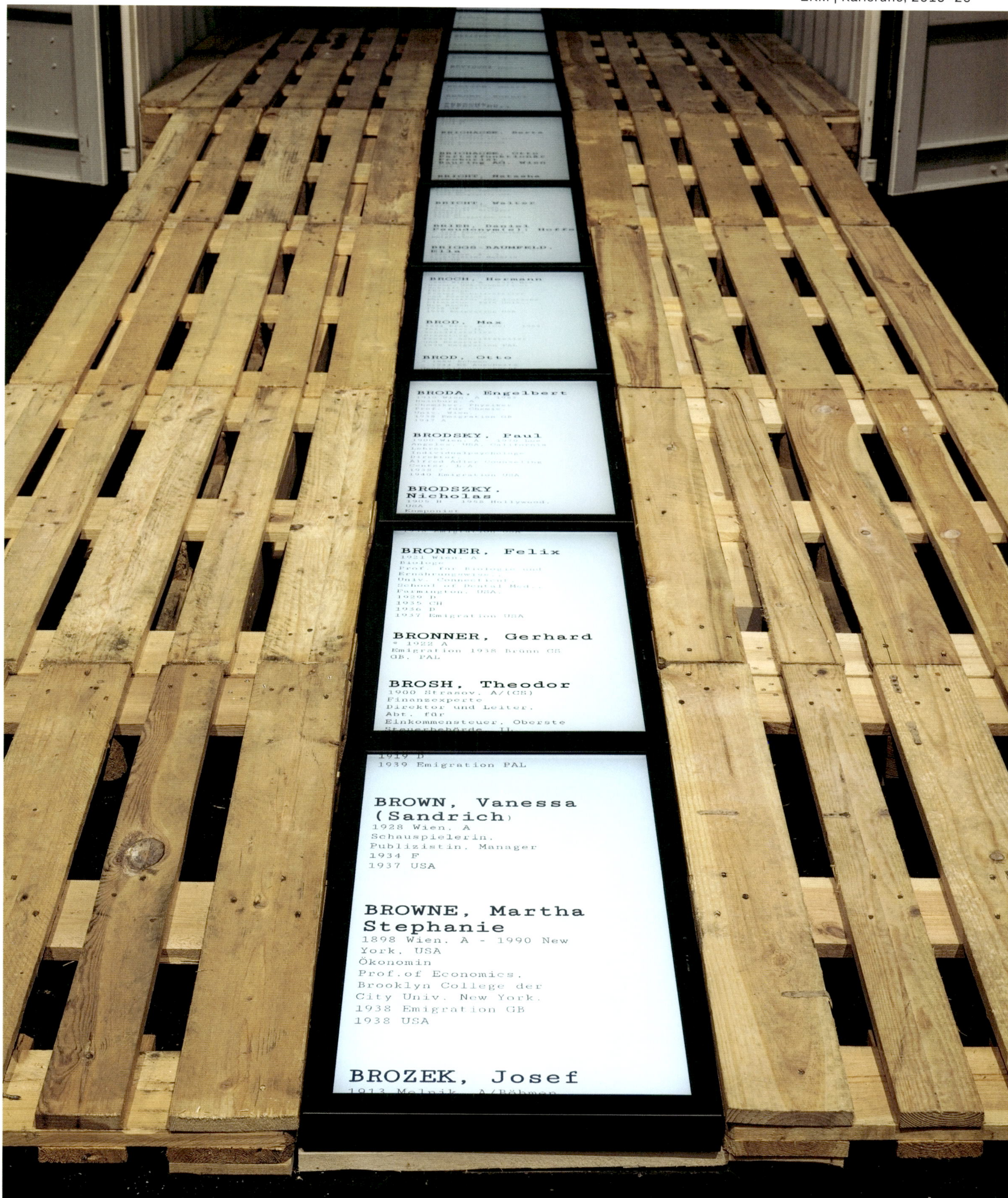

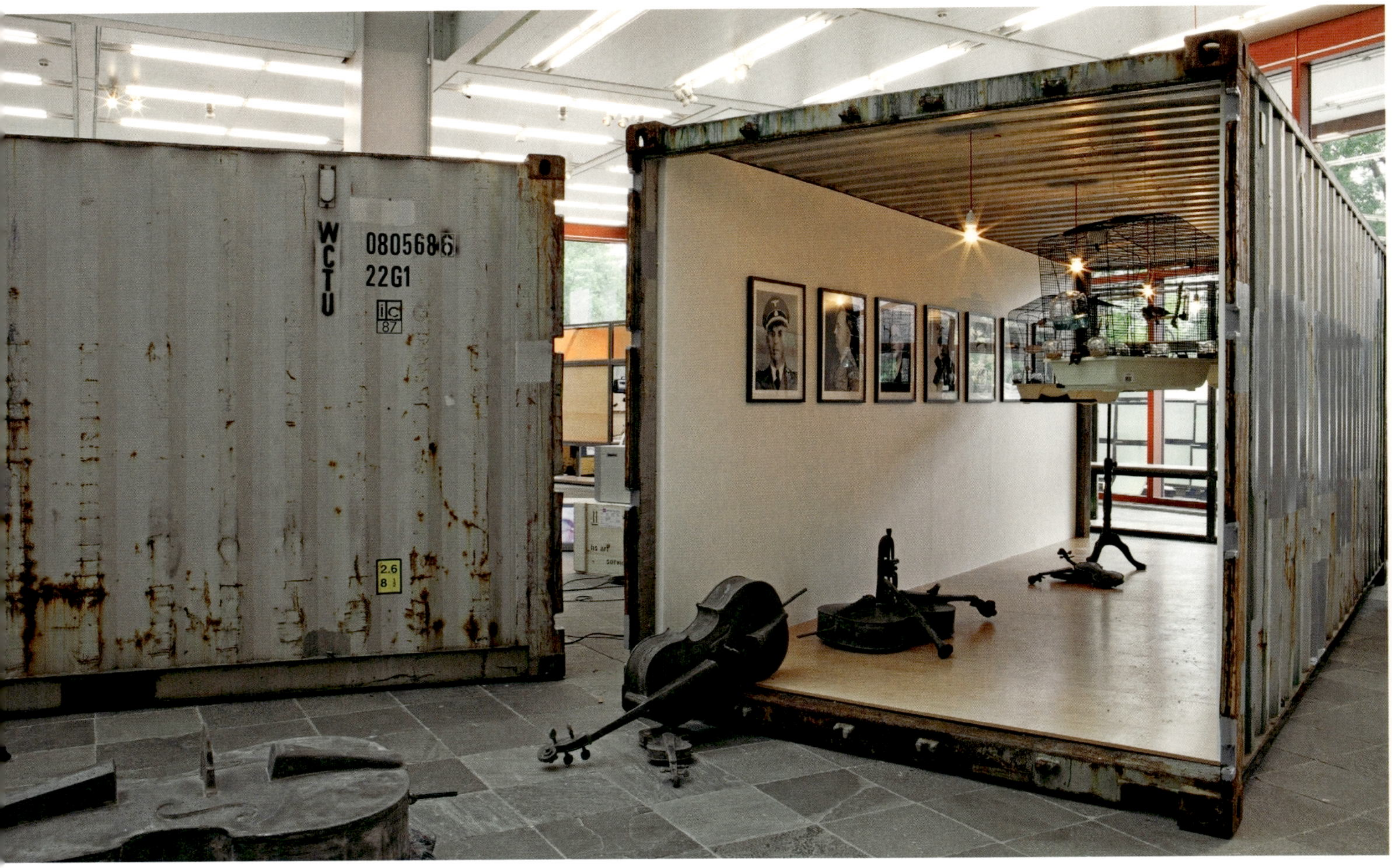

Installation view *Peter Weibel – MEDIA REBEL*, Belvedere, Vienna, 2014–15

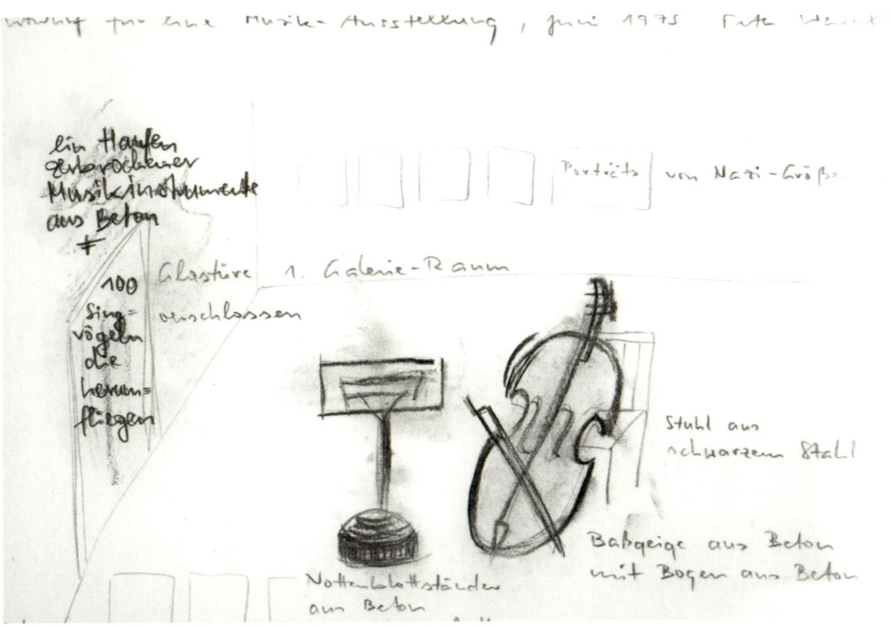

Conceptional drawing for *Music Exhibition*, 1975; 7 photographs of prominent Nazis, 7 photographs of victims, 100 birds behind a wall of glass, music stand and violin made of concrete, a heap of broken musical instruments made of concrete, steel chair

The installation *Music Exhibition*, conceived in 1975 and first realized in 2014 for his exhibition at the Belvedere in Vienna, consists of photographs of prominent Nazis, photographs of their victims, a music stand and several musical instruments made of concrete. Several songbirds in cages are part of the installation, which in the original concept for the installation would have flown freely around the gallery space. With the installation, the artist refers to the crimes of the Nazis that silenced the voice of humanity in the twentieth century with their barbaric acts. It also asks the question, why was culture unable to prevent this barbarism? Instruments of stone, petrified music, petrified conditions. The song of birds, the music of nature: An answer?

Music Exhibition / Musikausstellung

1975 (concept)/ 2014	musical instruments and music stands made of concrete, 14 b/w photographs, replicated songbirds in cages, dimensions variable, frame (14 photos, each): 60 × 50 × 2.5 cm

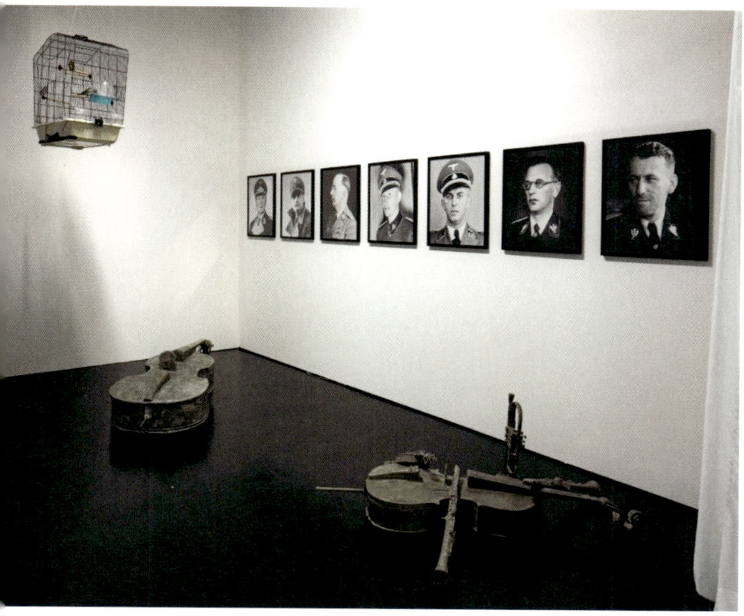

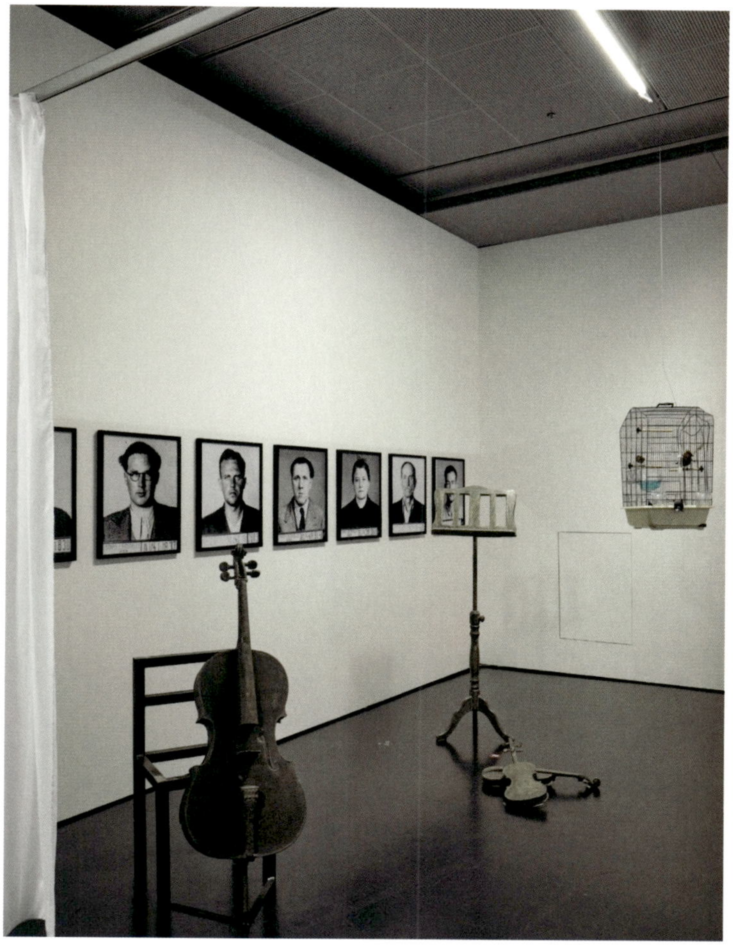

Installation views
respectively, Peter Weibel,
ZKM | Karlsruhe, 2019–20

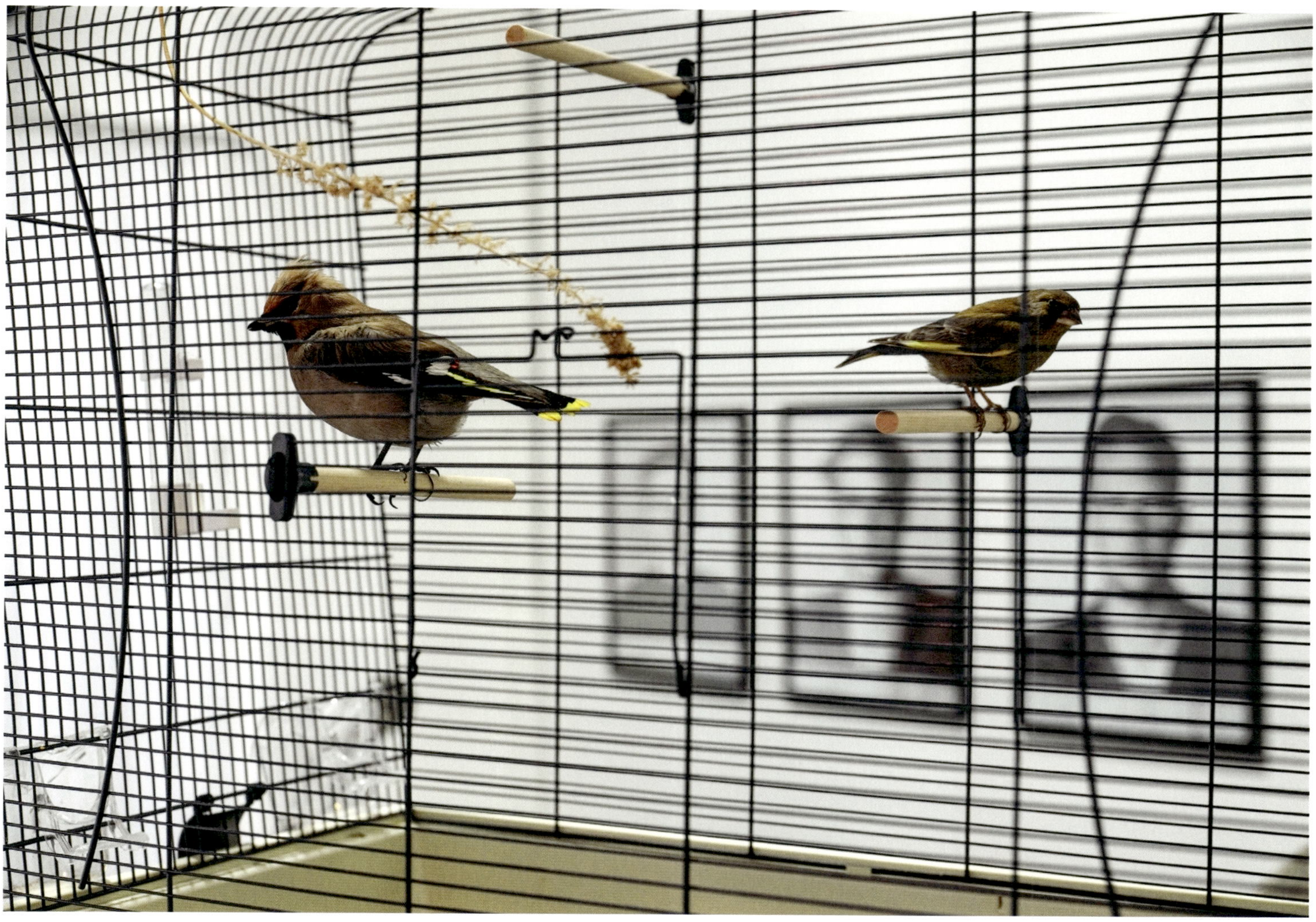

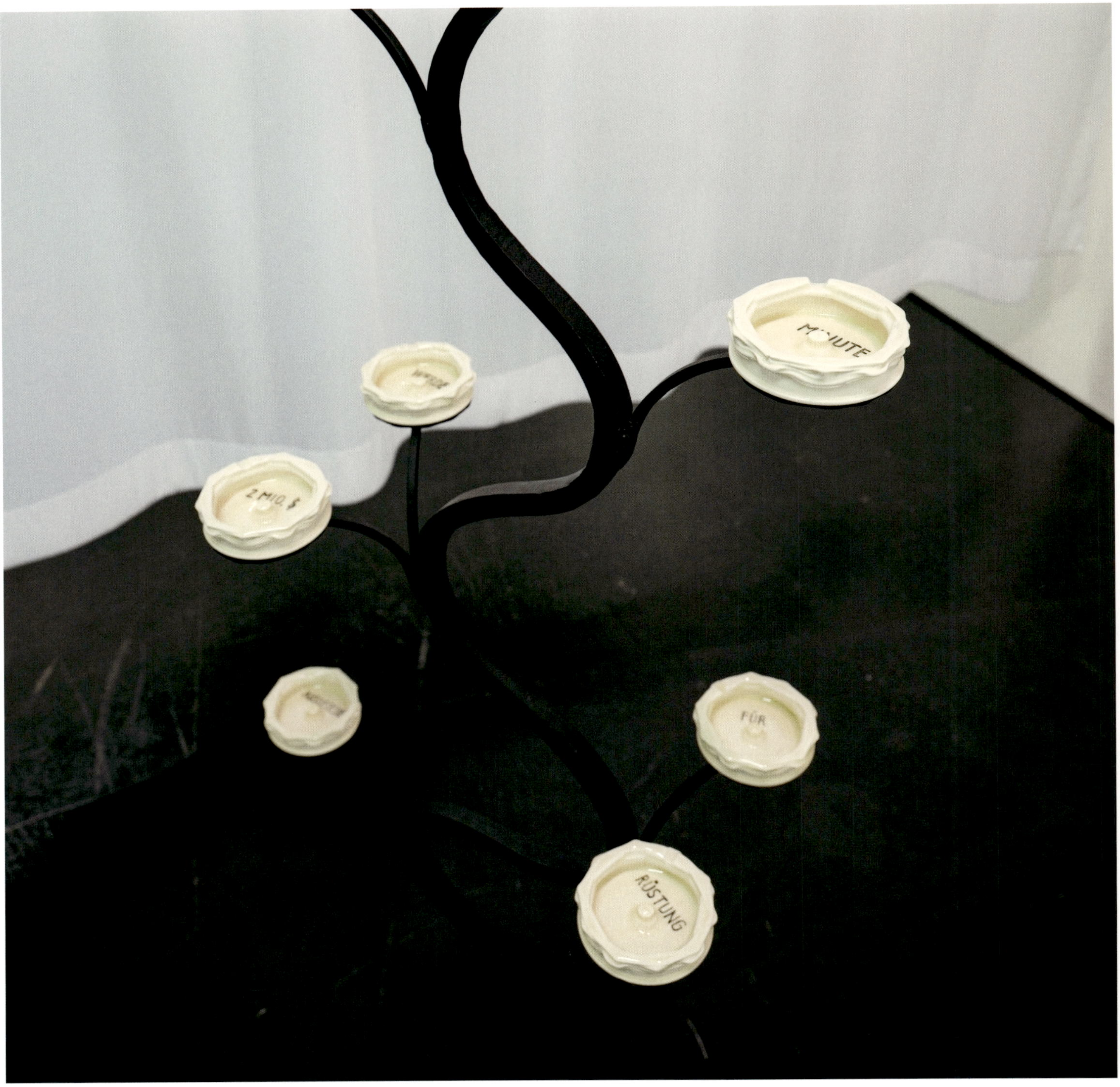

Installation views
respectively, Peter Weibel,
ZKM | Karlsruhe, 2019–20

The Living Room as a World Problem / The World as a Problem of the Living Room / Das Wohnzimmer als Weltproblem / Die Welt als Problem des Wohnzimmers

1990	wrought-iron flower rack, 7 ashtrays, white ceramic, glazed, cactus, blackboard with inscription in white chalk, 7 colored plastic balls with text (white adhesive lettering), dimensions variable Generali Foundation Collection–Permanent Loan to the Museum der Moderne Salzburg

In this installation, the furnishings of a Western, middle-class living room find themselves confronted with shocking realities. Conventional objects can be seen such as a wrought iron flower stand, which instead of plants holds ashtrays with the inscription *PRO/MINUTE/WERDEN/2 MIO.$/FÜR/RÜSTUNG/AUSGEGEBEN* (Every minute two million dollars are spent on armaments). Colored plastic balls are inscribed with the words *PRO/MINUTE/STERBEN/30/KINDER/AN/HUNGER* (Every minute thirty kids die of hunger). On a board a message in white chalk says: *WELTWEIT* (Worldwide). Through the metaphor of the living room, Peter Weibel, as a political artist, links these repressed realities with a critique of middle class lifestyle. How do we come to terms with the world we live in? Isn't the world the living room of all of us?

The owner of the installation is obliged to take care of the cactus, that is, to water it, as an activity of solidarity. The artwork as a living organism needs caretaking, just like the Earth does, to survive.

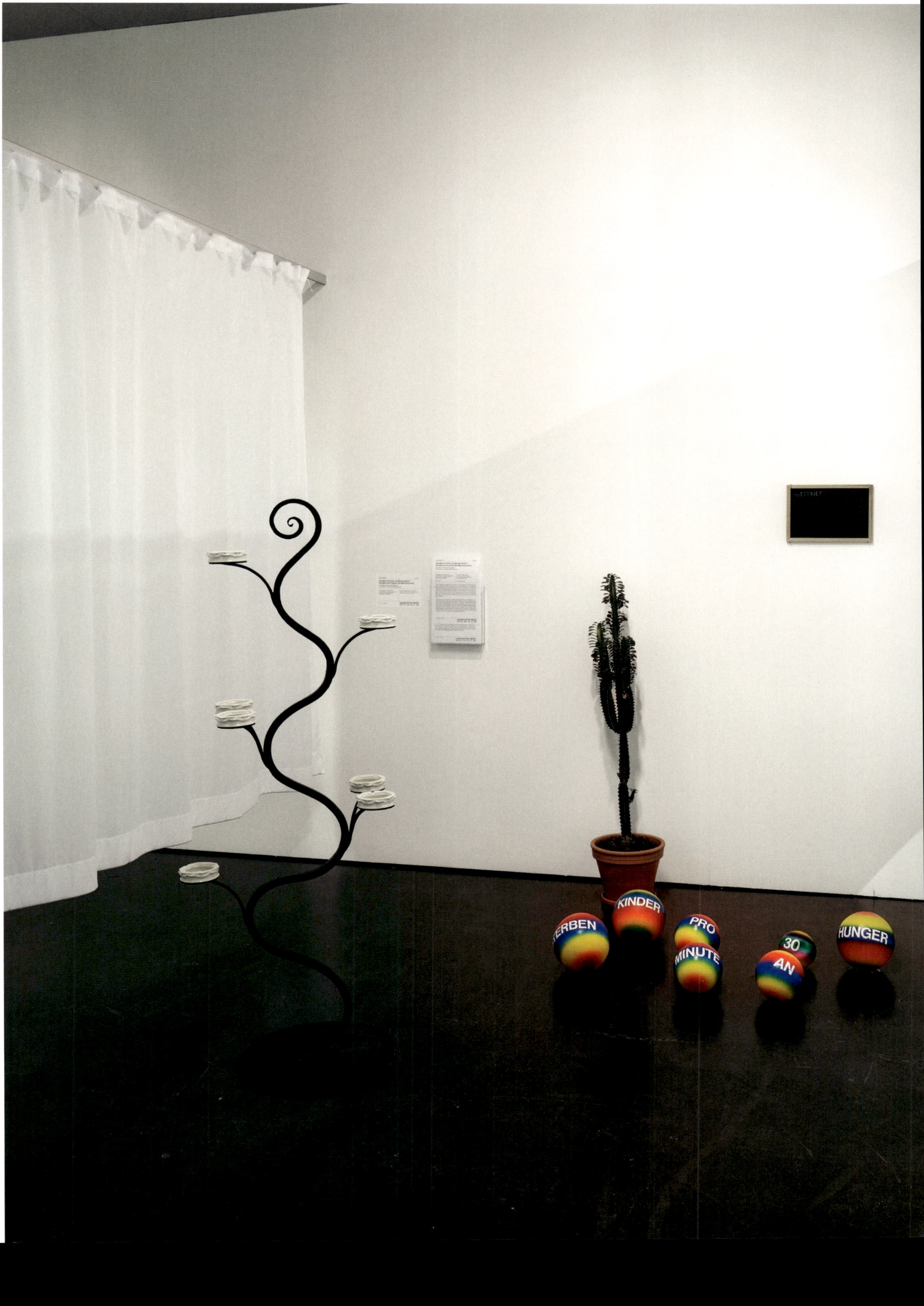
KINDER
PRO
MINUTE
30
AN
HUNGER

"The per capita gross national product of the sixty-one poorest states is 350 US dollars."

"The per capita gross national product of the fifty-four richest states is 25,890 US dollars."

"A millimeter is in the same proportion to a meter as a meter is to a kilometer. The size of a virus is the same proportion to a person as the size of a person is to the Earth. The diameter of the Earth is 107 meters, the diameter of a virus 10–7 meters. Compared to a virus, I am the size of the Earth. To the Earth, we are the same size that viruses are to us."

"It took humanity until the mid-nineteenth century to produce the first billion people. In 1930, only eighty years later, the world's population had already reached the second billion. And only fifteen years later, in 1975, we already had the fourth billion. Now we are approaching the eighth billion.

"About 130 million women and girls worldwide have been subjected to genital mutilation."

"Ninety-five percent of all anorexics and ninety percent of all bulimics are women."

"Each year, 200,000 women die of complications from an abortion."

"One percent of the world's population earns forty-two percent of all global profits."

"One billion people live off less than one dollar per day."

"130 million children worldwide do not attend school. Two thirds of them are girls."

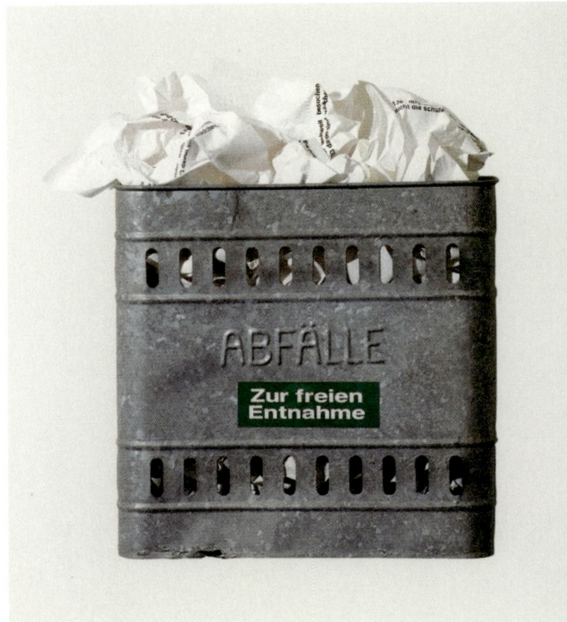

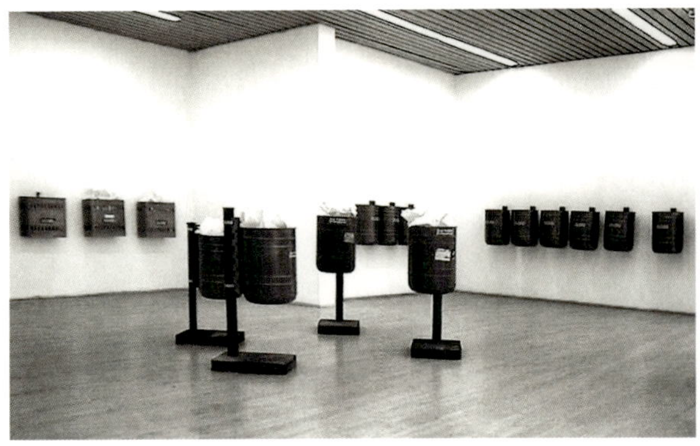

Installation view Galerie Mathias Kampl, Munich, 2000

Wastepaper baskets are filled with crumpled paper containing short messages. By seeing the garbage can as an everyday sculpture and the crumpled paper as the result of sculptural action, which is always defined by a change in form, Weibel is once again raising the question of how far the term "sculpture" goes. However, in so doing, he collides with conventional values, which are oriented on intact pieces of paper with legible messages and therefore interpret this deformation as the result of an act of destruction.
Peter Weibel transcends this intra-media discourse by turning to the social dimension, supplementing the baskets with a sign announcing that the paper can be taken away for free. Weibel adopts the language of advertising. The baskets impart the connotation of surplus, of that which has become unnecessary and unappetizing, whatever it may contain before it comes back to light. To members of an affluent society, taking something out of a public wastepaper bin is considered unseemly; but using it to dispose of what has become unlovely is seen as proof of good manners.
The waste in the baskets is of an intellectual origin. They are simple messages which Peter Weibel has termed global sentences. These sentences are not intended as an appeal; they produce a dismay which does not result in the development of approaches to solutions, but in the repression of any approach to the problem. This culminates in throwing away the message. The disposal is accepted as a communal necessity.

Throwaway Poems / Wegwerfgedichte

1999	15 street furniture wastepaper baskets (40.7 × 41 × 20 cm), PVC signs, paper waste

Note: Adapted from Arnulf Rohsmann, "peter weibel/wegwerfbotschaften", in *Peter Weibel: Globale Gier* (Klagenfurt: Kärntner Landesgalerie, 1999), 28. Translated from the German by Peter Dahm Robertson.

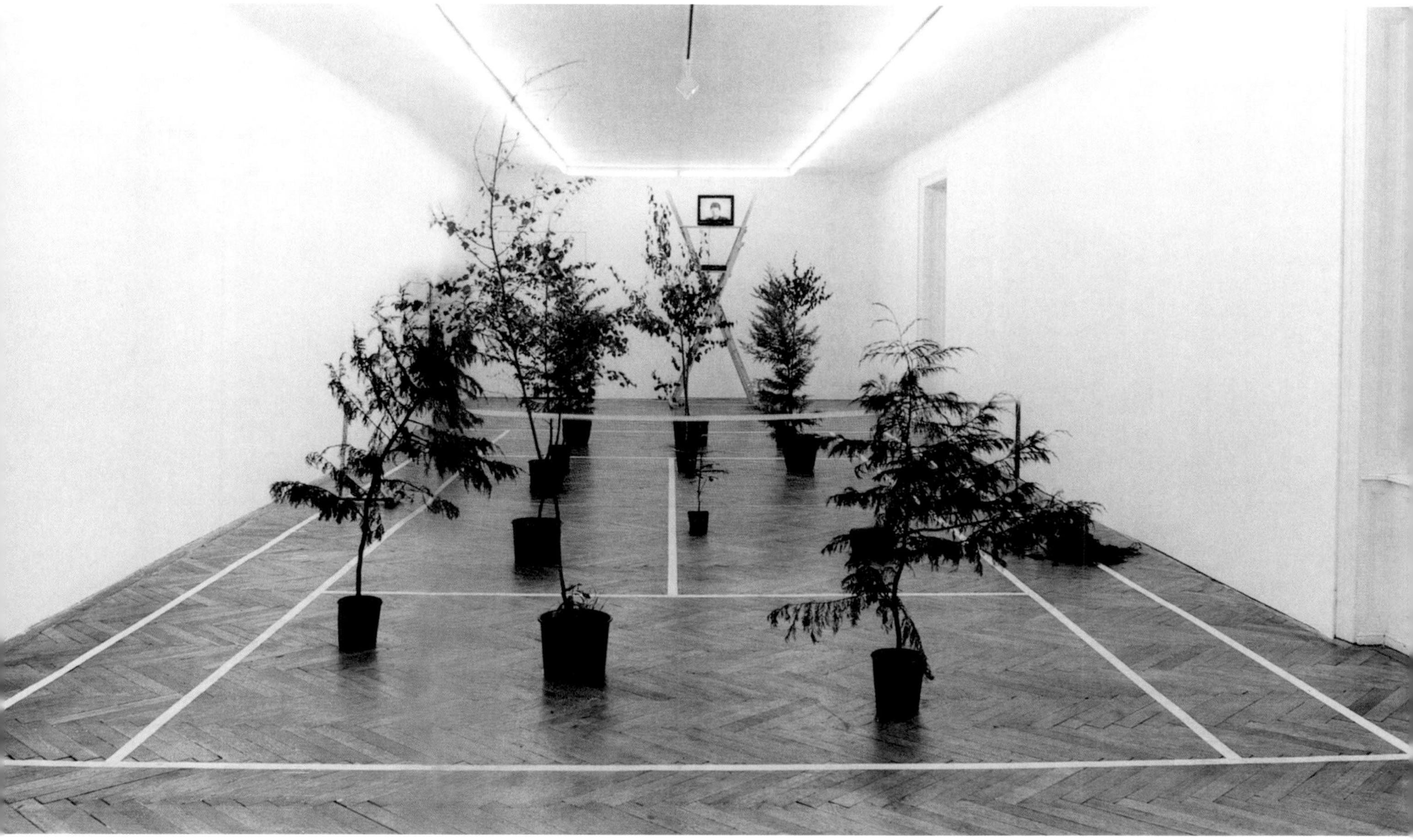

Installation view *Die Bitte-Gesellschaft*, Galerie Grita Insam, Vienna, 1995

Die selbstbewusste Nation (The self-confident nation) is the title of a collection of essays, published in 1995 and written by in some cases very well-known authors, in which the New Right stakes out its positions. The title refers to democracy's peculiar characteristic of offering few opportunities for identification, which many view as a weakness. Postcolonial dreams of a great nation and a self-confident people tap into these weaknesses. And indeed, we have not yet achieved the state of democracy. Citizens receive their due not because they deserve it, but because the state grants it to them. From childhood, the subject learns that it receives things only because it asks for them and not because it deserves them. From family to school, from training to job, citizens are continuously progressing through the institutions, commissions, councils, boards, etc. placed before them and to which the realization of their rightful desires is subject. Because the subject in administrative democracy is never released from a state of requesting, an ever greater potential accumulates, reaping populist profit from this real deficit of the contemporary form of democracy. Not the ostracization of those disappointed by democracy and of the enemies of the constitution, but only a radicalization of democracy and an admonition to fulfill its promises can help banish the populism of the New Right.

The installation poses anew fundamental questions of democracy, which as a social state must by definition be distinct from the state of nature. Democracy was, it might be said, invented to mitigate or avert what nature does to its subjects and what can be categorized by terms like contingency (chance, fate) and incommensurability (immeasurability). In democracy, comparison (of wages and all the way to positions) is meant to be possible; not chance and privilege are meant to rule but criteria of equality and justice. The game of tennis as a social game in which competition defines the winner is a modern metaphor for a competitive society as a complementary distortion of the request society. If trees play against trees, there can be no winner. Should we construct a society which undercuts this state of nature, rather than instead a society in which the disadvantaged have the same opportunities and rights and the weak are supported, in which thus all participate equally in the "wealth of nations" (Adam Smith)? —Peter Weibel, 1995

Translated from the German by Peter Dahm Robertson.

The Plea Society / Die Bitte-Gesellschaft

1995 | monitor, video player, ladder, pseudo-tennis court, trees

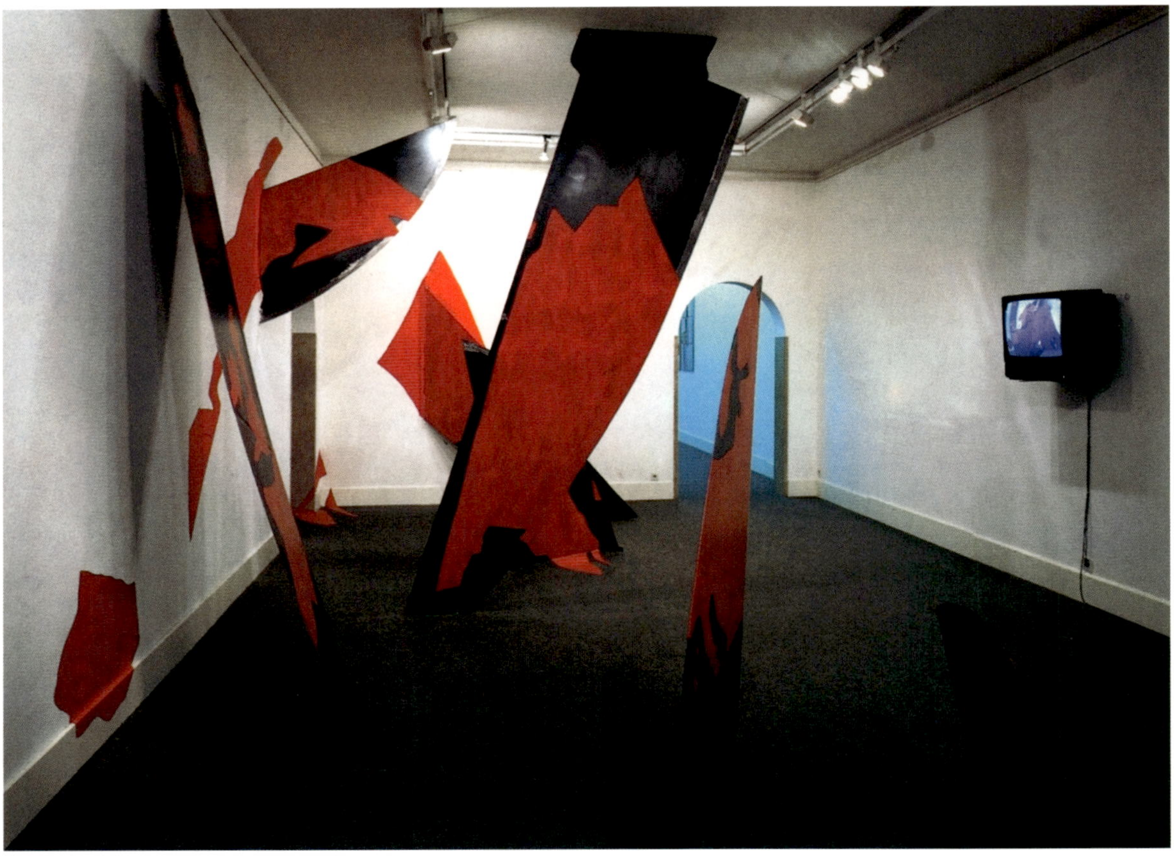

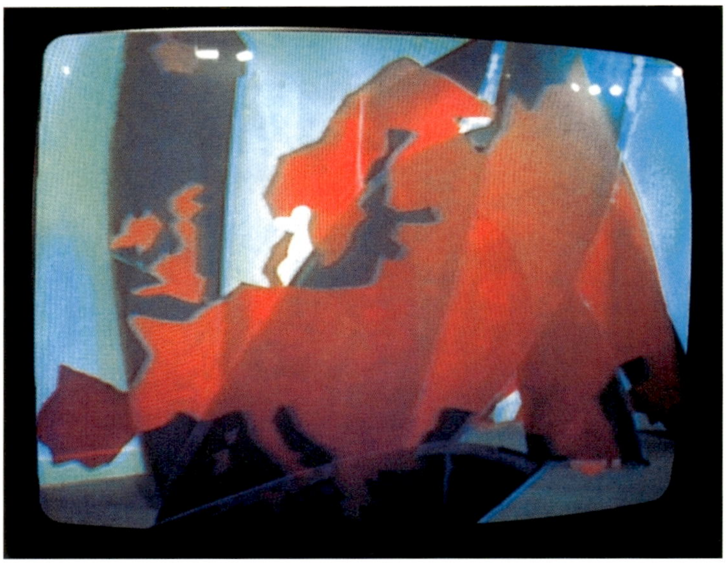

Installation views *aktuell '83*, Lenbachhaus, Munich, 1983

The German title *Europa(t)raum* playfully emanates from three nouns – Europe, dream (*traum*), and space (*raum*) – and manages to combine these into a single word, split up into two different meanings by a parenthesis in the middle. The inclusion or exclusion of these brackets and the letter "t" within them allow for a twofold reading and interpretation: first, as *Europatraum*, a dream about Europe; and second, as *Europaraum*, all of the European countries taken together and viewed as an interconnected space, a unified continent. Already with the choice of title, the installation demonstrates the difference between dream and reality, between the actual space and the symbolically charged knives cutting through the continent in conflict.

When visitors enter the space, they see a massacre: in the installation huge knife blades, which appear to be bloodstained, brutally penetrate the space. Blood red color is all over the wall, the floor, and the gigantic knives. However, from a fixed perspective and location, where the video camera is positioned to record the space and its contents, the various color fields within the space compose a simplified map of Europe without political boundaries on the screen. As the spectators move through the space, and thus through the map of Europe, they see themselves cut up on the screen by knives.

Almost ten years before the Yugoslav Wars, Peter Weibel portrayed Europe with this installation as a bleeding continent, fragmented by nationalism and separatism. A critical examination of the often bloody history of Europe can also be found in works like *45 European Pedestals with Recessed Sculptures* (*45 europäische Sockel mit nach innen versenkten Skulpturen*) from 1989 (see pp. 474–75 in this volume) and *Fortress Europe* (*Festung Europa*, see page opposite) from 1994. The latter is a large-scale installation with black flags, suitcases, and textiles that sketches Europe as a forest against migration and refugees, a bureaucratic-economic idea with a dark and inhumane history and present. In view of current political trends in Europe, these works are once again highly topical.

Europa(t)raum

1983	closed-circuit video installation; steel plates, paint, monitor, video camera, dimensions variable

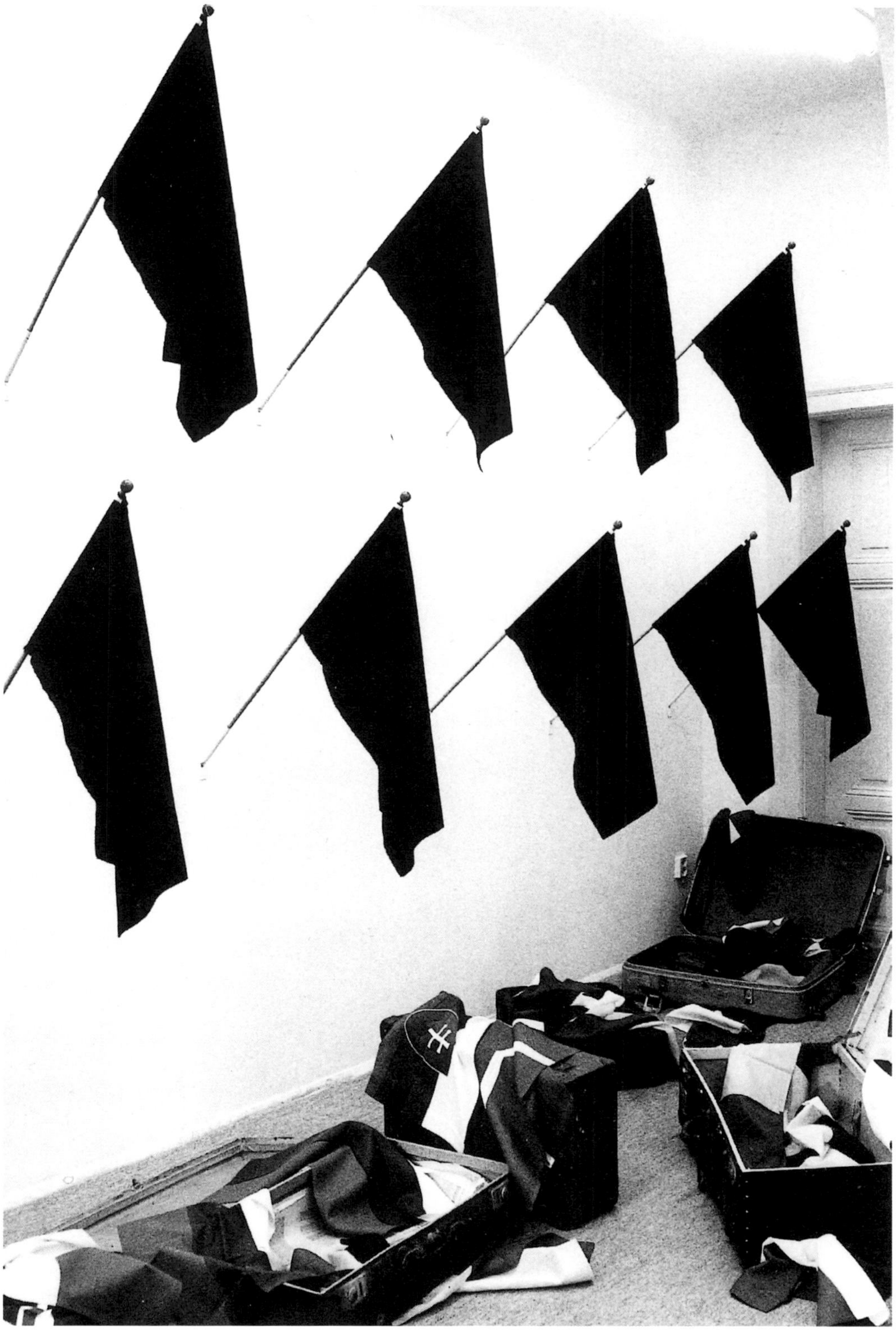

Installation view
Spielverderber, Forum
Stadtpark, Prague,
May 1994

Europe is a mask. Europe lives in Brussels and Maastricht. Europe is a bureaucratic and economic idea. Europe dies in Temelín and Sarajevo. Europe is not a political and humane idea. Horror, blood, hatred, and death rise up from the depths of Europe's past. Genocide is Europe's agenda. Europe exported colonialism and then reimported it. The sleep of Europe is the sleep that produces monsters. The Europe of Eastern Europe is not the same as the Europe of Western Europe. The two different Europes form a stereoscopic image of Europe. This image reveals Europe's underlying structure: plunder.

—Peter Weibel, 1994

Fortress Europe / Festung Europa

1994	black flags, flags of European Countries, suitcases

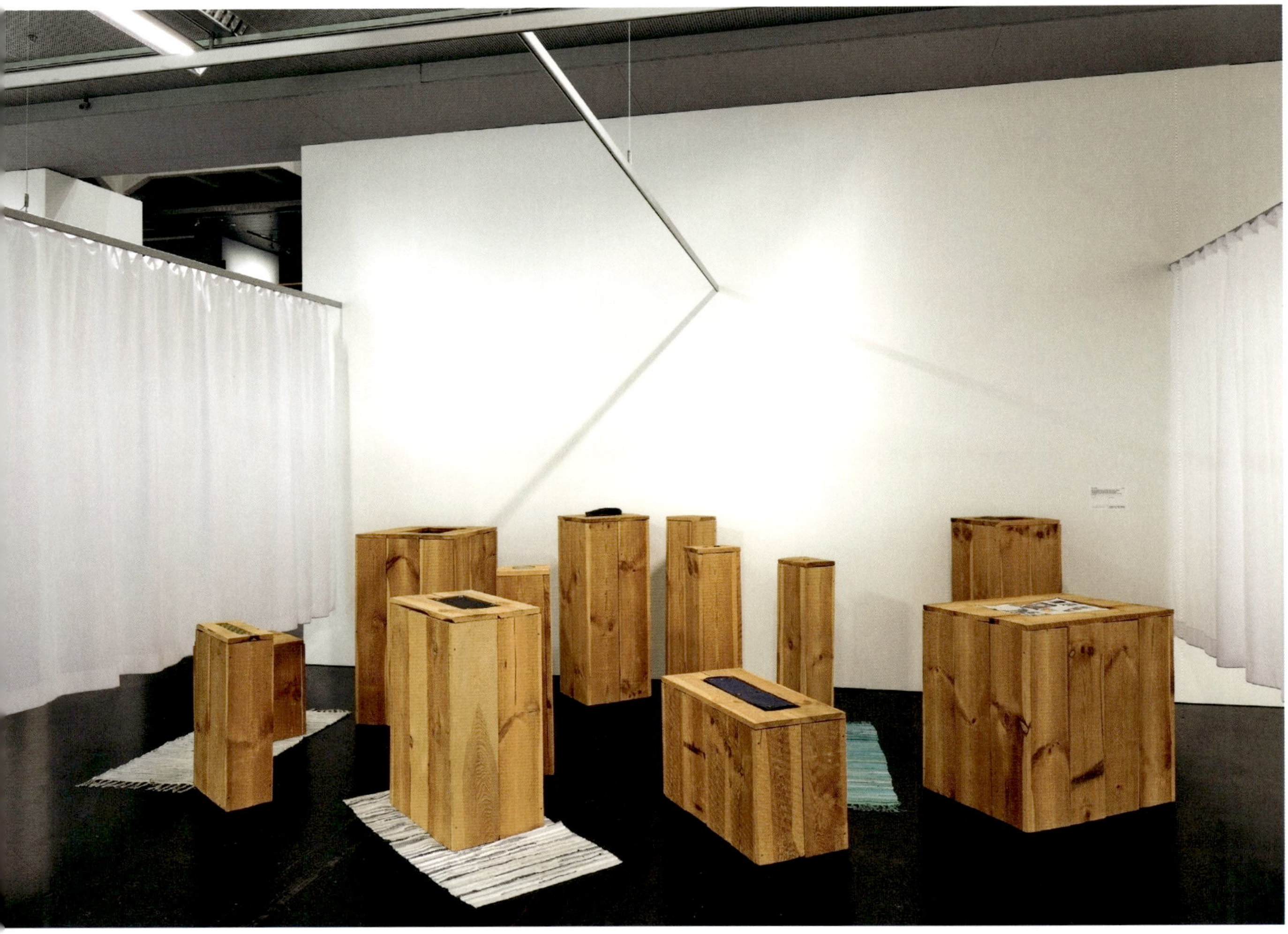

Installation views
respectively, Peter Weibel,
ZKM | Karlsruhe, 2019–20

Created in 1998, the installation *45 European Pedestals with Recessed Sculptures* consists of forty-five wooden pedestals of different shapes and heights, with various objects associated with the topic of migration inset into them. These pedestals are exhibited on colorful rag rugs in various groupings and configurations; a selection of twelve were included in the exhibition at the ZKM | Karlsruhe.

With this installation, Peter Weibel is operating on two different levels of meaning. On the one hand, he is questioning the classic relationship between sculpture and pedestal: the historical function of the pedestal is to isolate and elevate the sculpture; the insertion of objects into the pedestal inverts this relationship. The sculpture disappears inside the pedestal, becoming invisible to the viewer while the pedestal's presentational function is nullified: a double negation of classical sculpture. This aspect is further emphasized by the use of simple rag rugs, which signify the antithesis of the presentational function of the classic red carpet. On the other hand, the reference to Europe in the title makes clear that this array represents a political structure. As in a Tachist painting, a new image of Europe is being conceptualized here on rag rugs – on a patchwork of scraps. For Europe does not exist; it must first be invented. The map of Europe is like a patchwork that is kept constantly in flux by ethnic, linguistic, political, religious, and economic factors.

The objects selected constitute a report on the state of this flux, on Europe as a project of perpetual revision, a space of migration. In its political associations and its questioning of the essence of Europe, *45 European Pedestals with Recessed Sculptures* can be linked to works such as Weibel's 1983 closed-circuit video installation *Europa(t)raum*, his 1994 installation *Fortress Europe* (*Festung Europa*, see p. 473 in this volume), and his 2020 exhibition *(Post)Europa?* – three works that have been rendered distressingly topical by the rise of separatist movements in Europe.

Translated from the German by Patrick Hubenthal.

45 European Pedestals with Recessed Sculptures / 45 europäische Sockel mit nach innen versenkten Skulpturen

1998	wood, textile, objects, dimensions variable

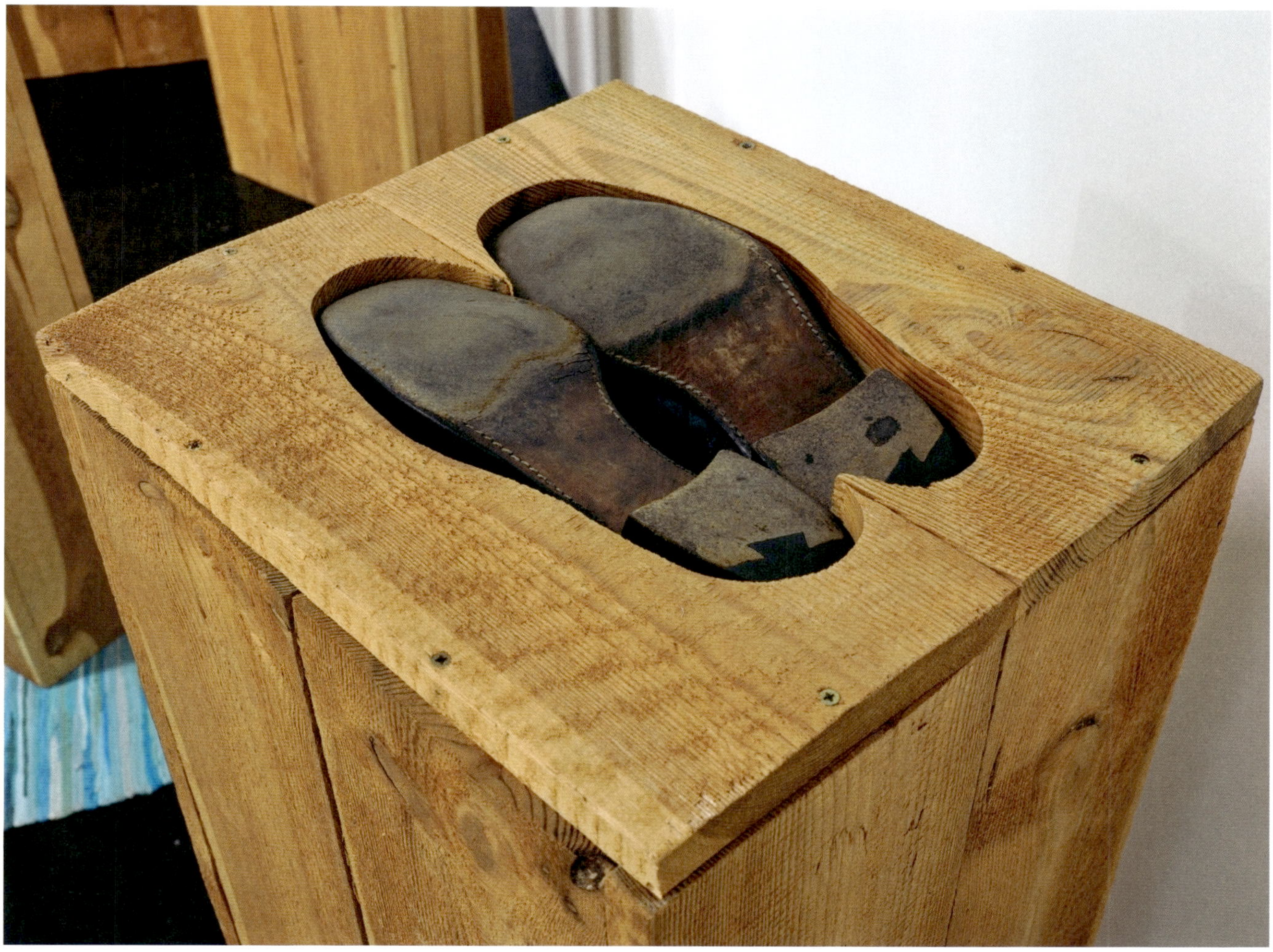

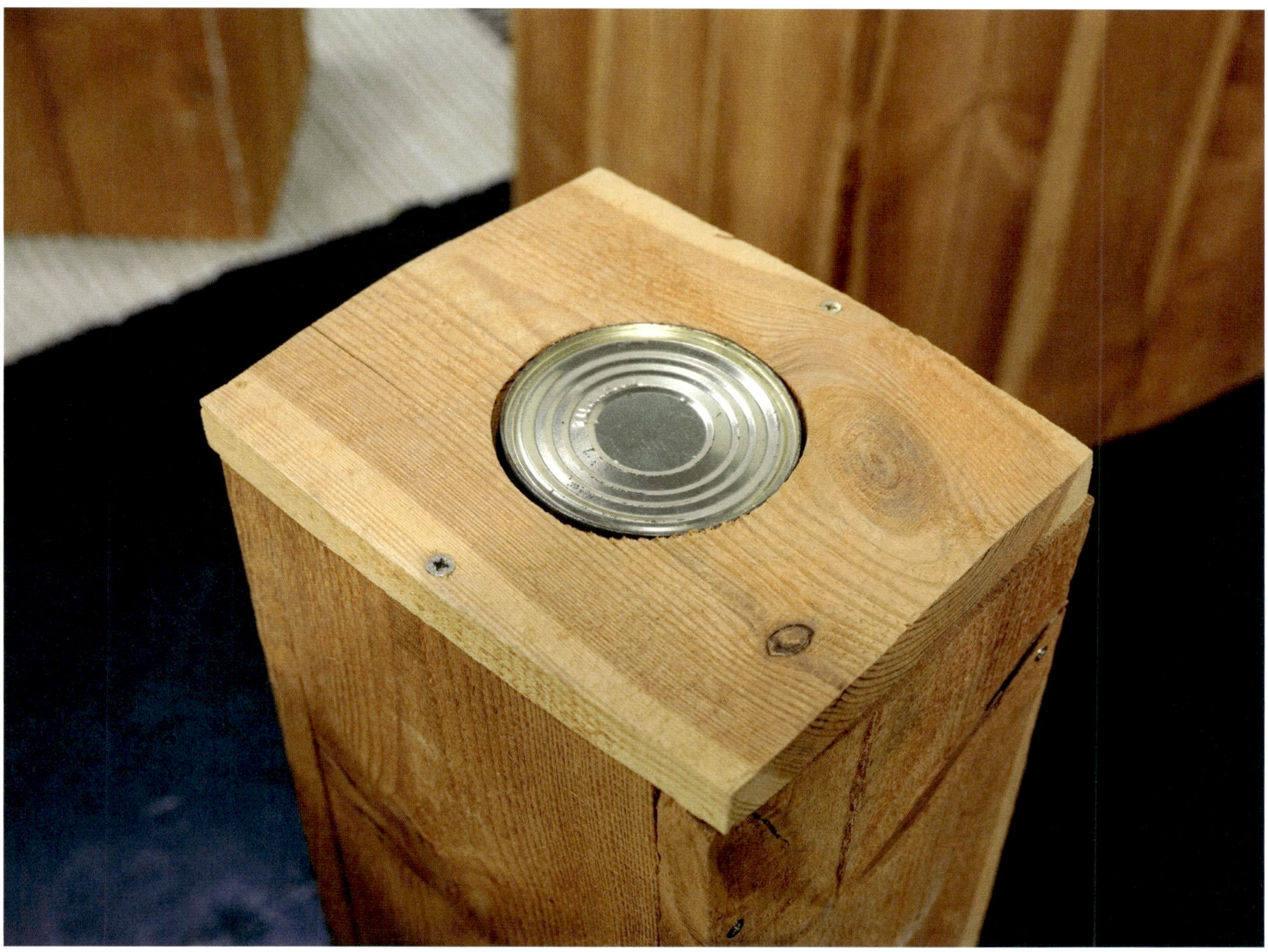

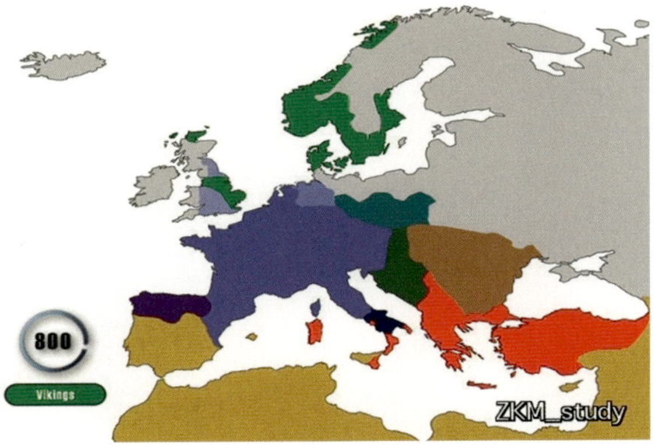

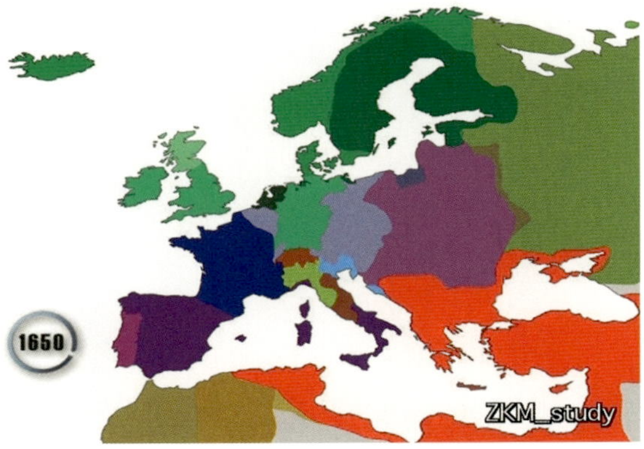

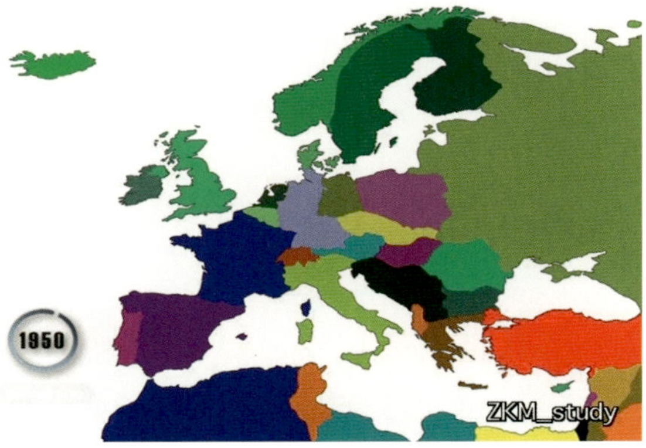

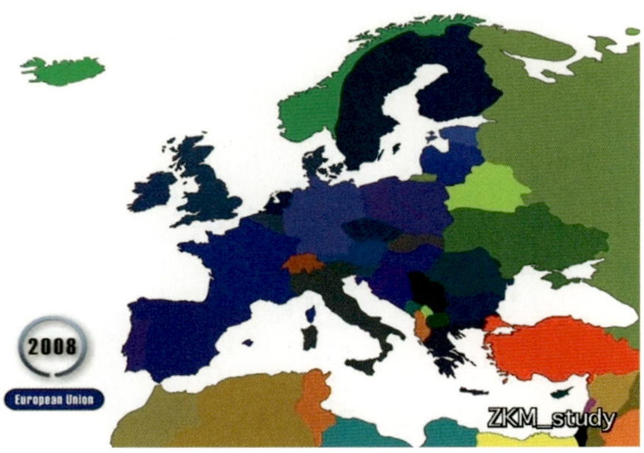

In different ways, both *Europe: A Rewriting Program* and *Is the House of Europe on Fire?* grapple with the European idea in times of upheaval, turning their gaze to Europe's past, present, and future.

Europe has always been in flux: a variety of founding myths, from Europa's abduction by Zeus to Aeneas's escape from the fall of Troy, define flight, migration, war, and violence as fundamental elements of European history. The key to Europe's success lies in the fact that it has always shown a willingness to engage with these processes of change, or "rewriting programs," as Peter Weibel calls them: "The European program ... was and is a rewriting program. ... The true history of Europe is a history of endless rewriting of texts and technologies, rites and myths, ideas and languages."[1] *Europe: A Rewriting Program* is a digital visualization that shows how the European continent's national borders have changed from the time of the Celts to today.

The title *Is the House of Europe on Fire?* evokes the words of Mikhail Gorbachev, the former president of the Soviet Union, who referred to Europe as "our common house" – a political metaphor that sought to bring about the peaceful coexistence of various parties under a single roof during the Cold War era. Europe as a habitable home where we can live together, a house that thrives on economic, ecological, cultural, and scientific interchange. This piece asks the question: What has become of the house of Europe? In light of the widening gap between rich and poor, the resurgence of the radical right, the ongoing erosion of democracy, the polarization of public opinion, and finally the wars in Yugoslavia and Ukraine, the political and economic failures of the recent past have become plain to see. *Is the House of Europe on Fire?* holds a literal mirror up to its viewers: everyone is responsible for Europe's future.

Translated from the German by Patrick Hubenthal.

Europe: A Rewriting Program /
Europa – ein Umschreibeprogramm

2010 | computer animation, color, stereo, 8:51 min.

[1]
Peter Weibel, "Europa as a Common Home or Post-Europe?," interview by Agnes Tieze, in *Peter Weibel: (Post-)Europa? Lovis-Corinth-Preis 2020*, ed. Agnes Tieze (Berlin: Hatje Cantz, 2020), 105–17 (quotation on 109).

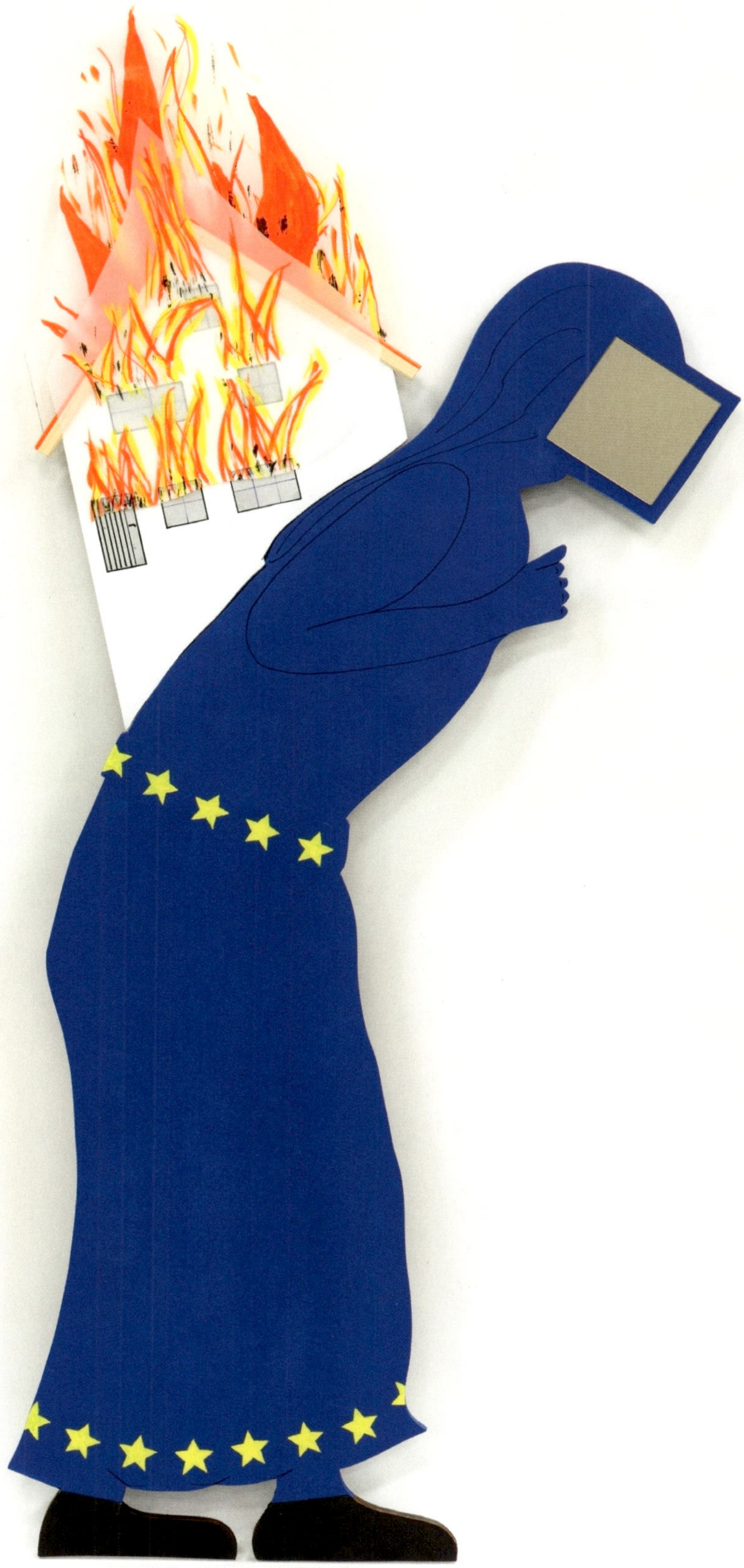

Is the House of Europe on Fire? / Brennt das Haus Europa?

2020 | sawed silk screen plate; paint

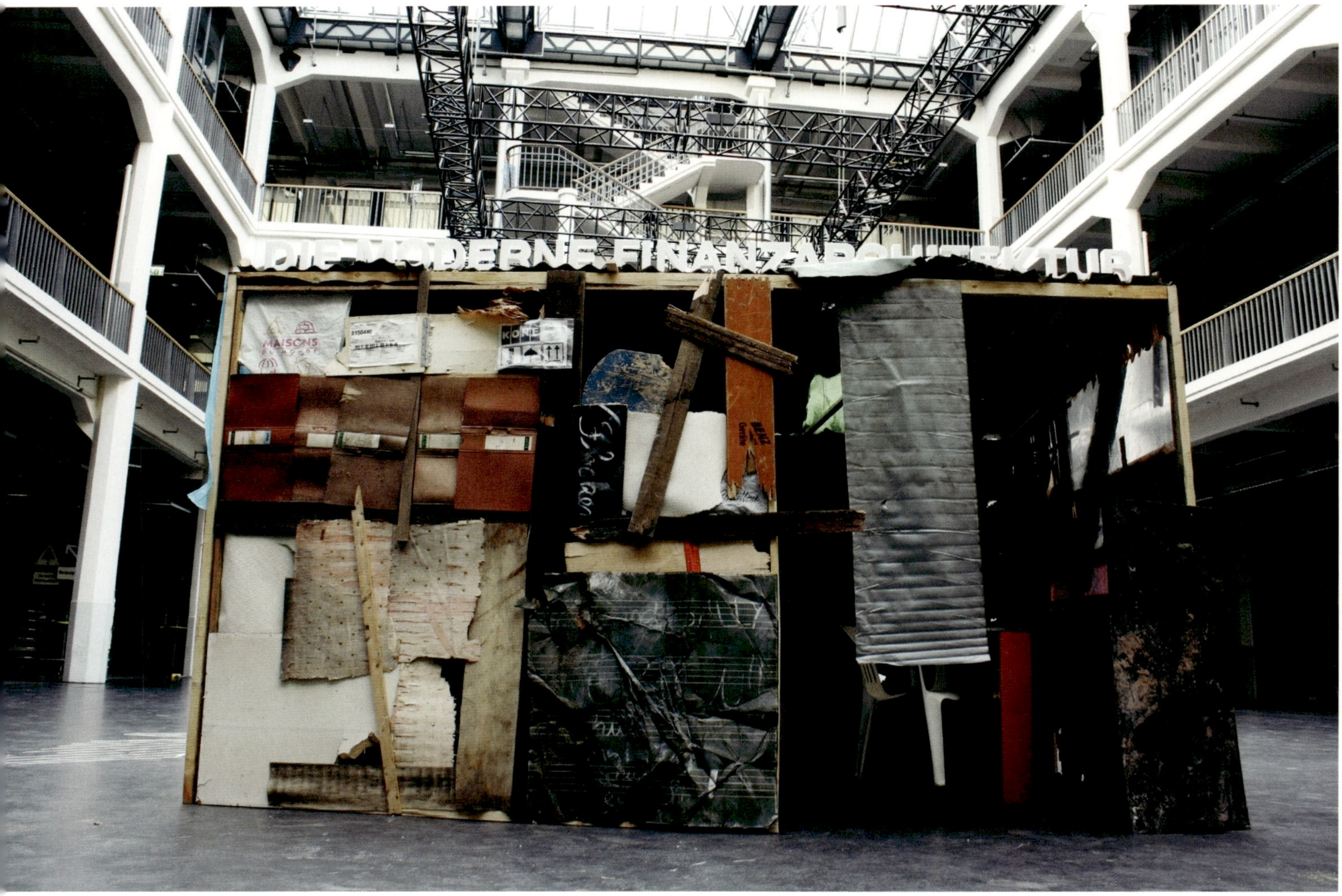

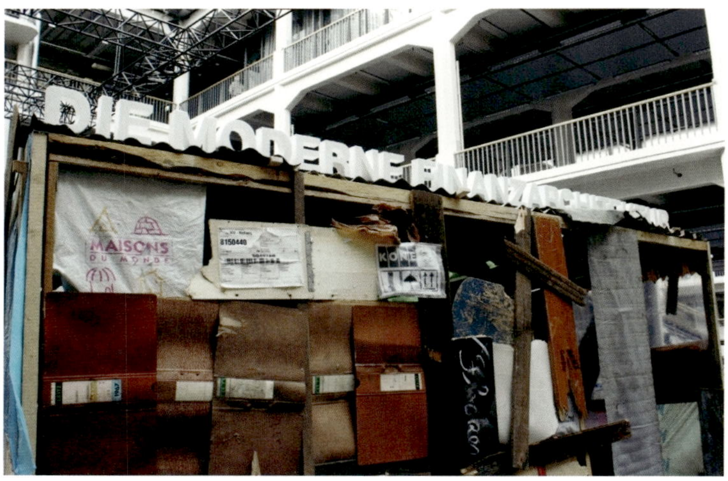

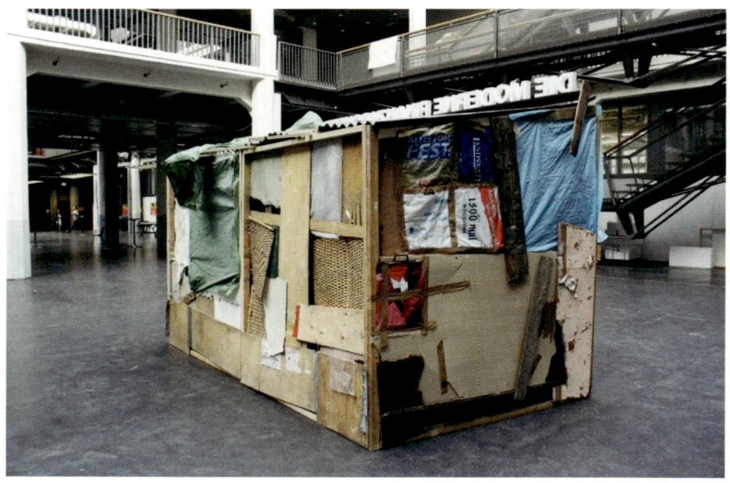

Installation views
Karlsruhe University of Arts and Design (Hfg), Karlsruhe, 2009

Modern Financial Architecture / Die moderne Finanzarchitektur

2009	hut, materials variable, with furniture, writing from styrofoam, ca. 200 × 400 × 240 cm

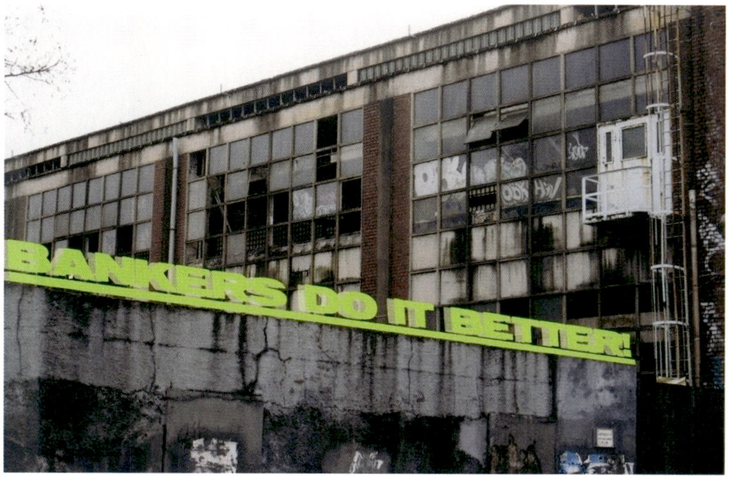

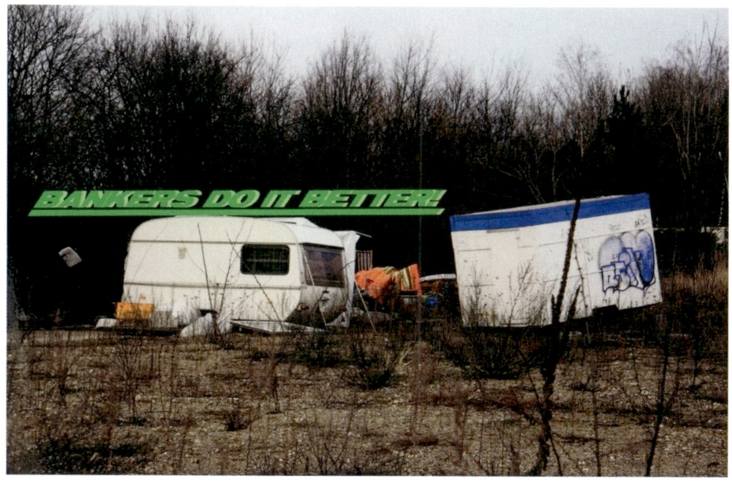

Bankers Do It Better

2009 — photographs on Alu-Dibond, sealed with UV-protective film, 2 parts, each 40 × 60 cm

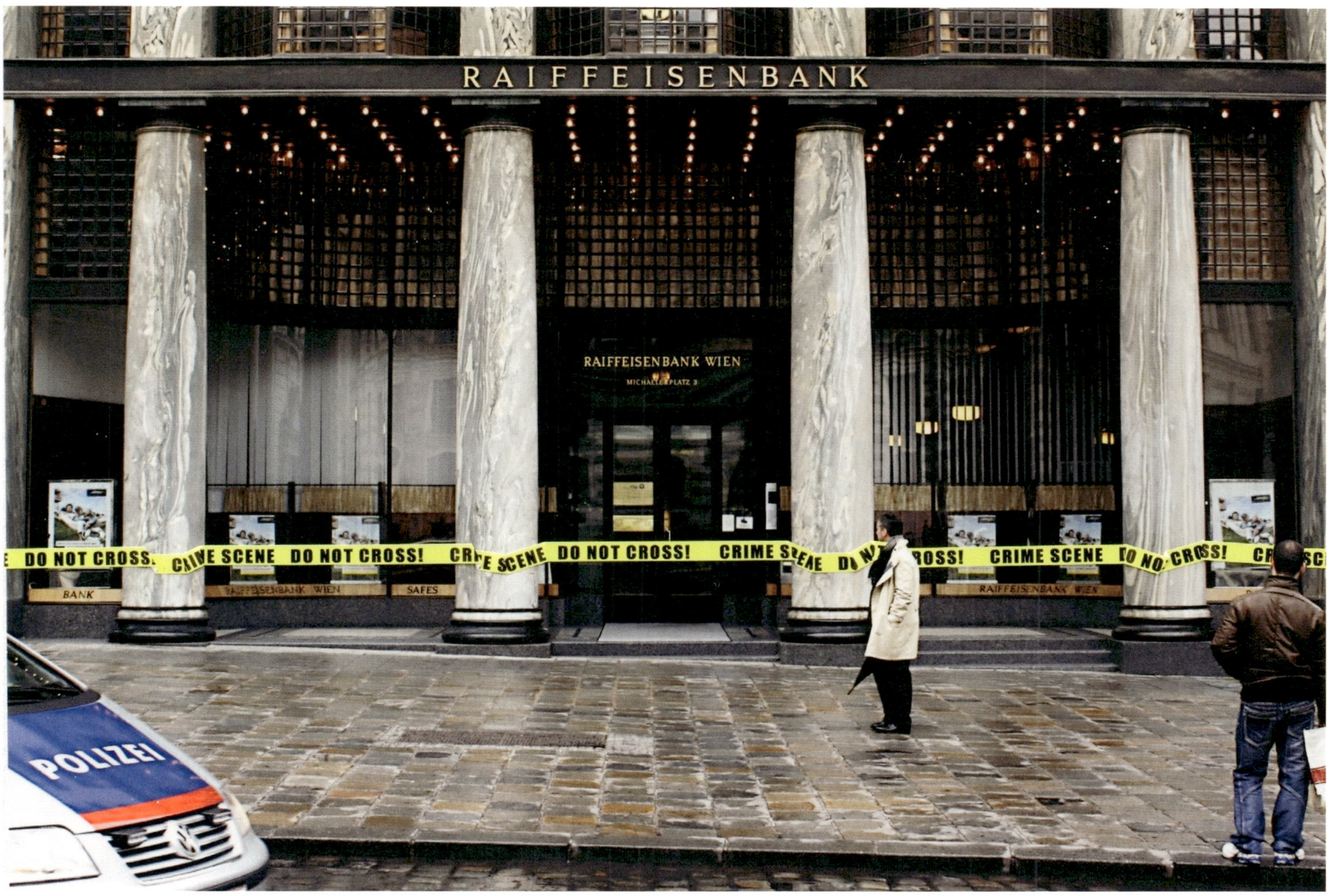

Crime Scene

2009 — photograph on Alu-Dibond, with UV-protective film and adhesive barrier tape, 21 × 29.7 cm

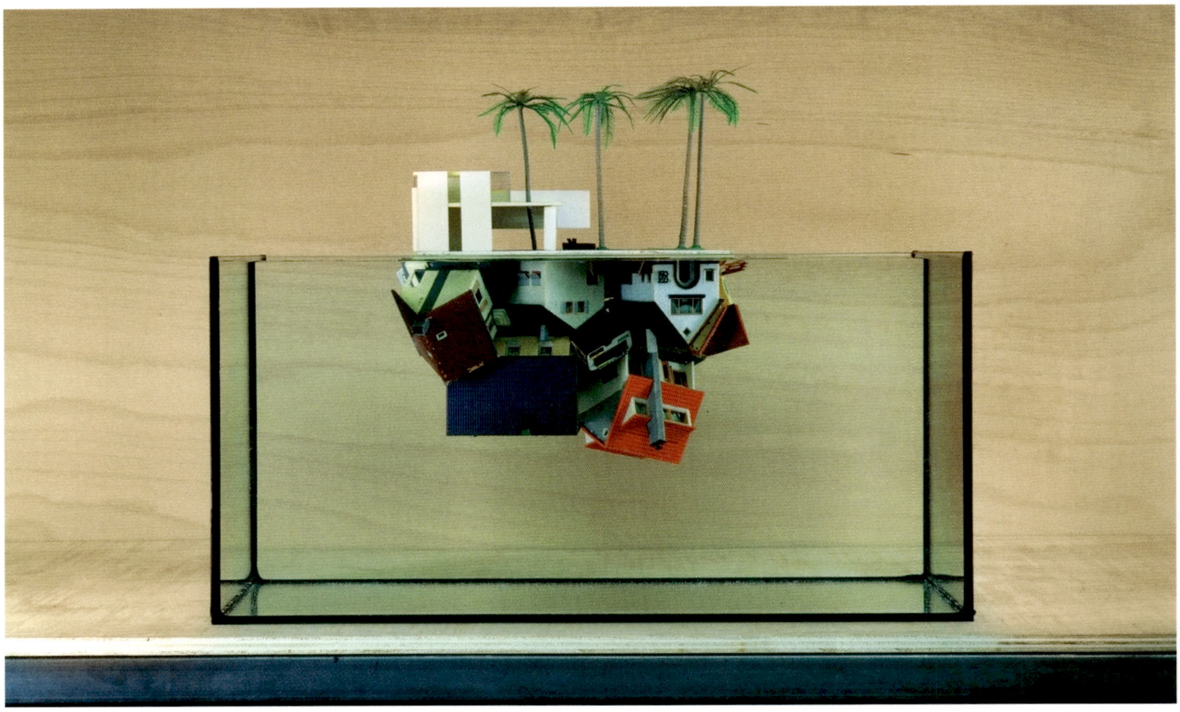

Installation view *Peter Weibel – MEDIA REBEL*, Belvedere, Vienna, 2014–15

Housing Crisis

2009	toy houses, glass basin, water, 30 × 60 × 30 cm

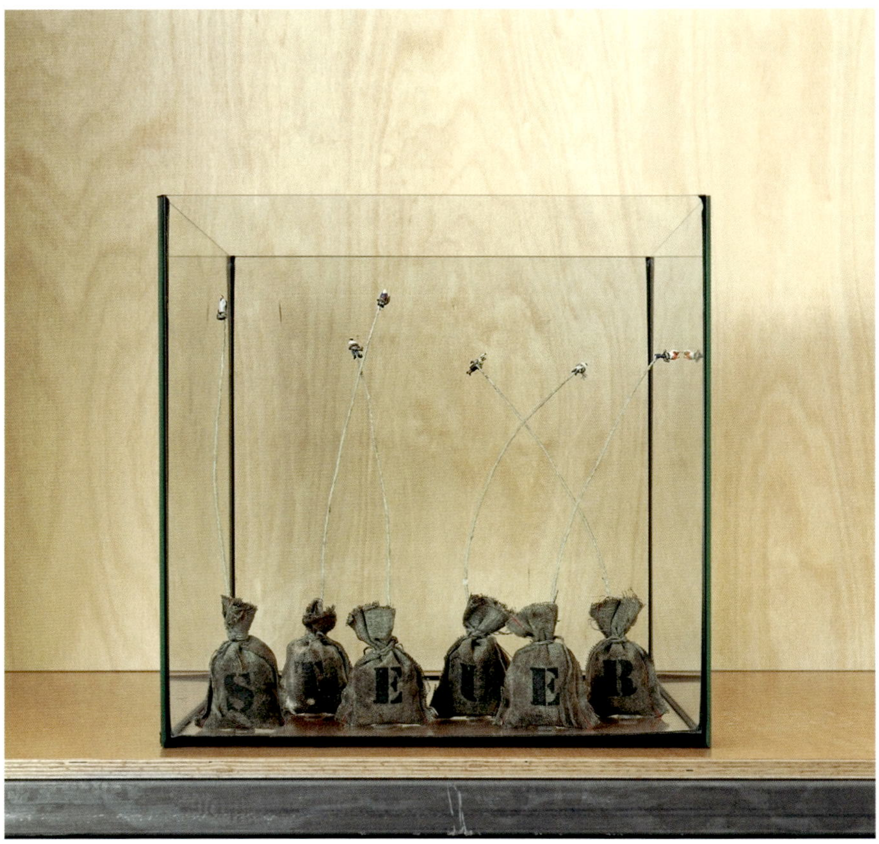

Installation view *Peter Weibel – MEDIA REBEL*, Belvedere, Vienna, 2014–15

TAX(payer) / STEUER(zahler)

2009	6 jute bags, yarn, toy figurines, glass basin, water

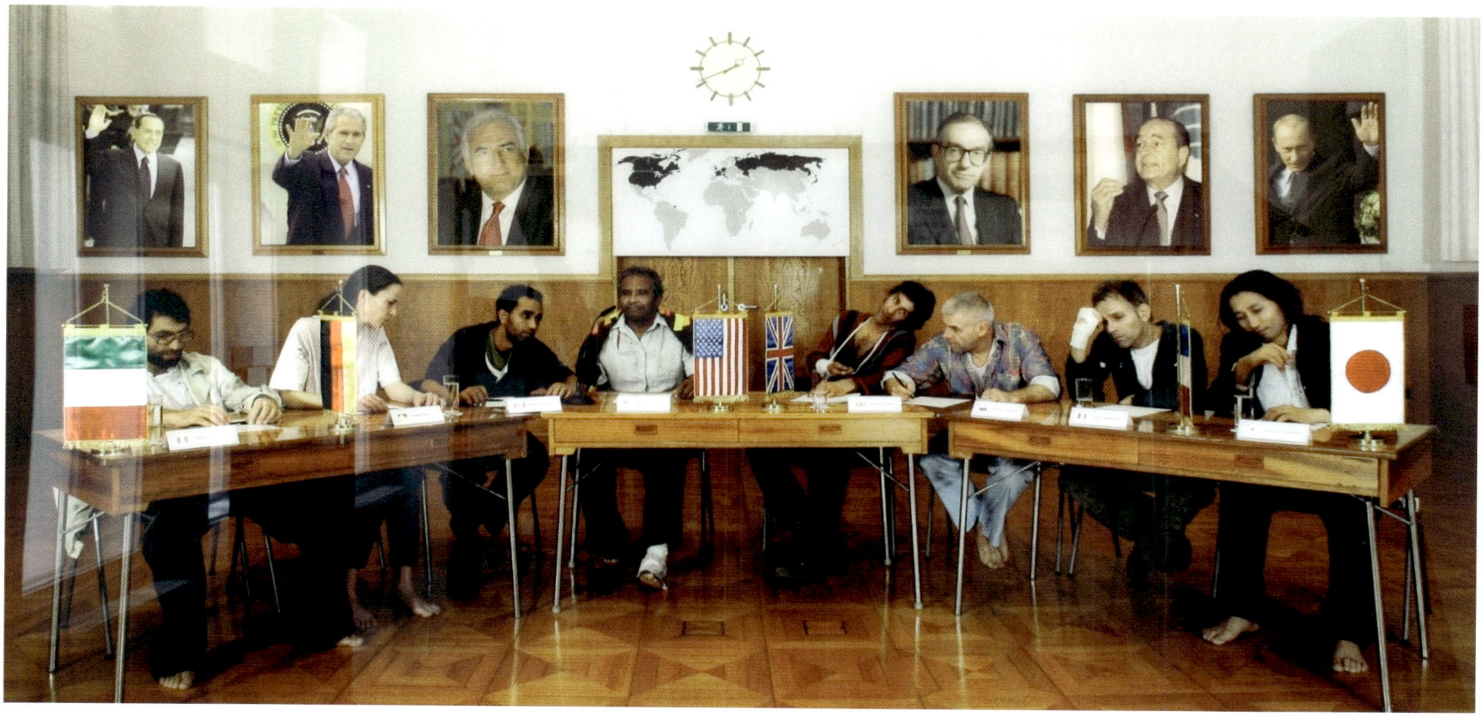

G 8 Summit / G-8 Gipfel

2009 — photograph on Alu-Dibond, with UV-protective film, behind glass, 100 × 200 × 0.5 cm

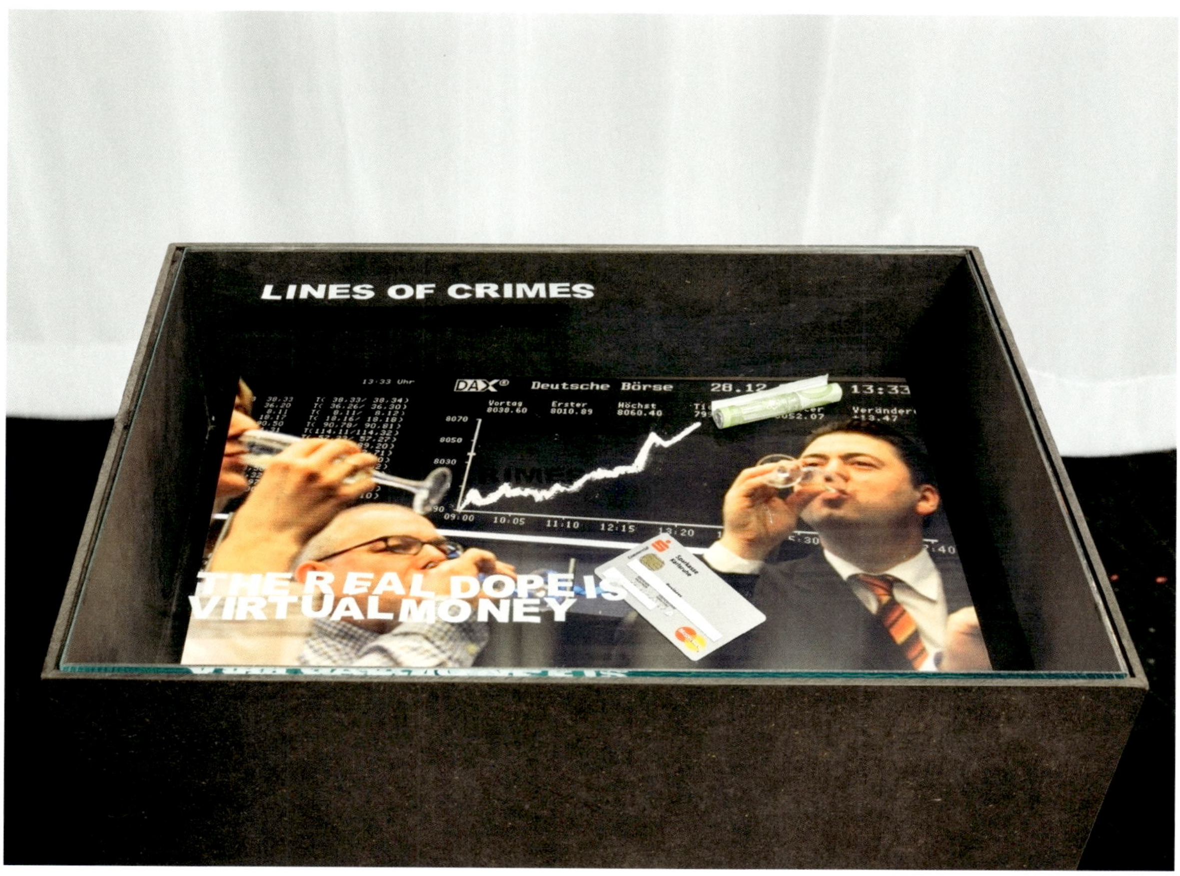

Lines of Crimes: The Real Dope Is Virtual Money

2009 — digital print, polystyrene block, powder, credit card, banknote, glass plate, pedestal

Installation view
respectively, Peter Weibel,
ZKM | Karlsruhe, 2019–20

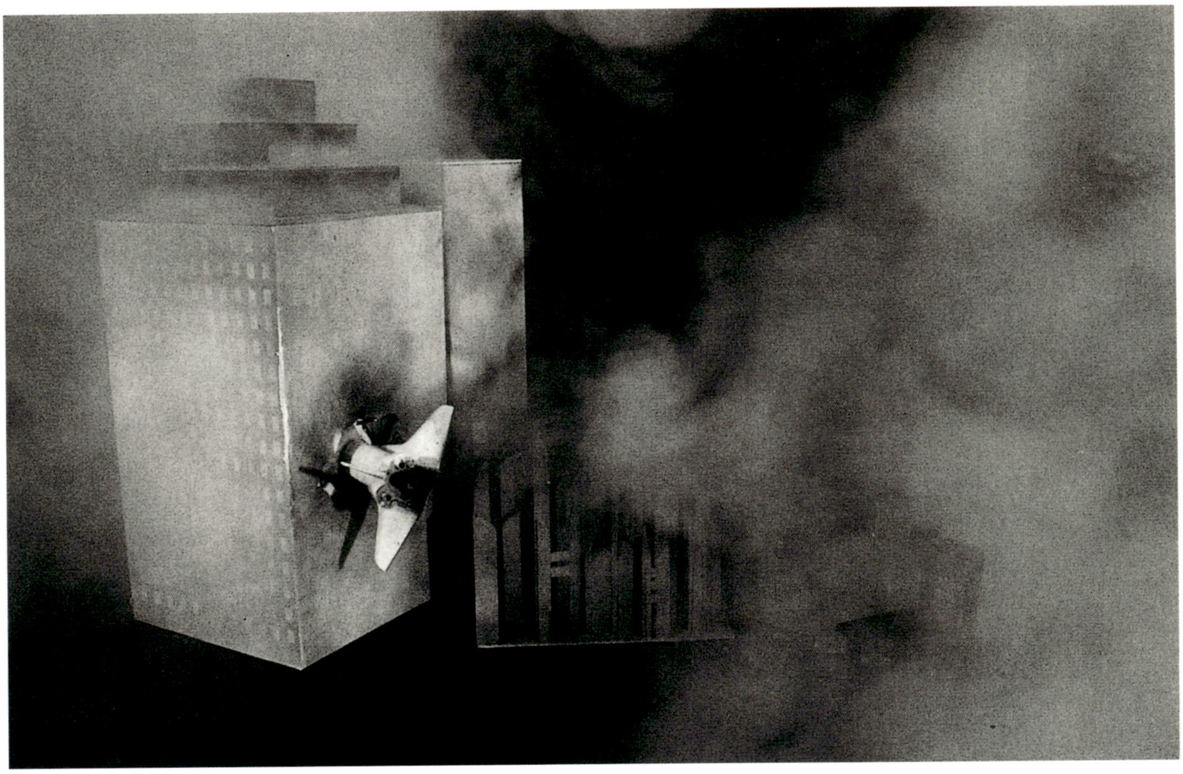

This motif was also used in the film *Chants of the Pluriverse* (1986–88, see pp. 304–307 in this volume).

Installation view *respectively, Peter Weibel*, ZKM | Karlsruhe, 2019–20

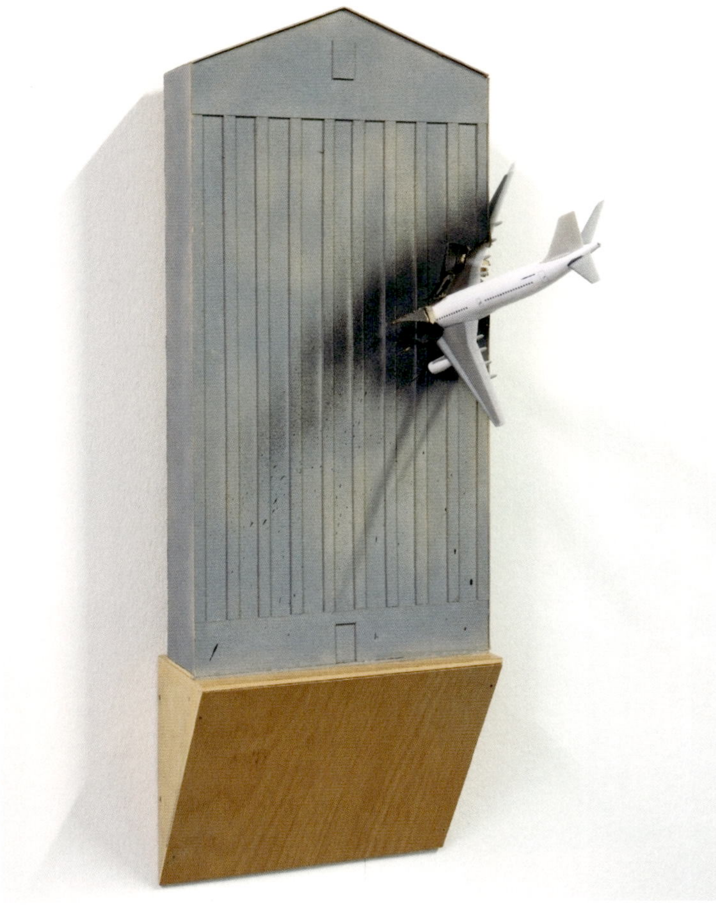

Sculpture in Public Space / Skulptur im öffentlichen Raum

1984	model, toy airplane, wood, paint, 80 × 40 × 40 cm

Peter Weibel's *Sculpture in Public Space* depicts an airplane crumpled against a skyscraper, as though it had flown into it. The idea for the work, which exists only as a model, arose out of Weibel's observations of air traffic over several decades and his conviction that someday, intentionally or not, an airplane would fly into a tall building. Especially in the context of the many hijackings that took place between 1967 and 1976, the artist intuited that a deliberate crash into a building – as a form of extortion, for example – was highly likely.

The piece was created in 1984 as a submission to a public art competition. Weibel's participation in this competition may be put down to his rejection of the rituals surrounding sculpture in public space: even in the early 1980s, his frame of reference for thinking about public sculpture was not merely local, but global and orbital, as can be seen in works such as *The Globe as Suitcase* (*Die Erdkugel als Koffer*, 2004, see p. 435 in this volume). This is the context in which *Sculpture in Public Space* may be understood.

Against the backdrop of the terror attacks of September 11, 2001, in which two airplanes were flown into the World Trade Center's twin towers, the work can also be regarded as a shocking foreshadowing of the iconic media images of that event.

Translated from the German by Patrick Hubenthal.

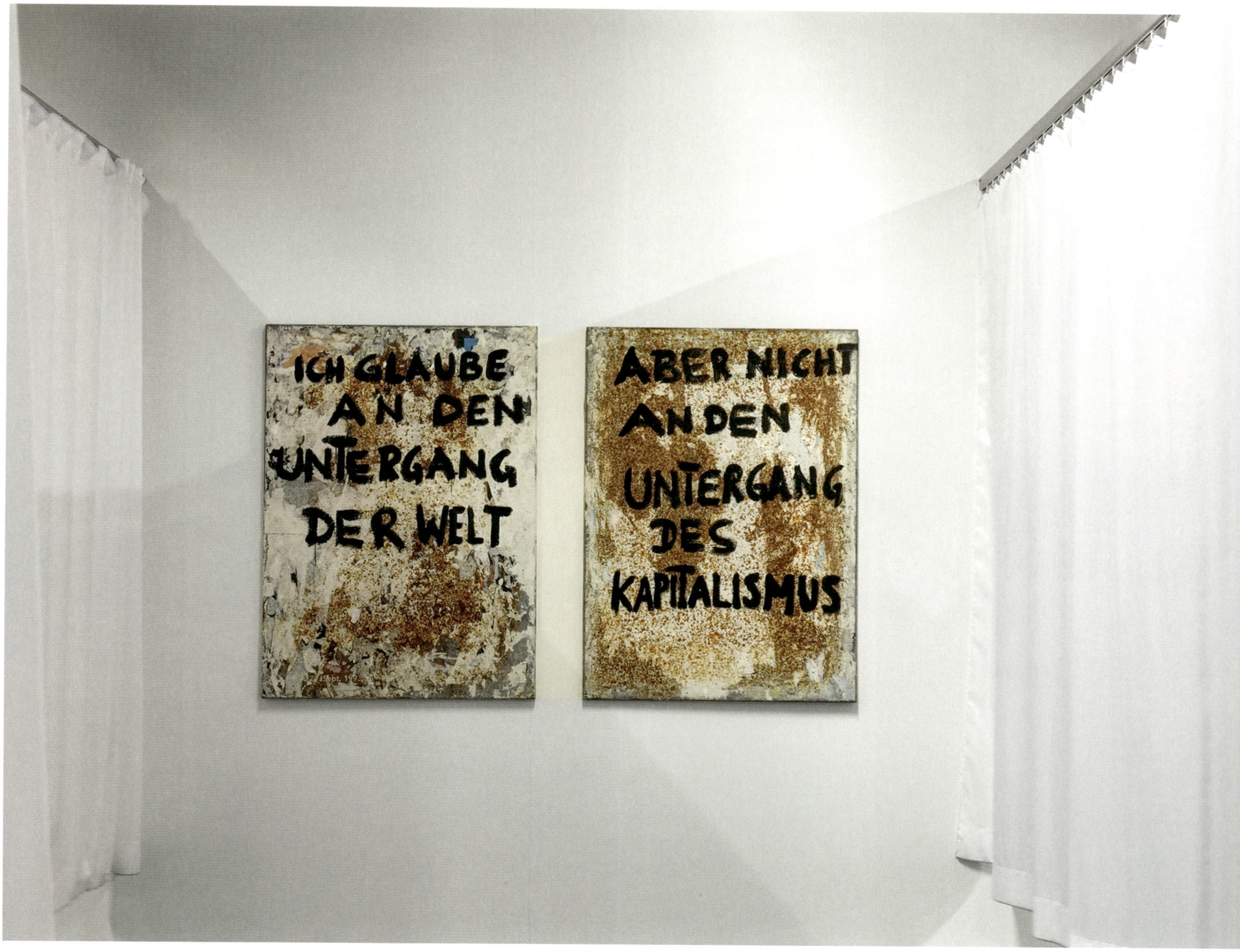

Installation view
respectively, Peter Weibel,
ZKM | Karlsruhe, 2019–20

I Believe In the End of the World ... ... But Not the Downfall of Capitalism / Ich glaube an den Untergang der Welt ... ... Aber nicht an den Untergang des Kapitalismus

1993 | 2 iron plates, paint, 120 × 84.5 cm

This work is a paraphrase of two poems by Girolamo Savonarola. The first was *De ruina mundi or Finis mundi* (1472). This poem about the end of the world was loved and praised by the Catholic Church, because apocalyptic visions drive people into the arms of the Church and today voters into the arms of political parties. The end of the world was welcome. But three years later Savonarola published a second poem under the title *De ruina ecclesiae or Finis ecclesiae* (1975), that is, about the end of the Church. This was too much darkness and apocalypse. The world may end, but never the Church or capitalism. Savonarola was publicly burned at the stake.

We still have no true democracy, but that is no reason to give up on it. We must first reach it, fight for it. At the moment, news bulletins of catastrophes are judged only by their entertainment value. Ideologies are the new religions, questions of faith. That is why in 1993 I created a two-part art work. *Ich glaube an den Untergang der Welt* ("I believe in the end of the world") is written on one large, rusty plate of iron. The second plate reads: *Aber nicht in den Untergang des Kapitalismus* ("But not in the end of capitalism"). People believe that we can save the human community without changing the structures and systems which underlie it. —Peter Weibel, 2022

Translated from the German by Peter Dahm Robertson.

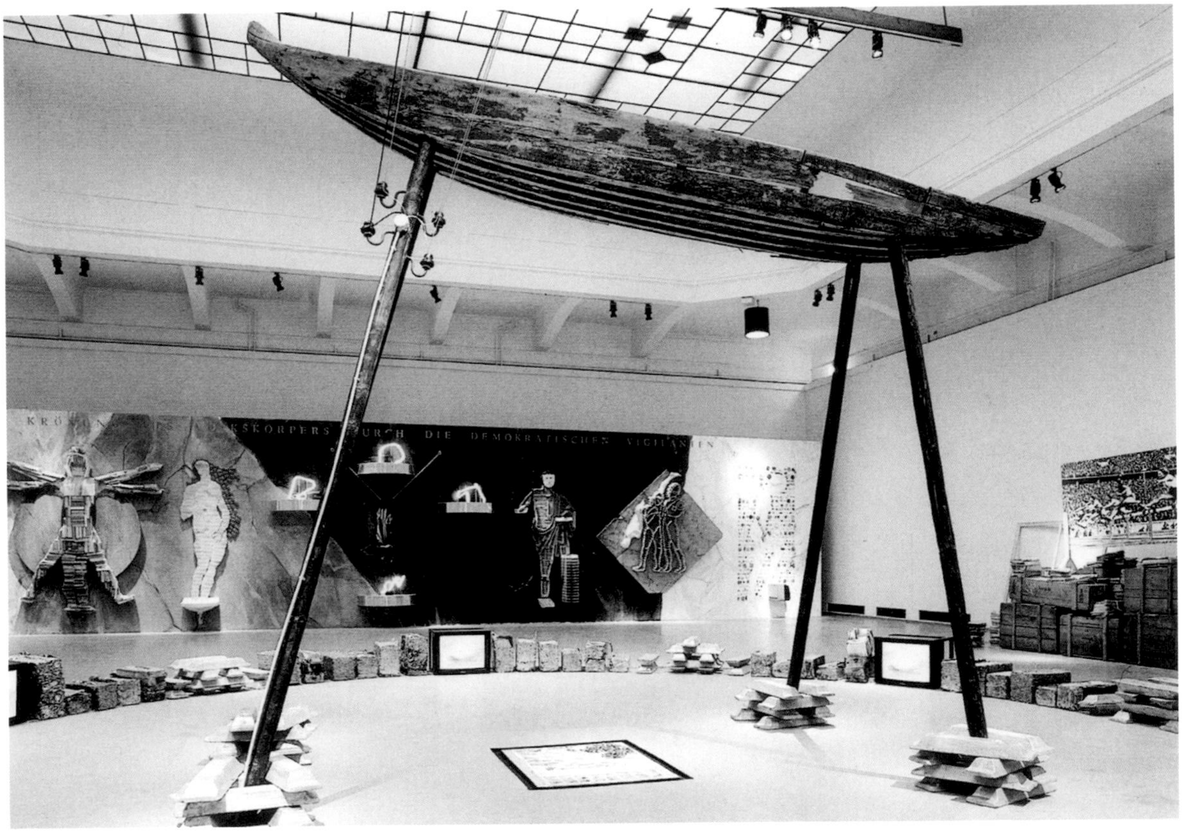

Installation views
Inszenierte Kunstgeschichte, MAK – Museum of Applied Arts, Vienna, 1988–89

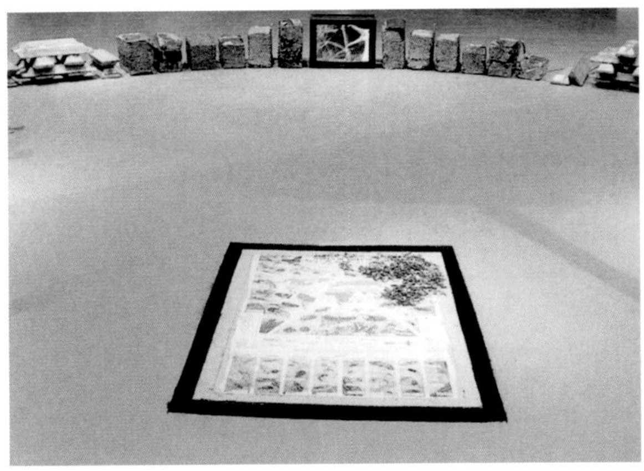

The energetic costs of civilization and its impending collapse as a result of the ecological crisis are visualized in the form of a boat balanced on telephone poles. Monitors set into a circle of aluminum replicas of railroad ties show videotapes of animal species threatened by extinction. The map beneath the raft is surrounded by a frame etched into the stone floor. This frame is filled with water, creating the impression of an island. Humanity is floating on a raft, but we don't realize it.

—Peter Weibel, 2022

Raft / Floß

1988 | boat mounted on telephone poles, wires, aluminum, map, water, monitors

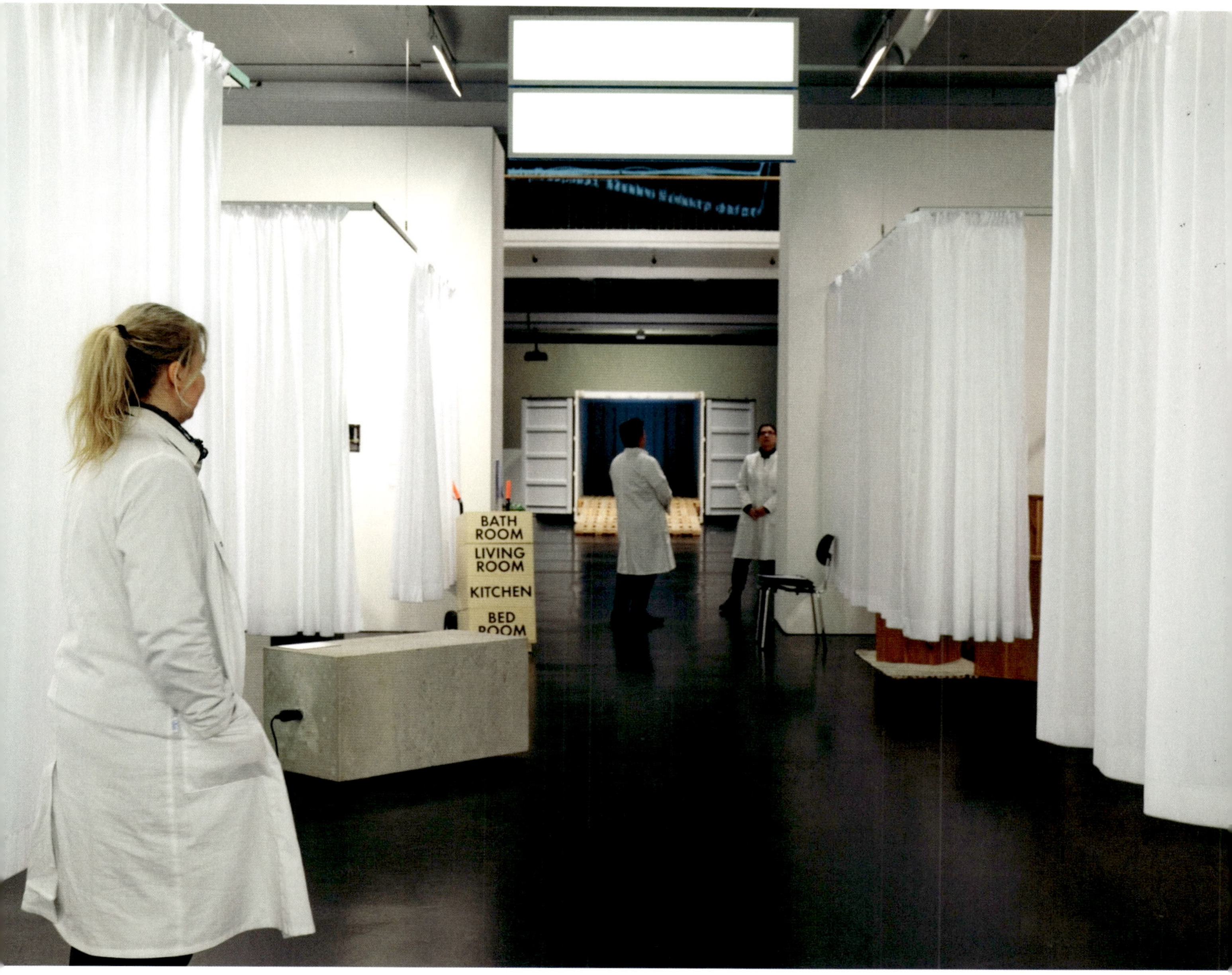

Installation view
respectively, Peter Weibel,
ZKM | Karlsruhe, 2019–20

Peter Weibel's installation *Ward W: The World's a Hospital* was created during preparations for *respectively, Peter Weibel* (ZKM | Karlsruhe, 2019–20). It is an illuminated sign that hangs from the ceiling and consists of two light boxes. The boxes have aluminum frames and are covered in a green plastic film with the words "Station W" and "Die Welt ein Krankenhaus" cut out of it. The work thus calls to mind exit signs and informational signage in hospital wards.

The work is part of a scenographic context depicting the world as a hospital: the space in which it was shown resembled a field hospital, with gallery attendants dressed in white lab coats and white curtains separating the different areas. This "infirmary" held up a mirror to a world unsettled by crises, fixed in the artist's diagnostic gaze. The socially critical works that appeared in this exhibition alongside *Ward W* functioned as examining rooms for the current state of the world, dealing with themes such as genocide and the climate crisis. In view of the coronavirus outbreak, which grew into a global pandemic during the exhibition's run, triggering Germany's first lockdown just a week after the show closed, the conceptual framework of the hospital also became immediately relevant and thus intensely topical; looking back, it now takes on the character of a prophecy.

Translated from the German by Patrick Hubenthal.

Ward W: The World's a Hospital / Station W: Die Welt ein Krankenhaus

2019	2 LED light panels, lettering, steel mounting brackets, steel cable, ca. 60 × 120 × 10 cm, each panel 30 × 120 × 2 cm

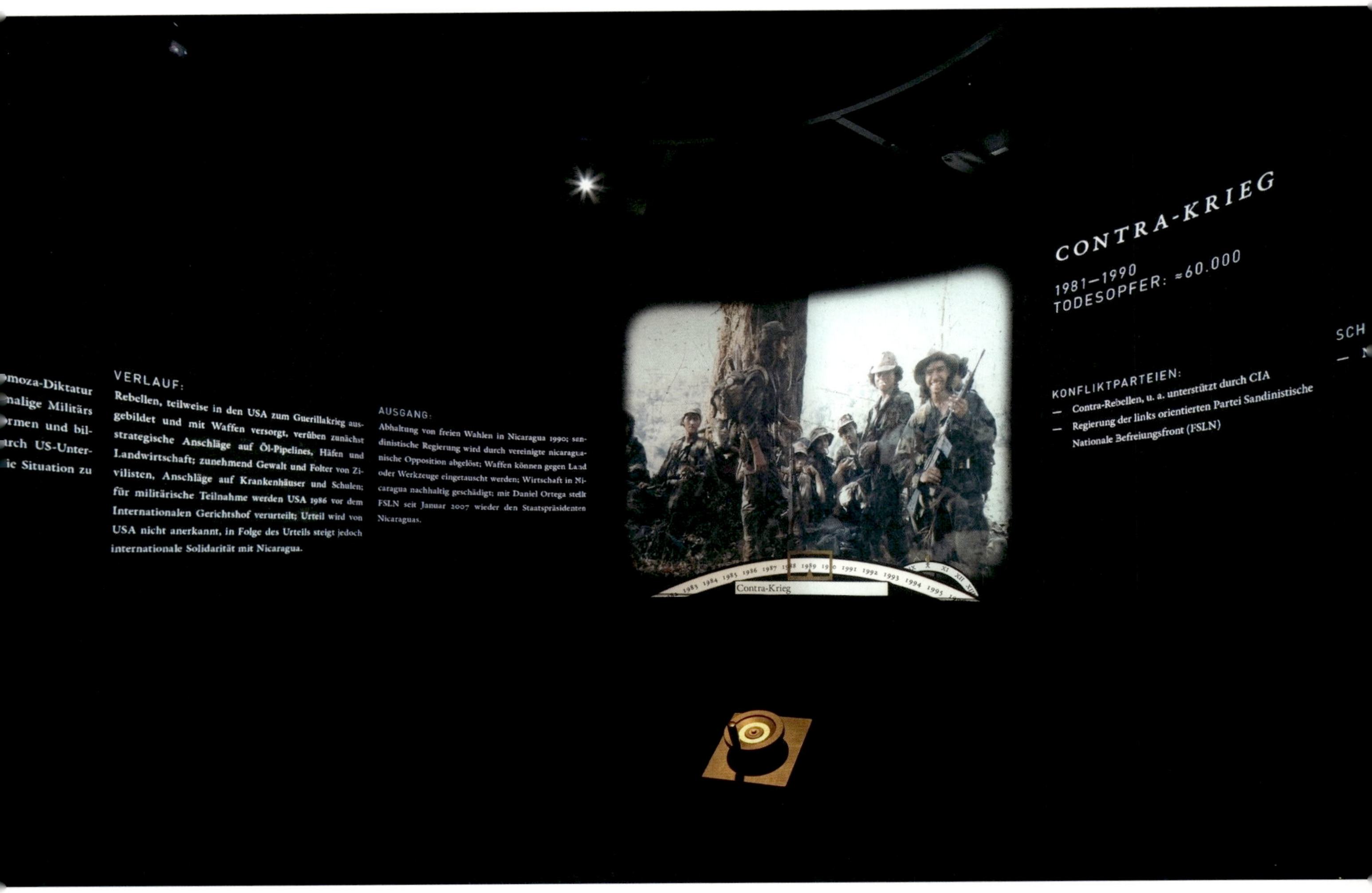

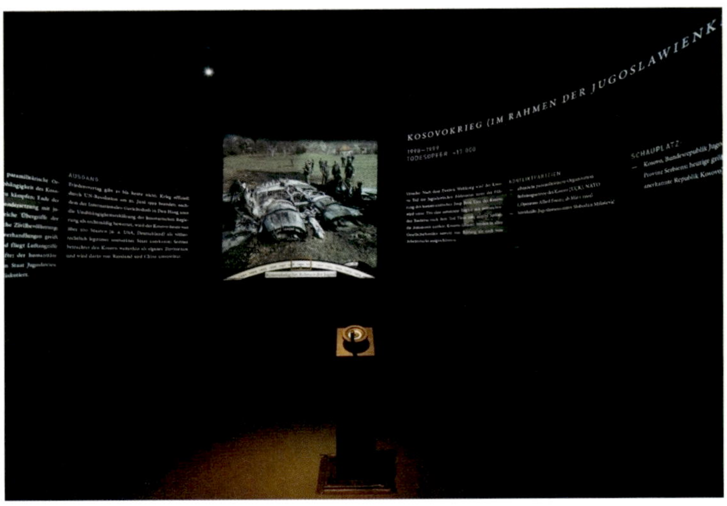

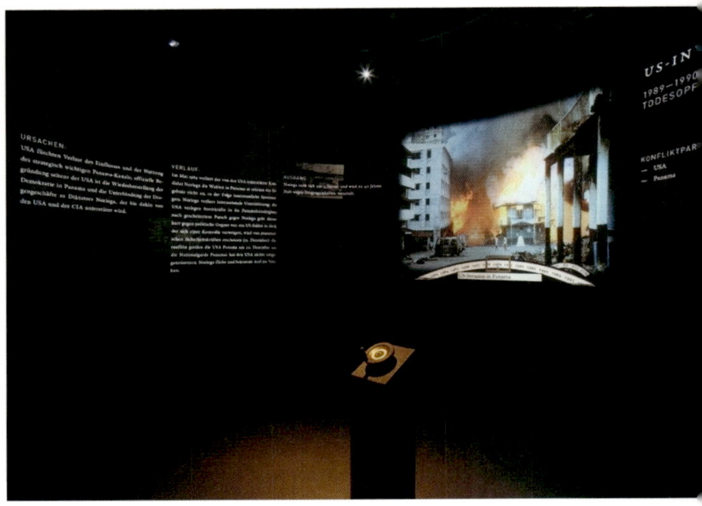

The Coldest Planet in the Universe: The Human Heart. Violence and Genocide in the Twentieth Century / Der kälteste Planet des Universums: Das menschliche Herz. Gewalt und Genozide im 20. Jahrhundert

2015	interactive media installation; panorama screen, panorama display software, crank, LED spotlight research: Antonia Marten, Vera Dietel, Anja Folberth, Adam Rafinski; copy editing: Antonia Marten; programming and graphic design: Nikolaus Völzow, Oliver Wrobel; technique Panorama-Lab: Bernd Lintermann

The installation *The Coldest Planet in the Universe: The Human Heart* w created by Peter Weibel for the ZKM's Panorama Lab as part of ZKM project *GLOBALE* (2015–16) which featured several exhibitions, conce performances, and conferences dealing with the cultural effects of d tization and globalization. The work shows images and texts of arou 250 million political murders (genocides, world wars, massacres, ci wars, assassinations, and terror attacks) from the twentieth century. T chronological overview of bloody events offers the opportunity to beco informed, year after year, about generally known crimes, forgotten crim and repressed memories of crimes and to reflect on how they have be dealt with in politics and society. An interactive control system, a dial o disk, enables the user to scroll through the history of the twentieth centu and call up facts in a panoramic view about the place and time of the eve the parties involved in the conflict, the number of deaths and weapons well as a brief description of the causes and course of events of the crim

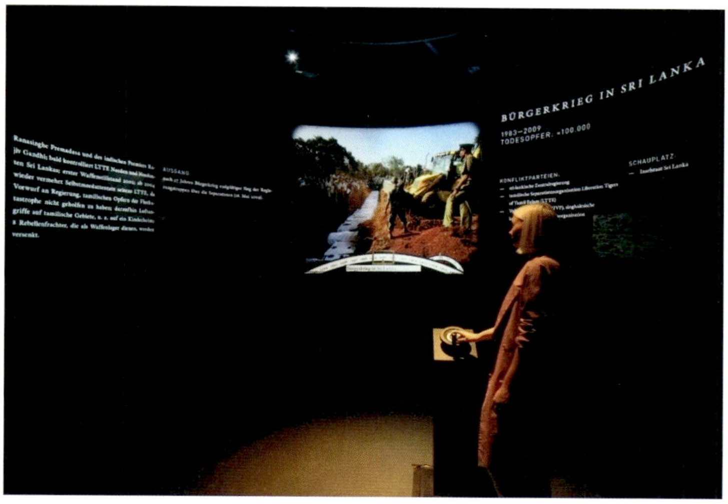

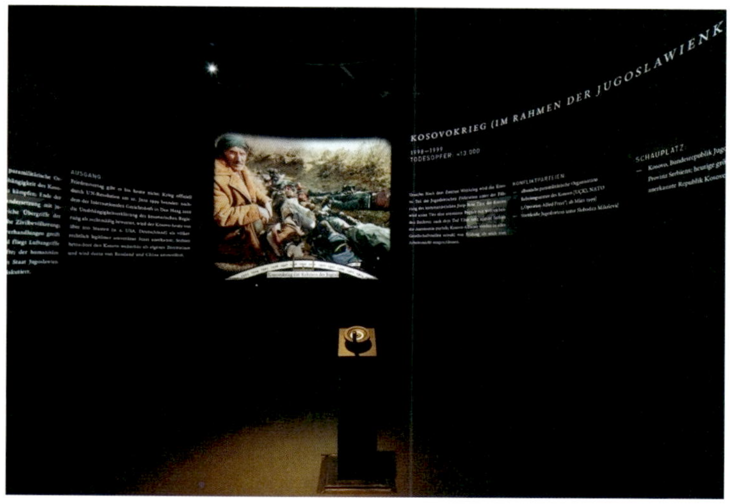

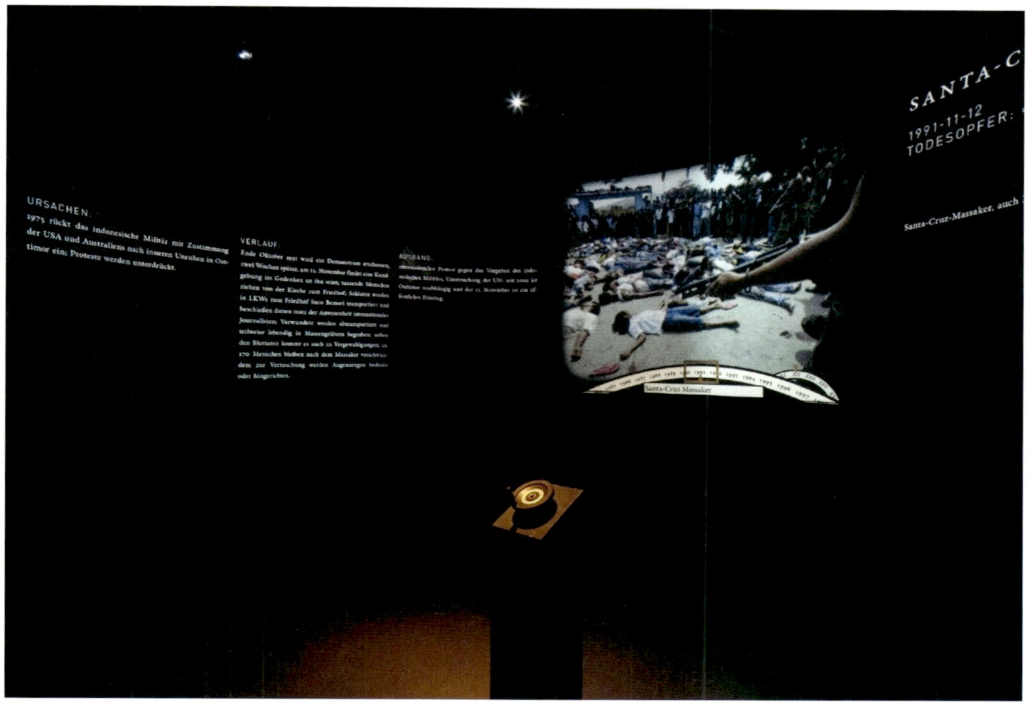

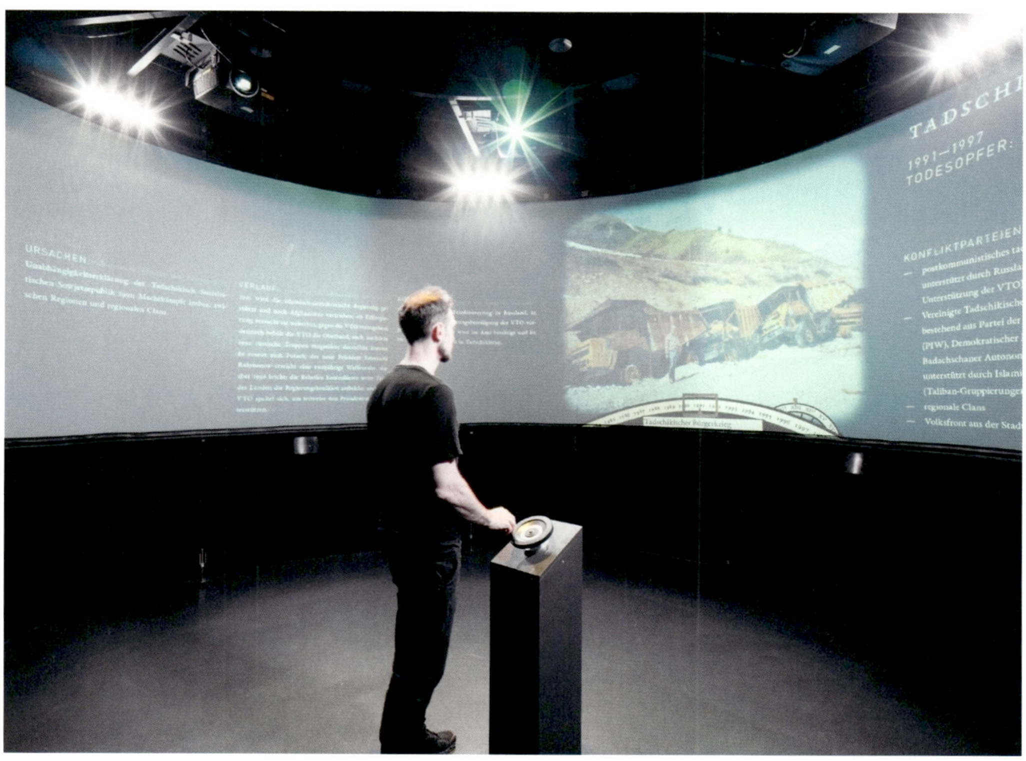

Installation views
respectively, Peter Weibel,
ZKM | Karlsruhe, 2019–20

Actions and Performances

Peter Weibel

Most performance artists developed their actions from painting, or to be more precise, from action painting, to which they also remained visually and conceptually bound. My performances differ from other actions performed in the 1960s through their relation to media and secondly, through their relation to politics. Photography, film and video are media and apparatus that enable a new écriture *corporelle*, a new body politics that is simultaneously a critique of traditional forms of representation as well as identity politics. The body image realized in my actions (in front of the audience or for photography and video) documents the body in the age of its constructability via media and gene technology – not as a matter of nature but as a cultural product – for example, in questions of gender. The body is an alphabet; its organs constitute its letters. In a continuation of Antonin Artaud's differentiation between bodies without organs and organs without bodies, my body image is formulated in a way that later becomes theoretically elaborated in the deconstructive philosophy of Gilles Deleuze and Félix Guattari. My body actions interpret the body against a horizon of transitional objects (Donald Winnicott) and partial drives. As a partisan of partial drives, unlike classical body artists I did not only search for the emancipation of the body, for instance through the sexual revolution. Above all, my idea was not liberation of the body, but liberation from the body through media and technology. My goal was emancipation from the prison of the body, the prison of a society that colonizes the body and holds it hostage. Through my body configurations and body constellations, I intended to transform the natural morphology of vertebrates and their organs into a logical calculus of variables to ultimately deliver humanity from the bondage of the body and of nature in general, precisely through the liberation from the body that appears to us as pure nature.

Note: Adapted from Peter Weibel's jacket text for the audiovisual publication *INDEX EDITION 043: Peter Weibel. Körperaktionen Bodyworks 1967–2003* (Vienna: sixpackfilm, 2018), 20-page booklet and DVD, 91 min.

Peter Weibel: The Master Artificer

Bazon Brock

Georg Wilhelm Friedrich Hegel (1770–1831) introduced the term *Werkmeister* (master artificer) in his *The Phenomenology of the Spirit* (*Phänomenologie des Geistes*, 1807) in order to distinguish himself, with his analysis of the modern version of creation as work, from the fuss about geniuses among the rhymers and rhetoricians of early Romanticism. The master artificer is characterized, on the one hand, by a connection to the creative activists of all earlier cultures, who did not as yet know the role of the autonomous artist and scientist, but relied on *techné*, that is, the knowledge of how something is accomplished. In the ancient world, the patroness of artisans and builders, of writers and sculptors, of preservers of regional heritage and civil diplomats, was therefore known as Pallas Athene in Greece and as Minerva in the Roman Empire. She lent people the strength to finish a task. That is difficult to understand today, when every jack-of-all-trades believes he knows how to rise above Aristotelian dramaturgy. In truth, these pseudo-modernists have merely lost the energy and ability to bring every beginning to an inevitable conclusion that results from it; indeed, it is wonderfully convenient to pass off every inability as disinclination and to attribute haphazard sloppiness to a program of openness and equality. To thunderous applause, the open work of art becomes the postulate, like cosmopolitanism today, for which, revealingly, there is no single, intelligible definition of the term.

The master artificer simply knows what will happen; for instance, if the car mechanic gets a whim to tinker with an electric motor or if mathematicians or physicists justify their work with explanations based on private mythology.

The master artificer is characterized, on the other hand, by legitimization through procedure while at the same time accepting violations of the procedure as an extension of the previously usual process – as far as they are successful. In Heinrich von Kleist's (1777–1811) *The Prince of Homburg* (*Der Prinz von Homburg*, 1822), which was written around the same time as Hegel's *The Phenomenology of the Spirit*, the aristocratic convention of sheer positivity initially seems to be victorious, before, finally, bourgeois morality – as an end that justifies the means – is praised from the depths of a woman's heart.

The master artificer exposes such escapades as a form of intellectual kitsch in which ideologues indulge (Napoleon dismissed intellectuals – that is, Hegelian conceptual workers – as ideologues because they did not bend to his will for power).

The master artificer Weibel is oriented in equal measure around technical ability and the replacement of the theologically based concept of creation by means of organized work. Where is there still room for the new, for the development or progress beyond fate and necessity? The master artificer Weibel knows, not only from his own experience, that one cannot step into the same river twice. At least every thirty years – that is, with every new generation – there is a new take on the old Albrecht Dürer (1471–1528) and the old Rembrandt (1606–69), on Gottfried Wilhelm Leibniz (1646–1716) and Gottlob Frege (1848–1925), on Hermann von Helmholtz (1821–94) and Henri Bergson (1859–1941). The work of these gentlemen has not changed one iota in its material composition, and yet, from time to time, it seems completely new. However, every master artificer knows that, logically, the newest new as something completely unknown is only operative in that it changes our view of the old; that it only gains in importance through its power to change the way we view traditions. This is in line with the media theory of Marshall McLuhan (1911–80), proposed in the 1960s, which states it is only the latest new media that enable us to appreciate the supposedly passé earlier media.

The invention of photography did not make painting obsolete, but it did change our awareness of the specific capability of painting. The invention of film – even of the "talking pictures" – forced us to reevaluate the aesthetic of the still image of photography. Our understanding of electromagnetism increases our respect for the pre-Socratic claim that "The whole world is full of gods," and computer-generated artificial intelligence will clarify the old question of how the physically existing brain can produce ideas (specifically, metaphysics) that can, in turn, determine how the form and hence the achievement of the materially existing world change.

The master artificer Peter Weibel also agrees with other master artificers on how this vast feedback of ideas on physics can be understood as a source of theoretical objects. For the sphere of artistic work, this results in the field of theoretical art. Therein, the master artificer assembles everything that has previously been grasped as a medium in its own autonomous, ontological determination with regard to form and shape; that is, as the nutrient fluid of "autopoiesis," so to speak. However, the theoretical objects are also characterized by "allopoiesis"; that is, the creation of reciprocal connections of meaning that are alien to each other. In the art world, "allopoiesis" has been defined at least since Marcel Duchamp's (1887–1968) lecture "The Creative Act" in Houston, 1957. Duchamp was aiming for the participation of the recipients, the observers, the viewers, and the listeners in the creation of works. "Allo," the Greek word for "other," refers to these other participants.

Theoretical art thus defines these efficacies called "work" that are not owed to the creative artist's central will, but rather result from participation, assistance, cooperation, helping, partaking, and involvement, and are, for that reason, constantly changing. The word "theory" conceals within it the ancient Greek understanding of theater viewers as "theoreticians," and indeed they had to play that role for only their perception and its intellectual processing produced a meaningful coherence of the events onstage. We are all necessarily theoreticians when we grasp in thought a material, physically existing occasion for perception, whether on a stage or on a gallery wall. It is precisely in this context that the phenomenon of *metaphysics* plays a decisive role, which has constituted an integral part of Peter Weibel's work ever since the early 1960s.

In our times, after the experiences of the past and their enduring impact, we should not fall prey again to the myth of unequivocal truths and ideologies promising the cure of all woes. Therefore, since existentially we are thrown back on ourselves and our own, individual interpretation of things, the conception and the understanding of what individuality is attains key significance: it is the prerequisite for the authority of statements by artists and scientists whose arguments are based on thought constructs and findings they have developed themselves, and which are not – as in the past – the result of the decisions of a collective or of invoking some higher power. Without constantly engaging intellectually with the reflecting, acting individuals in the outside world, their statements would be devoid of all validity and their claim of the authority of their authorship would be forfeit.

Metaphysics is in the tradition of conveying this understanding. In its core it addresses the relation between the *res extensa*, as René Descartes (1596–1650) expressed it during the Thirty Years' War, that is, the real, physical world, and the *res cogitans*, our world of ideas.[1] Metaphysics explores the question of how the immaterial sphere (Greek *meta*) impacts the material sphere (Greek *physis*). Both in his artistic and theoretical work Peter Weibel draws on these fundamental considerations, the essential features of which were already tabled by the Pre-Socratics in ca. 500 BCE. If we think back about our life, we notice that our life and our experiences have been affected to a far greater extent by thoughts than by *physis*. Thought processes actually provide our orientation in day-to-day life and are necessary in order to determine the boundaries of what is possible in the context of the general, externally given conditions, so that we can experience our limitations in view of the feedback between the two spheres.

Starting from this, Peter Weibel developed the concept of the *medium*, and confirmed the Ancient World's tenet that the world is full of gods – in the sense that through the proof of the physical existence of ubiquitous, electromagnetic waves, the terms "god" and "artwork" are filled with a new reality that they did not possess before. Weibel configures the space of the invisible and enables us to experience it as verifiably and comprehensibly visible in its role as a medium in his works, frequently with the aid of technological devices – metaphysics par excellence, a translation process of thought into reality. Facilitated by media, in this way the intellectual efforts of single individuals can be linked and assembled into a new collective experience, which everyone – without possessing artistic talent or prior knowledge – can understand and have. The metaphysics of Peter Weibel thus opens doors that did not exist before to a new collective, which unlike in previous centuries is not dependent on banks, churches, or rulers, for instance, but is based on the intellectual engagement and intellectual achievements of individuals. Thus, one could say that, ironically, it is through media art that the old theological justification for the idea of God has been helped to acquire new relevance on the basis of scientific findings pertaining to research on electromagnetic waves.

What links us to our fellow human beings is our ability to engage with things that we do not understand, day in, day out, to have confidence in the world and assert ourselves in it. Even though at times we feel somewhat helpless, fearful, and with no firm ground under our feet when confronted by virtual phenomena and increasingly opaque conditions. Whereas in the past we experienced art in terms of "that painting, hanging on the wall, is a masterpiece and we, as irresponsible minors, must accept this without any further thought," with Weibel, we as the viewers, are given a completely different function: works by the metaphysician Weibel represent for us opportunities for perception, which make considerable demands on our intellectual capabilities because we have to process the statements and methods ourselves. The works challenge us to draw our own insights from them and to define the concept of art ourselves. If we are willing to engage with the works, they offer us the opportunity to become aware of our own cognitive ability, receive confirmation of our individuality, and confidently insist upon it; to realize that the power of our thinking matters; to understand that how what we cannot understand about the world can be utilized in a way that generates communication with others, who like us are confronted with the same uncertainties, anxieties, and doubts. It is negative states such as these as well as the ability to control emotions and endure them that are the decisive qualities of individuals, artists, and scientists, in their role of models instead of shining examples. In the ability not to give way to despair, not to give up, and instead to develop strength, a metaphysical feedback of the *physis* with its own world of ideas, yes, to develop stability as an attitude: in this sense everyone is an artist.

Peter Weibel is the last and first metaphysician whose works seek to serve as the occasion for us to think about and engage with our inner and external worlds, with the connections between different spheres. He is able to process and realize the invisible for us, to generate practically a "real virtual reality" so to speak, in which the virtual, the mental, and the conceptual take on form, and which we can comprehend as viewers and "partners" of the metaphysician, as "theoreticians" – without any prior knowledge. Through Weibel's interdisciplinary work and through confronting the blind spots and imperfections of our cognitive ability, we are enabled to recognize false as false and for a moment be amused by the initially successful deception, by the realization of our own capacity for error. For this demonstration and for imparting these insights we owe an immense debt of gratitude to the metaphysician, the medium Peter Weibel.

Weibel's importance for the establishment of the status of a work as a *fake*, both in everyday subjects and those pertaining to art, science, politics, and economics, is evidenced by very many of his early actions, interventions, and demonstrations. These are designed as exemplary forms of work on the problems of the *verum falsum*, the true and the false. As Karl Popper (1902–94) recognized nearly a century ago, work is only serious work if it tackles what is *a priori* considered as the falsification (not the verification) of imaginative, speculative hypotheses, which are furthermore characterized by the desire to be recognized and assert themselves.[2] As an example of Weibel's artistic practice in this theoretical field, I would like to mention his *Attacks* (*Anschläge*) of 1970–75 (see pp. 536–37 in this volume). When Weibel, for instance, holds up the sign "LÜGT" (LIES) under a police station's sign in the manner of how a referee would do during a sporting event, this is not a work of art in the traditional sense, but it reveals the artist as a medium of connection between two units of meaning: "police" and "lies." By momentarily attaching signs to police stations – or other public institutions as in different actions of this kind – Weibel is committed to finding the truth by detecting the false.

In connection, however, with what the signs can actually accomplish and achieve in politics, the economy, or society, they can be seen as theoretical objects that sometimes, as in the case of protest signs by Joseph Beuys (1921–86) – such as the one stating "*Dürer, ich führe persönlich Baader + Meinhof durch die Dokumenta V*" (Dürer, I will personally guide Baader + Meinhof through Documenta V) of 1972 (see fig. 1) – end up in collections of theoretical art.

These Weibelian actions with their suggested demonstration objects are early criticisms of the misunderstandings over the alleged relation between the sign and denotation, or between the signified and signifier that have been rampant since Ferdinand de Saussure[3] (1857–1913). Signifier and signified can only be distinguished in the unity of the sign. Sign operations, whether as design or constellating order, must therefore be evaluated according to how clearly they are able to differentiate between our intrapsychic operations of perception, of imagination, of classification, of reflecting on our environment, and the outside world. To put it differently: according to how strongly sign operations are capable of bringing the aesthetic difference between consciousness and language/communication, the differences of appearances and characters, or of contents and forms into play. The peculiar status of the demonstration material, which is usually referred to as "model," "teaching material," or "toy" in simulative or experimental processes, becomes evident here. The material, arranged in a constellation formation, becomes the signifier of the "difference," the indicator of the aesthetic, ethical, and epistemological difference. These differences between the intra and extrapsychic world are dictated to us innately, with which phenomena such as optical illusions, ambivalence, ambiguity, as well as the diversity of interpretation such as in the case of confusion of ideas and misunderstandings are also predetermined. They force us, for instance, to ask the question of what is true before eating wild mushrooms, even if we are tempted to do so, for people with experience in this field tell us that, despite an outward similarity between a champignon and a death cap mushroom, only one of the two is edible. We therefore have to be prepared for the epistemological difference between character and appearance at all times. In the mimicry of various creatures, their ability to adjust to different environments, sometimes to such an extent that they become unrecognizable, nature forces us to recognize the ethical difference: the possibility of a deliberately wrong, "lying" allocation of consciousness and language. We have all experienced how thoughts or ideas change as soon as we try to offer them to the perception of others by linguistic con-

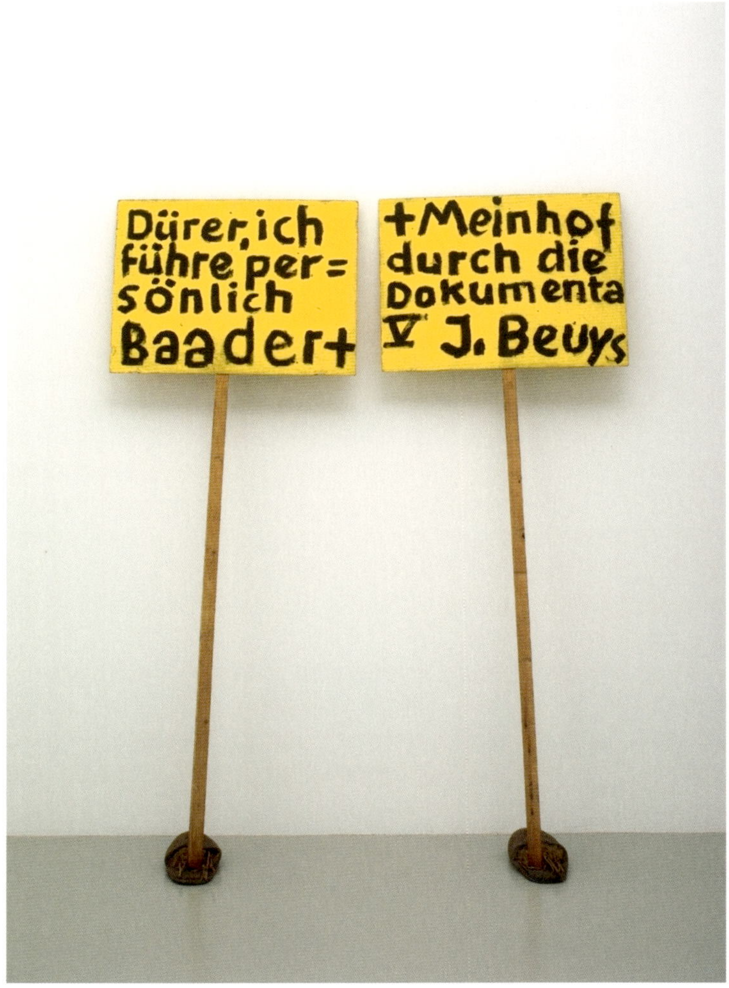

Fig. 1:
Joseph Beuys, *Dürer, ich führe persönlich Baader + Meinhof durch die Dokumenta V*, 1972

cretion. In addition, writers and artists such as Kleist or Picasso have shown us that thoughts and ideas can only gradually form themselves during the process of concretizing them as language, through speaking or scribbling; that thoughts change in the process of their formulation, or metamorphose into completely new thoughts which we only hit upon after several unsatisfactory attempts at formulation. Nor can we ever state explicitly what we feel we know so precisely (not even the texts of laws can be written in a sufficiently unequivocally explicit manner for all those working within the judicature to understand and interpret the law in precisely the same way). Furthermore, our assumed "understanding" is always transformed by other people's "strange" reactions to the very same texts.

There always remains a gap of inadequacy, of ambiguity, of polyvalence, between what we hope to impart to others by putting our ideas into language, and the manner in which these others actually seem to understand us. It is this aesthetic difference that is essential for any communication which refers to ourselves or to others, auto- and allocommunication, to which we innately, on a neurophysiological, biochemical, genetic, and sociobiological level, owe our ability to learn and to adjust.

Weibel repeatedly manifests explicitly these differences (Greek *chorismos*, English *gap*), like when he engraved the word "gemeinsam" (together) on a block of stone, similar to the engraving on a gravestone, in the Viennese gallery Galerie nächst St. Stephan in 1975, but then broke the block in the middle, so that the word "together" was broken in two to visualize the concept of "disunion": the disunion between visual perception and cognitive conception, and thus between language and consciousness (see p. 79 in this volume).

It is a characteristic feature of Weibel's figures of thought that they not only attempt to open up the idea of perceptual concepts to the public within the field of art, but that Weibel also tries to offer the analyses of this fact. Thus showing that the perception of ourselves and of others has always been prestructured by categories which have developed in line with the evolution of the machine responsible for our worldview (brain), and according to which our mind/consciousness is often bewitched by the linguistic output, and we lose ourselves in hopeless paradoxes, tautologies, and literal naiveties.

Since the beginning of the 1960s, formulations of aesthetic, ethical, and epistemological differences have proved to be the most sustainable which use the Latin *inter* as a prefix, as in intermedia, intertextuality, interpretation, intercultural, and of course also interest. *Inter* marks the opening up of relations from differences, for instance, in the form of spacing (such as between figure and ground), as indifference zones (such as in Marcel Duchamp's objects), or as "no-man's-lands" (like the internet as a public space). A good example for such relations resulting from differences that kept occurring in Weibel's work during the 1960s and 1970s was the *Tele-Action The Invisible Borders* (*Die unsichtbaren Grenzen*, 1970, see p. 450 in this volume). In this action, Weibel focuses on the pattern of the boundary present within traffic zones at airports, marks of difference such as the transit area, the duty-free zone, extraterritorial zones, the border police area, passport control, the area marked "Wait here" intended to protect people's privacy, etc. Weibel shows that invisible boundaries also have tangible consequences, and refers to our daily bewilderment in face of the fact that the merely virtual characteristics of what is legal and illegal, what is native and what is alien, what is "we" and what is "they," what is mine and what is yours, by no means remain virtual, but develop a claim to reality, or a power to disturb our intentions that we usually only grant to walls and electric fences.

These irritations occur in daily life because we hardly ever take into account the good old structuralist principle according to which commonality between different parties is produced by their insistence on this difference, and their differences are based on common codes, and organizational or existential forms, for instance, as cultures. Cultures that are proud of their singularity, of their unmistakable characteristics, have in common this claim to their respective singularity. Opposing armies have the no-man's-land between them in common because neither party owns it – in fact, it is precisely this territory that is the object of the war.

Weibel's work in the context of intermedia does not consider the relations between individual media – programmatic at the ZKM | Center for Art and Media Karlsruhe – as a childish "all and more," as an increase through additive accumulation, but neither is it considered a progress-oriented replacement of one medium by another, or as an entertaining motley assortment. Nowadays, we find commercial firms boasting "money and more," "toys and more," "food and more," "travel and more," in every high street. Apparently, the offered products or services are supposed to be more than, or different from, what they are or can be, as they are advertised as toys, food, travels, or banking services. In contrast, it is precisely the fundamental idea of an art or media center to emphasize the individual peculiarities of one medium by comparing it with other media.

Markus Lüpertz (b. 1941) poetically named the peculiar status of a representation of a relation-creating difference *Zwischenraumgespenst* (phantom of intermediate space).[4] Weibel's experimental modulations, his hypothetical architectures and simulations of art works, until recently also struck his contemporaries as such phantoms of intermediate space of relations between arts and sciences, poetry and thinking, humans and gods, past and present, being and having. In its most glorious moments, in moments of far-reaching visibility, not only Weibel, but the entire ZKM | Karlsruhe is perceived as a utopian, achronological phantom of intermediate space.

For the enlightened scientist today, evidence is the only admissible characteristic of truth as a claim to factuality. Those who will not or cannot submit to its claim to truth are thus forced to practice a critique of evidence. This is the true meaning of the *iconic turn*: to take into account the conditions of their reflexive self-criticism even during the stage of image creation.

Weibel appears at his most enlightened in his role of a friend of analysis, when he, as in the video poem *Tritity* (*Tritität*) of 1974 (see see pp. 236–37 in this volume), conveys to the viewers that knowledge gained from intermediality (a painting by Piero della Francesca (ca. 1415–92), a portrait of Nikolaus Lenau (1820–50), and Peter Weibel's video mask), results from the fact that the messages of the individuals relativize each other.

I would like, at this point, to draw special attention to Weibel's intermedium *The Magic Eye* (*Das Magische Auge*, 1969/2019, see pp. 178–79 in this volume). This "magic eye," which splits different levels of sign and meaning, shows a test person with a prototypical expression of a sense of self through awareness of others, exactly as we were shown in the images of torture taken in the U.S. military detainment camp Guantánamo and leaked to the public between 2004 and 2006. The still from Weibel's action shows a man standing with raised hands, which are connected to a circuit board via electric wires; the media splitting of *brain*, *image*, and *person* through diverse media reflections leads, as already implied in the still, to a hood-like shape over the head with signs of "nameless" suffering. The similarity between the Weibelian document of a media-enforced splitting of the person (torture focuses on precisely this destruction, or at least the threat of the destruction of the sense of self) and the so-called new statue of liberty from the torture prison of a U.S. military police brigade is particularly important because of the time difference between the images. Most likely, Weibel's performance at the time was a reaction to the publication of "Vietnam photos" in the tenor of an "infatuated connection between capital and media" that was then quoted everywhere. As Weibel, in 1969, had succeeded in presenting this connection as a function logic adopted by the media in order to produce evidence (thus not blaming it on the will of a few fat-necked, greedy, bloody-minded capitalists), particularly today *The Magic Eye* easily stands the test of providing criticism of the evidence of the images of torture from Iraq. The point is that these are not the cruel, bloodthirsty games of a few manic, antisocial individuals. Based on Weibel's model, the individuals produce those media facts – their dissemination via the internet and discussions in talk shows included – as docile agents of the function logics pertaining to a market economy that is governed by the advertising industry. Those who, for economic reasons, offer reality shows on all afternoon TV channels and even during prime time, would probably view reality shows from the torture cellars of the U.S. military police as a desirable new twist on the program format. After all, the disgust that was proclaimed worldwide unfortunately did not focus on the victims, but was a manifestation of the voyeuristic enjoyment of self-righteous indignation.

The Magic Eye was about the separation of photographic images and sound in a medium after the invention of optical sound. Based on the logic of this procedure, which produces sound from light, Weibel extended the film event in a unity of production and reception by converting light events from the performance room of the gallery into sound, especially those triggered by the audience. This corresponds to the method of the American military police who let their own media amusement become part of torture images. The separation of the image level "torture" through the magic eye of one's own little image machines from the real experience of destruction of all sense of self as an act of psychophysical torture produced a "silent scream," which was apparently heard all the more loudly because it could not be recorded, but only imagined as an experience of self-awareness. *The Magic Eye* awakens the echoes of these silent screams, this "eloquent silence," in the viewers themselves; it becomes the eye of the media monster Polyphem (the specter of multi = polymedia), confronted with whom we can only confess to being a "nobody."

In a major exhibition titled *Inszenierte Kunstgeschichte* (Staged art history) at MAK – Museum of Applied Arts, Vienna (1988), in another action Weibel presented a "great painter who has, however, received hardly any attention so far, because the overpoweringly great and new always receives little attention at first." With institutional and curatorial authority, Weibel compelled the public to follow his guidelines. After he succeeded in directing attention to this new painter, Weibel confessed, without any scorn or mockery, that the work presented was his own, but it was not exhibited in recognition of himself as a great painter; rather, he wanted to draw attention to the institutional, curatorial, and publication methods that the audience traditionally trusts.

Weibel destroys this trust, not as a bad-boy prank, though, but in order to raise a new, much more important question for all interested parties, namely: Why should the exhibited paintings suddenly no longer have any significance as great paintings just because a nonpainter, without further ado, simulated it? How can the knowledge of a deliberate forgery change the significance of a thing if it has not changed in between its presentation and the discovery of the simulation? I refer to this attitude towards the artistic sphere – which I share – as "always fishing for complications."

Working on questions like these with the audience – that is, the theoreticians – thus creates a work of theoretical art. In that sense, such forgeries are among the most important works of theoretical art today; they are not recognized by "experts" as forgeries, but, on the contrary, acknowledged as originals, and were only presented to us as fakes because the forgers revealed themselves as their makers. Forgers marketed their Georges Braques (1882–1963), Pablo Picassos (1881–1973), Henri Matisses (1869–1954), René Magrittes (1898–1967), or Salvador Dalís (1904–89) so brilliantly that no expert could detect the forgeries; these artifacts were openly labeled as "fake" because they were in fact ingenious paintings in the spirit of Braque, Picasso, and so on.

For decades, so-called appropriation artists around Elaine Sturtevant (1924–2014) have worked on such issues as painters. Until their arrival, an original painting seemed to possess an aura of uniqueness because its pictorial qualities could not be rendered by any reproduction technique. The appropriationists stripped or dispossessed these unique paintings of their aura, thereby raising a number of fundamental questions: Why is a work regarded as a forgery if it cannot be recognized as one even by experts, and was only exposed either by accident, or by the forgers themselves? Should not those forgeries that were declared masterpieces receive the highest recognition, and fetch the highest prices? What does it mean that, for instance, our prevailing perception of Greek art has to be recognized objectively as a forgery? (The fact that sculptures and architecture were originally colored even in the classic age of the fifth and fourth centuries can no longer be denied; however, due to our understanding of the classics to date, we still perceive this fact as "wrong.")

Unfortunately, this movement has so far had no influence on the market-driven terminological teeter-tottering and phrasemongering in the traditional art business. Theoretical physicists and theoretical mathematicians have, admittedly, gotten much further with this. The experimenting practitioners have granted them to assess whether the conclusions drawn from experiments are sensible or tenable at all, or if they are purely evidence-based, terminological sorcery that, at best, stimulates criticism of the contradiction between what one knows and what one sees as cognitive dissonance. In that respect, it is certainly understandable and acceptable that the master artificer Weibel is interested in mathematics as well as in physics and electronics, in genetics as well as in research on artificial intelligence. In those fields, nothing is considered great simply because it claims the label "scientific:" a large number of the scandals around fraud and cheating in science today were – and are – launched by the scientists themselves. Equally, in the field of the arts, something can no more assert a claim to validity simply because it bears the propagandistic label "art."

For a long time, artists have been orienting the form of their works towards initiating communicative processes. The material work as an abandoned appliance for the production of distinctions in the constellating order as shaping of material, preserves the possibility of updating the rites and liturgies at any time. The predominance of the art forms performance, action teaching, happening, agitation, and actionist demonstration proves the ritualization of artists' work, which commenced at the latest with Dada and the Surrealists' soirées. In this context, the concept of *Gesamtkunstwerk* in the Bayreuth vein was brought back to life, which consolidated all media and all forms of perception with the aim of endowing the most comprehensive liturgy of all historical times, together with the institution initiated by the artist himself. The practice of faith became the practice of art. Artists who had been molded by a strict Catholic upbringing knew how to present art as the civil religion of society that the latter had been searching for – also in the field of culture – since the age of Enlightenment and the process of secularization.

Weibel, for one, is certainly not planning to stop at the mere conversion of churchly activated sacraments into those that are commonly administered by museums, for instance, on the occasion of exhibition openings, or by scientific institutions when they award honorary doctorates. *Poetae docti* and their useful findings in the fields of technology, architecture, medicine, the military, and historiography, have been known in great quantities since the Renaissance. They all pride themselves on at least denouncing, if not defeating, *pro bonum contra malum*, the wrong and the evil, the warlike, the addictive, and are rewarded for it. *Pro verum contra falsum*, they exercise the criticism of ideology as an ideology; hunt down the fascist, racist, anti-Semite in everyone else, and are committed to the truth, and allegedly not to other interests; they are only keen on basic research, not on profane advantages. Faced with all this self-elevation of representatives of the truth, it is high time to prove the value of criticizing truth, especially criticizing truth that is fundamentalist in origin and enforced by totalitarian means. Not to shy away from the truth – that is easy for its agents. However, to criticize it, and not give in to it even when it appears in the shape of an evolutionary law, as a finding of geneticists, brain scientists, or sociobiologists, when its origins lie in consensus, and when it can be demonstrated with undeniable evidence – that is Weibel's role, that is the role of all of us.

He demonstrates that he is at the top of his game by managing to portray "science" and "art," the alleged seals of quality, as similarly meaningless or deceptive as many shoppers in supermarkets perceive the labels "organic" or "regional" to be. In this way he promotes clarification as the liberation from naive credulity.

The master artificer Weibel – as a proponent of theoretical art – works at his preeminent institution ZKM towards the goal of, on the one hand, guiding product liability out of the realm of goods, and establishing it in the realms of culture and those of art and science, on the other. For this highly commendable effort, he deserves prizes and awards of the kind that consumer protection and product testing now enjoy. Artists' tinkering with labels, galleries and museums committing prospectus fraud, and the ambition of curators promoted as service to the public, should no longer be tolerated. Thinkers at your service, master artificers at work – with Weibel first in line.

Note: This text is a composite of several texts and orations by Bazon Brock between 2006 and 2020, including "Ex opere operato: The Working Activist and Friend of Analysis Relies on Ingenuity Instead of Creation, on Falsification Instead of Proof of Truth, on Invention as Discovery Instead of Presumptuous Artistic Creation," in Peter Weibel and Harald Falckenberg, eds., *Peter Weibel: Das offene Werk 1964–1979*, exh. cat. Neue Galerie Graz, Universalmuseum Joanneum (Ostfildern: Hatje Cantz, 2006), 953–60, and "Peter Weibel, der Werkmeister," in *Peter Weibel: (Post-)Europe? Lovis-Corinth-Preis 2020*, exh. cat. Kunstforum Ostdeutsche Galerie, Regensburg, ed. Agnes Tieze (Berlin: Hatje Cantz, 2020), 102–4. The latter catalog accompanied the exhibition *Peter Weibel: (Post-)Europe? Lovis-Corinth-Prize 2020*, which was on view at Kunstforum Ostdeutsche Galerie Regensburg from October 3, 2020 to January 31, 2021. Translated from the German by Steven Lindberg, Y'plus, and Gloria Custance.

[1]
See René Descartes, *Principles of Philosophy*, trans. R. P. Miller (Dordrecht:Springer Netherlands, 1982). First published as *Principia philosophiae* (Amsterdam: L. Elzevir, 1644).

[2]
See Karl Popper, *The Logic of Scientific Discovery: On the Epistemology of Modern Natural Science* (Berlin: Springer, 1959). First published as *Logik der Forschung: Zur Erkenntnistheorie der modernen Naturwissenschaft* (Vienna: Springer, 1934).

[3]
See Ferdinand de Saussure, *Course in General Linguistics*, ed. Charles Bally and Albert Sechehaye, subsequent eds. Perry Meisel and Haun Saussy, trans. Wade Baskin (New York: Columbia University Press, 2011). First published as *Cours de linguistique générale* (Lausanne: Payot, 1916).

[4]
Zwischenraumgespenster is also the title of a publication about Lüpertz's series of 10 lithographies of the same name, see Markus Lüpertz, *Zwischenraumgespenster* (Berlin: Galerie Springer, 1986).

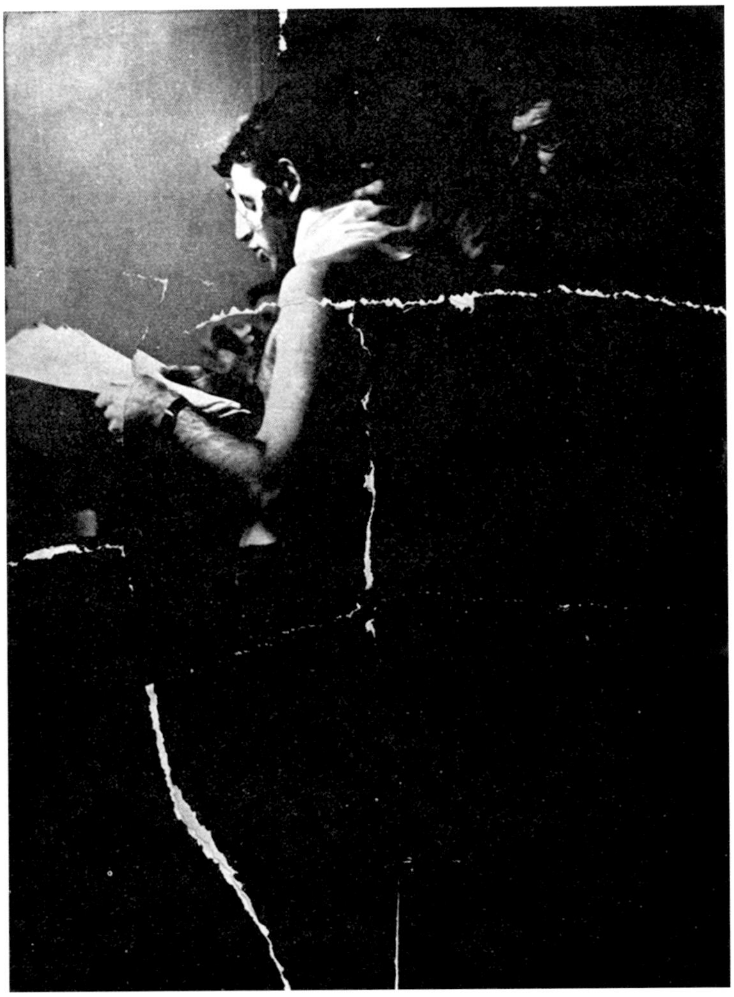

Peter Weibel and Otto Muehl during the performance

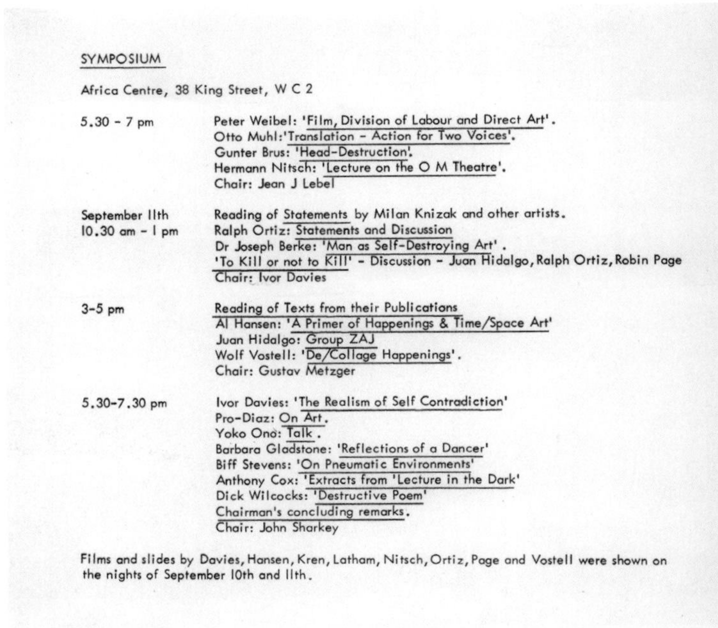

SYMPOSIUM

Africa Centre, 38 King Street, W C 2

5.30 - 7 pm	Peter Weibel: 'Film, Division of Labour and Direct Art'. Otto Muhl:'Translation - Action for Two Voices'. Gunter Brus: 'Head-Destruction'. Hermann Nitsch: 'Lecture on the O M Theatre'. Chair: Jean J Lebel
September 11th 10.30 am - 1 pm	Reading of Statements by Milan Knizak and other artists. Ralph Ortiz: Statements and Discussion Dr Joseph Berke: 'Man as Self-Destroying Art' . 'To Kill or not to Kill' - Discussion - Juan Hidalgo, Ralph Ortiz, Robin Page Chair: Ivor Davies
3-5 pm	Reading of Texts from their Publications Al Hansen: 'A Primer of Happenings & Time/Space Art' Juan Hidalgo: Group ZAJ Wolf Vostell: 'De/Collage Happenings'. Chair: Gustav Metzger
5.30-7.30 pm	Ivor Davies: 'The Realism of Self Contradiction' Pro-Diaz: On Art. Yoko Ono: Talk . Barbara Gladstone: 'Reflections of a Dancer' Biff Stevens: 'On Pneumatic Environments' Anthony Cox: 'Extracts from 'Lecture in the Dark' Dick Wilcocks: 'Destructive Poem' Chairman's concluding remarks. Chair: John Sharkey

Films and slides by Davies, Hansen, Kren, Latham, Nitsch, Ortiz, Page and Vostell were shown on the nights of September 10th and 11th.

On June 11, 1965, the International Poetry Incarnation, billed as "Poets of the World – Poets of Our Time," took place at the Royal Albert Hall in London, organized in part by the staff of the bookstore Better Books, Britain's hub for underground and experimental literature, film, and happenings. Readings were given by Allen Ginsberg, Alex Trocchi, Lawrence Ferlinghetti, Pete Brown, Gregory Corso, Simon Vinkenoog – and the Austrian concrete poet Ernst Jandl. A huge popular success, drawing an audience of about seven thousand, the event was documented by Peter Whitehead in his 1966 film *Wholly Communion*. Jandl was already highly regarded as a poet in England thanks to the publication of *mai hart lieb zapfen eibe hold* (London: Writers Forum, 1965) and had become well known through his readings at Better Books.

When Gustav Metzger and others were organizing the Destruction in Art Symposium (DIAS), to be held in London on September 9–15, 1966, they asked Jandl about others similar to himself in terms of radicality. As a close friend of mine, Jandl recommended me, and I recommended my comrades at the Institut für direkte Kunst (Institute for Direct Art), which had been founded that June by Otto Muehl and Günter Brus. And so Brus, Muehl, Kurt Kren, Hermann Nitsch, and I traveled to the now-legendary symposium, whose other participants included Al Hansen and Yoko Ono.

On September 10, I gave the lecture "Film, Division of Labour and Direct Art" at the Africa Centre. On September 15, a "Simultaneous Happening" took place there, featuring Brus, Hansen, Nitsch, Muehl, and myself. My contribution was *Action Lecture No. 1*: while pop music issued from a tape player and a record player, and various movies – including my film *Welcome* (1965, see p. 168 in this volume) as well as a porno flick and footage of an operation – were projected from portable projectors into the audience, into the room, and onto my half-naked body, I read aloud from a handwritten English manifesto titled "Proposals on Non-Affirmative Art." Meanwhile Muehl was at work on my lecture as well, tearing the pages out of my hand as I read them. He also removed more and more of my clothes and bandaged me with a roll of paper, creating montage effects in real time. At intervals I tossed the crumpled pages into the audience, while Muehl gradually dragged me to the floor. As I read the last word on the last page while lying on the floor, he concluded his bandaging by taping my mouth shut.

—Peter Weibel

Translated from the German by Patrick Hubenthal.

Action Lecture No. 1: Proposals on Non-Affirmative Art

1966	multimedia event, Expanded Cinema, with the participation of Otto Muehl, DIAS, London, September 15, 1966 See also pp. 168–69 in this volume

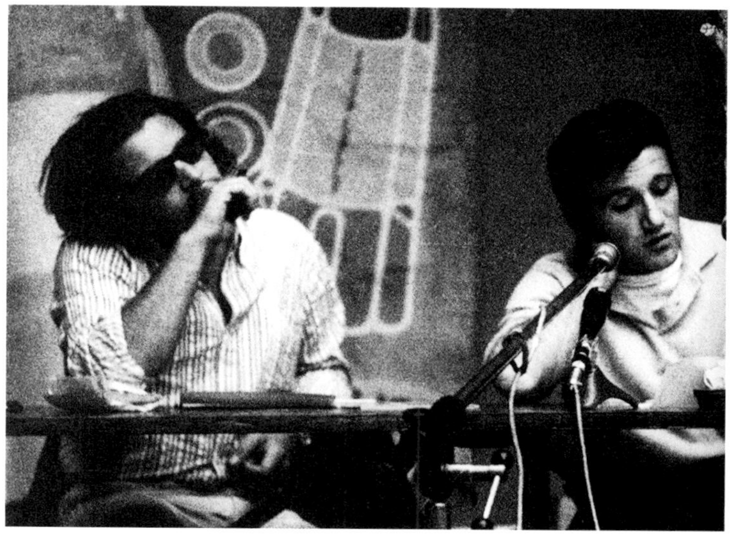

Peter Weibel (right) during his lecture "Film, Division of Labour and Direct Art" with Jean-Jacques Lebel (left)

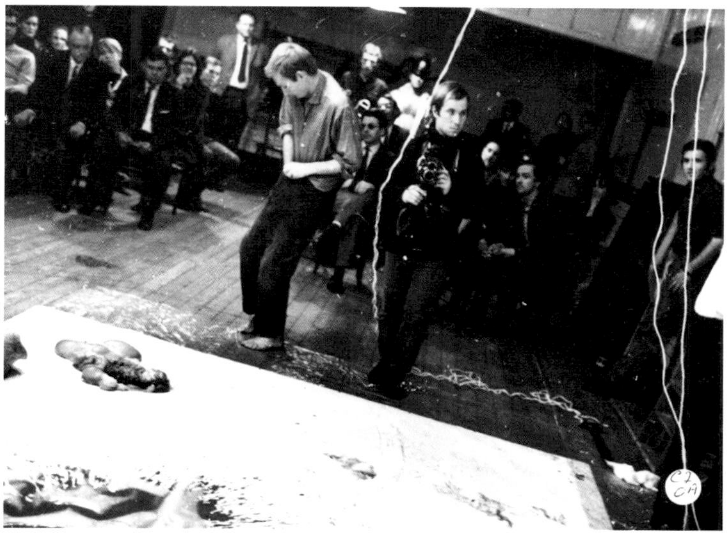

Herman Nitsch, *Crucifixion of a Lamb*, action at St. Brides Institute, Sept. 16, 1966; from left: [unidentified], Günter Brus, Peter Weibel (leaning against the wall)

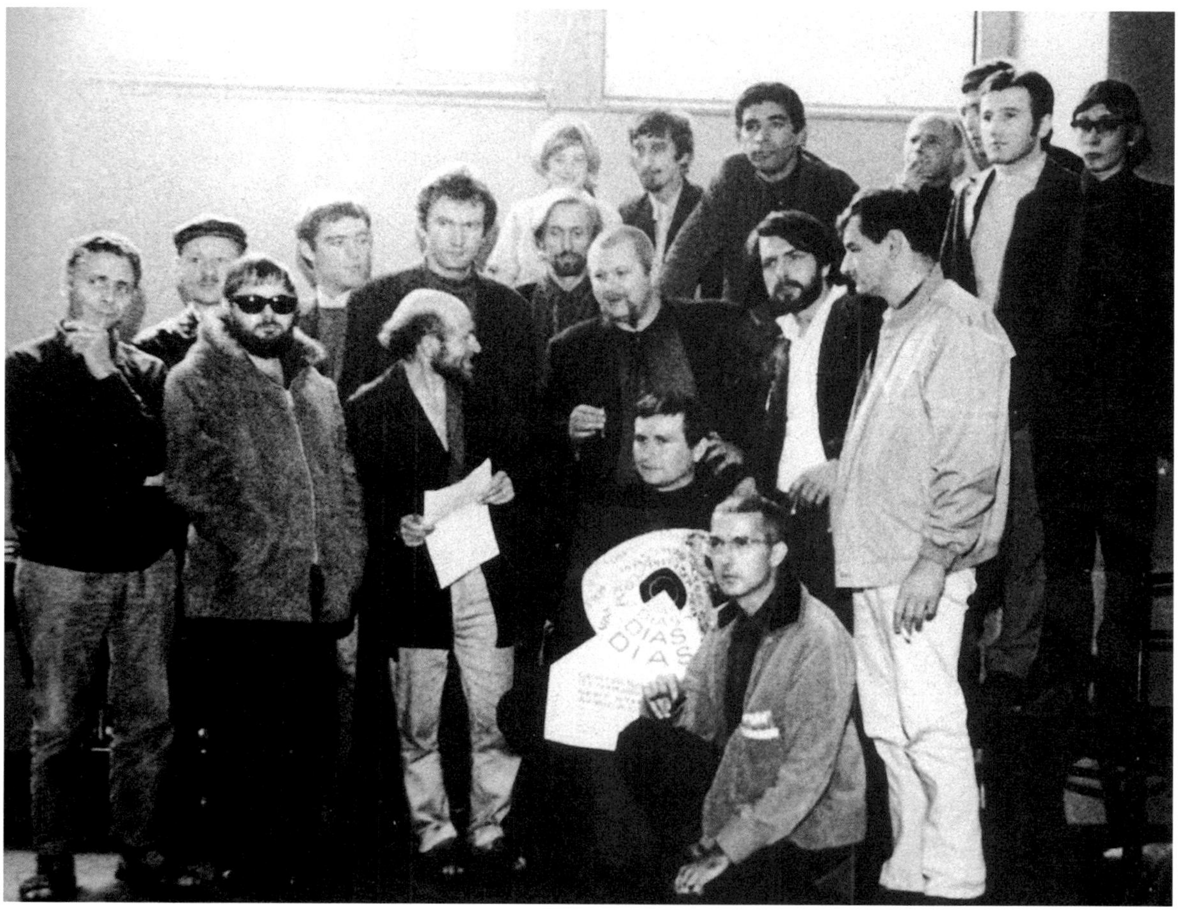

Participants of the Destruction in Art Symposium, from left to right:
first row: Henri Chopin, Jean Toche, Gustav Metzger, Wolf Vostell, Hermann Nitsch (with DIAS-poster), Juan Hidalgo (kneeling), Robin Page, Otto Muehl;
second row: Werner Schreib, John J. Sharkey, Ivor Davies, John Latham;
third row: Susan Cahn, John Sexton, Rafael Montañez Ortiz, Kurt Kren, Ener Donagh, Peter Weibel, Bryant Patterson

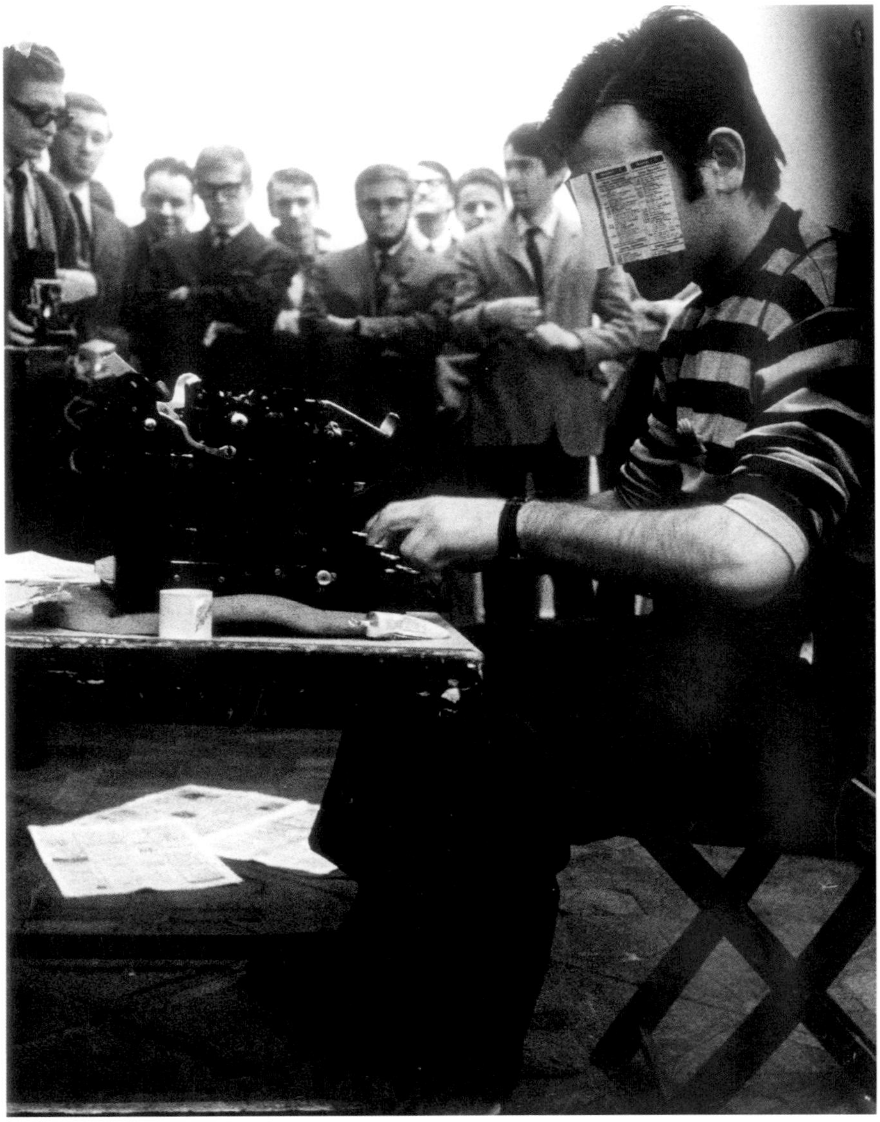

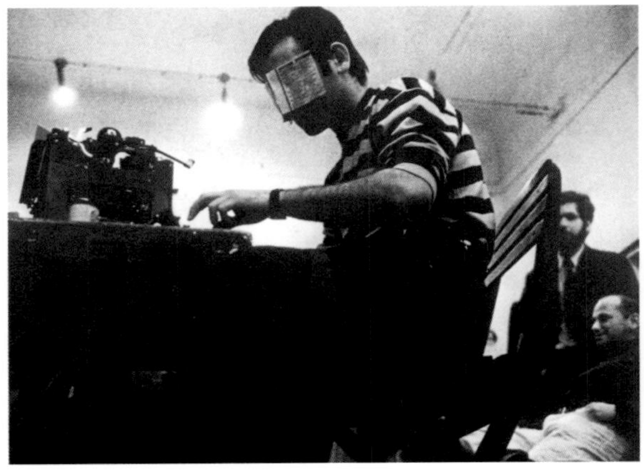

Rudolf Schwarzkogler (standing) and Kurt Kren (sitting) in the back

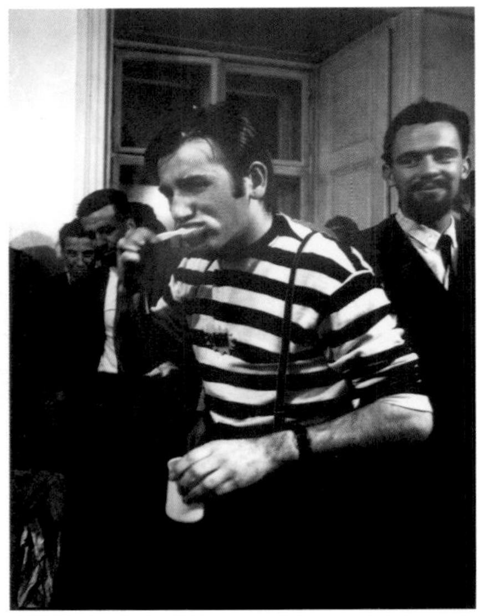

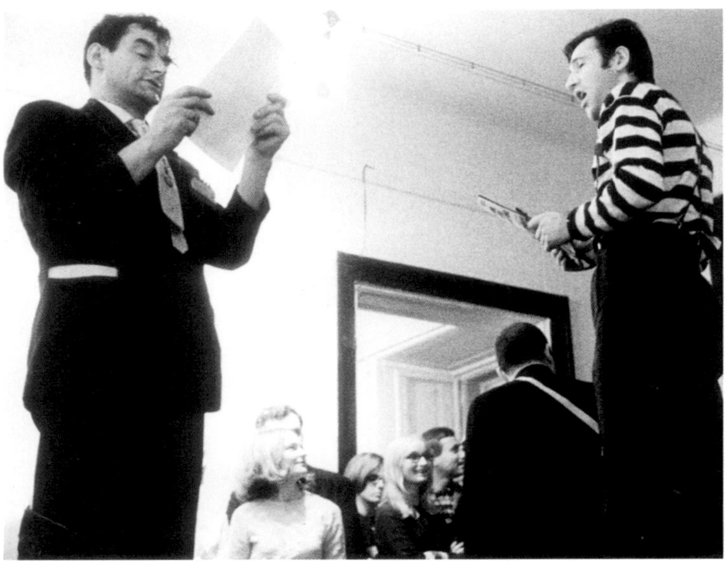

Otto Muehl and Peter Weibel reading the result of his blind typing

Critique of Art as Language, Blind Typing / Kritik der Kunst als Sprache, Blindschreiben

1966	performance at the *Action concert for Al Hansen*, Galerie nächst St. Stephan, Vienna, October 1966

Peter Weibel's performance *Critique of Art as Language*, with the addendum *Blind Typing* in its title, was staged in October 1966 at Galerie nächst St. Stephan in Vienna as part of a "performance concert" for Al Hansen. Besides Weibel, the other participants in this event in honor of the American Fluxus artist, which was instigated by Otto Muehl, included the Vienna Actionists Hermann Nitsch, Rudolf Schwarzkogler, and Oswald Wiener. For his "performance text," Weibel covered his eyes with pieces of newspaper and typed out a lecture about the critical potential of art on a typewriter. An alarm clock rang at regular intervals. Since the artist was not remotely well-versed in the ten-finger touch-typing system, this writing process produced a mishmash of meaningless letter combinations. Weibel and Muehl then read the result of his literal "blind typing" aloud as a lecture, demonstrating his scepticism toward the critical function of art.

As so often, Weibel gives precedence, here, not to the completed result, but rather to the medium itself – in this case, the typewriter – and to the process of writing, detached from fixed definitions of meaning. What is important is the primacy of the idea, the concept, the processual undertaking, not the finished product. He employs the typewriter as an instrument that can be used to produce not just written texts but also visual and auditory ones. He enters into an actual dialogue with the machine, a human-machine interaction, in which the machine is responsible for a significant portion of the artistic production.

Translated from the German by Patrick Hubenthal.

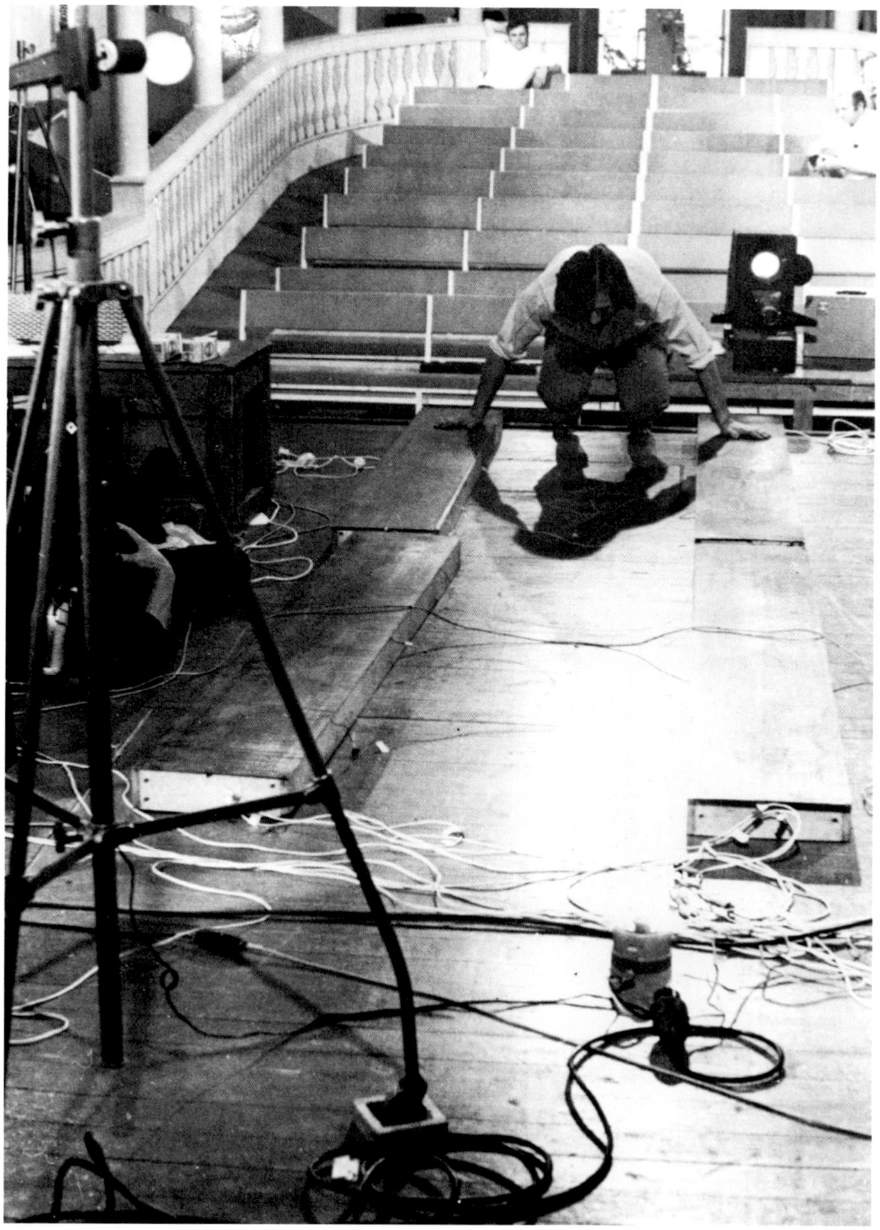

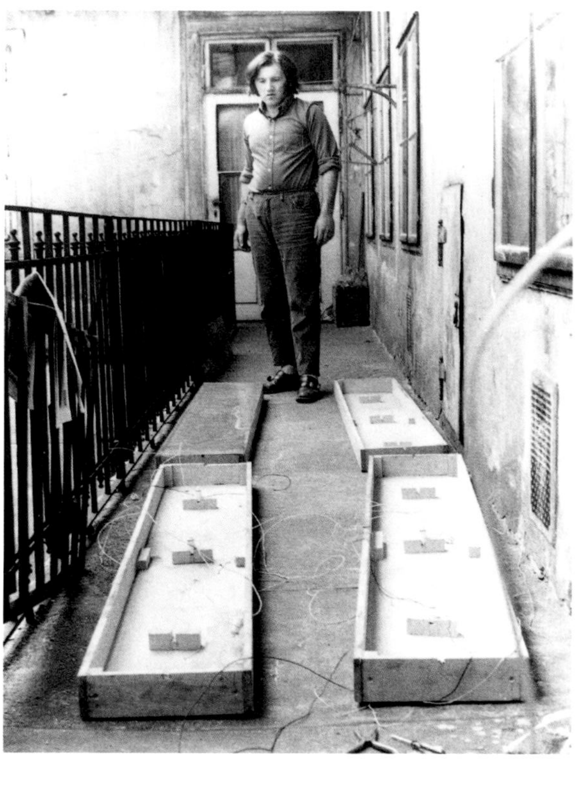

The interactive installation *The Long March* – whose title refers to "the long march through the institutions," an entrist political strategy initiated by the German left in the late 1960s – consists of parallel rows of steel plates covering six pairs of push-button switches, another variation of a keyboard. Each pair of switches is connected to a tape player set up to play two political programs, one on each of its two tracks. Pressing two switches at the same time starts one of the tracks playing. The six pairs of switches can thus be used to activate a total of six programs (three tape players with two tracks each). Each program consists of three concise demands:

abolish capital letters
abolish ticket inspectors
abolish the police

abolish tuition fees
abolish universities
abolish the legal system

abolish cunt and cock
abolish the constitution
abolish identity

abolish art
abolish work
abolish meaning

abolish language
abolish communication
abolish space and time

abolish society
abolish humanity
abolish nature

The interactor can crawl between the plates, pressing the buttons with his or her hands. Each time a button is pressed, however, the voltage running through the plates increases. This increase in voltage is indicated by a light bulb connected to the plates, which grows steadily brighter. As the interactor advances and the programs playing from the tape players become more radical, the way forward becomes more and more painful, until at last the interactor is forced to stop.

—Peter Weibel, 2022

Translated from the German by Patrick Hubenthal.

The Long March / Der lange Marsch

1969/71	interactive installation; steel plates, electrical wire, 12 push-button switches, 3 two-track tape players, speakers, light bulb, performed at *multi media 1*, Galerie Junge Generation, Vienna, April, 1969, recreated for Festival Konkrete Poesie, Hanau, May 1971

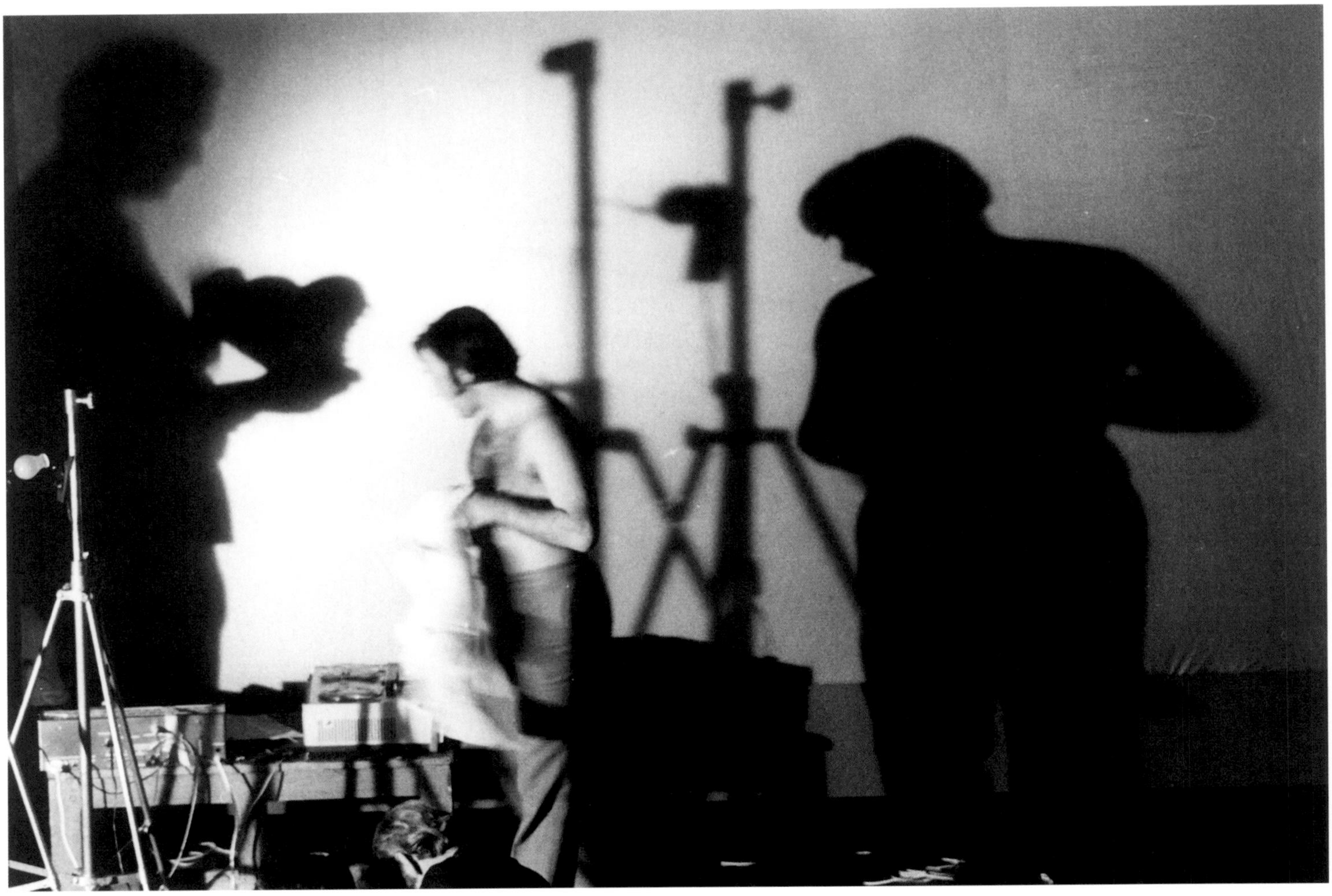

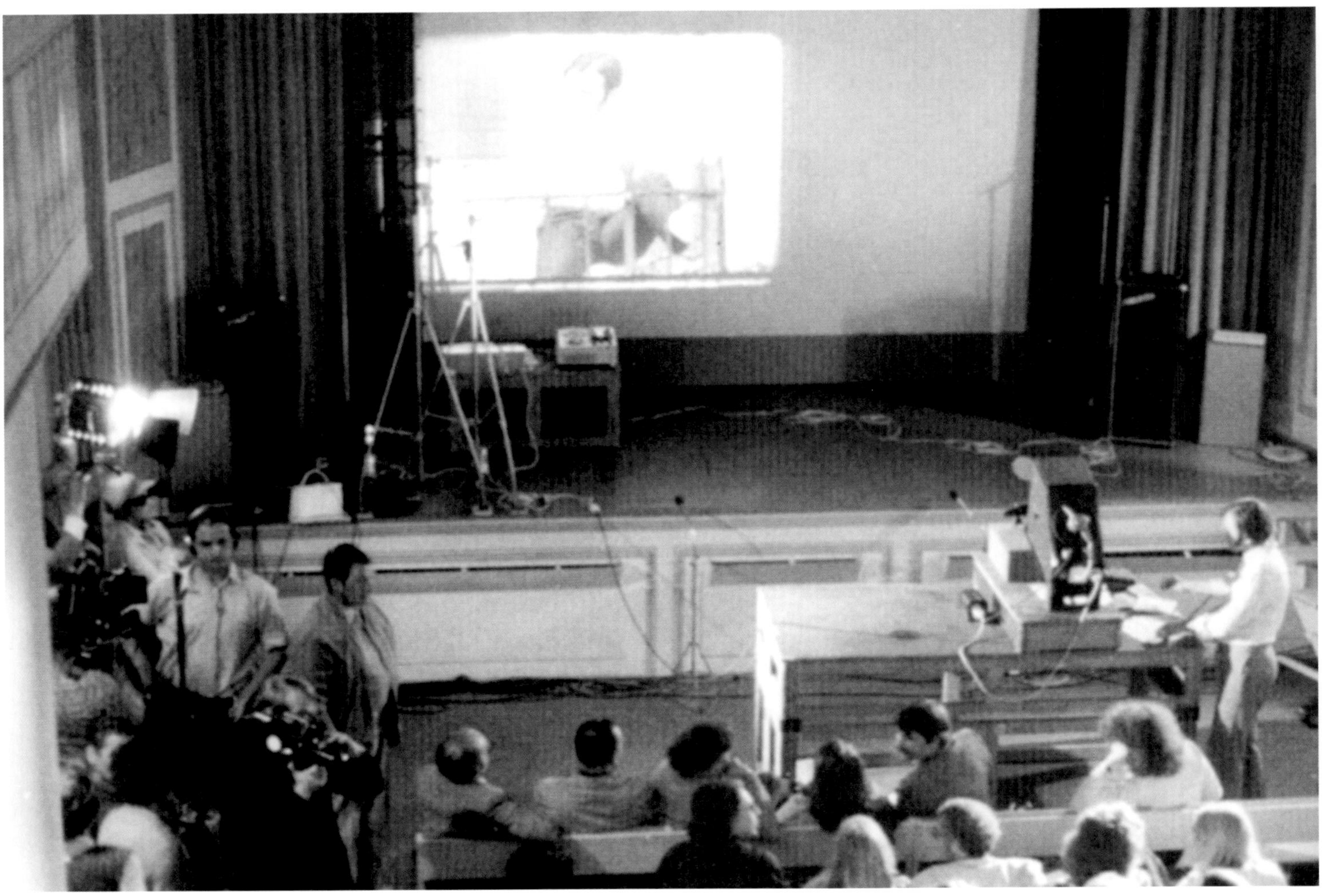

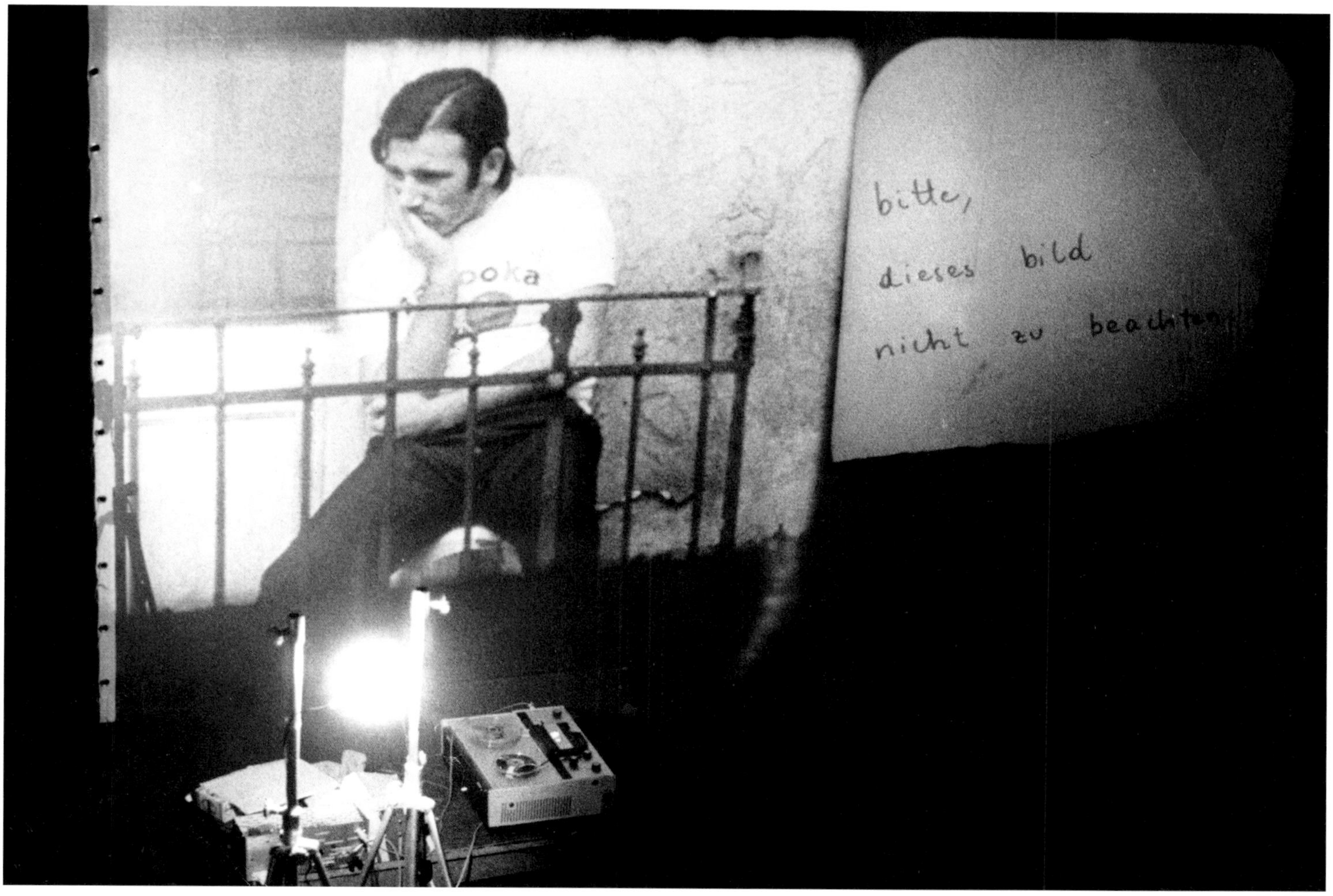

Communication Breakdown

1971 | performance; 8 mm films, *Denkakt* (Act of thinking) by Ernst Schmidt Jr., projectors, tape player, spotlights, microphone, loudspeakers, Festival Konkrete Poesie, Hanau, May 1971

The audio and video components of this performance are only activated at a certain level. When the noise made by the audience exceeds a certain volume, a light is switched on via a microphone. A tape player and a 16 mm projector turn on at the same time as the light. The light interferes with the projection, and the audience noise interferes with the speech on the tape. This means that the same condition (audience noise) that starts the projection and the tape simultaneously interferes with the projection machinery. The audience noise makes the speech impossible to hear, and the light makes the picture hard to see. If the audience quiets down, they hear and see nothing. Above a certain volume, however, an automatic control loop is created by the speakers: continual interference and self-interference keep the machinery running.

—Peter Weibel, 1971

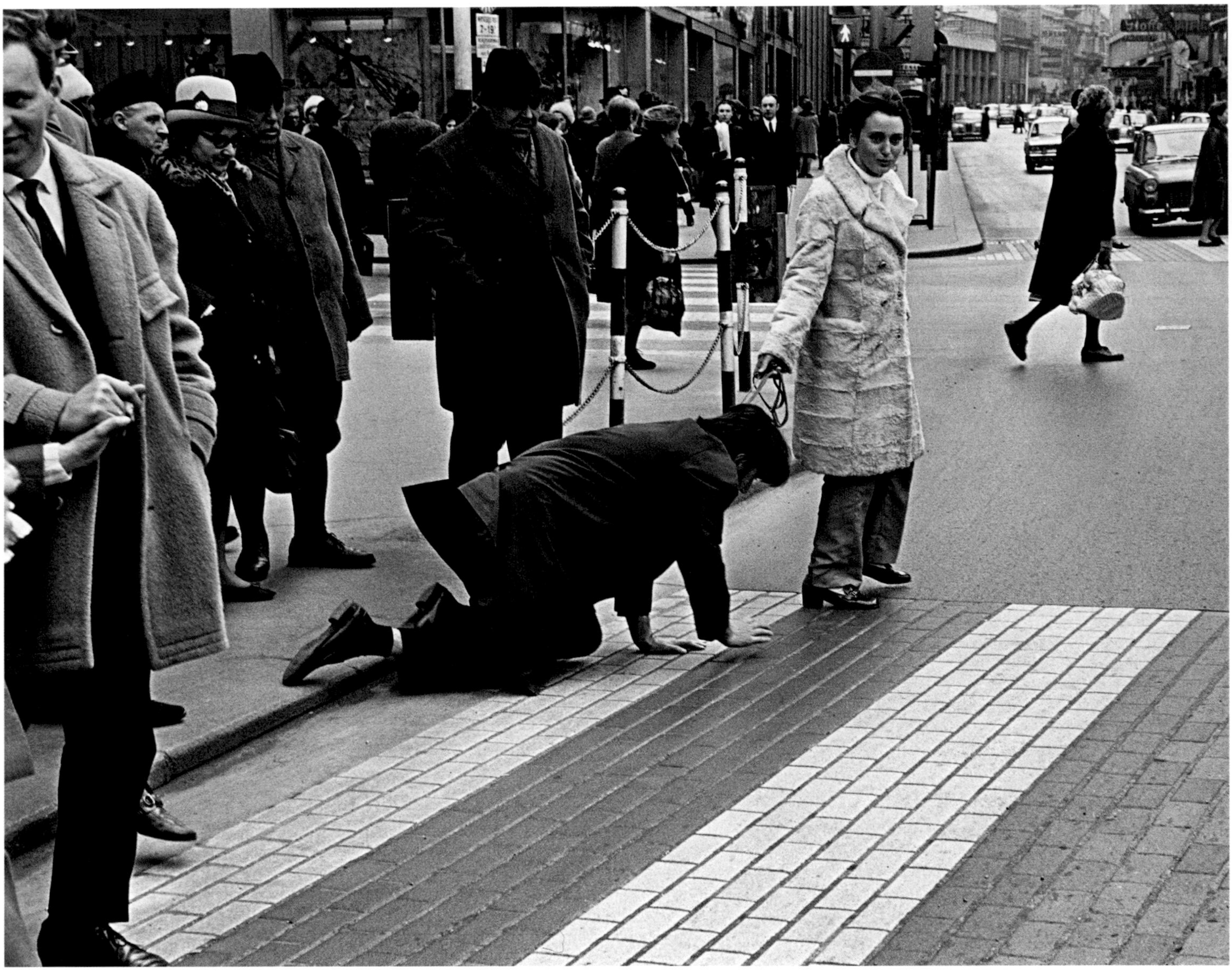

The action *From the Portfolio of Doggedness* by Peter Weibel and VALIE EXPORT was presented in Vienna in 1968 and aroused considerable public attention. VALIE EXPORT took Peter Weibel, who crawled on all fours, on a dog leash to an opening at the Galerie nächst St. Stephan, and a week later across the busy Kärntnerstrasse. Like all their actions at this time, such as *Touch and Feel Cinema* (*Tapp- und Tastkino*, 1968, see pp. 180–81 in this volume), this one was not announced as art and did not take place in a gallery. Not protected by the art context, they hit reality. During the public action, astonishing incidents occurred, the greatest irritation and anxiety being caused by the sadomasochistic dimension of the action.
The title is an ironic variation of "From the Portfolio of Humanity." This was the name of a pamphlet of the Red Cross or a similar humanitarian society that was distributed weekly in Austrian churches. To Peter Weibel, the state of society in the 1960s was far away from human dignity because the fascistic generation was still in power in Austria. The utopia of a "aufrechter Gang" (upright walk), articulated from Johann Gottfried Herder to Ernst Bloch, had not yet arrived. It was a critique of society, but also of representation, not only about gender etc., but also about visual culture in general. Weibel substituted representation with reality in many pieces to demonstrate the illusory and ideological character of representation. In this case, he turned animated cartoons à la Walt Disney featuring speaking animals (as humans) into reality: a human as a speaking animal. The motif of the "artist as dog" had already been the subject of a poem by Peter Weibel written in 1963, and had also been taken up in the photo booth picture series *Evolution of the Poet* (1967, see p. 112 in this volume), a title referring both to the novel by James Joyce and the short stories of Dylan Thomas. In this photo booth pictures series the artist reveals the constructedness of photographic self-representation, and its potential for slipping into fictitious, changing roles. Further, he points out the machine character of photography, which is based on a technical apparatus whose traits are automatic and anonymous.

From the Portfolio of Doggedness /
Aus der Mappe der Hundigkeit

1968	VALIE EXPORT and Peter Weibel action; Vienna, February 1968

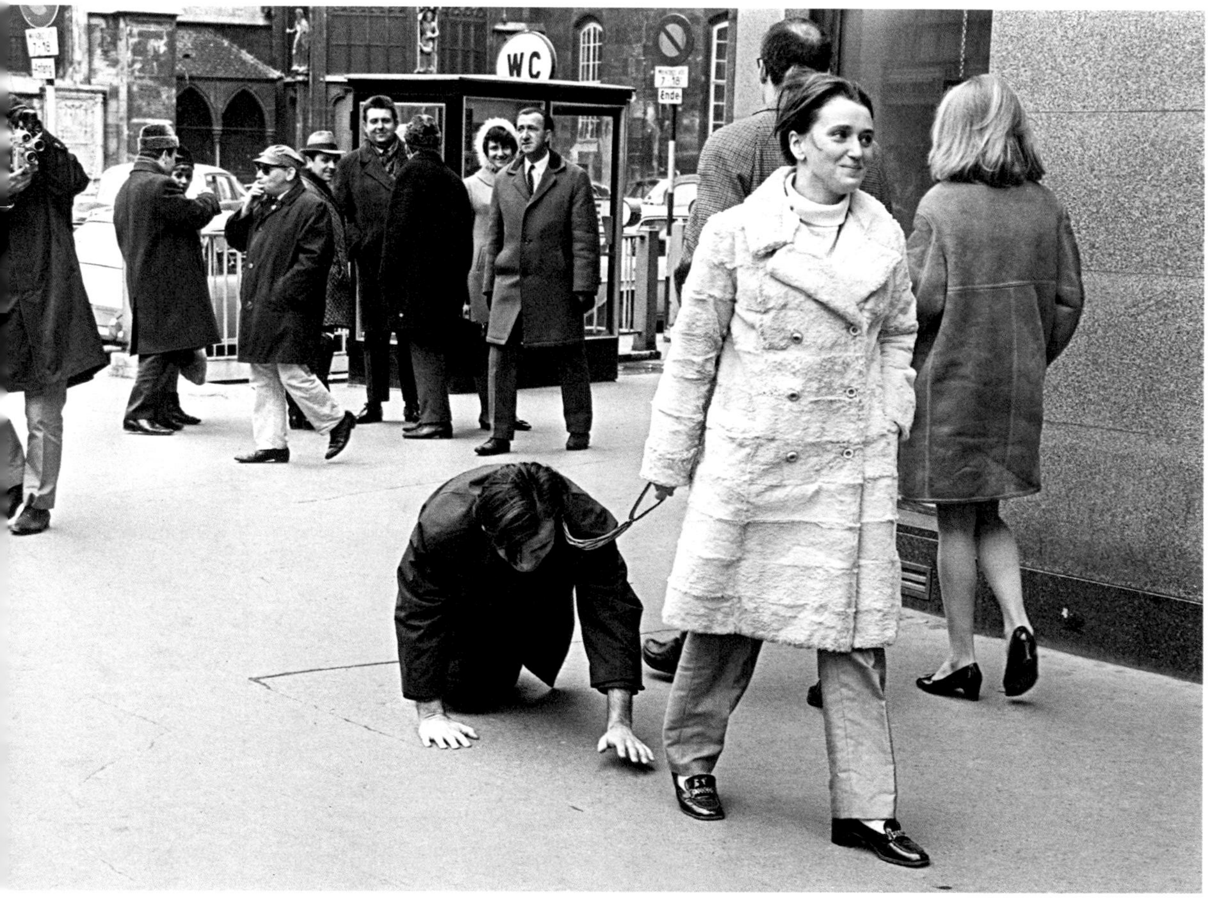

Hans Scheugl (with camera) and Kurt Kren (with sunglasses and cigarette) in the background

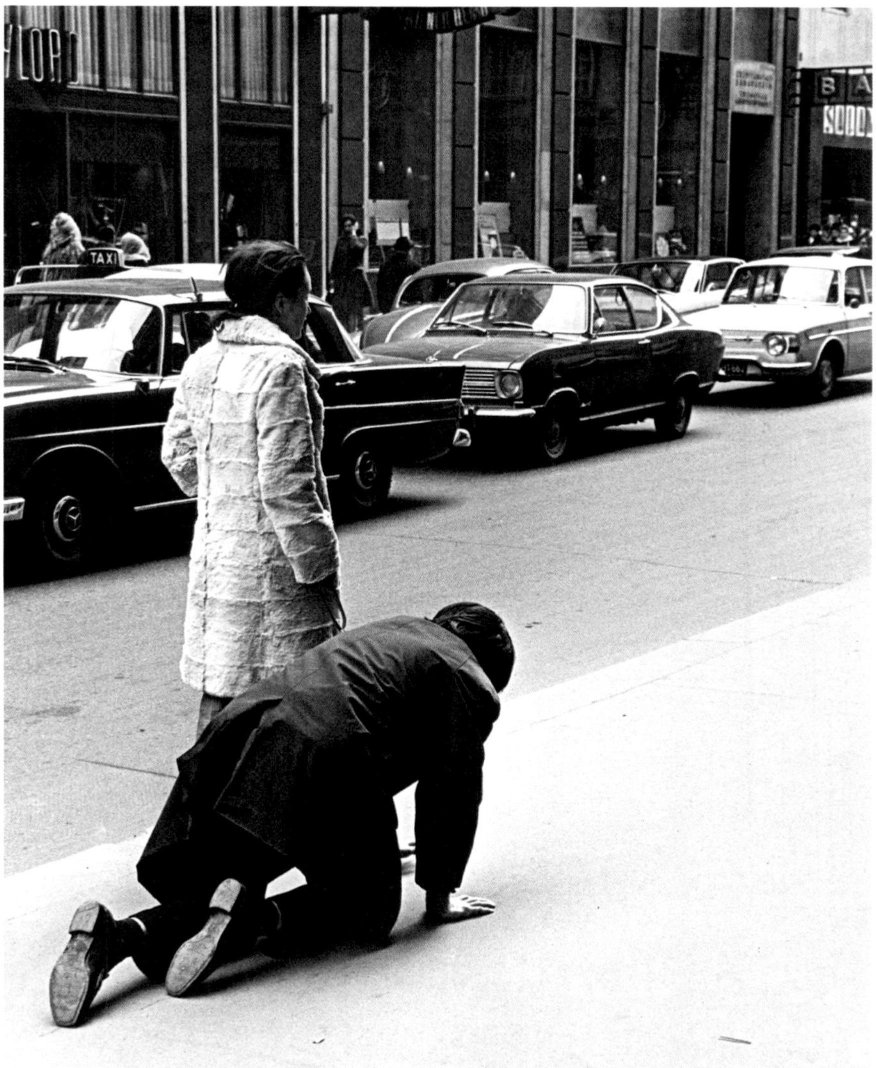

Documentation of the action, 5 b/w photographs, vintage prints,40.3 × 50.3 cm / 50 × 40.3 cm each Generali Foundation Collection–Permanent Loan to the Museum der Moderne Salzburg

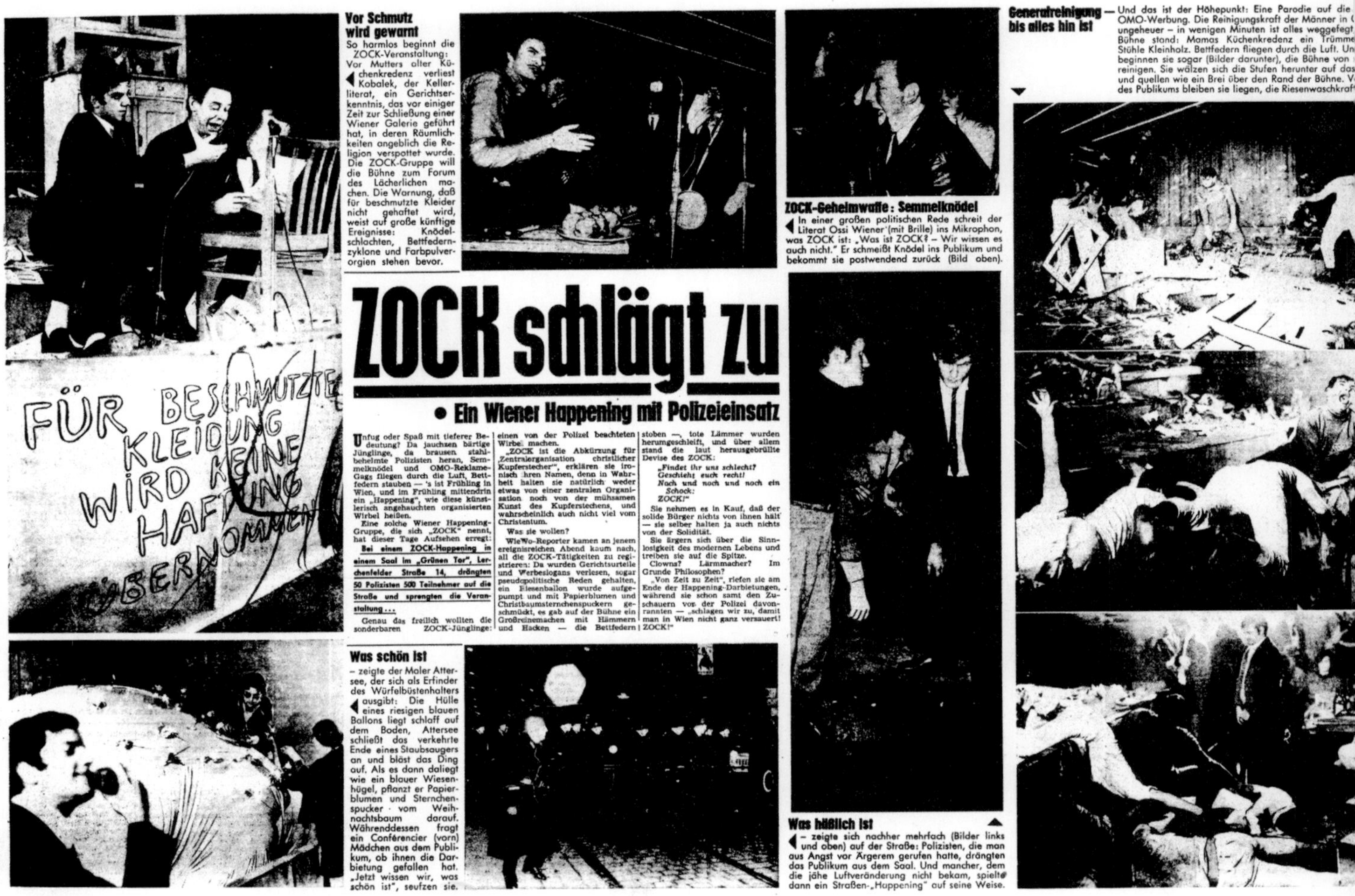

Vor Schmutz wird gewarnt

So harmlos beginnt die ZOCK-Veranstaltung: Vor Mutters alter Küchenkredenz verliest Kobalek, der Kellerliterat, ein Gerichtserkenntnis, das vor einiger Zeit zur Schließung einer Wiener Galerie geführt hat, in deren Räumlichkeiten angeblich die Religion verspottet wurde. Die ZOCK-Gruppe will die Bühne zum Forum des Lächerlichen machen. Die Warnung, daß für beschmutzte Kleider nicht gehaftet wird, weist auf große künftige Ereignisse: Knödelschlachten, Bettfedernzyklone und Farbpulverorgien stehen bevor.

ZOCK schlägt zu

• Ein Wiener Happening mit Polizeieinsatz

Unfug oder Spaß mit tieferer Bedeutung? Da jauchzen bärtige Jünglinge, da brausen stahlbehelmte Polizisten heran, Semmelknödel und OMO-Reklame-Gags fliegen durch die Luft, Bettfedern stauben — 's ist Frühling in Wien, und im Frühling mittendrin ein „Happening", wie diese künstlerisch angehauchten organisierten Wirbel heißen.

Eine solche Wiener Happening-Gruppe, die sich „ZOCK" nennt, hat dieser Tage Aufsehen erregt:

Bei einem ZOCK-Happening in einem Saal im „Grünen Tor", Lerchenfelder Straße 14, drängten 50 Polizisten 500 Teilnehmer auf die Straße und sprengten die Veranstaltung ...

Genau das freilich wollten die sonderbaren ZOCK-Jünglinge: einen von der Polizei beachteten Wirbel machen.

„ZOCK ist die Abkürzung für ‚Zentralorganisation christlicher Kupferstecher'", erklären sie ironisch ihren Namen, denn in Wahrheit halten sie natürlich weder etwas von einer zentralen Organisation noch von der mühsamen Kunst des Kupferstechens, und wahrscheinlich auch nicht viel vom Christentum.

Was sie wollen?

WieWo-Reporter kamen an jenem ereignisreichen Abend kaum nach, all die ZOCK-Tätigkeiten zu registrieren: Da wurden Gerichtsurteile und Werbeslogans verlesen, sogar pseudopolitische Reden gehalten, ein Riesenballon wurde aufgepumpt und mit Papierblumen und Christbaumsternchenspuckern geschmückt, es gab auf der Bühne ein Großreinemachen mit Hämmern und Hacken — die Bettfedern stoben —, tote Lämmer wurden herumgeschleift, und über allem stand die laut herausgebrüllte Devise des ZOCK:

„Findet ihr uns schlecht?
Geschieht euch recht!
Noch und noch und noch ein
Schock:
ZOCK!"

Sie nehmen es in Kauf, daß der solide Bürger nichts von ihnen hält — sie selber halten ja auch nichts von der Solidität.

Sie ärgern sich über die Sinnlosigkeit des modernen Lebens und treiben sie auf die Spitze.

Clowns? Lärmmacher? Im Grunde Philosophen?

„Von Zeit zu Zeit", riefen sie am Ende der Happening-Darbietungen, während sie schon samt den Zuschauern vor der Polizei davonrannten — „schlagen wir zu, damit man in Wien nicht ganz versauert! ZOCK!"

ZOCK-Geheimwaffe: Semmelknödel

In einer großen politischen Rede schreit der Literat Ossi Wiener (mit Brille) ins Mikrophon, was ZOCK ist: „Was ist ZOCK? – Wir wissen es auch nicht." Er schmeißt Knödel ins Publikum und bekommt sie postwendend zurück (Bild oben).

Generalreinigung bis alles hin ist — Und das ist der Höhepunkt: Eine Parodie auf die OMO-Werbung. Die Reinigungskraft der Männer in ungeheuer – in wenigen Minuten ist alles weggefegt, Bühne stand: Mamas Küchenkredenz ein Trümme Stühle Kleinholz. Bettfedern fliegen durch die Luft. Un beginnen sie sogar (Bilder darunter), die Bühne von reinigen. Sie wälzen sich die Stufen herunter auf das und quellen wie ein Brei über den Rand der Bühne. V des Publikums bleiben sie liegen, die Riesenwaschkraf

Was schön ist

– zeigte der Maler Attersee, der sich als Erfinder des Würfelbüstenhalters ausgibt: Die Hülle eines riesigen blauen Ballons liegt schlaff auf dem Boden, Attersee schließt das verkehrte Ende eines Staubsaugers an und bläst das Ding auf. Als es dann daliegt wie ein blauer Wiesenhügel, pflanzt er Papierblumen und Sternchenspucker vom Weihnachtsbaum darauf. Währenddessen fragt ein Conférencier (vorn) Mädchen aus dem Publikum, ob ihnen die Darbietung gefallen hat. „Jetzt wissen wir, was schön ist", seufzen sie.

Was häßlich ist

– zeigte sich nachher mehrfach (Bilder links und oben) auf der Straße: Polizisten, die man aus Angst vor Ärgerem gerufen hatte, drängten das Publikum aus dem Saal. Und mancher, dem die jähe Luftveränderung nicht bekam, spielte dann ein Straßen-„Happening" auf seine Weise.

164

Wiener Wochenblatt, May 6, 1967. Top left: Würthle and Kobalek, bottom left: Attersee and Schuppan, front: Würthle; top center: Steiger, Wiener, center: Kalb, supported by friends; bottom right: Omo Super Group, Kalb, urinating, and VALIE EXPORT, sitting next to him

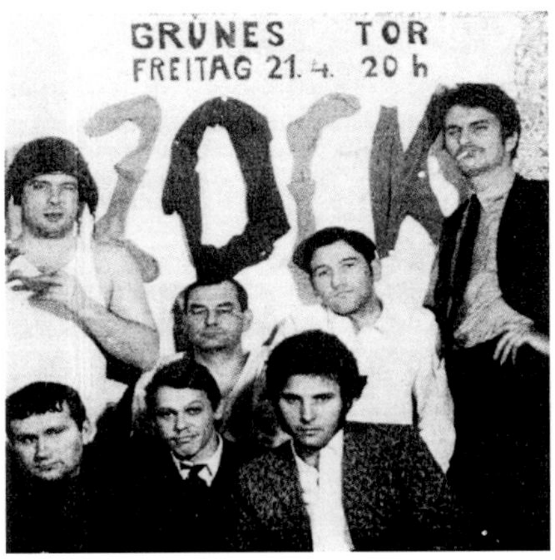

Slide for advertising in Vienna's cinemas; first row, from the left: Hermann Nitsch, Reinhard Priessnitz, Attersee; second row, from the left: Oswald Wiener, Otto Muehl, Peter Weibel, Dominik Steiger

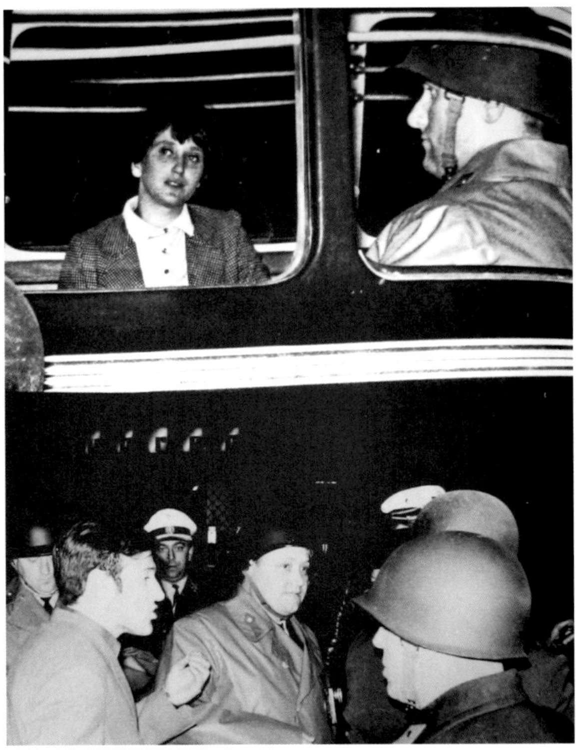

Top: VALIE EXPORT in a police car; bottom: Peter Weibel discusses with a police officer about the interrogation of VALIE EXPORT after the *Zock-Fest* to bail her out

Zock-Fest

1967	series of happenings; Galerie nächst St. Stephan, Vienna, April 17, 1967; Gasthaus Grünes Tor, April 21, 1967

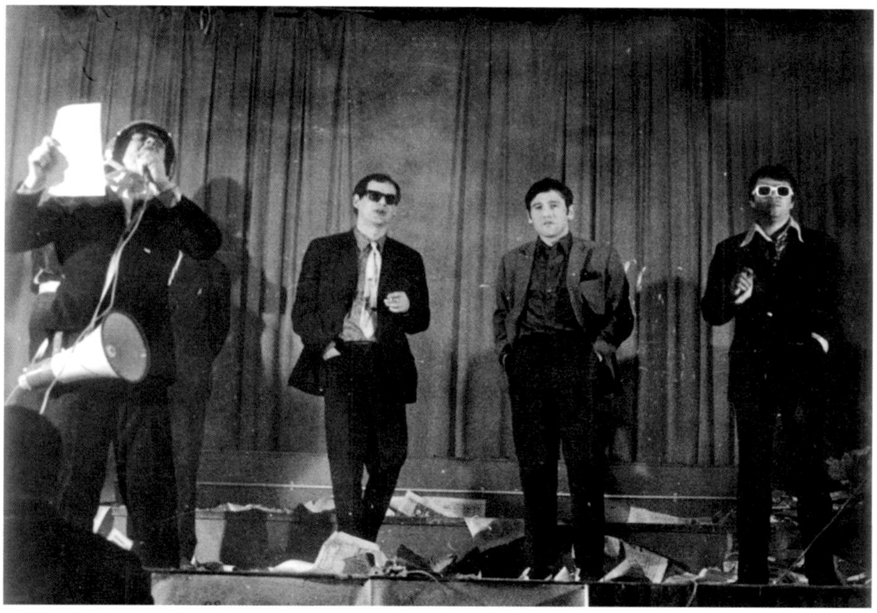

From the left: Oswald Wiener, Gunter Falk, Peter Weibel, Wolfgang Bauer

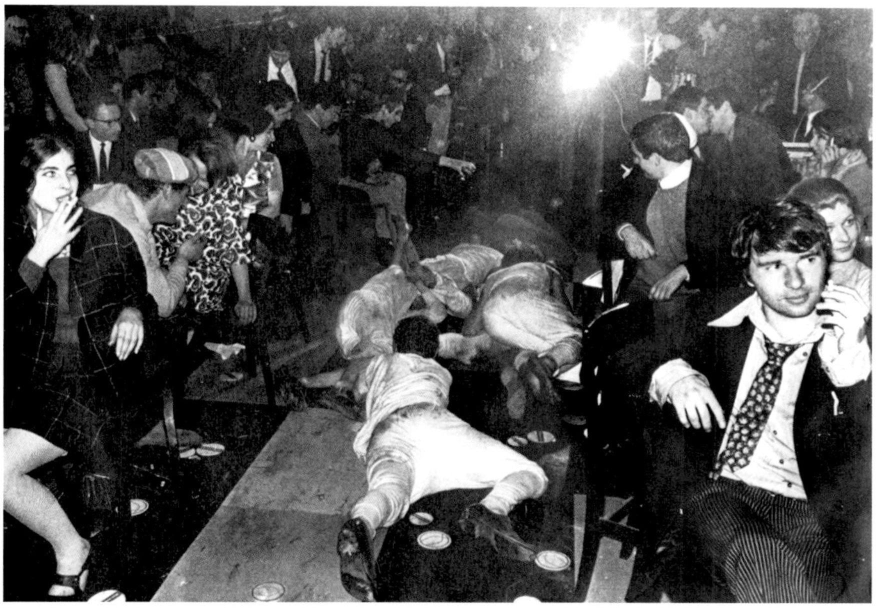

Omo Super Group (Otto Muehl, crawling); left: Ingrid Schuppan, behind her Ernst Jandl and Otto Breicha; fourth row, first seat: Elfriede Gerstl; right: Wolfgang Bauer

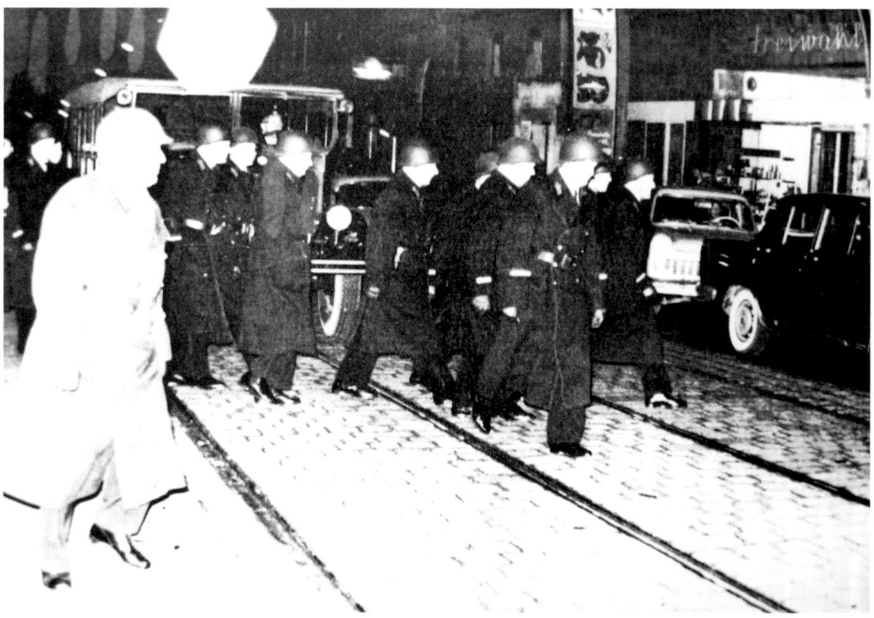

Police clearing the road, a total of 200 officers with police dogs and steel helmets

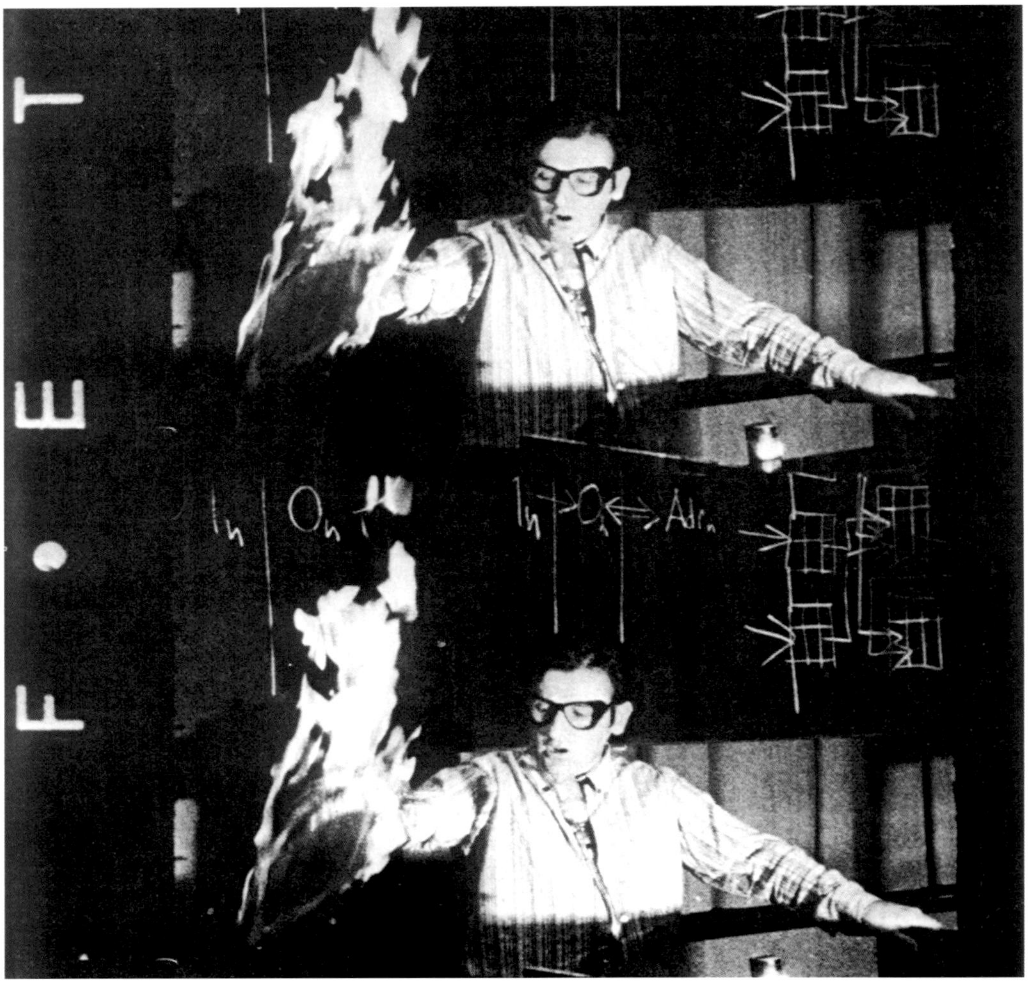

Peter Weibel during his incendiary speech, filmed by Ernst Schmidt Jr.

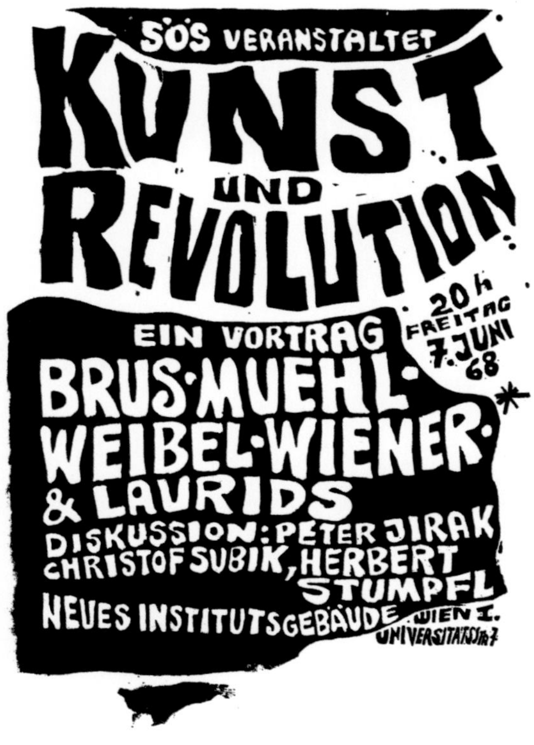

Poster for the action

Incendiary Speech / Brandrede

1968	action, University of Vienna, June 7, 1968

As part of the legendary action *Art and Revolution* (*Kunst und Revolution*) at the University of Vienna on June 7, 1968, Peter Weibel gave a literally incendiary speech with a burning hand, symbolizing the violence and real impact of language. During his speech entitled "Was tun?" (What to do?), paraphrasing Lenin and Nikolay Chernyshevsky, that denounced the failures of the government of the time, despite the protective goggles and asbestos glove, individual drops of gasoline got under the prepared glove and the fire spread to his bare arm. Weibel had to end the speech and threw the burning glove into a waiting bucket of water.

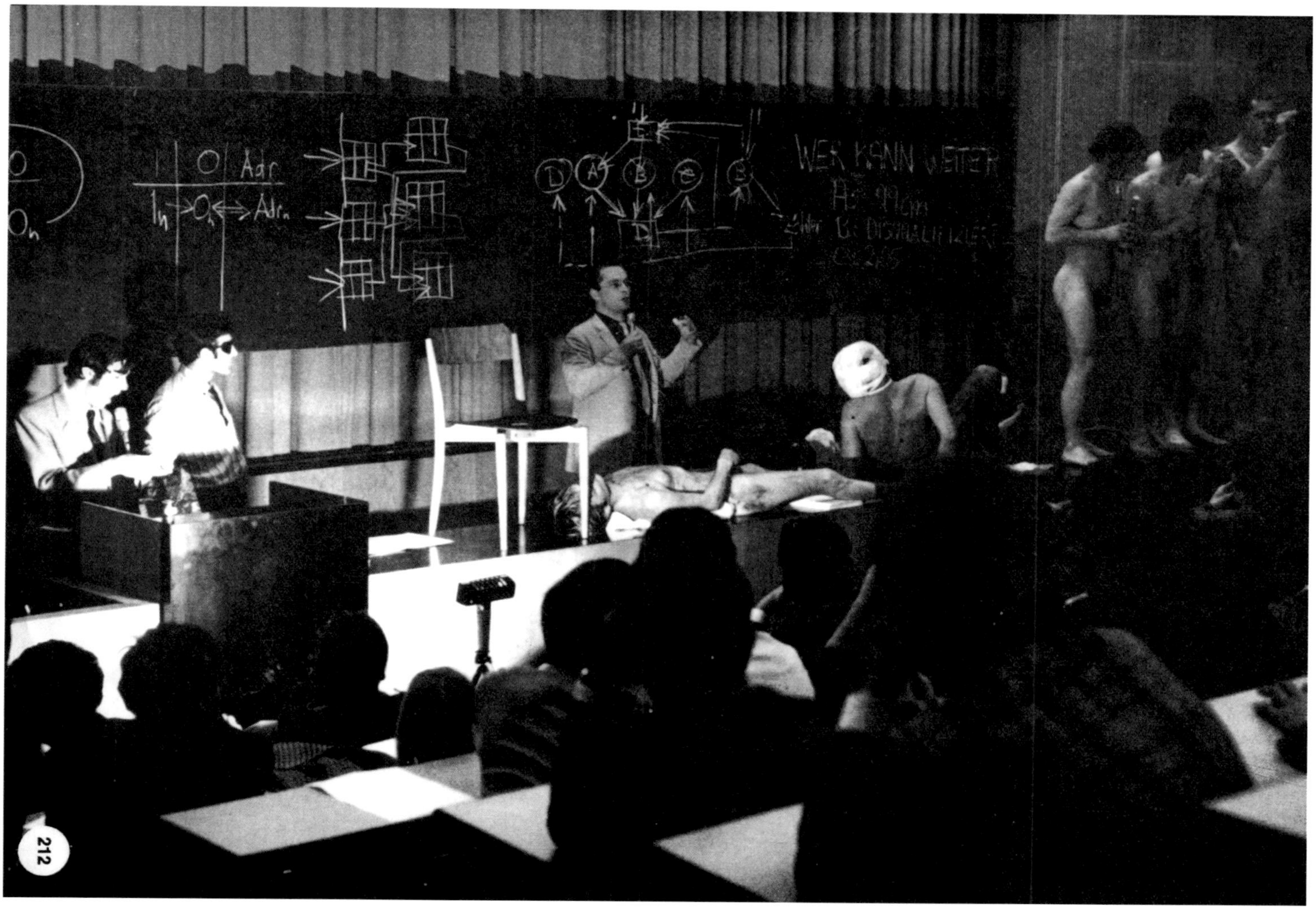

Peter Weibel (with sunglasses) next to Franz Kaltenbeck (left) during his lecture at the University of Vienna as part of *Art and Revolution*, 1968; speaking in the middle: Oswald Wiener; in the front, lying and masturbating: Günter Brus; man with bandaged head: Laurids; right: Otto Muehl and his assistants after his performance

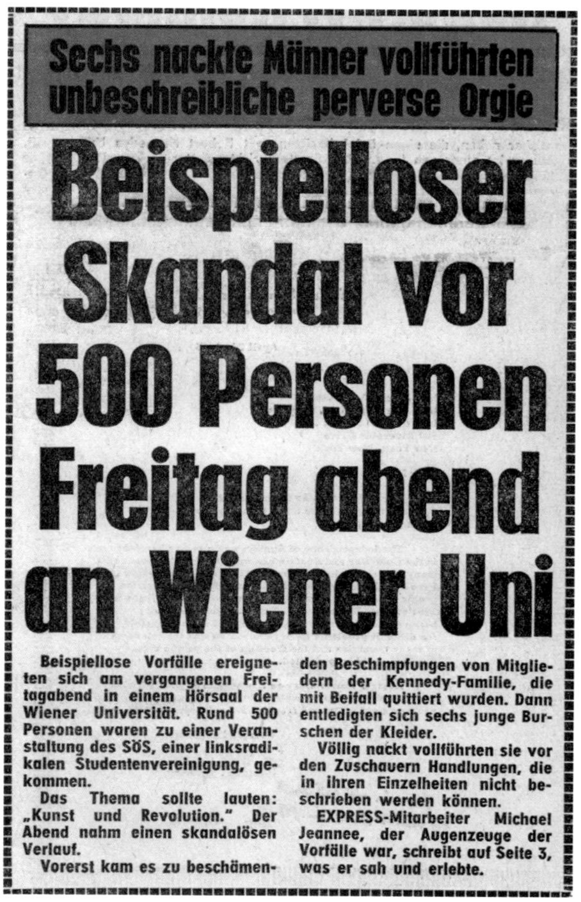

Sechs nackte Männer vollführten unbeschreibliche perverse Orgie

Beispielloser Skandal vor 500 Personen Freitag abend an Wiener Uni

Beispiellose Vorfälle ereigneten sich am vergangenen Freitagabend in einem Hörsaal der Wiener Universität. Rund 500 Personen waren zu einer Veranstaltung des SÖS, einer linksradikalen Studentenvereinigung, gekommen.

Das Thema sollte lauten: „Kunst und Revolution." Der Abend nahm einen skandalösen Verlauf.

Vorerst kam es zu beschämenden Beschimpfungen von Mitgliedern der Kennedy-Familie, die mit Beifall quittiert wurden. Dann entledigten sich sechs junge Burschen der Kleider.

Völlig nackt vollführten sie vor den Zuschauern Handlungen, die in ihren Einzelheiten nicht beschrieben werden können.

EXPRESS-Mitarbeiter Michael Jeannee, der Augenzeuge der Vorfälle war, schreibt auf Seite 3, was er sah und erlebte.

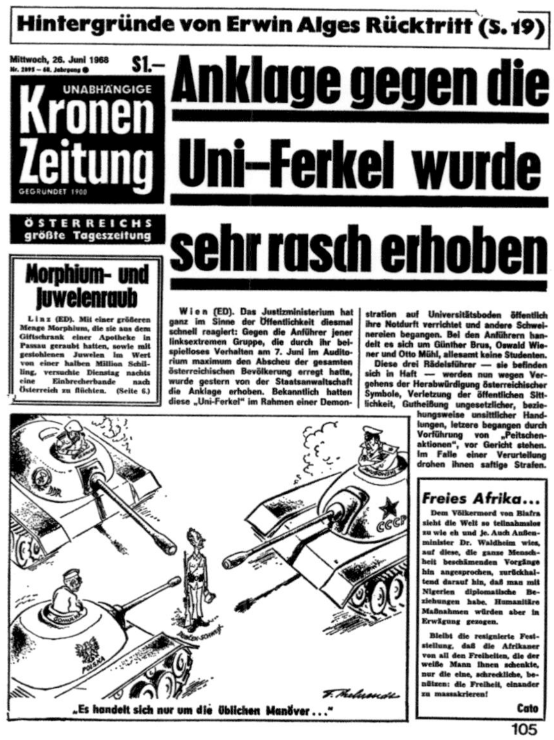

Hintergründe von Erwin Alges Rücktritt (S. 19)

Mittwoch, 26. Juni 1968 S1.–

UNABHÄNGIGE Kronen Zeitung GEGRUNDET 1900

ÖSTERREICHS größte Tageszeitung

Anklage gegen die Uni–Ferkel wurde sehr rasch erhoben

Wien (ED). Das Justizministerium hat ganz im Sinne der Öffentlichkeit diesmal schnell reagiert: Gegen die Anführer jener linksextremen Gruppe, die durch ihr beispielloses Verhalten am 7. Juni im Auditorium maximum den Abscheu der gesamten österreichischen Bevölkerung erregt hatte, wurde gestern von der Staatsanwaltschaft die Anklage erhoben. Bekanntlich hatten diese „Uni-Ferkel" im Rahmen einer Demonstration auf Universitätsboden öffentlich ihre Notdurft verrichtet und andere Schweinereien begangen. Bei den Anführern handelt es sich um Günther Brus, Oswald Wiener und Otto Mühl, allesamt keine Studenten.

Diese drei Rädelsführer — sie befinden sich in Haft — werden nun wegen Vergehens der Herabwürdigung österreichischer Symbole, Verletzung der öffentlichen Sittlichkeit, Gutheißung ungesetzlicher, beziehungsweise unsittlicher Handlungen, letzere begangen durch Vorführung von „Peitschenaktionen", vor Gericht stehen. Im Falle einer Verurteilung drohen ihnen saftige Strafen.

Morphium- und Juwelenraub

Linz (ED). Mit einer größeren Menge Morphium, die sie aus dem Giftschrank einer Apotheke in Passau geraubt hatten, sowie mit gestohlenen Juwelen im Wert von einer halben Million Schilling, versuchte Dienstag nachts eine Einbrecherbande nach Österreich zu flüchten. (Seite 6.)

„Es handelt sich nur um die üblichen Manöver ..."

Freies Afrika...

Dem Völkermord von Biafra sieht die Welt so teilnahmslos zu wie eh und je. Auch Außenminister Dr. Waldheim wies, auf diese, die ganze Menschheit beschämenden Vorgänge hin angesprochen, zurückhaltend darauf hin, daß man mit Nigerien diplomatische Beziehungen habe. Humanitäre Maßnahmen würden aber in Erwägung gezogen.

Bleibt die resignierte Feststellung, daß die Afrikaner von all den Freiheiten, die der weiße Mann ihnen schenkte, nur die eine, schreckliche, benützen: die Freiheit, einander zu massakrieren!

Cato

105

Headlines from Austrian newspapers *Express* (June 9, 1968) and *Kronenzeitung* (June 26, 1968) reading: "6 naked men performed an indescribable perverted orgy: unprecedented scandal in front of 500 people Friday evening at Vienna University" (left) and "charges against the university pigs were brought very quickly"(right).

Underground Explosion: W.A.R.

Wolfgang Kemp

With impetus from the protest movements of 1968, insulting the audience quickly became fashionable. It also became more physical.: for a 1969 performance staged at Munich's Zirkus Krone and in other cities, the Viennese Actionists VALIE EXPORT and Peter Weibel built an apparatus they called the *Audience Atomizer / Publikum(s) sprenger* – and the name was no exaggeration. "Weibel and EXPORT went into action again on another occasion. This time the audience was subjected to verbal abuse in the manner of Hitler and Goebbels from a lectern on the stage, which was surrounded by rolls of barbed wire. During the interaction with Weibel, EXPORT mimed fellatio, whereupon Weibel began to hurl the barbed wire into the audience while EXPORT laid into them with a whip (the theme here being 'war takes place in the auditorium,' as a counterpoint to 'art takes place in the auditorium'). This was no simulation; it was deadly serious – the Vietnam War protests made tangible. 'We're beating citizens into humans,' Weibel bellowed into the microphone. What he and EXPORT were interested in was the very real, directly felt experience of war and violence, not a simulation of them. At least that was the theory. The audience was thrown into a panic. Furious spectators stormed the stage and blasted Weibel with his own homemade water cannon."[1] The *Attack Lecture* ended with a counterattack.

Dialogue, which had only recently been taken up by reception aesthetics as a model for the relationship between artwork and viewer, underwent a change, turning first into a monologue and then into a battle, into thematic and physical violence. That same year, 1969, the term "structural violence" was coined, and with it a license to engage in unstructured counterviolence. EXPORT and Weibel began by arming themselves with the following manifesto: "any art that wishes to survive as a site of utopia, to be concerned and brass-knuckled for the future, any art that wishes to provide for survival, must become a paramilitary action, sweeping through the fortified territory of 'order' and leaving a trail of liberty in its wake."[2] The great violent excesses of the twentieth century were not yet far behind us, in 1969, yet as a structural principle violence was suddenly new again, and of course there was a timely reason for that: violence was not only being perpetrated by a world power in faraway Vietnam, but also closer to home in Berlin, where the student Benno Ohnesorg had been shot dead by a police officer. And that is to mention only two instances of the West's use of violence; the true empire of structural violence lay in the East, though hardly anyone in Actionist circles was interested in that.

In the cultural sphere, the spaghetti Western set new standards for on-screen brutality in the 1960s, taking a classic genre and heightening it. In the same way, the Actionists took the audience's disapproval or scorn, which in classical modernism had traditionally been the hallmark of truth, and elevated it into a physical altercation. In doing so they were also underscoring their predilection for action, which had been a major, albeit still somewhat latent aspect of artistic practice since the 1950s – think of action painting, or the early Happenings. But the factor that truly pointed the way forward was the transformation of the observer into an active participant.

Note: Adapted from Wolfgang Kemp, *Der explizite Betrachter* (Konstanz: Konstanz University Press, 2015), 58–59. Translated from the German by Patrick Hubenthal.

[1]
Christoph Wagner, "Der Krieg findet im Saal statt," *Die Wochenzeitung* (Zurich), Apr. 23, 2009, https://www.woz.ch/0917/underground-explosion/der-krieg-findet-im-Saal-statt. Wagner is referring to the Zurich showing of the performance, which premiered in Munich.

[2]
Peter Weibel and VALIE EXPORT, "W.I.R. sind W.A.R.," in *Underground Explosion* (self-pub., 1969), program booklet for the underground art event *Underground Explosion*, Munich, 1969.

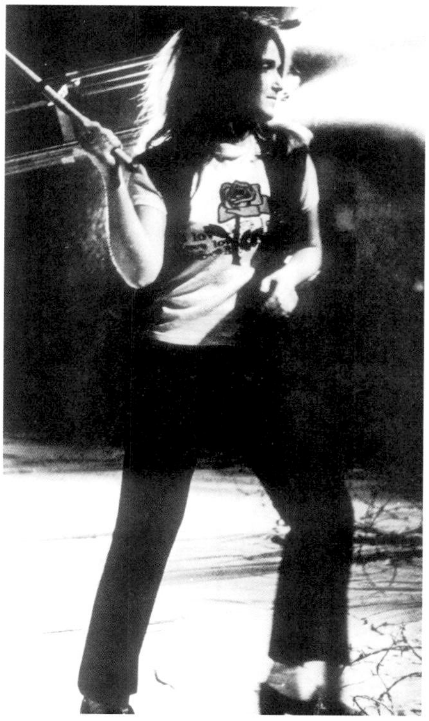

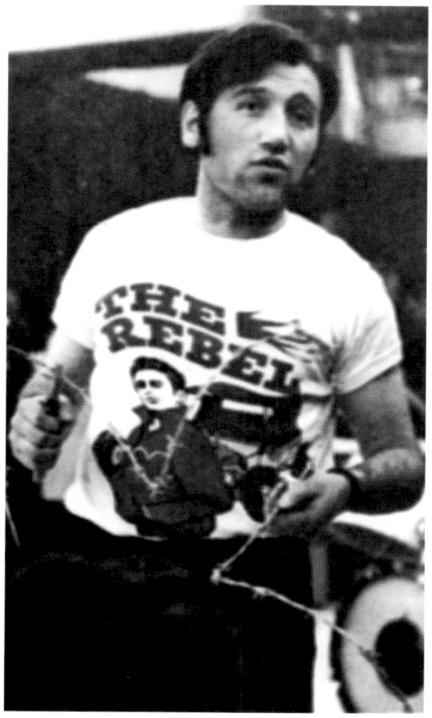

VALIE EXPORT and
Peter Weibel, photographs
of the performance at
Zirkus Krone, Munich,
Apr. 15, 1969

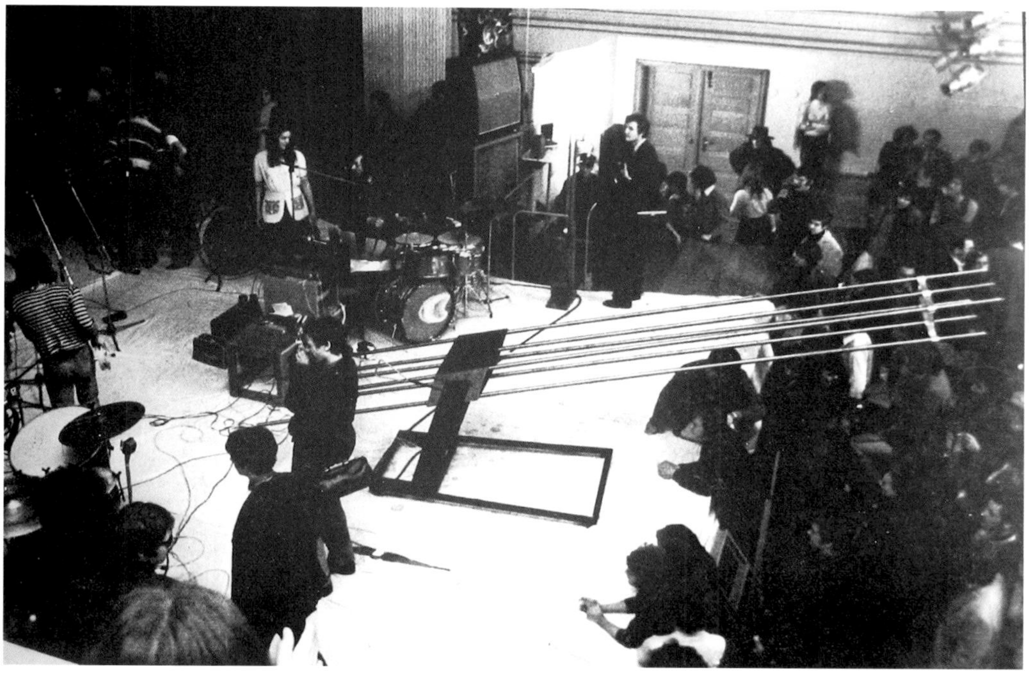

The *Audience Atomizer* at
Volkshaus Zürich,
Zurich, Apr. 18, 1969

Attack Lecture and Audience Atomizer / Attack Lecture und Publikum(s)sprenger

1969	Peter Weibel and VALIE EXPORT mobile sculpture for civilian combat, wheeled, with swivel arm, 4.5-meter steel pipes with water hoses idea: Peter Weibel; fabrication: Wolfgang Ernst

In April and May 1969, Peter Weibel and VALIE EXPORT were part of *Underground Explosion*, a tour with various groups from the theater and rock scenes that was organized by p.a.p (Karl-Heinz Hein and Dieter Meier, later famous as a member of the pop duo Yello) in collaboration with Paul and Limpe Fuchs, Amon Düül, Guru Guru Groove, Kinetic Lights, and the Wath-Tholl Theater. This tour, which drew thousands of spectators, traveled to Munich (Zirkus Krone, April 15), Zurich (Volkshaus, April 18), Cologne (Sporthalle, May 3), and Essen (May 6). The motto of the campaign was: "The audience as artwork, as victims of art, as guests at the wedding of Auschwitz!" And the program booklet proclaimed: "We are W.A.R., War Art Riot."

The two artists were seeking to create "a state of madness, anarchy, absolutism, anomie, aggression, perversion, ecstasy, discord, dissolution, transgression" with public fellatio, obscene speeches, etc. "A new form of communication, beyond the image and beyond state-sanctioned norms." They had not just declared war on the established concept of art, but on the audience itself. Their *Attack Lecture* can thus be seen as an extension and radicalization of the *Action Lectures* with which Weibel translated his theories into action.

Translated from the German by Patrick Hubenthal.

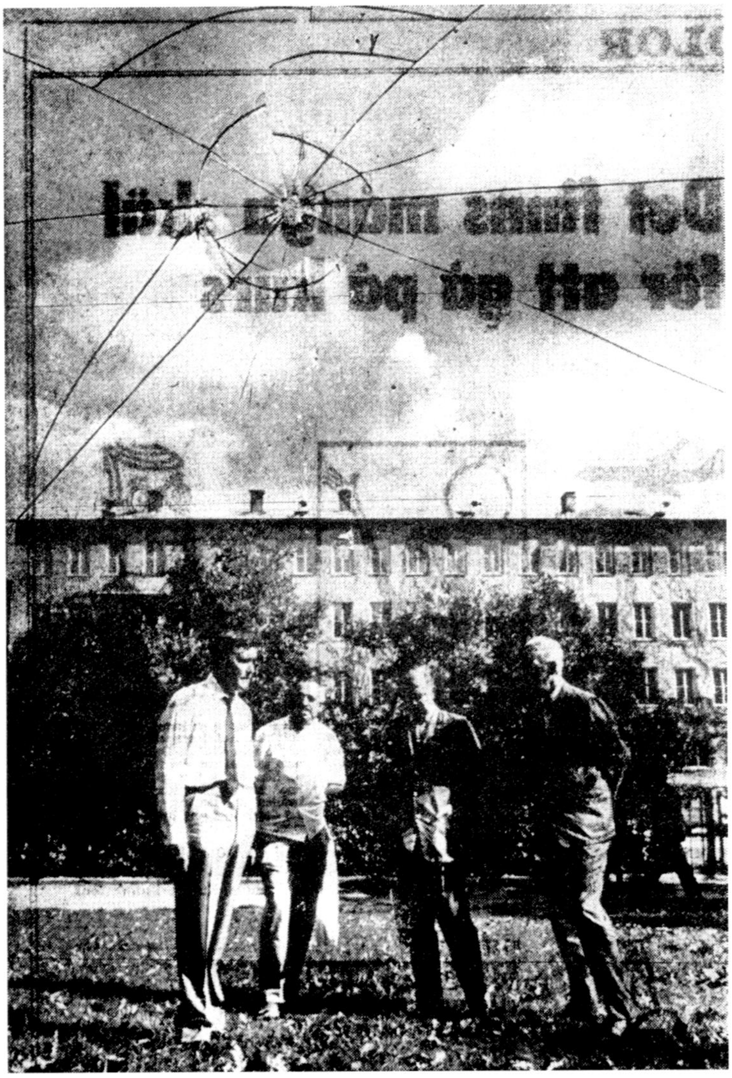

"Restaurant chef Uno Jansson (restaurant at the art museum), kitchen chef Dahl, Constable Holger Björkdahl, and Gunnar Johannsson in front of the art museum. Photographer Alf Weihed shot the photo from inside the museum through the broken window," caption from *Göteborgs Tidningen*, July 24, 1969

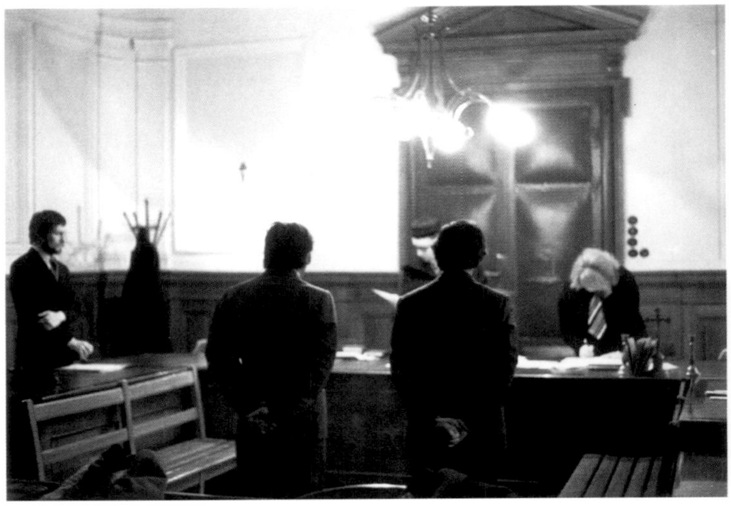

Trial for the action *Cryptic Message* in Vienna, Mar. 1972, for malicious damage to property

Cryptic Message / Kryptische Mitteilung

1969	Franz Kaltenbeck and Peter Weibel action; suction cups, lines, axe, 2 sledgehammers, iron wedge

On July 22, 1969, Franz Kaltenbeck and I had transformed a public building, the Gothenburg Art Museum in Sweden, into a private musical instrument, in a radicalization of Cage's and Fluxus's intentions according to which not only all sounds are music, but all objects can become musical instruments as well. Anonymously, at night, and without an audience, we used a civic institution as a private musical object. From the roof and using palm-sized suction cups capable of bearing loads of 25 kg, four strong lines were attached to four windows of a height of 6 m on the glass front of the Gothenburg Art Museum. On these lines, the following tools were hung: one axe, two heavy sledgehammers, one sharp-edged hard iron wedge. These were beaten against the glass front as keyboard at different time intervals according to specific principles. The smashing of the resonance chamber, or more accurately the museum; the beating of the hammers against the trigger wires of the alarm system behind the windows; and the concomitant siren as an aesthetic complement came together with the sirens of the police cars in a concert.

Two days later, on July 24 1969, the daily newspaper *Göteborgs Tidningen*, under the heading "Cryptic Message?" rightly asked: Was this deliberate vandalism or a cryptic message? Art museum and police presumed a break-in. It was in precisely this zone of ambivalence between cryptic artistic message and vandalism that the new territory of counterart set up camp.

Characteristically, the Austrian state, on the basis of the documentation of this artistic activity in the *Wien* book published in 1970, two years later brought charges for willful damage to property, although the Swedes, informed via Interpol, did not join the suit as private suitors but instead declared it lapsed.

To file the suit, the Austrian government made use of a law from the Metternich era. According to this law, an Austrian national could be tried for his conduct outside Austria according to Austrian law. The Austrian state in its fury even deemed the entire affair worth an expensive reconstruction of the museum's glass front with the help of the Technische Universität Wien, in order to test – using the same tools actually used during the act – whether a so-called *dolus eventualis* obtained; that is, whether the possibility of damage to the glass windows was rationally foreseeable. Upon this *dolus eventualis* the state sought to found its case, but the windows held, so no conviction was achieved. —Peter Weibel, 2006/22

Translated from the German by Peter Dahm Robertson.

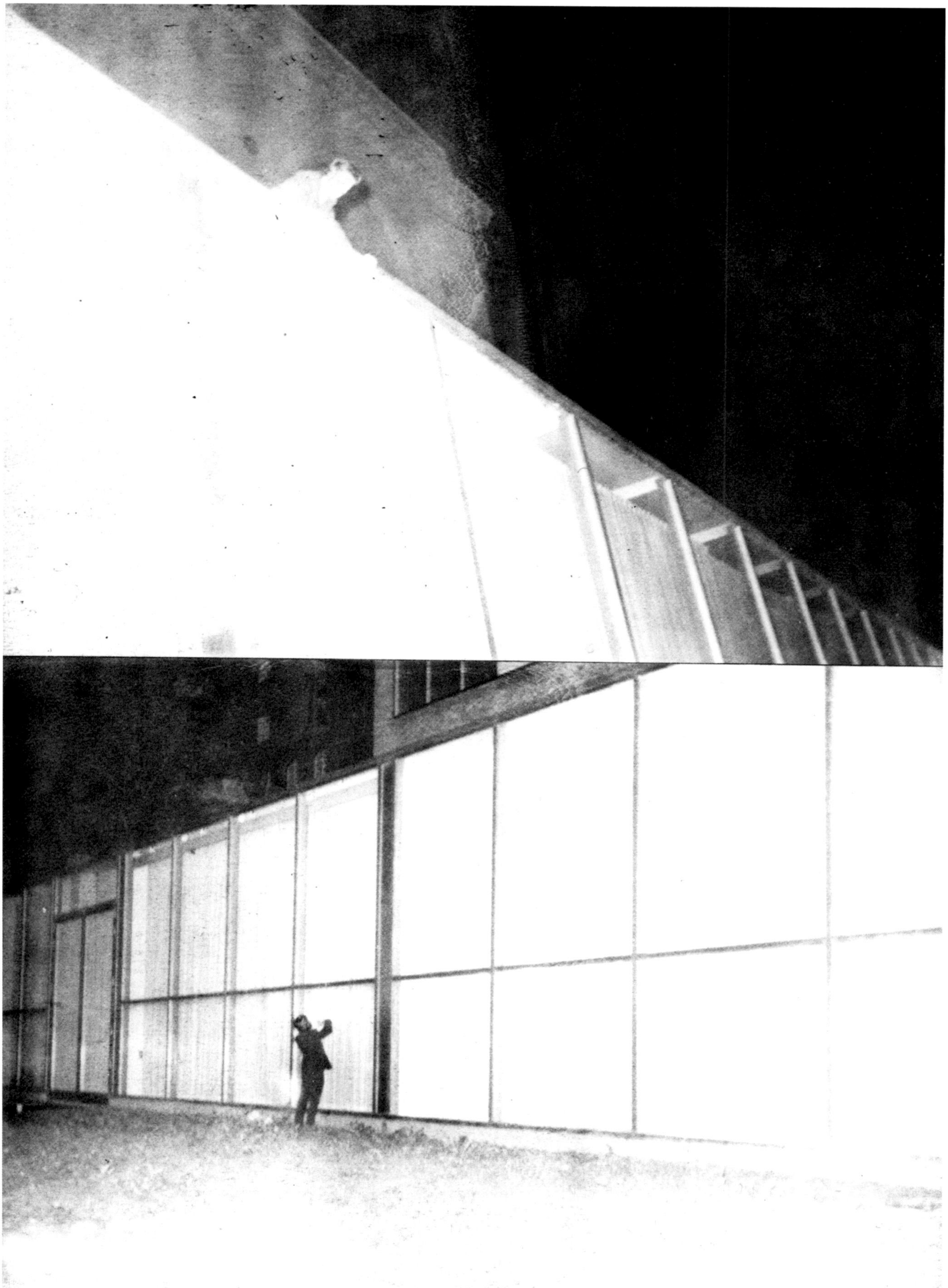

Franz Kaltenbeck fastens
the suction cups and the
different tools

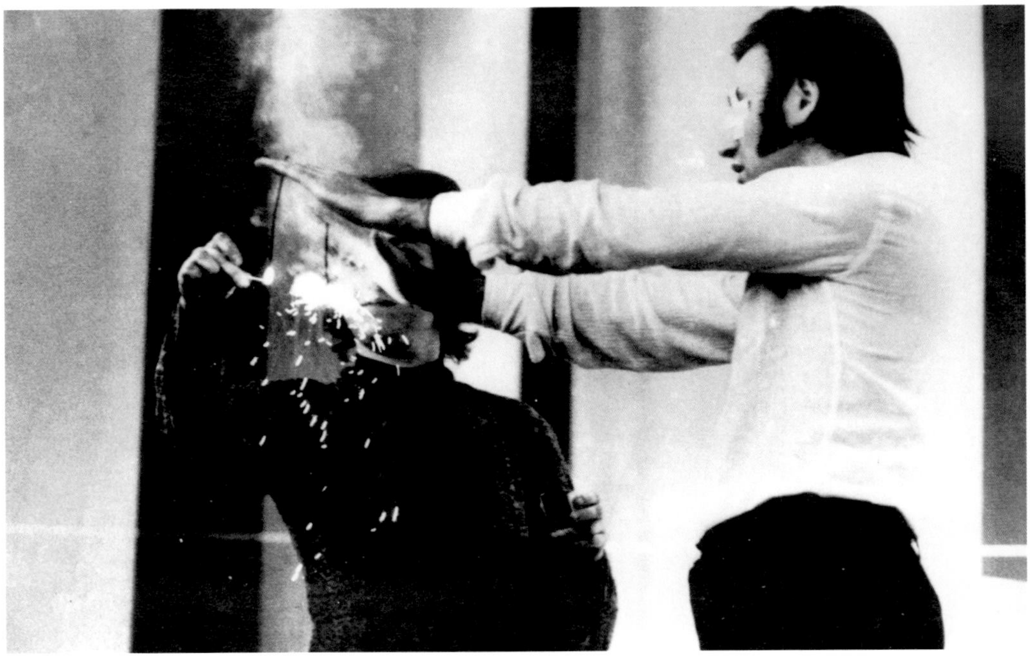

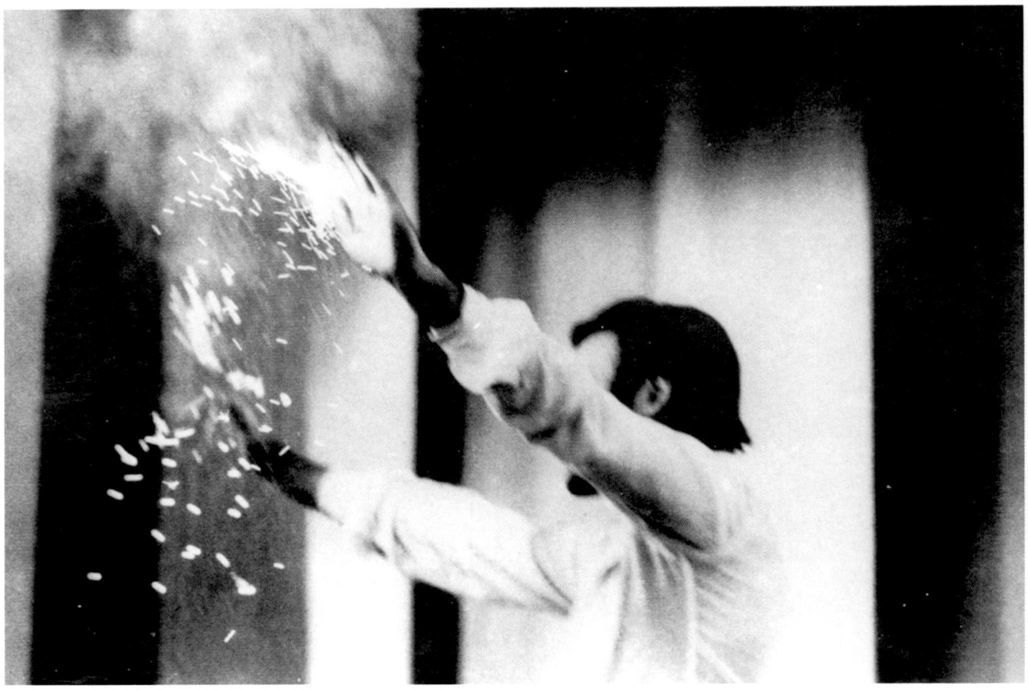

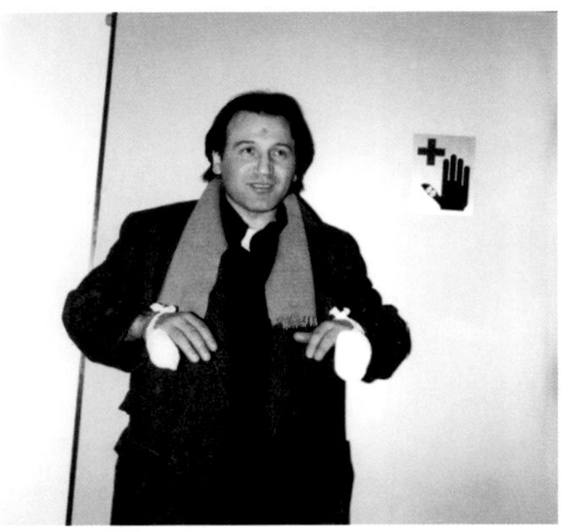

Peter Weibel after a performance at Forum Stadtpark, Graz, 1981

Initiation

1971 | performance; Experimenta 4, Frankfurt/Main, June 4, 1971

Eyes taped shut with crosses of medical tape, lengths of fuse in each hand. The fuses are lit and burn two lines in my index and middle fingers which are holding them. Whoever takes pain upon themselves while smiling initiates a journey that extinguishes the anxiety. One's own radiance undoes the repressive system.
The fingers require medical treatment.
After treatment at the hospital, the police call me and ask why I have "self-inflicted" burns on my fingers; whether I have been experimenting with explosives or prohibited chemicals. I offer the facts of the matter and their art-theoretical interpretation up for the record. They don't believe me.
Art – hospital – police: these are three stations of the repressive state system which I challenged.
The natural healing of the fingers as a process of art.
Process of body – process of art.
To show burning hands without any sign of pain demonstrates the resilience that the state cannot break your autonomy through torture of your body. —Peter Weibel, 1971/2022

Translated from the German by Peter Dahm Robertson.

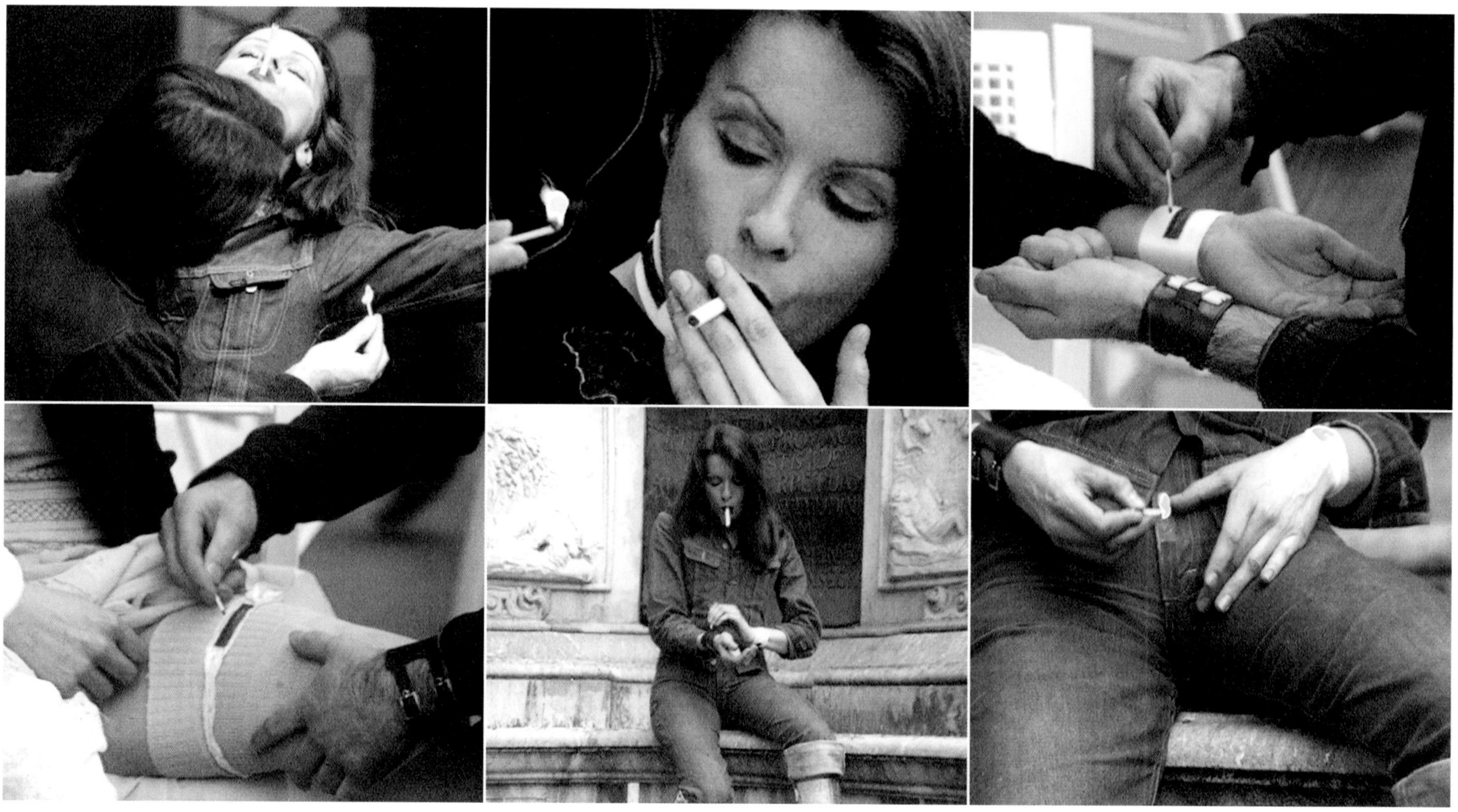

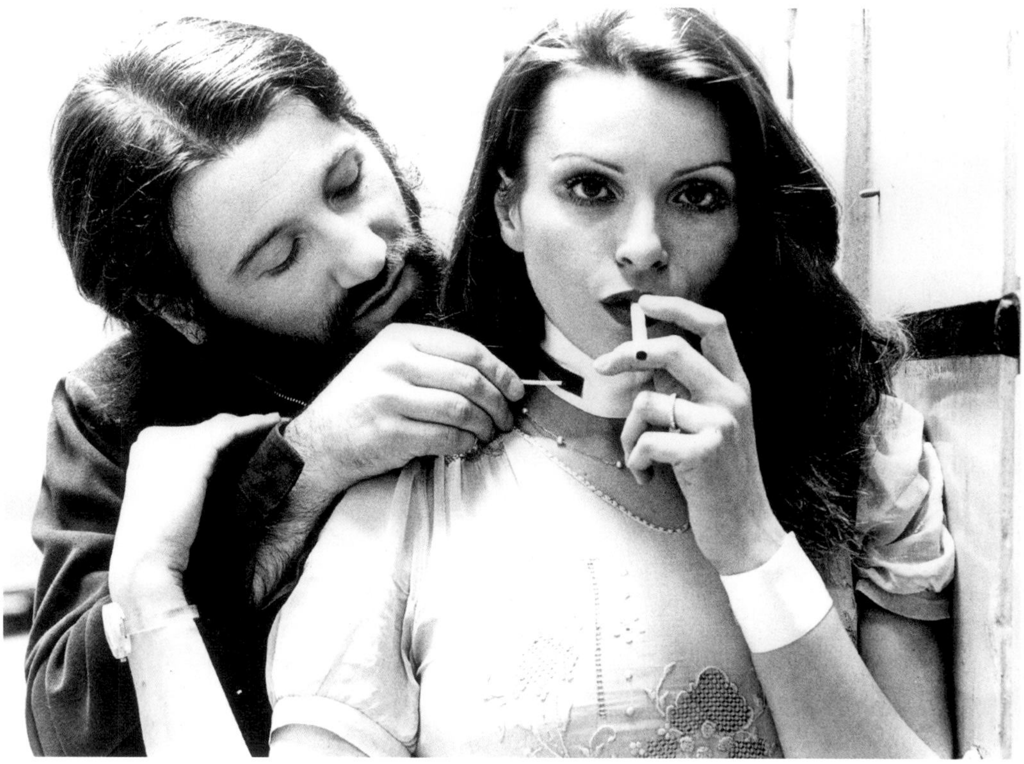

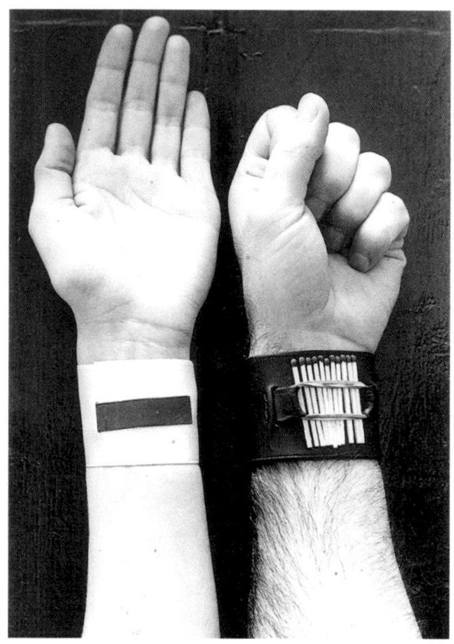

More Warmth among Human Beings / Mehr Wärme unter die Menschen

1972 — from the series *Tele-Actions*
first broadcast in *Impulse 7*, ORF,
June 29, 1972, 1:46 min.
actors: Susanne Widl, Peter Weibel

On the body of the actress (Susanne Widl) the match strike paper from matchboxes is fixed at different, especially erotically connotated body parts (neck, armpits, garter belt, genital area). The actor (Peter Weibel) lights a match on these friction surfaces and gives the actress a light. In the end, the actress lights her cigarette herself by striking the match on the strike paper in the genital area.

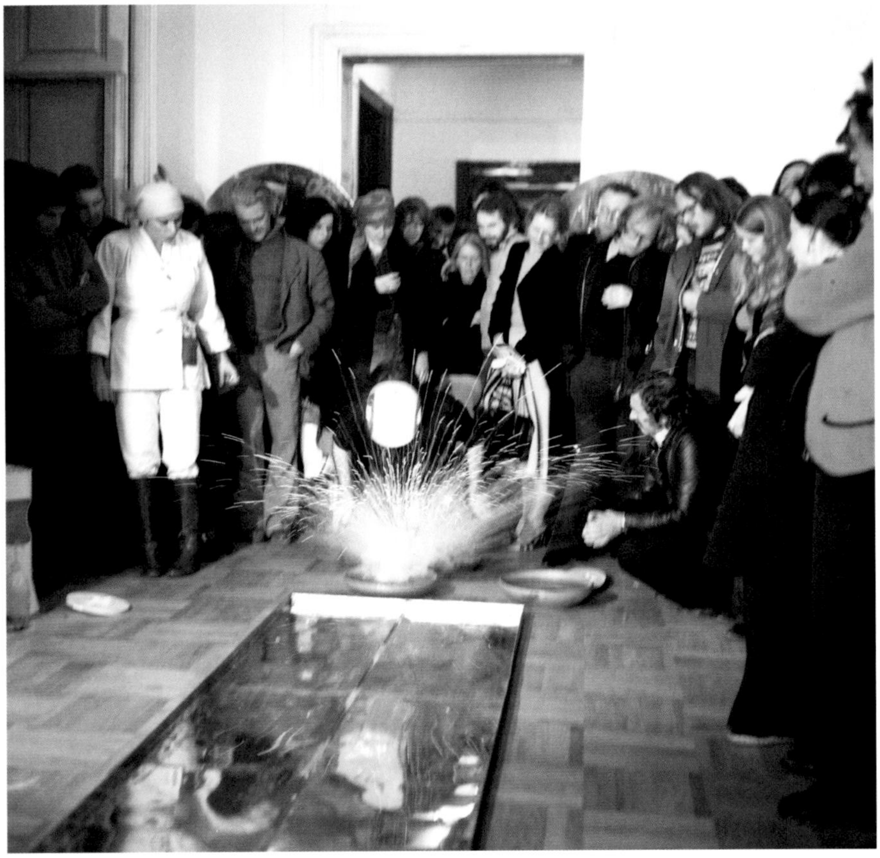

Performance at Galerie nächst St. Stephan, Vienna, Oct. 15, 1975

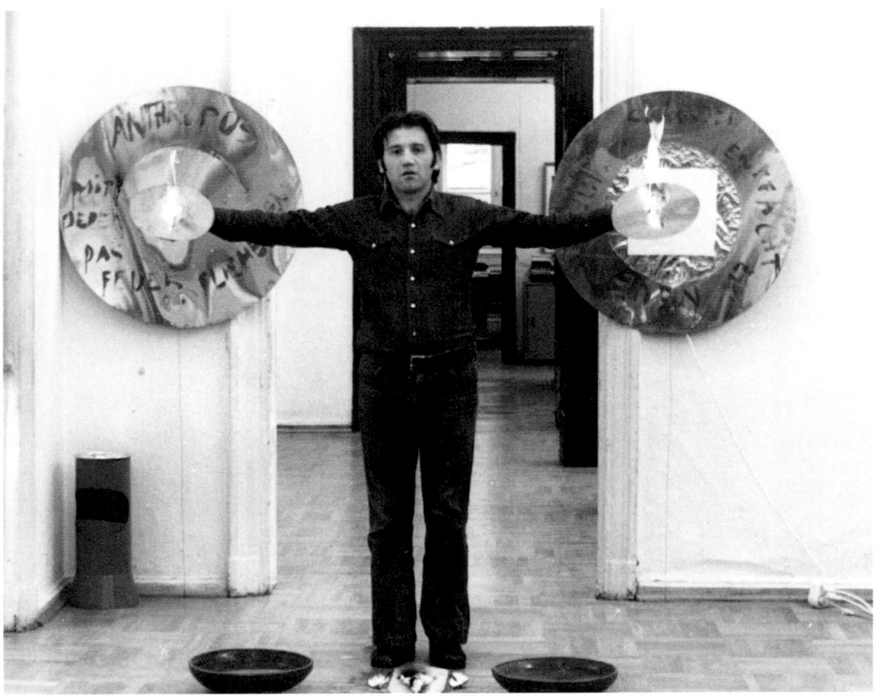

Fireflight – Flightfire /
Feuerflucht – Fluchtfeuer

1975	copper tub filled with water, 2 oval wooden discs, wooden mask, iron shoes, 2 clay discs filled with water

Copper tub, copper signs, ceramic basin, iron shoes, gloves, and face masks made of wood – all of which (paradoxically) began to burn on contact with water. The three ages: wood, clay, iron. The three *alchemia*: metal, water, fire. In the iron shoes, I walked over water and behind me fire burned; I dipped my hands into water and they burned, I immersed my face mask, on which a cross was carved, in water and the cross began to burn. The basis of the chemical reaction: the metal potassium ignites on contact with water. The situation of the *anthropos*: by fleeing the fire, their fleeing steps create new fire everywhere. The person who flees civilization and migrates into nature carries civilization into nature. —Peter Weibel, 2006

Translated from the German by Peter Dahm Robertson.

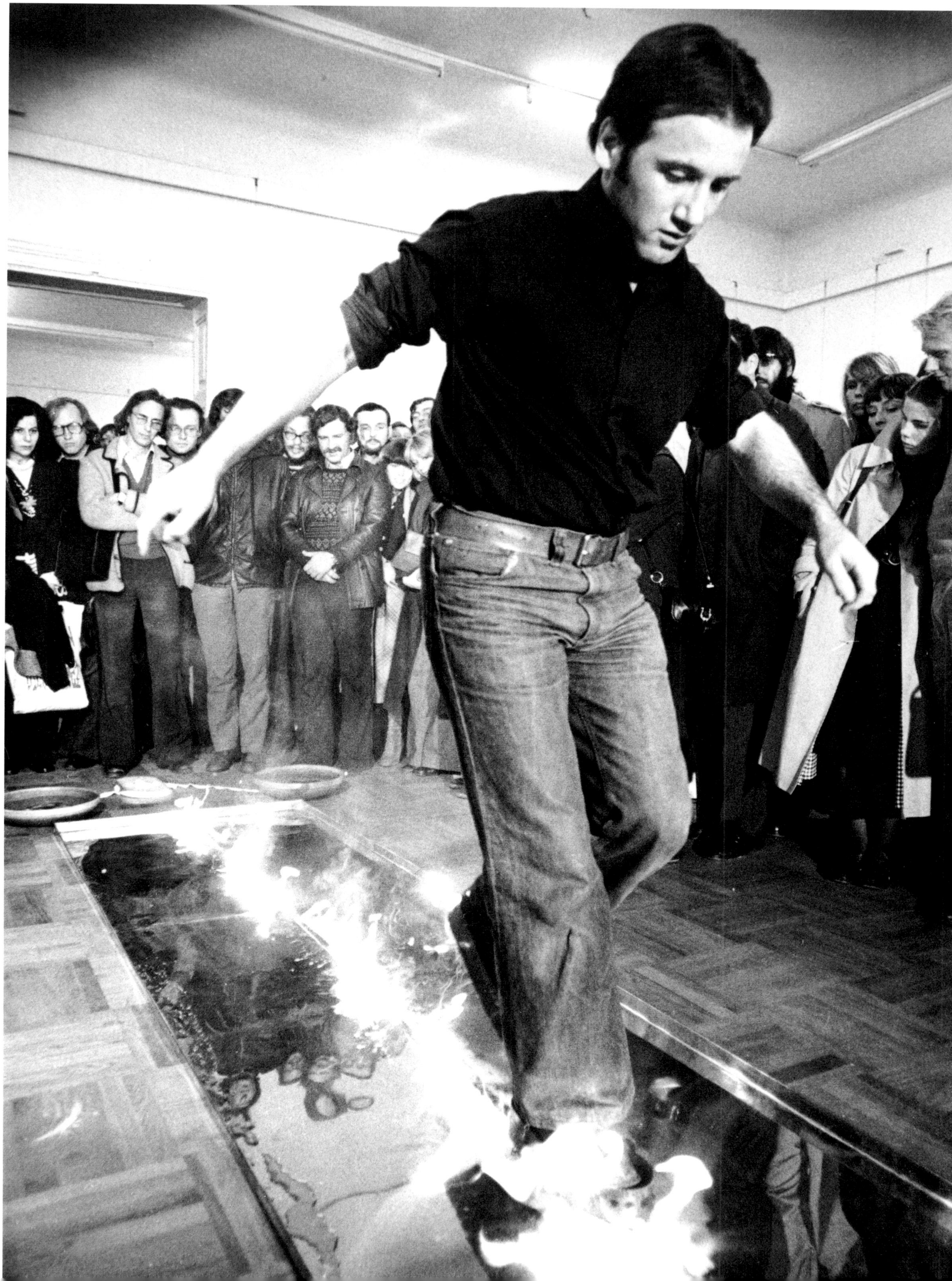

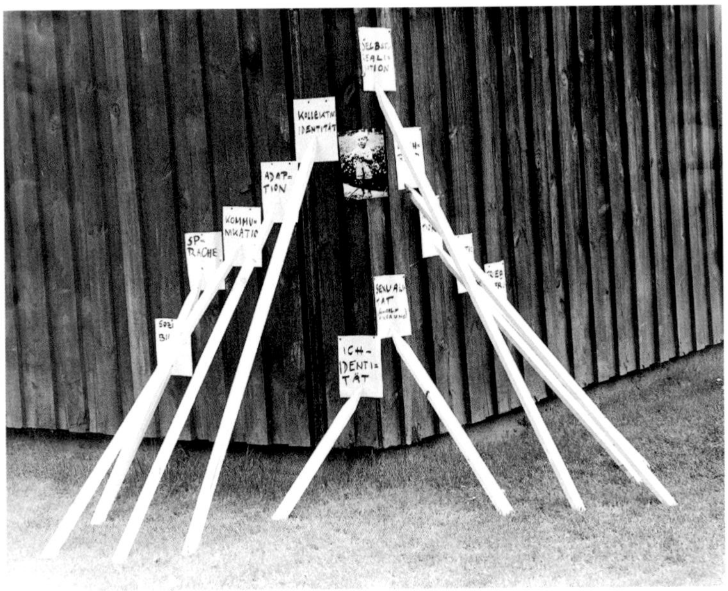

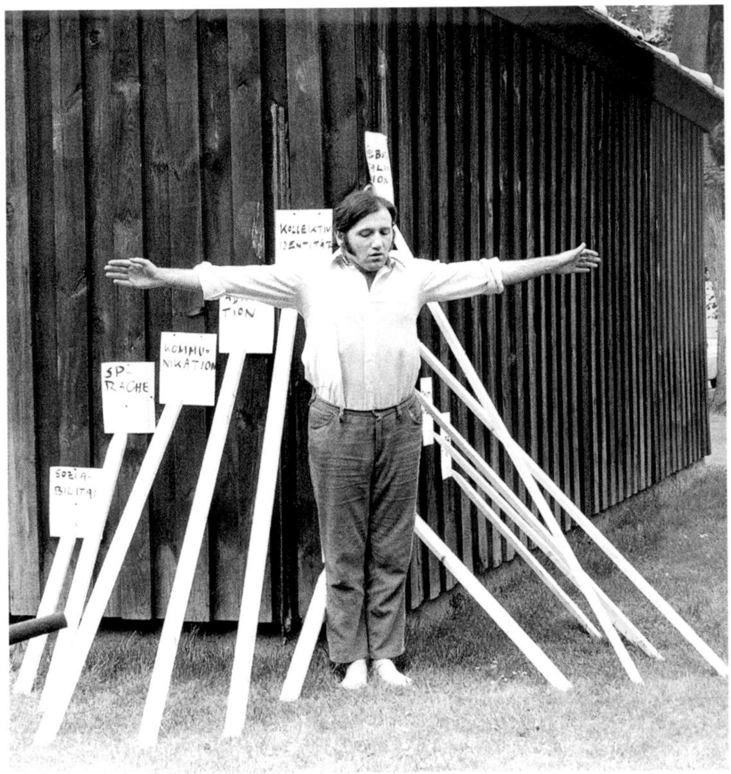

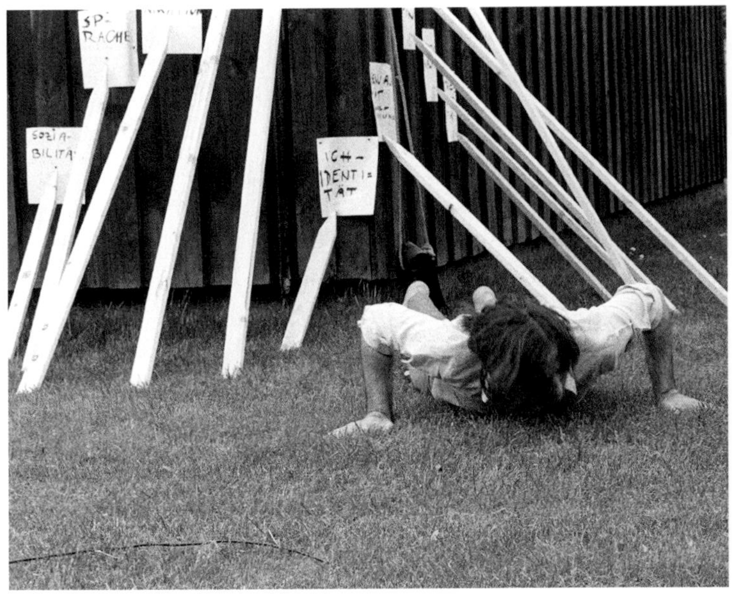

Social Matrix / Sozialmatrix

1971	text performance; wooden spears, socialization-related terms on paper, childhood photograph, scissors, fingernails, hair, skin, Festival Konkrete Poesie, Hanau, May 1971

Standing at the corner of a building, I spread my arms and fall through space to the ground. My shirt rips open, revealing a cross stuck to my back. I cut away everything on my body that grows and embodies time: hair, skin, nails. I bury my bodily residues in the shape of my body (legs vertical and arms horizontal), which also represents the four cardinal directions (N, S, E, W) and the shape of the cross.

The journey of life is an experience on the rails of concepts. The body, trapped in the corset of language, impaled on ideas, transfixed by words, embedded in time and space, nurtures the yearning of the sensory organs to be unbound. The language-ensheathed body interferes with the liberation of sensory data from their conceptual fetters. Language is the matrix of socialization. Linguistic conditioning begins in childhood and ends at the cross of death.

—Peter Weibel, 2006/22

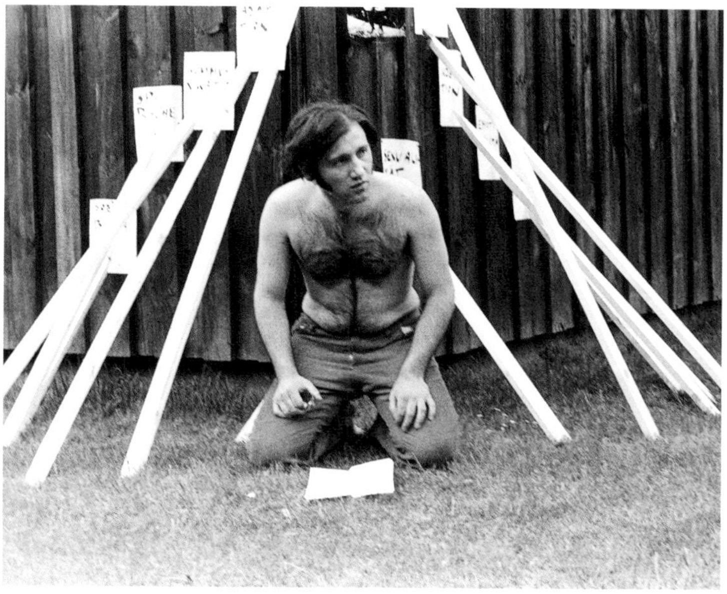

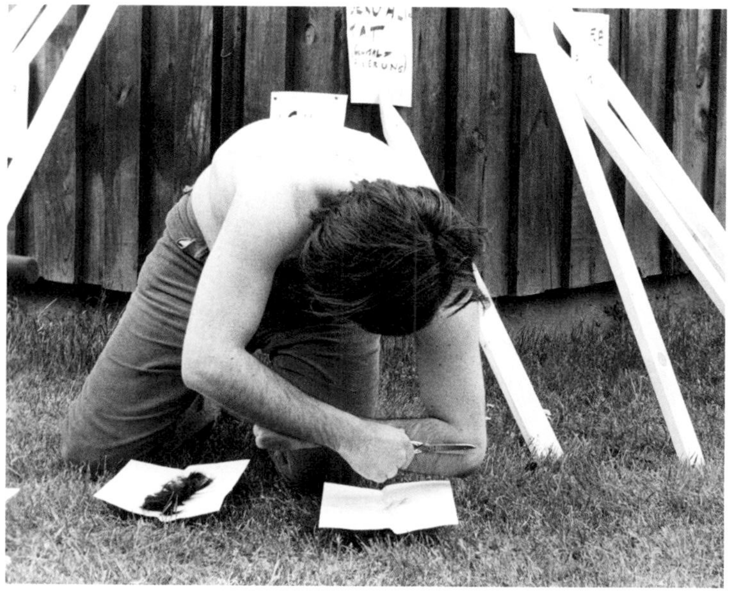

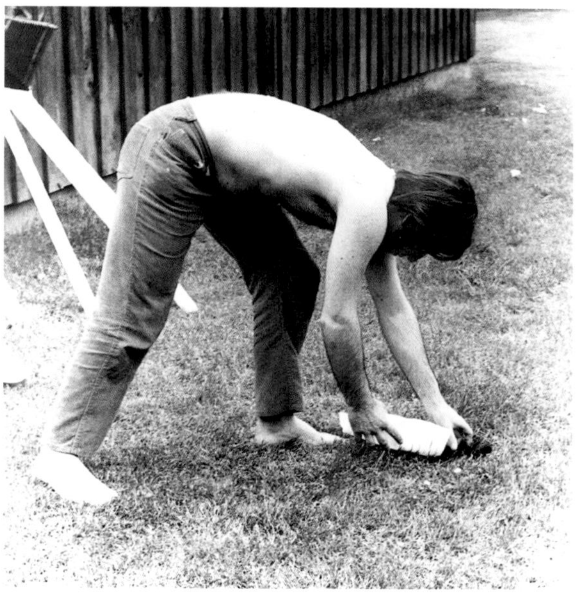

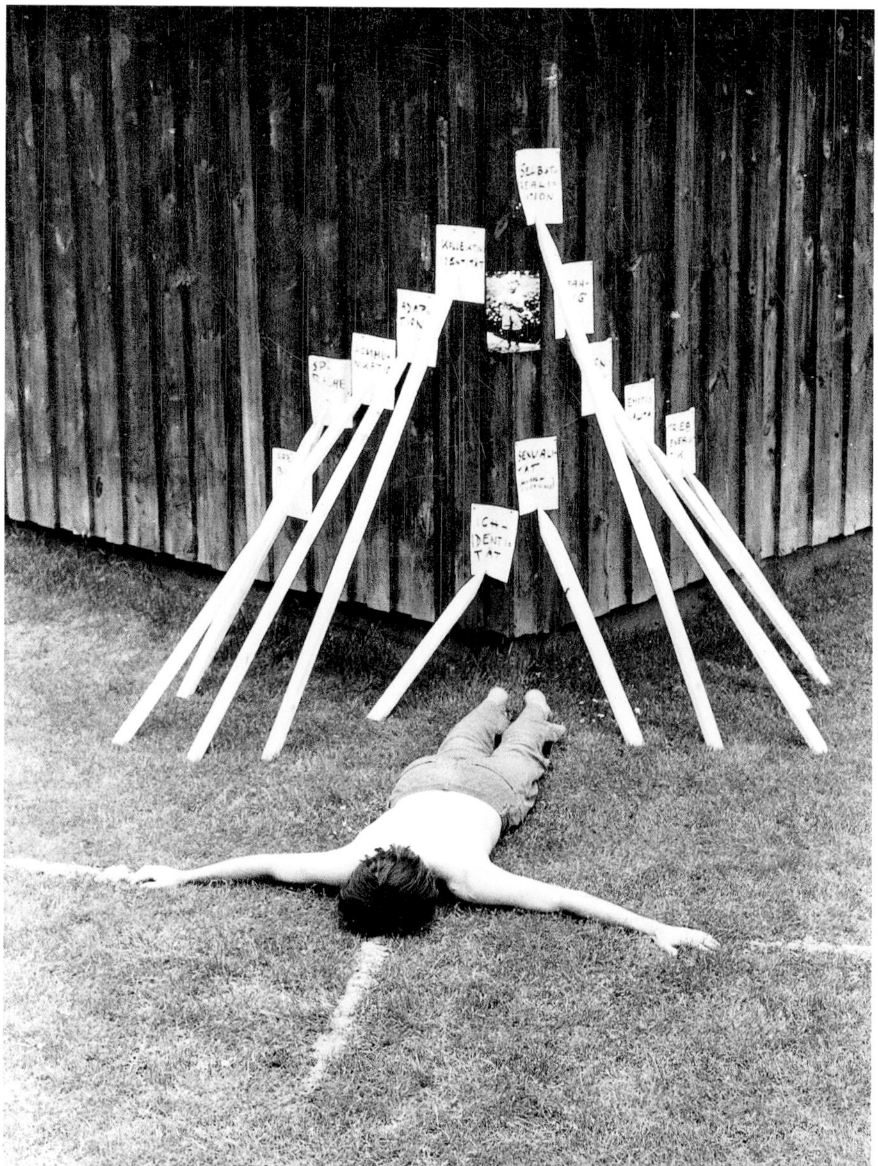
ADAP-
TION
ICH-
IDENT
TAT

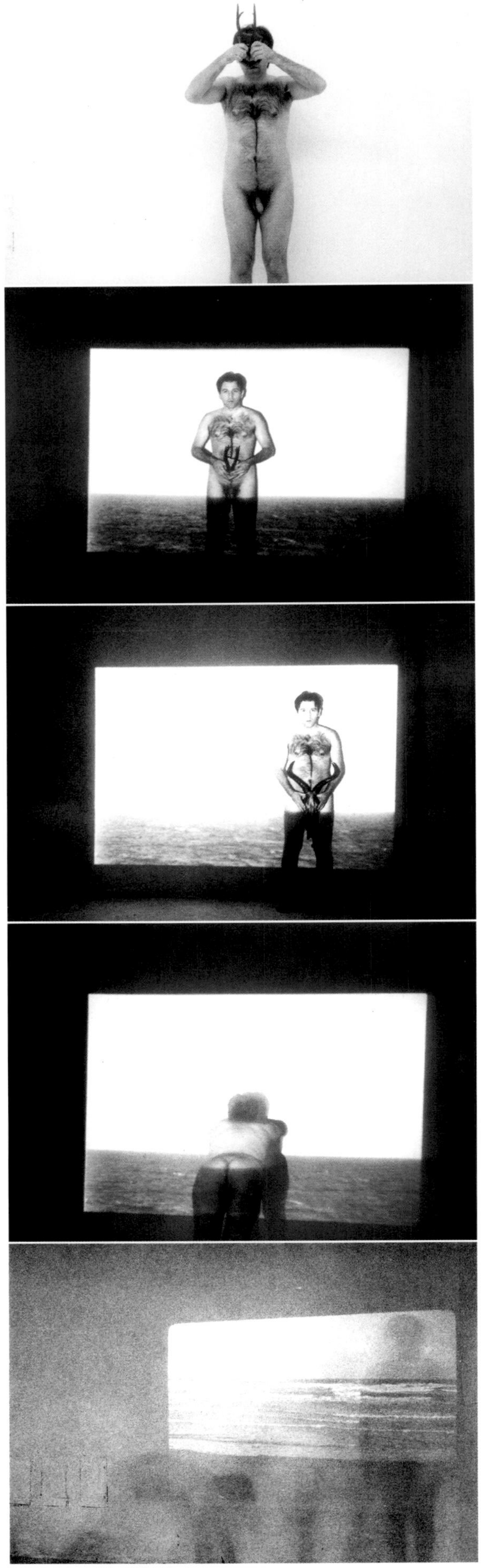

Against the backdrop of a slide depicting earth, air, and water, a presentation is given on the development of human primordia via fish, birds, and higher eutherians.
The descent of the gonads in male mammals plays a special role in this process, since their downward movement is accompanied by elaborations of the embryo's cranial and caudal ends that may be interpreted as a form of mammalian self-representation. Because the organs of spontaneous expression (involuntary utterances) are typically passed over in the study of human language, they are given priority here over controlled expression, with particular attention paid to intermediate forms such as laughing and crying. With the help of verbal and nonverbal interactions, an attempt is made to locate an anthropological basis for language in the principle of human ontogenesis. The preliminary stages of semantics and pragmatics are investigated. If language is, in one aspect, a social technology of behavior, and linguistic acts are a system of social interactions, then this comparative morphology is intended to bridge the gap between language and behavior, between verbal and nonverbal communication, and to highlight a possible common origin. —Peter Weibel, 2006

Translated from the German by Patrick Hubenthal.

Time Lung: Performances on the Morphology and Consciousness of a Vertebrate. On the Knowledge of the Body I / Zeit-Lunge. Aktionen zur Morphologie und zum Bewußtsein eines Wirbeltiers. Vom Wissen des Körpers I

1971 | performance; Voom Voom, Vienna, September 22, 1971

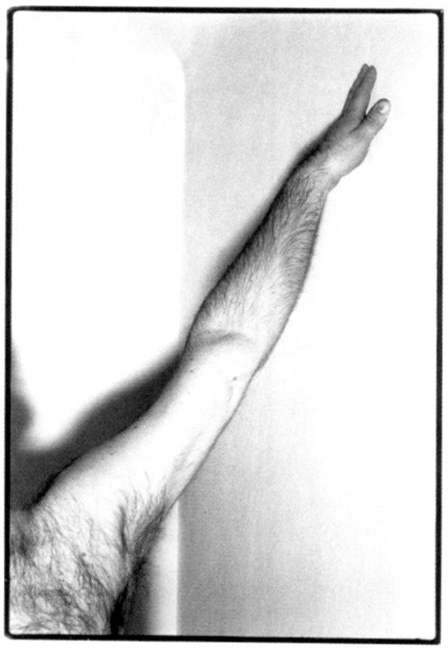
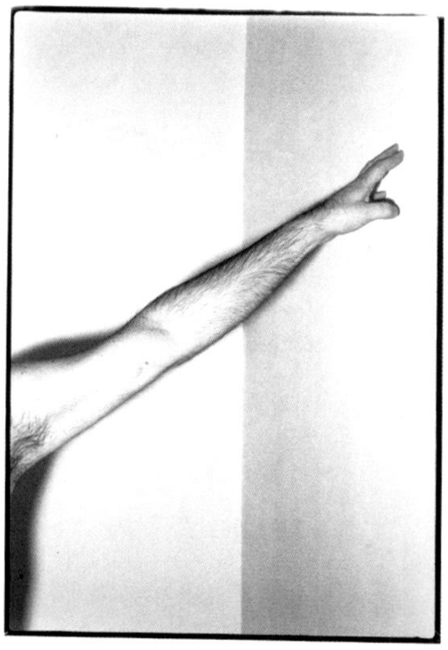
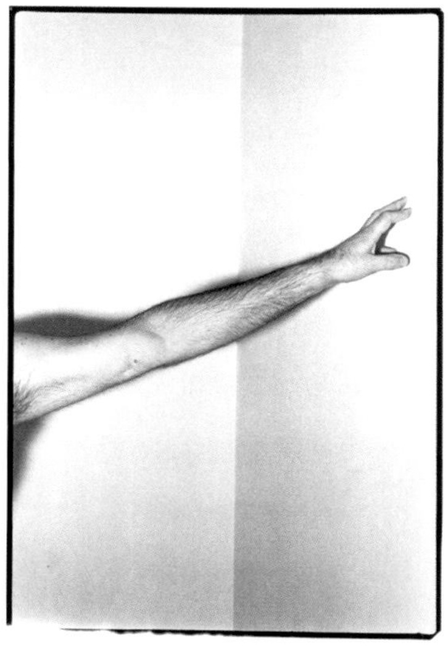
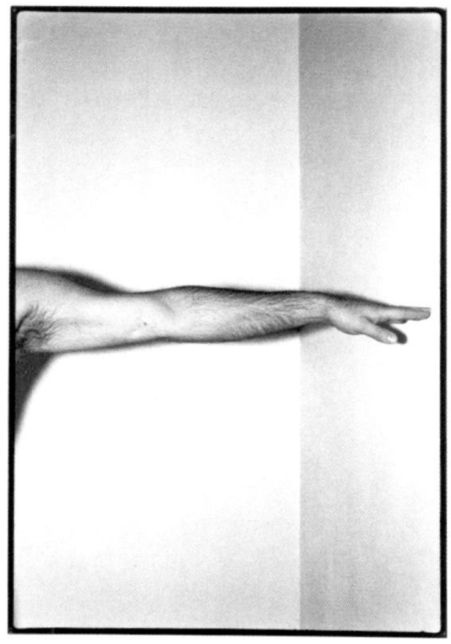
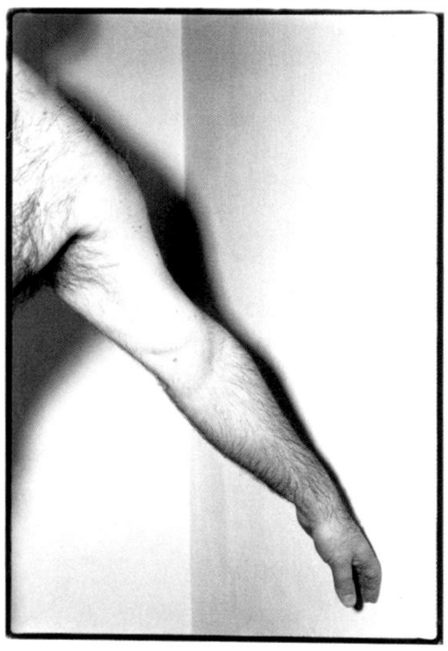
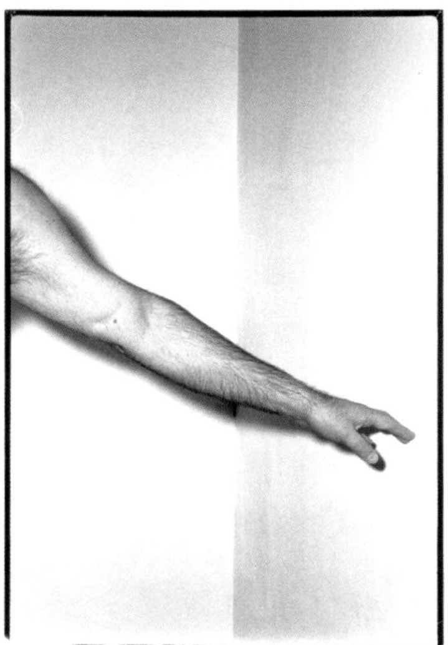
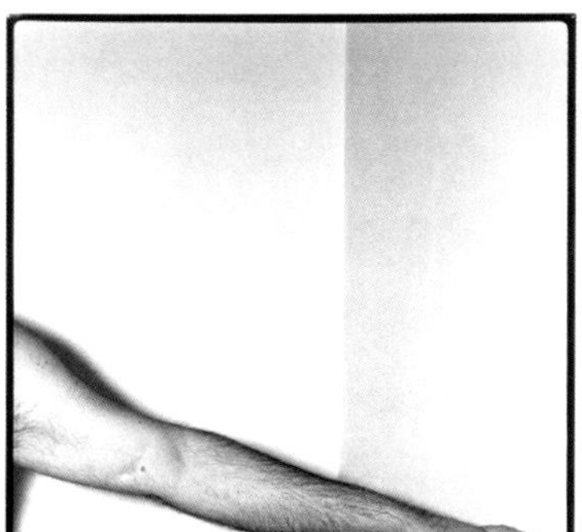

Time Bone – Space Muscle. On the Knowledge of the Body II / Zeitknochen – Raummuskel. Vom Wissen des Körpers II

1972 | 7 b/w photographs on hard foam boards, 78.3 × 53 cm each

In the seven-part work *Time Bone – Space Muscle. On the Knowledge of the Body II*, the movement of an outstretched arm is documented photographically. The photographs are arranged on the wall in such a way that together they form an arc that traces the movement – a cinematic process: photographs as film frames. Photography offers the possibility of immediately capturing movements, which the human eye can only perceive successively, and thus a photograph can highlight the anatomical precision of the unconscious process of a movement. For Weibel, who sees the body primarily as a storage medium for information, every movement of our fingers and arms reveals "the capital of evolution" unconscious knowledge that has poured into bones, muscles, and nerves over millions of years and constitutes a nonverbal language.

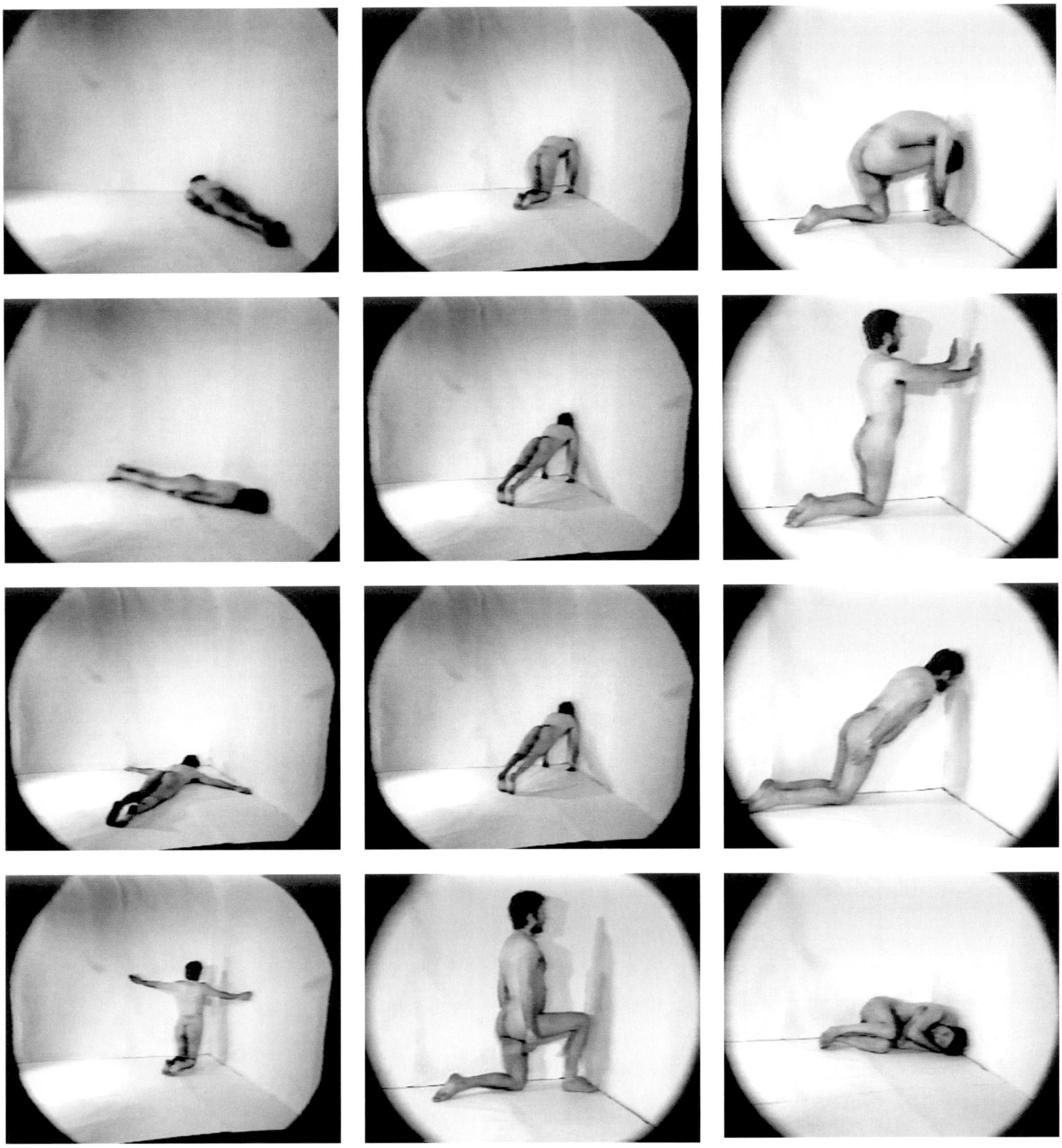

Volcanology of the Emotions / Vulkanologie der Emotionen

1971 (performance), 1973 (video)	body performance, video: 7:17 min.

The body performance *Volcanology of the Emotions* took place in Vienna in 1971. Here, Weibel deals with the body language of humans, their behavior in space, and the basic principles of human emotions that can be read from it. For Weibel, the actions of humans in space are a mirror of their emotional constitution. The more complex the geometric relationship of the body to space, the more expressive, or pathological, the respective emotional expression is perceived. A more complex geometry of body poses leads to more complex emotional expressions. Body poses in space can thus be understood as a direct expression of emotions. It is a nonverbal language created through the process of evolution, which can be analyzed according to criteria of language criticism and examined for the knowledge stored in it. Weibel developed a semantics of bodily expressions and emotions by means of a syntax of body positions in space. The emotions are only an effect of syntactic manoeuvres. A critique of Viennese Actionism and all other expressionisms.

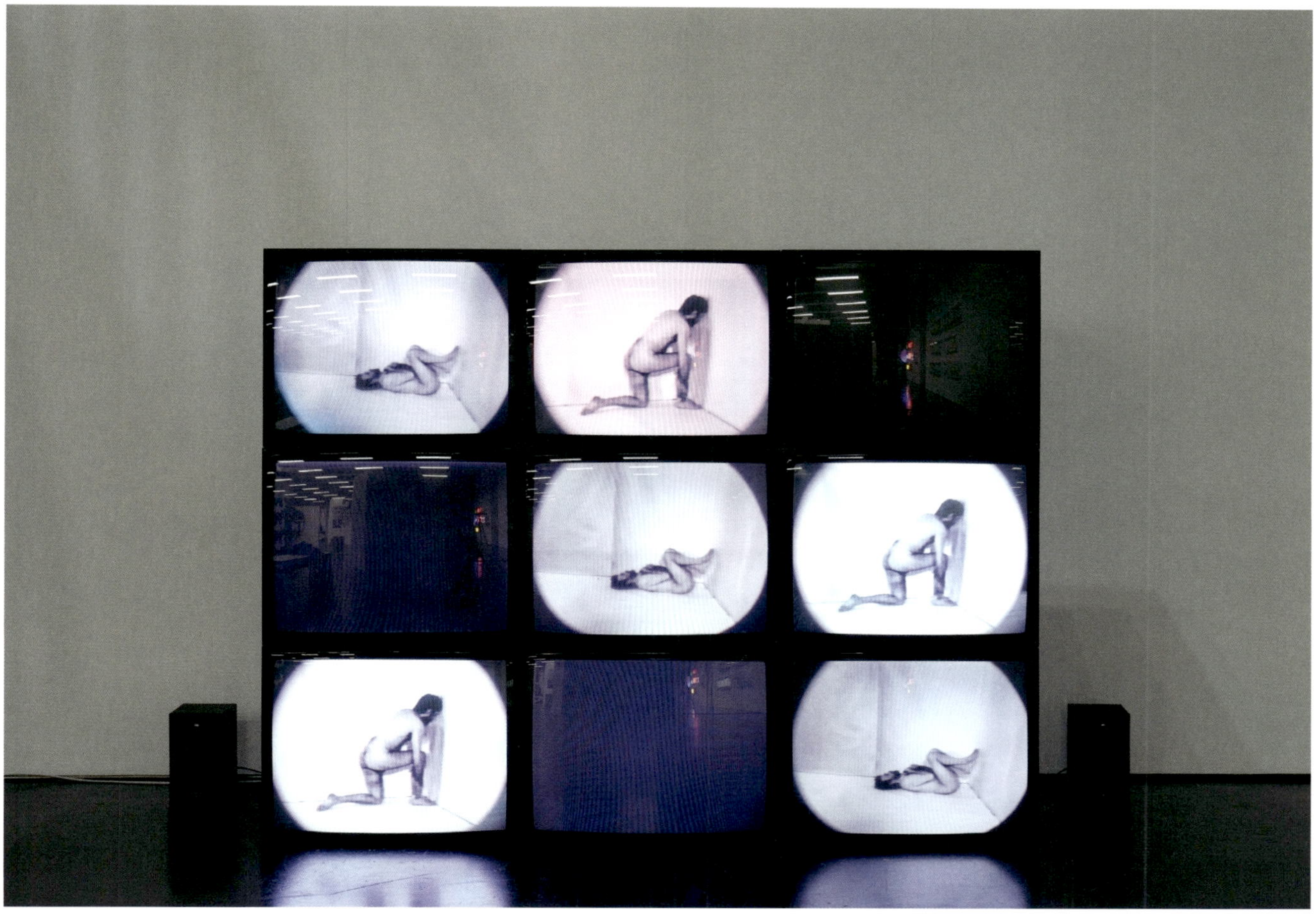

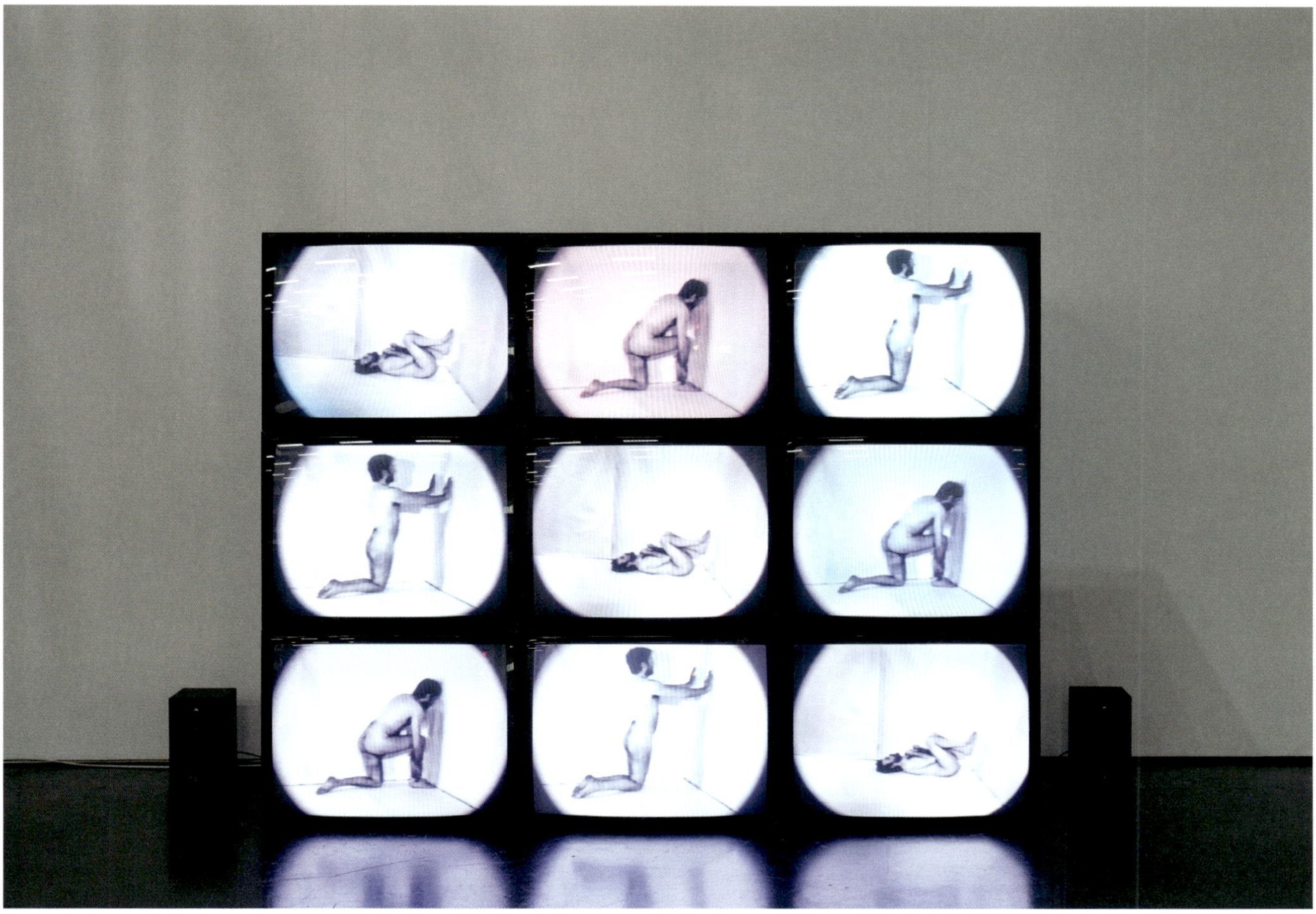

Installation views
respectively, Peter Weibel,
ZKM | Karlsruhe, 2019–20,
presented as a 3-channel
video installation on
a 9-part monitor wall

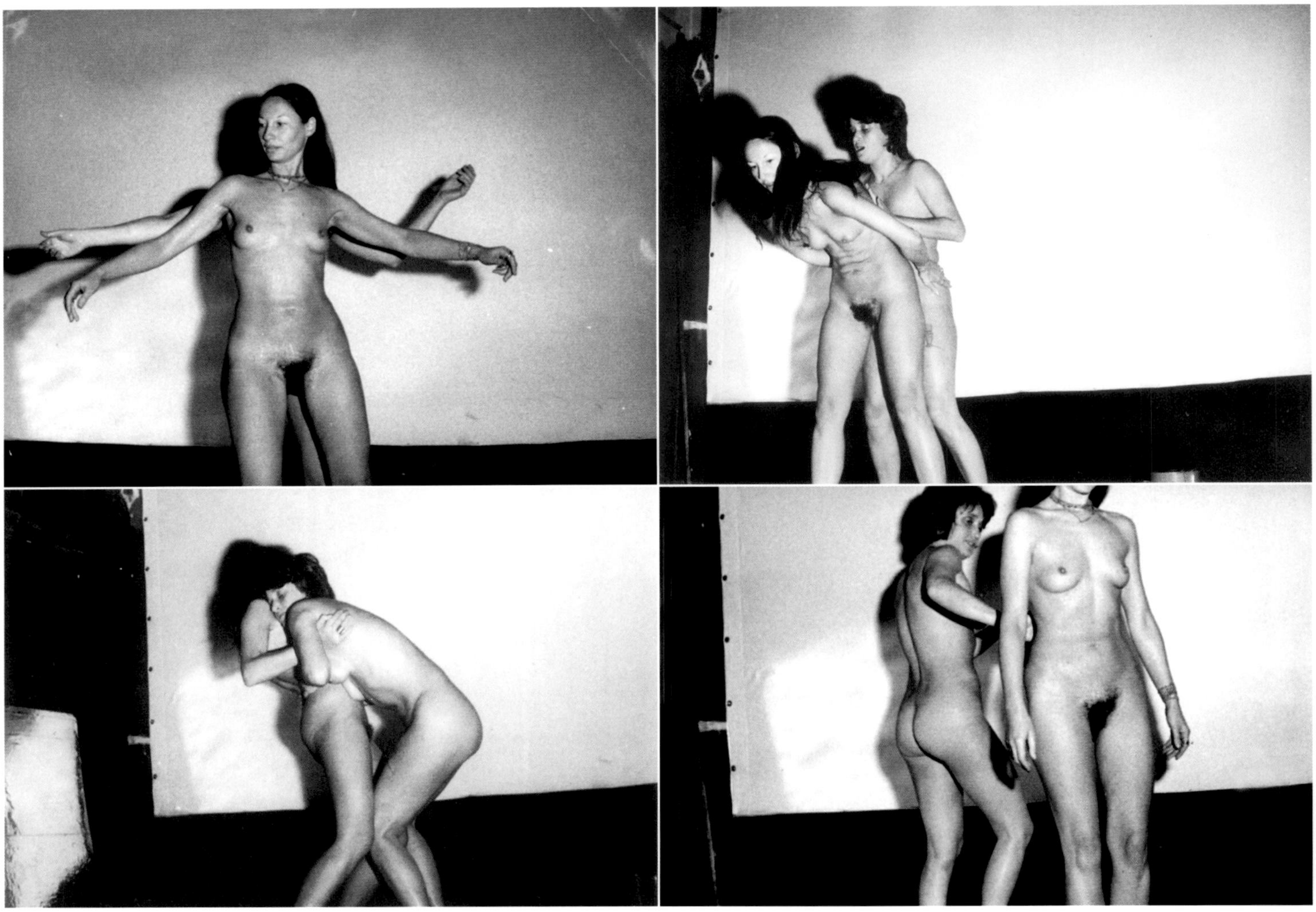

Energy Exchange / Energie Austausch

1971	delegated performance; Amsterdam 1971

Energy Exchange was performed in November 1971 by VALIE EXPORT and Erika Mies as Expanded Cinema, or rather body heat exchange performance in Amsterdam. One of the two women throws water over the other. The dry woman and the wet woman rub each other's skin on stage in front of a screen until both are dry. In contrast to most of the performances of that time, Weibel does not act on stage himself during the performance but assigns its realization to the two performers. He thus anticipates the genre of "delegated performance" that Gabriele Mackert introduced in 2002 and Claire Bishop in 2012 for Santiago Serra's works (since ca. 2000). As a film without film, *Energy Exchange* is one of the artist's early works in which he deals with the conventions of the classic film system – in this case the exposure of the female body as a voyeuristic object.

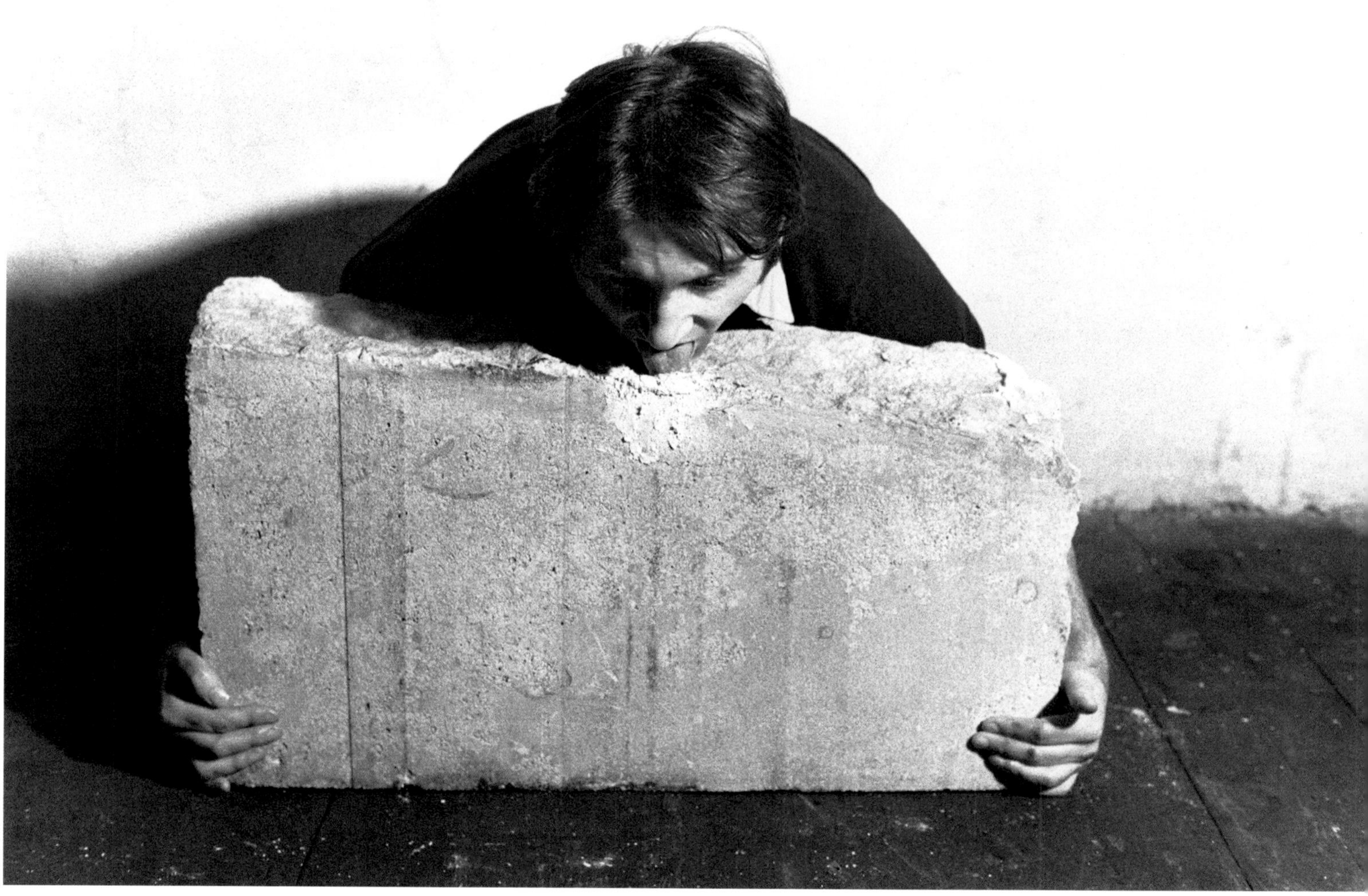

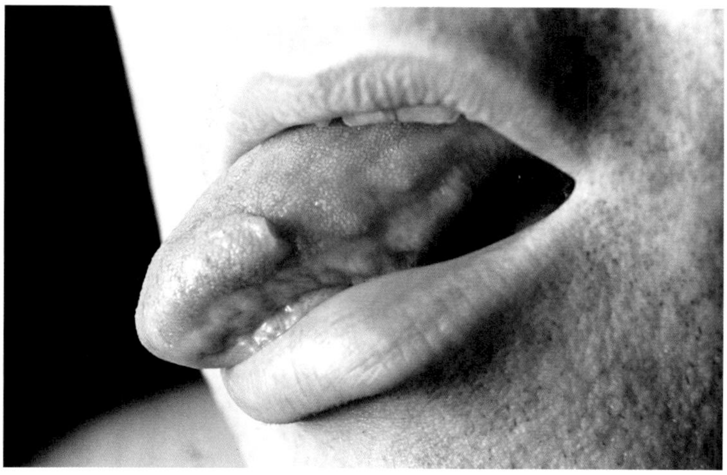

In the action *Space of Language*, Peter Weibel held a block of stone in which his tongue was set in concrete for hours so that he could not speak; a prisoner of language. The artist described the action as "sculpture with a living organism attached." When Weibel tried to separate himself from the stone after some hours had passed, it was not possible at first. The tongue was stuck in the concrete. Only with the help of tools could he finally free his tongue from the stone. During this process of liberation, the tongue was injured and violated. This was a metaphorical event to demonstrate the pains that need to be endured when people try to liberate themselves from the prison of language, of established codes. The basic principle of language is a spatial one: it makes air vibrate, and thus opens up the space in which it is heard. However, language is not only a phenomenon of immaterial vibrations, these have real, material consequences. Language is like a garment, a wall, a material space that surrounds and can constrict people. Sentences are not only verbal judgments, but can change, and even kill real lives. A mere spoken word, apparently only sound waves, can determine the freedom or imprisonment of a person in the courtroom. Where does this power of language and those who exercise it (such as judges, priests, or civil servants) come from? In order to soften the violence of language, the well-known separation of powers was introduced: Legislature, Judiciary, and Executive. But in vain. With this action, the artist renders the violence of language visible.

Space of Language / Raum der Sprache

1973 | action; Vienna 1973

abstract representation

$$w \subseteq w$$
$$a\ [\subseteq]\ w$$
$$b\ [\subseteq]\ f(a) \subseteq w$$
$$c\ [\subseteq]\ g(b) \subseteq g(f(a)) \subseteq w$$
$$d\ [\subseteq]\ h(c) \subseteq h(g(b)) \subseteq h(g(f(a))) \subseteq w$$
$$z_\infty\ [\subseteq]\quad \underbrace{\leftarrow \dots f_n \dots f_3, f_2, f_1(a)}_{f_\infty(a)} \subseteq w$$

table of symbols:

$\subseteq$	... inclusion relation, contained in subset of characteristics, in short: subcharacteristic
w	... weibel
a	... karl
b	... robert
c, d, etc.	... any other person
∞	... infinite
f, g, h	... functions

Transcript of the equation on the blackboard

For Peter Weibel, the crossing of boundaries represents a generative principle inherent to art. The accelerating convergence and fusion of the arts in the wake of Romanticism has been plain to see. At the same time, a broader view of the artist's materials has taken hold. The materials of art no longer refer solely to the external world – that is, they no longer manifest as imitations of the world around us, but have become self-referential. The material's identity has come to the fore, which allows for the concept of an autonomous, independent material and results in an expansion of what constitutes a material in art. Everything can serve as artistic material, everything is art – not just every material, but every activity as well – and every person is an artist; so goes the credo of the expanded arts movement.

In the light of these expansionistic tendencies in the arts, fundamental questions of identity become clearly discernible – especially in the case of the famous example provided by Marcel Duchamp in 1917 when he exhibited a urinal, a thoroughly unremarkable object, under the title *Fountain*. While the external form of the object remained intact, its meaning changed as a consequence of its new context. Duchamp had brought about a shift in identity, turning an ordinary toilet into a work of sculpture. The concept of the readymade was born. This transference of everyday things into the museum, into the context of art, thereby making artworks out of trivial objects, would subsequently be carried out on a massive scale, in every variation.

With *Identity Transfer Action*, a performance piece created for the trinational biennial Trigon '75 at the Künstlerhaus Graz, Peter Weibel took the technique of identity shifting to a new level, moving the focus from the identities of objects to those of people. As a first step, the artist chose another person to represent him before and during the exhibition, the theme of which was "Identity – Alternative Identity – Counteridentity": Karl Zuckriegl, a metalworker and former prison inmate, took on the communicative part of Weibel's identity, always appearing at his side and speaking for him. Weibel's communication took place via a double, who acted as a filter or intermedium. At the opening, which was attended by the president of Austria, he extended the identity transfer even further: Now Zuckriegl took over Weibel's role completely and was represented in turn by yet another speaker – namely, his own partner, Robert. Weibel was absent; only the installation and the two persons substituting for him were present.

The identity transfer Weibel initiated was based on the idea that we humans define ourselves as individuals by means of the characteristics that make up our identities. These characteristics, however, take the form of traits that can also be found in other people. For example, both Weibel and Zuckriegl speak German and belong, like Robert, to the male sex. And various characteristics of these three men can also be found in other people: for example, the fact that they possess human form. Human identity, then, is the sum of many constituent characteristics, and these, as this work demonstrates, can easily be transferred to other human beings. By an operation of subtraction, Weibel attempted to find out at what point one loses one's defining traits, one's identity, leaving nothing but an emptiness, devoid of the particular characteristics that actually constitute the individual. With *Identity Transfer Action*, Weibel demonstrated a fundamental principle of our culture and economy: identification (with a profession, a class, a gender, a nation, etc.). Society thus forms a continuum of identity transfers, substitutions, and identifications, in which personal freedom is an extremely relative construct.

Note: This text is a composite of Peter Weibel, "Grenzüberschreitungen in der Kunst der Gegenwart," in *Enzyklopädie der Medien*, vol. 3, *Kunst und Medien: Die Transformationen des Bildes in der Ära der Apparate*, 29–34 (Berlin: Hatje Cantz, 2019); and Karl Zuckriegl, "Mir fällt auf. Zu Peter Weibels und meinem Identitätstransfer," *protokolle: Zeitschrift für Literatur und Kunst* 11, no. 2 (1976). Translated from the German by Patrick Hubenthal.

Identity Transfer Action / Identität-Transfer-Aktion

1975	photographic documentation of a performance at Trigon '75, Künstlerhaus Graz; blackboard, table, chair, folder; 6 b/w photographs

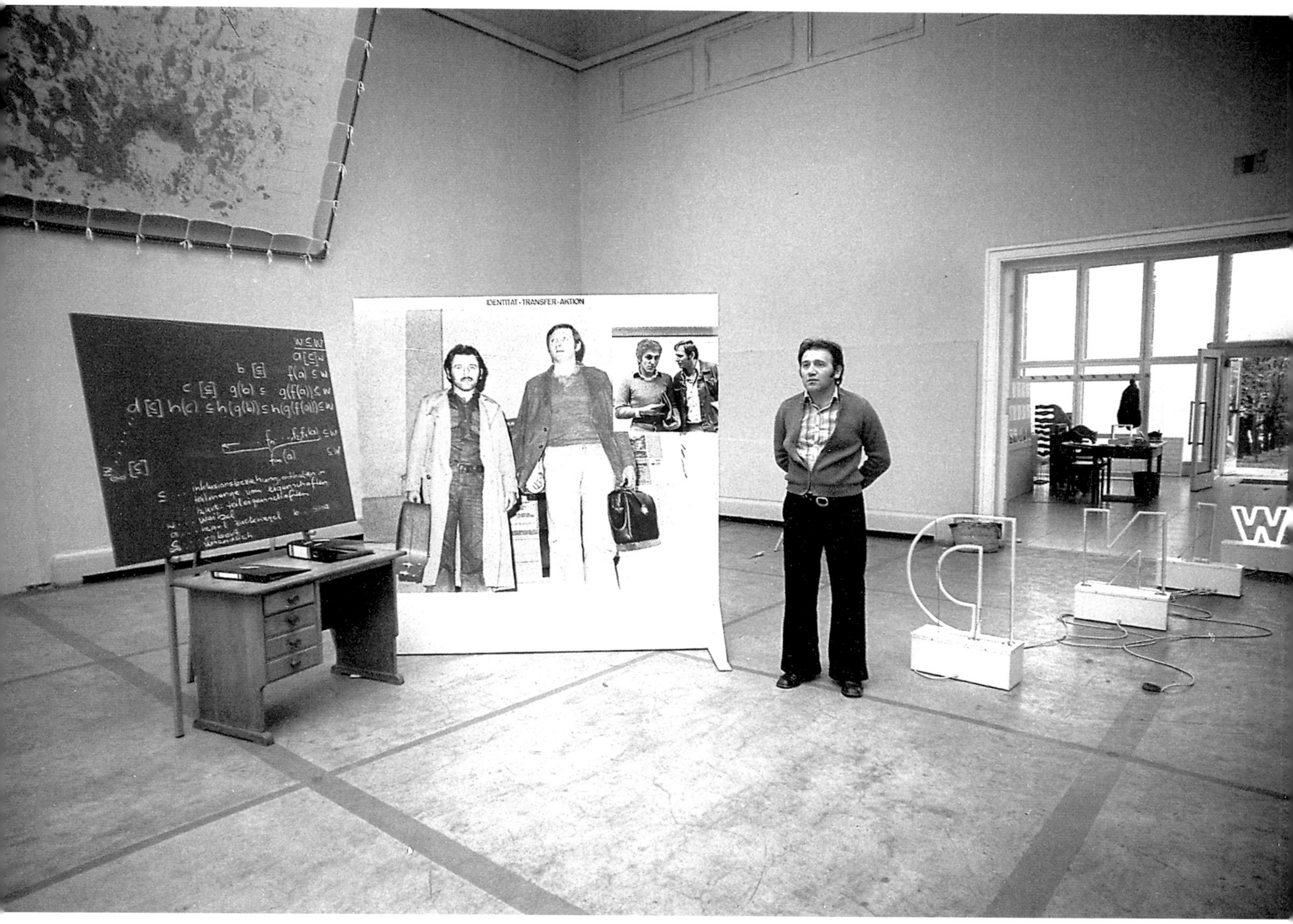

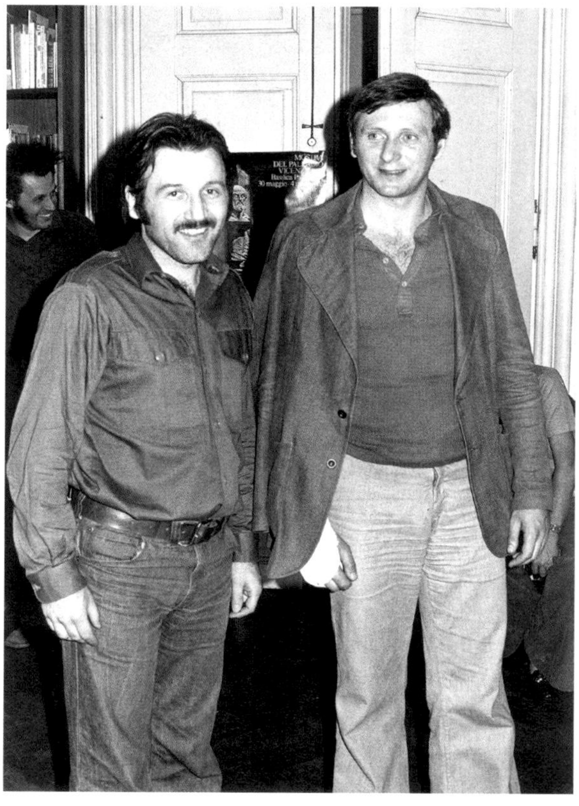

Peter Weibel and
Karl Zuckriegl

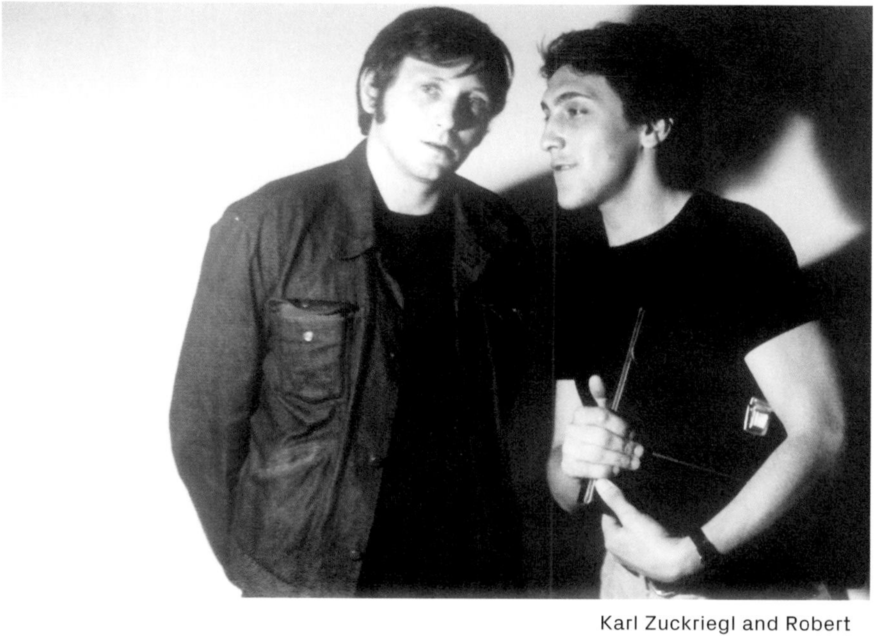

Karl Zuckriegl and Robert

I. Performance phase

Since the classic materials of representation like plaster, clay, and marble are dead, the artist substitutes himself for the material. Via the telephone and verbal instructions, the audience shapes the artist as a sculpture.
But the classic function of representation, which is homology (similarity of transfer), is dead as well. Therefore, the audience's instructions are performed in a nonhomologous, nonidentical way, specifically: as material it can only be shaped through the telephone calls. He enters the scenic space, picks up the receiver – but does not follow the instructions. Accordingly, there is a single homologous representation between audience and artist: my request to call me. Visually, this was expressed through the plaster hand (classic relic) – which I held in my hand during the entire performance. And in fact, one participant recognized this wound of identity that existed between his wish to call and my wish to pick up: he called several times and immediately after I picked up he hung up. This was the only audience instruction regarding my behavior that of necessity had to be followed during my artwork. The other telephone calls were sonorous input (of the imagination) for nonhomologous "sculptural identity." Apart from the electrically amplified telephone calls (amplifier, loudspeaker), the identity relation consisted of the following elements as well: A recorder storing the calls. The sentence "The site of the body is emptiness," written on the wall. On three walls (the fourth wall being the audience) hung three mirrors smeared with clay, reflecting one another and thus deprived of their own representational function (they became nonhomologous). In the center of the room stood a cube of clay on newspaper. From a clay bowl above water droplets dripped down onto this cube. A video camera recorded a close-up view of the drops falling onto the clay cube and then displayed it on a screen. This created a parallel between the dropping in of messages via the telephone and the falling drops of water: time in flux. The slow, gradual shaping of the material corresponds to a slow, persistent shaping of consciousness. Today, nonclassic materials such as telephone and newspaper shape people's brains and consciousness. The people themselves are the true material of nonhomologous "sculptural identity": the persistent shaping of consciousness through mass media is the true sculpture of today, built on the nonidentity or false identity of the people.

II. Installation phase

By using classic materials of representation such as clay and mirrors in a nonclassic, nonrepresentative way, by not following the audience's instructions, by not functioning as intended, by refusing to be the material of representation, I interrupted the chain of identifications on which the state and society are built. Nevertheless, the artist functions as a mirror once again: precisely by being the lacuna in the chain of identification, he mirrors the audience. The audience experiences itself as material, as its own material.
But actually what I did was collect the filtered calls and busied myself with something else in the gallery as the calls came in. Later, I left the residues of the action as an installation in the gallery and the wishes and commands of the audience were played back to the audience present at the exhibition the day after the performance. So the instructions for me became instructions for the audience by the audience.
On the day after the performance, the tape recorder, which now played the instructions of the audience, was installed in a room above the space of action. The three clay-smeared mirrors were now laid down on the floor horizontally. The drops fell on the clay cube again. With the exception of the video camera and the artist, it was the same installation as during the performance – only that now the visitors entering were greeted by the instructions of the audience. Demands and instructions were now directed at the visitors. The audience looked into the artist as a mirror and saw its own image.

—Peter Weibel, 1978

Translated from the German by Peter Dahm Robertson.

Sculptural Identity / Skulpturale Identität

1970 (idea)/ 1978	performance; De Appel, Amsterdam, March 24, 1978

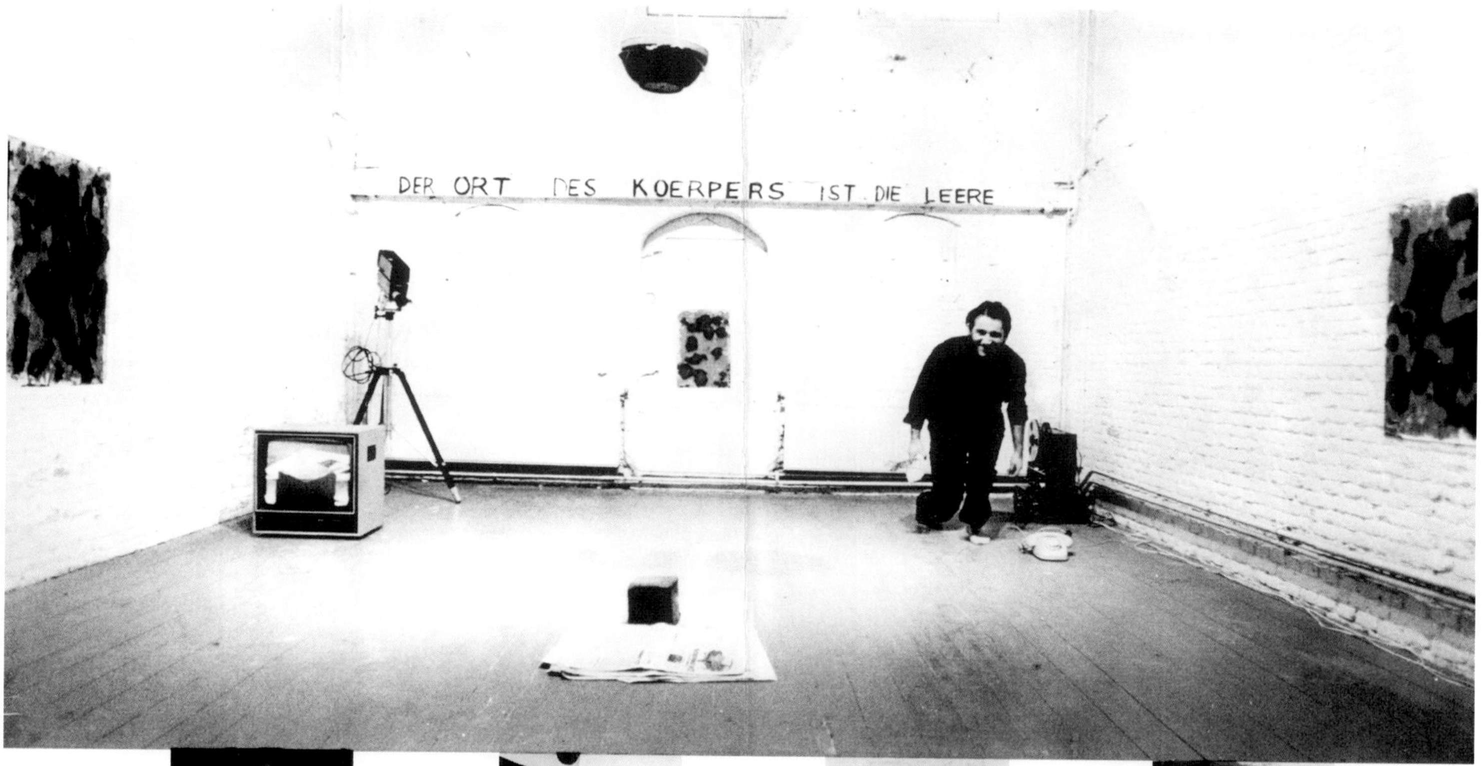

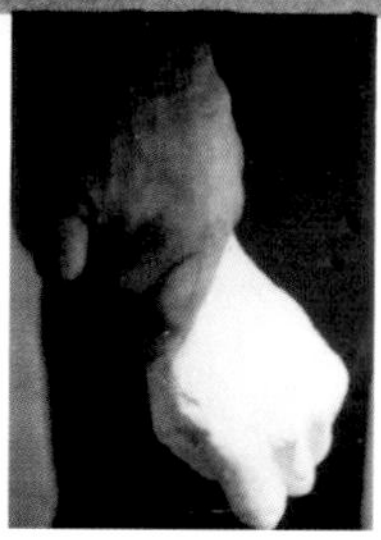

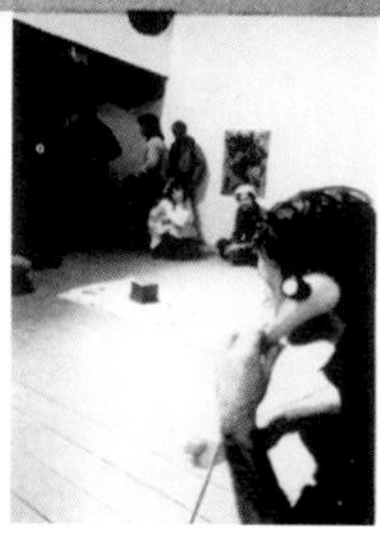

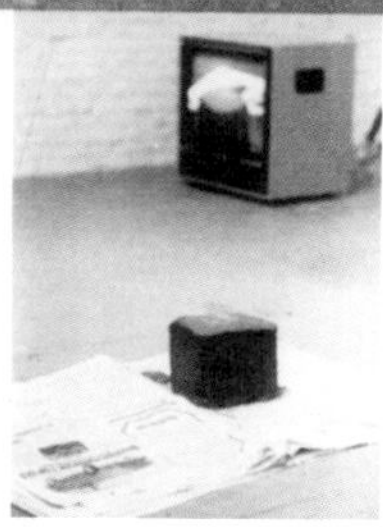

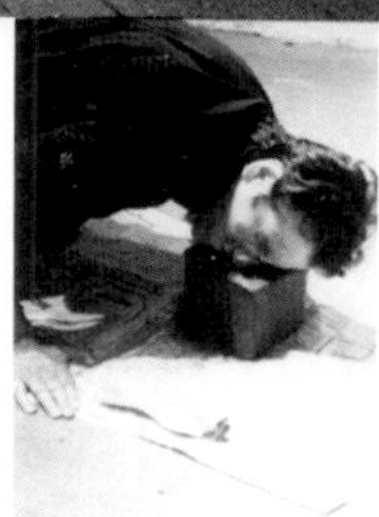

Friday, March 24, from 8 until 11 p.m.

Performance *Sculptural Identity*

The Austrian action-artist Peter Weibel places his body as plastic material at the public's disposal. On March 24, he will be present between 8 until 11 p.m. at de Appel; there he will be awaiting your telephone calls. You can reach him by dialing 020-255651 and give him your instructions as to what you want him to do with his body. He offers you his body for the free creation of environmental sculptures within a space; there is no limit as to the body articulations or body positions. He will obey all your orders given over the phone as long as they involve only himself, his body, the performance space, and dthe possibilities of the human body. Protected by anonymity, you can let him act out your most secret and intimate thoughts.

Phone and tell him what he should do with himself and his body.

The artist becomes the moulding paste of your deisres, the work result of your words.

The public present at de Appel will witness your sculptural creativity.

Announcement from de Appel of *Sculptural Identity*

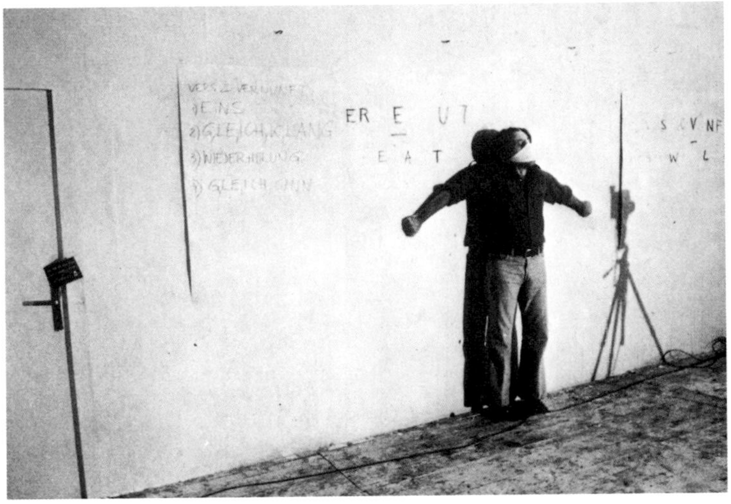

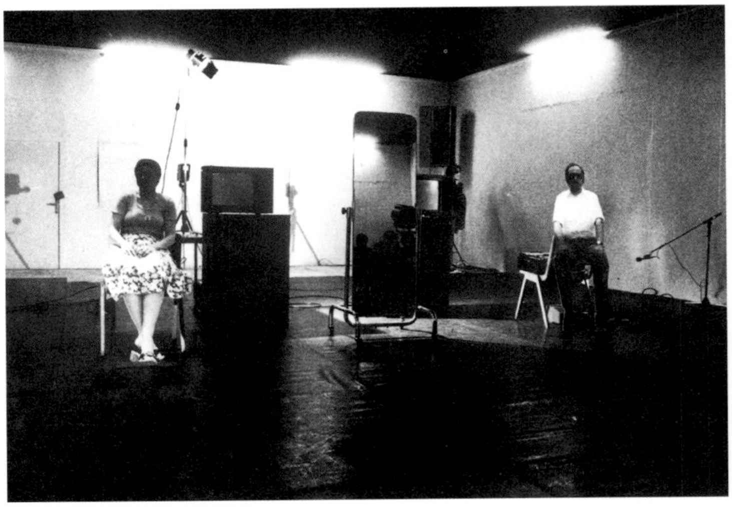

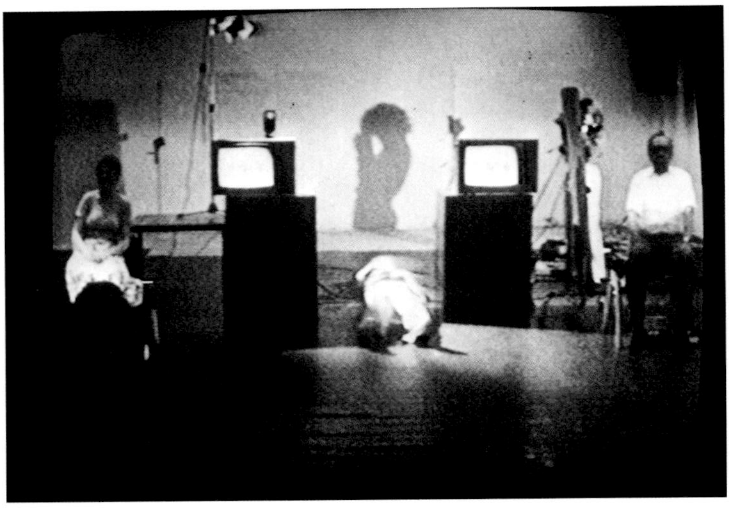

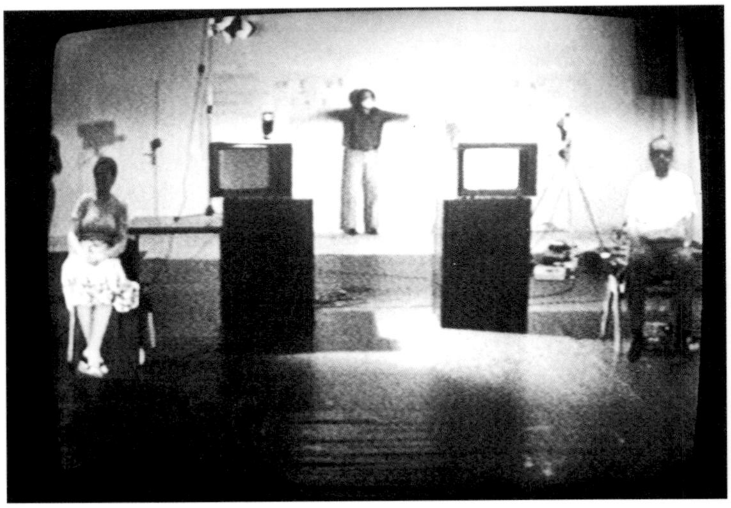

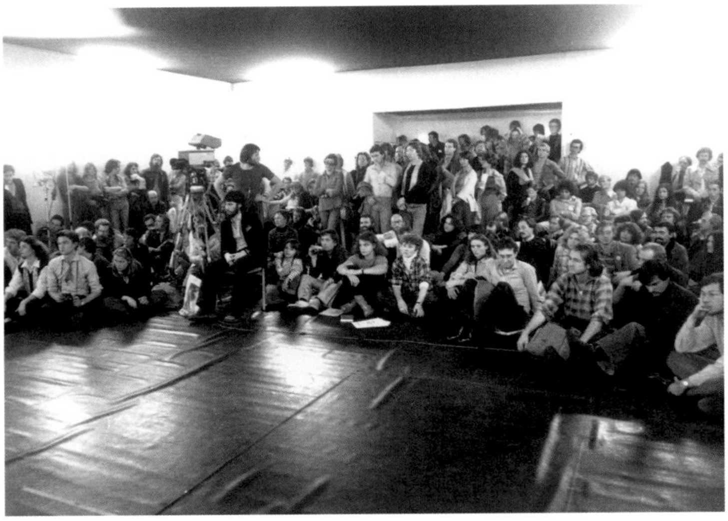

Rhyme and Reason I / Vers und Vernunft I

1977	performance; *Concept in Performance*, Internationaler Kunstmarkt Köln 1977, Cologne, October 26, 1977

Just as verse brings audible order, whose structural principles are merely mistaken for a corset, to the boundless chaos of the sound stream through rhyme and rhythm, reason brings objective law to the boundless chaos of wants and needs, whose terror is just as fallaciously seen as a private despotism. A performance on the theme of poetry and politics, presented through media and gesture, might make this discourse visible by turning concepts into actions.

By means of a mirror, the audience becomes part of the performance. It sees itself as a scene. Two photographic people watch the audience as it watches itself. Tape players play mirror texts, recursive texts. Two video cameras point at a wall, on which I write seemingly meaningless letters. By means of a mixer, the two cameras are united and display intelligible words on two monitors: *VERS + VERNUNFT = GEWALT* (rhyme + reason = violence). Wearing a light-dependent resistor, I roll around on the floor, producing sounds. I cover my mouth with a sail and start breathing more and more deeply until I collapse.

—Peter Weibel

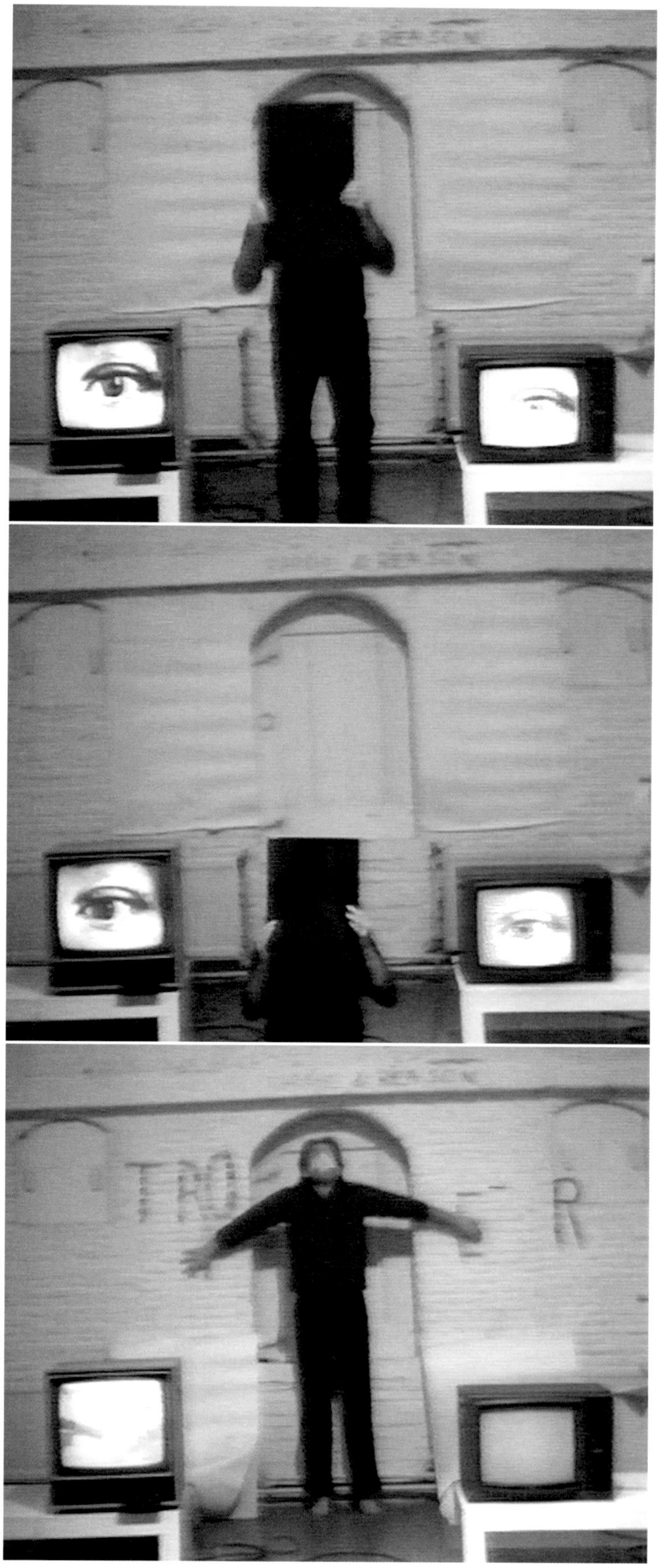

Rhyme and Reason II / Vers und Vernunft II

1978 | performance; De Appel, Amsterdam, March 25, 1978

Two eyes on two monitors blink for a while. I then start walking back and forth between two whiteboards, spray-painting letters on them. On the left-hand whiteboard: *V R + E N N T G W L*; on the right-hand one: *E S V R U F = E A T*. As I write, I maintain specific distances between the letters with the help of a control monitor. The seemingly nonsensical letters on the two whiteboards (reality) are superimposed on a monitor with the help of two cameras and a mixer (consciousness), where they form intelligible words: *VERS + VERNUNFT = GEWALT* (rhyme + reason = violence). A precise representational arrangement turns meaningless particles of reality into meaningful units. I begin rolling around on the floor between the two monitors, where light and shadow are separated by a sharply defined boundary. As I do this, a light-dependent resistor on my chest triggers a fluctuating tone. After a while, I start crawling on all fours around the monitors, whining and barking like a dog. The monitors display time-delayed images of me barking at them, alternating with images of two naked women barking at each other (the one on the left facing right and the one on the right facing left). Two cardboard cutouts (one of a man and one of a woman) have been sitting to the right and left of the monitors since the beginning. Now, when I place a mirror between them and the two monitors, the audience becomes part of this phalanx. I then roll around on the floor with the mirror (that is, the audience). I take a triangular white cloth on a pole that has been fluttering in front of a fan the whole time – a sail, filled by the wind of the world spirit – and tie it around my chin, like a robber or a doctor, and breathe heavily into a microphone, continuing until I pass out.

—Peter Weibel, 2006

Translated from the German by Patrick Hubenthal.

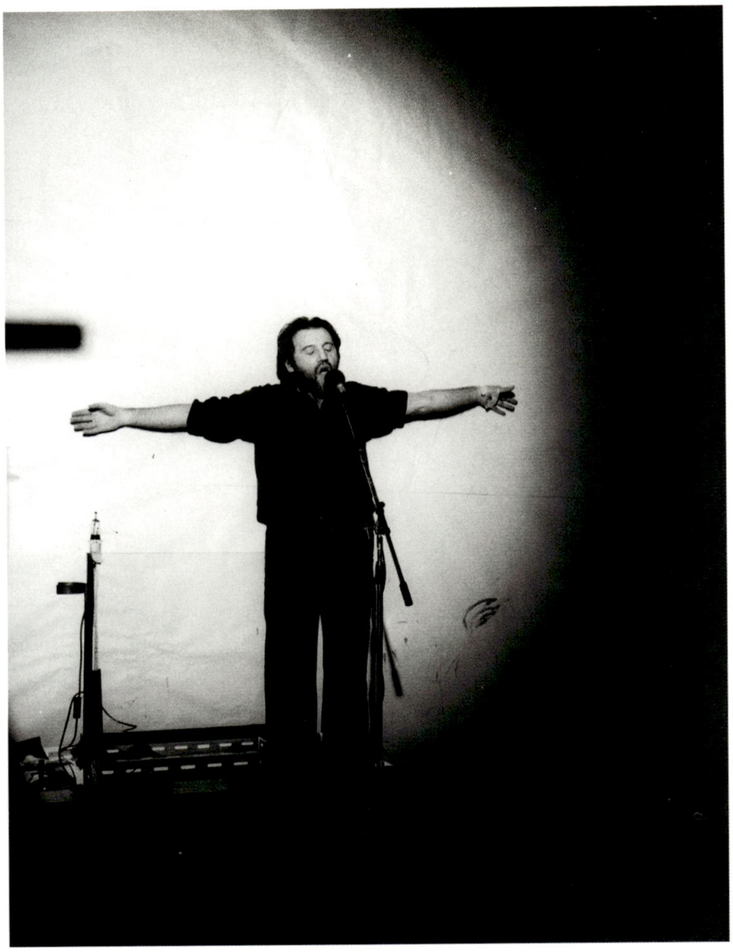

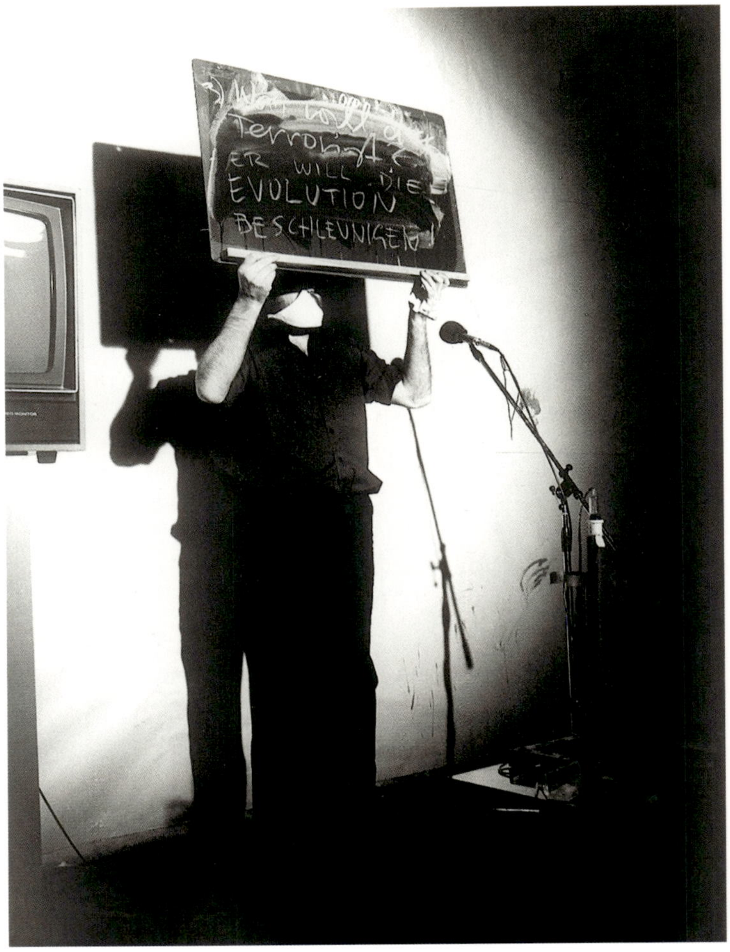

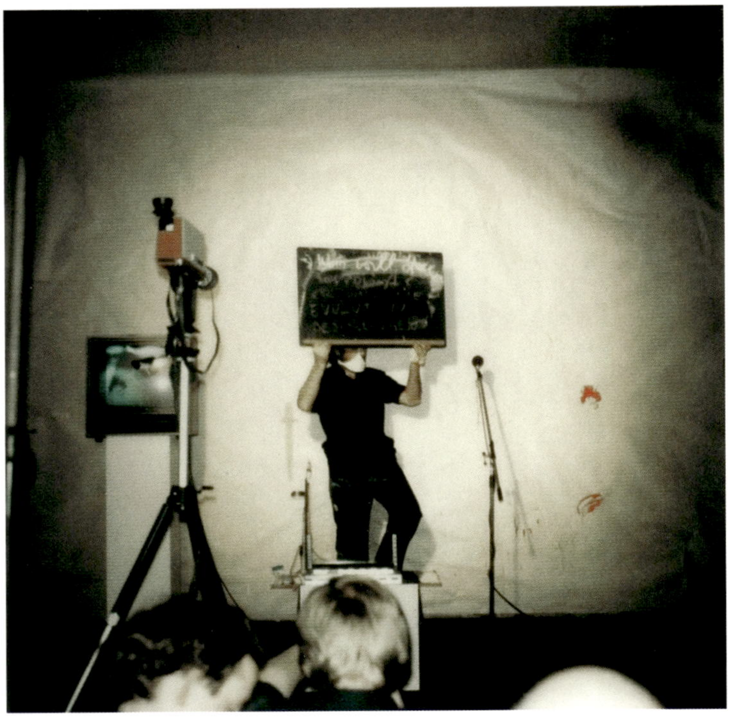

Rhyme and Reason III aka Time Jump / Vers und Vernunft III alias Zeitsprung

1978	performance; Internationales Performance Festival, Österreichischer Kunstverein, Vienna, April 27, 1978

I write the question "What does the alchemist want?" on the chalkboard and answer it orally: "To accelerate the evolution of nature." Then another written question: "What does the terrorist want?" My answer: "To accelerate the evolution of society." I then stand in front of two cameras with arms outstretched, making the shape of a cross. The right camera is pointed at my left arm, and the left camera is pointed at my right arm, so that on monitor 1, my two arms appear to be reaching toward each other to link hands in solidarity. I begin breathing rapidly and continue until I pass out. —Peter Weibel, 2006

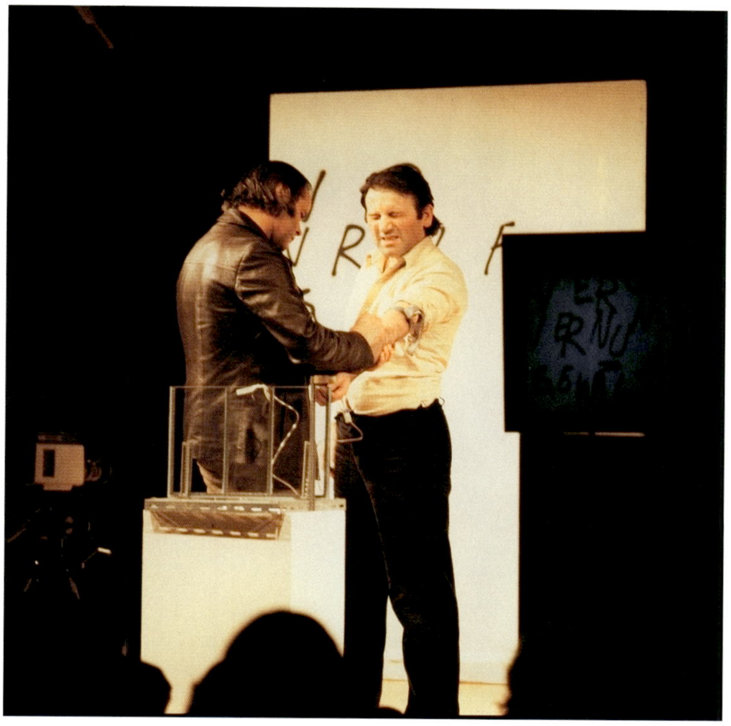

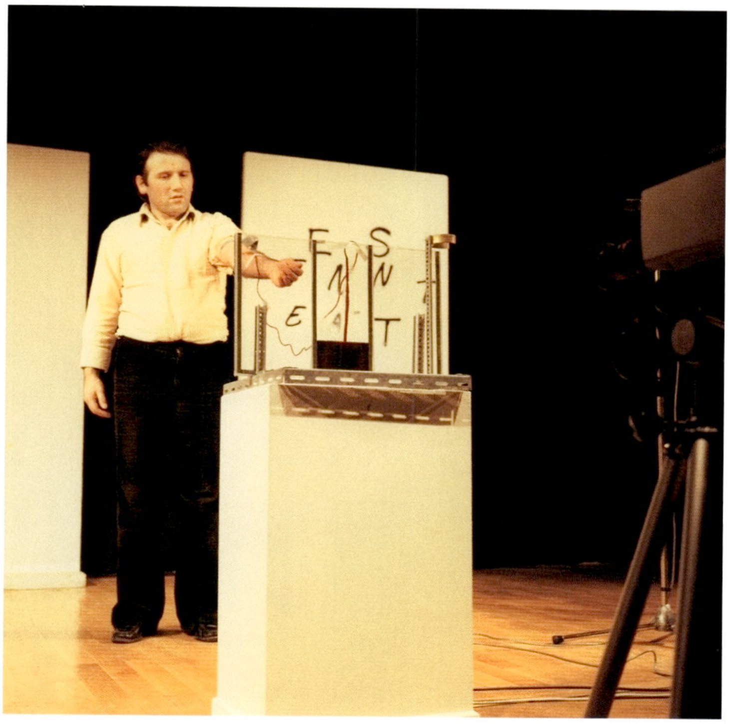

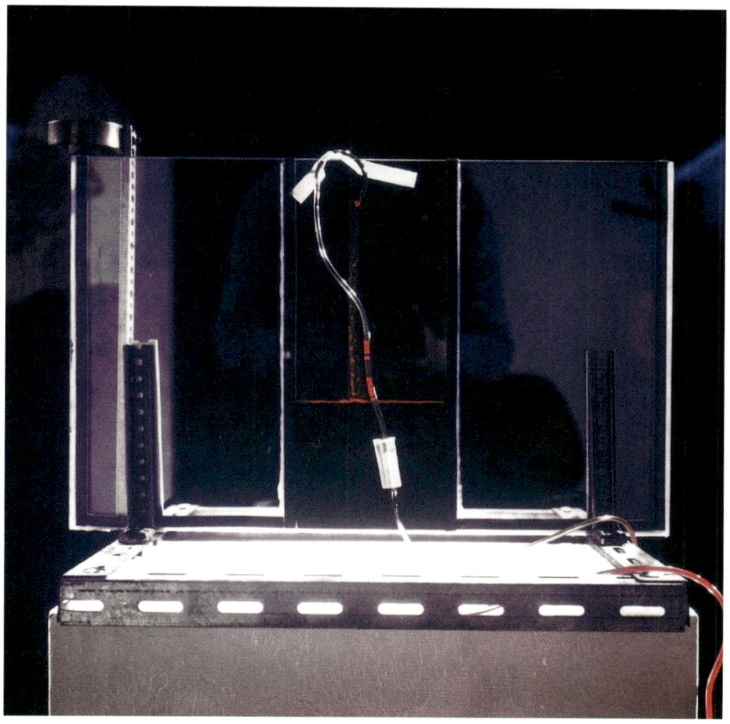

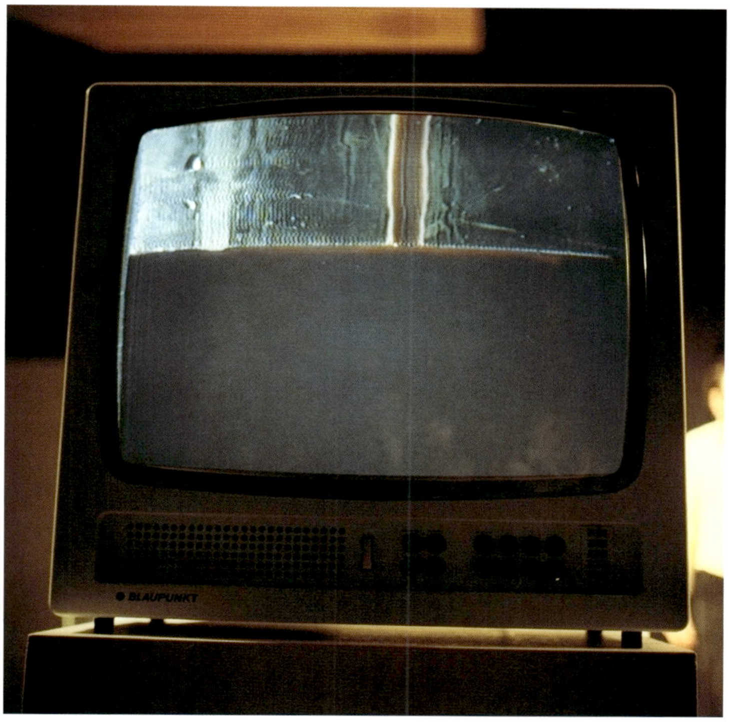

Blood Bell / Blutglocke

1979 — performance; *Performances '79*, Städtische Galerie, Lenbachhaus, Munich, February 2–18, 1979

Part 1: As in *Rhyme & Reason I*, meaningless word fragments on two spatially separated walls are recorded by two cameras and displayed on a single monitor, where they produce meaningful words.
Part 2: Display of my flowing blood that gradually obscures a video monitor (as in *Time Blood*)
Part 3: Breathing to exhaustion (as in *Rhyme and Reason III aka Time Jump*) —Peter Weibel, 2006

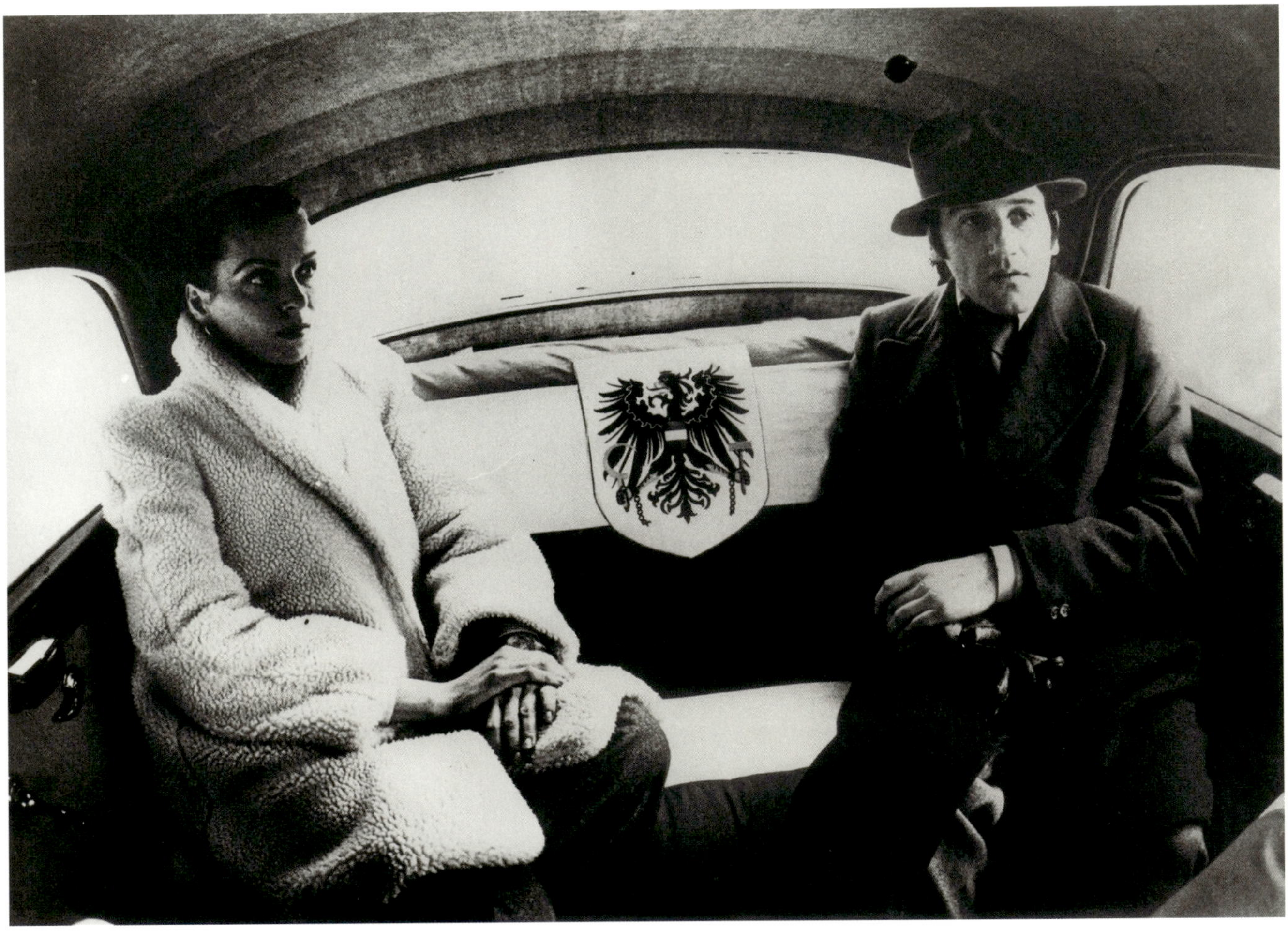

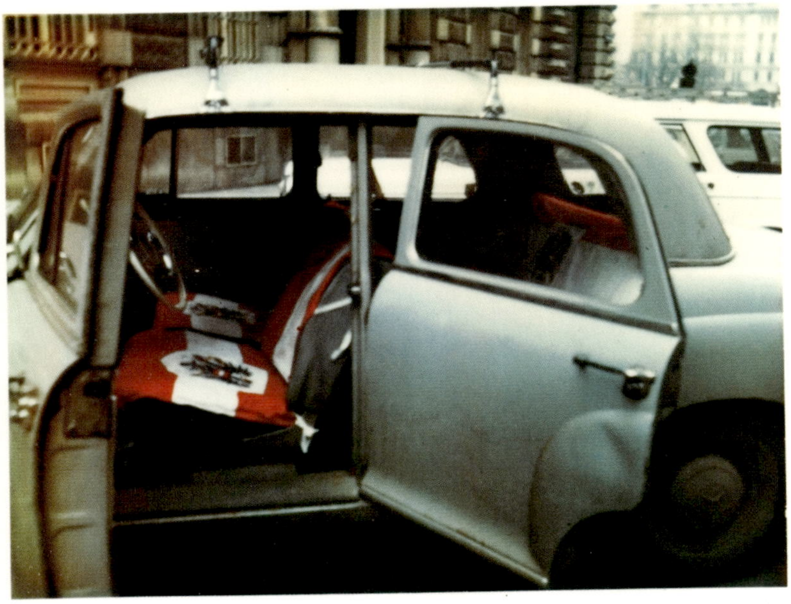

Private Vehicle as State Car / Privatauto als Staatskarosse

1972 | Susanne Widl and Peter Weibel performance

I had a very old and broken-down Mercedes. I covered it with the Austrian flag and national symbols. A private vehicle was decorated as an official vehicle. No sooner had Susanne Widl and I driven into downtown Vienna than we were stopped by the police, who charged me with "desecration of state symbols." The worst outrage was that I had placed the state seal so that it was underneath my anus – a visual metaphor that even the police could understand. —Peter Weibel, 2022

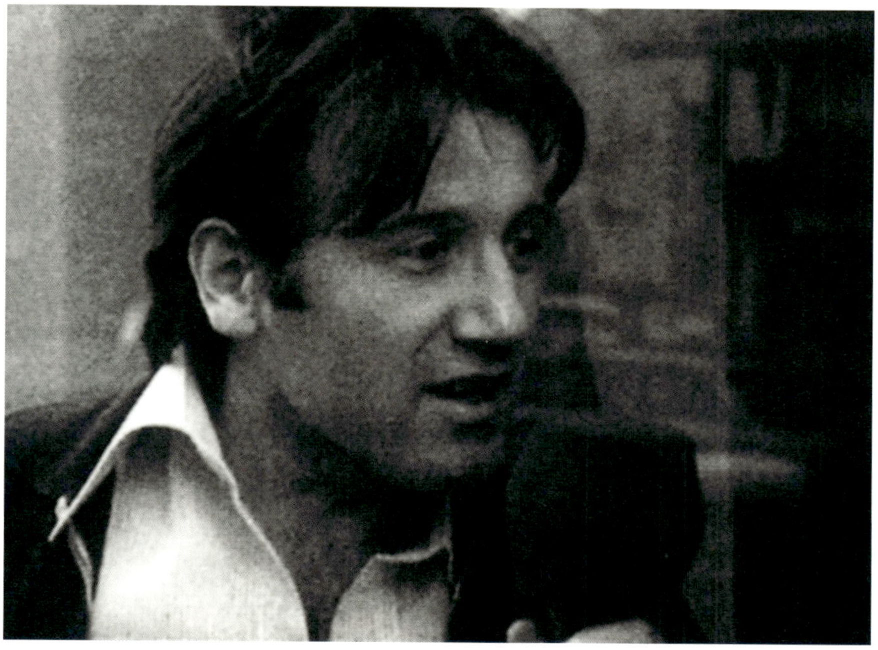

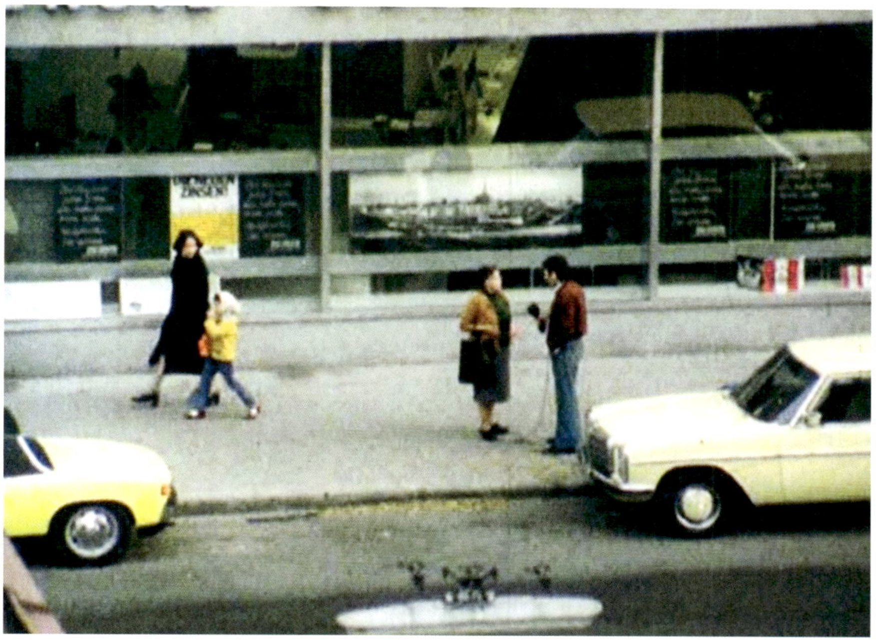

Countless election posters for Vienna's dominant political party, the Socialist Party of Austria, proclaimed that "Wien gehört den Wienern" – "Vienna belongs to the Viennese." I therefore asked passers-by: "Who does Vienna belong to?" At certain times, I read the following text from 1966 in public places:

"Vienna belongs to the Viennese. I'm Viennese. I go to Schmutz's bakery and say to the cashier: 'Two *Schwedenbomben* (literally Sweden bombs, chocolate-coated marshmallow confections) for Schumanngasse, 17th District.' I meet a guy from Linz on Kärntnerstraße. I inform him that Vienna belongs to the Viennese and I'm Viennese. I order him to get off Kärntnerstraße and leave the city. A guy from Deutsch-Wagram who overhears our conversation asks permission to take a walk on my street. I allow him to use Kärntnerstraße for two hundred Schillings an hour. When he wants to take his girlfriend along, I naturally forbid it. I also forbid him to use the sidewalks – their slipcovers are brand new. But in exchange for a nice tip, I do relieve him of a kerfuffle with the police. I cancel his half-hour promenade on Wiedner Hauptstraße and register him for one on Kärntnerstraße. By now it's getting cold. I go to Teller's clothes shop and ask for an overcoat for Favoritenstraße. They tell me it's taken, so I suggest Weihburggasse. That one's taken too. When Jägerstraße in the 20th District isn't mine either, I ask who it all belongs to. They tell me: 'Herr Mautner-Markhof.' To my astonishment, I realize that there are people who own nothing of Vienna even though they're Viennese."

—Peter Weibel, 2022

Who Does Vienna Belong To? / Wem gehört Wien?

1976/77	16 mm film; color and b/w, sound, 6:15 min. camera: Ernst Schmidt Jr.

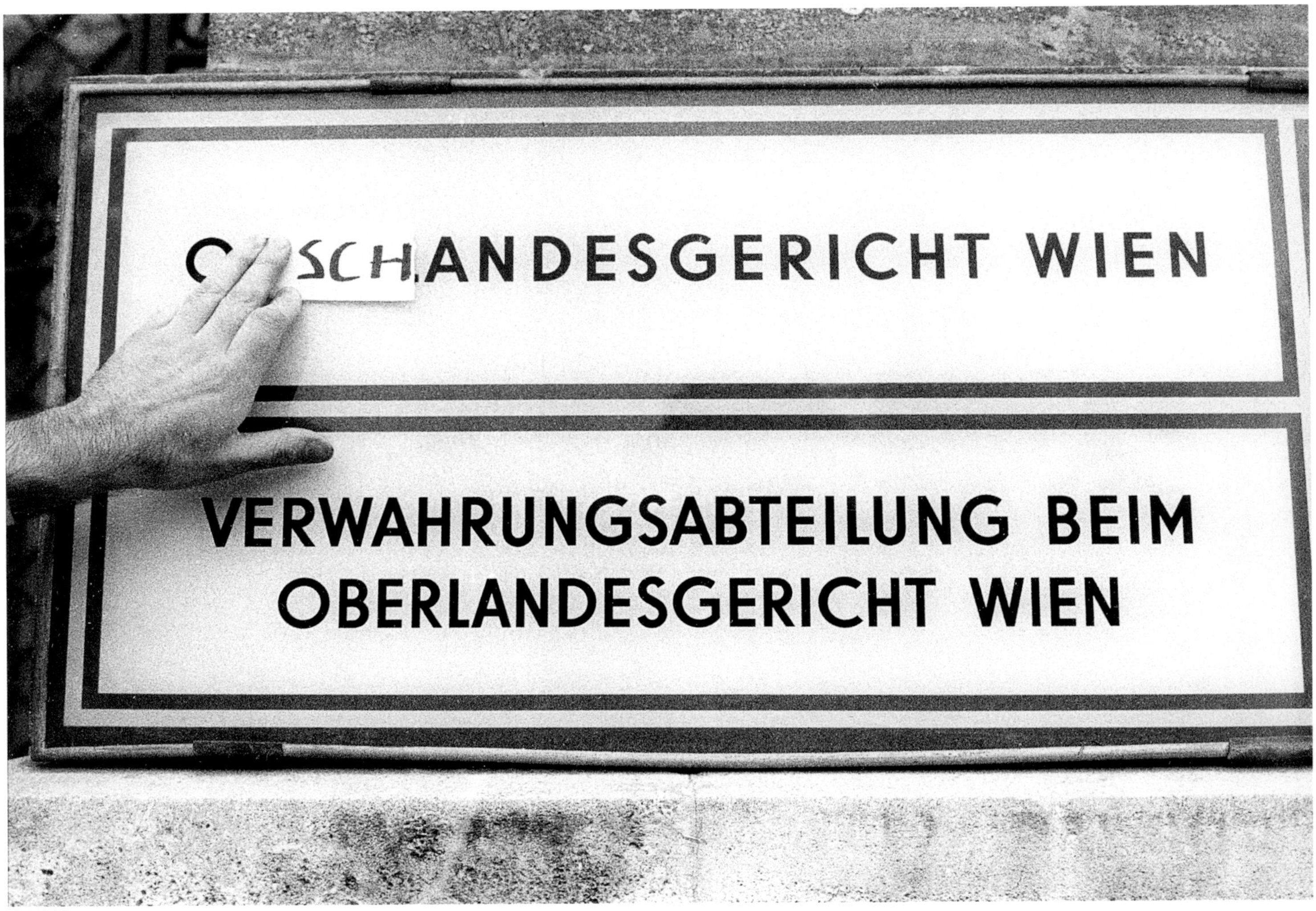

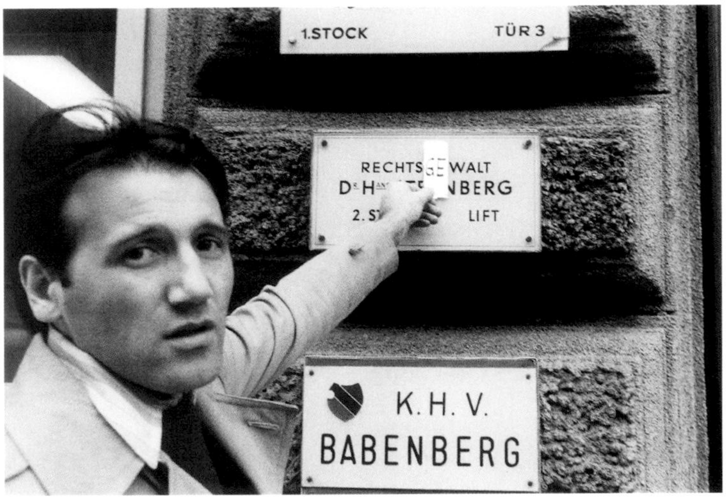

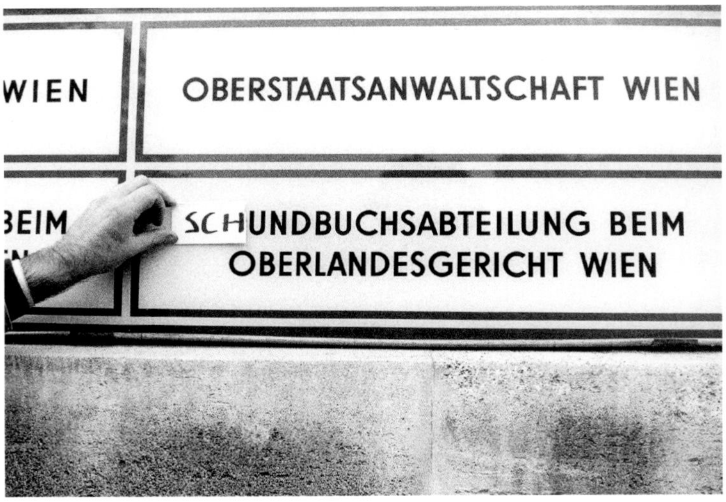

The word *Anschlag* is ambiguous. It can mean an official notice or announcement displayed in a public place, but it can also mean an attack – a semantic ambivalence exploited by Peter Weibel. In *Attacks*, a series of photographs created between 1970 and 1971, Weibel can be seen holding up letters and words in front of various public buildings and the signs identifying them – *Anschläge* in the official sense – thereby giving a different meaning to the existing labels. By changing *Rechtsanwalt* (attorney) to *Rechtsgewalt* (legal violence) and *Oberlandesgericht* (higher regional court) to *Schandesgericht* (disgraceful court), Weibel registered a protest against the state system, as he did in many of his works from the 1960s and 1970s. Other *Attacks* included changing *Grundbuchsabteilung* (land registry) to *Schundbuchsabteilung* (trash registry) and adding the words *verbricht* (commits crimes) and *Verbrechenshof* (court of crime) to the sign for Austria's Supreme Court of Justice, as well as the most famous image in the series, *Police Lie* (*Polizei lügt*). The places "attacked" were primarily public institutions and courts – locations where state power is exercised. With his linguistic interventions, Weibel commented on a system he perceived as oppressive and corrupt.

Translated from the German by Patrick Hubenthal.

Attacks / Anschläge

1970–71 | series of b/w photographs, various dimensions

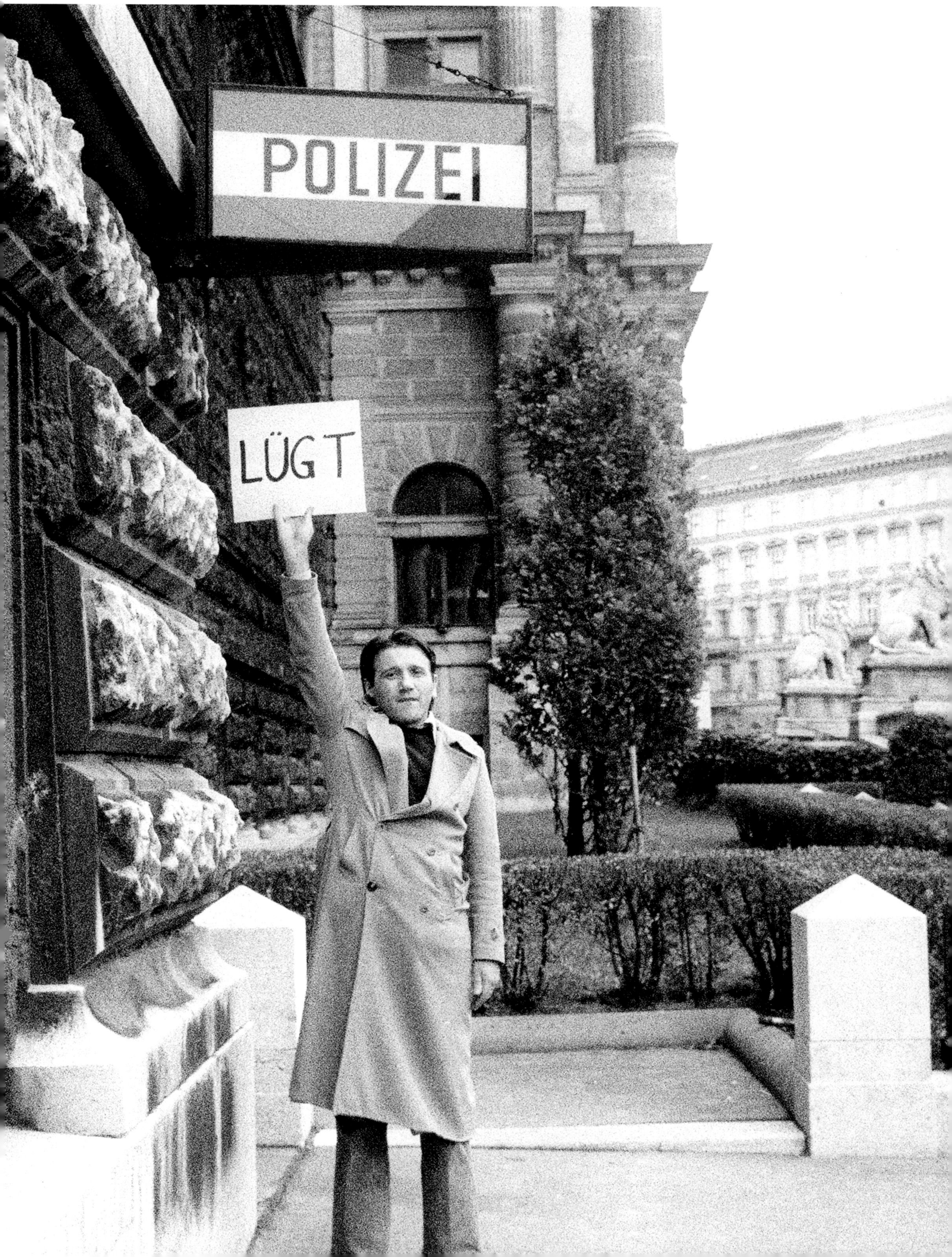
POLIZEI
LÜGT

Appendices

Peter Weibel's artistic work is genre-transcending in conception and exceedingly diverse in form. The order and structure in which it is presented in this book represents just one of many possibilities. The following thematic index of works therefore attempts to visually indicate recurring themes, leitmotifs, and technological parallels in his multifaceted oeuvre, revealing new relationships and affinities between individual works across boundaries of genre and medium.

Mirrors

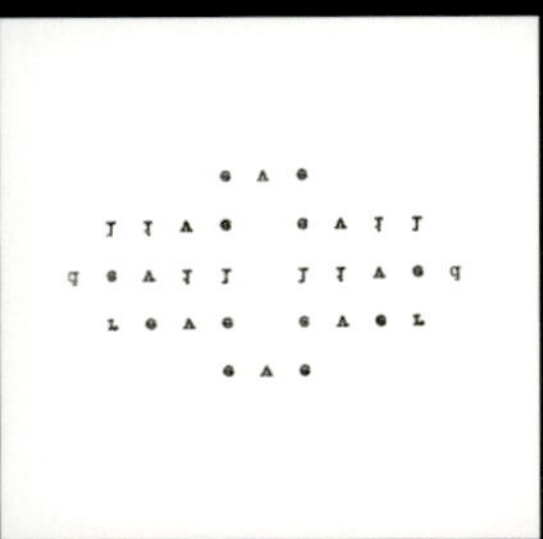

Mirror Eve, 1966 → p. 70

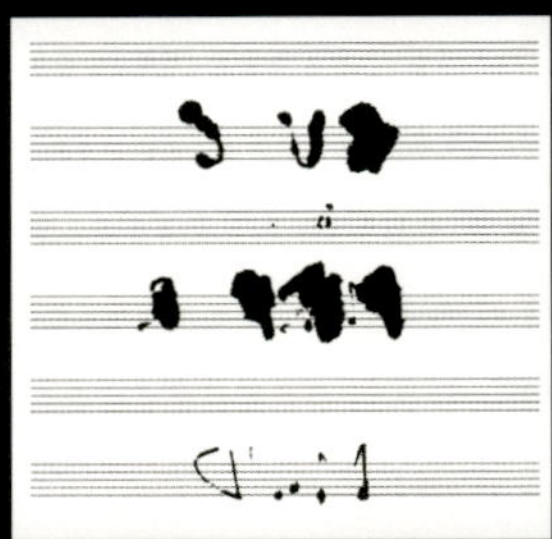

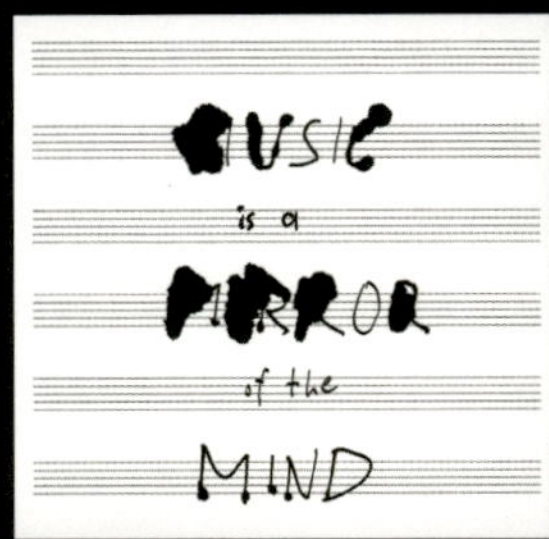

Music Is a Mirror of the Mind, 1967 → pp. 342–43

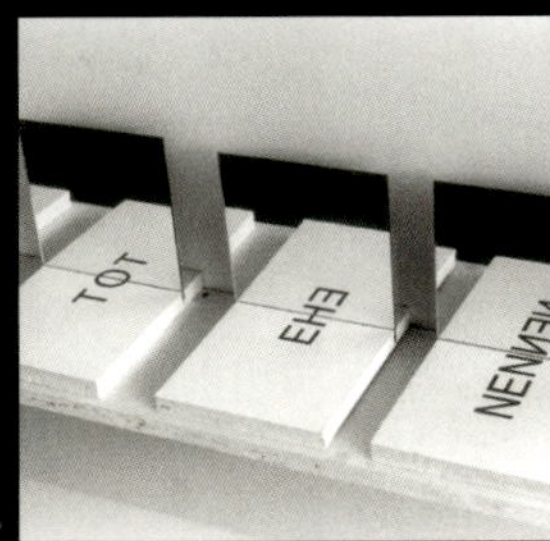

Mirror Texts, 1966 → p. 71

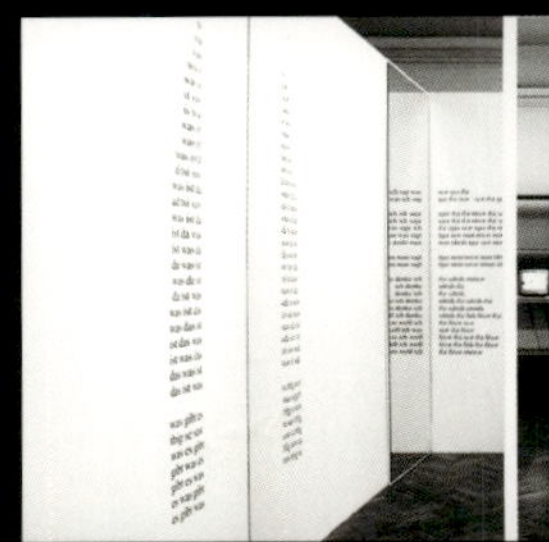

Soliloquy. Language Mirrorings, 1973 → pp. 96–97

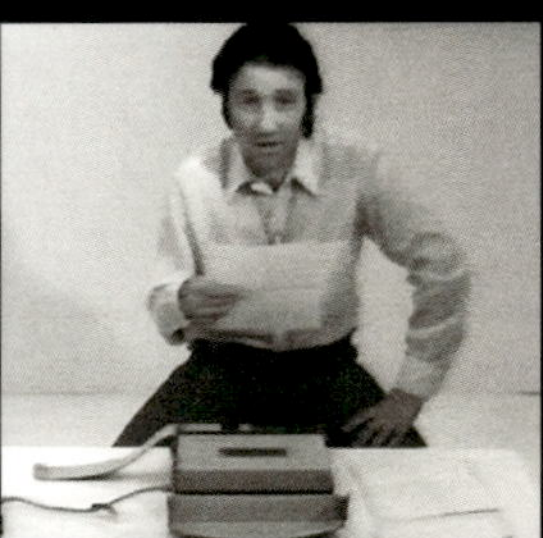

Soliloquy. Language Mirrorings, 1973 → pp. 96–97

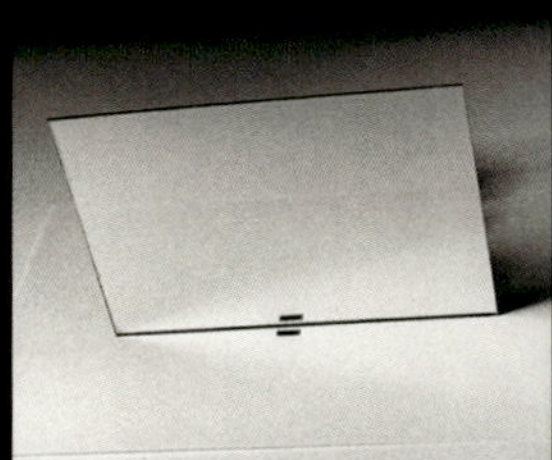

Reference Point for the Identity Sign, 1972 → p. 72

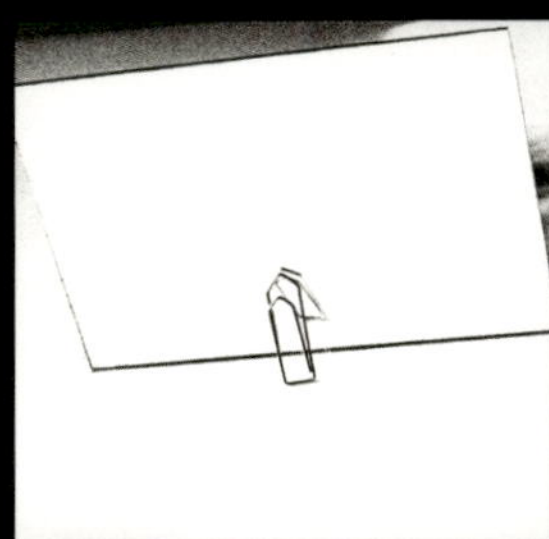

Refractions of Gravity, 1972 → p. 73

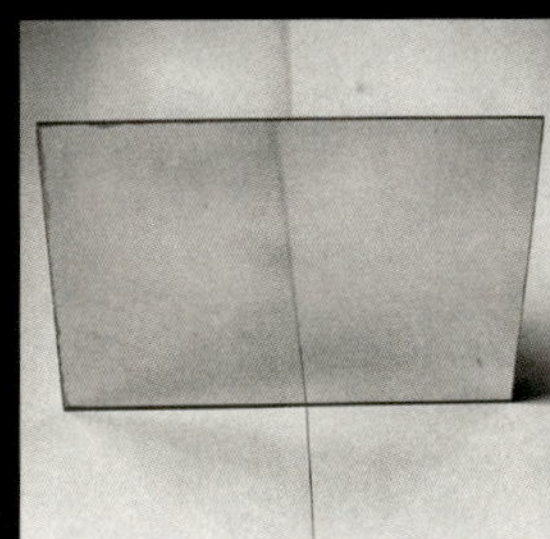

Iso- and Polymorphic Refractions, 1972 → p. 73

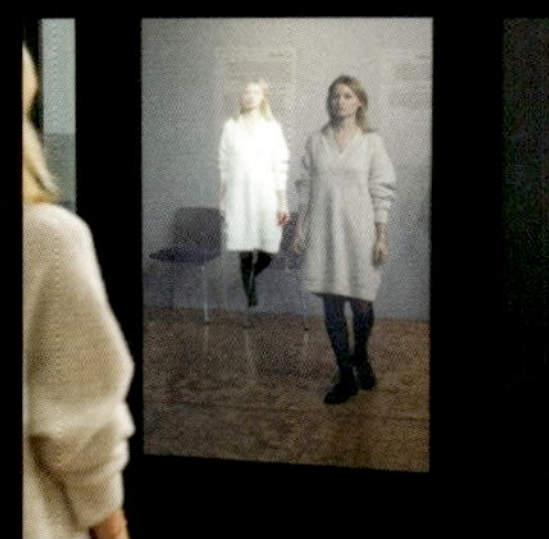

YOU:R:CODE, 2017 → pp. 322–25

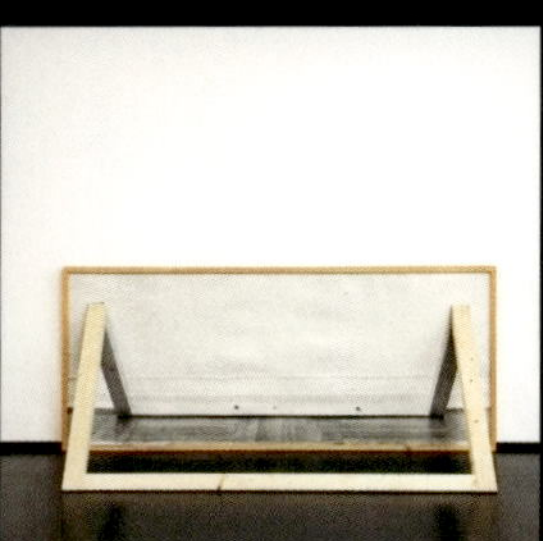

Mirror Square, 1977 → p. 391

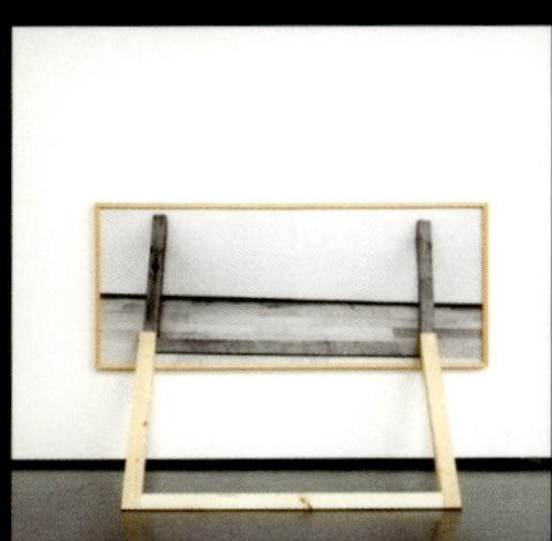

Halved Square, 1977 → p. 391

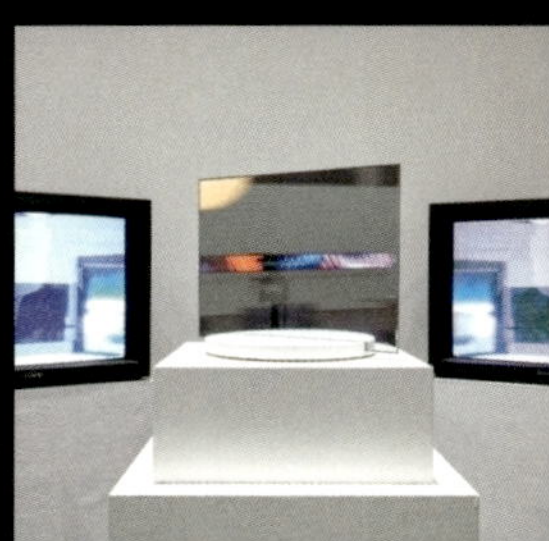

Video Corner, 1974/77 → p. 272

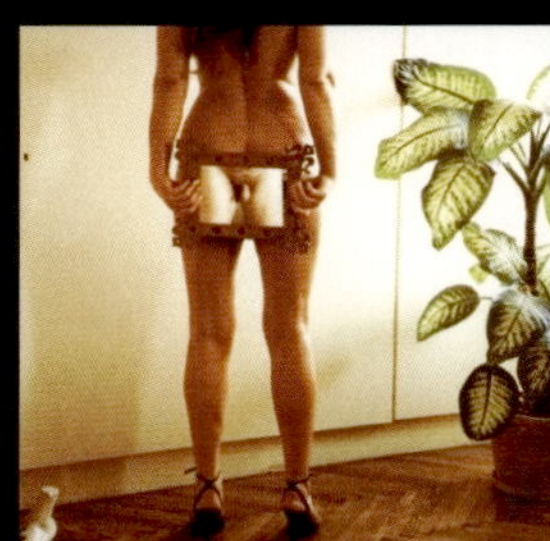

The Clinic of Two Mirrors, 1981 → pp. 132–33

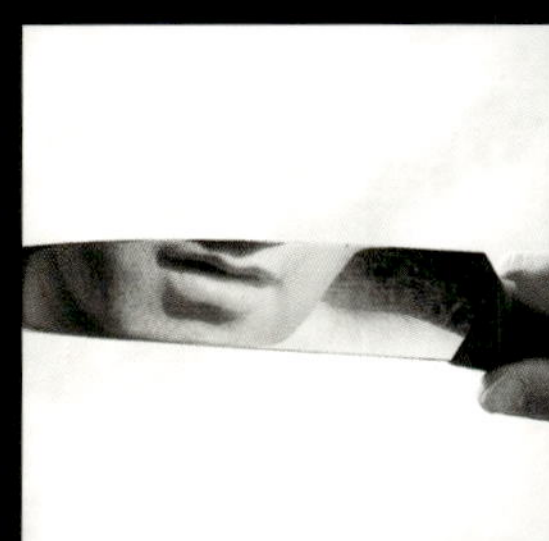

Self-Portrait (Knife as Mirror), 1975 → p. 116

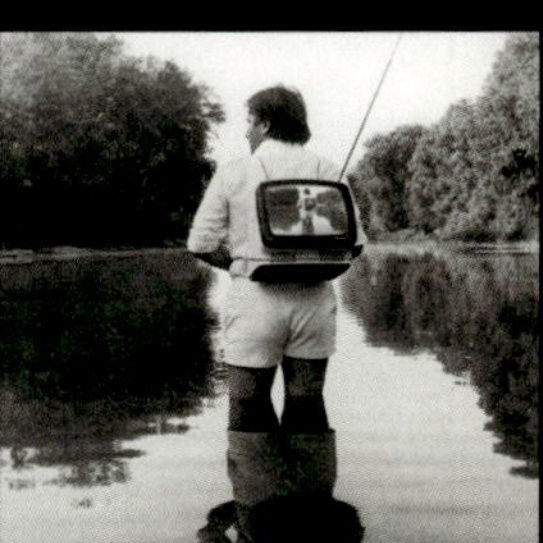

Infinite Contemplation, 1982 → p. 290

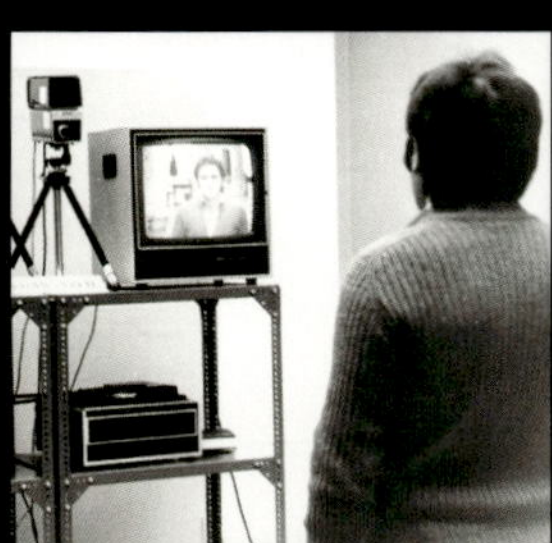

Self-Time –Time-Self, 1975 → pp. 266–67

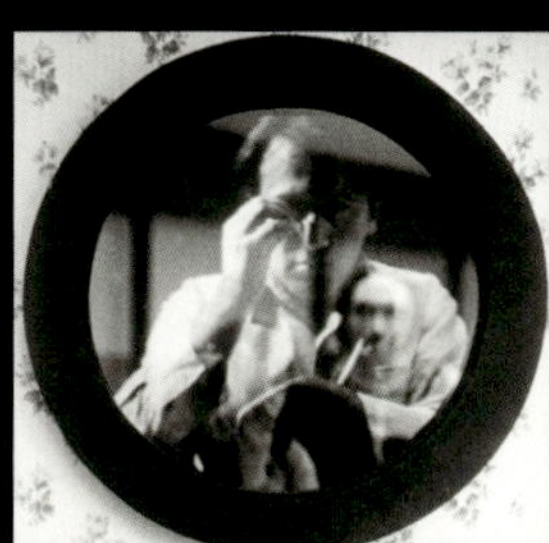

Mirror Check I, 1978 → p. 274

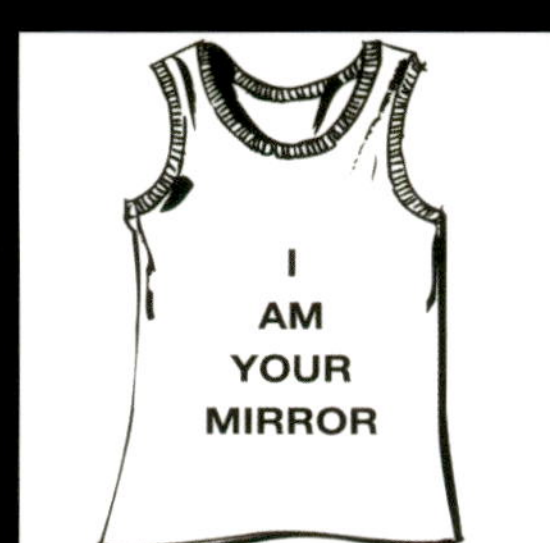

Shirts, 1999

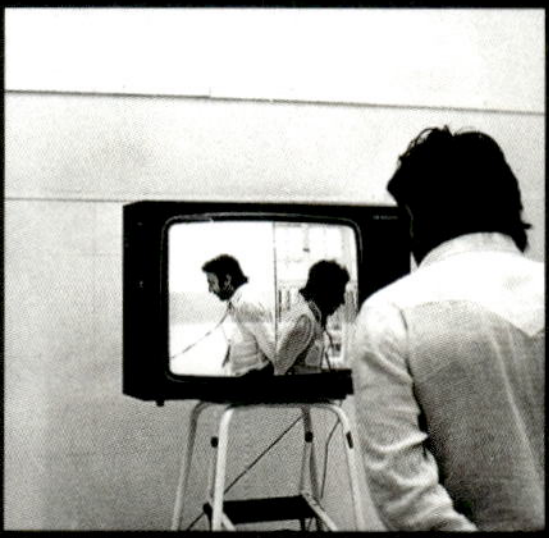
Doubles: Perspective Torsos, 1974 → p. 244

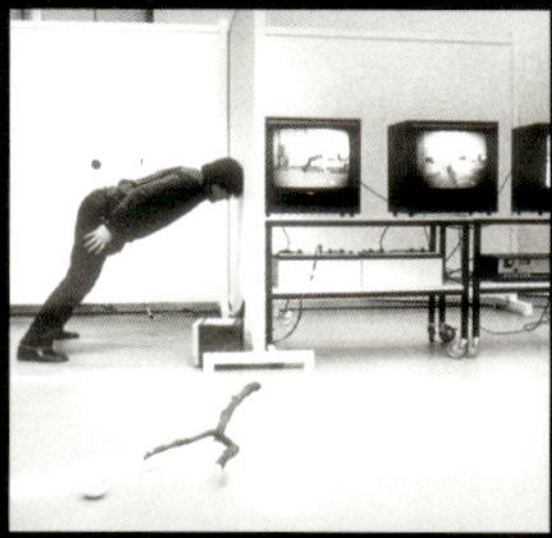
Bifurcation, 1979 → pp. 278–79

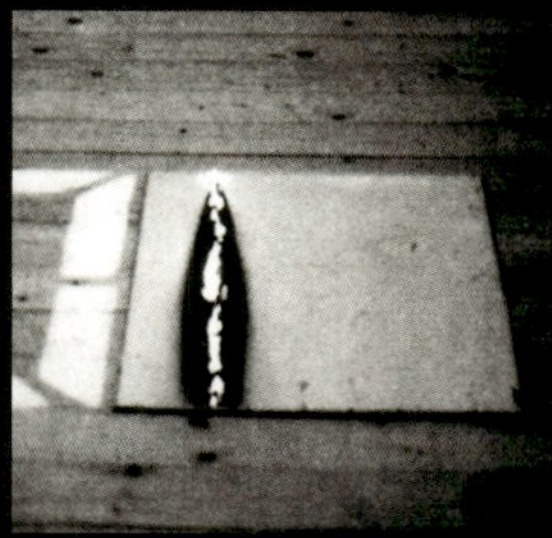
The Electric War, 1983 → pp. 458–59

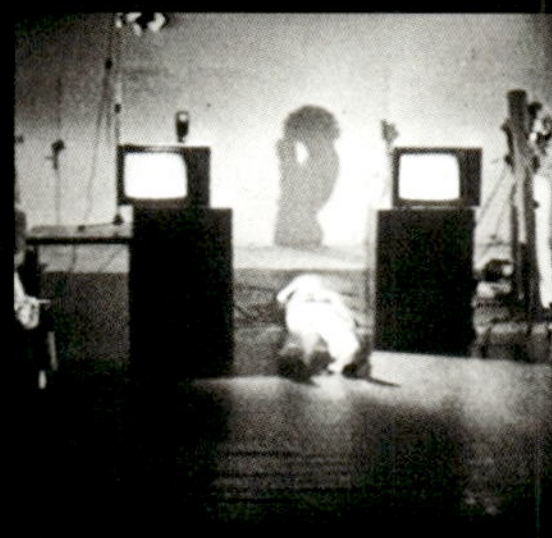
Rhyme and Reason I, 1977 → p. 530

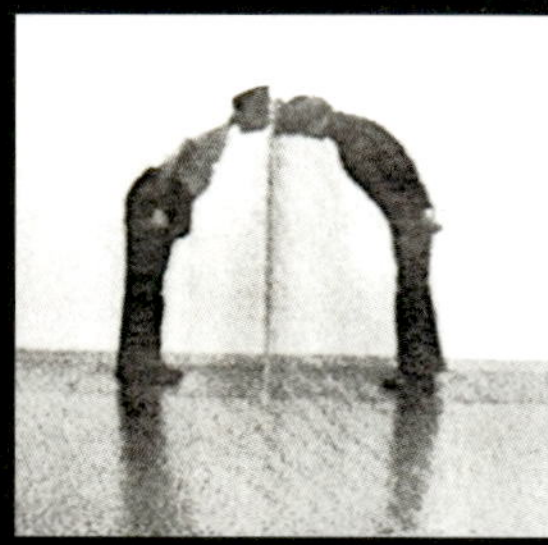
Symmys, 1978 (VALIE Export in collaboration with Peter Weibel)

Observation

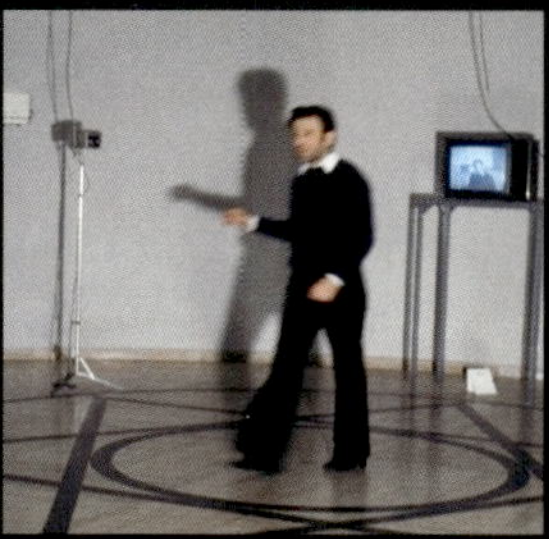
Observing Observation: Uncertainty, 1973 → p. 263

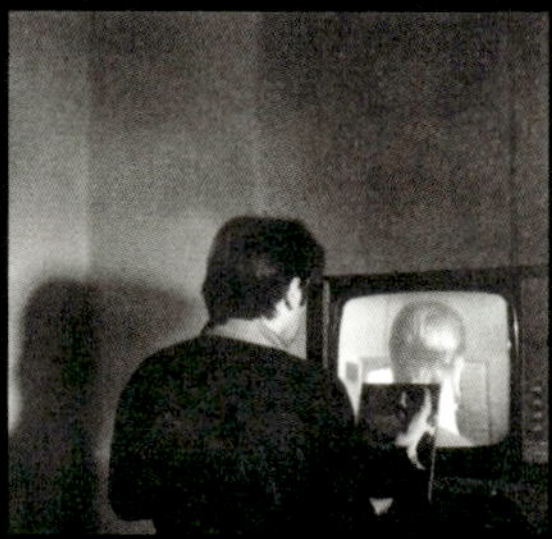
Reflection Permitted, 1977

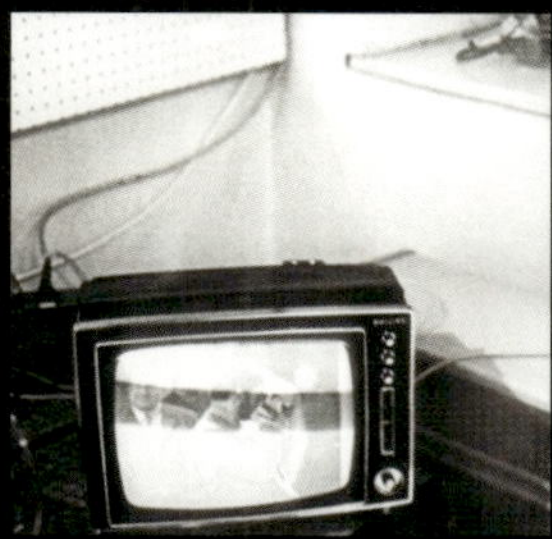
The Public as Exhibit (Audience Exhibited), 1969 → p. 41

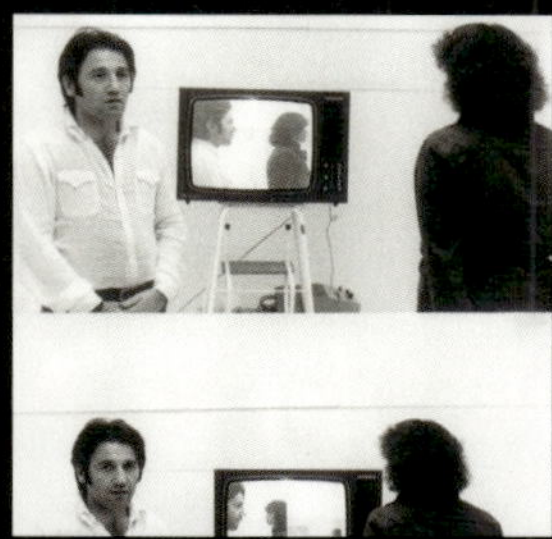
Paradoxical Encounters, 1974 → p. 245

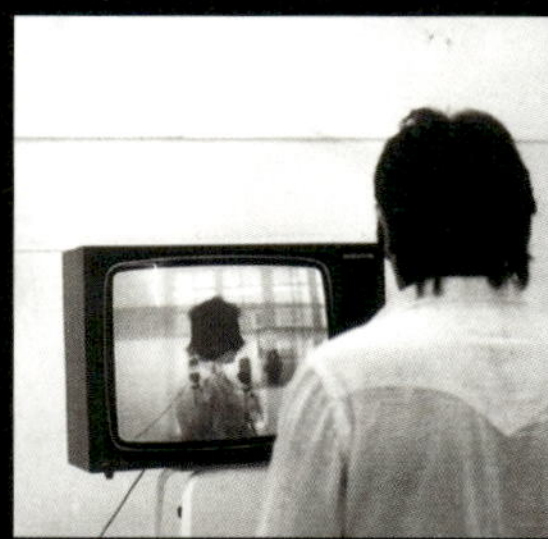
Self-Observation as Self-Obscuration as Self-Loss, 1974 → p. 243

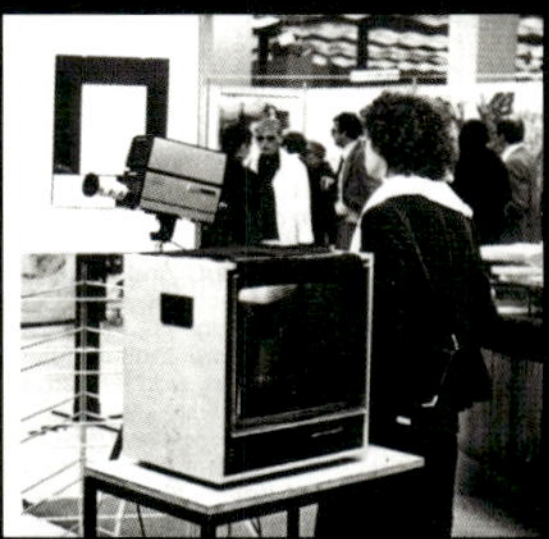
The Guard as Bandit, 1974 (idea)/1978 → p. 269

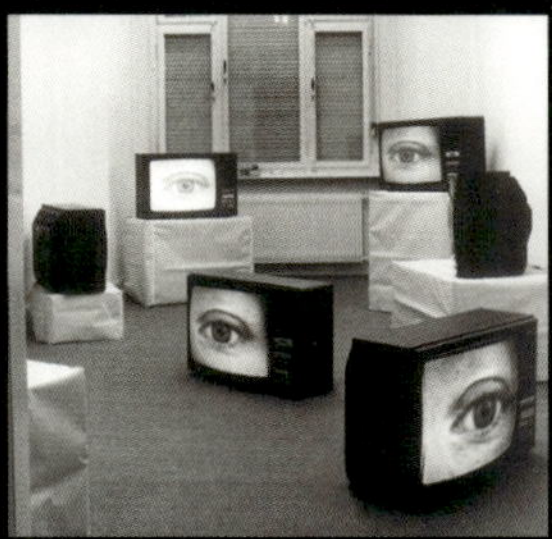
Video Lumina, 1972 (idea)/1977 → p. 268

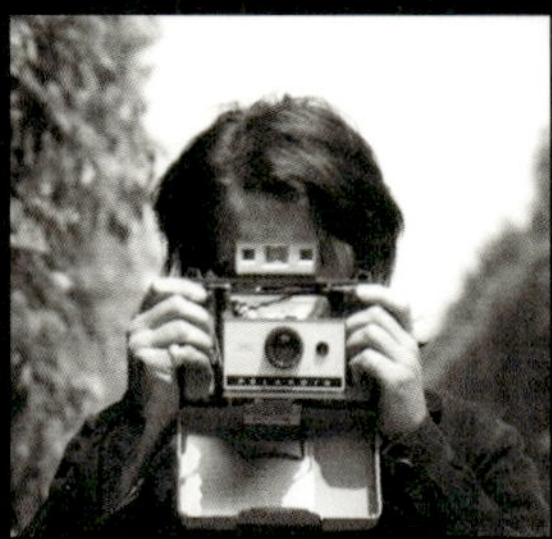
Depiction Is a Crime, 1970/71 → p. 249

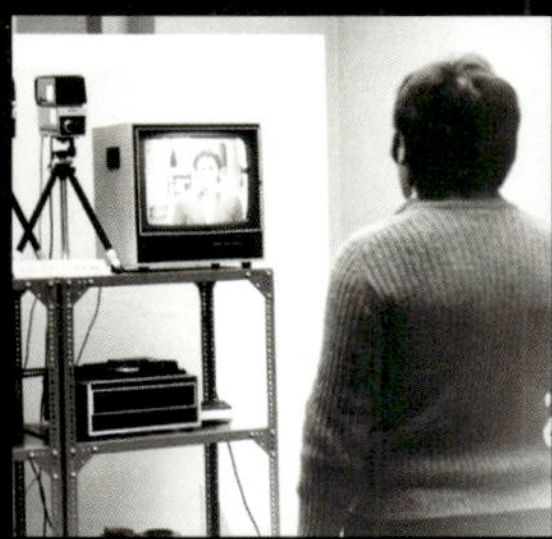
Self-Time –Time-Self, 1975 → pp. 266–67

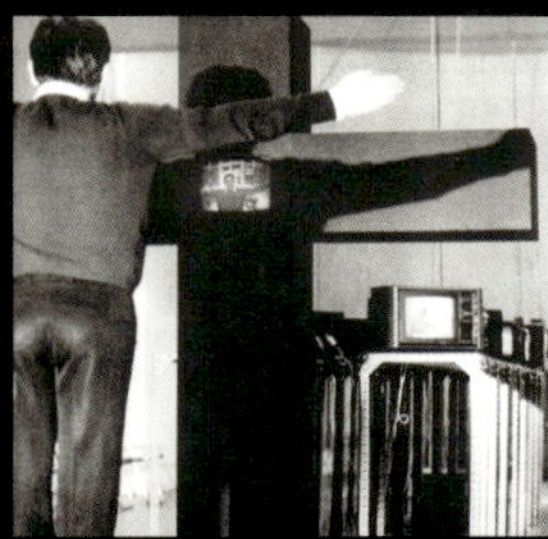
Crucifixion of Identity, 1973 → pp. 264–65

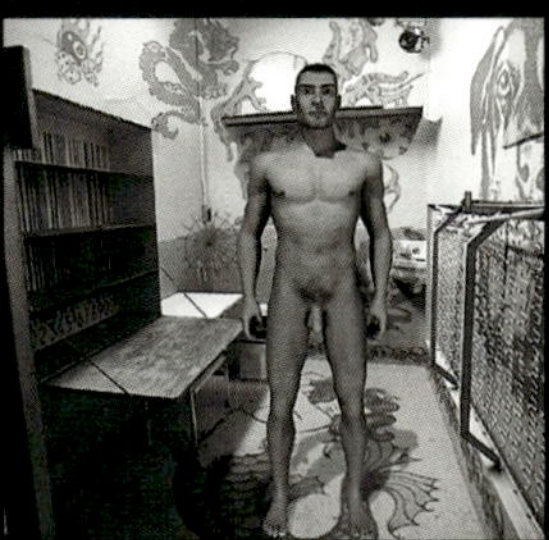
The Panoptic Society …, 2001 → p. 319

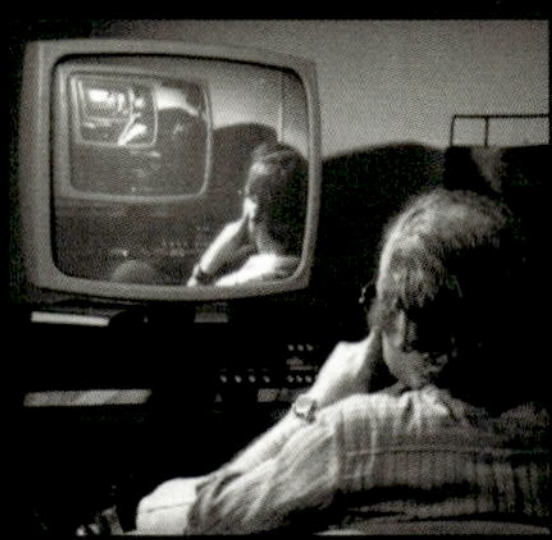
The Endless Sandwich, 1970/72 → p. 248

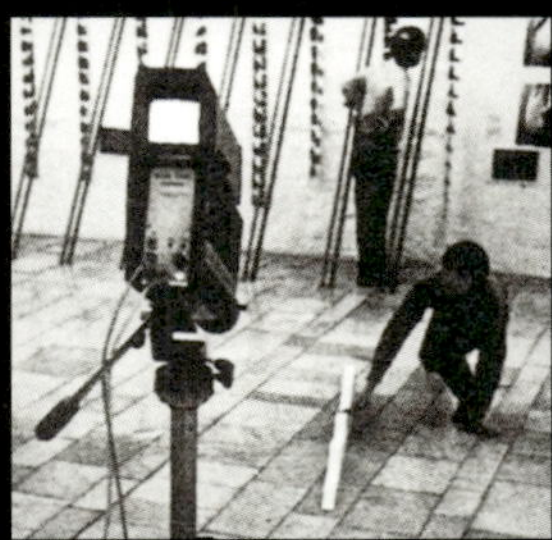
Straight and Wavy Line, 1981

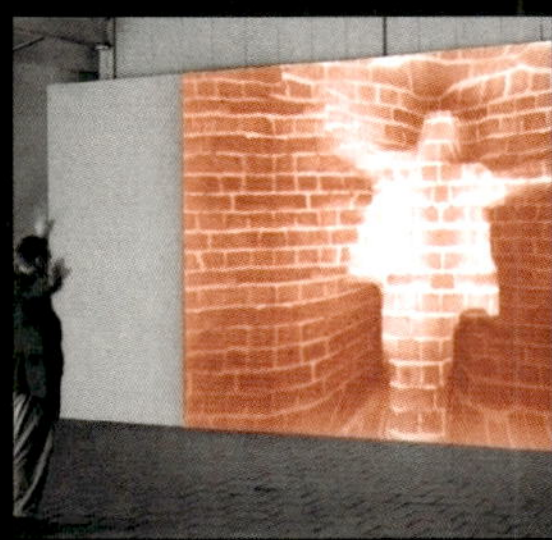
The Wall, the Curtain (Border, the) or, in Technical Terms: Lascaux, 1994 → p. 317

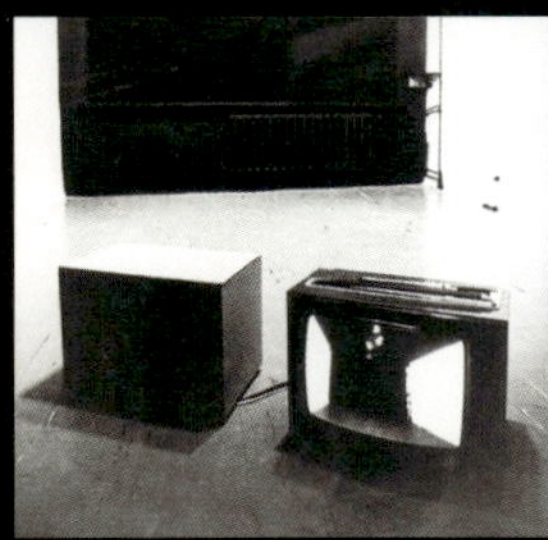
Inverse Box, 1972/80 → p. 273

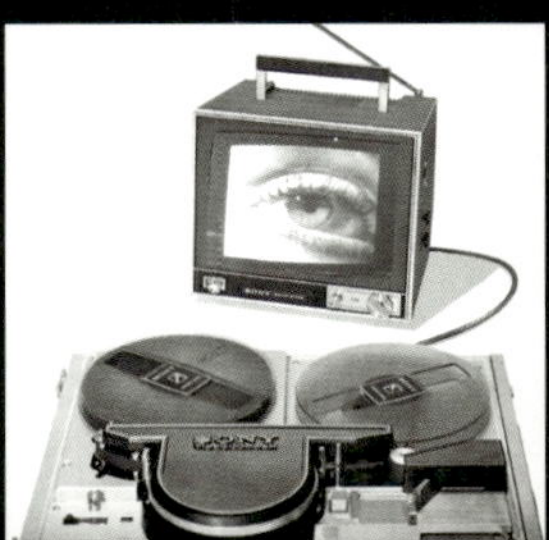
Monitoring, 1969 → p. 229

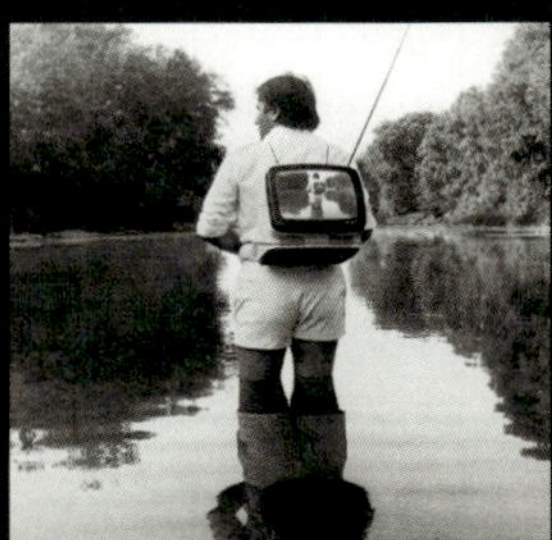
Infinite Contemplation, 1982 → p. 290

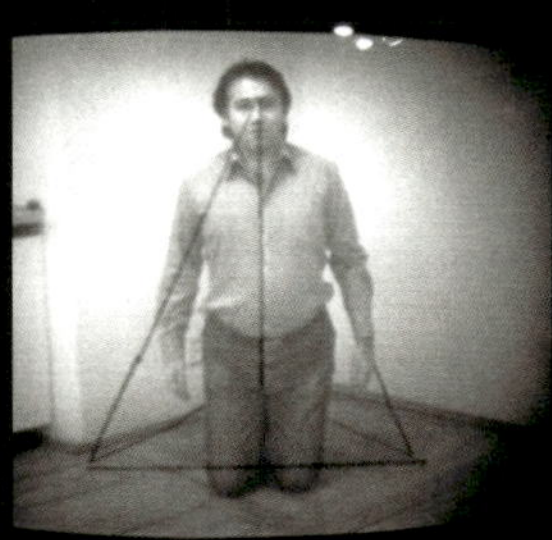
Virtual Tetrahedron, 1978 (concept)/1980 → p. 276

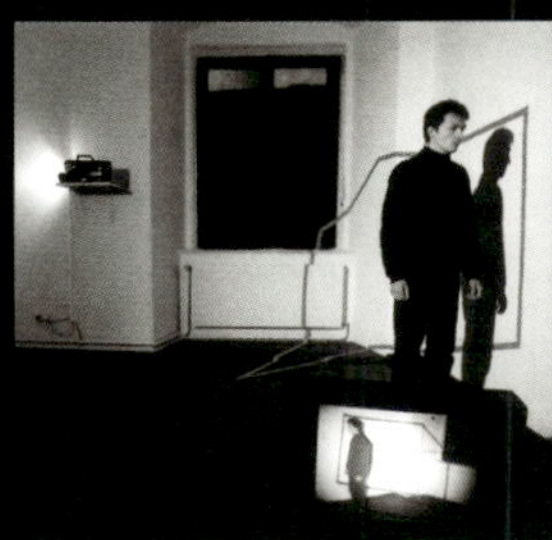
Virtual Cube, 1978/80 → p. 277

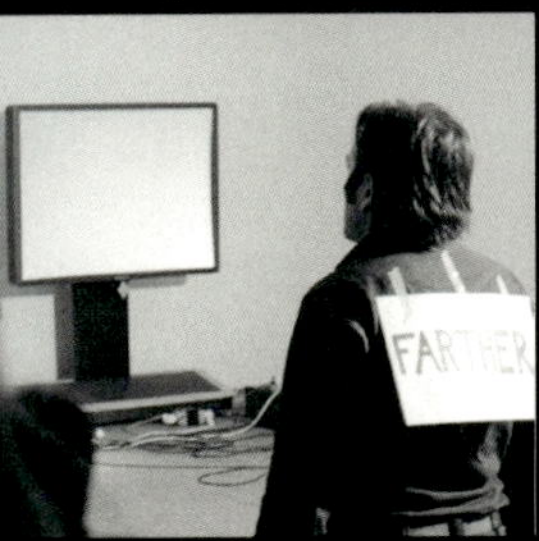

Investigation of Identity I, 1970/74 → p. 240

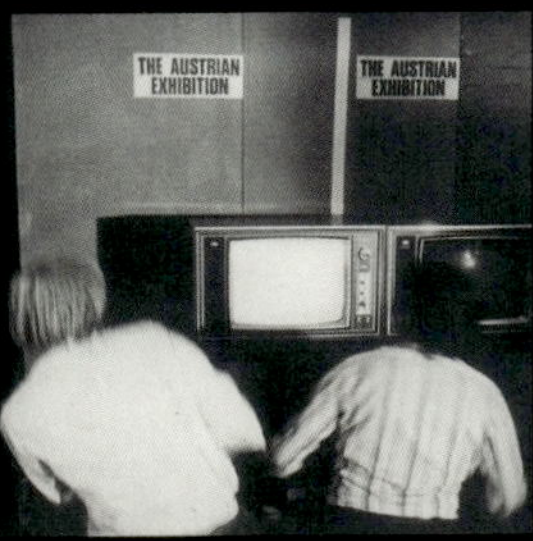

Investigation of Identity II, 1970/73 → p. 241

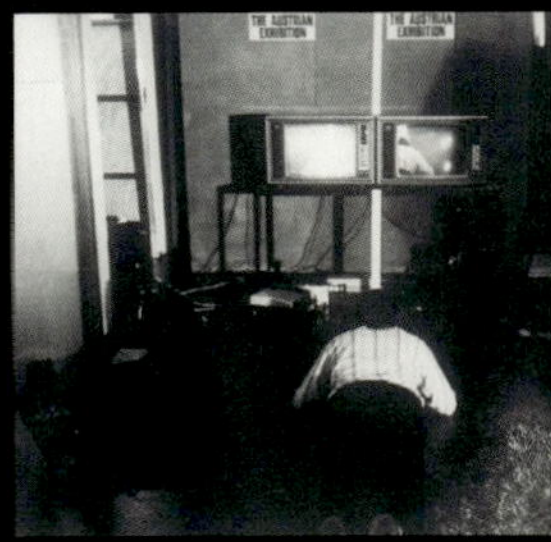
Investigation of Identity III, 1973 → p. 242

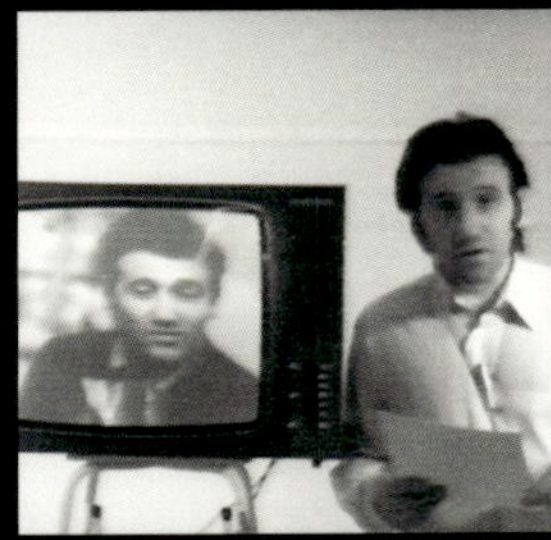
Parenthetical Identity, 1973 → p. 235

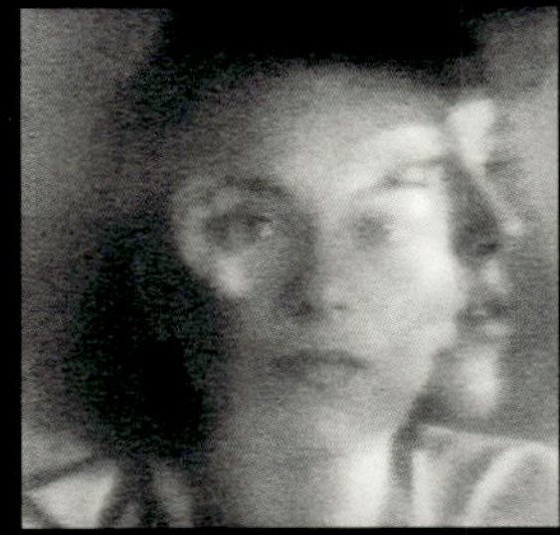
Switcher Sex, 1972 → p. 232

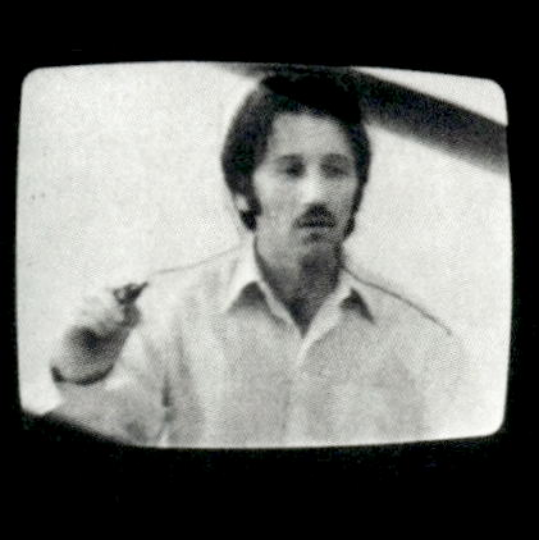
Self-Description, Unfinishable, 1973

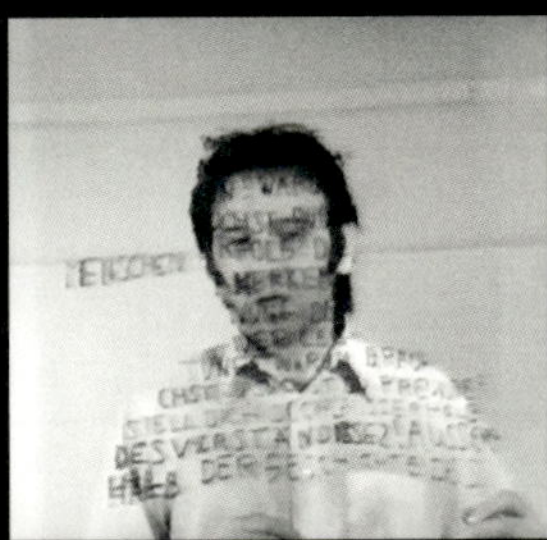
Self-Description, 1973 → p. 234

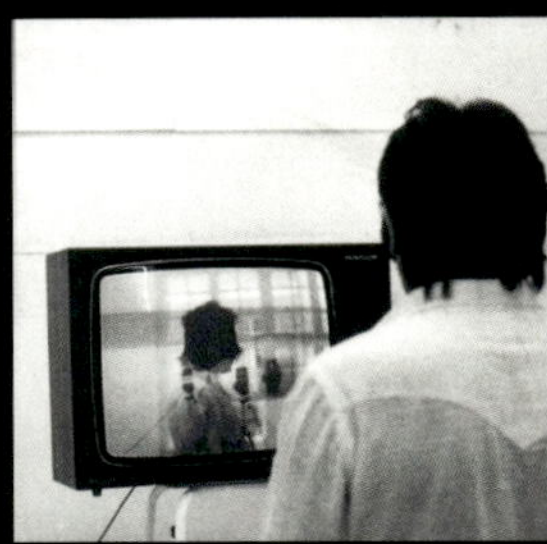
Self-Observation as Self-Obscuration as Self-Loss, 1974 → p. 243

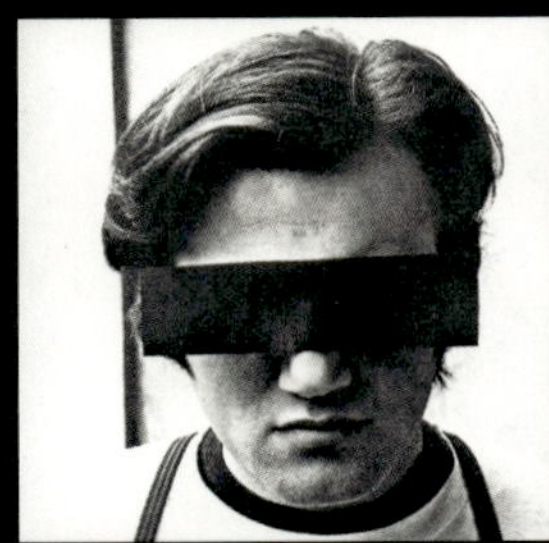
Self-Portrait as Anonymous, 1967 → p. 113

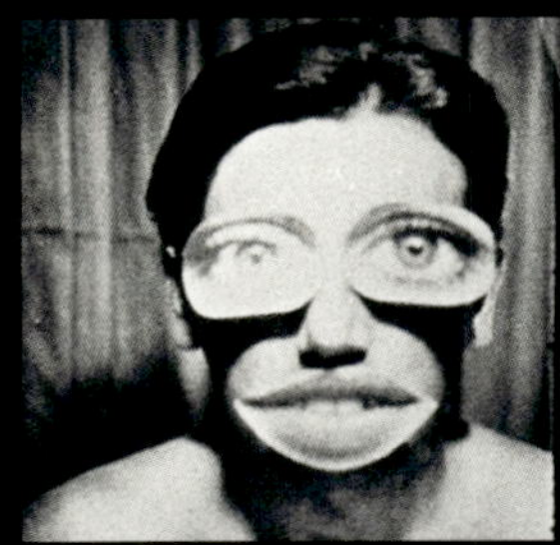
Self-Portrait as a Woman, 1967 → p. 113

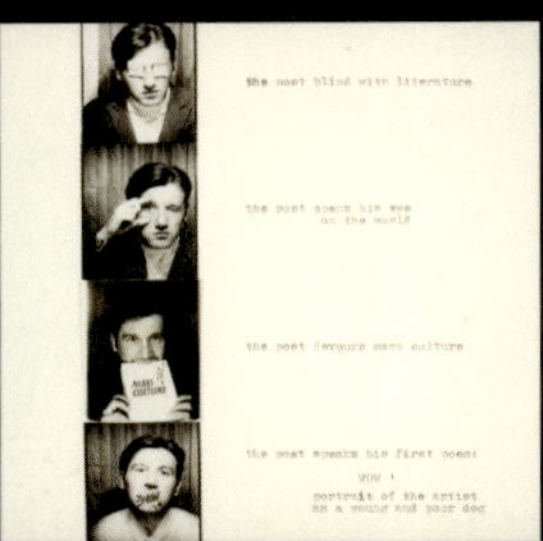
Evolution of the Poet, 1967 → p. 112

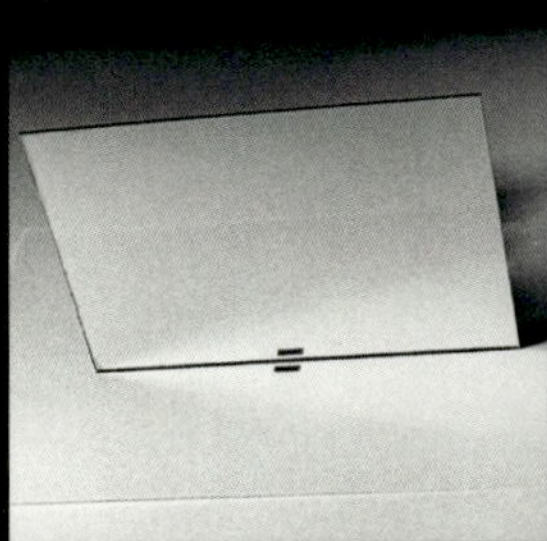
Reference Point for the Identity Sign, 1972 → p. 72

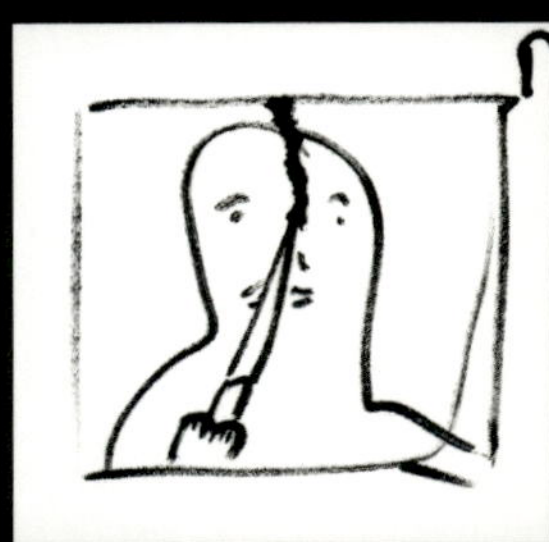
Schizoportrait, 1969

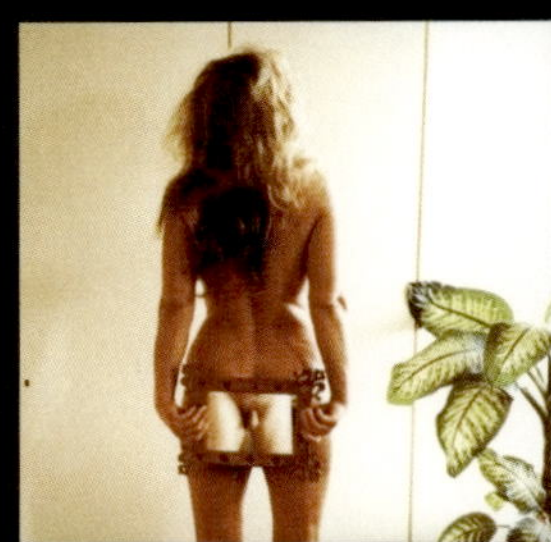
The Clinic of Two Mirrors, 1981 → pp. 132–33

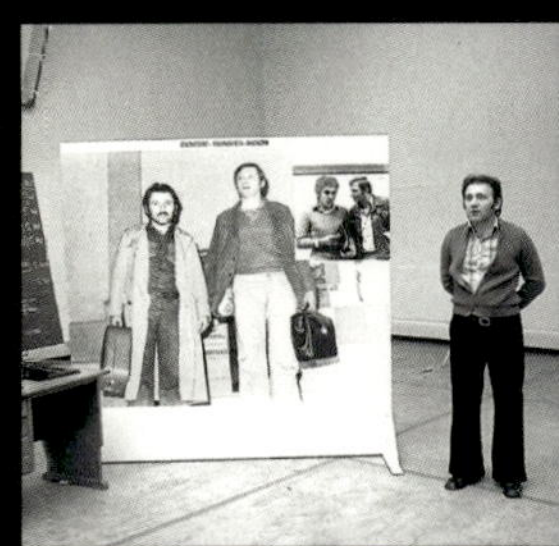
Identity Transfer Action, 1975 → p. 526

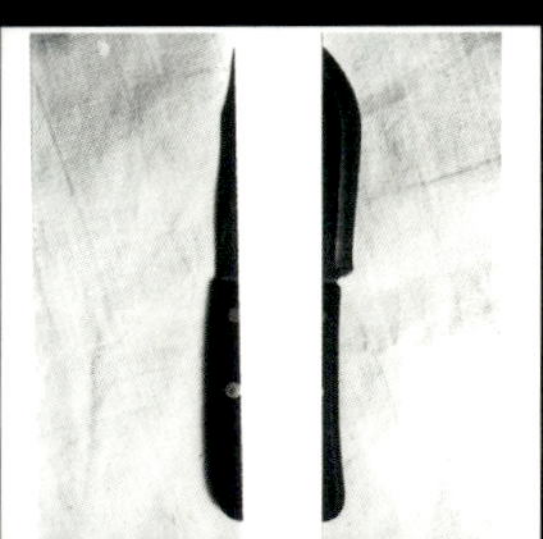
Identity of Action, 1975 → p. 145

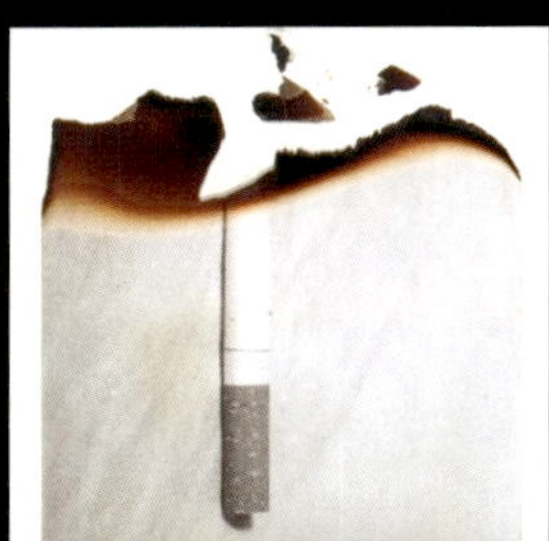
Identity of Action, 1975 → p. 145

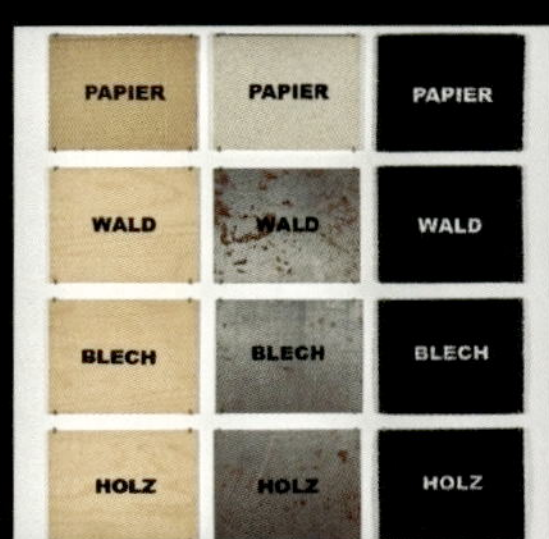

Iso- and Polymorphic Language of Materials, 1967 → p. 29

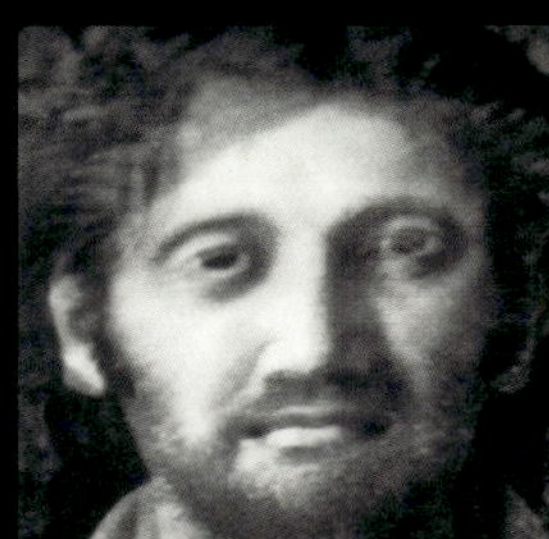
On the Theorem of Identity: Tritity, 1975 → pp. 236–37

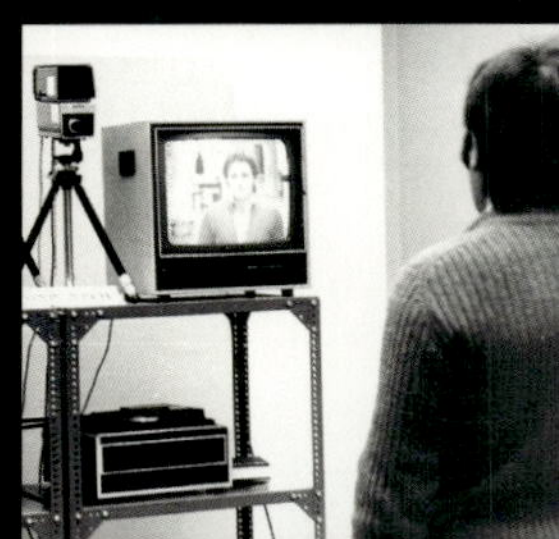
Self-Time –Time-Self, 1975 → pp. 266–67

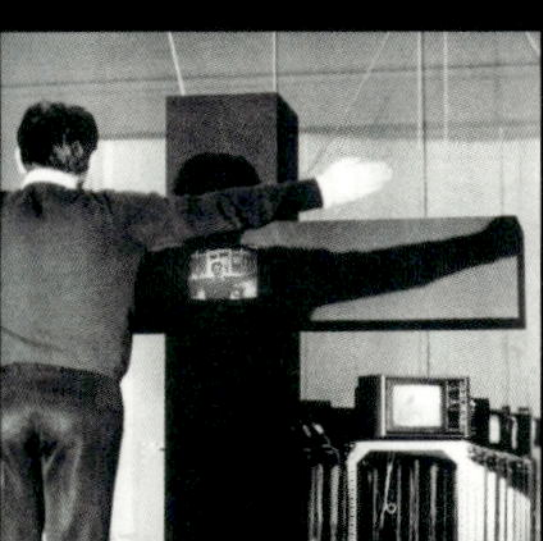
Crucifixion of Identity, 1973 → pp. 264–65

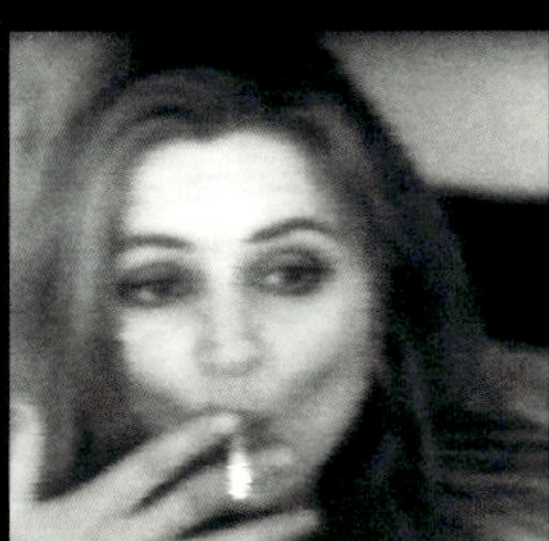
Review of the Nude, or Inverse Self-Portrait, 1975 → p. 233

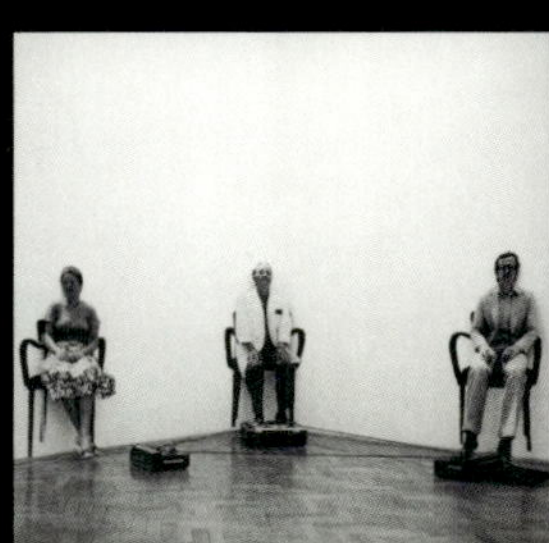
I Mass – Mass I, 1977 → p. 99

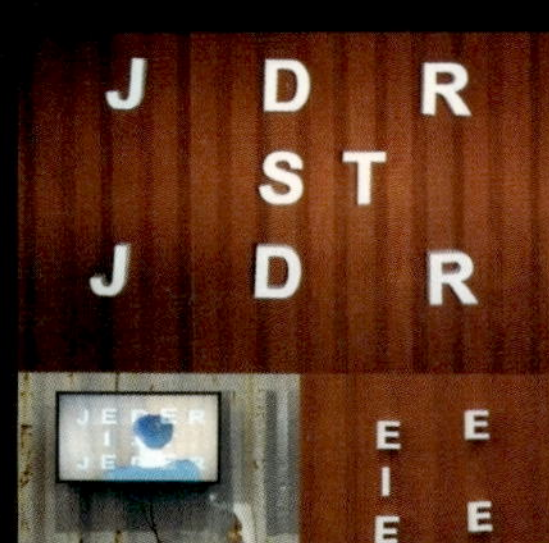

Everyone Is Everyone ..., 1993

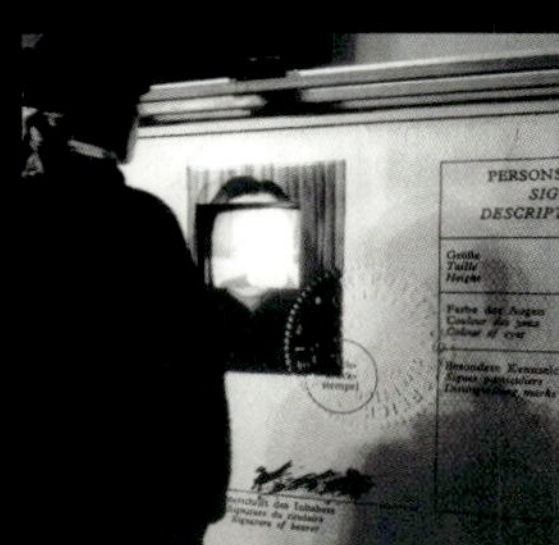
Against Institutional Identity, 1980/2014 → p. 275

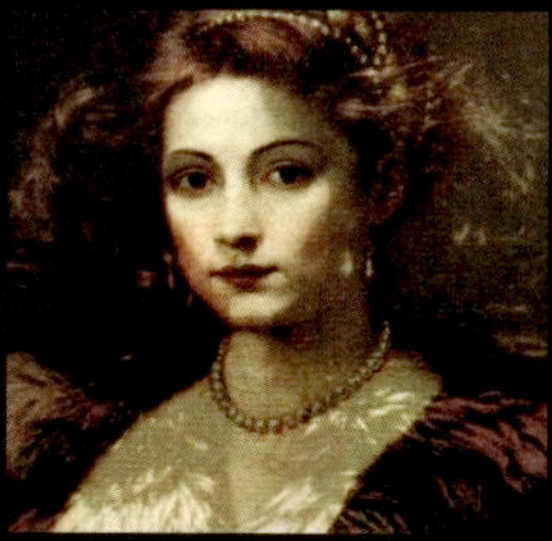

Venus in Furs, 2003 → p. 258

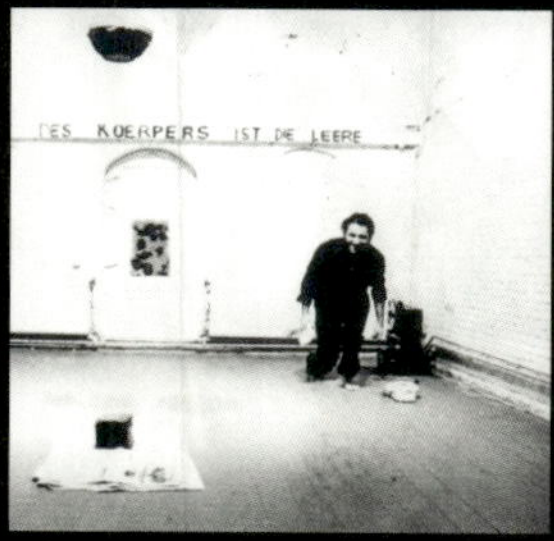

Sculptural Identity, 1970 (idea)/1978 → pp. 528–29

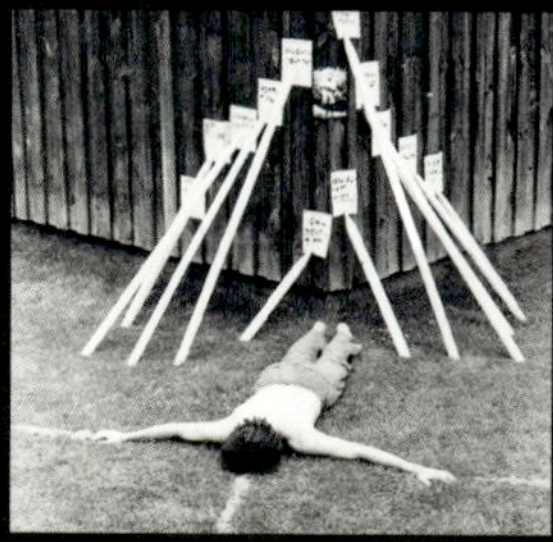

Social Matrix, 1971 → pp. 518–19

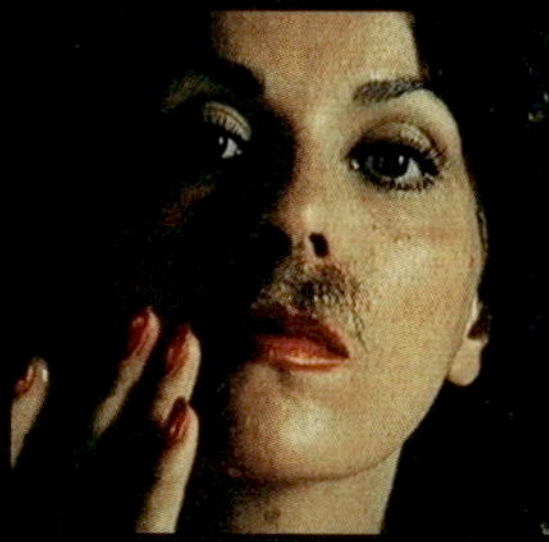

Scene from *Invisible Adversaries*, 1970 (idea)/1977 → pp. 134–35 and 202–203

Sex & Gender

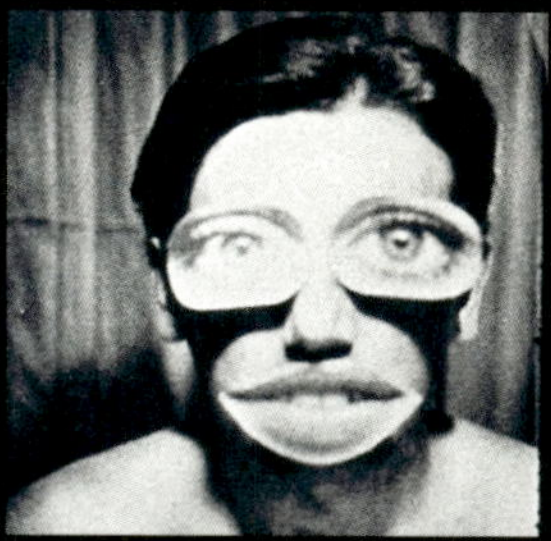

Self-Portrait as a Woman, 1967 → p. 113

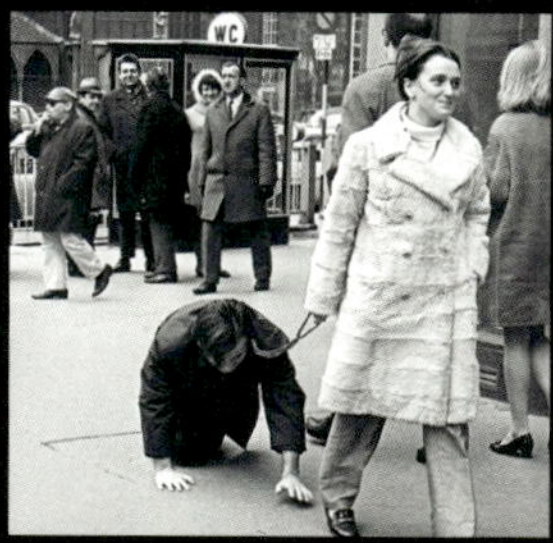

From the Portfolio of Doggedness, 1968 → pp. 504–505

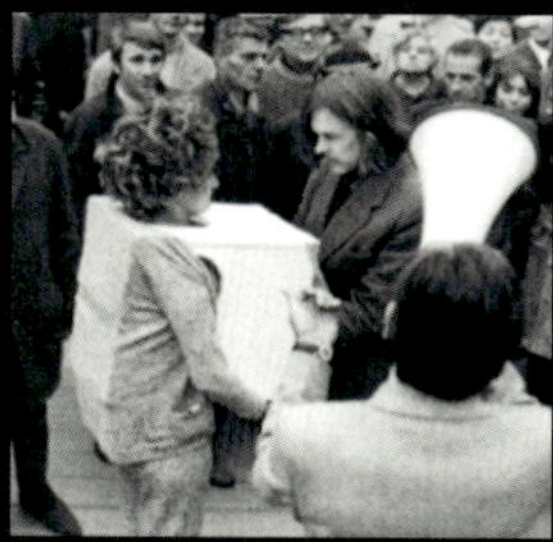

Touch and Feel Cinema, 1968 → pp. 180–81

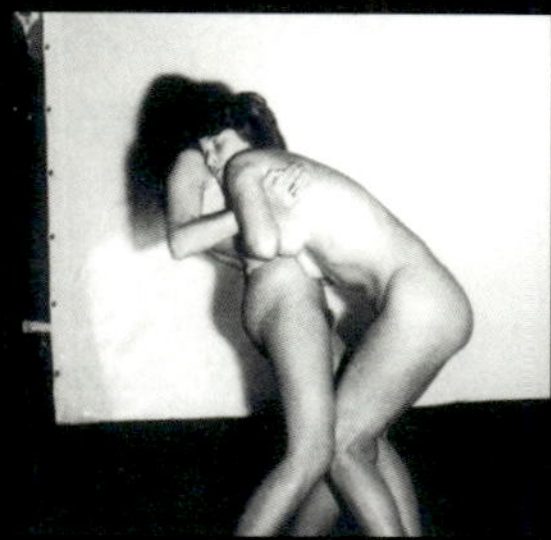

Energy Exchange, 1971 → p. 524

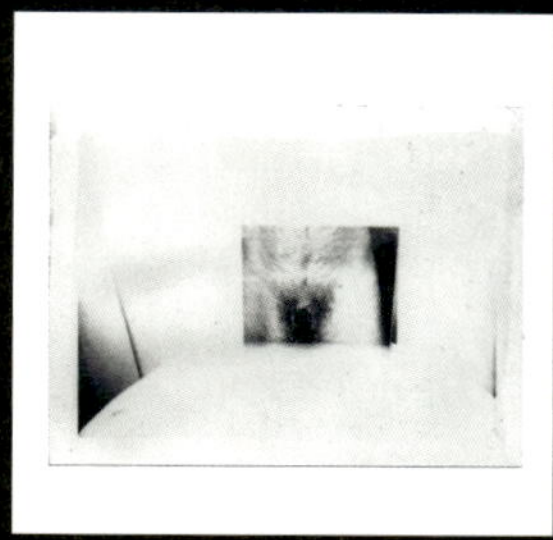

The Blind (Anonymous) Bandit, 1974 → p. 116

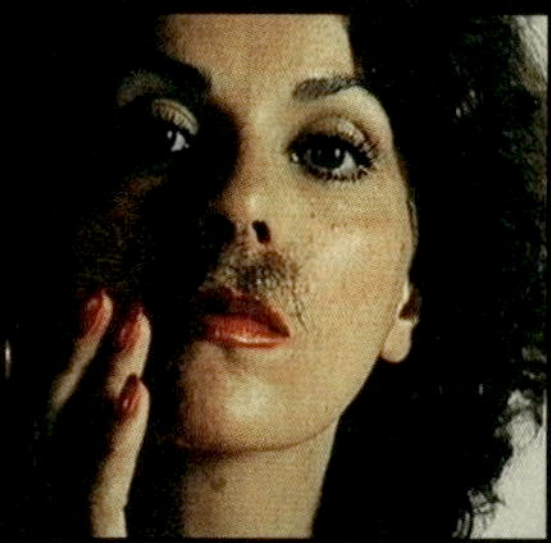

Scene from *Invisible Adversaries*, 1970 (idea)/1977 → pp. 134–35 and 202–203

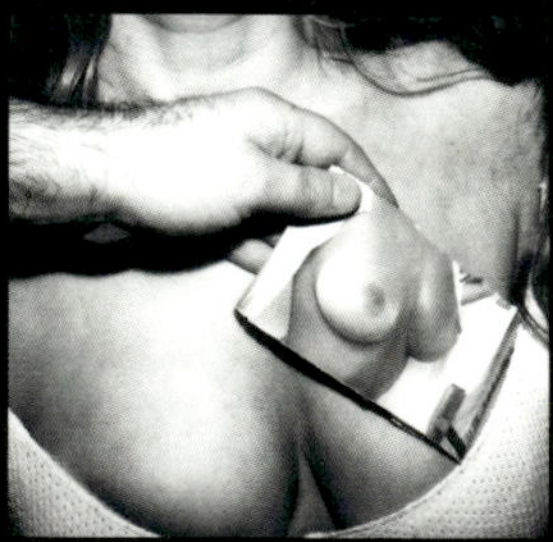

Partisan of Partial Drives, 1975 → p. 130

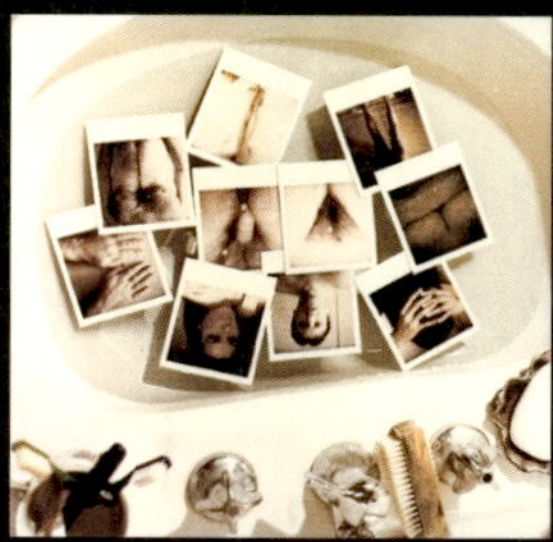

Anagrammar of Anatomy, 1972 → p. 131

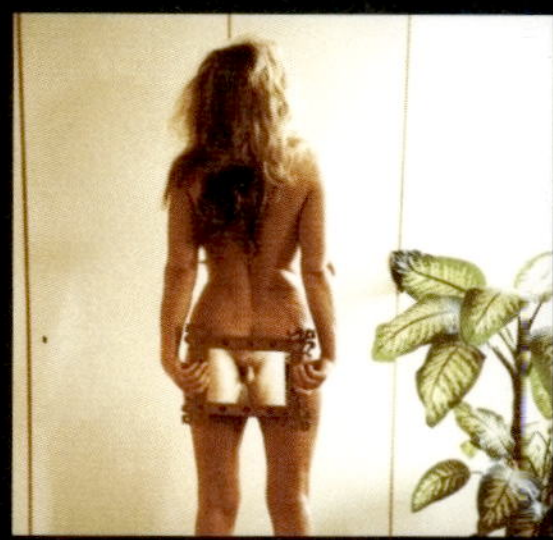

The Clinic of Two Mirrors, 1981 → pp. 132–33

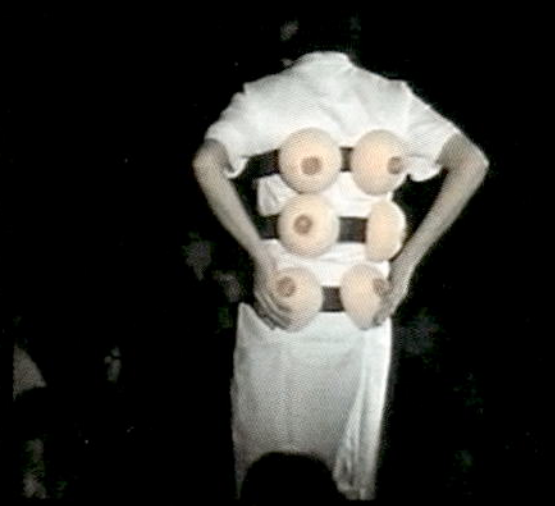

Scene from *Voices from an Inner Space*, 1988 → pp. 358–59

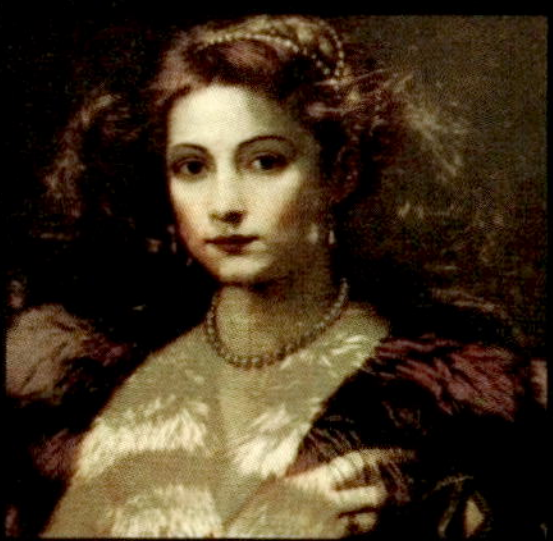

Venus in Furs, 2003 → p. 258

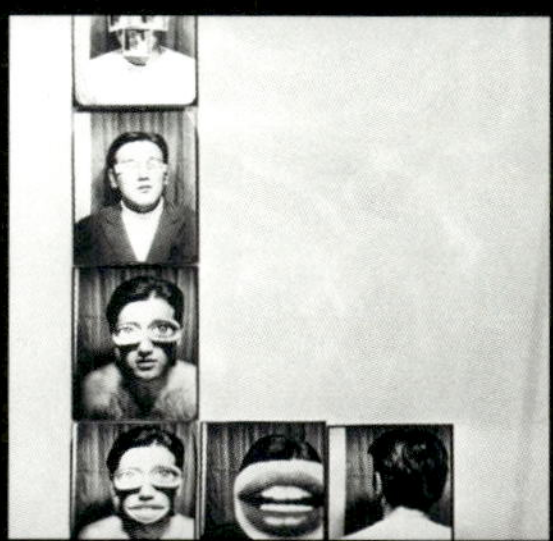

A Love Story, 1967

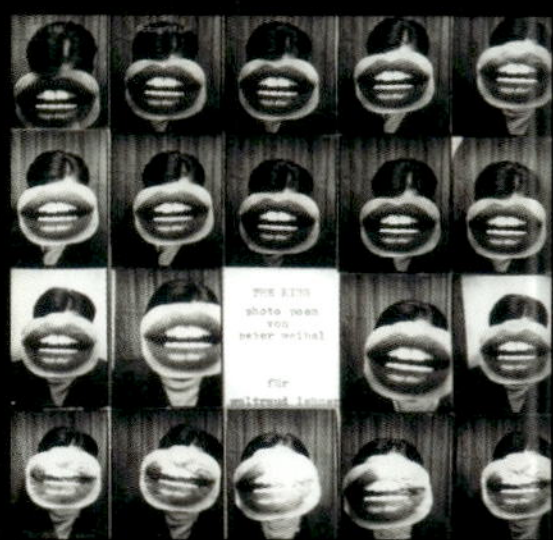

The Kiss, 1967

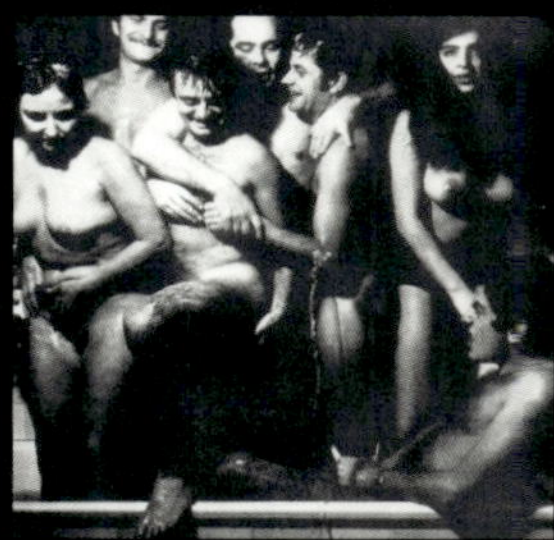

Announcement for the book *Wien* by Peter Weibel and VALIE EXPORT, 1970

Hair

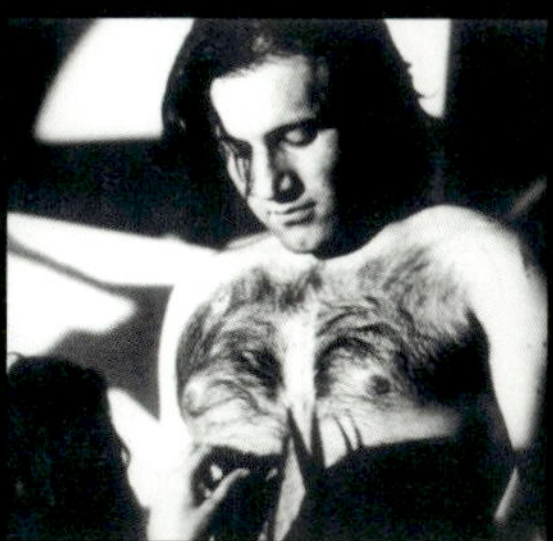

Cutting, 1967–68 → pp. 174–75

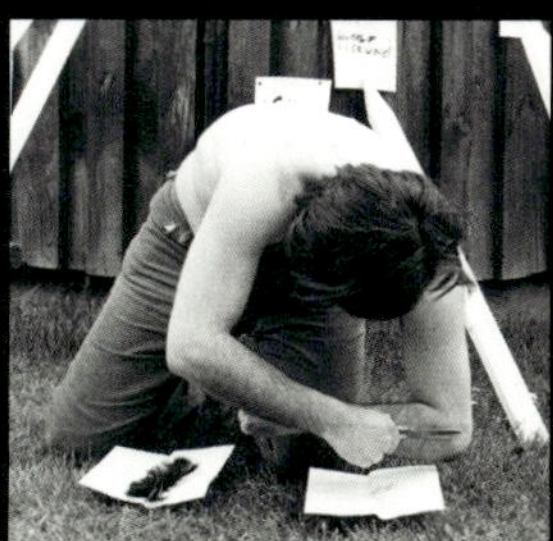

Social Matrix, 1971 → pp. 518–19

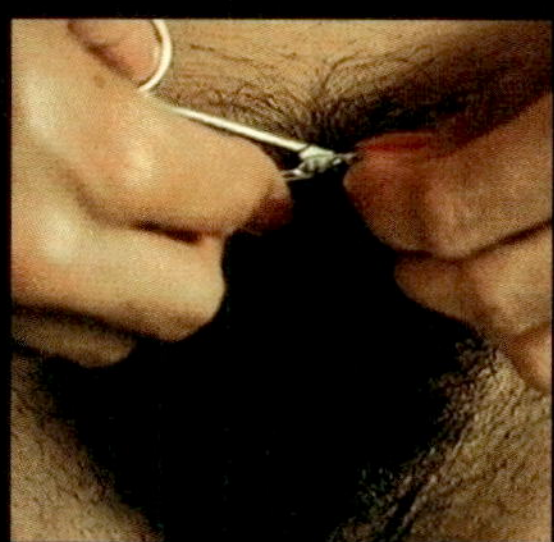

Scene from *Invisible Adversaries*, 1970 (idea)/1977 → pp. 134–35 and 202–203

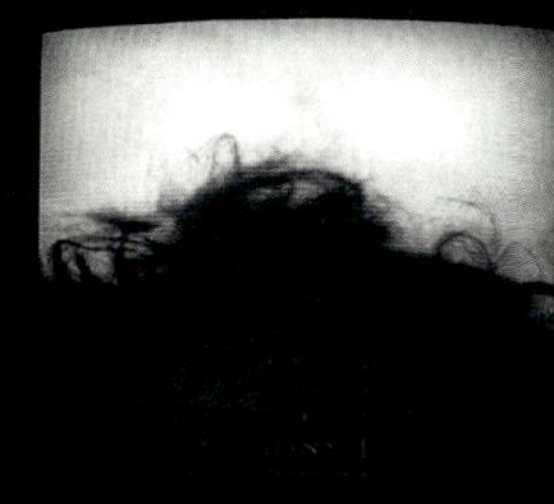

Solution of Fantasy, 1972 → p. 251

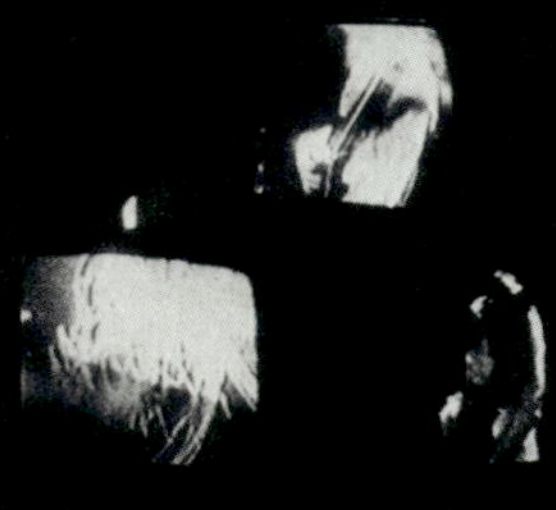

Video Music, 1978

Skin

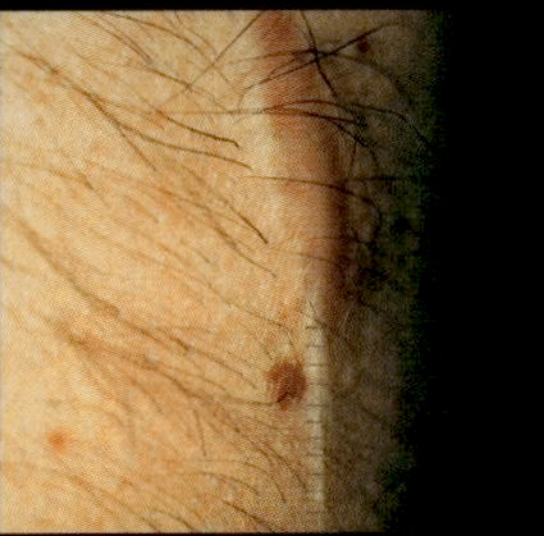

Skin Poems, 1968 → pp. 52–53

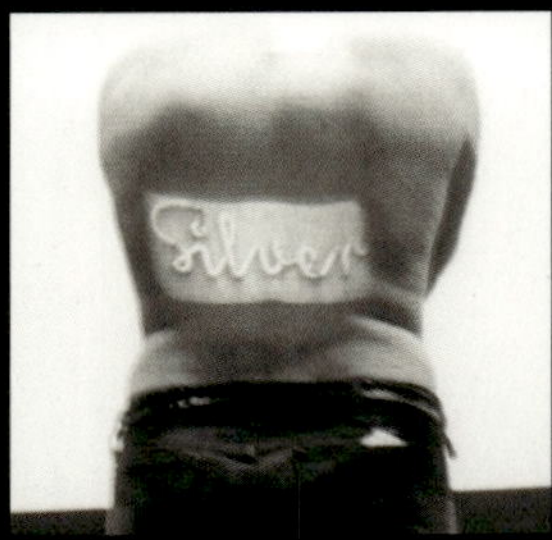

Action Lecture No. 1: Proposals on Non-Affirmative Art, 1966 → pp. 168–67 and 498–99

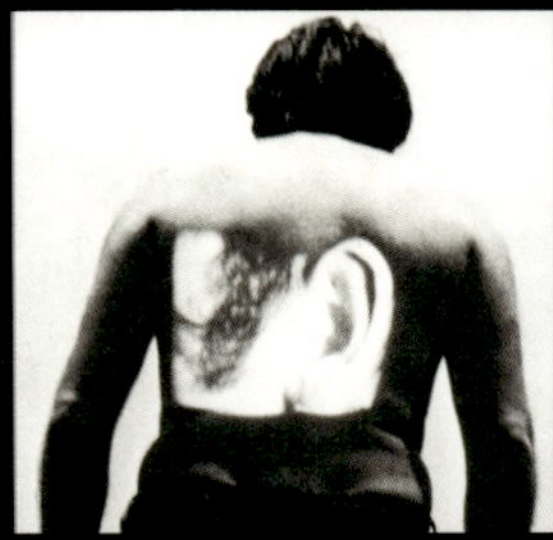

Listen, 1968 → p. 349

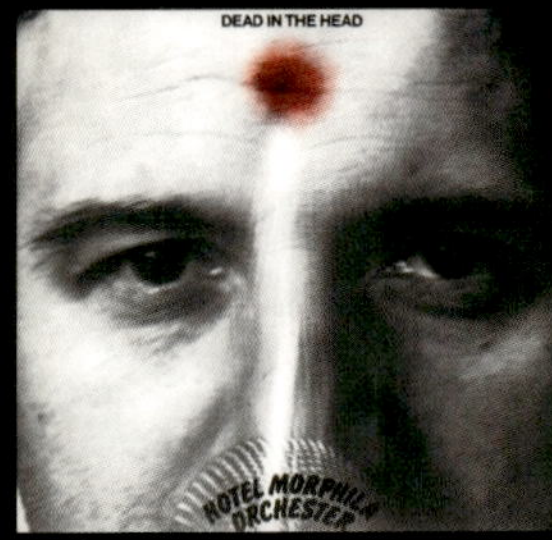

"Dead in the Head," 1982 → pp. 352–53

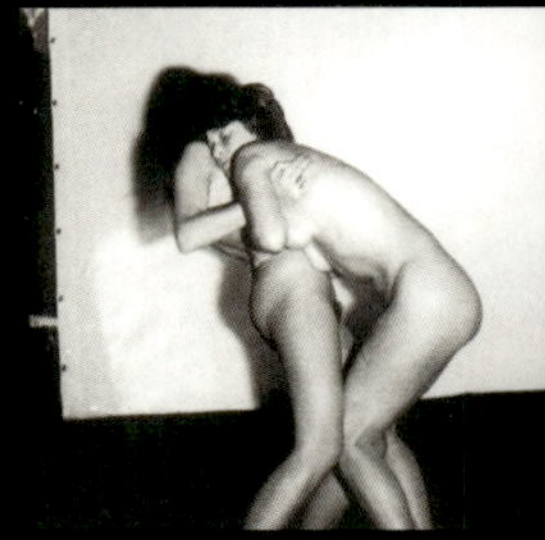

Energy Exchange, 1971 → p. 524

Blood

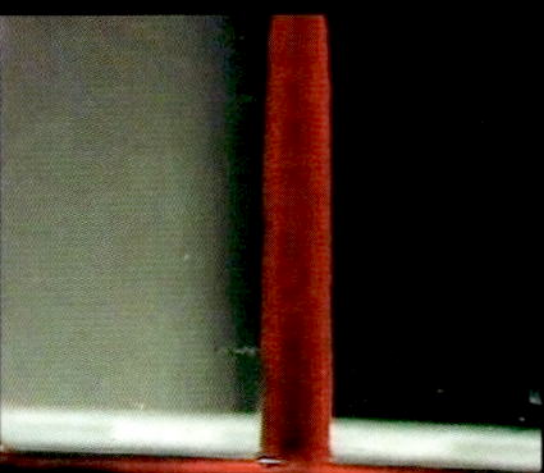

Time Blood, 1972/79 → pp. 254–55

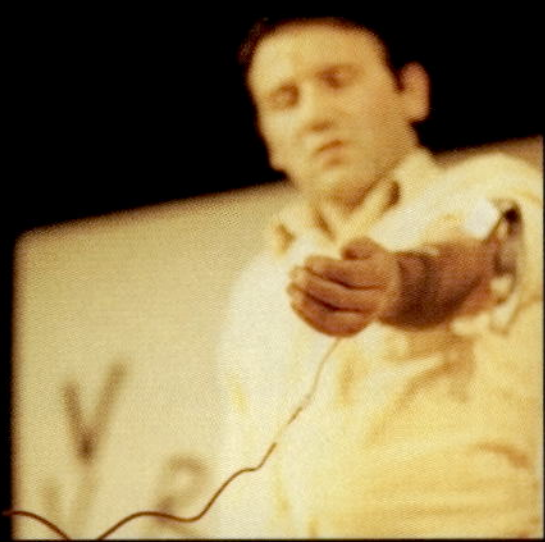

Blood Bell, 1979 → p. 533

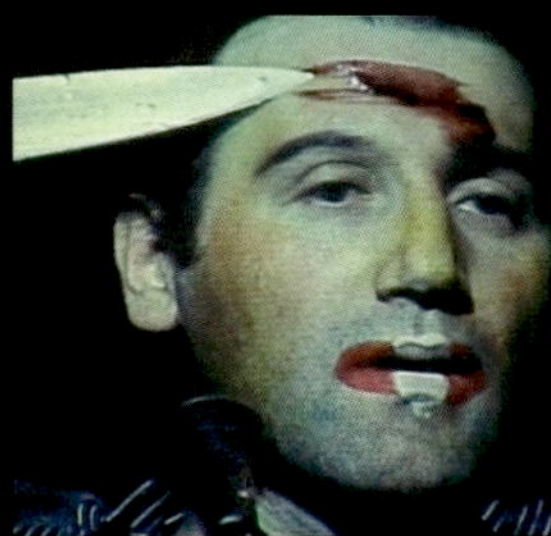

Performance by Hotel Morphila Orchester, 1980 → pp. 352–53

Eyes

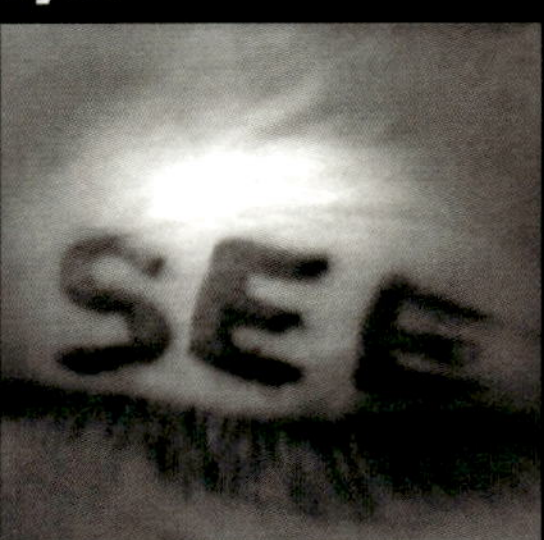

Eye Texts, 1974 → p. 56

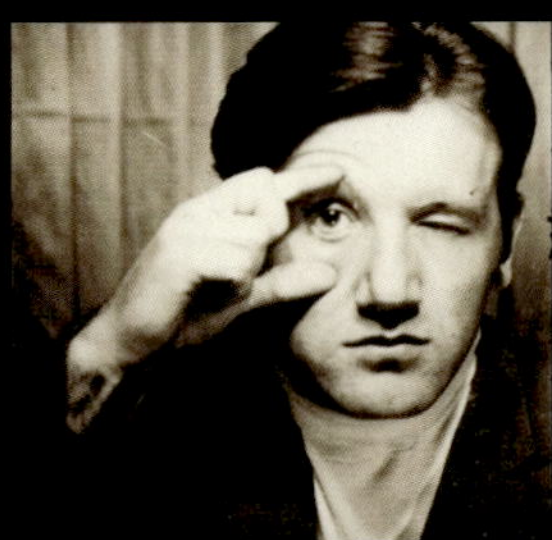

Evolution of the Poet, 1967 → p. 112

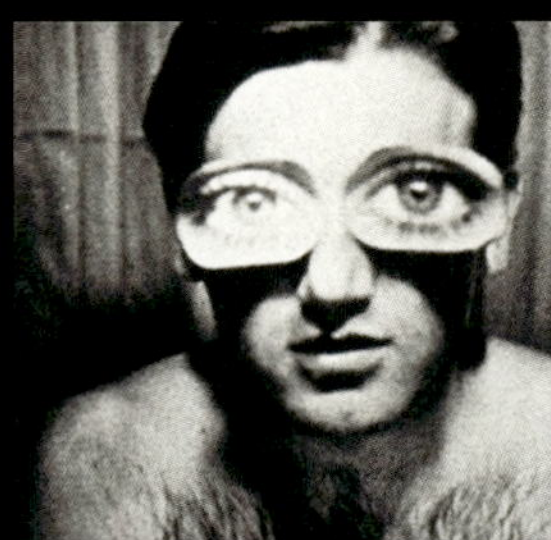

Self-Portrait as a Woman, 1967 → p. 113

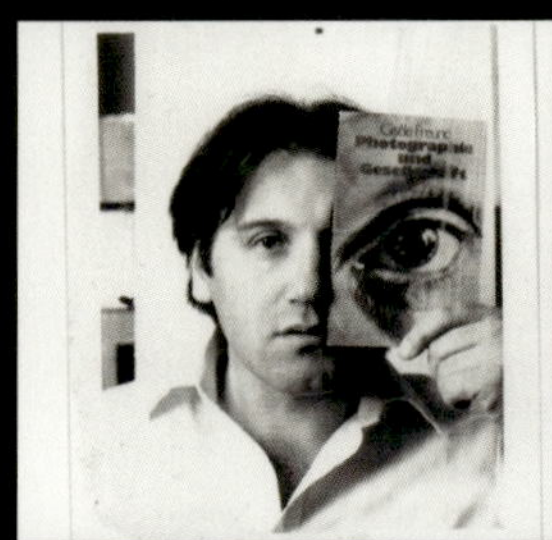

The Eye of Photography, 1977 → p. 117

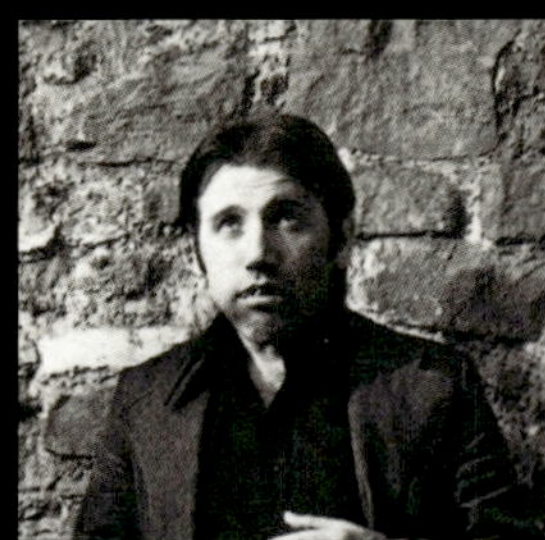

Blind, 1967/68 → p. 65

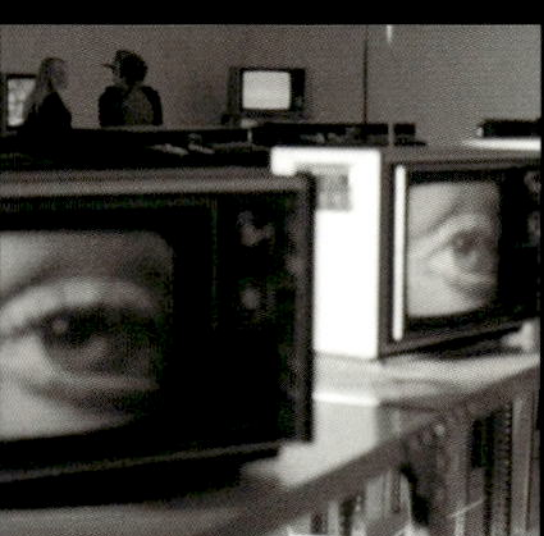

Eye, 1969

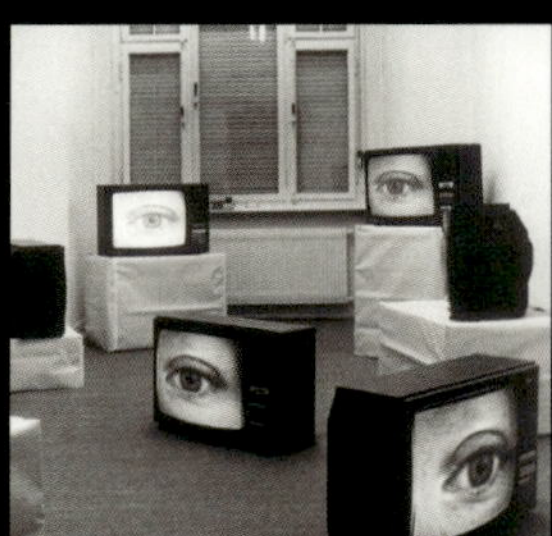

Video Lumina, 1972 (idea)/1977 → p. 268

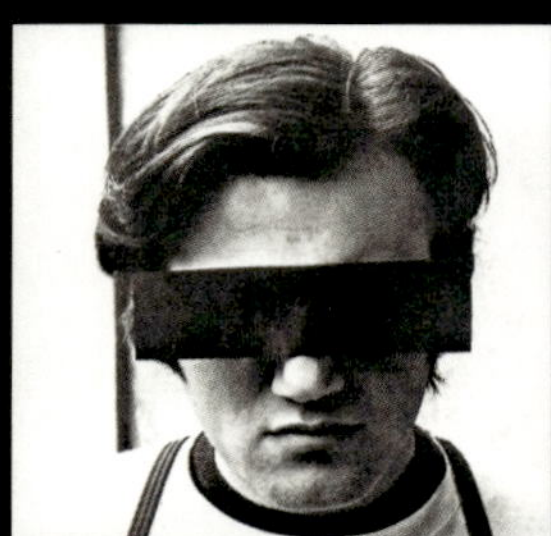

Self-Portrait as Anonymous, 1967 → p. 113

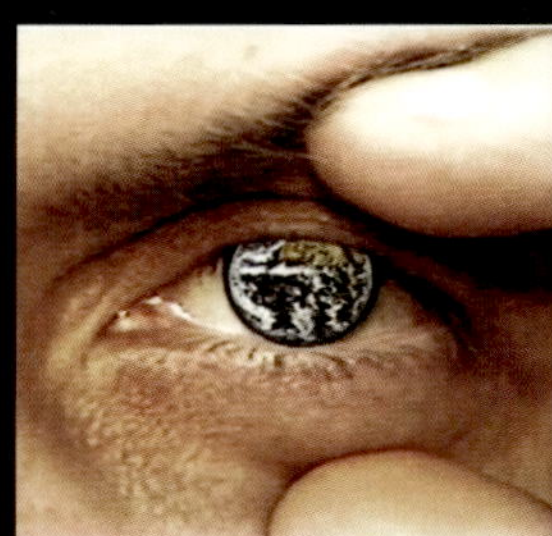

Scene from *The Artificial Will. Electronic Media Opera*, 1984 → pp. 355–57

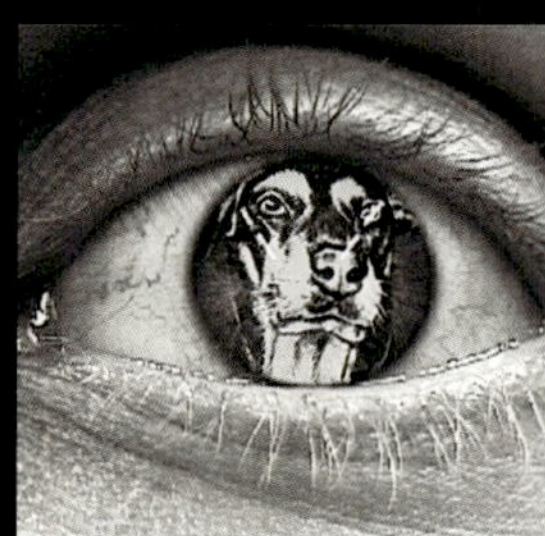

The Panoptic Society ..., 2001 → p. 319

Ears

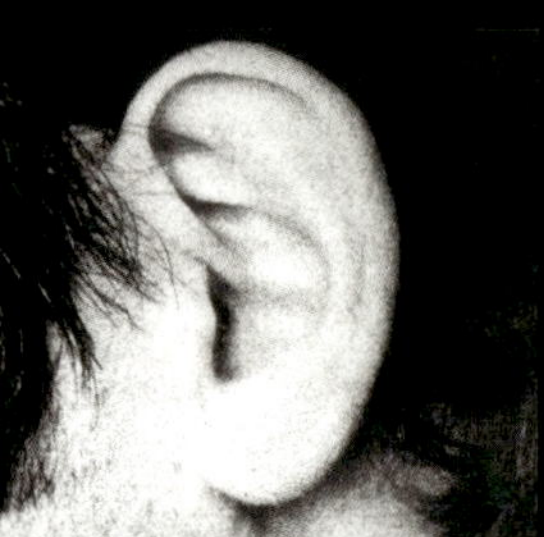

Transformational Music, 1972 → pp. 346–47

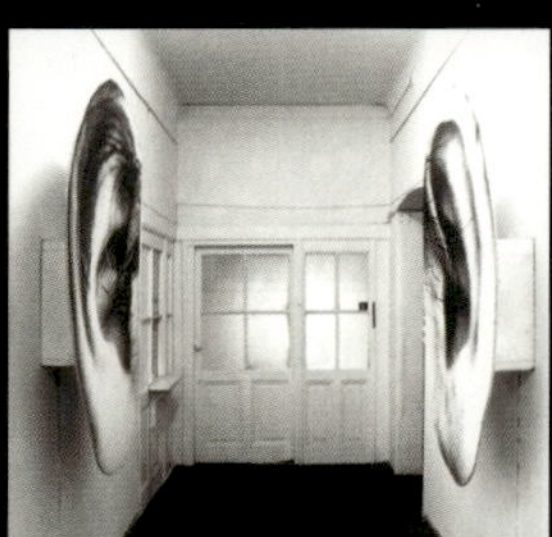

Auditory Canal, 1970/79 → pp. 350–51

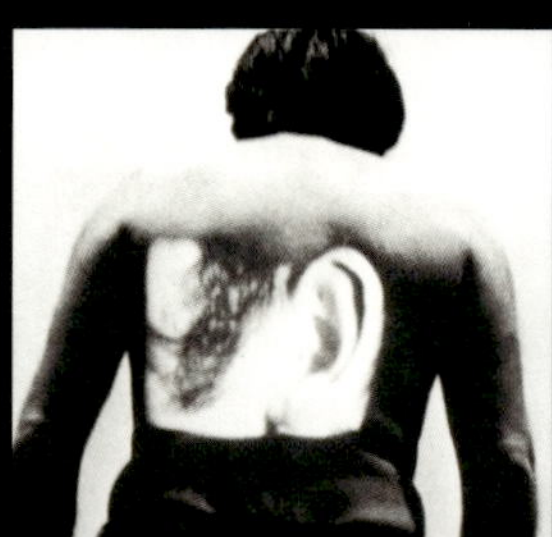

Listen, 1968 → p. 349

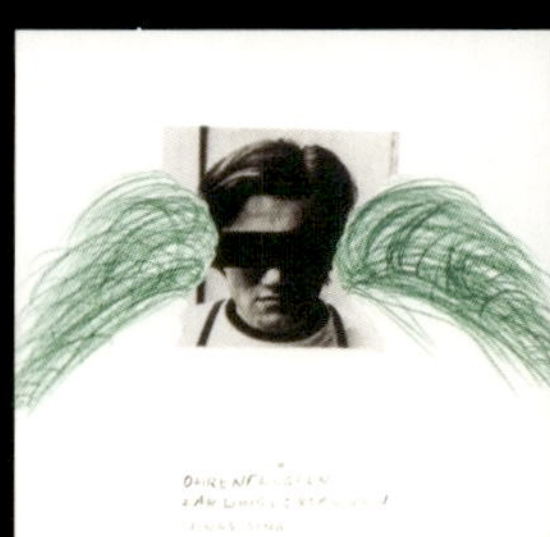

Ear Wings, 1978/2014 → p. 289

Mouth

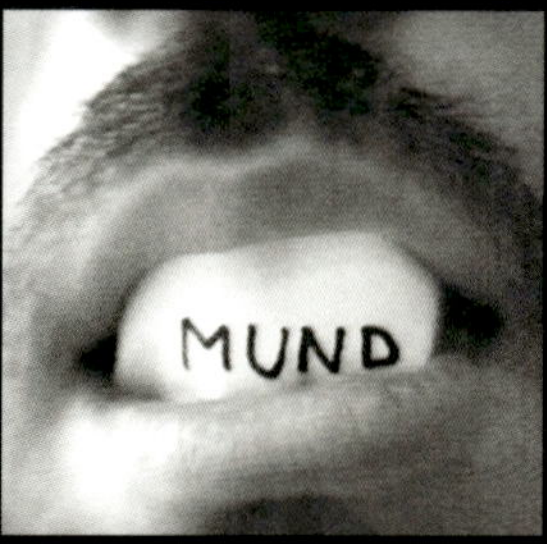

Mouth Texts, 1974 → p. 57

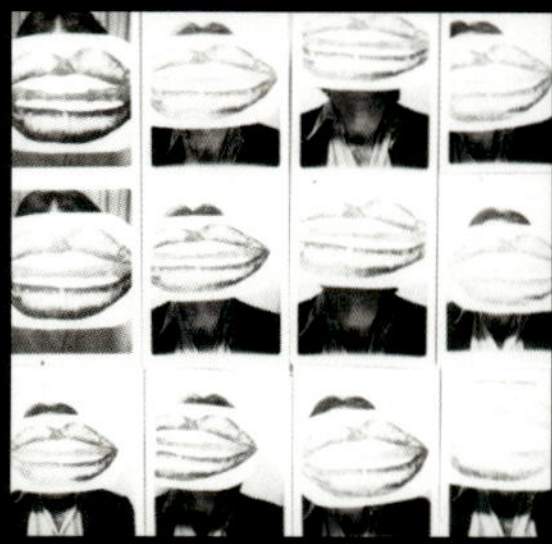
The World/Wilting Kiss, 1967 → p. 111

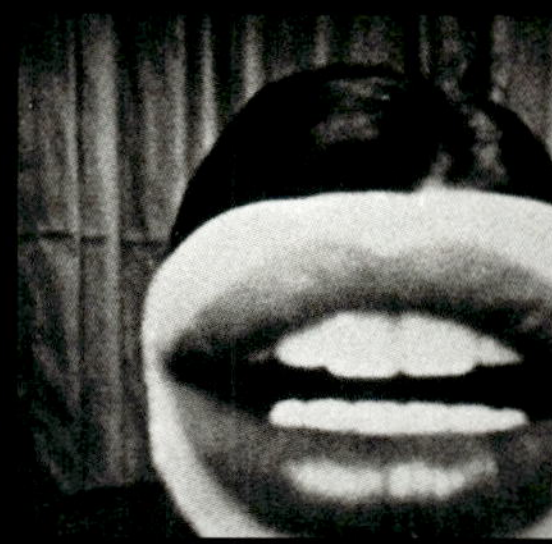
Self-Portrait as a Woman, 1967 → p. 113

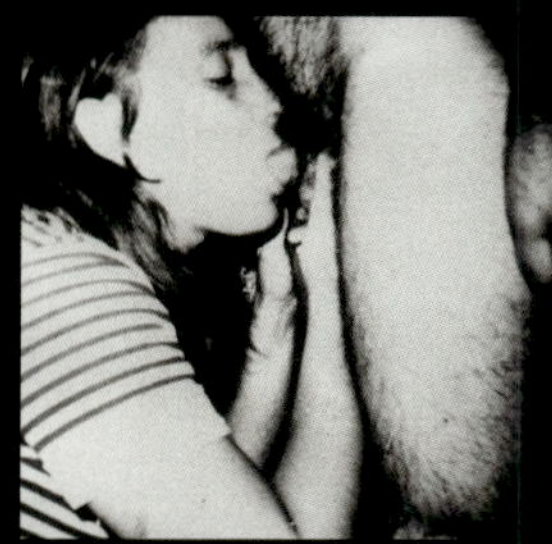
Cutting, 1967–68 → pp. 174–75

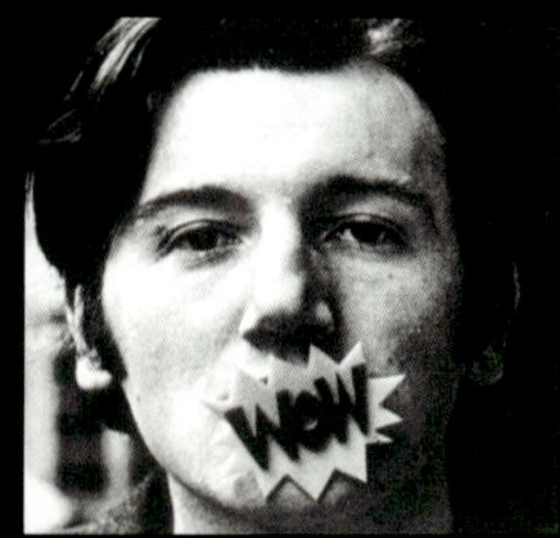

Self-Portrait of the Artist as a Young Dog, 1967

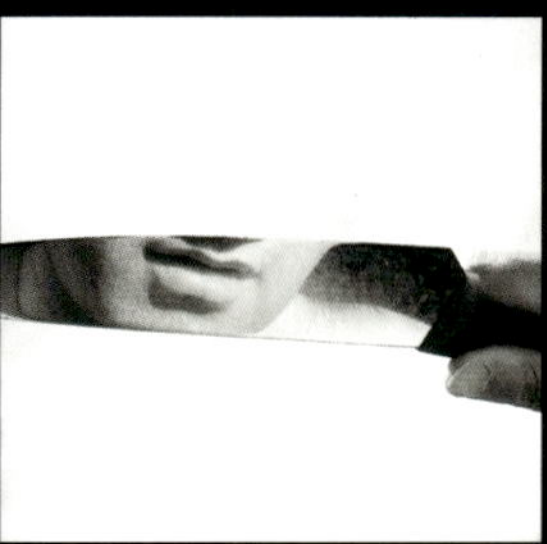
Self-Portrait (Knife as Mirror), 1975 → p. 116

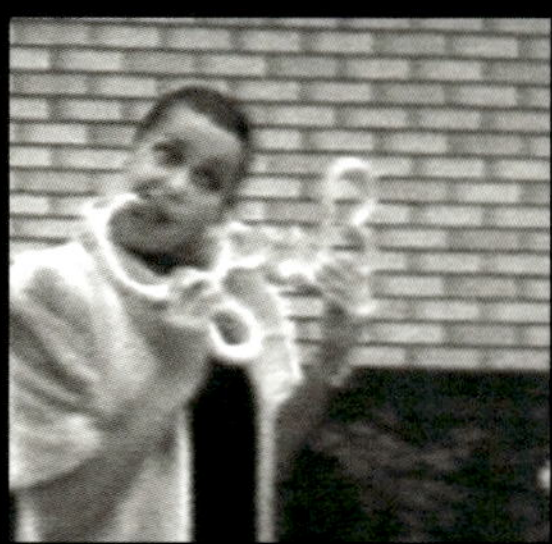
Greet God, 1967/72 → p. 64

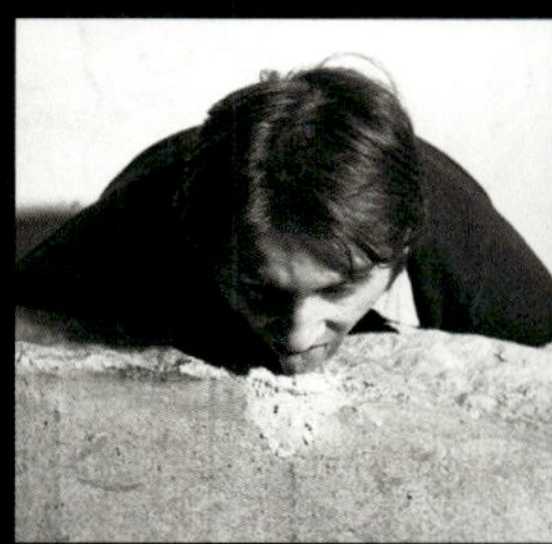
Space of Language, 1973

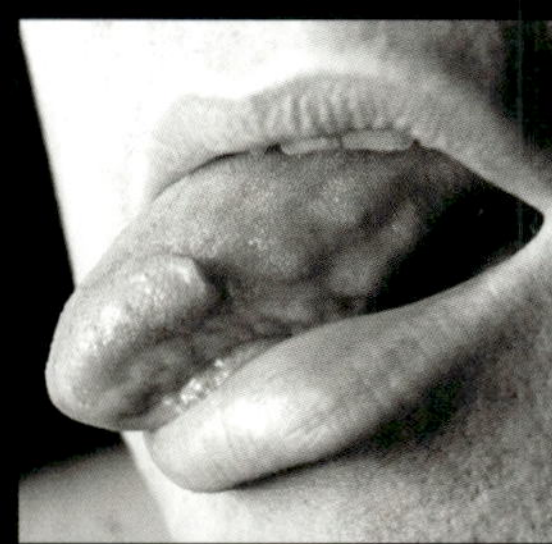
→ p. 525

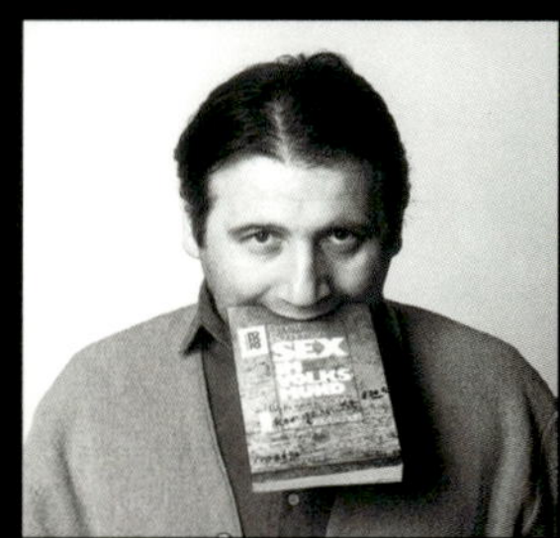

Sex on the Common Tongue, 1971 → p. 114

Hands & Fingers

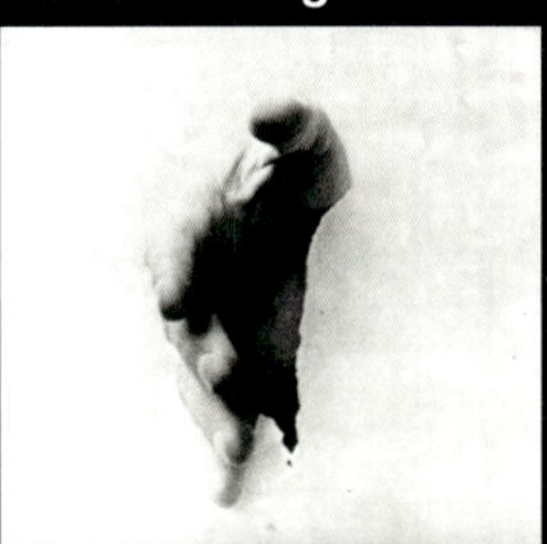
Rip, 1966 → front matter

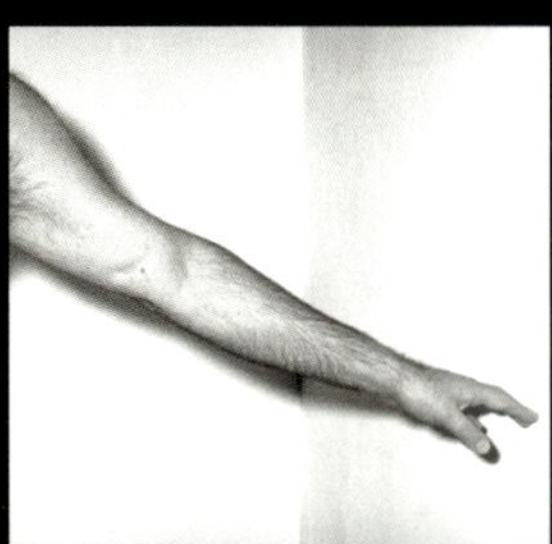
Time Bone – Space Muscle. On the Knowledge of the Body II, 1972 → p. 521

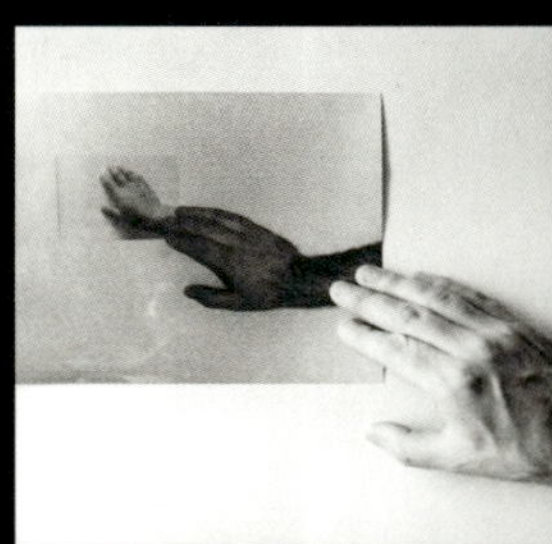
The Endless End → back matter

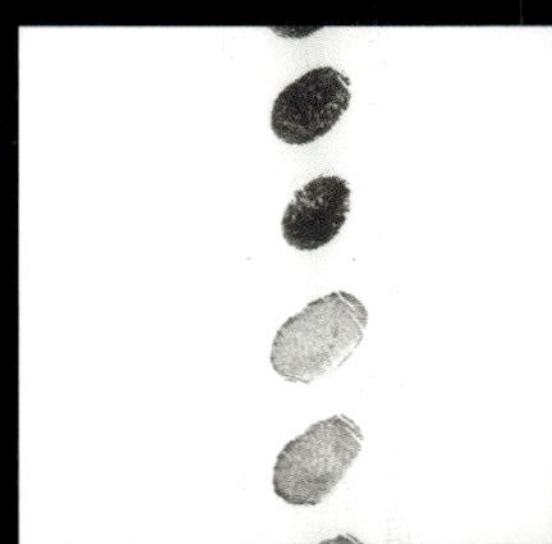
Fingerprint, 1968 → p. 171

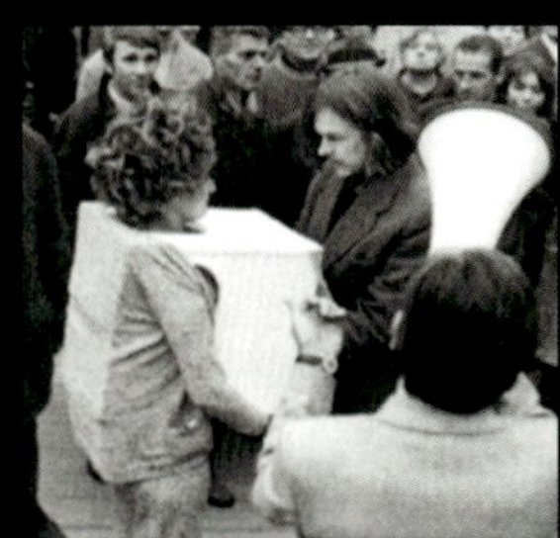
Touch and Feel Cinema, 1968 → pp. 180–81

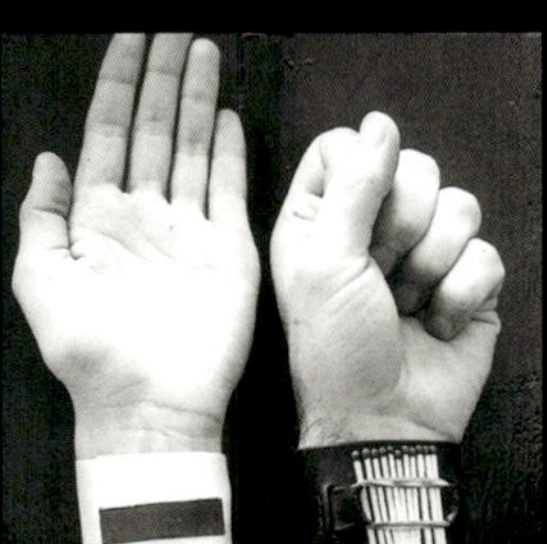
More Warmth among Human Beings, 1972 → p. 515

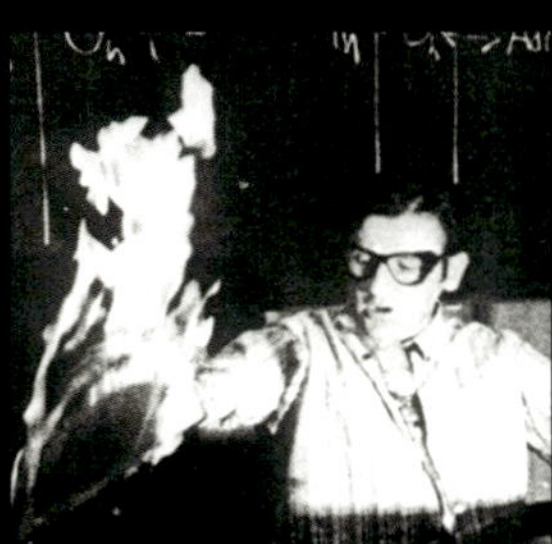
Incendiary Speech, 1968 → pp. 508–509

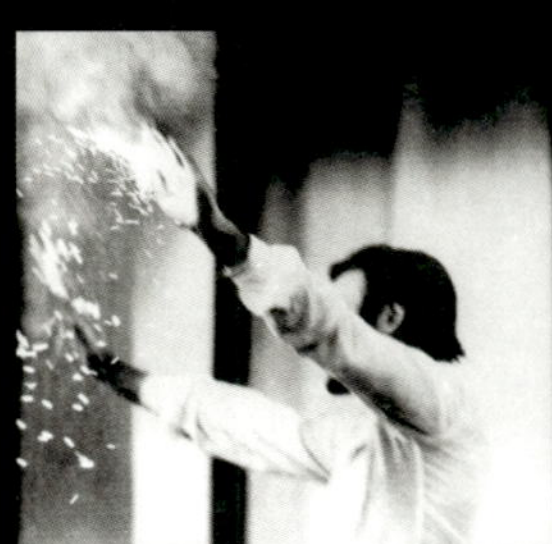
Initiation, 1971 → p. 514

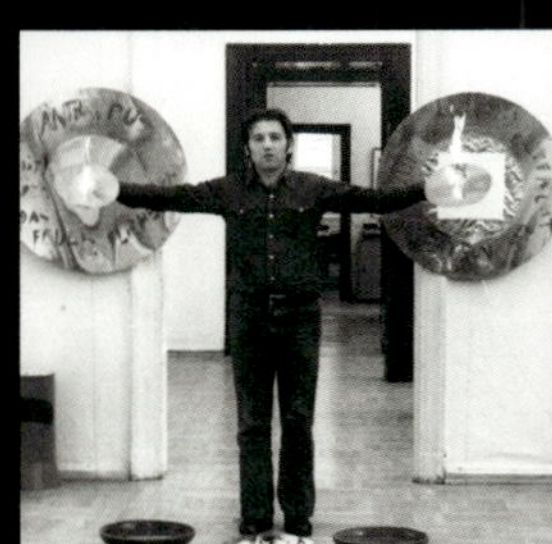
Fireflight – Flightfire, 1975 → pp. 516–17

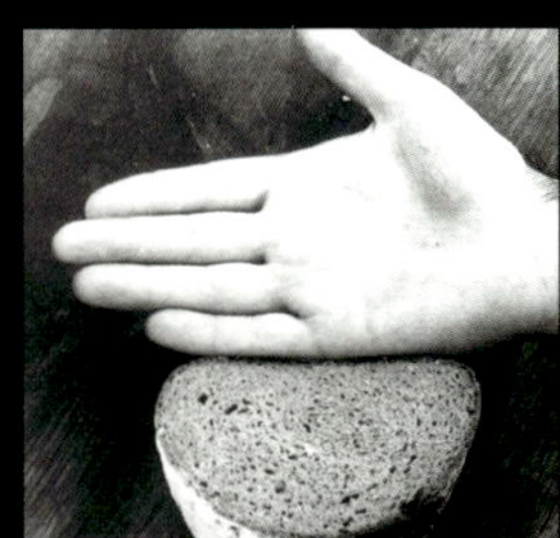
Invisible Force and Bodily Property: Body Relationships as Measures of Ownership, 1971

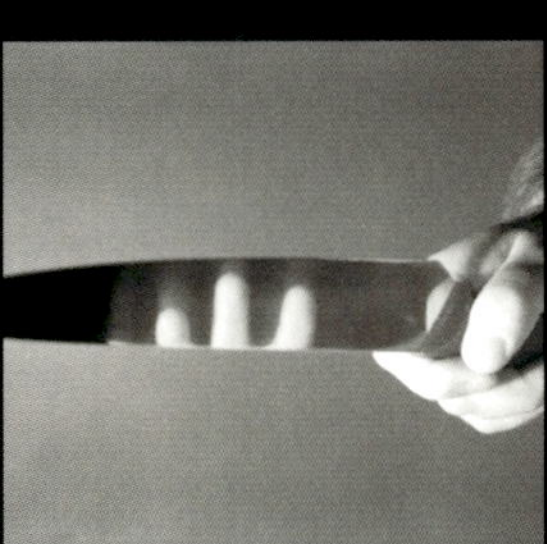
Self-Portrait (Knife as Mirror), 1975 → p. 116

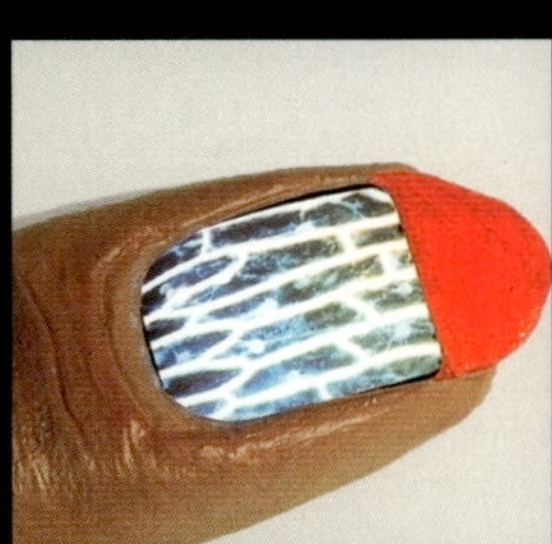
Orbital Sculpture V, 1988 → p. 434

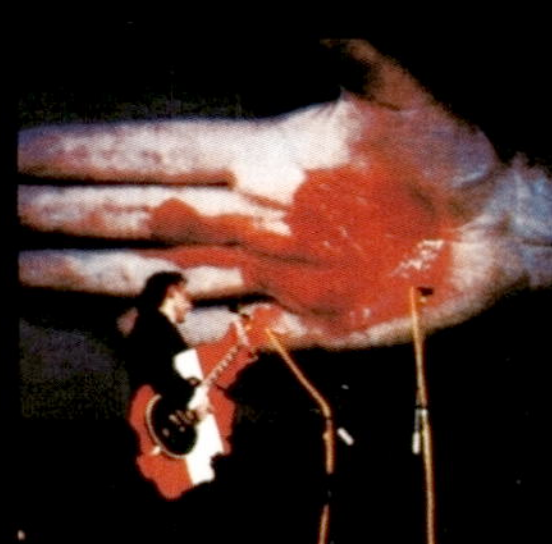
Scene from *The Artificial Will. Electronic Media Opera*, 1984 → pp. 355–57

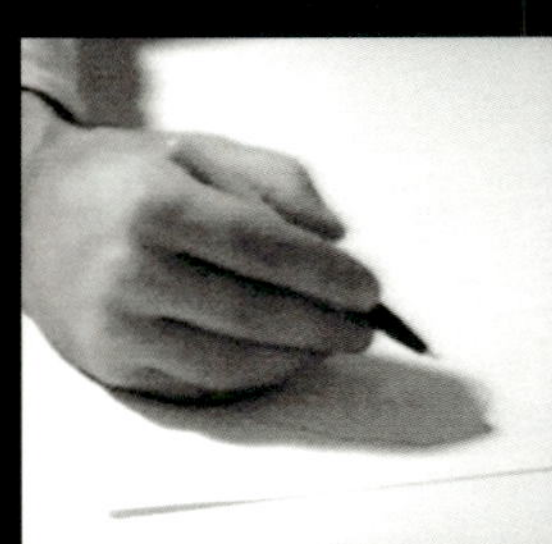
Self-Limitation, 1973

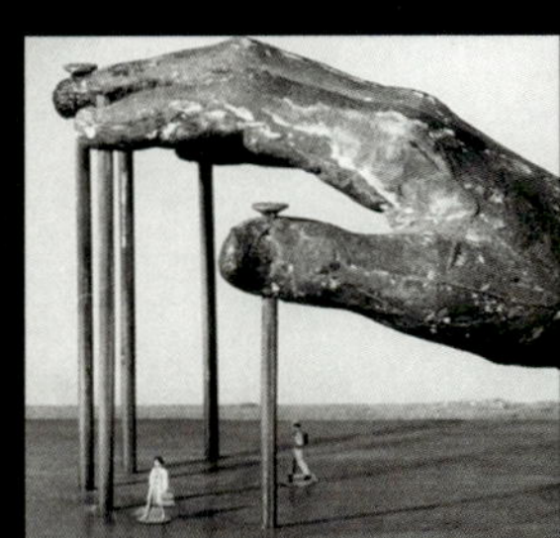
Place of Humanity, 1973

Stones

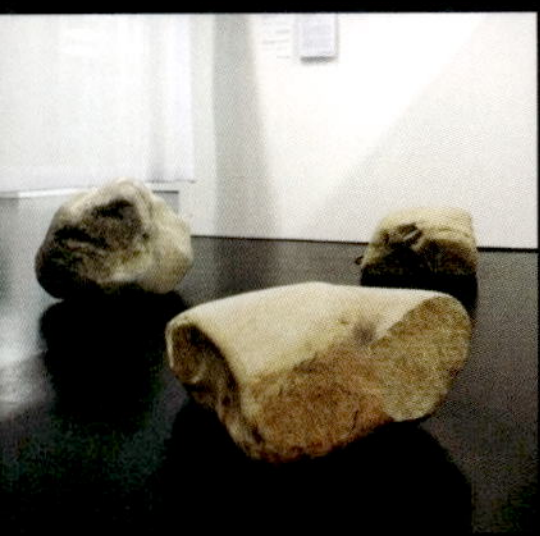

Moaning Stone. Nonhuman Poem, 1969 → pp. 100–101

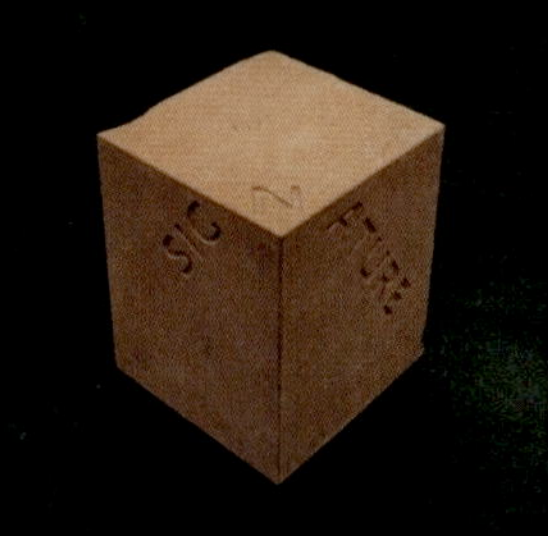

Sig/n/ature, 1967/75 → p. 77

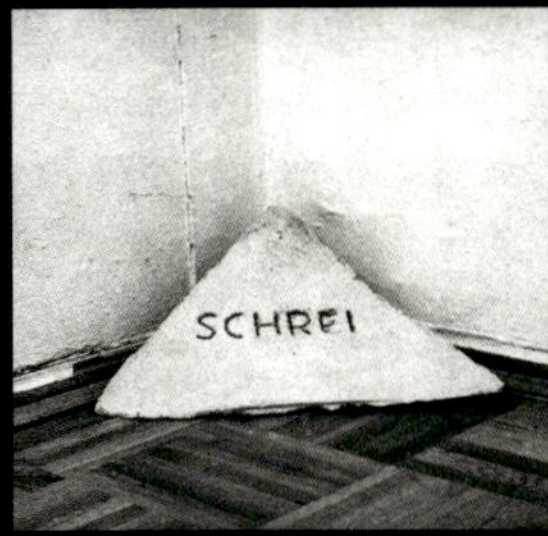

Scream, 1967/75 → p. 78

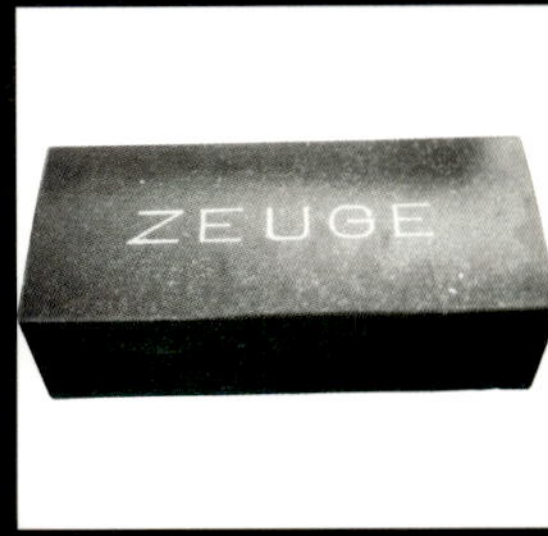

Silent Witness, 1966/75

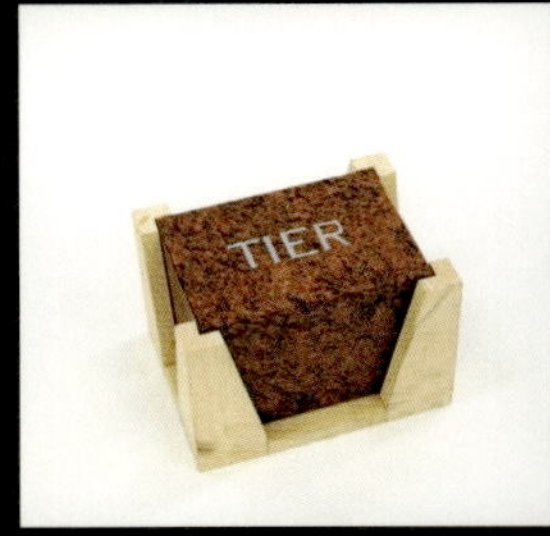

Cage, 1967/75 → p. 78

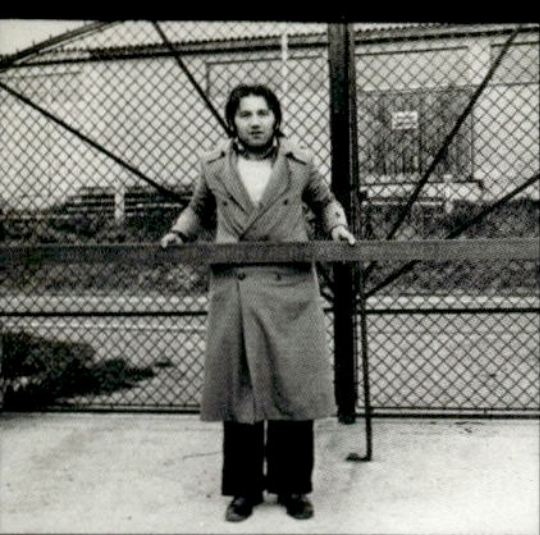

This Load Is Heavier Than I Can Bear, 1975 → p. 121

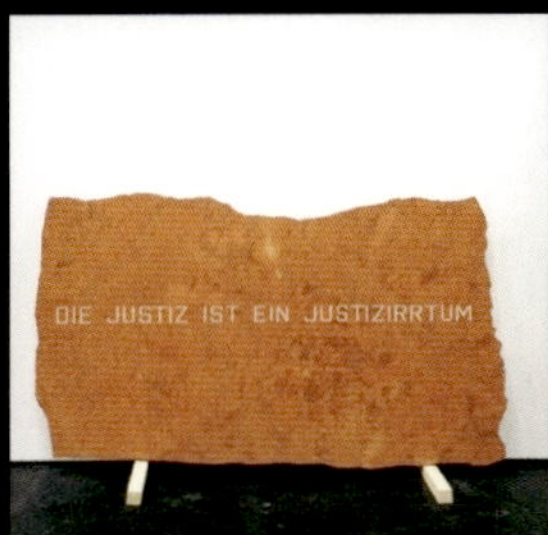

Justice Is a Miscarriage of Justice, 1967 → p. 76

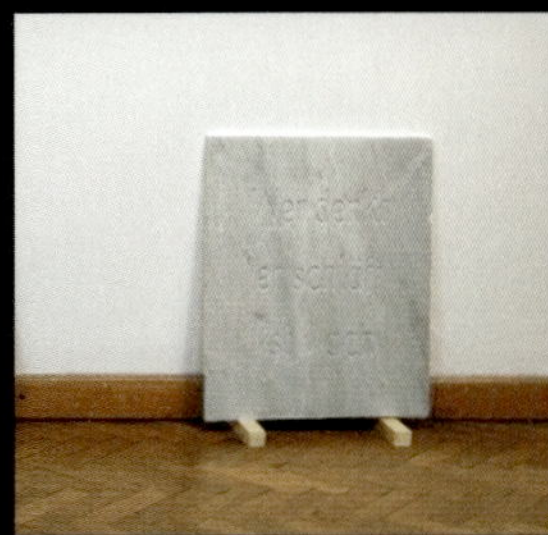

He Who Thinks He Sleeps Is Awake, 1967 (idea)/1974

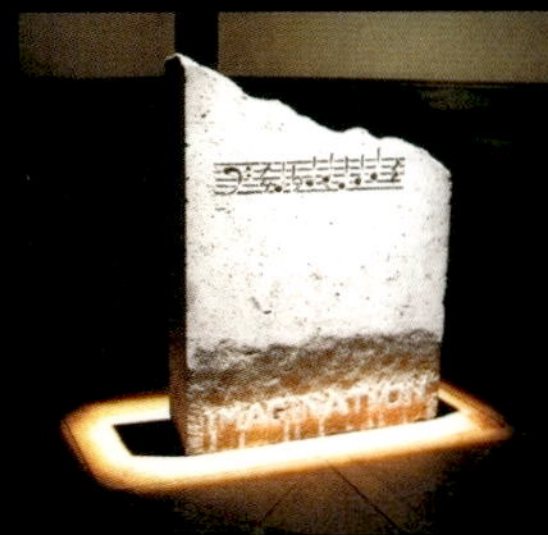

Imagination, 1974 → p. 83

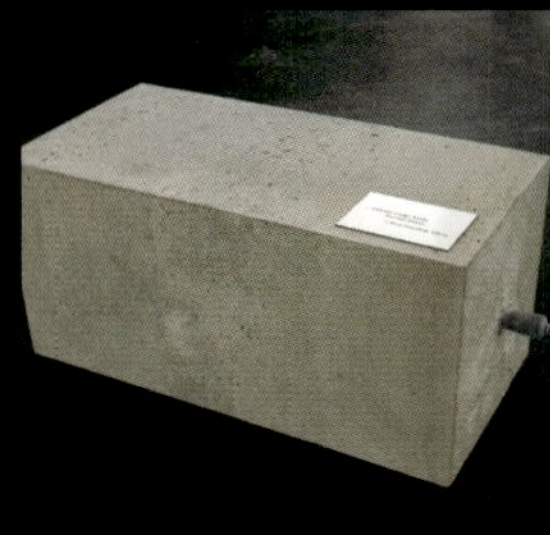

For the Everlasting Light Shall Shine upon You, 1974/2019 → p. 82

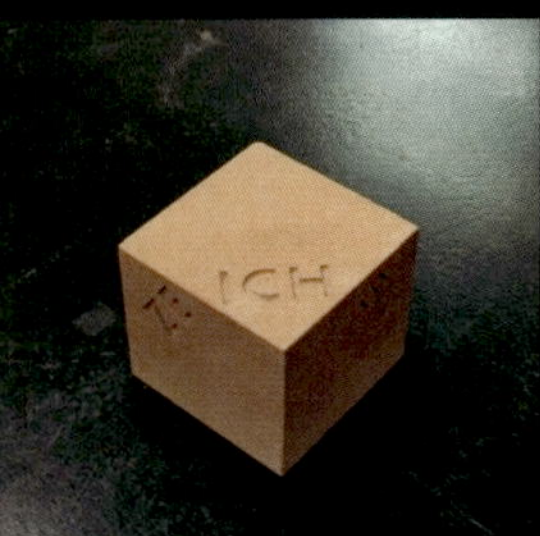

Sign, 1975

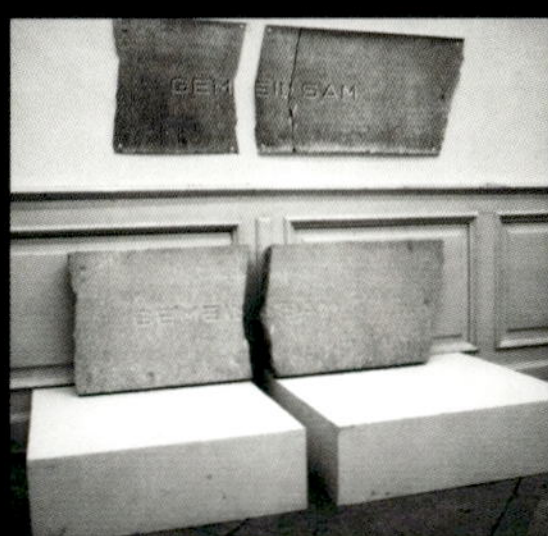

Gem-ein-sam, 1967 (idea)/1975 → p. 79

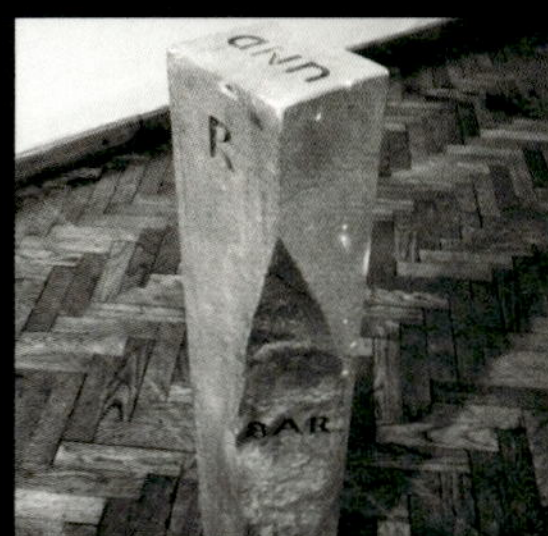

V.E.R.W.UND.BAR, 1974

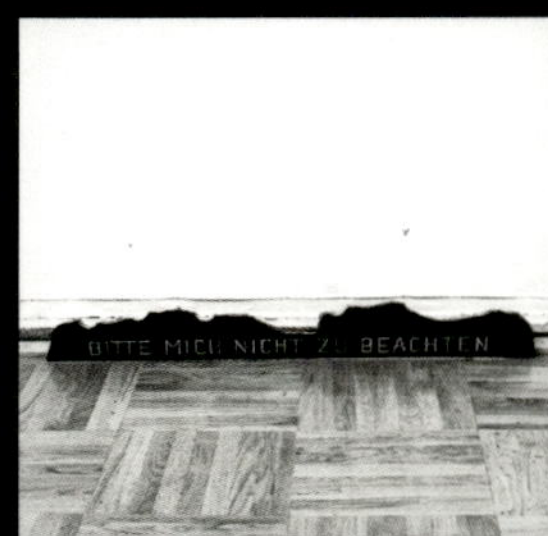

Please Do Not Pay Attention to Me, 1966/75 → p. 75

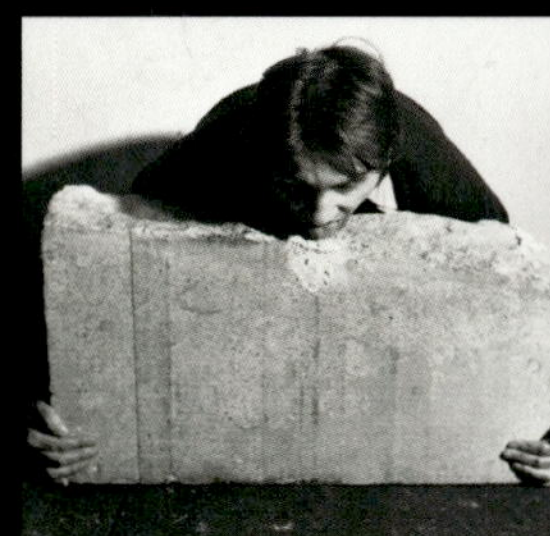

Space of Language, 1973 → p. 525

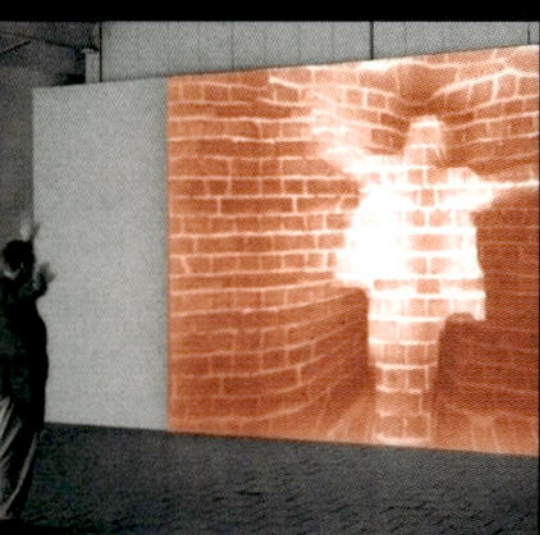

The Wall, the Curtain (Border, the) or, in Technical Terms: Lascaux, 1994 → p. 317

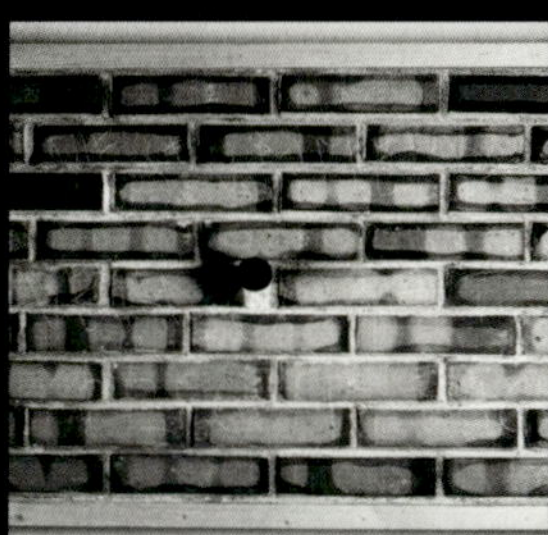

Mute Reason Speaks Playback, 1990/2014

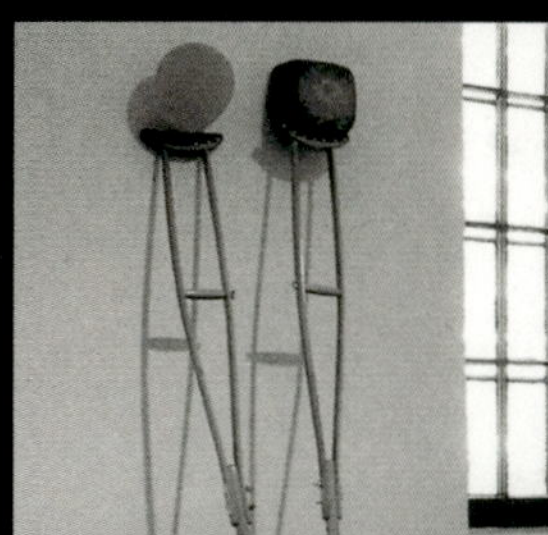

All Technology Is Crutch Technology and Overcomes the Pain of Absences and Distances, 1994 → p. 442

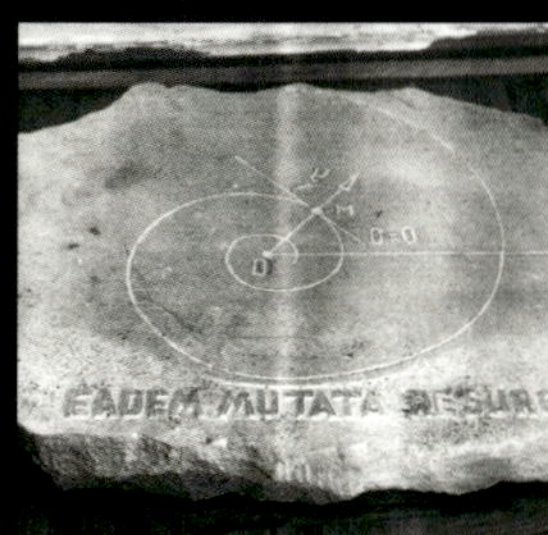

Logarithmic or Pascal Spiral, 1974 → pp. 80–81

Paper

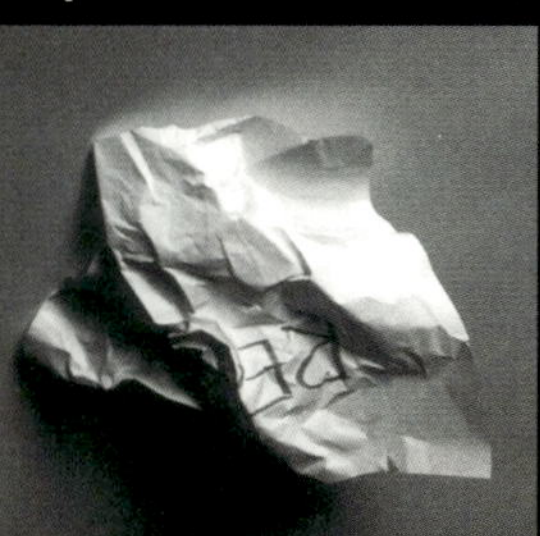

Paper Poems, 1965 → pp. 50–51

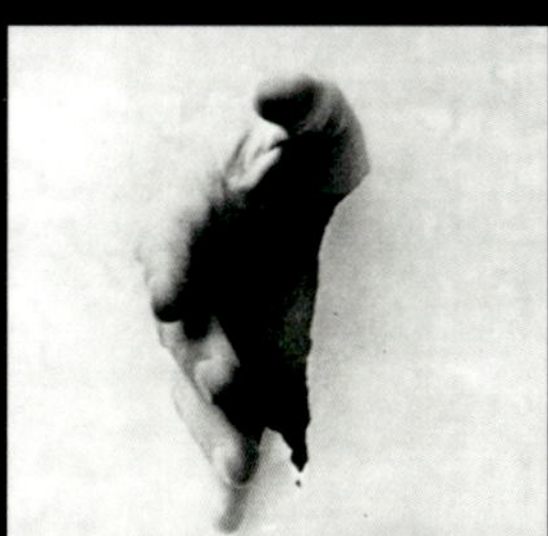

Rip, 1966 → front matter

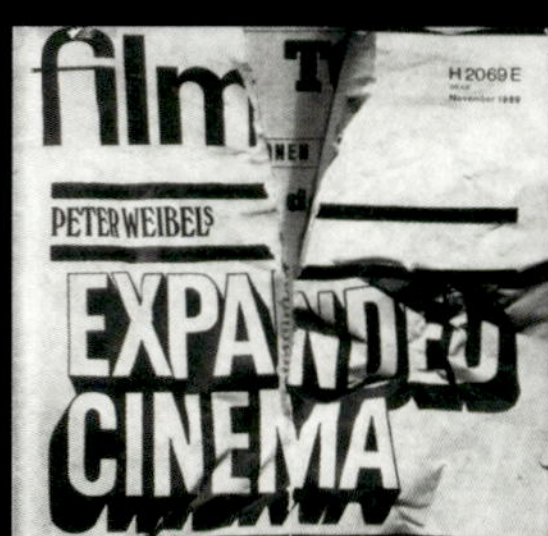

Expanded Cinema, Film no. 11, cover, 1969 → p. 167

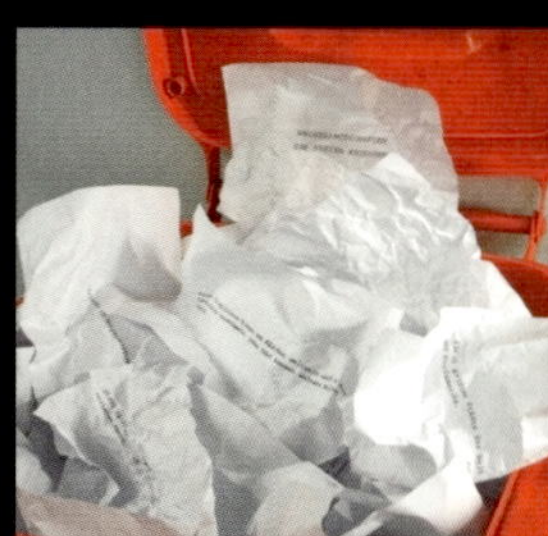

Throwaway Poems, 1999 → p. 470

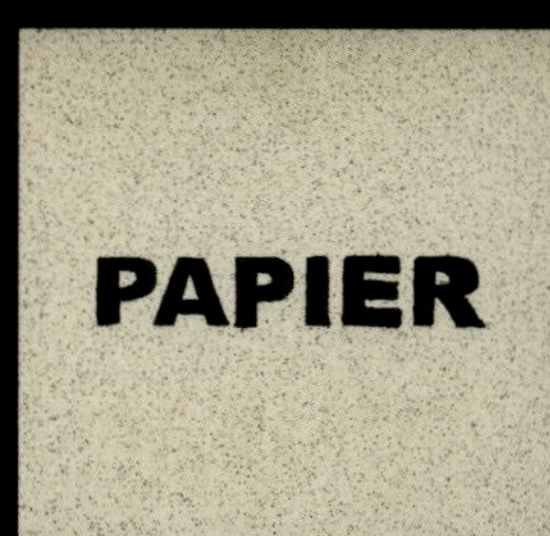

Iso- and Polymorphic Language of Materials, 1967 → p. 29

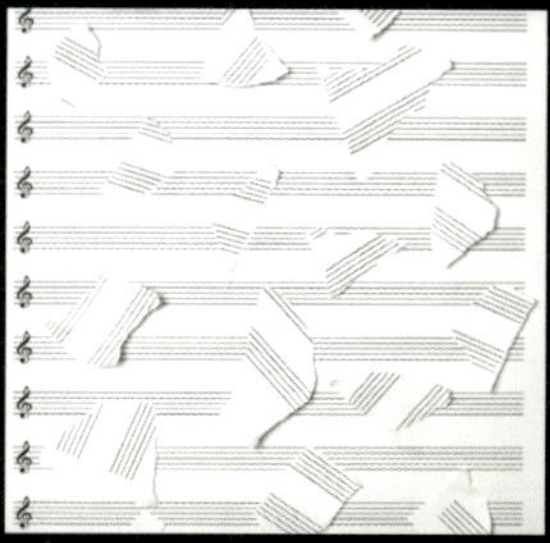
Self-Referential Music, 1968 → p. 344

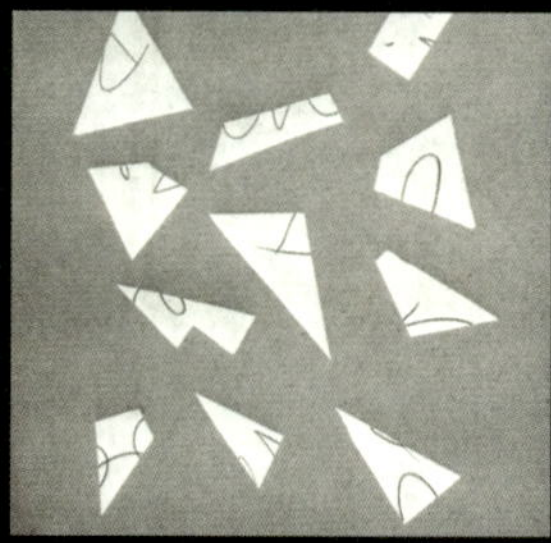
Proof, 1966

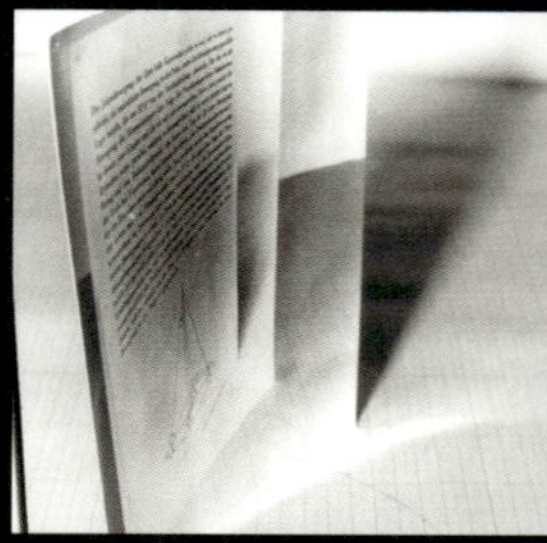
The Paradox of Perspective, 1973 → pp. 392–93

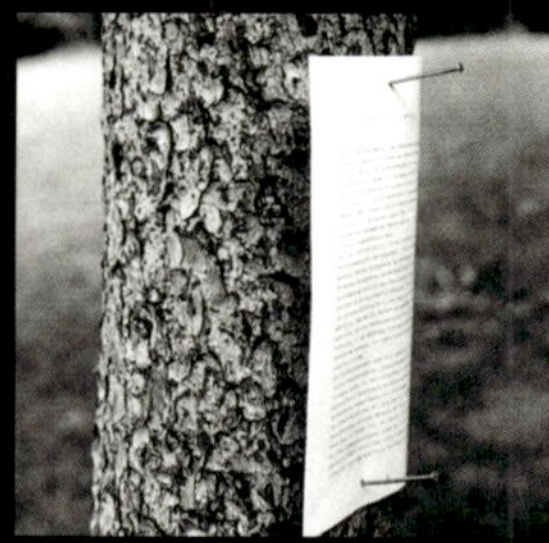
The Secret of Nature, 1967 → pp. 154–55

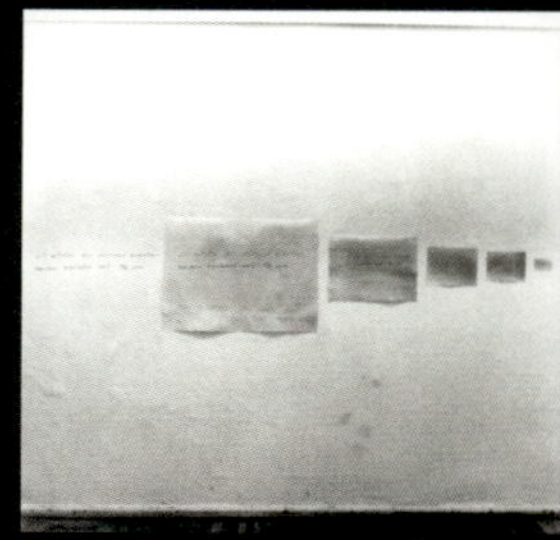
Estimation: The Paradox of the Reference System (Medium) II, 1975 → p. 156

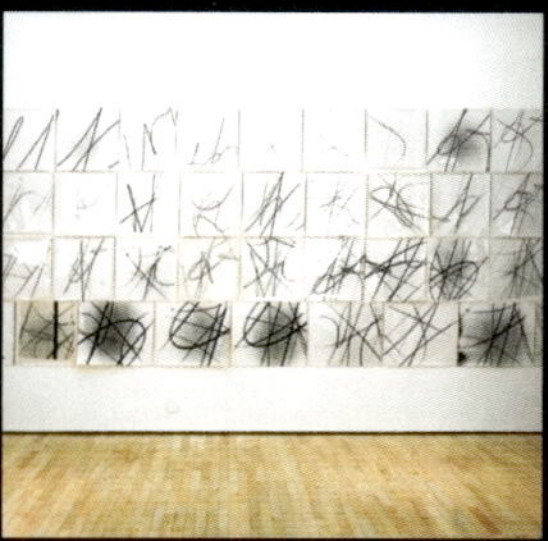
The Infinite One-Word Exhibition, 1977 → pp. 140–41

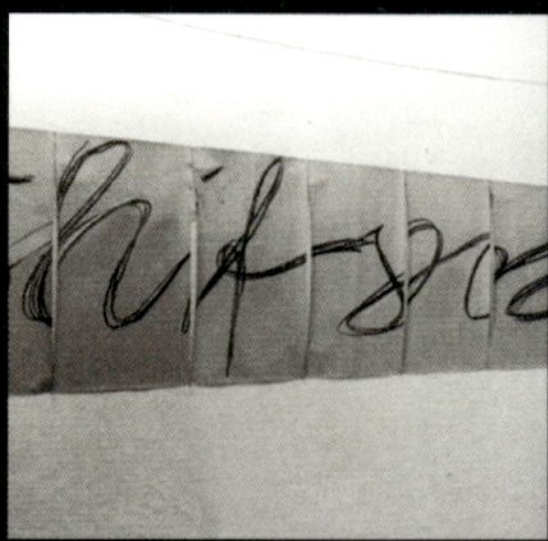
Parallel-Shifted Bijection, 1976

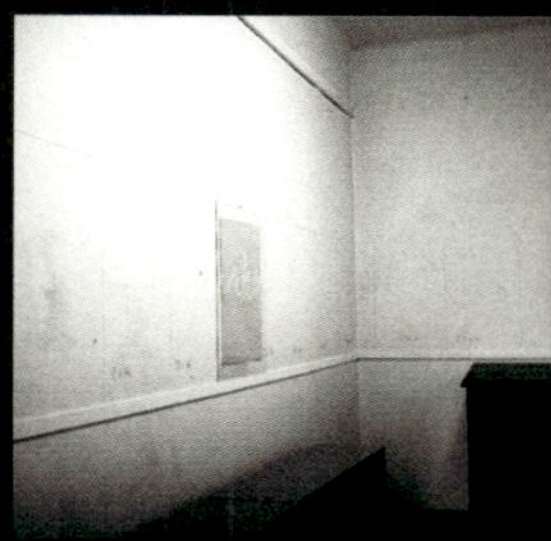
Elongation of a Sentence, 1975 (concept)/1976

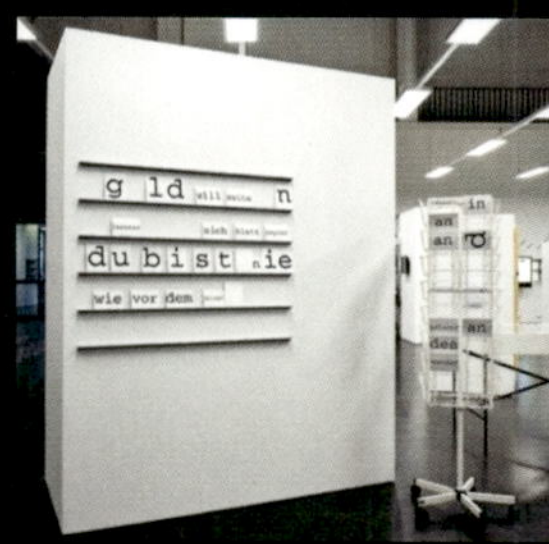
Mobile Texts, 1966 → p. 59

Notation

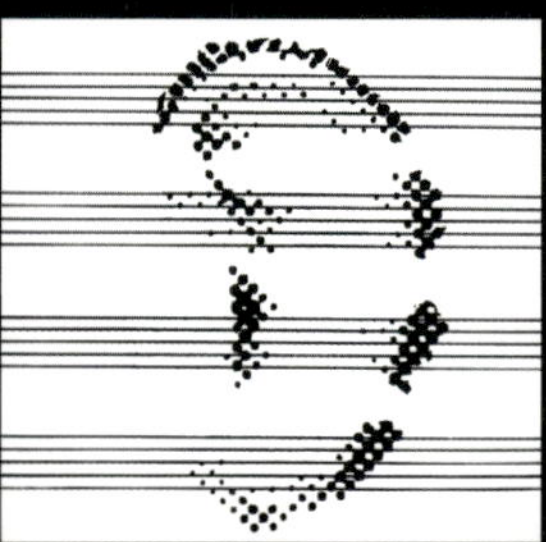
Transformational Music, 1972 → pp. 346–47

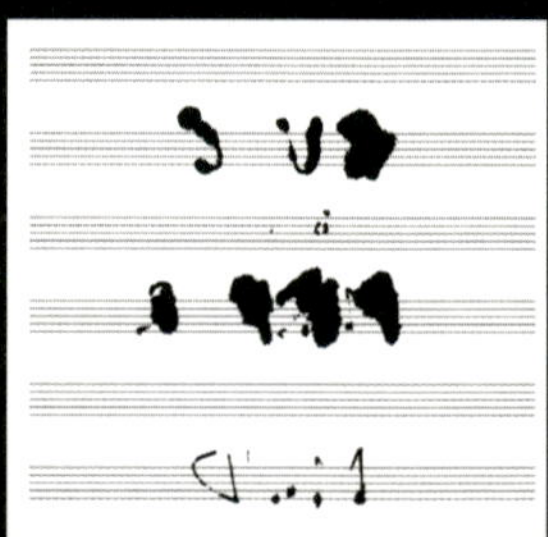
Music Is a Mirror of the Mind, 1967

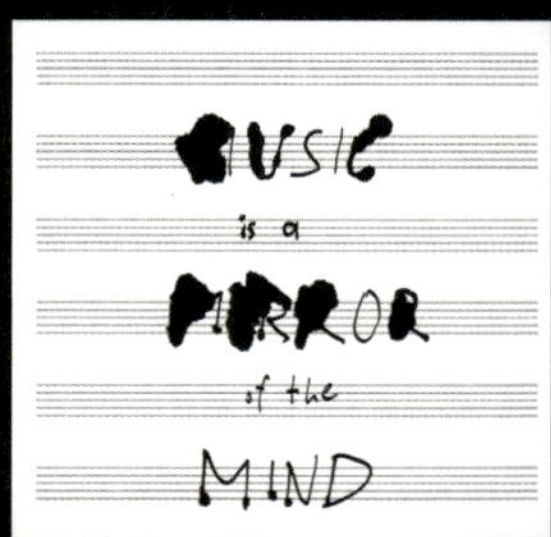

→ pp. 342–43

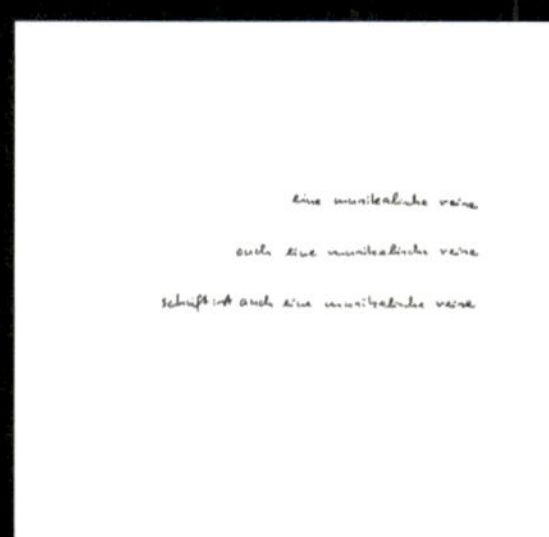
Writing-Music, 1967

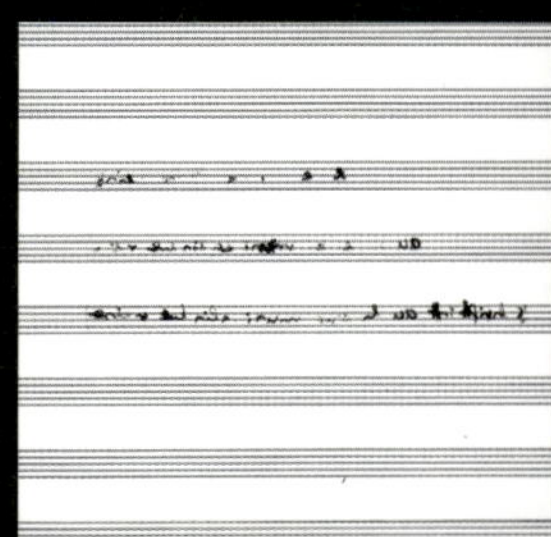

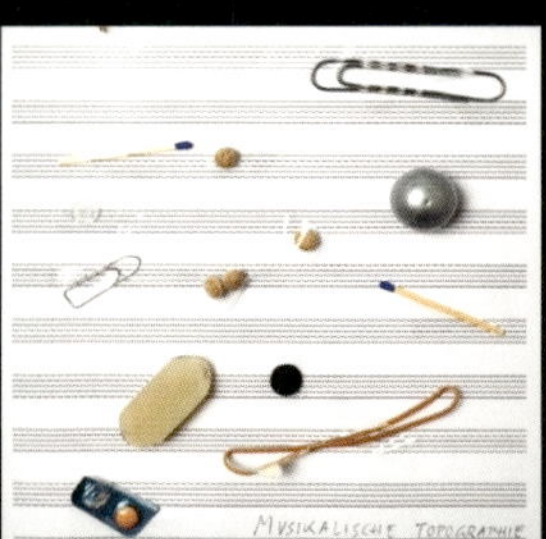

Musical Topography, 1968 → p. 345

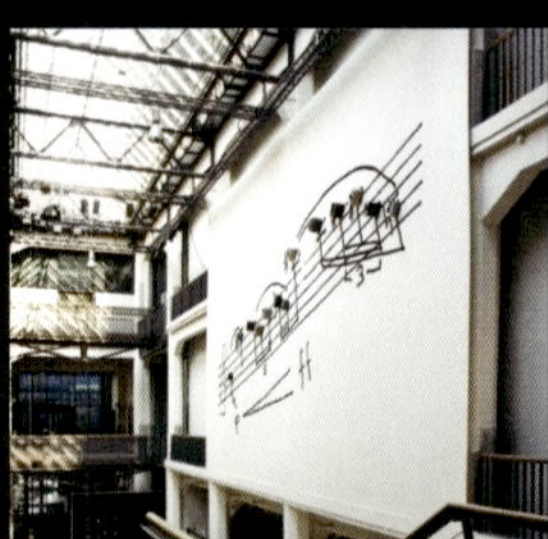
Loudspeakers as Notes, 2012

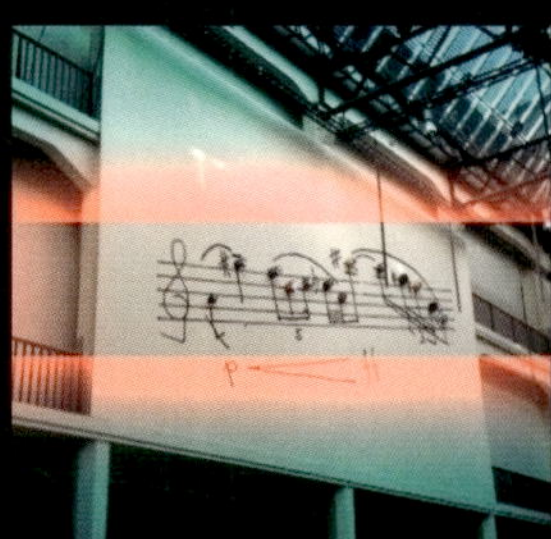
SynSeeThis, 2013 → p. 368

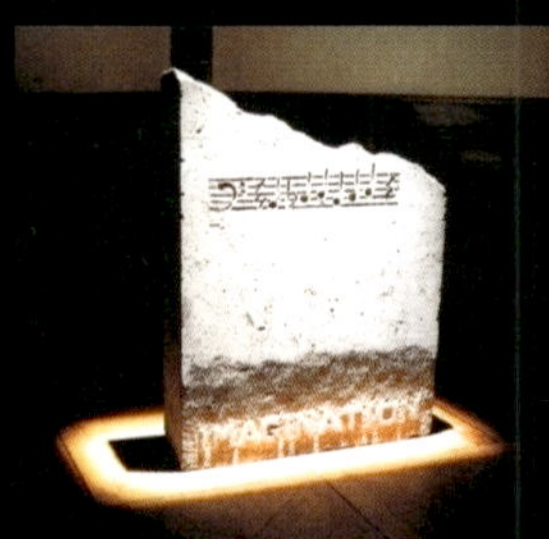

Imagination, 1974 → p. 83

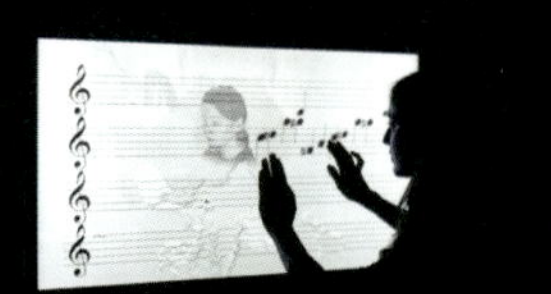
Music Passenger, 2019 → p. 373

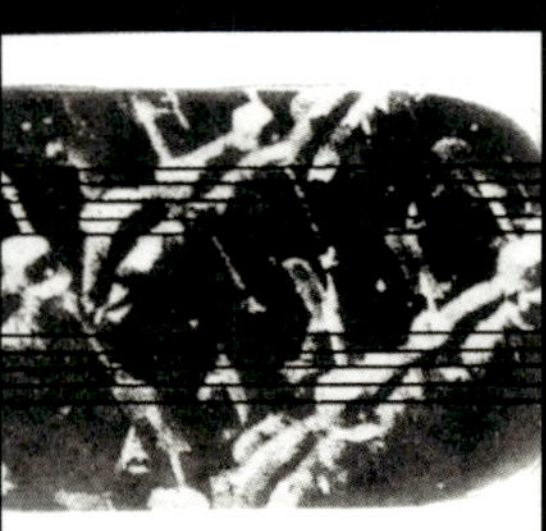
Totalitarian Symphony, 1971 → p. 253

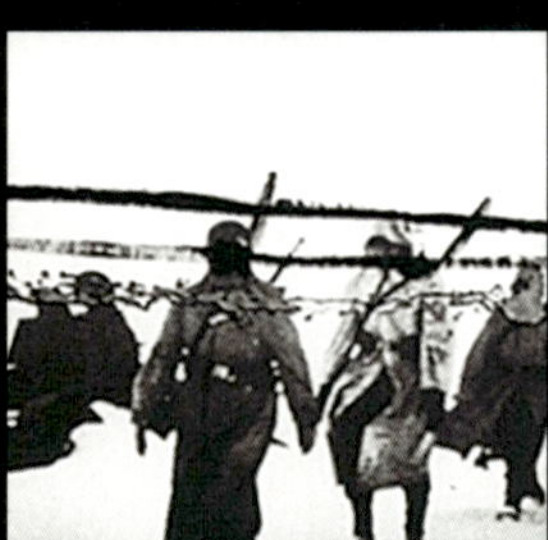
God's Memory, or the Arm of the Media, 1982

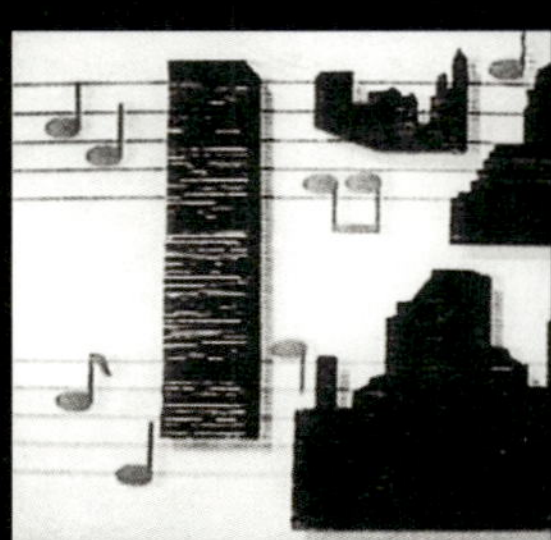
Scene from *Chants of the Pluriverse*, 1986–88 → pp. 304–307

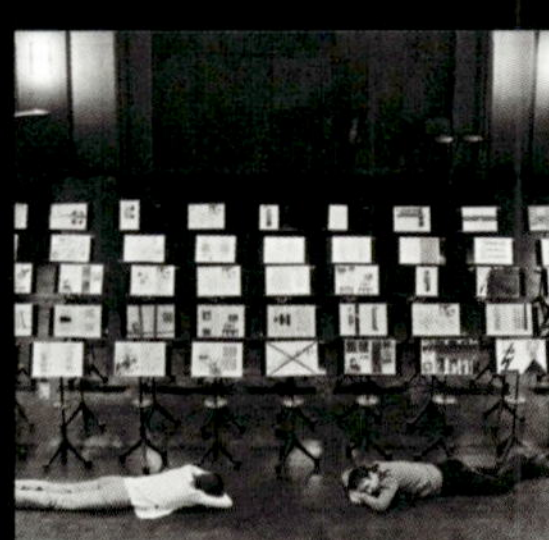
Bibliotheca Insomnia, 1979

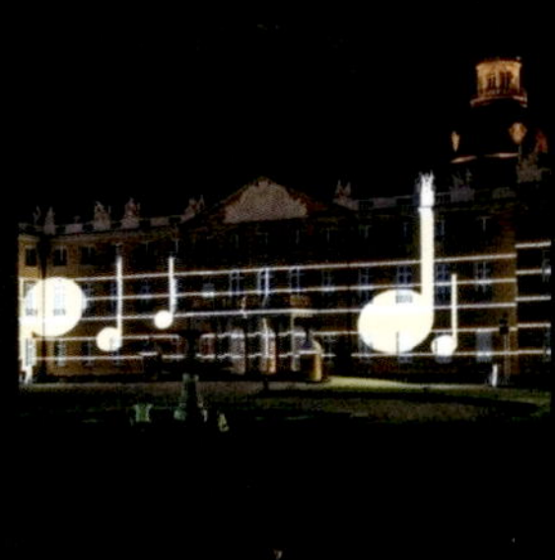
Scene from *Velografie für Drais*, Schlosslichtspiele 2017

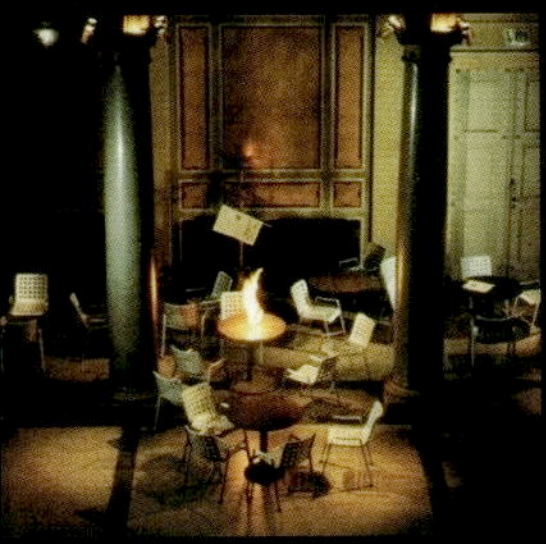
Burning Table, 1974/88 → p. 425

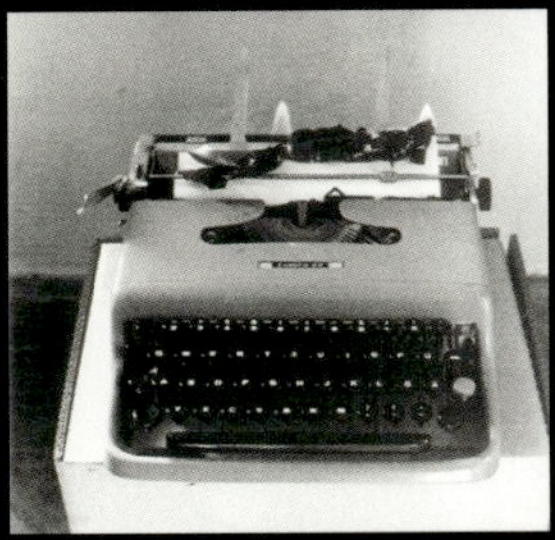
Principle of Hope, 1971 → p. 105

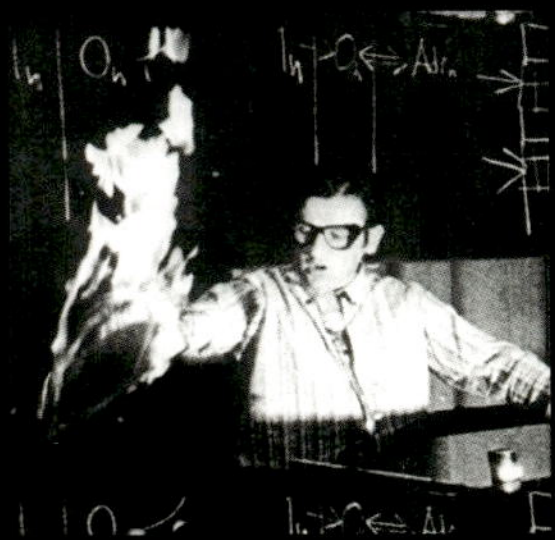
Incendiary Speech, 1968 → pp. 508–509

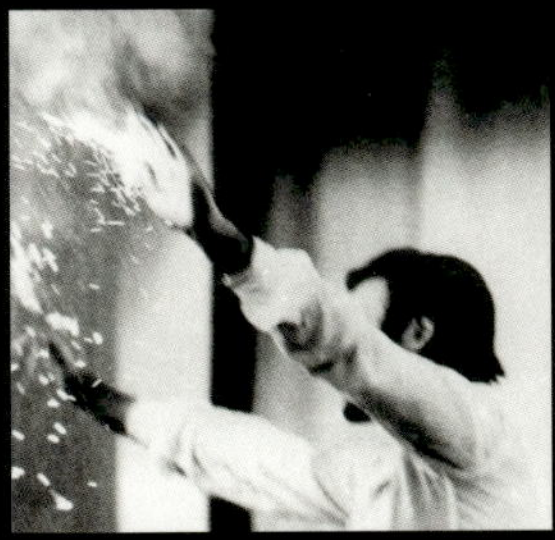
Initiation, 1971 → p. 514

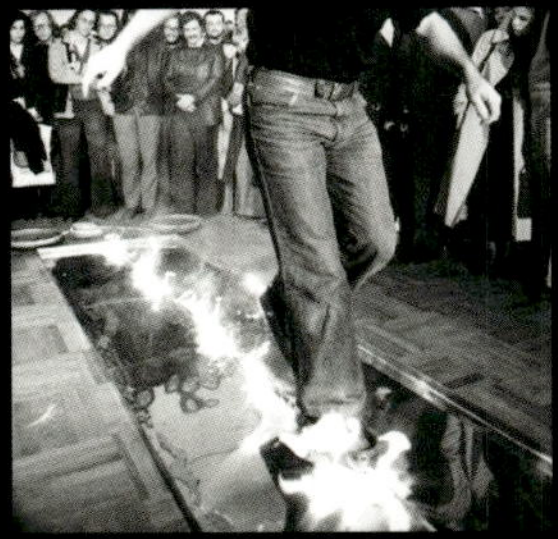
Fireflight – Flightfire, 1975 → pp. 516–17

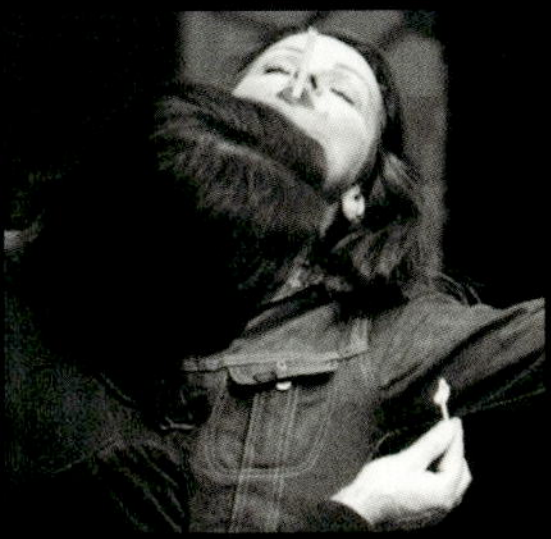
More Warmth among Human Beings, 1972 → p. 515

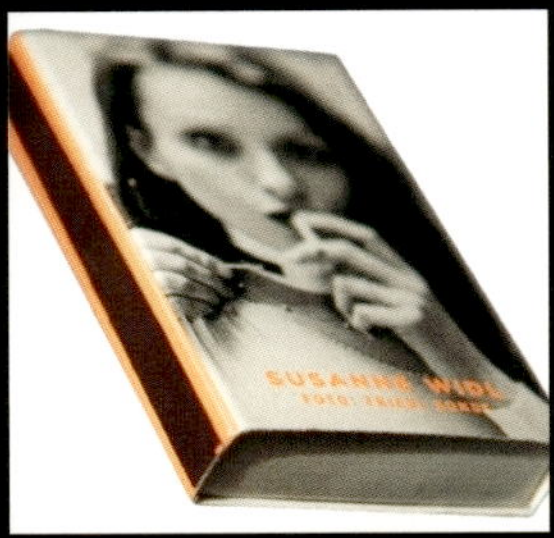
Matchbox for *More Warmth among Human Beings*, 1972 → p. 515

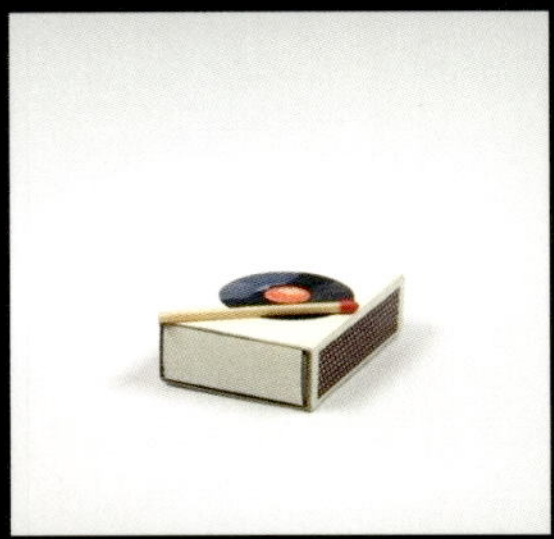
The Invention of Ignitive Sound, 1966 → p. 26

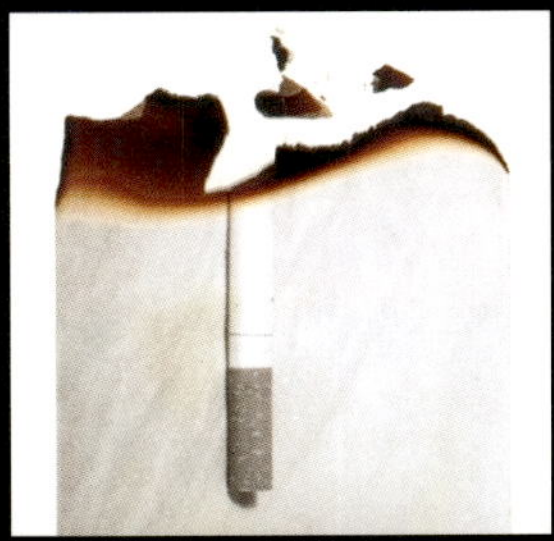
Identity of Action, 1975 → p. 145

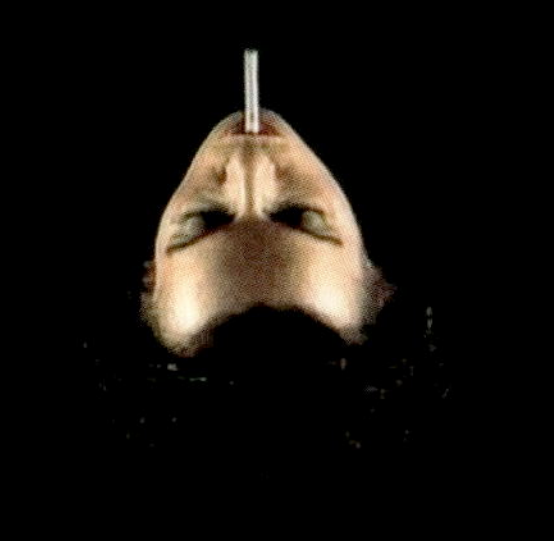
Scene from *Chants of the Pluriverse*, 1986–88 → pp. 304–307

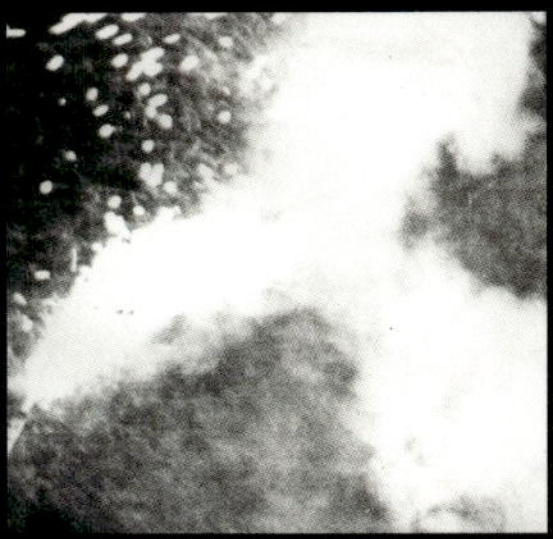
EXIT, 1968 → p. 182

Firewall, or More Traffic Deaths, Fewer Citizens, 1969

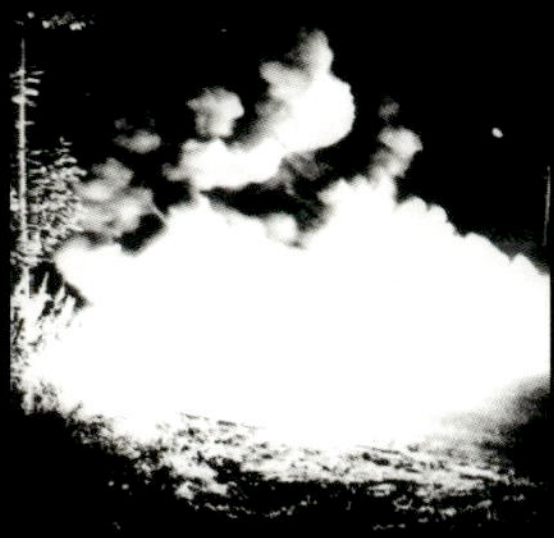
→ p. 212

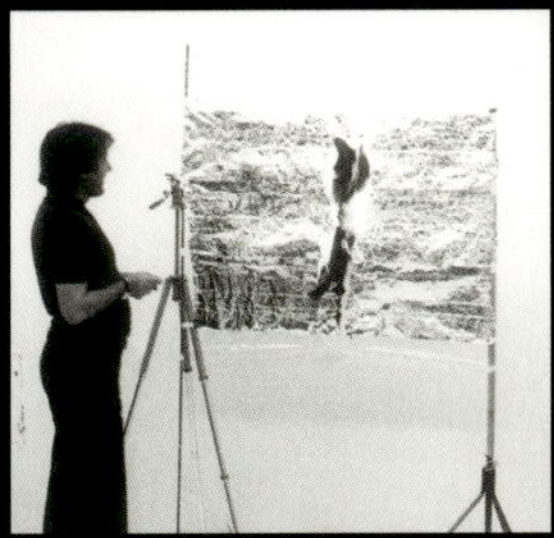
Got a Light?, 1968 → p. 208

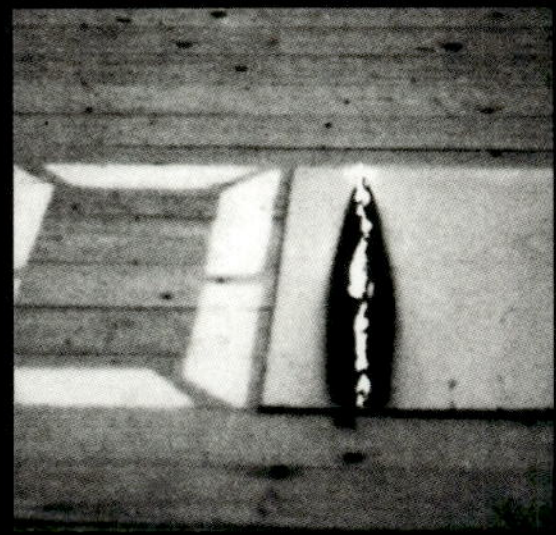
The Electric War, 1983 → pp. 458–59

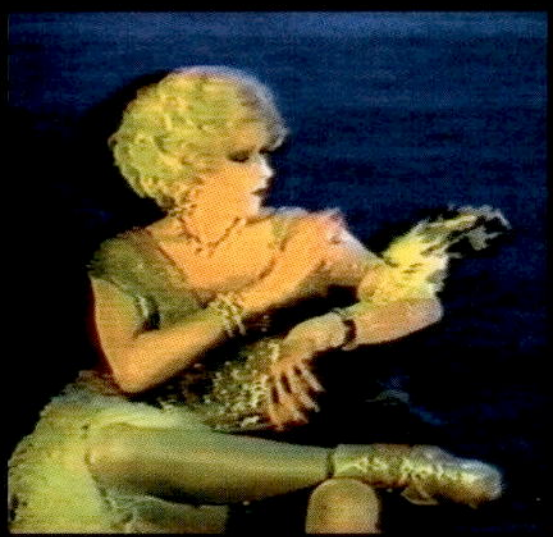
Scene from *Voices from an Inner Space*, 1988 → pp. 358–59

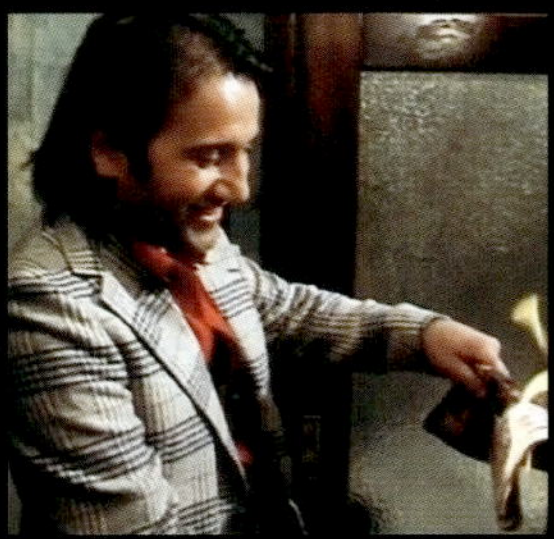
Scene from *Exit … Don't Panic!*, 1980

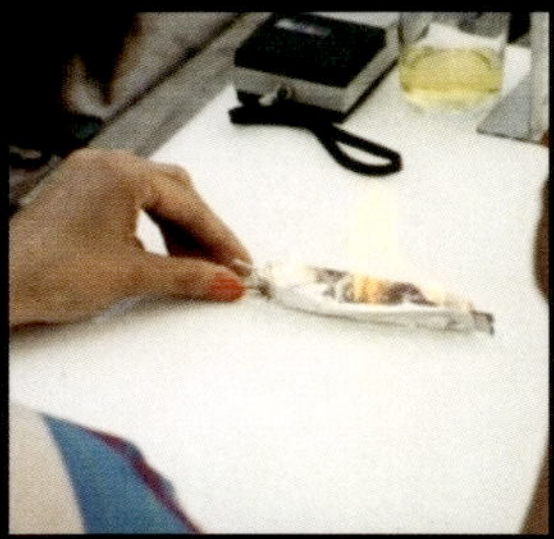
Scene from *Invisible Adversaries*, 1970 (idea)/1977 → pp. 134–35 and 202–203

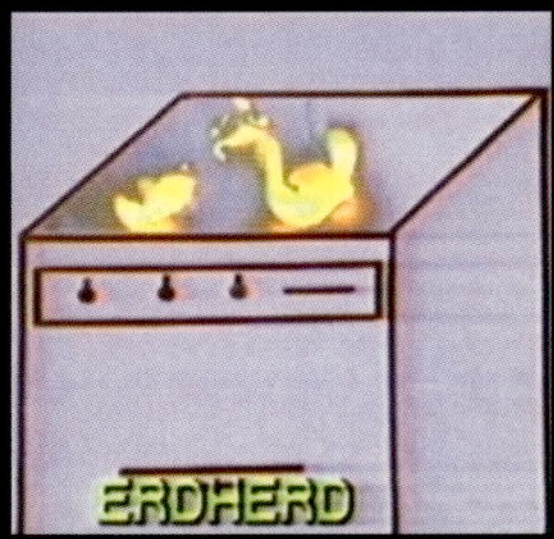

Scene from *The Artificial Will. Electronic Media Opera*, 1984 → pp. 355–57

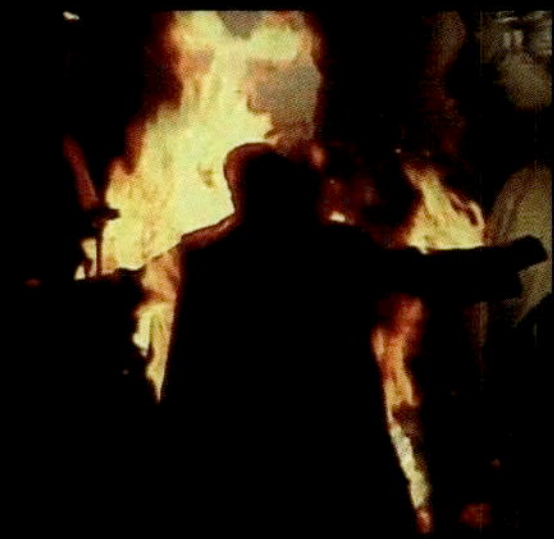
Performance by Hotel Morphila Orchester, 1980 → pp. 352–53

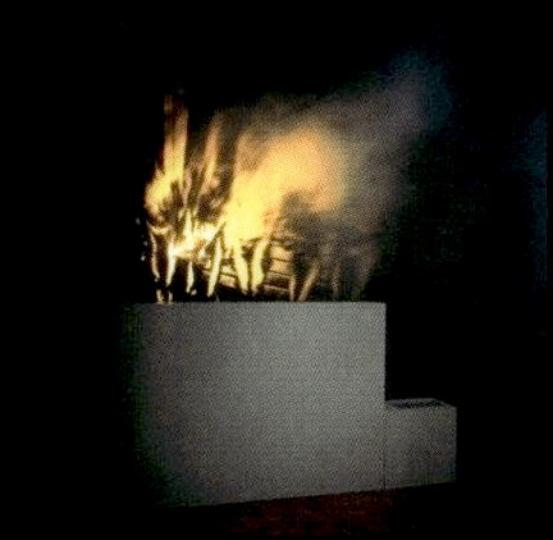
Dumb Show, 1969 → p. 183

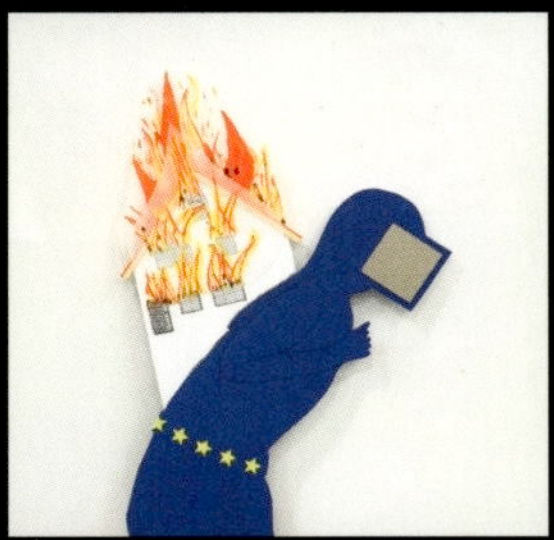
Is the House of Europe on Fire?, 2020 → p. 477

Water

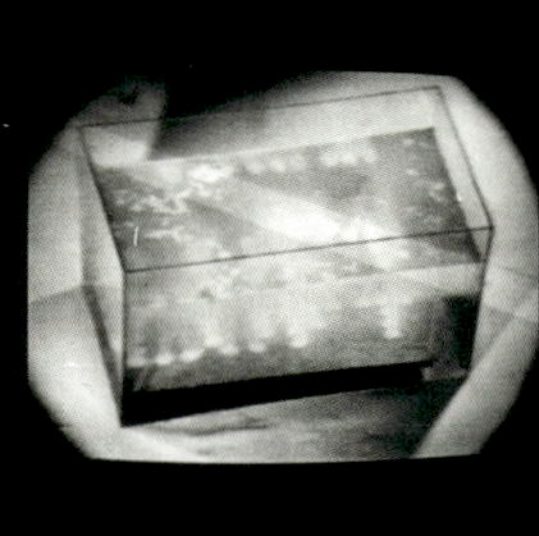
Airtext: The Product Is Extinguished in the Process, 1967–72 → p. 66

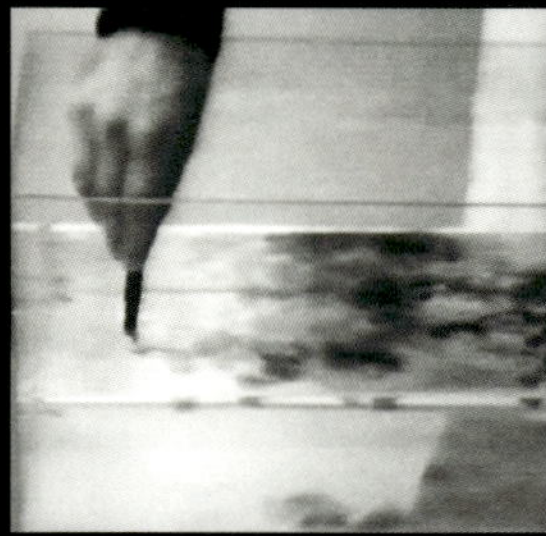
Sense, 1968 → p. 67

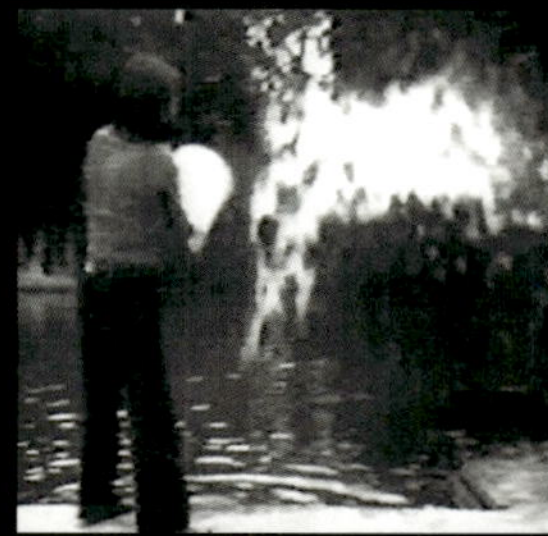
Virtual Water Sculpture, 1971 → p. 128

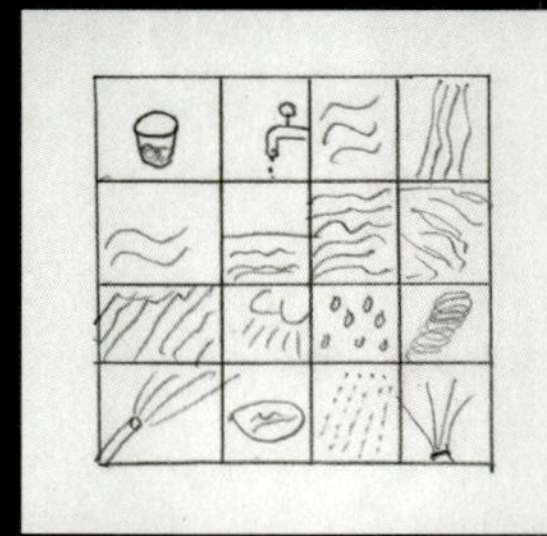
Concepts, 1967 → p. 123

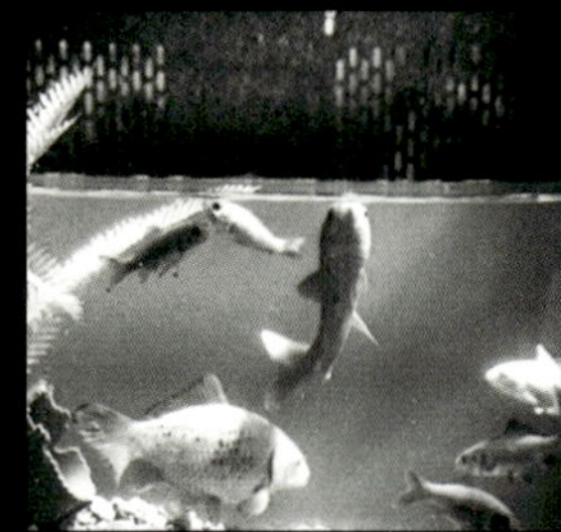
TV Aquarium (TV Death 1), 1970 (idea)/1972 → p. 250

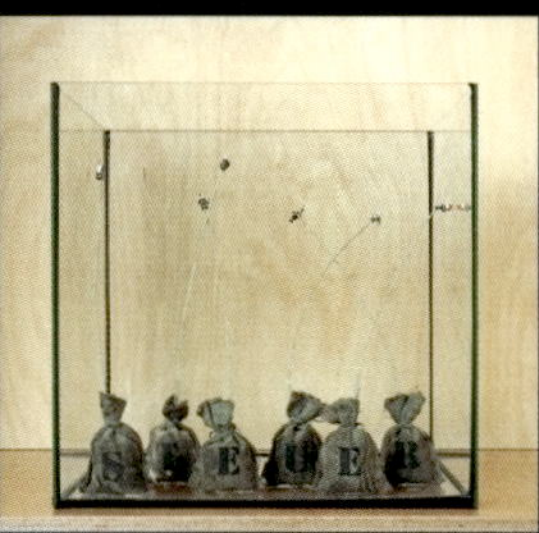
TAX(payer), 2009 → p. 480

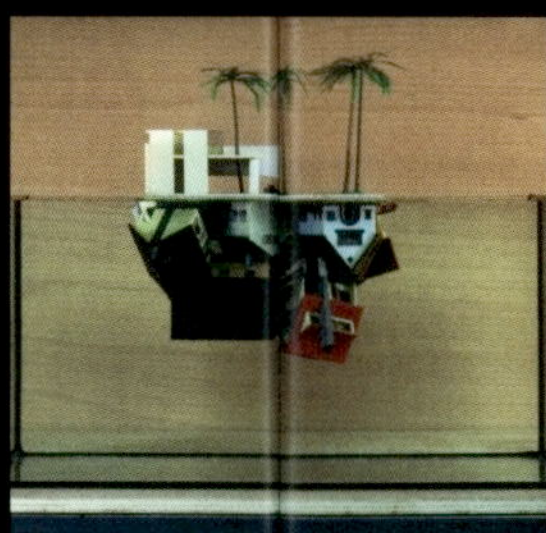
Housing Crisis, 2009 → p. 480

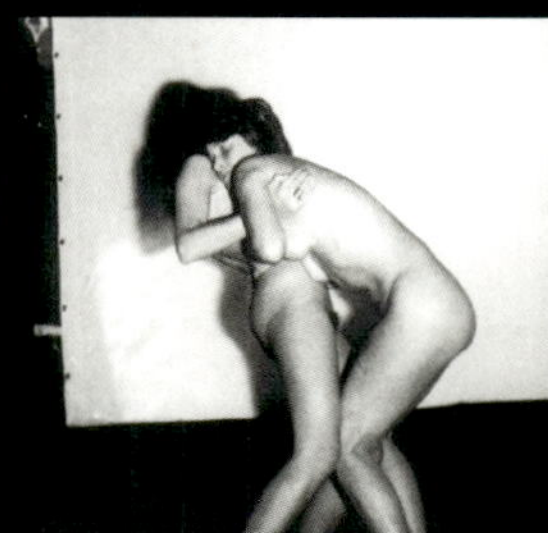
Energy Exchange, 1971 → p. 524

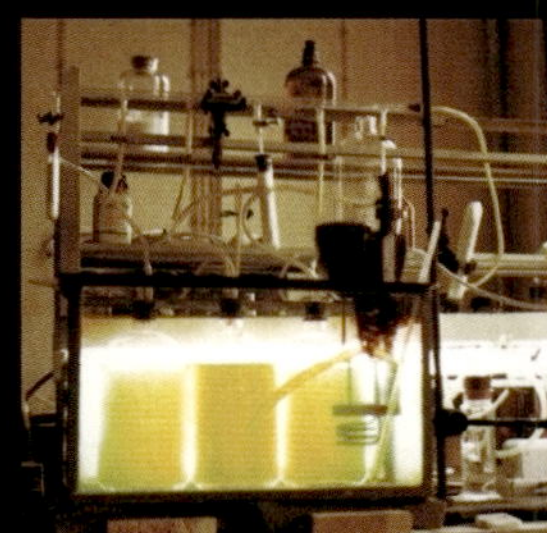
A Bioenergetic Paradox, 1975 → pp. 68–69

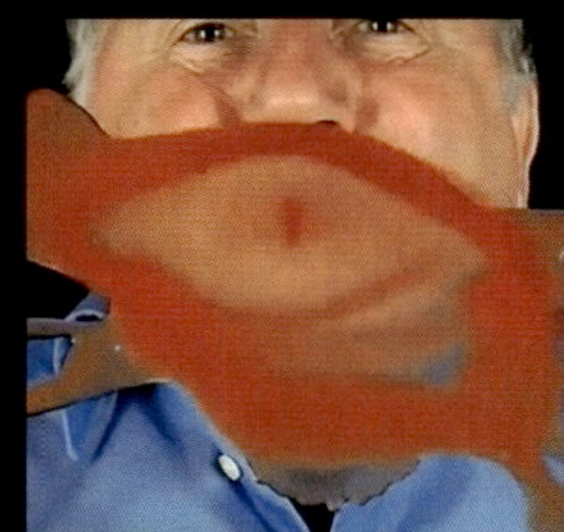
Spit Painting, 2009 → p. 259

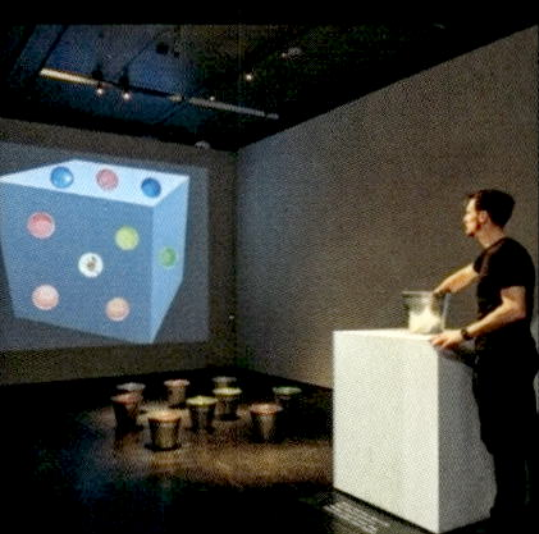
Water Dice, 1988 → p. 287

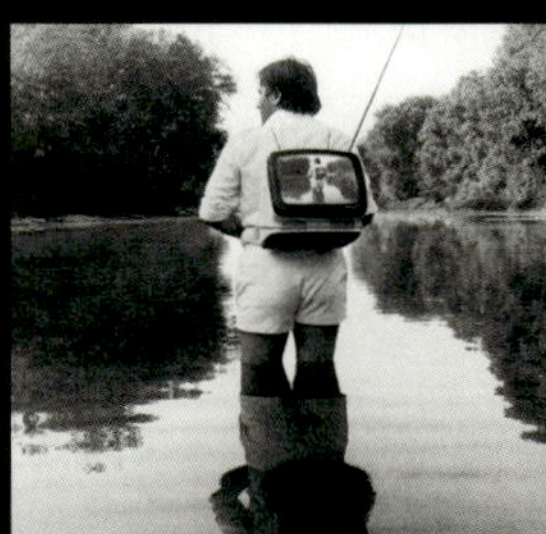
Infinite Contemplation, 1982 → p. 290

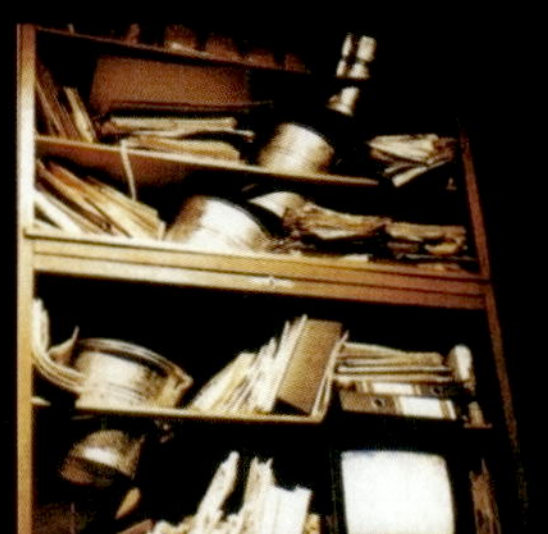
B.I.S. (Office Information System), 1989

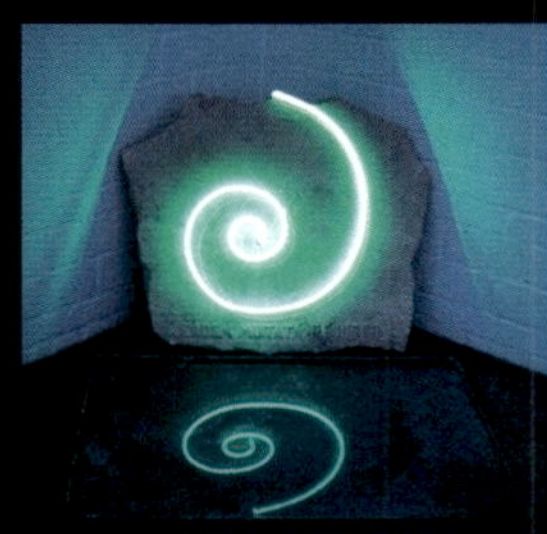
Logarithmic or Pascal Spiral, 1974 → pp. 80–81

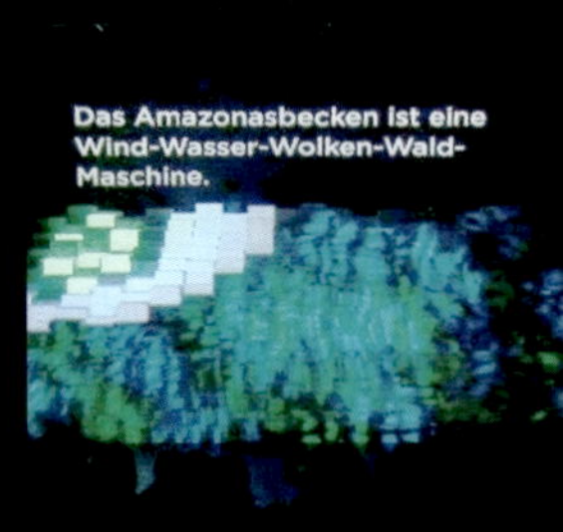

Amazon Conference ..., 2010 → pp. 362–63

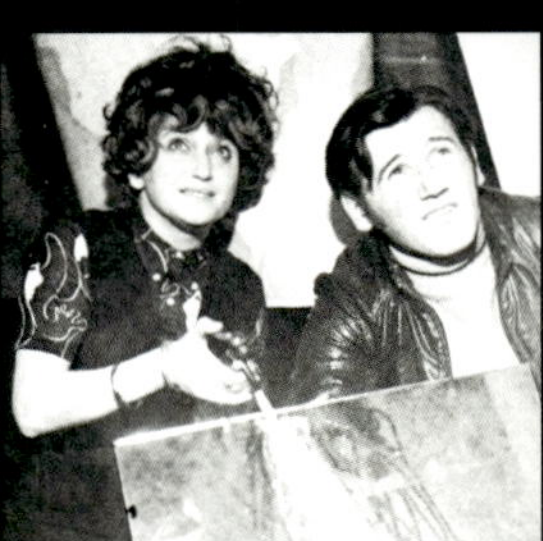
Abstract Film No. 1, 1968 → p. 207

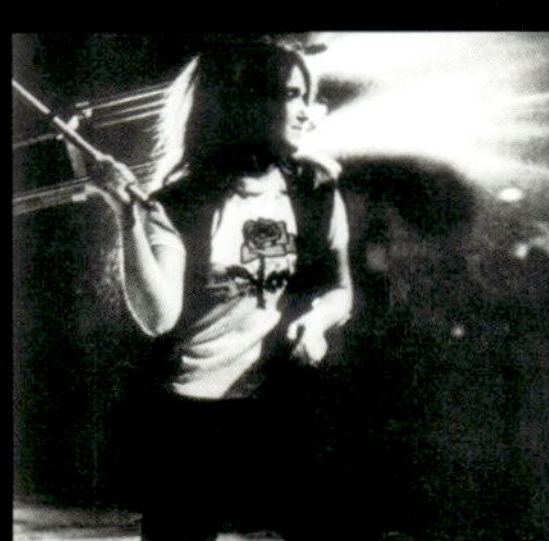
Audience Atomizer, 1969 → p. 511

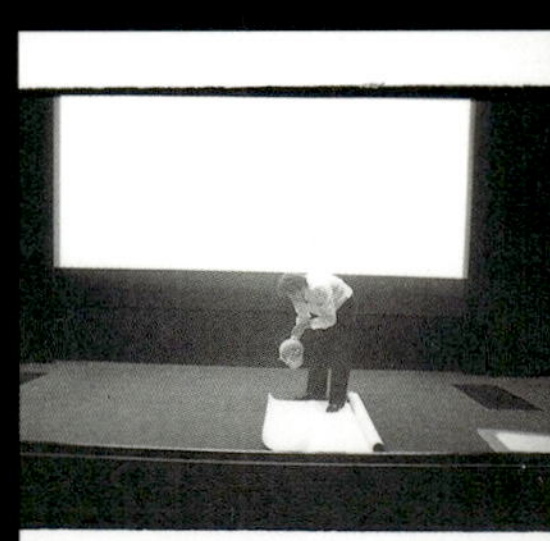
Pouring, 1967

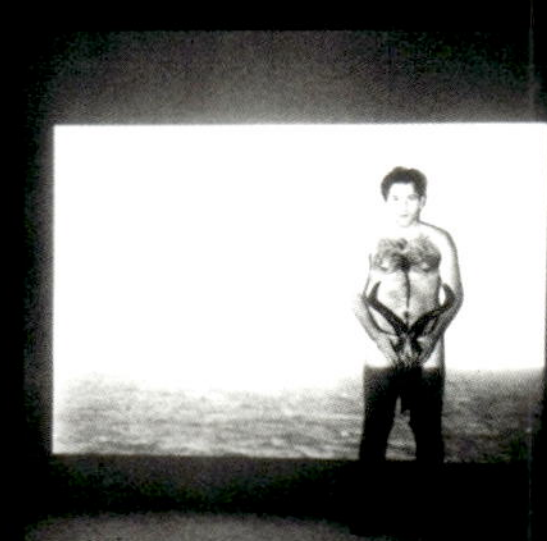
Time Lung: Performances on the Morphology and Consciousness of a Vertebrate ..., 1971 → p. 520

Electricity

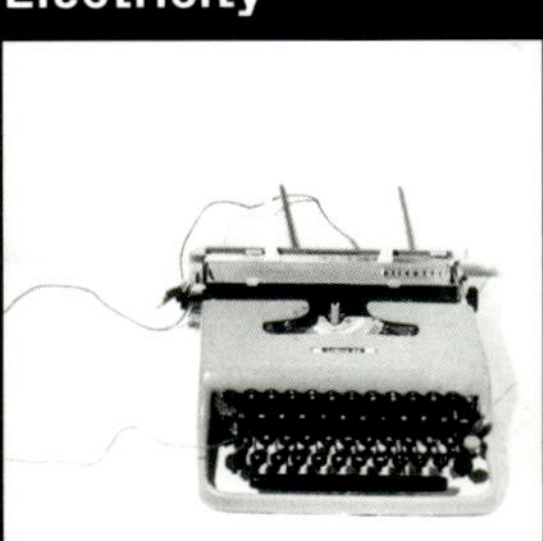
Principle of Scrip/tor/ture, 1971 → p. 104

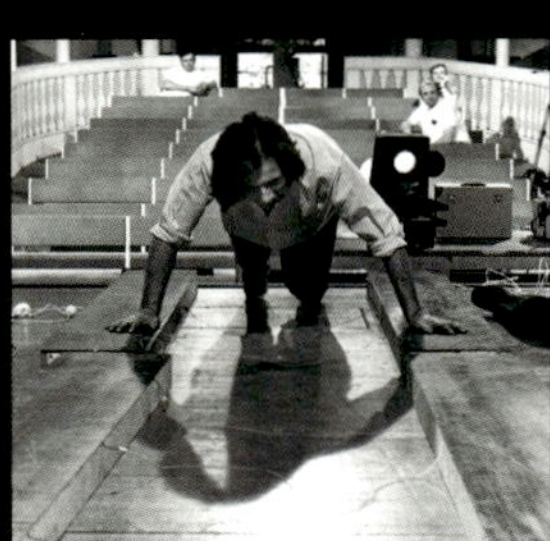
The Long March, 1968/71 → p. 501

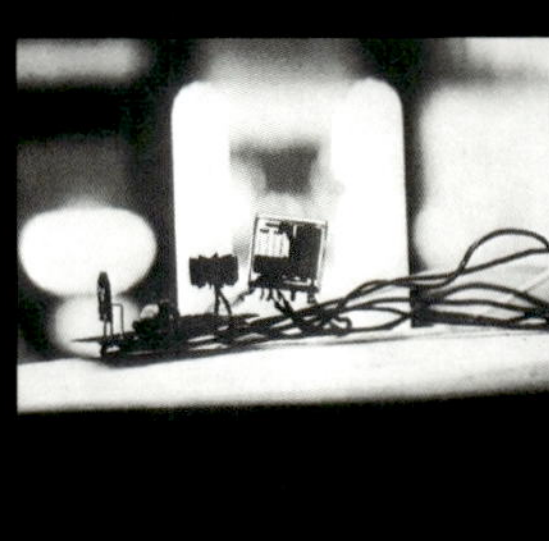
The Myth of the Twenty-First Century, 1967 → p. 33

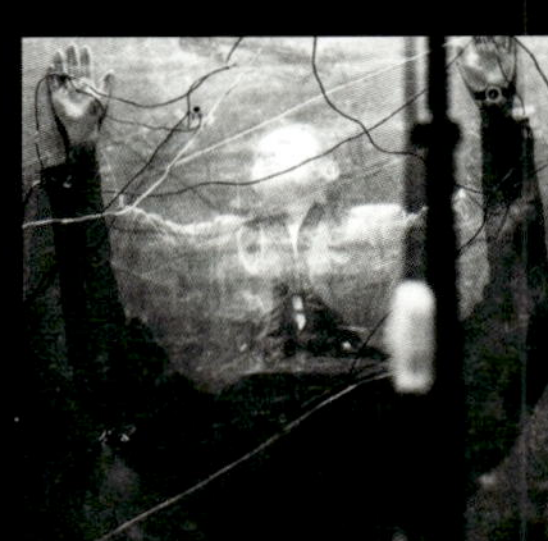
The Magic Eye, 1969 → pp. 178–79

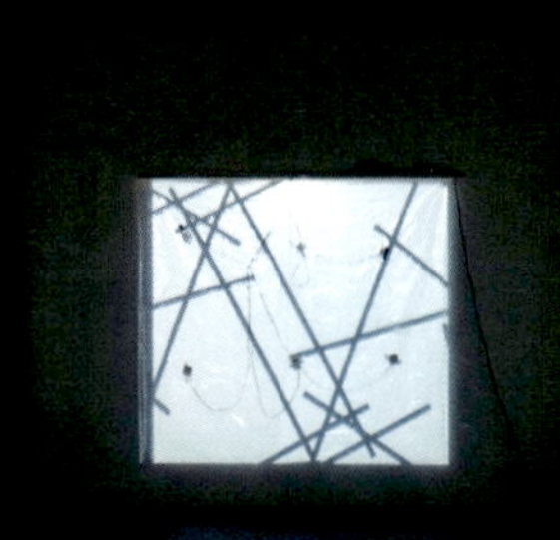
The Magic Eye, 2019 → pp. 178–79

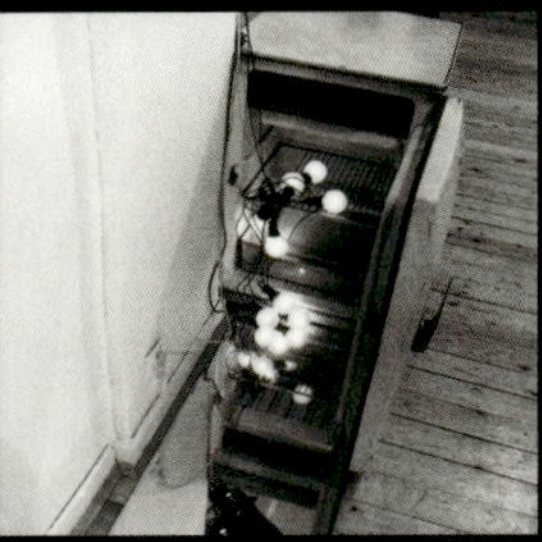

The Electric War, 1983 → pp. 458–59

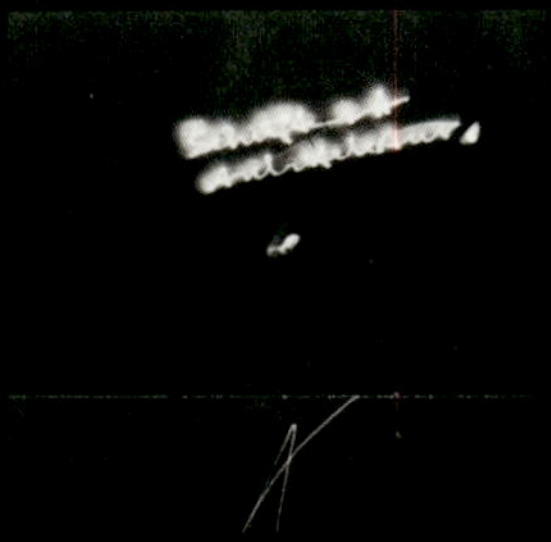

Writing Is Architecture, 1967 → p. 32

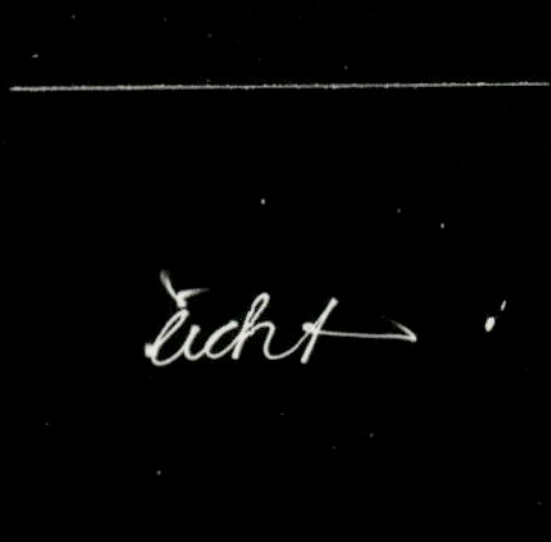

Light Text, 1967/75

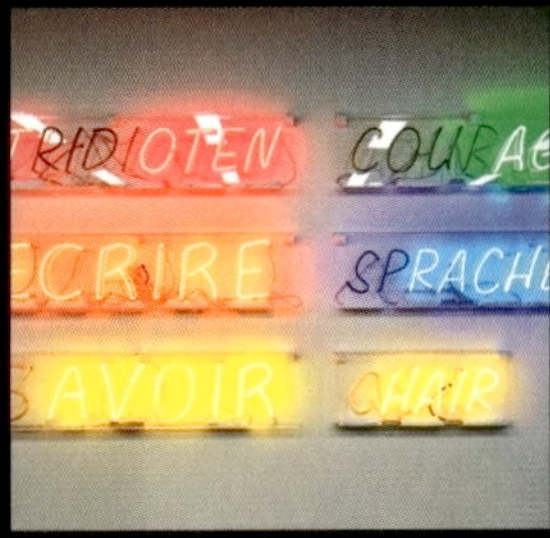

Motion Poems, 1966/2011/14 → pp. 31, 62, 63

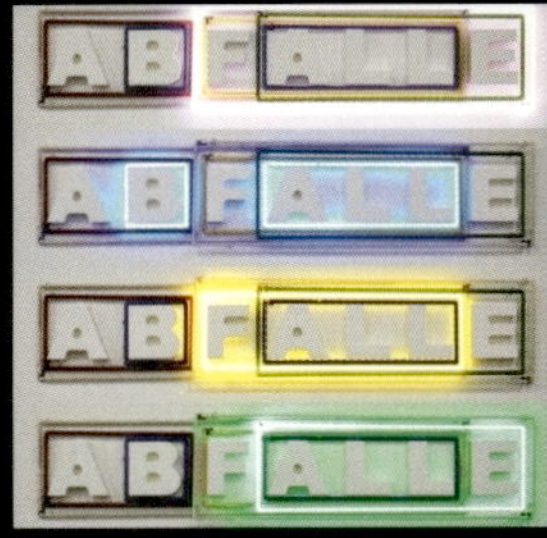

Motion Poem, 1966/2004

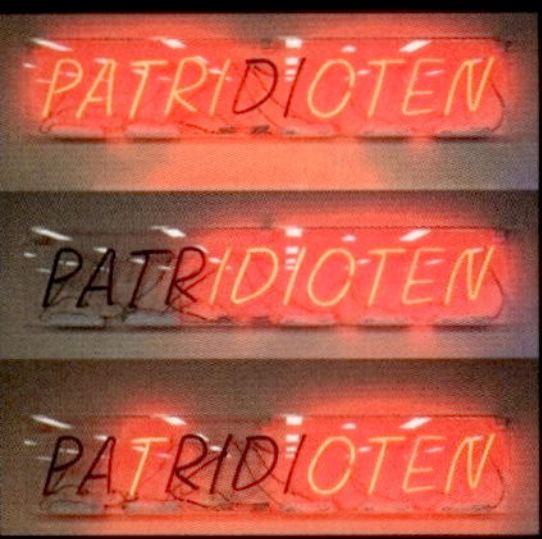

Patridioten, 1967/2019 → p. 31

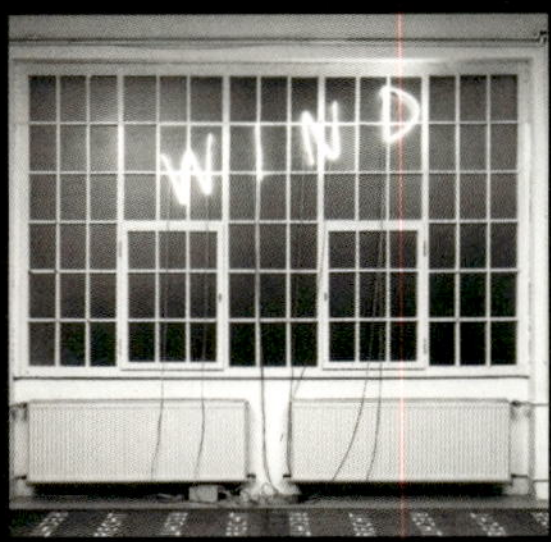

Wind: The Compulsion of Identity, 1966/75 → p. 87

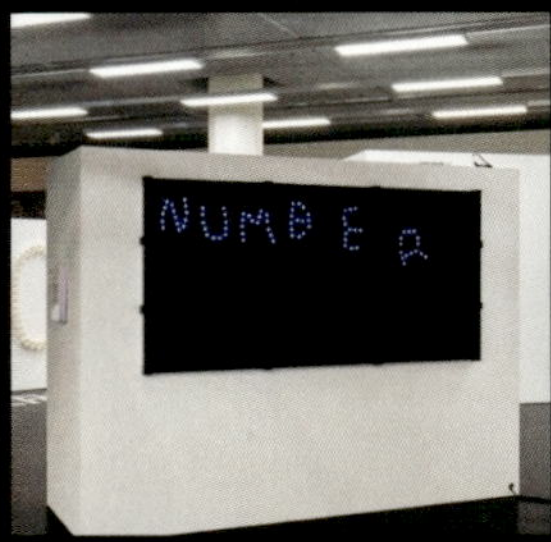

What Is a Number That a Man Can Recognize It …, 1992/2019 → pp. 328–29

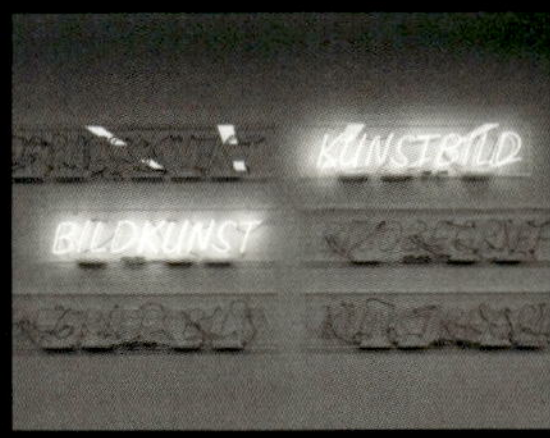

Concept Art, 1966/2014 → pp. 60–61

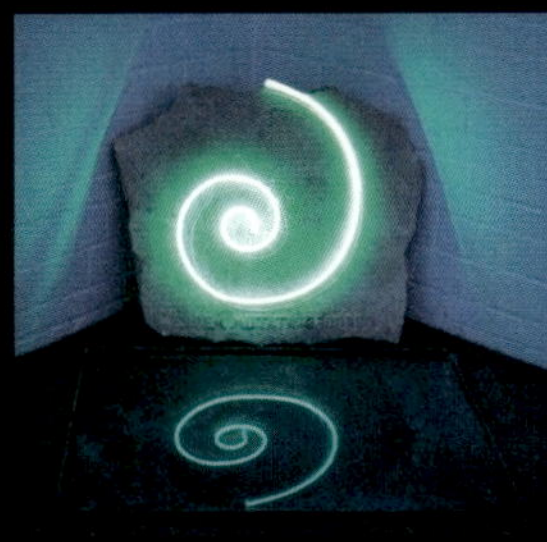

Logarithmic or Pascal Spiral, 1974 → pp. 80–81

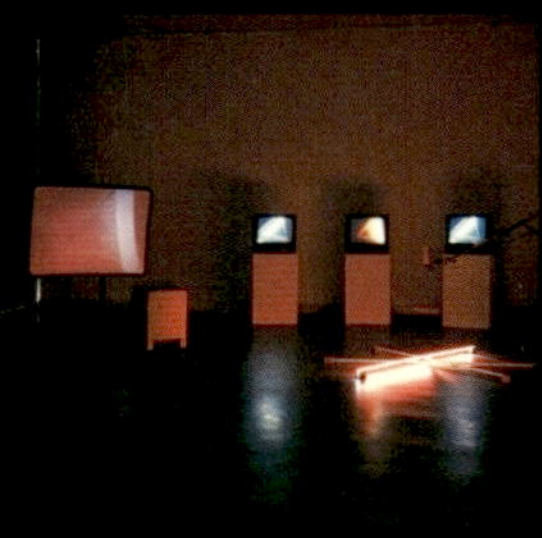

The Dream of Everyone Sharing the Same Consciousness, 1979 → pp. 280–83

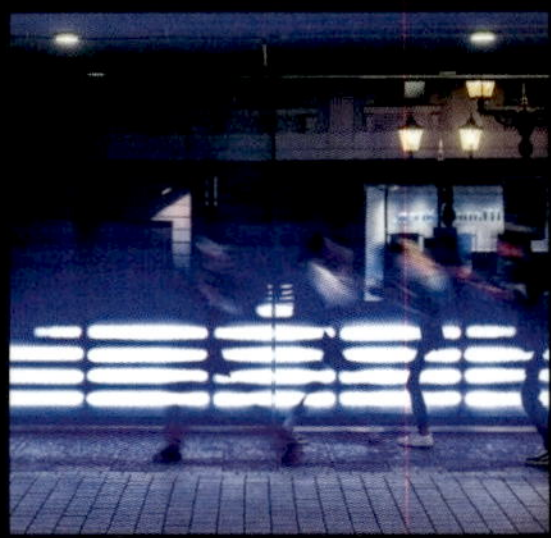

Sound Channel, 2012 → p. 372

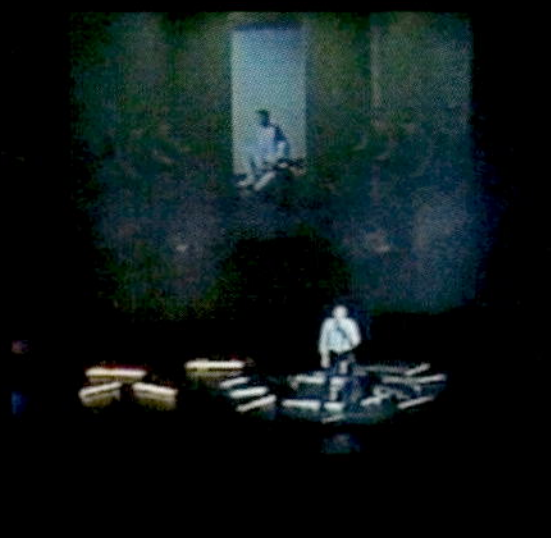

Scene from *The Artificial Will. Electronic Media Opera*, 1984 → pp. 355–57

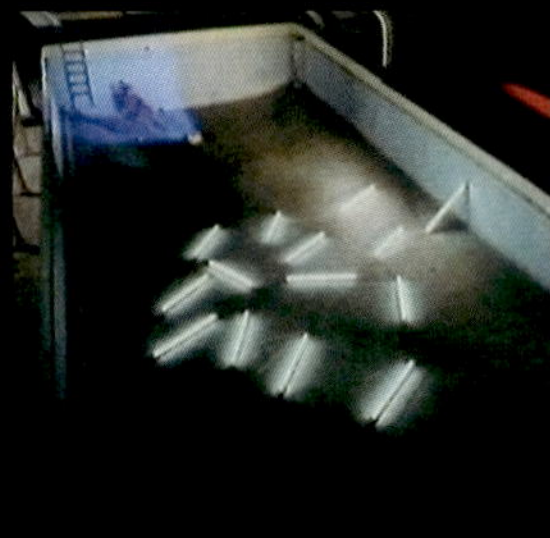

Scene from *Casablanca II*, 1983 → p. 302

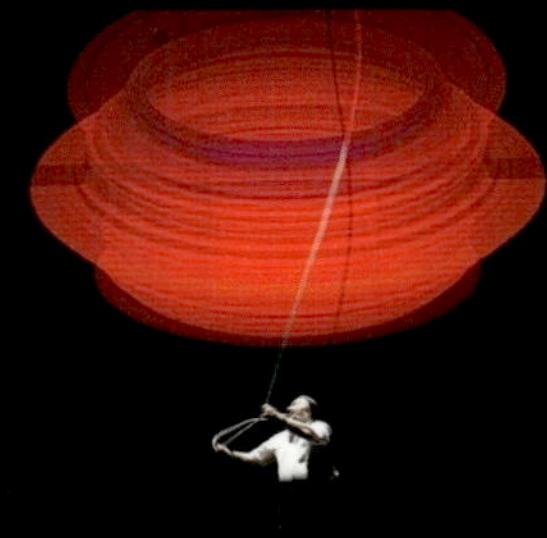

The Origin of Noise — The Noise of the Origin, 2013 → pp. 364–67

Light Bulbs

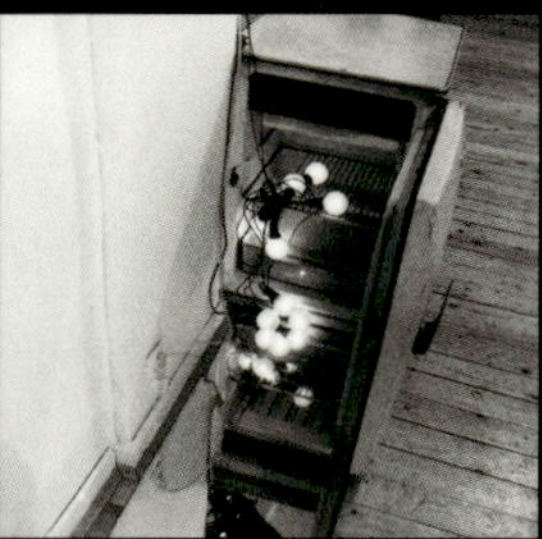

The Electric War, 1983 → pp. 458–59

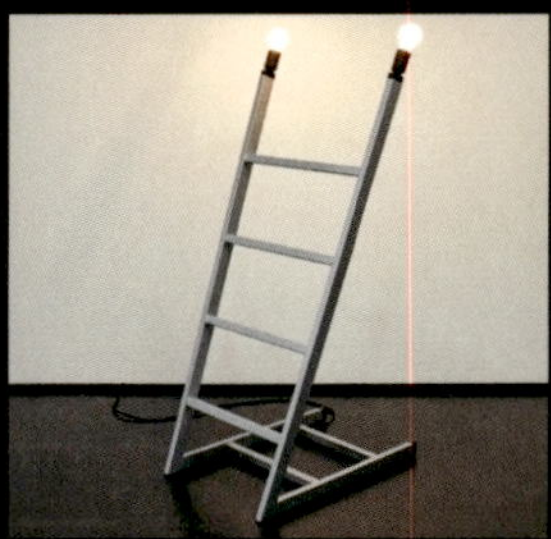

Ladder Lamp, 1988 → p. 422

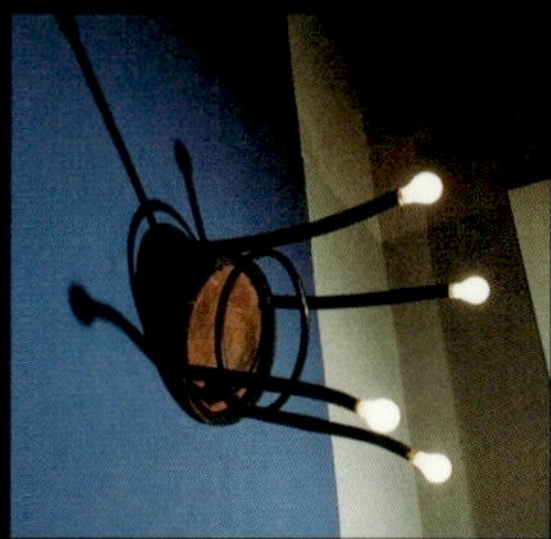

Chair Lamp, 1988 → p. 420

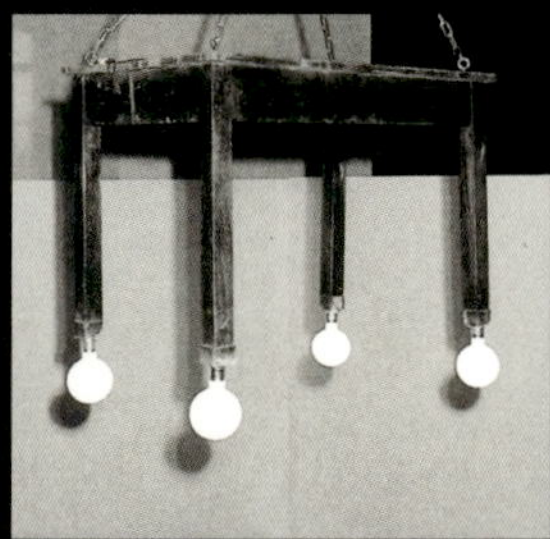

Table Lamp, 1988 → p. 420

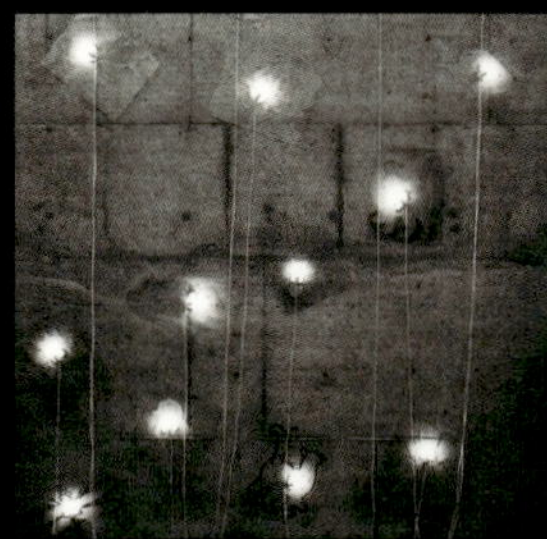

Wall Lights, 1982

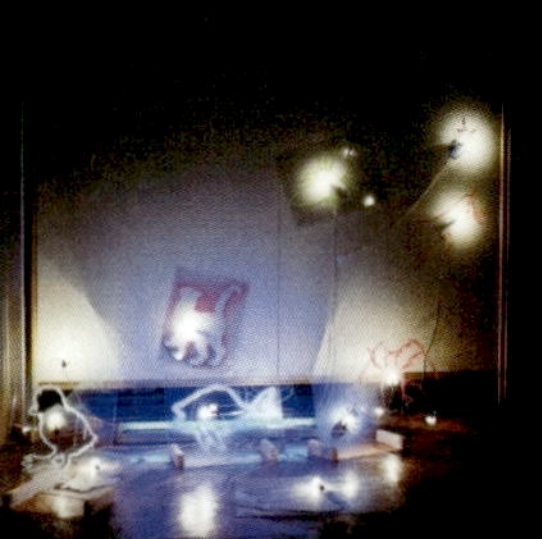

Electric Animals, 1982 → p. 456

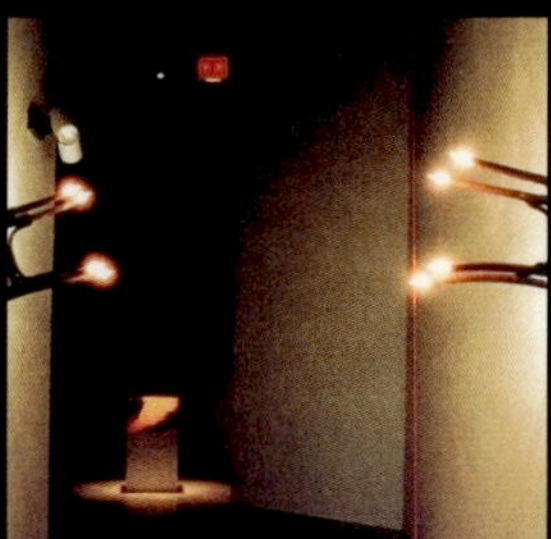

On the Dialectic of History. How Do History and I Meet in the Afternoon While Passing Stools, 1990 → p. 421

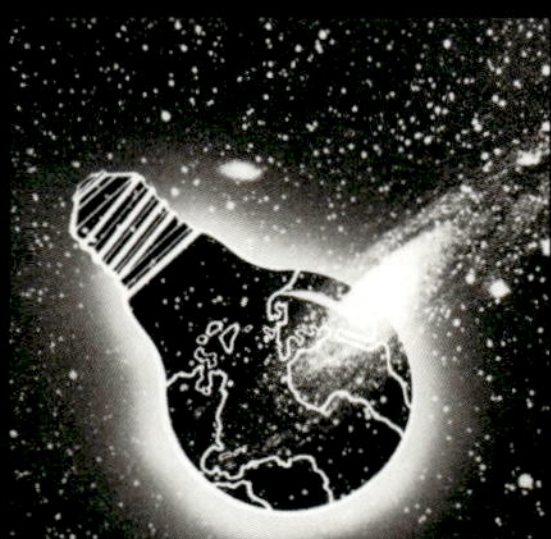

Scenes from *The Artificial Will. Electronic Media Opera*, 1984

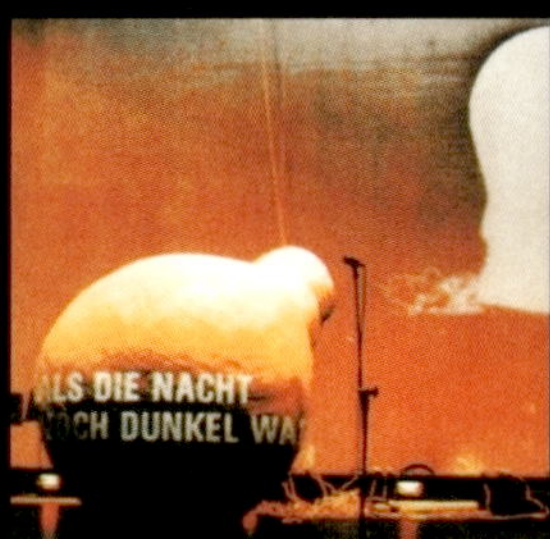

→ pp. 355–57

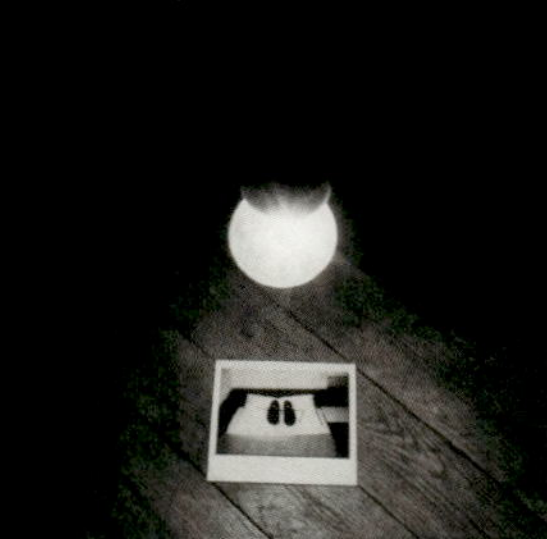

Untitled, 1991

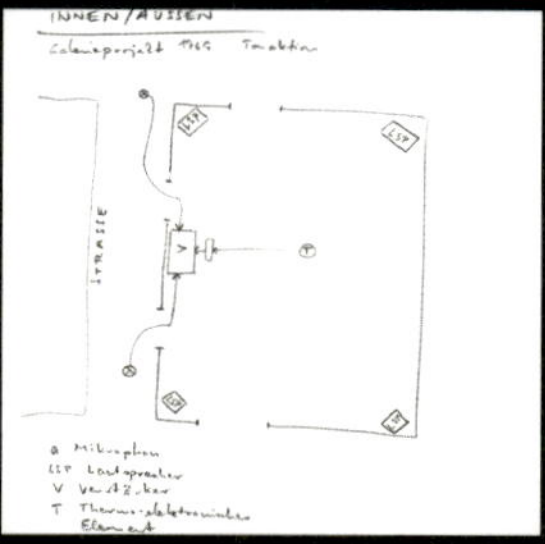
Inside/Outside, 1969 → p. 449

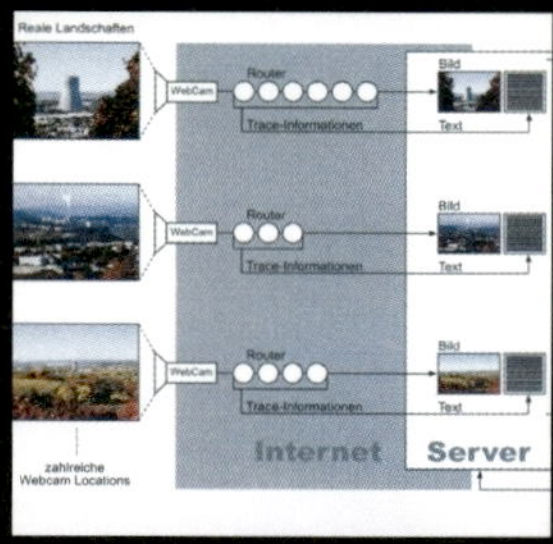

Islands of Nonlocality: From the Local and the Real to the Nonlocal and the Virtual, 1999 → p. 318

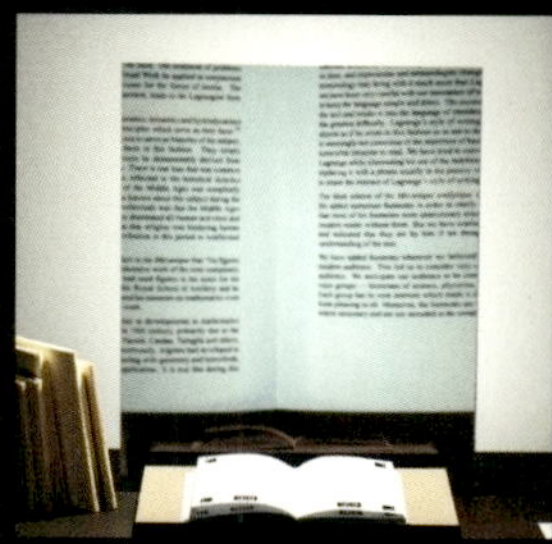
Bibliotheca Digitalis: Three Phases of Digitization, 2017 → pp. 326–27

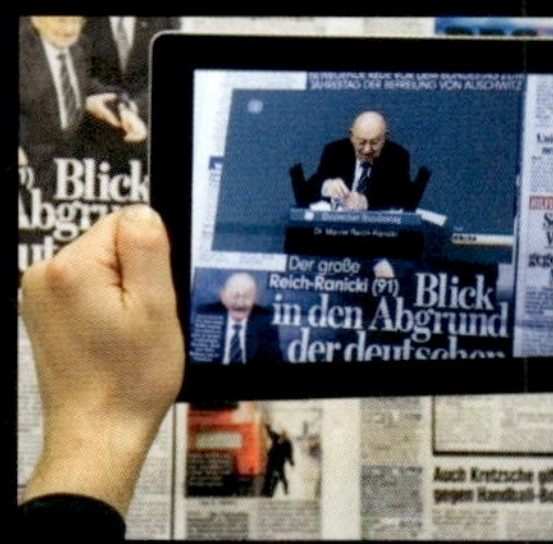

The Newspaper of the Future, 2011–12 → p. 332

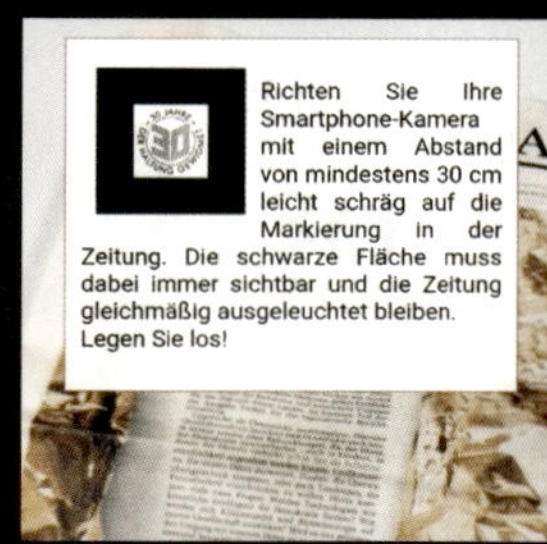

Word Art, 2018 → p. 333

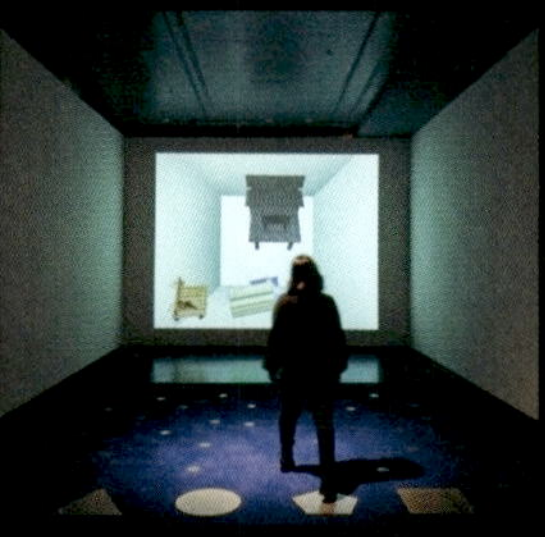
On Justifying the Hypothetical Nature of Art and the Non-Identicality within the Object World, 1992 → pp. 310–13

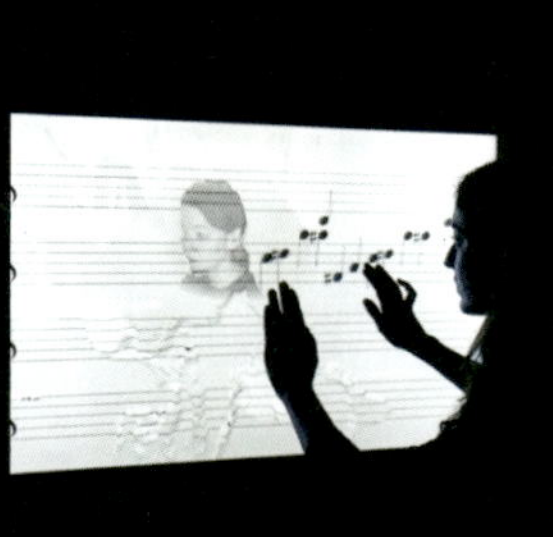
Music Passenger, 2019 → p. 373

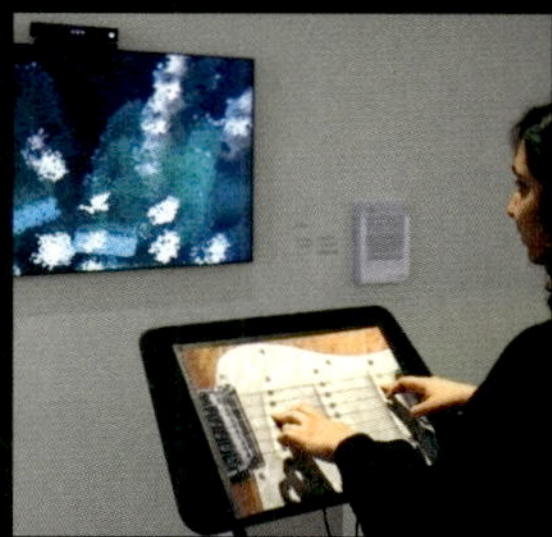
Data Music, 2016 → pp. 370–71

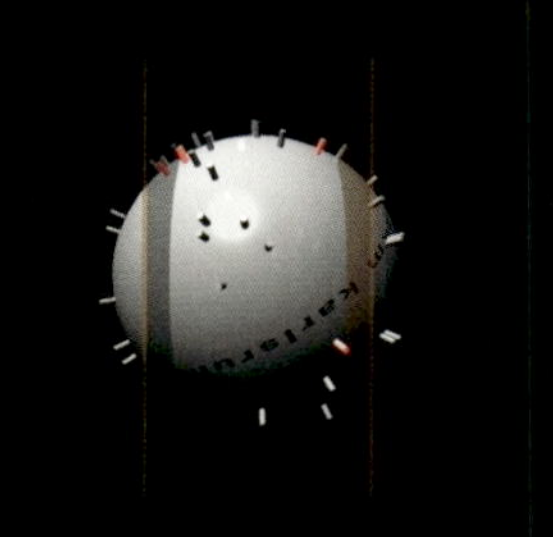
Sound Writer, 2016 → p. 375

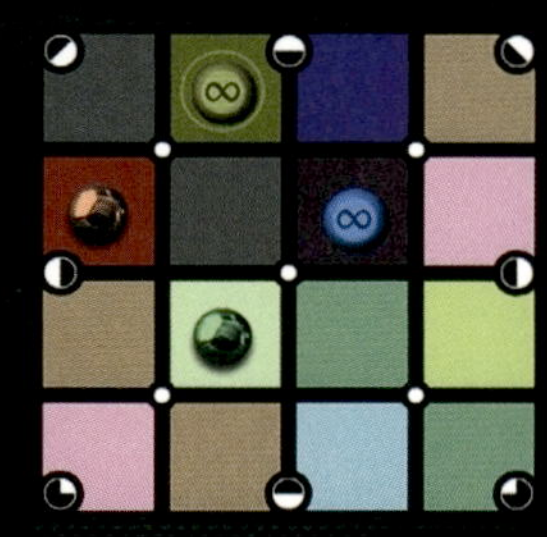
Music Board, 2011 → p. 374

Typewriters & Keyboards

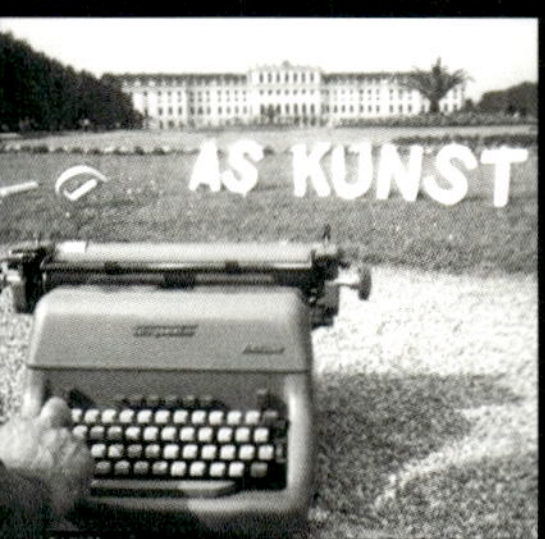

Is This Art?, 1972 → p. 102

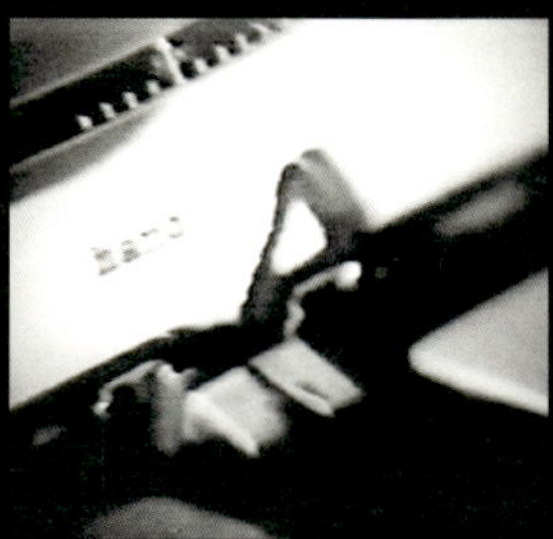
How Did Mathematics Evolve from Fish?, 1971 → p. 102

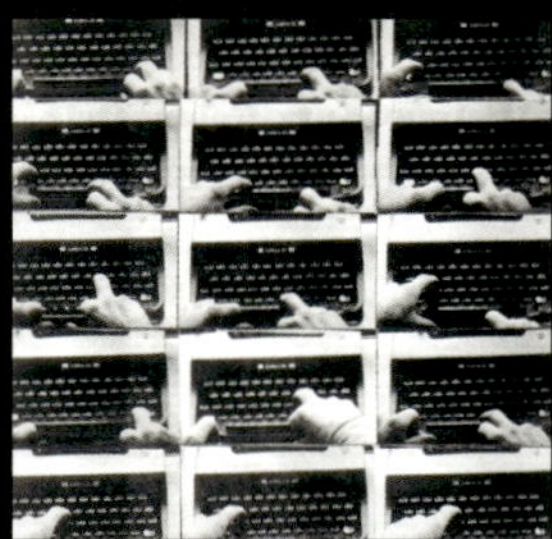
Metaphysical Poem for an Amphioxus, 1971 → p. 103

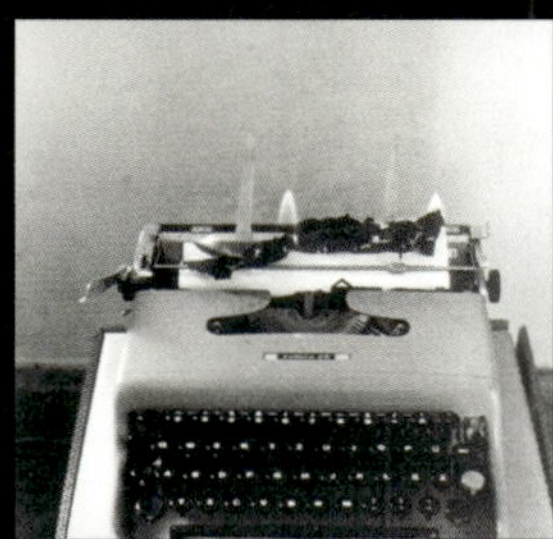
Principle of Hope, 1971 → p. 105

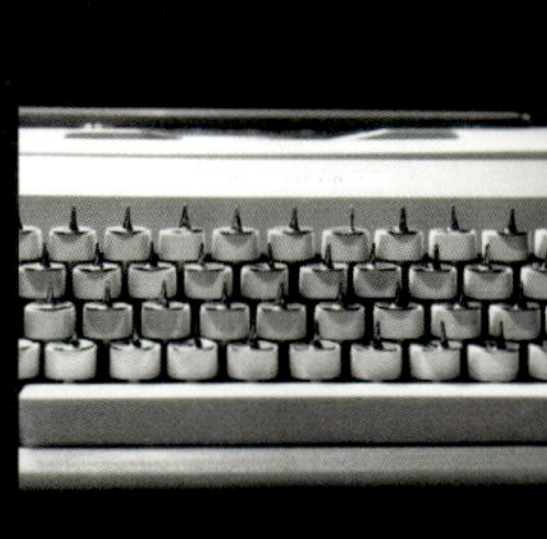
Principle of Resistance, 1971 → p. 104

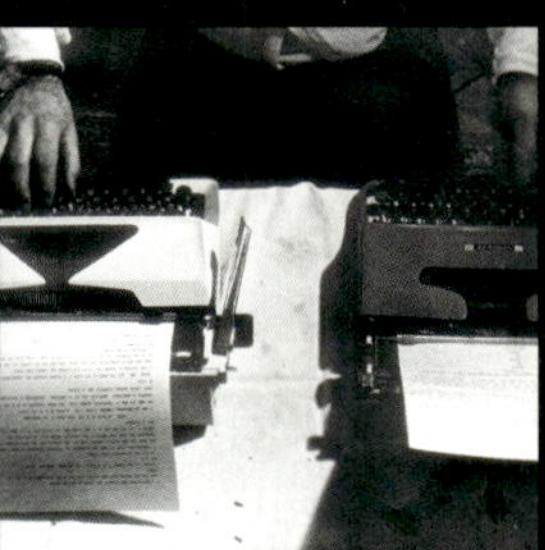
Simultaneous Split Medium, 1974 → pp. 106–107

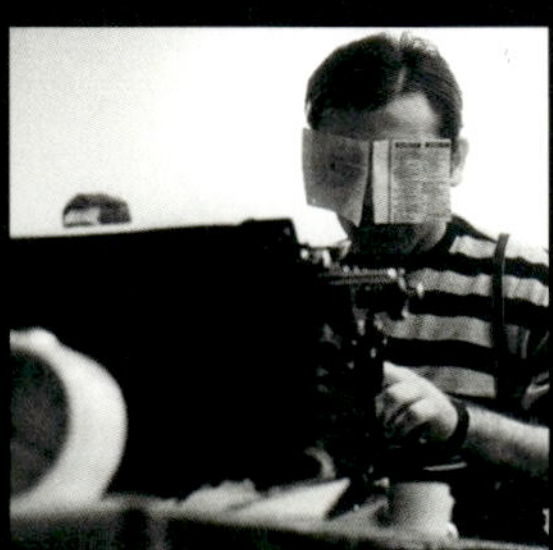
Critique of Art as Language, Blind Typing, 1966 → p. 500

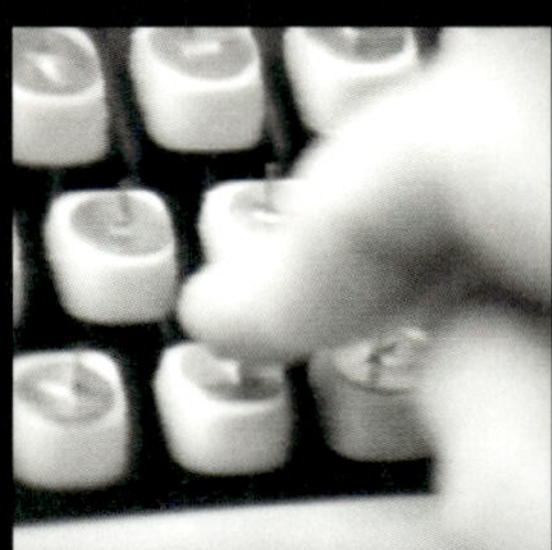
Scrip/tor/ture, 1975

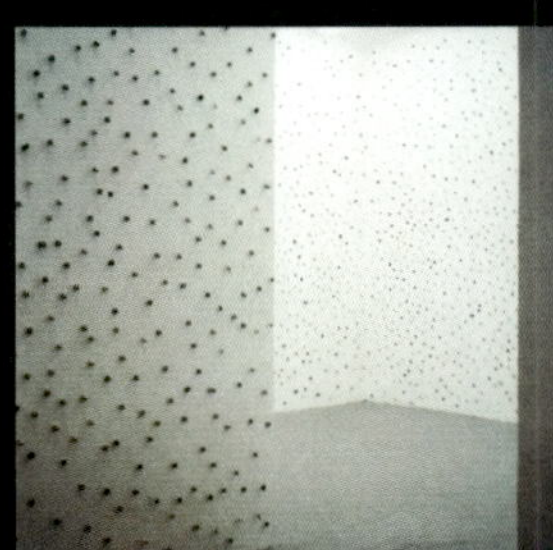
overproduction:only paranoiacs:survive, 1999

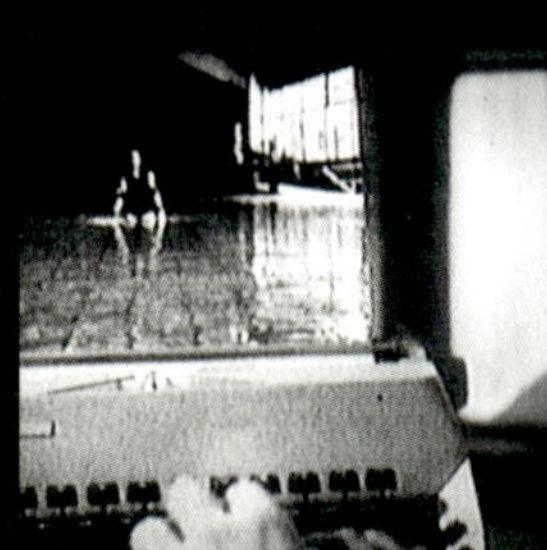
Scene from *Human Females*, 1980 → pp. 204–205

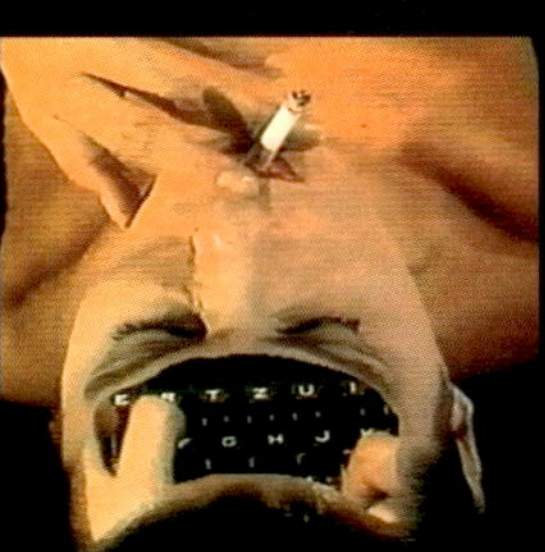
Scene from *Voices from an Inner Space*, 1988 → pp. 358–59

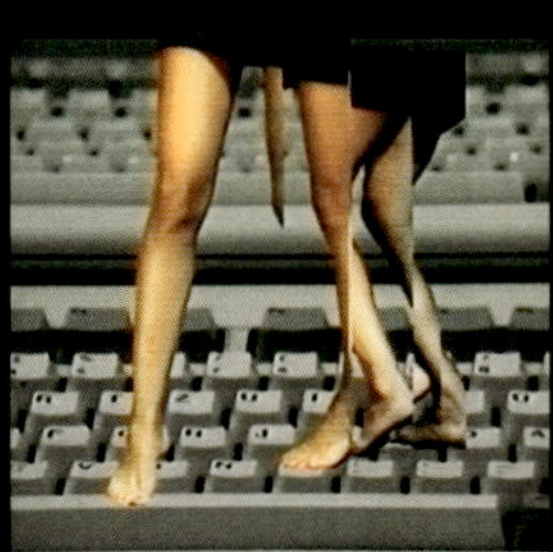
Scene from *Clip, Klapp, Bum …*, 1988

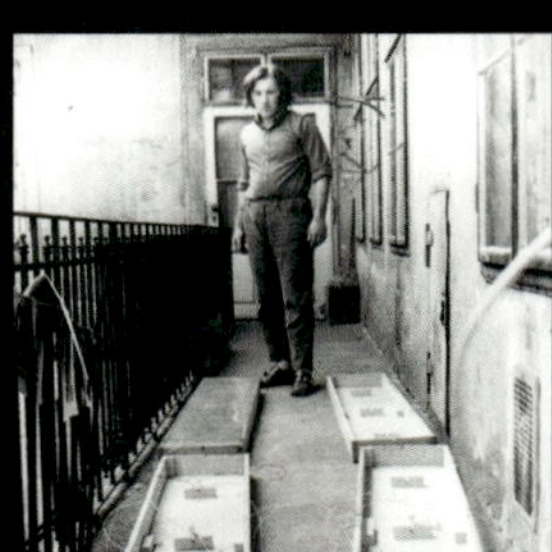
The Long March, 1969/71 → p. 501

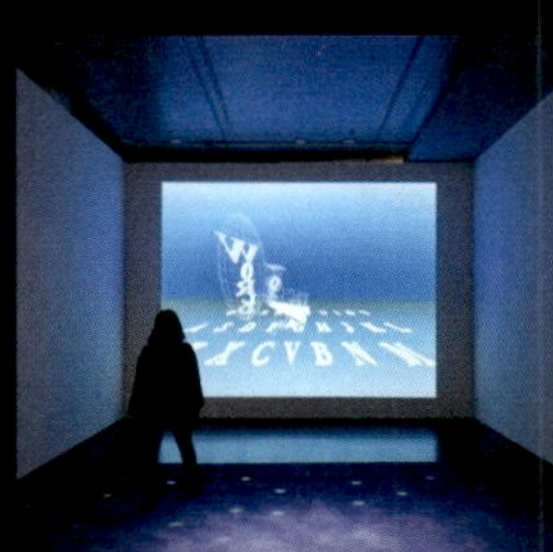
On Justifying the Hypothetical Nature of Art and the Non-Identicality within the Object World, 1992 → pp. 310–31

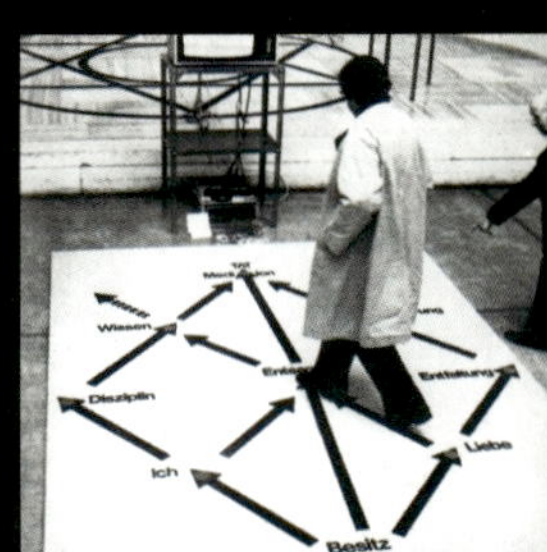

Autohypnosis, 1973 → p. 213

The Wheel

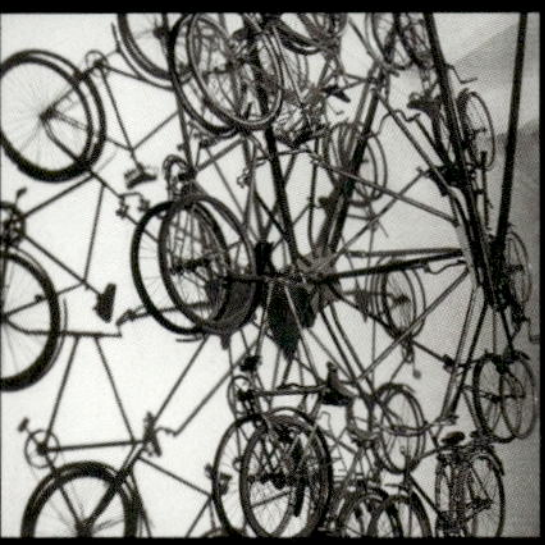
The Wheel of the Real, 1998/2014 → pp. 444–45

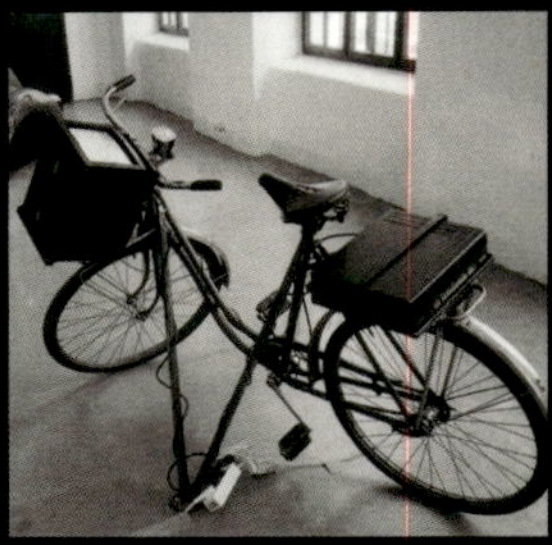
Vehicles and Bioprosthetics Are Intelligent Products …, 1994

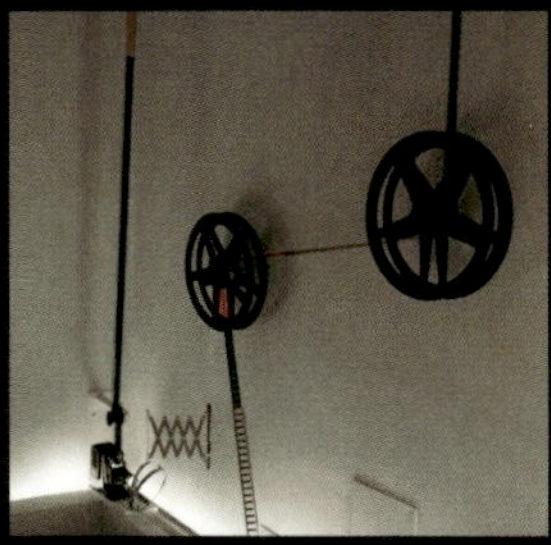
In the Shadow of the Cinema, 1966/2015 → pp. 172–73

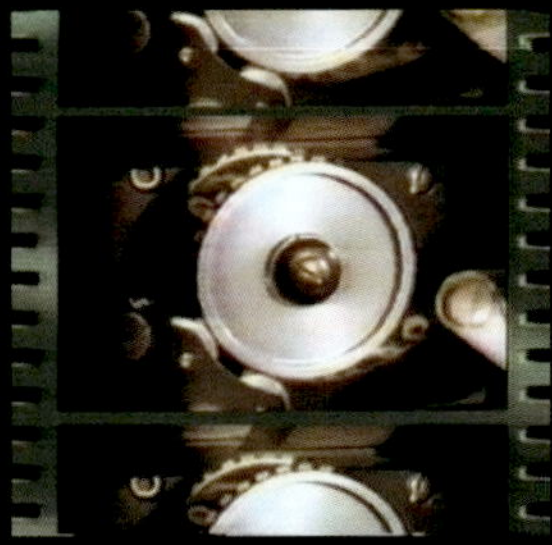
Scene from *Chants of the Pluriverse*, 1986–88 → pp. 304–307

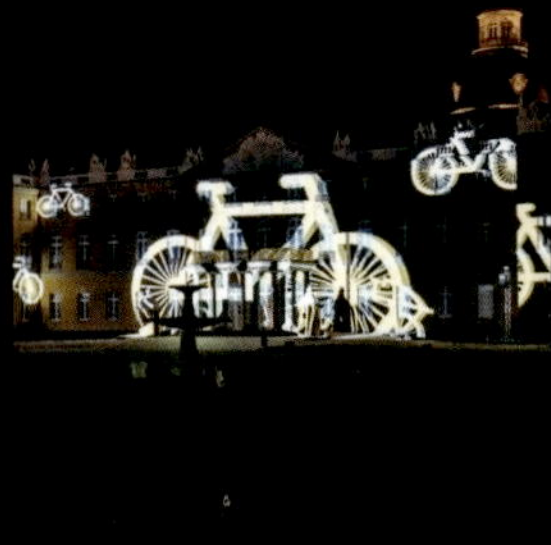
Scene from *Velografie für Drais*, Schlosslichtspiele 2017

Screens

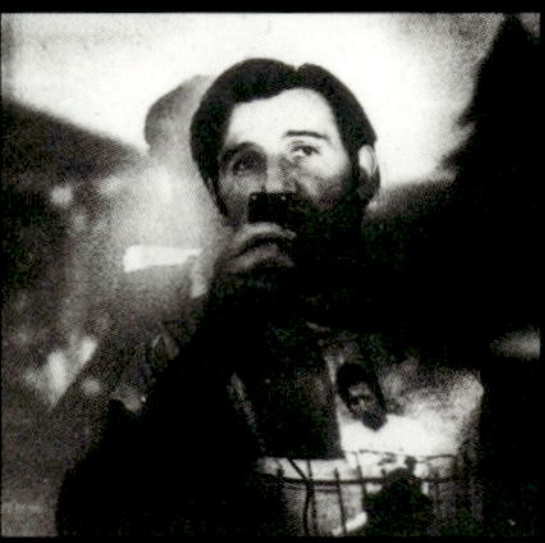
Action Lecture No. 4, 1968 → p. 176

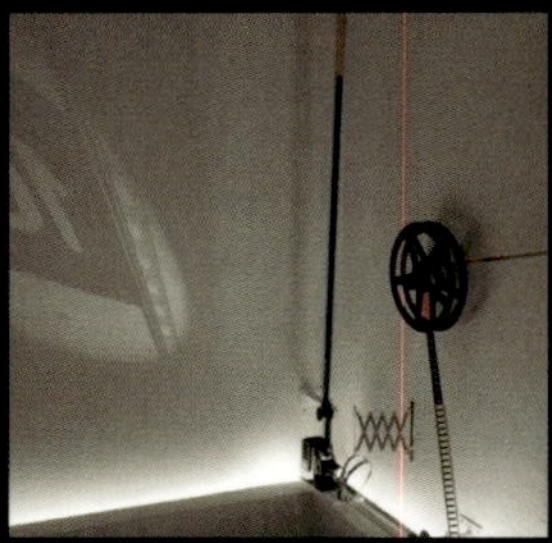
In the Shadow of the Cinema, 1966/2015 → pp. 172–73

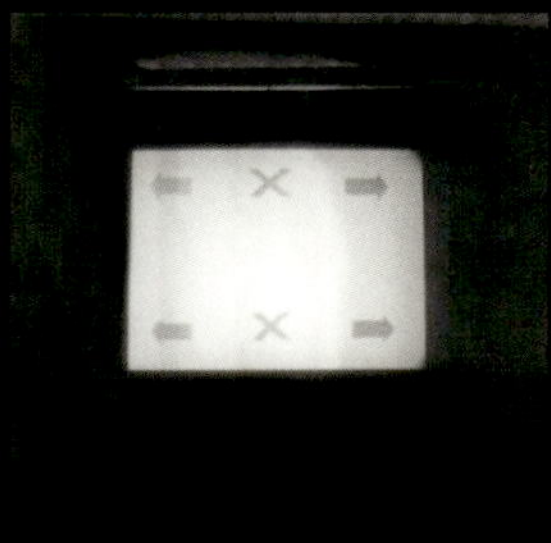
Light Beam : Screen, 1971 → p. 187

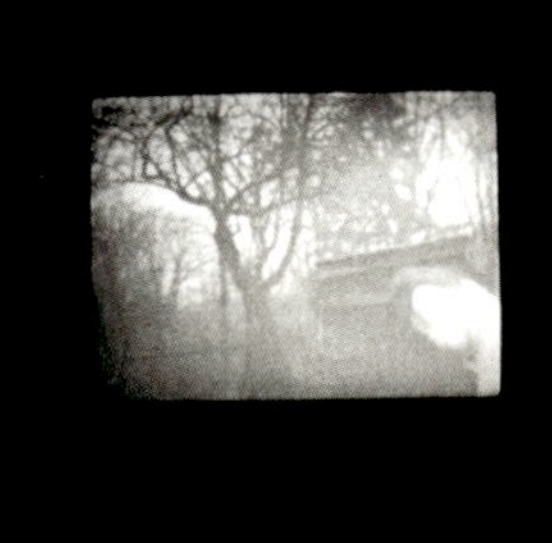
Storm Over ( ), 1969/70 → p. 186

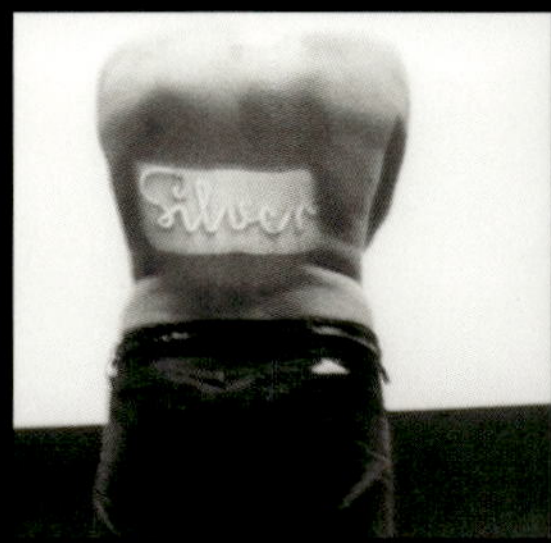
Action Lecture No. 1: Proposals on Non-Affirmative Art, 1966 → pp. 168–69 and 498–99

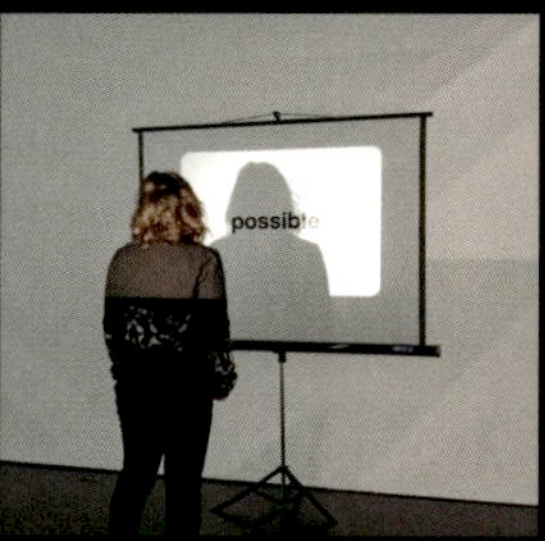

Possible, 1969 → pp. 188–89

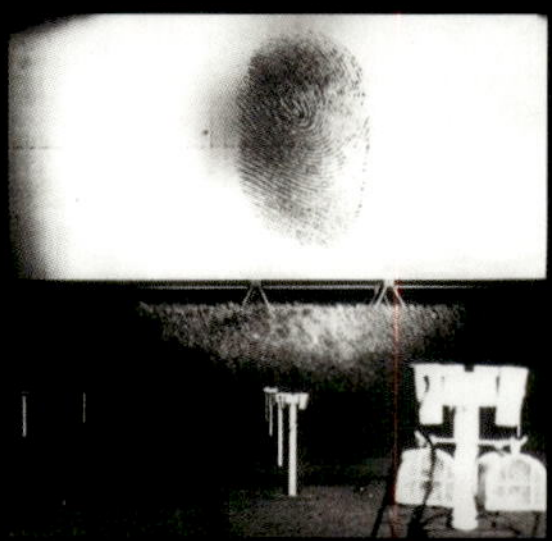
Fingerprint, 1968 → p. 171

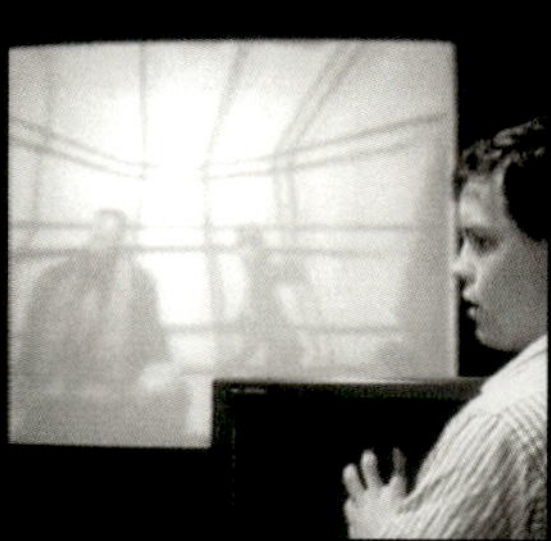
The Tangible Image, 1991/2019 → pp. 308–309

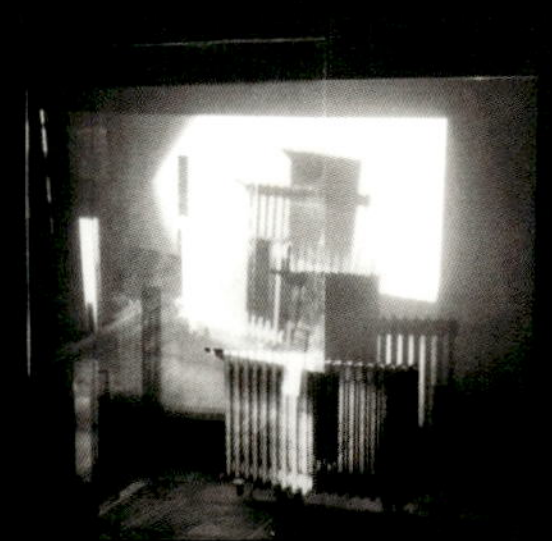
Object/Resistance Symphony: Four Gradual Displacements, 1982 → p. 136

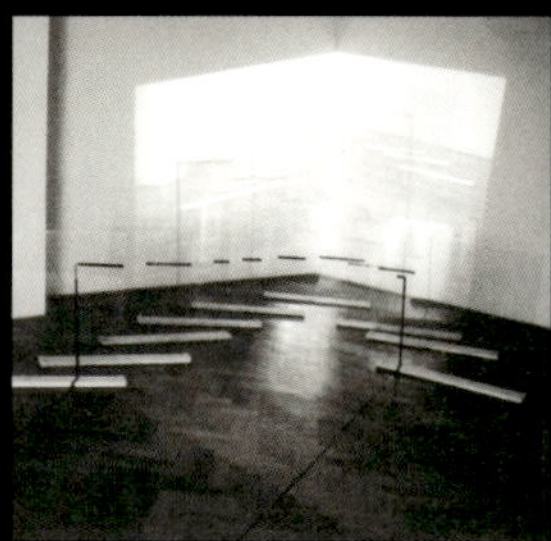
Three Degrees of Modality, 1982 → p. 137

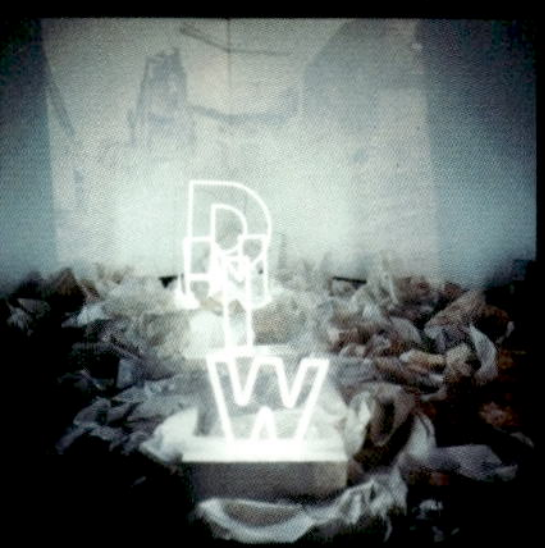
Wind: The Compulsion of Identity, 1966/75 → p. 87

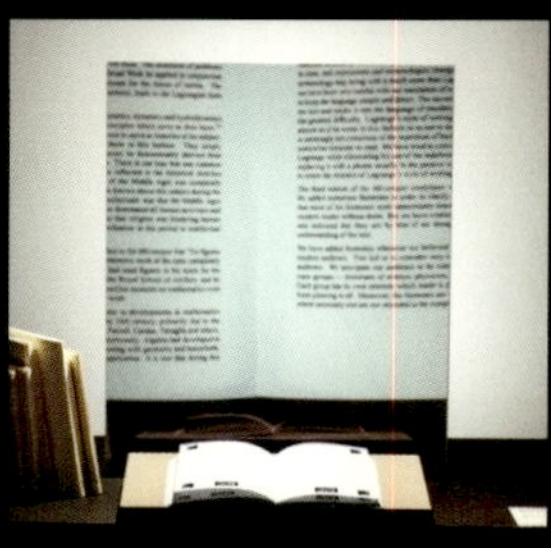
Bibliotheca Digitalis: Three Phases of Digitization, 2017 → pp. 326–27

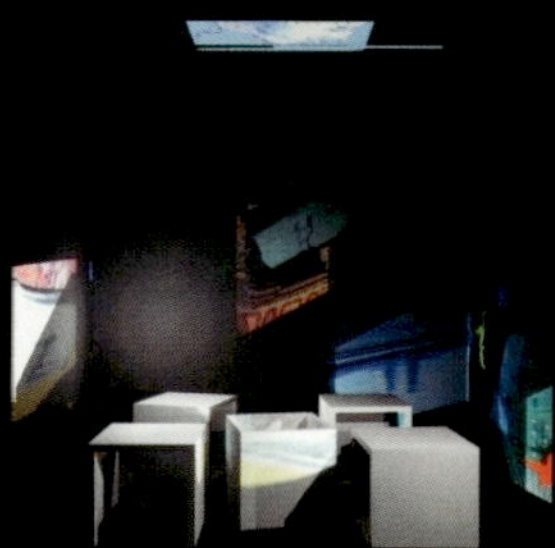
The Cycle of Boxes, 1999 → p. 293

Slides

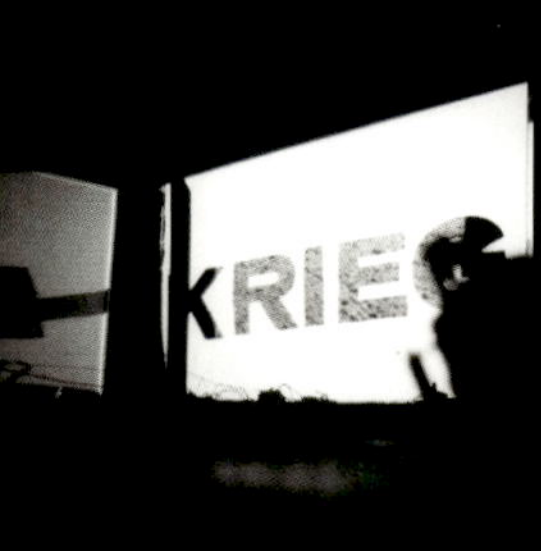

Semantic Study (on the Concept of Images and Image of the Concept), 1969/71

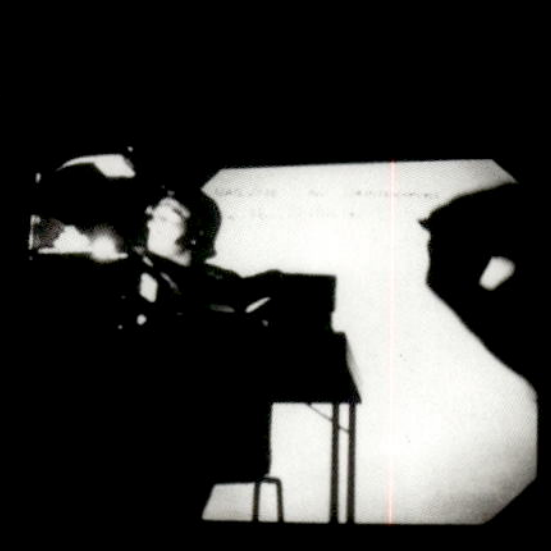
Precinema, Postcinema, Paracinema, 1971 → p. 191

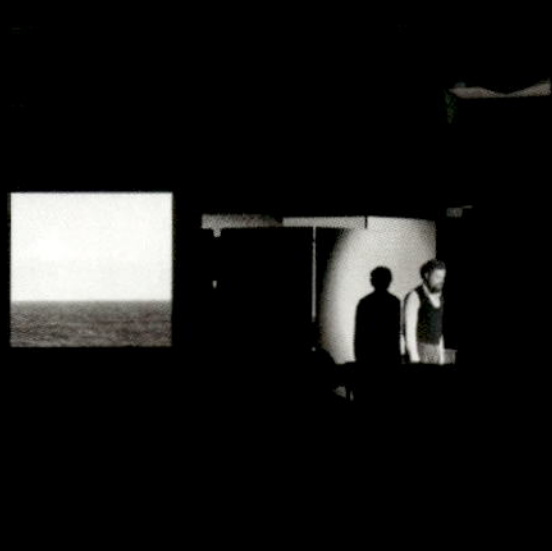
Zykan I, 1971 → pp. 192–93

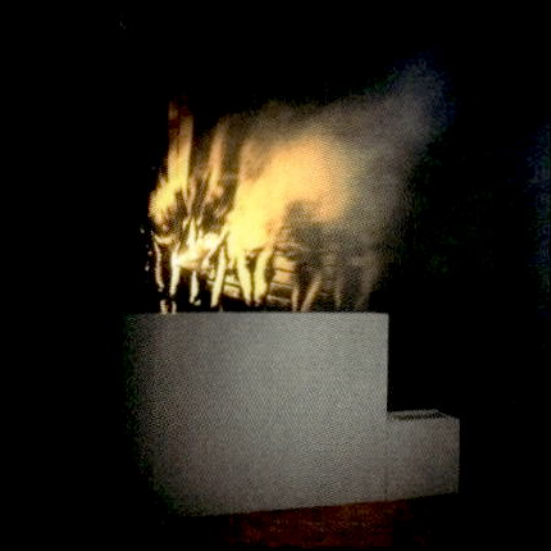
Dumb Show, 1969 → p. 183

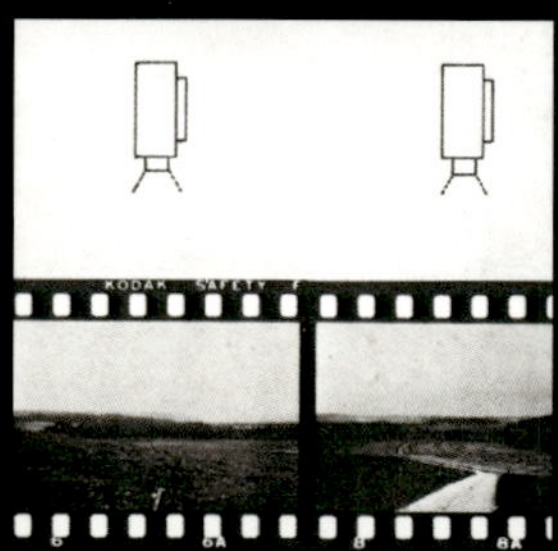

On the Nature of Technology: Landscape with Fine Grain, 1972 → pp. 194–95

Microphones

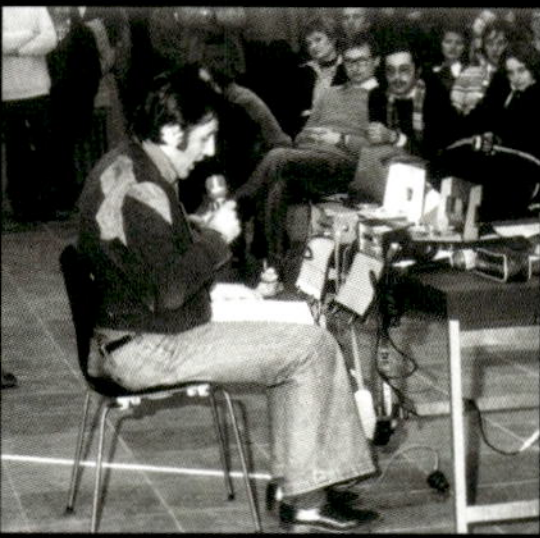
Recursive Texts, 1973 → p. 98

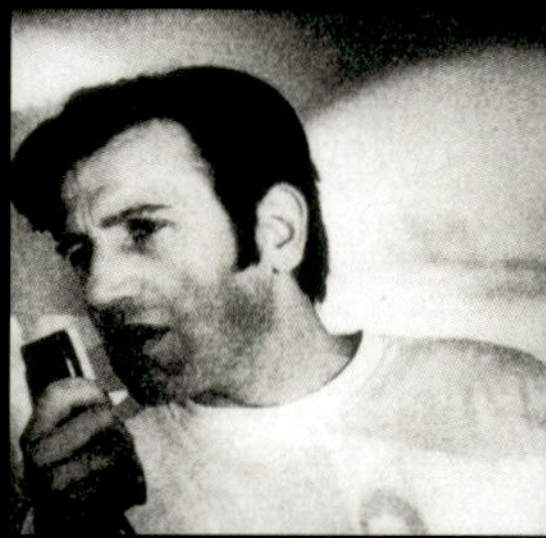
Action Lecture No. 4, 1968 → p. 176

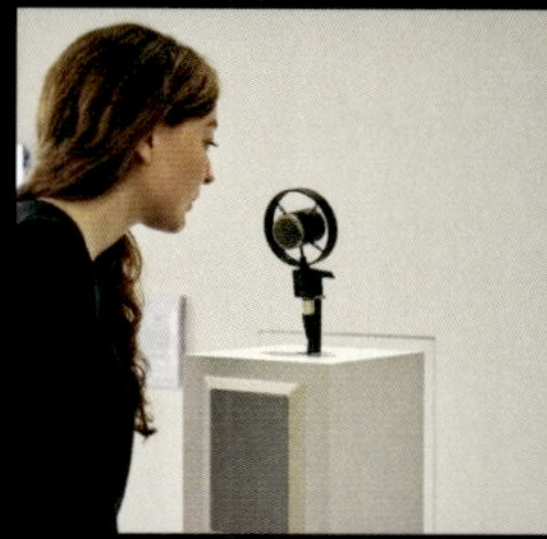
oh cet echo, 2012 → p. 369

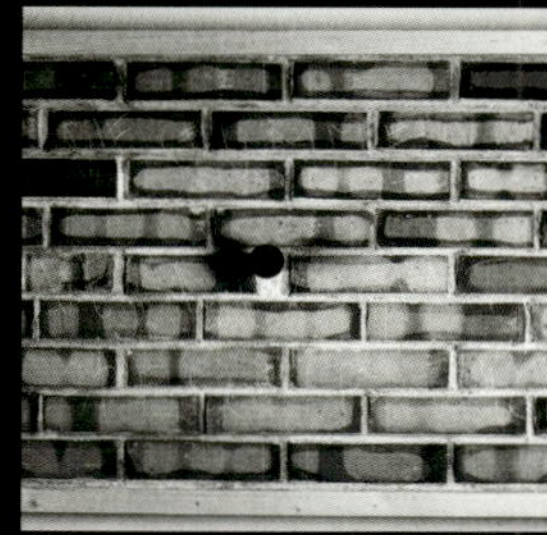
Mute Reason Speaks Playback, 1990/2014

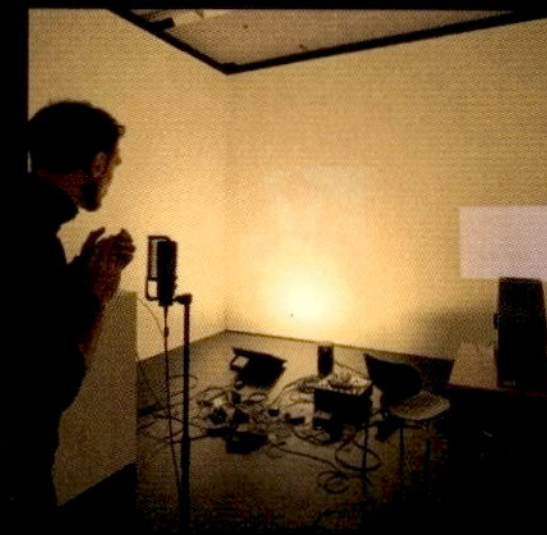
Communication Breakdown, 1971/2019 → pp. 502–503

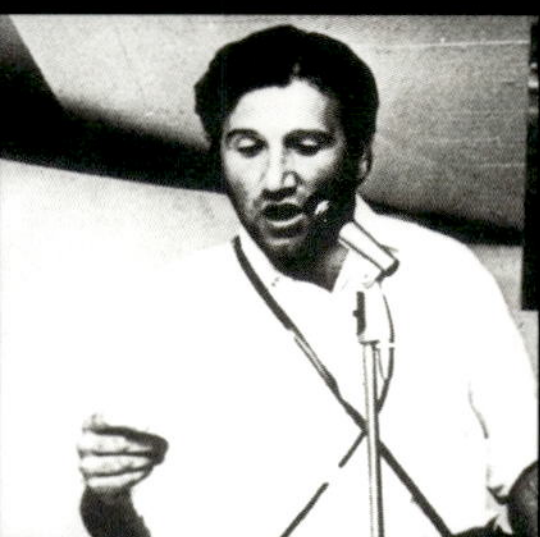
Action Lecture No. 5, 1968 → p. 177

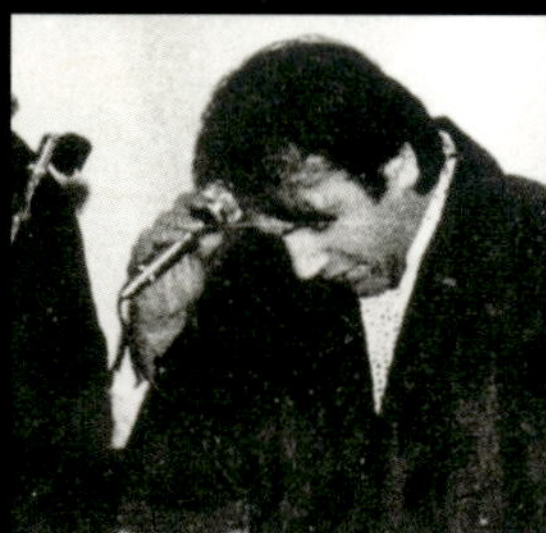
Performance by Hotel Morphila Orchester, 1979 → pp. 352–53

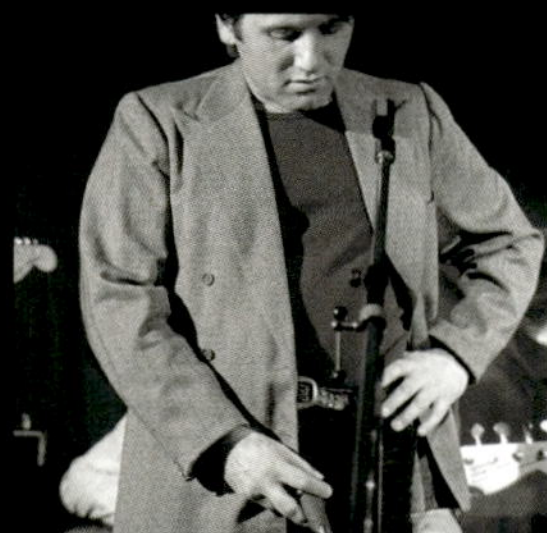
Performance by Hotel Morphila Orchester, 1979 → pp. 352–53

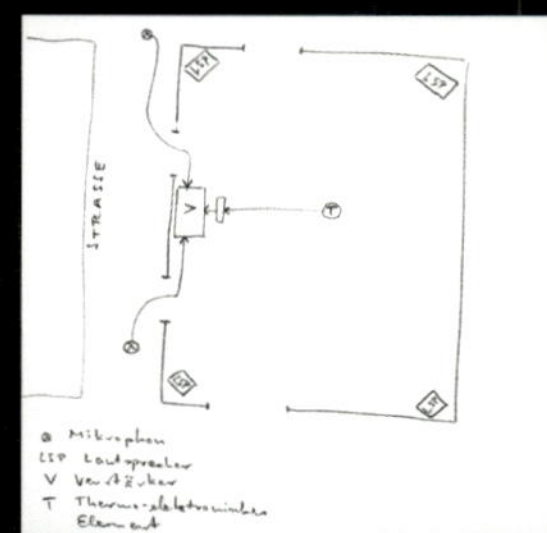

Inside/Outside, 1969 → p. 449

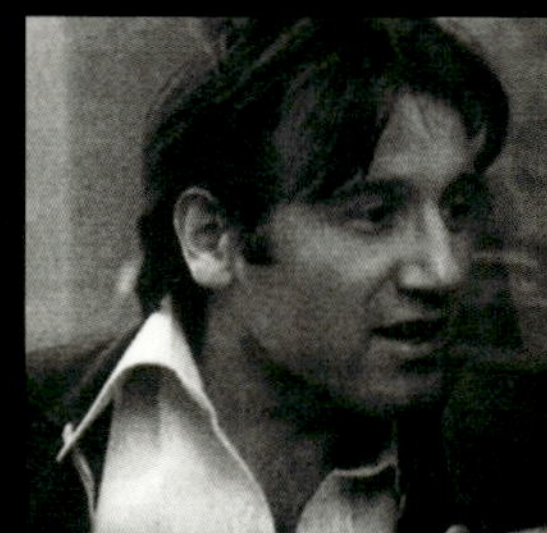
Who Does Vienna Belong To?, 1976/77 → p. 535

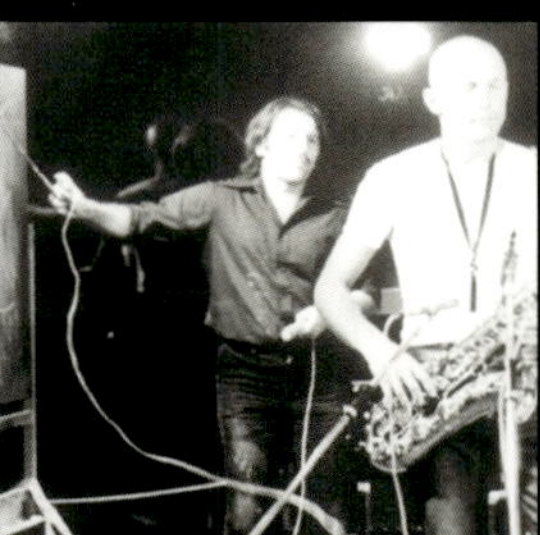
Video Music, 1978

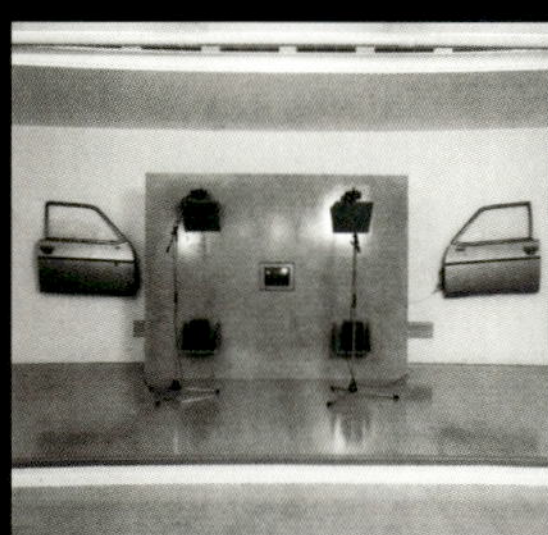
Lucifer's Dream: Non Serviam; The Revolt of the Automata; To the Self by Itself?, 1990 → p. 160

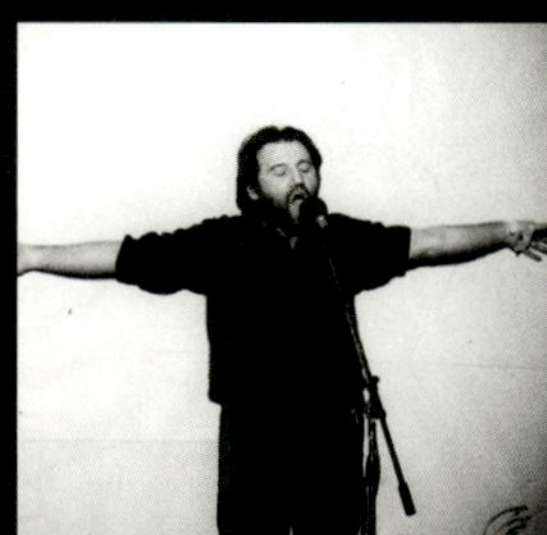
Rhyme and Reason III aka Time Jump, 1978 → p. 532

Telephones

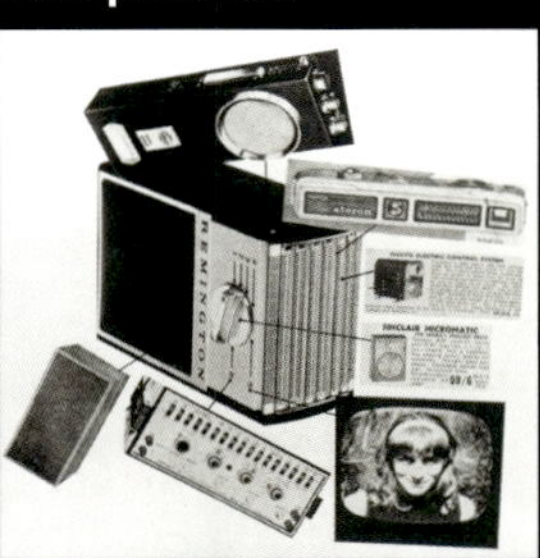
Information Unit, 1967 → p. 34

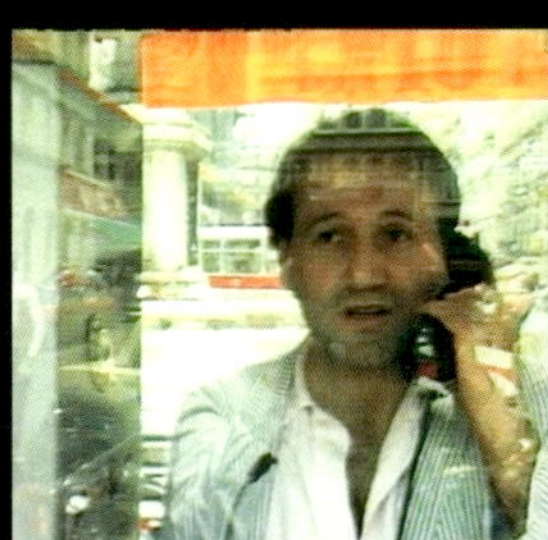
"Sex in der Stadt," 1982 → p. 353

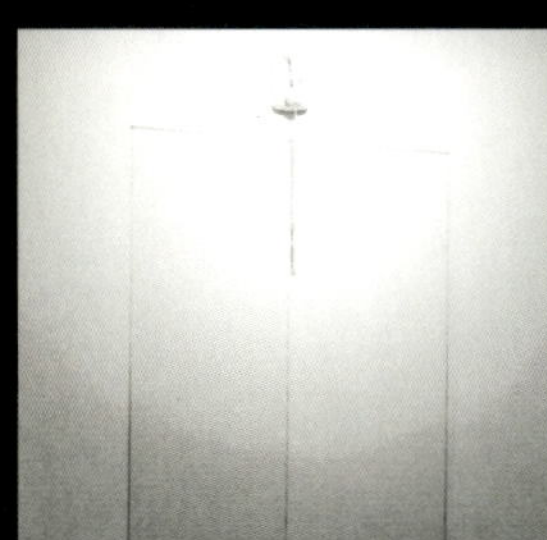
Untitled, 1988 → p. 385

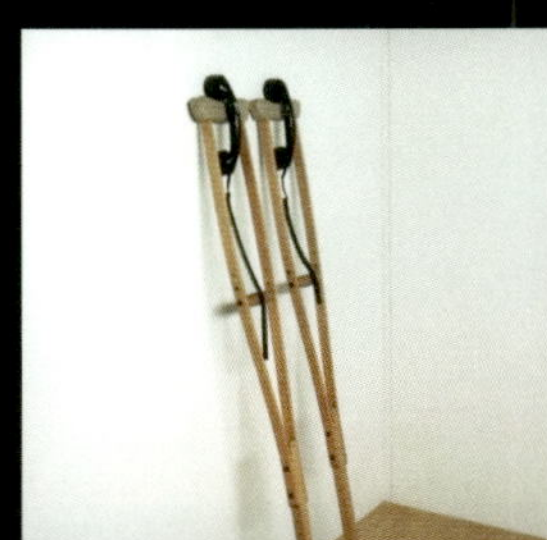
All Technology Is Teletechnology, All Teletechnology is Theotechnology, 1994 → p. 442

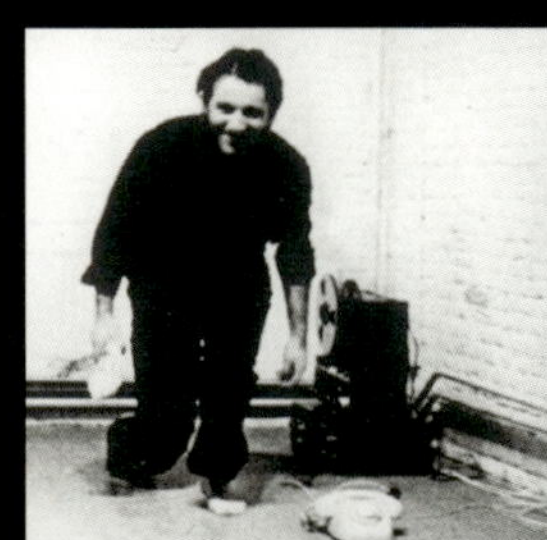
Sculptural Identity, idea (1970)/1978 → pp. 258–59

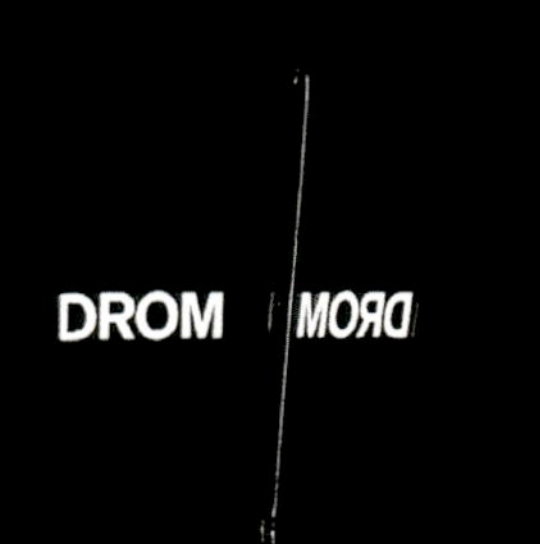

Monodrome, 1970/74 → p. 257

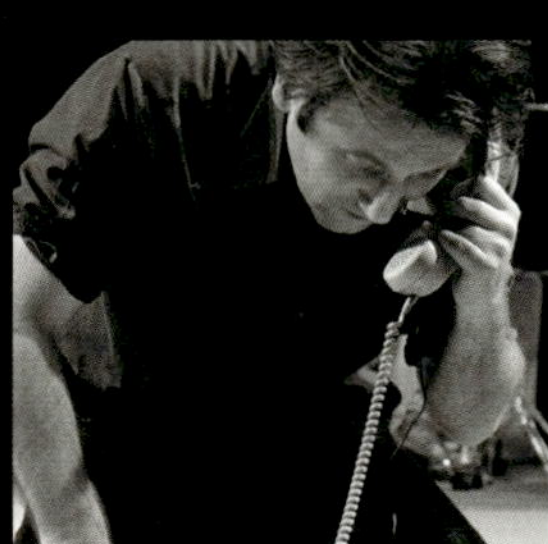
Performance by Hotel Morphila Orchester, 1979 → pp. 352–53

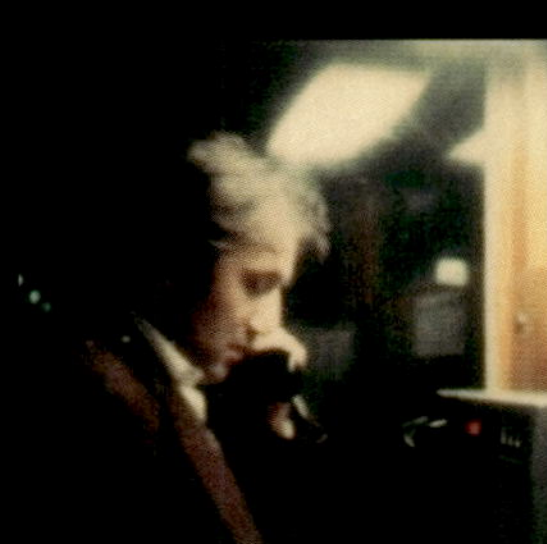
An Open Conversation, 1979

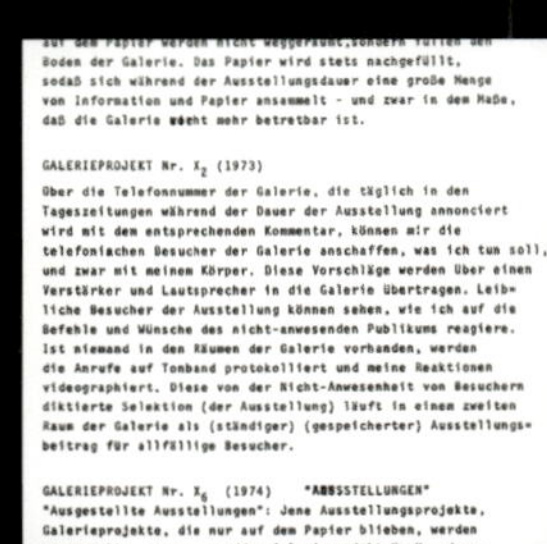
Boden der Galerie. Das Papier wird stets nachgefüllt, sodaß sich während der Ausstellungsdauer eine große Menge von Information und Papier ansammelt - und zwar in dem Maße, daß die Galerie nicht mehr betretbar ist.

GALERIEPROJEKT Nr. X_2 (1973)

Über die Telefonnummer der Galerie, die täglich in den Tageszeitungen während der Dauer der Ausstellung annonciert wird mit dem entsprechenden Kommentar, können mir die telefoniachen Besucher der Galerie anschaffen, was ich tun soll, und zwar mit meinem Körper. Diese Vorschläge werden über einen Verstärker und Lautsprecher in die Galerie übertragen. Leibliche Besucher der Ausstellung können sehen, wie ich auf die Befehle und Wünsche des nicht-anwesenden Publikums reagiere. Ist niemand in den Räumen der Galerie vorhanden, werden die Anrufe auf Tonband protokolliert und meine Reaktionen videographiert. Diese von der Nicht-Anwesenheit von Besuchern diktierte Selektion (der Ausstellung) läuft in einem zweiten Raum der Galerie als (ständiger) (gespeicherter) Ausstellungsbeitrag für allfällige Besucher.

GALERIEPROJEKT Nr. X_6 (1974) "AUSSTELLUNGEN"

"Ausgestellte Ausstellungen": Jene Ausstellungsprojekte, Galerieprojekte, die nur auf dem Papier blieben, werden ausgestellt, ausgenommen das Galerieprojekt Nr. X_6, da es

Gallery Project No. X2, 1973 → p. 448

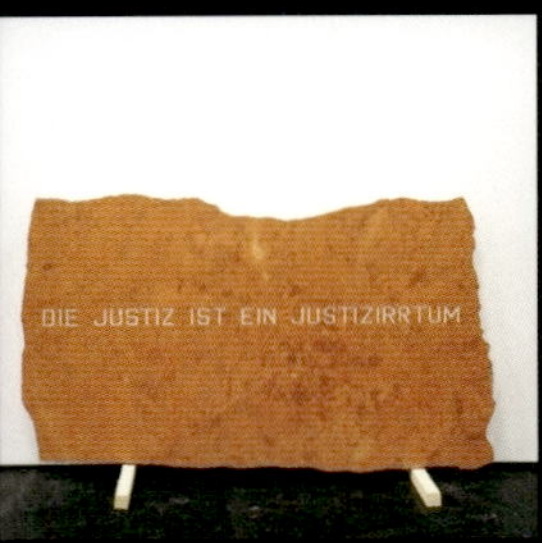

Justice Is a Miscarriage of Justice, 1967 → p. 76

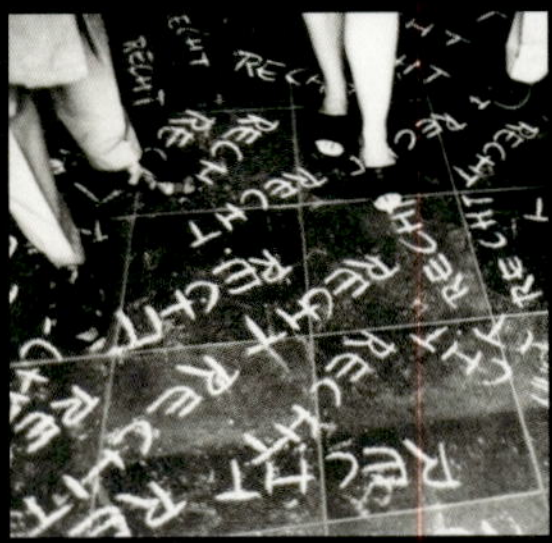

Trampling on Law, 1968 → p. 40

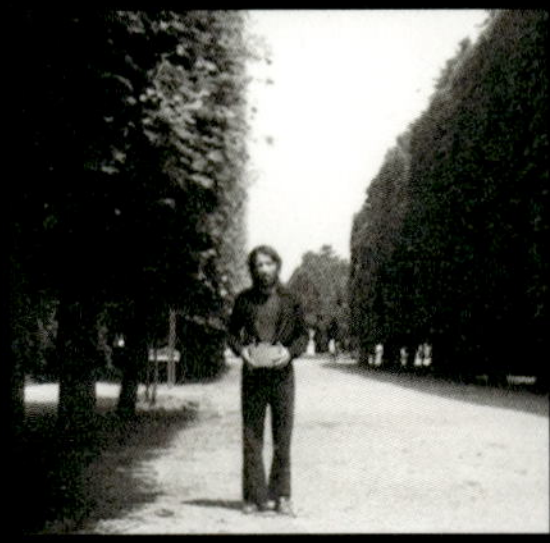

Depiction Is a Crime, 1970/71 → p. 249

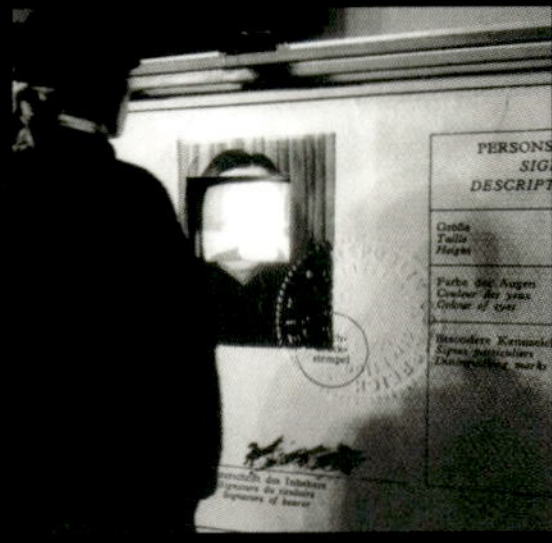

Against Institutional Identity, 1980/2014 → p. 275

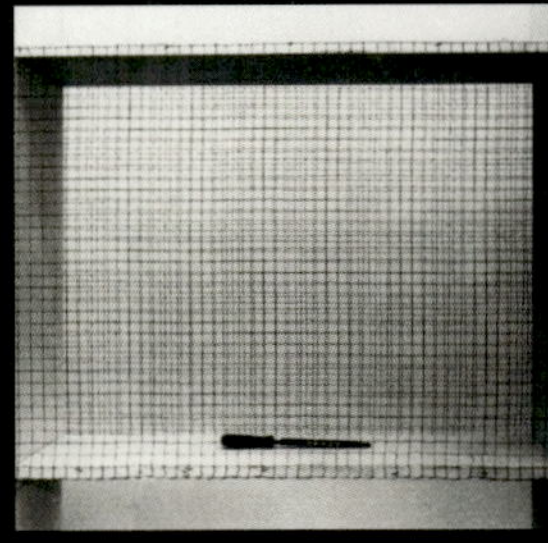

The World's a Prison, 1994

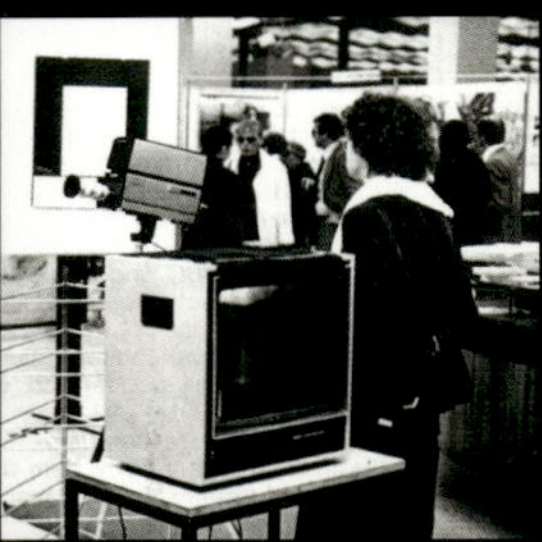

The Guard as Bandit, 1974 (idea)/1978 → p. 269

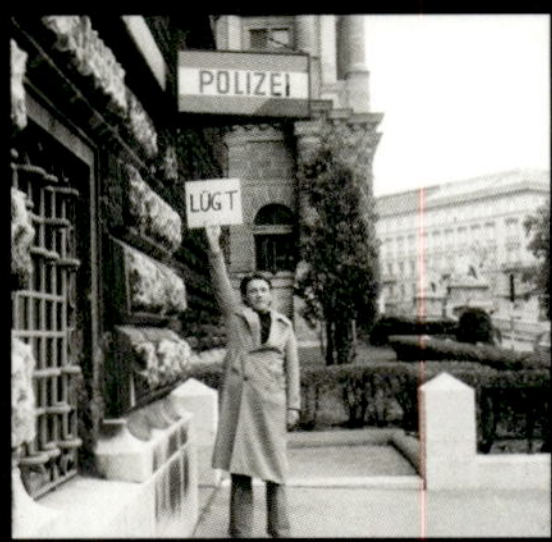

Attacks, 1970–71 → pp. 536–37

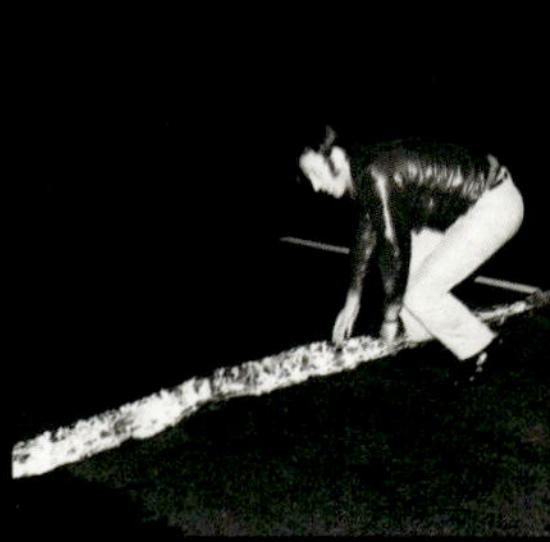

Firewall, or More Traffic Deaths, Fewer Citizens, 1969 → p. 212

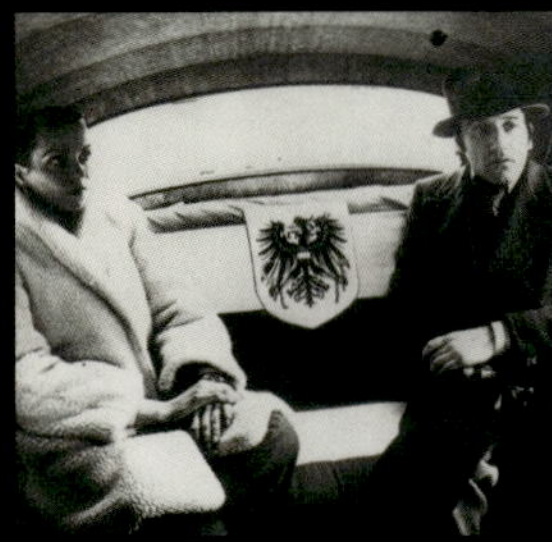

Private Vehicle as State Car, 1972 → p. 534

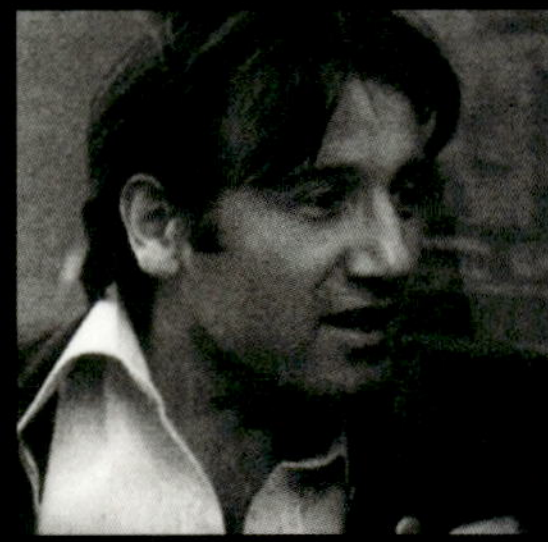

Who Does Vienna Belong To?, 1976/77 → p. 535

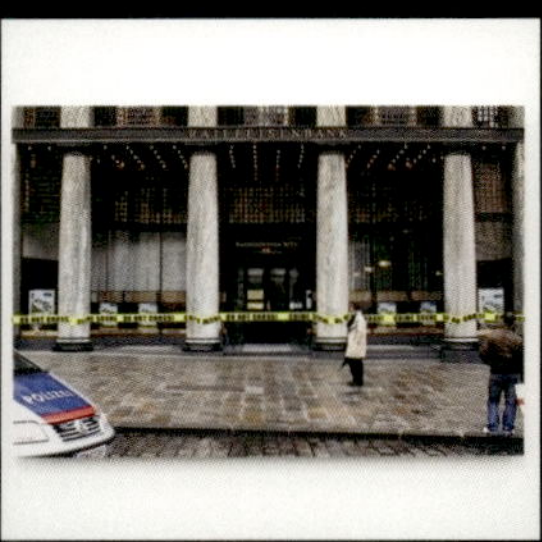

Crime Scene, 2009 → p. 479

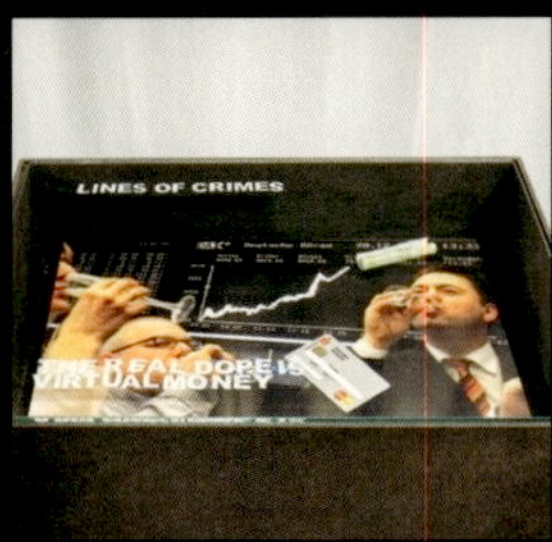

Lines of Crimes: The Real Dope Is Virtual Money, 2009 → p. 481

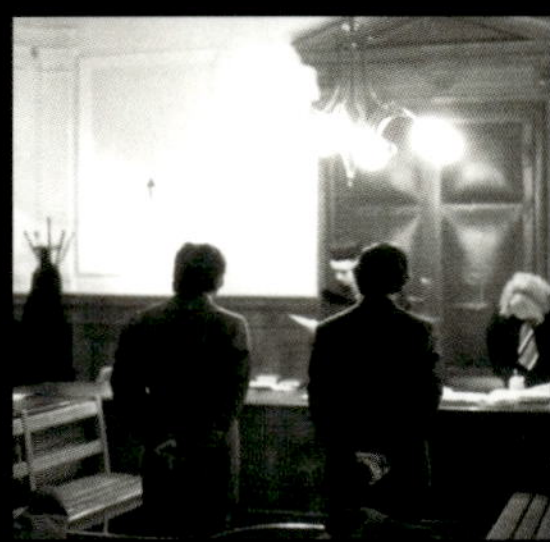

Cryptic Message, 1969 → pp. 512–13

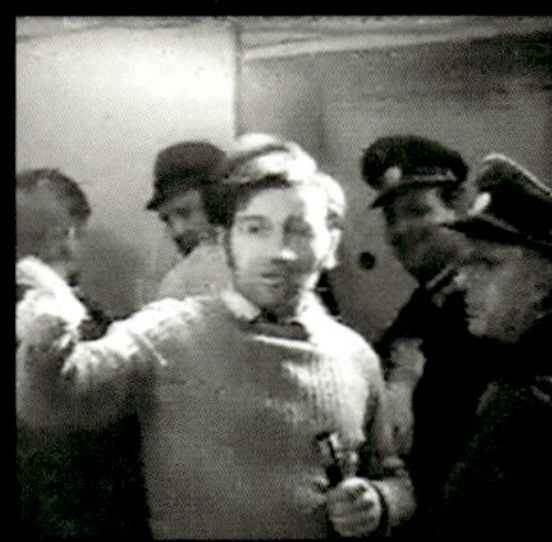

The Public as Exhibit (Audience Exhibited), 1969 → p. 41

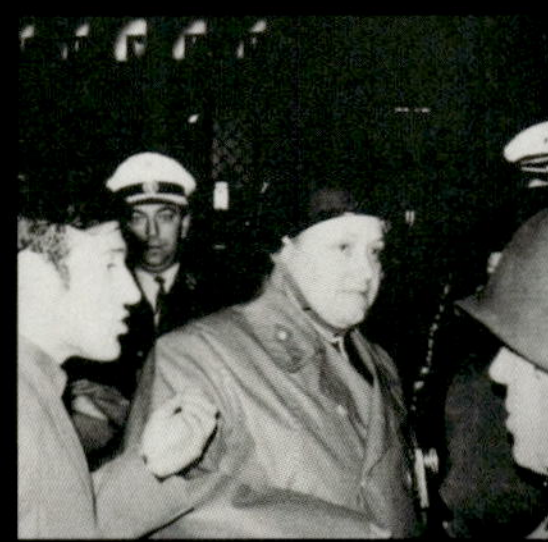

Zock-Fest, 1967 → pp. 506–507

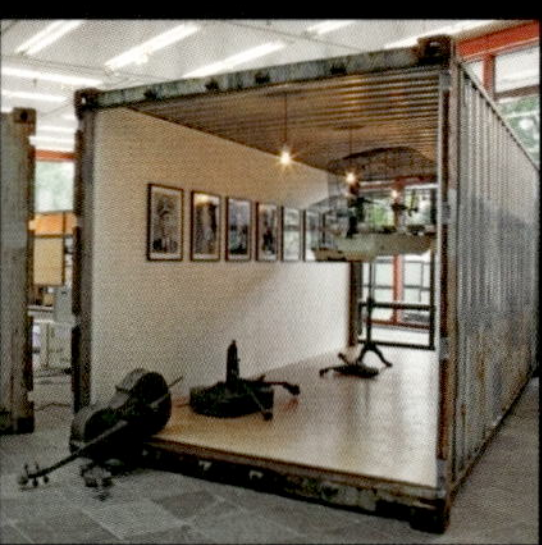

Music Exhibition, 1975 (concept)/2014 → pp. 466–67

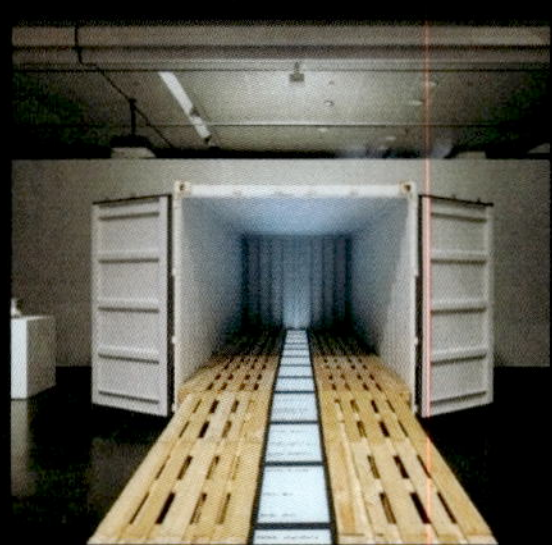

Expulsion of Reason, 1993 → pp. 464–65

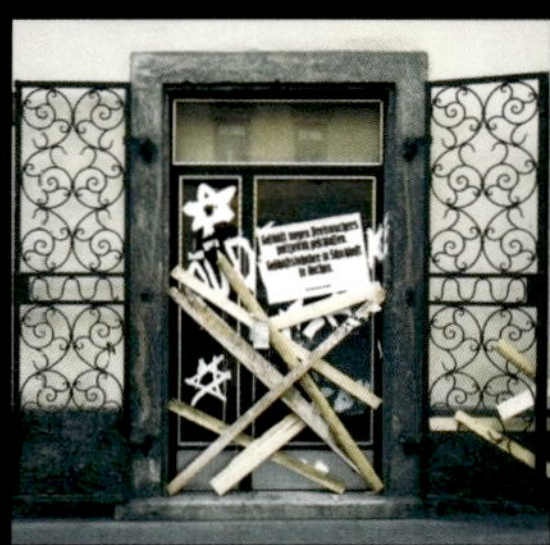

Time Display Window, 1970 (idea)/1979 → pp. 462–63

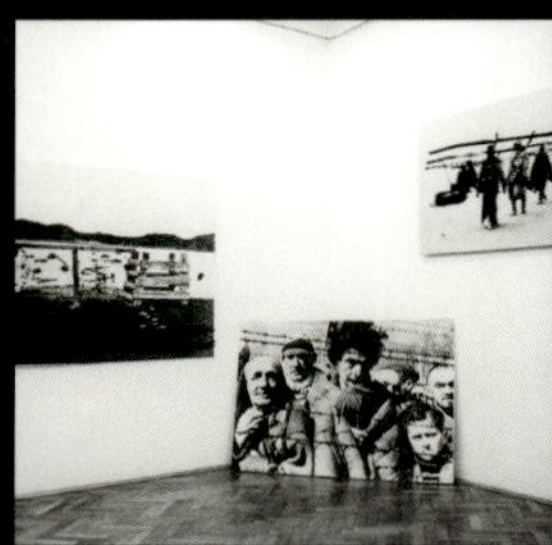

God's Memory, or the Arm of the Media, 1982

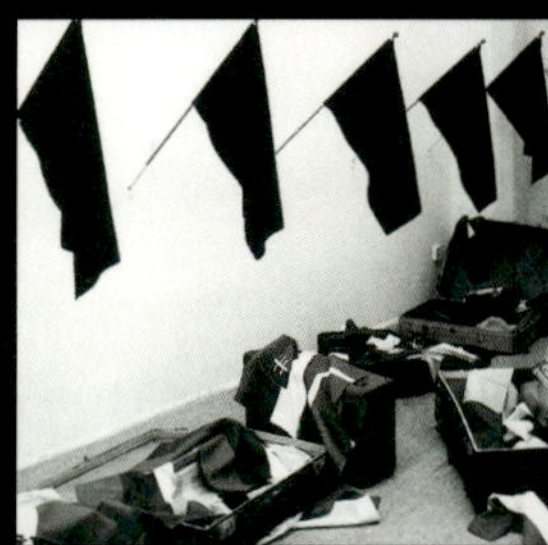

Fortress Europe, 1994 → pp. 473

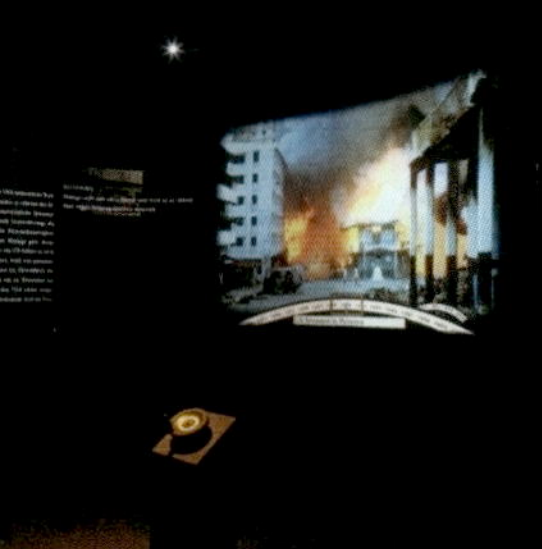

The Coldest Planet in the Universe: The Human Heart ..., 2015 → pp. 486–87

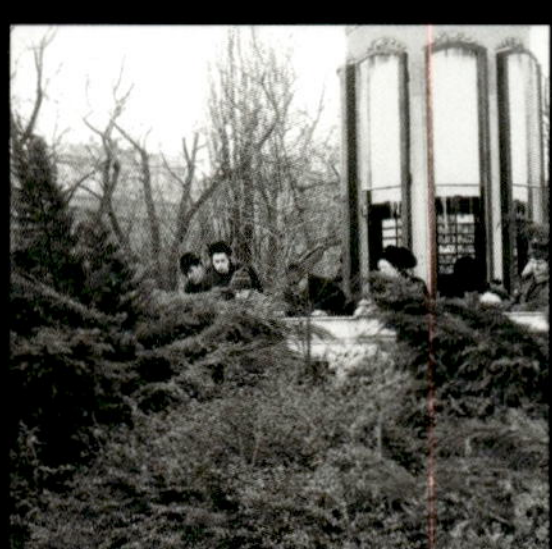

Moaning Stone. Nonhuman Poem, 1969 → pp. 100–101

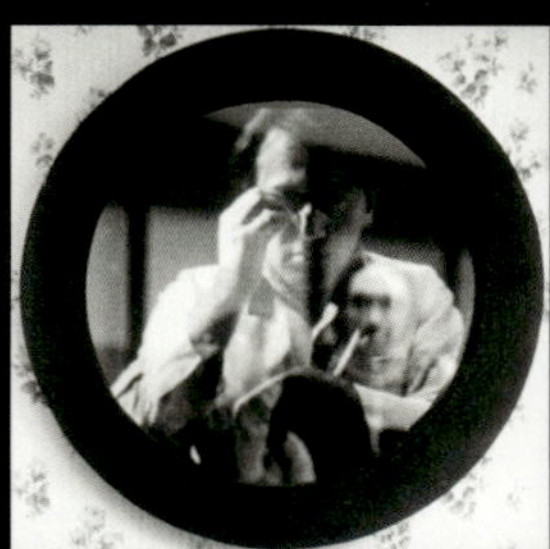

Mirror Check I, 1978 → p. 274

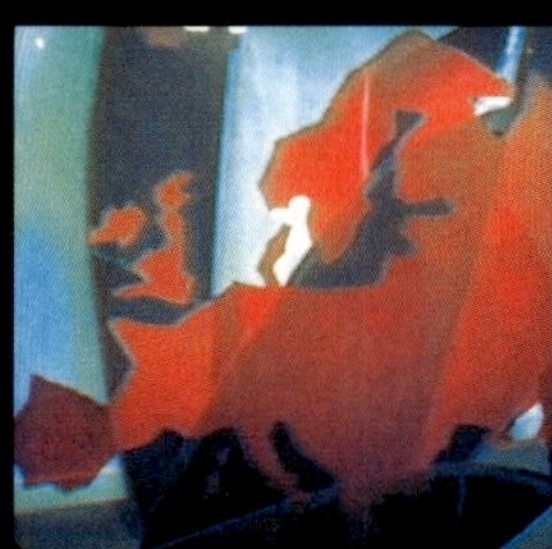

Europa(t)raum, 1983 → p. 472

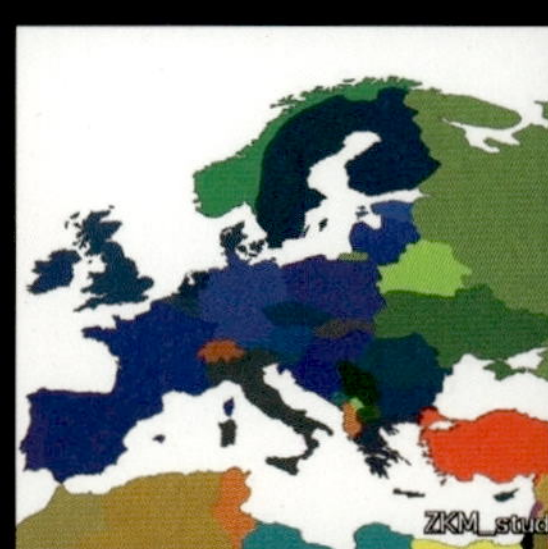

Europe: A Rewriting Program, 2010 → p. 476

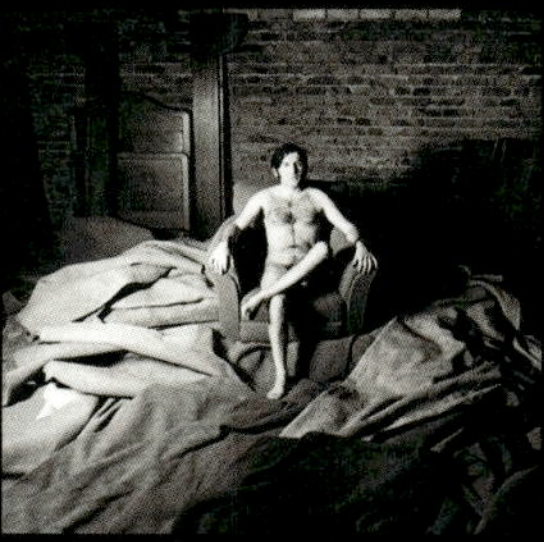
Prince of Darkness, 1970

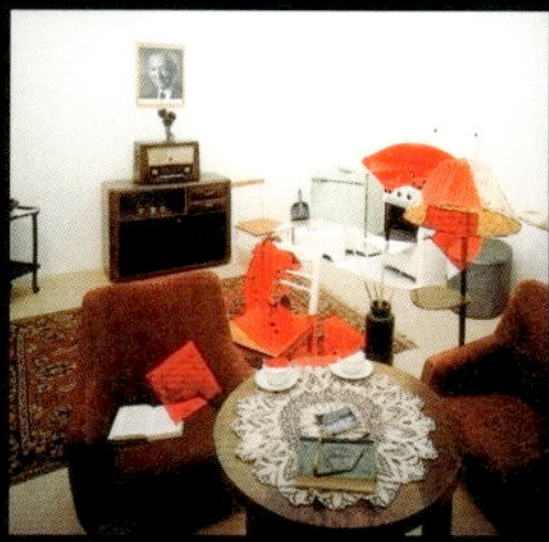
Austria Room (With Gas Heater for That Authentic Austrian Feeling), 1982 → pp. 284–85

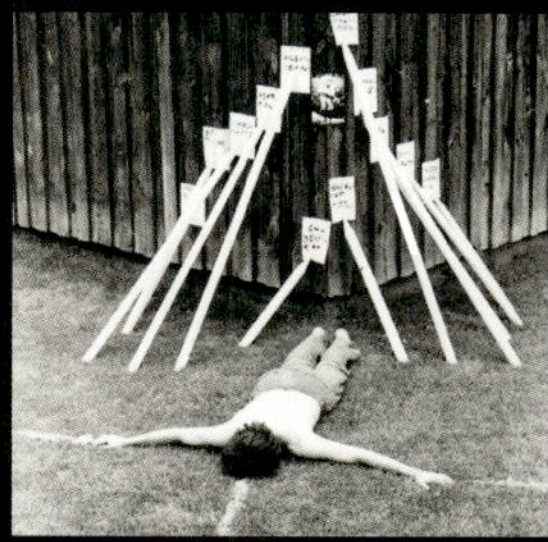
Social Matrix, 1971 → pp. 518–19

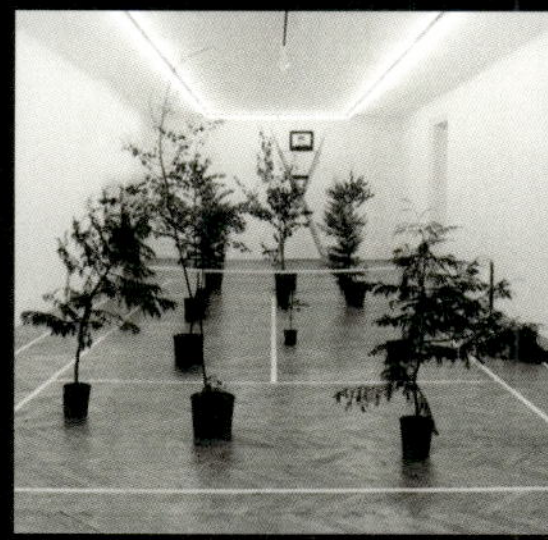
The Plea Society, 1995 → p. 471

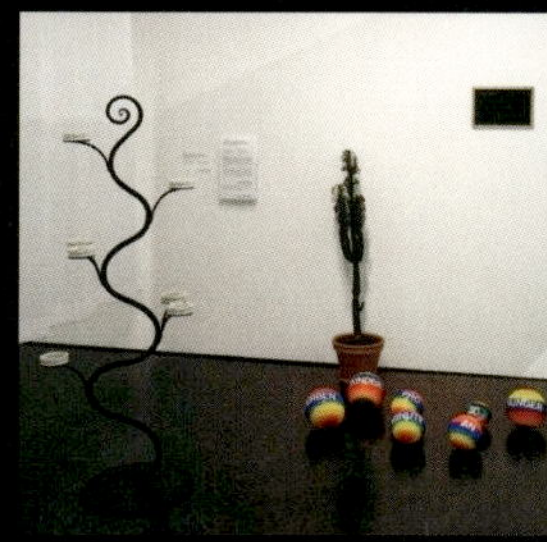
The Living Room as a World Problem / The World as a Problem of the Living Room, 1990 → pp. 468–69

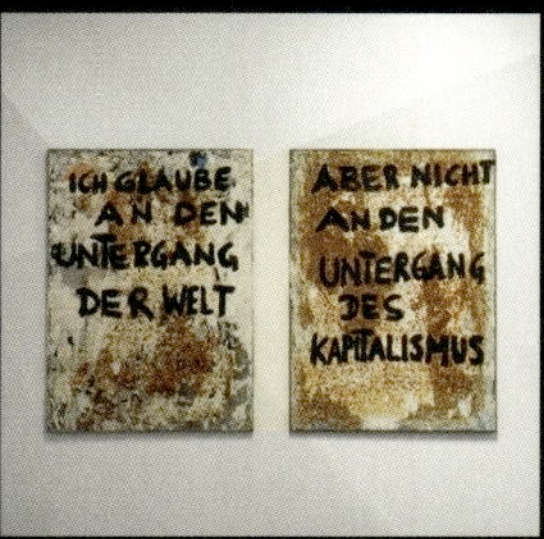

I Believe In the End of the World ... But Not the Downfall of Capitalism, 199 → p. 483

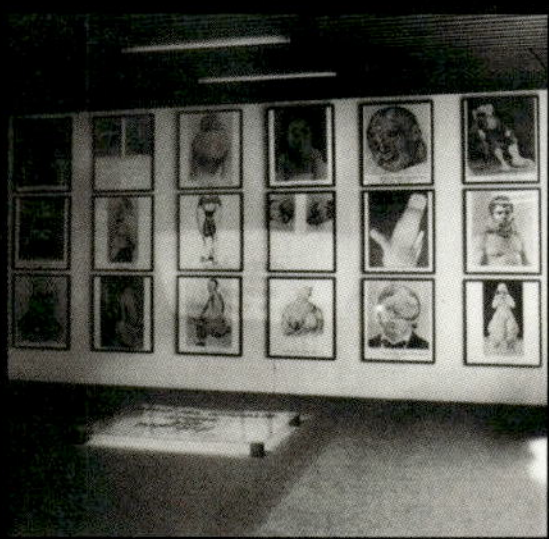
Music of the Anomaly – Voices of Man, 1982 → pp. 460–61

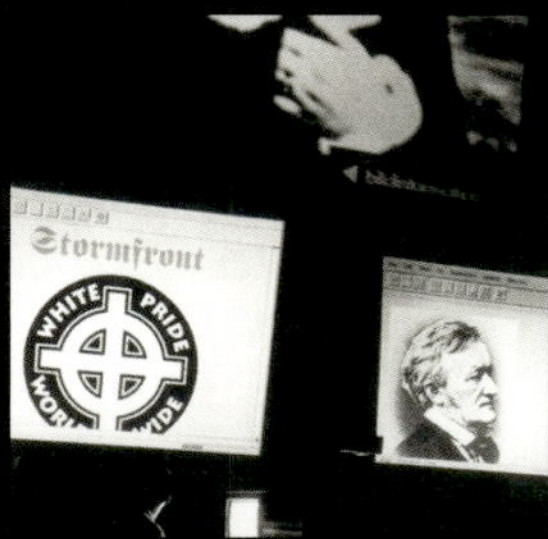

Wagner's Delusion, or the Holy Land of Capital, 1995 → pp. 360–61

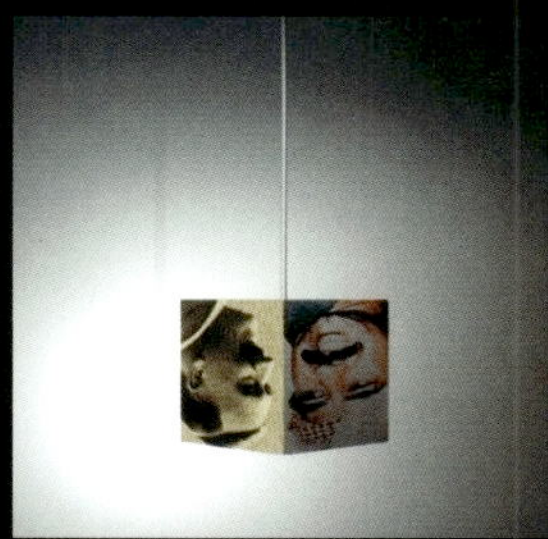
Crime as the Ornament of Politics ..., 1999

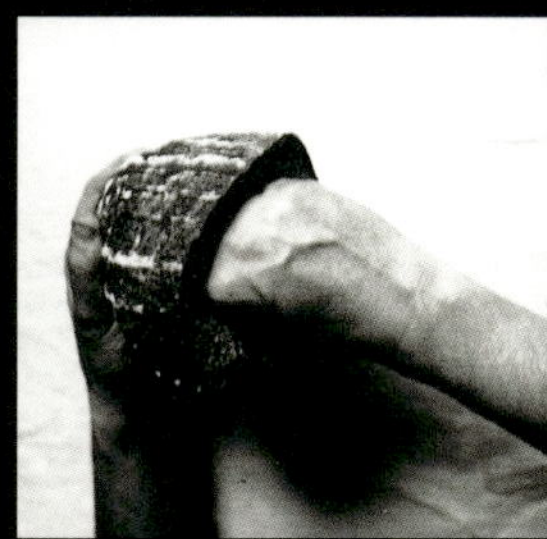
Sentencing, 1967

Religion

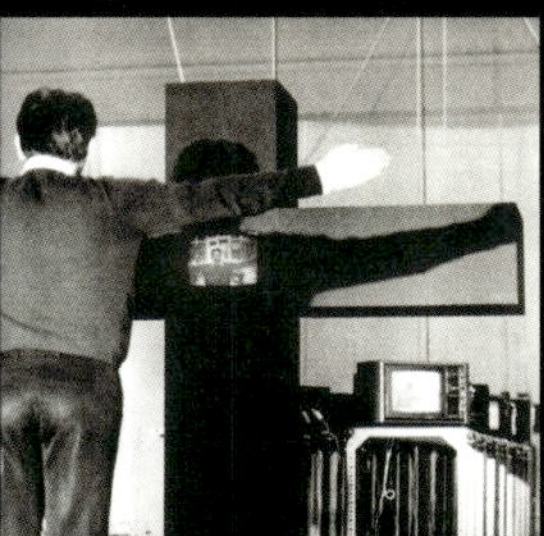
Crucifixion of Identity, 1973 → pp. 264–65

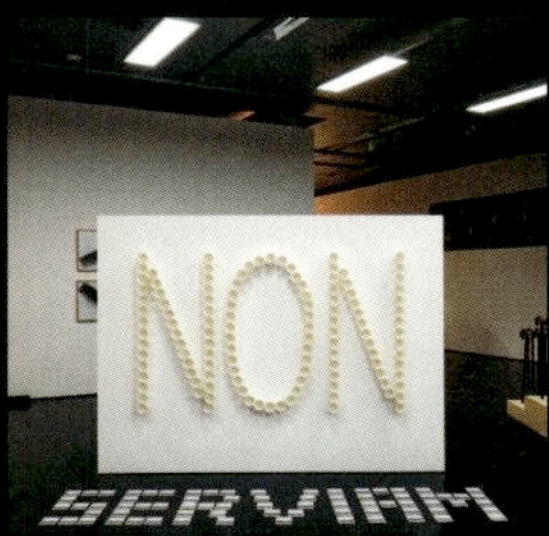

On the Problem Copy-Original: NON SERVIAM ..., 1991/2019 → p. 161

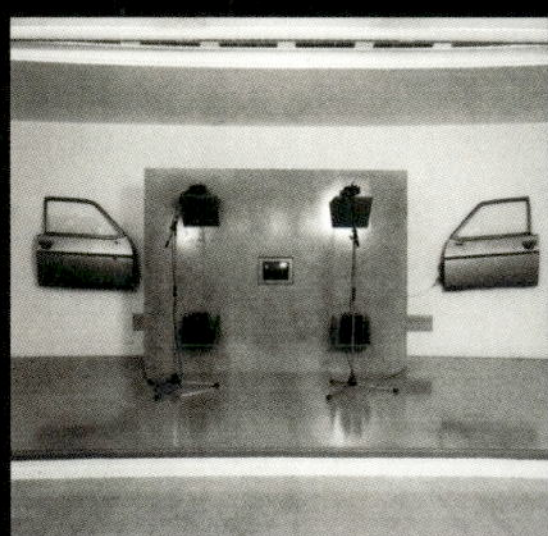
Lucifer's Dream: Non Serviam; The Revolt of the Automata; To the Self by Itself, 1990 → p. 160

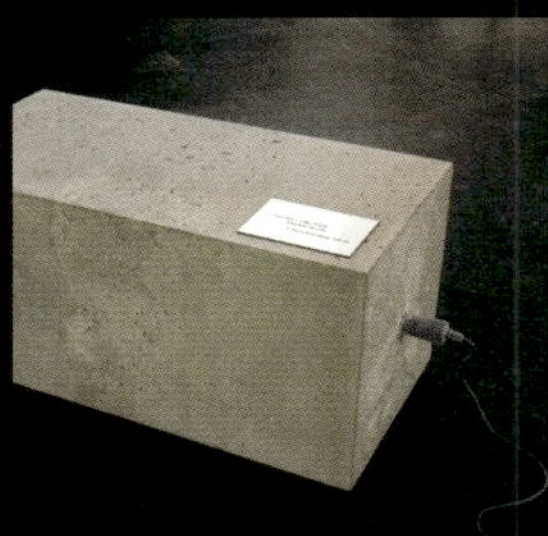
For the Everlasting Light Shall Shine upon You, 1974/2019 → p. 82

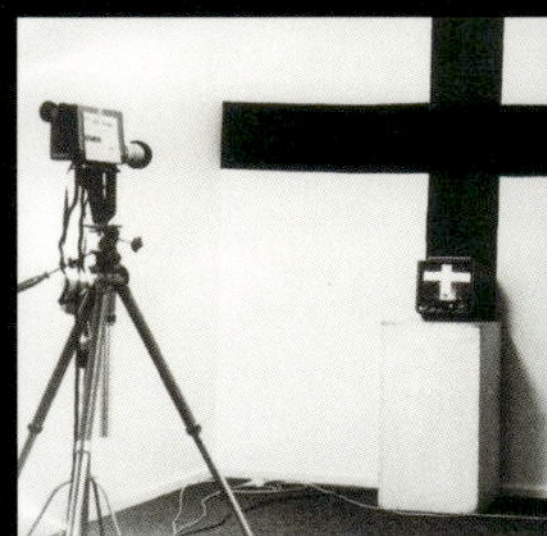
Transfinite Transference: The Cross, 1977 → p. 270

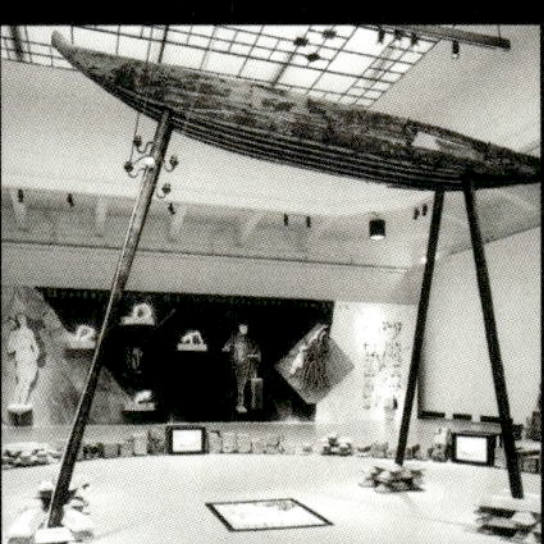
Raft, 1988 → p. 484

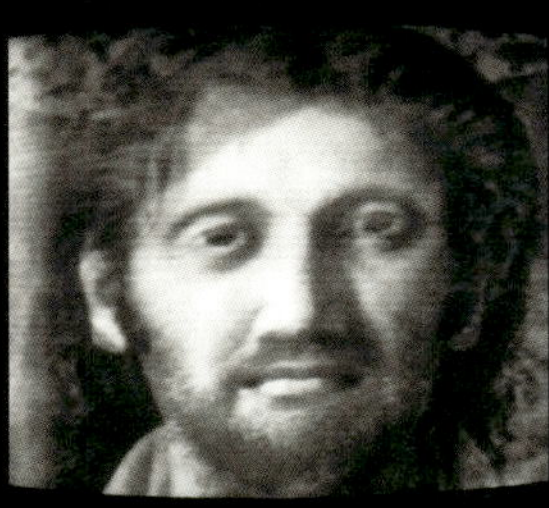
On the Theorem of Identity: Tritity, 1975 → pp. 236–37

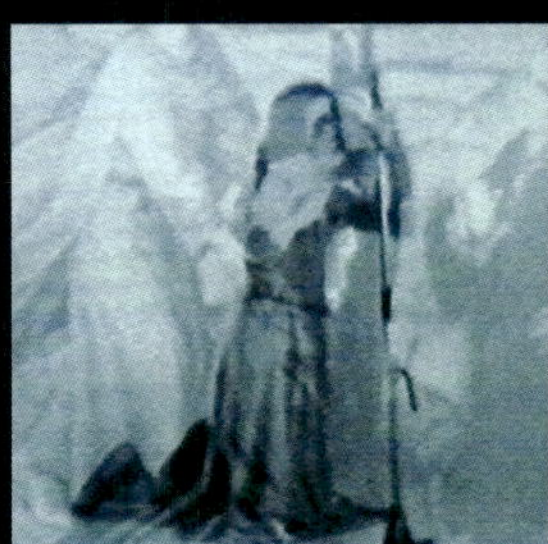
Scene from *Invisible Adversaries*, 1970 (idea)/1977 → pp. 134–35 and 202–203

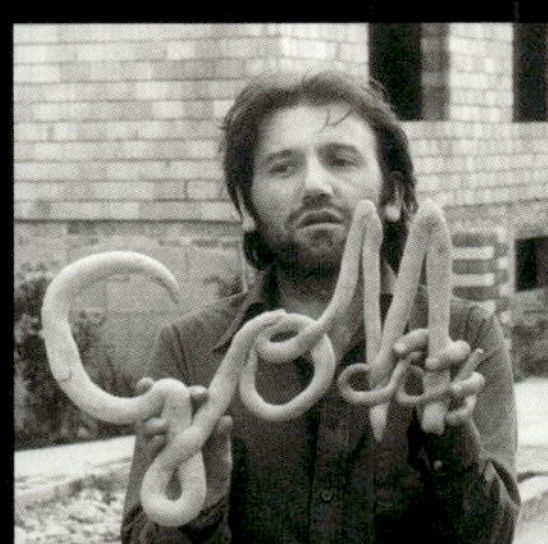
Greet God, 1967/72 → p. 64

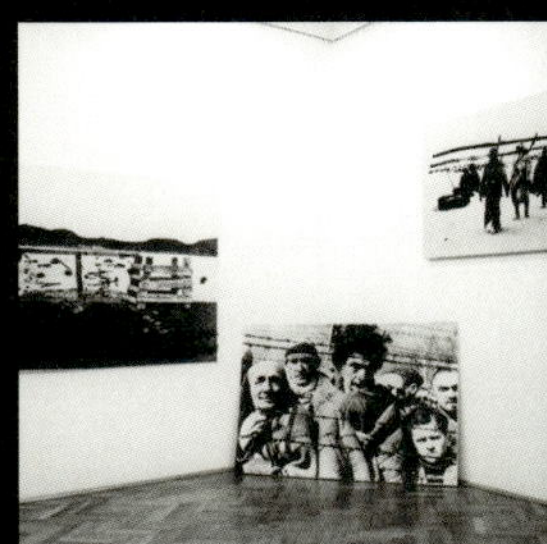
God's Memory, or the Arm of the Media, 1982

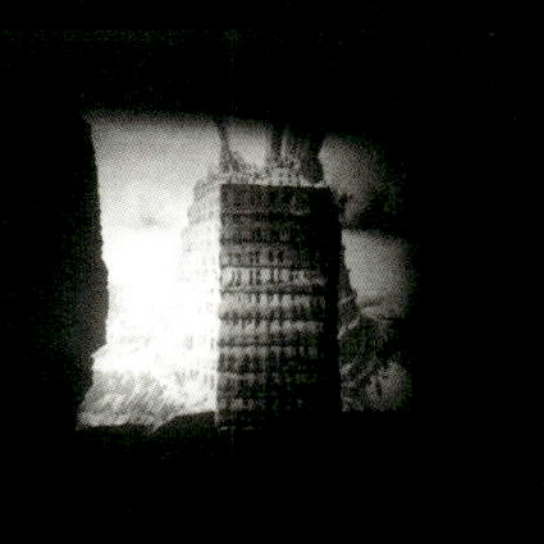
Everyone Is Everyone ..., 1993

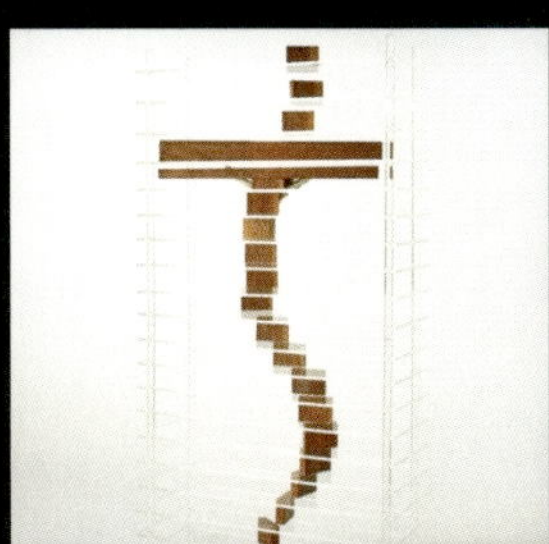
Scanned Cross, 1991 → p. 411

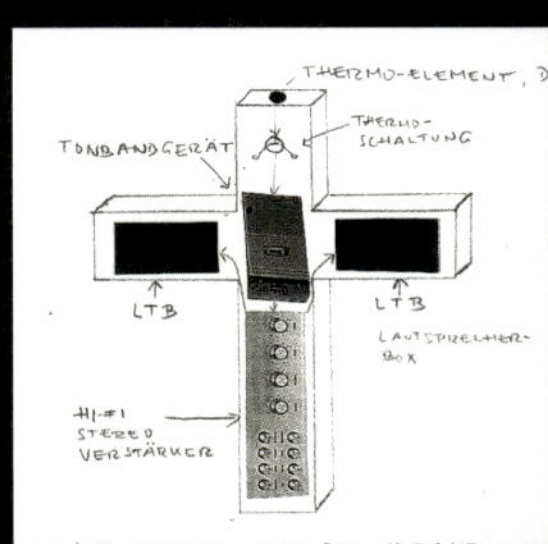

Project for a *War Art Campaign*, 1969

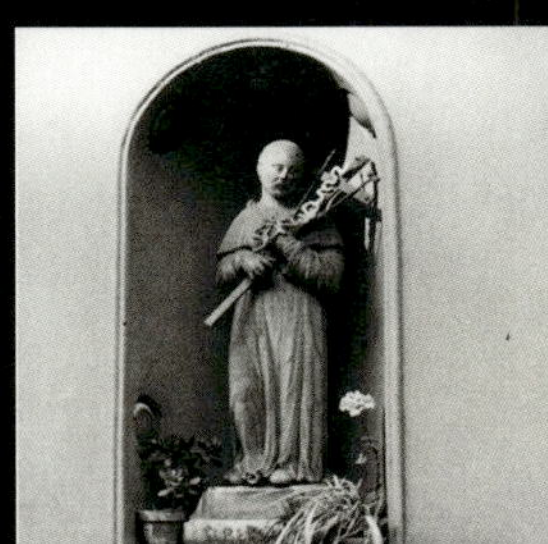
Sacred Phallus, 1972

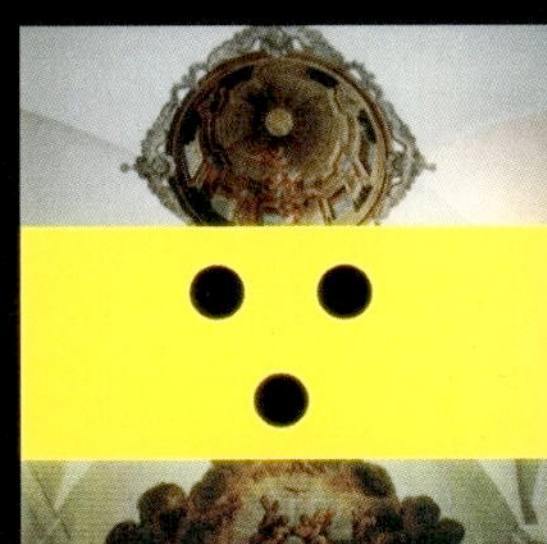
Heaven is Blind, 1999

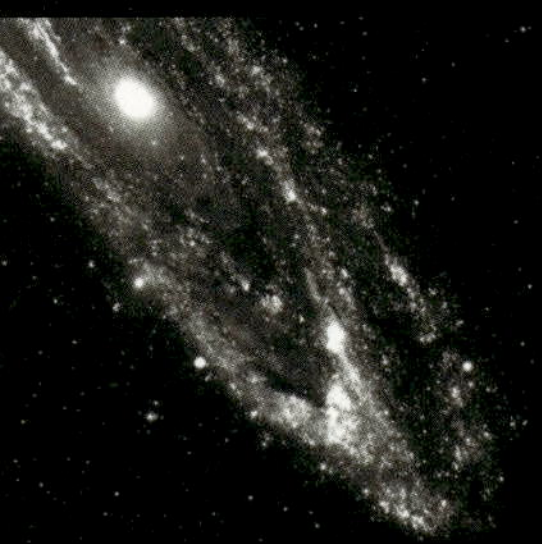

Navigation Systems, 1972
→ pp. 138–39

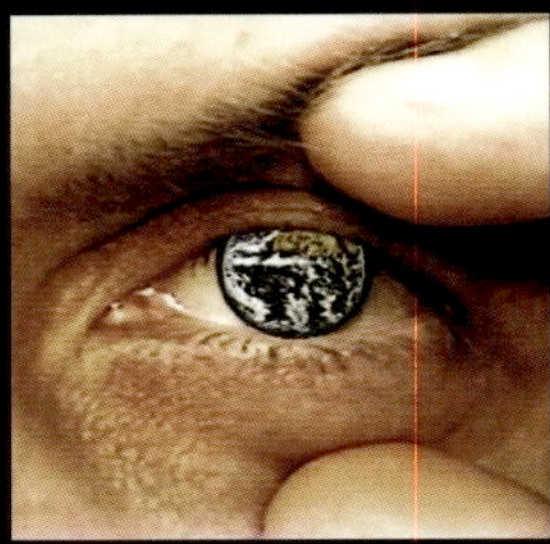

Scene from *The Artificial Will. Electronic Media Opera*, 1984
→ pp. 355–57

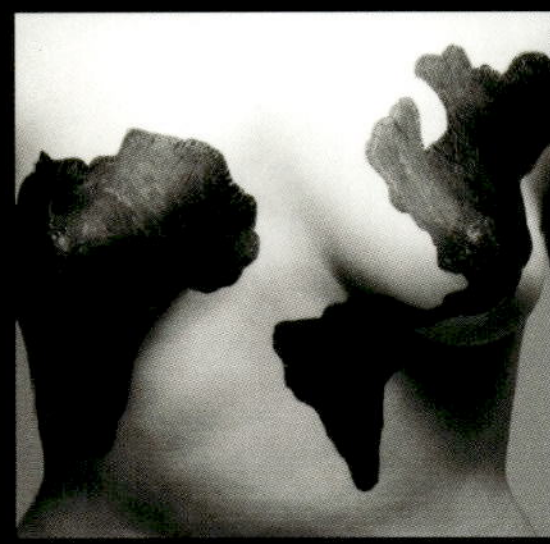

The Expanded Continent, 1971

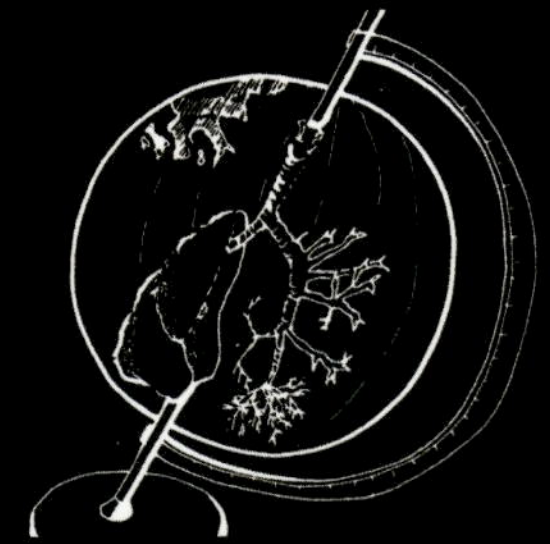

The Earth as Brainwork, 1985

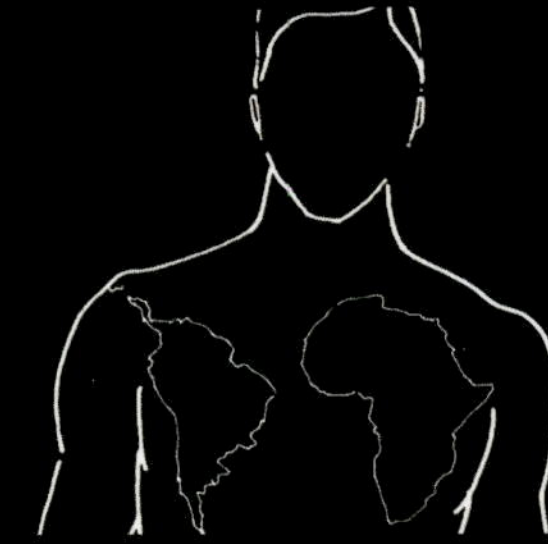

→ pp. 436–37

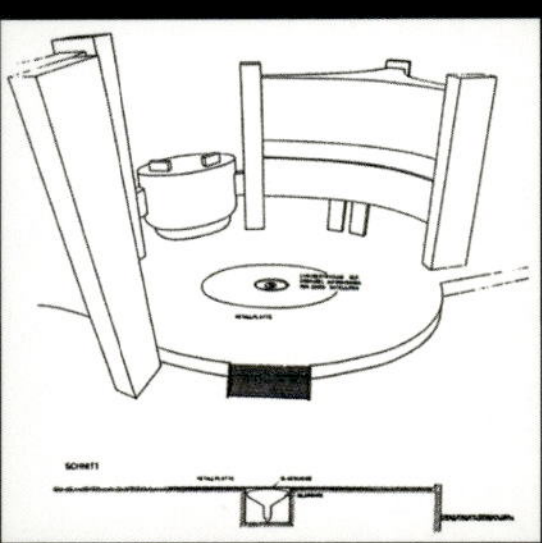

Orbital Sculpture I, 1976 → p. 430

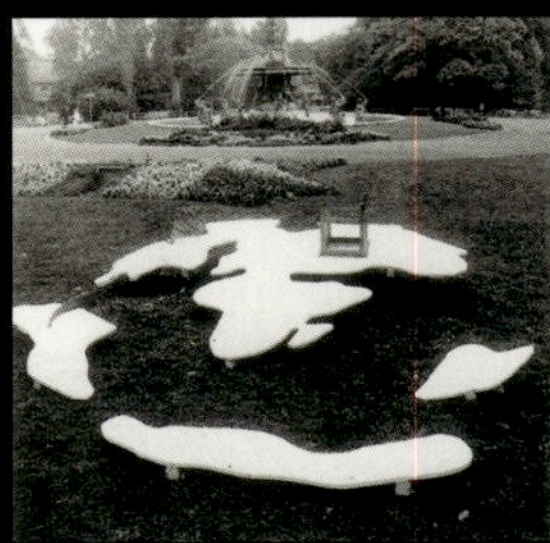

Orbital Sculpture II ..., 1985 → p. 431

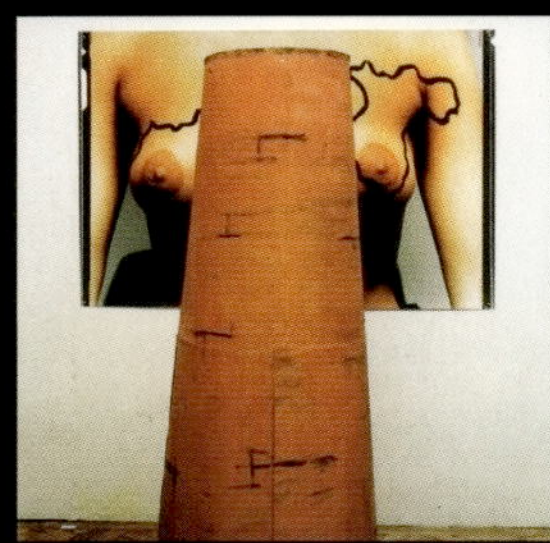

Orbital Sculpture III, 1985 → p. 432

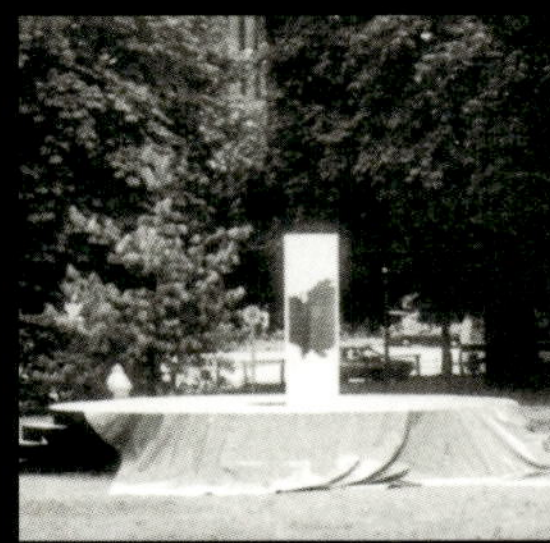

Orbital Sculpture IV: Spectral Music ..., 1976 → p. 433

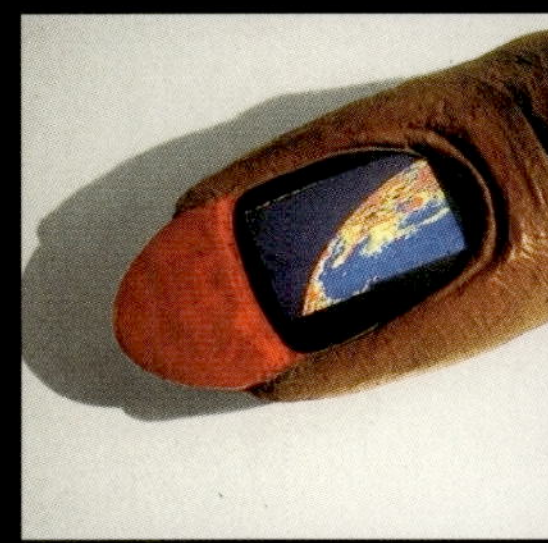

Orbital Sculpture V, 1988 → p. 434

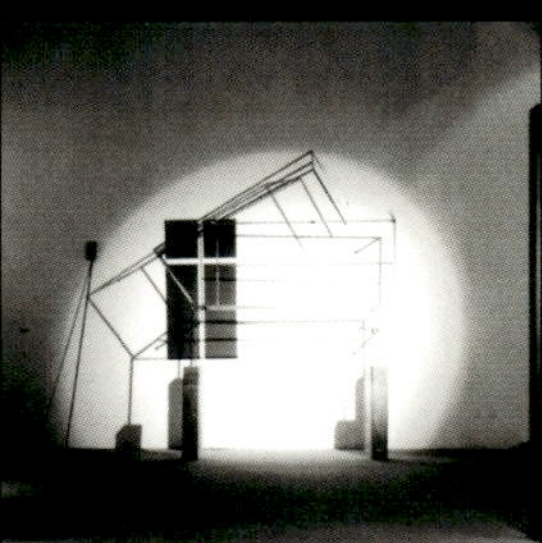

Oikos: Table as House, 1988
→ p. 426

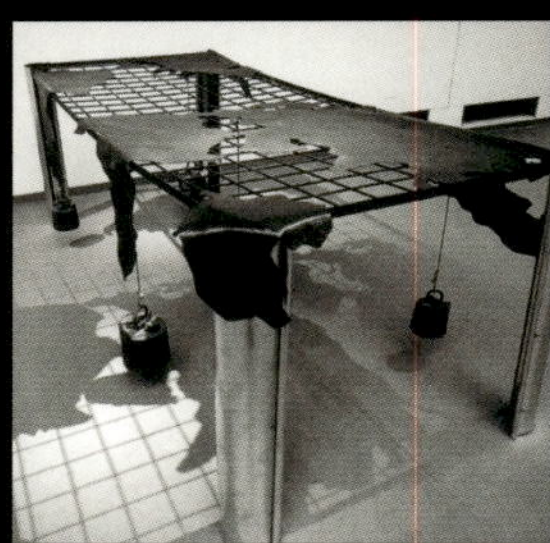

Roof of the World II, 1988 → p. 426

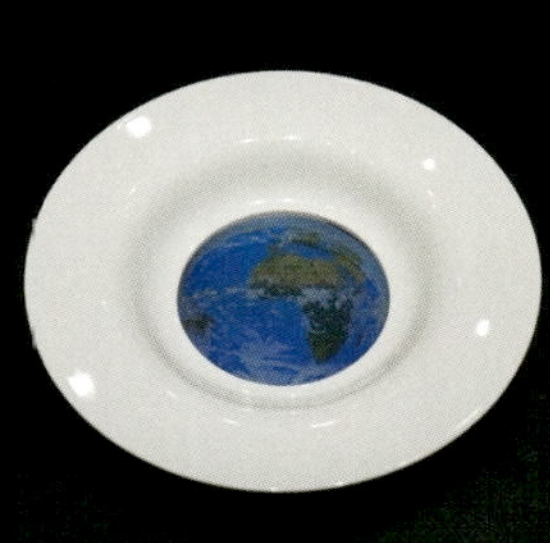

Roof of the World V, 1990/2003
→ pp. 428–29

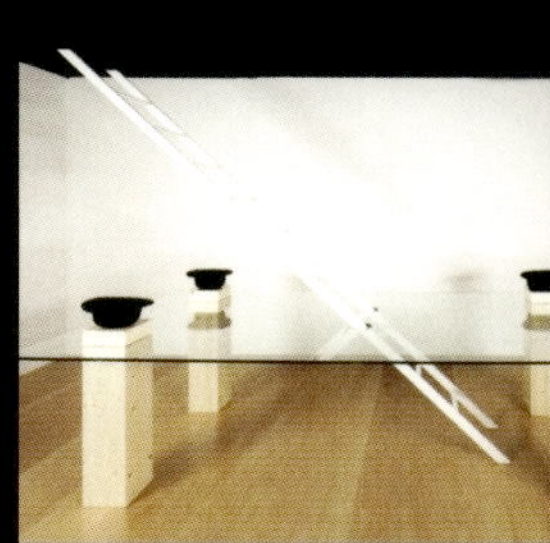

Roof of the World IV: Buy Nature, Buy Yourselves, 1990 → p. 427

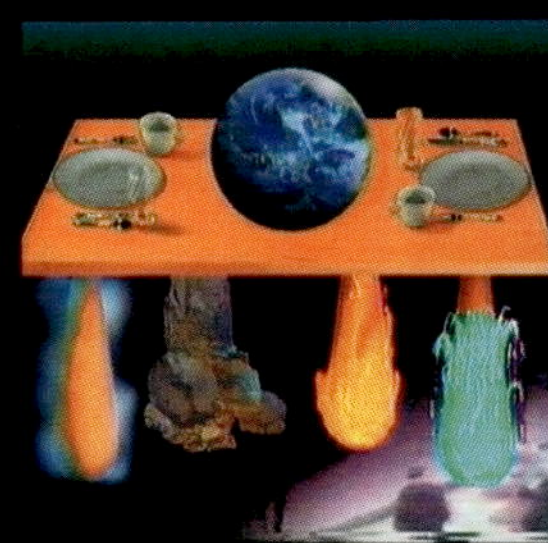

Scene from *Chants of the Pluriverse*, 1986–88 → pp. 304–307

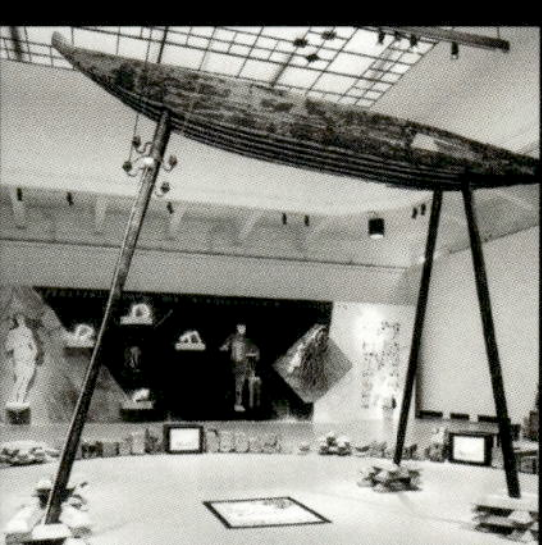

Raft, 1988 → p. 484

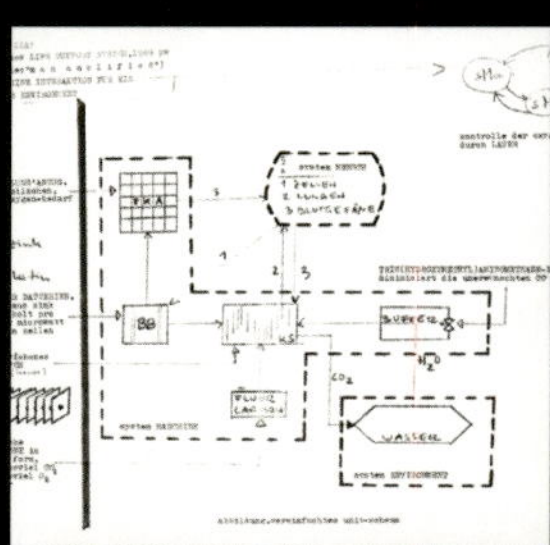

Back to the Sea!, 1969 → p. 42

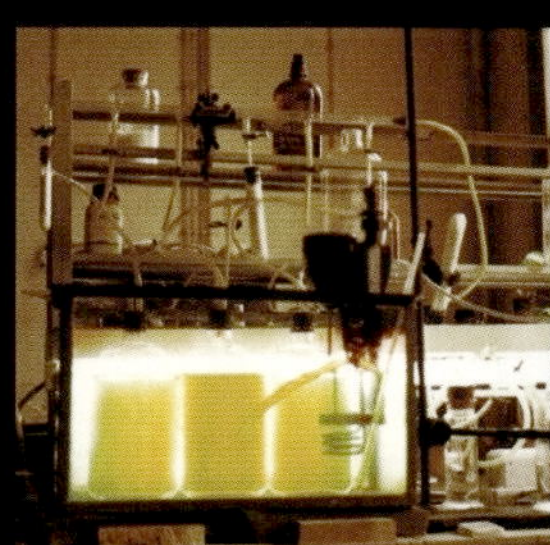

A Bioenergetic Paradox, 1975
→ pp. 68–69

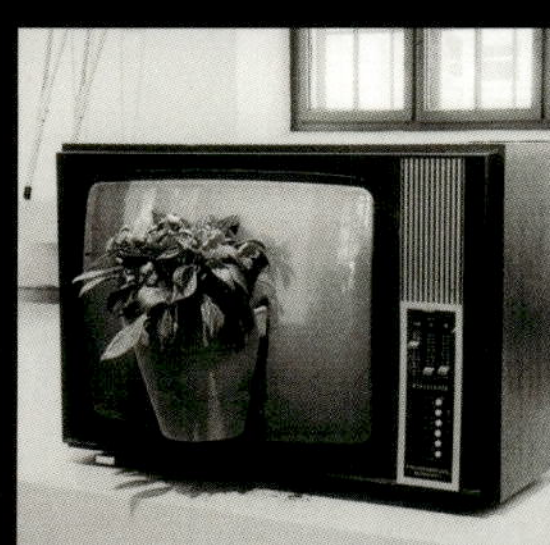

When Fuji Was Still a Mountain and Not a Company ..., 1990 → p. 291

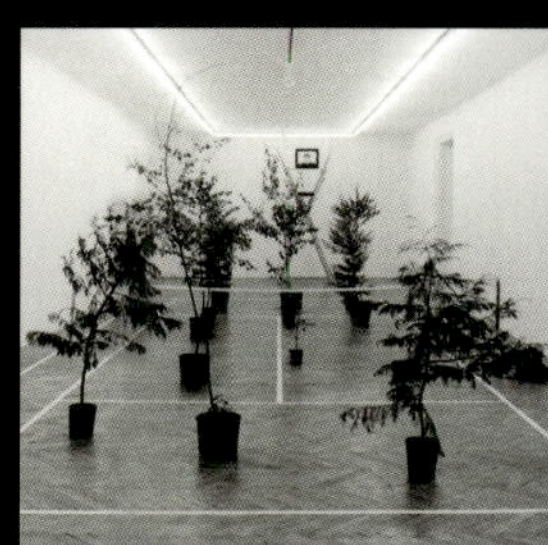

The Plea Society, 1995 → p. 471

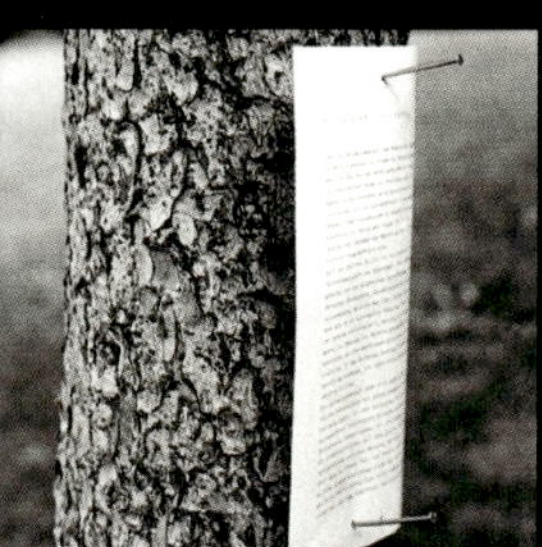

The Secret of Nature, 1967
→ pp. 154–55

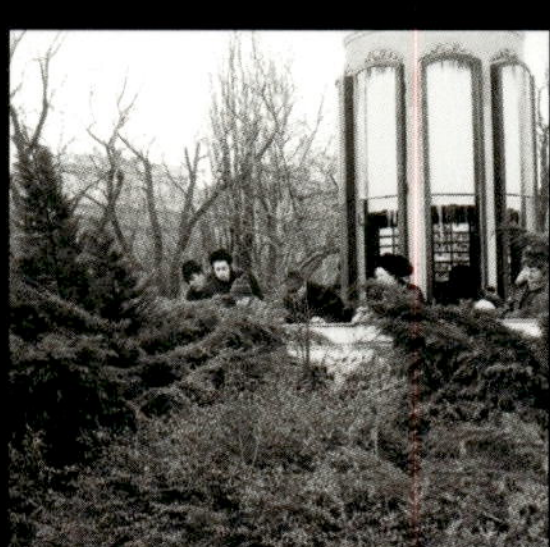

Moaning Stone. Nonhuman Poem, 1969 → pp. 100–101

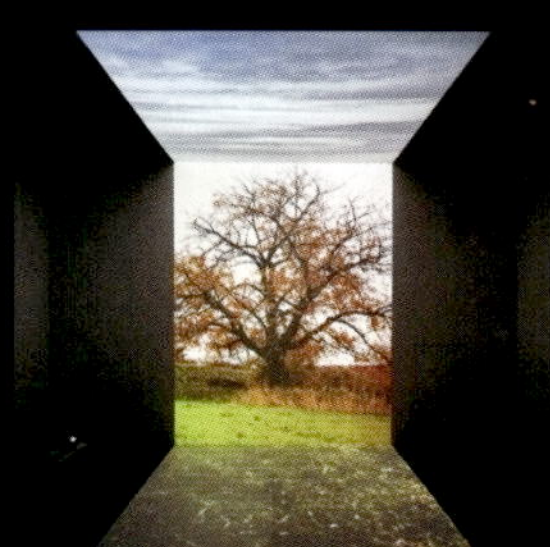

Way Way Way Out, 1968/2007
→ pp. 198–99

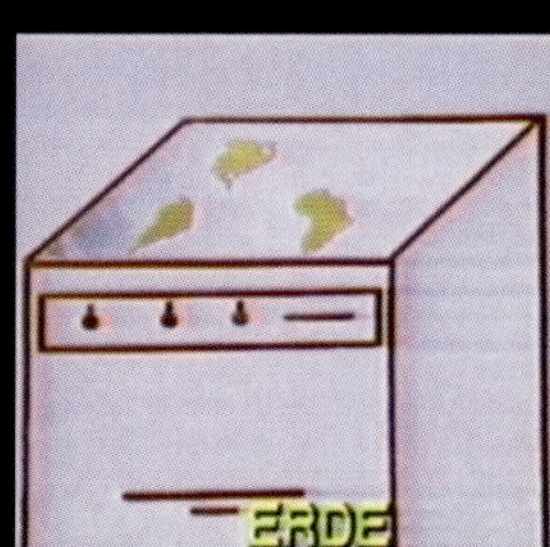

Scene from *The Artificial Will. Electronic Media Opera*, 1984
→ pp. 355–57

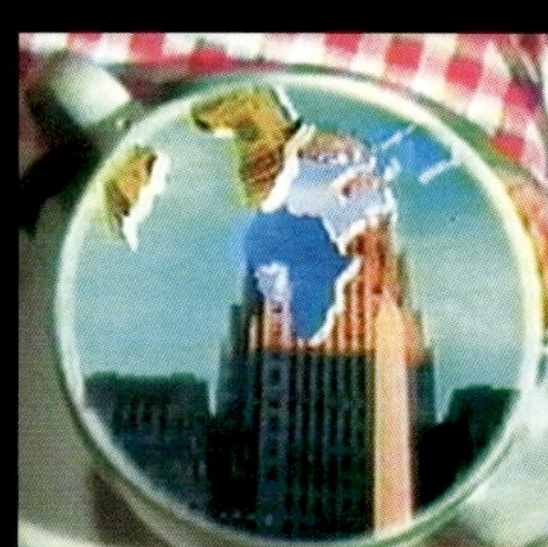

Scene from *Chants of the Pluriverse*, 1986–88 → pp. 304–307

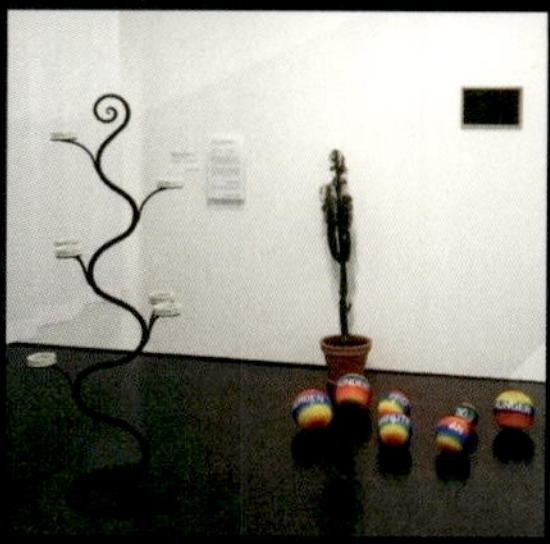
The Living Room as a World Problem / The World as a Problem of the Living Room, 1990 → pp. 468–69

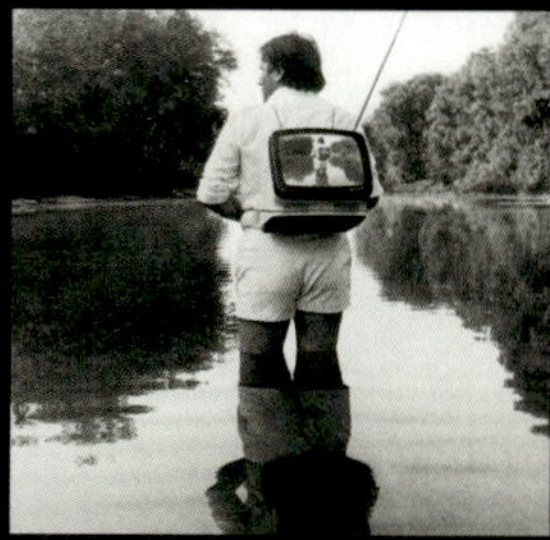
Infinite Contemplation, 1982 → p. 290

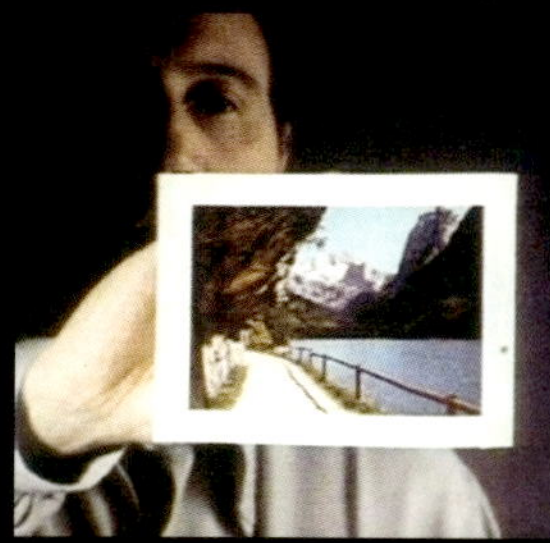
The Landscape of the 21st Century, 1984 → p. 129

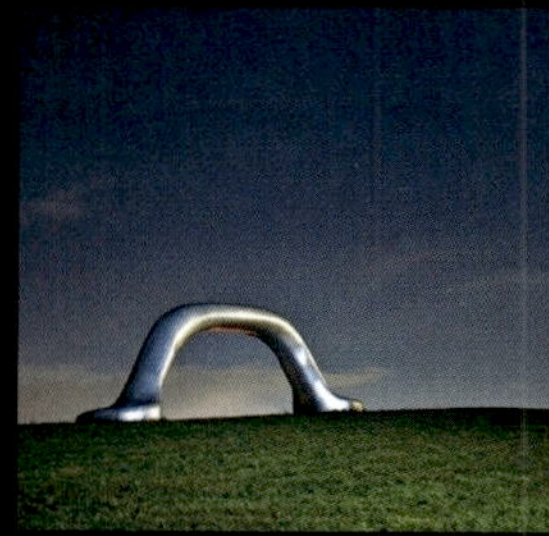
The Globe as Suitcase, 2004 → p. 435

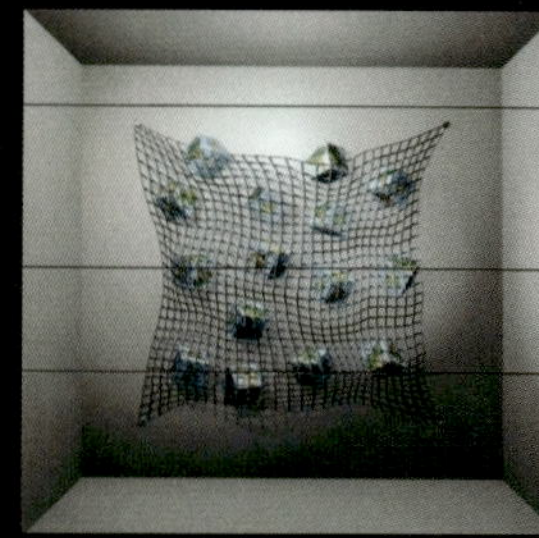
Global Greed ..., 1999

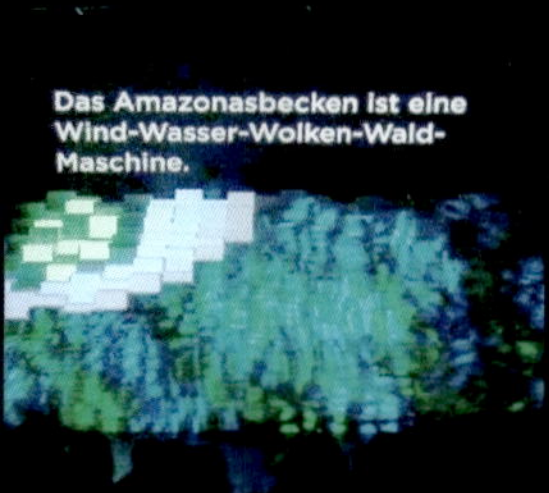

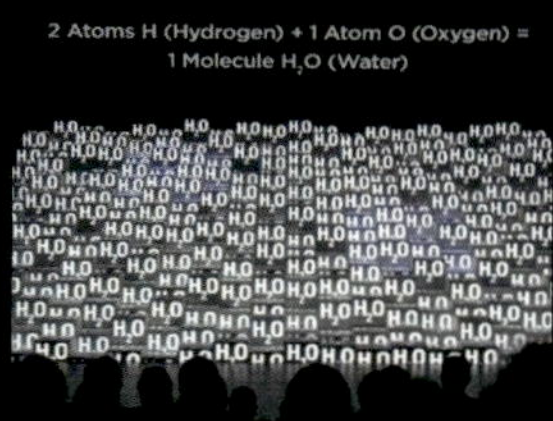

Amazon Conference ..., 2010 → pp. 362–63

Animals

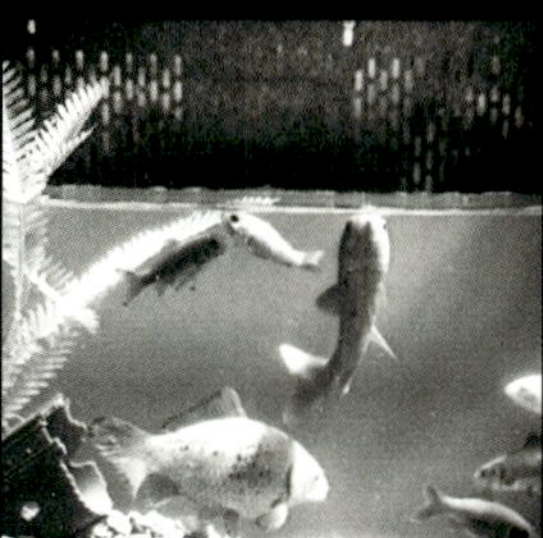
TV Aquarium (TV Death 1), 1970 (idea)/1972 → p. 250

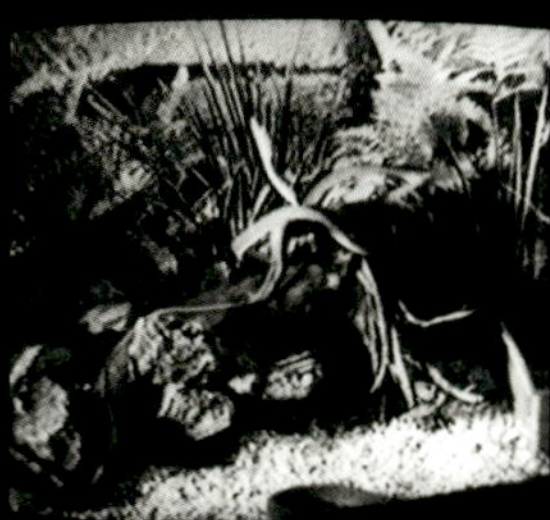
TV Terrarium (TV Death 3), 1970 (idea)/1974

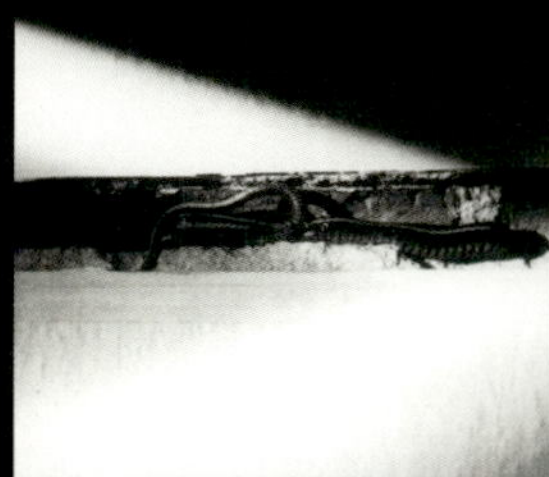
The Electric War, 1983 → pp. 458–59

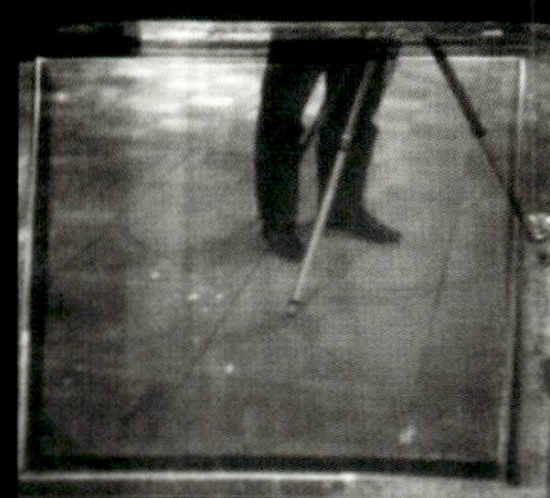
Video Mouse, 1982

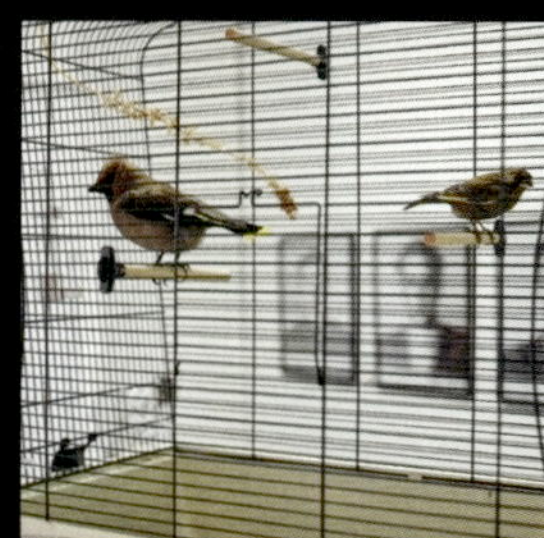
Music Exhibition, 1975 (concept) / 2014 → pp. 466–67

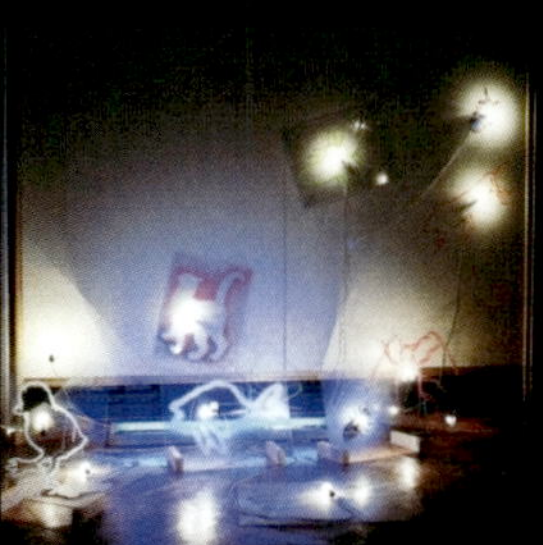
Electric Animals, 1982 → p. 456

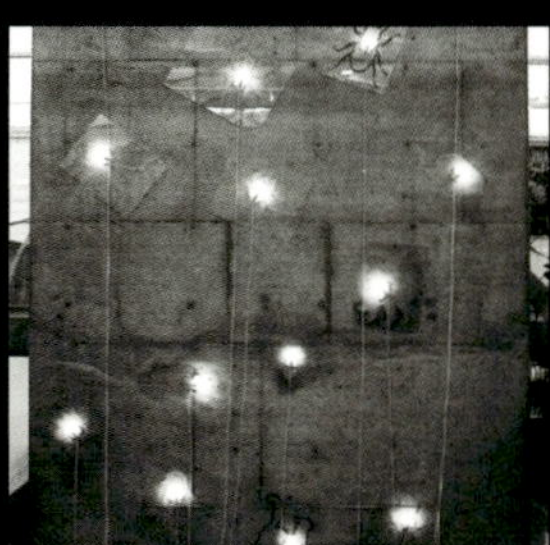
Wall Lights, 1982

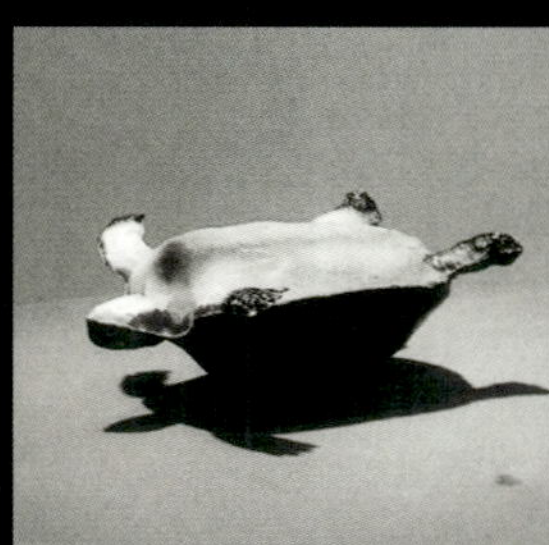
Mechanics of Organisms – Organics of Machines, 1994 → p. 443

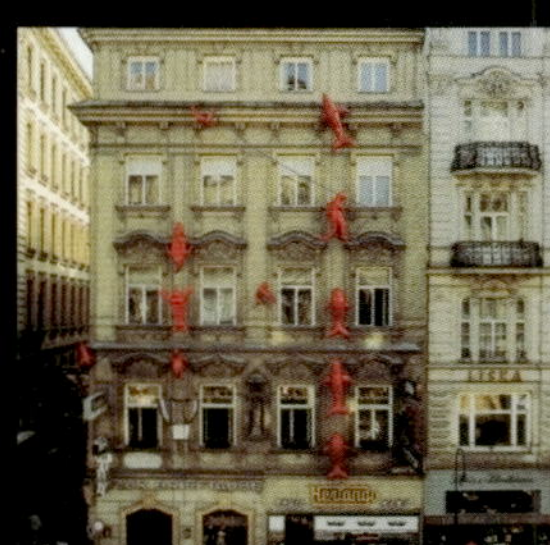
The Wrath of the Animals, 1988 → pp. 456–57

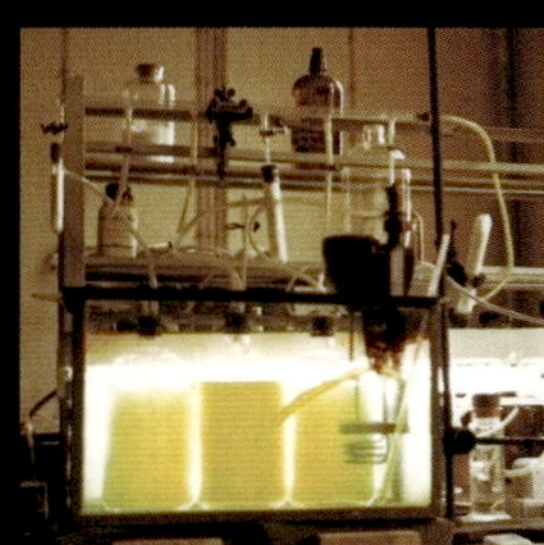
A Bioenergetic Paradox, 1975 → pp. 68–69

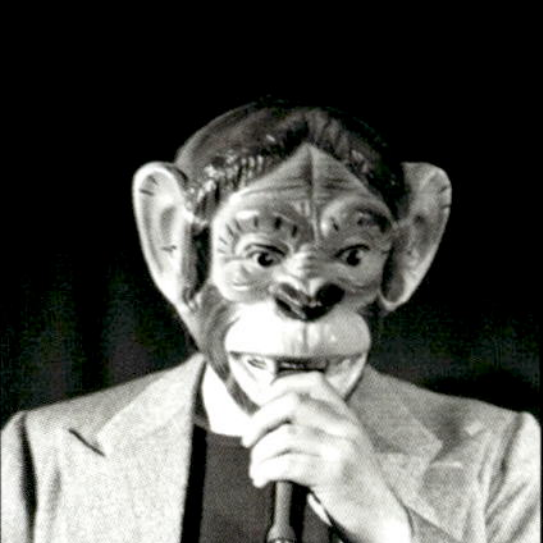
Performance by Hotel Morphila Orchester, 1979 → pp. 352–53

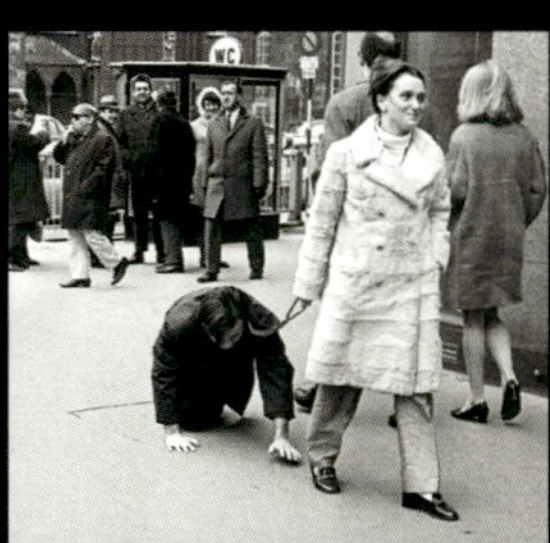
From the Portfolio of Dogedness, 1968 → pp. 504–505

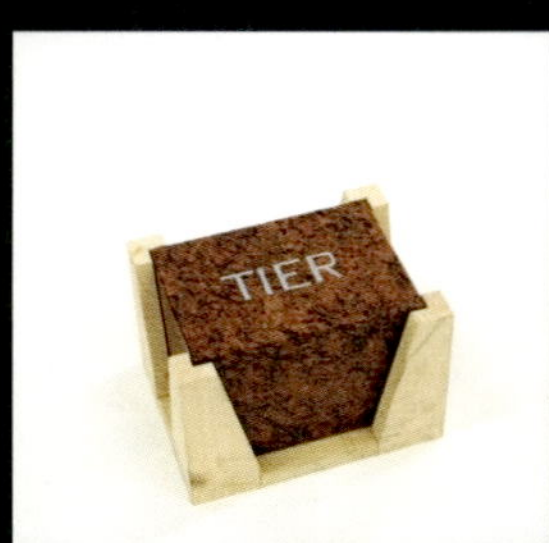

Cage, 1967/75 → p. 78

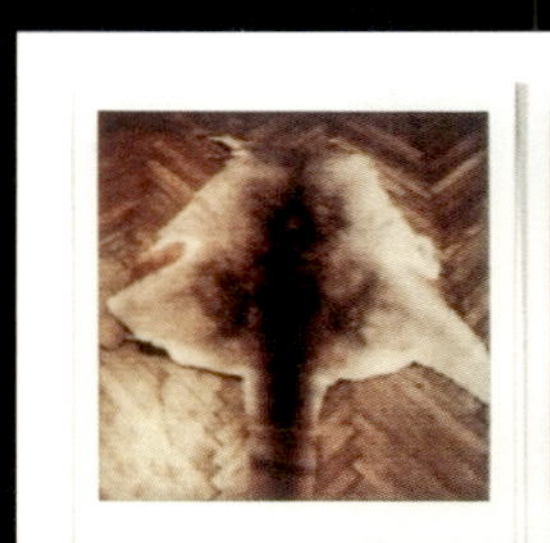
Honest Skin, 1979 → p. 127

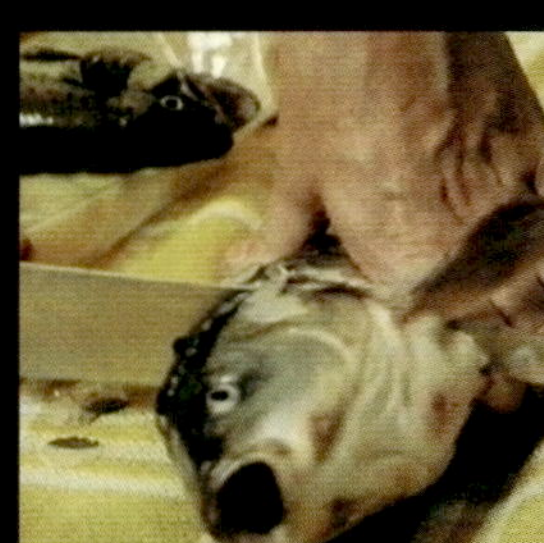
Scene from *Invisible Adversaries*, 1970 (idea)/1977 → pp. 134–35 and 202–203

1967

Der Mythos des 21. Jahrhunderts, Internationales Kulturzentrum, Vienna; Creamcheese, Düsseldorf (1968)

1975

Zur Kosmologie des Paradoxen, Galerie nächst St. Stephan, Vienna

1977

Video Lumina & Video Corner, Galerie Magers, Bonn

Video, Galerie Stampa, Basel

Die unendliche Ein-Wort-Ausstellung & Skulpturen, Galerie nächst St. Stephan, Vienna

Präkino-Parakino-Postkino 1968–76, Galerie nächst St. Stephan, Vienna

1978

Skulpturale Identität – Vers und Vernunft, de Appel, Amsterdam

1979

Loys Egg. Peter Weibel. Bibliotheca Insomnia, Galerie Heike Curtze, Vienna

Videoräume, Videowochen Essen '79, Folkwang Museum, Essen

1980

Imaginärer Raum, Galerie Stampa, Basel

Tribute to Tarski, Mala Galerija, Warsaw

Die Macht des Ausdrucks ist der Ausdruck meiner Ohnmacht, Galeria DESA, Krakow; Monte Video, Wuppertal

Wider die institutionelle Identität, Gesamthochschule Wuppertal

1982

Peter Weibel: Medienpoesie, Kulturhaus Graz, Graz

Medien und Realität, Kunsthalle Wilhelmshaven

1983

Der elektrische Krieg, Moltkerei Werkstatt, Cologne

1986

Video, Performance and Installation, Museum of Contemporary Art, Madrid

1988

Inszenierte Kunstgeschichte, MAK – Museum of Applied Arts, Vienna

1989

Computergraphiken, Galerie Grita Insam, Vienna

1990

Panische Objekte, Galerie Walcheturm, Zurich

1991

Scanned Objects, Galerie Grita Insam, Vienna; Gallery Insam-Gleicher, Chicago; Medien Galerie, Innsbruck (1992)

1992

Malerei zwischen Anarchie und Forschung, Neue Galerie Graz

Virtuelle Welten, Galerie Tanja Grunert, Cologne; Kunsthalle, Dorottya Galeria, Budapest (1993)

1993

Die Vertreibung der Vernunft. The Cultural Exodus from Austria, La Biennale di Venezia '93; Wiener Staatsoper, Vienna (1995); 565 Fifth Ave., New York (1995); Hochschule für Bildende Kunst–Städelschule, Frankfurt am Main (1995); Simon Wiesenthal Center, Los Angeles (1996); Forum Stadtpark, Graz (1996)

Peter Weibel. Jede(r) ist jede(r). Jede(r) ist der/die andere für jede(n) in der Heimat Babylon, Karmeliterkloster, Ausstellungshalle Frankfurt am Main

From Globe to Glove, USF Contemporary Art Museum, Tampa

Perzepte und Konzepte, Galerie Martin Detterer, Frankfurt am Main

1994

Hypothetische Produkte. Vom Elend der Privilegien, Prodomo Wien, Vienna

Pevnost Europa / Festung Europa, Forum Stadtpark, Prague

Peter Weibel / Lawrence Weiner, Galerie Tanja Grunert, Cologne

1995

Copyleft, Galleria Anna D'Ascanio, Rome

Die Bitte-Gesellschaft, Galerie Grita Insam, Vienna

1996

Die Symbol-Zeit ist abgelaufen, Galerie & Edition Artelier, Graz

Copyright, Galerie & Edition Artelier, Frankfurt am Main

Democrazia tra Telecrazia e Burocrazia, Galleria Giulia, Rome

1999

Videoretrospektive, Kunstverein Ludwigsburg

überproduktion:nur paranoiker:überleben, Galerie Hubert Winter, Vienna

Globale Gier, Kärntner Landesgalerie, Klagenfurt

2000

Wegwerfbotschaften. Zur freien Entnahme, Galerie Mathias Kampl, Munich

2003

B-Picture. Ein Film über Peter Weibel, Künstlerhaus Palais Thurn und Taxis, Bregenz

2004

Sozialmatrix. Werke 1965–1979, Galerie Meyer Riegger, Karlsruhe

Peter Weibel. Das Offene Werk 1964–1979, Neue Galerie Graz; Műcsarnok – Kunsthalle Budapest; Moderna Galerija, Ljubljana (2005); Sammlung Falckenberg / Phoenix Kulturstiftung, Hamburg (2006)

Rechtssubjekt statt Mensch, Artelier Contemporary, Graz

Käthe Kollwitz Prize, Akademie der Künste, Berlin

2009

Rewriter, Slought, University of Philadelphia

Zur modernen Finanzarchitektur. Bewohnbare Bibliotheken, Galerie Grita Insam, Vienna

2011

No Limits, Poésies visuelles, Médiathèque André Malraux, Strasbourg

No Limits, Réécrire l'Europe, Palais du Rhin, Strasbourg

No Limits, De la réalité virtuelle à la réalité augmentée, apollonia, Strasbourg

No Limits, Galerie Frank Pages, Baden-Baden

Peter Weibel – Mediapoet, Galerie Henze & Ketterer, Wichtrach/Bern

Die Kunst ist der Imker, nicht die Biene, Galerie Lisi Hämmerle, Bregenz

Konzeptuelle Fotografie 1965–1975, Galerie Mathias Kampl, Munich

Das Leben im 20. Jahrhundert, Artelier Contemporary, Graz

TONSPUR 42, Museums-Quartier Wien, Vienna

2012

Mobile Poeme, Kunstraum Tosterglope, Tosterglope

John Armleder versus Peter Weibel, Galerie Frank Pages, Geneva

2013

Politische Performance, Galerie Anita Beckers, Frankfurt am Main

2014

Peter Weibel – MEDIA REBEL. Warning! This exposition can change your life, 21er Haus, Belvedere, Vienna

2015

Scanned World, Artelier Contemporary, Graz

peter_weibel: technē_revolution, Moscow Museum of Modern Art (MMoMA), Moscow

2018

Loys Egg / Peter Weibel – Bibliotheca Insomnia, Galerie Kunst & Handel, Graz

2019

respectively, Peter Weibel (respektive Peter Weibel), ZKM | Center for Art and Media Karlsruhe

2020

Peter Weibel – (Post-) Europa? Lovis-Corinth-Preis 2020, Kunstforum Ostdeutsche Galerie Regensburg

2021

Ein Pluriversum – Künstlerporträt Peter Weibel, Media Art Corner Eisenach

2023

respectively, Peter Weibel. Art as an Act of Cognition, National Museum of Modern and Contemporary Art (MMCA), Seoul

1966

Destruction in Art Symposium (DIAS), Africa Centre, London

Action concert for Al Hansen, Galerie nächst St. Stephan, Vienna

1967

1. Filmschau der Wiener Filmgruppe, Palais Pálffy, Vienna

Zock-Fest, Galerie nächst St. Stephan, Vienna; Gasthaus Grünes Tor, Vienna

1968

Studententage, Melk Abbey, Melk

XSCREEN projection space, Cologne

U.I.E.A. (Internationale Union der Architekturstudenten), Vienna

Art and Revolution (*Kunst und Revolution*), University of Vienna

2. Maraisiade Junger Film 68, Vienna

1. Europäisches Treffen der unabhängigen Filmemacher, Munich

Capitol, Gothenburg

1969

Film Action Text, Buchhandlung Herrmann, Vienna

Underground Explosion, Zirkus Krone, Munich; Cologne; Stuttgart; Essen; Zurich

multi media I, Galerie Junge Generation, Vienna

1970

First International Underground Film Festival, National Film Theater, Arts Lab, London

Filmabend Peter Weibel, Viennale, Vienna

1971

Experimenta 4, Frankfurt am Main

Festival Konkrete Poesie, Hanau

1972

Musik / Film / Dia / Licht-Festival, Munich

Cinématheque Royale de Belgique, Palais de Beaux Arts, Brussels

London Avantgarde Film Festival 1972, London

Underground Film Festival, Forum für aktuelle Kunst, Innsbruck

1973

The Austrian Exhibition, Institute of Contemporary Arts (ICA), London; Richard Demarco Gallery, Edinburgh

Filme Video-Tapes, Städtische Galerie im Lenbachhaus, Munich

Audiovisuelle Botschaften, Trigon '73, Künstlerhaus Graz

1974

Flash Art Exhibition, Kölnischer Kunstverein; Salon international d'art contemporain, Paris

Projekt '74 – Aspekte internationaler Kunst am Anfang der 70er Jahre, Kölnischer Kunstverein, Cologne

Impact Art Video Art 74, Galerie Impact, Lausanne

Kulturforum, Bonn

Cinemonde, Munich

Experimentelle Epistemologie, Galerie Grünangergasse 12, Vienna

Independent International Experimental Film Festival (EXPRMNTL 5), Knokke-le-Zout

1975

Video Art, Institute of Contemporary Art, University of Philadelphia; The Contemporary Arts Center, Cincinnati, OH; Museum of Contemporary Art (MCA), Chicago, IL; Wadsworth Atheneum, Hartfort, CT

The Video Show, Serpentine Gallery, London

Kunst aus Sprache, Museum of the 20th Century, Vienna; Galerie im Taxispalais, Innsbruck (1976); Kulturhaus Graz (1976); Neue Galerie, Linz (1977)

Identität / Alternative-Identität / Gegenidentität, Trigon '75, Künstlerhaus Graz

Selbstportrait als Selbstdarstellung, Galerie im Taxispalais, Innsbruck; Galerie nächst St. Stephan, Vienna

Videotapes, Rencontre international ouvert de Video, Paris; Palais des Beaux Arts, Brussels

Herbstsalon, Galerie nächst St. Stephan, Vienna

1976

Schriftbild, Galerie nächst St. Stephan, Vienna

Narrative Art – Story Art, Galerie nächst St. Stephan, Vienna

Herbert Distel: The Museum of Drawers, Museum der Stadt Solothurn; Internationaal Cultureel Centrum (ICC), Antwerp; Museum Schwäbisch-Gmünd (1977); Städtische Kunsthalle, Düsseldorf (1977); Cooper-Hewitt Museum, New York (1978); Museum of Fine Arts, Dallas, TX (1978); New Orleans Museum of Art (NOMA) (1978); Wadsworth Atheneum, Hartford, CT (1978); Kunstmuseum Bern (1979); Kunsthaus Zürich, Zurich (1979); De Vleeshal, Middelburg (1981)

Video: Towards Defining an Aesthetic, Third Eye Centre, Glasgow

Buchobjekt – Objektbücher, Galerie nächst St. Stephan, Vienna

1977

Money Show, George Paton Gallery, Melbourne University

Projektionskunst, K45 – Internationale Kunstmesse, Künstlerhaus, Vienna

documenta 6, Kassel

Invisible Adversaries (*Unsichtbare Gegner*), The 27th Berlin International Film Festival (Berlinale)

Videotape, Galerie gm, Budapest

Film als Film. 1910 bis heute, Kölnischer Kunstverein, Cologne; Akademie der Künste, Berlin (1978); Folkwang Museum, Essen (1978); Württembergischer Kunstverein, Stuttgart (1978); Hayward Gallery & BFI South Bank, South Bank Centre, London (1979)

Locarno Film Festival (Locarno 30)

London Film Festival, National Film Theater, London

Photography as Art – Art as Photography, Fotoforum, Gesamthochschule Kassel; Centre for Contemporary Art, Warsaw; Institute of Contemporary Arts (ICA), London

Concept in Performance, Internationaler Kunstmarkt Köln 1977, Cologne

Das Sofortbild, Polaroid, Aktionsgalerie, Bern; Galerie Stadler, Paris; Canon Galerie, Geneva; Galerie AK, Frankfurt am Main

Vidéo, Museé d'art et d'histoire (MAH), Geneva

Kunstmesse Basel, Galerie Stampa, Basel

1978

Video Musik Performance, Kulturforum Bonn

Video Art 78, Herbert Art Gallery & Media Centre, Coventry

Photography as Tool, Video as Tool, Galeria Fotografii, Wroclaw

Logische Kunst: ist Kunst logisch?, Secession, Vienna

Wiener Aktionismus 1961–1971, Galerie nächst St. Stephan, Vienna

Internationales Performance Festival, Österreichischer Kunstverein, Vienna

Performance Festival, Belgrade

Screening of *Invisible Adversaries* (*Unsichtbare Gegner*), FILMEX International Film Exposition, Los Angeles

Europäisches Forum Alpbach, Vienna

Symmys (performance with Valie Export), Badischer Kunstverein, Karlsruhe

Paris–Berlin: Rapports et contrastes France-Allemagne, 1900–1933, Musée National d'Art Moderne, Centre Georges Pompidou, Paris

From Nature to Art, From Art to Nature, La Biennale di Venezia '78

Performance Art Festival, Beursschouwburg, Brussels

Loft, Munich

1979

Raumartikulationen, Forum Stadtpark, Graz

Performances'79: Körpersprache. Tanz-Musik. Video-Film. Theater, Städtische Galerie im Lenbachhaus, Munich

Video 1974–1979, Centre d'Art Contemporain, Geneva

Künstlerische Raum-Programme, Baack'scher Kunstraum, Cologne

3rd Biennale of Sydney, Art Gallery of New South Wales, Sydney

3rd International Avant-Garde Festival, British Film Institute (BFI), London

Photographie als Kunst 1879–1979, Kunst als Photographie 1949–1979, Tiroler Landesmuseum Ferdinandeum, Innsbruck; Neue Galerie am Wolfgang Gurlitt Museum, Linz (1980); Neue Galerie Graz (1980); Museum of the 20th Century, Vienna (1980); Salzburger Kunstverein, Salzburg (1980); Bregenzer Kunstverein, Bregenz (1980)

Sprachen jenseits von Dichtung, Westfälischer Kunstverein, Münster

Im Namen des Volkes: Das „gesunde Volksempfinden" als Kunstmaßstab, Wilhelm Lehmbruck Museum, Duisburg; Künstlerhaus Wien, Vienna; Forum Stadtpark, Graz (1981)

Audio Scene '79: Sound, Medium der bildenden Kunst, Modern Art Galerie (Grita Insam), Vienna

Ton in Dokumentation, Skulptur und Installation, Modern Art Galerie (Grita Insam), Vienna; Galerie Krinzinger, Innsbruck; Galerie Dany Keller, Munich; Gallerie Pellegrino, Bologna

Definition eines neuen Kunstbegriffes, Forum für aktuelle Kunst and Galerie Krinzinger, Innsbruck

Mail-Art, Künstlerhaus Stuttgart

Künstlerschaufenster, steirischer herbst '79, Graz

International Film Festival, Edinburgh

The International Festival of Independent Avantgarde Film, National Theatre, London

Bleecker Street Cinema, New York

1980

Human Females (*Menschenfrauen*), The 30th Berlin International Film Festival (Berlinale)

No Wave, Wiener Festwochen, Museum of the 20th Century, Vienna

Video made in Austria, Museum of the 20th Century, Vienna

Artists' Use of Telecommunication, Museum of the 20th Century, Vienna

Video, Photo, Performance, Goethe-Institute, Paris

Das Sofortbild – Entdeckung eines Mediums, Rheinisches Landesmuseum, Bonn; Frankfurter Kunstverein, Frankfurt am Main

Retrospektive Peter Weibel, Österreichische Filmtage, Kapfenberg

Österreichischer Avantgarde- und Undergroundfilm 1950–1980, Z-Club Alternativ, Österreichisches Filmarchiv, Wiener Festwochen, Vienna; Kunsthalle Basel

Künstlerfotografien in Österreich 1965 – 1980, Art 11, Basel Galerie Krinzinger and Galerie nächst St. Stephan, Vienna

Videos aus Österreich 1969–80, Galerie Stampa, Basel

Kunst im öffentlichen Raum, Galerie Grita Insam, Vienna

Rajz/Drawing '80, Pécsi Galéria, Pécs

1981

Kino der 70er Jahre, Kellerkino, Bern

Der Parkring: Ein Vitaskop. Ringpark-Projekt (with Werner DePauli-Schimanovich), Wiener Festwochen alternativ, Vienna

Gegeneinander – Miteinander, Galerie Grita Insam, Vienna

Video aus Österreich, Kunsthaus Zürich, Zurich

3. Symposium international d'art performance, Lyon

Vonal/Line, Pécsi Galéria, Pécs

Erweiterte Fotografie, 5. Wiener Internationale Biennale, Secession, Vienna

1982

Momentbild. Künstlerphotographie, Kestner Gesellschaft, Hanover

Metropoles Festival für Video und Filmexperiment, Alabamahalle, Munich

Videokunst Österreich, Tabakmuseum, Vienna

Einblicke, MAK – Museum of Applied Arts, Vienna

1983

3. Internationaler Experimentalfilm Workshop, Osnabrück

aktuell 83, Städtische Galerie im Lenbachhaus, Munich

Film und Video International, Galerie AK, Frankfurt am Main

Symbol Tier, Galerie Krinzinger, Innsbruck

Art Video, Palais des Beaux Arts, Charleroi

Geschichte der Fotografie in Österreich, Museum of the 20th Century, Vienna

1984

Arte Austriaca 1960–1984, Galleria Communale d'Arte Moderna, Bologna

Recent Acquisitions, Museum of Modern Art (MoMA), New York

Primer Festival Nacional de Vídeo, Círculo de Bellas Artes de Madrid

1984 – Orwell und die Gegenwart, Museum of the 20th Century, Vienna

39th Festival de Cannes

New Narration, American Film Institute, Los Angeles

Österreichische Filmtage, Wels

1984 National Video Festival, Los Angeles

1. Europäisches Musikvideo Festival, Vienna

Zur Geschichte und Ästhetik der digitalen Kunst, Ars Electronica '84, Brucknerhaus, Linz

1985

Zeit: Die vierte Dimension in der Kunst, mumok – Museum moderner Kunst Stiftung Ludwig Wien, Vienna

Schrägspur Videofestival, Graz

5th International Festival of Video Art, Locarno

Alles und noch viel mehr: Das poetische ABC, Kunsthalle Bern und Kunstmuseum Bern

40th Festival de Cannes

10th Toronto International Film Festival (TIFF)

Montreal World Film Festival 1985

Brainwork: Artificial Intelligence in the Arts, Municipal Art Gallery, Los Angeles; steirischer herbst '85, Graz

Encuentros en torno al vídeo, Círculo de Bellas Artes de Madrid

Die neue Dimension: Skulptur und Raum, Forum Stadtpark, Graz

Raum annehmen: Aspekte österreichischer Skulptur 1950 bis 1985, Galerie Grita Insam, Vienna

1986

Künstlerphilosophen, Kunsthaus Zürich, Zurich

Immagine elettronica, Galleria d'Arte Moderna, Bologna

The 36th Berlin International Film Festival (Berlinale)

1. Biennale de la Photographie d'Art, Paris

New Forms in Art, State University of San Francisco

Berner Filmfestival, Bern

ORF Videonale '86, Ars Electronica '86, Linz

Austrian Video, Long Beach Museum of Art, Los Angeles

Das Bild: Der Text, steirischer herbst '86, Forum Stadtpark, Graz

1987

Digital Visions: Computers and Art, IBM Gallery of Science and Art, New York; Everson Museum of Art, Syracuse, NY; Contemporary Arts Center, Cincinnati; Wright Art Gallery, University of California, Los Angeles

Digitale Bildwelten. Design, Kunst und Computer, Gesamthochschule Kassel

Freie Klänge – Offene Räume, Ars Electronica '87, Donaupark, Linz

National Video Festival, American Film Institute, Los Angeles

Animal Art, steirischer herbst '87, Graz

The Arts for Television, Stedelijk Museum, Amsterdam; The Museum of Contemporary Art (MoCA), Los Angeles; Institute of Contemporary Art (ICA), Boston, MA; Centro de Arte Reina Sofia, Madrid (1988); mumok – Museum moderner Kunst Stiftung Ludwig Wien, Vienna (1988); The Museum of Modern Art (MoMA), New York (1988); Musée National d'Art Moderne, Centre Georges Pompidou, Paris (1988); Tate Gallery, London (1988)

1988

Infermental 7, Hallwalls, Buffalo, NY

The 38th Berlin International Film Festival (Berlinale)

HKIFF12 – Hong Kong International Film Festival, Hongkong

North American New Music Festival, Slee Concert Hall, University at Buffalo, NY

Freizone Dorotheergasse, Galerie Metropol, Vienna

Das gläserne U-Boot, transart 1, Tabakfabrik, Krems an der Donau

The Dog Within, Galerie Hubert Winter, Vienna

Vor aller Augen. Kunst im öffentlichen Raum, Grazer Kunstverein, Graz

European Media Art Festival (EMAF), Osnabrück

Kunst der Szene, Ars Electronica '88, Linz

1989

Wien Wien – Vienna Vienna, 1960–1990, Museum für moderne Kunst Bolzano; Palazzo della Permanente, Milano (1990); Complesso Monumentale di San Michele, Rome (1990)

Fotografie, Wissenschaft und neue Technologien: Facetten der Interaktion, Kunstmuseum Düsseldorf

Wittgenstein. Das Spiel des Unsagbaren, Secession, Vienna; Palais des Beaux-Arts, Brussels

Das gequälte Quadrat, Schloß Buchberg, Gars/Kamp

Österreichischer Kunstsalon '89 – *Die Optik der Objekte*, Kunstsalon im Kulturhaus, Graz

Chaos & Ordnung, steirischer herbst '89, Graz

1990

Rencontres internationales Art cinéma/vidéo/ordinateur, Vidéothèque de Paris

Immagine elettronica, Palazzo del Diamante, Ferrara

Le désenchantement du monde, Villa Arson, Nice

Parterres, Galerie Martina Detterer, Frankfurt am Main

Elisabeth Schneider Preis 1990, Galerie Schneider, Freiburg im Breisgau

Zeichen im Fluss, Museum of the 20th Century, Vienna; Stadtmuseum, Prague; Museum moderner Kunst, Zagreb

Im Licht des Monitors, Kunstverein Horn; Stadtmuseum, Prague (1991); Kunstverein Graz, Künstlerhaus Graz (1991); Kunstraum Dornbirn (1991)

Österreichische Skulptur, Secession, Vienna

Raum annehmen III, Galerie Grita Insam, Vienna

1991

Das Belebte Bild, Art Frankfurt, Frankfurt am Main

Images du Futur '91, Montreal

Where are you? Where am I? Arbeiten mit Polaroid, Galerie Grita Insam, Vienna; Kunstraum Dornbirn, Dornbirn (1992)

Machines à communiquer, Atelier virtuel, La Villette, Paris

Fukui Video Biennale, Fukui

1992

Die Welt von innen – Endo und Nano, Ars Electronica '92, Oberösterreichisches Landesmuseum, Linz

Der entfesselte Blick, Kunstmuseum, Bern

Video der 80er Jahre, Deutsche Kunst- und Ausstellungshalle, Bonn

Identität: Differenz, Trigon '92, Neue Galerie Graz

Miteinander-Zueinander-Gegeneinander, steirischer herbst '92, Kulturhaus, Graz

1993

Die Sprache der Kunst, Frankfurter Kunstverein, Frankfurt am Main; Kunsthalle Wien, Vienna

Konzept 3, Karmeliterkloster, Frankfurt am Main

Genetische Kunst – Künstliches Leben, Ars Electronica '93, Oberösterreichisches Landesmuseum, Linz

Fontanelle – Kunst in (x) Zwischenfällen, Kunstspeicher Potsdam

Borderline, Postmuseum Frankfurt, Frankfurt am Main; Neuer Berliner Kunstverein (1994)

Montage 93: International Festival of the Image, Rochester, NY

First Europeans. Frühe Kulturen – moderne Visionen, Schloss Charlottenburg, Berlin

1994

Translucent Writings, Neuberger Museum of the Arts, Purchase College, State University of New York, NY

Lokalzeit: Wiener Material im Spiegel des Unbehagens, Kunstraum Strohal, Vienna; Moderna Galerija (MG), Ljubljana; Fondazione Querini stampaglia, Venice (1995)

Der Sprung ins Leere, Central'nyj Dom Chudoznika, Moscow

Körpernah, Galerie Krinzinger, Vienna; Galerie AK, Frankfurt am Main (1995)

Intelligente Ambiente, Ars Electronica '94, Linz

Images du Futur '94, Montreal

Exhibition, Museum für moderne Kunst Wien, Vienna

Oh Boy, It's a Girl!, Kunstverein München, Munich; Kunstraum Wien, Vienna

Spielverderber, steirischer herbst '94, Forum Stadtpark, Graz

The Media & Art Exhibition '94, Magic Media Company, Hürth

1995

Fisch & Fleisch: Photographie aus Österreich 1945–1995, Kunsthalle Krems

La nuova Europa, La Biennale di Venezia '95

Mythos Information. Welcome to the Wired World, Ars Electronica '95, Linz

Am Anfang war ..., Kunsthalle Tirol, Hall; Galerie Artelier, Frankfurt am Main

Falsch verbunden, Kunsthaus Hamburg

Self Construction, Museum of the 20th Century, Vienna

1996

Butterfly, Mücsarnok – Kunsthalle Budapest

White Cube / Black Box, Generali Foundation, Vienna

Reaching out, Galerie Insam, Vienna

Artisti austriaci a Roma, Palazzo Braschi, Rome

Interieur – Stillleben/ Lebensstil, Galerie Menotti, Baden/Vienna

Werkstatt Graz, Künstlerhaus Klagenfurt

Wunschmaschine Welterfindung: Eine Geschichte der Technikvision seit dem 18. Jahrhundert, Kunsthalle Wien, Vienna

Die Sechziger Jahre oder als alles möglich wurde. Kunst und Kultur in Österreich 1960–1970, Schloss Herberstein, St. Johann bei Herberstein

Vis À Vis, Galerie Krinzinger, Bregenz

Beyond Art (Jenseits von Kunst), Ludwig Museum – Museum of Contemporary Art, Budapest; Neue Galerie Graz; Musée des Beaux Arts, Antwerp

Elsewhere, John Hansard Gallery, Southampton

1997

L'empreinte, Musée National d'Art Moderne, Centre Georges Pompidou, Paris

Unimplosive Art, La Biennale di Venezia '97

1998

Out of Actions: Between Performance and the Object, 1949–1979, Museum of Contemporary Art (MoCA), Los Angeles

Crossings – Kunst zum Hören und Sehen, Kunsthalle Wien, Vienna

2001

Morphologies, Artspace Visual Arts Centre, Woolloomooloo

Ctrl [Space]. Rhetorics of Surveillance. From Bentham to Big Brother, ZKM | Center for Art and Media Karlsruhe

2002

Future Cinema. The Cinematic Imaginary after Film, ZKM | Center for Art and Media Karlsruhe

2004

Virtual Frame, Kunsthalle Wien, Vienna

2005

Open Systems: Rethinking Art c1970, Tate Modern, London

Light Art from Artificial Light. Light as a Medium in 20th and 21th Century Art, ZKM | Center for Art and Media Karlsruhe

2006

47th October Salon: Art, Life & Confusion, Belgrade Cultural Center

MindFrames. Media Study at Buffalo 1973-1990, ZKM | Center for Art and Media Karlsruhe

2007

True Romance, Kunsthalle Wien, Vienna; Villa Stuck, Munich (2008); Kunsthalle zu Kiel (2008)

2008

Hotelito San Rafael, Glasshill Projects, London

2009

1968. Die große Unschuld, Kunsthalle Bielefeld

Behind the Image / The Image Behind – on covering and uncovering media, Artefact Festival for Media Art, STUK Kunstcentrum, Leuven

under water / above water: From the Aquarium to the Video Image, Kunsthalle Wilhelmshaven

Imagining Media@ZKM, ZKM | Center for Art and Media Karlsruhe

Linie, 401contemporary, Berlin

We Interrupt This Program, Mercer Union – Centre for Contemporary Visual Art, Toronto

2010

Changing Channels. Art and Television 1963–1987, mumok – Museum moderner Kunst Stiftung Ludwig Wien, Vienna

Memento, 401contemporary, Berlin

Münchener Biennale 12, Reithalle München, Munich

Fremde Heimat – Kunst in Baden-Württemberg, Kunsthalle Mannheim

reich sein, Künstlerverein Walkmühle, Wiesbaden

Austria la vista, baby, TAF / The Art Foundation, Athens

Are You Ready for TV?, Museu d'Art Contemporani de Barcelona (MACBA)

2011

Shadows of the Bright, BERLINER KUNSTKONTAKTER and BERLIN ART PROJECTS, Berlin

Ich traue meinen Augen nicht, Karikaturmuseum Krems

Professorale, KISS – Kunst im Schloss Untergröningen, Abtsgmünd

no limits?, Haus der Wirtschaft, Stuttgart

Aufstand der Dinge, MIM & MORE Contemporary, Berlin

2012

MoMA – Media Lounge, The Museum of Modern Art (MoMA), New York

Happy Birthday Fluxus, zeitraumexit, Mannheim

Vidéo Vintage 1963–1983, Musée National d'Art Moderne, Centre Georges Pompidou, Paris; ZKM | Center for Art and Media Karlsruhe; Beirut Art Center, Beirut (2013); National Museum of Contemporary Art, Seoul (2013); Oi Futuro Cultural Center, Rio de Janeiro (2013)

The Immersion: Towards Haptic Cinema, MediaArtLab, Moscow

Faces. The Phenomenon of Face in Videoart, Galerie Rudolfinum, Prague

Sound Art: Sound as a Medium of Art, ZKM | Center for Art and Media, Karlsruhe

Poesie der Reduktion: Minimal, Concept, Land Art, mumok – Museum moderner Kunst Stiftung Ludwig Wien, Vienna

ARTandPRESS. Kunst. Wahrheit. Wirklichkeit., ZKM | Center for Art and Media Karlsruhe; Martin-Gropius-Bau, Berlin

Gnadenlos – Künstlerinnen und das Komische, KH/ Kunsthalle Vogelmann, Heilbronn; Paula Modersohn-Becker Museum, Bremen (2013)

Tiere gehen immer, Galerie Axel Obiger, Berlin

Zeit(lose) Zeichen. Gegenwartskunst in Referenz zu Otto Neurath, Künstlerhaus Wien, Vienna

2013

Transposition: Motion is Action – An International Media Art Exhibition, National Art Museum of China, Beijing

Elastic Video, Kunstraum Niederösterreich, Vienna

Through a Glass Darkly. Faces Past and Present, Ernst Museum Budapest

Fotos. Österreichische Fotografien von den 1930er bis heute, 21er Haus, Belvedere, Vienna

Body in Abstraction, ZKM | Center for Art and Media Karlsruhe; St John's College Oxford

Images for Images. Artists for Tichy – Tichy for Artists, Galerie Středočeského kraje – GASK, Kutná Hora

Interlace: Painting and Contemporary Media, Paço das Artes, São Paulo

EINE KLEINE MACHTMUSIK. Bericht aus dem Depot, Essl Museum, Klosterneuburg/Vienna

Teutloff Meets Ars Sacra, Salzburg Museum Neue Residenz, Salzburg

A Space Called Public / Hoffentlich Öffentlich, Landeshauptstadt München, Kulturreferat, Munich

Donaufestival Krèms Brûlée, Krems

TONSPUR 42, Canada Water Culture Space, London

Das schwache Geschlecht – Neue Mannsbilder in der Kunst, Kunstmuseum Bern

Weiße Normen der Macht, Graz Museum, Graz

electric parade. im fokus: medienkunst, Galerie Freihausgasse – Galerie der Stadt Villach, Villach

WeltKultur/GlobalCulture, Badisches Landesmuseum, Karlsruhe

Amazing! Clever! Linguistic! An Adventure in Conceptual Art, Generali Foundation, Vienna

Typemotion. Type as Image in Motion, ZKM | Center for Art and Media Karlsruhe; National Taiwan Museum of Fine Arts, Taichung (2015)

2014

Les Immatériaux, Kunstverein Düsseldorf

Underground, Kontur. Kunstverein Stuttgart, Stuttgart

Connecting Sound Etc. Cable Works, Cable Sounds, Cables Everywhere, frei_raum Q21 exhibition space, Vienna

Self-Timer Stories, Austrian Cultural Forum New York (ACFNY), New York

Electronic Sound in a Shifting Landscape, Galerie Dazibao, Montreal

30 Jahre – 30 Stimmen. Retrospektive anlässlich des 30-jährigen Jubiläums der Videonale Bonn, Bonner Kunstverein, Bonn

Close to Action: Performance Documentary 1961–1985, Galerie Angermeier, Winden

Time(less)Signs: Otto Neurath and Reflections in Austrian Contemporary Art, Austrian Cultural Forum London (ACFL), London

Graz – Offene Stadt. Ordnungspolitik und Möglichkeitsräume, Graz Museum, Graz

2D23D: photography as sculpture, sculpture as photography, OstLicht. Gallery for Photography, Vienna

Reines Wasser. Die kostbarste Ressource der Welt, LENTOS Kunstmuseum Linz

Body Talks – 100 Jahre BH, Museum für Kommunikation, Frankfurt am Main; Museum für Kommunikation, Nuremberg; Museum für Kommunikation, Berlin; Deutsches Bernstein-museum Ribnitz-Damgarten

Anarchy. Utopia. Revolution, Ludwig Museum – Museum of Contemporary Art, Budapest

Korrelation, Angewandte Interdisciplinary Lab (AIL), University of Applied Arts, Vienna

Damage Control. Body Art and Destruction 1968–1972, Neue Galerie Graz

Proudly Presenting: Sammlung Generali Foundation, Museum der Moderne Salzburg

2015

subversiv. Raum für Alternativen, Graz Museum, Graz

Gender in Art, Museum of Contemporary Art in Krakow (MOCAK), Krakow

Das ist Österreich!, vorarlberg museum, Bregenz

Landschaft: Transformation einer Idee, Neue Galerie Graz

body interference, Künstlerhaus Wien, Vienna

FLICK_KA, ZKM | Center for Art and Media Karlsruhe

Verkauf *in Nebenräumen / Sales in the Side Rooms*, Galerie Krinzinger, Vienna

VIDEODOX 2015, Galerie der Künstler, Munich

Bodenlos – Vilém Flusser und die Künste, ZKM | Center for Art and Media Karlsruhe; Akademie der Künste, Berlin

KINOMAGIE - Was geschah wirklich zwischen den Bildern?, METRO Kinokulturhaus, Vienna

2016

In Light of the 20th Century Avant-Garde, Atlas Sztuki, Łódz

Digital Synesthesia, Angewandte Interdisciplinary Laboratory (AIL), Vienna

Verfolgt, beraubt, vertrieben, Rotor Centre for Contemporary Art, Graz

Poesie der Veränderung, Museum der Moderne Salzburg

Rühm Weibel, Galerie Remixx, Graz

Deluge and Nebula, iCinema Scientia Facility, UNSW Kensington

Aus der Sammlung. Bild, Realität und Forschung von 1960 bis 1980, Neue Galerie Graz

Mikro Makro – Das Große im Kleinen entdecken, Nordrhein-Westfälische Akademie der Wissenschaften und der Künste, Düsseldorf

FLICK_EU, Galerie des Citoyens, Lyon

Interdependency – How to Picture Living Systems, Konrad Lorenz Institute (KLI), Klosterneuburg

2017

Under Construction, Schau Fenster, Berlin

Open Codes: Living in Digital Worlds, ZKM | Center for Art and Media Karlsruhe

Ästhetik der Veränderung – 150 Jahre Universität für Angewandte Kunst Wien, MAK – Museum of Applied Arts, Vienna

2018

Open Codes: The World as a Field of Data, ZKM | Center for Art and Media Karlsruhe

Ganz Wien. Eine Pop-Tour, Wien Museum, Vienna

Inside Out – Fotografie und Psychologie, DZ BANK Kunstsammlung, Frankfurt am Main

Flashes of the Future – Die Kunst der 68er oder die Macht der Ohnmächtigen, Ludwig Forum for International Art, Aachen

30 Jahre Generali Foundation. Im Dialog mit 1918 1938 1968, Museum der Moderne Salzburg

64. Internationale Kurzfilmtage Oberhausen

Schluss mit der Wirklichkeit! Avantgarde, Architektur, Revolution, 1968, Landesgalerie Linz, Oberösterreichisches Landesmuseum

ANiMAL – Art Science Nature Society, CityU Exhibition Gallery, Hong Kong

Scripted Reality: The Life and Art of Television, Fundación Jumex Arte Contemporáneo, Mexico City

Alternative Film/Video Festival 2019, Belgrade

2019

Wien 1900. Aufbruch in die Moderne, Leopold Museum, Vienna

Writing the History of the Future. The ZKM Collection, ZKM | Center for Art and Media Karlsruhe; Gwangju Museum of Art, Gwangju (2021)

Negative Space: Trajectories of Sculpture, ZKM | Center for Art and Media Karlsruhe

Über Grenzen, Kunsthalle Bahnitz

Sate of the Kunst, Art Bodensee, Dornbirn

Zoom auf die 1970er, Gdansk National Museum, Gdansk

Demoncracy, Alfons-Kern-Turm (A.K.T.), Pforzheim

2020

To Reach a Star, Gwangju Museum of Art, Gwangju

The Beginning. Kunst in Österreich 1945 bis 1980, Albertina Modern, Vienna

Berührend – Annäherung an ein wesentliches Bedürfnis, Museen Böttcherstraße, Bremen

In aller Munde, Kunstmuseum Wolfsburg

2021

Die Stadt als Datenfeld. Wie wir in Zukunft leben wollen, Graz Museum, Graz

Konkrete Poesie und so, kunsthaus muerz, Mürzzuschlag

Holobionti. Life Is Other, Magazin 4, Bregenz

Intopia – Welt ohne Lügen, Museum am Bach, Ruden

Infosphere, Galerie Zdeněk Sklenář, Prague

Ladies and Gentlemen. The Fragile Feminist We, Neue Galerie Graz

Ludwig Wittgenstein. Fotografie als analytische Praxis, Leopold Museum, Vienna

What If? Women in Film Festival 2021, blue Cinema Corso, Zurich

Seasons of Media Arts, Karlsruhe

2022

Kinetismus: 100 Years of Electricity in Art, Kunsthalle Praha, Prague

Macht! Licht!, Kunstmuseum Wolfsburg

Eye/View, eBoard Congress Center Basel

Kunst und Sprache, Galerie Artosphäre Schloss Puchenau, Puchenau

Kollaborationen, mumok – Museum moderner Kunst Stiftung Ludwig Wien, Vienna

Selbst/Self/I, Tichy Ocean Foundation, Zurich

Loys Egg / Peter Weibel – Bibliothec Insomnia Redux, Galerie Sommer, Graz

Faking the Real: Die Kunst der Verführung, Kunsthaus Graz

Das Tier in Dir, mumok – Museum moderner Kunst Stiftung Ludwig Wien, Vienna

Die (neue) Vermessung der Welt, Pochen Biennale 2022, Chemnitz

Medieninterventionen, Museum der Moderne Salzburg

Street Life. Die Straße in der Kunst von Kirchner bis Streuli, Wilhelm-Hack-Museum, Ludwigshafen

Im Detail. Die Welt der Konservierung und Restaurierung, Tiroler Landesmuseen Ferdinandeum, Innsbruck

Eye/View: Inter-Textualities, REX Box, Kino REX Bern

Awards & Distinctions

1974

Film Prize of the City of Berlin

1984

Award for Creative Excellence (Special Effects), Clio Awards, New York

Prix National, 39st Festival de Cannes

1985

Award for Creative Excellence (Special Effects), Clio Awards, New York

1986

Award for Creative Excellence (Special Effects), Clio Awards, New York

1988

Prix Ars Electronica '88 (Computer Animation), Linz

Special recognition, Locarno Film Festival (Locarno 43)

1991

Sculpture Prize of the EA-Generali Foundation, Vienna

1992

Prize for Fine Art of the City of Vienna

1993

Prix Ars Electronica '93 (Interactive Art), Linz

1997

Siemens Media Art Prize, awarded by the ZKM | Center for Art and Media Karlsruhe

2002

Grand Decoration of Honor for Services to the Republic of Austria

2004

Käthe Kollwitz Prize, awarded by the Akademie der Künste, Berlin

2005

Eyes & Ears Excellence Award, 7th International Eyes & Ears Awards, Munich

2007

Grand Decoration of Honor of the State of Styria, Graz

2008

Named Officier dans l'Ordre des Arts et des Lettres (Order of Arts and Literature), awarded by the French Ministry of Culture and Communication, Paris

2009

European Culture Prize, awarded by the European Cultural Foundation Pro Europa

Friedlieb Ferdinand Runge Award for unconventional approaches to publicising art, awarded by the Stiftung Preussische Seehandlung, Berlin

Order of Merit of Baden-Württemberg

2010

Austrian Decoration for Science and Art, First Class, awarded by the president of Austria, Vienna

2014

Oskar Kokoschka Prize, State Prize of the Austrian Federal Government, Vienna

2017

Austrian Art Prize – Media Art (Österreichischer Kunstpreis – Medienkunst), awarded by Federal Ministry of Arts, Culture, Civil Service and Sports of the Republic of Austria, Vienna

2020

International TREBBIA Award, awarded by the TREBBIA Foundation, Prague

Lovis Corinth Prize, awarded by the KünstlerGilde e.V. at Kunstforum Ostdeutsche Galerie Regensburg

Collections

BEEP Collection, Reus

Belvedere, Vienna

Bundesanstalt für Arbeitsschutz und Arbeitsmedizin, DASA Museum, Dortmund

Deutsches Museum Nürnberg (DMN)

Julia Stoschek Collection, Berlin

LENTOS Kunstmuseum Linz

MAK – Museum of Applied Arts, Vienna

Musée National d'Art Moderne, Centre Georges Pompidou, Paris

Museum der Moderne Salzburg

Museum of Modern Art (MoMA), New York

Muzeum Sztuki, Łódz

Neue Galerie Graz

The Pudil Family Foundation Art Collection, Prague

Ulsan Museum of Art, Ulsan

ZKM | Karlsruhe

Numerous private collections also hold works by Peter Weibel.

Peter Weibel as an Editor

1976

Oswald Oberhuber and Peter Weibel, eds., *Österreichs Avantgarde 1900–1938: ein unbekannter Aspekt. Bildende Kunst, Literatur, Wissenschaft*, exh. cat., Galerie nächst St. Stephan, Vienna / Galerie im Taxispalais, Innsbruck (Vienna: Oberhuber, 1976).

1979

Peter Weibel and Peter Pakesch, eds., *Künstlerschaufenster*, exh. cat. (Graz: steirischer herbst, 1979).

1981

Peter Weibel and Anna Auer, eds., *Erweiterte Fotografie*, exh. cat. (Vienna: Secession, 1981).

1985

Peter Weibel and Eva H. Plattner, eds., *Jaroslav Seifert: Auf den Wellen von TSF* (Vienna: Hora Verlagsgesellschaft, 1985).

1986

Peter Weibel, ed., *Martin Kippenberger. Zeichnungen über eine Ausstellung* (Vienna: Galerie Peter Pakesch, 1986).

1987

Peter Weibel and Veruschka Bódy, eds., *Clip, Klapp, Bum. Von der visuellen Musik zum Musikvideo* (Cologne: DuMont, 1987).

Peter Weibel, ed., *Jenseits der Erde: Kunst, Kommunikation, Gesellschaft im orbitalen Zeitalter* (Vienna: Hora Verlagsgesellschaft, 1987).

Peter Weibel, ed., *Freie Klänge – Offene Räume*, exh. cat. (Linz: Ars Electronica, 1987).

1988

Peter Weibel and Gottfried Hattinger, ed., *Kunst der Szene*, exh. cat. (Linz: Ars Electronica, 1988).

1989

Peter Weibel and Gerhard J. Lischka, eds., *Im Netz der Systeme: Eine Dokumentation anläßlich der Ars Electronica*, *Kunstforum*, no. 103 (1989).

1990

Peter Weibel and Edith Decker, eds., *Vom Verschwinden der Ferne. Telekommunikation und Kunst*, exh. cat., Deutsches Postmuseum, Frankfurt am Main (Cologne: DuMont, 1990).

Peter Weibel, ed., *Von der Bürokratie zur Telekratie. Rumänien im Fernsehen* (Berlin: Merve, 1990).

Peter Weibel and Georg Hattinger, eds., *Ars Electronica: Virtuelle Welten*, exh. cat. (Linz: Ars Electronica, 1990).

Peter Weibel and Georg Hattinger, eds., *Ars Electronica: Digitale Träume*, exh. cat. (Linz: Ars Electronica, 1990).

1991

Peter Weibel and Christian Meyer, eds., *Das Bild nach dem letzten Bild*, exh. cat., Galerie Metropol, Vienna (Cologne: Verlag der Buchhandlung Walther König, 1991).

Peter Weibel and Christian Meyer, eds., *Das Bild nach dem letzten Bild*, exh. cat., Galerie Metropol, Vienna (Cologne: Verlag der Buchhandlung Walther König, 1991).

Peter Weibel and Florian Rötzer, eds., *Strategien des Scheins: Kunst, Computer, Medien* (Munich: Boer Verlag, 1991).

Peter Weibel and Gerhard J. Lischka, eds., *Feminismus und Medien* (Wabern/Bern: Benteli, 1991).

Wolfgang Drechsler and Peter Weibel, eds., *Bildlicht. Malerei zwischen Material und Immaterialität*, exh. cat., Wiener Festwochen / Museum des 20. Jahrhunderts, Vienna (Vienna: Europaverlag, 1991).

1992

Götz Pochat, Christa Steinle and Peter Weibel, eds., *Kontinuität und Identität: Festschrift für Wilfried Skreiner* (Vienna/Cologne/Weimar: Böhlau, 1992).

Peter Weibel and Christa Steinle, eds., *Identität: Differenz. Tribüne Trigon, 1940–1990. Eine Topographie der Moderne*, exh. cat., steirischer herbst '92 / Neue Galerie Graz (Vienna/Cologne/Weimar: Böhlau, 1992).

Peter Weibel and Karl Gerbel, eds., *Ars Electronica: Die Welt von Innen – Endo und Nano*, exh. cat. (Linz: Ars Electronica, 1992).

Otto E. Rössler, *Endophysics: The World as an Interface*, ed. Peter Weibel (Singapore: World Scientific Publishing Company, 1998). Originally published as *Endophysik: Die Welt des inneren Beobachters*, ed. Peter Weibel (Berlin: Merve, 1992).

1993

Peter Weibel and Friedrich Stadler, eds., *Vertreibung der Vernunft. The Cultural Exodus*

from Austria, exh. cat. Venice Bienniale (Vienna: Löcker, 1993).

Peter Weibel and Christa Steinle, eds., *Räume für Kunst – Museumsmodelle: Europäische Museumsarchitektur der Gegenwart*, exh. cat., Stadtgeschichtliches Museum Leipzig / Kunst- und Ausstellungshalle der Bundesrepublik Deutschland, Bonn / Neue Galerie Graz / Groninger Museum, Groningen (Hanover: Kestner-Gesellschaft, 1993).

Florian Rötzer, *Cyberspace: Zum medialen Gesamtkunstwerk*, ed. Peter Weibel and Florian Rötzer (Munich: Boer Verlag, 1993).

Peter Weibel, ed., *Stellvertreter. Österreichs Beitrag zur 45. Biennale in Venedig*, exh. cat., La Biennale di Venezia (Cologne: Verlag der Buchhandlung Walther König, 1993).

Peter Weibel and Karl Gerbel, eds., *Ars Electronica: Genetische Kunst – Künstliches Leben*, exh. cat. (Linz: Ars Electronica, 1993).

1994

Peter Weibel, ed., *Kontext Kunst: The Art of the 90s*, exh. cat., Trigon '93, steirischer herbst '93, Neue Galerie Graz (Cologne: DuMont, 1994).

Peter Weibel, ed., *Felix Gonzales-Torres, Rudolf Stingel*, exh. cat., (Graz: Neue Galerie Graz, 1994).

Peter Weibel, ed., *Peter Fend. Ocean Earth. 1980 bis heute* (Stuttgart: Oktagon, 1994).

Peter Weibel and Karl Gerbel, eds., *Ars Electronica: Intelligente Ambiente*, exh. cat. (Linz: Ars Electronica, 1994).

Peter Weibel and Ulrike Lehmann, eds., *Ästhetik der Absenz. Bilder zwischen Anwesenheit und Abwesenheit* (Munich/Berlin: Klinckhardt & Biermann, 1994).

Peter Weibel and Carl Haenlein, eds., *Jon Kessler's Asia*, exh. cat., Neue Galerie, Graz (Hanover: Kestner-Gesellschaft, 1994).

1995

Peter Weibel, ed., *Pittura Immedia. Malerei in den 90er Jahren*, exh. cat., Neue Galerie Graz (Klagenfurt: Ritter Verlag, 1995).

Peter Weibel, ed., *The Media Pavilion: Art and Architecture in the Age of Cyberspace*, exh. cat., La Biennale di Venezia (Vienna: Springer, 1995).

Peter Weibel and Karl Gerbel, eds., *Ars Electronica: Mythos Information. Welcome to the wired world*, exh. cat., Ars Electronica, Linz (Vienna/New York: Springer, 1995).

Peter Weibel, Dietmar Kamper, Florian Rötzer et. al., eds., *Das Böse: Jenseits von Absichten und Tätern. Oder: Ist der Teufel ins System ausgewandert?*, exh. cat., Kunst- und Ausstellungshalle der Bundesrepublik Deutschland, Bonn (Göttingen: Steidl, 1995).

1996

Peter Weibel and Ingeborg Orthofer, eds., *Über Schwabtexte. Information. Beschreibung. Annäherungen*, exh. cat., Schwabsammlung, Graz/Vienna (Graz: Literaturverlag Droschl, 1996).

Peter Weibel, ed., *Günter Brus: Blitzartige Einfälle in vorgegebene Ideen. Bild-Imprimaturen und Trivialeums-Überzeichnungen von Günter Brus, auf den Weg gebracht und begleitet von Arnulf Meifert*, exh. cat., Neue Galerie Graz / Moderna Galerija Ljubljana (Ostfildern: Hatje Cantz, 1996).

Peter Weibel and Werner Hahn, eds., *Evolutionäre Symmetrietheorie. Selbstorganisation und dynamische Systeme* (Stuttgart: Wissenschaftliche Verlagsgesellschaft, 1996).

Peter Weibel, ed., *Quantum Daemon. Institutionen der Kunstgemeinschaft* (Vienna: Passagen Verlag, 1996).

Peter Weibel and Günther Holler-Schuster, eds., *Adrian Schiess* (Graz: Neue Galerie Graz, 1996).

Peter Weibel and Günther Holler-Schuster, eds., *"I started, for it amazed me that I had not seen him, since the reflection of the glass covered the whole room behind me": new paintings for the mirror room and archie in a studio off the courtyard by David Reed*, exh. cat. (Graz: Neue Galerie Graz, 1996).

Peter Weibel and Heinrich Klotz, eds., *Jeffrey Shaw: a User's Manual. From Expanded Cinema to Virtual Reality* (Ostfildern: Hatje Cantz, 1996).

1997

Peter Weibel, ed., *Inklusion : Exklusion. Versuch einer neuen Kartografie der Kunst im Zeitalter von Postkolonialismus und globaler Migration*, exh. cat., steirischer herbst '97, Graz (Cologne: DuMont, 1997).

Peter Weibel, ed., *Jenseits von Kunst*, exh. cat., Neue Galerie Graz (Vienna: Passagen Verlag, 1997).

Peter Weibel, ed., *Die Wiener Gruppe: ein Moment der Moderne 1954–1960. Die visuellen Arbeiten und die Aktionen* (Vienna/New York: Springer, 1997).

1998

Peter Weibel and Christa Steinle, eds., *Giulio Paolini. Von heute bis gestern, da oggi a ieri*, exh. cat., Neue Galerie Graz (Ostfildern: Hatje Cantz, 1998).

Peter Weibel, ed., *Die Meskalinzeichnungen von Henri Michaux: 1954–1959, 1966–1969*, exh. cat., Neue Galerie Graz (Cologne: Verlag der Buchhandlung Walther König, 1998).

Peter Weibel, ed., *Kunst ohne Unikat. Edition Artelier 1985–1998*, exh. cat., steirischer herbst '98, Graz (Cologne: Verlag der Buchhandlung Walther König, 1998).

Peter Weibel, Wolfgang Bauer and Jörg Schlick, eds.,*Martin*, exh. cat., steirischer herbst '98, Graz (Cologne: Verlag der Buchhandlung Walther König, 1998).

Peter Weibel, ed., *H. J. Painitz. Mehr als die Summe der Teile – 55 Stufen und 378 Bilder*, exh. cat. (Graz: Neue Galerie Graz, 1998).

1999

Peter Weibel, ed., *Jean Baudrillard. Fotografien, photographies, photographs: 1985–1998*, exh. cat., Neue Galerie Graz (Ostfildern: Hatje Cantz, 1999).

Peter Weibel, ed., *Offene Handlungsfelder – Open Practice – Prassi aperta*, exh. cat., La Biennale di Venezia (Cologne: DuMont, 1999).

Peter Weibel, ed., *Kunst ohne Original. Multiple und Sampling als Medium: Techno-Transformationen der Kunst*, exh. cat., Neue Galerie Graz (Cologne: Verlag der Buchhandlung Walther König, 1999).

Peter Weibel, ed., *KUNSTKALENDER 1999 Neue Galerie Graz. In der Rückschau 1998* (Graz: Neue Galerie Graz, 1999).

2000

Peter Weibel and Timothy Druckrey, eds., *Net_condition: Art and Global Media*, exh. cat., ZKM | Karlsruhe (Cambridge, MA: MIT Press, 2000).

Peter Weibel, Rainer E. Burkhard and Wolfgang Maas, eds., *Zur Kunst des formalen Denkens* (Vienna: Passagen Verlag, 2000).

Peter Weibel, ed., *Max Peintner – der Pilot als blinder Passagier: Wahrnehmung im technologischen Zeitalter*, exh. cat., Neue Galerie Graz (Ostfildern: Hatje Cantz, 2000).

Peter Weibel and Franz Kaltenbeck, eds., *Trauma und Erinnerung, Trauma and Memory: Cross-Cultural Perspectives*, exh. cat., steirischer herbst, Neue Galerie Graz (Vienna: Passagen Verlag, 2000).

Goethe-Institute and Peter Weibel, eds., *Update 2.0: Current media art from Germany* (Karlsruhe: ZKM | Karlsruhe, 2000).

2001

Peter Weibel and Zdenka Badovinac, eds., *2000+ Arteast Collection: The Art of Eastern Europe. A Selection of Works for the International and National Collections of Moderna Galerija Ljubljana*, exh. cat., Moderna galerija Ljubljana / Orangerie Congress, Innsbruck (Vienna: Folio, 2001).

Peter Weibel, ed. *Olafur Eliasson: Sourroundings Sourrounded. Essays on Space and Science*, exh. cat., ZKM | Karlsruhe / Neue Galerie Graz (Cambridge, MA: MIT Press, 2001).

Peter Weibel, ed., *Imbu chst aben feld: die Zukunft der Literatur*, exh. cat., Neue Galerie Graz (Graz: Literaturverlag Droschl, 2001).

Hans H. Diebner, Timothy Druckrey and Peter Weibel, eds., *Sciences of the Interface: Proceedings of the International Symposium Sciences of the Interface* (Tübingen: Genista Verlag, 2001).

Peter Weibel, ed., *Vom Tafelbild zum globalen Datenraum. Neue Möglichkeiten der Bildproduktion und bildgebender Verfahren* (Ostfildern: Hatje Cantz, 2001).

Peter Weibel and Günter Eisenhut, eds., *Moderne in dunkler Zeit. Widerstand, Verfolgung und Exil steirischer Künstlerinnen und Künstler 1933–1948*, exh. cat., Neue Galerie Graz (Graz: Literaturverlag Droschl, 2001).

2002

Peter Weibel and Bruno Latour, eds., *Iconoclash. Beyond the Image Wars in Science, Religion, and Art*, exh. cat., ZKM | Karlsruhe (Cambridge, MA: MIT Press, 2002).

Barbara Könches and Peter Weibel, eds., *Bilder Codes: Internationaler Medienkunstpreis* (Wabern/Bern: Benteli, 2002).

Thomas Y. Levin, Ursula Frohne and Peter Weibel, eds., *Ctrl [Space]. Rhetorics of Surveillance. From Bentham to Big Brother*, exh. cat., ZKM | Karlsruhe (Cambridge, MA: MIT Press, 2002).

Peter Weibel, ed., *Erwin Wurm*, exh. cat., Neue Galerie Graz / Centre national de la photographie, Paris / Galleria Moderna, Bologna / ZKM | Karlsruhe (Ostfildern: Hatje Cantz, 2002).

Roger Conover, Eda Čufer and Peter Weibel, eds., *In Search of Balkania: A User's Manual*, exh. cat. (Graz: Neue Galerie Graz, 2002).

Peter Weibel and Günther Holler-Schuster, eds., *Herbert Brandl: Chromophobie*, exh. cat., Neue Galerie im Künstlerhaus Graz / Neue Galerie Graz (Ostfildern: Hatje Cantz, 2002).

Peter Weibel and Christa Steinle, eds., *Die Cabinette des Dr. Czerny* (Cologne: Verlag der Buchhandlung Walther König, 2002).

Peter Weibel, Eckehart Köhler, Michael Stöltzner et al., eds., *Kurt Gödel: Wahrheit & Beweisbarkeit*, vol. 2, *Kompendium zum Werk* (Vienna: öbv & hpt Verlag, 2002).

Peter Weibel, ed., *Jordan Crandall. DRIVE*, exh. cat., Neue Galerie Graz (Ostfildern: Hantje Cantz, 2002).

Herbert Kapfer and Peter Weibel, eds., *intermedium 2: X-O Identities in the 21st Century* (Erding: Strunz, 2002).

2003

Werner DePauli-Schimanovich, *Europolis 01. Über die Zukunft der Verkehrs-Technik und Planung in Europa*, ed. Peter Weibel (Vienna: Passagen Verlag, 2003).

Jeffrey Shaw and Peter Weibel, eds. *Future Cinema. The Cinematic Imaginary after Film*, exh. cat., ZKM | Karlsruhe / Museum of Contemporary Art, Helsinki / NTT Intercommunication Center, Tokyo (Cambridge, MA: MIT Press, 2003).

Peter Weibel, ed., *Phantom der Lust: Visionen des Masochismus*, exh. cat., Neue Galerie Graz (Munich: Belleville, 2003).

Barbara Könches and Peter Weibel, eds., *Constructed Life: International Media Art Award 2003. Fiction Scenarios between Computer, Games, Cyber-Sex, Nanobytes and Robotic Arts*, exh. cat (Karlsruhe: ZKM | Karlsruhe, 2003).

Peter Weibel and Günther Holler-Schuster, eds., *M_ARS. Kunst und Krieg*, exh. cat., Neue Galerie Graz (Ostfildern: Hatje Cantz, 2003).

Peter Weibel and Gerhard J. Lischka, eds., *Regime des Image. Zwischen mimischem Display und Corporate Branding*, exh. cat., ZKM | Karlsruhe / Kornhaus, Bern (Wabern/Bern: Benteli, 2003).

2004

Roger Conover, Eda Čufer and Peter Weibel, eds., *Call me ISTANBUL ist mein Name. Kunst und urbane Visionen einer Metapolis*, exh. cat., ZKM | Karlsruhe (Tübingen/Berlin: Wasmuth, 2004).

Gerhard J. Lischka and Peter Weibel, eds., *Act! Handlungsformen in Kunst und Politik* (Wabern/Bern: Benteli, 2004).

Rolf Sachsse, *Wilhelm Ostwald: Farbsysteme. Das Gehirn der Welt*, ed. Peter Weibel (Ostfildern: Hatje Cantz, 2004).

Gerhard J. Lischka and Peter Weibel, eds., *Die Medien der Kunst, die Kunst der Medien* (Wabern/Bern: Benteli, 2004).

Werner DePauli-Schimanovich, *Europolis 02: Vom Bodensee zum Burgensee: Über die Zukunft des Verkehrs in Austria*, ed. Peter Weibel (Vienna: Passagen Verlag, 2004).

2005

Bruno Latour and Peter Weibel, eds., *Making Things Public. Atmospheres of Democracy*, exh. cat., ZKM | Karlsruhe (Cambridge, MA: MIT Press, 2005).

Peter Gente, Barbara Könches and Peter Weibel, eds., *Philosophie und Kunst: Jean Baudrillard. Eine Hommage zu seinem 75. Geburtstag* (Berlin: Merve, 2005).

Peter Weibel, ed., *Beyond Art: A Third Culture: A Comparative Study in Cultures, Art, and Science in 20th Century Austria and Hungary* (Vienna: Springer, 2005).

Peter Weibel and Andrea Buddensieg, eds., *Contemporary Art and the Museum: A Global Perspective* (Ostfildern: Hatje Cantz, 2007).

Boris Groys, Anne von der Heiden and Peter Weibel, eds., *Zurück aus der Zukunft. Osteuropäische Kulturen im Zeitalter des Postkommunismus* (Frankfurt am Main: Suhrkamp, 2005).

Georg Flachbart and Peter Weibel, eds., *Disappearing Architecture. From Real to Virtual to Quantum* (Basel: Birkhäuser, 2005).

Peter Weibel, Elisabeth Fiedler and Christa Steinle, eds., *Postmediale Kondition*, exh. Neue Galerie Graz / Centro Cultural Conde Duque, Medialab Center Madrid (Graz: Neue Galerie Graz, 2005).

Peter Weibel, Garnet Ord and Otto E. Rössler, eds., *Space Time Physics and Fractality: Festschrift in honour of Mohamed El Nashie on the occasion of his 60th birthday*, (Vienna/New York: Springer, 2005).

Hans Belting, *Szenarien der Moderne: Kunst und ihre offenen Grenzen*, FUNDUS vol. 164, ed. Peter Weibel (Hamburg: Philo & Philo Fine Arts, 2005).

Barbara Könches and Peter Weibel, eds., *UnSICHTBARes: Algorithmen als Schnittstelle zwischen Kunst und Wissenschaft*, Internationaler Medienkunstpreis 2004 (Wabern/Bern: Benteli, 2005).

2006

Nadja Rotter and Peter Weibel, eds., *Gego 1957–1988: Thinking the Line*, exh. cat., ZKM | Karlsruhe / Neue Galerie Graz / Miami Art Central (Ostfildern: Hatje Cantz, 2006).

Nadja Rotter and Peter Weibel, eds., *Ruth Vollmer 1961–1978: Thinking the Line*, exh. cat., ZKM | Karlsruhe / Neue Galerie Graz / Miami Art Central (Ostfildern: Hatje Cantz, 2006).

Barbara Könches and Peter Weibel, eds., *Stephan von Huene: Grenzgänger, Grenzverschieber*, exh. cat., ZKM | Karlsruhe (Heidelberg: Kehrer Verlag, 2006).

Marlene Angermeyer-Deubner and Peter Weibel, eds., *Syntax des Sehens. Die Videosammlung des ZKM* (Heidelberg: Kehrer Verlag, 2006).

Woody Vasulka and Peter Weibel, eds., *Buffalo Heads: Media Study, Media Practice, Media Pioneers, 1973–1990*, exh. cat., ZKM | Karlsruhe (Cambridge, MA: MIT Press, 2008).

Gregor Jansen and Peter Weibel, eds., *Faster! Bigger! Better!. Signetwerke der Sammlungen im ZKM | Museum für Neue Kunst* (Cologne: Verlag der Buchhandlung Walther König, 2006).

Peter Weibel, ed., *Wilhelm Ostwald: Farbsysteme. Das Gehirn der Welt* (Ostfildern: Hatje Cantz, 2006).

Peter Weibel, Gregor Jansen, eds., *Light Art from Artificial Light. Light as a Medium in 20th and 21th Century Art* (Ostfildern: Hatje Cantz, 2006).

Peter Weibel, eds., *The Role of Pictures in Society: New Ways of Using Images / Symposium 20–22 January, 2006* (Karlsruhe: ZKM | Karlsruhe, 2006).

Vanessa Diemand, Michael Mangold and Peter Weibel, eds., *Weblogs, Podcasting und Videojournalismus: Neue Medien zwischen demokratischen und ökonomischen Potenzialen* (Hanover: Heise, 2006.)

Peter Weibel, ed., *Ingrid Wiener: Träume*, exh. cat., Neue Galerie Graz (Cologne: Verlag der Buchhandlung Walther König, 2006).

2007

Peter Gente and Peter Weibel, eds., *Deleuze und die Künste* (Frankfurt am Main: Suhrkamp, 2007).

Bernhard Serexhe and Peter Weibel, eds., *Wolfgang von Kempelen. Mensch in der Maschine / Man in the Machine* (Berlin: Matthes & Seitz, 2007).

Wonil Rhee, Peter Weibel and Gregor Jansen, eds., *Thermocline of Art: New Asian Waves*, exh. cat., ZKM | Karlsruhe (Ostfildern: Hatje Cantz, 2007).

Jiří Ševčík and Peter Weibel, eds., *Utopien & Konflikte. Dokumente und Manifeste zur tschechischen Kunst 1938–1989* (Ostfildern: Hatje Cantz, 2007).

Peter Peer and Peter Weibel, eds., *Natalia Nikitin: Kunst-Orte*, exh. cat., Neue Galerie Graz (Graz: Neue Galerie Graz, 2007).

Michael Mangold, Peter Weibel and Julie Woletz, eds., *Vom Betrachter zum Gestalter. Neue Medien in Museen – Strategien, Beispiele und Perspektiven für die Bildung* (Baden-Baden: Nomos, 2007).

Peter Weibel, ed., Ecke Bonk: *Monte Carlo Methode: A Typosophical Manual*, exh. cat., Künstlerhaus Graz, steirischer herbst, Graz (Munich: Schirmer Mosel, 2007).

Peter Weibel and Anna Artaker, eds., *Marc Adrian* (Klagenfurt: Ritter Verlag, 2007).

Peter Weibel and Günther Holler-Schuster, eds., *Fluxus, Happening, Konzeptkunst: Aus der Sammlung der Neuen Galerie Graz*, exh. cat. (Graz: Neue Galerie Graz, 2007).

2008

Hubertus von Amelunxen, Dieter Appelt and Peter Weibel, eds., *Notation: Kalkül und Form in den Künsten*, exh. cat. (Berlin/Karlsruhe: Akademie der Künste / ZKM | Karlsruhe, 2008).

Harald Falckenberg and Peter Weibel, eds., *Paul Thek: artis's artist*, exh. cat., ZKM | Karlsruhe / Sammlung Falckenberg, Hamburg / Museo Nacional Centro de Arte Reina Sofía, Madrid (Cambridge, MA: MIT Press, 2008).

Peter Weibel, ed., *Bernhard Leitner: P.U.L.S.E. Räume der Zeit / Spaces in time* (Ostfildern: Hatje Cantz, 2008).

Gregor Jansen and Peter Weibel, eds., *Gerard Caris: Pentagonismus / Pentagonism* (Cologne: Verlag der Buchhandlung Walther König, 2008).

Holger Jost and Peter Weibel, eds., *Die Teile der Summe. Begegnungen mit Christian Chruxin. Visueller Gestalter 1937–2006* (Cologne: Verlag der Buchhandlung Walther König, 2008).

Peter Weibel and Peter Peer, eds., *Otto Beckmann – zwischen Mystik und Kalkül*, exh. cat., Neue Galerie Graz (Cologne: Verlag der Buchhandlung Walther König, 2008).

Boris Groys, *Die Kunst des Denkens*, FUNDUS vol. 169, ed. Peter Weibel (Hamburg: Philo Fine Arts, 2008).

Christian Eigner and Peter Weibel, eds., *Un/Fair Trade: Die Kunst der Gerechtigkeit*, exh. cat., Neue Galerie Graz (Vienna/ New York: Springer, 2008).

Morten Søndergaard and Peter Weibel, eds., *Thorbjørn Lausten: Magnet Visual Systems*, exh. cat., Museum of Contemporary Arts, Roskilde / Nordjyllands Kunstmuseum, Aalborg / ZKM | Karlsruhe (Heidelberg: Kehrer Verlag, 2008).

EnBW Energie Baden-Württemberg AG, ed., *Nam June Paik: Werke aus der Sammlung des ZKM*, broch. (Karlsruhe: EnBW and ZKM| Karlsruhe, 2008).

Peter Weibel, ed., *Giuseppe Uncini – Scultore/Bildhauer 1929–2008*, exh. cat., ZKM | Karlsruhe / Museo di Arte Moderna e Contemporanea di Trento e Rovereto, Rovereto / Neue Galerie Graz (Milan: Skira, 2008).

2009

Peter Weibel, ed., *Materialbild / Material Picture / Immagine Materiale. Italia 1950–1965*, exh. cat., ZKM | Karlsruhe (Milan: Silvana, 2009).

Götz Adriani and Peter Weibel, eds., *Just what is it ... 100 Jahre Kunst der Moderne aus privaten Sammlungen in Baden-Württemberg. 10 Jahre Museum für Neue Kunst*, exh. cat., ZKM | Karlsruhe (Ostfildern: Hatje Cantz, 2009).

Franz Kaltenbeck and Peter Weibel, eds., *Sigmund Freud: Immer noch Unbehagen in der Kultur?* (Zurich/Berlin: Diaphanes, 2009).

Peter Weibel, ed., *99,9% und mehr: Künstler-Gruppenprojekte „vor" dem ZKM* (Karlsruhe: ZKM | Karlsruhe, 2009).

Peter Weibel, ed., *Günter Brus – Offenlegungen. Ein Symposion zum 70.* (Cologne/Graz: Verlag der Buchhandlung Walther König, Bruseum, 2009).

2010

Peter Weibel and Slavoj Žižek, eds., *Inklusion: Exklusion. Probleme des Postkolonialismus und der globalen Migration* (Vienna: Passagen Verlag, 2010.)

Peter Weibel and Christoph Blase, eds., *Record Again! 40yearsvideoart.de – part 2*, exh. cat., ZKM | Karlsruhe / Ludwig Forum, Aachen / Kunsthaus, Dresden / Edith-Ruß-Haus für Medienkunst, Oldenburg (Ostfildern: Hatje Cantz, 2010).

Peter Weibel, ed., *rosalie: Lichtkunst light art. The universal theater of light* (Zurich: Scheidegger & Spiess, 2010).

Toni Stooss and Peter Weibel, eds., *Jürgen Klauke: Ästhetische Paranoia*, exh. cat., ZKM | Karlsruhe / Museum der Moderne Salzburg (Ostfildern: Hatje Cantz, 2010).

Peter Weibel, ed., *ZKM Guide: Archives and Collections, Exhibition and Events, Research and Production* (Karlsruhe: ZKM | Karlsruhe, 2010).

Peter Weibel, Friederike Fast and Martin Walde, eds., *Martin Walde: A Second Home for Schrödinger's Cat, exh. cat.*, ZKM | Karlsruhe / Neue Galerie Graz / MARTa, Herford (Ostfildern: Hatje Cantz, 2010).

2011

Peter Weibel, ed., *Car Culture. Medien der Mobilität*, exh. cat. (Karlsruhe: ZKM | Karlsruhe, 2011).

Boris Groys and Peter Weibel, eds., *Medium Religion: Faith. Geopolitics. Art*, exh. cat., ZKM | Karlsruhe (Cologne: Verlag der Buchhandlung Walther König, 2011).

Hans Belting, Jacob Birken, Andrea Buddensieg and Peter Weibel, eds., *Global Studies. Mapping Contemporary Art and Culture* (Ostfildern: Hatje Cantz, 2011).

Peter Weibel, ed., *Robert Wilson: Video Portraits*, exh. cat., ZKM | Karlsruhe (Cologne: Verlag der Buchhandlung Walther König, 2011).

Harald Falckenberg, Stefan Zweifel and Peter Weibel, eds., *Dieter Meier: Works 1968–2011 and the Yello Years*, exh. cat., ZKM | Karlsruhe (Cologne: Verlag der Buchhandlung Walther König, 2011).

Peter Weibel, ed., *Edmund Kuppel: Projektionen 1970–2010. Ausufernde Sehfelder*, exh. cat., ZKM | Karlsruhe (Cologne: Verlag der Buchhandlung Walther König, 2011).

Margit Rosen in collaboration with Peter Weibel, Darko Fitz, and Marija Gattin, eds., *A Little-Known Story About a Movement, a Magazine, and the Computer's Arrival in Art. New Tendencies and Bit International, 1961–1973*, exh. cat., Neue Galerie Graz (Cambridge, MA: MIT Press, 2011).

Siegfried Zielinski and Peter Weibel, eds., *Der Stand der Bilder: die Medien-pioniere Zbigniew Rybczyński und Gábor Bódy*, exh. cat., Akademie der Künste, Berlin / ZKM | Karlsruhe (Nuremberg: Verlag für Moderne Kunst, 2011).

Peter Weibel and Andreas F. Beiting, eds., *Elmgreen & Dragset: Trilogy*, exh. cat., ZKM | Karlsruhe (Cologne: Verlag der Buchhandlung Walther König, 2011).

Peter Weibel and Christa Steinle, eds., *Bruseum: Ein Museum für Günter Brus*, exh. cat., Neue Galerie Graz (Ostfildern: Hatje Cantz, 2011).

2012

Peter Weibel, ed., *Moderne: Selbstmord der Kunst? Im Spiegel der Sammlung der Neuen Galerie Graz*, exh. cat., Neue Galerie Graz (Ostfildern: Hatje Cantz, 2012).

Peter Weibel, ed., *Hans Hollein*, exh. cat., Neue Galerie Graz (Ostfildern: Hatje Cantz, 2012).

Boris Groys and Peter Weibel, eds., *Bilder eines Reiches: Leben im vorrevolutionären Russland* (Heidelberg: Kehrer Verlag, 2012).

Peter Weibel, ed., *Loys Egg: Zeichnungen/ Drawings. 1976–1982* (Klagenfurt: Ritter Verlag, 2012).

Peter Weibel and Andreas F. Beiting, eds., *Bernhard Sandfort: Produzentengalerie, Dialogische Malerei, Museum der Fragen*, exh. cat., ZKM | Karlsruhe (Heidelberg: Kehrer Verlag, 2012).

2013

Peter Weibel and Ljiljana Fruk, eds., *Molecular Aesthetics*, exh. cat., ZKM | Karlsruhe (Cambridge, MA: MIT Press, 2013).

Peter Weibel, Hans Belting and Andrea Buddensieg, eds., *The Global Contemporary and the Rise of the New Art Worlds*, exh. cat., ZKM | Karlsruhe (Cambridge, MA: MIT Press, 2013).

Peter Weibel, Sigrid Gareis and Georg Schöllhammer, eds., *Moments. Eine Geschichte der Performance in 10 Akten*, exh. cat., ZKM | Karlsruhe (Cologne: Verlag der Buchhandlung Walther König, 2013).

Peter Weibel, Andreas Beitin, Philipp Ziegler, eds., *Otto Piene Energiefelder*, exh. cat., ZKM | Karlsruhe (Nürnberg: Verlag für moderne Kunst, 2013).

Peter Weibel and Azzam Jaber, eds., *rosalie – LightScapes* (Ostfildern: Hatje Cantz, 2013).

Peter Weibel, ed., *Gemeine Wahrheiten – Werner Büttner*, exh. cat., ZKM | Karlsruhe / Weserburg Museum für moderne Kunst, Bremen (Ostfildern: Hatje Cantz, 2013).

ZKM | Karlsruhe, *Du Zhenjun Babel World*, exh. cat., ZKM | Karlsruhe (Karlsruhe: ZKM | Karlsruhe, 2013).

Peter Weibel, ed., *Jürgen Claus: Liebe die Kunst. eine Autobiografie in einundzwanzig Begegnungen* (Bielefeld: Kerber, 2013).

2014

Franz Erhard Walther, *Objekte, benutzen* [1968], ed. Peter Weibel (Cologne: Verlag der Buchhandlung Walther und Franz König, 2014).

Peter Weibel and Ulrike Gehring, eds., *Mapping Spaces: Networks of Knowledge in 17th Century Landscape Painting*, exh. cat., ZKM | Karlsruhe (Munich: Hirmer, 2014).

Peter Sloterdijk, *Der ästhetische Imperativ. Schriften zur Kunst*, ed. Peter Weibel (Hamburg: Philo & Philo Fine Arts, 2014).

Peter Weibel and Holger Jost, eds., *Claus Bremer: mitspiel. Die Aktivierung der Zuschauer: Vom Dynamischen Theater zur Theaterstadt. Dramaturgische Texte 1948–1971* (Berlin: Alexander Verlag, 2014).

Rolf Funck, Michael Heck and Peter Weibel, eds., *Das ZKM | Karlsruhe. Die Anfänge der Zukunft* (Paderborn: Wilhelm Fink Verlag, 2014).

Achille Bonito Oliva, Carla Subrizi, Peter Weibel et al., eds., *Baruchello: Certain Ideas* (Milano: Electa, 2014).

Christiane Riedel, Yoreme Waltz and Peter Weibel, eds., *Sasha Waltz. Installationen, Objekte, Performances*, exh. cat., ZKM | Karlsruhe (Ostfildern: Hatje Cantz, 2014).

Bernd Scheffer, Christine Stenzer, Peter Weibel et al., eds., *Schriftfilme. Schrift als Bild in Bewegung*, exh. cat., ZKM | Karlsruhe / FACT, Liverpool (Ostfildern: Hatje Cantz, 2014).

Peter Weibel, ed., *Kata Legrady: Smart Pistols*, exh. cat., ZKM | Karlsruhe (Berlin: Distanz Verlag, 2014).

2015

Peter Weibel, ed., *global aCtIVISm: Art and Conflict in the 21st Century*, exh. cat., ZKM | Karlsruhe (Cambridge, MA: MIT Press, 2015).

Peter Weibel and Siegfried Zielinski, eds., *Allah's Automata: Artifacts of the Arab-IslamicRenaissance (800–1200)*, exh. cat., ZKM | Karlsruhe (Ostfildern: Hatje Cantz, 2015).

Peter Weibel, Siegfried Zielinski and Daniel Irrgang, eds., *Flusseriana: An IntellectualToolbox* (Minneapolis, MN: Univocal Publishing, 2015).

Gerald Bast, Peter Weibel and Herwig Steiner, eds., *Performing Public Art*, exh. cat., University of Applied Arts, Vienna (Berlin/ Boston: De Gruyter, 2015).

Bernd Scheffer, Christine Stenzer, Peter Weibel et al., eds., *Typemotion: Type as Image in Motion*, exh. cat., ZKM | Karlsruhe / FACT, Liverpool (Ostfildern: Hatje Cantz, 2015).

Peter Weibel, ed., *Ryoji Ikeda: micro | macro*, exh. broch. (Karlsruhe: ZKM | Karlsruhe, 2015).

Peter Weibel, ed., *Transsolar + Tetsuo Kondo: Cloudscapes*, exh. broch. (Karlsruhe: ZKM | Karlsruhe, 2015).

Peter Weibel, ed., *Die Stadt ist der Star: Kunst an der Baustelle*, exh. broch. (Karlsruhe: ZKM, 2015).

Peter Weibel, ed., *Infosphäre / Infosphere*, exh. broch. (Karlsruhe: ZKM, 2015).

Peter Weibel, ed., *Exo-Evolution*, exh. broch. (Karlsruhe: ZKM, 2015).

Peter Weibel, ed., *ZKM AppArtAward 2015*, broch. (Karlsruhe: ZKM, 2015).

Peter Weibel, ed., *GLOBALE: Kurzführer / Short Guide*, exh. broch. (Karlsruhe: ZKM, 2015).

2016

Eckhart Gillen, Peter Weibel and Gunter De Wit, eds., *Facing the Future: Art in Europe 1945–1968*, exh. cat., ZKM | Karlsruhe / Palais des Beaux-Arts, Brussels / The Pushkin State Museum of Fine Arts, Moscow (Tielt: Lannoo, 2016).

Peter Weibel, ed., *Art in Europe 1945–1968: The Continent that the EU Does Not Know. Supplement to the exhibition catalogue*, exh. broch., (Karlsruhe: ZKM | Karlsruhe, 2016).

Noemi Smolik and Peter Weibel, eds., *Milan Grygar. Sound on Paper*, exh. cat. (Karlsruhe: ZKM | Karlsruhe, 2016).

Peter Weibel, ed., *Lynn Hershman Leeson: Civic Radar*, exh. cat., ZKM | Karlsruhe / Deichtorhallen Hamburg (Ostfildern: Hatje Cantz, 2016).

Peter Weibel, ed., *Beuys Brock Vostell. Aktion Demonstration Partizipation 1949–1983*, exh. cat., ZKM | Karlsruhe (Ostfildern: Hatje Cantz, 2016).

Peter Weibel, Andreas Beitin and Erec Gellautz, eds., *Albrecht Kunkel: QUEST. Fotografien 1989–2009 / Photographs 1989–2009* (Vienna: Verlag für moderne Kunst, 2016).

Peter Weibel, ed., *ZKM AppArtAward 2016*, broch. (Karlsruhe: ZKM, 2016).

ZKM | Karlsruhe, ed., *GLOBALE. Renaissance 2.0: Der Mensch als Möglichkeitswesen*, KUNSTFORUM International, no. 237 (Dec. 2015–Jan. 2016).

2017

Peter Weibel, ed., *Schlosslichtspiele 2015–2017* (Karlsruhe: ZKM | Karlsruhe, 2017).

Peter Weibel et al., *Leibniz' storehouse: photograms of a collected world*, exh. cat., ZKM | Karlsruhe (Berlin: Hatje Cantz, 2017).

Floris Neusüss and Renate Heyne, *Leibniz' storehouse: photograms of a collected world*, ed. Peter Weibel, exh. cat., ZKM | Karlsruhe (Ostfildern: Hatje Cantz, 2017).

Friedrich Kittler, Peter Berz, Peter Weibel and Joulia Strauss, eds., *Götter und Schriften rund ums Mittelmeer* (Paderborn: Wilhelm Fink, 2017).

Peter Weibel, ed., *Von Morgenröten, die noch nicht geleuchtet haben: Peter Sloterdijk. Ein Symposion zum 70. Geburtstag* (Karlsruhe: ZKM | Karlsruhe, 2017).

Peter Weibel, ed., *ZKM AppArtAward 2017*, broch. (Karlsruhe: ZKM | Karlsruhe, 2017).

Peter Weibel, ed., *Open Codes. Living in Digital Worlds*, exh. broch. (Karlsruhe: ZKM | Karlsruhe, 2017).

ZKM | Karlsruhe, Philipp Ziegler and Hanna Jurisch, eds., *Aldo Tambellini: Black Matters*, exh. broch. (Karlsruhe: ZKM | Karlsruhe, 2017).

2018

Amador Vega, Peter Weibel and Siegfried Zielinski, eds., *DIA-LOGOS: Ramon Llull's Method of Thought and Artistic Practice*, exh. cat., ZKM | Karlsruhe (Minneapolis: The University of Minnesota Press, 2018).

Peter Weibel, ed., *Ed Sommer: Planetare Allianz. Werke 1959–2015*, exh. cat., ZKM | Karlsruhe (Esslingen: Edition Cantz, 2018).

Rozemin Keshvani, Axel Heil and Peter Weibel, eds., *Better Books – Better Bookz: art, anarchy, apostasy, counter-culture & new avant-garde*, exh. cat., Trondheim Kunstmuseum / ZKM | Karlsruhe (London/ Cologne: Verlag der Buchhandlung Walther König, 2018).

Peter Weibel and Siegfried Zielinski, eds., *Kunst in Bewegung. 100 Meisterwerke mit und durch Medien*, exh. broch., (Karlsruhe: ZKM | Karlsruhe, 2018).

Loys Egg and Peter Weibel, eds., *Bibliotheca Insomnia, 1978–1979* (Vienna: Verlag für moderne Kunst, 2018).

Daniel Irrgang, Peter Weibel and Siegfried Zielinski, eds., *Political Economy of Data: On Psychometrics, Deep Learning and Net Populism*, exh. cat., ZKM | Karlsruhe / Karlsruhe University of Arts and Design (HfG) (Karlsruhe: ZKM | Karlsruhe, 2018).
ZKM | Karlsruhe, Sabiha Keyif and Bettina Korintenberg, eds., *The Art of Immersion*, exh. broch. (Karlsruhe: ZKM | Karlsruhe, 2018).

2019

Peter Weibel, ed., *Sound Art: Sound as a Medium of Art*, exh. cat., ZKM | Karlsruhe (Cambridge, MA: MIT Press, 2019).

Ulrich Arnswald, Friedrich Stadler and Peter Weibel, eds., *Der Wiener Kreis – Aktualität in Wissenschaft, Literatur, Architektur und Kunst* (Münster: LIT Verlag, 2019).

ZKM | Karlsruhe, Adam Mickiewicz Institute, Warsaw, and Muzeum Sztuki, Łódz, eds., *Ultra Sounds. The Sonic Art of Polish Radio Experimental Studio* (Heidelberg: Kehrer Verlag, 2019).

2020

Peter Weibel, Ludger Brümmer and Sharon Kanach, eds., *From Xenakis's UPIC to graphic notation today*, (Berlin: Hatje Cantz, 2020).

Bruno Latour and Peter Weibel, eds., *Critical Zones. The Science and Politics of Landing on Earth*, exh. cat., ZKM | Karlsruhe (Cambridge, MA: MIT Press, 2020).

Bruno Latour and Peter Weibel, eds., *Critical Zones. Die Wissenschaft und Politik des Landens auf der Erde*, exh. cat., (Karlsruhe: ZKM | Karlsruhe, 2021).

2021

Peter Weibel, *Negative Space: Trajectories of Sculpture in the 20th and 21st Centuries*, ed. Peter Weibel and Anett Holzheid, exh. cat., ZKM | Karlsruhe (Cambridge, MA: MIT Press, 2021).

Peter Weibel and Philipp Ziegler, eds., *Stephan von Huene. What's Wrong with Art?* (Munich: Hirmer, 2021).

2022

Karin Ohlenschläger, Peter Weibel, and Alfred Weidinger, eds., *Christa Sommerer & Laurent Mignonneau: The Artwork as a Living System 1992–2022*, exh. cat., ZKM | Karlsruhe (Cambridge, MA: MIT Press, 2022).

Peter Weibel and Zdeněk Sklenář, eds., *Michael Bielicky. Perpetuum Mobile* (Berlin: DCV, 2022).

Peter Weibel with Christelle Havranek, eds., *Kinetismus – 100 Years of Electricity in Art*, exh. cat., Kunsthalle Prague (Berlin: Hatje Cantz, 2022).

Peter Weibel, ed., *Joseph Beuys. Actions 1963–1986* (Cologne: Verlag der Buchhandlung Walther und Franz König, 2022).

2023

Peter Weibel, ed., *BioMedia: The Age of Media with Life-like Behavior*, exh. cat., ZKM | Karlsruhe (Leipzig: Spector Books, 2023).

Peter Weibel and Stephen Sarrazin, eds., *John Sanborn: Between Order and Entropy; Works 1976–2022* (Berlin: Hatje Cantz, 2023).

Peter Weibel and Siegfried Zielinski, eds., *Art in Motion: 100 Masterpieces with and through Media*, exh. cat. ZKM | Karlsruhe (Cologne: Verlag der Buchhandlung Walther und Franz König, 2023).

Audio-CDs, CD-ROMs and DVD-editions

1999

Peter Weibel, ed., *Dennis Del Favero: Cross currents* (Ostfildern: Hatje Cantz, 1999), CD-ROM.

2001

Peter Weibel, ed., *La nouvelle conception de l'homme. La construction de l'être humain. Podiumsdiskussion im ZKM 2000: Peter Weibel, Peter Sloterdijk; Michel Houellebecq, Alain Finkielkraut* (Karlsruhe: ZKM | Karlsruhe, 2001), audio-CDs.

2002

Gerhard J. Lischka and Peter Weibel, eds., *PVC3 – PerformanceVideo-ComputerClipClub: Elektronische Kunst, Projektion und Reflexion 1995/96* (Ostfildern: Hatje Cantz, 2002), DVD.

ZKM | Karlsruhe, ed., *the complete artintact komplett: Vols. 1–5, 1994–1999*, CD-ROMagazin interaktiver Kunst (Ostfildern: Hatje Cantz, 2002), DVD-ROM.

Barbara Könches, Bernhard Foos and Peter Weibel, eds., *Zeitraffer: Internationaler Medien-Kunst-Preis 1992–2002. 10 Jahre Medienkunst* (Stuttgart: SWR, 2002), DVD.

2003

Peter Weibel, ed., *Bernhard Leitner: Kopfräume / Headscapes* (Ostfildern: Hatje Cantz, 2003), audio-CD.

Peter Weibel, ed., *William Forsythe – Improvisation Technologies: A Tool for the Analytical Dance Eye*, 2nd ed. (Ostfildern: Hatje Cantz, 2003), CD-ROM.

2006

Gerhard J. Lischka and Peter Weibel, eds., *art_clips.ch.at.de: 90 Kurzvideos aus der Schweiz, Österreich und Deutschland* (Ostfildern, Hatje Cantz: 2006), DVD-edition.

2008

Dennis del Favero, Ursula Frohne and Peter Weibel, eds., *Un_imaginable* (Ostfildern: Hatje Cantz, 2008), CD-ROM/DVD.

Peter Weibel, ed., *Three Videos on Iconoclasm, Ritual and Immortality* (Ostfildern: Hatje Cantz, 2008), DVD-edition.

Peter Weibel, ed., *Jürgen Claus: Das audio-visuelle Zeitalter*, Edition ZKM (Ostfildern: Hatje Cantz, 2008), CD-ROM/DVD.

Boris Groys, *Thinking in Loop: Drei Videos über das Ikonoklastische, Rituelle und Unsterbliche*, ed. Peter Weibel (Ostfildern: Hatje Cantz, 2008), DVD.

2010

Peter Weibel and Christoph Blase, eds., *Record Again! 40yearsvideoart.de – part 2*, ZKM Digital Arts Edition (Karlsruhe: ZKM | Karlsruhe, 2010), DVD-edition.

2015

Bernd Scheffer, Christine Stenzer, Peter Weibel and Soenke Zehle, eds., *Schriftfilme Studienedition, Typemotion Study Edition*, ZKM digital arts edition (Karlsruhe: ZKM| Karlsruhe, 2015), DVD-edition.

2019

Rüdiger Schmidt-Grépály and Peter Weibel, eds., *Hans Heinz Holz: Spekulatives Denken Vorlesungen zur Geschichte der Philosophie von Parmenides bis Marx*, (Fridolfing: absolut MEDIEN, 2019), DVD.

Monographs and Essay Collections

1970

Peter Weibel in collaboration with VALIE EXPORT, *Wien: Bildkompendium Wiener Aktionismus und Film* (Frankfurt am Main: Kohlkunstverlag, 1970).

1973

Peter Weibel, *Kritik der Kunst, Kunst der Kritik: es says & I say, Texte 1965-1973* (Vienna/Munich: Jugend & Volk Verlagsgesellschaft, 1973).

1974

Claude Elwood Shannon and John McCarthy, eds., *Studien zur Theorie der Automaten = Automata Studies*, extended edition, trans. Franz Kaltenbeck and Peter Weibel (Munich: Rogner & Bernhard, 1974).

1977

Peter Weibel, *An Annotated Videography 1969–1976* (Innsbruck: Österreichische Hochschülerschaft Innsbruck, 1977).

1979

Gunter Rambow, Peter Weibel, Georg Nemec et al.,*"...das sind eben alles Bilder der Straße." Die Fotoaktion als sozialer Eingriff. Eine Dokumentation* (Frankfurt am Main: Syndikat, 1979).

1982

Peter Weibel, *Mediendichtung: Arbeiten in den Medien Sprache, Schrift, Papier, Stein, Foto, Ton, Film u. Video aus 20 Jahren*. Protokolle 80/2 (Vienna/ Munich: Jugend & Volk Verlagsgesellschaft, 1982).

1984

Peter Weibel, *Zur Geschichte und Ästhetik der digitalen Kunst*, supplementary volume of Ars Electronica '84 (Linz: Linzer Verlagsgesellschaft, 1984).

1987

Peter Weibel, *Die Beschleunigung der Bilder in der Chronokratie. Anhang. Das Reale und sein Double: Der Körper* (Wabern/Bern: Benteli, 1987).

1997

Peter Weibel and Gerhard J. Lischka, *Engagement: jetzt und hier: ein Gespräch*, ed. René Stettler (Basel/ Frankfurt am Main: Stroemfeld, 1997).

Peter Weibel and Werner DePauli-Schimanovich, *Kurt Gödel. Ein mathematischer Mythos* (Vienna: öbv & hpt Verlagsgesellschaft, 1997).

Peter Weibel and Otto E. Rössler, *Aussenwelt – Innenwelt – Überwelt: ein Gespräch.*, ed. René Stettler (Basel/ Frankfurt am Main: Stroemfeld, 1997).

2000

Peter Weibel, *Lebenssehnsucht und Sucht* (Berlin: Merve, 2000).

2004

Peter Weibel, *Gamma und Amplitude: Medien und kunsttheoretische Schriften*, ed. Rolf Sachsse (Berlin: Philo & Philo Fine Arts, 2004).

2006

Peter Weibel, *Time Slot: Geschichte und Zukunft der apparativen Wahrnehmung vom Phenakistiskop bis zum Quantenkino*, International Flusser Lecture (Cologne: Verlag der Buchhandlung Walther König, 2006).

2015

Peter Weibel, *Enzyklopädie der Medien*, vol. 1, *Architektur und Medien. Die Künste des Raumes im Zeitalter der Medien. Ausgewählte Schriften von Peter Weibel*, ed. University of Applied Arts, Vienna and ZKM | Karlsruhe (Ostfildern: Hatje Cantz, 2015).

2016

Peter Weibel, *Enzyklopädie der Medien*, vol. 2, *Musik und Medien. Vom Klang im technischen Zeitalter*, ed. University of Applied Arts, Vienna and ZKM | Karlsruhe (Berlin: Hatje Cantz, 2016).

2019

Peter Weibel, *Enzyklopädie der Medien*, vol. 3, *Kunst und Medien. Die Transformationen des Bildes in der Ära der Apparate*, ed. University of Applied Arts, Vienna and ZKM | Karlsruhe (Berlin: Hatje Cantz, 2019).

2021

Peter Weibel, *Enzyklopädie der Medien*, vol. 4, *Literatur und Medien. Expansion der Sprachkunst ins visuelle und technische Feld*, ed. University of Applied Arts, Vienna and ZKM | Karlsruhe (Berlin: Hatje Cantz, 2021).

2023

Peter Weibel, *Enzyklopädie der Medien*, vol. 5, *Politik und Medien. Von den Medien der Macht zur Macht der Medien*, ed. University of Applied Arts, Vienna and ZKM | Karlsruhe (Berlin: Hatje Cantz, 2023).

Peter Weibel, *Enzyklopädie der Medien*, vol. 6, *Theorie und Medien*, ed. University of Applied Arts, Vienna and ZKM | Karlsruhe (Berlin: Hatje Cantz, 2023).

Publications on Peter Weibel

1988

Peter Noever, ed., *Peter Weibel: Inszenierte Kunst-Geschichte. Mise-en-scène of Art History*, exh. cat., MAK – Museum of Applied Arts, Vienna (Vienna: University of Applied Arts, 1988).

1992

Andreas Hochegger, ed., *Peter Weibel – Malerei zwischen Anarchie und Forschung*, exh. cat. (Graz: Neue Galerie Graz, 1992).

Robert Fleck, ed., *Peter Weibel. Zur Rechtfertigung der hypothetischen Natur der Kunst und der Nicht-Identität in der Objektwelt*, exh. cat. (Cologne: Galerie Tanja Grunert, Cologne 1992).

1993

Peter Weibel. Jede(r) ist jede(r). Jede(r) ist der/die andere für jede(n) in der Heimat Babylon, exh. cat., Karmeliterkloster, Ausstellungshalle Frankfurt am Main (Mainz: 1993).

1994

Robert Fleck, ed., *Peter Weibel: Hypothetische Produkte. Vom Elend der Privilegien*, exh. cat. (Vienna: Prodomo, 1994).

1996

Romana Schuler, ed., *Peter Weibel: Bildwelten 1982–1996* (Vienna: Triton, 1996).

1999

Arnulf Rohsmann, ed., *Peter Weibel. Globale Gier*, exh. cat. (Klagenfurt: Kärntner Landesgalerie, 1999).

2004

Inge Zimmermann and Ingrid Krüger, eds., *Käthe-Kollwitz-Preis 2004. Peter Weibel* (Berlin: Akademie der Künste, 2004).

2006

Peter Weibel and Harald Falckenberg, eds., *Peter Weibel: Das offene Werk 1964–1979*, exh. cat., Neue Galerie Graz (Ostfildern: Hatje Cantz, 2006).

2009

Gregor Eichinger, *Bof!: Peter Weibel & Peter Sloterdijk* (Zurich: ETH, 2009).

2010

Werkstadt Graz (Joachim Baur and Barbara B. Edlinger), ed., *Peter Weibel: Europa – Ein Umschreibe-programm* (Graz: Edition Werkstadt Graz, 2010), DVD-Edition.

2015

Agnes Husslein-Arco and Alfred Weidinger, eds., *Peter Weibel. Medienrebell: Warnung! Diese Ausstellung kann ihr Leben verändern*, exh. cat., Belvedere, Vienna (Cologne: Verlag der Buchhandlung Walther König, 2015).

2020

Agnes Tieze, ed., *Peter Weibel: (Post-)Europa? Lovis-Corinth-Preis 2020*, exh. cat., Kunstforum Ostdeutsche Galerie Regensburg (Berlin: Hatje Cantz, 2020).

1944

Born in Odesa, Ukraine (then the Ukrainian Soviet Socialist Republic)

1945–49

In a camp in Ried im Innkreis, Austria, with "displaced person" status

1951–60

Childhood and youth in Austrian boarding schools (Gosau, Linz) and with foster parents

1960–2023

Working on experimental literature

1962–70

Working in Sweden during the summer months

1963–64

Studying French language and comparative literature in Paris, while visiting the Cinémathèque française almost daily

1964–67

Studying medicine at the University of Vienna

1964–2023

Working on a concept of expanded film and photography

1965–68

Actions with members of the Vienna Group (Wiener Gruppe) and Vienna Actionism (Wiener Aktionismus)

1967–74

Studying philosophy and mathematical logic at the University of Vienna

1967–2000

Artistic collaborations with VALIE EXPORT

1968–69

Cofounder and member of the Austria Filmmakers Cooperative (together with VALIE EXPORT, Kurt Kren, Hans Scheugl, Ernst Schmidt Jr., and Gottfried Schlemmer)

1971–74

Tele-Actions on Austrian Television (ORF)

1972–2023

Artistic collaborations with Susanne Widl

1974–2013

Production of five electronic media operas

1976–81

Teaching the morphology and theory of form at the University of Applied Arts Vienna

1977–80

Production of the feature films *Invisible Adversaries* (*Unsichtbare Gegner*, 1977) and *Human Females* (*Menschenfrauen*, 1980), in collaboration with VALIE EXPORT

1978

Cofounder of the band Hotel Morphila Orchester with Loys Egg

1979/80

Visiting professorship for media art at the Gesamthochschule Kassel

1981

Teaching theory of perception at the Gesamthochschule Kassel

1981–84

Visiting professorship for design theory and art education (Gestaltungslehre und Bildnerische Erziehung) at the University of Applied Arts Vienna

1982–85

Professor for photography at the University of Kassel

1984–89

Associate Professor for Video and Digital Arts an initiator of the Digital Arts Laboratory at the Center for Media Study, State University of New York at Buffalo

1984–2017

Professorship of visual media art at the University of Applied Arts, Vienna

1986–92

Artistic advisor of Ars Electronica, Linz

1989–95

Founding director of the Institute of New Media at the Hochschule für Bildende Künste–Städelschule, Frankfurt am Main

1992–95

Artistic Director of Ars Electronica, Linz

1993–2011

Head curator of the Neue Galerie Graz

1993–99

Austrian commissioner of La Biennale di Venezia

Early 1990s

First interactive computer-based installations

1998

Visiting professorship at the Nova Scotia College of Art and Design, Halifax

1999–2023

Chairman and CEO of the ZKM | Center for Art and Media Karlsruhe

2007

Honorary Doctorate from the University of Art and Design, Helsinki

2008

Artistic Director of the International Biennial of Contemporary Art of Seville (BIACS3)

2009–12

Visiting professorship at the University of New South Wales, Sydney

2009

Named full member of the Bavarian Academy of Fine Arts, Munich

Named full member of the Class for the Arts at the North Rhine-Westphalia Academy of the Sciences and the Arts, Düsseldorf

2011

Artistic Director of the 4th Moscow Biennale of Contemporary Art

2013

Honorary Doctorate from the University of Pécs

Named full member of the European Academy of Sciences and Arts, Salzburg

2015

Honorary member of the Russian Academy of the Arts, Moscow

2015–23

Curator of the *Schlosslichtspiele Light Festival Karlsruhe*

2015–17

Curator of *lichtsicht 5* and *6 – Projektions-Biennale*, Bad Rothenfelde

2017–23

Director of the Peter Weibel Research Institute for Digital Cultures at the University of Applied Arts, Vienna

2018

Visiting professorship at the Düsseldorf Academy of Fine Arts

2023

On March 1, Peter Weibel died in Karlsruhe

Hans Belting (1935–2023)

was a German art historian and media theorist. He is widely known for his contributions to the field of *Bildwissenschaft* ("image science"). He taught at numerous universities and colleges in Germany and abroad. Until his retirement in 2002, Belting was a professor at the Institute for Art History and Media Theory at the Karlsruhe University of Arts and Design (HfG). Pioneering in the development of a global perspective on art studies and museum practice was the research project GAM – Global Art and the Museum, which Belting initiated in 2006 with Peter Weibel and Andrea Buddensieg at the ZKM | Karlsruhe. The project, which ran until 2016, looked at new museum practices and the worldwide development of art biennials.

Bazon Brock (*1936)

is a German art theorist and critic, multimedia generalist, and artist. He is considered a member of the Fluxus movement. He was a professor of aesthetics at the Hochschule für bildende Künste Hamburg (HFBK), the University of Applied Arts Vienna, and the University of Wuppertal. He developed the method of "action teaching," and established the Documenta visitor schools in Kassel, Germany, which took place from 1968–92. In 2011, he founded the "Denkerei/Amt für Arbeit an unlösbaren Problemen und Massnahmen der hohen Hand" (Think tank/employment office on unsolvable problems and "high handed" measures) in Berlin.

Boris Groys (*1947)

is a philosopher, art critic, media theorist, essayist, and internationally recognized expert on the art and literature of the Soviet era, especially the Russian avant-garde. Groys teaches as Global Distinguished Professor of Russian and Slavic Studies at New York University, was Professor of Philosophy at the European Graduate School, and Senior Research Fellow at the Karlsruhe University of Arts and Design (HfG). Furthermore, he is a member of the Association Internationale des Critiques d'Art and a member of the Democracy in Europe 2025 (DiEM25) movement, which was founded in 2016.

Wolfgang Kemp (*1946)

is a German art historian, author, and professor of art history. He taught as a professor of art history at the Gesamthochschule Kassel, the University of Marburg and, until his retirement in 2011, at the University of Hamburg. In numerous publications, he has explored the history and theory of photography, reception aesthetics, and narrative research. He also held several visiting professorships, including at Harvard University, University of California Los Angeles (UCLA), the Wissenschaftskolleg zu Berlin, and the Getty Research Center in Los Angeles.

Jens Lutz (*1975)

has been head of the Publications Department at ZKM | Karlsruhe, Germany, since 2008. During this time, he has overseen the production of numerous books, brochures, DVD editions, and other publications in a wide variety of media, many of which have been published worldwide. He has also cocurated two exhibitions for the ZKM, *Bense and the Arts* (2010) and *Reinhard Döhl: Everything is Possible. Anything Goes* (2017).

Friederike Mayröcker (1924–2021)

was an Austrian writer. From the 1950s until her death, she published over 80 books; poetry, prose, children's books, stage texts, and radio plays. She experimented with language and was regarded as an avant-garde poet, and as one of the leading German-language authors. Mayröcker was awarded numerous German-language literary prizes and was frequently mentioned as a potential Nobel laureate. She was a close friend of Peter Weibel.

Hans Ulrich Obrist (*1968)

is artistic director of the Serpentine Galleries in London. Prior to this, he was the curator of the Musée d'Art Moderne de la Ville de Paris. Since his first show *World Soup (The Kitchen Show)* in 1991, he has curated more than 300 exhibitions. Obrist has lectured at international academic and art institutions, and is a contributing editor to *Artforum*, *AnOther Magazine*, *Cahiers D'Art*, and *032C*. He is the author of The Interview Project, an extensive, ongoing project of interviews. In 2011 Obrist received the CCS Bard Award for Curatorial Excellence, and in 2015 he was awarded the International Folkwang Prize for his commitment to the arts.

Pamela C. Scorzin (*1965)

is an art, design, and media theorist, and a professor of Art History and Visual Culture Studies at Dortmund University of Applied Sciences and Arts, Department of Design, Germany. Born in Vicenza, Italy, she studied European art history, philosophy, English and American literature, and history in Stuttgart and Heidelberg, Germany. Since 2005, she is a member of the German section of the International Association of Art Critics (AICA). She has published in German, English, French, and Polish on art-historical and cultural-historical topics from the seventeenth to the twenty-first century.

Peter Sloterdijk (*1947)

is a German philosopher, cultural theorist, and publicist. From 1992, Sloterdijk was professor of philosophy and aesthetics at Karlsruhe University of Arts and Design (HfG), and from 2001–15 its rector. From 1993–2008, he also headed the Institute for Cultural Philosophy and Media Theory at the Academy of Fine Arts in Vienna. Peter Sloterdijk has received numerous honorary doctorates and prizes (including the Ernst Robert Curtius Prize for Essay Writing, 1993; Sigmund Freud Prize, 2005; Helmuth Plessner Prize, 2017; European Prize for Political Culture, 2021). In 2013, he was listed as one of the "100 most influential thinkers in the world" by the Swiss Gottlieb Duttweiler Institute.

Terry Smith (*1944)

is an Australian art historian, art critic, and artist. Since 2001 he has been Andrew W. Mellon Professor of Contemporary Art History and Theory in the Department of the History of Art and Architecture at the University of Pittsburgh, USA. He also serves as a board member of The Andy Warhol Museum, Pittsburgh. His major research interests are contemporary art of the world, the histories of multiple modernities and modernisms, the history and theory of contemporaneity, and the historiography of art history and art criticism. He was the 2010 winner of the Franklin Jewett Mather Award for art criticism conferred by the College Art Association, USA, and in 2011 he received the Australia Council Visual Arts Laureate Award.

Philipp Ziegler (*1975)

is head of the Curatorial Department at ZKM | Karlsruhe, Germany. Since 2012, he has been responsible for numerous exhibitions as curator and cocurator. Besides his work at the ZKM, he has curated or cocurated several international projects; for example, he was one of the curators of the 6th Guangzhou Triennial at the Guangdong Museum of Art, Guangzhou, China in 2018, and in 2021 he was a member of the curatorial team of the exhibition *Machine Does Not Give Change* at Device_art festival in Zagreb, Croatia. He has published numerous articles and essays in magazines and catalogs.

Siegfried Zielinski (*1951)

is a German media theorist. He held the chair for Media Theory: Archaeology, and Variantology of the Media at Berlin University of the Arts, and is Michel Foucault Professor for Techno-Culture and Media Archaeology at the European Graduate School in Saas-Fee, Switzerland. He was founding rector (1994–2000) of the Academy of Media Arts Cologne, director of the Vilém Flusser Archive (1998–2016), and rector of the Karlsruhe University of Arts and Design (2016–18). Zielinski has published numerous books and essays mainly focusing on the archaeology and variantology of the relations between art and media. He has also curated several exhibitions, including at the ZKM. Zielinski is member of the Berlin Academy of Arts and the North Rhine-Westphalian Academy of Sciences, Humanities, and the Arts.

U

V

W

X

y

Z

Editors
Jens Lutz and Philipp Ziegler
With Clara Runge, Stephanie Stadler, and Patrick Trappendreher

Editorial staff, project management, image research
Stephanie Stadler (2021–23), Clara Runge, Patrick Trappendreher, Jens Lutz

Texts (if not stated otherwise)
ZKM | Curatorial Department (Judith Bihr, Svenja Clauss, Hanna Jurisch, Lena Reitschuster, Clara Runge, Leonard Sprüth, Philipp Ziegler), ZKM | Publications

Copy editing
Gloria Custance, Patrick Hubenthal

Translations (if not stated otherwise)
Patrick Hubenthal, Peter Dahm Robertson, Gloria Custance, Lauren K. Wolfe

Graphic design
2xGoldstein (Andrew Goldstein, Jeffrey Goldstein, Joshua Kaiss, Erik Schöfer)

Lithography
CoMYK Agentur für Grafik und Litho, Karlsruhe

Paper
GardaPat 13 KIARA 135 g/m²

This book was set in
Respektiva, Tobias, Px Grotesk

Printed and bound
Offizin Scheufele, Stuttgart, Germany

Project support
Christiane Riedel (former COO of ZKM)

Special thanks to
Karin Buol-Wischenau, Loys Egg, Sabine Grieb, Christian Haardt, Idis Hartmann, Ulrike Havemann, Anett Holzheid, Hartmut Jörg, Tobias Klingenmayer, Marianne Meister, Felix Mittelberger, Ulrike Rieger, Margit Rosen, Volker Sommerfeld, Miriam Stürner, Dominique Theise, Susanne Widl, Hannah-Maria Winters

Printed and bound in Germany.

ISBN 978-0-262-04913-9
Library of Congress Control Number: 2023945587

ZKM | Center for Art and Media Karlsruhe
Lorenzstraße 19
76135 Karlsruhe
Germany
Phone: +49 (0)721/8100-1200
info@zkm.de, www.zkm.de

zkm karlsruhe

respectively, Peter Weibel

ZKM | Center for Art and Media Karlsruhe
September 28, 2019–March 8, 2020

Curators
Philipp Ziegler, Judith Bihr, Clara Runge

Project assistance
Svenja Clauss, Hanna Jurisch, Lena Reitschuster, Leonard Sprüth

Head of technical museum and exhibition services
Martin Mangold

Exhibition graphic design
2xGoldstein

Scenography
Matthias Gommel

Technical project management
Andrea Hartinger

Construction team
Volker Becker, Claudius Böhm, Mirco Fraß, Rainer Gabler, Gregor Gaissmaier, Jan Gerik, Ronny Haas, Manfred Hauffen, Dirk Heesakker, Daniel Heiss, Christof Hierholzer, Werner Hutzenlaub, Gisbert Laaber, Christian Nainggolan, Marco Preitschopf, Martin Schläfke, Marc Schütze

Registrar
Marianne Meister, Juliane Peiser

Wissen – Collection, Archives & Research
Margit Rosen, Andreas Brehmer, Claudia Gehrig, Hartmut Jörg, Dorcas Müller, Sarah Donata Schneider, Morgan Stricot, Matthieu Vlaminck

Conservation team
Henrike Mall, Ellen Kotthaus, Jana Merseburg, Anna Virgin

Video studio
Christina Zartmann, Moritz Büchner, Güzide Coker, Andy Koch, Xenia Leidig, Peter Müller

Museum communication
Janine Burger, Banu Beyer, Sabine Faller, Regine Frisch, Barbara Zoé Kiolbassa, Kerstin Tscherbakova

Communication and marketing
Dominika Szope, Adamantia Goulandris, Regina Hock, Sabine Jäger, Alexa Knapp, Adrian Koop, Marie Schmidt

Library
Petra Zimmermann, Christiane Minter, Regina Strasser-Gnaedig

Lenders to the exhibition
Artelier Graz; Belvedere, Wien; Generali Foundation; Julia Stoschek Foundation; Kunsthaus Bregenz; Kunstsammlung der Landeshauptstadt Bregenz; LENTOS Kunstmuseum Linz; mumok – Museum moderner Kunst Stiftung Ludwig Wien; Neue Galerie Graz am Universalmuseum Joanneum; No-Collection Berlin; Oberösterreichisches Landesmuseum; Ostlicht Collection, Vienna; private collection, Berlin; Sammlung Museum der Moderne Salzburg; Dr. Steinek Heinrich; ZKM | Karlsruhe; and private collections

Special thanks to
Peter Weibel, Sascha Fronczek, Martin Häberle, Idis Hartmann, Robert O'Kane, Bernd Lintermann, Christian Lölkes, Roland Merz, Michael Truxa, Nikolaus Völzow, Dan Wilcox, and the entire team of the ZKM.

The exhibition was supported by:

Karlsruhe
vitra.

Founders of ZKM

Baden-Württemberg
MINISTERIUM FÜR WISSENSCHAFT, FORSCHUNG UND KUNST
Karlsruhe

Premium partner of ZKM

EnBW

Cover: © 2xGoldstein / 26: photo © ZKM | Karlsruhe; photo: Felix Mittelberger / 32: photos © Hermann Hendrich / 33: photo © Joseph Tandl / 41: photos © Bernd Junker / 49, top: photo © Wolfgang Woessner / 55: photo © ZKM | Karlsruhe; photo: Felix Mittelberger / 56, left: photos © Peter Hermann / 65: photos © VALIE EXPORT / 68–69: photos © VALIE EXPORT / 78, top: photo © VALIE EXPORT / 79: photo © VALIE EXPORT / 80: photo © VALIE EXPORT / 83: photo © Michael Schuster / 87: photo © Johann Koinegg / 88: photo © Kunsthalle Tirol / 93: photo © VALIE EXPORT / 99, bottom: photo © ZKM | Karlsruhe; photo: Steffen Harms / 100–101: photos © VALIE EXPORT / 102, right: photos © VALIE EXPORT / 103: photos © VALIE EXPORT / 106: photo © VALIE EXPORT / 111: photo © ZKM | Karlsruhe; photo: Franz Wamhof / 122: photo © VALIE EXPORT / 131: photo © Archiv Susanne Widl / 132–33: photos © Archiv Susanne Widl / 136: photos © Michael Schuster / 137: photos © Michael Schuster / 140: photo © Johann Koinegg / 148, fig. 1: photo © Artokoloro / Alamy Stock Photo / 150, figs. 4 a, b: photos © Library of Congress, Prints & Photographs Division, FSA/OWI Collection, LC-DIG-fsa 8b29516, Courtesy of Gendell Gallery / 151: photo © The Robert Capa and Cornell Capa Archive, Gift of Cornell and Edith Capa, 2010 / 152: photo © Ludwig Hoffenreich / 155: photo © VALIE EXPORT / 160: photo © Mark Gisler / 163, top: photo © ZKM | Karlsruhe / 163, bottom: photo © ZKM | Karlsruhe; photo: Manfred Hauffen / 164: photos © ZKM | Karlsruhe / 171, bottom: photo © Werner H. Mraz / 172, bottom: photo © Jan Gerigk / 173: photo © Jan Gerigk / 174, top and center (parts 1 and 2): photo © Peter Hassmann / 174, bottom (part 3): photo © Werner H. Mraz / 175: photos © Werner H. Mraz / 176: photos © Hans Scheugl / 177: photos © Werner Schulz / 179: photo © Werner Schulz / 180: photos © Werner Schulz / 181: photos © Hans Scheugl / 183: photo © ZKM | Karlsruhe; photo: Barbara von Woellwarth / 195: photo © Wolf Isser / 198: photos © ZKM | Karlsruhe; photo: Barbara von Woellwarth / 208, top: photo © Hermann Hendrich / 209, bottom: photo © Schmidt/ Pongracz / 211, bottom: photo © Gertrude Wolfschwenger / 212, top: photos © Franz Kaltenbeck / 219: photo © MONO-CHROM / 221, fig. 5: photo © Erik Kessels / 221, fig. 6: photo © ZKM | Karlsruhe; photo: Felix Grünschloß / 235: photos © VALIE EXPORT / 243, top: photo © VALIE EXPORT / 244: photos © VALIE EXPORT / 245: photos © VALIE EXPORT / 262, top: photo © Werner Schulz / 262, bottom: photo © VALIE EXPORT / 268: photo © F. Fischer / 271: photo © VALIE EXPORT / 272, top: photo © F. Fischer / 273: photos © Wojciech Bruszewski / 276: photos © Wojciech Bruszewski / 275: photo © Michael Schuster / 279: photos © Michael Schuster / 280–83: photos © Michael Schuster / 284: photo © Belvedere, Vienna / 285: photos © Michael Schuster / 286: photos © A. Palacios-Nunez / 287, bottom: photo © Michael Schuster / 288: photos © Walter Scholz-Ruhs / 289, bottom: photo © Belvedere, Vienna / 290: photo © Karl Heinz Koller / 291, top: photos © Gerhard Koller / 297: photo © ZKM | Karlsruhe; photo: Elias Siebert / 305: photo © MMCA, photo: Hong Cheolki / 312: photo © Galerie Tanja Grunert; photo: Michael Janssen / 314, top: photo © Teri Wehn-Damisch / 315: photos © Thomas Dreher / 317: photo © Galerie Tanja Grunert; photo: Michael Janssen / 321, top left and bottom: photo © ZKM | Karlsruhe / 321, top right: photo © ZKM | Karlsruhe; photo: Franz Wamhof / 322: photo © ZKM | Karlsruhe; photo: Jonas Zilius / 323, left: photo © ZKM | Karlsruhe; photo: Jonas Zilius / 323, right: photo © ZKM | Karlsruhe / 324: photo © ZKM | Karlsruhe; photo: Felix Grünschloß / 325: photos © ZKM | Karlsruhe / 326: photo © ZKM | Karlsruhe; photo: Jonas Zilius / 327: photo © ZKM | Karlsruhe; photo: Felix Grünschloß / 331: photo © ZKM | Karlsruhe; photo: Jonas Zilius / 348: photo © Johann Koinegg / 349: photo © VALIE EXPORT / 351: photo © Michael Schuster / 358: photo © Archiv Susanne Widl / 360: photos © Klomfar & Sengmüller / 361, top: photo © Klomfar & Sengmüller / 365: photo © Jan Gerigk / 366, top right: photo © Jan Gerigk / 366, bottom left: photo © ONUK / 367, top right: photo © Manfred Hauffen / 367, center right: photo © Jan Gerigk / 367, bottom left and right: photos © ONUK / 371, bottom: photo © ZKM | Karlsruhe; photo: Felix Grünschloß / 372, top: photo © ZKM | Karlsruhe; photo: Uli Deck / 379, top and bottom right: photos © Michael Schuster / 382: photo © Johann Koinegg / 383: photo © Gert Rappenecker / 389, bottom: photo © Michael Schuster / 394: photos © Michael Schuster / 396, bottom: photo © Johann Koinegg / 402, top left: photo © Michael Schuster / 402, bottom left: photo © Michael Schuster / 402, bottom right: © Bauhaus-Archiv Berlin / 403, top left: photo © Michael Schuster / 407, top right and bottom: photos © Norbert Truxa / 418: photo © Dieter Bogner / 421, bottom: photos © Belvedere, Vienna / 427: photo © Belvedere, Vienna / 428: photos © Jean Brasille / 431: photos © Peter Philipp / 432: photo © Wolfgang Woessner / 435, top: photo © Wolfgang Werner / 435, bottom left: photo © Bernhard Wimmer / 435, bottom right: photo © N. Lackner / 440: photos © Mark Gisler / 442, left: photo © Belvedere, Vienna / 442, right: photo © Gerhard Koller / 443: photos © Gerhard Koller / 452–53: photos © Johann Koinegg / 456: photo © Michael Schuster / 457: photo © Georg Kargl / 458–59: photos © Elisabeth Jappe / 461: photos © Michael Schuster / 464: photos © Gerhard Koller / 466, top: photo © Belvedere, Vienna / 468, 469: © Generali Foundation / 470: photos © Galerie Kampl, Munich / 471: photo © Werner Kaligofsky / 472: photos © Philipp Schönborn / 473: photo © Blanka Lamrová / 478: photos © Norbert Truxa / 480, top: photo © Belvedere, Vienna / 480, bottom: photo © Belvedere, Vienna / 481, top: photo © Norbert Truxa / 484: photos © Michael Schuster / 494: photo © Viehof Collection, formerly Speck Collection; photo: Henning Rogge / 498, left: photo © Michael Broome / 499, center: photo © Tom Picton, National Archives / 499, bottom: photo © Tate, London; photo: Tom Picton / 500: photos © Joseph Tandl / 501: photos © Lisl Ponger / 502: photos © Lisl Ponger / 503: photo © Inge Werth / 504–505: © Generali Foundation, photos: Josef Tandl / 506, bottom right: photos © Detlev Heiliger / 507: photos © Adolf Ljubic / 508, top: photos © Ernst Schmidt Jr. / 509, top: photo © Klein / 511: photos © Candid Lang / 515, bottom: photos © Friedl Bondy / 516–17: photos © Wolfgang Sos / 518–19: photos © Peter Zollna / 521: photos © VALIE EXPORT / 525, top: photo © VALIE EXPORT / 530: photos © VALIE EXPORT / 532: photos © Lotte Hendrich-Hassmann / 535: photos © Ernst Schmidt Jr. / 542, 5.4: photo © Belvedere, Vienna / 545, 4.5: photo © VALIE EXPORT / 547, 2.3: photo © VALIE EXPORT / 547, 4.2: photo © ZKM | Karlsruhe; photo: Steffen Harms / 548, 2.2: photo © Friedl Bondy / 548, 4.2: photo © Franz Novotny / 549, 3.3: photo © Angelika Gradwohl / 550, 2.2: photo © Johann Koinegg / 551, 4.4: photo © Galerie Hubert Winter, Vienna / 552, 4.3: photo © Michael Schuster / 552, 5.1: photo © Inge Werth / 554, 1.5: photo © Hans Weigand / 555, 2.5: photo © VALIE EXPORT / 555, 5.1: photo © Gert Rappenecker / 557, 3.2: photo © Lauda

Photos © ZKM | Karlsruhe; photos: Tobias Wootton

27; 29; 31; 59; 60–61; 62; 63; 75, top; 76; 77; 78, bottom; 82; 84; 85; 86; 112; 121; 123; 125, left; 125, right; 126, left; 126, right; 146, top; 146, bottom; 159; 161; 178; 189; 197; 199; 220, fig. 3; 258, right; 264; 272, bottom; 287, top; 291, bottom; 298; 309; 311; 313; 314, bottom; 328; 330; 369; 370, top; 371, top and center; 372, bottom; 373; 391, top; 391, bottom; 405; 406; 407, top left; 408–409; 410; 411 left; 411, right; 412; 413; 414; 415 top; 415, bottom; 416; 422, top; 429; 444, bottom; 445; 449, bottom; 465; 467; 468–69; 474–75; 477; 479 top; 479, bottom; 481, bottom; 482, bottom; 483; 485; 486–87; 523; 546, 1.1; 546, 5.4; 550, 1.5; 553, 1.5; 554, 4.2

Photos © Archive Peter Weibel

28; 30; 34; 35, bottom; 37, top; 37, bottom; 38; 39; 40; 42; 49, bottom; 50–51; 52–53; 56, right; 57; 58; 64; 66; 67; 70; 71; 72; 73, top; 73, bottom; 74; 75, bottom; 81; 94; 95; 96–97; 98; 99, top; 102, left; 104, top; 104, bottom; 105; 107; 113, left; 113, right; 114, top; 114, bottom; 115, top; 115, bottom; 116, top; 116, bottom; 117, left; 117, right; 118; 127, left; 127, right; 128; 129; 130; 138–39; 141; 142; 145; 156; 157; 167; 168; 170; 171, top; 172, top; 184; 186; 187; 190; 191; 192–93; 196; 200; 201; 202; 206; 207, top; 207, center; 207, bottom; 208, center; 208, bottom; 209, top; 209, center; 210, bottom; 211, top; 212, center; 212, bottom, 213, top; 213, bottom; 228; 229, top; 229, bottom; 230; 232; 233; 234, left; 234, right; 237; 238; 239; 240; 241; 242; 243, bottom; 248; 249, left; 250, left; 250, right; 251, left; 251, right; 252, top; 253, left; 253, right; 254–55; 256; 257; 258, left; 259; 265; 266–67; 269; 270, top; 270, bottom; 274; 277; 289, top; 292; 293; 301, left; 301, right; 302, top; 302, bottom; 303, top; 303, bottom; 306–307; 308; 310; 316; 318; 319; 329; 342–43; 344; 345; 346–47; 352–53; 354; 355–57; 370, bottom; 374; 375; 379, bottom left; 380; 381; 384; 385; 388–89; 390; 392–93; 395; 396, top; 397; 400; 401; 403, bottom; 404; 420, top; 420, bottom; 421, top; 422, bottom; 423; 424; 425; 426, top; 426, bottom; 430; 433; 434; 437; 438; 444, top; 449, top; 450; 451; 454; 455; 460; 462–63; 466, bottom; 476; 482, top; 498, right; 499, top; 506 top and bottom left; 508, bottom (poster design Günter Brus); 509 bottom; 512–13; 514; 515, top; 520; 522; 524; 525, bottom; 527; 529; 531; 533; 534; 536–37; 540, 4.4; 541, 1.5; 541, 2.3; 541, 4.4; 541, 5.5; 542, 2.1; 542, 3.3; 543, 4.2; 543, 4.3; 543; 4.4; 544. 2.3; 544, 3.2; 544, 4.1; 545, 3.3; 546, 1.4; 546, 2.3; 546, 3.3; 546, 4.3; 547, 1.2; 547, 3.4–5; 547, 5.1; 547, 5.2; 547, 4.3; 547, 4.4; 548, 4.4; 549, 4.3; 550, 3.3; 550, 3.4; 550, 4.5; 551, 4.3; 553, 2.3–4; 553, 3.2; 553, 4.2; 553, 5.3; 553, 5.4; 555, 1.1; 555, 2.4; 555, 5.3; 555, 5.5; 556, 5.5; 557, 5.1, endpapers and spine

© VG Bild-Kunst, Bonn 2024

Joseph Beuys, Giorgio de Chirico, VALIE EXPORT, Joseph Kosuth, Le Corbusier, Hermann Nitsch, Franz Wamhof

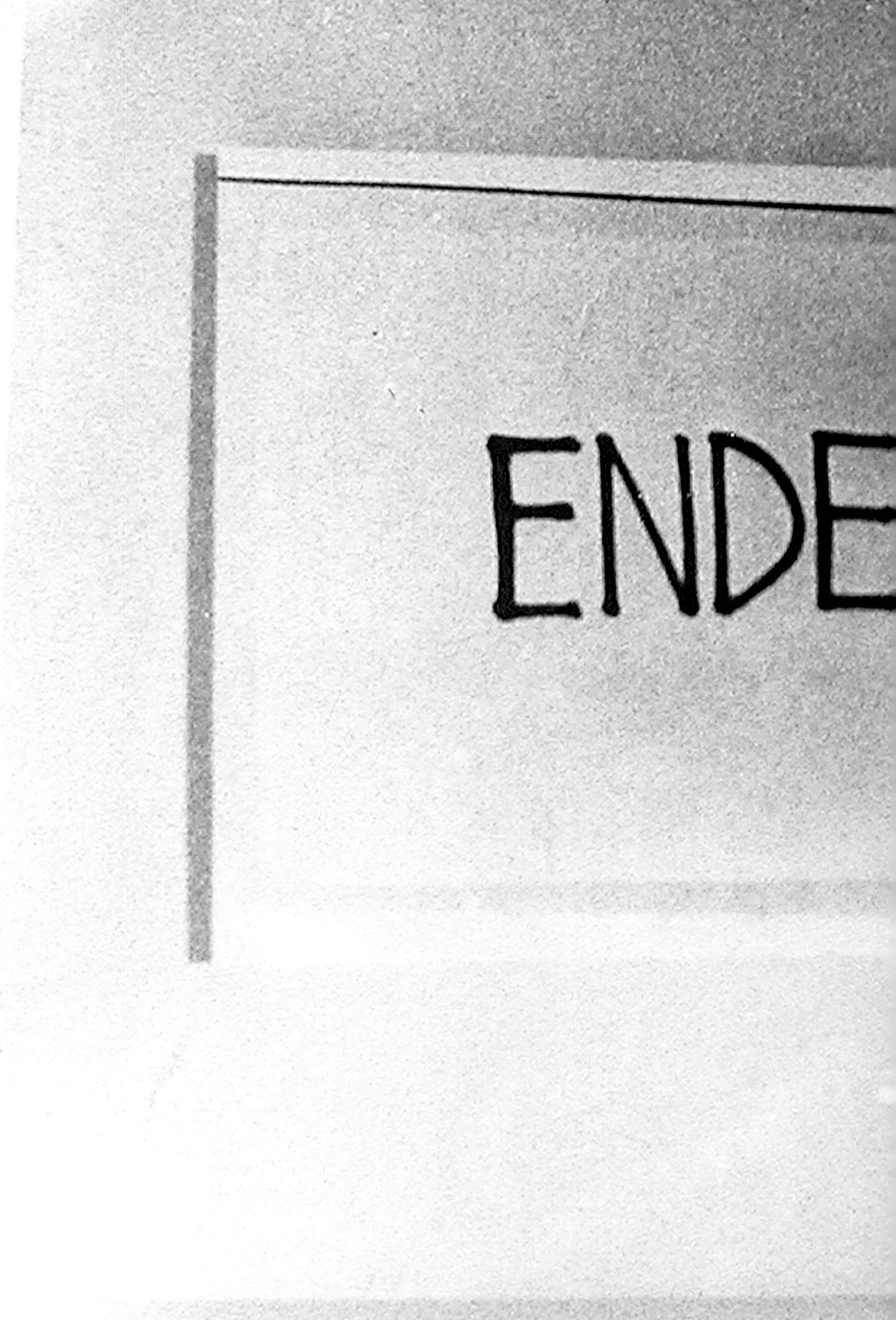

The Endless End / Das Endlose Ende

1969 | photo poem

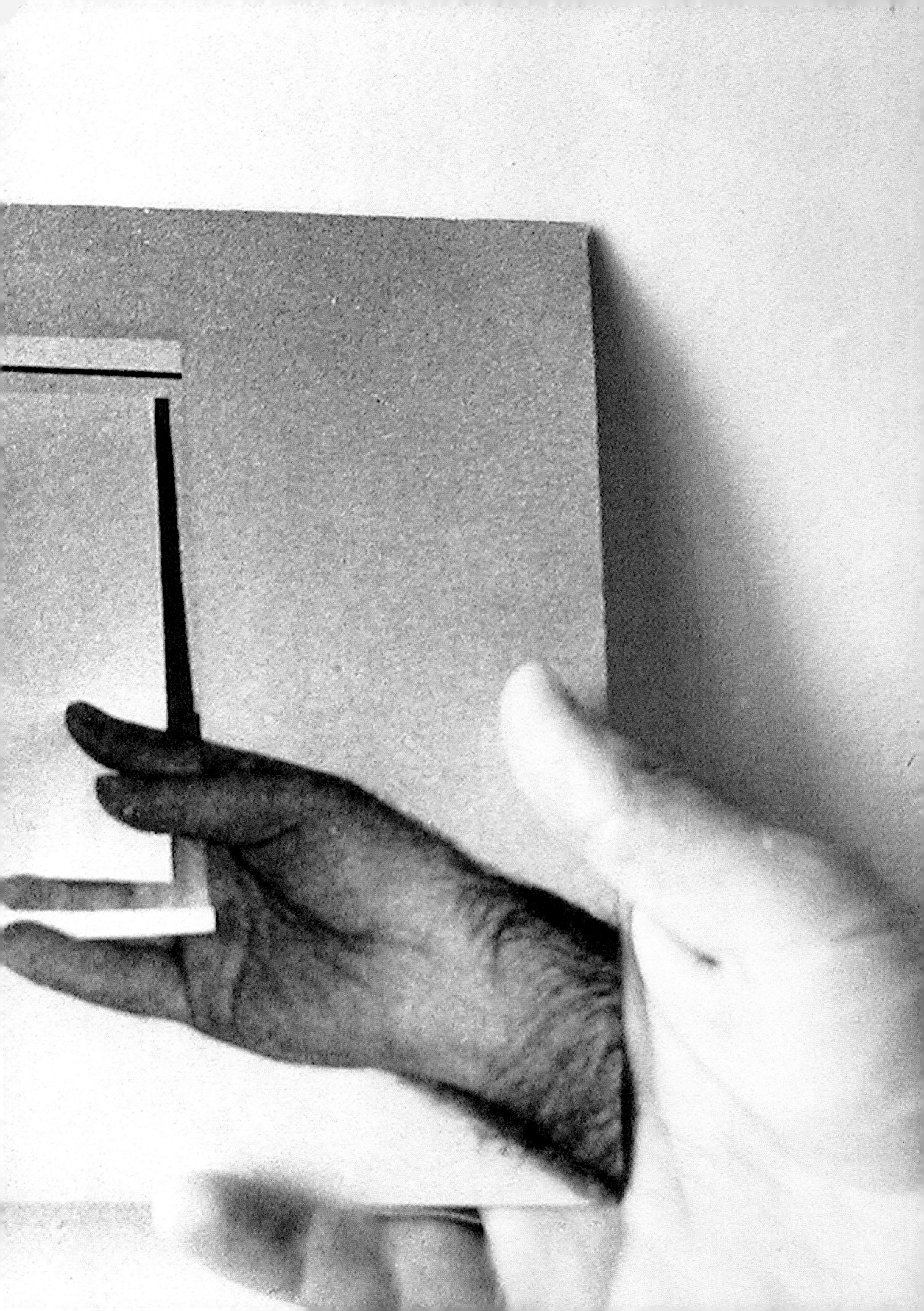